# Understanding Food

## Principles and Preparation

### Fourth Edition

## Amy Brown
*University of Hawaii at Manoa*

WADSWORTH
CENGAGE Learning

Australia • Brazil • Japan • Korea • Mexico • Singapore • Spain • United Kingdom • United States

## WADSWORTH
CENGAGE Learning™

*Understanding Food: Principles and Preparation,* **Fourth Edition**
Amy Brown

Senior Acquisitions Editor: Peggy Williams

Developmental Editor: Elesha Feldman

Editorial Assistant: Alexis Glubka

Media Editor: Miriam Myers

Senior Marketing Manager: Laura McGinn

Marketing Assistant: Elizabeth Wong

Senior Marketing Communications Manager: Linda Yip

Content Project Management: Pre-PressPMG

Creative Director: Rob Hugel

Senior Art Director: John Walker

Print Buyer: Linda Hsu

Rights Acquisitions Account Manager, Text: Roberta Broyer

Rights Acquisitions Account Manager, Image: Dean Dauphinais

Production Service: Pre-PressPMG

Photo Researcher: Pre-PressPMG

Cover Designer: Riezebos/Holzbaur: Bill Alexander

Cover Image: © Taillard/photocuisine/Corbis

Compositor: Pre-PressPMG

For product information and technology assistance, contact us at **Cengage Learning Customer & Sales Support, 1-800-354-9706.**

For permission to use material from this text or product, submit all requests online at **www.cengage.com/permissions.** Further permissions questions can be e-mailed to **permissionrequest@cengage.com.**

Library of Congress Control Number: 2010926934

ISBN-13: 978-0-538-73498-1

ISBN-10: 0-538-73498-1

**Wadsworth**
20 Davis Drive
Belmont, CA 94002-3098
USA

Cengage Learning is a leading provider of customized learning solutions with office locations around the globe, including Singapore, the United Kingdom, Australia, Mexico, Brazil, and Japan. Locate your local office at **www.cengage.com/global.**

Cengage Learning products are represented in Canada by Nelson Education, Ltd.

To learn more about Wadsworth, visit **www.cengage.com/wadsworth**

Purchase any of our products at your local college store or at our preferred online store **www.cengagebrain.com.**

Printed in the United States of America
2 3 4 5 14 13 12 11

# Brief Contents

# Contents

# PART II   FOOD SERVICE

## 4 Food Safety    64

## 5 Food Preparation Basics    99

# PART III   FOODS
## PROTEIN—MEAT, POULTRY, FISH, DAIRY, & EGGS

# Preface

Comprehensive is the word that describes *Understanding Food*. It brings together the most current information in food science, nutrition, and food service. Founded on research from more than 35 journals covering these disciplines, the text incorporates the very latest information on food—its science and its application. *Understanding Food*, 4th edition, provides students with a broad foundation to launch a career in any of these food-related fields.

## ORGANIZATION OF CONTENT

*Understanding Food* is organized according to the various food disciplines. Part I represents information related to food science and nutrition, such as food selection, sensory and physical evaluation, and food chemistry. Part II covers aspects of food service from food safety, food preparation basics, and meal management. Part III covers all of the standard food items arranged into protein (meat, poultry, fish, dairy, and eggs); phytochemicals (vegetables, fruits, soups, salads, and gelatins); complex carbohydrates (cereals, flour, breads); refined carbohydrates and fat (sweeteners, fats and oils, cakes and cookies, pies and pastries, candy, and frozen desserts); and water (beverages) groupings. Part IV relates to the food industry in terms of food preservation, government food regulations, and food careers. The chapter on food careers introduces students to the many careers associated with a basic foods course. In addition, the Professional Profile feature, found in many chapters, spotlights individuals working in various aspects in the food industry, so students really get a hands-on understanding of various career opportunities. Extensive appendixes provide additional key information, including approximate food measurements, weights and measures, storage temperatures, ingredient substitutions, flavorings and seasonings, and more.

## NEW TO THIS EDITION

- **Calorie Control** is a new feature teaching students where the calories are in foods, how many daily calories are recommended, and quick pointers on how to control calories within each food group. Obesity is at epidemic proportions in the United States and yet the public and many health professionals remain *calorie challenged*. An introductory food textbook is the perfect place to provide this information for future food and nutrition professionals.
- **New Food Evaluation Chapter** responds to reviewers' requests to create a separate chapter on this topic. Now this topic can be included as part of the basic food course or for a more advanced food course.
- **Revised Food Safety Chapter** responds to readers wanting a more "applied" hands-on approach to food safety. The content was totally revised to teach students practical tips on preventing foodborne illness that follow food flow—purchasing, storage, preparation, cooking, holding, cooling, reheating, and sanitation.
- **Temperature Danger Zone** updated to include both FDA and USDA recommendations.
- **Updated Information Added** on "gluten free" definitions and labeling regulations, stevia sweeteners, irradiation research, and new functional foods. Website links to the latest information have been incorporated, and more opportunities in food service careers have been listed.
- **Updated Art and Photos** providing new and intriguing ways to better illustrate concepts in the book.
- **Updated Chemistry Corners** and **How & Why Features** expanding on two popular features already in the book.

## OTHER FEATURES

The unique features of this text allow flexibility in teaching and create a dynamic learning environment for students.

- **Professional Profile** features provide interviews with people in the food arena and give advice to students to help them on their career path.
- **How and Why** inserts answer the questions most frequently asked by students. They are used to spark natural curiosity, trigger inquisitive thought patterns, and exercise the mind's ability to answer.
- **Chemist's Corner** features provide information on food chemistry in boxes within the chapters for those students and instructors who wish to further explore the chemistry of food. These Chemist's Corners create a book with two chemistry levels, allowing for flexibility based on the chemistry requirements of the individual course.
- **Nutrient Content** boxes in each of the foods chapters provide an overview of the nutritional composition of the foods, reflecting the increased emphasis in the food industry on food as a means for health promotion and disease prevention.
- **Food Additive information** has been incorporated throughout the book responding to students' requests to learn more about this topic.
- **Pictorial Summaries** at the end of every chapter are a proven favorite

with readers. Instead of a standard narrative summary, these pictorial chapter summaries use a combination of art and narrative text to encapsulate the key concepts in each chapter for student review.

- **Key terms,** boldfaced in the text, are defined in boxes on the same page to allow for quick review of the essential vocabulary in each chapter. A glossary at the end of the book assembles all of the key terms in the chapters in one place.
- **Functions of ingredients** are highlighted in the introduction to each chapter to aid students in successful food product development and food preparation. They introduce a focus of the food industry that is often missing in other books.
- **Problems and causes tables** in various food chapters summarize the problems that may occur when preparing specific food products and describe the possible causes, providing students with a handy reference tool for deciphering "what went wrong."
- **Numerous illustrations** placed throughout the text enhance students' understanding of the principles and techniques discussed.
- **A 16-page full-color insert** displays exotic varieties of fruits and vegetables, salad greens, flowers used in salads, traditional cuts of meats (including the lowest-fat meat cuts), and much more, all with detailed captions describing use and preparation tips.
- **Chapter review questions** were changed from 5 to 7 questions at the end of each chapter responding to requests to help prepare students for their class exams and also to help prepare them for the American Dietetic Association Registration Examination.

The dynamic world of food changes rapidly as new research constantly adds to its ever-expanding knowledge base. *Understanding Food: Principles and Preparation,* 4th edition, is designed to meet the needs of this evolving and expanding discipline, and to provide students with a strong foundation in any food-related discipline that they select.

## ANCILLARY MATERIALS

An assortment of student and instructor support materials, thoroughly updated for the fourth edition, are available:

- The print **Lab Manual,** revised by Janelle M. Walter (Baylor University), presents food experiments and recipes to demonstrate the principles discussed in the text. Pretest questions and materials/time needed information for instructors enhance the lab units, which parallel the organization and content of the text.
- The **Instructor's Resource CD-ROM** delivers several key instructor tools.
- **PowerPoint**® resources include JPEGs of text figures and ready-to-use (or modify) lecture presentations.
- An expanded **Test Bank** by Joan Aronson (New York University) provides multiple-choice, true/false, matching, and discussion/essay items.
- The **Instructor's Manual,** by Joan Aronson and Cheryl Houston (Fontbonne University), features engaging classroom activities, objectives, recommendations, and lecture outlines.
- The text's **Companion Website** offers various test preparation exercises for students, including quizzes, and instructor downloads.

## ACKNOWLEDGMENTS

Many individuals assisted me in the development of this textbook. First and foremost I thank Peter Marshall, Publisher, without whose knowledge and experience this book would never have come to be. I also thank Peggy Williams, who masterfully brought this book to the completion of its fourth edition.

I also extend my thanks to the outstanding members of the Cengage nutrition team: Elesha Feldman, Developmental Editor, for helping me revise and enhance the fourth edition; Elizabeth Howe, second edition Developmental Editor, for her excellent skills in working with me to create a well-organized manuscript; and Laura McGinn, Marketing Manager, who understands the process of book publishing and marketing to such a high degree that her presence alone is invaluable. My thanks to Yolanda Cossio, Publisher; Alexis Glubka, Editorial Assistant; and Bob Kauser and Dean Dauphinais, Permissions Editors. A thank you also goes to Elizabeth Wong, Marketing for getting the word out about this text. I also thank the tremendous production staff at Pre-PressPMG who worked miracles on this book, especially Kristin Ruscetta, Antonina Smith, and Catherine Schnurr.

I gratefully acknowledge Eleanor Whitney and Sharon Rolfes for contributing the Basic Chemistry Concepts appendix in this text.

A special thanks goes to the person who kindled my writing career, Nackey Loeb, Publisher of *The Union Leader.* Your early support and encouragement did far more than you will ever know.

Many colleagues have contributed to the development of this text. Their thoughtful comments provided me with valuable guidance at all stages of the writing process. I offer them my heartfelt thanks for generously sharing their time and expertise.

They are:

Dorothy Addario,
*College of St. Elizabeth*

Koushik Adhikari,
*Kansas State University*

Gertrude Armbruster (retired),
*Cornell University*

Mike Artlip,
*Kendall College*

Hea Ran-Ashraf,
*Southern Illinois University*

Mia Barker,
*Indiana University of Pennsylvania*

Nancy Berkoff,
*Art Institute of Los Angeles*

Margaret Briley,
*University of Texas*

Helen C. Brittin,
*Texas Tech University*

Mildred M. Cody,
*Georgia State University*

Carol A. Costello,
*University of Tennessee*

Barbara Denkins,
*University of Pittsburgh*

Nikhil V. Dhurandhar,
*Wayne State University*

Joannie Dobbs,
*University of Hawaii/Manoa*

Linda Garrow,
*University of Illinois/Urbana*

Natholyn D. Harris,
*Florida State University*

Sylvia Holman,
*California State University/Northridge*

Zoe Ann Holmes,
*Oregon State University*

Alvin Huang,
*University of Hawaii*

Wendy T. Hunt,
*American River College*

Karen Jameson,
*Purdue University*

Faye Johnson,
*California State University/Chico*

Nancy A. Johnson,
*Michigan State University*

Mary Kelsey,
*Oregon State University*

Elena Kissick,
*California State University/Fresno*

Patti Landers,
*University of Oklahoma*

Deirdre M. Larkin,
*California State University/Northridge*

Colette Leistner,
*Nicholls State University*

Lisa McKee,
*New Mexico State University*

Marilyn Mook,
*Michigan State University*

Martha N. O'Gorman,
*Northern Illinois University*

Polly Popovich,
*Auburn University*

Rose Tindall Postel,
*East Carolina University*

Beth Reutler,
*University of Illinois*

Susan Rippy,
*Eastern Illinois University*

Janet M. Sass,
*Northern Virginia Community College*

Anne-Marie Scott,
*University of North Carolina*

Sarah Short,
*Syracuse University*

Sherri Stastny,
*North Dakota State University*

Darcel Swanson,
*Washington State University*

Ruthann B. Swanson,
*University of Georgia*

M. K. (Suzy) Weems,
*Stephen F. Austin University*

Finally, I wish to express my appreciation to the students. Were it not for them, I would not have taken pen to paper. I am grateful to be part of your academic journey.

**Amy Christine Brown, Ph.D., R.D.**
**University of Hawaii at Manoa**
**amybrown@hawaii.edu**

# About the Author

© 2004 Carl Shaneff

**Amy Christine Brown, Ph.D., R.D.,** received her Ph.D. from Virginia Polytechnic Institute and State University in 1986 in the field of Human Nutrition and Foods. She has been a college professor and a registered dietitian with the American Dietetic Association since 1986. Dr. Brown currently teaches at the University of Hawaii's John A. Burns School of Medicine in the Department of Complementary and Alternative Medicine. Her research interests are in the area of bioactive plant substances beneficial to health and medical nutrition therapy. Some of the studies she has conducted include "Diet and Crohn's disease," "Potentially harmful herbal supplements," "Kava beverage consumption and the effect on liver function tests," and "The effectiveness of kukui nut oil in treating psoriasis." Selected research journal publications include: "Position of the American Dietetic Association: functional foods" (*Journal of the American Dietetic Association*); "The Hawaii Diet: Ad libitum high carbohydrate, low fat multi-cultural diet for the reduction of chronic disease risk factors" (*Hawaii Medical Journal*); "Lupus erythematosus and nutrition: A review" (*Journal of Renal Nutrition*); "Dietary survey of Hopi elementary school students" (*Journal of the American Dietetic Association*); "Serum cholesterol levels of nondiabetic and streptozotocin-diabetic rats" (*Artery*); "Infant feeding practices of migrant farm laborers in northern Colorado" (*Journal of the American Dietetic Association*); "Body mass index and perceived weight status in young adults" (*Journal of Community Health*); "Dietary intake and body composition of Mike Pigg—1988 Triathlete of the Year" (*Clinical Sports Medicine*); and numerous newspaper nutrition columns.

Feedback welcome, contact: amybrown@hawaii.edu

## To Jeffery Blanton

*To the person who saw me through four years of writing the first edition.*
*Four years, four thousand laughs, and only one you.*

*Always Grateful,*

*Amy Christine Brown*

RapidEye/istockphoto.com

# 1 Food Selection

ot too long ago, meats, milk, grains, nuts, vegetables, and fruits were the only foods available for consumption. Today food companies offer thousands of prepared and packaged foods, many of which are mixtures of these basic foods and often include artificial ingredients. The number of different foods now available can make it more difficult, rather than easier, to plan a nutritious diet. Food companies compete fiercely to develop ever newer and more attractive products for consumers to buy. This competition makes food scientists focus on why people eat what they eat, and what it is about a food or beverage that causes them to choose one over another.

People choose food and beverage based on several factors: how foods look and taste, health, cultural and religious values, psychological and social needs, and budgetary concerns (21). The factors influencing consumer food selection are the focus of this chapter, and each of the food selection criteria is addressed in detail.

## SENSORY CRITERIA

When people choose a particular food, they evaluate it consciously or unconsciously, primarily by how it looks, smells, tastes, feels, and even sounds (Figure 1-1). These sensory criteria are discussed first because of their strong influence on food selection. How a food or beverage affects the senses is more important to most consumers than other criteria when it comes to what a person chooses to eat or drink. The sensory criteria of sight, odor, and taste are evaluated below.

### Sight

The eyes receive the first impression of foods: the shapes, colors, consistency, serving size, and the presence of any outward defects. Color can denote the ripeness, strength of dilution, and even degree to which the food was heated. Black bananas, barely yellow lemonade, and scorched macaroni send visual signals that may alter a person's choices. Color can be deceiving; if the colors of two identical fruit-flavored beverages are different, people often perceive them as tasting different even though they are exactly the same (87). People may judge milk's fat content by its color. For instance, if the color, but not the fat, is improved in reduced-fat (2%) milk, it is often judged to be higher in fat content, smoother in texture, and better in flavor than the reduced-fat milk with its original color (71).

The color palette of foods on a plate also contributes to or detracts from

**FIGURE 1-1**    Sensory impressions of food provided by the five senses.

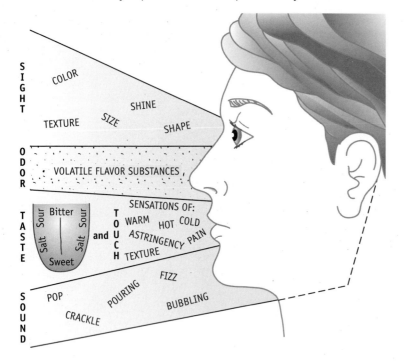

rabbit (18), reflecting the difference in importance of the sense of smell between people and rabbits. The exact function of these specialized cells in the sense of smell is not well understood.

Who has not experienced the feeling of bubbles tingling in the nose brought on by drinking a carbonated beverage while simultaneously being made to laugh unexpectedly? This illustrates how the mouth and nose are connected and how molecules can reach the olfactory epithelium by either pathway.

their appeal. Imagine a plate containing baked flounder, mashed potatoes, boiled cabbage, and vanilla ice cream, and then compare it to one that contains a nicely browned chicken breast, orange sweet potatoes, green peas, and blueberry cobbler. Based on eye appeal alone, most people would prefer the latter.

## Odor

Smell is almost as important as appearance when people evaluate a food item for quality and desirability. Although the sense of smell is not as acute in human beings as it is in many other mammals, most people can differentiate between 2,000 to 4,000 odors, whereas some highly trained individuals can distinguish as many as 10,000 (8).

### Classification of Odors

Naming each of these thousands of odors separately would tax even the

most fertile imagination; researchers have categorized them into major groups. One classification system recognizes six groups of odors: spicy, flowery, fruity, resinous (eucalyptus), burnt, and foul. The other widely used grouping scheme consists of four categories: fragrant (sweet), acid (sour), burnt, and caprylic (goaty) (8).

### Detecting Odors

Regardless of the classifications, most odors are detected at very low concentrations. Vanillin can be smelled at $2 \times 10^{-10}$ (0.0000000002) mg per liter of air (18). The ability to distinguish between various odors diminishes over the time of exposure to the smells; this perception of a continuously present smell gradually decreasing over time is called *adaptation*. People living near a noxious-smelling paint factory will, over time, come not to notice it, whereas visitors to the area may be taken aback by the odor.

We are able to detect odors when **volatile molecules** travel through the air and some of them reach the yellowish-colored **olfactory** epithelium, an area the size of a quarter located inside the upper part of the nasal cavity. This region is supplied with olfactory cells that number from 10 to 20 million in a human and about 100 million in a

**Volatile molecules** Molecules capable of evaporating like a gas into the air.

**Olfactory** Relating to the sense of smell.

### ? How & Why?

*Imagine the scent of chocolate chip cookies wafting through the house as they bake. How does this smell get carried to people? Why is the odor of something baking more intense than the odor of cold items like ice cream or frozen peaches?*

Heat converts many substances into their volatile form. Because only volatile molecules in the form of gas carry odor, it is easier to smell hot foods than cold ones. Hot coffee is much easier to detect than cold coffee. Relatively large molecules such as proteins, starches, fats, and sugars are too heavy to be airborne, so their odors are not easily noticed. Lighter molecules capable of becoming volatile are physically detected by the olfactory epithelium by one of two pathways: (1) directly through the nose and/or (2) during eating when they enter the mouth and flow retronasally, or toward the back of the throat and up into the nasal cavity (Figure 1-2) (69).

## Taste

Taste is usually the most influential factor in people's selection of foods (25). Taste buds—so named because the arrangement of their cells is similar to the shape of a flower—are located primarily on the tongue, but are also found on the mouth palates and in the pharynx. Taste buds are not found on the flat, central surface of the tongue, but rather on the tongue's underside, sides, and tip.

**FIGURE 1-2**   Detecting aroma, mouthfeel, and taste.

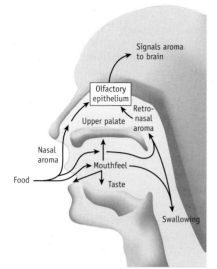

Stockbyte/Jupiter Images

## Mechanism of Taste

What is actually being tasted? Many tasted substances are a combination of nonvolatile and volatile compounds. In order for a substance to be tasted, it must be dissolved in liquid or saliva, which is 99.5% water. In the middle of each taste bud is a pore, similar to a little pool, where saliva collects.

When food comes into the mouth, bits of it are dissolved in the saliva pools and they come into contact with the cilia, small hair-like projections from the **gustatory** cells. The gustatory cells relay a message to the brain via one of the cranial nerves (facial, vagus, and glossopharyngeal). The brain, in turn, translates the nervous electrical impulses into a sensation that people recognize as "taste." As people age, the original 9,000 to 10,000 taste buds begin to diminish in number, so people over 45 often find themselves using more salt, spices, and sugar in their food. Another important factor influencing the ability of a person to taste is the degree to which a compound can dissolve (61). The more moisture or liquid, the more the molecules triggering flavor can dissolve and spread over the tongue, coming in contact with the taste buds (37).

## The Five Taste Stimuli

Different areas on the tongue are associated with the five basic types of tastes: sweet, sour, bitter, salty, and savory (*umami*, a Japanese word meaning "delicious") (65). The fifth taste stimulus,

savory (*umami*), is found in certain amino acids. The tip of the tongue is more sensitive to sweet and sour tastes, the sides are sensitive to salty and sour sensations, and the back is sensitive to bitter taste perceptions. The time it takes to detect each of these taste stimuli varies from a split second for salt to a full second for bitter substances (18). Bitter tastes, therefore, have a tendency to linger. The chemical basis of these five categories of taste is as follows:

- The sweetness of sugar comes from the chemical configuration of its molecule. A long list of substances yield the sweet taste, including sugars, glycols, alcohols, and aldehydes. Little is known, however, about the sweet taste receptor and how "sweetness" actually occurs (37, 88).
- Sour taste comes from the acids found in food. It is related to the concentration of hydrogen ions ($H^+$), which are found in the natural acids of fruits, vinegar, and certain vegetables.
- Bitterness is imparted by compounds such as caffeine (tea, coffee), theobromine (chocolate), and phenolic compounds (grapefruit). The ability to taste bitterness can act as a warning system to prevent us from ingesting toxins. Among the many

substances that can yield bitter tastes are the alkaloids that are often found in poisonous plants (11).
- Salty taste comes from ionized salts such as the salt ions ($Na^+$) in sodium chloride (NaCl) or other salts found naturally in some foods.
- Savory (*umami*) taste was first identified in 1908 by researchers at Tokyo Imperial University. *Umami* is actually glutamate, an amino acid that imparts the taste of beef broth, but without the salt. Some people can detect monosodium glutamate (MSG) in foods because it contains glutamate.

## Taste Interactions

Each item used in food preparation contains several compounds, and bringing these items together creates new tastes when all their compounds interact. Salt sprinkled on grapefruit or added to fruit pies tends to decrease tartness and enhance sweetness. Conversely, acids in subthreshold concentrations, which are present but not yet detectable, increase saltiness. Adding sugar to the point that it is not yet tasted decreases

**Gustatory**   Relating to the sense of taste.

## PROFESSIONAL PROFILE

### Ben Cohen and Jerry Greenfield— Cofounders of Ben & Jerry's

It's hard to believe, but some people taste food for a living. "Taste testers" have such sensitive taste buds or olfactory detection that they are hired by food companies to taste new products being developed. Food companies need to be sure that the absolute best product is being produced for consumers. John Harrison, the official taste tester for Dreyer's and Edy's ice creams, has tasted samples from over 100 million gallons of ice cream, and his taste buds are insured for $1 million. Derek Spors, who obtained his bachelor's degree in food science, works in the same capacity at Ben & Jerry's as a "product developer." The first taste tester at Ben & Jerry's Ice Cream was the company's cofounder, Ben Cohen. However, he had such weak taste buds that he kept asking the flavor developers for more sugar, salt, chocolate cookies, or caramel. His challenged taste buds are what caused Ben & Jerry's to produce ice creams

**Ben Cohen and Jerry Greenfield**

that are now famous for their very intense flavors.

Ben & Jerry's started small. Ben Cohen and Jerry Greenfield were high school friends on Long Island, New York, in the late 1960s. During high school, Ben drove an ice cream truck, selling ice cream pops to kids. He went off to college, only to drop out after a year and to return to his ice cream job. Enrolling in another college, he studied pottery and jewelry, before dropping out again to teach crafts in a residential school for emotionally disturbed children. It was there that he began to experiment with ice cream–making as an activity to do with the students. In 1977, Ben left the school and decided to sell ice cream with Jerry (who had applied to medical school and had been rejected twice). They opened up "Ben & Jerry's Homemade Ice Cream Parlor" in a renovated gas station in Vermont. The rest is ice cream history. You can visit the company website at www.benjerry.com.

---

salt concentration and also makes acids less sour and coffee and tea less bitter. Some compounds, like monosodium glutamate, often used in Chinese cooking, actually improve the taste of meat and other foods by making them sweeter (55, 69).

### Factors Affecting Taste

Not everyone perceives the taste of apple pie the same way. There is considerable genetic variation among individuals in sensitivity to basic tastes (70). Tasting abilities may also vary within the individual, depending on a number of outside influences (67). One such factor affecting taste is the temperature of a food or beverage. As food or beverage temperatures go below 68°F (20°C) or above 86°F (30°C), it becomes harder to distinguish their tastes accurately. For example, very hot coffee tastes less bitter, whereas slightly melted ice cream tastes sweeter. Other factors influencing taste include the color of the food; the time of day it is eaten; and the age, gender, and degree of hunger of the taster (38).

Variety in available food choices also affects taste. This can be seen when

the "taste," or appetite, for a food eaten day after day starts to diminish. Even favorite foods consumed everyday can lose their appeal after a while. Some weight-reducing fad diets are based on this principle, banking on the idea that people will get tired of eating just one type of food and therefore will come to eat less of it. A routine of grapefruit for breakfast, grapefruit for lunch, and grapefruit for dinner quickly becomes boring and unappetizing.

### Definition of Flavor

In examining the factors affecting taste, it is important to distinguish between taste and **flavor.** Taste relies on the taste buds' connection to the brain via nerve cells, which signal the sensations of sour, salt, sweet, bitter, and savory. Flavor is a broader concept than either taste or aroma; aroma provides about 75% of the impression of flavor (23, 78). To get some idea of how the ability to smell affects flavor perception, think of having a cold with a badly stuffed-up nose. Everything tastes different. The nasal congestion interferes with the function of the olfactory sense, impairing the ability to detect the aromas contributing to the perception of flavor. Some people apply this principle to their advantage by pinching their nostrils shut to lessen the bad flavor of a disagreeable medicine they must swallow.

**Flavor** The combined sense of taste, odor, and mouthfeel.

### ? How & Why?

Flavors, regardless of the medium in which they are dissolved, do not stay at the same intensity day after day, but diminish over time. Sensory chemists and flavor technologists know that one way to keep the food products sold by manufacturers from losing their appeal is to prevent the volatile compounds responsible for flavor from deteriorating, escaping, or reacting with other substances. They look at methods in processing, storage, and cooking, all of which affect the volatile flavor compounds, to devise strategies against these occurrences. One of the major functions of protective packaging is to retain a food's flavor. Packaging guards flavor in several ways. It protects against vaporization of the volatile compounds and against physical damage that could expose food to the air and result in off odors. It keeps unpleasant odors from the outside from attaching to the food. It also prevents "flavor scalping," or the migration of flavor compounds from the packaging (sealers, solvents, etc.) to the food or vice versa (55).

Whether in a package or on a plate, a commercial food's flavor is the single most important factor determining its success in the marketplace (26). There are thousands of ways to prepare foods and beverages, but the method or chef that yields the best flavor will most likely be best received by consumers whose number-one selection criterion is how something tastes.

## Touch

The sense of touch, whether it operates inside the mouth or through the fingers, conveys to us a food's texture, consistency, astringency, and temperature.

Texture is a combination of perceptions, with the eyes giving the first clue. The second comes at the touch of fingers and eating utensils, and the third is mouthfeel, as detected by the teeth and the tactile nerve cells in the mouth, located on the tongue and palate. Textural or structural qualities are especially obvious in foods such as apples, popcorn, liver, crackers, potato chips, tapioca pudding, cereals, and celery, to name just a few. Textures felt in the mouth can be described as coarse (grainy, sandy, mealy), crisp, fine, dry, moist, greasy, smooth (creamy, velvety), lumpy, rough, sticky, solid, porous, bubbly, or flat. Tenderness, which is somewhat dependent on texture, is judged by how easily the food gives way to the pressure of the teeth.

**Consistency** is only slightly different from tenderness, and is expressed in terms of brittleness, chewiness, viscosity, thickness, thinness, and elasticity (rubbery, gummy).

**Astringency,** which causes puckering of the mouth, is possibly due to the drawing out of proteins naturally found in the mouth's saliva and mucous membranes (18). Foods such as cranberries, lemon juice, and vinegar have astringent qualities.

Another term used in the sensory perception of foods is **chemethesis.** Chemethesis defines how certain foods that are not physically hot or cold appear to give the impression of being "hot" (hot salsa) or "cooling" (cucumbers) when placed on the tongue (39). Although extremely hot temperatures can literally burn the taste buds (they later regenerate), the other kind of "hot" that may be experienced with food is the kind generated by "hot" peppers (Chemist's Corner 1-1). The hotness in peppers is produced by a chemical called capsaicin (cap-SAY-iss-in). Many people enjoy the sensation of capsaicin in moderation, but it can cause real pain because it is a powerful chemical that irritates nerves in the nose and mouth. In fact, this compound is so caustic when concentrated that it is now used by many law enforcement agencies in place of the mace-like sprays.

## Hearing

The sounds associated with foods can play a role in evaluating their quality. How often have you seen someone tapping a melon to determine if it is ripe? Sounds like sizzling, crunching, popping, bubbling, swirling, pouring, squeaking, dripping, exploding (think of an egg yolk in a microwave), and crackling can communicate a great deal about a food while it is being prepared, poured, or chewed. Most of these sounds are affected by water content, and their characteristics thus give clues to a food's freshness and/or doneness.

# NUTRITIONAL CRITERIA

Over the past several decades, emerging scientific evidence about health and nutrition has resulted in changing food consumption patterns in the United States (12). Past surveys reveal at least half of all consumers reportedly making a major change in their diets, with nutrition being second only to taste in importance to shoppers (48, 93). The changing food habits are related to the increased awareness that diet can be related to some of the leading causes of death—heart disease, cancer, and diabetes—as well as to other common health conditions such as osteoporosis, diverticulosis, and obesity (40).

Obesity has reached epidemic proportions in the United States (24) and is a risk factor for heart disease, cancer, diabetes, and other health conditions. Health care costs are higher in people who are obese compared to people of normal weight (3). Being

**Consistency** Describes a food's firmness or thickness.

**Astringency** A sensory phenomenon characterized by a dry, puckery feeling in the mouth.

**Chemethesis** The ability to feel a food's chemical properties (e.g., cool mints or hot chili peppers).

overweight is one of the biggest and costliest health problems in the United States (3). Also, some $33 billion are spent annually by 65 million Americans on "quick fix" weight loss solutions, most of which achieve no permanent results.

To reduce dietary risk factors for some of the major health conditions affecting Americans, the U.S. government published several diet-planning guides. Two important ones are the Dietary Guidelines for Americans and MyPyramid (www.mypyramid.gov).

## Dietary Guidelines for Americans

The emphasis on adjusting fat and other dietary factors in the diet was reinforced by the Dietary Guidelines, which have been published every five years since 1980 by the United States Department of Agriculture (USDA) and the United States Department of Health and Human Services (DHHS) (92). As of this writing, the latest Dietary Guidelines for healthy adults are available in 2010. They encourage people to follow the recommendations available at www.mypyramid.gov (select Dietary Guidelines).

**Food group plan**  A diet-planning tool that "groups" foods together based on nutrient and calorie (kcal) content and then specifies the amount of servings a person should have based on their recommended calorie (kcal) intake.

**Antioxidant**  A compound that inhibits oxidation, which can cause deterioration and rancidity.

**Nutraceutical**  A bioactive compound (nutrients and nonnutrients) that has health benefits.

**Functional food**  A food or beverage that imparts physiological benefits that enhance overall health, prevents or treats a disease or condition, and/or improves physical/mental performance.

## MyPyramid

MyPyramid is a visual **food group plan** developed by the USDA to illustrate the concepts of the Dietary Guidelines. The www.mypyramid.gov website is an interactive site designed to assist people in creating a personal food plan and making better food and lifestyle choices by taking small steps toward those goals each day. The Food Guide Pyramid, MyPyramid's predecessor, was first developed in 1992 to encourage Americans to improve their diets and to replace the basic four food groups of milk, meat, vegetable/fruit, and bread/cereal (34, 91). Other countries have their own versions of this type of guideline; Canada's version is available online (see websites at end of chapter) (74). The easy-to-comprehend visual concept of MyPyramid organizes foods into six food groups. The lower-fat, complex carbohydrate foods such as grains are emphasized, followed by vegetables, fruits, and milk, meat and beans, and oil. The narrow slivers of color at the top imply moderation in foods rich in solid fats and added sugar.

Other U.S. pyramids exist and include the Mediterranean, Asian, Latin American, and vegetarian pyramids. Regarding this last pyramid, the American Dietetic Association suggested that properly planned vegetarian diets may reduce the risk of certain chronic, degenerative diseases and conditions including heart disease, some cancers, diabetes mellitus, obesity, and high blood pressure (2). Other factors may, however, contribute to the decreased morbidity and mortality from these diseases among vegetarians. These include positive lifestyle differences such as lower rates of smoking and drinking. Nevertheless, vegetarian diet benefits probably come from lower intakes of fat, saturated fat, cholesterol, and animal protein, balanced by higher levels of phytochemicals, fiber, complex carbohydrates, **antioxidants** such as vitamins C and E, carotenoids, and folate (a B vitamin) (60).

## Consumer Dietary Changes

As a result of these dietary guidelines and other influences, consumers have shifted their dietary concerns and intakes, and more people are reading the Nutrition Facts on food labels to understand what they are consuming (Chapter 29). Throughout the 1990s, consumers reported their biggest nutritional concern to be fat; this exceeded concerns about salt, cholesterol, sugar, and even calories (kcal) (84). Today, Americans are ingesting less red meat and whole milk, and more poultry, reduced-fat (2%) milk, fresh fruits, fresh vegetables, pasta, and rice. As a result, fat consumption has dropped from 42% of calories (kcal) in the mid-1960s to less than 33% today (16).

### Health Focus

The focus of the Dietary Guidelines since their 2010 revision has been to promote the best diet for reducing the risk of chronic disease often resulting from excess consumption of the wrong kinds of foods and to boost the intake of nutrients such as calcium, fiber, vitamin A, vitamin C, and potassium. According to an International Food Information Council survey, the majority of Americans believe that some foods can have health benefits besides their nutritive value and can delay the onset, or reduce the risk, of serious and chronic diseases (www.health.gov/DietaryGuidelines/). The vegetarian movement is gaining ground; approximately 15% of college students define themselves as vegetarians. About 33% of Americans have used herbs or herb products medicinally, and about 60% take a multivitamin supplement. The concept that "food is medicine" is common to many cultures, and the shift from treating an established disease to preventing its occurrence is slowly gaining ground. Overall, more people view foods as an integral part of maintaining their health (1).

### Complementary and Alternative Medicine

Another influence on consumer dietary changes is complementary and alternative medicine (CAM), which is making permanent inroads in the U.S. medical landscape. Terms such as **nutraceuticals** and **functional foods** (described more fully following this section) are becoming commonplace. In the United States, *nutraceuticals* is a term often used to refer to dietary supplements, while the official

## CALORIE CONTROL
### Calorie Balance

One out of every four people in the United States was classified as "obese" in 2007 according to the Centers for Disease Control and Prevention (CDC) (14). A website link at the end of this chapter allows each person to calculate his/her body mass index (BMI). This number is a ratio based on a person's weight to height that classifies him or her as underweight, normal weight, overweight, or obese.

The CDC is concerned about obesity because of health consequences, high health care costs, increased absenteeism, and work-related injuries (52). Although many other factors such as environment, genetics, disease, and drugs may contribute to obesity, this book focuses on the primary cause of obesity—too many calories (Figure 1-3).

The purpose of the "Calorie Control" sections in this book is to address the obesity epidemic by providing readers with calories found in foods and healthful ways to modify their diets. Specific topics to be included are: (1) average daily caloric intakes by Americans (see below), (2) calorie sources (see Chapter 3 on Food Composition), (3) the average number of calories found in foods (see individual chapters), (4) suggestions for practicing portion control (see Chapter 5 on Food Preparation Basics and various individual chapters), and (5) healthful preparation methods (various chapters).

### How Many Calories Do People Consume Each DAY?

The Dietary Reference Intakes (DRI) for calories (2403 per day for women and 3067 for men) exceed those reported by the National Health and Nutrition Examination Survey (1999–2000), which measures the actual caloric intakes (1,833 for women and 2,475 for men) of a population in which one fourth are obese (15). Although it's best for people to determine their caloric and nutrient needs by seeing a registered dietitian (RD), the approximate calories needed each day by healthy adults to "reach" and "maintain" a healthy goal weight are suggested below:

| | |
|---|---|
| Women | approximately 1,600 calories for each day |
| Men | approximately 2,400–2,600 calories for each day |

This estimate includes exercising three times a week for at least 20 minutes each session. A person will need more calories if they exercise more than three times a week—approximately 300–600 calories for each hour of aerobic exercise. The exceptions are active (athletes) and larger people, who need more calories; sedate and shorter people, who need fewer calories; and older people, who need fewer calories (after 40, people need 100 fewer calories for each 10 years of age) (92).

**FIGURE 1-3** Caloric balance is like a scale. To remain in balance and maintain your body weight, the calories consumed (from foods) must be balanced by the calories used (in normal body functions, daily activities, and exercise).

| If you are: | Your caloric balance is: |
|---|---|
| Maintaining Weight | CALORIES IN Food Beverages / CALORIES OUT Body functions Physical activity | • "IN BALANCE" <br> • You are eating roughly the same number of calories that your body is using. <br> • Your weight will remain stable. |
| Gaining Weight | CALORIES IN Food Beverages / CALORIES OUT Body functions Physical activity | • "IN CALORIC EXCESS" <br> • You are eating more calories than your body is using. <br> • You will store these extra calories as fat and you'll gain weight. |
| Losing Weight | CALORIES OUT Body functions Physical activity / CALORIES IN Food Beverages | • "IN CALORIE DEFICIT" <br> • You are eating fewer calories than you are using. <br> • Your body is pulling from its fat storage cells for energy, so your weight is decreasing. |

*(Continued)*

### How Many Calories for Each MEAL?

Because it's challenging to count total daily calories, the easier method is to break it down for each "meal." This is done by taking the average number of calories needed for women (1,600) and dividing it by 4—three 400-calorie meals plus one 400-calorie snack (or two 200-calorie snacks). The snacks are best eaten mid-morning and mid-afternoon, but can be taken in any combination of calories during any part of the day and even as part of a meal. For men, the average 2,400 calories a day is divided by four to equal three 600-calorie meals plus one 600-calorie snack (or two 300-calorie snacks).

### Starving Is a Bad Idea

About two thirds of a person's calories are used to sustain life: heart beating, lungs breathing, body temperature at 97.6°F/36.4°C, and other bodily functions. Most of the remaining 30% of calories are burned by activity.

The bottom line is that based on gender, a person should not consume less than the following amount of daily calories:

Women   1,200 calories (about 70% of 1,600)

Men       1,600 calories (about 70% of 2,400)

### How to Gain or Lose Weight

The recommended method of gaining or losing weight is to either increase or decrease caloric intake, respectively, by at least 500 calories a day. This should result in a weekly 1-pound weight gain or loss, respectively.

### How Many Calories Equal a Pound?

3,500 calories = 1 pound

> To <u>lose</u> 1 pound = Consume 3,500 calories less and/or burn it off with exercise
>
> To <u>gain</u> 1 pound = Consume 3,500 calories over what your body burns

### Combination of Diet and Exercise

If a person can achieve a deficit of 500 calories per day through diet and/or exercise, they will lose approximately 1 pound a week.

### Successful Weight Loss Is Usually Slow

Consistency is the goal. The slower you lose the weight, the more likely it will stay off.

1 pound a week for 1 month = 4 weeks = 4 pounds

1 pound a week for 1 year = 52 weeks = 52 pounds

© 2010 Amy Brown

---

definition in Canada is "a product isolated or purified from foods, and generally sold in medicinal forms not usually associated with food and demonstrated to have a physiological benefit or provide protection against chronic disease" (44).

Europe and Japan lead the way in complementary medicine. In Germany, the E Commission was created in 1978 to ensure product standardization and safe use of herbs and phytomedicines. Composed of a body of experts from the medical and pharmacology professions, the pharmaceutical industry, and laypersons, the German E Commission studies the scientific literature for research data on herbs based on clinical trials, field studies, and case studies. It has created a collection of **monographs** representing the most accurate information available in the world on the safety and efficacy (power to produce effects or "does it work?") of herbal products. Germany defines herbal remedies in the same manner as it does drugs, because its physicians, and others in Europe, often prescribe herbal remedies that are paid for by government health insurance.

### Functional Foods

Known in Japan as "Foods for Specified Health Use," functional foods are those produced, selected, or consumed for reasons beyond basic caloric and nutrient content. The functional food concept first developed in Japan in the late 1980s. Both Japan and Europe appear to surpass the United States in their interest in how foods can benefit health beyond providing carbohydrates, protein, fat, and vitamins/minerals. In fact, Japan is the only country that recognizes functional foods as a distinct category, and the Japanese functional food market is now one of the most advanced in the world (95). Purported

**Monograph** A summary sheet (fact sheet) describing a substance in terms of name (common and scientific), chemical constituents, functional uses (medical and common), dosage, side effects, drug interactions, and references.

imagebroker/Alamy

uses for which functional foods have been manufactured include, but are not limited to, maintenance of gastro-intestinal health, blood pressure, and blood cholesterol levels (4). Japan has imported record shipments of blueberries from the United States because the blueberry's blue pigment, anthocyanin, is a powerful antioxidant thought to possibly benefit eyesight (47).

The United States has no official definition existing for "functional foods" because they are not recognized as a regulatory category by the Food and Drug Administration. Moreover, the largest organization of food and nutritional professionals in the United States, The American Dietetic Association (ADA), classifies all foods as functional at some physiological level because they provide nutrients or other substances that furnish energy, sustain growth, and/or maintain and repair the body. However, functional foods move beyond basic survival needs by providing additional health benefits that may reduce disease risk and/or promote optimal health. Specifically, the ADA defines functional foods as including conventional foods, modified foods (fortified, enriched, or enhanced), medical foods, and foods for special dietary use (Table 1-1) (43). These functional food categories—published in a 2009 ADA Position Paper on functional foods (co-published by the author)—are now explained:

1. *Conventional Foods.* Unmodified whole foods or conventional foods such as fruits and vegetables are the simplest functional foods. For example, tomatoes, raspberries, kale, or broccoli are considered functional foods because they are rich in such bioactive components as lycopene, ellagic acid, lutein, and sulforaphane, respectively. A few of the many examples of health benefits linked to conventional foods by emerging evidence include:

Cancer Risk Reduction
- Cruciferous vegetables reduce risk of several types of cancers (67).
- Tomato products rich in lycopene may reduce the risk of prostate, ovarian, gastric, and pancreatic cancers (54).
- Citrus fruit may reduce the risk of stomach cancer (6).

Heart Health
- Dark chocolate reduces high blood pressure (29).
- Tree nuts and peanuts reduce the risk of sudden cardiac death (59).

Intestinal Health Maintenance
- Fermented dairy products (probiotics) may reduce irritable bowel syndrome symptoms (79).

Urinary Tract Function
- Cranberry juice reduces bacterial concentrations in the urine (64).

Some other health conditions affected by conventional foods include osteoporosis, diabetes, arthritis, brain health (mood, memory, depression, insomnia, stress, anxiety, and alertness), weight (appetite, weight loss or gain), eyesight, and enhanced energy and sports performance (64).

2. *Modified Foods.* Functional foods can also include those that have been modified through fortification, enrichment, or enhancement. These include calcium-fortified orange juice (for bone health), folate-enriched breads (for proper fetal development), or foods enhanced with bioactive components, such as margarines containing plant stanol or sterol esters (for cholesterol lowering), and beverages enhanced with "energy-promoting" ingredients such as ginseng, guaraná, or taurine. Modifying foods through biotechnology to improve their nutritional value and health attributes may also bring new functional foods, such as omega-3 fatty acid-enhanced or trans fat-free oils, to the marketplace (76), although the topic remains controversial.

3. *Medical Foods.* The term *medical food*, as defined by the Orphan Drug Act, is "a food which is formulated to be consumed or administered internally? under the supervision of a physician and which is intended for the specific dietary management of a disease or condition for which distinctive nutritional requirements, based on recognized scientific principles, are established by medical evaluation" (33). Examples of medical foods include oral supplements in the form of phenylketonuria (PKU) formulas free of phenylalanine, and diabetic, renal, and liver formulations. How the product is marketed to consumers determines the regulatory status of a product. For example, a canned or bottled oral supplement used under medical supervision is a medical food; however, it becomes a food for special dietary use when sold to consumers at the retail level.

4. *Foods for Special Dietary Use.* The Federal Food, Drug, and Cosmetic

**TABLE 1-1** Functional Food[1] Categories and Selected Food Examples

| Functional food category | Selected functional food examples |
| --- | --- |
| **Conventional foods (whole foods)** | Garlic |
| | Nuts |
| **Modified foods** | |
| Fortified | Calcium-fortified orange juice |
| | Iodized salt |
| Enriched | Folate-enriched breads |
| Enhanced | Enhanced energy bars, snacks, yogurts, teas, bottled water, and other functional foods formulated with bioactive components such as lutein, fish oils, ginkgo biloba, St. John's wort, saw palmetto, and/or assorted amino acids |
| **Medical foods** | Phenylketonuria (PKU) formulas free of phenylalanine |
| **Foods for special dietary use** | Infant foods |
| | Hypoallergenic foods such as gluten-free foods, lactose-free foods |
| | Weight-loss foods |

[1]As defined by the American Dietetic Association

Act [Section 411(c)(3)] defines *special dietary use* as "a particular use for which a food purports or is represented to be used, including but not limited to the following: (1) supplying a special dietary need that exists by reason of a physical, physiological, pathological, or other condition . . .; (2) supplying a vitamin, mineral, or other ingredient for use by humans to supplement the diet by increasing the total dietary intake, and (3) supplying a special dietary need by reason of being a food for use as the sole item of the diet. . . ." (33). Examples of such foods would include infant formulas, and hypoallergenic foods such as gluten-free foods, lactose-free foods, and foods offered for reducing weight.

Although functional foods are emerging as one of the latest trends in food and nutrition, this concept is not entirely new; about 2,500 years ago, Hippocrates said, "Let food be thy medicine, and medicine be thy food" (9).

### Nutrigenomics

Someday people might receive diet plans tailored to their genes thanks to **nutrigenomics,** which first appeared in the scientific literature in 2001 (75). Before the term was coined, nutrigenomics existed undefined within the study of metabolic disorders (inborn errors of metabolism). These genetic errors often occur due to the lack of an enzyme within a biochemical pathway resulting in a need for dietary intervention, as in the case with phenylketonuria (PKU). Nutrigenomics includes not only these diseases, but all others where less dramatic genetic differences result in different dietary needs—such as heart disease, diabetes (types 1 and 2), osteoporosis, rheumatoid arthritis, hypertension, bipolar

disorder, and a myriad of inflammatory disorders—and any other disease or condition with a genetic link that may be improved by dietary modification (20). Nutrigenomics relies on nutritional biochemistry to explain why differences in genes cause variations in absorption, circulation, or metabolism of essential nutrients. This knowledge enables people to select certain foods for optimal health or reduced risk of chronic disease (30). Some have suggested that this science is still in its infancy, so its contribution to human health may take some time (68).

## CULTURAL CRITERIA

**Culture** is another factor influencing food choice. Culture influences food habits by dictating what is or is not acceptable to eat. Foods that are relished in one part of the world may be spurned in another. Grubs, which are a good protein source, are acceptable to the aborigines of Australia. Whale blubber is used in many ways in the arctic region, where the extremely cold weather makes a high-fat diet essential. Dog is considered a delicacy in some Asian countries. Escargots (snails) are a favorite in France. Sashimi (raw fish) is a Japanese tradition that has been fairly well accepted in the United States. Locusts, another source of protein, are considered choice items in the Middle East. Octopus, once thought unusual, now appears on many American menus.

## Ethnic Influences

Ethnic minorities comprise at least 25% of the U.S. population of approximately 300 million people, with the four major groups being African, Other (includes two or more races), Asian/Native Hawaiian/Pacific Islanders, and Native/Alaskan Americans (Figure 1-4). The U.S. Census does not classify "Hispanic" or "Latino" as a race. Rather, those taking the survey are asked whether or not they are "Spanish/Hispanic/Latino" and to select the race to which they identify. The belief is that people from this group may be of any race, but this makes the overall percentage "picture" a little confusing. The latest U.S. Census reported 13% of the

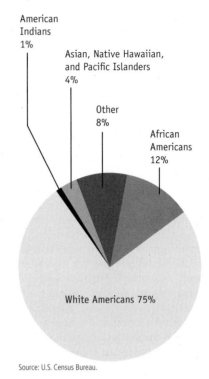

**FIGURE 1-4**   Percent distribution of racial/ethnic groups in the United States. Fifteen percent of Americans report themselves to be "Spanish/Hispanic/Latino."

American Indians 1%

Asian, Native Hawaiian, and Pacific Islanders 4%

Other 8%

African Americans 12%

White Americans 75%

Source: U.S. Census Bureau.

American population being of "Hispanic or Latino" descent. An increasingly diverse population in the United States, accompanied by people traveling more and communicating over longer distances, has contributed to a more worldwide community, and a food industry that continues to "go global" (80). Within the boundaries of the United States alone, many foods once considered ethnic are now commonplace: pizza, tacos, beef teriyaki, pastas, and gyros. More recently arrived ethnic foods, such as Thai, Indian, Moroccan, and Vietnamese, are constantly being added to the mix to meet the escalating demands for meals providing more variety, stronger flavors, novel visual appeal, and less fat (85).

## Place of Birth

Birthplace influences the foods that a person will be exposed to, and helps to shape the dietary patterns that are often followed for life. Salsa varies in flavor, texture, and color depending on whether it was prepared in Mexico,

**Nutrigenomics** A field of study focusing on genetically-determined biochemical pathways linking specific dietary substances with health and disease.

**Culture** The ideas, customs, skills, and art of a group of people in a given period of civilization.

Guatemala, Puerto Rico, or Peru. Curry blends differ drastically depending on where in the world the recipe evolved. In Mexican cuisine, the same dish may taste different in different states.

## Geography and Climate

Not so long ago, geography and climate were the main determinants of what foods were available to be chosen. People ate foods that were grown close to where they lived and very rarely were presented with the possibility of eating those of a more exotic nature. For example, guava fruit grown in tropical regions was not even a consideration in an area such as Greenland. Now the wide distribution of formerly "local" foods throughout the world provides many people with an incredible variety of food choices.

## Cultural Influences on Manners

Culture not only influences what types of foods are chosen, but also the way they are consumed and the behavior surrounding their consumption. In some parts of India, for example, only the right hand is used for eating and manipulating utensils; the left hand is reserved for restroom duties. Foods may be served on banana leaves or wrapped in cornhusks. It may be eaten with chopsticks, as is the custom throughout Asia, or with spoons, forks, and knives as in Europe and the Americas. It is considered impolite in China not to provide your guest with a bountiful meal, so an unusually large number of food courses is served when guests are present.

# RELIGIOUS CRITERIA

Religion is another important influence on food choices. Religious beliefs affect the diets of many by declaring which foods are acceptable and unacceptable and by specifying preparation procedures. By designating certain foods for specific occasions and assigning symbolic value to some, religious principles wield further influence (28).

More than 85% of the American population claims to be Christian, and the bread (wafers) and wine served by many denominations during communion symbolize the body and blood of Christ. A traditional holiday meal with a turkey or ham as the main entrée is usually served at Christmas and/or Easter. The eggs used at Easter symbolize new life and were originally painted red to represent Christ's blood. Early Christians exchanged these eggs to recognize each other. Another food used among Christians is fish, which for many Catholics, until recently, was served on Fridays instead of meat.

Some of the food practices of Buddhists, Hindus, Seventh-Day Adventists, Mormons, Jews, and Muslims are explored in further detail below.

## Buddhism

There are over 100 million Buddhists in China and 300 million worldwide. Buddhists believe in *karuna,* which is compassion, and karma, a concept that implies that "good is rewarded with good; evil is rewarded with evil; and the rewarding of good and evil is only a matter of time" (50). Many Buddhists consider it uncompassionate to eat the flesh of another living creature, so vegetarianism is often followed; however, not all Buddhists are vegetarian. Whether Buddhists are vegetarian depends on their personal choice, the religious sect to which they belong, and the country where they live (27).

## Hinduism

Most of the 930 million followers of Hinduism live in India, and the Hindu American Foundation estimates that there are 2 million Hindus in the United States. Like Buddhism, Hinduism also promotes vegetarianism among some, but not all, of its followers (57). Buddhism actually originated in India before being disseminated to Asia and surrounding areas. The goal of both Hinduism and Buddhism is to reach "enlightenment" or "nirvana," in which the soul transcends "individual" ego and unites with the cosmos' higher state of consciousness (sometimes described as One, Supreme God). It is believed that souls who do not reach this state on earth are reincarnated. As a result,

some Hindus believe that the soul is all-important, uniting all beings as one, so it is against their beliefs to injure or kill a person or an animal. Thus strict Hindus reject poultry, eggs, and the flesh of any animal. The cow is not considered sacred among Hindus as is widely believed, but it is an animal so it is not slaughtered for food. However, dairy products from cattle are acceptable and even considered spiritually pure (27). Coconut and *ghee,* or clarified butter, are also accorded sacred status, but may be consumed after a fast. Some strict Hindus do not eat garlic, onions, mushrooms, turnips, lentils, or tomatoes.

## Seventh-Day Adventist Church

A vegetarian diet is recommended but not required for members of the Seventh-Day Adventist Church. About 40% of its members are vegetarians, the majority of them lacto-ovo-vegetarians, meaning that they allow milk and egg products (42). Consumption of between-meal snacks, hot spices, alcohol, tea, and coffee is discouraged (10).

## Church of Jesus Christ of Latter-Day Saints (Mormon Church)

The Church of Jesus Christ of Latter-Day Saints discourages the consumption of alcohol, coffee, and tea. Section 89:12 of the Doctrine and Covenants written in 1833 states, "Yea, flesh also of beasts and of the fowls of the air . . . they are to be used sparingly." Although not all Mormons follow these lifestyle recommendations, several studies suggest that they are healthier as a group compared to the average American. A significant number of Mormons live in Utah, and several studies have shown that the death rate attributed to specific diseases for Utah residents is 40% below the average U.S. rate because of lower rates of heart disease and cancer. Other factors possibly affecting the death rate are the discouragement of smoking and using illegal drugs, the recommendations of regular physical activity and proper sleep, and a positive religious outlook (77). The lower fat content of some vegetarian diets and the strength of Utah's health care system also cannot be ignored as possible contributing factors.

## Judaism

The *kashruth* is the list of dietary laws adhered to by orthodox Jews. Kosher dietary laws focus on three major issues (81):

1. Kosher animals allowed
2. Blood not allowed
3. Mixing of milk and meat not allowed

Foods are sorted into one of three groups: meat, dairy, or *pareve* (containing neither meat nor dairy) (58). Milk and meat cannot be prepared together or consumed in the same meal. In fact, separate sets of dishes and utensils are used to prepare and serve them, and a specified amount of time (1 to 6 hours) must pass between the consumption of milk and meat. Foods considered **kosher** include fruits, vegetables, grain products, and with some exceptions during Passover, tea, coffee, and dairy products from kosher animals as long as they are not eaten simultaneously with meat or fowl (82). Kosher animals are ruminants, such as cattle, sheep, and goats that have split hooves and chew their cud. Other meats that are considered kosher are chicken, turkey, goose, and certain ducks.

Orthodox Jews are not allowed to eat nonkosher foods such as carnivorous animals, birds of prey, pork (bacon, ham), fish without scales or fins (shark, eel, and shellfish such as shrimp, lobster, and crab), sturgeon, catfish, swordfish, underwater mammals, reptiles, or egg yolk containing any blood. These foods are considered unclean or *treif*. Even the meat from allowed animals is not considered kosher unless the animals have been slaughtered under the supervision of a rabbi or other authorized individual who ensures that the blood

**FIGURE 1-5** Examples of kosher and halal food symbols.

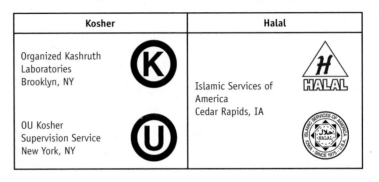

| Kosher | | Halal | |
|---|---|---|---|
| Organized Kashruth Laboratories Brooklyn, NY | Ⓚ | Islamic Services of America Cedar Rapids, IA | |
| OU Kosher Supervision Service New York, NY | Ⓤ | | |

has been properly removed. Foods that are tainted with blood, a substance considered by Jews to be synonymous with life, are forbidden (27).

Kosher foods are labeled with a logo such as those of the kosher-certifying agencies shown in Figure 1-5. Manufacturing facilities are inspected by a rabbi before a kosher certification can be given for a food (13). People other than Jews who often purchase kosher foods include Moslems, Seventh-Day Adventists, vegetarians, individuals with allergies (shellfish) or food intolerances (milk), and anyone who perceives kosher foods as being of higher quality (58).

Food figures prominently in the celebration of the major Jewish holidays. Rosh Hashanah, the Jewish New Year, is celebrated in part with a large meal. Yom Kippur, or the Day of Atonement, requires a day of fasting preceded by a bland evening meal the night before. Passover, which is an eight-day celebration marking the Exodus from Egypt, is commemorated in part by a meal whose components represent different aspects of the historic event. The Jews left Egypt without enough time for their bread to leaven (rise); to commemorate this event, leavened bread is prohibited during the Passover celebration. As a result, the five prohibited grains are wheat, rye, oats, barley, and spelt. The only grain allowed during Passover is unleavened bread made from wheat (matzo).

## Islam

Worldwide, there are over 1 billion Muslims compared to 13 million Jews (27). The Koran, the divine book of Islam, contains the **halal** dietary food laws recommended for Muslims that

describe halal (permitted) or haram (prohibited) foods (82). The five major areas addressed by the halal follow:

1. Animals not allowed
2. Blood not allowed
3. Improper slaughtering method
4. Carrion (decaying carcass) not allowed
5. Intoxicants not allowed

Many of the halal dietary food laws are similar to the food laws of Judaism, and like kosher foods, halal foods are identified with symbols (Figure 1-5). However, the most striking similarity is that the kosher meat consumed by Jews is permitted for Muslims because the animal has been slaughtered in a manner that allows the blood to be fully drained. Halal meat is also permitted and defined as any meat from approved animals processed according to Muslim guidelines. Most meat is allowed except pork, carnivorous animals with fangs (lions, wolves, tigers, dogs, etc.), birds with sharp claws (falcons, eagles, owls, vultures, etc.), land animals without ears (frogs, snakes, etc.), shark, and products containing pork or gelatin made from the horns or hooves of cattle (19). Alcohol and products containing alcohol in any form, including vanillin and wine vinegar, are forbidden. Stimulants such as tea and coffee are also discouraged.

Ramadan is a time of the year that significantly affects diet for Muslims. Islam teaches that the ninth month of the lunar calendar is the month in which the Prophet Muhammad received the revelation of the Muslim scripture, the Koran. This month, which depends on the sighting of the new moon, is a time of religious observances that include fasting from dawn to sunset.

**Kosher** From Hebrew, food that is "fit, right, proper" to be eaten according to Jewish dietary laws.

**Halal** An Arabic word meaning "permissible." Usually refers to permissible foods under Islamic law.

# PSYCHOLOGICAL AND SOCIOLOGICAL CRITERIA

Social and psychological factors strongly influence food habits. For most people, the knowledge that food is readily available provides a sense of security. The aim of every food company's advertising is to develop a sense of security among consumers about its products. A soft drink held in the hand of an athlete, the cereal touted by a child's favorite cartoon character, and diet foods offered by slim, vivacious spokespeople create positive associations in people's minds for these products and assure them of their quality. Social conscience and peer pressure sometimes influence food choices. One recent trend has seen consumers moving toward more environmentally sound purchases. At a buffet, the presence of other people may influence a person's choice of food and beverages.

Psychological needs intertwine with social factors when foods are used more for a display of hospitality or status than for mere nourishment. Caviar is just fish eggs, but is esteemed by many as a delicacy. Beer tastes terrible to most people when they try it for the first time, but the social surroundings and pressures may cause it to become an acquired taste. Several studies have shown that information influences expectations, and expectations mold choices (21), so it is no surprise that consumers report that television is their predominant source of information about nutrition (42%), followed by magazines (39%) and newspapers (19%) (9, 66).

Psychological factors also influence people's response to three relatively recent additions to the food market: genetically modified, organic, and natural foods.

## Bioengineering

Psychological and social factors are involved in the formation of public attitudes toward the **biotechnology** of foods, a term preferred over genetically engineered foods (31). The resulting **genetically modified organisms (GMOs)** are slowly gaining ground, but not everyone is knowledgeable about or accepting of the new foods (2).

Iconica/Getty Images

### History of Biotechnology

In the past, it took years to accomplish hybridization, or crossbreeding, by matching "the best to the best" in the plant, livestock, and fishery worlds to achieve the desired results. Cattle, corn, and even dogs were bred this way to yield desirable results. Dogs would not look the way they do without humans' modifying their genes through many years of selective breeding. Depending on the desired results, it could take decades or even centuries to develop a certain "look" and/or function in an animal or plant. Traditional ways of breeding to combine the genes of two species in order to obtain a specific trait were thus time consuming, cumbersome, and unpredictable (17).

Along came the age of food biotechnology (genetic engineering), which began in the early 1970s when DNA was isolated from a bacterium, duplicated, and inserted into another bacterium. The resulting DNA, known as recombinant deoxyribonucleic acid

(rDNA), allows researchers to transfer genetic material from one organism to another (51). Instead of crossbreeding for years, researchers can now identify the **genes** responsible for a desired trait and reorganize or insert them from

---

**Biotechnology** Previously called genetic engineering, this term describes the alteration of a gene in a bacterium, plant, or animal for the purpose of changing one or more of its characteristics.

**Genetically modified organisms (GMOs)** Plants, animals, or microorganisms that have had their genes altered through genetic engineering using the application of recombinant deoxyribonucleic acid (rDNA) technology.

**Gene** A unit of genetic information in the chromosome.

the cells of one bacterium, plant, or animal into the cells of other bacteria, plants, or animals (2). The goal of this process is to produce new species or improved versions of existing ones. The U.S. Department of Agriculture envisions food biotechnology being used to increase production potential, improve resistance to pests and disease, and develop more nutritious plant and animal products (83).

## Foods Created with Biotechnology

Food biotechnology has so far resulted in benefits that increase the food's resistance to the following (51):

- Pests (less pesticide required)
- Disease (lower crop losses)
- Harsh growing conditions (drought, salty soil, climate extremes)
- Transport damage (less bruising allows more produce to make it to market)
- Spoilage (longer shelf life)

Foods using biotechnology can be categorized as one of the following (2):

1. Actual food (e.g., corn)
2. Foods derived from or containing ingredients of actual food (e.g., cornmeal)
3. Foods containing single ingredients or additives from GMOs (e.g., amino acids, vitamins, colors)
4. Foods containing ingredients obtained from enzymes produced through GMO foods

What actual foods have been produced using biotechnology? Some examples of GMO foods include ripening-delayed fruits, grains with a higher protein content, potatoes that absorb less fat when fried, insect-resistant apples, and more than 50 other plant products. The first genetically engineered food to be approved by the Food and Drug Administration was Calgene's FlavrSavr™ tomato (Figure 1-6). Introduced to the consumers in the mid-1990s, this tomato was resistant to some common tomato crop diseases; it could also be left on the vine until fully ripened and flavorful, yet withstand the hardship of shipping without bruising (45). The FlavrSavr tomato softens at a slower rate because of food biotechnology that reduces the activity of an enzyme responsible for breaking down the cell wall during ripening (94). Conventional tomatoes are picked while they are green to prevent damage during transport; after shipping they are typically ripened by exposure to ethylene gas. The FlavrSavr™ was taken off the market after a few years as some consumers objected to genetic modification and others complained that it did not taste good. Many consumers still seek the succulence of a vine-ripened tomato.

After the FlavrSavr, other genetically engineered foods have been introduced including celery without strings, squash that is resistant to a common plant virus, presweetened melons, and tomatoes resistant to damage from both cold and hot temperatures. Genes have also been reorganized in strawberries to increase their natural sweetness. Possible genetically engineered foods of the future include cow's milk with some of the immune benefits of human milk (49), fruits containing higher amounts of vitamins A and C, fats and oils containing more omega-3 fatty acids (62), foods that generate proteins that could be used as oral vaccines (5), and soybeans providing a more complete source of protein (22). Currently, the three most popular GMOs among U.S. farmers are soybeans, corn, and cotton, with GM soybeans representing 92% of planted acres in 2008. These soybeans infiltrate the food supply because so many processed foods contain their oil (90).

**Clones**   Some people may think that cloned animals or plants are genetically engineered; however, that is not the case. The Biotechnology Industry Organization describes cloning as a "breeding method that does not manipulate the animal's genetic make-up nor change an animal's DNA; it is simply another form of sophisticated assisted reproduction. Cloning allows livestock breeders to create a genetic copy of an existing animal—essentially an identical twin. Animal clones are not 'biotech' or 'genetically engineered' animals; and their offspring are considered 'conventional' animals."

## Concerns About Food Biotechnology

Some consumers view genetic engineering as an invasion of nature's domain, and fear that scientists are treading on dangerous ground. Their concerns include allergies, gene contamination, and religious/cultural objections.

- *Allergies.* The concern most commonly expressed to the Food and Drug Administration by consumers was the possibility that the proteins produced by these new genes could cause allergic reactions. In one study, soy was infused with a gene from Brazil nuts, a known allergen, or allergy-causing substance (56). Some people participating in the experiment became ill, but this was a preliminary research study and the modified soy never reached the market (71). Researchers would be prudent to avoid food allergens in the process of genetically engineering foods because, even though protein food allergies affect only a small percentage of the population, they still exist and can cause problems (35).
- *Gene Contamination.* Another concern is that genetically engineered plants might "escape" into the wild, take over, and change the environment. Scientists assure us, however, that such plants are no more dangerous than traditionally bred crops. The greatest fear for some is that food biotechnology will lead to researchers using this type of biotechnology to try to "improve" the human race (56).
- *Religious/Cultural Concerns.* Some people, for religious or cultural reasons, do not want certain animal genes appearing in plant foods. For example, if swine genes were inserted into vegetables for some purpose, those vegetables would not be considered kosher. In one instance, a group of chefs refused to use a genetically engineered tomato when they found out that its disease resistance was obtained from a mouse gene. Vegetarians may object to a fish gene being placed in a tomato to provide resistance to freezing (45). Hawaiians objected when researchers tried to modify the gene sequence of their sacred taro plant, which is commonly used to make poi (a starchy paste made from the plant's corm, its thickened underground stem).

## Acceptance/Rejection of Genetically Engineered Foods

Despite the controversy over animal genes being inserted in plant foods,

**FIGURE 1-6**   Genetically engineering a tomato to soften more slowly.

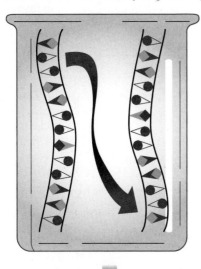

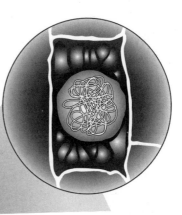

**1.** Ripe tomatoes contain an enzyme, polygalacturonase (PG), that causes them to soften. The PG gene that forms this enzyme is isolated and cloned. Scientists reverse the PG gene sequence and place it into bacteria.

**2.** The bacteria are grown in a petri dish filled with cut tomato leaves. The leaves' edges absorb the bacteria and the PG gene becomes part of the tomato plant cells' genetic material.

**3.** Tomato plants are regenerated from leaf cuttings containing the reversed PG gene.

**4.** The genetically engineered tomatoes can now ripen more fully on the vine prior to harvest, and be transported later with less concern for rotting due to softening.

SOURCE: Adapted from Ref 45.

the line between "plant genes" and "animal genes" is already blurred. Bacteria, plants, and animals share a large number of the more than 100,000 genes found in higher organisms. Nevertheless, research into people's attitudes about food repeatedly reveals that consumers are more likely to accept biotechnology conducted on plants rather than on animals or fish (46).

The Food and Drug Administration accepts genetically engineered foods as posing no risk to health or safety, and for this reason it does not require mandatory labeling, unless the foods contain new allergens, have modified nutritional profiles, or represent a new plant (17). The National Academy of Sciences has stated that genetic transfers between unrelated organisms do not pose hazards or risks different from those encountered by natural selection or crossbreeding. Currently, there is no evidence that transferring genes will convert a harmless organism into a hazardous one (51). People who wish to avoid GMOs can ensure that their foods are free of this type of genetic modification by purchasing organic foods. It's also a way to avoid foods produced with the use of antibiotics, hormones, or pesticides (36, 63).

## Organic Foods

Some people prefer to select *organic foods,* a term that had no official definition until 2002, following the Organic Foods Production Act of 1990 (31). This act created the United States National Organic Standards Board (NOSB), which in turn makes organic definition recommendations to the National Organic Program (located in the USDA's Agricultural Marketing Service). Terms commonly used in the marketplace that do not have official definitions or certification by the government include "free-range," "hormone-free," "natural," "organically produced," "pesticide free," "raised without antibiotics," or even "certified organic." Prior to 2002, products were often labeled "organic" by growers without any real certification, or they were certified by private agencies according to a patchwork of regulations that varied from state to state. Now, for a food to be labeled "organic" it

has to fit one of the four official definitions listed by the U.S. Department of Agriculture (USDA) and shown in Table 1-2. The USDA's definition of what is organic goes beyond just describing foods that are not sprayed with chemicals. The word *organic* now refers to food products that have been produced without most synthetic pesticides and fertilizers (including sewage sludge), crops that have not been genetically modified (no GMOs or bioengineering), food products not exposed to irradiation, and livestock produced without antibiotics or hormones, raised on 100% organic feed, and grazing on pasture at least four months of the year, and 30% of their feed must come from grazing (86).

A few concerns exist for organic foods. One is the cost to consumers, as they are typically more expensive than their conventionally grown counterparts. Another is that some proponents mislead the public with the "fear factor" of suggesting that only organic foods and beverages are healthy or safe. On the side of organic advocates is the concern that in 2007 the USDA allowed

38 nonorganic ingredients into foods that were 95% or less organic. This list includes whey protein concentrate, gelatin, 19 food colorings, two starches, unmodified rice starch, sweet potato starch, konjac flour, intestinal casings for hot dogs, fish oil, Wakame seaweed, fructoligosaccharides, and some flavor-providing items such as chipotle chili peppers, celery powder, dill weed oil, chia, frozen lemon grass, Turkish bay leaves, unbleached orange shellac, frozen galangal (citrus flavor from a rhizome), and hops for beer (86).

### Organic Certification

The government agency certifying that a food is organic is the USDA. A USDA certifier inspects the facilities where the food is grown, determines if organic standards were met, and then labels such food products with the organic seal shown in Figure 1-7. Both 100% and 95% organic products may use the USDA organic seal, while those made with less than 95% organic ingredients are limited in what they may place on the label. USDA agents determine if food is organic

| **TABLE 1-2** The U.S. Government's Criteria for Defining Organic Food Products | | |
|---|---|---|
| **What does the label mean? The USDA Organic seal tells you a product is at least 95% organic.** | | |
| **Organic Term** | **Definition** | **Labeling Allowed** |
| 100% Organic | All ingredients of the finished product are certified 100% organic. | USDA ORGANIC |
| Organic | 95% of finished product ingredients meet organic criteria. | USDA ORGANIC |
| Made with Organic Ingredients | 70% of finished product ingredients meet organic criteria. | "Made with Organic Ingredients" |
| Contains Organic Ingredients | Less than 70% of finished product ingredients meet criteria. | May only list organic ingredients on the information panel. |

**FIGURE 1-7** USDA's official organic seal.

USDA

by following the guidelines set by the USDA's Agricultural Marketing Service (AMS), published as the National Organic Program (NOP) in the Federal Register (December 21, 2000). Producers selling less than $5,000 in organic products are exempt from certification, but they must still follow the standards (86). Only those food products that were organically grown or processed and certified by an accredited USDA organic–certifying agent can carry the organic seal. Violators making false claims can be fined $10,000 per offense.

## Natural Foods

The word "natural" on a food label may or may not mean anything about how the food was produced or what it contains. This is because no official, United States FDA definition exists for "natural" foods except for meat, poultry, and eggs overseen by the USDA. *Natural* is defined by the USDA for its products by answering two questions: (1) Does the product contain an artificial ingredient, a chemical preservative, or any other synthetic or artificial ingredient? (2) Are the product and its ingredients only minimally processed? The product is natural if the answer to the first question is no, and the answer to the second question is yes (32). The lack of an official definition for "natural" for all other foods and beverages can lead to inconsistent claims and consumer confusion.

# BUDGETARY CRITERIA

Cost is a very important limiting factor in food purchasing. In fact, debit cards obtained through the U.S. Department of Agriculture's Supplemental Nutrition Assistance Program (SNAP; formerly the Food Stamp Program) are limited by the "Thrifty Food Plan" that calculates what an average family needs to spend on food (89).

Cost helps determine the types of foods and brands that are bought and the frequency of restaurant patronage. People feeling financial strain may still eat beef, but they may choose ground beef over prime rib. "Can I afford this?" is a question that also applies to time, which can make convenience foods effectively more economical, even if the dollar price is higher. Budgeting and time management are discussed in greater detail in Chapter 6.

# PICTORIAL SUMMARY / 1: Food Selection

People choose foods that satisfy their senses of sight, smell, taste, touch, and hearing, their nutrient needs, cultural and religious values, psychological and social influences, and budget.

## FOOD SELECTION CRITERIA

**Sensory Criteria.** When most people choose a particular food, they evaluate it using the sensory reactions illustrated below rather than by considering its nutritional content.

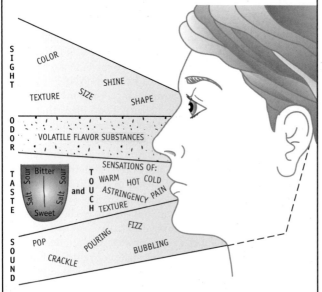

**SIGHT:** COLOR, SHINE, TEXTURE, SIZE, SHAPE

**ODOR:** VOLATILE FLAVOR SUBSTANCES

**TASTE:** Sour, Bitter, Salt, Salt, Sweet

**TOUCH** and SENSATIONS OF: WARM, HOT, COLD, ASTRINGENCY, PAIN, TEXTURE

**SOUND:** POP, CRACKLE, POURING, FIZZ, BUBBLING

 **Nutritional Criteria.** Over the last several decades, emerging awareness of health and nutrition has resulted in six out of ten consumers making a major change in their diets. Guidelines that reinforce an emphasis on better health through nutrition include the U.S. Government's Dietary Guidelines and the MyPyramid food guide. Portion control starts with understanding average daily caloric intakes. Although no official U.S. regulatory definition for functional food exists, the ADA defines them as conventional foods, modified foods (fortified, enriched, or enhanced), medical foods, and foods for special dietary use.

## MyPyramid

The colors of the pyramid illustrate variety: each color represents one of the five food groups, plus one for oils. Different band widths suggest the proportional contribution of each food group to a healthy diet.

A person climbing steps reminds consumers to be physically active each day.

The narrow slivers of color at the top imply moderation in foods rich in solid fats and added sugars.

The broad bases at the bottom represent nutrient-dense foods that should make up the bulk of the diet.

Greater intakes of grains, vegetables, fruit, and milk are encouraged by the broad bases of orange, green, red, and blue.

GRAINS    VEGETABLES    FRUITS    MILK    MEAT & BEANS

 **Cultural and Religious Criteria.** An increasingly diverse population, with greater access to travel and expanded global communication, has resulted in a huge increase in the variety of foods that are available in the United States today. Familiar taste preferences acquired in childhood as well as religious tenets affect many people's food habits throughout their lives.

**Psychological and Sociological Criteria.** Advertising, social conscience, and peer pressure can all play a part in an individual's food choices. The controversies surrounding genetically engineered, organic, and natural foods are examples of how food products can be affected by these criteria.

 **Budgetary Criteria.** Cost helps determine the types of foods and brands that are bought and the frequency of restaurant patronage. A shortage of time for food preparation or eating out can result in greater use of convenience foods and "fast foods," even if they are often more expensive and less nutritious.

# CHAPTER REVIEW AND EXAM PREP

## Multiple Choice*

1. The word *olfactory* is most closely related to which of the following senses?
   a. taste
   b. smell
   c. touch
   d. sight

2. Total daily calorie (kcal) needs for adults decrease by _____ calories (kcals) for every decade in people after age 40.
   a. 100
   b. 200
   c. 300
   d. 400

3. Which of the following religions encourage(s) a vegetarian diet?
   a. Buddhism
   b. Hinduism
   c. Seventh-Day Adventist
   d. all of these

4. Identify the correct statement about genetically engineered foods.
   a. Genes are programmed by sequencing the amino acids.
   b. Food can be genetically engineered to delay ripening.
   c. In the United States, all genetically engineered foods must be labeled.
   d. Not a single genetically engineered food has been approved by the FDA.

5. Which of the following functional food examples is categorized as a medical food?
   a. Tomatoes rich in lycopene
   b. Gluten-free foods
   c. Folate-enriched breads
   d. Phenylketonuria (PKU) formulas free of phenylalanine

6. How many calories (kcal) equal a pound of body weight?
   a. 500
   b. 1000
   c. 2500
   d. 3500

7. A produce grower may place the USDA Organic label on a product if _____ or more of the ingredients are organic.
   a. 50%
   b. 75%
   c. 85%
   d. 95%

## Short Answer/Essay

1. List the five taste stimuli and the proposed mechanism of taste for each.
2. Why is the odor of just-baked bread more intense than the odor of cold foods such as ice cream?
3. Give two examples of taste interactions.
4. How does taste differ from flavor?
5. Obesity is a rising problem. Discuss the basics of the energy balance equation.
6. List four categories of functional foods as defined by the American Dietetic Association.
7. Discuss three examples of cultural influences on food intake.
8. Discuss the possible influences that religions such as Buddhism, Judaism, and Islam may have on food intake.
9. Describe the process of producing a genetically engineered food. Discuss the pros and cons of this process.
10. Describe the four categories of organic food and the labeling allowed for each category.

*See p. AK-1 for answers to multiple choice questions.

# REFERENCES

1. *2008 Food & Health Survey: Consumer Attitudes toward food, nutrition & health.* International Food Information Council Foundation. www.ific.org/research/foodandhealthsurvey.cfm. Accessed 4/22/09.
2. American Dietetic Association. Position of the American Association: Agricultural and Food Biotechnology. *Journal of the American Dietetic Association* 106(2):285–293, 2006.
3. Andreyeva T, R Sturm, and JS Ringel. Moderate and severe obesity have large differences in health care costs. *Obesity Research* 12(12):1936–1943, 2004.
4. Arai S, et al. Recent trends in functional food science and the industry in Japan. *Bioscience, Biotechnology, and Biochemistry* 66:2017–2029, 2002.
5. Armtzen DJ. Edible vaccines. *Public Health Reports* 112:190–197, 1997.
6. Bae J-M, AJ Lee, and G Guyatt. Citrus fruit intake and stomach cancer risk: A quantitative systematic review. *Gastric Cancer* 11:23–32, 2008.

7. Bialostosky K, and ST St. Jeor. The 1995 Dietary Guidelines for Americans. *Nutrition Today* 31(1):6–11, 1996.

8. Bodyfelt FW, J Tobias, and GM Trout. *The Sensory Evaluation of Dairy Products.* Van Nostrand Reinhold, 1988.

9. Borra ST, R Earl, and EH Hogan. Paucity of nutrition and food safety "news you can use" reveals opportunity for dietetics practitioners. *Journal of the American Dietetic Association* 98:190–193, 1998.

10. Bosley GC, and MG Hardinge. Seventh-Day Adventists: Dietary standards and concerns. *Food Technology* 46(10):112–113, 1992.

11. Breslin PAS, and AC Spector. Mammalian taste perception. *Current Biology* 18(4):R148–R155, 2008.

12. Camire ME, et al. IFT research needs report: Diet and health research needs. *Food Technology* 55(5):189–191, 2001.

13. Caranfa M. Kosher goes mainstream: Consumers outside the Jewish faith increase demand in supermarkets and restaurants. *ADA Times* 5(4):9, 2008.

14. Centers for Disease Control and Prevention. U.S. Obesity Trends 1985–2007. www.cdc.gov/nccdphp/dnpa/obesity/trend/maps/. Accessed 4/7/09.

15. Centers for Disease Control and Prevention. National Health and Nutrition Examination Survey. Intake of calories and selected nutrients for the United States population, 1999–2000. www.cdc.gov/nchs/data/nhanes/databriefs/calories.pdf. Accessed 4/7/09.

16. Chanmugam P, JF Guthie, S Cecilio, JF Morton, PP Basiotis, and R Anand. Did fat intake in the United States really decline between 1989–1991 and 1994–1996? *Journal of the American Dietetic Association* 103:867–872, 2003.

17. Chapman N. Developing new foods through biotechnology. *Prepared Foods* 166(2):34, 1997.

18. Charley H. *Food Science.* Prentice-Hall, 1995.

19. Chaudry MM. Islamic food laws: Philosophical basis and practical implications. *Food Technology* 46(10):92–93, 1992.

20. DeBusk R. Diet-related disease, nutritional genomics, and food and nutrition professionals. *Journal of the American Dietetic Association* 109(3):410–413, 2009.

21. Deliza R, HJH MacFie, and D Hedderley. Information affects consumer assessment of sweet and bitter solutions. *Journal of Food Science* 61(5):1080–1084, 1996.

22. De Lumen BO, DC Krenz, and MJ Revilleza. Molecular strategies to improve the protein quality of legumes. *Food Technology* 51(5):67–70, 1997.

23. De Roos KB. How lipids influence food flavor. *Food Technology* 51(5):60–62, 1997.

24. Dietz WH, DE Benken, and AS Hunter. Public health law and the prevention and control of obesity. *Milbank Q* 87(1):215–227, 2009.

25. Drewnowski A. Why do we like fat? *Journal of the American Dietetic Association* 97(7):S58–S62, 1997.

26. Duxbury D. Sensory evaluation provides value. *Food Technology* 59(5):68, 2005.

27. Eliasi JR, and JT Dwyer. Kosher and halal: Religious observances affecting dietary intakes. *Journal of the American Dietetic Association* 101(7):911–913, 2002.

28. Ensminger AH, et al. *Foods and Nutrition Encyclopedia.* CRC Press, 1994.

29. Faridi Z, VY Njike, S Dutta, S Ali, and DL Katz. Acute dark chocolate and cocoa ingestion and endothelial function: A randomized controlled crossover trial. *American Journal of Clinical Nutrition* 88:58–63, 2008.

30. Ferguson LR. Nutrigenomics approaches to functional foods. *Journal of the American Dietetic Association* 109(3):452–458, 2009.

31. Finucane ML, and JL Holup. Psychosocial and cultural factors affecting the perceived risk of genetically modified food: An overview of the literature. *Social Science and Medicine* 60(7):1603–1612, 2005.

32. Fisher C, and R Carvajal. What is natural? *Food Technology* 62(11):24–31, 2008.

33. Food and Drug Administration. Orphan Drug Act (as amended). www.fda.gov/orphan/oda.htm. Accessed 1/9/09.

34. Food guide pyramid replaces the Basic 4 Circle. *Food Technology* 46(7):64–67, 1992.

35. Fuchs RL, and JD Astwood. Allergenicity assessment of foods derived from genetically modified plants. *Food Technology* 50(2):83–88, 1996.

36. Giese J. Pesticide residue analysis in foods. *Food Technology* 56(9):93, 2002.

37. Godshall MA. How carbohydrates influence food flavor. *Food Technology* 51(1):63–67, 1997.

38. Goff S. Preferred flavors. *Nutraceuticals World* 10(7):46–48, 2007.

39. Green BG, M Alvarez–Reeves, P George, and C Akirav. Chemesthesis and taste: Evidence of independent processing of sensation intensity. *Physiology and Behavior* 86(4):526–537, 2005.

40. Guh DP, W Zhang, N Bansback, Z Amarsi, CL Birmingham, and AH Anis. The incidence of comorbidities related to obesity and overweight: A systematic review and meta-analysis. *BMC Public Health* 9(1):88, 2009.

41. Hamilton EMN, and SAS Gropper. *The Biochemistry of Nutrition.* West, 1987.

42. Harding TB, and LR Davis. Organic foods: Manufacturing and marketing. *Food Technology* 59(1):41–46, 2005.

43. Hasler CM, and AC Brown. Position of the American Dietetic Association: Functional foods. *Journal of the American Dietetic Association* 109(4):735–746, 2009.

44. Health Canada. Policy paper: Nutraceuticals/functional foods and health claims on foods. Health Canada website. www.hc-sc.gc.ca/fn-an/label-etiquet/claims-reclam/nutra-funct_foods-nutra-fonct_aliment-eng.php. Accessed 1/9/09.

45. Henkel J. Genetic engineering: Fast-forwarding to future foods. *FDA Consumer* 29(3):6–11, 1995.

46. Hoban TJ. How Japanese consumers view biotechnology. *Food Technology* 96(5):85–88, 1996.

47. Hollingsworth P. Growing nutra-ceuticals. *Food Technology* 55(9):22, 2001.

48. *How Are Americans Making Food Choices?* 1994 Update. The American Dietetic Association and International Food Information Council, 1994.

49. How Food Technology covered biotechnology over the years. *Food Technology* 51(4):68, 1997.

50. Huang Y, and CYW Ang. Vegetarian foods for Chinese Buddhists. *Food Technology* 46(10):105–108, 1992.

51. IFT Backgrounder: Genetically modified organisms (GMOs). *Food Technology* 54(1):42–45, 2000.

52. Jordan T. Lite'n Up program. *Journal of the Kentucky Medical Association* 106(3):114–117, 2008.

53. Katz F. The changing role of water binding. *Food Technology* 51(10):64–66, 1997.

54. Kavanaugh CJ, PR Trumbo, and KC Ellwood. The U.S. Food and Drug Administration's evidence-based review for qualified health claims: Tomatoes, lycopene, and cancer. *Journal of the National Cancer Institute* 99:1074–1085, 2007.

55. Kemp SE, and GK Beauchamp. Flavor modification by sodium chloride and monosodium glutamate. *Journal of Food Science* 59(3):682–686, 1994.

56. Kendall P. Food biotechnology: Boon or threat? *Journal of Nutrition Education* 29:112–115, 1997.

57. Kilara A, and KK Iya. Food and dietary habits of the Hindu. *Food Technology* 46(10):94–102, 1992.

58. Kosher certification a marketing plus. *Food Technology* 51(4):67, 1997.

59. Kris-Etherton PM, FB Hu, E Ros, and J Sabate. The role of tree nuts and peanuts in the prevention of coronary heart disease: Multiple potential mechanisms. *Journal of Nutrition* 38(suppl):1746S–1751S, 2008.

60. Leitzmann C. Vegetarian diets: What are the advantages? *Forum Nutrition* 57:147–156, 2005.

61. Leland JV. Flavor interactions: The greater whole. *Food Technology* 51(1):75–80, 1997.

62. Liu K, and EA Brown. Enhancing vegetable oil quality through plant breeding and genetic engineering. *Food Technology* 50(11):67–71, 1996.

63. Lu C, K Toepel, R Irish, RA Fenske, DB Barr, and R Bravo. Organic diets significantly lower children's dietary exposure to organophosphorus pesticides. *Environmental Health Perspectives* 114(2):260–263, 2006.

64. Maletto P. Functional food formulation. *Food Technology* 11(6):38–44, 2008.

65. Marcus JB. Culinary applications of umami. *Food Technology* 59(5):24–30, 2005.

66. McMahon KE. Consumer nutrition and food safety trends 1996. *Nutrition Today* 31(1):19–23, 1996.

67. Moriarty RM, R Naithani, and B Surve. Orgaosulfur compounds in cancer chemoprevention. *Mini Reviews in Medicinal Chemistry* 7:827–838, 2007.

68. Müller M, and S Kersten. Nutrigenomics: Goals and perspectives. *Nature Reviews Genetics* 4:315–322, 2003.

69. Nagodawithana T. Flavor enhancers: Their probable mode of action. *Food Technology* 48(4):79–85, 1994.

70. Navarro-Allende A, N Khataan, and A El-Sohemy. Impact of genetic and environmental determinants of taste with food preferences in older adults. *Journal of Nutrition for the Elderly* 27(3–4):267–276, 2008.

71. Nettleton JA. Warning: This food has been genetically engineered. *Food Technology* 51(3):20, 1997.

72. Nowack R. Cranberry juice—A well-characterized folk-remedy against bacterial urinary tract infection. *Wiener medizinische Wochenschrift* 157:325–330, 2007.

73. Onyenekwe PC, and GH Ogbadu. Radiation sterilization of red chili pepper (*Capsicum frutescens*). *Journal of Food Biochemistry* 19:121–137, 1995.

74. Painter J, JH Rah, and YK Lee. Comparison of international food guide pictorial representations. *Journal of the American Dietetic Association* 102(4):483–489, 2002.

75. Peregrin T. The new frontier of nutrition science: Nutrigenomics. *Journal of the American Dietetic Association* 101(11):1306, 2001.

76. Pew Initiative on Food and Biotechnology. Application of biotechnology for functional foods. Pew Charitable Trusts website. http://pewagbiotech.org/research/functionalfoods/.

77. Pike OA. The Church of Jesus Christ of Latter-Day Saints: Dietary practices and health. *Food Technology* 46(10):118–121, 1992.

78. Porzio MA. Flavor delivery and product development. *Food Technology* 61(1):22–29, 2007.

79. Quigley EM. Bacteria: A new player in gastrointestinal motility disorders: Infections, bacterial overgrowth, and probiotics. *Gastroenterology Clinics of North America* 6:735–748, 2007.

80. Raghavan S. Developing ethnic foods and ethnic flair with spices. *Food Technology* 58(8):35–42, 2004.

81. Regenstein JM, MM Chaudry, and CE Regenstein. The kosher and halal food laws. *Comprehensive Reviews in Food Science and Food Safety* 2:111–127, 2003.

82. Regenstein JM, and CE Regenstein. The kosher food market in the 1990s—A legal view. *Food Technology* 46(10):122–124, 1992.

83. Saleh-Lakha S, and BR Glick. Is the battle over genetically modified foods finally over? *Biotechnology Advances* 23(2):93–96, 2005.

84. Schwartz NE, and ST Borra. What do consumers really think about dietary fat? *Journal of the American Dietetic Association* 97(suppl):S73–S75, 1997.

85. Sloan AE. Ethnic foods in the decade ahead. *Food Technology* 55(10):18, 2001.

86. Spano M. Organics in overdrive. The explosion of natural food products. *Today's Dietitian* 9(10):66–69, 2007.

87. Stillman JA. Color influences flavor identification in fruit-flavored beverages. *Journal of Food Science* 58(4):810–812, 1993.

88. Turning sugar into sand and sweet citric acid. *Prepared Foods* 163(11):63–66, 1994.

89. U.S. Department of Agriculture. Center for Nutrition, Policy, and Promotion. *Official USDA food plans: Cost of food at home at four levels. U.S. averages.* Available: www.cnpp. usda.gov/usdafoodplanscostoffood. htm. Accessed 4/7/09.

90. U.S. Department of Agriculture. *Adoption of genetically engineered crops in the U.S.* www.ers.usda.gov/ Data/BiotechCrops. Accessed 4/22/09.

91. U.S. Department of Agriculture. Human Nutrition Information Service. www.MyPyramid.gov. Washington, D.C.: Government Printing Office, 2005.

92. U.S. Department of Agriculture and U.S. Department of Health and Human Services. *Dietary Guidelines for Americans,* 6th ed. Washington, D.C.: US Government Printing Office, 2005.

93. Vartanian LR, CP Herman, and B Wansink. Are we aware of the external factors that influence our food intake? *Health Psychol* 27(5):533–538, 2008.

94. Wilkinson JQ. Biotech plants: From lab bench to supermarket shelf. *Food Technology* 51(12):37–42, 1997.

95. Yamaguchi P. *Japan's Nutraceuticals Today: Functional Foods Japan 2006.* NPIcenter website. www.npicenter. com/anm/templates/newsATemp. aspx?articleid=15160&zoneid=45. Accessed 4/7/09.

# WEBSITES

Find more information on the USDA's Dietary Guidelines:

**www.health.gov/DietaryGuidelines/**

Find more information on the USDA's Food Pyramid:

**www.mypyramid.gov**

Canada's Food Guide at:

**http://www.hc-sc.gc.ca/fn-an/food-guide-aliment/index-eng.php**

Find details about the USDA's Thrifty Food Plan:

**www.cnpp.usda.gov/ usdafoodplanscostoffood.htm**

Calculate your body mass index (BMI) at:

**www.cdc.gov/healthyweight/assessing/ bmi/adult_bmi/english_bmi_ calculator/bmi_calculator.html**

Learn about the statistics on different ethnic groups in the United States and your state:

**www.census.gov**

Find the latest obesity trends state by state from the Centers for Disease Control and Prevention (CDC) at:

**www.cdc.gov/nccdphp/dnpa/obesity/ trend/maps/**

Discover more about the National Organic Program (NOP) from the USDA's website on the subject:

**www.ams.usda.gov/nop**

Attend free flavor classes (travel and lodging not included) from FONA International at:

**www.fona.com/FlavorUniversity.html**

Find more information about food and nutrition from the USDA's Food and Nutrition Information Center (FNIC) located at the National Agricultural Library (NAL):

**www.nal.usda.gov/fnic**

Find information about complementary and alternative medicine from the National Institutes of Health:

**http://nccam.nih.gov/**

Learn more about herbal products from the Memorial Sloan-Kettering Cancer Center:

**www.mskcc.org/mskcc/html/11570.cfm**

Explore the USDA site on Plants and Crops: Biotechnology, Genetics, and Breeding:

**http://riley.nal.usda.gov/nal_display/ index.php?info_center=8&tax_ level=2&tax_subject=7&topic_ id=1058&&placement_default=0**

Note: Website page links frequently change, so if a particular URL does not bring you to the desired website, just enter the key words from the description into a search engine.

# 2 Food Evaluation

The food industry uses an array of testing methods to measure the sensory factors related to food selection and to evaluate food quality. These tests are conducted for research and development (R&D), product improvement, sales and marketing, quality assurance, nutrient content analysis for labeling requirements (Nutrition Facts), and detecting contamination or adulteration (7). Food evaluation is accomplished using both **sensory (subjective)** and **objective tests** (5). The specific types of tests and tools that the food industry uses to evaluate the palatability of food among consumers are the focus of this chapter.

## SENSORY (SUBJECTIVE) EVALUATION

Sensory evaluation (or analysis) is the scientific discipline of measuring the responses of people to food products as perceived by their senses of sight, taste, touch, smell, and hearing (2). This type of testing is termed *subjective* because it relies on the opinions of selected individuals.

Although certain machines are designed to replicate the ability to perceive the five senses, very few succeed in matching a human being. As a result, sensory evaluation tests are often used by large food companies in their research and development departments for the purposes of evaluating potentially new and/or established consumer products. Human panels are required to evaluate the products through various types of established scientific sensory tests. The results are then statistically analyzed to determine consumer preference and/or acceptability.

## Two Types of Sensory Testing

There are two basic types of sensory tests: analytical (effective) and affective (1). Analytical tests are more objective and based on discernible differences, whereas affective tests are more subjective and based on individual preferences (Figure 2-1). In both types of testing,

**Sensory (subjective) tests** Evaluations of food quality based on sensory characteristics and personal preferences as perceived by the five senses.

**Objective tests** Evaluations of food quality that rely on numbers generated by laboratory instruments that are used to quantify the physical and chemical differences in foods.

**FIGURE 2-1**    Summary of subjective tests for food evaluation.

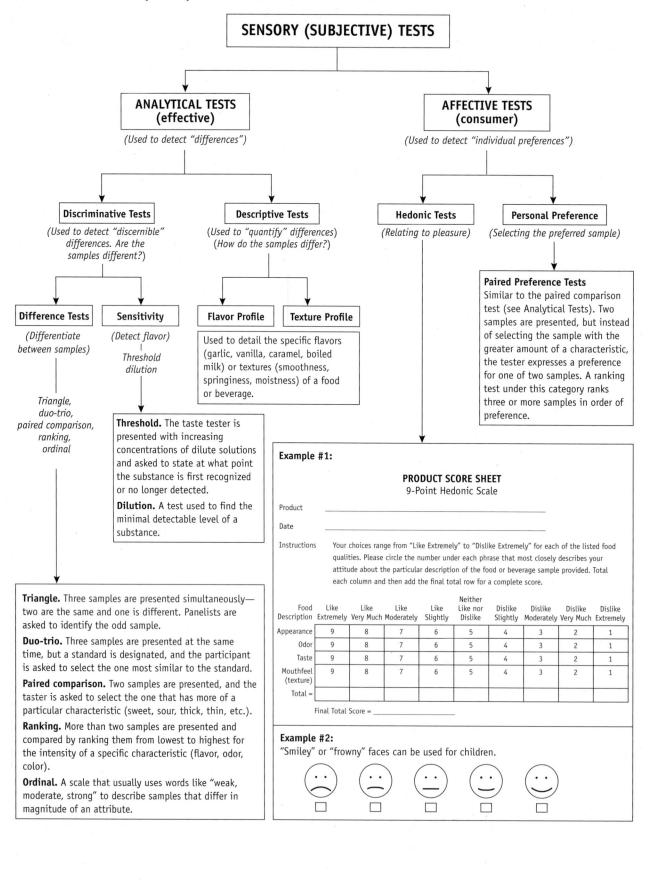

food samples are presented to taste panel participants, who evaluate the foods according to specific standards for appearance, odor, taste, texture, and sound.

### Analytical (Effective) Tests

The more objective analytical tests are usually conducted by a trained panel that evaluates food products through either discriminative (are the samples different?) or descriptive tests (how much do the samples differ?). The most common difference tests are the triangle and duo-trio tests in which the person compares three samples and has to determine if a difference exists (11). Descriptive tests rely on a trained panel to document a product's sensory characteristics (10).

### Affective Tests

Whether or not a person prefers a certain aspect of a food is the focus of affective or consumer testing. Because anyone can have an opinion, these types of tests are usually given to untrained consumers. The test instruments range from simple questions (which of the two samples do you prefer?) to complex 9-point hedonic product score sheets evaluating one or more factors of a food on a scale from "like extremely" to "dislike extremely" (Figure 2-1).

## Taste Panels

The individuals on a taste panel can range from randomly selected members of the population to experts who are highly trained in tasting a particular food or beverage (Figure 2-2). Vintners and brewers rely on the latter types of skilled tasters to evaluate the proper timing for each step in the process of making wine or beer (13). The ability to detect slight differences in specific foods is a sought-after trait, prized so much that the taste buds of one gourmet ice cream taste expert are insured for $1 million. General taste panels usually consist of at least five people who meet the following criteria: they are free of colds, chew no gum immediately before testing, have not ingested any other food for at least 1 hour before testing, are nonsmokers, are not color blind, and have no strong likes or dislikes for the food to be tested. An equal distribution in gender is preferred, because

**FIGURE 2-2** Taste test panel at NASA.

NASA

women can usually detect sweetness better than men can. Age distribution of the panel is also considered, because it may affect test results.

## Sample Preparation

The environment in which the taste panel evaluates foods or beverages is also carefully controlled (8). Panelists may be seated at tables, cubicles (Figure 2-3), or booths, and the food is presented in a uniform fashion. Food samples must be of the same size (enough for two bites), from the same portion of the food (middle vs. outside), equally fresh, at the same temperature, and presented in containers or plates that are of the same size, shape, and color. White or clear containers are usually chosen so as not to influence panelists' perceptions of the food's color. Care is taken that the lighting in the room is uniform and that the ambient temperature is comfortable

**FIGURE 2-3** Testers evaluating samples in private booths that minimize outside influences.

Al Behrman/AP Photo

and the surroundings quiet and odor-free. Mid-mornings or mid-afternoons are considered the best times for sampling, because at these times people are not usually overly hungry or full. Samples are randomly coded and are kept to a reasonable number to avoid "taste fatigue." Room-temperature water or plain bread is made available for panelists to eat between samples to prevent carryover tastes, and at least a 30-second rest period is scheduled between samples. Paper towels or napkins are provided, and, because swallowing the food or beverage influences the taste of subsequent samples, small containers into which samples may be spit are provided.

# OBJECTIVE EVALUATION

In objective evaluations, laboratory instruments instead of humans are used to measure the characteristics of foods quantitatively. The two major types of objective evaluation tests, physical and chemical, sometimes attempt to mimic the five senses, and serve as the basis of most objective food testing. Despite the benefits of objective tests, they cannot substitute for sensory testing by real human beings, who ultimately decide which foods and beverages they will select. However, human senses cannot detect the quantitative measures determined by physical and chemical tests. These tests analyze for the presence of potentially harmful bacteria, yeast, and mold; create standards for maintaining quality control; and identify almost any chemical in foods used for nutrition fact labeling, moisture content analysis, and detecting the presence of allergens or toxins, to name just a few examples.

## Physical Tests

Physical tests measure certain observable aspects of food such as size, shape, weight, **volume, density,** moisture, texture, and **viscosity** (Chemist's Corner 2-1) (3). Table 2-1 lists some of the laboratory instruments used to measure the various physical aspects of foods (Chemist's Corner 2-2). Figure 2-4 shows an example of one such instrument. These and other instruments are usually purchased by companies that need to ensure that their products meet certain quality control guidelines.

**Volume**  A measurement of three-dimensional space that is often used to measure liquids.

**Density**  The concentration of matter measured by the amount of mass per unit volume. Objects with a higher density weigh more for their size.

**Viscosity**  The resistance of a fluid to flowing freely, caused by the friction of its molecules against a surface.

| TABLE 2-1 | Selected Physical Tests for Food Evaluation |
|---|---|
| **Visual Evaluation** | |
| Microscope | Used to observe microorganisms as well as starch granules, the grain in meats, the crystals of sugar and salt, the fiber in fruits and vegetables, and for any texture changes in processed foods. |
| Spectrophotometer | Measures color by detecting the amount and wavelength of light transmitted through a solution. Spectroscopy is based on the principle that the molecules in foods and beverages will absorb light at different wavelengths on the spectrum. The amount of absorption parallels the amount of substance found in the sample. Spectroscopy can be used to determine the amount of caffeine in coffee or the concentration of riboflavin (vitamin $B_2$) in milk. |
| **Weight/Volume Measurements** | |
| Weight | Weight is measured in pounds/ounces or milligrams/grams/kilograms. |
| Volume | Volume quantifies the area occupied by a mass, whereas density is the measure of mass (weight) in a given volume. Specific density relates a substance's density to an equal amount of water. |
| **Texture Measurements** | |
| Penetrometer | Simulates teeth biting into a food to measure its tenderness. |
| Warner-Bratzler Shear | Evaluates meat and baked product tenderness by measuring the force required to cut through a cylindrical sample. |
| Shortometer | Measures tenderness by determining the resistance of baked goods, such as cookies, pastries, and crackers, to breakage. Puncture testing evaluates the firmness of fruit or vegetable tissue. |
| **Viscosity Measurements** | |
| Line-spread test | Measures the consistency of batters and other viscous foods. Food is placed in a hollow cylinder in the middle of the spread sheet; the cylinder is then lifted, allowing the food to spread, and the spreading distance is measured in centimeters. |
| Viscometer (or viscosimeter) | Measures the viscosity of food such as pudding, sour cream, salad dressing, sauces, cream fillings, cake batters, and ketchup. |
| **Concentration Measurements** | |
| Polarimeter | Measures the concentration of various organic compounds, especially sugars, in solution by determining the angle (refractive index) of polarized light passed through the solution. Refractometers are commonly used to measure sugar concentrations in soft drinks. The Brix/acid ratio is used to measure the palatability of fruit juices that depends on the delicate balance between sweetness (sugars) and tartness (acid). This ratio is obtained by measuring the degrees Brix (determined by the use of a refractometer) divided by the total acid concentration (determined by acid titration) (9). |
| Atomic absorption | Used to measure mineral content. |

## Chemical Tests

The number of chemical tests available for use on foods is almost limitless, but Table 2-2 lists some of these tests. Many are based on the work of the Association of Official Analytical Chemists (AOAC) International, which publishes a book on chemical tests, including those for determining various nutrient and nonnutrient substances in foods. Using instruments to evaluate foods provides more objective data than does sensory testing, and is less costly and time consuming.

### Commercial Laboratories

Chemical tests can be conducted within a corporation, but they can also

**FIGURE 2-4** Texture analyzer.

Food Technology Corporation

| TABLE 2-2 | Selected Chemical Tests for Food Evaluation |
|---|---|
| Benedict and Fehling tests | Determine the presence of sugars (reducing) such as lactose and maltose, which are more likely to be involved in a chemical reaction that turns food brown. |
| Chromatography | Identifies the presence of various compounds, especially those associated with flavor. |
| Electrophoresis | Specific proteins are characterized by passing an electrical field through a gel containing proteins and measuring the rates at which they migrate. |
| Enzyme tests | The peroxidase 1 test evaluates peroxidase enzyme activity in pasteurized foods: if the heat of pasteurization is adequate to destroy harmful bacteria, it should also inactivate the peroxidase enzyme. The effectiveness of briefly boiling food to destroy the enzymes responsible for vegetable deterioration can be determined by measuring the catalase enzyme activity. |
| Fuchsin test | Detects aldehydes in fats and oils. |
| Iodine value test | Measures the degree of unsaturation in fats. |
| Peroxide value test | Measures the extent of oxidation that has occurred in a fat. |
| pH meter | Detects the amount of acidity or alkalinity in food mixtures or beverages. |
| Proximate analysis | A sequence of chemical tests to determine the macronutrient (protein, fat, carbohydrate) content of food. |

**Source:** USDA.

be sent out to commercial food testing companies. These food laboratories specialize in certain tests, such as microbial evaluation, which is often necessary for food safety testing. Many companies analyze their food products for certain bacteria, yeast, and/or molds. Food testing companies can provide corporations with Nutrition Facts labels by analyzing the nutrients in a new food product. They also can conduct a variety of chemical tests, some of which are listed in Table 2-3. The potential for contamination and adulteration is another reason why a food company or government agency tests a food; pesticides, herbicides, and industrial residues are just a few of the chemicals that can be analyzed. After melamine was illegally used in pet food by certain Chinese manufacturers to elevate protein content, the FDA and some private manufacturers began analyzing for melamine (12).

**TABLE 2-3**    An Example of Chemical Tests Conducted by a Food Testing Company

**Chemistry Analyses**

| | |
|---|---|
| Allergens | Maximum internal temperature |
| Ammonia | Moisture/protein ratio |
| Ash | Nitrate |
| Calcium | Nutritional analysis and labeling |
| Calories (by calculation) | Percent bone |
| Calories from fat | Oxidative rancidity |
| Carbohydrates (by calculation) | Pesticide residue |
| Collagen | pH |
| Crude fiber | Phosphate |
| Fat (Soxhlet) | Protein |
| Fatty acid content (saturated, unsaturated, trans) | Salt |
| Iron | Sodium nitrite |
| Heavy metals | Soy protein concentrate |
| Hydroxyproline | Thiobarbituric acid reactive substances (TBA) |
| Moisture (water) | Unknown compound identification |

**Source:** http://www.abcr.com/ana_meat.asp

Chemical testing opens up a world of exploration in food evaluation that is a rich array of both sensory and objective testing. Food corporations, certain government agencies, nutrition and food university departments, culinary schools, food marketing companies, and even individuals deciding daily which foods to eat all incorporate the principles of food evaluation.

## PICTORIAL SUMMARY / 2: Food Evaluation

Food manufacturers use both sensory (subjective) and objective evaluation methods to help in determining consumer acceptance of new products in research and development (R&D), product improvement, sales and marketing, quality assurance, analyzing nutrient content for labeling requirements (Nutrition Facts), and detecting contamination or adulteration.

**Sensory (subjective) tests** evaluate food quality by relying on the sensory characteristics and personal preferences of selected individuals. Taste panels, consisting of either randomly chosen members of the population or experts trained in tasting a particular product, are used to conduct subjective tests:

- *Analytical tests* are based on discernible differences.
- *Affective tests* are based on individual preferences.

NASA

**Objective tests** rely on laboratory methods and equipment to evaluate foods through physical and chemical tests.

- *Physical tests* measure certain observable aspects of food such as size, shape, weight, volume, density, moisture, texture, and viscosity.
- *Chemical tests* are used to determine the various nutrient and nonnutrient substances in foods.

# CHAPTER REVIEW AND EXAM PREP

## Multiple Choice*

1. Which of the following is an example of a subjective food evaluation test?
   a. Triangle test
   b. Spectrophotometry
   c. Measuring volume
   d. Measuring weight

2. Which of the following is an example of a chemical test?
   a. Duo-trio
   b. Threshold
   c. Nutrient analysis
   d. Shortometer

3. In what analytical sensory test is a tester asked to find the minimal detectable level of a substance?
   a. Triangle
   b. Flavor profile
   c. Hedonic
   d. Dilution

4. A line-spread test is a physical test for measuring
   a. sweetness.
   b. meat and baked product tenderness.
   c. consistency of batters and other viscous foods.
   d. degree of unsaturation in fats.

5. A food evaluation containing a range of from "like extremely" to "dislike extremely" is best described as a _____ test.
   a. discriminative
   b. descriptive
   c. hedonic
   d. personal preference

6. Which chemical test evaluates the degree of unsaturation in fats?
   a. Electrophoresis
   b. Enzyme test
   c. Iodine value test
   d. Proximate analysis

7. What is the chemical method of determining the protein, fat, and carbohydrate content of foods called?
   a. Proximate analysis
   b. Fuchsin test
   c. Benedict and Fehling tests
   d. Peroxide value test

## Short Answer/Essay

1. Describe the difference between sensory (subjective) and objective evaluation of foods.
2. How do discriminative and descriptive tests differ from each other?
3. Create a product score evaluation sheet for a cookie based on a 9-point hedonic scale.
4. Describe the general requirements for setting up a taste panel and the process of preparing samples to be subjectively tested.
5. List and describe three examples of difference testing.
6. List and describe three examples of physical testing.
7. List and describe three examples of chemical testing.
8. Describe viscosity and how it is used to test food quality.
9. How is chromatography used to evaluate certain foods?
10. Describe how a polarimeter and the Brix/acid ratio are used to test sweetness in liquids.

*See p. AK-1 for answers to multiple choice questions.

# REFERENCES

1. Drake MA. Invited review: Sensory analysis of dairy foods. *Journal of Dairy Science* 90:4925–4937, 2007.

2. Duxbury D. Sensory evaluation provides value. *Food Technology* 59(5):68, 2005.

3. Finucane ML, and JL Holup. Psychosocial and cultural factors affecting the perceived risk of genetically modified food: An overview of the literature. *Social Science and Medicine* 60(7):1603–1612, 2005.

4. Giese J. Instruments for food chemistry. *Food Technology* 50(2):72–77, 1996.

5. Giese J. Measuring physical properties of foods. *Food Technology* 49(2):54–63, 1995.

6. Henshall A. Analysis of starch and other complex carbohydrates by liquid chromatography. *Cereal Foods World* 41(5):419, 1996.

7. Hollingsworth P. Sensory testing and the language of the consumer. *Food Technology* 50(2):65–69, 1996.

8. IFT Sensory Evaluation Division. Sensory evaluation guide for testing food and beverage products. *Food Technology* 35(11):50–59, 1981.

9. Jordan RB, RJ Seely, and VA McGlone. A sensory-based alternative to Brix/acid ratio. *Food Technology* 55(6):36–44, 2001.

10. Leake LL. Sensory evaluation techniques. *Food Technology* 61(8): 80–83, 2007.

11. Meilgaard, MM, GV Civille, and T Carr. Overall difference tests: Does a sensory difference exist between samples? *Sensory Evaluation Techniques,* 4th ed. CRC Press, New York, NY, page 63–104, 2007.

12. Mermelstein NH. Analyzing for melamine. *Food Technology* 63(2):70–75, 2009.

13. Moskowitz HR. Experts versus consumers: A comparison. *Journal of Sensory Studies* 11:19–37, 1996.

## WEBSITES

The Sensory Science Laboratory at Oregon State University is just one example of a university food science department providing sensory testing services to the food industry.
**http://oregonstate.edu/dept/sensory/**

AOAC International (formerly the Association of Official Analytical Chemists) is a nonprofit association founded in 1884 in part to provide consensus on chemical analysis methods.
**http://www.aoac.org/**

Complimentary classes on flavors are offered by Fona International.
**http://www.fona.com/flavorUniversity. html**

A specific gravity to Brix table:
**http://www.fermsoft.com/gravbrix.php**

# 3 Chemistry of Food Composition

"**Y**ou are what you eat." When the 19th-century German philosopher Ludwig Feuerbach coined this phrase, he probably did not realize himself how true it was. Foods and people are composed of the same chemical materials, and there was a time when people served as nourishment to other animals in the food chain. All foods, including people, consist of six basic nutrient groups: water, carbohydrates, lipids, protein, vitamins, and minerals (Figure 3-1). Foods consist of varying amounts of these nutrients. For example, milk is 80 percent water, meats serve as primary sources of protein, potatoes and grains are rich in carbohydrates, and nuts are almost all fat. Actually, most foods contain a combination of the six major nutrient groups. Figure 3-2 shows the proportion of these six nutrients in humans.

**FIGURE 3-1**  Nutrient groups.

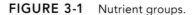
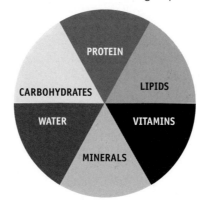

Because people literally are what they eat, the main purpose of eating and drinking is to replace those nutrients used up in the body's maintenance, repair, and growth, and to obtain the calories (kcal) necessary for energy. Calories are fuel to the body, as gas is fuel to a car.

## BASIC FOOD CHEMISTRY

The body benefits from the energy and nutrients in foods at the cellular level. To comprehend how this occurs, it is necessary to know some biochemistry. Although this seems a daunting task, biochemistry is simply the study of the chemistry that occurs within living organisms. Knowing something about biochemistry helps explain how nutrients from foods and beverages are assimilated in living systems.

**FIGURE 3-2**    Approximate proportion of nutrients in the human body. Differences occur due to age, gender, and condition. The proportion shown represents percent by weight. Vitamins and carbohydrates contribute a minimal amount.

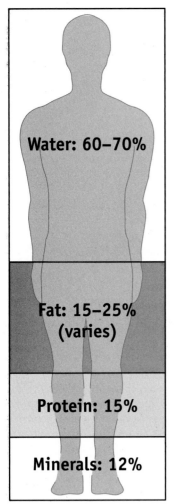

Water: 60–70%

Fat: 15–25% (varies)

Protein: 15%

Minerals: 12%

# Six Key Atoms— CHNOPS

A basic principle of biochemistry is that all living things contain six key elements (or **atoms**): carbon, hydrogen, nitrogen, oxygen, phosphorus, and sulfur (CHNOPS) (Chemist's Corner 3-1). These are the building blocks of organic material, carbon-containing compounds that are often living material. All the elements have the capacity to join together with similar or different elements to produce **molecules** or **compounds,** which then combine to create all the substances on earth, including the focus of this book—foods and beverages.

This chapter focuses on both organic and inorganic compounds by covering the six nutrient groups found in food and people: water, carbohydrates, fats, proteins, vitamins, and minerals. These **nutrients** serve as the foundation underlying all the principles in food and nutrition. They are discussed in this chapter with attention to what foods contain them, their chemical composition, and their functions in foods. Sugar (a form of carbohydrate) is discussed in greater detail in Chapter 21, and fat is covered further in Chapter 22.

**Atoms**    The basic building blocks of matter; individual elements found on the periodic table.

**Molecule**    A unit composed of one or more types of atoms held together by chemical bonds.

**Compound**    A substance whose molecules consist of unlike atoms.

**Nutrients**    Food components that nourish the body to provide growth, maintenance, and repair.

**Gram**    A metric unit of weight. One gram (g) is equal to the weight of 1 cubic centimeter (cc) or milliliter (mL) of water (at a specific temperature and pressure).

## CHEMIST'S CORNER 3-1

### Atomic Structure

Everything physical in the universe is made from atoms, some of the smallest particles in existence. How are these smallest units of matter identified? The number of protons and electrons that they contain identifies them. Protons are positively charged particles in the atom's nucleus, and electrons are negatively charged particles surrounding the nucleus like the rings around Saturn. The number of electrons on the outside ring of an atom dictates how many bonds that particular atom can form, and, therefore, what kind of substances it forms. For example, the carbon (C) atom, the backbone of carbohydrates, fats, and proteins, usually forms four bonds. Nitrogen (N) is capable of forming three bonds, whereas oxygen (O) can form two, and hydrogen (H) only one (Figure 3-3). The bond holding atoms together through the sharing of electrons is called a covalent bond.

## ✹ CALORIE CONTROL
### Where Do Calories Come From?

Some organic compounds can be broken down by the body to release the energy, in the form of calories, needed to sustain life. Carbohydrates, fats, proteins, and alcohol are the only sources of calories from the diet. No calories are obtained from vitamins, minerals, or water. Both water and minerals are inorganic compounds, substances that do not contain carbon and cannot provide calories. The following table shows the calories (kcal) provided per **gram.**

#### Calorie (kcal) Sources

| Yes: | No: |
|---|---|
| • Carbohydrate = 4 kcal/gram | • Vitamins |
| • Protein       = 4 kcal/gram | • Minerals |
| • Fat           = 9 kcal/gram | • Water |
| • Alcohol       = 7 kcal/gram | • Fiber[1] |

[1]The carbohydrates in fiber are not digested, so they are not absorbed to provide calories.

**Conversion factors:**  5 grams      = 1 teaspoon
28.35 grams = 1 ounce
100 grams   = ½ cup liquid

© 2010 Amy Brown

**FIGURE 3-3** The number of bonds that selected atoms can form with other atoms.

$$H — \quad — O — \quad — \overset{\displaystyle |}{\underset{\displaystyle |}{N}} — \quad — \overset{\displaystyle |}{\underset{\displaystyle |}{C}} —$$

1      2      3      4

# WATER

Water is the simplest of all nutrients, yet it is the most important (23). Without it life could not exist. Life probably began in water billions of years ago, and it is still essential at every stage of growth and development. Water brings to each living cell the ingredients that it requires and carries away the end products of its life-sustaining reactions. The life functions of assimilating, digesting, absorbing, transporting, metabolizing, and excreting nutrients and their by-products all rely on water. The body's cells are filled with water and bathed in it. The human body averages 60 to 70 percent water, and losing as little as 10 percent can result in death. Water balance is maintained by drinking fluids and by eating foods, all of which naturally contain at least some water. A small portion is also obtained through metabolic processes.

## Water Content in Foods

People get the water they need from foods and beverages. Although it may not always be apparent, many foods contain more water than any other nutrient. Foods have a water content of 0 to 95+ percent (Figure 3-4). Those that yield the most water are fruits and vegetables (70 to 95 percent), whole milk (over 80 percent), and most meats (average just under 70 percent). The foods with the least water are vegetable oils and dried foods such as grains and beans.

## Free or Bound Water

The water in foods may be in either free or bound form. Free water, the largest amount of water present in foods, is easily separated from the food, whereas

**FIGURE 3-4** The percent water content of certain foods.

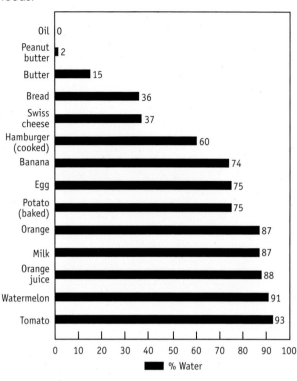

■ % Water

bound water is incorporated into the chemical structure of other nutrients such as carbohydrates, fats, and proteins. Examples would be the free water found in fruit and the bound water found in bread. Bound water is not easily removed and is resistant to freezing or drying. It also is not readily available to act as a medium for dissolving salts, acids, or sugars.

## Composition of Water

Whether bound or free, water's chemical formula remains the same. Water is a very small molecule consisting of three atoms—one oxygen atom flanked by two hydrogen atoms ($H_2O$) (Chemist's Corner 3-2).

## Measuring Calories

It takes heat, or its loss, to move the molecules of water through their different states, and this heat is commonly measured in the form of calories. This unit of measurement when expressed with an uppercase "C," is equal to the amount of energy required to raise 1 kilogram of water 1° Celsius (measured between 14.5°C and 15.5°C at normal atmospheric pressure). The energy

> **CHEMIST'S CORNER 3-2**

**The Chemical Structure of Water**

Water has an overall neutral charge. This "neutrality" is derived from the combination of its two hydrogen ($H^+$) atoms, each with one positive charge, being balanced by the two negative charges of water's one oxygen ($O^{-2}$) atom. Overall, this gives water a neutral charge. However, it is not completely neutral in the sense that the water molecule has a negative pole and a positive pole, making it dipolar (Figure 3-5). Dipolar molecules have poles with partial charges that oppose each other, and this dynamic contributes to some of water's very unique properties.

**FIGURE 3-5** Two atoms of hydrogen combine with one oxygen atom, creating a dipolar molecule.

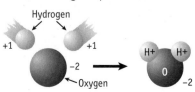

values of foods are actually measured in thousands of calories, more accurately expressed as *kilocalories* and represented by the terms *kcal* or *Calories* with an uppercase C. One kilocalorie (kcal) equals 1,000 calories.

In theory, the small c calorie is used by chemists, whereas the large C Calorie, or, more commonly, kilocalorie (kcal), is the more accurate term for referring to the energy value of foods. In practice, however, calorie with a lowercase c is often used, especially with the general public, and it is assumed that those in the food field know that what is really meant is kilocalorie (kcal) or Calorie. Throughout this book, *calorie (kcal)* is used to represent the unit of measurement of the energy derived from food.

## Specific Heat

It takes more energy to heat water than it does any other substance now known. Water's high **specific heat** makes it unique compared to other compounds on earth. Given the same amount of heat, a metal pan or the oil in it will become burning hot, whereas water will become only lukewarm. This important characteristic of water enables animals, including people, with a high water content to withstand the very hot or cold temperatures sometimes found on earth. Water's specific heat of 1.00 (1 calorie will raise 1 gram of water 1°C) is used as the measure against which all other substances are compared. Similarly, water differs from other compounds in the

amount of energy it takes to reach its specific freezing, melting, and boiling points.

### How & Why?

*If the atoms in $H_2O$ do not change, how is water able to exist as a gas (steam or humidity), liquid, or solid (ice)?*

The distance between the molecules determines these differences, and the distances are influenced by temperature. At very low temperatures, ice forms as the water molecules line up very close together. Elevating the temperature increases the movement of the water molecules against each other, pushing them farther away from each other. When enough heat is applied, ice melts into a liquid. Continued heating transforms liquid water into a gas (steam) by giving the molecules freedom to move even farther apart (Figure 3-6). The variations of water from solid to liquid to gas are called changes in state. In spite of the obvious differences in these states, they do not involve any changes in the structure of the water molecule.

## Freezing Point

In many parts of the world, winter temperatures can turn water into ice when its **freezing point** is reached. People living in snow country are particularly aware that when the temperature drops

to freezing (32°F/0°C at normal atmospheric pressure), ice can form on the roads. The lower temperature decreases water's kinetic energy, or the energy associated with motion, which slows the movement of the water molecules until they finally set into a compact configuration. **Heat of solidification** occurs when at least 80 calories (0.08 kcal) of heat are lost per gram of water.

Unlike other substances, water expands and becomes less dense when completely frozen, which is why ice floats. The expansion of frozen water ruptures pipes and containers filled with water. It should come as no surprise, then, that it also ruptures the cells in plants and meats, diminishing the potential food's textural quality. Pure water freezes at 32°F (0°C), but adding anything else to the water changes its freezing point. The addition of **solutes** such as salt or sugar to water lowers the freezing point. Adding too much, however, slows the freezing process. Thus, frozen desserts made with large quantities of sugar take extra time to freeze.

## Melting Point

Just as removing heat from water causes it to turn into ice, returning the same 80 calories (0.08 kcal) of heat to a gram of ice will cause it to reach its **melting point** and turn it back into water. While the ice absorbs the 80 calories (0.08 kcal) of heat, there is no rise in temperature. This **latent heat** does not register because it was put to work in moving the

---

**Specific heat** The amount of heat required to raise the temperature of 1 gram of a substance 1°C.

**Freezing point** The temperature at which a liquid changes to a solid.

**Heat of solidification** The temperature at which a substance converts from a liquid to a solid state.

**Solute** Solid, liquid, or gas compound dissolved in another substance.

**Melting point** The temperature at which a solid changes to a liquid state (liquid/solid/gas).

**Latent heat** The amount of energy in calories (kcal) per gram absorbed or emitted as a substance undergoes a change in state (liquid/solid/gas).

---

**FIGURE 3-6** Molecular movement dictates whether a substance is a solid, liquid, or gas.

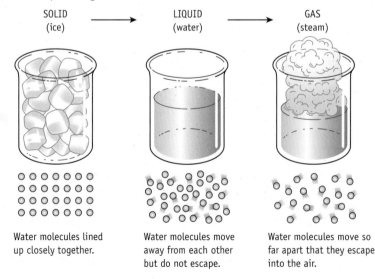

SOLID (ice) → LIQUID (water) → GAS (steam)

Water molecules lined up closely together.

Water molecules move away from each other but do not escape.

Water molecules move so far apart that they escape into the air.

molecules of water far enough apart to change the physical structure of the solid ice into liquid water.

## Boiling Point

Bubbles start to break the surface of water when it reaches 212°F (100°C) at sea level. This is its **boiling point** (further explained in Chapter 5). The water will not get any hotter, nor will the food cook faster, no matter how much more heat is added, and this is why a slow rolling boil is often recommended. Keeping the temperature at a slow rolling boil is also more gentle on the foods and results in less evaporation.

The point at which water boils is reached when the pressure produced by steam, called *vapor pressure,* equals the pressure of the atmosphere pushing down on the earth. At this point the natural pressures of the atmosphere are not strong enough to push back the expanding gases of boiling water. Water requires 540 calories (0.54 kcal) of energy per gram to boil and vaporize.

This **heat of vaporization** is quite a bit higher than the 80 calories (0.08 kcal) needed to melt ice. Serious burns can result from human exposure to steam because the amount of heat required to produce it is so high.

### Elevation and Boiling Point

Increasing the elevation decreases the boiling point of water. At sea level, water boils at 212°F (100°C), but this drops 1°F for every 500-foot increase in altitude (an increase of 960 feet in elevation decreases water's boiling point by 1°C). Water boils in the mountains at lower temperatures than it does at sea level because there is less air and atmospheric pressure pushing down on the earth's surface. Steam has less resistance to overcome, and therefore occurs at lower temperatures. For example, at 7,000 feet water boils at 198°F (92°C). People at even higher elevations, such as on Mount Everest, could put a hand in a pan of boiling water and find it quite comfortable. Recipes are usually modified for elevations above 3,000 feet because the lower boiling temperature might affect ingredient actions and reactions.

Artificial changes in atmospheric pressure can be achieved by pressure cookers as well as by special equipment used only in the commercial food industry. A pressure cooker speeds up heating time by increasing atmospheric pressure to 15 pounds; thus, water temperatures up to 240°F (116°C) can be achieved.

## Hard vs. Soft Water

Most water is not pure water, but contains dissolved gases, organic materials, and mineral salts from the air and soil. The minerals in water determine whether it is hard or soft water. Hard water contains a greater concentration of calcium and magnesium compounds, whereas soft water has a higher sodium concentration. The temperatures at which water freezes, melts, or boils remain constant regardless of whether it is hard or soft water.

| TABLE 3-1 | Functional Properties of Water in Food | |
|---|---|---|
| **Heat Transfer** | **Universal Solvent** | **Chemical Reactions** |
| Moist Heating of Foods | Solution | Ionization |
| Boiling | Colloidal dispersion | Changes in pH |
| Simmering | Suspension | Salt formation |
| Steaming | Emulsion | Hydrolysis |
| Stewing | | $CO_2$ release |
| Braising | | Food preservation |

## Functions of Water in Food

Water is the most abundant and versatile substance on earth. Among its many uses in food preparation, its two most important functions are as a transfer medium for heat and as a universal solvent. In addition, it is important as an agent in chemical reactions, and is a factor in the perishability and preservation of foods (Table 3-1).

### Heat Transfer

Water both transfers and moderates the effects of heat. A potato heated by itself in a pan will burn, but surrounding that same potato with water ensures that the heat will be evenly distributed. Water also transfers heat more efficiently, which explains why a potato heats faster in boiling water than in the oven. Because water has a higher specific heat than other substances, it buffers changes in temperature. More energy is needed to increase the temperature of 1 gram of water than 1 gram of fat. For example, the specific heat of oil is 0.5; thus, it heats twice as fast as water when given the same amount of heat.

### *Moist-Heat Cooking Methods*

Almost half of the methods used to prepare foods rely on water to transfer heat, and these are known collectively as *moist-heat methods.* The major moist-heat methods discussed in this book are boiling, simmering, steaming, stewing, and braising. Dry-heat methods use

---

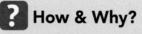 **? How & Why?**

**How can you tell if water is hard or soft?**

Hard water leaves a ring in the bathtub, a grayish sediment on the bottom of pans, and a grayish cast in washed whites. Although permanently hard water cannot be softened by boiling or distilling, it can be converted by a water softener, which works by exchanging sodium for calcium and magnesium. Another way to determine if water is hard or soft is to call the local water department and ask how much calcium carbonate (in ppm—parts per million) is in the water. The following breakdown defines whether it is hard or soft:

| Water Hardness | Calcium Carbonate (ppm) |
|---|---|
| Soft | 0–50 |
| Medium | 50–100 |
| Hard | 100–200 |
| Very Hard | 200+ |

---

**Boiling point** The temperature at which a heated liquid begins to boil and changes to a gas.

**Heat of vaporization** The amount of heat required to convert a liquid to a gas.

heat in the form of radiation and include baking, grilling, broiling, and frying. Microwaving uses both dry- and moist-heat methods; microwaves are actually a form of radiation that heats the water molecules in foods, which then heat the food itself. Microwaving techniques are discussed throughout this book under moist-heat preparation methods.

### Universal Solvent

The many biochemical interactions occurring in living organisms—human, animal, and plant—could not occur in the absence of a **solvent** environment. Water is considered to be the earth's

---

**Solvent** A substance, usually a liquid, in which another substance is dissolved.

**Solubility** The ability of one substance to blend uniformly with another.

**Solution** A completely homogeneous mixture of a solute (usually a solid) dissolved in a solvent (usually a liquid).

**Precipitate** To separate or settle out of a solution.

**Distillation** A procedure in which pure liquid is obtained from a solution by boiling, condensation, and collection of the condensed liquid in a separate container.

**Saturated solution** A solution holding the maximum amount of dissolved solute at room temperature.

**Supersaturated solution** An unstable solution created when more than the maximum solute is dissolved in solution.

**Colloidal dispersion** A solvent containing particles that are too large to go into solution, but not large enough to precipitate out.

**Suspension** A mixture in which particles too large to go into solution remain suspended in the solvent.

**Emulsion** A liquid dispersed in another liquid with which it is usually immiscible (incapable of being mixed).

**Flocculation** A partial gel in which only some of the solid particles colloidally dispersed in a liquid have solidified.

---

universal solvent. The fluid substance, mostly water, within and around the cell is a solvent that contains many dissolved solutes.

Combining a solvent and a solute results in a solution, a colloidal dispersion, a suspension, or an emulsion. These mixtures differ from each other based on the size or **solubility** of their solutes.

*Solutions*  In a **solution**, the molecules of the solute are so small that they completely dissolve and will not **precipitate** from their fluid medium. They cannot be separated by filtering, but can sometimes be removed by **distillation**. If a substance is able to enter into a solution by dissolving, it is considered to be soluble.

Much of what people perceive as the taste of foods depends on the formation of solutions with solutes in foods such as sugars, salts, acids, and other flavor compounds, and their resulting enhanced ability to attach to flavor receptors. Water also forms solutions with minerals and water-soluble vitamins (B complex and C). This increases the likelihood that these minerals and vitamins may leach out of foods into cooking water, which is often discarded, causing nutrients to be lost. To the delight of tea and coffee lovers, water can also dissolve caffeine and other flavorful compounds from tea leaves and coffee beans. Higher temperatures increase the amount of solute that will dissolve in the solvent, which explains why very hot water is used for making coffee and tea.

The solubility of a substance is measured by the amount in grams that will dissolve in 100 mL of solvent. Raising the temperature allows more solute to dissolve in the solvent, creating a **saturated solution**. Increasing the temperature even higher sets the stage for a **supersaturated solution**, which is very unstable and must be cooled very slowly to avoid having the solute precipitate out or crystallize. Many candies, including fudge, rely on the creation of supersaturated solutions.

*Colloidal Dispersions*  Not all particles dissolve readily or homogeneously. Some particles, called colloids (e.g., proteins, starches, and fats), never truly dissolve in a solvent, but remain in an unstable **colloidal dispersion**. Unlike solutes in solutions, which completely dissolve, colloids, because

of their large size, do not completely dissolve, but neither do they noticeably change the dispersion's freezing or boiling point. Examples of different types of dispersions include a solid in a liquid, a liquid in another liquid (salad dressing) or solid (jam, gelatin, cheese, butter), and a gas that can be incorporated into either a liquid (egg white or whipped cream foams) or a solid (marshmallow). Two types of dispersions are **suspensions** and **emulsions**.

- *Suspension.* Mixing cornstarch and water results in a suspension in which the starch grains float within the liquid. Cornstarch suspensions are often used in Chinese cooking and give Chinese food its particular shiny appearance and smooth mouthfeel.

- *Emulsion.* Another type of colloidal dispersion involves water-in-oil (w/o) or oil-in-water (o/w) emulsions. Neither water nor fats will dissolve in each other, but they may become dispersed in each other, creating an emulsion. Examples of food emulsions include milk, cream, ice cream, egg yolk, mayonnaise, gravy, sauces, and salad dressings (6). These and other emulsions can be separated by freezing, high temperatures, agitation, and/or exposure to air (10). Emulsions are discussed in more detail in Chapter 22.

Colloidal dispersions, which are unstable by nature, can be purposely or accidentally destabilized, causing the dispersed particles to aggregate out into a partial or full gel, a more-or-less rigid protein structure. An example of this is seen when milk is heated; its unstable water-soluble milk proteins precipitate out and end up coating the bottom of the pot, creating a **flocculation**. Full gels such as yogurt and cheese are also made possible by the colloidal nature of milk.

## Chemical Reactions

Water makes possible a vast number of chemical reactions that are important in foods. These include ionization, pH changes, salt formation, hydrolysis, and the release of carbon dioxide.

### Ionization

When particles dissolve in a solvent, the solution is either molecular or ionic in nature. Molecular solutions are those in

**FIGURE 3-7** Hydrolysis of sucrose to glucose and fructose.

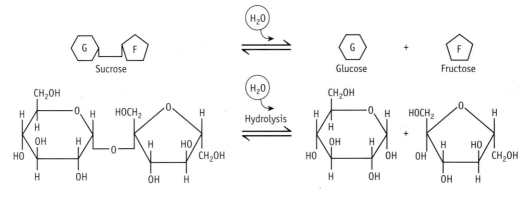

which the dissolved particles remain "as is" in their molecular form. An example would be the dissolving of a flavored sugar mix in water to make a beverage. The sugar molecules remain unchanged in solution. Ionic solutions occur when the solute molecules **ionize** into electrically charged ions or **electrolytes.** When salt, or sodium chloride (NaCl), is dissolved in water, it ionizes into the individual ions of sodium ($Na^+$) and chloride ($Cl^-$).

This chemical reaction is written:

$$NaCl \rightarrow Na^+ + Cl^-$$

### Changes in pH—Acids and Bases

Acids are substances that donate hydrogen ($H^+$) ions, and bases provide hydroxyl ($OH^-$) ions. Another defining difference between acids and bases is that acids are proton donors, whereas bases are proton receptors. The **pH scale** (*pH* stands for *power* and *hydrogen*) is a numerical representation of the hydrogen ($H^+$) ion concentration in a liquid. A solution with a pH of under 7 is considered acidic, whereas anything over 7 is alkaline or basic. A pH of 7 indicates that the solution is neutral, containing equal concentrations of hydrogen ($H^+$) ions and hydroxyl ($OH^-$) ions. Each number on the scale represents a tenfold change in degree of acidity (Chemist's Corner 3-3).

Water is naturally neutral, but tap water is normally adjusted to be slightly alkaline (pH 7.5 to 8.5), because acidic water causes pipe corrosion. Overly alkaline water, however, results in deposits of carbonates that may block water pipes. Many coffee connoisseurs prefer distilled water for making coffee because of its neutral nature.

### Salt Formation

The nature of water as a universal solvent makes it possible to form salts. This occurs when a positive ion combines with a negative ion, as long as neither is a hydrogen ($H^+$) nor hydroxyl ($OH^-$) ion. The primary example is sodium chloride ($Na^+Cl^-$), resulting from sodium ($Na^+$) combining with

chloride ($Cl^-$). Salts can also be formed by combining an acid and a base, or a metal and a nonmetal. Metal salts include potassium fluoride ($K^+F^-$) and lithium bromide ($Li^+Br^-$).

### Hydrolysis

Countless chemical reactions rely on **hydrolysis.** Figure 3-7 shows an example of how a water molecule is used to break a sugar into smaller molecules, and the hydrolysis of a lipid is illustrated in Figures 3-8 and 3-9. Just a few of the hydrolysis applications used in the food industry include breaking down cornstarch to yield corn syrup; dividing table sugar into its smaller components to create another sugar helpful in the manufacture of some candies (see Chapter 21); and creating protein hydrosylates, smaller molecules derived from protein hydrolysis, to add to foods to improve flavor, texture, foaming abilities, and nutrient content.

> **CHEMIST'S CORNER 3-3**
>
> **The Logarithmic pH Scale**
>
> The concentrations of ions in water are so small that it is awkward to speak or write about these concentrations using ordinary words. For example, water has a hydrogen ion concentration of 0.0000001 mole per liter, which is translated in terms of the negative logarithm of the hydrogen ion concentration (18):
>
> $$pH = -\log(H^+)$$
>
> pH is also understood in a scale of 1 to 7. The expression of 0.0000001 mole per liter using a decimal can also be written in its exponential form of $1 \times 10^{-7}$ and then placed into the negative logarithm to yield a pH of 7:
>
> $$\begin{aligned} pH &= -\log(1 \times 10^{-7}) \\ &= \log 1/(1 \times 10^{-7}) \\ &= \log(1 \times 10^7) = 7 \end{aligned}$$
>
> These single-digit numbers are much easier to fathom as long as it is understood that each number in the pH scale represents a tenfold change in the degree of acidity.

**Ionize** To separate a neutral molecule into electrically charged ions.

**Electrolyte** An electrically charged ion in a solution.

**pH scale** Measures the degree of acidity or alkalinity of a substance, with 1 the most acidic, 14 the most alkaline, and 7 neutral.

**Hydrolysis** A chemical reaction in which water (*hydro*) breaks (*lysis*) a chemical bond in another substance, splitting it into two or more new substances.

**FIGURE 3-8**   Hydrolysis of a triglyceride to glycerol and three fatty acids.

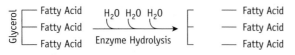

Triglyceride                Glycerol   +   3 Fatty Acids

**FIGURE 3-9**   Hydrolysis of a monoglyceride to glycerol and one fatty acid.

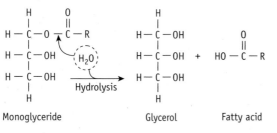

Monoglyceride              Glycerol        Fatty acid

## Carbon Dioxide Release

Many baked products are allowed to rise before baking. One of the agents making this possible is baking powder, which is a combination of baking soda and acid. It is only when baking powder is combined with water that the gas carbon dioxide is released, which causes baked products to rise. The chemical reaction is a two-step process:

*Step 1:*

$$NaHCO_3 + HX \xrightarrow{\text{water}} NaX + H_2CO_3$$
baking soda + acid ⟶ salt + carbonic acid

*Step 2:*

$$H_2CO_3 \longrightarrow H_2O + CO_2$$
carbonic acid ⟶ water + carbon dioxide

# Food Preservation

While water is essential to the chemical reactions on which living things and many foods depend, it is also important for the life of **microorganisms** such as bacteria, molds, and yeasts. The actions of these microorganisms on food cause deterioration and decay. Atmospheric humidity alone increases the likelihood of foods degenerating. For example, a relative humidity of 75 percent or more, especially if combined with warm temperatures, encourages the growth of microorganisms. Thus, removing water from fruits, vegetables, meats, and herbs was among the earliest forms of food preservation. Without water, microorganisms cannot survive, so limiting the amount of water available to them inhibits their growth. Conversely, water in a cool environment helps preserve the freshness of fruits and vegetables by preventing dehydration—hence those artificial "rain" showers we see in supermarket produce displays.

Removing dirt and other debris from fruits and vegetables by rinsing them in water or even washing them with detergent eliminates many microorganisms. Detergents lower the surface tension of water, which improves its ability to act as a cleansing agent.

## Water Activity

A food's **water activity (a_w)** or water availability determines its perishability. Bacteria need water to grow, and most foods do not support bacterial growth if their water activity ($a_w$) is below 0.85 (pure water has a water activity of 1.00) (36). Thus, foods high in water content, such as milk, meat, vegetables, and fruits, are much more prone to microbial spoilage than drier foods such as grains, nuts, dried milk, dried beans, or dried fruits (45). Moreover, once deterioration sets in, the putrefying food itself releases water, which fuels the further growth of microorganisms. One way to lower the water activity ($a_w$) of even pure water below 1.00 is to add other substances (Chemist's Corner 3-4) (44). Water molecules orient themselves around any added solute, making them unavailable for microbial growth. Solutes such as sugar and salt added to jams and cured meats inhibit microbial growth by lowering water activity. The food industry makes water unavailable to microorganisms by using solutes such as salt, sugars, glycerol, propylene glycol, and modified corn syrups (2).

## Osmosis and Osmotic Pressure

Salting has been used as a method of preserving foods for thousands of years because salt draws water out of foods and to itself. Of course, ancient peoples did not understand the process of **osmosis,** which causes water to be drawn to solutes; all they knew was that salting kept their food edible for long periods of time. Part of this process depends on the fact that water passes through membranes freely, but most solutes do not (Figure 3-10). The side of the membrane with more solute has more **osmotic pressure** and draws the necessary water to that side to dilute its solute concentration. Any bacteria contacting heavily salted food lose their water by the

---

**Microorganism** Plant or animal organism that can only be observed under the microscope—e.g., bacteria, mold, yeast, virus, or animal parasite.

**Water activity (a_w)** Measures the amount of available (free) water in foods. Water activity ranges from 0 to the highest value of 1.00, which is pure water.

**Osmosis** The movement of a solvent through a semipermeable membrane to the side with the higher solute concentration, equalizing solute concentration on both sides of the membrane.

**Osmotic pressure** The pressure or pull that develops when two solutions of different solute concentration are on either side of a permeable membrane.

---

> **CHEMIST'S CORNER 3-4**
>
> **Measuring Water Activity**
>
> As free water decreases, so does the water activity. Water activity is measured by dividing the vapor pressure exerted by the water in food (in solution) by the vapor pressure of pure water ($P_w$), which is equal to 1.00 (2).

**FIGURE 3-10** Water flows in the direction of the higher concentration of solute.

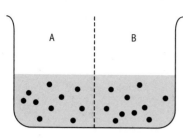 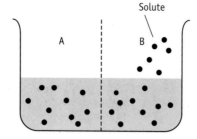 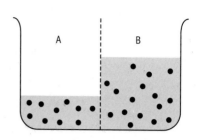

1. With equal numbers of solute particles on both sides, the concentrations are equal, and the tendency of water to move in either direction is about the same.

2. Now additional solute is added to side B. Solute cannot flow across the divider (in the case of a cell, its membrane).

3. Water can flow both ways across the divider, but has a greater tendency to move from side A to side B, where there is a greater concentration of solute. The volume of water becomes greater on side B, and the concentrations on sides A and B become equal.

**SOURCE:** Whitney and Rolfes, Understanding Nutrition, 11th edition, Cengage, 2007.

same process and die by dehydration. Beef jerky is the result of the combined processes of salting, smoking, and drying of meats. The high sugar concentration of jams and jellies acts to preserve them in the same way as salt on meats.

# CARBOHYDRATES
## Foods High in Carbohydrates

Carbohydrates are the sugars, starches, and fibers found in foods. Plants are the primary source of carbohydrates, with the exception of milk, which contains a sugar called lactose. The muscles from animals can also contain some carbohydrate in the form of glycogen, but much of this is converted to a substance called lactic acid during slaughter. Most carbohydrates are stored in the seeds, roots, stems, and fruit of plants. Common food sources for carbohydrates include grains such as rice, wheat, rye, barley, and corn; legumes such as beans, peas, and lentils; fruits; and some vegetables, such as carrots, potatoes, and beets. Sugar cane and sugar beets provide table sugar, whereas honey is derived from the nectar of flowers.

## Composition of Carbohydrates

Carbohydrates have been described in the media as good, bad, simple, complex, and even as net carbs (17), but regardless

of how they are described, the simplest elements making up carbohydrates are carbon (C), hydrogen (H), and oxygen (O). The word *carbohydrate* can be broken down into *carbon* (C) and *hydrate* ($H_2O$). This leads to the basic chemical formula of carbohydrates, which is $C_n(H_2O)_n$, where *n* stands for a number ranging from 2 into the thousands. Carbohydrates are found primarily in green plants, where they are synthesized through the process of photosynthesis. The chemical reaction of photosynthesis is written:

carbon dioxide + water + sun energy
$\longrightarrow$ glucose + oxygen

$$6CO_2 + H_2O + \text{sun energy} \longrightarrow C_6H_{12}O_6 + 6O_2$$

The carbon, hydrogen, and oxygen atoms making up carbohydrates are arranged in a basic unit called a *saccharide*. Carbohydrates are classified into monosaccharides, disaccharides, oligosaccharides, and polysaccharides, depending on the type and number of saccharide units they contain (Figure 3-11).

- Monosaccharides (one saccharide)
- Disaccharides (two monosaccharides linked together)
- Oligosaccharides (*few*—three to ten—monosaccharides linked together; these are not as common in foods as either monosaccharides or disaccharides)
- Polysaccharides (*many* monosaccharides linked together in long chains; these include starch and fibers)

**FIGURE 3-11** Classification of carbohydrates.

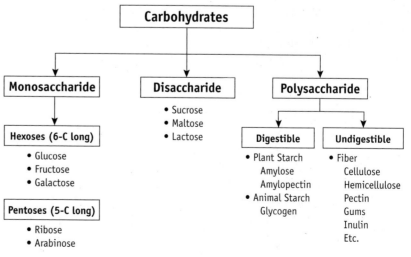

# Monosaccharides

The simplest sugars, monosaccharides, are classified by the number of carbons in the saccharide unit—triose (three carbons), tetrose (four carbons), pentose (five carbons), and hexose (six carbons). The chemical names of many of the carbohydrates end in -ose, which means sugar (Chemist's Corner 3-5). Pentose and hexose sugars are more common in foods, the main pentoses being ribose and arabinose, and the three most predominant hexoses being glucose, fructose, and galactose (Figure 3-12).

## Ribose and Arabinose

Ribose is an extremely important component of nucleosides, compounds that are part of the genetic material deoxyribonucleic acid (DNA), ribonucleic acid (RNA), and the energy-yielding adenosine triphosphate (ATP). Ribose also plays an important role as part of vitamin B$_2$ (riboflavin). Arabinose contributes to the structure of many vegetable gums and fibers.

## Glucose

Glucose is the most common hexose found in foods and the major sugar in the blood. It is present in its free form

> ### CHEMIST'S CORNER 3-5
>
> **D and L Sugars**
>
> Saccharide nomenclature uses D or L (or alpha or beta) to describe the chemical spatial arrangement of certain saccharides. The designations D and L allude to two series of sugars. Most natural sugars belong to the D series, in which the highest-numbered asymmetric carbon has the hydroxyl group pointed to the right (Figure 3-13). L-series sugars point to the left. The prefix *alpha* or *beta* can also be used to describe whether the hydroxyl group points to the right (alpha) or left (beta) of the saccharide.

in fruits, honey, corn syrup, and some vegetables. It also exists as the repeating saccharide unit in starch and glycogen, and is incorporated into many fibers. Refined glucose, called dextrose in the food industry, is used in the production of candies, beverages, baked goods, canned fruits, and alcoholic beverages. Glucose is also the major ingredient of corn syrup, which

**FIGURE 3-13**    The D or L system of nomenclature describes the right or left chemical configuration of a molecule.

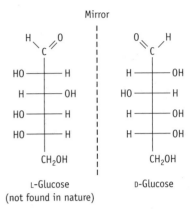

L-Glucose (not found in nature)    D-Glucose

is made commercially by hydrolyzing cornstarch.

## Fructose

Also called fruit sugar or levulose, fructose is found primarily in fruits and honey. Fructose is the sweetest of all sugars, yet it is seldom used in its pure form in food preparation because it can cause excessive stickiness in candies, overbrowning in baked products, and lower freezing temperatures in ice cream. High-fructose corn syrup, however, is the preferred and predominant sweetening agent used in soft drinks.

## Galactose

Seldom found free in nature, galactose is part of lactose, the sugar found in milk. A derivative of galactose, galacturonic acid, is a component of pectin, which is very important in the ripening of fruits and the gelling of jams.

# Disaccharides

Combining two saccharides results in a disaccharide. The three most common disaccharides are sucrose, lactose, and maltose (Figure 3-14).

## Sucrose

Sucrose is table sugar, the product most people think of when they use the word *sugar*. Chemically, sucrose is one glucose molecule and one

**FIGURE 3-12**    Monosaccharides—hexoses.

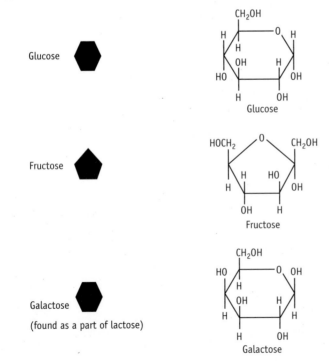

Glucose

Fructose

Galactose (found as a part of lactose)

**FIGURE 3-14** Disaccharides.

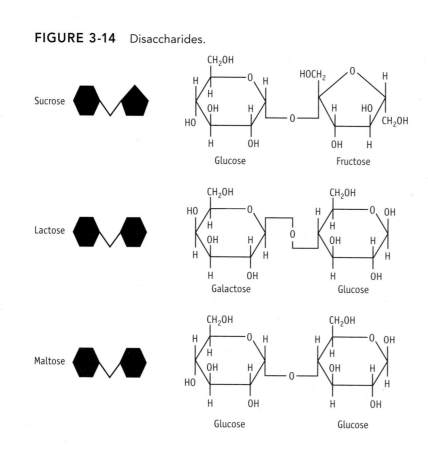

Sucrose

Glucose Fructose

Lactose

Galactose Glucose

Maltose

Glucose Glucose

fructose molecule linked together. The types of sugars derived from sucrose are explained in Chapter 21.

### Lactose

A glucose molecule bound to a galactose molecule forms lactose, one of the few saccharides derived from an animal source. About 5 percent of fluid milk is lactose, or milk sugar. Some people are unable to digest lactose to its monosaccharides because they lack sufficient lactase, the **enzyme** responsible for breaking down milk sugar into glucose and galactose. The symptoms of lactase deficiency, or lactose intolerance, include gas, bloating, and abdominal pain caused by the disaccharides not being properly digested. In some cheeses, yogurt, and other fermented dairy products, bacteria break down the lactose to lactic acid, which can usually be tolerated by lactase-deficient individuals.

### Maltose

Two glucose molecules linked together create maltose, or malt sugar. Maltose is primarily used in the production of beer and breakfast cereals and in some

infant formulas. This saccharide is produced whenever starch breaks down, for example, in germinating seeds and in human beings during starch digestion.

## Oligosaccharides

*Oligo* means "few" in Greek, so compounds made up of three to 10 monosaccharides are called oligosaccharides. The two most common are raffinose (three monosaccharides) and stachyose (four monosaccharides). These saccharides, found in dried beans, are not well digested in the human digestive tract, but intestinal bacteria do break them down, forming undesirable gas as a by-product. There are twelve classes of food-grade oligosaccharides in commercial production. These are either extracted directly from soybeans or synthesized by building up disaccharides or breaking down starch.

### Food Industry Uses

The food industry can use oligosaccharides for bulking agents in low-calorie diet foods such as confections,

beverages, and yogurt, and as fat replacers in beverages (7). One benefit of oligosaccharides is that they are not cariogenic, or cavity producing, as are many of the disaccharides.

## Polysaccharides

Starch, glycogen, and fiber are the polysaccharides most commonly found in foods. Polysaccharides contain many monosaccharides linked together and are divided into two major groups: digestible (starch and glycogen) and indigestible (fiber).

### Starch—Digestible Polysaccharide from Plant Sources

The glucose derived from photosynthesis in plants is stored as starch. As a plant matures, it not only provides energy for its immediate needs, but also stores energy for future use in starch granules. Microscopic starch granules are found in various foods such as rice, tapioca, wheat, and potato. A cubic inch of food may contain as many as a million starch molecules (50). Amylose and amylopectin are the two major forms of starch found in these granules. The glucose units in both of these starch molecules are joined together with a glycosidic bond (alpha-1, 4) that is capable of being digested by human enzymes. Amylose is a straight-chain structure of repeating glucose molecules, whereas amylopectin is highly branched with alpha-1, 6 bonds (every 15 to 30 glucose units) (Figure 3-15). The majority of starchy foods in their natural state usually contain a mixture of about 75 percent amylopectin and 25 percent amylose. These two starches are further explained and illustrated in Chapter 18.

The body can break starch down during digestion into its individual glucose units for absorption (48). In foods, heat, enzymes, and acid are used to break starches down into smaller, sweeter segments called *dextrins*. The sweeter taste of toasted bread, compared to its untoasted counterpart, comes from the dextrins formed in the toaster.

---

**Enzyme** A protein that catalyzes (causes) a chemical reaction without itself being altered in the process.

**FIGURE 3-15**   Starch consists of a mixture of amylose (straight chain) and amylopectin (branched chain). Each G represents a glucose molecule.

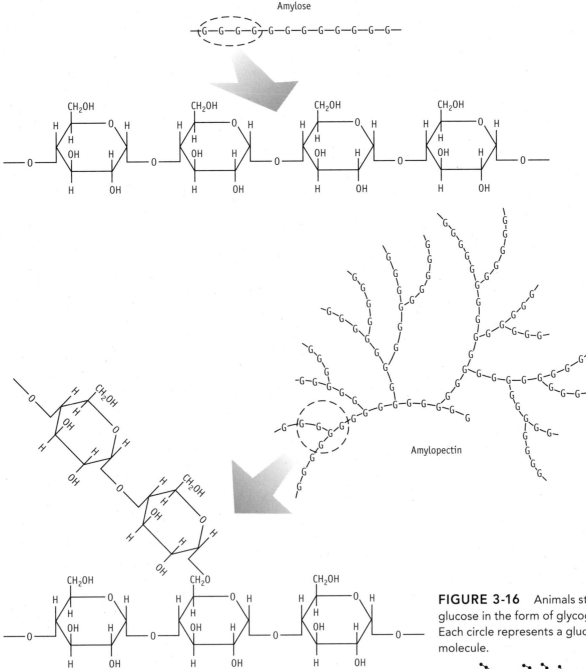

## Glycogen—Digestible Polysaccharide from Animal Sources

Glycogen, or animal starch, is one of the few digestible carbohydrates found in animals. It is located only in the liver and muscles. Just as glucose is stored by plants as starch, it is stored by animal bodies in long chains of glycogen (Figure 3-16). It is a highly branched arrangement of glucose molecules, and serves as a reserve of energy. Glycogen can be quickly hydrolyzed by an animal's enzymes to release the glucose needed to maintain blood glucose levels. The glycogen in meat is converted to lactic acid during slaughtering and so is not present by the time it reaches the table. Shellfish such as scallops and oysters provide a minuscule amount of glycogen, which is why they tend to taste slightly sweet compared to other fish and plants.

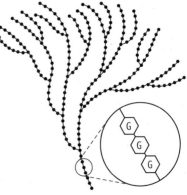

**FIGURE 3-16**   Animals store glucose in the form of glycogen. Each circle represents a glucose molecule.

## Fiber—Undigestible Polysaccharide

Fiber, also known as roughage or bulk, describes a group of indigestible polysaccharides. Unlike those in starch, the sugar units in fibers are held together by bonds that the human digestive enzymes cannot break down. Most fibers, therefore, pass through the human body without providing energy. Fiber is found only in foods of plant origin, especially certain cereals, vegetables, and fruits. Plant cells rely on the fiber between their cell walls for structural strength.

A variety of fiber definitions are accepted by food and nutrition professionals. Some definitions are based on analytical methods while others are based on physiological effects (43).

### Dietary Fiber vs. Crude Fiber

Several different laboratory methods are used to measure the amount of fiber in foods. The older technique consisted of treating a food with strong acid to simulate the environment of the stomach, and then treating it with a base to parallel the experience in the small intestine. The remaining weight of undigested fiber was measured as "crude fiber" and was listed in most food composition tables as "fiber" (24). This rather imprecise method has been largely replaced by an analytical method approved by the Association of Official Analytical Chemists International (AOAC) that measures **dietary fiber.** For every 1 gram of crude fiber, there are about 2 to 3 grams of dietary fiber.

### Soluble vs. Insoluble Fiber

Chemists classify fibers according to how readily they dissolve in water: soluble fibers dissolve in water, whereas insoluble fibers do not. The insoluble fibers of foods act as a sponge in the intestine by soaking up water. This increases the softness and bulk of the stool and may thereby decrease the risk of constipation, **diverticulosis,** and possibly colon cancer (27). Scientists have also suggested that soluble fibers may benefit health by lowering high blood cholesterol levels and reducing high blood glucose in certain kinds of diabetics (22). Foods containing fiber usually have a mixture of both soluble and insoluble fiber. Foods high in soluble fiber include dried beans, peas, lentils, oats, rice bran, barley, and oranges. Insoluble fibers are found predominantly in whole wheat (wheat bran) and rye products, along with bananas.

The Institute of Medicine recommends that the terms soluble and insoluble fibers no longer be commonly used to classify dietary fibers even though they may still appear on some food labels. Both are considered *functional fibers*—nondigestible carbohydrates that have beneficial physiological effects in humans. Another new term is *total fiber,* which is the sum of dietary fiber and functional fiber (20).

### Common Fibers

The most common fibers are cellulose, hemicellulose, and pectic substances. A few other types of fiber include vegetable gums, inulin, beta-glucan, oligosaccharides, fructans, some resistant starches, and lignin—one of the few fibers that is not a carbohydrate (20).

**Cellulose** Cellulose is one of the most abundant compounds on earth. Every plant cell wall is partly composed of cellulose, long chains of repeating glucose molecules similar to starch. Unlike starch, however, the chains do not branch, and the bonds holding the glucose molecules together cannot be digested by human enzymes (Figure 3-17). As a result, the cellulose fiber is not absorbed, provides no calories, and simply passes through the digestive tract. The digestive systems of herbivores such as cattle, horses, goats, and sheep have the proper enzymes to digest cellulose, allowing them to use the energy from glucose found in grass and other plants.

**Hemicellulose** Hemicellulose is composed of a mixture of monosaccharides. The most common monosaccharides comprising the backbone of hemicelluloses are xylose, mannose, and galactose; the common side chains are arabinose, glucuronic acid, and galactose. Baking soda is sometimes added to the water in which green vegetables are boiled to maintain their color. Unfortunately, it breaks down the hemicellulose of the vegetables, causing them to become mushy.

**Pectic Substances** These polysaccharides found between and within the cell walls of fruit and vegetables include protopectin, pectin, and pectic acid.

**Food Industry Uses.** Pectic substances act as natural cementing agents, so they are extracted from their source foods by the food industry for use in thickening jams, jellies, and preserves; keeping salad dressings from separating; and controlling the texture and consistency of a variety of foods. Not all the pectic substances, however, can be used for gelling purposes, and the amounts that can be obtained vary depending on the ripeness of the fruit or vegetable. The pectin found in ripe, but not overripe, fruit is responsible for gel formation in jams. Protopectin and pectic acid are prevalent in unripe and overripe fruit, respectively, and are insufficient themselves to cause gel formation. Further details on the use of pectins by the food industry for the formation of jams, jellies, and preservers are covered in Chapter 14.

**Dietary fiber** The undigested portion of carbohydrates remaining in a food sample after exposure to digestive enzymes.

**Diverticulosis** An intestinal disorder characterized by pockets forming out from the digestive tract, especially the colon.

**FIGURE 3-17**    The bonds differ between the glucose molecules in starch (alpha-1,4) and cellulose (beta-1,4). The latter is not digestible to humans.

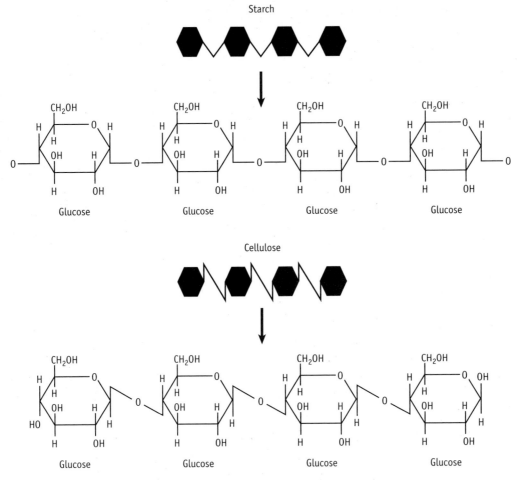

**Vegetable Gum**    Vegetable gums belong to a group of polysaccharides known as hydrocolloids.

***Food Industry Uses.*** Composed of simple sugars, gum fibers are used by the food industry to thicken, increase viscosity of, gel, stabilize, and/or emulsify certain processed foods. They impart body, texture, and mouthfeel to foods, while also making it less likely that dispersed ingredients will separate (40). In addition, the gums' "water-loving" nature combined with their ability to bind as much as 100 times their weight in water contribute a certain desirable appearance, texture, and

stability to food products (8). As shown in Figure 3-18, gums may be derived from plants or from a bacterium (10).

Vegetable gums are normally sold as a dry powder and are used extensively as stabilizers in the production of low-calorie salad dressings, confections, ice cream, puddings, and whipped cream. Gums are also used in many frozen products because they control crystal growth, yield optimal texture, and make the food more stable in the freezing and thawing process. Typical applications in the food industry of a gum, specifically carrageenan, are listed in Figure 3-18. Agar gum can be used for quick-drying frostings and to reduce chipping or cracking in glazed doughnuts (47).

**Inulin**    Inulin consists of repeating units of fructose with an end molecule of glucose. Although this fiber occurs naturally in over 30,000 plants, it is

most commonly found in asparagus, Jerusalem artichoke, and garlic.

***Food Industry Uses.*** Commercial processors extract inulin from the chicory root (31), and this soluble fiber is used by the food industry to impart a creamy texture to frozen dairy products such as no-fat or no-sugar ice cream, improve the textures of margarine spreads, and develop no-fat icings, fillings, and whipped toppings.

**Lignin**    Lignin is the one fiber that is not a carbohydrate. Instead of saccharides, it consists of long chains of **phenolic** alcohols linked together into a large, complex molecule. As plants mature, their cell walls increase in lignin concentration, resulting in a tough, stringy texture. This partially explains why celery and carrots get tougher as they age. Boiling water does not dissolve or even soften the lignin.

**Phenolic**    A chemical term to describe an aromatic (circular) ring attached to one or more hydroxyl (−OH) groups.

**FIGURE 3-18** Natural and synthetic gums used by the food industry.

| Gum Source | Examples | Structure | Selected Function[1] | Selected Foods[1] |
|---|---|---|---|---|
| Plant exudates[2]  | Gum Arabic | Galactose, arabinose, rhamnose, glucuronic acid, arabino-galactan-protein complex (AGP) | Emulsifier, flavor stabilizer, film former, enhances mouthfeel in beverages | Flavor emulsions, candy coating, icing, beverages |
| | Gum karaya | Galacturonic acid, galactose, rhamnose, glucuronic acid | Gelling | Bakery goods, French dressing, ice pops (prevents ice crystals, bleeding of free water), ground meats, meringues |
| | Gum tragacanth | Polymer of galacturonic acid, galactose, arabinose, xylose | Emulsifier, thickener | Dressings, sauces, tomato-based products, milkshake |
| Seeds  | Guar gum | Mannose and galactose | Provides mouthfeel to dairy, thickens, stabilizer, moisture retention | Ice cream, breads, dressings, sauces, baked goods |
| | Locust (carob) bean gum | Galactose, mannose | Provides heat/shock resistance to ice cream, prevents whey off in dairy, texture | Sauces and gravy, ice cream, cottage cheese, frozen foods |
| Seaweeds  | Agar-agar | Alpha-D-galactopyranosyl and 3,6-anhydro-alpha-L-glactopyranosyl | Gelling agent, suspends particulates, melt point management | Yogurt, bakery glazes, icings, confections |
| | Alginates | D-mannuronic acids and L-glucuronic acids | Gelling, controls syneresis, improves cooking yield and texture in meat | Frozen desserts, pie filling, meat sauce, dairy, beverages |
| | Carrageenan | Sulfated linear polysaccharide of D-galactose | Improves mouthfeel, controls syneresis suspension, binder gelation | Sauces, pudding, whipped toppings, chocolate milk, meat, pie filling ice cream |
| | Furcellaran (Danish agar) | Sulfated polysaccharide of D-galactose | Gelling | Jams & jellies |
| Microbial[3] 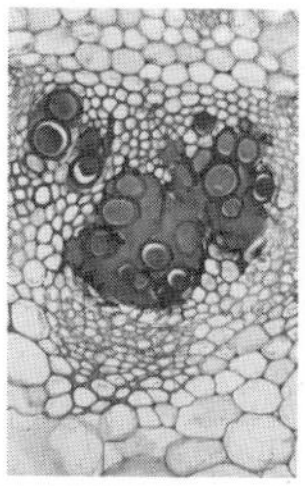 | Xanthan gum | D-glucose, D-mannose, and D-glucuronic acids | Stabilizes and emulsifies salad dressings and sauces, improves cling, suspends particulates, controls moisture in refrigerated dough | Salad dressings & sauces, refrigerated dough, dry mixes, bakery emulsions |
| Plant fibers | Cellulose gum | Glucose | Stabilizer, thickener, moisture retention | Egg (frozen/dried), syrup, milk, cakes |
| | Konjac (elephant yam) | Glucose and mannose (40% glucomannan) with some acetyl substituted groups | Viscosity, forms strong films, gelling agent | Jellies, diet food |
| Synthesized | Carboxymethyl-cellulose (CMC) | Glucose | Provides mouthfeel to beverages, provides structure and freeze/thaw stability, thickener, ice crystal growth management, thickener | Dietetic foods, bakery goods, syrups, frozen dairy, puddings, ice cream |
| | Hydroxypropyl methylcellulose | Glucose | Firm texture, stabilize foam | Soy burgers, whipped cream |
| | Hydroxypropyl cellulose | Glucose | Thickener, emulsion stabilizer | Dairy, batter and breaded coatings |

[1]Only a few examples of gums many functions and food uses were selected

[2]Sap-like oozings from injured plants

[3]Produced from fermentation by the *Xanthomonas campestris* bacterium

**Sources:** Adapted by Amy Brown jorgegonzalez/istockphoto.com from www.ticgums.com, www.gumtech.com, www.foodadditives.org, and www.marcelcarrageenan.com

redbrickstock.com/Alamy

Grant Heilman Photography/Alamy

gary corbett/Alamy

Biodisc/Visuals Unlimited/Corbis

# Functions of Carbohydrates in Foods

Carbohydrates play numerous roles in foods. Saccharides or sugars contribute to sweetness, solubility, crystallization, color (through browning), moisture absorption, texture, fermentation, and even preservation. These functions are described in Chapter 21. Starches also have many roles in food, including thickening agent, edible film, and sweetener source (syrups) (see Chapter 18).

# LIPIDS OR FATS

The fats and oils in foods belong to a group called lipids. Lipids are commonly called "fats" when their content in foods is under discussion; although this terminology is not precisely accurate, this textbook will follow this generally accepted practice. Fats and oils are differentiated in two ways: (1) fats are solid at room temperature, whereas oils are liquid and (2) fats are usually derived from animal sources, whereas oils are derived predominantly from plants. Three exceptions are coconut and palm oils, which are solid at room temperature, and fish oils, which, at the same temperature, are liquid.

## Foods High in Lipids

The foods that are high in fats from animal sources include meats, poultry, and dairy products. Plant food sources high in fat include nuts, seeds, avocado, olives, and coconut. Most fruits, vegetables, and grains, however, contain little, if any, fat. Invisible fats are those not easily observed in foods, such as the marbling in meat. Visible fats, such as the white striations found in bacon and the outside trim on meats, are easily seen. Vegetable oils, butter, margarine, shortening, lard, and tallow are also obvious visible fats.

## Composition of Lipids

Lipids, like carbohydrates, are composed of carbon, hydrogen, and oxygen atoms, but in differing proportions. One way to determine if a substance is a lipid is to test whether it will dissolve in water. Lipids are not water soluble, but can be dissolved in organic solvents not used in food preparation, such as benzene, chloroform, ether, and acetone. Acetic acid, which is responsible for the sour taste of vinegar, is the one lipid that will dissolve in water because its molecule is so small. Edible lipids are divided into three major groups: triglycerides (fats and oils), phospholipids, and sterols.

## Triglycerides

About 95 percent of all lipids are triglycerides, which consist of three ("tri") fatty acids attached to a glycerol molecule (Figure 3-19) (Chemist's Corner 3-6). (Two fatty acids linked to the glycerol molecule form a diglyceride, whereas one fatty acid linked to glycerol is a monoglyceride.) The fatty acids on the glycerol can be identical (simple triglyceride) or different (mixed triglyceride).

## Fatty Acid Structure

Fatty acids differ from one another in two major ways: (1) their length, which is determined by the number of carbon atoms and (2) their degree of *saturation,* which is determined by the number of double bonds between carbon atoms.

**FIGURE 3-19**    A triglyceride (fat) is made by combining three fatty acids with a glycerol molecule. Water is released as a by-product.

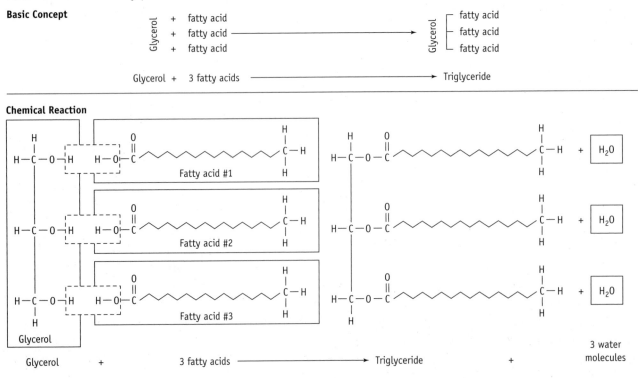

**Basic Concept**

**Chemical Reaction**

Glycerol    +    3 fatty acids ⟶ Triglyceride    +    3 water molecules

**Chemical Formation of Triglycerides**

Few fatty acids occur free in foods, but rather are incorporated into triglycerides. Each fatty acid consists of an acid group (–COOH) on one end and a methyl group (–CH³) on the other end. The fatty acids are attached to the glycerol molecule by a condensation reaction: the hydrogen atom (H) from the glycerol and a hydroxyl (–OH) group from a fatty acid form a molecule of water (Figure 3-19). When a fatty acid reacts like this with an alcohol such as glycerol, the resulting compound is called an ester. Because *acyl* refers to the fatty acid part of an ester, what is called *triglyceride* should actually be named *triacylglycerol* (11).

The number of carbons can range from 2 to 22, with the number usually being even. A fatty acid is said to be saturated if there are no double bonds between carbons—every carbon on the chain is bonded to two hydrogens and therefore fully loaded. If one hydrogen from two adjacent carbons is missing, the carbons form double bonds with one another and form a point of unsaturation. A fatty acid with one double bond present is called a monounsaturated fatty acid. If there are two or more double bonds in the carbon chain of a fatty acid, the fatty acid is called polyunsaturated (Figure 3-20).

The degree of unsaturation of the fatty acids in a fat affects the temperature at which the molecule melts. Generally, the more unsaturated a fat, the more liquid it remains at room temperature. In contrast, the more saturated a fat, the firmer its consistency. Thus, vegetable oils containing mostly mono- or polyunsaturated fatty acids are generally liquid at room temperature, whereas largely saturated animal fats are solid.

## Fatty Acids in Foods

Most foods contain all three types of fatty acids—saturated, monounsaturated, and polyunsaturated—but one type usually predominates (Figure 3-21). Generally speaking, most vegetable and fish oils are high in polyunsaturates, whereas olive and canola oils are rich in monounsaturates. The animal fats, as well as coconut and palm oils, are more saturated. Overall, foods of animal origin usually contain approximately a 50:50 **P/S ratio** of polyunsaturated and saturated fats, whereas for plant foods the ratio is usually 85/15. The higher the P/S ratio, the more polyunsaturated fats the food contains.

## Fatty Acid Nomenclature

Each fatty acid is identified by a common name, systematic name (Chemist's Corner 3-7), chemical configuration (Chemist's Corner 3-8), or numerical ratio (Table 3-2). Usually a fatty acid is referred to by its common name, whereas the systematic name is used when a more formal or correct chemical nomenclature is required. The long number of carbons is abbreviated in a type of chemical shorthand that conveys the length and saturation of fatty acids in a numerical ratio. For example, palmitic acid is a saturated fatty acid that is represented by 16:0, meaning that it is sixteen carbons long with zero double bonds. Approximately 40 fatty acids are found in nature. Some of the more common fatty acids are butyric acid, found in butter, and the two fatty acids that are **essential nutrients**—linoleic and linolenic.

## Phospholipids

Phospholipids are similar to triglycerides in structure, in that fatty acids are attached to the glycerol molecule.

**FIGURE 3-20** Fatty acids differ in their degree of saturation.

**Saturated**

**Primarily Animal Sources:** Meat, Poultry, Milk/Butter/Cheese, Egg Yolk, Lard
**Plant Sources:** Chocolate, Coconut/Coconut Oil, Palm Oil, Vegetable Shortening

(a)

**Monounsaturated**

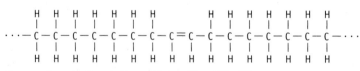

**Sources:** Avocado, Peanuts/Peanut Butter, Olives/Olive Oil

(b)

**Polyunsaturated**

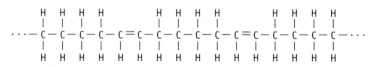

**Primarily Plant Sources:** Vegetable Oils (Corn, Safflower, Soybean, Sunflower, Canola, etc.)
Margarine (most), Mayonnaise, Certain Nuts (Almonds, Filberts, Pecans, Walnuts)
**Animal sources:** Fish

(c)

**P/S ratio** The ratio of polyunsaturated fats to saturated fats. The higher the P/S ratio, the more polyunsaturated fats the food contains.

**Essential nutrients** Nutrients that the body cannot synthesize at all or in necessary amounts to meet the body's needs.

**FIGURE 3-21**    Classification of fatty acids.

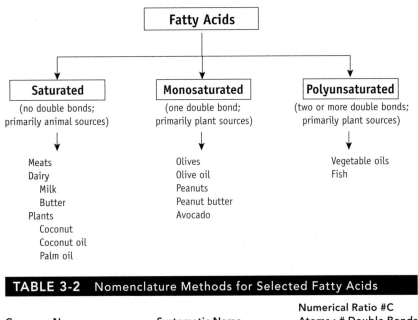

**TABLE 3-2**    Nomenclature Methods for Selected Fatty Acids

| Common Name | Systematic Name | Numerical Ratio #C Atoms : # Double Bonds |
| --- | --- | --- |
| Butyric | Butanoic | 4:0 |
| Linoleic | 9, 12-Octadecadienoic | 18:2 |
| Linolenic | 9, 12, 15-Octadecatrienoic | 18:3 |

The difference is that one of the fatty acids is replaced by a compound containing phosphorus, which makes the phospholipid soluble in water, whereas its fatty acid components are soluble in fat (Figure 3-22). Phospholipids are very important in the body as a component of cell membranes, where they assist in moving fat-soluble vitamins and hormones in and out of the cells. Foods that naturally contain phospholipids include egg yolks, liver, soybeans, wheat germ, and peanuts.

**Emulsifier**  A compound that possesses both water-loving (hydrophilic) and water-fearing (hydrophobic) properties so that it disperses in either water or oil.

**Hydrophobic**  A term describing "water-fearing" or non-water-soluble substances.

**Hydrophilic**  A term describing "water-loving" or water-soluble substances.

**Bile**  A digestive juice made by the liver from cholesterol and stored in the gall bladder.

## Food Industry Uses

The dual nature of phospholipids makes them ideal **emulsifiers**. The best-known phospholipid is lecithin, which is found in egg yolks. Lecithin acts as an emulsifying agent that allows **hydrophobic** and **hydrophilic** compounds to mix. This natural function of phospholipids results in them being widely used by the food industry as emulsifiers in such products as beverages, baked goods, mayonnaise, and candy bars.

## Sterols

Sterols are large, intricate molecules consisting of interconnected rings of carbon atoms with a variety of side chains attached. Many compounds important in maintaining the human body are sterols, including cholesterol, **bile,** both sex (testosterone, estrogen) and adrenal (cortisol) hormones, and vitamin D. The sterol of most significance in foods is cholesterol (Figure 3-23). Although both animal and plant foods contain sterols, cholesterol is found only in foods of animal origin such as meat, poultry, fish, fish roe (caviar), organ meats (liver, brains, kidneys), dairy products, and egg yolks. Plants do not contain cholesterol, but they may contain other types of sterols.

### CHEMIST'S CORNER 3-7

**Naming Chemical Compounds**

Billions of compounds exist in nature. This vast number does not even include the compounds that are synthesized in a laboratory. At first people named compounds after people, places, and things, but that proved too cumbersome. Scientists then developed a system to name chemical substances known as *chemical nomenclature,* and this is defined as the *systematic* naming of chemical compounds. In chemistry, the systematic name literally describes the chemical construction of the compound. The International Union of Pure and Applied Chemistry (IUPAC) is the official organization responsible for mandating the nomenclature of all chemical compounds. Learn more about the IUPAC through their website at www.iupac.org

### CHEMIST'S CORNER 3-8

**Cis, Trans, and Omega Fatty Acids**

Other notations that are frequently encountered when discussing fatty acids are cis or trans, and omega-3 or omega-6. The terms *cis* and *trans* describe the geometric shape of the fatty acid. A cis fatty acid has the hydrogens on the same side as the double bond, causing it to fold into a U-like formation. A trans fatty acid has the hydrogens on either side of the double bond, creating a linear configuration (Figure 3-24). Most of the fatty acids in nature are in the cis or slightly V-shaped configuration, whereas trans fatty acids often result from hydrogenation. The difference between the omega-3 and omega-6 fatty acids is the location of the double bond between the third and fourth, or the sixth and seventh, carbon from the left of the fatty acid molecule, respectively (Figure 3-25).

**FIGURE 3-22**   Phospholipids, such as this lecithin molecule, have a phosphorous-containing compound that replaces one of the fatty acids on a triglyceride. Each compound in the lecithin molecule is derived from simpler molecules (see arrows).

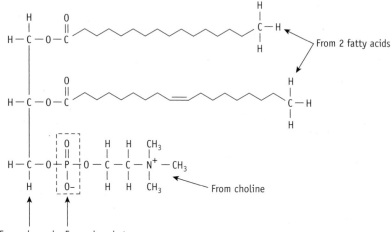

From glycerol   From phosphate

**FIGURE 3-23**   Cholesterol is a lipid found in foods of animal origin.

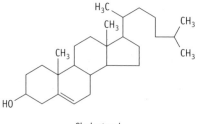

Cholesterol

### Plant Sterols

Soybeans are not the only plants containing sterols; in fact, they are found in small amounts in many fruits, vegetables, nuts, seeds, cereals, legumes, and other plant sources (46). In 2000, the FDA allowed the use of health claims regarding foods containing these substances with regard to the role of plant sterols or plant sterol esters in reducing the risk for coronary heart disease. To qualify for a claim, the food must contain at least 0.65 gram of plant sterol esters or 1.7 grams of **plant stanol esters** per serving. It also must not contain over 13 grams of total fat per serving and per 50 grams. Spreads and salad dressings are exempted if the label refers the consumer to the product's Nutrition Facts panel (32).

## Functions of Lipids in Foods

Chapter 22 discusses the roles of lipids in foods, which include heat transfer during food preparation, contributing to the tenderness, mixing (emulsifying), texture (melting, plasticity, solubility), and flavor of foods, and increasing one's feeling of fullness after eating (satiety).

## PROTEINS

Proteins derive their name from the Greek word *proteos*, of "prime importance." The body can manufacture most of the necessary carbohydrates (except fiber) and lipids (except a few essential fatty acids) it needs, but when it comes to protein, the body can synthesize only about half of the compounds it requires in order to manufacture the proteins needed for the body. These substances needed for protein manufacture are called amino acids. Of the 22 amino acids, 9 are essential nutrients and thus must be obtained daily from the diet (Table 3-3).

**FIGURE 3-24**   *Cis* or *trans* fatty acids are defined by their chemical configuration at the double bonds.

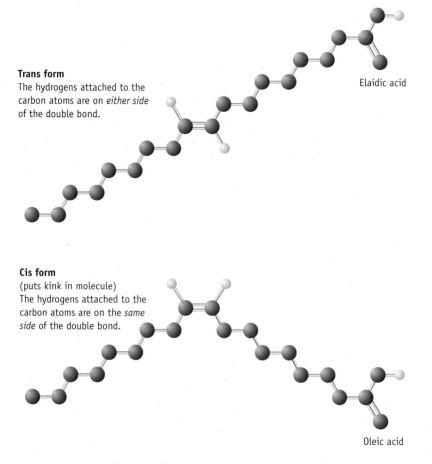

**Trans form**
The hydrogens attached to the carbon atoms are on *either side* of the double bond.

Elaidic acid

**Cis form**
(puts kink in molecule)
The hydrogens attached to the carbon atoms are on the *same side* of the double bond.

Oleic acid

**Plant stanol esters** Naturally occurring substances in plants that help block absorption of cholesterol from the digestive tract.

**FIGURE 3-25**    Omega-3 or omega-6 fatty acids are defined by the location of the first double bond.

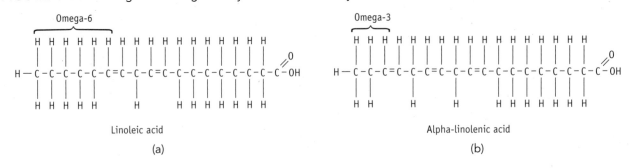

Linoleic acid
(a)

Alpha-linolenic acid
(b)

## Protein Quality in Foods

Foods vary in their protein quantity and quality. Most protein from animal sources—meat, poultry, fish and shellfish, milk (cheese, yogurt, etc.), and eggs—is **complete protein.** Gelatin is one of the few animal proteins that is not complete. Plant protein, with the exception of that from soybeans and certain grains (quinoa and amaranth), is **incomplete protein** and will support maintenance, but not growth. In order to obtain complete protein primarily from plant sources, it is necessary to practice the strategy of **protein complementation.** The best sources of protein from plants are the legumes—beans, peas, and lentils—which are often combined with grains.

## Composition of Proteins

One key way in which proteins differ from carbohydrates and lipids is that proteins contain nitrogen atoms, whereas carbohydrates and lipids contain only carbon, hydrogen, and oxygen atoms. These nitrogen atoms give the name "amino," meaning "nitrogen containing," to the amino acids of which protein is made. Protein molecules resemble linked chains, with the links being amino acids joined by **peptide bonds.** A protein strand does not remain in a straight chain, however. The amino acids at different points along the strand are attracted to each other, and this pull causes some segments of the strand to coil, somewhat like a metal spring. Also, each spot along the coiled strand is attracted to, or repelled from, other spots along its length. This causes the entire coil to fold this way and that, forming a globular or fibrous structure.

## Amino Acids

Each protein has its own specific sequence of amino acids. The 22 amino acids that exist in nature are like an alphabet, forming the "letters" of the "words"—proteins—that make up the language of life itself. All amino acids have the same basic structure—a carbon with three groups attached to it: an amine group ($-NH_2$), an acid group ($-COOH$), and a hydrogen atom (H). Attached to the carbon at the fourth bond is a side chain called an R group (Figure 3-26). It is this fourth attachment, the side chain, different for each amino acid, that gives the amino acid its unique identity and chemical nature (Figure 3-27). The simplest amino acid is glycine, with only a hydrogen for the R group. In other amino acids, the R group may consist of carbon chains or cyclic structures. Amino acids that are acidic contain more acid groups ($-COOH$) than amine groups ($-NH_2$), whereas alkaline amino acids contain more amine than acid groups.

**Complete protein**  A protein, usually from animal sources, that contains all the essential amino acids in sufficient amounts for the body's maintenance and growth.

**Incomplete protein**  A protein, usually from plant sources, that does not provide all the essential amino acids.

**Protein complementation**  Two incomplete-protein foods, each of which supplies the amino acids missing in the other, combined to yield a complete protein profile.

**Peptide bond**  The chemical bond between two amino acids.

| TABLE 3-3    Essential and Nonessential Amino Acids | |
|---|---|
| **Classification** | **Amino Acid** |
| Essential for all humans | Histidine |
| | Isoleucine |
| | Leucine |
| | Lysine |
| | Methionine |
| | Phenylalanine |
| | Threonine |
| | Tryptophan |
| | Valine |
| Nonessential | Alanine |
| | Arginine |
| | Asparagine |
| | Aspartic acid |
| | Cysteine |
| | Glutamic acid |
| | Glutamine |
| | Glycine |
| | Proline |
| | Serine |
| | Tyrosine |
| Related compounds sometimes classified as amino acids | Carnitine |
| | Cystine |
| | Hydroxyglutamic acid |
| | Hydroxylysine |
| | Hydroxyproline |
| | Norleucine |
| | Taurine |
| | Thyroxine |

**FIGURE 3-26**    The structure of an amino acid.

$$NH_2$$
$$|$$
$$R - C - H$$
$$|$$
$$COOH$$

The R represents different groups that can attach here. This group makes each amino acid different.

**FIGURE 3-27**   An amino acid's chemical nature is determined by its side group.

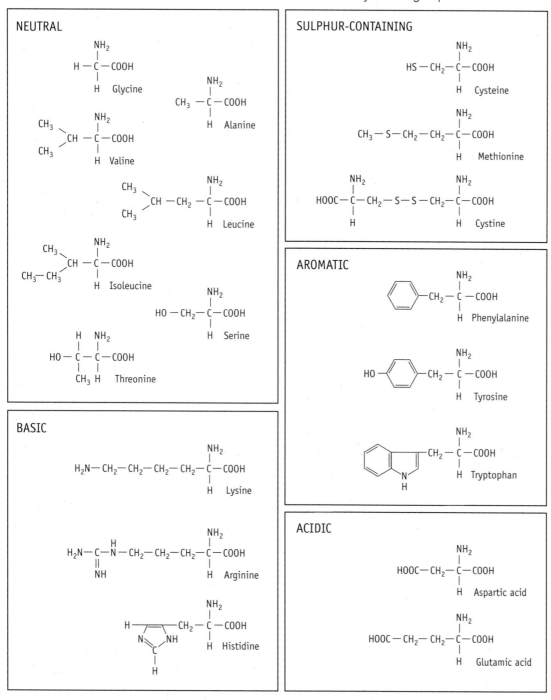

# Functions of Proteins in Food

The proteins in foods allow several important reactions to occur during food preparation:

- Hydration
- Denaturation/coagulation
- Enzymatic reactions
- Buffering
- Browning

## Hydration

The ability of proteins to dissolve in and attract water, a process called hydration, allows them to play several important roles in foods. One of these is the capability to form a gel, an intricate network of protein strands that trap water, resulting in a firm structure. Another is to aid in dough formation to produce numerous bread products.

***Food Industry Uses***   Proteins from milk, meat, egg, and soy are used in a variety of gels (Chemist's Corner 3-9) (26, 30). The gelling ability of proteins allows them to be used as binders (5), stabilizers (51), and thickeners in a variety of foods such as preserves, confectioneries (gums, marshmallows), and desserts (ice cream, puddings, custards, pie fillings, mousses, and gelatins). Sausages and gelled fish products also rely on the ability of proteins to gel.

## CHEMIST'S CORNER 3-9

### Protein Gels

It is the different types, concentrations, and interactions of proteins in any given food that determine the overall resulting gel strength. Myosin has the highest gel-forming ability of all the muscle proteins, but the whole myofibrillar protein fraction, sarcoplasmic proteins, and connective tissue proteins can also gel (42).

See Chapter 15 for the use of animal protein in the formation of gelatin products.

Another example of protein's hydrating ability in food preparation is in bread making. Water or milk is combined with yeast and the two major proteins of wheat—gliadin and glutenin—through the process of kneading to yield the protein gluten, whose elastic qualities allow it to stretch with the carbon dioxide gas produced by the yeast during fermentation. This is how bread rises, and without protein's ability to hydrate, rising would not take place.

## Denaturation/Coagulation

Large protein molecules are sensitive to their surroundings. When subjected to heat, pH extremes, alcohol, and physical or chemical disturbances, proteins undergo **denaturation.** Denaturation can result in **coagulation,** which is described as a curdling or congealing of the proteins. Both denaturation and coagulation are irreversible in most proteins. Examples include the hardening of egg whites with heating, the formation of yogurt as bacteria convert lactose to lactic acid and lower the pH, and the stiffening of egg whites when they are whipped (3).

**Denaturation** The irreversible process in which the structure of a protein is disrupted, resulting in partial or complete loss of function.

**Coagulation** The clotting or precipitation of protein in a liquid into a semisolid compound.

**Substrate** A substance that is acted upon, such as by an enzyme.

***Food Industry Uses*** Foam formation depends on denaturing the protein found in eggs through mechanical disruption. Once the foam forms, adding sugar to beaten egg whites helps to stabilize the delicate denatured proteins; therefore, sugar is often added near the end of whipping, just before the egg whites have reached their optimal consistency. Foams produced by milk and eggs are discussed in Chapters 10 and 12, respectively.

Cheese production (discussed in Chapter 11) also relies on the coagulation of proteins, which is speeded up by adding salt. Cheese is made by creating a curd (hard mass) composed of denatured milk proteins that collapse together. The enzymes that break down milk proteins are enhanced by adding salt, which is why salt is frequently used by cheese makers to help produce a firm curd.

## Enzymatic Reactions

Enzymes (or biocatalysts) are one of the most important proteins formed within living cells because they act as biological catalysts to speed up chemical reactions (Chemist's Corner 3-10). Thousands of enzymes reside in a single cell, each one a catalyst that facilitates a specific chemical reaction. Without enzymes, reactions would occur in a random and indiscriminate manner. The lock-and-key concept describes enzyme action (Figure 3-28). An enzyme combines with a substance, called a **substrate,** catalyzing or speeding up a reaction, which releases a product. The enzyme is freed unchanged after the reaction and is able to react with another substrate, yielding another product.

***Enzyme Nomenclature*** The names of most enzymes end in *-ase.* Enzymes are usually named after the substrate they act upon or the resulting type of chemical reaction. For example, sucrase is the enzyme that acts on sucrose, and lactase breaks down lactose to glucose and galactose. This general nomenclature rule does not always apply; the enzyme papain is named after papaya, from which it is derived, and ficin gets its name from figs. These enzymes, obtained from fruits, are used in meat tenderizers to break down meat's surface proteins.

***Structure of Enzymes*** The overall structure of an enzyme, called the

## CHEMIST'S CORNER 3-10

### Enzyme Classification

Most enzymes are grouped into one of six different classes according to the type of reaction they catalyze (49). Hydrolases are the most common enzymes used by the food industry; they catalyze hydrolysis reactions (37). These hydrolytic enzymes break, or cleave, a chemical bond within a molecule by adding a molecule of water. Water actually is broken apart as its two hydrogens and oxygen become part of the two new molecules formed. Examples of hydrolases include lipases that hydrolyze lipids, proteases that hydrolyze protein, and amylases that hydrolyze starch.

Another type of enzyme, oxidoreductase, catalyzes oxidation-reduction reactions. This type includes dehydrogenases, which act by removing hydrogen, and oxidases, which add oxygen.

Lyases assist in breaking away a smaller molecule, such as water, from a larger substrate. Transferases, as their name implies, transfer a group from one substrate to another. Ligases catalyze the bonding of two molecules. The last type, isomerases, transfer groups within molecules to yield isomeric forms.

holoenzyme, contains both a protein and nonprotein portion. Most of the enzyme is protein, but the nonprotein portion, which is necessary for activity, is either a coenzyme (usually a vitamin) or a cofactor (usually a mineral).

***Factors Influencing Enzyme Activity*** Enzymes are readily inactivated and will only operate under mild conditions of pH and temperature. Because enzymes are primarily protein, they are subject to denaturation caused by extremes in temperature or pH or even by physical and/or chemical influences. Every enzyme has an optimal temperature and pH for its operation, but most do best in the 95°F to 104°F (35°C to 40°C) range and with a pH near neutral.

***Food Industry Uses*** Many foods would not be on the market if it were

**FIGURE 3-28** Lactase enzyme hydrolyzing lactose to glucose and galactose.

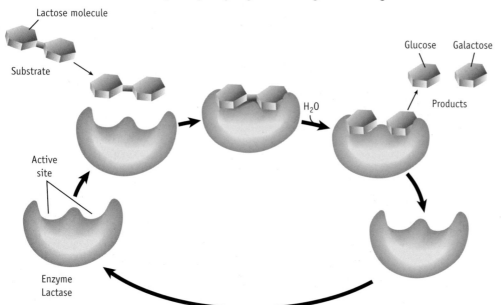

Lactose molecule

Substrate

Glucose   Galactose

$H_2O$

Products

Active site

Enzyme Lactase

Enzymes are like 3-D puzzle pieces that fit alone or with the help of a coenzyme (usually a vitamin) or a cofactor (usually a mineral).

Enzymes catalyze thousands of reactions per second without being changed themselves.

not for certain enzymes. Foods that can be manufactured with the aid of enzymes include wines, cheeses, corn syrups, yogurt, cottage cheese, baked goods, sausages, juices, egg white replacers, the artificial sweetener aspartame, and various Asian foods relying on molds (37, 38). Examples include:

- Rennin, also known as chymosin, aids in cheese production by converting milk to a curd.
- Meats can be tenderized with the enzymes of papain, bromelain, and/or ficin.
- Phenol oxidase imparts the characteristic dark hue to tea, cocoa, coffee, and raisins.
- Glucose oxidase has been used for decades in the desugaring of eggs, flour, and potatoes, and in the preparation of salad dressings.
- Manufacturers of baked products use enzymes to retard staling, improve flour and dough quality, bleach flour, and enhance crust color.
- Enzymes can also be used in improving the filtration of beer (4).

Fruit juice processors use enzymes to increase juice yields, enhance juice clarity, improve filtration, reduce bitterness, and speed fruit dehydration. The enzymes most commonly used by fruit juice processors are pectinase, cellulase,

hemicellulase, amylase, and arabinase. The bitter compounds in grapefruit juice—naringin and limonin—can be hydrolyzed with naringinase and limonase, respectively (21).

Sometimes the food industry is more interested in inhibiting the action of enzymes. This is the case for lipoxygenase activity in milk, which produces off-flavors (2). The vulnerability of enzymes to high temperatures makes it easy to destroy enzymes that cause the spoilage of fruits and vegetables. Briefly submerging foods (usually vegetables) in boiling water denatures the enzymes that contribute to deterioration. Pasteurization of milk, which is intended to kill harmful bacteria, also halts enzyme activity.

Another major use of enzymes by the food industry is in quality testing of a variety of food products (Table 3-4).

**TABLE 3-4** Use of Enzymes by the Food Industry to Test for Food Quality

| For This Food | Use This Enzyme | To Test for |
|---|---|---|
| Fruits and vegetables | Peroxidase | Proper heat treatment |
| Milk, dairy products | Alkaline phosphatase | |
| Eggs | ß-Acetylglucosaminidase | |
| Oysters | Malic enzyme | Freezing and thawing |
| Meat | Glutamate oxaloacetate | |
| Meat, eggs | Acid phosphatase | Bacterial contamination |
| Beans | Catalase | |
| Potatoes | Sucrose synthetase | Maturity |
| Pears | Pectinase | |
| Fish | Lysolecithinase | Freshness |
| | Xanthine oxidase | Hypoxanthine content |
| Flour | Amylase | Sprouting |
| Wheat | Peroxidase | |
| Coffee, wheat | Polyphenol oxidase | Color |
| Meat | Succinic dehydrogenase | |

A test for ensuring that adequate pasteurization temperatures have been reached is to measure the activity of the phosphatase enzyme that naturally exists in milk. Lack of phosphatase activity indicates that sufficient heat was applied to destroy harmful microorganisms. Fish quality can be measured by using xanthine oxidase, which acts on hypoxanthine, a compound that increases as the fish spoils (21). A strip of absorbent paper soaked partially in xanthine oxidase can be used aboard ships, dockside, or in a food processing plant. The strip of paper is moistened in fish extracts and then observed for color intensity, which is correlated to freshness. Enzymes can also be used to detect bacterial contamination in meat, poultry, fish, and dairy products.

## Buffering

Proteins have the unique ability to behave as buffers, compounds that resist extreme shifts in pH (Chemist's Corner 3-11). The buffering capacity of proteins is facilitated by their **amphoteric** nature.

## Browning

Proteins play a very important role in the browning of foods through two chemical reactions: the Maillard reaction and enzymatic browning.

*Maillard Reaction*   The brown color produced during the heating of many different foods comes, in part, from the **Maillard reaction.** This reaction contributes to the golden crust of baked products, the browning of meats, and the

dark color of roasted coffee. Temperatures most conducive to the Maillard reaction are those reaching at least 194°F (90°C), but browning can occur at lower temperatures, as seen in dried milk that has been stored too long.

*Enzymatic Browning*   Enzymatic browning is the result of an entirely different mechanism than the Maillard reaction. It requires the presence of three substances: oxygen, an enzyme (polyphenolase), and a phenolic compound (Figure 3-29).

Another type of enzymatic browning occurs when the enzyme tyrosinase oxidizes the amino acid tyrosine to result in dark-colored melanin compounds such as those observed in browning mushrooms (21). Although the browning from either phenolase or tyrosinase is unappealing in itself, it is harmless.

> ### ? How & Why?
>
> *Why does an apple turn brown when you take a bite out of it and then let it sit?*
>
> Enzymatic browning is responsible for the discoloration seen on the cut surface of certain fruits and vegetables. Normally, the cell structure separates the enzymes from the phenolic compounds in the fruit. When the vegetable or fruit is cut or bruised, however, the phenols and enzymes, thus exposed to oxygen, react in its presence to produce brown-colored products. Not all fruits and vegetables contain phenolic compounds, but sliced apples, pears, bananas, and eggplants turn brown rather rapidly after cutting. Potatoes turn slightly pink or gray.

**Amphoteric** Capable of acting chemically as either acid or base.

**Maillard reaction** The reaction between a sugar (typically reducing sugars such as glucose/dextrose, fructose, lactose, or maltose) and a protein (specifically the nitrogen in an amino acid), resulting in the formation of brown complexes.

**Enzymatic browning** A reaction in which an enzyme acts on a phenolic compound in the presence of oxygen to produce brown-colored products.

**FIGURE 3-29**   Enzymatic browning.

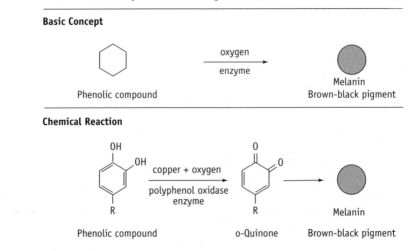

**Basic Concept**

Phenolic compound →(oxygen / enzyme)→ Melanin Brown-black pigment

**Chemical Reaction**

Phenolic compound →(copper + oxygen / polyphenol oxidase enzyme)→ o-Quinone → Melanin Brown-black pigment

# VITAMINS AND MINERALS

## Foods High in Vitamins and Minerals

Most foods contain some vitamins and minerals. Vitamins can be categorized into two major groups: fat-soluble (A, D, E, and K) or water-soluble (B complex and vitamin C). Minerals may be termed either macro or micro (Table 3-5). Meats are good sources of B vitamins, iron (Fe), and zinc (Zn). Dairy foods provide about 80 percent of the average American's daily calcium (Ca). Vitamin C (ascorbic acid) is found only in plants. All the fat-soluble vitamins (A, D, E, and K) are found in an egg yolk. Vitamin $B_{12}$ is found naturally only in foods of animal origin or fermented foods such as tempeh, tofu, and miso, which contain bacteria that produce vitamin $B_{12}$ as a by-product. The two major sources of sodium (Na) in the diet are processed foods and the saltshaker.

## Composition of Vitamins and Minerals

Vitamins are organic (carbon-containing) compounds, each with a unique chemical composition. Minerals are inorganic elements and are depicted in the periodic table. Unlike vitamins, minerals cannot be destroyed by heat, light, or oxygen. Vitamins and minerals do not provide calories.

## Functions of Vitamins and Minerals in Food

Vitamins and minerals regulate metabolic functions. Because of this, they are often added to foods.

### Enrichment and Fortification

Because of the vital role vitamins and minerals play in the body's processes, many foods are now **enriched** or **fortified** with additional vitamins and minerals. During processing and preparation, foods such as wheat and rice may lose some of their vitamin or mineral content. Some of the nutrients, such as vitamin $B_1$ (thiamin), vitamin $B_2$ (riboflavin), niacin, and iron (calcium optional), may be added back to the processed food (enrichment). Fortification is intended to deliver nutrients to the general public in an effort to deter certain nutrient deficiencies (28). In 1922, salt became the first food ever to be fortified, with the addition of iodine. Iodine deficiencies were resulting in goiter (enlarged thyroid gland) and cretinism (dwarfism, mental retardation) in children born of mothers who had not ingested sufficient iodine. Other nutrients that are used to fortify foods include vitamins A and D (milk), calcium (orange juice), and/or folate, a B vitamin (cereal products) (25, 52).

The decision to fortify with a particular nutrient is a complex one. It starts with the realization that a significant number of people are not obtaining desirable levels of a specific nutrient and the determination that the food to be fortified makes an appreciable contribution to the diet. It must be further ascertained that the fortification will not result in an essential nutrient imbalance, that the nutrient is stable under storage and capable of being absorbed from the food, and that toxicity from excessive intakes will generally not occur (39).

### Antioxidants

Certain nutrients, especially vitamins A, C, and E and the mineral selenium, may also be added to foods to act as **antioxidants** (15, 16).

***Food Industry Uses*** These compounds neutralize **free radicals** (Figure 3-30), leading to an increased shelf life (9). Foods to which antioxidants are commonly added include dry cereals, crackers, nuts, chips, and flour mixes. Consumer interest in antioxidants and health has also caused manufacturers to add additional amounts of these nutrients to other food products.

### Sodium

Another compound in the vitamin/mineral category that is used to preserve foods is salt, the only mineral directly consumed by people.

---

**TABLE 3-5**   Major Vitamins and Minerals in Foods

### Vitamins

| Water Soluble | Fat Soluble |
| --- | --- |
| B complex: | Vitamin A |
|   Thiamin ($B_1$) | Vitamin D |
|   Riboflavin ($B_2$) | Vitamin E |
|   Vitamin $B_6$ (pyridoxine) | Vitamin K |
|   Vitamin $B_{12}$ (cobalamin) | |
|   Niacin | |
|   Folate | |
|   Pantothenic acid | |
|   Biotin | |
| Ascorbic acid (vitamin C) | |

### Minerals

| Macrominerals (minerals present in the body in relatively large amounts) | Microminerals (minerals present in the body in relatively small amounts) |
| --- | --- |
| Calcium (Ca) | Iron (Fe) |
| Phosphorous (P) | Zinc (Zn) |
| Potassium (P) | Selenium (Se) |
| Sulfur (S) | Manganese (Mn) |
| Sodium (Na) | Copper (Cu) |
| Chlorine (Cl) | Iodine (I) |
| Magnesium (Mg) | Molybdenum (Mo) |
| | Chromium (Cr) |
| | Fluorine (F) |

**Enriched**  Foods that have had certain nutrients, which were lost through processing, added back to levels established by federal standards.

**Fortified**  Foods that have had nutrients added that were not present in the original food.

**Antioxidant**  A compound that inhibits oxidation, which can cause deterioration and rancidity.

**Free radical**  An unstable molecule that is extremely reactive and that can damage cells.

**FIGURE 3-30**   Antioxidant action of vitamin E.

**1.** A chemically reactive oxygen free radical attacks fatty acids, forming other free radicals.

**2.** This initiates a rapid, destructive chain reaction.

**3.** The result is deterioration of foods containing unsaturated fatty acids.

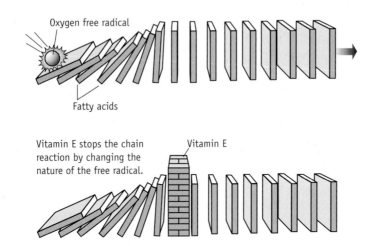

Oxygen free radical

Fatty acids

Vitamin E stops the chain reaction by changing the nature of the free radical.

Vitamin E

*Food Industry Uses*   Salt is used for more than preservation of certain foods. It provides flavor to so many processed foods that it is now the second most common food additive by weight, after sugar, in the United States. Salt in a variety of forms can be purchased and added to products by the food industry: topping salt for saltine crackers, breadsticks, and snack crackers; surface-salting for pretzels, soft pretzels, and bagels; fine crystalline salt for potato chips, corn chips, and similar snacks; blending/dough salts for flour and cake mixes; light salt (potassium chloride) for reducing sodium levels; and encapsulated salt for frozen doughs (29).

# NONNUTRITIVE FOOD COMPONENTS

Not everything found in food can be classified within the six basic nutrient groups. All sorts of substances—natural, intentional, and/or unintentional—may be present. Natural compounds, whether beneficial or harmful, are discussed in this chapter. The intentional and unintentional compounds are food additives discussed here and in Chapter 29. Among the beneficial compounds naturally found in foods are those that provide color and flavor, along with certain plant compounds.

# Food Additives

According to the FDA, a food additive is any substance added to food (Figure 3-31). The more precise legal definition is "any substance the intended use of which results or may reasonably be expected to result—directly or indirectly—in its becoming a component or otherwise affecting the characteristics of any food (FDA)." Over 3,000 food additives exist, but the three most common, by weight, are salt, sugar, and corn syrup. Others that are commonly used include citric acid (from oranges and lemons), baking soda, vegetable colors, mustard, and pepper. These food additives account by weight for 98 percent of all food additives listed on packaged food labels (26).

## Everything Added to Food in the United States (EAFUS)

EAFUS is an online list of food additives used in U.S. foods that is maintained by the FDA. It contains information on over 3,000 substances added directly to food that are approved by the FDA or listed or affirmed as Generally Recognized As Safe (GRAS) (13).

# Purposes of Food Additives

The purposes of food additives, according to FDA regulations, are to meet one or more of the four objectives listed below:

1. Improve the appeal of foods by improving their flavor, smell, texture, or color
2. Extend storage life
3. Maximize performance
4. Protect nutrient value

Many types of ingredients are used to meet these objectives, and a few examples are listed in Table 3-6. The purposes of food additives are now addressed in more detail.

## Improve Appeal

Food appeal can be improved by the addition of coloring, flavoring, and texture-enhancing agents.

**FIGURE 3-31**   Food additives are available in powder, granule, liquid, or other forms.

jorgegonzalez/istockphoto.com

**TABLE 3-6**  Types of Food Additives

**The purpose, use, and names of common food additives.**

| Types of Ingredients | What They Do | Examples of Uses | Names on Labels |
|---|---|---|---|
| Preservatives | Prevent food spoilage from bacteria, molds, fungi, or yeast (antimicrobials); slow or prevent changes in color, flavor, or texture; maintain freshness | Fruit sauces and jellies, beverages, baked goods, cured meats, oils and margarines, cereals, dressings, snack foods, fruits, and vegetables | Ascorbic acid, citric acid, sodium benzoate, calcium propionate, sodium erythorbate, sodium nitrite, calcium sorbate, potassium sorbate, BHA, BHT, EDTA, tocopherols (vitamin E) |
| Sweeteners | Add sweetness with or without adding calories | Beverages, baked goods, confections, table-top sugar, substitutes, many processed foods | Sucrose (sugar), glucose, fructose, sorbitol, mannitol, corn syrup, high-fructose corn syrup, saccharin, aspartame, sucralose, acesulfame potassium (acesulfame-K), neotame |
| Color Additives | Offset color loss due to exposure to light, air, temperature extremes, moisture and storage conditions; correct natural color variations; enhance natural colors; provide color to colorless and "fun" foods | Many processed foods (candies, snack foods, margarine, cheese, soft drinks, jams/jellies, gelatins, pudding and pie fillings) | FD&C Blue Nos. 1 and 2, FD&C Green No. 3, FD&C Red Nos. 3 and 40, FD&C Yellow Nos. 5 and 6, Orange B, Citrus Red No. 2, annatto extract, beta-carotene, grape skin extract, cochineal extract or carmine, paprika oleoresin, caramel color, fruit and vegetable juices, saffron (Exempt color additives are not required to be declared by name on labels but may be declared simply as colorings or color added) |
| Flavors and Spices | Add specific flavors (natural and synthetic) | Pudding and pie fillings, gelatin dessert mixes, cake mixes, salad dressings, candies, soft drinks, ice cream, BBQ sauce | Natural flavoring, artificial flavor, and spices |
| Flavor Enhancers | Enhance flavors already present in foods | Many processed foods | Monosodium glutamate (MSG), hydrolyzed soy protein, autolyzed yeast extract, disodium guanylate or inosinate |
| Fat Replacers (and components of formulations used to replace fats) | Provide creamy mouthfeel in reduced-fat foods | Baked goods, dressings, frozen desserts, confections, cake and dessert mixes, dairy products | Olestra, cellulose gel, carrageenan, polydextrose, modified food starch, microparticulated egg white protein, guar gum, xanthan gum, whey protein concentrate |
| Nutrients | Replace vitamins and minerals lost in processing (enrichment), add nutrients that may be lacking in the diet (fortification) | Flour, breads, cereals, rice, macaroni, margarine, salt, milk, fruit beverages, energy bars, instant breakfast drinks | Thiamine hydrochloride, riboflavin (vitamin $B_2$), niacin, niacinamide, folate or folic acid, beta-carotene, potassium iodide, iron or ferrous sulfate, alpha-tocopherols, ascorbic acid, vitamin D, amino acids (L-tryptophan, L-lysine, L-leucine, L-methionine) |
| Emulsifiers | Allow smooth mixing of ingredients, prevent separation. Keep mixed products stable, reduce stickiness, control crystallization, keep ingredients dispersed, and help products dissolve more easily | Salad dressings, peanut butter, chocolate, margarine, frozen desserts | Soy lecithin, mono- and diglycerides, egg yolks, polysorbates, sorbitan monostearate |
| Stabilizers and Thickeners, Binders, Texturizers | Produce uniform texture, improve mouthfeel | Frozen desserts, dairy products, cakes, pudding and gelatin mixes, dressings, jams and jellies, sauces | Gelatin, pectin, guar gum, carrageenan, xanthan gum, whey |
| pH Control Agents and Acidulants | Control acidity and alkalinity, prevent spoilage | Beverages, frozen desserts, chocolate, low-acid canned foods, baking powder | Lactic acid, citric acid, ammonium hydroxide, sodium carbonate |
| Leavening Agents | Promote rising of baked goods | Breads and other baked goods | Baking soda, monocalcium phosphate, calcium carbonate |
| Anti-Caking Agents | Keep powdered foods free-flowing, prevent moisture absorption | Salt, baking powder, confectioner's sugar | Calcium silicate, iron ammonium citrate, silicon dioxide |
| Humectants | Retain moisture | Shredded coconut, marshmallows, soft candies, confections | Glycerin, sorbitol |

*(continued)*

**TABLE 3-6**   Types of Food Additives (*continued*)

The purpose, use, and names of common food additives.

| Types of Ingredients | What They Do | Examples of Uses | Names on Labels |
|---|---|---|---|
| Yeast Nutrients | Promote growth of yeast | Breads and other baked goods | Calcium sulfate, ammonium phosphate |
| Dough Strengtheners and Conditioners | Produce more stable dough | Breads and other baked goods | Ammonium sulfate, azodicarbonamide, L-cysteine |
| Firming Agents | Maintain crispness and firmness | Processed fruits and vegetables | Calcium chloride, calcium lactate |
| Enzyme Preparations | Modify proteins, polysaccharides, and fats | Cheese, dairy products, meat | Enzymes, lactase, papain, rennet, chymosin |
| Gases | Serve as propellant, aerate, or create carbonation | Oil cooking spray, whipped cream, carbonated beverages | Carbon dioxide, nitrous oxide |

**Source:** Adapted by Amy Brown from International Food Information Council (IFIC) and U.S. Food and Drug Administration. "Food Ingredients & Colors." http://www.fda.gov/Food/FoodIngredientsPackaging/ucm094211.htm. Accessed 3/27/10.

*Color Compounds*   Food additives in this category can be any dye, pigment, or substance that imparts color. Color additives are used to offset color loss due to exposure to air, moisture, temperature changes, light, or storage conditions; enhance natural colors already existing within the food; keep colors that may naturally vary consistent; and provide color to colorless or "fun" foods (14). Foods would be dramatically different without added colors: imagine clear colas, white margarines, and nongreen mint ice-cream. Colors added to foods may be synthetic (certified) or natural.

**Certified Colors**   Certified colors are synthetically produced (manufactured) to create colors that are uniform in hue and blend more easily than natural colors. Examples of certified colors include nine Food, Drug, and Cosmetic (FD&C) colors: Blue No. 1 and 2, Green No. 3, Yellow No. 5 and 6, Red No. 3 and 40, Orange B, and Citrus Red No. 2. FD&C Yellow No. 5 might cause hives in a few people (less than one out of 10,000), so the law requires that it be listed as an ingredient.

**? How & Why?**

*What is the difference between dyes and lake color additives?*

Dyes and lakes are both certified colors, but dyes dissolve in water, while lakes do not. As a result, dyes are less stable than the lakes that are often used in fat-based products (FDA).

**Natural Colors**   Colors exempt from certification include natural colors that are obtained from vegetable, animal, or mineral sources (or synthetic replicas of them) (14). Examples of natural colors include annatto extract (yellow), dehydrated beets (bluish-red to brown), caramel (yellow to tan), beta-carotene (yellow to orange), and grape skin extract (red, green).

**Use of Colors in Foods**   Food is made more appetizing and interesting to behold by the wide spectrum of colors made possible through pigments. One way these pigments are classified by food scientists is according to the following four categories: (1) shiny (diffuse reflection), (2) glossy (specular or mirror-like reflection), (3) opaque or cloudy (diffuse transmission), or (4) translucent (specular transmission) (16). Most of the natural pigments contributing to color are found in fruits and vegetables. The colors of foods from animal products and grains are less varied and bright. The three dominant color pigments found in most plants are carotenoids (orange-yellow), chlorophyll, and flavonoids (blue, cream, red). These plant pigments, their colors, and food examples are explained in more detail in Chapter 13 on vegetables.

Although foods of animal origin are less colorful, even meat varies in color depending on its stage of maturity (33). When first sliced with a knife, a cut of beef is purplish red from the presence of a pigment called myoglobin. As it is exposed to air, the myoglobin combines with oxygen to turn the meat a bright red color. The meat then turns grayish brown during cooking when the protein holding the pigment becomes denatured.

Cured meats present an altogether different scenario as added nitrites, compounds that are used as a preservative, react with the myoglobin to cause the meat to be a red color, which converts to pink (denatured protein) when cooked. Meat pigments are covered in more detail in Chapter 7 on meats.

Milk appears white as light reflects off the colloidal dispersion of milk protein. The yellowish hue of cream comes from carotene and riboflavin (vitamin $B_2$). Carotene, a fat-soluble pigment, is also the substance that gives butter its yellow color.

*Flavor Compounds*   The flavors in foods are usually mixtures of substances derived from both nutrient and nonnutrient compounds (46). These are sometimes too numerous to track as the source of a specific flavor. Like colors, flavoring agents can also be synthetic, like saccharin, or natural, like fruit extracts, juice concentrates, processed fruits, fruit purées, spice resins, and monosodium glutamate (MSG) (30).

Organic acids found naturally in foods, or sometimes added, determine whether the food is acidic or basic. An acidic pH in foods not only contributes to a sour taste, but also the color of fruit juices, the hue of chocolate in baked products, and the release of carbon dioxide in a flour mixture. An alkaline pH contributes a bitter taste and soapy mouthfeel to foods.

*Texture Compounds*   Food additives are also used to add body and texture to foods. For example, thickeners generate a smooth texture in ice cream by preventing ice crystals from forming.

## Extend Storage Life

By reducing the rancidity of fats, food storage life can be extended with additives such as butylated hydroxyanisole (BHA) and butylated hydroxytoluene (BHT), which slow or prevent food deterioration.

## Maximize Performance

Emulsifiers, stabilizers, and other additives maximize the performance of foods. Emulsifiers make it possible to distribute tiny particles of one liquid into another, thereby preventing immiscible liquids from separating. For example, emulsifiers prevent the oil from separating out of peanut butter. Stabilizers and thickeners give milkshakes body and a smoother feel in the mouth. If the need exists to alter the pH of a food, certain compounds can be added to achieve the necessary acidity or alkalinity. Some additives retain moisture, whereas anticaking agents prevent moisture from lumping up powdered sugar or other finely ground powdered substances.

## Protect Nutrient Value

Food additives that protect nutrient value include vitamins and minerals that are added to enrich or fortify foods (28). Enriched foods have the vitamins thiamin ($B_1$), riboflavin ($B_2$), folate, and niacin, and the minerals iron and sometimes calcium added back to levels established by federal standards for breads and cereals. Most table salt is fortified with iodine to help prevent goiter. Milk is fortified with vitamin D to help prevent rickets. Many fruit drinks are fortified with vitamin C, which tends to be missing in the diets of people who do not consume sufficient amounts of fruits and vegetables.

# Plant Compounds

In addition to color and flavor compounds, some plants contain other nonnutritive substances that, when ingested, may have either beneficial or harmful effects.

## Beneficial

Many of the possible anticarcinogens, or compounds that inhibit cancer, come from plants (1, 34). In particular, phytochemicals, a special group of substances found in plants, appear to

**TABLE 3-7** Phytochemicals: Potential Cancer Protectors

| Phytochemical Family | Major Food Sources |
|---|---|
| Allyl sulfides | Onions, garlic, leeks, chives |
| Carotenoids | Yellow and orange vegetables and fruits; dark green, leafy vegetables |
| Flavonoids | Most fruits and vegetables |
| Indoles | Cruciferous vegetables (broccoli, cabbage, kale, cauliflower, etc.) |
| Isoflavones | Soybeans (tofu, soy milk) |
| Isothiocyanates | Cruciferous vegetables |
| Limonoids | Citrus fruits |
| Lycopene | Tomatoes, red grapefruit |
| Phenols | Nearly all fruits and vegetables |
| Phenolic acids | Tomatoes, citrus fruits, carrots, whole grains, nuts |
| Plant sterols | Broccoli, cabbage, cucumbers, squash, yams, tomatoes, eggplant, peppers, soy products, whole grains |
| Polyphenols | Green tea, grapes, wine |
| Saponins | Beans and legumes |
| Terpenes | Cherries, citrus fruit peel |

have a protective effect against cancer (35). One class of these phytochemicals, called indoles, is found in vegetables such as cabbage, cauliflower, kale, kohlrabi, mustard greens, swiss chard, and collards. Laboratory animals given indoles and then exposed to carcinogens developed fewer tumors than animals exposed to the same carcinogens, but not given indoles. A few of the plant substances that appear to protect against cancer are listed in Table 3-7.

## Harmful

There are several potentially harmful substances in plants (19, 34). The U.S. Department of Health and Human Services has gone so far as to say that natural toxins are so widespread that the only way to avoid them completely is to stop eating. Other substances, although not strictly toxins, can cause problems for certain people if ingested in excess. One such substance is caffeine (41).

*Caffeine*   Caffeine is a natural stimulant that belongs to a group of compounds called methylxanthines. The most widely used sources of caffeine include coffee beans, tea leaves, cocoa beans, and cola nuts. Caffeine is also found in the leaves of some plants, where its high concentrations in the leaves provide protection from insects seeking something to eat.

Caffeine ingested at high concentrations may temporarily increase heart rate, basal metabolic rate, secretion of stomach acid, and urination. The increased secretion of stomach acids may cause problems for people with ulcers. In healthy adults, however, a moderate intake of caffeine does not appear to cause health problems. Individuals who habitually drink a lot of caffeine-containing beverages may, however, experience withdrawal headaches and irritability if they stop drinking the beverage. Another possible side effect in sensitive individuals is fibrocystic conditions in the breast, which are painful but usually harmless lumps within the breasts and under the arms (12).

Excessive caffeine intake, defined as more than five 5-ounce cups of strong, brewed coffee daily, can also cause "coffee nerves." *The Diagnostic and Statistical Manual of Mental Disorders* published by the American Psychiatric Association defines caffeine intoxication as exhibiting at least five of the following symptoms: nervousness, agitation, restlessness, insomnia, frequent urination, gastrointestinal disturbance, muscle twitching, rambling thought and speech, periods of exhaustion, irregular or rapid heartbeat, and psychomotor agitation. Infants who ingest caffeine through their mother's milk may also get the "jitters." Because infants are unable to metabolize caffeine efficiently, the compound may stay in their system up to a week, compared to about 12 hours in a healthy adult.

# PICTORIAL SUMMARY / 3: Chemistry of Food Composition

Truly, we are what we eat. Food provides energy (kilocalories) and nutrients, which are needed for the maintenance, repair, and growth of cells. Understanding food chemistry is important in planning good nutrition.

## BASIC FOOD CHEMISTRY

The six major nutrient groups are:
- Water
- Carbohydrates
- Proteins
- Lipids
- Vitamins
- Minerals

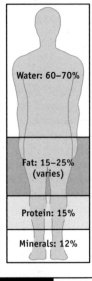

Water: 60–70%

Fat: 15–25% (varies)

Protein: 15%

Minerals: 12%

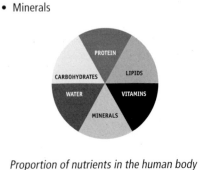

*Proportion of nutrients in the human body*
- Most foods contain a combination of these six groups. The main purpose of eating and drinking is to provide calories and to replace those nutrients used up in the body's maintenance, repair, and growth.

## WATER

Water (H$_2$O) is the simplest but most important of all nutrients. The human body contains about 60 to 70 percent water, and water concentration in foods ranges from 70 to 95 percent in fruits, vegetables, and meats to less than 15 percent in grains, dried beans, and fats. In food preparation, water acts as a heat-transferring agent and a universal solvent, plays a crucial role in preservation, and is involved in the formation of numerous solutions and colloidal dispersions, suspensions, and emulsions.

## OTHER FOOD COMPONENTS

Other components in foods include beneficial or harmful plant compounds and food additives used to (1) improve appeal, (2) extend storage life, (3) maximize performance, and (4) protect nutrient value. Both natural and synthetic food additives are available; two examples are color compounds (carotenoids, chlorophylls, flavonoids) and flavor compounds.

## PROTEINS

Proteins are essential to proper growth and maintenance. They differ from carbohydrates and lipids in that they contain nitrogen. Proteins consist of amino acids linked together by peptide bonds. Plant proteins, with the exception of the protein in legumes (beans, peas, and lentils), lack all the essential amino acids and are therefore "incomplete."

## LIPIDS OR FATS

Lipids, which are derived from both plant and animal sources, include fats and oils. Foods that are high in fat:

**Animal sources**
- Meats
- Poultry
- Dairy products

**Plant sources**
- Seeds and nuts
- Vegetable oils
- Avocados, olives, coconut

Three major lipid groups:
- Triglycerides
- Phospholipids
- Sterols

## CARBOHYDRATES

Carbohydrates are the sugars, starches, and fibers found primarily in plants. The basic building block of carbohydrates is the saccharide, which is composed of carbon, hydrogen, and oxygen. Starches consist of amylose and amylopectin. Fiber can be characterized in several ways, including crude or dietary, soluble or insoluble, functional, or total.

## VITAMINS AND MINERALS

Both vitamins and minerals function at the cellular level. Neither provides calories and both are found to some degree in most foods. Some foods are enriched (nutrients added that were lost in processing) or fortified (nutrients added that were not originally in the food) with vitamins and/or minerals. Salt is one of the few minerals used for a functional purpose in foods, specifically its ability to act as a preservative.

## NONNUTRITIVE FOOD COMPONENTS

Nonnutritive components in foods include color compounds (carotenoids, chlorophylls, flavonoids), flavor compounds, and beneficial or harmful plant compounds.

# CHAPTER REVIEW AND EXAM PREP

## Multiple Choice*

1. What is the word used to describe the process by which water splits a larger molecule into two smaller molecules?
   a. Hydrolysis
   b. Fission
   c. Hydration
   d. Fusion

2. Sucrose is an example of a(n) _____.
   a. monosaccharide
   b. disaccharide
   c. oligosaccharide
   d. polysaccharide

3. Which of the following groups is a good source of polyunsaturated fatty acids?
   a. Meats
   b. Fruits
   c. Dairy products
   d. Most vegetable oils

4. Orange juice that has had calcium added during processing is considered to be _____.
   a. hydrolyzed
   b. fortified
   c. enriched
   d. purified

5. Which of the following words is used to describe the group of nonnutritive plant compounds that may have protective effects against certain types of cancer?
   a. Triglycerides
   b. Vitamins
   c. Minerals
   d. Phytochemicals

*See p. AK-1 for answers to multiple choice questions.

6. Which of the following foods is an example of a complete protein source?
   a. Egg
   b. Spinach
   c. Jell-O
   d. Carrot

7. Which of the following foods is an example of a monosaturated fat source?
   a. Butter
   b. Steak
   c. Fish
   d. Olive

## Short Answer/Essay

1. List the six basic nutrient groups and describe the basic composition of each group. What is the caloric contribution (kcal) of each group to the diet?
2. Does the addition of solutes to water increase or decrease its freezing point?
3. Define *solute, solvent,* and *solution,* and explain how they differ from colloidal dispersions.
4. Define *precipitate, electrolyte, emulsion, hydrolysis,* and *water activity.*
5. List and describe the most common monosaccharides, disaccharides, oligosaccharides, and polysaccharides.
6. Describe the basic structure of a triglyceride, and explain the two primary ways in which fatty acids differ.
7. Describe the basic structure of an amino acid. What is the difference between complete and incomplete protein?
8. Discuss the various functions of protein in foods.
9. What is the difference between a food being enriched and its being fortified?
10. List and describe three nonnutritive compounds found in food.

# REFERENCES

1. Ahmad N, and H Muck. Green tea polyphones and cancer: Biological mechanisms and practical implications. *Nutrition Reviews* 57(3):78–83, 1999.

2. Ashie INA, BK Simpson, and JP Smith. Mechanisms for controlling enzymatic reactions in foods. *Critical Reviews in Food Science and Nutrition* 36(1&2):1–30, 1996.

3. Baniel A, A Fains, and Y Popineau. Foaming properties of egg albumen with a bubbling apparatus compared with whipping. *Journal of Food Science* 62(2):377–381, 1997.

4. Berne S, and CD O'Donnell. Filtration systems and enzymes: A tangled web. *Prepared Foods* 165(9):95–96, 1996.

5. Cai R, and SD Arntfield. Thermal gelation in relation to binding of bovine serum albumin-polysaccharide systems. *Journal of Food Science* 62(6):1129–1134, 1997.

6. Cornec M, et al. Emulsion stability as affected by competitive absorption between an oil-soluble emulsifier and milk proteins at the interface. *Journal of Food Science* 63(1):39–43, 1998.

7. Crittenden RG. Functional polymers for the next millennia. *Prepared Foods* 166:123–124, 1997.

8. Dartey CK, and GR Sanderson. Use of gums in low-fat spreads. *Inform* 7(6):630–634, 1996.

9. Elliot JG. Application of antioxidant vitamins in foods and beverages. *Food Technology* 53(2):46–48, 1999.

10. Ensminger AH. *Foods and Nutrition Encyclopedia*. CRC Press, 1994.

11. Fats and fattening: Fooling the body. *Prepared Foods* 166(7):51–53, 1997.

12. Ferrini RL, and E Barrett-Connor. Caffeine intake and endogenous sex steroid levels in post-menopausal women. *American Journal of Epidemiology* 144(7):642–644, 1996.

13. Food and Drug Administration. *EAFUS: A food additive database*. www.cfsan.fda.gov/~dms/eafus. html. Accessed 4/15/09.

14. Food and Drug Administration. *Food ingredients & colors*. 2004. http://www.fda.gov/ForConsumers/ ConsumerUpdates/ucm048951.htm. Accessed 02/26/10.

15. Giese J. Antioxidants: Tools for preventing lipid oxidation. *Food Technology* 50(11):73–78, 1996.

16. Giese J. Color measurement in foods quality parameter. *Food Technology* 54(2):62, 2000.

17. Grabitske HA, and L Slavin. Low-digestible carbohydrates in practice. *Journal of the American Dietetic Association* 108(10):1677–1681, 2008.

18. Hamilton EMN, and SAS Gropper. *The Biochemistry of Human Nutrition*. West, 1987.

19. Horn S, et al. End-stage renal failure from mushroom poisoning with *Cortinarius orellanus*: Report of four cases and review of the literature. *American Journal of Kidney Diseases* 30(2):282–286, 1997.

20. Institute of Medicine and Food and Nutrition Board. *Dietary Reference Intakes for Energy, Carbohydrate, Fiber, Fat, Fatty Acids, Cholesterol, Protein, and Amino Acids (Macronutrients)*. The National Academies Press, 2005.

21. James J, and B Simpson. Application of enzymes in food processing. *Critical Reviews in Food Science and Nutrition* 36(5):437–463, 1996.

22. Jenkins AL, et al. Comparable postprandial glucose reductions with viscous fiber blend enriched biscuits in healthy subjects and patients with diabetes mellitus: Acute randomized controlled clinical trial. *Croatian Medical Journal* 49(6):772–782, 2008.

23. Kleiner SM. Water: An essential but overlooked nutrient. *Journal of the American Dietetic Association* 99(2):200–206, 1999.

24. Kritchevsky D. Dietary fiber. *Annual Reviews in Nutrition* 8:301–328, 1988.

25. Labin-Godscher R, and S Edelstein. Calcium citrate: A revised look at calcium fortification. *Food Technology* 50(6):96–97, 1996.

26. Lehmann P. More than you ever thought you would know about food additives. *FDA Consumer* 13(3):10–16, 1979.

27. Marcason W. What is the latest research regarding the avoidance of nuts, seeds, corn, and popcorn in diverticular disease? *Journal of the American Dietetic Association* 108(11):1956, 2008.

28. Mertz W. Food fortification in the United States. *Nutrition Reviews* 55(2):44–49, 1997.

29. Niman S. Using one of the oldest food ingredients—salt. *Cereal Foods World* 41(9):728–731, 1996.

30. O'Donnell CD. Proteins and gums: The ties that bind. *Prepared Foods* 165(4):50–51, 1996.

31. Ohr L. Nutraceuticals: Delving into dietary fiber. *Food Technology* 63(8):63–70, 2009.

32. Ohr L. Nutraceuticals and functional foods. *Food Technology* 56(6): 109–115, 2002.

33. Pegg RB, and F Shahidi. Unraveling the chemical identity of meat pigments. *Critical Reviews in Food Science and Nutrition* 37(6):561–589, 1997.

34. Phillips BJ. A study of the toxic hazard that might be associated with the consumption of green potato tops. *Food and Chemical Toxicology* 34(5):439–448, 1996.

35. Potter JD. Reconciling the epidemiology, physiology, and molecular biology of colon cancer. *Journal of the American Medical Association* 268:1573–1577, 1992.

36. Powitz RW. Water activity: A new food safety tool. *Food Safety Magazine* 24, 2007.

37. Pszczola DE. Enzymes: Making things happen. *Food Technology* 53(2):74–79, 2001.

38. Pszczola DE. From soybeans to spaghetti: The broadening use of enzymes. *Food Technology* 55:11, 2001.

39. Reilly C. Too much of a good thing? The problem of trace elements fortification of foods. *Trends in Food Science & Technology* 7(4):139–142, 1996.

40. Sanderson GR. Gums and their use in food systems. *Food Technology* 50(3):81–84, 1996.

41. Scientific status summary: Evaluation of caffeine safety. *Food Technology* 41(6):105–113, 1987.

42. Sikorski ZE (ed). *Chemical and Functional Properties of Food Proteins*. CRC Press, Danvers, MA, 2001.

43. Slavin JL. Positions of the American Dietetic Association: Health implications of dietary fiber. *Journal of the American Dietetic Association* 108(10):1716–1731, 2008.

44. U.S. Dept of Health and Human Services. Food and Drug Administration. Center for Food Safety and Applied Nutrition. Factors affecting growth of microorganisms in foods. *Foodborne Pathogenic Microorganisms and Natural Toxins Handbook*. 1997.

45. Van Garde SJ, and M Woodburn. *Food Preservation and Safety: Principles and Practice*. Iowa State University Press, 1994.

46. Waddell WJ, et al. Gras flavoring substances 23. *Food Technology* 61(8):22–49, 2007.

47. Ward FM. Hydrocolloid systems as fat memetics in bakery products: Icings, glazes and fillings. *Cereal Foods World* 42(5):386–390, 1997.

48. Wheeler ML, and X Pi-Sunyer. Carbohydrate issues: Type and amount. *Journal of the American Dietetic Association* 108:S34–S39, 2008.

49. Whitehead IM. Challenges to biocatalysts from flavor chemistry. *Food Technology* 52(2):40–46, 1998.

50. Whitney EN, CB Cataldo, and SR Rolfes. *Understanding Normal and Clinical Nutrition*. Wadsworth/ West Publishing, 2009.

51. Xie YR, and NS Hettiarachchy. Xanthan gum effects on solubility and emulsification properties of soy protein isolate. *Journal of Food Science* 62(6):1101–1104, 1997.

52. Yetley EA, and JI Rader. Folate fortification of cereal-grain products: FDA policies and actions. *Cereal Foods World* 40(2):67–72, 1995.

# WEBSITES

This FDA site providing links related to food chemistry:

**http://www.fda.gov/Food/ FoodIngredientsPackaging/ ucm113293.htm**

Learn about scientific food composition from the Journal of Food Composition and Analysis:

**http://www.elsevier.com/wps/find/ journaldescription.cws_home/622878/ description#description**

Food chemists publish their research articles in numerous journals and one of these is *Food Chemistry*, which can be found at:

**www.elsevier.com/wps/find/ journaldescription.cws_home/405857/ description#description**
(click on full journal articles)

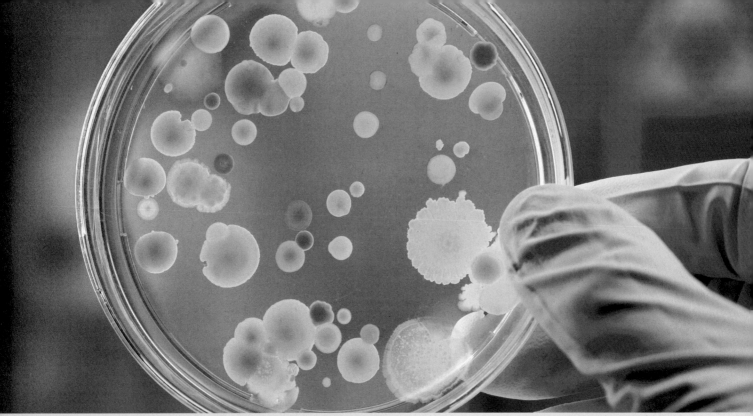

# 4 Food Safety

**Outbreak** Defined by the CDC as
the occurrence of two or more cases
of a similar illness resulting from the
ingestion of a common food.

**Foodborne illness** An illness
transmitted to humans by food.

The United States food supply is
probably the safest in the world
(13). Why is this so? Food safety is
primarily achieved by controlling contamination at the food source (44).
Federal and state regulations, along
with local inspections, require vigi-
lance at all levels of food production
and distribution. (Chapter 29 covers
government regulations in detail.)
Additionally, the Centers for Disease
Control and Prevention (CDC) track
down causal factors when even as few
as one or two **outbreaks** of **foodborne
illness** occur. The U.S. Food and Drug
Administration (FDA) maintains the
following public website with updates
on recent outbreaks:

www.fda.gov/oc/opacom/hottopics.

Finally, food manufacturers and
distributors are motivated to avoid law-
suits brought against them as a result of
negligence. Therefore, many follow food
safety practices and programs to ensure
that their food products are safe.

In spite of all these preventive measures, people still get sick from food
and beverages. Although it is difficult
to assess the total number of people
afflicted with foodborne illness each
year, the General Accounting Office estimates that as many as 76 million illnesses, 325,000 hospitalizations, and
up to 5,000 deaths can be traced to contaminated foods (81). Approximately
80 percent of these foodborne illnesses
originate at restaurants and other food
service establishments, and most of the
rest are traced to errors at home (80).
Control and prevention of restaurant-
related contamination is inconsistent
due to complicated regulatory jurisdiction and a fragmented food safety
system with uneven enforcement on
the part of various agencies (70). **Relating food manufacturers is similarly
problematic.** President Barack Obama

pointed out in a weekly White House address in March 2009 that "many of the laws and regulations governing food safety in America have not been updated since they were in written in the time of Teddy Roosevelt . . ." and the FDA has the staffing capability to inspect "just 7,000 of our 150,000 food processing plants and warehouses each year. That means roughly 95% of them go uninspected." (See www.whitehouse.gov/the_press_office/Weekly-Address-President-Barack-Obama-Announces-Key-FDA-Appointments-and-Tougher-F/.)

In some instances, criminal charges have been brought against negligent food manufacturers. In 1996, a juice company was fined when its contaminated apple juice resulted in the death of an infant. In 2001, a major food corporation was fined when the deaths of 15 people were linked to its contaminated hot dogs and lunchmeat (12).

There is also the perceived threat of intentional *food supply terrorism,* a concern that arose after September 11, 2001. The concept of *food defense,* dealing with intentional food contamination, has grown in recent years (89). Accidental contaminants are usually associated with microorganisms innate to the local environment, while intentional contaminants are more likely to be rare agents associated with very high mortality rates (76). Food biosecurity aims to keep the food supply free from planned contamination with biological, chemical, or physical hazards due to malicious and/or criminal intent (73). The Bioterrorism Act of 2002 expanded government authority to prevent, detect, and respond to potential threats. This newfound interest in food defense has resulted in the call for additional action in the form of a multidisciplinary advisory board to address risks and the development of new biosensors for mass screening of the food supply (5).

Before developing an irrational fear of foodborne illnesses, whatever their cause, it is necessary to consider the relative risk. More than 273 billion meals and an inestimable number of snacks are consumed each year in the United States. When compared with these numbers, the number of illnesses resulting from food contamination in this country is minimal. This chapter defines foodborne illness and discusses its causes and methods of prevention.

# WHAT IS A FOODBORNE ILLNESS?

Many people have suffered the unpleasant experience of a foodborne illness. Symptoms of foodborne illness include inflammation of the gastrointestinal tract lining (gastroenteritis), nausea, abdominal cramps, diarrhea, and vomiting. About one third of all reported diarrhea cases in the United States have been linked to foodborne illnesses (57). The severity of diarrhea or any of the other symptoms varies depending on the type of causative agent, the amount of the agent consumed, and the age and the susceptibility of the immune system of the affected individual. Those most seriously affected by foodborne illness are the very young, the old, and those with immune systems compromised by diseases such as AIDS or cancer (10). Mild cases of foodborne illness usually subside with time. Dehydration resulting from diarrhea and vomiting can be treated by the consumption of electrolyte-rich liquids. Still, severe cases may result in hospitalization or even death.

## What Causes Foodborne Illness?

People get sick from food that has been contaminated by one of three types of food hazards: (1) biological, (2) chemical, and (3) physical (Table 4-1). Biological hazards are living organisms or organic material that includes bacteria, molds, viruses, and parasites. Some of these hazards are so small that they cannot be seen except with the aid of a microscope. Bacteria, molds, viruses, and some parasites are examples of these kinds of microorganisms (*micro* means small), and are the topics of an introductory microbiology class. Chemical hazards are chemical substances that can harm living systems. These range from agricultural and industrial contaminants (including cleaners and sanitizers) to plant and animal toxins. Physical hazards include foreign material such as glass, metal, stones, and wood that could cause harm if ingested. Other common

**TABLE 4-1** Types of Foodborne Hazards

| Biological | Chemical | Physical |
|---|---|---|
| Bacteria | Plant toxins | Glass |
| Molds | Animal toxins | Bone |
| Viruses | Agricultural chemicals | Metal |
| Parasites | | Plastic |
| Prions | Industrial chemicals | |

substances following these top four include jewelry, insects, insulation, bone, and plastic (16).

# BIOLOGICAL HAZARDS— LIVING CULPRITS

Foodborne biological hazards are organisms such as bacteria, molds, viruses, and parasites. The seriousness of these biological hazards varies greatly (Table 4-2). It is difficult to avoid microorganisms because they are everywhere. However, most biological hazards are inactivated or killed by adequate cooking and/or their numbers are kept to a minimum by sufficient cooling.

## Bacteria: Number-One Cause of Foodborne Illness

More than 90 percent of foodborne illnesses are caused by **bacteria,** but only about 4 percent of identified bacteria are **pathogenic,** that is, can cause illnesses. The remaining 96 percent are benign (harmless). Some are even used to produce foods such as cheese, yogurt, soy sauce, butter, sour cream, buttermilk, cured meats, sourdough bread, and fermented foods such as pickles, beer, and sauerkraut. Beneficial and pathogenic bacteria are everywhere. Pathogenic bacteria cause three types of foodborne

**Bacteria** One-celled microorganisms abundant in the air, soil, water, and/or organic matter (i.e., the bodies of plants and animals).

**Pathogenic** Causing or capable of causing disease.

## TABLE 4-2    Biological Hazards Grouped According to Severity of Risk

| Severe Hazards | Moderate Hazards: Potentially Extensive Spread* | Moderate Hazards: Limited Spread |
|---|---|---|
| *Clostridium botulinum* | *Listeria monocytogenes* | *Bacillus cereus* |
| *Shigella dysenteriae* | *Salmonella* | *Campylobacter jejuni* |
| *Salmonella typhi* | *Shigella* | *Clostridium perfringens* |
| Hepatitis A and E | Enterovirulent *Escherichia coli* (EEC) | *Staphylococcus aureus* |
| *Brucella abortus* | *Streptococcus pyogenes* | *Vibrio cholerae* (non-01) |
| *Vibrio cholerae* (01) | Rotavirus | *Vibrio parahaemolyticus* |
| *Vibrio vulnificus* | Norwalk virus group | *Yersinia enterocolitica* |
| | | *Giardia lamblia* |

*Although classified as moderate hazards, complications and aftereffects may be severe in certain susceptible populations.

**Source:** Adapted from International Commission on Microbiological Specifications for Food (ICMSF) (1986); Pierson and Corlett, eds., *HACCP Principles and Applications* (New York: Chapman & Hall, 1992).

## TABLE 4-3    Bacterial Food Infections

| Causative Agent | Latency Period (duration) | Principal Symptoms | Typical Foods | Mode of Contamination | Prevention of Disease |
|---|---|---|---|---|---|
| *Listeria monocytogenes* | 3–70 days (up to 1 month) | Meningoencephalitis; stillbirths; septicemia or meningitis in newborns | Raw milk, cheese, and vegetables | Soil or infected animals, directly or via manure | Pasteurization of milk; cooking |
| *Salmonella* species | 12–36 hr (2–7 days) | Diarrhea, abdominal pain, chills, fever, vomiting, dehydration | Raw, undercooked eggs; raw milk, meat, and poultry | Infected food-source animals; human feces | Cook eggs, meat, and poultry thoroughly; pasteurize milk; irradiate chickens |
| *Shigella* species | 12–48 hr (4–7 days) | Diarrhea, fever, nausea; sometimes vomiting, cramps | Raw foods | Human fecal contamination, direct or via water | General sanitation; cook foods thoroughly |
| *Streptococcus pyogenes* | 1–3 days (varies) | Various, including sore throat; erysipelas, scarlet fever | Raw milk, deviled eggs | Handlers with sore throats, other "strep" infections | General sanitation; pasteurize milk |
| *Yersinia enterocolitica* | 3–7 days (2–3 weeks) | Diarrhea, pains, mimicking appendicitis, fever, vomiting, etc. | Raw or undercooked pork and beef; tofu packed in spring water | Infected animals, especially swine; contaminated water | Cook meats thoroughly; chlorinate water |

illnesses: (1) infection, (2) intoxication or poisoning, and (3) toxin-mediated infection. These are summarized below and explained in detail in the shaded section.

**Food infection**  An illness resulting from ingestion of food containing large numbers of living bacteria or other microorganisms.

**Food intoxication**  An illness resulting from ingestion of food containing a toxin.

### Food Infections

About 80 percent of bacterial foodborne illnesses are due to **food infections.** These foodborne illnesses are caused by ingesting bacteria that grow in the host's intestine, replicate, and create an infection through their colonization. Table 4-3 lists the bacteria primarily responsible for food infections.

### Food Intoxication

Foodborne illnesses can also be the result of **food intoxication** or poisoning. Bacteria grow on the food and release toxins that cause illness in the person consuming the toxin-laden food or beverage. Certain plants and animals produce toxins, but the most common food intoxicants originate from bacteria. Table 4-4 identifies bacteria most often associated with food intoxication.

### Toxin-Mediated Infection

This type of foodborne illness occurs when bacteria enter the intestinal tract and *then* start to produce the toxin in the intestine (Table 4-5).

**TABLE 4-4** Bacterial Food Intoxicants

| Causative Agent | Latency Period (duration) | Principal Symptoms | Typical Foods | Mode of Contamination | Prevention of Disease |
|---|---|---|---|---|---|
| *Clostridium botulinum* (botulism) | 12–36 hr (months) | Fatigue, weakness, double vision, slurred speech, respiratory failure, sometimes death | Types A & B: vegetables, fruits; meat, fish, and poultry products; condiments; Type E: fish and fish products | Types A & B: soil or dust; Type E: water and sediments | Thorough heating and rapid cooling of foods |
| *Clostridium botulinum* (infant botulism) | Unknown | Constipation, weakness, respiratory failure, sometimes death | Honey, soil | Ingested spores from soil or dust or honey colonize intestine | Do not feed honey to infants—will not prevent all |
| *Clostridium perfringens* | 8–24 hr (12–24 hr) | Diarrhea, cramps, rarely nausea and vomiting | Cooked meat and poultry | Soil, raw foods | Thorough heating and rapid cooling of foods |
| *Bacillus cereus* (diarrheal) | 6–15 hr (12–24 hr) | Diarrhea, cramps, occasional vomiting | Meat products, soups, sauces, vegetables | From soil or dust | Thorough heating and rapid cooling of foods |
| *Bacillus cereus* (emetic) | 1/2–6 hr (5–24 hr) | Nausea, vomiting, sometimes diarrhea and cramps | Cooked rice and pasta | From soil or dust | Thorough heating and rapid cooling of foods |
| *Staphylococcus aureus* | 1/2–8 hr (6–48 hr) | Nausea, vomiting, diarrhea, cramps | Ham, meat, poultry products, cream-filled pastries, whipped butter, cheese | Handlers with colds, sore throats or infected cuts; food slicers | Thorough heating and rapid cooling of foods |

**TABLE 4-5** Bacterial Toxin-Mediated Infections

| Causative Agent | Latency Period (duration) | Principal Symptoms | Typical Foods | Mode of Contamination | Prevention of Disease |
|---|---|---|---|---|---|
| *Campylobacter jejuni* | 2–5 days (2–10 days) | Diarrhea, abdominal pain, fever, nausea, vomiting | Infected food-source animals | Chicken, raw milk | Cook chicken thoroughly; avoid cross-contamination; irradiate chickens; pasteurize milk |
| *Vibrio cholerae* (cholera) | 2–3 days (up to 7 days) | Profuse, watery stools; sometimes vomiting; dehydration; often rapidly fatal if untreated | Raw or undercooked seafood | Human feces in marine environment | Cook seafood thoroughly; general sanitation |
| *Escherichia coli* (enterohemorrhagic) | 12–60 hr (2–9 days) | Watery, bloody diarrhea | Raw or undercooked beef, raw milk | Infected cattle | Cook beef thoroughly |
| *Escherichia coli* (enteroinvasive) | at least 18 hr (uncertain) | Cramps, diarrhea, fever, dysentery | Raw foods | Human fecal contamination, direct or via water | Cook foods thoroughly; general sanitation |
| *Escherichia coli* (enterotoxigenic) | 10–72 hr (3–5 days) | Profuse watery diarrhea; sometimes cramps, vomiting | Raw foods | Human fecal contamination, direct or via water | Cook foods thoroughly; general sanitation |
| *Vibrio parahaemolyticus* | 12–24 hr (4–7 days) | Diarrhea, cramps; sometimes nausea, vomiting, fever; headache | Fish and seafood | Marine coastal environment | Cook fish and seafood thoroughly |
| *Vibrio vulnificus* | In persons with high serum iron (1 day) | Chills, fever, prostration, often death | Raw oysters and clams | Marine coastal environment | Cook shellfish thoroughly |

# BACTERIAL FOOD INFECTIONS

The main bacteria that cause food infections via colonization in the intestinal tract are *Salmonella, Listeria monocytogenes, Yersinia enterocolitica,* and *Shigella* (Table 4-3).

## Salmonella

*Salmonella* is one of the most common causes of illnesses traced to contaminated foods and water (Figure 4-1). Foods most susceptible to *Salmonella* contamination are meat, fish, poultry, eggs (especially eggnog or Caesar salad made with raw egg), and dairy products (especially custard fillings, cream, ice cream, sauces, dressings, and raw or unpasteurized milk).

Poultry is particularly vulnerable to *Salmonella* contamination (49). If birds are to be stuffed, this should be done just prior to cooking, and the stuffing removed from the cavity immediately after cooking and refrigerated as soon as possible. If reheated, it should be brought to a temperature of at least 165°F (74°C) prior to consumption. Current recommendations suggest that large birds should not be stuffed at all.

Eggs are also at risk for *Salmonella* (*S. enteritidis*). Any crack or hole in an egg allows bacterial contamination to occur, so any damaged eggs in a carton should be discarded. Research suggests that *Salmonella enteritidis* can even be

**FIGURE 4-1**    Salmonella.

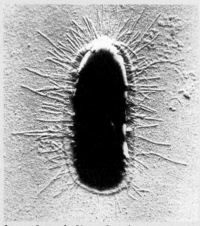

**Source:** Centers for Disease Control

transmitted from infected hens to the eggs they lay (6). Consequently, some states have laws prohibiting the use of raw eggs in institutional settings. An FDA regulation now requires a printed warning on egg cartons for consumers regarding the risk of undercooked eggs— "To prevent illness from bacteria: Keep eggs refrigerated, cook eggs until yolks are firm, and cook foods containing eggs thoroughly" (Chemist's Corner 4-1).

In 2008, a peanut plant was the source of a *Salmonella* outbreak that sickened more than 400 people and was suspected in 6 deaths (11). The plant manufactured peanut products that were used in the production of crackers, cookies, and a variety of other foods. During this outbreak, the FDA reported that, although *Salmonella* is a heat-sensitive bacteria, it could become heat resistant in low-water activity conditions such as in the production of peanut products. Failing to monitor foods for bacterial counts and/or allowing contaminated products out to market is a potential cause of outbreaks for all products (27).

Yet another source of *Salmonella* contamination is pet turtles, iguanas, and other reptiles, so hand washing is essential after handling such pets.

## Listeria monocytogenes

*Listeria monocytogenes* causes the second most costly foodborne illness, after *Salmonella* (47). Listeriosis can have

serious consequences. The fatality rates are as high as 20 to 35 percent of those infected. The CDC records about 500 U.S. deaths annually from *Listeria* (22). *Listeria* infection may also cause pneumonia, septicemia, urethritis, meningitis, and spontaneous abortion (71). *Listeria monocytogenes* is unique for several reasons. It is a facultative bacterium (capable of growing with or without oxygen); it can survive in a wide pH range (from 4.8 to 9.0); and it grows in a wide temperature range (39°F to 113°F/4°C to 45°C). It is one of the few bacteria that can thrive at refrigerator temperatures (79), and frozen dairy desserts have been implicated in some cases of *Listeria monocytogenes* contamination (18). Other foods associated with *Listeria* outbreaks are contaminated cabbage, pasteurized milk, luncheon meats, and Mexican-style soft cheese (see Chemist's Corner 4-2) (4). The principle source of *Listeria* in ready-to-eat foods is recontamination from the processing environment (47).

## Yersinia enterocolitica

*Yersinia enterocolitica* is destroyed by heat, but, like *Listeria,* can grow in a wide temperature range (32°F to 106°F/0°C to 41°C). The ability of this bacterium to grow at refrigerator temperatures makes it all the more hazardous. Yersiniosis infection commonly occurs in children, resulting in gastrointestinal upset, fever, and appendicitis-like symptoms. In one

**FIGURE 4-2**

"It's the waiter at the restaurant where we ate tonight. He wants to know if everything is still all right."

Dean Vietor/Cartoonbank.com

**FIGURE 4-3** Staphylococcus aureus.

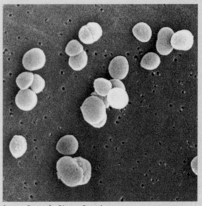

Source: Centers for Disease Control

**FIGURE 4-4** Clostridium botulinum.

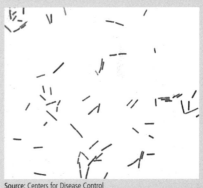

Source: Centers for Disease Control

outbreak, 36 children were hospitalized with apparent acute appendicitis and several underwent appendectomies before health officials determined that they had been infected with *Yersinia enterocolitica* by drinking contaminated chocolate milk (7). Yersiniosis infection can occasionally also cause septicemia (bacteria in the blood), meningitis (inflammation of the spinal cord or brain membranes), and arthritis-like symptoms.

## Shigella

Poor personal hygiene by food handlers is the number-one cause of *Shigella* infection. *Shigella* is carried in the intestinal tract and transferred to the hands of food service personnel who visit the restroom and do not wash their hands (Figure 4-2).

# BACTERIAL FOOD INTOXICATIONS

Food intoxication or poisoning occurs when a food that contains a toxin produced by bacteria such as *Staphylococcus aureus* or *Clostridium botulinum* is consumed (Table 4-4).

## Staphylococcus aureus

A major cause of foodborne illness, *Staphylococcus aureus* is ubiquitous (found everywhere) (Figure 4-3). Up to half of all healthy humans carry it, and it is a common cause of sinus infections and infected pimples and boils. It lives in the throat and nasal passages and in small cuts, so it is easily transmitted to foods through sneezing, coughing, and hand contact (80).

## Clostridium botulinum

The *Clostridium botulinum* toxin causes botulism, one of the deadliest, but fortunately rarest, forms of foodborne illness (Figure 4-4). Less than half a cup of *botulinum* toxin is enough to poison every person on earth. Medical advances, including the development of an antitoxin, have contributed to reducing the death rate from botulism to less than 2 percent (85).

The most common cause of botulism is improperly home-canned food (see Chapter 28). Cans that are dented, have leaky seals, or bulge (indicating the presence of the gas produced by the

bacterium) should be discarded. A foul odor or milky liquid in any can is also a sign of contamination.

# BACTERIAL TOXIN-MEDIATED INFECTIONS

Common examples of bacteria causing toxin-mediated infections include *Escherichia coli, Campylobacter jejuni,* and *Vibrio* (Table 4-5).

## Escherichia coli

*Escherichia coli (E. coli)* is a normal inhabitant of the digestive tract in both humans and animals; in its normal environment, it prevents the growth of more dangerous bacteria. However, when it contaminates food sources, it can produce serious illness. In 2006, raw spinach was contaminated with *E. coli*, resulting in 200 cases of infection (87). In this outbreak, over one half of infected individuals were hospitalized, and three deaths occurred. Eleven percent of the infections occurred in children under the age of 5. The source of the contaminant was found to be animals, including cattle and a wild boar, located near the spinach fields.

*E. coli* is particularly dangerous in children. The CDC estimates that between 7,600 and 20,400 people become ill and 120 to 360 people die each year from *E. coli* (12). The main concern for children is that certain strains of *E. coli* cause infant diarrhea, traveler's diarrhea, and bloody diarrhea. *E. coli* may also cause hemorrhagic colitis—severe abdominal cramps, vomiting, diarrhea, and a short-lived fever followed by watery, bloody diarrhea (10). A potentially deadly condition caused by *E. coli* is hemolytic uremic syndrome (HUS), which is the leading cause of acute renal (kidney) failure in children (36). Of those developing HUS, about 5 percent may die (61).

*E. coli* is naturally found in the intestinal tract, and causes problems only when fecal matter gets into the food or water supply (14). Most infections have been linked to undercooked meat, because contamination often occurs during the butchering of a carcass when the meat comes into contact with the animal's intestinal tract (Figure 4-5) (19). Undercooked hamburger is the most common meat source of *E. coli* contamination.

Scientists began identifying *E. coli* 0157:H7 in the early 1980s, but it was not until an outbreak in 1993 in the Pacific Northwest that national attention and preventive efforts were focused on the problem. The outbreak, which was caused by the consumption of undercooked ground beef in a fast-food restaurant, brought about a prompt response from the U.S. government. A new safety program was instituted, featuring more rapid testing of ground beef, tighter controls at slaughterhouses and processing plants, and the labeling of fresh meat products with instructions for safe handling and preparation, including increasing final cooking temperature of ground meats from 140°F to 160°F/60°C to 71°C (42). Health

**FIGURE 4-5**    How *E. coli* can cause serious health problems.

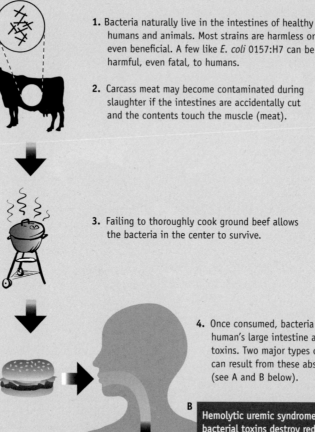

1. Bacteria naturally live in the intestines of healthy humans and animals. Most strains are harmless or even beneficial. A few like *E. coli* 0157:H7 can be harmful, even fatal, to humans.

2. Carcass meat may become contaminated during slaughter if the intestines are accidentally cut and the contents touch the muscle (meat).

3. Failing to thoroughly cook ground beef allows the bacteria in the center to survive.

4. Once consumed, bacteria live in the human's large intestine and produce toxins. Two major types of symptoms can result from these absorbed toxins (see A and B below).

B. Hemolytic uremic syndrome (HUS) – bacterial toxins destroy red blood cells, which may clot the kidney's small blood vessels. *E. coli* 0157:H7 is the leading cause of acute renal failure in children; it can even result in death.

A. Hemolytic colitis – severe abdominal cramps, vomiting, brief fever, and bloody diarrhea.

Stomach
Kidneys
Large intestine
Bacteria toxin absorbed
Small intestine

officials recommend that ground beef be thoroughly cooked so that no pink remains.

Undercooked hamburgers are not the only sources of *E. coli* infection. Other foods or food-handling practices implicated in *E. coli* 0157:H7 outbreaks include unpasteurized (raw) milk, unpasteurized apple juice or cider, raw sprouts (30), dry cured salami, fresh produce (especially manure fertilized) (Chemist's

Corner 4-3), yogurt, sandwiches, and water (78).

*E. coli* may also be transmitted to children in child-care centers because of poor hand washing after diaper changing. Ideally, those who change diapers should not be the same people who prepare food (80). Other outbreaks have been reported from an improperly chlorinated swimming pool and contaminated water systems.

## Campylobacter jejuni

The number of people infected with *Campylobacter jejuni* now equals or exceeds those affected by *Salmonella* (40). *Campylobacter* species are responsible for more than 14 percent of the estimated annual food-related illnesses and deaths attributed to foodborne pathogens. Although the largest foodborne disease outbreak was traced to a municipal water supply, most other cases are linked to raw meat, undercooked poultry, unpasteurized milk, and untreated water.

## Vibrio

Seafood is the major carrier of *Vibrio* infection. *V. parahaemolyticus* is the most common cause of foodborne illness in Japan (90). Cholera (*V. cholera*) is responsible for thousands of deaths each year in Asia. Poor sanitary conditions contaminate water supplies and usually account for the deaths reported in other countries (64). In the United States, very few cases are reported and they are usually associated with raw oyster consumption (20). The bacteria can also be transmitted through skin wounds during the cleaning or harvesting of shellfish, or if seawater washes over a preexisting wound. The Food and Drug Administration estimates that about 5 to 10 percent of raw shellfish on the market are contaminated.

## Molds

**Molds** produce **mycotoxins,** which can cause food intoxication. Over 300 mycotoxins have been identified, most of which do not present a significant food safety risk (59). However, some are carcinogenic (cancer causing) (56). Aflatoxin, a carcinogenic toxin made by the mold *Aspergillus flavus,* is the most potent liver carcinogen known. Foods infected with *Aspergillus flavus* are most likely to be peanuts and grains; it has also been identified in corn, cottonseed, Brazil nuts, pistachios, spices, figs, and dried coconut.

Patulin is a toxin produced by both *Aspergillus* and *Penicillium* that can contaminate fruits and cereals (59). Fumonisins are produced by *Fusarium verticillioides* and *Fusarium proliferatum* and are associated with corn.

Unlike bacteria, molds are visible, exhibiting **bloom** on affected foods. They also thrive at room temperature and need less moisture than bacteria do. Foods susceptible to molds are breads, jams and jellies, and salty meats such as ham, bacon, and salami.

Black spots in the refrigerator, often called mildew, are actually molds that can be cleaned by washing the surface with a solution of 1 tablespoon of baking soda dissolved in 1 quart of water. Musty-smelling dishcloths, sponges, and mops should be thoroughly cleaned or replaced, because such odors indicate that mold has taken root.

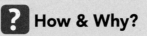

### ? How & Why?

***Why are some molds all right on foods whereas others are not?***

As a rule, foods that show signs of mold should not be eaten. The exceptions are cheeses (such as Roquefort, bleu, Brie, and Camembert), whose flavor, texture, and color depend on specific safe molds. Other foods relying on molds during processing include soy sauce, tempeh, and certain types of Italian-style salami that are coated in a thin, white mold. Cheeses such as cheddars and Swiss that become moldy can safely be trimmed 1 inch away from the mold. Soft cheeses such as cottage cheese and cream cheese that have become moldy, however, should be thrown out, because the mold may penetrate through the cheese.

## Viruses

**Viruses** are one of nature's simplest organisms. Unlike bacteria, which can exist independently, a virus needs a living cell in order to multiply. These microorganisms have been identified as causal agents in about 3 to 10 percent of foodborne illnesses (Table 4-6) (38). All foodborne viruses are transmitted via the oral-fecal route, that is, from contaminated feces to the mouth. They may be passed from person to person, or through carriers such as flies, soiled diapers, water, and food. Two of the most common viruses known to cause foodborne illnesses are the hepatitis A virus and the Norwalk virus.

### Hepatitis A Virus

Hepatitis A infection occurs most frequently after food is contaminated with fecal matter. (This differs from the hepatitis B virus, which is transmitted through body fluids and not through food.) Another common source of hepatitis A is polluted shellfish beds and vegetable fields (65). Shellfish are a source of hepatitis A infection because they are eaten with their digestive tracts intact. Another possible source of hepatitis A

**Mold**  A fungus (a plant that lacks chlorophyll) that produces a furry growth on organic matter.

**Mycotoxin**  A toxin produced by a mold.

**Bloom**  Cottony, fuzzy growth of molds.

**Virus**  An infectious microorganism consisting of RNA or DNA that reproduces only in living cells.

**TABLE 4-6**   Viruses Causing Foodborne Illness

| Causative Agent | Onset (duration) | Principal Symptoms | Typical Foods | Mode of Contamination | Prevention of Disease |
|---|---|---|---|---|---|
| Hepatitis A virus | 10–50 days (2 weeks to 6 months) | Fever, weakness, discomfort, nausea; often jaundice | Raw or undercooked shellfish; sandwiches, salads, etc. | Human fecal contamination, via water or direct | Cook shellfish thoroughly; general sanitation |
| Norwalk-like viruses | 1–2 hours (1–2 days) | Nausea, vomiting, diarrhea, pains, headache, mild fever | Raw or undercooked shellfish; sandwiches, salads, etc. | Human fecal contamination, via water or direct | Cook shellfish thoroughly; general sanitation |
| Rotaviruses | 1–3 days (4–6 days) | Diarrhea, especially in infants and young children | Raw or mishandled foods | Probably human fecal contamination | General sanitation |

contamination is child-care centers where diaper changing occurs. A vaccine is available that is 95 percent effective against the virus.

## Norwalk Virus

Norovirus is also known as Norwalk virus (after the town in Ohio where the first identified outbreak occurred) (87). Infection is most common in the summer months and is commonly referred to as the "stomach flu." Norwalk virus is the most common cause of gastroenteritis in the United States. In fact, it is the second most common viral infection after the common cold.

The Norwalk virus is spread via contaminated shellfish, food handlers, and water containing raw sewage. Heating will destroy this virus, but freezing will not. Norwalk virus infection outbreaks can be large, as in the case of a Minnesota restaurant in which two salad makers contaminated the food and infected over 2,000 people (80).

## Parasites

**Parasites** need a host to survive. They infect people in many parts of the world, but in the United States fewer than 500 cases of parasitic infection are reported each year. Two of the more common foodborne parasites are roundworms and protozoa (Table 4-7).

**Parasite** An organism that lives on or within another organism at the host's expense without any useful return.

## Roundworms

Roundworm infections can result from eating undercooked pork or uncooked or undercooked fish.

***Trichinella spiralis***   The *Trichinella spiralis* roundworm is probably the most common parasite carried in food and is responsible for causing trichinosis. Pork products are the primary source of infection, with 1 out of every 100 swine in the United States infected. There are now relatively few cases of trichinosis each year, and most infections are thought to be contracted through the consumption of raw or improperly cooked pork, especially sausage (51). Heating pork to an internal temperature of 137°F (58°C) will kill the *T. spiralis* larvae, but the National Livestock and Meat Board recommends a final internal temperature of 160°F (71°C) as a safety margin. Microwave cooking of pork is not recommended because of uneven heating.

***Herring Worms*** (Anisakis simplex) ***and Codworms*** (Pseudoterranova dicipiens)   Japanese cooks preparing sushi (a Japanese dish of thin raw fish slices or seaweed over a cake of cooked rice) inspect fish for these tiny white worms, but because the worms are no wider than a thread, some may be missed. Therefore, not all raw fish dishes are guaranteed to be worm-free. There is no commercial method to detect all parasites. Even candling, which involves placing a fillet over a lighted translucent surface, finds only 60 to 70 percent of the worms (45).

People who consume raw or undercooked fish containing the live worms may experience a tingling throat sensation caused by the worm wriggling as it is swallowed. Other symptoms usually appear within an hour after ingestion, but may show up as much as two weeks later. In serious cases, the worm penetrates through the stomach or intestinal wall, causing severe abdominal pain, nausea, vomiting, or diarrhea. Symptoms often continue for several days and have been misdiagnosed as appendicitis, gastric ulcer, Crohn's disease, and gastrointestinal cancer (84). After several weeks, the worm dies, or it may be coughed or vomited up by the host. It also may be removed by a physician using a fiber-optic device equipped with mechanical forceps. Despite these dramatic problems arising from worm infection from contaminated fish, only about ten cases are reported every year. The number of actual cases, however, may be significantly higher because of underreporting.

## Protozoa

Protozoa are animals consisting of just one cell. They most frequently infect humans through contaminated water. Only 3 out of about 30 types of protozoa are related to food safety: *Giardia*, *Cryptosporidium*, and *Cyclospora*. The most common of these is *Giardia*.

***Giardia lamblia***   *Giardia lamblia* is primarily transmitted through surface streams and lakes that have been contaminated with the feces of infected livestock and other animals. This protozoan is responsible for the most common parasitic infection in the world, and is most frequently associated with the consumption of

**TABLE 4-7**   Parasites Causing Foodborne Illness

| Causative Agent | Onset (duration) | Principal Symptoms | Typical Foods | Mode of Contamination | Prevention of Disease |
|---|---|---|---|---|---|
| **Roundworms (Nematodes)** | | | | | |
| *Trichinella spiralis* | 8–15 days (weeks, months) | Muscle pain, swollen eyelids, fever; sometimes death | Raw or undercooked pork or meat of carnivorous animals (e.g., bears) | Larvae encysted in animal's muscles | Thorough cooking of meat; freezing pork at 5°F for 30 days; irradiation |
| *Ascaris lumbricoides* | 10 days–8 weeks (1–2 years) | Sometimes pneumonitis, bowel obstructions | Raw fruits or vegetables that grow in or near soil | Eggs in soil from human feces | Sanitary disposal of feces; cooking food |
| *Anisakis simplex, Pseudoterranova decipiens* | Hours to weeks (varies) | Abdominal cramps, nausea, vomiting | Raw or undercooked marine fish, squid or octopus | Larvae occur naturally in edible parts of seafoods | Cook fish thoroughly or freeze at −4°F for 30 days |
| **Protozoa** | | | | | |
| *Giardia lamblia* | 5–25 days (varies) | Diarrhea with greasy stools, cramps, bloating | Mishandled foods | Cysts in human and animal feces, directly or via water | General sanitation; thorough cooking |
| *Cryptosporidium parvum* | 2–3 days (2–3 weeks) | Diarrhea; sometimes fever, nausea, and vomiting | Mishandled foods | Oocysts in human feces | General sanitation; thorough cooking |
| *Entamoeba histolytica* (amebic dysentery) | 2–4 weeks (varies) | Dysentery, fever, chills, sometimes liver abscess | Raw or mishandled foods | Cysts in human feces | General sanitation; thorough cooking |
| *Toxoplasma gondi* | 10–23 days (varies) | Resembles mononucleosis; fetal abnormality or death | Raw or undercooked meats; raw milk; mishandled foods | Cysts in pork or mutton, rarely beef; oocysts in cat feces | Cook meat thoroughly; pasteurize milk; general sanitation |
| **Tapeworms (Cestodes)** | | | | | |
| *Taenia saginata* (beef tapeworm) | 10–14 weeks (20–30 years) | Worm segments in stool; sometimes digestive disturbances | Raw or undercooked beef | "Cysticerol" in beef muscle | Cook beef thoroughly or freeze below 23°F |
| *Diphylliobothrium latum* (fish tapeworm) | 3–5 weeks (years) | Limited; sometimes vitamin $B_{12}$ deficiency | Raw or undercooked freshwater fish | "Plerocercoids" in fish muscle | Heat fish 5 minutes at 133°F or freeze 24 hours at 0°F |
| *Taenia sollium* (pork tapeworm) | 8 weeks–10 years (20–30 years) | Worm segments in stool, sometimes "cysticercosis" of muscles, organs, heart, or brain | Raw or undercooked pork; any food mishandled by a *T. serum* carrier | "Cysticerol" in pork muscle; any food—human feces with *T. serum* eggs | Cook pork thoroughly or freeze below 23°F; general sanitation |

contaminated water. Another common source of infection is child-care centers. Approximately 2 percent of the population in the United States is infected (84). Infection with this organism causes recurring attacks of diarrhea and the passage of stools containing large amounts of unabsorbed fats and yellow mucus. When a *Giardia* infection is contracted, medications can be taken for the symptoms.

# Prions—Mad Cow Disease

**Prions** are related to mad cow disease, or bovine spongiform encephalopathy (BSE) (86). It is a type of transmissible spongiform encephalopathy (TSE) that riddles the brain with holes, making it look like a sponge. TSE is a group of diseases that affect the brain, resulting in symptoms that range from loss of coordination to convulsions and ultimately death. BSE is a disease affecting cattle. When humans ingest beef from cattle with BSE, they may develop what is called variant Creutzfeldt-Jakob disease (vCJD). TSEs other than

**Prion** An infectious protein particle that does not contain DNA or RNA.

mad cow disease include Creutzfeldt-Jakob disease (CJD; a disease causing dementia and not related to BSE), Kuru ("the laughing death," a disease formerly found in New Guinea when cannibalism was practiced there), and scrapie (a disease causing coordination loss and itching/scraping in goats and sheep).

In mad cow disease, prions, or viruses producing prions (identity remains controversial), travel up the spinal cord to the brain. This virus-related protein material incorporates itself into the brain, causing chain reactions that create holes in the brain. The incubation period between infection and manifestation can be months, years, or decades.

Controversy exists concerning how prions are transmitted from food to humans. The foods most often believed to be linked to these prions have been cattle and sheep in Great Britain. Prior to the understanding of prions, the practice for some livestock growers was to kill sickly animals and feed the remains to other cattle. It is speculated that healthy cattle being fed rendered livestock would then become infected and, when slaughtered for their meat, would potentially spread this disease to the consumer.

Over 95 percent of all BSE cases have occurred in Great Britain (37). In 2003, the first case of BSE in a native-born North American cow was identified in Canada (67). This prompted fear that BSE could be spread to the United States because of the frequent trade of cattle and cattle products across the Canadian-U.S. border. Less than a year after the initial case of BSE in Canada was reported, a cow in the United States, which was born in Canada, was found to have BSE. In response to this occurrence, the USDA implemented tighter regulations to prevent the ingestion of contaminated beef products. These included complete removal of high-risk material (central nervous tissue and small intestines) from all cattle products and the introduction of a national identification system to track animals throughout the food supply chain. The USDA also banned the use of rendered carcasses as feed for other livestock (53), and the livestock industry, interested in keeping its food products safe, has exhibited

strong compliance with this ban. Previous measures to keep livestock safe in the United States included a 1989 USDA ban on importation of livestock from Great Britain, which was extended to Europe in 1997. Although BSE prevalence in the United States remains extremely low, some countries have banned beef imports from the United States (bans are sometimes temporarily lifted).

Additional attempts to prevent the contamination of the food supply with infectious forms of BSE include the development of rapid bioassays to detect and remove prions and new methods to screen the meat supply for the presence of nervous tissue or prions (23).

To date, over 150 cases of vCJD have been diagnosed in the United Kingdom, and epidemiological evidence strongly suggests BSE-contaminated beef as the causative factor. Three cases of vCJD have been diagnosed in the U.S. (9). Two of these individuals are thought to have been exposed to BSE in Great Britain, and the third in Saudi Arabia. In August of 2001, the Department of Health and Human Services introduced the BSE/TSE Action Plan to further improve the scientific understanding and control of BSE and other TSEs (9). Components of the action plan include surveillance for human disease, protection, research, and oversight. Each component is assigned to a specific government agency, such as the CDC, FDA, or the National Institutes of Health (NIH).

## New Virulent Biological Hazards

It is not unusual for microorganisms that were relatively unheard of to emerge with a new virulence (actively harmful), making them a public threat (60). Examples of pathogens not previously recognized as a serious cause of foodborne illness are the Norwalk virus, *Campylobacter jejuni, Listeria monocytogenes, Vibrio vulnificus, Vibrio cholera,* and *Yersinia enterocolitica* (81).

Pathogens are living organisms that rapidly evolve (13). Bacteria are constantly appearing as potential hazards to public health, so health departments and government agencies must be vigilant. Several serious

outbreaks resulting from "new" microorganisms led the CDC to implement Emerging Infections Programs (EIPs) in state health departments in 1994 (32).

## Advanced Techniques for Detecting Contamination

The conventional method of confirming food contamination by microorganisms is culture. This involves growing the organism in the lab (in a petri dish) until it can be identified visually or by additional tests. Culture is a highly accurate method for identification of pathogens, but can take a long time—up to several months for some species of mold—and is labor intensive. Culture and other "gold standard" methods for microbe identification are published in the FDA's *Bacterial Analytical Manual* (available online at www.cfsan.fda.gov/~ebam/bam-toc.html) (52). New techniques allow for rapid testing that can be performed in a much shorter time, are very reliable, and can make an accurate identification in 90 percent to 95 percent of cases (52). Some rapid tests use conventional culture to grow suspected *E. coli,* but in a medium that produces specific colors based on the presence of the pathogen, simplifying identification (63). New, faster testing methods for *Salmonella* are also available (48).

Many of the rapid tests available require 8 to 24 hours to complete, so even faster techniques are still needed. The most time-consuming step is growing the microbe to levels that can be detected by the test that identifies it (48).

In 1996, the CDC created PulseNet, a national network of food safety testing and regulatory agencies. Participants perform DNA fingerprinting of food contaminants to identify specific bacterial strains (50). They upload this data to the CDC PulseNet database, where participants all over the nation can view outbreaks and determine if the same bacterial strain is responsible for multiple occurrences in different locations. PulseNet can be accessed online at www.cdc.gov/pulsenet.

# CHEMICAL HAZARDS— HARMFUL CHEMICALS IN FOOD

Chemical hazards are any chemical substance hazardous to health. Harmful chemicals can come from additives (unintentional), plant toxins, animal toxins, or certain metals (Table 4-8). Contamination can also be deliberate, as was the case in 2007, when pet foods manufactured in China were intentionally spiked with melamine, an industrial chemical, to artificially increase their protein content (see Chemist's Corner 4-4). This resulted in the death of several animals, and highlighted the possibility of a similar adulteration in the human food market (39).

Experts from the World Health Organization (WHO) called for food safety reform in China (54), and in 2008 the FDA announced the establishment of eight full-time FDA positions at the U.S. diplomatic posts in China (29).

Unlike melamine, most chemical toxins negatively affecting health are not added intentionally or are natural (Chemist's Corner 4-5). The National Academy of Sciences published the Food Chemicals Codex (FCC) in 1958 to help differentiate safe additives from

> ## CHEMIST'S CORNER 4-4
>
> ### How Melamine Falsely Elevates Protein Content
>
> The industrial chemical melamine is used in fabric, glues, resins, and fertilizers, but is very harmful when ingested. Dog food manufacturers in China added it to their products in 2006 in order to falsely increase the protein content, so the food would appear to have the necessary amount of dietary nutrients.
>
> The mechanism of the test performed to evaluate protein content allows melamine to be used as a substitute to "fool" the test. The level of nitrogen in a product is used as a surrogate marker of the amount of protein, because protein has many nitrogen molecules. Melamine also contains high levels of nitrogen, but no protein. When the dog food was evaluated for nitrogen content, the level appeared high. This adulteration was not discovered until animals became sick, and in some cases died, after ingesting the food (39).

harmful ones; however, the FCC has only sporadically enforced these guidelines (39). Technically, additives are considered "safe" when there is "reasonable certainty in the minds of competent

> ## CHEMIST'S CORNER 4-5
>
> ### Acrylamide
>
> Acrylamide is a chemical that in high dosages can cause cancer in laboratory animals, but has unknown effect in humans. Acrylamide is not naturally found in food, but is formed by a chemical reaction when certain foods such as french fries are exposed to high temperatures. The reaction requires sugar, the amino acid asparagine, and high heat (88).

scientists that a substance is not harmful under the intended conditions of use" (72). It is beyond the scope of this book to provide a complete list of harmful additives, but selected chemical toxins are presented in Table 4-8, and those from seafood are now further discussed.

## Seafood Toxins: Chemicals from Fish/Shellfish

Both fish and shellfish may harbor toxins, unaffected by cooking, that can cause foodborne illness (Figure 4-6).

### Ciguatera Fish Poisoning

Ciguatera fish poisoning is the most common toxin-related food poisoning in the United States. It is caused by eating fish, usually from tropical waters, that contain a ciguatoxin that is not destroyed by heating (33). Although less than 1 percent of fish found in tropical areas with coral reefs are actually contaminated, more than one third of Florida barracuda were found to contain ciguatera toxin, resulting in a ban on the sale of barracuda for human consumption (46).

### Histamine Food Poisoning

Excessive histamine accumulation in fish (especially tuna) may result in histamine food poisoning (scombrotoxism). Other popular fish affected include mackerel, sardines, and herrings (69). This is one of the most common forms of fish poisoning in the United States and occurs when the fish have not been chilled immediately after being caught. The fish

## TABLE 4-8  Potential Chemical Contaminants

| Food Additives (Unintentional) | Plant Toxins | Animal Toxins | Toxic Metals |
|---|---|---|---|
| • Pesticides<br>• Herbicides<br>• Fertilizers<br>• Pharmaceuticals in water supply (toilet disposal)<br>• Industrial chemicals<br>  acrylamide<br>  benzene<br>  perchlorate<br>• Pollutants<br>  ethers<br>  dioxins<br>  polybrominated diphenyl<br>  polycyclic aromatic hydrocarbons<br>• Cleaning products | • Raw or undercooked red kidney beans or fava beans<br>• Certain mushrooms<br>• Certain herbs<br>• Fruit pits<br>• Mold toxins (mycotoxins) | • Certain seafood toxins | • Heavy metals<br>  arsenic<br>  lead<br>  cadmium<br>  mercury<br>• Other metals<br>  antimony<br>  brass<br>  copper<br>  zinc |

**FIGURE 4-6** Sign warning unsafe to eat seafood collected from shore.

Chris Howes/Wild Places Photography/Alamy

become toxic when bacteria (such as *Morganella morganii*) produce histamine due to time-temperature abuse.

### Pufferfish Poisoning

One of the most violent poisonings originating from seafood occurs when the liver, gonads, intestines, and/or skin of the pufferfish are consumed. These organs contain tetrodotoxin, which if ingested results in a mortality rate of 50 percent. Only a few cases have been reported in the United States, but in Japan, where pufferfish, or *fugu*, is a traditional delicacy, 30 to 100 cases are reported each year (84). Most of these cases in Japan originate from people preparing the dish at home, rather than eating it at special restaurants ,where licensed chefs are specially trained in how to remove the poisonous viscera from the pufferfish.

### Red Tide

Red tide is the result of the rapid growth of a reddish marine alga, usually occurring during the summer or in tropical waters. Shellfish, such as mollusks, oysters, and clams, and certain fish that consume red tide algae become poisonous and should not be eaten until the red tide has disappeared.

**Cross-contamination** The transfer of bacteria or other microorganisms from one food to another.

# FOOD ALLERGY

A food allergy is an immune response to a specific protein within a food. The allergic reaction may result in a variety of symptoms, including a skin rash, swelling of the mouth or lips, nausea, abdominal pain, vomiting, diarrhea, runny nose (rhinitis), and difficulty breathing, which can be life threatening. Food allergy is different than food intolerance, which does not involve an immune response, but rather an inability to absorb or process a certain food due to an enzyme deficiency. The most common food intolerance is for lactose, found in dairy products.

## Prevention

Complete avoidance of the allergen is the best method of preventing an allergic reaction, although other options are being investigated (Chemist's Corner 4-6). Foods containing allergens are not always easy to recognize. For instance, peanuts may show up in chili, egg rolls, candies, and ice cream, and wheat is present in soy sauce. Milk is commonly present in deli meats, canned tuna, seafood, and frozen French fries.

## Most Common Food Allergens

While there are over 160 known food allergens, the vast majority of allergies are triggered by the top eight: wheat, eggs,

peanuts, tree nuts, fish, shellfish, soy, and milk. The Food Allergen Labeling and Consumer Protection Act (FALCPA) requires food manufacturers to appropriately label any product that contains a potential allergen. Products must be labeled with the statement "Contains" followed by a list of all allergens present, or with a parenthetical statement within the ingredient list (2).

Manufacturers are not required to include a FALCPA statement if they can prove their product does not contain any allergens. Allergen test kits that can detect 0.1 to 5 ppm (parts per million) of an antigen are available for milk, eggs, almonds, peanuts, hazelnuts, and soy flour. Rapid, on-site tests are available for detecting peanuts and gluten (68). Additional tests are available from specialized laboratories.

Internationally, regulations vary from country to country. Table 4-9 lists labeling requirements in the United States and several other countries.

## Cross-Contamination

Prevention of **cross-contamination** with food allergens is a critical step in preventing unwanted reactions in consumers. Because they are under the regulation of the USDA, meat,

**TABLE 4-9** Allergen labeling requirements in the U.S., European Union (EU), Japan, Australia, and New Zealand (NZ)

| US | EU | Japan | Australia/NZ |
|---|---|---|---|
| *Mandated:* | *Mandated:* | *Mandated:* | Voluntary incidental trace allergen labeling (VITAL) system with three action levels: |
| Tree nuts | Nuts | Peanuts | |
| Peanuts | Peanuts | Eggs | |
| Soy | Soy | Milk | |
| Egg | Eggs | Wheat | Level 1 (Green): Allergen does not need label. |
| Milk | Milk/dairy products | Buckwheat | |
| Fish | Fish | | Level 2 (Yellow): Allergen advisory label stating that allergen may be present. |
| Wheat | Cereals containing | *Labeled when possible:* | |
| Shellfish | gluten | | |
| | Crustaceans | Abalone | |
| | Celery | Squid | Level 3 (Red): Significant levels of allergen are likely, labeling advised. |
| | Mustard | Salmon roe | |
| | Sesame seeds | Shrimp/prawn | |
| | Sulfites | Oranges | |
| | | Crab | |
| | | Kiwifruit | |
| | | Beef | |
| | | Tree nuts | |
| | | Salmon | |
| | | Mackerel | |
| | | Soybeans | |
| | | Chicken | |
| | | Pork | |
| | | Mushrooms | |
| | | Peaches | |
| | | Yams | |
| | | Apples | |
| | | Gelatin | |

Source: adapted from Reference 2.

poultry, and egg products are not subject to FALCPA rules; however, the USDA requires producers to include allergen information in their HAACP plan (see below). Avoidance of cross-contamination must be addressed during storage, production, and cleaning. Control of allergens must be documented in the standard operating procedures of the manufacturer, and must include a label development procedure.

# PHYSICAL HAZARDS— OBJECTS IN FOOD

Physical hazards that can harm the consumer's health when found in food and beverages include glass, bone, metal (especially from opening cans), wood, stones, false fingernails, toothpicks, watches, jewelry, insects, staples from food boxes, and many other foreign items that have been known to find their way into the food supply.

# PREVENTING FOODBORNE ILLNESS

Because biological hazards (especially bacteria) are everywhere and many other chemical and physical food hazards can potentially enter the food supply, it is important to pursue foodborne illness prevention through food safety programs. In order for prevention to be successful, it's important to ask, "Where do most foodborne illnesses occur?"

## Location, Location, Location

The majority of foodborne illnesses originate at restaurants; only about 3 percent can be traced to food processing plants (3). Most outbreaks in food processing plants are caused by contamination of incoming foods, failure of pathogen-killing processes, or contamination of foods after sanitization. The low percentage of foodborne illnesses originating from food

manufacturers is due in part to the implementation of FDA's **Current Good Manufacturing Practices (cGMPs)** that provide guidelines to minimize the risk of contamination (available online at www.cfsan.fda.gov/guidance.html). The use of GMPs allows any product to be traced back to where it was manufactured in case of a recall.

Because most outbreaks occur in restaurants, the second half of this chapter discusses preventing foodborne illness in this environment by focusing on food service personnel, food flow (purchasing, storage, preparation, cooking, holding, cooling, reheating, and sanitation), and food safety monitoring programs (FDA Food Code, health department inspections, HACCP, and national surveillance). These topics are often covered in food safety courses for food service employees, who complete the course, take a test, and receive a certificate. According to an FDA report, the top three factors associated with foodborne illness are poor personal hygiene, cross-contamination, and time/temperature control (43).

## Personnel

Food service personnel are critical to food safety. In the early 20th century, a cook in New York named Mary Mallon appeared perfectly healthy, but infected about 50 people with typhoid fever. Not surprisingly, she came to be known as "Typhoid Mary." She believed that because she could not see germs, she did not have to wash her hands before cooking. As this story illustrates, a top priority for any food-serving establishment is that food workers be healthy and know how to handle food safely (Figure 4-7). Typhoid is far from the only illness that can be transferred through carelessness and poor hygiene. The common cold, mumps, measles, pneumonia, scarlet fever, tuberculosis, trench mouth, diphtheria, influenza, and whooping cough

**Current Good Manufacturing Practices (cGMPs)** A set of regulations, codes, and guidelines for the manufacture of food products, drugs, medical devices, diagnostic products, and active pharmaceutical ingredients (APIs).

**FIGURE 4-7**   Personal hygiene checklist.

**Self**
☐ Stay healthy. Maintain daily sleep, well-balanced diet, and relaxation.
☐ Report to supervisor if you are sick.
☐ Stay clean. Practice daily bathing, shampoo hair regularly, use deodorant, and take care of fingernails—they should be cleaned, trimmed, and free of polish and decorations.
☐ Wear only clothes that are new or have been washed. Shoes should cover the foot (no sandals, open toe) and have nonskid soles.
☐ Wear caps or hairnets.
☐ Avoid items that may fall into food/beverages: hairpins, jewelry, false nails, nail polish, nail decorations, bandages on hand (cover with plastic gloves), handkerchiefs.

**Food Handling**
☐ Avoid handling food; use serving spoons, scoopers, dippers, tongs, and ladles.
☐ Cover all exposed food with lids, plastic wrap, or aluminum wrap.
☐ Taste food with clean spoon and do not reuse.
☐ If gloves are used, change them between food and nonfood handling.

**Kitchen**
☐ Wash hands in the hand-washing sink before starting and after breaks/meals.
☐ Cover all coughs/sneezes and immediately wash hands in hand sink.
☐ No smoking or gum chewing.
☐ Keep all surfaces clean.
☐ Use potholders for pots and dish towels for dishes.
☐ Keep cleaning items away from foods/beverages.

**Serving**
☐ Hold plates without touching the surface.
☐ Carry silverware only by the handles.
☐ Handle glassware without touching the rim or the inside.

may also be spread this way. Only people who are free of colds, diarrhea, wounds, and illnesses should be working with food.

### Training

A Food Management Certificate obtained through a health department education class, commercial in-person or online class, or the National Restaurant Association Educational Foundation at www.nraef.org ensures that a food handler has learned safe food-handling techniques, which are often a job requirement for food service employees. Periodic retraining on sanitation techniques is available from local health departments, the National Restaurant Association, corporations offering food safety courses, or the CDC.

### Personal Hygiene Habits

Numerous health habits help keep contamination under control and a few of these are avoiding hand-to-mouth gestures, hand washing and clean uniforms.

- *Avoid hand-to-mouth.* Any hand-to-mouth movements should be discouraged, including even simple habits such as smoking, gum chewing, and eating in the food-handling areas. Sampling foods with fingers should not be permitted as this also transfers bacteria. Double-dipping with utensils is also not allowed. *Staphylococcus* can be transferred by workers who touch their mouth, nose, a pimple, or infected cut, and then handle food. Sneezing or coughing sends millions of microorganisms into the surrounding air to settle on food.

- *Hand washing.* Hands should be washed frequently, especially before handling food and after touching raw meat or eggs, using the restroom, sneezing, or handling garbage. Hot soapy water should be used for at least 20 seconds to scrub hands, in between fingers, wrists, and under fingernails. To ensure maximum effect from hand washing, the routine should consist of washing up to the elbow for at least 20 seconds,

using a nailbrush, and then drying with disposable towels or an air dryer (not a cloth towel). Disposable paper towels or air drying is preferred over cloth, which can harbor bacteria if used repeatedly. When food handlers touch a doorknob, handrail, telephone, counter, or any other surface that is frequently contacted by others, their hands should be washed again before food is touched. One of the most common causes of foodborne illness is failure of employees to wash their hands after using the restroom, particularly if they have diarrhea or are changing an infant's diaper. Hand sanitizers assist in reducing bacterial numbers, but should never substitute for hand washing. Food establishments are often required to have a separate sink strictly for hand washing (Figure 4-8). It should never be used for washing foods or utensils.

- *Uniforms.* Food service workers should clean their uniforms frequently, wear caps or hairnets, and avoid jewelry such as rings and bracelets that can collect minute particles of food and dust.

## Vulnerable Foods

Just like humans, bacteria need food to survive, so it's natural for them to break down organic material as part of the earth's ecology. Foods that best support their growth are known as *high-risk* foods because they contain large amounts of protein and water (Figure 4-9). Other factors making foods more prone toward bacterial contamination are low acid content and sufficient oxygen, although exceptions exist. These risk factors affecting bacterial growth are illustrated by the "Food Risk Road" in Figure 4-10, and are now further explained.

### High-Risk Foods

Meats, which top the high-risk foods list due to their protein and water content, are particularly susceptible to contamination because the digestive tracts of people and animals naturally contain bacteria. During the rendering of animals at the slaughterhouse, the digestive tract may be accidentally

**FIGURE 4-8** Hand-washing sinks are separated from food preparation sink(s).

Guy Croft Industrial/Alamy

**FIGURE 4-9** High-risk foods (high levels of protein/water) for disseminating foodborne illness.

- Meat—beef, pork, lamb
- Poultry
- Fish and shellfish
- Dairy
- Eggs
- Broth, stocks
- Gravies/sauces (meat-, milk-, or egg-based)
- Tofu and other soy foods
- Stuffings (when exposed to poultry cavity)

cut open or nicked, releasing bacteria that may then come in contact with meat (21). Other possible sources of contamination include any cuts, skin, feet, hair, hide, or feathers that can carry bacteria.

Other vulnerable foods include those made with the high-risk foods and extensively handled, such as meatloaf, hamburger, salads (coleslaw, and pasta, chicken, egg, and tuna salads), Chinese and Mexican dishes, some baked goods, and cream fillings. The juice from marinades used to coat raw meat, poultry, pork, or lamb is particularly full of bacteria. Egg dishes likely to become contaminated include baked or soft custard, French toast, quiches, hollandaise sauce, meringues, eggnog,

**FIGURE 4-10** Food Risk Road. Risk factors on the road to foodborne illness.

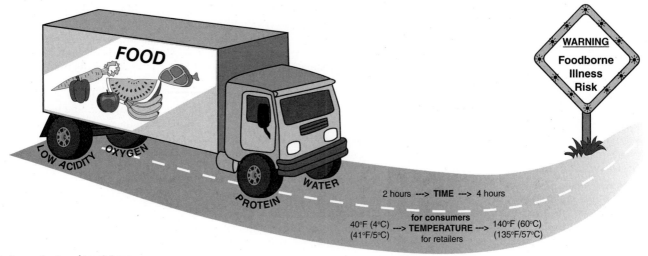

**Source:** Courtesy of Amy C. Brown

and mayonnaise. Damaged eggs are good vectors for organisms that cause foodborne illnesses and should be discarded. Food establishments have the safer option of using pasteurized eggs. Proper refrigeration or freezing is a must for high-risk foods or the dishes that contain them.

## Foods with High Water Activity

It is primarily foods with a high water activity ($a_w$) of 0.85–0.97 that are at risk (58). Bacteria cannot survive without water because they will dry out and die. This principle is why drying foods is one of the oldest methods of food preservation. Meat is highly susceptible to contamination, but beef jerky lasts for a long time, making it a perfect food to transport during long trips without refrigeration. Another way that food manufacturers reduce water activity and hence the risk of bacterial contamination is to add sugar or salt as a preservative. Water, including that within the bacteria, is pulled by osmosis toward the sugar and salt, causing the bacteria to die.

It is also important to keep the food facility as water free as possible (15). Contaminated water is a severe problem for food service establishments that can only use water from an approved water supply free of pathogenic microorganisms. In addition, only restaurants with adequate plumbing along and proper water waste and sewage disposal are allowed to serve food. Mechanisms must be in place to prevent backflow through pipes (sewer)—an occurrence that will cause the health department to immediately close a food service establishment.

## Foods with Low Acidity

The acidity or alkalinity of a substance often determines which bacteria, if any, will grow in a food. The FDA states that proper pH (the measure of acidity, less than 7, or alkalinity, greater than 7, of a solution) must be maintained in order to prevent the growth of harmful bacteria. Because most (but not all) bacteria do not grow well below the low pH of 4.6, acidification is one way to maintain safe pH levels and keep various foods safe from harmful bacteria (25). The food industry uses two methods to increase the acidic concentration of foods:

1. Acid is added to foods as a chemical
   - Acetic acid (vinegar, salad dressing)
   - Citric or lactic acid (candy, orange juice)
   - Phosphoric acid (cola soda)
2. Acid is produced through fermentation by microorganisms such as bacteria or yeast that are added to foods
   - Proprionic acid (Swiss cheese)
   - Lactic acid (yogurt, pepperoni, sauerkraut)

In general, foods that are naturally high in acid (such as limes) or are made more acidic are less likely to engender bacterial growth than low-acid foods. The pH sensitivity of bacteria means that fruits and other high-acid content (pH < 4.6) foods are less likely to be sources of microbial contamination than foods with a lower acid content (pH > 4.6), such as meats and vegetables (31). Exceptions always exist. For example, tomatoes were once considered high-acid foods and relatively safe, but some newer varieties are not so acidic.

## Exceptions to the High Protein/Water/pH Rules

In recent years, foodborne illnesses have been traced to foods that do not contain high levels of protein and water. Fresh fruits and vegetables, cooked rice, sliced fruits, sautéed onions, potatoes, garlic-in-oil combinations, and apple cider have all been implicated (81). According to the FDA, more than 75 percent of produce-related outbreaks can be traced to five groups: lettuce (30 percent), tomatoes (17 percent), cantaloupes (13 percent), herbs (11 percent), and green onions (5 percent) (35). Other implicated foods include raspberries, strawberries, and almonds.

Illnesses from fresh-pressed apple cider, affecting hundreds of people and resulting in at least one death, have been reported over the last 20 years (71). At one time it was thought that the acidity of fresh apple cider (pH < 4.0) was protective, but it is now known that one strain of *E. coli* can survive in fresh cider with a pH of 3.7. This conflicts with the FDA Food Code that has historically considered foods with a pH of less than 4.6 as generally safe (66). Because fruits and vegetables

may be contaminated by microorganisms in the soil or by manure used to fertilize crops, the FDA recommends pasteurization for juices and requires the fresh juice industry to place the following warning on untreated juice or cider:

WARNING: This product has not been pasteurized and therefore may contain harmful bacteria that can cause serious illness in children, the elderly, and persons with weakened immune systems (28).

## Oxygen and Food

Many microorganisms need oxygen to grow, but others prefer oxygen-free environments such as cans, foil-wrapped baked potatoes, or untreated garlic-in-oil mixtures (58). A common method of reducing the risk of bacterial growth is to remove oxygen from a food product bag through vacuum packaging.

# Purchasing

Food should be purchased from safe sources.

## Written Specifications

A quality control program in a food establishment often ensures that only foods that meet written specifications are purchased. Foods should be purchased from reputable vendors, meet temperature and humidity requirements, show no evidence of being refrozen (such as frozen fluid lining the bottom of the food container or large ice crystals on the food's surface), be received in undamaged containers, and meet specifications, such as USDA grade, weight, size, and form (e.g., fresh or frozen; see Chapter 6). Suspect cans (dented, bloated, or showing signs of leakage) and foods in unmarked containers should be discarded. All foods should be in their original containers or clearly labeled if they have been transferred to another receptacle.

# Inspection

Shipments should be inspected upon delivery by trained employees that can evaluate the quantity and quality of the food (each food-related chapter in this book covers quality under the selection section). Temperatures should be checked, and then the products correctly stored as soon as possible after receipt. Employees then document whether or the shipment was accepted, rejected, or corrected.

## Storage

Stored food slowly deteriorates, making it vulnerable to microbrial contamination. So during storage or at any other time, two main risk factors on the "Risk Road" to foodborne illnesses are (1) temperature and (2) the amount of time that food stays within a certain temperature (Figure 4-10).

For this reason, after delivery perishable foods should be immediately stored in one of three types of storage—(1) refrigerator, (2) freezer, or (3) under dry conditions—according to the following temperatures:

1. Refrigerator: USDA = 40°F (4°C) or below for consumers
   • FDA = 41°F (5°C) or below for retailers
2. Freezer: below 0°F (−18°C)
3. Dry storage: 60–70°F (15–21°C) canned goods
   • 50–70°F (10–21°C) root vegetables (potatoes, onions) and whole citrus, eggplant, and squash (hard-rind)

The optimal type of storage and the storage temperatures recommended for foods vary; the inside back cover of this book along with the Storage sections of the food-related chapters describe these differences. For instance, some fruit is better left to ripen at room temperatures, especially bananas, tomatoes, avocados, and pears.

In any case, storage areas should be properly maintained and kept clean to prevent contamination of foods during storage. Microorganisms by their nature are ubiquitous and lodge themselves in air filters, drains, equipment, floor cracks, food scraps, and even dust.

### Temperature

Proper storage is important because bacteria grow rapidly in the **temperature danger zone** and this includes the human body temperature of 98.6°F (37°C) (Figure 4-11). The goal is to keep foods

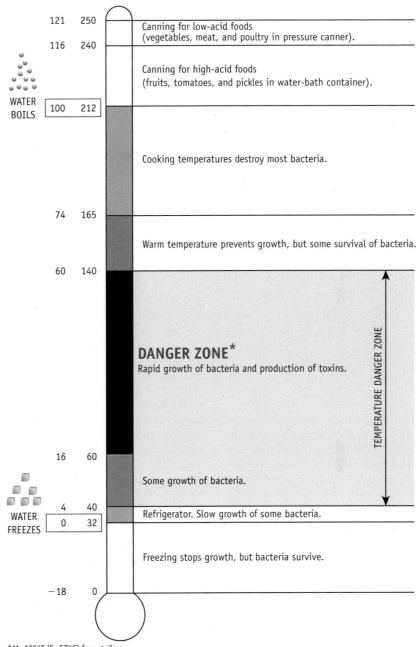

**FIGURE 4-11**   The temperature danger zone and its surrounding temperatures.

| °C | °F | |
|---|---|---|
| 121 | 250 | Canning for low-acid foods (vegetables, meat, and poultry in pressure canner). |
| 116 | 240 | Canning for high-acid foods (fruits, tomatoes, and pickles in water-bath container). |
| 100 | 212 | WATER BOILS |
| | | Cooking temperatures destroy most bacteria. |
| 74 | 165 | Warm temperature prevents growth, but some survival of bacteria. |
| 60 | 140 | **DANGER ZONE*** Rapid growth of bacteria and production of toxins. |
| 16 | 60 | Some growth of bacteria. |
| 4 | 40 | Refrigerator. Slow growth of some bacteria. |
| 0 | 32 | WATER FREEZES |
| | | Freezing stops growth, but bacteria survive. |
| −18 | 0 | |

TEMPERATURE DANGER ZONE

*41–135°F (5–57°C) for retailers

out of the temperature danger zone. It is important to recognize that temperatures suggested by the USDA primarily for consumers and those recommended by the FDA Food Code for retailers may differ. The USDA suggests storing cold foods under 40°F (4°C), and hot foods above 140°F (60°C). The FDA Food Code recommendation is different for retailers (restaurants, etc.), which have a slightly modified temperature danger zone of 41°F (5°C) to 135°F (57°C). Bacteria normally do not

survive temperature extremes, although a few bacteria survive below freezing (32°F/0°C).

---

**Temperature danger zone** The temperature range—considered to be 40°F–140°F (4°C–60°C) by the USDA for consumers, and 41°F–135°F (5°C–57°C) according to the FDA Food Code for retailers—that is ideal for bacterial growth.

**FIGURE 4-12**    Bacteria divide to reproduce, resulting in billions from just one cell in less than one day.

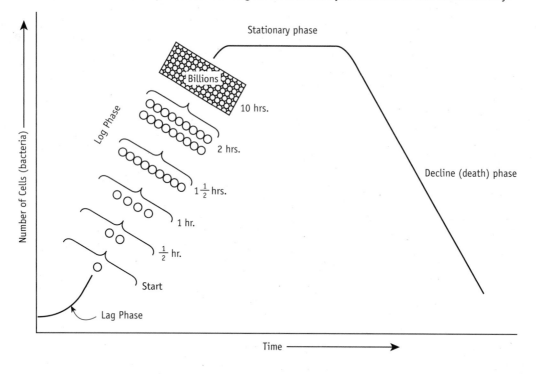

Despite precautions, some bacteria may survive environmental stresses in **spore** form. Spores are very resistant to drying and heating, and the bacteria may remain in this dormant state for long periods until their environment becomes more hospitable. In addition, food processing plants are vulnerable when instruments and appliances used in pathogen-killing processes, such as temperature gauges, heaters, seals, and refrigeration units, fail to work properly.

***Proper Refrigerator and Freezer Use***    Refrigerators and freezers, especially in a commercial food establishment, should be opened only when absolutely necessary because frequent door opening decreases their temperature efficiency. Refrigerator temperatures should be checked regularly to ensure that they are being maintained correctly. Studies show that 10 to 20 percent of home refrigerators are above 50°F (10°C), a temperature that cannot ensure safety (17). The refrigerator or freezer should also not be overloaded to avoid compromising the capacity of the unit to maintain correct temperatures. In addition, the practice of placing large amounts of hot food in a cold refrigerator will cause temperatures to fluctuate dangerously.

### Storage Times

Any perishable food exposed to danger zone temperatures for more than 2 hours of actual time, or *4 hours* of cumulative time, should be discarded. Cumulative time includes the time from the truck to the store, the store to the freezer, and the freezer to the kitchen, and the time on the counter where the food is being prepared. Microbial growth occurs exponentially; the number of bacteria can grow from harmless to staggering in a relatively short time (Figure 4-12).

There is also a limit as to how long perishable foods can be held safely in the refrigerator or freezer, or under dry conditions. Storage limits are provided on the inside back cover of this book; however, recommendations for maximum storage times are not exact and vary from source to source. A general rule to follow is 3 days maximum for fresh meats and high-water-content fruits and vegetables. The "first in, first out" (FIFO) rule should also be followed—foods brought into the storage area at an earlier date are used before those purchased later. This is especially true for the high-risk foods such as meat, dairy, and eggs. Each food service establishment should follow a schedule for reviewing all perishable foods so that those passing their expiration dates are discarded.

## Preparation

The various steps of food preparation—pre-preparation, cooking, holding, cooling, reheating, and serving—are vulnerable to creating conditions for foodborne illness. These individual steps of food preparation are now discussed in terms of applying food safety measures.

**Spore**  Encapsulated, dormant form assumed by some microorganisms that is resistant to environmental factors that would normally result in its death.

### Pre-Preparation

Two important pre-preparation steps related to food safety are now discussed—thawing and cross-contamination.

*Thawing*   For safe thawing, only one of the methods listed below should be used (Figure 4-13).

- Refrigerator, on the bottom shelf to avoid contaminating other foods with any drippings. Although this method requires planning, thawing frozen meat at room temperature is considered an unsafe practice.
- Submerged under running water. However, running cold water over meat wrapped in protective plastic, or placing it in a bath of ice water and frequently replacing the water, are not as safe as defrosting in a refrigerator.
- Microwave oven followed by immediate cooking will work for smaller items, but not large roasts or turkey.
- As part of the cooking process, but be sure to check final internal temperatures.

*Cross-Contamination*   Cross-contamination is one of the most common causes of foodborne illness during the summer months when backyard barbecuing is popular. People carry raw meat on a plate to the barbecue, cook the meat, and place it back on the same plate, which is contaminated with raw meat juices. Even using sponges may be a problem because they can harbor bacteria even after they dry (62). An example of commercial cross-contamination occurred when at least 224,000 people became ill from eating ice cream contaminated with *Salmonella*. The bacteria was traced to the trucking company that had transported the pasteurized ice cream premix in trailers that had previously carried *Salmonella*-contaminated liquid eggs (41).

To prevent cross-contamination, food should never touch contaminated surfaces unless it is to be thoroughly cooked. Surfaces should also be regularly washed and sanitized, especially after coming into contact with raw food. Particularly susceptible surfaces include hands, utensils, tabletops, cutting boards, and slicers, as well as aprons, cleaning cloths, and sponges. Using separate chopping surfaces for raw meat and salad vegetables is ideal. Dust and soil should be washed off the tops of cans before they are opened. Raw meats should never be stored in the refrigerator above cooked or ready-to-eat foods where they may drip onto the food below.

Foodservice workers should not touch the surfaces of food-serving utensils. The forks, knives, and spoons for customer use are always placed head down in serving canisters. The same rule applies to the ice scoop handle, which should never come in contact with ice after touching an employee's hands or an unclean surface. The biggest mistake to avoid is double-dipping when taste-testing food. A tasting utensil used once is best. Liquids can be poured into a separate, small cup.

### Cooking (Heating)

Temperature-time abuse contributes to most foodborne outbreaks in the United States (9). The best temperature for microorganisms to grow rapidly is in the middle of the temperature danger zone, so the three most common mistakes to avoid when heating food are:

1. Failing to heat food to its minimum internal temperature
2. Failing to cool food properly
3. Failing to reheat food to its minimum internal temperature

*Minimum Internal Temperatures*
Heat often destroys bacteria. It is well known that bacteria die when exposed to at least 10 minutes of boiling (212°F or 100°C) and that freezing (32°F or 0°C) slows their growth. However, not all food can be boiled, so the goal is to keep foods out of the temperature danger zone and to heat foods to be cooked to their minimum internal temperatures to destroy any microorganisms that may be present. Certain foods, especially meat, poultry, and fish, need to reach higher temperatures than the minimum consumer standard of 140°F (60°C), or 135°F (57°C) for retailers, during cooking in order to ensure safety (Table 4-10).

**FIGURE 4-13**  Thawing foods—four acceptable methods.

## Refrigerator

## Submerged under running water

## Part of the cooking process

## Microwave oven (cook right after)

**TABLE 4-10** Minimum Internal Temperatures (for 15 seconds)

*(Temperatures vary depending on the source [FDA Food Code and the USDA], and individual states [health departments] are not required to adopt either of these recommendations, so some temperatures will vary.)*

| Food | Temperature |
|---|---|
| Poultry (includes ground poultry) | 165°F (74°C)[1] |
| Stuffing, sauces, gravies, soups | 165°F (74°C) |
| Reheated or microwaved foods, leftovers, casseroles, hot dogs | 165°F (74°C) |
| Ground meats: beef, pork, veal, lamb | USDA = 160°F (71°C) or FDA = 155°F (68°C) |
| Pork (fresh or raw) | USDA = 160°F (71°C) or FDA = 145°F (63°C) |
| Precooked ham | 140°F (60°C) |
| Eggs | USDA = 160°F (71°C) or FDA = 145°F (63°C) |
| Beef, veal, lamb, fish | 145°F (63°C) for medium rare[2] |
| Vegetables and fruits | 135°F (57°C) |
| Commercially processed ready-to-eat foods | 135°F (57°C) |

[1]In 2006, the USDA selected 165°F (74°C) to be the single safe minimum endpoint temperature. However, consumers can choose to cook poultry to higher temperatures. It is recommended to check the temperature in whole birds at three locations—thigh (deep crevice), wing joint, and breast.
[2]160°F (71°C) for medium and 170°F (78°C) for well done.

**Source:** www.fsis.usda.gov/Fact_Sheets/Use_a_Food_Thermometer/index.asp.

*Thermometers* The only way to be sure about a food's temperature—and whether or not it is in the temperature danger zone—is to use a thermometer. Both Fahrenheit (nonmetric) and Celsius centigrade (metric) scales are usually displayed together. Standard bulb thermometers work on the principle that mercury expands when heated and contracts when cooled. Mercury in the bulb at the bottom of an extended glass tube expands and contracts along a marked graduated scale. However, glass thermometers with mercury should never be used in commercial food operations because they can break. Numerous types of other thermometers are now available, including those that record the results and/or operate remotely.

# PROPER USE OF THERMOMETERS

## Types of Thermometers

As shown in Figure 4-14, an array of food thermometers that include dial, digital, disposable, appliance, and others are available for different purposes in commercial food preparation (83). Among the most popular are the pocket-size, instant-read (dial or digital) thermometers that can be used to check foods being held on steam tables in food service establishments. These are portable, much like a pen is, and give a reading a few seconds after being inserted into the food.

## How to Use a Thermometer

Correct temperatures rely on the use of selecting the right thermometer, testing foods correctly, the proper care of thermometers.

### Selecting a Thermometer

It's important to choose the right thermometer from among the different types discussed here. Thermometers differ not only in style but also in temperature ranges, cost, ability to be calibrated, and use. For instance, a meat thermometer has a short rod for insertion into the meat and usually has an upper limit of 185°F (85°C). A candy thermometer has an upper range of 325°F (163°C), and a deep-frying thermometer goes up to at least 500°F (260°C).

### Testing Temperatures

Only a few thermometers can be left in the food or appliance for checking temperatures continuously. Most are used at the end of preparation (cooking or cooling, or while being held) by testing the food in two or three different areas to see if it has reached a safe temperature.

- *Instant Readings.* Instant-read dial thermometers are inserted straight or at an angle at least 2 inches into the thickest part of the food without touching fat or bone and given 10–20 seconds before reading. Thin pieces can be measured by inserting the probe into the side of foods such as hamburgers or pork chops. Instant-read digital thermometers only need to be inserted about a half inch into the thickest part of the food and only take 2–10 seconds to register (83).

- *Dial Readings.* Noninstant read dial thermometers take longer than their digital counterparts to read temperatures—at least 1–2 minutes.

### Care of Thermometers

Instant-read thermometers must be cleaned between uses (with alcohol wipes or washed with hot soapy water by hand, but never immersed in water) or contamination may result. It's also best to leave the sharp probe in the sheath and turn off the device between uses because they are battery operated. Store them away from heat or cold.

## Calibration of Thermometers

Thermometers have a tendency to malfunction when dropped or jarred. They may also simply drift off in their readings, so thermometers need to be tested frequently for accuracy against a known standard by testing them in either freezing (ice water method) or boiling water (boiling water method). The more common ice water method consists of blending a cup of crushed ice in a cup of water (or 50 percent ice and 50 percent water at least 2 inches deep),

**FIGURE 4-14** Different types of food thermometers.

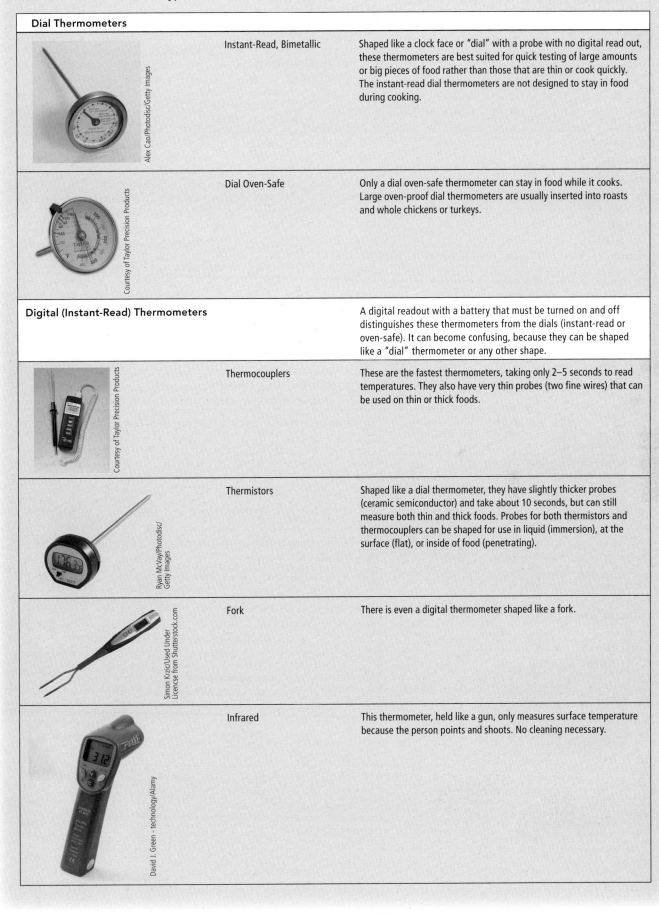

| Dial Thermometers | | |
|---|---|---|
| | Instant-Read, Bimetallic | Shaped like a clock face or "dial" with a probe with no digital read out, these thermometers are best suited for quick testing of large amounts or big pieces of food rather than those that are thin or cook quickly. The instant-read dial thermometers are not designed to stay in food during cooking. |
| | Dial Oven-Safe | Only a dial oven-safe thermometer can stay in food while it cooks. Large oven-proof dial thermometers are usually inserted into roasts and whole chickens or turkeys. |
| **Digital (Instant-Read) Thermometers** | | A digital readout with a battery that must be turned on and off distinguishes these thermometers from the dials (instant-read or oven-safe). It can become confusing, because they can be shaped like a "dial" thermometer or any other shape. |
| | Thermocouplers | These are the fastest thermometers, taking only 2–5 seconds to read temperatures. They also have very thin probes (two fine wires) that can be used on thin or thick foods. |
| | Thermistors | Shaped like a dial thermometer, they have slightly thicker probes (ceramic semiconductor) and take about 10 seconds, but can still measure both thin and thick foods. Probes for both thermistors and thermocouplers can be shaped for use in liquid (immersion), at the surface (flat), or inside of food (penetrating). |
| | Fork | There is even a digital thermometer shaped like a fork. |
| | Infrared | This thermometer, held like a gun, only measures surface temperature because the person points and shoots. No cleaning necessary. |

| Disposable Temperature Indicators (Single-Use) | | |
|---|---|---|
| <br>*Courtesy of Jaccard Corporation* | Pop-Up | Commonly used with whole chickens, turkeys, or roasts, but it's best to double-check the results against a conventional thermometer. The indicator pops up when the recommended temperature is reached because this is the heat required to melt the firing material which releases the spring in the nylon device. |
| <br>*Courtesy of Taylor Precision Products* | Disposable | Designed for just one use. |
| <br>*Courtesy of 3M* | Time Temperature Indicators (TTI) | These small monitors about the size of two stamps have self-adhesive on the back so they can stick to food products in transit or storage. They irreversibly record the high or low temperature so that any temperature abuse can be detected. |

| Appliance Thermometers | | Specialized thermometers can be IN or OUTSIDE ovens, refrigerators, or freezers (dial or digital) to check the accuracy of the equipments' thermostats. |
|---|---|---|
| <br>*JOSH51699/istockphoto.com* | Dial Oven-Safe (Bi-Metallic) | Thermometers that can measure temperatures from 100 to 600°F (38–316°C). |
| <br>*Courtesy of Taylor Precision Products* | Refrigerator/ Freezer | Refrigerators need to be at or below 40°F (4°C), while freezers need to be below 32°F (0°C). |
| <br>*Per Karlsson – BKWine.com/Alamy* | Panel | Many units now have the internal temperatures digitally displayed on the outside or have panel thermometers built onto the outside of the unit to minimize temperature fluctuations caused by opening the door. |
| <br>*Freddy Eliasson/Used Under License from Shutterstock.com* | Oven Cord Thermometer | A thermometer probe inserted into the cooking food within the oven is connected by a cord to the base unit on the counter, maintaining a continuous temperature check. |

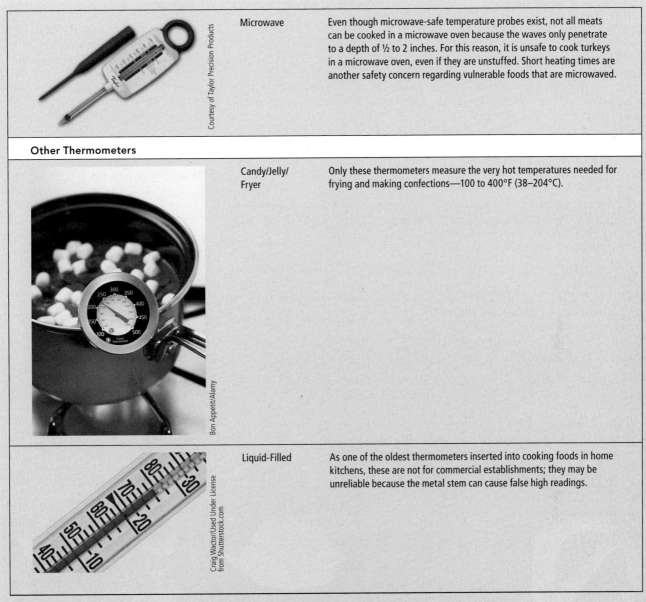

| | Microwave | Even though microwave-safe temperature probes exist, not all meats can be cooked in a microwave oven because the waves only penetrate to a depth of ½ to 2 inches. For this reason, it is unsafe to cook turkeys in a microwave oven, even if they are unstuffed. Short heating times are another safety concern regarding vulnerable foods that are microwaved. |
|---|---|---|

*Courtesy of Taylor Precision Products*

**Other Thermometers**

| | Candy/Jelly/ Fryer | Only these thermometers measure the very hot temperatures needed for frying and making confections—100 to 400°F (38–204°C). |
|---|---|---|

*Bon Appetit/Alamy*

| | Liquid-Filled | As one of the oldest thermometers inserted into cooking foods in home kitchens, these are not for commercial establishments; they may be unreliable because the metal stem can cause false high readings. |
|---|---|---|

*Craig Wactor/Used Under License from Shutterstock.com*

Adapted from reference 82.

dipping the point into the stirred slurry for 30 seconds, and seeing if it reads 32°F (0°C). Detailed instructions are provided by one of the website links at the end of this chapter. The boiling-point method involves dipping the point of a bulb thermometer into boiling water for 30 seconds without touching the sides or bottom of the pan. Digital and dial thermometers are checked by placing them 2 inches above boiling water to obtain the reading of 212°F (100°C). The reading should be 212°F (100°C) at sea level (with 1°F [2.2°C] subtracted for every 500 feet of increase in altitude).

Some dial thermometers have a calibration nut underneath their face and can be recalibrated by gripping the hex nut with pliers and twisting the dial face until it reads the correct temperature (81). Digital thermometers are slightly more expensive, but more accurate because they contain an electronic sensor near the tip that is effective when inserted only ½ inch into food. If they drift off from calibration, they need to be replaced.

Exceptions to this are battery-operated digital thermometers that can be sent back to the manufacturer or independent testing facility for recalibration, or that are equipped with a recalibration knob.

Ideally, thermometers would be calibrated daily before the work shift begins or at least weekly or monthly, and definitely after each time they are dropped, jarred, or exposed to extreme temperatures. Not all food thermometers can be calibrated, so it's best to determine this prior to purchase.

***Temperature-Time Monitoring Program***    Every restaurant or food-related establishment should have an organized plan written into its standard operating procedures that states what, when, and who (the person responsible) will monitor in terms of the temperature and times for food preparation. Employees should be given their own thermometers, times to check certain foods or equipment (refrigerators, freezers, fryers, serving areas), methods to log the temperature/time, and an established set of corrective action procedures. This information should then be filed in the organization's records. An active monitoring program is essential for the prevention of foodborne illnesses.

## Holding

Foods being held before or after serving must also be kept out of the temperature danger zone. Thermometers should be used to ensure that the following temperatures are reached:

- Hot foods
  FDA   = at least 135°F (57°C)
  USDA = at least 140°F (60°C)

- Cold foods
  FDA   = under 41°F (5°C)
  USDA = under 40°F (4°C)

**Time Limit**    Held food must be sold, served, or discarded within 4 hours (58). Food temperatures should be checked in the beginning and at least every 2 hours to allow for corrective action. After 4 hours the food has to be discarded if it falls outside these minimal internal temperatures. It's easier to stay within recommended temperatures by distributing evenly by stirring and preparing foods in smaller batches.

**Protective Barriers**    Held food should be covered as much as possible or blocked from people by plastic sneeze guards.

## Cooling

Inappropriately cooled foods are another major cause of foodborne illnesses. Larger, denser foods are the most vulnerable. According to the USDA, foods should be cooled to below 40°F (4°C) within 4 hours of removal from cooking or they pose a danger to consumers. The FDA Food Code suggests that the cooling of hot foods occur in two stages: 135°F–70°F (57°C–21°C) in the first 2 hours, and then 135°F–41°F (57°C–5°C) within 6 hours or less. Food not reaching these temperatures within 6 hours should be discarded.

Four methods exist to speed the cooling of foods before they are placed in the refrigerator or freezer (Figure 4-15) (58). Liquid foods should be placed in shallow pans less than 3 inches deep to cool, and thicker foods in pans less than 2 inches deep (Figure 4-15). Again, stirring distributes the heat more quickly.

## Reheating

Within 2 hours before being served, all hot foods must be reheated to at least 165°F (74°C) for 15 seconds. In a food service establishment, untouched leftovers are sometimes discarded, because

**FIGURE 4-15**    Cooling foods—four safe methods

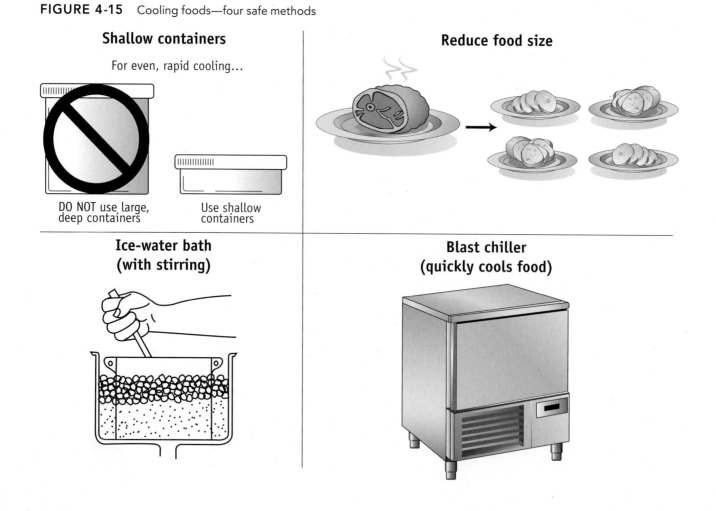

**Shallow containers**

For even, rapid cooling...

DO NOT use large, deep containers

Use shallow containers

**Reduce food size**

**Ice-water bath (with stirring)**

**Blast chiller (quickly cools food)**

**FIGURE 4-16**   Three-compartment sink.

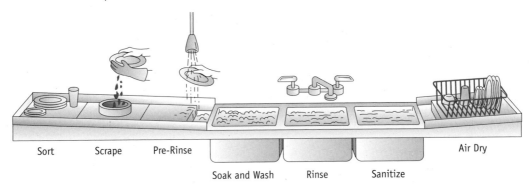

Sort      Scrape      Pre-Rinse            Air Dry

Soak and Wash      Rinse      Sanitize

they are a potential source of microbial contamination.

### Serving

Serving is also a vulnerable point for food contamination. Good personal hygiene on the part of food service employees is essential to the safety of the foods and beverages being served. Even when serving, the USDA recommends that the 140°F (60°C) and 40°F (4°C) boundaries continue to be observed, while the FDA Food Code suggests ≤41°F (5°C) and ≥135°F (57°C) for retailers. See the previous sections on personnel hygiene and cross-contamination for specific methods of ensuring food safety during serving. The safest practice is to make sure that bare hands never touch food or surfaces that will touch food.

## Sanitation

Sanitation of surfaces coming in contact with food, whether dishes, utensils, counters, equipment, floors, or other parts of the foodservice area, can be achieved either through (1) heating and/or (2) chemicals.

### Dishes

Dishes require specific processes to meet sanitation requirement for cleaning and drying.

*Heating*   Dishes in a food service establishment may be hand or machine washed. Whatever method is used, the process must meet certain sanitation guidelines to pass a health department food inspection. In order to kill pathogens, dishwashing temperatures should be between 140°F and 160°F (60°C and 71°C), and rinse temperatures must be at least 180°F (82°C) for 10 seconds or 170°F (77°C) for 30 seconds. Despite concerns that dishwashing failures result in foodborne illness, only 5 percent of sanitation failures can be traced to faulty equipment; the remaining 95 percent are a result of human error.

*Three-Compartment Sink*   Manual washing in a food establishment requires a **three-compartment sink** for washing pots and pans, dishes, glasses, cutlery, and tools (Figure 4-16). The first compartment is used for soaking and washing items in water heated to 110°F (43°C). The second compartment is for rinsing, and the third area is for sanitation. The last compartment can sterilize items with either hot water or chemical sanitizers. If water is used, then temperatures must reach at least 180°F (82°C) for 1 minute.

*Chemicals*   Food establishments may also sanitize with the chemical sanitizers shown in Table 4-11.

*Drying*   Items should always be air or heat dried. Damp cloth towels should be avoided because they serve as an ideal medium for microorganism growth.

### Scheduling

Schedules for cleaning should be posted and followed scrupulously to maintain a sanitary work environment. Floors, walls, windows, lights, and equipment should all be included in the frequent cleanup routine. Sanitation guidelines involving cleanup, personnel, equipment, facilities, pest control, and water could be set and routinely followed by establishing cleaning schedules to be checked off on predetermined dates.

### Equipment

The National Sanitation Foundation (NSF) sets standards for equipment to be used in food service establishments. Equipment should be as free of crevices as possible, so it is best to buy equipment

| TABLE 4-11   Chemical Sanitizers Used in Commercial Food Establishments | |
| --- | --- |
| Sanitizer* | How to Use (at 75°F or 24°C) |
| Chlorine | 50 ppm** for water and equipment or 1 teaspoon for each quart of warm water or 1 tablespoon for each gallon of warm water |
| Iodine | 12.5-25 ppm for hand washing and equipment |
| Quaternary ammonium compounds | 200 ppm for walls; 500 ppm for floors |
| Organic acids (lactic, acetic, propionic) | 130 ppm for stainless steel surfaces |

*If manufacturer's instructions differ from this table, follow manufacturer's recommendations

**ppm = parts per million

**Three-compartment sink**   A sink divided into three sections, the first for soaking and washing, the second for rinsing, and the third for sanitizing.

with rounded junctions. Wooden cutting boards are more prone to nicks and crevices than are those made of plastic or marble, and it once was thought that they were more prone to microbial contamination. Studies now show that once they are cleaned, bacterial levels do not differ significantly among the various types of boards (55). All equipment, utensils, containers, and meat grinders and slicers should be thoroughly cleaned after each use. Only trained employees should clean meat slicers, because these instruments are the number-one cause of cuts in a food service organization.

After cleaning, utensils are best stored covered, and glasses and cups should always be stored upside down. Disposable utensils such as plastic cups, forks, knives, and plates should never be washed and reused. Freezers and refrigerators should have at least 6 inches (10 inches is preferred) between the bottom storage shelf and the floor; this allows for adequate cleaning. They should also contain thermometers to determine that correct temperatures are being maintained.

### Facilities

In order to remain sanitary, a food service establishment should be designed and maintained in ways that promote cleanliness. Floors, walls, and ceilings should have adequate ventilation. Materials used in construction should allow for easy cleaning. Garbage should be discarded in covered, pest-proof containers that are frequently cleaned and free of litter. Lighting fixtures must be covered to prevent dust and insects from collecting on light bulbs and falling on the food. Food service organizations are required to have dressing rooms, restrooms, and hand-washing sinks available to the employees. Having unclean restrooms is the most common complaint against eating establishments that consumers file with health departments.

### Pest Control

Food naturally draws living creatures, from bacteria to bears. Even the cleanest facility can be put at risk of transmitting foodborne illness by the presence of insects, rodents, birds, turtles, or other animals. Rodents such as mice and rats can carry *Salmonella*, typhus, and the bubonic plague. Insects and

**FIGURE 4-17**   Common pests that may transmit foodborne illness.

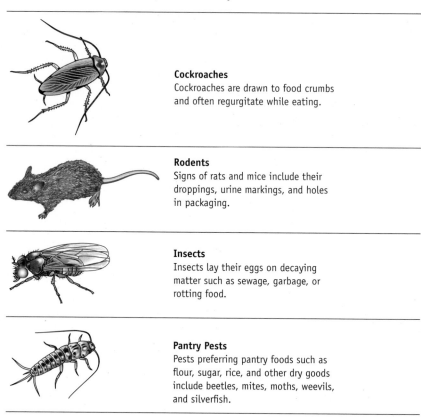

**Cockroaches**
Cockroaches are drawn to food crumbs and often regurgitate while eating.

**Rodents**
Signs of rats and mice include their droppings, urine markings, and holes in packaging.

**Insects**
Insects lay their eggs on decaying matter such as sewage, garbage, or rotting food.

**Pantry Pests**
Pests preferring pantry foods such as flour, sugar, rice, and other dry goods include beetles, mites, moths, weevils, and silverfish.

cockroaches transfer microorganisms by landing, walking, and regurgitating their stomach contents on foods when feeding. Figure 4-17 shows some common pests. To discourage pests from taking up residence, it's best to (1) block pests from entering the establishment, (2) block pests from all food, and (3) maintain a pest control program.

***Blocking Entrances***   All entrances should be screened (at least 16 mesh/square inch) or sealed. Doors and windows should be self-closing, closed, or covered with air curtains. Although not always possible, maintaining a low humidity (under 50 percent) through dehumidifiers and/or ventilation keeps cockroach eggs from hatching (58).

***Blocking Access to Food***   Keep counters, floors, and all surfaces as food-free as possible. Clean up all spills immediately and thoroughly. Even a crumb is a large meal for an insect. Keep all garbage covered and remove it frequently. All food should be securely stored in pest-proof containers. Sometimes flours and grains already contain pests that then multiply within the bags

and these should be immediately discarded. It may be necessary in humid climates to refrigerate grains, nuts, or any powdered food such as flour and dried milk. Water is critical to pest survival, so standing water should be removed and all drains should be working properly.

***Pest Control Program***   A regular maintenance program is key to controlling pests. It's important to block all entrances and access to food, but pests who manage to surmount these obstacles need to be removed or eliminated. Immediate removal prevents breeding that would result in exponential numbers of pests. Even though it might be tempting to spray insects, only a licensed pest control operator (PCO) should apply insecticides.

## Food Safety Monitoring

Food needs to be inspected to ensure that the nation's supply is safe. First and foremost are the internal food inspections that start on site with the foodservice employees because no government agency can check every establishment every day. The frontline agencies responsible for food inspection are local state health departments. On the national level, the CDC receives reports from these health departments through FoodNet. Manufacturers and other pre-consumer food-related establishments (but not restaurants) can also be inspected by the FDA or the USDA.

## FDA's Food Code

The FDA's "Food Code" (24) became available in 1993 to serve as a foundation for all U.S. food safety programs. This list of government food safety recommendations is updated every several years. Although not actual law, it serves as a guide for state health departments conducting inspections. As a result, minimum temperatures and other aspects of food safety lack uniformity across states, but the variations are often minor.

A survey of restaurant workers revealed that the main barriers to implementing safe food practices were time constraints, inconvenience, inadequate training, and inadequate resources (43). These barriers need to be overcome in order to ensure food safety for consumers and to pass health department inspections. An additional measure to ensure consumer safety is the Hazard Analysis and Critical Control Point or **HACCP** (pronounced hass-ip) system, a seven-step program primarily utilized by large institutions and national surveillance. Food safety monitoring through state health department inspections and HACCP is now discussed in detail.

## Health Department Inspection

Proactive foodservice organizations already perform their own internal inspections, and thus are prepared to pass unannounced health department inspections. Too many violations can result in suspension or revocation of an establishment's license. A temporary suspension is served when there is an imminent hazard to health or when there has been a failure to comply with an earlier inspection's findings.

A revocation occurs with more serious or repeated violations.

As shown in Figure 4-18, a sample health department inspection form assigns point values to various safety practices based on their importance. Points for any practices that are not being demonstrably followed are subtracted from 100 percent to obtain the score. The most common violations found by health inspectors involve food-holding temperatures, improper refrigeration, and improper cooling of cooked foods. Because food safety regulations vary from state to state, it is wise to obtain a copy of the state inspection form in order to be absolutely sure of what is inspected by the local health department.

## HACCP

A more formal approach for larger institutions is to incorporate **HACCP.** The FDA recommends the HACCP system be incorporated by the food industry, including processing plants, food service establishments, and food corporations. Previously, the only real standard method of ensuring food safety compliance was health department inspections. Having a customized, written HACCP plan now shifts the emphasis from policing to prevention. It is also based on the flow of food rather than meeting minimum sanitary recommendations.

*History of HACCP*   HACCP celebrated its 50th anniversary in 2009 (74). The program was initially implemented in the 1960s by the Pillsbury Company with the assistance of the U.S. Army Natick Research Laboratories and the National Aeronautic and Space Administration (NASA). NASA wanted to provide food that approached a 100 percent assurance against foodborne illness to astronauts during space flights. At its inception, HACCP included three principles, which were later expanded to seven. When the National Advisory Committee on Microbiological Criteria for Foods (NACMCF) published these seven principles in 1989, the HACCP guidelines became the international standard for food safety (75).

*The Seven HACCP Principles*
HACCP is a step-by-step food safety

program based on the seven principles shown in Figure 4-19 (57). All seven steps in a HACCP plan are written by the food-related establishment's management into a customized plan that follows the food flow of their organization. As a result, the HACCP program can be somewhat complex and time-consuming. It involves (1) assessing (identifying) potential hazards (biological, chemical, and physical); (2) identifying **critical control points (CCPs)**; (3) establishing quantifiable limits such as temperatures for each CCP; (4) monitoring CCPs to make sure that they stay within the recommended limits; (5) taking corrective action if they do not; (6) verifying through regularly evaluating records; and (7) documenting through record keeping. To assist with ensuring that food safety practices are incorporated, Table 4-12 lists some of the records that could be included in the HACCP plan or any other food safety program.

## Food Surveillance

Food surveillance is provided nationally by FoodNet and globally the World Health Organization (WHO) wants to institute a system.

*FoodNet*   The Foodborne Diseases Active Surveillance Network (FoodNet) was established in 1995 and is a collaborative project among local health departments, the CDC, the FDA, and the USDA Food Safety and Inspection Service. The Network attempts to provide active surveillance of foodborne outbreaks by keeping national estimates of the number of reported foodborne illnesses (26). The specific biological hazards that are monitored include: *Campylobacter, Escherichia coli* O157:H7, *Listeria, Salmonella,*

---

**HACCP** Hazard Analysis and Critical Control Point System, a systematized approach to preventing foodborne illness during the production and preparation of food.

**Critical control point (CCP)** A point in the HACCP process that must be controlled to ensure the safety of the food.

**FIGURE 4-18**   Health department inspection form.

| FOOD | | POINTS |
|---|---|---|
| *01 | Source; sound condition, no spoilage | 5 |
| 02 | Original container; properly labeled | 1 |

**FOOD PROTECTION**

| | | |
|---|---|---|
| *03 | Potentially hazardous food meets temperature requirements during storage, preparation, display, service transportation | 5 |
| 04 | Facilities to maintain product temperature | 4 |
| 05 | Thermometers provided and conspicuous | 1 |
| 06 | Potentially hazardous food properly thawed | 2 |
| *07 | Unwrapped and potentially hazardous food not re-served | 4 |
| *08 | Food protection during storage, preparation, display, service, transportation | 2 |
| 09 | Handling of food (ice) minimized | 2 |
| 10 | In use, food (ice) dispensing utensils properly stored | 1 |

**PERSONNEL**

| | | |
|---|---|---|
| *11 | Personnel with infections restricted | 5 |
| *12 | Hands washed and clean, good hygienic practices | 5 |
| 13 | Clean clothes, hair restraints | 1 |

**FOOD EQUIPMENT AND UTENSILS**

| | | |
|---|---|---|
| 14 | Food (ice) contact surfaces: designed, constructed, maintained, installed, located | 2 |
| 15 | Non-food contact surfaces: designed, constructed, maintained, installed, located | 1 |
| 16 | Dishwashing facilities: designed, constructed, maintained, installed, located, operated | 2 |
| 17 | Accurate thermometers, chemical test kits provided, gauge cock (1/4" IPS valve) | 1 |
| 18 | Pre-flushed, scraped, soaked | 1 |
| 19 | Wash, rinse water: clean, proper temperature | 2 |
| *20 | Sanitization rinse: clean, temperature, concentration, exposure time; equipment, utensils sanitized | 4 |
| 21 | Wiping cloths; clean, use restricted | 1 |
| 22 | Food-contact surfaces of equipment and utensils clean, free of abrasives, detergents | 2 |
| 23 | Non-food contact surfaces of equipment and utensils clean | 1 |
| 24 | Storage, handling of clean equipment/utensils | 1 |
| 25 | Single-service articles, storage, dispensing | 1 |
| 26 | No re-use of single service articles | 2 |

**WATER**

| | | |
|---|---|---|
| *27 | Water source, safe: hot and cold under pressure | 5 |

| SEWAGE | | POINTS |
|---|---|---|
| *28 | Sewage and waste water disposal | 4 |

**PLUMBING**

| | | |
|---|---|---|
| 29 | Installed, maintained | 1 |
| *30 | Cross-connection, back siphonage, backflow | 5 |

**TOILET AND HANDWASHING FACILITIES**

| | | |
|---|---|---|
| *31 | Number, convenient, accessible, designed, installed | 4 |
| 32 | Toilet rooms enclosed, self-closing doors, fixtures, good repair, clean; hand cleaner, sanitary towels/tissues/hand-drying devices provided, proper waste receptacles | 2 |

**GARBAGE AND REFUSE DISPOSAL**

| | | |
|---|---|---|
| 33 | Containers or receptacles, covered: adequate number insect/rodent proof, frequency, clean | 2 |
| 34 | Outside storage area enclosures properly constructed, clean; controlled incineration | 1 |

**INSECT, RODENT, ANIMAL CONTROL**

| | | |
|---|---|---|
| *35 | Presence of insects/rodents — outer openings protected, no birds, turtles, other animals | 4 |

**FLOORS, WALLS, AND CEILINGS**

| | | |
|---|---|---|
| 36 | Floors, constructed, drained, clean, good repair, covering installation, dustless cleaning methods | 1 |
| 37 | Walls, ceiling, attached equipment: constructed, good repair, clean, surfaces, dustless cleaning methods | 1 |

**LIGHTING**

| | | |
|---|---|---|
| 38 | Lighting provided as required, fixtures shielded | 1 |

**VENTILATION**

| | | |
|---|---|---|
| 39 | Rooms and equipment — vented as required | 1 |

**DRESSING ROOMS**

| | | |
|---|---|---|
| 40 | Rooms clean, lockers provided, facilities clean, located | 1 |

**OTHER OPERATIONS**

| | | |
|---|---|---|
| *41 | Toxic items properly stored, labeled, used | 5 |
| 42 | Premises maintained free of litter, unnecessary articles, cleaning maintenance equipment properly stored. Authorized personnel | 1 |
| 43 | Complete separation from living/sleeping quarters. Laundry. | 1 |
| 44 | Clean, soiled linen properly stored | 1 |

*Critical items requiring immediate attention.

Score = 100 – ☐ pts. = ☐

**Source:** Adapted from Texas Department of Health's Food Service Establishment Inspection Report.

*Shigella, Yersinia, Vibrio,* and parasites, such as *Cryptosporidium* and *Cyclospora.*

National surveillance occurs when physicians and coroners notify local health departments of certain disease cases as required by law. These reports are then sent to the state public health epidemiology office, where their laboratories may also receive food samples to test. Eventually this information is sent to federal offices such as the CDC. Unfortunately, not every person who reports a foodborne illness provides a sample, so only about 38 percent of all recognized outbreaks reported to the CDC ever have their cause identified (41).

**WHO**   Globally, the World Health Organization (WHO) has proposed the development of a food safety plan to detect food hazards around the world (34).

**TABLE 4-12**  Food Safety Records to be Maintained by Food Manufacturers

### Registration

Registration under the Bioterrorism Act
Registration with FDA (required for low-acid and acidified food processors)
FDA must be given advanced notice on shipments of imported food
Filed processes and letters from processing authority (required for low-acid and acidified food processors)
Supplemental processes from equipment manufacturers

### Management and Personnel

Compliance regulation assigned to a qualified supervisor
Assignment of qualified supervisor(s) for overall sanitation
Supervisor and employee training documenting competency in identifying sanitation failures or food contamination
Certification of supervisors' completion of Good Manufacturing Practice (GMP) schools under 21 *CFR* 108 (required for low-acid and acidified food processors)

### Procedures

Hygiene and proper food handling
Cleaning and sanitization
Quality control
Allergen control
*Listeria monocytogenes* control plan
Recall
Hazard Analysis and Critical Control Points Plan

### Supplier Guarantees

Raw materials, packaging materials, and ingredients
Cleaning and sanitization materials
Pest control measures and treatments
Water safety and testing
Boiler and gasses indirect additives

### Quality Control and Testing

Documented compliance with procedures
Instrument calibration
Inspection of incoming materials
Sanitation and allergen testing
Aflatoxin and other naturally occurring toxin testing (if applicable)
Unavoidable defects testing (if applicable)
Inspection and control of physical hazards
Validation of preventative measures
Monitoring of final control parameters (e.g., pH, $a_w$)

### Production

Batch control records identifying ingredients
Batch, lot, and coding records
In-process controls of critical parameters (e.g., temperature)
Reconditioning and rework
Distribution
Pre-distribution record verification
Non-carrier source of ingredients and transport
Transportation and initial distribution of final product

**Source:** Adapted from reference 73.

**FIGURE 4-19**  The seven principles of the Hazard Analysis and Critical Control Point (HACCP) system.

1. Assess the hazards

2. Identify the critical control points

3. Establish limits at each critical control point

4. Monitor critical control points

5. Take corrective action

6. Verification

7. Documentation

# PICTORIAL SUMMARY / 4: Food Safety

Federal and state regulations, along with regular inspections throughout the food industry, ensure that the food supply in the United States is the safest in the world.

## FOODBORNE ILLNESS

Symptoms of an illness transmitted through food often include inflammation of the gastrointestinal tract lining, nausea, abdominal cramps, diarrhea, and/or vomiting.

The most common causes of foodborne illness are:

**Biological**
- Microorganisms
  - Bacteria
  - Molds
  - Viruses
- Animal parasites
- Prions

**Chemical**
- Plants
- Seafood toxins
- Agricultural/industrial

**Physical**
Foreign objects in food

## BIOLOGICAL HAZARDS

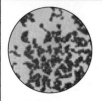

Biological hazards in food include bacteria, molds, viruses, parasites, and prions. The most common cause of foodborne illness is bacteria via infection or intoxication (poisoning). Food infections, responsible for about 80 percent of foodborne illnesses, occur when a person consumes a food or beverage containing large numbers of bacteria. Food intoxication and toxin-mediated infection occur when a person consumes a toxin from bacteria growing on food. Other foodborne illnesses are caused by mycotoxins in molds, viruses such as the hepatitis A virus, parasites such as worms and protozoa, and protein particles called prions, which are responsible for mad cow disease.

## CHEMICAL HAZARDS

Chemicals that are hazardous to a person's health can come from plants such as poisonous mushrooms, agricultural or industrial chemicals that are included in food unintentionally, or fish or shellfish that harbor dangerous toxins. Four examples of toxic seafood are ciguatera fish poisoning (the most common in the United States), histamine food poisoning, pufferfish poisoning, and red tide.

## PHYSICAL HAZARDS

Foreign objects that inadvertently turn up in food and beverage products can threaten the consumer's health.

## PREVENTING FOODBORNE ILLNESS

Food safety management is maintained by food-related establishments following food safety principles according to the flow of food.

### Personnel
Food Management Certificates ensure that employees are trained in personal hygiene habits such as avoiding hand to mouth movements, using hand-washing sinks frequently, wearing clean uniforms, and other safe work-related food techniques.

### Purchasing
Vulnerable, *high-risk* foods prone to microbial contamination are those high in protein and water—meat, poultry, fish, dairy, egg, broth, stocks, sauces, tofu, soy foods, and stuffing. Other factors important for microbial growth include the acidity (> pH 4.6) and oxygen exposure. Written specifications and inspections upon delivery by trained personnel help reduce problems.

### Storage
Temperature and time (4 hours maximum) are two additional key factors important in preventing microbial growth. Food should be stored at correct temperatures: USDA = 40°F (4°C) in the refrigerator (FDA = 41°F [5°C]), 0°F (−18°C) in the freezer, 60–70°F (15–21°C) dry storage for canned goods, and 50–70°F (10–21°C) dry storage for root vegetables, citrus, eggplant, and hard-rind squash.

### Preparation
**Pre-Preparation**    Thawing—The four safe methods for thawing are in the refrigerator, submerged under running water, microwaving followed by cooking, and as part of the cooking process.

Cross-contamination—Be sure to avoid having raw, *high-risk* vulnerable foods come in contact with cooked food or food that will not be cooked. Surfaces (hand, counter, cutting boards) should be carefully monitored.

**Cooking (Heating)**    Minimal internal temperatures needed to destroy microorganisms in different foods.

| Food | Temperature |
|---|---|
| Poultry (includes ground poultry) | 165°F (74°C) |
| Stuffing, sauces, gravies, soups | 165°F (74°C) |
| Reheated or microwaved foods, leftovers, casseroles, hot dogs | 165°F (74°C) |
| Ground meats: beef, pork, veal, lamb | USDA = 160°F (71°C) or FDA = 155°F (68°C) |
| Pork (fresh or raw) | USDA = 160°F (71°C) or FDA = 145°F (63°C) |
| Precooked ham | 140°F (60°C) |
| Eggs | USDA = 160°F (71°C) or FDA = 145°F (63°C) |
| Beef, veal, lamb, fish | 145°F (63°C) for medium rare |
| Poultry (includes ground poultry) | 165°F (74°C)[1] |
| Stuffing, sauces, gravies, soups | 165°F (74°C) |
| Vegetables and fruits | 135°F (57°C) |
| Commercially processed ready-to-eat foods | 135°F (57°C) |

[1]Figure 4-14 shows that the main types of thermometers are dial, digital, disposable, and other.

**Holding and Serving**    Food must be held and served outside the temperature danger zone which, according to the USDA, is 40°F–140°F (4°C–60°C) for consumers. The FDA Food Code suggests a slightly different temperature for retailers—above 135°F (57°C) for hot foods and below 41°F (5°C) for cold foods.

**Cooling**    The four safe methods to cool food are placing in shallow containers, reducing food size, placing in an ice-water bath, and using a blast chiller. The USDA suggests foods be cooled to

## PICTORIAL SUMMARY / 4: Food Safety (contineud)

below 40°F (4°C) within 4 hours. The FDA Food Code suggests two cooling stages: 135°F –70°F (57°C to 21°C) in the first 2 hours, and then 135°F to 41°F (57°C–5°C) within 6 hours or less.

**Sanitation**
Regularly scheduled cleaning, safe equipment, and a clean facility are all part of sanitation.

**Washing**   Three-compartment sink temperatures should reach 110°F (43°C) during soaking and at least 180°F (82°C) for 1 minute during final rinsing. Dishwashing machine temperatures should be between 140°F and 160°F (60°C and 71°C), and rinse temperatures

at least 180°F (82°C) for 10 seconds or 170°F (77°C) for 30 seconds to kill most pathogens.

**Drying**   Only air or heat drying is considered safe.

**Pest Control**   Pests should be blocked from entering food establishments, prevented from accessing stored food, and eliminated by a pest control operator (PCO).

**Food Safety Monitoring**
This may involve FDA's Food Code, State Health Department inspections, HACCP (hass-sip) (seven steps), and national surveillance (FoodNet, CDC).

# CHAPTER REVIEW AND EXAM PREP

## Multiple Choice*

1. What is the name for any illness transmitted to humans by food?
   a. Food outbreak
   b. Food intoxication
   c. Foodborne illness
   d. Food infection

2. HACCP is a(n) _____ designed to prevent _____ from occurring.
   a. inspection program, food outbreaks
   b. mandatory program, food outbreaks
   c. inspection program, foodborne illnesses
   d. food-safety program, foodborne illnesses

3. Which temperature range represents what is called "the temperature danger zone" for consumers and retailers, respectively, in food safety?
   a. between 20°F/–7°C and 80°F/27°C; between 21°F/–6°C and 75°F/24°C
   b. between 80°F/27°C and 180°F/82°C; between 81°F/27°C and 175°F/79°C
   c. between 40°F/4°C and 140°F/60°C; between 41°F/5°C and 135°F/57°C
   d. between 60°F/16°C and 160°F/71°C; between 61°F/16°C and 155°F/68°C

4. *Salmonella* bacteria may be found in which type of food?
   a. Poultry products
   b. Vegetable products
   c. Beef products
   d. All of the above

5. Which of the following is the safest method for thawing frozen foods?
   a. At room temperature overnight
   b. In the sun
   c. In a refrigerator
   d. In a sink overnight

*See p. AK-1 for answers to multiple choice questions.

6. What is the most common fish toxin-related food poisoning in the United States?
   a. Red tide
   b. Pufferfish
   c. Ciguatera
   d. Histamine

7. Foods considered high-risk for contributing to foodborne illness include those that are _____
   a. high in protein and acidity.
   b. high in protein and water.
   c. low in protein and acidity.
   d. low in water and fat.

## Short Answer/Essay

1. Discuss the difference between food infection and toxin-mediated infection. Give examples of each.
2. List and briefly explain the seven principles of the HACCP system.
3. Identify the major hazards to food safety.
4. Define the following: *pathogenic, temperature danger zone, spore,* and *cross-contamination*. What are the recommended temperatures for refrigerators, freezers, and dry storage?
5. For the agents of foodborne illness listed below, indicate: (a) whether it causes an infection or intoxication; (b) what the general symptoms of infection are; and (c) the foods that are most commonly associated with the illness. *V. vulnificus, Salmonella, E. coli, Staphylococcus aureus, Listeria monocytogenes, Hepatitis A virus*
6. List and briefly describe three parasites that can cause foodborne illness.
7. List and briefly describe three natural toxicants that can cause foodborne illness.
8. Which nine food categories are particularly prone to bacterial contamination?
9. To what internal temperatures should the following be heated for safe consumption: poultry, pork, beef, fish, ground meats, reheated foods, and stuffing?
10. Discuss the process of a general county health department inspection of a food service establishment.

# REFERENCES

1. Anonymous. Children can complete treatment for peanut allergies and achieve long-term tolerance, studies suggest. *Science Daily* March 16, 2009.

2. Baldus K, K Werronen, and V Deibel. Adverse reactions to allergens: The importance of labeling and cross-contamination control. *Food Safety Magazine* 14(6):16–22, 2009.

3. Bernard DT, and VN Scott. *Listeria monocytogenes* in meats: New strategies are needed. *Food Technology* 53(3):124, 1999.

4. Berrang ME, JF Frank, RJ Buhr, JS Bailey, and NA Cox. Eggshell membrane structure and penetration by *Salmonella typhimurium*. *Journal of Food Protection* 62(1):73–76, 1999.

5. Bhunia AK. Biosensors and bio-based methods for the separation and detection of foodborne pathogens. *Advances in Food and Nutrition Research* 54:1–44, 2008.

6. Black RE, et al. Epidemic *Yersinia enterocolitica* infection due to contaminated chocolate milk. *New England Journal of Medicine* 298:76, 1978.

7. Bolton DJ, AH Oser, GJ Cocoma, SA Palumbo, and AJ Miller. Integrating HACCP and TQM reduces pork carcass contamination. *Food Technology* 53(4):40–43, 1999.

8. Bren L. Bacteria-eating virus approved as food additive. *FDA Consumer News* January–February 2007.

9. Centers for Disease Control and Prevention. Fact sheet: Variant Creutzfeldt-Jakob disease. www.cdc.gov/ncidod/dvrd/vcjd/factsheet_nvcjd.htm. Accessed 4/20/09.

10. Centers for Disease Control and Prevention. Food Safety Office. Foods and high risk groups. www.cdc.gov/foodsafety/foods.htm. Accessed 4/20/09.

11. Centers for Disease Control and Prevention. Multistate outbreak of Salmonella infections associated with peanut butter and peanut butter-containing products—United States, 2008–2009. *MMWR Weekly* 58(4):85–90, 2009.

12. Centers for Disease Control and Prevention. Outbreak of Escherichia coli O157:H7 infections associated with drinking unpasteurized commercial apple juice—British Columbia, California, Colorado, and Washington, October 1996. *MMWR Weekly* 45(44):975, 1996.

13. Centers for Disease Control and Prevention. Preliminary FoodNet Data on the incidence of infection with pathogens transmitted commonly through food—10 States, 2008. *MMWR Weekly* 10;58(13):333–337, 2009.

14. Chapman PA, CA Siddons, ATC Malo, and MA Harkin. A 1-year study of *Escherichia coli* 0157 in cattle, sheep, pigs, and poultry. *Epidemiology & Infection* 119(2):245–250, 1997.

15. Chirife J, and M del Pilar Buera. Water activity, water glass dynamics, and the control of microbiological growth in foods. *Critical Reviews in Food Science and Nutrition* 36(5):465–513, 1996.

16. Clark JP. Drying still actively researched. *Food Technology* 56(9):76, 2002.

17. Daniels RW. Applying HACCP to new-generation refrigerated foods at retail and beyond. *Food Technology* 45(6):122–124, 1991.

18. Dean JP, and EA Zottola. Use of nisin in ice cream and effect on the survival of *Listeria monocytogenes*. *Journal of Food Protection* 59(5):476–480, 1996.

19. Delmore LRG, JN Sofos, JO Reagan, and GC Smith. Hot-water rinsing and trimming/washing of beef carcasses to reduce physical and microbiological contamination. *Journal of Food Science* 62(2):379–376, 1997.

20. Duan J, and Y Su. Occurrence of *Vibrio parahaemolyticus* in two Oregon oyster-growing bays. *Journal of Food Science* 70(1):58–63, 2005.

21. DuPont HL. The growing threat of foodborne bacterial enteropathogens of animal origin. *Clinical Infectious Disease* 45(10):1353–1361, 2007.

22. Duxbury D. Keeping tabs on *Listeria*. *Food Technology* 58(7):74–80, 2004.

23. Edgeworth JA, GS Jackson, AR Clarke, C Weissmann, and J Colling. Highly sensitive, quantitative cell-based assay for prions adsorbed to solid surfaces. *Proceedings of the National Academy of Sciences USA* 106(9):3479–3483, 2009.

24. Food and Drug Administration. *FDA Food Code*. www.foodsafety.gov/~dms/foodcode.html. Accessed 4/26/09.

25. Food and Drug Administration. *The FDA list of terms: Acidification*. www.cfsan.fda.gov/~dms/a2z-a.html#acidification. Accessed 4/28/09.

26. Food and Drug Administration. *The FDA list of terms: F (comprehensive)*. www.cfsan.fda.gov/~dms/a2z-f.html. Accessed 4/28/09.

27. Food and Drug Administration. CFSAN/Retail Food Protection Team. *Peanut-derived products used as ingredients*. www.cfsan.fda.gov. Accessed 4/20/09.

28. Food and Drug Administration. *Talking about juice safety. What you need to know*. June 2006. www.cfsan.fda.gov/~dms/juicsaf3.html. Accessed 5/21/09.

29. Food and Drug Administration takes next step in establishing overseas presence. *Food Safety Magazine* 14(3):8, 2008.

30. Ferguson-Smith DD, and JA Richt. Temporally distinct *Escherichia coli* 0157 outbreaks associated with alfalfa sprouts linked to a common seed source—Colorado and Minnesota, 2003. *Epidemiology and Infection* 133(3):439–447, 2005.

31. Food microbiology: Controlling the greater risk. *Medallion Laboratories Analytical Progress* 4(1):1–8, 1988.

32. Foodborne diseases active surveillance network. *MMWR Weekly* 46(12):258–261, 1996.

33. Friedman MA, et al. Ciguatera fish poisoning: Treatment, prevention and management. *Marine Drugs* 6(3):456–479, 2008.

34. Galanis E, et al. Web-based surveillance and global *Salmonella* distribution, 2000–2002. *Emerging Infectious Diseases* 12(3):381–388, 2006.

35. Garrett E. Meeting the challenges of produce supply chain safety.

*Food Safety Magazine* 13(6):42–44, 2008.

36. George JN. The thrombotic thrombocytopenic purpura and hemolytic uremic syndromes: Overview of pathogenesis (Experience of the Oklahoma TTP-HUS Registry, 1989–2007). *Kidney International* S112:S8–S10, 2009.

37. Giese J. It's a mad, mad, mad, mad cow test. *Food Technology* 55(6):60, 2001.

38. Greig JD, ECD Todd, CA Bartleson, and BS Michaels. Outbreaks where food workers have been implicated in the spread of foodborne disease. Part 1. Description of the problem, methods, and agents involved. *Journal of Food Protection* 70(7):1752–1761, 2007.

39. Griffith J, and D Abernethy. Improving food safety through public standards for food ingredients. *Food Safety Magazine* 14(4):12–14, 2008.

40. Grigoriadis SG, PA Koidis, KP Vareltzis, and CA Batzios. Survival of *Campylobacter jejuni* innoculated in fresh and frozen beef hamburgers stored under various temperatures and atmospheres. *Journal of Food Protection* 60(8):903–907, 1997.

41. Hennessy TW, et al. A national outbreak of *Salmonella enteritidis* infections from ice cream. *New England Journal of Medicine* 334(20):1281–1286, 1996.

42. Hoch GJ. New rapid detection test kits speed up food safety. *Food Processing* 58(11):35–39, 1997.

43. Howells AD, KR Roberts, CW Shanklin, VK Pilling, LA Brannon, and BB Barrett. Restaurant employees' perceptions of barriers to three food safety practices. *Journal of the American Dietetic Association* 108(8):1345–1349, 2008.

44. Jackus LA, et al. Managing food safety: A systematic approach. *Food Technology* 58(10):37–39, 2004.

45. Jenks WG, CG Bublitz, GS Choudhury, YP Ma, and JP Wikswo. Detection of parasites in fish by superconduction quantum interference device magnetometry. *Journal of Food Science* 61(5):865–869, 1996.

46. Juranovic LR, and DL Park. Foodborne toxins of marine origin: *Ciguatera*. *Reviews of Environmental Contamination and Toxicology* 117:51–94, 1991.

47. Kornacki JL. Controlling *Listeria* in the food processing environment. *Food Technology* 59(11):36–42, 2005.

48. Leake LL. Advancing rapid microbial testing. *Food Technology* 60(6):68–72, 2006.

49. Li Y, MF Slavik, JT Walker, and H Xiong. Pre-chill of chicken carcasses to reduce *Salmonella typhimurium*. *Journal of Food Science* 62(3):605–607, 1997.

50. Mann J. Good hand washing is management 101. *Food Safety Magazine* 13(6):18–21, 2008.

51. McAuley JB, MK Michelson, and PM Schantz. *Trichinosis* surveillance, United States, 1987–1990. *MMWR Weekly* 40(3):35–42, 1991.

52. Mermelstein NH. Testing for *Salmonella*. *Food Technology* 62(10):87, 2008.

53. Mermelstein NH. Comprehensive BSE risk study released. *Food Technology* 56(1):75, 2002.

54. Merrett N. China must step up food safety reform, claims expert. May17, 2007. www.ap-foodtechnology.com. Accessed 5/30/07.

55. Miller AJ, RC Whiting, and JL Smith. Use of risk assessment to reduce *listeriosis* incidence. *Food Technology* 51(4):100–103, 1997.

56. Moss MO. Fungi, quality and safety issues in fresh fruits and vegetables. *Journal of Applied Microbiology* 104(5):1239–1243, 2008.

57. National Advisory Committee on Microbiological Criteria for Foods. Hazard analysis and critical control point principles and application guidelines. *Journal of Food Protection* 61(6):762–775, 1998.

58. National Restaurant Association. *ServSafe®* (5th ed.). Prentice Hall, 2008.

59. Newsome R. Understanding mycotoxins. *Food Technology* 60(6):50–58, 2006.

60. Nwachcuku H, and CP Gerba. Emerging waterborne pathogens: Can we kill them all? *Current Opinion in Biotechnology* 15(3):175–80, 2004.

61. Omaye ST. Shiga-toxin-producing *Escherichia coli*: Another concern. *Food Technology* 55(5):26, 2001.

62. Powitz RW. Creating a great cutting board and wipe rags program. *Food Safety Magazine* 13(4):20–25, 2007.

63. Rapid *E. coli* test speeds lab work for dietary supplement maker. *Food Safety Magazine* 13(4):38, 2007.

64. Risks misjudged in cholera epidemic. *Food Insight* January/February 2, 1992.

65. Sánchez G, A Bosch, and RM Pintó. Hepatitis A virus detection in food: Current and future prospects. *Letters in Applied Microbiology* 45(1):1–5, 2007.

66. Sado PN, KC Jinneman, GJ Husby, SM Sorg, and CJ Omiecinski. Identification of *Listeria monocytogenes* from unpasteurized apple juice using rapid test kits. *Journal of Food Protection* 61(9):1199–1202, 1998.

67. Sakaguchi S. Prospects for preventative vaccines against prion diseases. *Protein Peptide Letters* 16(3):260–270, 2009.

68. Schubert-Ullrich P, et al. Commercialized rapid immunoanalytical tests for determination of allergenic food proteins: An overview. *Analytical and Bioanalytical Chemistry* 395(1):69–81, 2009.

69. Shakila J, G Jeyasekara, SA Vyla, and RS Kumar. Effect of delayed processing on changes in histamine and other quality characteristics of 3 commercially canned fishes. *Journal of Food Science* 70(1):24–29, 2005.

70. Shames LR. The food safety system needs restructuring. *Food Technology* 61(5):96, 2007.

71. Silk TM, ET Ryser, and CW Donnelly. Comparison of methods for determining coliform and *Escherichia coli* levels in apple cider. *Journal of Food Protection* 60(11):1302–1305, 1997.

72. Sotomayor RE, KB Arvidson, JN Mayer, AJ McDougal, and C Sheu. Assessing the safety of food contact substances. *Food Safety Magazine* 13(4):26–30, 2007.

73. Stearns D. Intentional contamination: The legal risks and responsibilities. *Journal of Environmental Health* 70(6):58–59, 2008.

74. Stier RF, and JG Surak. Evolution of HACCP: A natural progression to ISO 22000. *Food Safety Magazine* 14(4):17–22, 2008.

75. Surak JG. A global standard puzzle solved? How the ISO 22000 Food Safety Management System integrates HACCP and more. *Food Safety Magazine* 11(6):52–80, 2006.

76. Takhistov P, and CM Bryant. Protecting the food supply. *Food Technology* 60(7):34–44, 2006.

77. Tarver T. Biofilms: A threat to food safety. *Food Technology* 63(2):46–52, 2009.

78. Tauxe RV. Emerging foodborne diseases: An evolving public heatlh challenge. *Emerging Infectious Diseases* 3(4):425–434, 1997.

79. Tinney KS, MF Miller, CB Ramsey, LD Thompson, and MA Carr. Reduction of microorganisms on beef surfaces with electricity and acetic acid. *Journal of Food Protection* 60(60):625–628, 1997.

80. Todd EC, JD Greig, CA Bartleson, and BS Michaels. Outbreaks where food workers have been implicated in the spread of foodborne disease. Part 5. Sources of contamination and pathogen excretion from infected persons. *Journal of Food Protection* 71(12):2582–2595, 2008.

81. Todd EC, JD Greig, CA Bartleson, and BS Michaels. Outbreaks where food workers have been implicated in the spread of foodborne disease. Part 2. Description of outbreaks by size, severity, and settings. *Journal of Food Protection* 70(8):1975–1993, 2008.

82. United States Department of Agriculture. Food Safety and Inspection Service (FSIS). *Fact sheets. Appliances and thermometers.* May 16, 2008. www.fsis.usda.gov/Fact_Sheets/Kitchen_Thermometers/index.asp. Accessed 5/22/09.

83. United States Department of Agriculture. Food Safety and Inspection Service (FSIS). *Fact sheets. Thermy. Types of food thermometers.* www.fsis.usda.gov/food_safety_education/Types_of_Food_Thermometers/index.asp.

84. U.S. Dept. of Health and Human Services. Food and Drug Administration. Center for Food Safety and Applied Nutrition. *Foodborne Pathogenic Microorganisms and Natural Toxins Handbook* (the "Bad Bug Book"), 1997. www.foodsafety.gov/~mow/intro.html. Accessed 4/24/09.

85. Vangelova L. *Botulinum* toxin: A poison that can heal. *FDA Consumer* 29(10):16–19, 1995.

86. Watts JC, A Balachandran, and D Westaway. The expanding universe of prion diseases. *PloS Pathogens* 2(3):e26, 2006.

87. Wilkins J. Microorganisms that make us worry. *Today's Dietician* 9(6):10–16, 2007.

88. Winter C. Acrylamide: The issue that refuses to die. *Food Safety Magazine* 14(3):46–50, 2008.

89. Yoe C, M Parish, D Eddy, DK Lei, B Paleg, and JG Schwarz. The value of a food defense plan. *Food Safety Magazine* 14(2):16–21, 62, 2008.

90. Yoneyama N, Y Hara-Kudo, and S Kumagai. Effects of heat-degraded sugars on survival and growth of Vibrio parahaemolyticus and other bacteria. *Journal of Food Protection* 70(2):373–377, 2007.

# WEBSITES

The Council to Improve Foodborne Outbreak Response (CIFOR) provides an online set of "Guidelines for Foodborne Disease Outbreak Response" at:
**www.cifor.us/**

The American Medical Association's website provides information on foodborne illness:
**www.ama-assn.org/go/foodborne**

The federal government's food safety website is located at:
**www.foodsafety.gov**

The Centers for Disease Control and Prevention (CDC) maintains a website on food safety:
**www.cdc.gov/foodsafety/cme.htm**

The USDA's food safety and inspection website:
**www.fsis.usda.gov/**

More information about the FDA's Center for Food Safety & Applied Nutrition can be found at its website:
**www.cfsan.fda.gov**

Mad cow disease information can be found at these websites:
USDA: **www.fsis.usda.gov/Fact_Sheets/Bovine_Spongiform_Encephalopathy_BSE/index.asp**

BSE information provided by the US Beef Industry
**www.BSEinfo.org**

The FDA's "Bad Bug Book" clearly describes numerous pathogenic microorganisms and toxins:
**www.cfsan.fda.gov/~mow/intro.html**

Calibrate a bimetallic stemmed thermometer using the ice-point method:
**www.johnson-county.com/publichealth/pdf/foodsafety/CalibrateThermometer.pdf**

FDA's HACCP website:
**www.cfsan.fda.gov/~lrd/haccp.html**

FDA's Information about Food Allergens website:
**www.cfsan.fda.gov/~dms/wh-alrgy.html**

FDA's list of press releases on product recalls:
**www.cfsan.fda.gov/~lrd/press.html**

FDA's bioterrorism websites:
**www.cfsan.fda.gov/~dms/defterr.html**
**www.cfsan.fda.gov/~dms/fsbtact.html**
**www.cfsan.fda.gov/~dms/fsbtbook.html**

Generic HACCP plans from the University of California at Davis are found at its website:
**http://seafood.ucdavis.edu/haccp/plans.htm**

# 5 Food Preparation Basics

**C**ooking is a craft which can rise, on occasion, to an art," said Arno Schmidt, once an executive chef of the Waldorf-Astoria Hotel in New York City (19). Understanding the basics of food preparation is essential to getting a meal together, but because it is not an exact science, no matter how knowledgeable and careful the food preparer is results vary from meal to meal. So as Schmidt also said, it is "no wonder that seemingly similar foods taste and act differently depending upon endless factors."

The factors that Schmidt is referring to include the type of heat used, the cooking utensils (Appendix A gives a summary of the equipment used in food preparation), the amount of food prepared, and the fact that a cup of, say, fresh leeks tastes more potent than twice the quantity of dried leeks. Add to the equation the foibles of human nature and the unique tastes and preferences of individuals, and it is easy to see how two chefs following the same recipe could come up with different products.

Food preparation most definitely approaches art at times, but until its basic techniques are learned and mastered, the results will more nearly resemble preschool finger painting than the work of the Great Masters. The purpose of this chapter is to describe the basic heating methods in food preparation, cutlery techniques, measuring and mixing techniques, the proper use of seasonings and flavorings, and the guidelines of food presentation.

## HEATING FOODS

Heating not only destroys microorganisms that cause illness but also changes the molecular structure of foods, altering their texture, taste, odor, and appearance. During food preparation, heat is transferred by either moist- or dry-heat methods. Regardless of the method used, food should never be left unattended while it is cooking because that is the number-one cause of kitchen fires.

### Moist-Heat Preparation

**Moist-heat preparation** techniques include scalding, poaching, simmering, stewing, braising, boiling, parboiling, blanching, and steaming. In

> **Moist-heat preparation**
> A method of cooking in which heat is transferred by water, any water-based liquid, or steam.

99

these methods, liquids not only heat the food, but may also contribute flavor, color, texture, and appearance to the final product. This is especially the case if broth and mixtures containing herbs, spices, and seasonings have been added. Moist-heat preparation helps to soften the fibrous protein in meats and the cellulose in plants, making it more tender. Liquids generated from heating foods can also be used as a flavorful stock to make soups or sauces. One possible drawback to moist-heat methods is that color, flavor compounds, vitamins, and minerals may leach out and be lost in the liquid. However, using this liquid in a sauce or gravy retains them in the diet.

The various moist-heat preparation methods are presented below in order of increasing heat requirements, ranging from a low heat of 150°F (66°C) for scalding water to a high heat of 240°F (116°C) for pressure steaming.

# TYPES OF MOIST-HEAT PREPARATION

## Scalding

Scalding water reaches a temperature of 150°F (66°C). It is indicated by the appearance of large, but relatively still, bubbles on the bottom and sides of the pan. This process was most frequently used with milk to improve its function in recipes and to destroy bacteria. Pasteurized milk does not need to be scalded, even though many older recipes call for scalded milk. Recipes now use scalded milk to speed the combination of ingredients; in hot milk sugar dissolves more readily, butter and chocolate melt more easily, and flour mixes in more evenly without creating lumps (25).

## Poaching

Water heated to a temperature of 160–180°F (71–82°C) is used for poaching, in which the food is either partially or totally immersed. The water is hotter than it is at scalding, but has not yet reached the point of actually bubbling, although small, relatively motionless bubbles appear on the bottom of the pan. Poaching is used to prepare delicate foods, like fish and eggs, which could break apart under the more vigorous action of boiling.

## Simmering

Water simmers at just below the boiling point, never less than 180°F (82°C). Simmering is characterized by gently rising bubbles that barely break the surface. Many food dishes, especially rice, soups, and stews, are first brought to a boil and then simmered for the remainder of the heating time. Simmering is preferred over boiling in many cases because it is more gentle and will usually not physically damage the food, and foods will not overcook as quickly as they do when boiled. The lower heat of a simmer is essential when cooking tough cuts of meat that require gentle cooking in order to become tender.

## Stewing

Stewing refers to simmering ingredients in a small to moderate amount of liquid, which often becomes a sauce as the food cooks. Most stew dishes consist of chopped ingredients such as meat (often browned first) and vegetables placed in a large casserole or stock pot with some water, stock, or other liquid. The pot is covered and the food simmered for some time on the range or in a moderate oven. Stews often taste better the day after their initial preparation, because the overnight rest deepens their flavors.

## Braising

Braising is similar to stewing in that food is simmered in a small amount of liquid in a covered casserole or pot. The liquid may be the food's own juices, fat, soup stock, and/or wine. Flavors blend and intensify as foods are slowly braised on top of the range or in an oven (1).

In order to generate a browner color and better flavor, meats are frequently browned with a dry-heat method such as sautéing before being braised. Frequently, meats are braised, and then the vegetables are added during the final cooking to preserve some of their texture and flavor.

**Parboil** To partially boil, but not fully cook, a food.

> **? How & Why?**
>
> *Why are stewing and braising called by different names if both entail simmering food in a small amount of liquid?*
>
> The primary difference between stewing and braising is that stewing generally refers to smaller pieces of meat, whereas braising entails larger cuts. Stews are also most often made with more liquid and served in their sauce.

## Boiling

In order to boil, water must reach 212°F (100°C) at sea level, a temperature at which water bubbles rapidly. The difference in the bubbles between poaching, simmering, and boiling is shown in Figure 5-1. The high temperature and agitation of boiling water are reserved for the tougher-textured vegetables and for dried pastas and beans. A common technique is to bring a liquid to a rolling boil, gradually add the food, distributing it evenly, and then bring the liquid back to a full boil before reducing the heat so that boiling becomes gentle. A lid on the pot or pan will bring the liquid to a boil more quickly by increasing the pressure. It is always recommended to reduce the heat setting once a boil has been reached, because food will not cook any faster at a higher setting than at the one required to maintain a gentle boil. Spillovers, burns, and loss of cooking liquid through evaporation all can be avoided if a gentle boil is used.

Food may also be **parboiled** in boiling water, after which it is removed and its cooking completed either at a later time or by a different heating method. Parboiling is used frequently in restaurant service when food must

**FIGURE 5-1**   Bubble size and movement differ during poaching, simmering, and boiling.

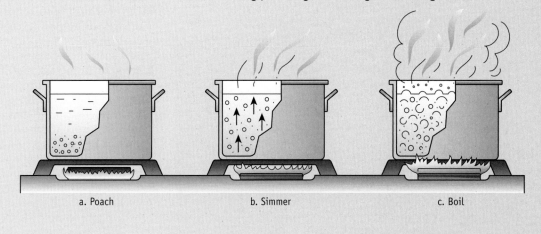

a. Poach          b. Simmer          c. Boil

be prepared in advance and finished to order. Another use for boiling water is for **blanching**, which sets the color of green vegetables; loosens the skins of fruits, vegetables, and nuts for peeling; and destroys enzymes that contribute to deterioration. Foods are often blanched before being canned or frozen.

## Steaming

Any food heated by direct contact with the steam generated by boiling water has been steamed. Cooked vegetables are at their best when steamed, because this method helps to retain texture, color, taste, and nutrients. A common method for steaming is to place food in a rack or steamer basket above boiling water and to cover the pot or pan with a lid in order to trap the steam. An indirect technique, called *en papillote* (on pap-ee-yote), is to wrap the food in foil or parchment paper before it is baked or grilled. Then, in an oven or over the grill, the food cooks by the steam of its own juices, which are trapped in the packet. In a microwave oven, covering foods with plastic wrap facilitates steaming. Pressure cookers heat food by holding steam in an enclosed container under pressure. The temperature increases with increasing pounds of pressure per square inch.

## Microwaving

Although microwaving is listed under moist-heat preparation, it actually belongs in an entirely separate category because it incorporates both dry- (radiation) and moist-heat preparation methods. Microwaves are a form of radiation aimed at the water in the food or beverage. The specifics of preparing food using a microwave oven are discussed in Appendix A as well as in chapters on specific foods.

## Dry-Heat Preparation

Examples of **dry-heat preparation** include baking, roasting, broiling, grilling, barbequing, and frying. Higher temperatures are reached in dry-heat preparation than they are in moist-heat methods, because water can heat only to its boiling point of 212°F (100°C) or slightly higher under pressure, whereas ovens can reach up to 500°F (260°C).

### Baking

Baking is the heating of food by hot air in an oven. The average baking temperature is 350°F (177°C), although temperatures may range from 300°F to 425°F (149°C to 219°C). Baking results can be affected by rack position and the color of the pan.

*Rack Position*   For the best outcome, the food should be placed in the middle of the center rack (Figure 5-2). Foods placed on the uppermost rack may brown excessively on their top surface, whereas on the lowest rack foods are prone to burning on the bottom. It is also best to position foods using only one rack; if this is not possible, the foods should be staggered so that they are not directly over each other in order to allow hot air to flow more freely through the oven. At least 2 inches should be left between pans and between the pans and the oven walls. If these guidelines are ignored, the resulting inadequate air circulation may cause uneven browning, and food may not be thoroughly cooked.

*Pan Color*   In addition to rack position and placement of pans, the cooking pan material will affect the baking outcome. Shiny metal pans reflect heat and are best for cakes or cookies, in which only light browning and a soft crust are desired. The darker, duller metal pans (including anodized and satin-finish) tend to absorb heat, resulting in browner,

**Blanch**   To dip a food briefly into boiling water.

**Dry-heat preparation**   A method of cooking in which heat is transferred by air, radiation, fat, or metal.

**FIGURE 5-2**   Oven rack position.

| POSITION: | USED FOR: |
|-----------|-----------|
| RACK 5 | Toasting bread, or for two-rack baking. |
| RACK 4 | Most broiling and two-rack baking. |
| RACK 3 | Most baked goods on a cookie sheet or jelly roll pan, or frozen convenience foods, or for two-rack baking. |
| RACK 2 | Roasting large cuts of meat and large poultry, pies, soufflés, or for two-rack baking. |
| RACK 1 | Roasting large cuts of meat and large poultry, pies, soufflés, or for two-rack baking. |

Digital Works

crisper crusts ideal for pies or bread baking. Glass pans require that oven temperatures be reduced by 25°F (4°C), because food tends to heat more quickly in glass (exceptions are pies and bread).

Because baking times are dependent on many factors, it is important to check the food's progress at the suggested minimum baking time and then at intervals after that until the food is done. This must be done judiciously, however, because checking too soon or too frequently will allow heat and/or steam to escape from the oven, adversely affecting the baking outcome.

# TYPES OF DRY-HEAT PREPARATION

## Roasting

Roasting is similar to baking except that the term is usually applied to meats and poultry. Roasted meats are often **basted** every 20 minutes or so to prevent the food from drying out. Some roasted meats are initially **seared** at 400–450°F (200–230°C) for about 15 minutes before reducing the heat to normal roasting temperatures. Although searing adds a desirable texture, color, and flavor to the meat's outer surface, roasts cooked at lower temperatures are juicier, shrink less, and are easier to carve than those that are seared.

**Baked or Roasted?** The word *roasting* can also refer to cooking on an open fire, as with roasted marshmallows and vegetables, and to cooking with a rotisserie. To make things even more confusing, meats such as ham, meat loaf, and fish are often referred to as *baked*. Chicken may be described as either baked or roasted.

**Baste**  To add a liquid, such as drippings, melted fat, sauce, fruit juice, or water, to the surface of food (usually roasting meat) to help prevent drying.

**Sear**  To brown the surface of meat by brief exposure to high heat.

## Broiling

To broil is to cook foods under an intense heat source. The high temperatures of broiling cook foods in approximately 5 to 10 minutes, so only tender meats, poultry, and fish are broiled; tougher foods require longer heating times. Temperature is controlled by moving the rack closer to or farther away from the heat source. Thicker cuts are broiled farther from the heat, thinner ones closer—on the fourth or fifth rack (up from the bottom) of a home oven. Foods are often slightly oiled to prevent drying and sticking, placed under the broiler only after it has been preheated to its full heat, and then turned over only once. Food service operations often employ a *salamander,* also called a cheese-melter, a low-intensity broiler used just prior to serving to melt or brown the top layer of a dish.

## Grilling

Grilling is the reverse of broiling, in that food is cooked above, rather than below, an intense heat source. The grill may be a rack or a flat surface on a stove. *Grilled* can also refer to foods that are seared on a grill over direct heat (22).

## Barbecuing

Barbecuing and grilling are no longer used to refer to the same heating method. Barbecuing was once synonymous with grilling over a pit, but has since assumed a unique and more specific meaning. Barbecuing now refers to foods being slow-cooked, usually covered in a zesty sauce, over a longer period of time (22). The temperature in barbecuing is regulated by adjusting the intensity of the heat source (charcoal, wood, gas, or electric); adjusting the distance between the food and the heat source; and moving the food to different places on the grill.

## Frying

Frying is heating foods in fat. Oils used in frying serve to transfer heat, act as a lubricant to prevent sticking, and contribute to flavor, browning, and a crisp outside texture (28). Although oils are liquid, frying is a method of dry-heat preparation because pure fat contains no water. Types of frying—sautéing, stir-frying, pan-broiling, pan-frying, and deep-frying—are distinguished by the amount of fat used, ranging from a thin sheet to complete submersion. Temperatures vary among the different methods: sautéing, stir-frying, and pan-frying require only a medium or high heat—lower heat results in higher fat absorption—whereas deep-frying temperatures range from 350°F to 450°F (177°F to 232°C). Chapter 22 discusses frying with fats in greater detail.

### Sautéing and Stir-Frying

These methods use the least amount of fat to heat the food. Stir-frying is predominantly used in Asian cooking; the pan is held stationary while the food is stirred and turned over very quickly with utensils. Sautéing is done in a frying pan, a special sauté pan, or on a griddle. The foods most frequently prepared on a griddle with a little fat are eggs, pancakes, and hamburgers (with the fat derived from the meat itself).

### Pan-Broiling and Pan-Frying

Pan-broiling refers to placing food, usually meat, in a very hot frying pan with no added fat and pouring off fat as it accumulates. If the fat is not poured off, pan-broiling becomes pan-frying, which uses a moderate amount of fat (up to ½-inch deep), but not enough to completely cover the food.

### Deep-Frying

In deep-frying, the food is completely covered with fat. Many deep-fried foods are first coated with breading or batter to enhance moisture retention, flavor development, tenderness, browning, crispness, and overall appearance. The characteristics of the coating influence a fried food's final outcome (18, 21). A fine-crumb breading absorbs less fat, but a coarser grain produces a crisper texture. Sugar in the coating speeds up browning, but this is undesirable if the outside browns and appears done while the inside remains uncooked (Figure 5-3). Although the breading or batter protects the food from absorbing too much fat, it can also simultaneously protect the deep-frying oil from the deterioration that occurs when it contacts the food's natural moisture and salt content.

Because deep-frying requires high temperatures, it is best to rely on the fryer's thermostat to obtain the desired temperature. If this is not possible, another method is to place a 1-inch cube of white bread into the oil and time how long it takes to turn golden brown (Table 5-1).

**FIGURE 5-3** Sugar in the batter darkens fried food, as seen in the fritter on the right.

| TABLE 5-1 Deep-Frying Based on Color | |
|---|---|
| Temperature of Fat | Approximate Number of Seconds to Turn a 1-Inch Bread Cube Golden Brown |
| 385°F–395°F (196°C–201°C) | 20 |
| 375°F –385°F (190°C–196°C) | 40 |
| 365°F –375°F (185°C–190°C) | 50 |
| 355°F –365°F (179°C–185°C) | 60 |

# Types of Heat Transfer

Energy sources for heating foods are usually electricity and gas (natural or butane), but secondary sources such as wood, coal, and charcoal may also be used for heating. All of these produce heat energy that can be transferred through conduction, convection, radiation, or induction (Figure 5-4).

## Conduction

In preparing foods on the range or in the fryer, heat is transferred by **conduction.** Heat from the electric coil or gas flame is conducted to the pan or fryer and then to its contents (5). Conduction is based on the principle that adding heat to molecules increases their energy (kinetic) and hence their ability to transfer heat to neighboring molecules.

The material and size of the pan greatly affect the speed and efficiency of heat transfer. Copper is an excellent heat conductor and is often used to line the bottom of stainless steel pans. Dark, dull surfaces absorb heat more readily, which shortens the baking time. Tempered glass conducts heat in such a manner that baking temperatures should be reduced by 25°F (4°C) when it is used.

## Convection

**Convection** relies on the principle that heated air or liquid expands, becomes less dense, and rises to the surface. The cooler, heavier air or liquid originally on top moves to the bottom, where it is heated, thus creating continuous circular currents. A common example of the use of convection in cooking is baking in an oven. Baked goods rely on convection to allow the hot air to rise. Having the heating unit located at the bottom of the oven contributes to the rising of hot air. Convection ovens, which are more common in food service institutions than in home kitchens, have an air-circulating system, whereas standard ovens do not. Fans move the air more quickly and evenly around the food, which speeds up baking times.

Convection ovens do have drawbacks, however: The moving air causes foods to lose moisture, and cake batters are more prone to develop uneven tops. Injecting steam into a convection oven helps to reduce the drying and shrinking effects.

Other examples of convection cooking are simmering, steaming, and deep-frying. The use of water and fat to heat food relies on both conduction and convection. For example, once the heat from convection begins to heat a baked potato, conduction takes over when the heat penetrates the potato's water molecules and is transferred to the center of

**Conduction** The direct transfer of heat from one substance to another that it is contacting.

**Convection** The transfer of heat by moving air or liquid (water/fat) currents through and/or around food.

**FIGURE 5-4**    Four types of heat transfer: Conduction, convection, radiation, and induction.

**FIGURE 5-5**    The three main scales used to measure heat intensity.

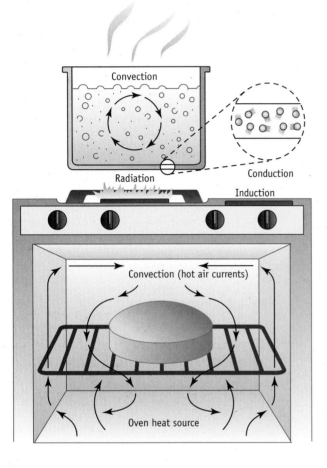

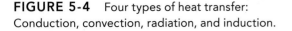

the potato. Because water conducts heat more efficiently than air does, it takes less time to boil than to bake a potato.

### Radiant Heat

Heat is transferred by **radiation** in broiling, grilling, and microwaving. The short electromagnetic waves that are generated by microwave ovens can pass through glass, paper, and most plastic. Infrared heat lamps and ovens are other

**Radiation**  The transfer of heat energy in the form of waves of particles moving outward from their source.

**Induction**  The transfer of heat energy to a neighboring material without contact.

**Kinetic energy**  Energy associated with motion.

heat sources that use electromagnetic waves for heat. These are usually found in restaurants and institutional kitchens where they are used to keep foods warm and to prepare frozen foods.

### Induction

Flat-surfaced ranges that have the coils buried underneath conduct heat through **induction.** The cooktop consists of a smooth, ceramic surface that allows the transfer of heat from the coiled electrical apparatus below. Because no coils are exposed on the surface, cleaning is easy.

## Measuring Heat

Food preparation relies on accurate heat measurements. The three basic temperature scales, their freezing and boiling points, and how calories are related to energy are now discussed.

### Temperature Scales

Heat is a form of energy that can be measured. The three main scales used to measure heat intensity are Fahrenheit (°F), Celsius or centigrade (°C), and Kelvin (K) (Figure 5-5). The last is used primarily in scientific research and will not be discussed here.

Freezing and boiling are extremes in the range in temperatures encountered in food preparation that owe their effects to changes in the **kinetic energy** of molecules. The molecules in living organisms always have some motion; heat speeds up that motion, whereas cold slows it down. Heating or freezing foods is accomplished by increasing or decreasing, respectively, the movement of molecules. Figure 5-6 summarizes various temperatures used in food preparation.

***Freezing and Boiling Points***    The freezing point of water is 32° on the

**FIGURE 5-6**  Temperatures important in preparing foods.

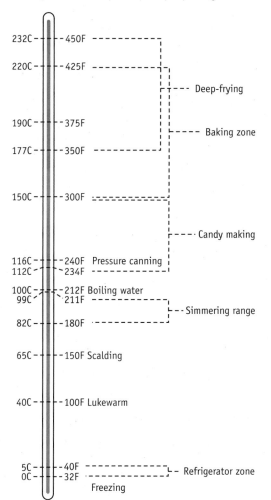

Fahrenheit scale and 0° on the Celsius scale. Water boils (boiling point) at 212° on the Fahrenheit scale and 100° on the Celsius scale. The boiling point changes slightly with altitude; 1°F must be subtracted for every 500-foot increase in elevation up from sea level (an increase of 960 feet in elevation decreases water's boiling point by 1°C). Other compounds in the water, such as sugar or salt, influence its boiling and freezing temperatures, so all three scales pertain to pure water. Other materials have their own freezing and boiling points.

### Calories

Energy can be correlated to heat and is measured in units called **calories**. For everyday use, it is more common to refer to the kilocalorie (1,000 calories), abbreviated kcal, which is the amount of energy required to raise 1 kilogram of water 1 degree Celsius (see Chemist's Corner 5-1). Calories are discussed in more detail in Chapters 1 (see Calorie Control) and 3.

Regardless of which term is used to quantify dietary calories (kcal), it is more accurate to speak of energy rather than calories, unless a specific amount is being discussed.

# CUTLERY TECHNIQUES

It's important to know about the various aspects of heating foods, yet often food must be cut into smaller pieces before it can be heated. Thus, another basic pillar of food preparation is the knowledge and use of cutlery. Selecting and caring for knives are discussed in Appendix A. The following sections cover their handling and the styles of cutting food. The techniques vary according to the type of knife selected, and selection depends on the task to be performed.

## Handling Knives

The most frequently used knife is the chef's or French knife. The positioning of the grip and of the food under the blade both influence the degree of control and leverage a person has over the knife (Figure 5-7). A chef's knife should be firmly held with the base of the blade between the thumb and forefinger and the other fingers wrapped around the handle. While one hand grips the knife, the other hand must hold the food and guide it toward the blade. Curling the fingers of the guiding hand under while holding the food allows the knuckles to act as a protective shield and keeps the fingertips away from the cutting edge. It is best to allow at least a half-inch barrier of food between the blade and the fingers holding the food.

**Calorie (kcal)** The amount of energy required to raise 1 kilogram of water 1°C (measured between 14.5°C and 15.5°C at normal atmospheric pressure).

**FIGURE 5-7** Technique of holding a chef's knife.

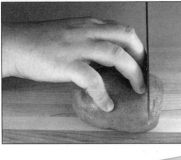

Curl fingers of guiding hand inward for protection.

Different sections of a blade are used for different tasks (Figure 5-8). Light tasks such as cutting out the stem end of a tomato can usually be accomplished with the tip of the blade, or, even better, with a knife more suitable to small tasks, such as a paring knife. Heavy duties such as chopping off the tough base of a bunch of celery are better accomplished by making use of the weight and thickness found at the base of the blade. Most other cutting tasks are carried out using the center of the blade.

## Cutting Styles

Uniformity is the usual goal in cutting food. It allows for even heating and gives food an appetizing appearance. Cutting styles include slicing, shredding, dicing (cubing), mincing, and peeling.

- *Slice.* To move the food under the blade while keeping the point of the blade firmly on the cutting board. The base of the knife is lifted up and down with a forward and backward motion (Figure 5-9).
- *Julienne.* Sliced food can be further cut up, or **julienned,** resulting in delicate sticks that are usually 1 to 3 inches long and only 1/16 to 1/8 of an inch thick (Figure 5-10).
- *Shred.* To cut leaf vegetables into thin strips. This may be done by first rolling the leaves into cigar-like shapes and then cutting them into

**Julienne** To cut food lengthwise into very thin, stick-like shapes.

**FIGURE 5-8** Blade position determined by the cutting task.

Tip—delicate tender work.

Center—all-purpose work.

Heel—heavy work.

**FIGURE 5-9** Slicing technique.

Start with the blade tip down.

Press down and forward simultaneously to slice.

As soon as the blade heel touches the board, it is moved up and over (keep tip down) for the next slice.

**FIGURE 5-10** Julienned slices.

shreds. Hand shredders and food processors with different sizes of shredding blades may also be used.

- *Dice.* To cut food into even-size cubes.
- *Mince.* To chop food into very fine pieces. This is done by placing the holding hand on the tip of the knife

**FIGURE 5-11**  Dicing an onion; further cuts result in mincing.

First vertical cuts
(do not go all the way through onion)

Second vertical cuts
(go entirely through onion)

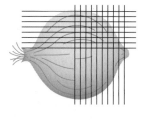

**FIGURE 5-12**  Peeling with a paring knife.

Digital Works

and rocking the base up and down in short strokes while moving it across the food several times, and then repeating as necessary. Figure 5-11 illustrates how to dice and mince an onion.

- *Peel.* To remove the skin. The peel and rind can be cut from an orange or any thick-skinned fruit by first cutting off in a circular fashion the top of the fruit's skin, then scoring the skin through to the flesh of the fruit in four places. The skin can then be peeled in segments down from the top. Fruits can also be peeled directly with a paring knife (Figure 5-12). Avocados can be

**FIGURE 5-13**  How to cut, seed, and peel an avocado.

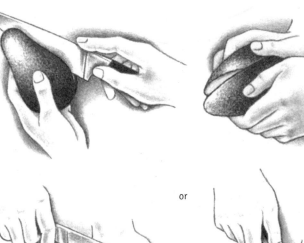

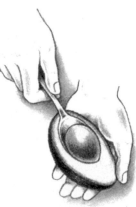

or

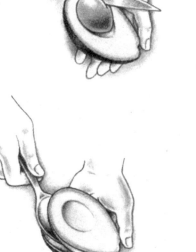

stripped of their peel by cutting the avocado from stem to stern through to the pit. Each half is cupped in the hands and twisted gently to separate the halves. The seed (nut) can be removed with the fingers or the tip of a sharp knife. At this point the avocado can be scooped out with a large serving spoon or peeled and sliced (Figure 5-13).

# MEASURING INGREDIENTS

Correct measuring is another essential aspect of basic food preparation. The three major steps in measuring are:

1. Approximating the amount of required food (e.g., 4 ounces of cheese yields 1 cup shredded)
2. Selecting the right measuring utensil
3. Using an accurate measuring technique

## Approximating the Amount of Required Food

Appendix B provides common food purchase quantities and their approximate yields. For example, one pound of all-purpose flour yields 4 cups (sifted), while one pound of granulated sugar provides 2–2¼ cups.

| TABLE 5-2 | Basic Measuring Equivalents | | | | | |
|---|---|---|---|---|---|---|
| **Teaspoon (tsp)** | **Tablespoon (T )** | **Ounce (oz)** | **Cup (C)** | **Pint** | **Quart** | **Gallon** |
| ⅛ tsp (dash) | | | | | | |
| ¼ tsp | | | | | | |
| 1 tsp | | | | | | |
| 3 tsp    =    1 T | | | | | | |
| | 2 T    =    1 oz | | | | | |
| | | 2 oz   = | ¼ C | | | |
| | | 4 oz   = | ½ C | | | |
| | | 8 oz   = | 1 C | | | |
| | | 16 oz   = | 2 C   =   1 pint | | | |
| | | | 4 C   = | 2 pints   =   1 qt | | |
| | | | 16 C   = | 8 pints   = | 4 qt   =   1 gallon | |

## Selecting the Right Measuring Utensil

Whether an ingredient is liquid or dry determines the kind of measuring utensil that will be used. Appendix A describes these in detail. A graduated measuring cup with a lip for pouring is best for measuring liquid ingredients. Sets of flat-topped measuring cups are reserved for measuring dry ingredients. All dry ingredients are best measured by first stirring them to eliminate any packing or lumps. Amounts less than ¼ cup should be measured with measuring spoons. Sifting flour with dry ingredients such as baking soda or salt is an efficient way to blend and distribute the ingredients evenly.

It is important to be able to use measuring utensils interchangeably whether using liquid or dry measuring cups. This is easy if a few basic equivalents are remembered (see Table 5-2, and the inside back cover of this book).

1 teaspoon    = about 5 grams*
1 tablespoon  = 3 teaspoons
2 tablespoons = 1 fluid ounce or 28.35 grams
¼ cup         = 2 fluid ounces
½ cup         = 4 fluid ounces
1 cup         = 8 fluid ounces, 16 tablespoons, or 48 teaspoons

1 pint    = 2 cups, or 16 fluid ounces (1 pound)
1 quart   = 2 pints or 4 cups
1 gallon  = 4 quarts, 8 pints, or 16 cups
1 pound   = 16 ounces
1 kilogram = 2.2 pounds

*When weighing water

Knowing the general units used in measuring allows for the next step required for accuracy—using the largest measuring device possible. For example, 3 teaspoons of sugar should be measured using 1 tablespoon; ¾ cup should be measured using ½ cup plus a ¼ cup. Accuracy is also achieved by using the guide provided in Table 5-3 for rounding off weights and measures.

***Volume vs. Weight***   Volume measures the space filled by an ingredient. Weight measures the heaviness of an ingredient. It's important to be aware of

| TABLE 5-3 | Guide to Rounding Off Weights and Measures |
|---|---|
| **If the total amount of an ingredient is:** | **Round it to:** |
| ***WEIGHTS*** | |
| Less than 2 oz | Measure unless weight is ¼-, ½ -, or ¾-oz amounts |
| 2–10 oz | Closest ¼ oz or convert to measure |
| More than 10 oz but less than 2 lb 8 oz | Closest ½ oz |
| 2 lb 8 oz–5 lb | Closest full ounce |
| More than 5 lb | Closest ¼ lb |
| ***MEASURES*** | |
| Less than 1 T | Closest ⅛ tsp |
| More than 1 T but less than 3 T | Closest ¼ tsp |
| 3 T–½ cup | Closest ½ tsp or convert to weight |
| More than ½ cup but less than ¾ cup | Closest full tsp or convert to weight |
| More than ¾ cup but less than 2 cups | Closest full tsp or convert to weight |
| 2 cups–2 qt | Nearest ¼ cup |
| More than 2 qt but less than 4 qt | Nearest ½ cup |
| 1–2 gal | Nearest full cup or ¼ qt |
| More than 2 gal but less than 10 gal* | Nearest full quart |
| More than 10 gal but less than 20 gal* | Closest ½ gal |
| More than 20 gal* | Closest full gallon |

*For baked goods or products in which accurate ratios are critical, always round to the nearest full cup or ¼ qt.

this distinction and its implications; for example, 1 cup of powdered sugar does not weigh the same as 1 cup of honey. Also, a fluid ounce only measures *volume*, whereas an ounce measures *weight*. They are only equal when measuring water.

*Using Scales*   For even better accuracy, different types of scales shown in Appendix A may be used to measure ingredients. Scales are used by commercial operations because they use weight to measure ingredients. Businesses cannot afford incorrect measurements that result in a loss of food, time, and money.

# Using an Accurate Measuring Technique

Specific volume-measuring techniques for liquids, eggs, fat, sugar, and flour are discussed below.

## Liquids

Only transparent graduated measuring cups with pouring lips should be used to measure liquids. The cup should sit on a flat surface and all measuring be done at eye level in order to accurately read the line at the bottom of the meniscus (the arc formed by the liquid's surface; see Appendix A, Figure C-21). The exception is milk, which is read at the top of the meniscus. Viscous liquids, such as honey, oil, syrup, and molasses, have a tendency to stick to the sides as they are poured, so the amount measured can be diminished by the amount that sticks to the sides. Should this happen, a rubber scraper can be used to remove the remaining contents.

## Eggs

Eggs range in size from *pee wee* to *jumbo*, but most standard recipes are based on *large* size eggs, if not specified. When half an egg or less is called for, it can be measured by beating a whole egg into a homogeneous liquid, which can then be divided in half or smaller increments. When measuring eggs, it is helpful to remember the following volume equivalents:

- 1 large egg = 2 ounces
- 4 large eggs = 7 ounces (just under 1 cup)
- 8 to 10 egg whites, or 12 to 14 yolks = 1 cup

## Fat

Manufacturers of butter and margarine have made it easy to measure their products. Both usually come in 1-pound packages that contain four ¼-pound sticks, with each stick equivalent to ½ cup. Thus, 1 pound of butter is equivalent to approximately 2 cups. The same weight of vegetable shortening, on the other hand, is equivalent to 2½ cups by volume. The wrappings of the ¼-pound sticks are usually further marked into eight 1-tablespoon segments.

Different methods are used to measure liquid and solid fats. Liquid fats such as oil and melted butter are measured in glass measuring cups. Solid fats such as lard, shortening, butter, and margarine should be removed from the refrigerator and allowed to become **plastic** at room temperature. Once pliable and soft, they can be pressed into a fractional metal measuring cup with a rubber scraper. The fat should be pressed down firmly to remove any air bubbles and the top of the cup leveled with the straight edge of a spatula. As with liquids, amounts under ¼ cup should be measured with measuring spoons.

Solid fats may also be measured by using the water-displacement method. For example, if ½ cup of fat is required, a 1-cup liquid measuring cup is filled with cold water to ½ cup. The fat is added and pressed below the water line until the water line reaches the 1-cup measuring line. The colder the water, the easier the cleanup will be, because cold fat is less likely to stick to the sides of the cup. Some water may cling to the fat and should be shaken free or patted away lightly with a paper towel.

## Sugar

The amount of sugar needed depends on its type—granulated white sugar, brown sugar, or confectioners' sugar (powdered or icing). Measuring methods differ among these sugars, because 1 pound of each yields 2, 2¼, and 4½ (sifted) cups, respectively.

White granulated sugar is usually poured into fractional measuring cups and leveled with a spatula. If it becomes lumpy, it can be mashed and sifted before measuring. Brown sugar has a tendency to pack down and become hard because it contains 2% moisture, which has a tendency to evaporate. Lumping

can be prevented by placing the brown sugar in an airtight container and storing it in the refrigerator or freezer. Hardened brown sugar can be softened by placing it in a microwave oven for a few seconds, or in a conventional oven set at about 200°F (93°C) for a few minutes. Brown sugar is best measured by pressing it firmly into a fractional metal measuring cup and leveling it. The packing should be firm enough that the brown sugar retains the shape of the measuring cup when it is turned out. Lump-free or free-flowing brown sugar, which weighs 25% less than regular brown sugar, is measured in the same manner as granulated white sugar.

*Measuring Confectioners' Sugar*   Confectioners' sugar must be sifted before measuring to break up any existing lumps. The light, airy nature of confectioners' sugar causes it to have a greater volume than the same amount of granulated sugar, which is why 1¾ cups of confectioners' sugar is equal to the weight of 1 cup of granulated sugar. After sifting, confectioners' sugar is measured the same way as granulated sugar.

## Flour

White flour is one of the more difficult ingredients to measure accurately by volume, because its tiny particles not only vary in shape and size, but also have a tendency to pack. In addition, the various white flours differ in density, ranging from 88 grams per cup in soy flour to approximately 132 grams per cup in wheat flour. This influences the number of cups obtained from various flours of the same weight (Table 5-4). Although there is no standard weight for a cup of flour, 1 pound of all-purpose flour averages 4 cups. Professional bakers and chefs avoid the discrepancy in volume measurement by always weighing the flour.

White flour should be sifted before being lightly spooned into a fractional measuring cup and leveled with a spatula. The cup should never be tapped or shaken down, because doing so can pack the flour particles tightly, which may result in too much flour being

**Plasticity**  The ability of a fat to be shaped or molded.

**TABLE 5-4**    One Pound of Flour Varies in Volume (# cups) and Weight (grams) Depending on the Flour

| Flour (1 lb) | Volume (approximate cups) | Weight (per cup) | |
|---|---|---|---|
| | | g | oz |
| All purpose (sifted) | 4 | 115 | 4.2 |
| Cake (sifted) | 4½ | 96 | 3.4 |
| Rice (sifted) | 3½ | 126 | 4.4 |
| Rye (sifted–light) | 5 | 88 | 3.2 |
| Rye (sifted–dark) | 3½ | 128 | 4.6 |
| Soy (low-fat) | 5½ | 84 | 3.0 |
| Whole wheat (stirred) | 3⅓ | 132 | 4.8 |

used. To avoid sifting and still get consistent baking results with regular white flour, one technique is to remove 2 tablespoons from each cup of unsifted flour (15).

Not all flours are sifted prior to being used. Whole-grain and graham flours and meal should not be sifted, because sifting will remove the bran particles. These flours should simply be lightly stirred before being scooped into a fractional measuring cup. Presifted or instantized flours (discussed in Chapter 17) have already been processed into uniform particles and should not be sifted. Instantized flour should not be used in baked products.

### Other Ingredients and Substitutions

Some foods to be measured do not fall into the basic categories described above. The methods for measuring foods such as cheese, nuts, chocolate, and garlic depend on their form—whole, cubed, shredded, minced, and so forth. Foods cut into pieces tend to occupy a greater volume. For example, 1 pound of cheese is equivalent to approximately 2 cups, but 1 pound of grated cheese is equivalent to approximately 4 cups. The following basic recipe amounts are helpful:

- Three apples usually equal about 1 pound, and six apples are needed for the average apple pie.
- A medium orange or lemon yields up to ½ cup of juice.
- A medium orange yields approximately 1 to 2 tablespoons of grated rind (zest), whereas a lemon yields only ½ to 1 tablespoon zest. Appendix B gives the volume-to-

weight equivalents of different types of foods.

- Another helpful reference is the number of cups found in common can sizes (inside back cover of this book).

Often a basic ingredient turns up missing during the preparation of a dish. This can put a halt to food preparation, but in some situations, a substitution may save the day. Appendix C lists some substitutions that can be made for standard ingredients.

# MIXING TECHNIQUES

Once the ingredients have been selected and measured, the next step is often to mix them all together. *Mixing* is a general term that includes stirring, beating, blending, binding, creaming, whipping, and folding. In mixing, two or more ingredients are evenly dispersed in one another until they become one product. In general, the other processes accomplish the same thing, but there are distinctions.

- *Stir.* This method is the simplest, as it involves mixing all the ingredients together with a utensil (usually a spoon) using a circular motion.
- *Beat.* The ingredients are moved vigorously in a back-and-forth, up-and-down, and around-and-around motion until they are smooth. An electric mixer is often used to beat ingredients together.
- *Blend.* Ingredients are mixed so thoroughly that they become one.
- *Bind.* Ingredients adhere to each other, as when breading is bound to fish.

- *Cream.* Fat and sugar are beaten together until they take on a light, airy texture.
- *Whip or whisk.* Air is incorporated into such foods as whipping cream and egg whites through very vigorous mixing, usually with an electric mixer or a whisk.
- *Fold.* One ingredient is gently incorporated into another by hand with a large spoon or spatula.

There are many methods for combining the ingredients of cakes and other baked products, but the most commonly used are the conventional (creaming), conventional sponge, single-stage (quick-mix), pastry-blend, biscuit, and muffin methods. Methods for mixing yeast bread doughs are discussed in Chapter 20.

## Conventional (Creaming) Method

The conventional method, also known as the creaming or cake method, is the most time consuming, and is the method most frequently used for mixing cake ingredients. It produces a fine-grained, velvety texture. The three basic steps are:

1. Creaming
2. Egg incorporation
3. Alternate addition of the dry and moist ingredients

The fat and sugar are creamed together by working the fat until it is light and foamy and then gradually adding small portions of sugar until all of it is well blended. A well-creamed combination of fat and sugar incorporates air while suspending sugar crystals and air bubbles in the fat. As the fat melts during baking, it creates air cells that migrate toward the liquid, resulting in a very fine-grained texture (2).

The eggs or egg yolks are then added one at a time to the creamed fat and sugar. An alternative method is to whip the egg whites separately and fold them into the cake batter after all the other ingredients have been mixed.

Finally, flour, baking powder or soda, and salt are sifted together with other dry ingredients such as cocoa in order to distribute the leavening agent evenly. The sifted dry ingredients, divided into three or five portions, are then added alternately with a liquid (usually milk) into the fat, sugar, and egg base. After one portion of dry ingredients has been incorporated, a portion of liquid is added and stirred or beaten until well blended. The process begins and ends with the dry ingredients.

### Stirring Too Little or Too Much

As with any type of mixing method, too much or too little stirring can cause problems. Overstirring a cake batter creates such a viscous mass that the cake may not be able to rise during baking, and the texture will tend to be fine but compact or lower in volume, full of tunnels, and have a peaked instead of a rounded top. Too little stirring can also result in a low-volume cake from an uneven distribution of the baking powder or soda. The texture of an understirred cake tends to contain large pores, have a crumbly grain, and brown excessively.

## Conventional Sponge Method

The conventional sponge method, also known as the conventional meringue method, is identical to the creaming method except that a portion of the sugar is mixed in with the beaten egg or egg white, and the egg foam is folded into the batter in the end. The conventional sponge method is preferred for foam or sponge cakes because it contributes volume, and for baked goods made with soft fats whose creamed foam breaks and releases much of its incorporated air when egg yolks are added. In either case, the air in the foam that is folded in during the last stage increases volume.

## Single-Stage Method

In the single-stage method, also known as the quick-mix, one-bowl, or dump method, all the dry and liquid ingredients are mixed together at once. Packaged mixes for cakes, biscuits, and other baked goods rely on the single-stage method. Only baked products containing higher proportions of sugar, liquid, and possibly an emulsifier in the shortening can be mixed by this method. Starting with the dry ingredients in a bowl, the fat (usually

vegetable oil), part of the milk, and the flavoring are added and stirred for a specified number of strokes or amount of time (if an electric mixer is being used). The eggs and remaining liquid are then added, and the batter is mixed again for a specified period of time. The sequence and mixing of ingredients is important, because creaming is not a part of the process. To attain a uniform blend, the bottom and sides of the bowl should be scraped frequently. Quick-mix batters are more fluid than conventional batters.

## Pastry-Blend Method

Fat is first cut into flour with a pastry blender, or with two knives crisscrossed against each other in a scissor-like fashion, to form a mealy fat-flour mixture. Half the milk and all of the sugar, baking powder, and salt are then blended into the fat-flour mixture. Last, eggs and more milk may then be blended into the mixture.

## Biscuit Method

This method is similar to the pastry method except that all the dry ingredients—flour, salt, and leavening—are first combined. The fat is then cut into the flour mixture until it resembles coarse cornmeal. Liquid is added last. The dough is mixed just until moistened and not more or the biscuits will be tough.

## Muffin Method

This is a simple, two-stage mixing method. The dry and moist ingredients are mixed separately and then combined and blended until the dry ingredients just become moist. Over-mixing will result in a tough baked product riddled with tunnels.

# SEASONINGS AND FLAVORINGS

The most nutritious and beautifully presented meal in the world cannot be enjoyed unless it tastes good. Enhancing the flavor of foods is an art that is

critical to the acceptability of foods, and a restaurant can succeed or fail depending on how that art is practiced. The most common reason for consumers to reject food is unacceptable flavor (17). **Seasonings** and **flavorings** help food taste its best.

They are rarely, however, capable of redeeming foods that are not of good quality to start with or of rejuvenating foods that have lost their quality during preparation. No amount of cinnamon will raise the flavor quality of an apple pie made from frozen apple slices to the level of one made from fresh, juicy apples.

## Types of Seasonings and Flavorings

The number and variety of seasonings and flavorings available from all over the world would be nearly impossible to catalog, so this chapter focuses on the basics: salt, pepper, herbs and spices, flavor enhancers, oil extracts, marinades, batters, and condiments.

### Salt

Salt was esteemed so highly in ancient times that the word *salary* is derived from *salt*. Salt, or sodium chloride (NaCl = 40% Na, 60% Cl), is the second most frequently used food additive by weight. (Sugar, which is fully discussed in Chapter 21, is first.) Salt was originally introduced into foods as a preservative; salting, or curing, meat and fish was the only way to preserve food prior to refrigerators, freezers, or canning. The function of salt in foods expanded to those shown in Table 5-5. Salt in its most common form is a crystalline seasoning that may or may not be iodized and combined with an anticaking material.

*Types of Salt* A variety of salts may be purchased, including sea salt, rock salt, kosher salt, and a number of flavored salts, the most common being

---

**Seasoning** Any compound that enhances the flavor already found naturally in a food.

**Flavoring** Substance that adds a new flavor to food.

**TABLE 5-5**    Functions of Salt in Foods

| Function | Description |
|---|---|
| Flavor Enhancer | Salt's best-known function is to enhance the flavor of foods. People like salt because our bodies need sodium chloride (14). Breads are less bland, cheeses are not as bitter, and processed meats are more flavorful in part because of the addition of salt. |
| Preservative | Salt cures and has been used for thousands of years to preserve foods. Refrigerators and freezers have not always been around, and drying out the moisture in foods with salt prevented bacteria from being able to live on the food. |
| Binder | Food manufacturers use salt to help form a gel in sausage and other smoked meat products. |
| Texture Enhancer | Salt contributes to the texture of ham, processed meats, bread, and certain cheeses. |
| Color Aid | The color of processed meats such as ham, bacon, hot dogs, and sausage is partially due to salt. |
| Control Agent | Bacteria and yeast are sensitive to salt concentrations and so salt is used to control their growth during fermentation in such foods as bread, cheese, pickles, sauerkraut, and sausage. |

**Source:** Salt Institute.

Very low salt = 35 mg or less per serving
Low salt (sodium) = 140 mg or less per serving
Reduced salt = at least a 75% reduction compared to original food
Unsalted = no salt added
Salt-free = 5 mg or less per serving

***Other Dietary Salt Sources***    Sources of sodium to watch out for when preparing food are seasoning salts (garlic, parsley, onion, celery); meat tenderizers; meat flavorings (barbecue sauce, smoke-flavored products); salad dressings; molasses; party spreads; dips; condiments (mayonnaise, mustard, ketchup, tartar sauce, chili sauce, soy sauce, relish, horseradish, Worcestershire sauce, and steak sauces); monosodium glutamate (MSG); and bouillon cubes.

garlic, onion, and celery (Figure 5-14) (24). There are also some expensive and rare sea salts known as *fleur de sel* and *sel gris* that are used only in the finest restaurants.

***Adding Salt in Food Preparation***    Regardless of the type, salt should be added in small increments because of its potential to overwhelm the taste buds when too much is added. The preparer should also be aware that any liquid, such as a sauce or soup, that will be reduced should be only lightly salted because the salt becomes even more concentrated as the volume of the liquid diminishes. Although removing excess salt is almost impossible, salty soup may be partially neutralized by adding a touch of sugar or by dropping in a raw, peeled potato to absorb some of the salt.

***Processed Foods***    Foods that are canned, frozen, cured, or pickled provide more than 75% of all the sodium ingested (16). Because high sodium intake has become a health concern, food companies have developed lower-sodium versions of many processed food products. To make it easier to look for these lower-sodium products, food labels carry the following terms describing sodium/salt content:

**FIGURE 5-14**    Different types of salt used in food preparation.

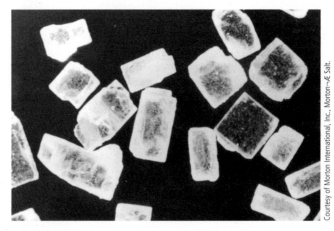

Courtesy of Morton International, Inc., Morton-Æ Salt.

**Salt Varieties**

| | |
|---|---|
| Sea salt — | Obtained from evaporated seawater. Sea salt is more costly than other salts, yet often preferred because it has a pure, mineral-like taste. |
| Rock salt — | Derived from ancient sea beds that have long dried up and are underground. |
| Table salt — | Refined rock salt that is often fortified with iodine and contains additives to prevent caking. |
| Kosher salt — | Rock salt with no additives. Preferred by professional chefs because of its large, flaky crystals that are picked up easily with fingers (use one-third more if substituting for table salt and vice versa). |
| Flavored salts — | Garlic, onion, and celery salt mixtures. |

## NUTRIENT CONTENT

*Reducing Salt* Sodium is the portion of salt that has raised concerns because of its possible connection to high blood pressure, or hypertension. Because table salt is 40% sodium, one teaspoon contains 2,300 milligrams of sodium. Not everyone is susceptible to high blood pressure caused by a high-sodium diet, but about 15 to 25% of Americans are genetically prone to developing the condition. The average North American diet is high in sodium, which automatically puts this subgroup at risk.

To safeguard these sodium-sensitive individuals, the surgeon general recommends that all Americans lower their sodium intake. The concern is that high blood pressure, regardless of its cause, is a known risk factor for heart disease, kidney disease, and strokes (6).

How much should sodium intake be reduced? Most Americans ingest about 1¾ to 2 teaspoons of salt each day (this amount contains 4,000 to 5,000 mg of sodium). The Dietary Guidelines suggest that amount be lowered to approximately 1 teaspoon of salt (2,300 mg of sodium). Taking three steps helps to achieve this goal: (1) cutting back on high-sodium sources such as processed foods, (2) not adding salt during food preparation, and (3) removing the saltshaker from the table.

**Salt Substitutes** Finally, the salt added at the table should also be reduced. Salt substitutes are an option for some people. Many salt substitutes, however, contain potassium, which may also be inappropriate for people with kidney, heart, or liver problems. Calcium chloride is another salt substitute. "Low sodium" salt, which contains half the sodium of regular table salt, is not considered a salt substitute. "Lite" salt products sold to replace the saltshaker should be avoided, because they also contain some sodium. Ultimately, the craving for salt is an acquired taste, so gradually cutting back on salt will eventually decrease cravings.

### Pepper

Pepper is just behind salt in popularity as a seasoning. Pepper is added most frequently to meats, soups, sauces, and salads. Ground black or white pepper comes from the berries of a tropical climbing shrub. The color of pepper depends on the berry's ripeness. Black pepper is from the dried, unripe berry, while white pepper is from the ripe berry from which the dark outer skin has been removed. Green peppercorns, a less common variety, are from underripe berries that are preserved in brine or freeze dried.

Peppercorns belong to an entirely different genus than the *Capsicum* family of chili peppers, which are classified as vegetables. Many varieties of *Capsicum* peppers are dried and used in chili powder, cayenne pepper, and paprika.

### Herbs and Spices

Foods would be boring if salt and pepper were the only two methods to season foods. A variety of herbs and spices from all over the world exist to improve food flavors.

**Herbs** Herbs were described by Charlemagne as "a friend of physicians and the praise of cooks" (10). The Food and Drug Administration groups culinary herbs and spices together and considers them both to be spices. Regardless of how they are defined, herbs are well known for their seasoning capabilities in food preparation (Appendix D). The best-known seasoning herbs include basil, sage, thyme, oregano, bay leaves, cilantro, dill, marjoram, mint, parsley, tarragon, rosemary, and savory. For the best in flavor and texture, fresh herbs are generally preferred over dried.

**Spices** Spices are distinguished from herbs by their origin: herbs are derived from leaves, spices from other parts of the plant. Some examples include:

- Allspice (from a fruit)
- Saffron (flower)
- Cinnamon (bark)
- Anise, caraway, celery, cumin, fennel, mustard, poppy, and sesame (seeds)
- Ginger and turmeric (**rhizomes**)

Although garlic, onions, and shallots can serve as a spice, they are officially recognized as vegetables. Appendix D also lists some of the more common spices.

History records a time when spices were greater in value than gold. In fact, they have been called "vegetable gold" and were once used as currency. A Goth leader once demanded 3,000 pounds of pepper as a partial ransom for calling off his siege of Rome (12). The search for these flavoring ingredients resulted in the carving of trade routes between countries, the founding of wealthy empires, and the exploration of far-off lands. Their value now rests in their unique ability to add a flavorful difference to dishes. The various world cuisines owe their distinctiveness to the unique combinations of spices in foods. Thai food relies heavily on hot peppers, and Central American dishes are distinguished by their use of chili peppers or powder. Mexican meals often incorporate cumin, coriander, paprika, pepper, and cilantro. Indian dishes are enhanced with curry mixtures, which are combinations of spices whose exact ingredients and proportions can be closely guarded family secrets.

**Purchasing Herbs and Spices** Herbs and spices can be purchased in whole, crushed, or ground form. Whole herbs retain their freshness longer than crushed, which in turn keep longer than ground. Whole seeds and leaves provide a visual and textural appeal, although the flavor release may be slow and unevenly distributed. Ground spices provide a quick infusion of flavor that is more uniform, but their aromas are easily lost when exposed to oxygen (oxidized) during storage. The natural antioxidant properties of certain herbs are also lost when they are exposed to oxygen (20).

**Storing Herbs and Spices** According to the American Spice Trade Association, dried spices and herbs should be kept below 60°F (16°C) for optimal potency and replaced every 12 months. Herbs and spices deteriorate rapidly

**Herb** A plant leaf valued for its flavor or scent.

**Spice** A seasoning or flavoring added to food that is derived from the fruit, flowers, bark, seeds, or roots of a plant.

**Rhizome** An underground (usually) stem that generates (1) shoots that rise up and/or horizontally to propagate new plants, and (2) roots that grow down to the ground.

when exposed to air, light, and heat. They keep best in airtight, opaque containers stored in cool, dry places. Green herbs such as chives and parsley are light sensitive and will fade if exposed to light (4). The freshness of a particular spice or herb is tested by crushing it in the palm of the hand and then sniffing it to detect its intensity. The full-bodied aroma of fresh herbs becomes weak and barely detectable over time. If an herb or spice is to be used only occasionally, it is best to buy it in small quantities.

## Flavor Enhancers

Monosodium glutamate (MSG) is a compound that does not fit into any particular category. It influences flavor without contributing any flavor of its own. Hundreds of years ago in Asia, people found that food cooked in a seaweed-based soup stock had a unique flavor. In 1909, this compound was isolated from seaweed by a Japanese scientist and called *umami,* meaning "delicious" (see Chapter 1 defining *umami* as the fifth taste) (27). Its scientific name, *monosodium glutamate,* comes from glutamic acid, an amino acid found in seaweed. It is now widely used in processed foods, including canned/dried soups, spaghetti sauces, sausages, and frozen meat dishes. It has been implicated in "Chinese Restaurant Syndrome," in which MSG-sensitive people experience nausea, diarrhea, dizziness, grogginess, sleepiness, warmth, headache, chest pain, and arthritis-like symptoms from consuming MSG (11).

## Oil Extracts

Oil extracts can be used as food flavorings. These essential oils are obtained from natural sources such as flowers (orange), fruits (oranges, lemons), leaves (peppermint), bulbs (garlic), bark (cinnamon), buds (clove), and nuts (almonds). The flavor in essential oils is so concentrated that only a small amount is required for flavoring purposes. Oil extracts are primarily used to flavor puddings, candy, ice cream, cakes, and cookies.

Vanilla beans from the cured pod of a tropical orchid provide the purest, most intense vanilla flavor. The small black specks in vanilla sauces and ice cream are the seeds of the pod. To obtain pure vanilla extract, cured vanilla beans are steeped in alcohol. The Food and Drug Administration defines "pure vanilla extract" as containing 35% alcohol by volume, while those of lesser content are labeled "pure vanilla flavor." Vanilla/vanillin blends or imitation versions should be avoided, because they contribute an artificial flavor to foods.

Extracts are made by steam-distilling the oils from various plant sources and blending them with ethyl alcohol, which can evaporate. For that reason they should be stored in a cool, dark place and used within a year to retain maximum flavor.

## Marinades

Marinades are seasoned liquids that flavor and tenderize foods, usually meats, poultry, and fish. A vinaigrette is a marinade used for vegetables served cold. The basic marinade consists of one or more of the following ingredients: oil, acid (lemon juice, vinegar, wine), and flavorings (herbs, spices). The food is completely submerged in the marinade and refrigerated from a few minutes to several days. The food should be turned occasionally in order to evenly distribute the marinade. Meat, fish, and poultry marinades should be discarded after use and never served raw with the cooked food.

## Breading and Batters

Breading and batters enhance the flavor and moisture retention of many foods. Most foods coated in this manner are deep-fried, pan-fried, or sautéed to give them a browned, crisp outer texture.

***Breadings***    The flours most frequently used for breading are either wheat- or corn-based. Coating the food lightly in flour, called dredging or *à la meuniere* (ala moon-yare), results in a light, golden crust. Crumb coatings differ in that they are applied in three steps (Figure 5-15). First the food is dredged lightly in flour to seal in moisture and provide a base for the next step. The flour-coated food is then dipped quickly in an egg wash consisting of beaten eggs plus a tablespoon of water or milk. (Substituting oil for the water or milk results in a richer, more tender coating.) The proteins in the eggs or milk act as binding agents to "glue" the breading to the surface of the food (13). Finally, the sticky-coated food is placed in a bowl of crumbs for the final coating. Seasoned breadcrumbs, cracker crumbs, cornmeal, or cereal (cornflakes) may be used to coat foods. Smaller, more delicate foods such as

**FIGURE 5-15**    Breading—application of a crumb coating.

Roll in flour and shake off excess.

Dip floured piece in egg wash.

Dip in bowl of crumbs and toss more crumbs on top.

mushrooms require finer-grained breadings. Seasonings or flavorings such as salt, pepper, rosemary, thyme, sage, or others can be added at any of the three steps of breading, although mixing them into the egg wash ensures they are evenly distributed (23). Sugar can also be added, but be aware that it results in a browner product (8).

***Batters***    Another way to coat foods is through the use of batters, which are wet flour mixtures containing water, starch, and seasonings into which foods are dipped prior to being fried. Commercial batters that require simply adding water are available. There is no one recipe for a batter, and formulas can be extremely flexible (26). The addition of eggs to the batter will produce a darker coating because of the yolk content. Commercial batters often have added ingredients such as gums for viscosity and starches to increase adhesion by the swelling of their granules. Shortening or oils contribute to overall flavor and mouthfeel (11). Figure 5-16 shows the basic differences between using a breading process and a batter process to coat foods.

## Condiments

Condiments are seasonings or prepared relishes used in cooking or at the table. Some of the most common condiments are mustard, ketchup, mayonnaise, relish, tartar sauce, salsa, barbecue sauce, chili sauce, soy sauce, horseradish, Worcestershire sauce, chutney, and steak sauce.

# Adding Seasonings and Flavorings to Food

Knowing how much and when to add the different seasonings and flavorings to foods is essential to food preparation.

## How Much to Add?

If tested recipes are available, they should be followed. If there is no recipe, start by adding ¼ teaspoon of spice (or ⅛ teaspoon for chili, cayenne, or garlic powder) for every pound of meat or pint of liquid (soup, sauce). Flavor-test and add more seasonings as desired. It is always easier to add than to subtract, and because it is important not to overpower other ingredients in a dish, it pays to be cautious. There really is no easy set rule or formula for adding seasoning and flavoring to foods. The freshness of herbs and spices will influence how much should be added, and evaporation of liquid during heating will concentrate what is already present. Successfully prepared foods have well-balanced flavors that are complementary.

> ## ? How & Why?
>
> **How much dried herbs should one use in place of fresh herbs?**
>
> When substituting dried herbs for fresh, the general rule is to use about one third as much as you would fresh herbs, because the flavor of dried herbs that have not become stale is generally more intense.

## When to Add?

Seasonings should be added to prepared foods early enough in the cooking process to release their flavor, but not so soon that their flavor is lost. Most seasonings (especially ground) are added near the end of the heating period, whereas a few (whole or lightly crushed) need more time to release their flavors and aromas to blend with the other ingredients. Foods tend to better retain the flavor of seasonings and flavorings if their surfaces are partially cooked and therefore permeable to what is added. This stage is commonly referred to by professional chefs as **sweating** (3). Delaying the addition of seasonings and flavorings is particularly important for salts, which tend to shrink meats if they are added too soon. However, some chefs salt during the entire process of cooking to allow time for the salt to disperse and interact with the food (14). Flavor retention is influenced by the length of the heating and the final temperature attained. Experience may well be the best teacher.

## Food Industry Uses

Food corporations sometimes add seasonings to their products with the aid of rotating metal drums. These are equipped with spray bars that can lightly coat the chips, pretzels, or other snacks with an oil so that the seasoning will stick. The dry seasoning mix such as salt, powdered cheese, or garlic is then released

**FIGURE 5-16** The difference between using breadings and batters.

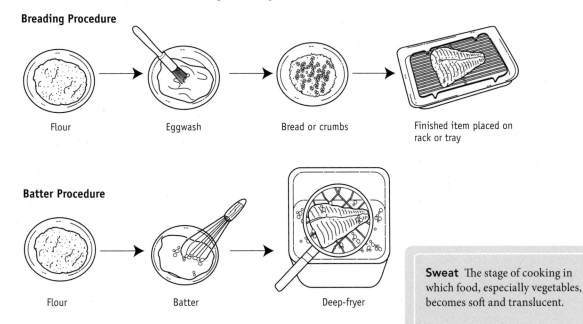

**Breading Procedure**

Flour → Eggwash → Bread or crumbs → Finished item placed on rack or tray

**Batter Procedure**

Flour → Batter → Deep-fryer

**Sweat** The stage of cooking in which food, especially vegetables, becomes soft and translucent.

through a feeder inserted into the drum. The drums are only 30% full to prevent pieces from falling out. Another option is to spray the snacks moving on a flat belt as they exit the oven (7).

# FOOD PRESENTATION

The highest quality, best-prepared food is shortchanged if the plate presentation has not achieved or surpassed the same level of quality. An artistic layout of food items on the plate plays a very important role in winning over and satisfying the customer, whose first impression is based largely on sight.

## Plate Presentation

The first impression of food is how it is laid out on a plate. When plating food, the top consideration is coordination of colors, shapes, sizes, textures, and flavors. Following are some guidelines to help in achieving this coordinated balance.

First, a hot plate is selected for hot foods, whereas a cold plate is reserved for cold foods. The size of the plate should be sufficient so that food is not crowded, but not so large that the amount of food looks meager on it. Items are placed on the plate to achieve balance. The main food item, often the meat, is set in front of the guest with the best part forward, and any fat or bone facing away. The plate should not have to be turned in order for the main entrée to be consumed. Accompanying items are plated around the main entrée, and garnishes may be added to contribute to balance (Figure 5-17). Space should be kept between each item on the plate, with the border of the plate serving as the frame. The border should never be covered with food; any food that does spill over onto the edges should be wiped clean. The exception is when the plate rim is dusted with chopped herbs, spices, or other decorative touches.

## Garnishes

Garnishing adds color and design to a plate, making it more attractive to the eye. Garnishes are edible items used to decorate food and should generally reflect the flavors of the dish being served.

---

> ### ◆ CALORIE CONTROL
> #### Calorie Control: Lowering Fat
> Avoiding fried foods is one of the most effective ways to reduce fat intake and excess calories. A baked potato has 145 calories (kcal) compared to the 750 calories (kcal) from that same potato prepared as French fries (9). All of the moist and dry heat preparations discussed earlier, with the exception of frying (pan and deep), are healthful options for preparing food (that includes sautéing). Note that many pan- and deep-fried foods are breaded and the added flour contributes even more calories.

For example, a rosemary sprig would be appropriate for a rosemary-scented meat sauce. Other possible garnishes, depending on what is being served, may include:

- Leaves, such as parsley sprigs, or mint leaves in iced tea
- Fruit, such as pineapple sticks; kiwifruit slices; olives; or lemon, lime, or orange wedges
- Vegetables, such as cucumbers, tomatoes, green peppers, radishes, or onions
- Pickled items, such as olives, pickles, or pimentos
- Nuts, croutons, crackers
- Hard-boiled egg slices or halves

Only fresh, high-quality foods should be used for making garnishes. Using garnishes adds balance; if the items on a plate are already harmonized, a garnish is not necessary. Plate garnishes are best when they are colorful, contrasting but not clashing, and compatible with the food being served in terms of flavor, size, and shape. Garnishes should not crowd the dish, and an odd number tends to be more visually appealing. For example, three slices of apple on a plate look better than two or four slices. To prevent any possible injuries, unfrilled toothpicks and other hard inedible items should be avoided.

**FIGURE 5-17**    Plate presentation.

Jancouver/istockphoto.com

# PICTORIAL SUMMARY / 5: Food Preparation Basics

Mastering the basics of food preparation is essential to putting a good meal together, but food preparation is not an exact science. Understanding and adjusting for the many variables at play in preparing even the simplest recipe can elevate food preparation from a craft to an art form.

## METHODS OF HEATING FOODS

**Moist-heat preparation:** Heat is transferred by water, water-based liquid, or steam.

**Dry-heat preparation:** Heat is transferred by air, radiation, fat, or metal.

**Microwaving:** Usually listed as moist-heat, microwaving actually incorporates both dry-(radiation) and moist-heat methods.

| Moist Heat | | Dry Heat | |
|---|---|---|---|
| Scalding | Poaching | Baking | Roasting |
| Simmering | Stewing | Broiling | Grilling |
| Braising | Boiling | Frying | Sautéing |
| Parboiling | Blanching | Stir-Frying | Pan-Broiling/Frying |
| Steaming | | Deep-Frying | |

Poach     Simmer     Boil

**Temperature scales** used to measure heat in cooking are available in two commonly used scales, Fahrenheit and Celsius or centigrade. These scales are based on the freezing and boiling temperatures of pure water at sea level (32°F/0°C and 212°F/100°C, respectively).

## HEAT TRANSFER METHODS

**Conduction:** The direct transfer of heat from one substance to another by direct contact, for example, heat from a gas flame warms the pot on the stove and then its contents.

**Convection:** Air or liquid expands and rises as it heats up, creating a circular current. Oven baking, simmering, steaming, and deep-frying are all examples of convection cooking. The use of water and fat to heat foods combines both conduction and convection.

**Induction:** Transferring heat energy to adjacent material without contact.

**Radiation:** Radiant heat in the form of particle waves moving outward is generated by broiling, grilling, and microwaving. Infrared heat lamps and ovens use electromagnetic waves to keep foods warm and to prepare frozen foods.

## MEASURING INGREDIENTS

The three major steps in correct measuring:
- Approximating the amount of required food (e.g., 4 ounces of cheese yields 1 cup shredded)

- Selecting the right measuring utensil
  - *Wet ingredients:* transparent, graduated cup with pour spout
  - *Dry ingredients:* flat-topped measuring cups for leveling
- Using accurate measuring technique

**Know your substitutions:** Sometimes knowing what item can replace a missing ingredient can save the day!

## MIXING TECHNIQUES

Mixing is a general term describing stirring, beating, blending, binding, whipping (whisking), and folding. The ingredients for baked goods can be mixed in several different ways, but the most common methods are the conventional (creaming), conventional sponge, single-stage (quick-mix), pastry-blend, biscuit, and muffin.

## SEASONINGS AND FLAVORINGS

**Seasoning:** Any compound that *enhances the flavor* already found naturally in a food.

**Flavoring:** An addition that *adds a new flavor* to a food.

The major seasonings/flavorings are:
- Salt and pepper
- Herbs and spices
- Oil extracts
- Flavor enhancers
- Marinades
- Breadings and batters
- Condiments

The freshness of herbs and spices will influence how much should be added, and evaporation of the liquid during heating will concentrate the effect of the flavoring/seasoning added.

It is always easier to add than to subtract! Flavor-test and add more seasoning as desired.

Most seasonings are added near the end of cooking time.

## CUTLERY TECHNIQUES

Knowing knives and how to use them is essential to basic food preparation.

**Holding a chef's knife.**

It is important to know how to hold the knife; the different sections of the blade assigned to various tasks; and the differences among slicing, shredding, dicing, mincing, chopping, and peeling.

## FOOD PRESENTATION

A restaurant customer's first impression is based largely on sight, and an artful arrangement of food on a plate contributes a great deal to the dining experience.

# CHAPTER REVIEW AND EXAM PREP

## Multiple Choice*

1. In which cooking method is food simmered in a small amount of liquid in a covered pot or casserole dish?
   a. Braising
   b. Scalding
   c. Simmering
   d. Steaming

2. Food that is cut into very thin, stick-like shapes has been _____.
   a. chopped
   b. minced
   c. peeled
   d. julienned

3. Three teaspoons is equivalent to _____.
   a. 1 tablespoon
   b. 1½ tablespoons
   c. 2 tablespoons
   d. 2 ⅓ tablespoons

4. When using the muffin method of mixing, over-mixing will produce what type of unwanted result?
   a. Tender
   b. Tough
   c. Overly moist
   d. Overly dry

5. What is the best environment for storing herbs and spices?
   a. A moist place
   b. A warm, bright place
   c. A cool, dry place
   d. A hot, moist place

6. Which of the following heating methods is based on the principle of conduction?
   a. Heat from an electrical coil or gas flame touches a pan that heats the pan's contents
   b. Heated air in an oven rises and cooler air moves downward
   c. Heat is transferred by waves during grilling, broiling, and microwaving
   d. Heat is moved from the coils under a flat-surfaced range to the pot

7. The correct sequence for breading is _____.
   a. egg wash, flour, crumbs
   b. flour, crumbs, egg wash
   c. egg wash, crumbs, flour
   d. flour, egg wash, crumbs

## Short Answer/Essay

1. Briefly describe each of the following moist heat preparation methods: *scalding, poaching, simmering, stewing, braising, boiling,* and *steaming.*
2. Explain the difference between dry-heat and moist-heat preparation. Define the following dry-heat preparation methods, providing examples for each: *baking, roasting, broiling, grilling,* and *frying.*
3. Define the following cutting styles: *slicing* (include *julienning*), *shredding, dicing, mincing,* and *peeling.*
4. Fill in the following equivalent measurements:
   _____ tsp = _____ T    _____ oz = _____ 1 lb
   _____ oz = _____ 1 C    _____ grams = 1 tsp
   _____ C = _____ pint    _____ pints = _____ quart(s)
5. How would you measure each of the following items: *liquid, eggs, fat, sugar,* and *flour?*
6. List and briefly describe three mixing methods.
7. Explain the difference between seasonings and flavorings, herbs and spices, and breadings and batters.
8. What is a marinade, and how is it used?
9. What are the main factors to be considered in plate presentation?
10. List six basic types of garnishes that can add color and design to a meal.

*See p. AK-1 for answers to multiple choice questions.

# REFERENCES

1. Adams SJ. Slow cooking enhances winter vegetables. *Fine Cooking* 18:40–43, 1997.
2. Anonymous. Reduce fat, maintain function. *Prepared Foods* 177(1):151, 2008.
3. Anonymous. "Sweating" vegetables coaxes out flavor. *Fine Cooking* 25:76, 1998.
4. ASTA. *The Foodservice and Industrial Spice Manual.* American Spice Trade Association, 1990.
5. Banooni S, SM Hosseinalipour, AS Mujumdar, E Taheran, M Bahiraei, and P Taherkhani. Baking of flat bread in an impingement oven: An experimental study of heat transfer and quality aspects. *Drying Technology* 26(7):902–909, 2008.
6. Centers for Disease Control and Prevention. Application of lower sodium intake recommendations to adults—United States, 1999–2006. *MMWR Weekly* 58(11):281–283, 2009.
7. Clark JP. Applying seasonings and coatings. *Food Technology* 61(11):72–74, 2007.
8. Corriher SO. Taking the fear out of frying. *Fine Cooking* 16:78–79, 1996.

9. Deppe M. Half the fat. Your cooking method makes all the difference. *Today's Diet & Nutrition* 4(1):7, 2007.

10. Farrell KT. *Spices, Condiments, and Seasonings.* Avi, 1985.

11. Freeman M. Reconsidering the effects of monosodium glutamate: A literature review. *Journal of the American Academy of Nurse Practitioners* 18(10):482–486, 2006.

12. Giese J. Spices and seasoning blends: A taste for all seasons. *Food Technology* 48(4):88–98, 1994.

13. Loewe R. Role of ingredients in batter systems. *Cereal Foods World* 38(9):673–677, 1993.

14. Masibay KY. Salt makes everything taste better. *Fine Cooking* 91:80, 2008.

15. Mathews RH, and R Batcher. Sifted versus unsifted flour. *Journal of Home Economics* 55:123, 1963.

16. Mattes RD. Discretionary salt use. *American Journal of Clinical Nutrition* 51:519, 1990.

17. Ponte JG. Sugar in baking foods. In *Sugar: A User's Guide to Sucrose,* eds. NL Pennington and CW Baker. Van Nostrand Reinhold, 1990.

18. Saguy IS, and EJ Pinthus. Oil uptake during deep-frying: Factors and mechanism. *Food Technology* 49(4):142–145, 1995.

19. Scott ML. *Mastering Microwave Cooking.* Bantam, 1976.

20. Shahin SA, et al. Indian medicinal herbs as sources of antioxidants. *Food Research International* 41(1):1–15, 2008.

21. Singh RP. Heat and mass transfer in foods during deep-fat frying. *Food Technology* 49(4):134–137, 1995.

22. Sloan AE. Grilling and slow cooking are gaining. *Food Technology* 53(6):28, 1999.

23. Stevens M. Coating food for a golden crisp crust. *Fine Cooking* 20:74, 1997.

24. Stevens M. What's the difference in salt? *Fine Cooking* 24:10, 1998.

25. Stevens M. Why scald milk? *Fine Cooking* 27:12, 1998.

26. Suderman DR. Selecting flavorings and seasonings for batter and breading systems. *Cereal Foods World* 38(9):689–694, 1993.

27. van Wassenaar PD, AHA van den Oord, and WMM Schaaper. Taste of 'delicious' beefy meaty peptide. *Journal of Agricultural and Food Chemistry* 43(11):2828–2832, 1995.

28. Varela G, and B Ruiz-Roso. Some effects of deep-frying on dietary fat intake. *Nutrition Reviews* 50(90):256–262, 1992.

# WEBSITES

The Culinary Institute of America, a leading chef school in the United States, has two useful websites: **www.ciachef.edu** and **www.ciaprochef.com**

Find information in this food preparation encyclopedia: **http://allrecipes.com**

# 6  Meal Management

A successful meal is both psychologically and physiologically satisfying. Planning such meals requires a basic knowledge of food preparation, nutrition, and presentation strategies.

**Job description** An organized list of duties used for finding qualified applicants, training, performance appraisal, defining authority and responsibility, and determining salary.

**Organizational chart** A descriptive diagram showing the administrative structure of an organization.

Effective meal planning and preparation, whether for a household, an institution, or a restaurant chain, are made possible by the efficient management of money, time, and energy. These resources are used in the various steps of meal production: food procurement, storage, preparation, serving, and cleanup. All these steps require organization on the part of the individual or of the food service manager. In the case of the latter, good organization involves people working together toward the common goal of preparing and serving attractive, tasty, nutritious, and profitable meals. This chapter covers how food service establishments are organized, meal planning, purchasing, time management, types of meal service, and table settings.

## FOOD SERVICE ORGANIZATION

At the core of every food service operation is an organization with a structure set up to achieve specific goals. Management determines the objectives necessary to reach those goals and then mobilizes people toward meeting them. This entails the division of work, which necessitates clear and effective **job descriptions.** Positions are often described more fully in a job description than they are in a newspaper ad for that job, or even in the verbal description an employer provides when interviewing or training a new employee. Anyone applying for employment in such an operation should ask the employer to provide the formal job description in writing and check that it matches the job duties, performance evaluations, and salary as explained by the person doing the hiring. If there's a performance evaluation form, ask to see it so that you can attempt to meet the performance criteria. It's also a good idea to get a copy of the annual evaluation form when you start the job. An **organizational chart** is also helpful to a new employee. Figure 6-1 compares the organizational charts between a hospital dietary department and a restaurant.

**FIGURE 6-1** Organizational chart comparison of a hospital and a restaurant.

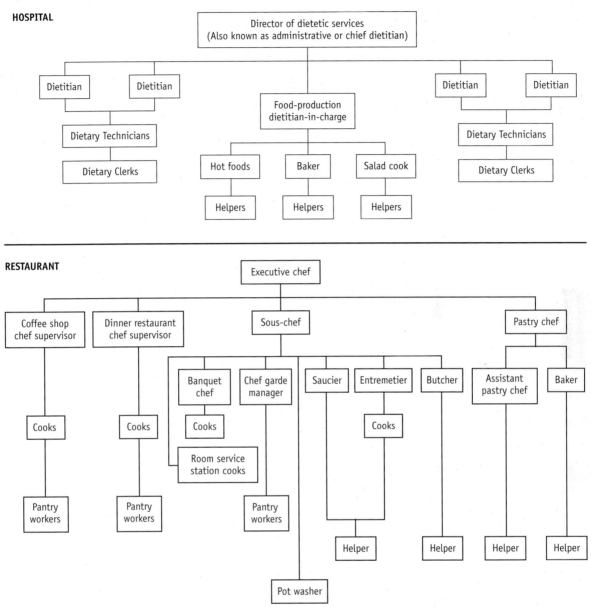

**Source:** Adapted from Mizer et al., *Food Preparation for the Professional* (John Wiley & Sons, 1998).

# Commercial Food Service Organization

Large food service organizations usually follow a historical structure that was pioneered by George Auguste Escoffier (1847–1935). Escoffier, called the father of 20th-century cookery, created stations for particular areas of food production. Escoffier's system of dividing up large kitchens into various preparation areas led to the creation of jobs requiring specific skills. The kitchen team of employees under this type of food service organization is called the *brigade de cuisine* (bree-gahd-de-kweé-zeen).

## Escoffier's System of Organization via Stations

At the head of each station in the kitchen are station chefs or heads with a particular area of expertise:

- Sauce chef/*saucier* (so-see-ay). The highest position among the stations. This chef specializes in the production of sauces, sauce-related dishes, hot hors d'oeuvres, stews, and sautéed foods.
- Fish cook/*poissonier* (pwah-so-nee-ay). Sometimes this station is covered by the sauce chef.
- Vegetable cook/*entremetier* (on-tramet-ee-ay). Prepares vegetables.

- Soup cook/*potager* (poh-ta-zhay). Prepares soups and stocks.
- Roast cook/*rotisseur* (ro-tee-sur). Responsible for meat dishes, particularly if they are roasted or braised.
- Broiler cook/*grillardin* (gree-yar-dan). Specializes in preparing grilled, broiled, or deep-fried meats, poultry, and seafood.
- Pantry chef/*chef garde manger* (guard-mon-zhay). Prepares all nondessert cold foods such as salads and cold hors d'oeuvres.
- Pastry chef/*patissier* (pa-tees-see-ay). Prepares baked goods—pastries, desserts, and breads.

| | | | |
|---|---|---|---|
| **TABLE 6-1** | Brief Summary of Hospital Diets | | |
| **Diet** | **Other Common Names** | **Definition** | **Purposes** |
| **General** | Routine, House, or Selective Diet | For adults who are not on any dietary restriction | Maintain optimal nutritional status |
| **Prescribed** | Diabetic, Low-Sodium, Renal Diet, etc. | Prescribed by physician as medical nutrition therapy for patient | For people with specific medical conditions (diabetes, renal disease, heart disease, etc.) |
| **Modified Consistency** | Clear Liquid — | Foods that are clear liquids containing minimal fiber | Preparation for and recovery from abdominal surgery, etc. |
| | Full Liquid   Surgical Liquid Diet | Similar to clear liquid except permits strained items such as milk, juice, and eggs (in drinks/custards) | Follows clear liquid diet |
| | Blenderized Liquid   Pureed Diet | Foods blenderized to liquid form | For those unable to tolerate solid food following oral or plastic surgery; chewing or swallowing problems; wired jaws |
| | Mechanically Altered   Surgical Soft, Mechanical Soft Diet | Food modified only in texture—blended, chopped, ground | Promote ease of chewing following head and neck surgery or radiation; swallowing problems; oral or dental problems |
| | Soft Diet   Bland, Low-Fiber, Low-Residue Diet | Low fiber, little seasoning, smooth texture, low on fried and strong-flavored foods or gas-forming vegetables | Transition between liquid and general; post-operation to prevent nausea and vomiting, gas and distention from anesthesia and gastrointestinal immobility |
| | High-Fiber   High-Residue Diet | Increased fiber (25–35 g/day), gradual, plenty of fluid | For gastrointestinal problems—constipation, diverticulosis (if not inflamed), irritable bowel syndrome, hemorrhoids, colon cancer, heart disease, diabetes mellitus, obesity |

- Relief, swing, or rounds cook/*tournant* (tour-non). Capable of handling any station in order to relieve one of the other chefs.

In smaller kitchens, there may be only two stations, one for hot foods and one for cold foods. In such kitchens the pantry chef is a major position. Each station head has cooks and helpers or assistants that aid with that area of food production. The first step on the ladder of jobs in a large kitchen is the entry-level position of helper or assistant, which requires virtually no skills. Once the skills are acquired, however, the person may be promoted to cook. People can sign on as apprentices and receive formal training in food service by rotating through each of the kitchen stations.

## Administrative Positions

In general, after about 5 to 10 years of experience in all stations as a chef, a person can be promoted to administrative positions in the kitchen:

- Executive chef/*chef executif*. The person in charge of the entire operation, including kitchen administration, hiring, budgeting, purchasing, work scheduling, menu planning, and more.
- Production manager/*sous* (soo) chef. The second-highest position in the kitchen. The sous chef is in charge of all areas of production and supervision of the staff.

## Hospital Food Service Organization

Patients in a hospital have to be fed and the dietary department is responsible for taking care of people that need regular or specialized meals (Table 6-1). Many larger hospital food operations are professionally managed by Sodexo Corporation. Working for this or a similar company allows the hospital food service employees more flexibility in shifting positions geographically. Hospital food service operations differ slightly from large hotel or restaurant establishments in that the person in charge is the food service director, or administrative or chief **dietitian**. Working for the chief dietitian are other dietitians, their numbers depending on the size of the hospital. The *number of beds* is a common phrase used to describe the size of a hospital. Assisting the dietitians are the dietetic technicians, who hold at least a 2-year degree. The entry-level employees in a hospital dietary department,

**Dietitian (registered dietitian or RD)** A health professional who counsels people about their medical nutrition therapy (diabetic, low cholesterol, low sodium, etc.). Registration requirements consist of completing an approved four-year college degree, exam, internship, and ongoing continuing education.

who may start out relatively untrained, are the dietary aides. People working in this position assist the dietitians and dietetic technicians, often filling out diet orders, taking care of paperwork, and sometimes working the tray line where meals are put together before being delivered to hospital patients.

# MEAL PLANNING

The ultimate goal of a food service organization is to plan, prepare, and serve meals. Food production begins with planning the menu. The menu dictates all other actions that will follow, such as purchasing, choosing equipment to use, scheduling labor, and serving.

## USDA Menu Patterns

What is typically served for breakfast, lunch, and dinner? The meal pattern varies tremendously from country to country. In North America the standard fare can be divided into the selections shown in the United States Department of Agriculture's (USDA) Adult Care Meal Pattern (Figure 6-2). A similar meal pattern for infants (birth–1 year) and children (1–2 years; 3–5 years; 6–12 years) is called the USDA's Child Care Meal Pattern and is available at the USDA's website: www.fns.usda.gov/cnd/care/ProgramBasics/Meals/Meal_Patterns.htm.

The USDA also distributes menu guidelines for its School Lunch Programs. These guidelines offer a choice between two meal patterns: Traditional Food-Based Menu Planning and Nutrient Standard Menu Planning (NuMenues).

### Traditional Food-Based Menu Planning

This approach creates a menu of five food items from four food components: meat/meat alternate, vegetables and/or fruits, grains/breads, and milk. Portion sizes depend on the children's ages and grade groups.

### Nutrient Standard Menu Planning (NuMenues)

The trend toward healthier eating led the USDA to develop NuMenues based on nutrient content. Instead of

**FIGURE 6-2** USDA's Adult Care Meal Pattern.

**Adult Care Meal Pattern**

**Breakfast for Adults**
Select All Three Components for a Reimbursable Meal

| | | |
|---|---|---|
| 1 milk | 1 cup | fluid milk |
| 1 fruit/vegetable | 1/2 cup | juice,[1] fruit and/or vegetable |
| 1 grains/bread[2] | 2 slices | bread or |
| | 2 servings | cornbread or biscuit or roll or muffin or |
| | 1 1/2 cups | cold dry cereal or |
| | 1 cup | hot cooked cereal or |
| | 1 cup | pasta or noodles or grains |

**Lunch for Adults**
Select All Four Components for a Reimbursable Meal

| | | |
|---|---|---|
| 1 milk | 1 cup | fluid milk |
| 2 fruits/vegetables | 1 cup | juice,[1] fruit and/or vegetable |
| 1 grains/bread[2] | 2 slices | bread or |
| | 2 servings | cornbread or biscuit or roll or muffin or |
| | 1 1/2 cups | cold dry cereal or |
| | 1 cup | hot cooked cereal or |
| | 1 cup | pasta or noodles or grains |
| 1 meat/meat alternate | 2 oz. | lean meat or poultry or fish[3] or |
| | 2 oz. | alternate protein product or |
| | 2 oz. | cheese or |
| | 1 | egg or |
| | 1/2 cup | cooked dry beans or peas or |
| | 4 Tbsp. | peanut or other nut or seed butter or |
| | 1 oz. | nuts and/or seeds[4] or |
| | 8 oz. | yogurt[5] |

**Dinner for Adults**
Select All Three Components for a Reimbursable Meal

| | | |
|---|---|---|
| 2 fruits/vegetables | 1 cup | juice,[1] fruit and/or vegetable |
| 1 grains/bread[2] | 2 slices | bread or |
| | 2 servings | cornbread or biscuit or roll or muffin or |
| | 1 1/2 cups | cold dry cereal or |
| | 1 cup | hot cooked cereal or |
| | 1 cup | pasta or noodles or grains |
| 1 meat/meat alternate | 2 oz. | lean meat or poultry or fish[3] or |
| | 2 oz. | alternate protein product or |
| | 2 oz. | cheese or |
| | 1 | egg or |
| | 1/2 cup | cooked dry beans or peas or |
| | 4 Tbsp. | peanut or other nut or seed butter or |
| | 1 oz. | nuts and/or seeds[4] or |
| | 8 oz. | yogurt[5] |

[1] Fruit or vegetable juice must be full-strength.

[2] Breads and grains must be made from whole-grain or enriched meal or flour. Cereal must be whole-grain or enriched or fortified.

[3] A serving consists of the edible portion of cooked lean meat or poultry or fish.

[4] Nuts and seeds may meet only one-half of the total meat/meat alternate serving and must be combined with another meat/meat alternate to fulfill the lunch requirement.

[5] Yogurt may be plain or flavored, unsweetened or sweetened.

**Source:** Food & Nutrition Service of the USDA.

using food groups to create a menu, NuMenues use nutrition-analysis computer software to design a healthy meal. This is currently defined as one meeting the Dietary Guidelines for Americans; limiting calories (which may still be too high; e.g., 825 calories/kcal for a 7–12th grader's lunch); and providing at least one-third of the daily Recommended Dietary Allowances for protein, iron, calcium, and vitamins A and C. Percentage fat limitations depend on the grade (e.g., 7–12th graders should not have more than 30% fat in the course of one week).

In the same spirit of altering eating habits to meet healthier goals, Figure 6-3 suggests an Eating Right Menu in which complex carbohydrates comprise the bulk of a meal, followed by fruits, vegetables, dairy, and, last, meat.

## Hospital Menu Patterns

Patients in a hospital have various dietary needs. They may be on a regular or general diet, a modified-consistency diet, or a prescribed diet, depending on their current health condition (Table 6-1). It is the responsibility of the hospital's dietary department to ensure that all patients are being provided the appropriate diet. Dietitians also provide dietary counseling to patients if a specific medical nutrition therapy (MNT) is prescribed by their physicians.

## Creating the Menu

Regardless of what menu pattern is followed, planning menus is the essential first step of food production. The first decision concerns what type of menu will be used: no choice, limited choice, or choice menu. The menu can reflect what is offered daily, weekly, or for several weeks (**cycle menus**).

**FIGURE 6-3**   Eating Right Menu: Selected suggestions for breakfast, lunch, and dinner.

**Breakfast –**   Eggs (limit 4/week)
Cereal (High fiber—at least 3gm/serving)
  + milk (nonfat or lowfat)
Pancakes
Waffles
French toast
Bagel
Muffin
Scone
Cottage cheese
Yogurt
Smoothie
+ fruit (whole or juice)
+ high fiber cracker

**Lunch –**   Sandwich (all have tomato and/or lettuce)
  (bread = whole wheat, whole grain, low calorie, pita)
  (meat if included not to exceed 2 oz)
  Examples: Tuna/sunflower/dill/mustard
       Grilled tuna/melted cheese
       Turkey/cheese/cranberry sauce
       Cheese/mustard
       Grilled cheese/mustard
       Peanut butter + jelly
       Grilled cheese
Soup/salad
  Examples: Pasta/rice/bean
       Tuna/chicken/shrimp
       Potato
       Greens
Baked potato
Pasta
Rice/beans
Fruit/cheese/crackers + yogurt
+ vegetable  (3 servings/day) (vitamin A & C containing
  at least 4x/week)
+ fruit (2 servings/day) (vitamin A & C containing
  at least 4x/week)
+ bread (at least 1 gm fiber/serving)

**Dinner –**   Any lunch entrée
Meat (3 oz)
  Lean meat (beef, pork, lamb)
  Poultry (no skin)
  Fish
Pasta
  Examples: Spaghetti
       Macaroni
       Lasagna
Rice/bean
+ vegetable (vitamin A & C containing at least 4x/week)
+ fruit (vitamin A & C containing at least 4x/week)
+ bread (at least 1 gm fiber/serving)

**Source:** Mizer et al., *Food Preparation for the Professional* (John Wiley & Sons, 1998).

**Cycle menu**  A menu that consists of two or more weeks, usually three or four, that cycles through a certain order of meals. Cycle menus offer a combination of variety and controlled costs.

Because dinners are usually the most prominent meal, most people plan dinner first. Meals then are built around the main entrée, and the remaining items are usually decided upon in the following order: vegetable, starch, salad, bread, soup, appetizer, and dessert (Figure 6-4). Some provision must also be made for breakfasts, lunches, beverages, and snacks. Fruit is recommended twice a day, and it can be incorporated into the menu as a salad, dessert, snack,

**FIGURE 6-4** Checklist for organizing a menu.

❑ **Main Entrée** – Meat (beef, pork, lamb, fish/shellfish, poultry)
  Cereal (rice, wheat, oat, rye, barley, breakfast cereals)
  Beans (red, kidney, pinto, lima, etc.)
  Pasta (lasagna, macaroni, spaghetti, etc.)
  Eggs (fried, scrambled, omelet, shirred, poached, etc.)

❑ **Vegetable** – (3 servings/day)         ❑ **Appetizer**
  Vitamin C and A containing
  at least 4x/week each

❑ **Starch**                      ❑ **Dessert**

❑ **Salad**                       ❑ **Breakfasts**

❑ **Bread** –      At least 1 g fiber/serving   ❑ **Beverages** – Water, lowfat or
                                                        nonfat milk, juice,
                                                        soda, coffee, tea

❑ **Soup**                        ❑ **Snacks**

* Fruit (2 servings/day) can be incorporated into breakfasts, snacks, salads, desserts, and/or beverages.

**FIGURE 6-5** An example of a one-week (minus Friday) cycle menu.

**Month 1. Week 1**

|  | Breakfast | Lunch | Dinner |
|---|---|---|---|
| **MONDAY** | Cereals<br>Scrambled Eggs<br>Grapefruit<br>Wheat Toast/Jam | Spaghetti<br>Garlic Bread<br>Caesar Salad<br>Juices | Roast Beef<br>Baked Potato<br>Glazed Carrots<br>Garden Salad |
| **TUESDAY** | Cereals<br>Pancakes<br>Apricot Cup<br>Wheat Toast/Jam | Roast Beef Sandwiches<br>Apple or Banana<br>Coleslaw | Lasagna<br>Garlic Bread<br>Tossed Salad<br>Fruit Pizza Dessert |
| **WEDNESDAY** | Cereals<br>French Toast<br>Fruit Salad<br>Wheat Toast/Jam | Rice/Bean Combo<br>Steamed Broccoli<br>Whole Grain Bread<br>Fruit Salad | Baked Fish<br>Red Potatoes<br>Green Beans<br>Apple Pie (low-fat) |
| **THURSDAY** | Poached Eggs<br>Orange Juice<br>Scones<br>Wheat Toast/Jam | Baked Chicken<br>Mashed Potatoes<br>Cranberry Salad<br>Corn | Angel Hair Pasta<br>Garlic Bread<br>Broccoli/Cauliflower<br>Peach Cobbler |

**Source:** Ashley S. Anderson, *Catering for Large Numbers* (Reed International Books, 1995).

breakfast, and/or a beverage. At least three vegetable servings a day should also be part of the menu.

## Cycle Menus

Creating several weekly menus in a row sets up a menu cycle (Figure 6-5). This is a common practice for food service institutions, especially schools. Three-week cycles improve cost control, but four-week cycles are less monotonous, and longer cycles are preferred for people who are unable to eat elsewhere, such as those residing in nursing homes or other long-term care institutions. After deciding on the number of weeks in the cycle menu, the contents of each week are then planned.

Planning a weekly menu cycle is usually the responsibility of the food service establishment management team, which attempts to balance numerous factors, such as those itemized in the checklist in Figure 6-6.

## Planning Healthful Meals

The caloric and nutritive value of meals is another responsibility of the food service manager or director. Fortunately, the trend toward healthful eating is spurred on by numerous chefs publishing culinary books on how to prepare healthful foods (4). A new generation of student chefs is entering the culinary world with many believing that nutrition is an important part of the meal (13).

### ? How & Why?

*Why are obesity rates increasing in the United States if fat intake has decreased?*

Despite the successful decrease in percentage of calorie intake from fat, Americans continue to increase their overall consumption of calories (6). Excess calorie intake from any food source results in weight gain.

The U.S. government also encourages citizens to learn about calories and improve their diets through the USDA's Center for Nutrition Policy and Promotion. The center offers the MyPyramid Tracker at www.mypyramid.gov so that people can use it to analyze their own diets. This diet self-assessment tool allows users to input their daily food intake and receive an evaluation of their overall diet. Continually improving implementation of the Dietary Guidelines will benefit the nutrient intake of North Americans and possibly reduce the dietary risk factors for degenerative diseases. These guidelines assist the more than half of all shoppers motivated by health reasons to make major modifications in their diets (5). The "Calorie Control" section that follows explains

**FIGURE 6-6**    Menu cycle evaluation checklist.

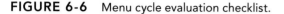

❑ **Clientele**
   Age
   Religion
   Cultural preferences
   Regional differences

❑ **Cost**

❑ **Taste**
   Does the entrée selection include meals that taste better than the
   competition?

❑ **Holiday meals**

❑ **Seasonal availability** (fruits and seafoods)

❑ **Nutrition guidelines**
   Risk factors for diseases
   Fat — 20-35% calories
   Complex carbohydrates — 45-65% calories
   Cholesterol — less than 300 mg/day
   Fiber — at least 25 g/day for women, 38 g/day for men
   Vitamins — A- and C-rich vegetables/fruits at least 4x/week
   Minerals — avoid excess sodium
   National Cancer Institute recommendations
   Exchange list
   Dietary guidelines

❑ **Appealing menu items**
   Flavor/color/texture/shapes (diced/strips/chopped)/temperature variation
   Type of preparation (fried/baked/broiled/sauced/plain)
   Are records of consumption/popularity incorporated?
   Garnishes

❑ **Equipment use balanced**

❑ **Workload/schedules balanced** (broiling, frying, microwave, oven, etc.)

❑ **Cycle/day sequence**
   Is the end of the cycle different from the beginning?
   Are the day's options for breakfast/lunch/dinner different?
   Is any one item repeated too frequently in the cycle?

❑ **Descriptive menu**
   Steak — Sizzling Swiss Steak
   Peas — Buttered Peas and Mushrooms
   Potatoes — Boiled New Potatoes
   Salad — Fresh Garden Salad
   Brownies — Chewy Fudge Brownies

**Food cost** The cost of foods purchased, often expressed as a percentage obtained by dividing the raw food cost by the menu price.

**Forecast** A predicted amount of food that will be needed for a food service operation within a given time period.

**Specifications** Descriptive information used in food purchasing that defines the minimum and maximum levels of acceptable quality or quantity (i.e., U.S. grade, weight, size, fresh or frozen).

how to ensure that meals contain the recommended amount of calories and nutrients.

# PURCHASING

A third aspect of meal planning is deciding how much of each food to buy. Budget limitations determine both the types and amounts of food to be purchased. A further consideration in food service establishments is that the **food cost** accounts for about half of all costs, with the majority of the other half

incurred by labor (labor cost). These two costs are primary concerns to any food service manager who knows that the bottom line is of paramount importance (9).

## Buyers

Larger food service operations usually have a buyer, a purchasing department, or a cooperative arrangement with other institutions to purchase foods according to the **forecasted** menu requirements. The buyer or purchasing department determines food needs, selects vendors, bargains for price, and negotiates contracts.

Food service purchasing may be formal or informal. Informal purchasing, or open-market buying, consists of ordering food supplies from vendors on a daily, weekly, or monthly basis. Formal purchasing, or competitive-bid buying, occurs when the buyer sends vendors an invitation to quote prices on a needed food item (14). **Specifications** describe in detail the food items to be purchased and may be developed by either the buyer or the seller (12). Deadline dates are given in formal purchasing, and bids are placed in a sealed envelope that is not opened until all the qualified vendors' bids have been submitted. The lowest bid is awarded the purchasing contract.

## Food Stores and Vendors/Suppliers

The cost of anything, including food, depends in part on where it is purchased. Understanding the differences among the types of retail and wholesale food supply sources allows buyers to select the ones that will give them the most for their money. The variety of food stores available to consumers includes supermarkets; warehouse stores; co-ops; farmers' markets; and convenience, specialty, and health food stores. Food service establishments rely on large food distribution centers to obtain their supplies.

### Supermarkets

A century or so ago, consumers went to the local grocery store and gave a storekeeper or clerk a list of what they wished to purchase. Then they waited while all the products on the list were collected for them. Although this was an accepted part of life in the community, it could be tedious and time consuming, and it prevented customers from browsing and selecting items at their leisure.

In the early 1900s, large city grocery stores began allowing retailers of individual products to sell from booths inside the stores. This opened up the market for different kinds of foods, increased consumer choices, and made shopping faster and more convenient. Eventually, this arrangement developed into the modern supermarket, where the major departments include meats, produce, dairy, bakery, frozen, canned, and otherwise processed foods, as well as nonfood items such as cleaning, beauty, and even car care supplies.

The easy availability of items is a major factor in consumers' selection of a supermarket. A marketing company that polled consumers nationwide, however, found that the most important consideration in this selection was the cleanliness of the store, followed by the convenience of its location, the courtesy of its clerks, its prices, and speedy checkout service. Less important were attractive displays, baggers, weekly specials, and store coupons. Those polled also indicated that they would appreciate a checkout lane for those with a quantity of groceries in between that accepted in the express and regular lanes.

### Warehouse Stores

Although supermarkets are undoubtedly the most popular avenue for purchasing food, there are other options. Warehouse stores are less expensive than supermarkets because they offer the basic foods with little glitz. Food is often found on the shelves in the original shipping containers, and shoppers may find themselves bagging their own purchases.

### Co-ops

The food cooperative (co-op) is a membership arrangement that cuts out the middle, retail level by purchasing foods in bulk at wholesale prices to sell to members and, in some instances, the public. Any profits are divided among the co-op members. Some co-ops expect the members to put in several

hours per week helping with the operation of the co-op, whereas others hire the necessary help. Co-ops have some disadvantages. They tend to offer limited choices, their management tends to be top-heavy and suffer from inexperience, and they are unable to offer specials that can compete with supermarkets.

### Smaller Outlets

The smaller food outlets include convenience, specialty, and health food stores, and farmers' markets. Convenience stores are a mini-version of the supermarket, with easily accessible foods and fast service being the keys to their success. They are the closest thing to the old-fashioned corner grocery store, but their prices are considerably higher and they carry only the fastest moving items. Specialty stores include bakeries, delicatessens, butcher shops, and ethnic food, cookie, candy, and ice cream stores. Although specialty stores are usually more expensive, they offer unique items that may not be found at the supermarket. Health or natural food stores offer a wide selection of foods, many of which have been produced without chemical pesticides, herbicides, fertilizers, hormones, or antibiotics. Bulk items, herbs, fermented milks, soy milk, food supplements, and natural cosmetics may be bought in such stores. Finally, farmers' markets and roadside stands offer fresh produce straight from the grower's fields.

### Food Service Vendors

Food service distribution centers and vendors supply the food for food service establishments. These large food warehouses obtain food directly from the food companies and deliver it to various private and public food service organizations. Vendors, the purveyors or food suppliers, usually specialize in a given product or category of products, such as produce, meat, or dairy.

# Keeping Food Costs Down

Although the budget is certainly not the only consideration in making a purchase, it is a vital factor, and there are

several methods for keeping food costs down.

### Meats

The biggest expense in the food budget is meat. A money-saving and nutrition-conscious step would be to reduce daily meat intake to no more than 5 ounces per person. A 12-ounce steak is not the most healthy or economical serving, and cutting back on meat is a major move toward saving money. The less tender cuts of meat are just as nutritious, often less fatty, and less expensive than are tender cuts. In addition, it is generally more economical to buy a large piece of meat and cut it up than it is to buy meat already cut up.

The many nonmeat substitutes available provide inexpensive protein options. They include dried beans (including soybeans and the tofu made from them), peas, and lentils. These legumes are high in complex carbohydrates and fiber, and the best source of plant protein. Eggs are another nutritious and inexpensive protein source.

### Fish

Frozen or canned fish is often less expensive than fresh fish. Lobster, crab, and jumbo shrimp are usually more costly than other protein sources, so they are best saved for special occasions.

### Dairy

The least expensive form of milk is nonfat dried milk. If the taste and texture are unacceptable, it can be mixed with fluid milk, or a teaspoon of vanilla flavoring can be added for each gallon of reconstituted nonfat dried milk. Cheese varies widely in cost, with presliced or shredded cheeses tending to be more expensive than those sold in block form.

### Bread/Grain

Creating a diet based on pasta and grains subjects the budget to automatic belt-tightening. The more processing that is involved in a product, the higher the cost will be. Prepared baked goods such as cakes and cookies are more expensive than those made from scratch. Ready-to-eat cereals cost quite a bit more than uncooked cereals and grains.

# ✸ CALORIE CONTROL

## Balanced Calorie and Nutrient Intake

### Calorie Sources

A growing national concern among North Americans is the increase in the obesity rate and the resulting health problems. Excess calories are a major contributor to obesity, and it would help many "calorie challenged" consumers to learn (1) how many calories they actually need (see Chapter 1 Calorie Control) and (2) how many calories are found in foods and beverages. Several methods exist to help consumers regulate caloric intakes: Dietary Guidelines (see Chapter 1), MyPyramid (see Chapter 1), and the Exchange Lists. The first two were already discussed and the Exchange Lists are found in a booklet entitled *Choose Your Foods: Exchange Lists for Diabetes* and jointly published by the American Diabetes Association and the American Dietetic Association (see website section). Although developed to help clients with diabetes manage blood glucose by keeping carbohydrate intake consistent, they can also be used by anyone wishing to lose weight. Once the daily caloric level is known, a certain number of servings from each food list are selected to create a healthful meal plan. The person can then use the booklet to determine appropriate portion sizes of each type of food so they stay within their daily calorie goal. Using the Exchange Lists method is a good way for people to (1) realize how many grams of carbohydrate, protein, and fat are found in foods and (2) control their food intake based on their personal daily calorie recommendations.

### Planning Healthful Meals

A dietary counseling session with a registered dietitian will provide the actual amounts from each food list recommended depending on that person's gender, age, activity level, and desired weight. According to *Choose Your Foods: Exchange Lists for Diabetes,* a healthy daily meal plan starts with at least 6 servings of starch, 2 servings of fruit, 2 servings of milk, 2–3 servings of nonstarchy vegetables, and 6 ounces of meat/substitutes. These add up to approximately 1,325 calories, none of which are derived from fat (45 calories for each teaspoon), sweets (includes soft drinks or sodas), or alcohol. Considering these latter sources of calories and the fact that many people still consume more servings than recommended, it's easy to see why so many North Americans are overweight or obese.

### Meal Plans

The next step is to convert the number of servings into a dietary meal plan, which can be (1) pre-planned or (2) customized. Figure 6-7 shows how to choose a more healthful diet based on a given number of daily exchanges selected from a pre-planned list of foods. Diet B based on the Exchange Lists provides more food volume, but delivers at least 2,000 fewer calories (kcal) than the fast-food diet. Large portions (increasing your number of exchanges) or sugary and/or fatty food choices make it easy to go over the daily calorie cap, and an excess of only 500 calories a day leads to a weekly gain of 1 pound (500 calories × 7 days = 3,500 calories). A customized meal plan can be created by an individual using her/his daily calorie cap, the number of exchanges from each food list, and the booklet of Exchange Lists showing the serving sizes for each food (*Choose Your Foods: Exchange Lists for Diabetes*).

### Healthful Portions

Another important part of calorie control is overcoming portion distortion. A standard "serving size" is not equal to what is served on the plate. In addition, a serving size is not equal to an "exchange." When using the Dietary Guidelines, MyPyramid, the Exchange Lists, or any other means of limiting food intake, it's important to realize how servings are defined (because they are not always the same) and how big (or small) these servings really are. Figure 6-8 provides a practical visual translation of how much ½ cup, 1 cup, and 1 or 3 ounces of anything really adds up to, and it's not very much. A 12-ounce glass is really 1 ½ cups. A 12-ounce steak is double the 6 ounces of meat recommended each day—3 ounces is equivalent to a deck of cards. A 12-ounce regular soft drink (soda) often delivers approximately 150 calories. Excessive "serving sizes" require that people downsize their ideas of how much food they need, and if they are overweight or obese, it's often less than what they are currently eating.

**FIGURE 6-7**    Choosing a Healthier Diet

| Diet A | | | Diet B ("B" for "better") | | |
|---|---|---|---|---|---|
| **Breakfast** | | | **Breakfast** | | **Exchanges (for Diet B)** |
| 1 | | donut | 1 C | bran cereal | 2 starch |
| 1 C | | coffee | 1 C | 2% milk | 1 milk |
| 1 tbsp | | cream | 1 C | cantaloupe | 1 fruit |
| **Lunch** | | | **Lunch** | | |
| 1 | | burger (deluxe) | 3 oz | baked chicken | 3 meat |
| 1 | | fries (medium) | ½ large | potato with skin | 2 starch |
| 1 | | milk shake (medium) | ½ C | carrots | 1 vegetable |
| | | | 1 C | spinach | 1 vegetable |
| | | | 1 tsp | margarine | 1 fat |
| **Dinner** | | | **Dinner** | | |
| 1 | | fried chicken breast | 3 oz | round steak | 3 meat |
| 1 C | | coleslaw | ½ C | broccoli | 1 vegetable |
| 20 | | potato chips | 2/3 C | brown rice | 2 starch |
| 1 | | pickle | 1 tsp | margarine | 1 fat |
| **Snack** | | | **Snack** | | |
| 1 C | | ice cream | 1 C | frozen yogurt | 1 milk |

*(continued)*

**Nutrient Analysis[1]**

| | Diet A | Diet B |
|---|---|---|
| Calories | 3,350 (high) | 1,300 |
| % Carbohydrate | 39% (low) | 55% |
| % Fat | 49% (high) | 22% |
| % Protein | 11% | 23% |
| Protein (g) | 95 | 85 |
| Fat (g) | 183 (high) | 34 |
| Fiber (g) | 16 (low) | 33 |
| Cholesterol (mg) | 502 (high) | 180 |
| Vitamins | 8+ (low) | 2 (low) |
| Minerals | 4+ (low) | within normal limits |
| Sodium (mg) | 5,150 (high) | 1,012 |

[1] Analyzed using FoodProSQL (Esha Research; www.esha.com)
© 2010 Amy Brown.

## Nutrients

After establishing a healthful calorie balance goal coupled with realistic portion sizes, the next step is to make sure the diet is balanced in terms of the nutrients it provides: macronutrients (carbohydrates, fat, and protein), cholesterol, fiber, and micronutrients (vitamins and minerals).

## Macronutrients

The overall percentage of calories (kcal) from carbohydrates should be at least 45–65%, with fat providing 20–35%, and protein approximately 10–35% (7).

**Carbohydrates**  An Institute of Medicine publication recommends an intake of at least 130 grams of carbohydrate daily—this is the average minimum amount of glucose the brain needs. Most people exceed this amount; the median intake averages 180–230 grams per day for women and 220–330 grams a day for men, and falls within the recommended range of 45–65% of energy (7).

**Carbohydrate Sources**  About 15 grams of carbohydrates are found in an Exchange List serving of starches (bread, cereal, grain, and some starchy vegetables), fruit, or sweets. A cup of milk contains almost the same amount of carbohydrate (12 grams); however, it is in the form of lactose. Nonstarchy vegetables also contain some carbohydrates (5 grams). Meat, fats (including nuts), and water contain very little to no carbohydrate.

**FIGURE 6-8**  Size-Up Your Food.

**Three Methods for Measuring Portion Sizes**
1. Hand
2. Common Items
3. Measuring Utensils (See measuring ingredients in Chapter 5.)

**1. Using "Hand" Measurements (portion sizes based on using different parts of your hand)**

| | | |
|---|---|---|
| Fist | = | 1 cup (8 ounces) |
| | | Used to measure beverages, grains, pasta, cereal, mashed potatoes, vegetables, ice cream, yogurt, etc. |
| ½ Fist | = | ½ cup (4 ounces) |
| Palm | = | 3 ounces (oz) |
| | | Used to measure an average meat serving |
| Closed Palm | = | 1 ounce (oz) held within space under closed fingers (fingers flat, not curled as in fist) |
| Whole Thumb | = | 1 ounce (oz) |
| | | Used to measure cheese, nuts |
| Thumb Joint | = | 1 teaspoon (tsp) |
| | | Used to measure fats |
| 3 Thumb Joints | = | 1 tablespoon (T) or ½ ounce (oz) |

**2. Using "Common Items" Measurements**

| | | |
|---|---|---|
| Soda Can | = | 12 ounces (oz) |
| Baseball (not softball) | = | Average fruit or 1 cup of anything |
| Yogurt Container | = | 6 ounces (oz) |
| Tennis Ball | = | Medium/small fruit |
| | | ½ cup of anything (ice cream, yogurt, pudding, vegetable) |
| Card Deck | = | 3 ounces (oz) (meat servings such as chicken breast or beef patty) |
| Checkbook | = | 3 ounces (oz) (grilled fish, pasta) |
| Ice Cream Scoop | = | 1/3 cup if rounded (rice serving) |
| Golf Ball | = | ¼ cup (2 ounces) |

© 2010 Amy Brown

Complex carbohydrates are preferred and can be consumed by choosing whole-grain breads, cereals, grains, pasta, legumes (dried beans, peas, and lentils), and starchy vegetables such as peas, potatoes, and corn.

**Recommended Fiber Intake**   Fiber is a carbohydrate that is not digested or absorbed, so it contributes few if any calories. The Institute of Medicine recommends that healthy adults consume:

- 25 grams of daily fiber for women
- 38 grams of daily fiber for men

Fiber recommendations for children and elderly are 14 grams of fiber for every 1000 calories (kcal) consumed. The average United States intake of only 15 grams of dietary fiber per day falls short of these goals (7). Good food sources of fiber are provided in Chapters 13, 14, 16, and 20 (vegetables, fruits, grains, and bread products, respectively).

One of the ways to increase both complex carbohydrates and fiber is by offering a vegetarian menu option, which, as restaurateurs are now realizing, is no longer just a passing fad (11).

**Fat**   Foods high in total fat, saturated fat, trans fat, and/or cholesterol should be minimized. The trend toward healthier eating has even led to the suggestion of a national "fat tax" aimed at fast-food restaurants delivering super-sized, high-fat meals (10).

**Fat Sources**   The main sources of fat in the diet, according to the Exchange Lists, are primarily animal foods such as whole milk (8 grams/cup), high-fat meat (8 grams/ounce), and fat itself (5 grams/teaspoon). Plant sources of fat include coconut, avocado, and vegetable oils. Saturated fat is derived primarily from animal sources—milk and meat food products. The only plant products high in saturated fat are coconut and oils derived from coconut or palm.

**Cholesterol**   Although cholesterol is technically not a fat, the American Heart Association recommends limiting daily dietary intake to 300 mg for adults without heart disease (200 mg with heart disease) (1). Sources high in dietary cholesterol include eggs (213 mg/large egg yolk), shrimp, caviar, and organ meats such as liver, heart, kidney, and sweet breads (brains). (Note: Dietary cholesterol is different from blood cholesterol, which is measured in mg/dL. Only animal foods can contain cholesterol, which is made by a liver. Plants do not have livers, so even "oily" foods such as corn oil or peanut butter contain zero cholesterol.)

**Protein Sources**   The primary sources of protein in the diet are milk/dairy foods (8 grams per cup), meat and meat substitutes (7 grams per ounce), and eggs (7 grams each). Small amounts of protein are available from starch (3 grams per serving) and vegetables (2 grams per serving). Little to no protein is found in fruits or fats.

**Recommended Protein Intake**   An Institute of Medicine report states 0.8 grams/day of protein are recommended for every kilogram body weight (or 0.36 grams for every pound) in healthy adults. In practical terms, that equals about 46 grams a day for women (19+ years) and 56 grams for men (19+ years) (7).

### Micronutrients

Vitamins and minerals are obtained from various foods, with the most common sources (especially for vitamins A and C) being certain vegetables and fruits (Chapters 13 and 14). Approximately 80% of dietary calcium is derived from dairy ingredients (Chapter 10). Iron, zinc, and many B vitamins are found primarily in meats (Chapter 7).

### Nutrient Analysis

Actual caloric and nutrient intake is best determined through a nutrient analysis using one of the many professional or consumer computer software programs available. After the person types in a profile (gender, age, height, desired weight, activity level) along with the type and amounts of foods consumed in a day(s), the program generates a nutrient analysis providing a summary of total calories and amounts (grams, milligrams) of macronutrients, cholesterol, fiber, and micronutrients. These are often presented as a percentage of the government standard, which in the United States is the Dietary Reference Intakes for calories and nutrients based on a person's gender, age, pregnancy, or lactation (7).

Individual foods can be analyzed for free using the USDA's "National Nutrient Database for Standard Reference," and diets can be evaluated using one of the many free online tools (see websites). Both consumer and professional nutrient analysis software applications are available (2).

Cereals offered in mini-packages or single portions also come at a premium price. Seasoned grain and pasta mixtures cost more than plain pasta and grains to which seasonings are added during cooking. A wide variety of grains are sold in bulk at very reasonable prices.

### Fruits and Vegetables

Savings can be achieved in the purchase of fruits and vegetables by determining the cost of the fresh form against the processed versions. Seasonal availability, brand, grade, and added ingredients all factor into the equation when comparing the price of fresh, dried, canned, or frozen fruits and vegetables.

### Price Comparisons

Comparing prices is accomplished by calculating the cost per serving. This is easily done when the price per unit (ounce, pound, count, etc.) is given by the supplier. It pays to check this price and not be deceived by a product's packaging, shape, or size.

Prices differ not only among brands, but among forms as well—fresh, dried, canned, or frozen. Convenience foods are almost guaranteed to be more costly, as shown in Figure 6-9. Prices escalate whenever produce has been trimmed or cut, making it more expensive to purchase shredded cabbage, diced carrots, watermelon cut in portions, or strawberries and pineapple sold ready to be served in plastic containers. Pound for pound, frozen vegetables and fruits are often more expensive than fresh unless they are out of season.

**FIGURE 6-9** Price comparison.

Instant oatmeal:      .40 per oz
Old Fashioned Oatmeal:    .17 per oz

## Reading Label Product Codes

Familiarity with the various types of dating on some packaged items helps consumers select the freshest available products (Figure 6-10). Code dates (not shown) are useful in the event of a recall because they identify the manufacturer and/or packer of the food.

## Reducing Waste Saves Costs

Careful purchasing avoids waste that can commonly occur in the following areas (3):

- Overpurchasing perishable produce or other foods. Fresh produce losses can be avoided by following the Three-Day Rule: never buy more than what can be consumed in three days.
- Losses resulting from food preparation (peeling, coring, trimming of fat, deboning).
- Losses from shrinkage during cooking.
- Losses from *plate waste,* that is, food left on the plate because of too-large portion sizes or poor food quality.

### As Purchased vs. Edible Portion

There can be a great disparity between the amount of food purchased and what ends up on the table. These different

**FIGURE 6-10** Dates on labels and what they mean.

**Freshness or quality assurance date**—The last day the product will be of optimum quality. Often preceded by "best when used by."

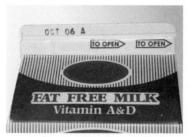

**Pull date**—The last day a store will sell an item, even though the food may be safe for consumption for a little while longer. Dairy and other perishable and semi-perishable items have a pull date that indicates the last day a store should sell the item. Such items are often priced very low and are a good buy if used within a short period of time.

**Expiration date**—The last day a food should be consumed. Certain products that will "expire" such as baking powders, yeast packages, and refrigerated doughs, need to show expiration dates to let consumers know whether or not they are still capable of making baked products rise.

**Pack date**—The date the food was packed at the processing plant. Canned, bottled, or frozen goods have pack dates that inform consumers how old the food is when purchased. It is often used by stores, which need to know when to rotate stock.

quantities are termed **as purchased (AP)** and **edible portion (EP).** Extra quantities of food must sometimes be purchased to make up for losses incurred during preparation, especially when buying meats, fruits, and vegetables.

## Percentage Yield

In terms of waste resulting from pre-preparation, the **percentage yield** gives an estimate of how much edible food will remain after peeling and trimming (Table 6-2). The two steps to determining how much food to purchase according to percentage yield are:

1. Determine the edible portion by multiplying the number of servings by the serving size.
2. Determine the amount to purchase by dividing the edible portion by the percentage yield for that particular food (Table 6-3).

For example, a meal of boneless, skinless chicken breasts for five people, each having a 3-ounce serving, calls for a total edible portion of 15 ounces (step 1). The percentage yield for chicken breast with ribs and skin is 66%, so 15 ounces divided by 0.66 results in 22.7, or approximately 23, ounces (step 2). Thus, for each guest to receive 3 ounces of edible chicken breast, a total of 23 ounces of chicken breast with ribs and skin will need to be purchased.

## Portion Control

One of the most important aspects of controlling the food budget of a food service organization is portion control. Food cost is a major expense in running a food service establishment, so it is crucial to adhere to set serving sizes. If 300 people are served 4 ounces of roast beef instead of the planned 3 ounces, this results in the consumption of almost 19 pounds

---

**As purchased (AP)** The total amount of food purchased prior to any preparation.

**Edible portion (EP)** Food in its raw state, minus that which is discarded—bones, fat, skins, and/or seeds.

**Percentage yield** The ratio of edible food to food as purchased. Edible food = food as purchased minus inedible or wasted food.

---

**TABLE 6-2**  Percentage Yield: Approximate Edible Portion (EP) Yield per Pound of Selected Foods as Purchased (AP)

| Food Items | % Yield Pounds of EP | Food Items | % Yield Pounds of EP |
|---|---|---|---|
| *Meat, Poultry, Fish (Cooked)* | | *Vegetables (Cooked)* | |
| Beef, ground (no more than 30% fat) | 40 | Beans, green | 88 |
| Beef, chuck roast (boneless) | 60 | Carrots | 60 |
| Beef, round (boneless)* | 60 | Corn on the cob | 55 |
| Beef, stew meat* | 55 | Potato, baked with skin | 80 |
| Lamb, leg (boneless)* | 60 | Sweet potato, baked with skin | 60 |
| Lamb, stew* | 65 | *Vegetables (Raw)* | |
| Pork, loin chops* | 40 | Asparagus | 55 |
| Pork, loin roast (boneless)* | 55 | Beets | 45 |
| Pork, spareribs | 45 | Broccoli | 80 |
| Ham (boneless)* | 65 | Cabbage | 90 |
| Chicken, breast (rib and skin) | 66 | Carrots | 70 |
| Chicken, drumstick (skin) | 50 | Cauliflower | 55 |
| Turkey, whole (skin) | 55 | Celery | 75 |
| Turkey, fr. rolls | 65 | Leeks | 50 |
| Fish, fr. portions (raw breaded)† | 60 | Lettuce, head | 75 |
| Fish, fr. sticks (raw breaded)† | 60 | Mushrooms | 95 |
| *Fruits (Raw)* | | Onions | 90 |
| Apples | 75 | Parsley | 85 |
| Apricots | 94 | Peppers | 82 |
| Avocados | 75 | Radishes | 90 |
| Berries | 95 | Spinach | 80 |
| Bananas, with peel | 65 | Squash (summer) | 90 |
| Cantaloupe | 50 | Squash (winter) | 75 |
| Coconut | 50 | Tomato | 99 |
| Grapefruit | 50 | Turnips | 80 |
| Grapes | 90 | | |
| Kiwi | 80 | | |
| Lemons | 45 | | |
| Mangoes | 75 | | |
| Oranges | 65 | | |
| Papayas | 65 | | |
| Peaches | 75 | | |
| Pears, pared | 80 | | |
| Pineapple | 55 | | |
| Plums | 95 | | |
| Watermelon | 55 | | |

*Lean Meat
†75% fish

---

of roast beef beyond what was calculated into the budget. It may also leave customer number 301 without any meat and dissatisfied. The three ways of measuring portions are by:

1. Weight (ounce, pound) or volume (cup, pint)
2. Number (five olives, one ear of corn, two dinner rolls)
3. Size (1.8-inch slice of cake, 2 × 2-inch brownie)

Careless measuring is ultimately reflected in the final yield of the food for the diner and the cost to the establishment. Portion guides are available to help keep food costs under control (Table 6-3).

# TIME MANAGEMENT

Quality meals rely on timing. Food is usually best when it is prepared as soon as possible after purchase, and served immediately after preparation. Different foods are prepared at different rates, so coordination is key to an organization's providing timely meal service.

## Estimating Time

The preparer gains control by beginning with a realistic assessment of available time and energy. The stress of planning a

**TABLE 6-3**   Portion Guide, Common Serving Sizes*

| Breakfast | | | |
|---|---|---|---|
| Eggs | 2–4 oz (1–2 eggs) | 50–125 g | |
| Meat | 2–4 oz | 50–125 g | |
| Fruit | ½ C | 125 mL | |
| Cereal | 3/4 C | 175 mL | |
| Juice | ½–¾ C | 125–175 mL | |
| Bread | 1–2 slices (1–2 oz) | 30–60 g | |

| Lunch | | |
|---|---|---|
| Soup | 4–6 oz | 125–175 mL |
| Salad | 4–8 oz | 125–250 g |
| Salad dressing | 1–2 oz | 25–50 mL |
| Main dish | 4–6 oz | 125–175 g |
| Starch | 2–3 oz | 50–100 g |
| Vegetable | 2–3 oz | 50–100 g |
| Sauce | 1–2 oz | 25–50 mL |
| Bread | 1–2 oz | 30–60 g |
| Dessert | 2–4 oz | 50–125 g |

| Dinner | | |
|---|---|---|
| Soup | 6–8 oz | 175–250 mL |
| Salad | 4–8 oz | 125–250 g |
| Salad dressing | 1–2 oz | 25–50 mL |
| Main dish | 6–8 oz | 175–250 g |
| Sauce | 1–2 oz | 25–50 mL |
| Starch | 2–3 oz | 50–75 g |
| Vegetable | 2–3 oz | 50–75 g |
| Bread | 1–2 oz | 30–60 g |
| Dessert | 2–4 oz | 50–125 g |

| Hors d'ouevres and Canapes[†] | |
|---|---|
| Lunch, with meal | 2–4 per person |
| Lunchtime, without meal | 4–6 per person |
| Dinner, with meal | 4–6 per person |
| Before-dinner reception | 6–8 per person |
| After-dinner reception | 4–6 per person |

*Quantities given reflect general practice. Specific needs will vary.

[†]Average size is 1 oz (30 g). Total number reflects combined hot and cold items.

**Source:** Based on Mizer D. A., M. Porter, B. Sonnier, and K. E. Drummond. Food Preparation for the Professional (3rd Ed.), 1999.

meal can be minimized by logging how long it will take for each menu item to be prepared, estimating the time at which the meal will be served, and working backward to determine at what time the preparation should be started in order to serve the meal on time. When the meal is prepared, items are usually prepared in descending order of time required.

## How & Why?

**How many pounds of carrots should you buy if you need 4 ounces per serving for 50 people?**

First multiply 4 ounces per serving by 50 servings, for a total of 200 ounces. You know that carrots are sold by the pound, so you have to convert ounces to pounds. Divide the 200 ounces by 16 ounces per pound to get 12.5 pounds. However, 12.5 pounds of carrots would not be enough because Table 6-3 shows that the percent yield for carrots is 70%. This means that you have to divide 12.5 pounds of carrots by 0.70 to obtain the amount you need to buy, which is 17.86 pounds. A buyer would probably round that off to a 20-pound purchase of carrots because vendors often do not sell by the half pound or even under 5- or 10-pound increments.

# Efficient Meal Preparation

Effective management of time can improve the efficiency of all the steps of meal preparation, which include:

1. Planning the menu
2. Developing a purchase list
3. Purchasing the food
4. Storing the food
5. Planning the order in which the menu items will be prepared
6. Preparing the food
7. Preparing the table
8. Serving
9. Cleaning up

The preparer can increase efficiency through menu planning and wise purchasing as described above, and through recipe consultation.

## Recipes

There are four styles of recipe writing: the descriptive, standard, action, and narrative forms (Figure 6-11). The ingredients in the descriptive method are listed in the sequence in which they are used. This method displays the ingredient, amount, and directions in three columns, which makes it easy to read. The standard recipe style lists all ingredients and amounts with the instructions in numerical order. A modification of that form is the action recipe, which gives the instruction followed by the ingredients for that step only. Probably the most tedious to decipher is the narrative form, which reads like an essay, explaining ingredients, amounts, and preparation methods in text form.

Food service establishments rely on **standardized recipes** that have been tested and adapted for serving a large number of people (48 to 500 servings). Standardized recipes, which frequently follow the descriptive style, record ingredients, proportions, and procedures so the number of servings can easily be increased or decreased. When standardized recipes are stored in a computer, changing the number of servings automatically changes the amount of each ingredient needed. Standardized recipes are repeatedly tested and adapted to suit a particular food service operation.

**Standardized recipe** A food service recipe that is a set of instructions describing how a particular dish is prepared by a specific establishment. It ensures consistent food quality and quantity, the latter of which provides portion/cost control.

**FIGURE 6-11** The four different styles of recipes: Descriptive, standard, action, and narrative.

**DESCRIPTIVE**

**Texas Chocolate Cake**

Desserts: C-33　　　　　　　　　Oven Temperature: 350°F
Portion: 16 servings　　　　　　　Time: 20 minutes

| Ingredients | Amount | Procedure |
|---|---|---|
| All-purpose flour | 1 C | Sift flour and sugar together into large mixing bowl. |
| Sugar | 1 C | |
| Water | 1 C | Melt margarine in sauce pan and add water and cocoa. Bring it to a boil while stirring constantly. Take off heat and pour into flour/sugar mixture. |
| Margarine | 2 sticks | |
| Cocoa | 4 tbs | |
| Eggs | 2 | In a separate bowl, beat eggs slightly and add to milk, baking soda, and flavorings. Add gradually to creamed mixture and mix with spoon until just blended smooth. Pour batter into a 13 × 9 greased baking pan. Immediately place on center rack of preheated oven. Bake until toothpick comes out clean. Cool briefly in pan on rack. |
| Sour milk* | ½ C | |
| Baking soda | 1 tsp | |
| Cinnamon | 1 tsp | |
| Vanilla | 1 tsp | |

* or ½ C milk + 1 tbs lemon juice or vinegar

**STANDARD**

**Texas Chocolate Cake**

Preparation Time: 15 minutes
Cooking Time:　　 20 minutes
Yield:　　　　　　 16 servings

Ingredients:　　1 C All-purpose flour
　　　　　　　　1 C Sugar
　　　　　　　　1 C Water
　　　　　　　　2 Sticks Margarine
　　　　　　　　4 tbs Cocoa
　　　　　　　　2 Eggs
　　　　　　　　½ C Sour milk
　　　　　　　　or ½ C milk + 1 tbs lemon juice or vinegar
　　　　　　　　1 tsp Baking soda
　　　　　　　　1 tsp Cinnamon
　　　　　　　　1 tsp Vanilla

Directions:　　1) Preheat oven to 350°F and grease sides and bottom of 13 × 9 pan.
　　　　　　　　2) Sift together flour and sugar into large mixing bowl.
　　　　　　　　3) Melt margarine in sauce pan and add water and cocoa. Bring it to a boil while stirring constantly. Take off heat and pour into flour/sugar mixture.
　　　　　　　　4) In a separate bowl, beat eggs slightly and add to milk, baking soda, and flavorings. Add gradually to creamed mixture with spoon until just blended smooth.
　　　　　　　　5) Pour batter into greased baking pan. Immediately place on center rack of preheated oven.
　　　　　　　　6) Bake until toothpick comes out clean. Cool briefly in pan on rack.

**ACTION**

**Texas Chocolate Cake**

Preheat oven to 350°F and grease sides and bottom of 13 × 9 pan.

Sift together flour and sugar into large mixing bowl.
　　　　1 C All-purpose flour
　　　　1 C Sugar

Melt margarine in sauce pan and add water and cocoa. Bring it to a boil while stirring constantly. Take off heat and pour into flour/sugar mixture.
　　　　2 Sticks Margarine
　　　　1 C Water
　　　　4 tbs Cocoa

In a separate bowl, blend ingredients below, add gradually to creamed mixture, and mix with spoon until just blended smooth.
　　　　2 Eggs (slightly beaten)
　　　　½ C Sour milk
　　　　or ½ C milk + 1 tbs lemon juice or vinegar
　　　　1 tsp Baking soda
　　　　1 tsp Cinnamon
　　　　1 tsp Vanilla

Pour batter into greased baking pan. Immediately place on center rack of preheated 350°F oven. Bake until toothpick comes out clean. Cool briefly in pan on rack.

**NARRATIVE**

**Texas Chocolate Cake**

*Cake.* Preheat oven to 350°F and grease sides/bottom of 13 × 9 pan. Sift together 1 C all-purpose flour and 1 C sugar into large mixing bowl. Melt 2 sticks of margarine in sauce pan and add 1 C water and 4 tbs cocoa. Bring to boil, stir stirring constantly, and pour liquid over the flour/sugar mixture. In a separate bowl, blend together 2 eggs (slightly beaten), ½ C sour milk (or ½ C milk + 1 tbs lemon juice or vinegar), 1 tsp baking soda (more soda results in a more cake-like cake, less produces a more brownie-like product), 1 tsp cinnamon, and 1 tsp vanilla. Add this gradually to the creamed mixture and mix with spoon until just blended smooth. Pour batter into greased baking pan. Immediately place on center rack of preheated oven and bake for 20 minutes or until toothpick comes out clean. Cool briefly in pan on rack.

*Icing.* Bring to boil ½ C margarine and 4 tbs cocoa. Stir constantly until mixed and remove from heat. Pour into mixing bowl and add 1 box of powdered sugar, blend with mixer on medium speed. If not blending smoothly, then add 1 tbs of milk at a time until it does. Too much milk makes runny icing, which is corrected by adding more powdered sugar. Cakes are usually cooled completely on racks before frosting however, frosting only 5–10 minutes later makes the cake more moist. Covering the sheet pan with aluminum foil further traps in moisture.

# TYPES OF MEAL SERVICE

There are six basic types of meal service in North America. In descending order of formality they are: Russian, French, English, American, family, and buffet. Not only do the table settings for each differ, but the manner in which the food and beverages are placed on and removed from the table differs as well (8). No matter which type of service is employed, dessert is served only after the table has been cleared of all extraneous items, including salt and pepper shakers, all condiments, and unnecessary flatware.

## Russian Service

The most formal type of meal service is Russian, also known as European, Continental, or formal service. The entire meal is served by well-trained waiters. Normally, the waiting staff serves and clears food items from the left with the left hand, whereas beverages are always served and removed from the right. The guest to the right, or the host or hostess, is served first, with the rest of the diners being served in a counterclockwise direction. Service plates, or place plates, sometimes made of or embossed with silver or gold, are part of each place setting and serve as underliners; the food is never placed directly on them.

The meal is served in courses, starting with the appetizer, and then the soup. Each of these courses has its own underliner plates, which go on top of the place plates. A fish course may follow the soup. Prior to the introduction of the main entrée, a miniature serving of chilled sorbet is provided to clear the taste buds of any lingering flavors. The place plate is then removed and the main entrée served from a platter or on plates placed before guests. Salad is served and consumed before or after the main entrée has been removed. Once the diners have finished their salads, the waiters remove all flatware, tableware, glassware, and condiments and finish clearing the table with a procedure known as **crumbing**. Filling all glasses and/or coffee or tea cups before serving dessert is a common policy, regardless of the type of meal service. After the dessert is finished and removed, a finger bowl containing cool water and, usually, a lemon slice is provided to each guest. The fingertips only are dipped into the water and dried on a napkin.

## French Service

Another very formal type of meal service is French, or cart, service. The food is brought out on a cart (*guerdon*) to the table, where it is cooked or has its cooking completed in a small heater (*rechaud*) by the *chef de rang* (chief or experienced waiter) and *commis de rang* (assistant waiter). The French method is expensive, requires skilled personnel, and results in slower service. It tends to be reserved for elegant French restaurants or those found in Europe.

## English Service

In English service, waiters bring in the various courses, clear the table at the appropriate times, and may take servings dished out by the host and hostess to the individual guests. Frequently, when it is a family or small gathering, the host serves the meat to the guests, who pass their warmed plates to the hostess, who serves the vegetables. The necessary maneuvering of dishes between host or hostess and guests makes this type of service useful for no more than about six or eight people.

## American Service

American service is that in which the meal is placed on the plates in the kitchen and then brought out to the table. This type of service is useful in smaller spaces, allowing for faster service, hotter food, and fewer dishes to wash.

## Family Service

Family service allows the guests to serve themselves from serving platters and bowls brought to the table and passed counterclockwise among the diners. The main meal-serving dishes, condiment containers, and all accompanying flatware, dinnerware, and glassware are removed after completion of the main entrée. Dessert is served at the table by the hostess or brought to the table in individual portions.

## Buffet Service

The buffet service allows guests to walk to a separate buffet table from which they serve themselves. If the group of people is large, it is preferable to set the table up in such a way that guests can pass down both sides simultaneously.

To avoid "the line," people can be divided alphabetically (A–O, P–Z) or by any other method and sent to the buffet in 10-minute intervals.

The sequence of items on a buffet table varies considerably, but it usually starts with the plates and is followed by the vegetables, salad, bread, main entrée, condiments, beverages, flatware, and napkins. Hot items are usually kept warm in a chafing dish with a burner underneath, whereas chilled foods are placed on a bed of ice.

The guests may sit down at a table, hold the plates in their laps, or use collapsible TV trays. Regardless of where they end up, they should never have to set glasses on the floor. Also, unless a table is available, runny menu items such as soups and sauces should not be offered to avoid possible spills and embarrassment. In the same vein, if paper plates are used, they should be extra sturdy and resistant to foods soaking through.

# TABLE SETTINGS

A well-prepared meal deserves to be enhanced by aesthetically pleasing table settings and surroundings. Expensive silver and china do not make up for grease marks on the utensils or an improperly set table. This section focuses on the correct presentation of the cover and linens, flatware, dinnerware, glassware, and accessories.

## Cover and Linens

The table arrangement focuses on a **cover,** or table setting, for each individual. Each place setting should occupy from 20 to 24 inches to give diners elbow room. The table linen can be a tablecloth, place mats, or a combination of the two, along with a napkin. For formal service, the most common choice is

**Crumbing** A ceremonious procedure of Russian service in which a waiter, using a napkin or silver crumber, brushes crumbs off the tablecloth into a small container resembling a tiny dust pan.

**Cover** The table setting, including the place mat, flatware, dishes, and glasses.

a white damask cloth with a felt cloth or other **silence cloth** placed underneath.

When a tablecloth is used, it should be centered on the table with an 8- to 12-inch overhang. Most people do not like to sit at a table at which the tablecloth is any longer than lap length. Place mats, either alone or in addition to a tablecloth, are arranged so that each table setting is clearly distanced from the one(s) next to it. The napkin is placed to the left of the fork with the open edge facing the plate and its open corner at its own lower left. This makes it easy to grasp with the fingers of the right hand so it can be brought across the lap with a single motion. For formal service, the napkins are often folded attractively and placed in the center of the service plate or in the glassware. Napkins may be of linen or paper and vary in size—large for dinner, medium for lunch, and small for tea or cocktails. Paper napkins are acceptable for most meals, but linen or other cloth napkins should be used for more formal occasions.

## Flatware/Dinnerware/ Glassware

**Flatware** has assigned positions on the table setting, depending on the type of meal being served. A standard placement of flatware is shown in Figure 6-12, but most everyday and restaurant meals do not include two sets of forks, knives, and spoons. Most restaurants position their flatware according to the general rules shown in Figure 6-12, but creativity or necessity can influence the final arrangement.

Plates, saucers, bowls, and other dinnerware also have generally assigned positions, as shown in Figure 6-13. Dinnerware does not have to match, but patterns should harmonize. It should also be in balance with the position of the glassware.

## Accessories

Accessory items that can be distributed attractively on the table include the

**FIGURE 6-12**    Standard placement of flatware at the start of service.

Dinner forks                    Dinner knife Teaspoons

**FIGURE 6-13**    Standard placement of dinnerware and glassware.

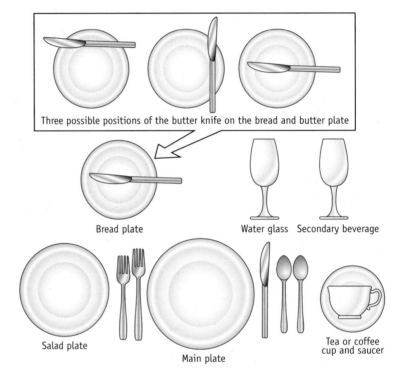

Three possible positions of the butter knife on the bread and butter plate

Bread plate                    Water glass   Secondary beverage

Salad plate            Main plate            Tea or coffee cup and saucer

salt and pepper shakers, sugar bowl with spoon, cream pitcher, butter dish with butter knife, bread baskets, any decorations, and condiments removed from their containers and displayed in attractive serving dishes. Items with the potential to drip are always placed on underliners. Salt and pepper shakers are usually placed near the center of the table, with the salt placed to the right of the pepper. Individual salt and pepper are placed just above or slightly to the left on each cover. Depending on the type of service, serving dishes and serving utensils may or may not be on the table.

## Centerpieces

Centerpieces and other table decorations should be given special attention, with simple elegance the rule, unless the meal has a specific theme that requires something more distinctive. Centerpieces should be in scale with the table, should not be overpowering, and should not keep guests from being able to see each other across the table. Candle flames should be kept below eye level to avoid the problem of people having to bend right or left to talk to someone across the table. Many a host or hostess failing to observe these guidelines ends up removing a centerpiece after guests are seated.

**Silence cloth**  A piece of fabric placed between the table and the tablecloth to protect the table, quiet the placement of dishes and utensils, and keep the tablecloth from slipping.

**Flatware**  Eating and serving utensils (e.g., knives, forks, and spoons).

# PICTORIAL SUMMARY / 6: Meal Management

Planning meals that are both psychologically and physiologically satisfying requires a basic knowledge of food preparation, nutrition, and presentation strategies. Effective meal management, whether for a household, an institution, or a restaurant, involves efficient management of resources such as money, time, and energy.

## FOOD SERVICE ORGANIZATION

Food service management must set goals, determine what is needed to achieve them, and mobilize people toward these goals. Division of labor is central to good organization. Brigade de cuisine is a system of dividing a kitchen into stations supervised by chefs with expertise in specific areas:

- Sauce
- Fish
- Vegetable
- Soup
- Roast
- Pantry
- Pastry
- Relief

## MEAL PLANNING

Planning menus and setting up menu cycles helps to control costs as well as balance nutrition. Some points to consider:

- Calorie and nutrient recommendations: Including correct serving sizes.
- Individual preferences and needs: Based on age and religious, cultural, ethnic, and regional differences.

- Costs: Planning nutritious, flavorful, and appealing meals within the available budget.
- Food preparation methods: Alternating oven-baked, boiled, and fried foods for optimal nutrition and to avoid monotony.
- Seasonal factors: Availability of fresh food products, method of preparation, and temperature at which food is served.

## PURCHASING

Budget limitations determine both the types of food to be purchased and the amounts. Careful control of foods and labor costs is critical in food service establishments and waste must be avoided. Food bills can be reduced through organized purchasing, comparison pricing, and controlling portions.

## TIME MANANGEMENT

Food is usually best when it is prepared as soon as possible after purchase and served immediately after preparation.

Coordinating the preparation of different foods to be served at the same time is crucial, but equally important is an awareness of the time involved in menu planning, purchasing, preparing the table, serving, and cleaning up.

Efficiency is increased with careful menu planning and recipe consultation.

## TYPES OF MEAL SERVICE

The six basic types of meal service differ in the manner in which the table is set and the food served. They include, in descending order of formality,

- Russian: Most formal; entire meal is served by waiters.
- French: Food is served and or prepared from a cart brought to the table by specially trained chef/staff.
- English: Host participates in serving guests; waiters assist.
- American: Served on plates in kitchen and brought to table.
- Family: Guests serve themselves from platters brought to table and passed counterclockwise among the diners.
- Buffet: Guests serve themselves from a central buffet table.

## TABLE SETTINGS

A cover or place setting should be laid for each individual. There are specific customs for the placement of linens (tablecloth and/or place mat, and napkin), flatware, dinnerware, glassware, and accessories such as decorations and condiments.

**Three possible positions of the butter knife on the bread and butter plate.**

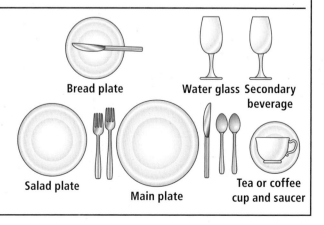

Bread plate

Water glass  Secondary beverage

Salad plate

Main plate

Tea or coffee cup and saucer

# CHAPTER REVIEW AND EXAM PREP

## Multiple Choice*

1. Both one ounce of meat and one egg contain _____ grams of protein.
   a. 3
   b. 5
   c. 7
   d. 9

2. One teaspoon of fat contains _____ grams of fat and 45 calories.
   a. 5
   b. 15
   c. 25
   d. 30

3. AP stands for _____ and EP stands for _____ .
   a. as planned, eating plan
   b. amount purchased, edible product
   c. as purchased, edible portion
   d. as planned, eating production

4. The last day a food product will be of optimal quality is referred to as the _____ .
   a. freshness/quality assurance date
   b. pull date
   c. expiration date
   d. code date

5. The table setting, including the place mat, flatware, dishes, and glasses, is referred to as the _____ .
   a. placement
   b. coding
   c. coverage
   d. cover

*See p. AK-1 for answers to multiple choice questions.

6. The second highest chef position in the kitchen after the executive chef is _____ chef.
   a. sauce
   b. pantry
   c. pastry
   d. *sous*

7. Which of the following basic types of meal service is the most formal?
   a. French
   b. Russian
   c. Family
   d. American

## Short Answer/Essay

1. Describe the Exchange Lists and list how many grams of carbohydrate, fat, and protein are in each Food List. Include the primary food sources for carbohydrate, protein, and fat.
2. Explain the basic serving sizes and calories provided by each Food List and why it is easy to go over the daily calorie limit.
3. Discuss the difference between AP and EP.
4. How many ounces of roast beef do you need to purchase if the roast weighs 5 pounds, the percentage yield is 60%, and there are 100 people to serve?
5. Define each of the following label dates: freshness/quality assurance date, pull date, expiration date, pack date, and code date.
6. Define what standardized recipes are and explain their function.
7. Briefly outline the *brigade de cuisine* of kitchen organization.
8. Diagram and label a standard lunch cover, including flatware, dinnerware, and glassware.
9. Diagram three positions for the butter knife on the bread and butter plate.
10. Briefly describe the most formal type of meal service and how food is served in French service.

# REFERENCES

1. American Heart Association. *AHA Scientific Position. Cholesterol.* http://books.nap.edu/catalog.php?record_id=10490. Accessed 06/07/09.

2. Arnson D. Before you buy, get the lowdown on nutrition software. *ADA Times* 5(4)1:22, 2008.

3. Ashley S, and S Anderson. *Catering for Large Numbers.* Reed International Books, 1993.

4. Culinary Institute of America. *Techniques of Healthy Cooking.* The Culinary Institute of America (CIA), 2007.

5. Harrison GG. Reducing dietary fat: Putting theory into practice—conference summary. *Journal of the American Dietetic Association* 97(7):S93–S96?, 1997.

6. Hollingsworth P. Burgers or biscotti? The fast-food market is changing. *Food Technology* 56(9):20, 2002.

7. Institute of Medicine. *Dietary Reference Intakes for Energy, Carbohydrate, Fiber, Fat, Fatty Acids, Cholesterol, Protein, and Amino Acids (Macronutrients),* Food and Nutrition Board, The National Academies Press, Washington, D.C., 2005. http://books.nap.edu/catalog.php?record_id=10490. Accessed 06/06/09.

8. Kinder F, NR Green, and N Harris. *Meal Management.* Macmillan, New York, 1984.

9. Lafferty L, and RA Dowling. American Dietetic Association: Position of the American Dietetic Association: Management of health care food and nutrition services. *Journal of the American Dietetic Association* 97(12):1427–1430, 1997.

10. Laurance J. Time for a fat tax? *Lancet* 373(9675):1597, 2009.

11. McTiernan A, et al. Low-fat, increased fruit, vegetable, and grain dietary pattern, fractures, and bone mineral density: The Women's Health Initiative Dietary Modification Trial. *American Journal of Clinical Nutrition* 89(6):1864–1876?, 2009.

12. Payne-Palacio J, and T Monica. *Introduction to Foodservice* (10th ed.). Pearson/Prentice Hall, Upper Saddle River, NJ, 2005.

13. Reichler G, and S Dalton. Chefs' attitudes toward healthful food preparation are more positive than their food science knowledge and practices. *Journal of the American Dietetic Association* 98(2):165–169, 1998.

14. Warfel MC, and FH Waskey. *The Professional Food Buyer: Standards, Principles, and Procedures.* McCutchan, 1979.

# WEBSITES

The National Restaurant Association, the restaurant industry's number one association, provides links to numerous websites for other major food service organizations and associations:

**www.restaurant.org/**

The USDA's Center for Nutrition Policy and promotion offers nutrition information for planning healthy menus:

**www.usda.gov/cnpp**

The Exchange Lists are found in the *Choose Your Foods: Exchange Lists for Diabetes* booklet published by the American Dietetic Association/American Diabetes Association ($3.25 for each booklet + $7.50 shipping/handling).

**http://www.eatright.org/Shop/Product.aspx?id=4962**

Analyze your own diet for calories and nutrients at:

**www.nat.uiuc.edu** (click on "NAT Tool Version 2.0")

The USDA Agricultural Research Service's Nutrient Data Laboratory has an online data resource for nutrient-conscious consumers. Check out how many calories and nutrients are in different foods using the USDA's database:

**www.nal.usda.gov/fnic/foodcomp/search**

PhotoDisc/Getty Images

# 7 Meat

In North America and Europe, meat from herbivores such as beef cattle, sheep, and swine serves as an important source of complete protein. Meat from other animals, such as goat, rabbit, deer, elk, moose, horse, possum, and squirrel, is less commonly eaten. Significant sources of meat in other countries include the camel in the Middle East, the llama in Peru, the kangaroo in Australia, and the dog in some parts of the Far East. This chapter's content is confined to the meat from cattle, sheep, and swine.

Meat is generally defined as the muscles of animals, but in a broader sense it also covers the organs and glands obtained from the animal. Although the word *meat* includes the flesh of poultry and fish, these are each covered separately in the next two chapters.

The focus of this chapter is to briefly describe the different types of meat (beef, lamb, mutton, and pork); their composition (muscle, connective, and fatty tissues; bone; pigments; extractives); the various considerations involved in purchasing meat (inspection, grading, tenderness, fresh cuts, and processed meats); their preparation (heating changes, determining doneness, dry and moist preparation, carving); and their storage.

## TYPES OF MEATS

### Beef

The ancestor of beef cattle was a type of wild ox domesticated in ancient Greece and Turkey during the Stone Age (around 10,000 BC). Since that time, hundreds of breeding lines have been specially developed to provide cattle that serve as abundant sources of good quality beef. Red meat consumption continues to increase among North American consumers (62). Beef originates from cattle that are classified according to age and gender.

- **Steers.** Male cattle that are castrated while young so that they will gain weight quickly.
- **Bulls.** Consumers often do not see the tougher meat from bulls. These older uncastrated males that provide *stag meat* are usually used for breeding and then later for processed meats and pet foods.
- **Heifers and cows.** Heifers, females that have not borne a calf, are also used for meat. The meat from cows, female cattle that have borne calves, is less desirable than that from steers or heifers.
- **Calves.** Calves 3 to 8 months old are too old for veal and too young for beef. If they go to market between 8 and 12 months, their meat is referred to as *baby beef*.

### Veal

Veal comes from the young calves of beef cattle, either male or female, between the ages of 3 weeks and 3 months. These very young animals are fed a milk-based diet or formula and have their movements greatly restricted, resulting in meat with an exceptionally milky flavor, pale color, and tender texture. Some retailers have stopped selling veal, however, because of possible consumer objections over what is perceived as the inhumane treatment of these animals. The meat from calves allowed to roam in a pasture is called *free-range* veal and it is slightly less tender than traditionally fed veal.

## Lamb and Mutton

Lamb and mutton are the meat of sheep. The primary difference between the two is the age of the animal from which they come: in general, lamb comes from sheep less than 14 months old, and mutton from those over 14 months. Further confirmation of whether one is dealing with lamb or mutton may be found in the position where the lower leg of a carcass will snap. Lamb breaks off above the joint, whereas mutton will break in the joint. Mutton is also darker and tougher than lamb and has a stronger flavor, which grows even more pronounced as the animal matures.

## Pork

Most pork is derived from young swine of either gender slaughtered at between 5½ and 7 months of age. Technically, pigs are less than 4 months old, whereas hogs are older than 4 months, although the terms are often used interchangeably. In recent times, pork has been bred to be leaner and more tender. Over the last 30 years, this has resulted in a 50% increase in the amount of lean meat yielded per animal. About one third of all pork is sold fresh, whereas the rest is cured and provided to consumers as ham, sausage, luncheon meats, and bacon. Salt pork and fat back are cuts of fatty tissue from pigs that are used as flavoring agents, for example, in Boston baked beans or to wrap cuts of meat prior to cooking.

# COMPOSITION OF MEATS

## Structure of Meat

Meats are composed of a combination of water, muscle, connective tissue, adipose (fatty) tissue, and often bone. The proportions of these elements vary according to the animal and the part of its anatomy represented by the cut of meat.

### Muscle Tissue

Most of the protein in animals is found in their muscles, which serve as the main sources of dietary meat. The characteristics of muscles are an important consideration in deciding how the resulting meat should be prepared.

Muscles are made up of a collection of individual muscle cells, called muscle fibers, that are each surrounded by an outer membrane called the sarcolemma (Figure 7-1). Each muscle fiber is further filled with cell fluid (sarcoplasm) in which there are about 2,000 smaller muscle fibrils serving as the contractile components of the muscle fiber. If the muscle fibrils are small, the result is finer muscle bundles, which give the meat a very delicate, velvety consistency.

**FIGURE 7-1** Muscles are composed of bundles of muscle cells (fibers). Each of these muscle cells (fibers) is a bundle of fibrils. The individual fibrils are responsible for muscle contraction/relaxation.

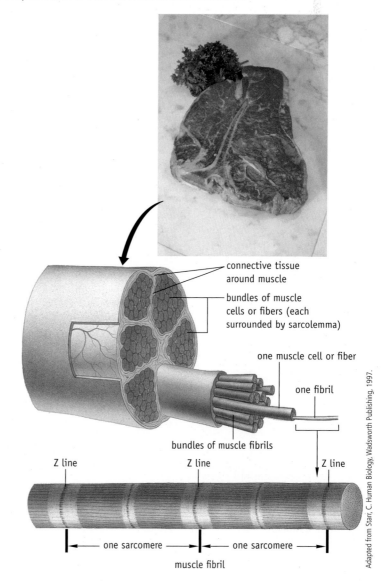

connective tissue around muscle

bundles of muscle cells or fibers (each surrounded by sarcolemma)

one muscle cell or fiber

one fibril

bundles of muscle fibrils

Z line        Z line        Z line

— one sarcomere —        — one sarcomere —

muscle fibril

Adapted from Starr, C. Human Biology, Wadsworth Publishing, 1997.

**FIGURE 7-2** Muscle contraction and relaxation. The sliding filament theory states that sarcomere units consist of two protein filaments, actin (thin) and myosin (thick), that interact with each other to form actinomyosin, which shortens (contracts) the sarcomere.

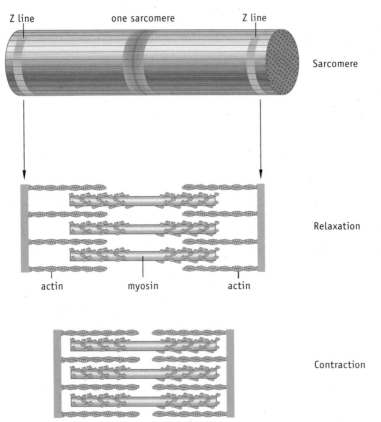

Z line          one sarcomere          Z line

Sarcomere

Relaxation

actin          myosin          actin

Contraction

### Muscle Contraction and Relaxation

Muscle fibrils play an important role in muscle contraction and relaxation. The muscle fibril is separated into segments called sarcomeres, which are bordered by dark bands called Z lines. The sarcomeres contain two proteins, actin (thin) and myosin (thick), that are alternately aligned. It is thought that muscle contraction occurs when the sarcomeres shorten as the thick and thin filaments "slide" past each other, forming another protein called actinomyosin (Figure 7-2) The energy for muscle contraction is provided by **adenosine triphosphate (ATP)**.

### Connective Tissue

**Connective tissue** is a part of ligaments and tendons, and also acts as the "glue" that holds muscle cells together. It is composed primarily of a mixture of proteins and mucopolysaccharides (a type of polysaccharide). The most abundant protein in connective tissue is **collagen** (Chemist's Corner 7-1). It is tough and fibrous, but converts to a gel when exposed to moist heat. The other two main types of connective tissue proteins are elastin and reticulin.

**Adenosine triphosphate (ATP)** A universal energy compound in cells obtained from the metabolism of carbohydrate, fat, or protein. The energy of ATP, which is located in high-energy phosphate bonds, fuels chemical work at the cellular level.

**Connective tissue** A protein structure that surrounds living cells, giving them structure and adhesiveness within themselves and to adjacent tissues.

**Collagen** A pearly white, tough, and fibrous protein that provides support to muscle and prevents it from over stretching. It is the primary protein in connective tissue.

Elastin, as the name implies, has elastic qualities, and reticulin consists of very small fibers of connective tissue that form a delicate interlace around muscle cells. Reticulin fibers create a fine meshwork that supports tissues such as the bone marrow, liver, and lymphatic system.

### Effect of Collagen on Tenderness

The type and amount of connective tissue found in a meat cut determines its tenderness or toughness and the best type of cooking method. Cuts high in connective tissue are naturally tough and need to be properly prepared in order to become more tender. Muscles used for movement, such as those found in the neck, shoulders, legs, and flank, contain more collagen and tend to be tougher than muscles from the loin, or lower back, and rib areas, which get less exercise.

### Effect of Age on Tenderness

Collagen concentration also increases as animals age, which is why meat from older animals is tougher. These usually less expensive, tougher cuts require slow, moist heating at low temperatures to convert, or hydrolyze, the tough connective tissue to softer gelatin. Conversely, the tougher cuts have more flavor than the more tender ones.

### Effect of Elastin on Tenderness

The other two components of connective tissue have less effect on meats when they are cooked. Elastin, which is yellowish, rubbery, and often referred to as *silver skin,* does not soften with heating, so it should be removed before

## Figure 1

# Beef

Shown here are selected retail beef cuts derived from the rib, brisket, and short loin wholesale cuts, as well as suggested preparation methods for each. Some of the preferred beef cuts for **broiling** are filet mignon steak, strip loin steak, Delmonico steak, rib-eye steak, top butt sirloin steak, chuck tender steak, and top round steak.

## Rib

**Rib eye steak**
*Broil, pan-broil, pan-fry*

**Rib steak, small end**
*Broil, pan-broil, pan-fry*

**Rib eye roast**
*Roast*

**Rib roast, small end**
*Roast*

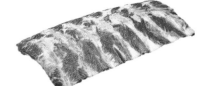

**Back ribs**
*Braise, cook in liquid*

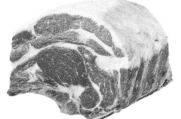

**Rib roast, large end**
*Roast*

## Brisket

**Corned brisket, point half**
*Braise*

**Brisket, whole**
*Braise, cook in liquid*

## Short Loin

**T-bone steak**
*Roast, broil*

**Tenderloin roast**
*Roast, broil*

**Boneless tenderloin steak**
*Broil, pan-broil, pan-fry*

**Porterhouse steak**
*Broil, pan-broil, pan-fry*

**Tenderloin steak**
*Broil, pan-broil, pan-fry*

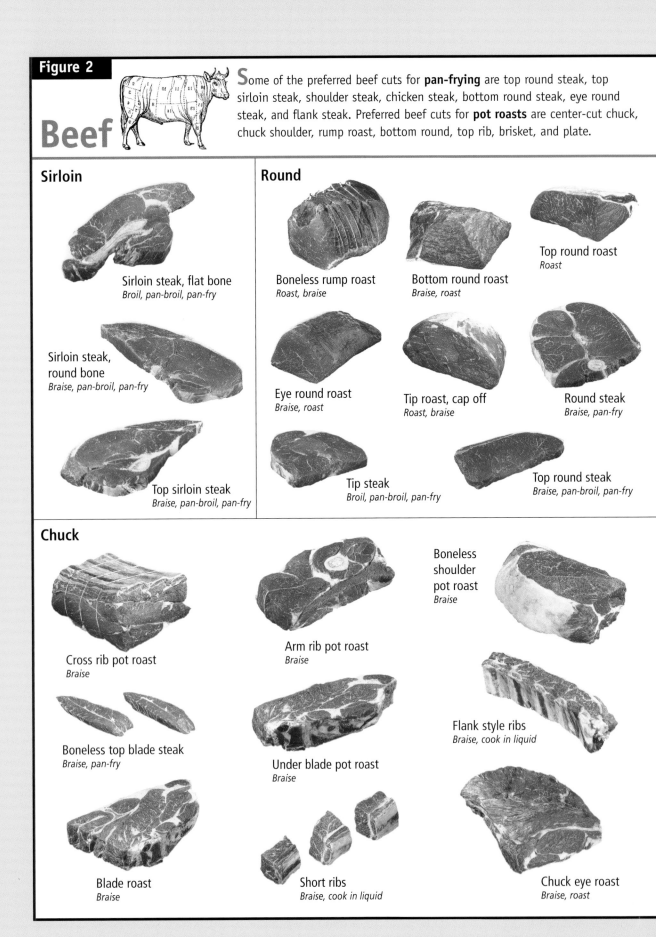

**Figure 2**

# Beef

Some of the preferred beef cuts for **pan-frying** are top round steak, top sirloin steak, shoulder steak, chicken steak, bottom round steak, eye round steak, and flank steak. Preferred beef cuts for **pot roasts** are center-cut chuck, chuck shoulder, rump roast, bottom round, top rib, brisket, and plate.

## Sirloin

**Sirloin steak, flat bone**
*Broil, pan-broil, pan-fry*

**Sirloin steak, round bone**
*Braise, pan-broil, pan-fry*

**Top sirloin steak**
*Braise, pan-broil, pan-fry*

## Round

**Boneless rump roast**
*Roast, braise*

**Bottom round roast**
*Braise, roast*

**Top round roast**
*Roast*

**Eye round roast**
*Braise, roast*

**Tip roast, cap off**
*Roast, braise*

**Round steak**
*Braise, pan-fry*

**Tip steak**
*Broil, pan-broil, pan-fry*

**Top round steak**
*Braise, pan-broil, pan-fry*

## Chuck

**Cross rib pot roast**
*Braise*

**Boneless top blade steak**
*Braise, pan-fry*

**Blade roast**
*Braise*

**Arm rib pot roast**
*Braise*

**Under blade pot roast**
*Braise*

**Short ribs**
*Braise, cook in liquid*

**Boneless shoulder pot roast**
*Braise*

**Flank style ribs**
*Braise, cook in liquid*

**Chuck eye roast**
*Braise, roast*

USDA

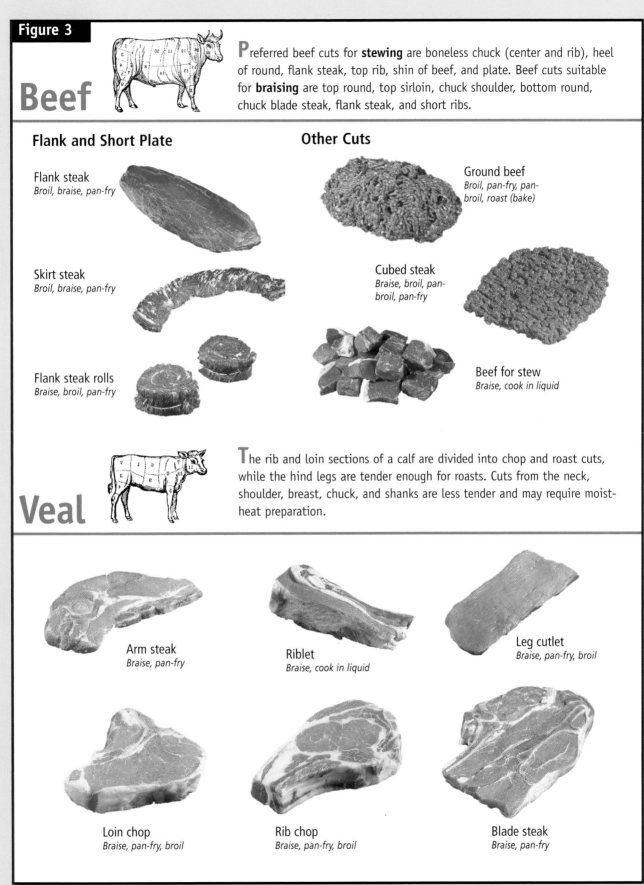

## Figure 3

### Beef

Preferred beef cuts for **stewing** are boneless chuck (center and rib), heel of round, flank steak, top rib, shin of beef, and plate. Beef cuts suitable for **braising** are top round, top sirloin, chuck shoulder, bottom round, chuck blade steak, flank steak, and short ribs.

### Flank and Short Plate

**Flank steak**
*Broil, braise, pan-fry*

**Skirt steak**
*Broil, braise, pan-fry*

**Flank steak rolls**
*Braise, broil, pan-fry*

### Other Cuts

**Ground beef**
*Broil, pan-fry, pan-broil, roast (bake)*

**Cubed steak**
*Braise, broil, pan-broil, pan-fry*

**Beef for stew**
*Braise, cook in liquid*

### Veal

The rib and loin sections of a calf are divided into chop and roast cuts, while the hind legs are tender enough for roasts. Cuts from the neck, shoulder, breast, chuck, and shanks are less tender and may require moist-heat preparation.

**Arm steak**
*Braise, pan-fry*

**Riblet**
*Braise, cook in liquid*

**Leg cutlet**
*Braise, pan-fry, broil*

**Loin chop**
*Braise, pan-fry, broil*

**Rib chop**
*Braise, pan-fry, broil*

**Blade steak**
*Braise, pan-fry*

USDA

Figure 4

# Pork

**P**referred pork steaks and chops suitable for **pan-frying** are center-cut loin chop, center-cut rib chop, loin end chop, fresh ham steak, shoulder arm steak, and blade pork steak.

## Chops

Sirloin chop

Center-cut loin chop

Blade steak

Chops are one of the most familiar pork cuts. Chops can be prepared by pan-broiling, grilling, broiling, roasting, sautéing, or braising.

Thin chops (3/8-inch) are best quickly sautéed. Thicker chops (3/4-inch to 1 1/2 inches) can be grilled, roasted, braised, or pan-broiled.

Pork rib chop

Pork loin chop

Boneless pork sirloin chop

## Tenderloin

Pork tenderloin

Pork tenderloins are among the leanest cuts of pork. A pork tenderloin has only 4.1 grams of fat and 141 calories per 3-ounce roasted, trimmed serving.

## Ribs

Back ribs

Ribs are commonly used for barbecue meals. Slow-roasting or braising yields tender, flavorful results.

## Roasts

A roast is a large cut of pork from the loin, leg, shoulder, or tenderloin. It can be roasted in the oven, barbecued over indirect heat, or braised in the oven.

Boneless blade roast

Bone-in blade roast

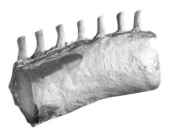

Center rib roast (rack of pork leg)

**Figure 5**

# Lamb

Lamb is traditional in Middle Eastern and Navajo cuisine. In North America, it is often served with mint sauce or jelly.

## METHODS FOR COOKING LAMB

### PAN-FRYING (if cut thin enough)

Loin chops

Ground lamb

Rib chop

Sirloin chops

### BRAISING, STEWING, AND MOIST COOKING

Shanks

Center leg

Shoulder chops

Half of leg

### ROASTING

Boneless lamb shoulder

Boneless loin roast

PhotoDisc

USDA

PhotoDisc

## Figure 6

# Lower-Fat Meats

More people are achieving the goal of deriving less than 30 percent of their calories from fat. Choosing lean meat cuts and following the tips for reduced fat cooking listed below are some of the steps that can be taken toward achieving a healthy, balanced diet.

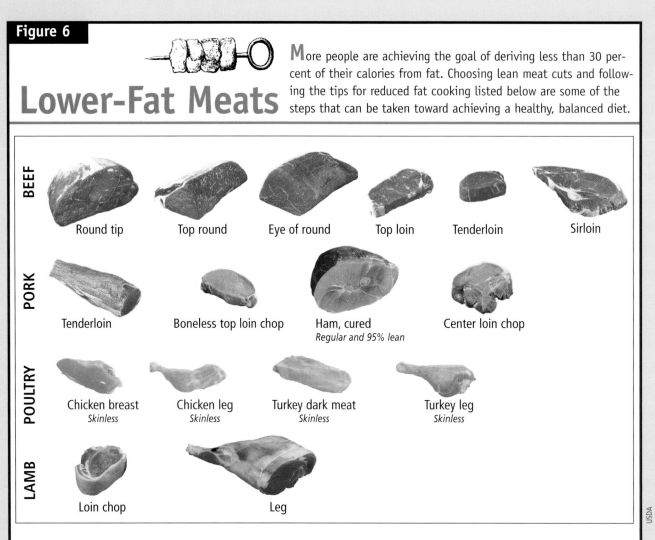

**BEEF**

Round tip — Top round — Eye of round — Top loin — Tenderloin — Sirloin

**PORK**

Tenderloin — Boneless top loin chop — Ham, cured *Regular and 95% lean* — Center loin chop

**POULTRY**

Chicken breast *Skinless* — Chicken leg *Skinless* — Turkey dark meat *Skinless* — Turkey leg *Skinless*

**LAMB**

Loin chop — Leg

USDA

## TIPS FOR REDUCED FAT COOKING

☑ Choose 3-ounce servings (for a total of 6 ounces per day). Start with 4 ounces of raw meat to end up with a 3-ounce cooked serving. This will account for cooking losses.

☑ Look for beef labeled with the "USDA Select" grade. It's lower in fat and calories than "Choice" or "Prime." Marbling (the flecks of fat in the lean) makes the difference.

☑ Use the "loin/round" rule of thumb for beef and "loin/leg" for pork, lamb, and veal. Cuts with these words on the label will be lean choices.

☑ Tenderize lean cuts of meat by cooking them slowly in liquid or marinating them before cooking. Pounding, grinding, and slicing across the grain can also help.

☑ Keep your meat selections lean. Trim all visible fat and let the remainder drip off during cooking. When you prepare meat, broil, grill, bake, roast on a rack, or microwave. Buy skinless poultry or remove skin before cooking and you will reduce fat content by about half.

☑ Remove fat from stews, soups, and casseroles by chilling them and skimming the hardened fat from the top. If you're pressed for time, use a baster to remove it.

☑ Don't fry. The batter or breading on fried chicken, for example, acts like a sponge—soaking up fat. And after frying, you're less likely to remove the coating and skin before you eat the meat. Also skip the heavy gravies and rich sauces. Even the butter or margarine you use on broiled food makes the fat add up fast.

# Figure 7

## Seafood

Consuming fish twice a week, especially those high in omega-3 fatty acids, has been reported to lower the risk for heart disease. Most fish—except mackerel, shark, herring, and eel—also contain fewer than 160 calories (kcal) per three-ounce cooked serving.

PhotoDisc

Lois Frank

Lois Frank

## Figure 8

# Onions

The type of onion chosen depends on how it will be used in food preparation. Yellow onions are all-purpose, white onions are the most pungent, red onions lend themselves to certain salads, the smaller pearl onions are preferred for soups and casseroles, and vidalias yield a sweeter flavor.

## VARIETIES OF ONIONS

**WHITE**
- Pungent odor
- Sharp flavor

**YELLOW**
- All-purpose
- Medium-strong flavor

**RED**
- Best used raw
- Tangy flavor

**VIDALIA**
- Salads
- Sweet flavor

**PEARL ONIONS**
- Soups and casseroles
- Regular onions whose growth has been stunted

Lois Frank

Figure 9

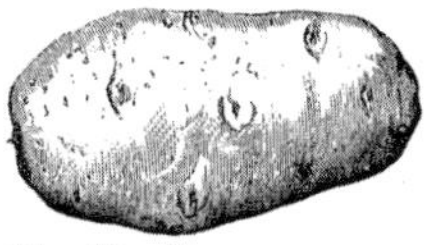

# Potatoes

Potatoes differ in their moisture and starch content. Russet/Idaho potatoes are high in starch and best for baking, while reds or waxy potatoes have the least starch and are best for boiling and microwaving.

## SELECTING THE RIGHT TYPE OF POTATO

**RED POTATOES**
- Boiling
- Microwaving
- Salads
- Soups
- Steaming

**RUSSET/IDAHO POTATOES**
- Baking
- Boiling
- Casseroles
- Mashing
- Microwaving
- Pan-frying
- Sautéing
- Frying

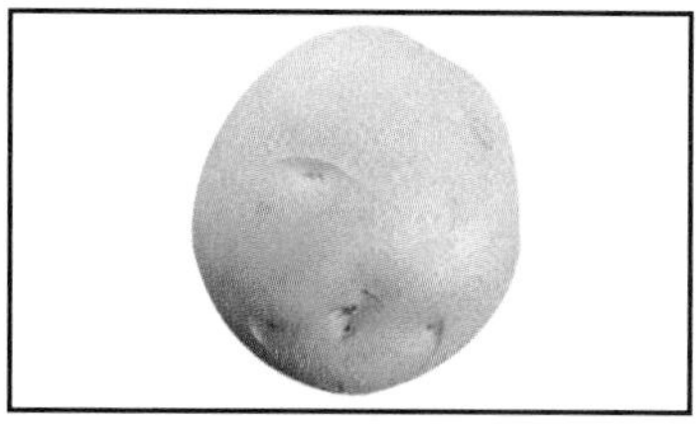

**WHITE POTATOES**
- Braising
- Boiling
- Casseroles
- Frying
- Mashing
- Microwaving
- Pan-frying
- Salads
- Sautéing

## POTATOES

| Bintje | Austrian Crescent | Blue Pride | Ozette Indian | Ruby Crescent |

| Red Thumb | Rote Erstling | French Fingerling | Russian Banana | Yukon Gold |

Vincent Lee

PhotoDisc

Vincent Lee

**Figure 10**

# Squash

Summer squashes are harvested in the summer, usually elongated, and can be left unpeeled and cooked whole, sliced, cubed, or grated. Winter squashes, harvested in the fall, usually have hard rinds that are cut in half to remove their fibrous matter and seeds before being baked, broiled, or steamed.

## SOME POPULAR VARIETIES OF SQUASH

Banana

White Acorn

Australian Blue

Spaghetti

Acorn

Butternut

White Acorn

Golden Nugget

Yellow Crookneck

Zucchini

# Figure 11

## Melons

These round to oblong fruits grow on vines. The skin on melons is actually a rind that can be smooth, netted, ridged, wrinkled, or warty. Inside, the edible pulp varies in color and can be white, yellow, pink, green, or red.

## SOME VARIETIES OF MELON

Persian

Santa Claus

Casaba

Crenshaw

Juan Canary

Vincent Lee

As the global economy has expanded and transportation methods have improved, the selection of available fruits and vegetables includes produce from around the world.

# Exotic Vegetables and Fruits

## Figure 12   VEGETABLES

**Chinese long beans**
- Grow up to 18" long
- Steam or stir-fry

**Jerusalem artichoke**
- Root of sunflower plant
- Nutty, sweet, mild flavor

**Breadfruit**
- Not used to make bread
- Used like a potato

**Kohlrabi**
- Sweeter and crisper than turnips
- Flavor of broccoli stems

**Jicama**
- Pronounced "hee-ka-ma"
- Sweet, starchy taste

**Belgian endive**
- Mild, bitter flavor
- Used in salads or soups

**Calabaza**
- Variety of squash
- Dark orange flesh

**Chayote**
- Pronounced "chy-o-tay"
- Flavor similar to zucchini/cucumber

## Figure 13   FRUITS

**Kumquats**
- Tart pulp

**Cherimoya**
- Sweet, custard-like flavor

**Plantains**
- Served as vegetable
- Starchy, no banana flavor

**Passion fruit**
- Lemony, tart flavor
- Many small black seeds

**Lychee**
- Grape-like flesh

**Red banana**
- Maroon when ripe
- Tangy-sweet flavor

**Pummelos**
- Largest of citrus fruits

**Guavas**
- High vitamin C content

**Figure 14**

# Apples

Over 7,500 varieties of apples are grown worldwide, but only about 18–25 comprise the majority of the North American commercial crop. While many apples can be used for both eating and cooking, tart varieties with a high acid content and firm texture are best for baking.

## Some Apple Favorites

### Red Delicious
- Bright to dark red, sometimes striped
- Favorite eating apple
- Mildly sweet, juicy
- Available year-round

### Winesap
- Dark red
- Appropriate for cider, snacking, and cooking
- Spicy, slightly tart
- Available October to August

### Rome Beauty
- Brilliant red, round
- Great for baked apples; holds shape well when cooked
- Available October to June

### Golden Delicious
- Yellow-green
- All-purpose apple for baking, salads, and fresh eating; flesh stays white longer than other apples
- Mellow, sweet
- Available year-round

### Criterion
- Sweet, yellow, often with red blush
- Wonderful eaten fresh, in salads, or baked; flesh stays white longer than other apples
- Available October to Spring

### Gala
- Yellow to red
- Appropriate for cider, snacking, and cooking
- Spicy, slightly tart
- Available October to August

### Granny Smith
- Green
- Excellent for cooking, salads, fresh eating
- Tart, crisp, juicy
- Available year-round

### Fuji
- Ranges from yellow-green with red highlights to very red
- Excellent for eating or applesauce
- Sweet, spicy, crisp
- Available year-round

Washington State Apple Commission

**Figure 15**

# Greens

Iceberg, butterhead, romaine, and loose-leaf lettuce are the greens most commonly used in salads, but a variety of other greens are also available.

When selecting greens, look for clean, crisp, tender leaves free of "tipburn"—the ragged brown borders that can appear on a leaf's edge.

**Bibb Lettuce (Butterhead)**

**Boston Lettuce**

**Belgian Endive**

**Green Cabbage**

**Radicchio**

**Chicory**

**Savoy**

**Escarole**

**Watercress**

**Red Leaf Lettuce**

**Swiss Chard**

**Green Leaf Lettuce**

**Tarragon**

**Savory**

**Romaine lettuce**

**Cilantro**

**Flat Parsley**

**Oregano**

Rosemary

**Figure 16** EDIBLE FLOWERS

One of the newest trends in gourmet produce is edible flowers. A colorful, peppery-tasting addition to salads can be made by adding a sprinkle of nasturtium flowers or calendula petals. Daylily, squash, and pumpkin blossoms are delicious dipped in tempura batter and quickly deep-fried. Lavender and many geranium blossoms add a perfumy, herbal scent to beverages and desserts. Viola, pansy, and violet blossoms can be candied and used as edible decorations for cakes and other desserts.

Digital Works

Digital Works

Richard Brewer

Lois Frank

**Figure 17**

Desserts not only satisfy the sweet tooth but are sometimes identified with certain meals or occasions—fortune cookies following a Chinese meal, marshmallow eggs at Easter, birthday cakes, pumpkin pie at Thanksgiving, Christmas cookies and fruitcakes during Christmas, and complimentary mints at some restaurants.

# Desserts

## COOKIES

| **Bar** | **Dropped** | **Pressed** | **Rolled** | **Molded** |
|---------|-------------|-------------|------------|------------|
| Brownies | Chocolate Chip | Tea | Sugar | Peanut Butter |
| | Oatmeal Raisin | Lady Fingers | | Shortbread |
| | | Coconut Macaroons | | |

## PIES

Cream
Fruit
Ice Cream
Chiffon
Custard
Meringue

## CAKES

| **Shortened** | **Unshortened** | **Chiffon** |
|---------------|-----------------|-------------|
| White | Angel Food | Lemon Chiffon |
| Yellow | Sponge | Chocolate Chiffon |
| Chocolate | | |
| Spice | | |
| Fruit | | |

## CANDY

## PASTRIES

| **Blitz/Puff Pastry** | **Phyllo** |
|-----------------------|------------|
| Napoleons | Baklava |
| Tart Shells | |

| **Strudel** | **French** |
|-------------|------------|
| **Danish** | **Pâte à Choux** |
| Eclairs | Cream Puffs |

## FROZEN DESSERTS

Ice Cream
Imitation Ice Cream
Sherbet
Sorbets
Water Ices
Frozen Yogurt
Still-Frozen
—Mousses
—Bombes
—Parfaits

**FIGURE 7-3**   Collagen molecule.

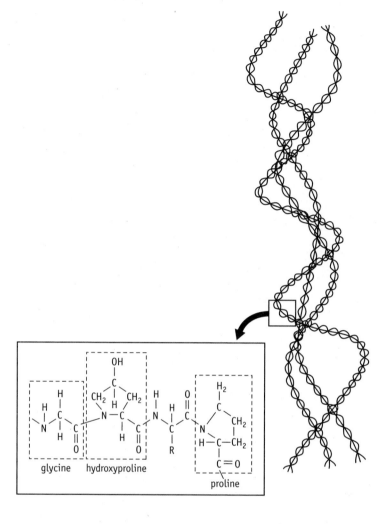

glycine   hydroxyproline

proline

**FIGURE 7-4**   A polyunsaturated diet will yield pork that is higher in polyunsaturated fat.

Wolfgang Kaehler/Encyclopedia/CORBIS

preparation if possible. There is very little elastin in meats, except in cuts from the neck and shoulder, so it is less likely to affect tenderness.

### Adipose (Fatty) Tissue

Adipose tissue is, simply, fat, which serves as insulation under the skin (subcutaneous) and as padding in the abdominal cavity for sensitive internal organs. When it appears on the outside of meat, this fat is known as cover fat.

Cover fat helps retain the moisture of meats, but this separable fat is often trimmed from meats prior to preparation. Fat found within muscles is called intramuscular fat or **marbling.** Fat content varies widely among meats and is dependent on the source animal's genetics, age, diet, and exercise, and on the cut of the meat.

Well-marbled beef fetches a higher price, so many cattle ranchers, in an attempt to improve marbling, feed cattle

richer grain during the last weeks or months before slaughter. Paradoxically, however, because of the recent consumer trend away from fatty meats, some ranchers are raising lower-fat beef to meet these consumer demands. Some livestock are being bred to average 25 to 30% fat (26). Similarly, a recent technique in swine livestock management is the use of a growth hormone, somatotropin, which results in a leaner animal (41). Conversely, certain hormones (medroxyprogesterone acetate) are sometimes given to animals to speed up fattening (20).

*Fat Color and Texture*   The animal's age, diet, and species affect the color and texture of fat. It is white in younger animals, and turns progressively more yellow as the animals age because of the presence of carotenoid pigments in the feed. Feeding-lot practices that provide swine with fats

that are primarily saturated will yield pork fat that is more saturated and hard (Figure 7-4). Conversely, including more polyunsaturated fatty acids in the animal's diet will make its fat softer. The species and breed of the animal also influence the softness of fat; beef fat, for example, is very different from the hard, more brittle and dense fat observed in lamb.

### ? How & Why?

*Why is marbling desirable in meat?*

When meat is cooked, the intramuscular fat deposits melt and contribute to perceived flavor and juiciness. For this reason, the more marbling in a cut of beef, the higher the grade (see the section "Purchasing Meats").

**Marbling**   Fat deposited in the muscle that can be seen as little white streaks or drops.

## Bone

Bones are used as landmarks for identifying the various meat cuts from a carcass (Figure 7-5). When buying meat, keep in mind that bone weighs more than meat and that the higher the proportion of bone there is to meat, the less the meat yield and the more the cost of the edible portion will be.

*Marrow*    Marrow is the soft, fatty material in the center of most large bones. The marrow found within the bone will generally be of two different types: (1) yellow marrow, found in the long bones, and (2) red marrow—red because it is supplied with many blood vessels—in the spongy center of other bones. Marrow is a valued food in many cultures and can provide much of the flavor in soups. (See Chapter 15 for more on how bones are used in soups.)

# Antibiotics and Hormones

Why are antibiotics and hormones given to livestock? Each year, more than 20 million pounds of antibiotics are given to animals raised for meat to shield them from disease and promote growth. In contrast, about 3 million pounds of antibiotics are given to humans. Regulations require that the drugs be withheld prior to slaughter so that any remaining residues fall below federal limits. Some people believe that antibiotic use in animals possibly contributes to the growth and spread of drug-resistant bacteria (31). The World Health Organization (WHO) recommended that nations phase out the use of antibiotic growth promoters in animal feed in order to preserve the effectiveness of medicinal antibiotics (See Chemist's Corner 7-2 and Chemist's Corner 7-3).

Not all countries permit the use of hormones in livestock to encourage rapid weight gain (increased production by 15%), help them reach market weight sooner, and reduce the production cost (hormone-treated animals gain more on less feed) (36). The United States Department of Agriculture (USDA) has allowed the use of hormones in raising cattle and sheep (but not swine or poultry) for almost half a century. Since 1988, the European Union has banned meat imported from countries that permit hormone use in livestock. The six hormones approved in the United States include three natural hormones (testosterone, progesterone, and estradiol), and three synthetic hormones (trenbolone acetate, which mimics testosterone; melengestrol acetate, which mimics progesterone; and zeranol, which mimics estradiol). The hormones are delivered to the animal through its feed or an ear implant (removed at slaughter).

Advocates state that the hormone levels in such beef are within the natural

**FIGURE 7-5**    Bones identify retail cuts of meat.

| CUT | BONE | | |
|---|---|---|---|
| SHOULDER ARM | Arm bone | | |
| SHOULDER BLADE | Blade bone (near neck) | Blade bone (center cuts) | Blade bone (near rib) |
| RIB | Back bone and rib bone | | |
| SHORT LOIN | Back bone (T-shape) T-bone | | |
| HIP (SIRLOIN) | Pin bone (near short loin) | Flat bone (center cuts) | Wedge bone (near round) |
| LEG OR ROUND | Leg or round bone | | |
| BREAST OR BRISKET | Breast and rib bones | | |

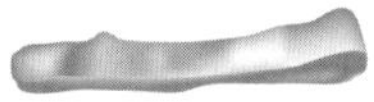

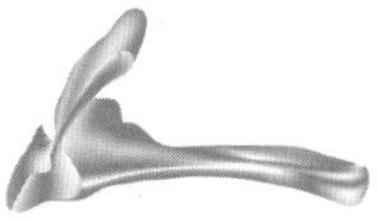

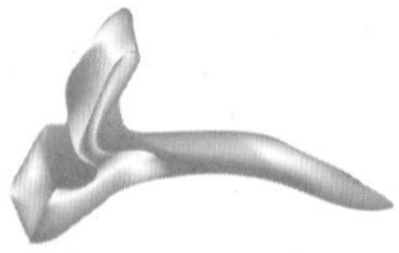

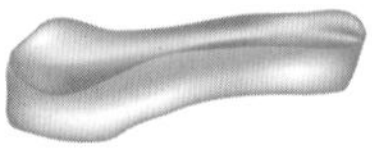

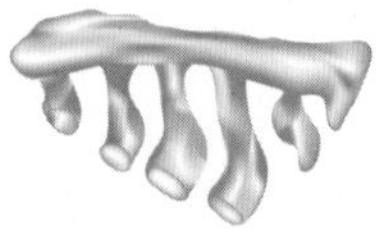

> ## CHEMIST'S CORNER 7-2
>
> ### Antibiotic Resistance
>
> Bacteria and other microorganisms can develop antibiotic resistance—the ability to survive in the presence of a medication that once killed them. Antibiotic resistance can develop sporadically, due to a genetic mutation, or can be a process of adaptation to the constant presence of an antibiotic or disinfectants (3, 24). In both humans and animals, long-term or frequent use of antibiotics and disinfectants has resulted in resistant bacteria that can cause infections that have the potential to be unresponsive to these antibiotic medications.

levels of hormones found in the animal. European beef is primarily from bulls (high in testosterone), whereas beef comes predominantly from steers (castrated bulls, which are low in testosterone) in North America. Also, eggs and milk that are not treated with hormones naturally have higher hormone levels than meat from steers treated with hormones.

Critics raise the issue of whether or not hormone trace residues in the meat have serious human-health consequences. They also question the effect on the environment of those hormones eliminated through the feces, especially as these hormones may affect the water supply. Some have attributed changes in the reproductive cycle of fish to water runoff from feedlots and slaughterhouses (40). Adverse effects of these hormones in humans remains to be proven.

Over three fourths of North American cattle are treated with hormones. Although much more expensive, organic beef ensures the consumer that neither hormones nor antibiotics were used. Such beef is sometimes labeled "no hormone administered." However, just because a meat is labeled "natural" does not mean that hormones were not used. Any meat without added ingredients can be called natural.

## CHEMIST'S CORNER 7-3

### Superbugs in Food

*Clostridium difficile,* commonly called *C. diff.,* is a bacteria most commonly associated with diarrhea in hospitalized patients. While *C. diff.* causes only mild diarrhea in some patients, others develop recurring, severe diarrhea that requires the use of powerful antibiotics. Infection with *C. diff.* is often attributed to antibiotic use—antibiotics kill off the normal bacteria in the gut, allowing overgrowth of *C. diff. C. diff.* has recently been identified in commercial meat (35). It is still unknown whether *C. diff.* can be transmitted from animals to humans and whether the infection can be contracted by eating tainted meat.

# Pigments

Many people evaluate a meat's color when deciding whether or not to purchase a particular meat cut. The color of meat is derived from pigment-containing proteins, chiefly myoglobin and, to a lesser extent, hemoglobin. The so-called red meats—beef, pork, sheep, and lamb—have more of these pigments than poultry or fish do. Myoglobin receives oxygen from the blood and stores it in the muscles, whereas hemoglobin transports oxygen throughout the body and is found primarily in the bloodstream.

The higher the concentration of myoglobin in raw meat, the more intense is its bright red color (see Chemist's Corner 7-4). Several factors influence the concentration of myoglobin. Heavily exercised muscle has a higher demand for oxygen, so it is higher in myoglobin and therefore redder than the less exercised muscles. The red color of meat also increases as the animal ages, which is why beef is redder than veal, and mutton is darker than the pink hue of lamb. Meat color also varies from species to species. Beef is darker than lamb, which, in turn, is darker than pork, a meat that is on the pink side with no visible red.

## Effect of Oxygen on Color

Exposure of meat to oxygen changes the color of myoglobin, and therefore the meat. After slaughter, meat undergoes

## CHEMIST'S CORNER 7-4

### Meat Pigments

Each pigment-containing compound in meat consists of two parts: a protein (globin), and a nonprotein pigment (heme). The heme is an atom of iron surrounded by four connecting pyrrole rings. The difference between myoglobin and hemoglobin is that the simpler myoglobin molecule consists of one protein polymer strand and one heme (molecular weight = about 17,000), whereas the larger hemoglobin molecule is made of four protein polymer strands and four hemes (molecular weight = about 68,000) (42).

several changes in color over time due to modifications in the molecular structure of myoglobin and/or hemoglobin (Chemist's Corners 7-4, 7-5, and 7-7, on pages 145, 146, and 175). Myoglobin within the meat is purplish red, but once exposed to oxygen (when meat is cut), it becomes bright red—a color indicating freshness and so desired by consumers. A meat's color is the number-one factor influencing consumers when they are purchasing meats (45). After a while, meats left in storage may be exposed to bacteria or less oxygen, and/or kept under fluorescent or incandescent lights, all of which turn the meat brownish-red (38). Using plastic wrap that is permeable to oxygen allows meat retailers to maintain the bright red color for a longer period of time, whereas vacuum wrap, which eliminates the oxygen, causes the meat to appear purplish-red.

## Effect of Heat on Color

Cooking meat initially converts the color of raw meat to bright red, but then the denaturing of the pigment-containing proteins yields the classic color of well-done meat—grayish brown. Storing cooked meat too long causes the denatured protein to further break down, causing the meat to turn yellow, green, or faded.

# Extractives

Meat derives some of its flavor from nitrogen compounds called **extractives.** The most common extractives are creatine and creatinine, but urea, uric acid, and other compounds also contribute to the flavor of meat. The meat from older animals contains more connective tissue and extractives, and therefore yields more flavor than that from younger livestock. Extractives are water soluble, so some of the flavor of boiled or simmered beef may be lost in the cooking water, but the flavor can be recaptured by using the cooking liquids in the preparation of soup or gravy.

**Extractives** Flavor compounds consisting of nonprotein, nitrogen substances that are end-products of protein metabolism.

## CHEMIST'S CORNER 7-5

### Color Changes in Meat

The molecular changes in the pigment-containing proteins determine the color of meat from slaughter to consumption. Color changes are dependent on the oxidation or reduction of the iron in the heme. Initially, the internal color of meat is purplish-red, because slaughter depletes oxygen concentrations in the meat. As soon as meat is cut from the carcass and exposed to the oxygen in the air, a bright red compound known as oxymyoglobin forms (Figure 7-6). Over time, the bright red oxymyoglobin is further oxidized to metmyoglobin, which is a brownish-red color (65). Older meat cuts look browner because myoglobin or oxymyoglobin is converted to metmyoglobin as the iron in the pigment is oxidized from its ferrous (+2) to ferric (+3) state. This usually occurs during storage when the meat continues to be exposed too long to bacteria, oxygen, or light (fluorescent and incandescent). The brownish-red color resulting from metmyoglobin is undesirable to retailers (46).

**FIGURE 7-6**    Color changes in cooked meats.

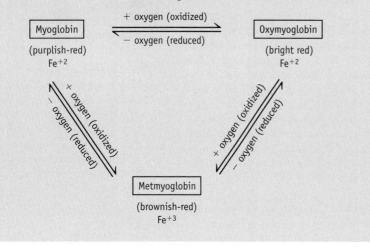

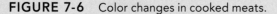

Myoglobin
(purplish-red)
$Fe^{+2}$

$+$ oxygen (oxidized)
$-$ oxygen (reduced)

Oxymyoglobin
(bright red)
$Fe^{+2}$

$+$ oxygen (oxidized)
$-$ oxygen (reduced)

$+$ oxygen (oxidized)
$-$ oxygen (reduced)

Metmyoglobin
(brownish-red)
$Fe^{+3}$

# PURCHASING MEATS

To ensure that consumers are purchasing meat that is safe, federal laws require the inspection of animal carcasses. In addition to this mandatory inspection for safety, meat may also be assigned yield grades and then later quality grades to assist consumers in selection. Meat processors submit to the grading system voluntarily.

**Quality grades** The USDA standards for beef, veal, lamb, and mutton.

## Inspection

The Federal Meat Inspection Act of 1906 made inspection mandatory for all meat crossing state lines or entering the United States through foreign commerce. Inspections are the responsibility of the USDA Food Safety and Inspection Service. This inspection is a guarantee of only wholesomeness and does not ensure the quality or tenderness of the meat itself. A meat inspection is conducted by licensed veterinarians or by specially trained, supervised inspectors.

The inspectors or veterinarians examine live animals prior to slaughter as well as animal carcasses, observe the meat at various processing stages, monitor temperatures and additives, review packaging materials and labels, determine employee and facility hygiene, and check imported meat. Meat that passes this federal inspection is marked with an inspection stamp (Figure 7-7) to distinguish it from meats that are diseased or slaughtered in unsanitary conditions. The exception is inspection for *Trichinella spiralis*, because visual inspection does not assure the absence of this small parasite (see Chapter 4 for information on *Trichinella spiralis* and other diseases).

Other laws passed since 1906 further protect the meat supply. The USDA can oversee only those meats that are transported between states, so the Wholesale Meat Act of 1967 was passed to require that meat sold within the states meet requirements equal to the federal standards. The most recent law was the implementation in 1997 of new USDA inspection regulations incorporating hazard analysis and critical control points (HACCP) within meat and poultry slaughterhouses, along with mandatory testing for *E. coli* (22).

## Grading

The grading of meat is not under government mandate or control, but is a strictly voluntary procedure that the meat packer or distributor may have done under contract with the USDA (39). For purposes of grading, a cut is made between the twelfth and thirteenth rib in order to expose the rib muscle.

### Quality

The **quality grades** for the different types of meat are shown in Table 7-2. Factors considered in grading are color,

**FIGURE 7-7**    USDA meat inspection stamp. The number is assigned to the meat processing plant. Consumers rarely see the stamp, which is placed on larger wholesale cuts.

## NUTRIENT CONTENT

Meat consists primarily of water (75%), protein (20%), and fat (varies), with a few minerals and some B vitamins (46). It contains very little to no carbohydrates (liver is the richest source), no fiber, and no vitamin C.

*Protein.* According to the Exchange Lists, meat delivers 7 grams of protein per ounce. Because current recommendations suggest a meat intake of 4 to 6 ounces a day, this means that meat provides a majority of our recommended daily protein intake—28 to 42 grams. The Dietary Reference Intake for protein for adults is approximately 50 grams per day (46 grams for women and 56 grams for men), and meat is an excellent source of this nutrient. The remaining protein need is usually met by consuming foods from other food groups—especially dairy (milk, cheese, etc.), which provides 8 grams for each cup of milk or ounce of cheese.

*Fat.* Fat content can vary widely, according to the grade of meat and its cut. Several cuts of beef are lower in fat than an equal amount of some poultry choices, yet consumers often select poultry over beef, thinking it is lower in fat (58). See Calorie Control for more information on how to choose lower-fat meats.

*Carbohydrate.* Meat contains very little carbohydrate. Glycogen, found in liver and muscle tissue, is present when the animal is alive, but the glucose that makes up the glycogen is broken down to lactic acid during and after slaughter.

*Vitamins.* Meat is an excellent source of certain B vitamins—thiamin ($B_1$), riboflavin ($B_2$), pyridoxine ($B_6$), vitamin $B_{12}$, niacin, and some folate. Niacin is obtained from tryptophan, an amino acid plentiful in meats and milk. Lean pork is an excellent source of thiamin ($B_1$). Vitamin $B_{12}$ and fat-soluble vitamins, especially vitamin A, are found in liver. Vitamin loss during meat preparation depends on the temperature, the time exposed to the heat source, and the cooking method. Thiamin ($B_1$) is especially sensitive to heat, so levels of this vitamin are somewhat reduced in canned meats, which have undergone high-heat processing (62). Water-soluble B vitamins can be leached from meat into cooking liquid, but can be recaptured by making gravy or soup from that liquid.

*Minerals.* Meat is an excellent source of iron, zinc, copper, phosphorus, and a few other trace minerals; liver is especially rich in iron. Minerals are stable when heated, and although they can be lost in cooking water, retention of most minerals in cooked meat ranges from 80 to 100%.

grain, surface texture, and fat distribution. Unfortunately, this system is not used uniformly by retailers. Instead of Prime, Choice, Select, and Standard, retailers frequently designate the quality of their meat with a descriptive word or phrase such as 5 Star, Blue Ribbon, or Supreme. This is purely a marketing strategy and leaves it up to the consumer to determine the validity, or lack thereof, of the designation. Although a large percentage of meat sold is graded, the term *no roll* is used to indicate ungraded meat.

Any judgment of quality must be somewhat subjective, but several identifiable factors separate a poor cut of meat from one that is excellent. Top cuts of meat have the optimum color for their type and fine-grained, smooth surfaces that are velvety, silky, or satiny to the touch. They contain fat that is evenly distributed, white or creamy-white rather than yellow in color, and firm instead of brittle or runny. These factors contribute to tenderness, which is never directly measured in grading, although it remains a top concern among both retailers and restaurateurs (10).

### Influence of Fat Content on Grading

Fat, especially in the form of marbling, melts during heating, thereby increasing the flavor and perceived tenderness of the meat. USDA quality grades of beef reflect this marbling. Prime, the top USDA grade, contains the most marbling and is the most expensive (Figure 7-8). The marbling and any fat trim of a beef steak being examined for possible purchase should be cream colored. If the fat is yellowish, the meat may be from an older animal and therefore may be tough. However, when retailers trim the fat to 1/8 inch around the meat, it makes it difficult to judge the fat's actual color and texture. The differences in USDA quality grades for beef are described below.

*Beef USDA Quality Grades* There are eight USDA quality grades for beef, with the top three—Prime, Choice, and Select—being of most concern to consumers (Table 7-3). Prime accounts for less than one fifth of the beef marketed. It is usually sold to restaurants, because the price is not competitive enough for the average supermarket consumer. Choice and Select are the grades most commonly purchased by consumers in the supermarket. Select cuts contain 5 to 20% less fat than Choice cuts, and 40% less fat than Prime. Standard and Commercial USDA grades are not seen at the retail level, because they are usually from older, more mature, and therefore less tender cattle. USDA grades identified as Utility, Cutter, and Canner are usually used in processed foods such as canned meats, sausages, and pet foods.

### Yield

*Yield grade* is the other main factor determining the grade of a meat cut (Figure 7-9). The evaluation for yield does not occur at the consumer level, but rather at the slaughterhouse, often referred to as a rendering plant. Carcasses of beef, lamb, and mutton are rated at yields of 1 to 5, with 1 providing the highest yield and 5 the lowest (Table 7-4). Pork is yield-graded from 1 to 4. Veal is not yield-graded because it contains so little fat. Although 4 ounces of raw meat with little or no bone generally constitutes one serving, ¾ to 1 pound of meat per serving may be needed if the meat contains high amounts of refuse.

> **Yield grade** The amount of lean meat on the carcass in proportion to fat, bone, and other inedible parts.

# Tenderness of Meats

Tender meat generally is preferred by consumers. In fact, tenderness and flavor are the two most important factors affecting consumer acceptance of cooked beef (17). It's difficult to select tender meat, because a top quality grade does not guarantee its tenderness. The only real test is how easily the meat gives way to the teeth. Extreme variations of tenderness exist in beef, even within different areas of a single meat cut, but overall, natural meat tenderness is due to factors such as the cut, age, and fat content. Meats can also be treated to make them more tender by adding enzymes, salts, and acids, or by subjecting them to mechanical or electrical treatments. Preparation temperatures and times also have an influence on tenderness.

## Natural Tenderizing

The particular cut of the meat, the animal's age at slaughter (connective tissue concentration), the animal's heredity and diet, the meat's marbling, slaughtering conditions, and aging all play a part in determining tenderness (57).

*Cut* The most important influence on the tenderness of meat is the location, on the animal's body, of the muscle from which it came. Muscles that are exercised are tougher than those that are not, due to higher concentrations of connective tissue (9). Meat cuts such as chuck and round from the shoulder and hindquarters come from muscles that are used for locomotion and are therefore tougher than those from the loin (lower back) and rib areas of the animal. The least tender cuts are flank steak, short plate, shank, short ribs, rump roast, and brisket from the legs and underside of the animal. The most tender cuts of the carcass, such as sirloin, tenderloin, and rib eye, are found in the loin and rib areas where the muscles are used less and thus develop less connective tissue. Other tender cuts of meat include strip steak, strip loin, T-bone steak, and standing rib roast. Cuts of beef with intermediate degrees of tenderness include flank, chuck, top blade, and skirt steaks (9).

*Animal's Age* An animal's age at the time of slaughter contributes to tenderness, and top USDA grades usually come from relatively young animals. As muscles age, the diameter of

---

## ✦ CALORIE CONTROL
### Choosing Meats and Meat Substitutes

Portion control starts with recognizing how many calories are being consumed for each meal, which may average 400 calories (kcal) for women and 600 calories (kcal) for men (see Chapters 1 and 6). Meat and meat substitutes contribute to these calories because they are a main source of daily dietary protein. Calories from meats are controlled by (1) limiting the number of ounces or servings and (2) leaning more toward the lean meats.

- **Limit meat to 5–6 ounces a day.** Five to six ounces of meat/meat substitutes are recommended by MyPyramid and the Exchange Lists, respectively. Visually, this is not much more than two decks of cards or the palm of an average hand. In terms of the Exchange Lists, a serving size for meat is 1 ounce, so if the average meat portion on the plate weighs 3 ounces, that means it provides about 250 calories/kcal (75 calories/ounce × 3 ounces = 225 calories, rounded to the nearest 50).
- **Lean Meats.** The less fat, the fewer calories.
  - The general rule of thumb is that beef cuts from the loin or round, and veal and lamb cuts from the loin or leg, are the leanest choices. Examples of lean beef cuts include sirloin, tenderloin, top loin, top round, and eye round (see Figure 7 in the color insert).
  - Avoid processed meats, such as hot dogs, bologna, and sausage as they average 30 to 50% fat (exceptions are the lower-fat alternatives).
  - Wild game tends to be lower in fat compared to beef, as shown in Table 7-1.

Approximate calories (from low to high) and fat content in meat and meat substitute servings can be found at http://www.nal.usda.gov/fnic/foodcomp/search/. Meats also include poultry, fish, cheese, and eggs, but these are listed under their respective chapters.

**TABLE 7-1** Fat Content of Wild Game Compared to Beef*

| Species | Fat (grams) | Calories |
|---|---|---|
| Beef, T-bone, USDA choice | 10 | 214 |
| Antelope | 4 | 165 |
| Bison (buffalo) | 2 | 143 |
| Deer (venison) | 3 | 158 |
| Duck (skinless) | 11 | 201 |
| Elk | 2 | 146 |
| Moose | 1 | 134 |
| Pheasant, breast without skin | 3 | 133 |
| Rabbit | 8 | 197 |
| Rabbit (wild) | 4 | 173 |
| Squirrel | 5 | 173 |

*All values shown are for a 100 gram (about 3 oz) cooked portion, with visible fat removed unless noted.

© 2010 Amy Brown

---

**TABLE 7-2** USDA Quality Grades for Beef, Veal, Lamb, and Mutton from Highest to Lowest*

| Beef | Veal | Lamb | Mutton |
|---|---|---|---|
| Prime | Prime | Prime | Choice |
| Choice | Choice | Choice | Select |
| Select | Select | Select | Commercial |
| Standard | Standard | Commercial | Utility |
| Commercial | Commercial | Utility | Cull |
| Utility | Utility | Cull | |
| Cutter | Cull | | |
| Canner | | | |

*There are no quality grades for pork.

**FIGURE 7-8**  USDA grades related to marbling.

**USDA Prime:** Very heavy marbling that looks like snowflakes and is evenly distributed.

**USDA Choice:** Moderate marbling in delicate lacy streaks that is less evenly distributed than in Prime.

**USDA Select:** Spotty marbling scattered like rice grains.

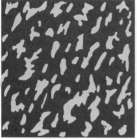 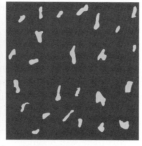

| TABLE 7-4 | Percentage of Lean Meat Required for USDA Yield Grades |
|---|---|
| Yield Grade | Usable Carcass Meat* |
| 1 | 74% or more |
| 2 | 71% |
| 3 | 68% |
| 4 | 64% |
| 5 | 61% or less |

*Percentages vary depending on the range within each yield grade and the way the carcass is cut.

the muscle fibers increases and more connective tissue develops, resulting in toughening of the meat (32).

***Heredity***  Cuts of meat will vary in tenderness because of genetic factors. For example, beef from Black Angus cattle, which are bred to be heavily muscled and marbled, will be very different from meat obtained from dairy cattle or from one of the other, larger breeds of cattle.

***Diet***  The type of diet fed to the animal directly influences its fat accumulation, which is one of the factors affecting the tenderness and flavor of its meat. Ranchers have long known that grain-fed cattle yield ground beef that is more tender and better flavored than that from cattle fed hay or left to feed on the range (62).

***Marbling***  Fattening animals before slaughter is thought to increase tenderness by increasing marbling and the development of subcutaneous fat. The amount of subcutaneous fat on the carcass contributes to tenderness by delaying the speed at which the carcass chills

| TABLE 7-3 | Top Three USDA Quality Grades for Beef |
|---|---|
| USDA Grade | What the Grade Means |
|  | Very tender, juicy; flavorful; the greatest degree of marbling. The most expensive of the grades, Prime is sold to finer restaurants and some meat stores. |
|  | Quite tender and juicy, good flavor; slightly less marbling than Prime. The grade most frequently found in retail stores. |
| USDA SELECT | Fairly tender; not as juicy and flavorful as Prime and Choice; has least marbling of the three, and is generally lower in price. |

**FIGURE 7-9**  USDA yield grades. The amount of lean muscle is compared in a ratio to the non-meat portion—fat, bone, and inedible material.

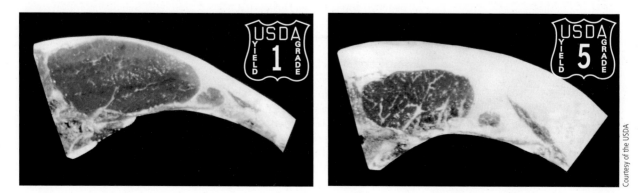

when refrigerated. When choosing meat cuts, consumers seem to prefer lean-looking meats over more marbled ones, but usually reverse their choices in a taste test after those same cuts are prepared.

*Rigor Mortis*   Within 6 to 24 hours after slaughter, the muscles of livestock enter the state of **rigor mortis.** This condition reverses naturally 1 or 2 days after slaughter.

The way that meat is handled during this period is important, because it can affect pH, which influences meat quality. The perception of a meat's juiciness or dryness depends on the binding of water to muscle proteins, and this is influenced by pH. Water-holding capacity is best in meats with a pH of 5.8 (44). A pH that is too low or too high results in less than desirable meat.

Meat pH changes during rigor mortis because the oxygen-deprived cells switch to glycogen as an energy source. Glycogen is converted to lactic acid and the increase in acidity causes the pH to fall from approximately 7.0 to 5.8, which is a desirable pH for meat quality.

*Slaughtering Conditions*   Both the conditions preceding slaughter and the handling of the carcass immediately afterward can result in several problems affecting the meat's quality: poor quality; dark-cutting beef; pale, soft, and exudative (PSE) pork; thaw rigor; cold shortening; and green meat. The first three problems are all related to pH. Under ideal conditions, the process of rigor mortis results in meat with a pH of 5.8; however, the pH can drop either too much or not enough depending on the timing and duration of stress to the animal.

- *Poor Quality.* A poor-quality meat will result if all the glycogen has already been converted to lactic acid prior to slaughter. This causes

the pH to drop too low and occurs when the animal has already used up all of its glycogen as a result of stress from fear, fasting, temperature extremes, and/or exercising.

- *Dark-cutting beef.* If glycogen stores are depleted before death because the animal is exercised or stressed, insufficient lactic acid will be produced during rigor mortis. The resulting higher pH (above 5.8) of the meat will result in a deep-purple brown meat known as dark-cutting beef, which has a sticky texture that is unacceptable to consumers (42).
- *PSE pork.* Pale, soft, and exudative (PSE) pork results if the pH drops too low, causing the meat to become very dry when cooked. A low pH—under 5.1, or even up to 5.4—can cause the pork to become extremely pale, mushy, slimy, flavorless, and full of excess drip (16). Japanese export buyers evaluate pork based on its pH (8). Because a low pH can damage meat proteins and alter color to a pale tan, they watch for this color change and select darker-colored meats with a higher pH, and thus better flavor. Some meat packing companies even measure the pH of their meat products.
- *Thaw rigor.* Freezing meat before it undergoes rigor mortis can cause thaw rigor, a phenomenon in which the meat shrinks violently by almost 50% when thawed.
- *Cold shortening.* A kind of thaw rigor occurs, although to a lesser degree, when meat has been chilled too rapidly before rigor mortis, called cold shortening. In both cases, the meat will be tougher. Neither thaw rigor nor cold shortening meat is allowed to be sold at the consumer level.
- *Green meat.* Meat cooked while in a state of rigor mortis (before the muscles have had time to relax) will be tough and tasteless. However, it can be quite tender if prepared before stiffening begins.

*Aging*   Aging meats improves their juiciness, tenderness, flavor, color, and ability to brown during heating. This treatment pertains primarily to beef (64). All fresh beef is aged for at least a few days and may be aged up to several weeks. Enzymes naturally found in the meat break down the muscle tissue, improving its texture and flavor. Hanging the carcass also aids in the aging process

by stretching the muscles (Chemist's Corner 7-6). The animal's species, size, age, and activity before slaughter influence how long rigor mortis lasts. Beef takes about 10 days to age, which is about the same amount of time it takes for meat to be transported, packaged, and sold to the consumer. Top-quality beef is often aged longer, up to 6 weeks. Mutton is sometimes aged, but pork and veal come from such young animals that aging is not required to increase tenderness. The fat in pork tends to go rancid quickly, and veal's lack of protective fat covering causes it to dry out too quickly—further reasons these meats are not routinely aged.

Meats are aged in one of several ways. The time required for aging depends on the method used.

- *Dry aging.* Carcasses are hung in refrigeration units at 34°F–38°F (1°C to 3°C) with low (70 to 75%) or high (85 to 90%) humidity for 1½ to 6 weeks. Specialty steak houses and fine restaurants usually purchase dry-aged meat. The meat is more expensive than other types of aged beef because the exposure to air can cause it to lose up to 20% of its original weight. The carcass weight is further reduced because the dry exterior layer must then be trimmed. The advantage of dry aging is that the dehydration concentrates the meat's flavor, making it more succulent and mellow (5).
- *Fast or wet aging.* Most beef is aged in plastic shrink-wrap. Warmer temperatures of 70°F (21°C) with a high humidity of 85 to 90% lower the aging time to 2 days, but additional aging will occur during the 10 or so days it takes the meat to reach the consumer. Ultraviolet lights are used to inhibit microbial growth. Most retail meat is fast aged.

**Rigor mortis**   From the Latin for "stiffness of death," the temporary stiff state following death as muscles contract.

**Aging**   Holding meat after slaughter to improve texture and tenderness. A ripening that occurs when carcasses are hung in refrigeration units for longer periods than that required for the reversal of rigor mortis.

> **CHEMIST'S CORNER 7-6**

**Passing of Rigor Mortis**

The passing of rigor occurs when the muscles gradually extend again. This is facilitated by the proteases that hydrolyze proteins and disrupt the Z bands. As a result, the actin and myosin release from each other, causing the muscles to relax (29).

- *Vacuum-packed aging.* Less weight loss and spoilage occur in meats that are aged by vacuum packing (cryovacing). During this process, meat carcasses are divided into smaller cuts, vacuum packed in moisture- and vapor-proof plastic bags, and then aged under refrigeration.

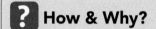

### ? How & Why?

#### Why does a carcass stiffen?

Rigor mortis is caused by a cascade of events that take place at the cellular level. Death interrupts the blood flow and prevents oxygen from reaching the cells. Changes then occur within the cells of the muscles, causing them to contract and stiffen. Muscles stay relaxed in the presence of ATP (adenosine triphosphate), but once it is used up through glycolysis, the lack of ATP causes actin to bind irreversibly with myosin. The muscles then contract. This rigidity of the muscles in rigor mortis occurs because the cross-links between the actin and myosin filaments overlap and cause the sarcomeres to shorten. The automatic contraction of fibrils in the muscle cells causes the characteristic muscle stiffness.

### Artificial Tenderizing

External treatments can be applied to meats to increase their tenderness. These include the use of enzymes, salts, acids, and mechanical methods such as grinding or pounding.

*Enzymes*  One of the reasons that contracted muscles begin to "relax" toward the end of rigor mortis is that proteolytic enzymes work internally to break down the proteins within the muscle fibrils (29). A more even distribution of enzymes may be achieved by injecting a tenderizing solution of papain, or some other proteolytic enzyme, into the bloodstream of animals 10 minutes before slaughter. This optional treatment sends enzymes traveling to all the muscles through the circulatory system, but they are not activated until meat from the animal is exposed to heat during preparation. This process not only increases tenderness, but shortens the time of rigor mortis and aging as well (47).

Commercial meat tenderizers containing enzymes are available for consumers to use, but they are effective only on fairly thin cuts of meat because they penetrate to a depth of only 1/2 to 2 millimeters. They are ineffective on larger cuts such as roasts. Tenderizers are sold as a salt or liquid mixture and differ in the proteolytic enzymes they contain: papain from papayas, bromelin from pineapples, ficin from figs, trypsin from the pancreases of animals, and rhyozyme P-11 from fungi.

The enzymes are not active at room temperature. The optimal activity temperature (highest rate of activity) for papain, the most common tenderizing enzyme, is about 131°F–170°F (55°C–76°C), which is reached only during heating. Exceeding 185°F (85°C) denatures the enzyme, thus inhibiting its activity. Uniform distribution is hard to achieve with the use of commercial tenderizers, and any attempt to get more of the enzyme to penetrate by adding excessive amounts of it can cause the meat to have an unappetizing, mealy, mushy texture.

### ? How & Why?

#### How Do Meat Tenderizers Work?

Meat tenderizers contain enzymes that break down muscle proteins. They are sprinkled on meat, which is then pierced with a fork to drive the enzymes below the surface, where they hydrolyze muscle cell proteins and connective tissue when activated by the heat of preparation.

*Salts*  Tenderness can also be increased by the addition of salts in the form of potassium, calcium, or magnesium chlorides. These salts retain moisture and break down the component that surrounds the muscle fibers, resulting in the release of proteins. Polyphosphates are sometimes added to the salts to improve the meat's juiciness by increased water retention ability (49), and, if added to processed meats, they also increase firmness, emulsion stability, and antimicrobial activity (25). However, this increased water retention capacity is accompanied by an increase in sodium concentration.

*Acids*  Meats can be made more tender by applying marinades containing acids or alcohol, which break down the outside surface of the meat (56). The various acids found in marinades include vinegar; wine; and lemon, tomato, and other fruit juices. Not only do marinades tenderize the meat, but they increase flavor and also contribute to color. The maximum benefit of a marinade can be obtained by increasing the surface area of the meat. This may be done by cutting the meat into small pieces, such as teriyaki strips or kabob cubes. Marinades penetrate only the surface of the meat and are therefore not effective at tenderizing large cuts of meat or poultry. Generally, the acid in a marinade is responsible for tenderizing, although some marinades rely on added enzymes from certain tropical fruits such as papayas and pineapples. The meat is then allowed to soak in the marinade, in the refrigerator, from half an hour to overnight, or for several days for sauerbraten.

*Mechanical Tenderization*  Meat can be tenderized mechanically by several methods, including grinding, cubing, needling, and pounding. These actions physically break the muscle cells and connective tissue, making the meat easier to chew. Grinding and cubing meat simply increase the surface-area-to-volume ratio, causing the teeth to have less work to do. Needling uses a special piece of equipment to send numerous needle-like blades into the meat, separating the tissues. Because of the equipment required, needling is usually not practiced at the consumer level. Another method of mechanical tenderization more easily done in the home is simply pounding the meat with a special hammer that breaks apart its surface tissue.

*Electrical Stimulation*  The meat of beef cattle and sheep, but not swine, becomes more tender when a current of electricity is passed through the carcass after slaughter and before the onset of rigor mortis. Electrical stimulation speeds up rigor mortis by accelerating glycogen breakdown and enzyme activity, which disrupts protein structure, making the meat more tender (10). In this way, the meat can be immediately cut up without any loss of quality.

# Cuts of Meat

Consumer confusion abounds when it comes to deciphering the various meat cuts. Part of this confusion stems from the fact that there are two major types of meat cuts, wholesale and retail. Prior to reaching the supermarket and the consumer, a carcass is divided into about seven **wholesale** or **primal cuts.** Although the carcasses of each species are sectioned slightly differently, the basic wholesale cuts are similar to each other and are identified by the major muscles and by bone "landmarks" (see Figures 1 to 5 in the color insert). These wholesale cuts are then divided into the **retail cuts** purchased by consumers. Figure 7-10 shows how common retail cuts are obtained from the wholesale cut of a hindquarter.

## Terminology of Retail Cuts

The use of the standard system of naming retail cuts is not mandatory, so consumers often face additional confusion at the market. The same cut of meat may be called by different names, depending on the retailer or the part of the country in which it is sold. For example, beef chuck cross-rib pot roast is also known as Boston cut, bread and butter cut, cross-rib roast, English cut roast, and thick rib roast. Lack of a single title for each cut prevents consumers from catching on to a standardized nomenclature for cuts based on shear repetition.

Commercial establishments do have a system of standardized names out of necessity. The names and specifications for over 300 cuts of beef, veal, pork, and lamb are known as Institutional Meat Purchases Specifications (IMPS). The IMPS are listed in a booklet, "The Meat Buyers Guide," that can be purchased online. It serves as an industry reference for those responsible for the preconsumer purchase and the sale of meat. Most retail meat markets adhere to "The Meat Buyers Guide." Under this system, meat

**FIGURE 7-10**    How retail cuts are obtained from a hindquarter wholesale cut.

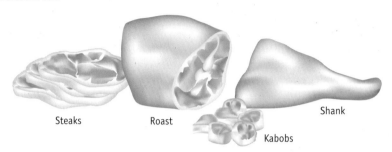

Steaks        Roast        Kabobs        Shank

**FIGURE 7-11**    Meat labeling.

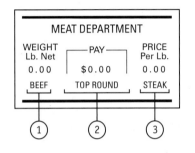

1. **The kind of meat**
   BEEF, VEAL, PORK or LAMB. It is listed first on every label.

2. **The primal (wholesale) cut**
   CHUCK, RIB, LOIN or ROUND. Tells where the meat comes from on the animal.

3. **The retail cut**
   BLADE ROAST, SPARERIBS, LOIN CHOPS, etc. Tells from what part of the primal cut the meat comes.

labels include the species (beef, veal, pork, or lamb), primal cut, and retail cut (Figure 7-11). Hence, rib eye steak would be labeled "Beef, rib, rib eye steak."

## Beef Retail Cuts

Rib, short loin, and sirloin wholesale cuts lie along the back of the animal and are usually the most tender and expensive cuts of beef (Figure 7-12). Rib roasts are the most tender roasts, and tenderloin the most tender steak. Filet mignon is the small end of the tenderloin, but some retailers incorrectly, perhaps deliberately, label any cut from the tenderloin as filet mignon. Although less tender, chuck (shoulder area) and round (rump area) wholesale cuts provide many popular retail cuts. The least tender wholesale cuts are flank, short plate, brisket, and foreshank (Table 7-5).

---

**Wholesale (primal) cuts** The large cuts of an animal carcass, which are further divided into retail cuts.

**Retail cuts** Smaller cuts of meat obtained from wholesale cuts and sold to the consumer.

---

**TABLE 7-5    Retail Cuts Obtained from the Primal Cuts of Beef**

**Some of the more tender retail cuts**

| Rib | Short Loin | Sirloin |
|---|---|---|
| Rib eye (Spencer) | Tenderloin | Sirloin steak (Delmonico) |
| Rib roast | Porterhouse steak | Top loin steak |
| Rib steak | T-bone steak | Sirloin tip roast |
| | | Tenderloin steak |

**Less tender but still popular retail cuts**

| Chuck | Round |
|---|---|
| Chuck roast | Top round steak or roast |
| Cross-rib roast | Eye of round steak or roast |
| Boneless chuck eye roast | Bottom round or roast |
| Blade roast or steak | Rump roast |
| Arm pot roast or steak | Heel of round |
| Boneless shoulder pot roast or steak | Cubed steak |
| Chuck short ribs | Ground beef |
| Stew meat | |
| Ground chuck | |

**The least tender cuts**

| Flank | Short Plate | Brisket | Foreshank |
|---|---|---|---|
| Flank steak | Skirt steak rolls | Brisket | Crosscut shank |
| | Short ribs | Corned beef | Stew meat |
| | Stew meat | Stew meat | |

**FIGURE 7-12** Wholesale and retail cuts of beef.

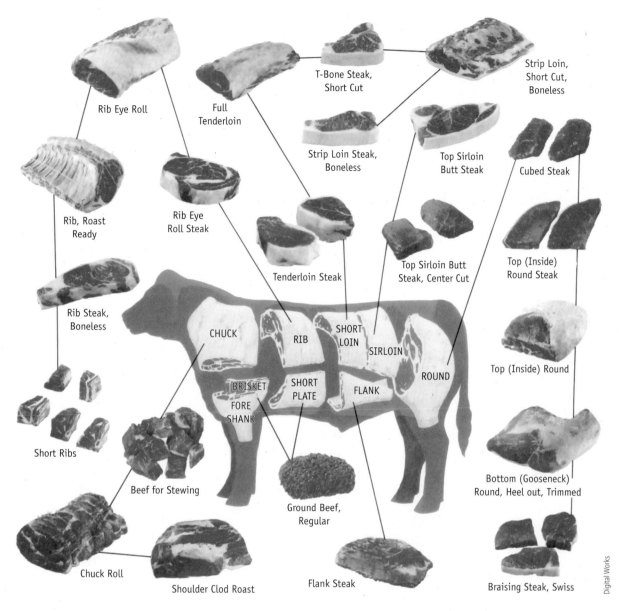

Rib Eye Roll

T-Bone Steak, Short Cut

Strip Loin, Short Cut, Boneless

Full Tenderloin

Strip Loin Steak, Boneless

Top Sirloin Butt Steak

Cubed Steak

Rib, Roast Ready

Rib Eye Roll Steak

Tenderloin Steak

Top Sirloin Butt Steak, Center Cut

Top (Inside) Round Steak

Rib Steak, Boneless

CHUCK

RIB

SHORT LOIN

SIRLOIN

ROUND

Top (Inside) Round

BRISKET

SHORT PLATE

FLANK

FORE SHANK

Short Ribs

Beef for Stewing

Ground Beef, Regular

Bottom (Gooseneck) Round, Heel out, Trimmed

Chuck Roll

Shoulder Clod Roast

Flank Steak

Braising Steak, Swiss

Digital Works

### Ground Beef

*Ground Beef* About 44% of all fresh beef is sold in the form of ground beef and used extensively in fast-food restaurants, schools, military programs, and homes (12). The terms *ground beef* and *hamburger* are often used interchangeably, but there is a difference. The USDA classifies ground beef as beef that has been ground. Hamburger is ground beef that is often combined with ground fat; seasonings may also be added. Neither ground beef nor hamburger may exceed 30% fat by weight. Regular ground beef contains 30% fat, lean ground beef about 23%, and extra lean ground beef does not exceed 15% fat. Draining the fat off hamburger or ground beef during and after cooking lowers the fat content appreciably. Consumer preference

studies have shown that ground beef containing 15 to 20% fat is preferred. Reducing the fat content below 20% decreases the flavor, tenderness, and juiciness of the product (13).

The fat in ground beef can be reduced by adding extenders such as nonfat dry milk solids, texturized vegetable protein (TVP), plant starches, soy proteins, oat bran or fiber, modified food starches, maltodextrins (starches), and vegetable gums (carrageenan) (21). Many of these extenders enhance the flavor as well as lower the fat content.

### Veal Retail Cuts

The retail cuts of veal, like those of pork and lamb described below, are fewer in number than those of beef because the

carcasses are smaller. The hind legs of these animals are suitable for roasts, but veal roasts are usually tender regardless of their wholesale cut origin.

### Pork Retail Cuts

Pork is usually tender, regardless of the cut, because it comes from animals under 1 year of age. When compared to beef, veal, or lamb wholesale cuts, the wholesale loin and spare rib cuts of pork are much longer because there is no separation of the rib and sirloin as in other carcasses (Figure 7-13). In addition, modern breeders have developed an even longer swine with fourteen ribs (as compared with thirteen in beef and lamb). The leg is the largest primal cut, representing about one fourth of the

**FIGURE 7-13**    Wholesale cuts of pork.

# PURCHASING PORK

# A Consumer Guide To Identifying Retail Pork Cuts.

Left: tenderloin
Right: Canadian-style bacon

TENDERLOIN & CANADIAN-STYLE BACON

CHOPS

## CHOPS

Upper row (l-r):
sirloin chop, rib chop, loin chop.
Lower row (l-r):
boneless rib end chop (Chef's Prime Filet™), boneless center loin chop (America's Cut™- 1 1/4-1 1/2″ thickness), butterfly chop.

RIBS

ROASTS

## ROASTS

Upper row (l-r):
center rib roast (Rack of Pork),
bone-in sirloin roast.
Middle: boneless center loin roast.
Lower row (l-r):
boneless rib end roast (Chef's Prime™),
boneless sirloin roast.

SHOULDER BUTT

**RIBS** Left: country-style ribs.
Right: back ribs.

## SHOULDER BUTT

Upper row (l-r):
bone-in blade roast, boneless blade roast.
Lower row (l-r): ground pork (The Other Burger®), sausage, blade steak.

PICNIC SHOULDER

## PICNIC SHOULDER

Upper row (l-r): smoked picnic, arm picnic roast.
Lower row: smoked hocks.

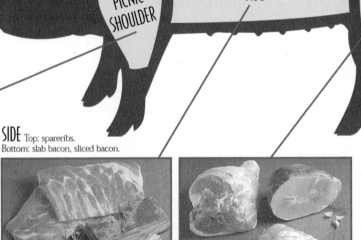

SIDE

LEG

**SIDE** Top: spareribs.
Bottom: slab bacon, sliced bacon.

## LEG

Upper row (l-r):
bone-in fresh ham, smoked ham. Lower row (l-r): leg cutlets, fresh boneless ham roast.

pork.

SHOULDER BUTT

LOIN

PICNIC SHOULDER

SIDE

LEG

carcass weight. The following wholesale cuts provide the majority of fresh pork retail cuts:

- **Loin.** Pork loin chop or roast, Canadian-style bacon, pork loin tenderloin
- **Spare rib.** Spare ribs, bacon, salt pork

Baby back ribs are cut from the backbone, close to where tenderloin and pork chops are cut (50). These ribs, also called top loin or loin back ribs, are smaller and more expensive than spareribs. Spareribs are cut from the belly, similar in location to bacon. Spareribs are fattier and meatier, but are also cheaper.

Thicker cuts of pork, up to 1½ inch, are juicier because they have a lower surface area to weight ratio. This results in better retention of their natural juices (55).

### Lamb Retail Cuts

Lambs are smaller than either cattle or swine, so the leg wholesale cuts are usually cut into roasts, with leg of lamb being the most common (Figure 7-14). A rack of lamb consists of seven or eight rib chops; the backbone is usually removed to make carving easier. A fancier cut is crown roast of lamb, which consists of two rib sections or racks attached to the backbone. Formed into a circle or crown, it can be stuffed and is often decorated just before serving by covering the bone tips with paper frills, making a very handsome main dish for any table. Lamb chops are frequently cut from the loin, rack (rib), or shoulder. Loin chops are the most tender.

### Variety Meats

**Variety meats,** also known as organ, offal, sundry, or specialty meats, can be divided into two categories: organ meats and muscle meats (Figure 7-15). Organ meats such as liver, kidneys, and brains from young animals are generally very soft, are extremely tender, and require only very short heating times. Sweetbreads can be obtained only from calves or young beef, because the thymus gland disappears as the animal matures.

The meat of heavily exercised muscles such as the tongue and heart is quite tough and requires long, slow

**FIGURE 7-14**   Wholesale cuts of lamb.

*Courtesy of the American Lamb Board*

cooking. Tripe, the inner lining of the stomach, can be smooth or honeycombed. Smooth tripe originates from the first stomach, and honeycombed tripe, which is more popular, comes from the second stomach. Both types are extremely tough and strong in flavor. As with the tongue and heart, they require long, slow cooking.

### Kosher Meats

Kosher meats are from certain animals (cattle, sheep, and goats, but not swine) designated as clean that have been slaughtered according to Jewish religious practices dating back more than 3,000 years (see Chapter 1). The animal must be slaughtered in the presence of a rabbi or other approved individual with a single stroke of a knife, be completely bled, and have all its arteries and veins removed. Blood must not be consumed because in the Jewish tradition it is synonymous with life. The hindquarter is rarely used for kosher meats because it is so difficult to remove the blood vessels in this area.

**Variety meats** The liver, sweetbreads (thymus), brain, kidneys, heart, tongue, tripe (stomach lining), and oxtail (tail of cattle).

**FIGURE 7-15** Variety meats.

Oxtail

Tongue

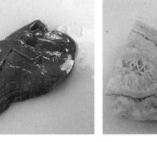
Heart

Kidney

Sweetbreads

Brains

Liver

Tripe

Digital Works

## Halal Meats

As discussed in Chapter 1, *halal* is defined as "permitted," and it often refers to meat. Most meat is allowed except pork and carnivorous animals with fangs (lions, wolves, tigers, dogs, etc.). Acceptable animals need to be sacrificed according to Muslim guidelines.

## Organic Meats

The demand for the more expensive organic meats is increasing. Organic beef standards were established in 2002 by the United States government. Organic meat is defined as being derived from cattle fed only milk, grasses, and grains from birth to slaughter.

# Processed Meats

About one third of all meat is processed, meaning it has been changed from its original fresh cut (Table 7-6). Ham, sausage, and bacon are the most popular processed meat products. Other examples of processed meats include salami, bologna, bratwurst, and pastrami.

## Processing Methods

Before the advent of refrigeration, meat was preserved by such processing

**TABLE 7-6** Examples of Processed Meats: The meats are grouped according to their major meat ingredient. Differences within each group are based on added ingredients and processing techniques

| Beef | Pork/Ham | Beef and Pork | Veal and Pork | Liver |
|------|----------|---------------|---------------|-------|
| Beef bologna | Blood sausage | Club bologna | Bockwurst | Braunschweiger |
| Beef salami | Bratwurst | Cervelat | Bratwurst | Liverwurst (pork) |
| Pastrami | Capacolla | Frankfurters* | Veal loaf | |
| | Chorizo | Honey loaf | Wiesswurst | |
| | Frizzies | Hot dog* | | |
| | Ham | Knockwurst | | |
| | Ham bologna | Luncheon meat | | |
| | Linguica | Mettwurst | | |
| | Lola/Lolita | Mortadella | | |
| | Luncheon meat | Olive loaf | | |
| | Lyons | Peppered loaf | | |
| | New England-style | Pimento loaf | | |
| | sausage | Salami | | |
| | Old-fashioned loaf | Smokies | | |
| | Pork sausage | Weiner* | | |
| | Prosciutto | Vienna sausage* | | |
| | Salsiccia | | | |
| | Scrapple | | | |
| | Thuringer | | | |

*Terms used interchangeably.

methods as curing, smoking, canning, and drying.

*Curing* Commonly cured meat products include ham, bacon, sausages, frankfurters, corned beef, and luncheon meats. Meat once was cured by saturating it with salt. Corned beef, a cured beef brisket, was so named because in the 16th century the word *corn* was used interchangeably with *grain,* so meat rubbed with coarse grains of salt was called *corned* (23). The term "cured" today is generally used to mean the addition of synthetic nitrates or nitrites, salt, and other preservatives (7). This mixture often includes sodium or potassium nitrate, sugar, spices, phosphates, dextrose, corn syrup, lactates, and seasonings (7). Nevertheless, salt remains one of the major flavoring agents of cured meat. The different proportions and combinations of ingredients used for curing contribute to the varying flavors of cured meats, which often garner additional flavor from being smoked.

### ? How & Why?

**How does nitrite cure meat?**

Nitrite or nitrate salts are used in most cured meat products. When added to meat, the nitrite molecule (-NO$_2$) chemically reacts with the meat to create nitric oxide, which is the agent responsible for curing the meat (7). Nitrate (-NO$_3$) will not cure meat until it is converted to nitrite; this can be accomplished before or after it is added to the meat.

There are several ways to cure meat. Dry curing consists of mixing the ingredients together and rubbing them into the surface of the meat, so that they can penetrate their way to the center. Another method involves brining the meat (soaking it in a salt solution) or immersing it in a pickling solution. Salt and other flavors migrate into the meat, making it more flavorful. Osmosis is the mechanism whereby fluids diffuse through cells into the meat to equalize the increased salt (ion) concentration. The greater salt concentration in the meat causes it to absorb some of the water of the brining solution (6 to 8% of the meat's original weight), resulting in a moister meat (6). Whole turkey carcasses are commonly brined; however, this process can be applied to any meat.

The most common commercial curing technique is one in which the curing solution is mechanically pumped or injected into the meat using a machine lined with needles. These injected curing solutions increase the meat's weight. If the meats are not shrunk back to their original weight through heating and/or smoking, and if they contain up to 10% added moisture, they must be labeled "Water Added."

Although the original purpose of salting foods was to keep them from spoiling, now that refrigeration is widely available, meats are no longer cured solely for preservation. The high sodium content of many cured meat products now serves several purposes: to provide flavor, to improve texture by facilitating the binding of proteins, and to increase the proteins' water-binding capacity, which reduces fluid loss within the packages. Nevertheless, lower-sodium processed meats are becoming increasingly available on the market.

*Smoking*   Most cured meats are also smoked and cooked. Smoke imparts flavor, aroma, and color to foods. Meats are placed in smokers, where they are exposed to the smoke of burning wood. In smoke houses, the intensity of the smoke, the humidity, and the temperature are all carefully regulated, and the type of sawdust or wood used to produce the smoke determines the product's resulting flavor. Sawdust is the most economical fuel and is often used by commercial processors, but other woods available for smoking include mesquite, hickory, oak, apple, and various combinations of them. In the late 1800s, a technique was developed to distill the smoke from burning wood to create liquid smoke, which could be spread on cured meats to achieve the flavor of meat smoked in a smoke house (43). Today the use of liquid smoke is more common, and it saves time and minimizes air pollution. Although the additional flavor provided by smoked meats is preferred by some consumers, there is some concern about its posing a possible cancer risk regardless of the type of smoking used.

*Canning*   Canned meats are processed through either pasteurization or sterilization. Pasteurized canned meats require refrigeration and are labeled "Perishable—Keep Refrigerated," whereas those that are sterilized do not need refrigeration as long as the can remains sealed.

*Drying*   Drying is not widely used for meats, but it has some applications for them. Certain types of sausage, including pepperoni, salami, and cervelat, are dried. They are cooked, sometimes smoked, and dried under specific conditions of humidity and temperature. Beef jerky, usually dried to a water activity of 0.7 to 0.85, is convenient, is ready to eat, and requires no refrigeration (52).

## Food Additives in Processed Meats

A variety of food additives are added to processed meats, as shown in Figure 7-16. Nitrite is a common food additive used by the food industry to keep processed meat from turning brown (Chemist's Corner 7-7). Nitrite and salts of nitrate are used as a preservative in approximately 7% of foods, particularly processed meats such as ham, hot dogs, bacon, sausage, bologna, salami, and other cold cuts (4). These food additives are responsible for keeping many packaged processed meats permanently pink, while simultaneously reducing the risk of botulism and creating a distinctive flavor (46). To maintain the pinker color in the grocery store, ham slices are often stored upside down with the label on the back of the Styrofoam board because grocery store lights promote undesirable color changes.

The safety of foods containing nitrites became an issue after the discovery that carcinogenic nitrosamines can form when nitrites combine with secondary amines in the stomach acid (1). This concern resulted in the lowering of nitrite levels used in processing, but not in their elimination, because of their role in preventing botulism poisoning. In fact, nitrites are also formed in the body, and are found naturally in such foods as cabbage, cauliflower, carrots, celery, lettuce, radishes, beets, and spinach. Nevertheless, antioxidants such as ascorbic acid (vitamin C) or vitamin E are now often added to cured meats to help reduce nitrite reactions.

Other additives used as preservatives in processed meats include BHA (butylated hydroxyanisole), BHT (butylated hydroxytoluene), citric acid, potassium nitrite, propyl gallate, and EDTA (ethylenedi-aminetetraacetic acid). Potassium sorbate and propylparaben are used as preservatives to prevent mold growth on sausage casings. Flavoring additives include sucrose, sodium, sorbitol, corn syrup, glucose (dextrose), hydrolyzed vegetable (plant) protein, and MSG (monosodium glutamate). Papain is sometimes added as a meat tenderizer.

**FIGURE 7-16**  Selected additives[1] used in processed meat and poultry products.

**Butylated Hydroxytoluene (BHT), Butylated Hydroxyanisole (BHA), Tocopherols (Vitamin E)**—antioxidants that help maintain the appeal and wholesome qualities of food by retarding rancidity in fats, sausages, and dried meats.

**Carrageenan**—seaweed is the source of this additive that may be used as a binder in meat products.

**Citric Acid**—widely distributed in nature in both plants and animals. It can be used as an additive to protect the fresh color of meat cuts during storage. Citric acid also helps protect flavor and increases the effectiveness of antioxidants.

**Corn Syrup**—sugar that is derived from the hydrolysis of cornstarch. Uses include flavoring agent and sweetener in meat and poultry products.

**Emulsifier**—substance added to products, such as meat spreads, to prevent separation of product components to ensure consistency. Examples of these types of additives include lecithin, and mono- and diglycerides.

**Gelatin**—thickener from collagen that is derived from the skin, tendons, ligaments, or bones of livestock. It may be used in canned hams or jellied meat products.

**Humectant**—substance added to foods to help retain moisture and soft texture. An example is glycerin, which may be used in dried meat snacks.

**Monosodium Glutamate (MSG)**—MSG is a flavor enhancer. It comes from a common amino acid, glutamic acid, and must be declared as monosodium glutamate on meat and poultry labels.

**Phosphates**—the two beneficial effects of phosphates in meat and poultry products are moisture retention and flavor protection. An example is the use of phosphates in the curing of ham in which approved additives are sodium or potassium salts of tripolyphosphate, hexametaphosphate, acid pyrophosphate, or orthophosphates, declared as phosphates on labels.

**Propyl Gallate**—used as an antioxidant to prevent rancidity in products such as rendered fats or pork sausage. It can be used in combination with antioxidants such as BHA and BHT.

**Sodium Caseinate**—used as a binder in products such as frankfurters and stews.

**Sodium Erythorbate**—the sodium salt of erythorbic acid, a highly refined food-grade chemical closely related to vitamin C, synthesized from sugar, and used as a color fixative in preparing cured meats.

**Sodium Nitrite**—used alone or in conjunction with sodium nitrate as a color fixative in cured meat and poultry products (bologna, hot dogs, bacon). Helps prevent growth of *Clostridium botulinum*, which can cause botulism in humans.

**Whey, Dried**—the dried form of a component of milk that remains after cheese making. Can be used as a binder or extender in various meat products, such as sausage and stews.

[1]See Chapter 3 for more information about food additives in processed meats.
**Source:** Food Safety and Inspection Service, United States Department of Agriculture.

Annatto, saffron, and tumeric are coloring agents used in sausage casings. Cochineal, derived from the dried female insect, *Coccus cacti*, generates a red color in some meat products.

"Natural" additives are sometimes used to accomplish the same goal as chemical additives. While "natural curing" is not a term recognized by the USDA, meats have been cured naturally for centuries in a wide variety of cultures, including those of ancient Egypt and Rome (7). Natural curing is sometimes used to refer to the use of microbes (typically bacteria) to convert nitrates in the environment to nitrites, which then react with meat to cure it. For example, the addition of harmless food-grade microorganisms and sea salt to meat can result in a natural curing process.

Examples of other additives considered as natural curing agents include carrageenan (seaweed), sodium bicarbonate (baking soda), vegetable and fruit juices, vinegar, honey, sugar, food start cultures, and spices.

Flavoring additives that are considered natural include celery, onion, garlic powder, and fruit or vegetable juices.

When meat is packaged and labeled as "natural," this term refers to the additives in the product, not the meat itself. This is because all meat is considered natural, irrespective of the antibiotics or hormones used in raising the animals (7).

## Types of Processed Meat

There are three types of meats that are commonly processed: ham, bacon, and sausage. In addition, lower-fat

processed meats are becoming popular with consumers.

**Ham**  Ham is cured pork, and according to USDA standards, only meat from the hind leg of a hog can be labeled ham. Several types of cooked ham products are available for purchase:

- *Canned ham.* Boneless, fully cooked ham that can be served cold or heated. Most are cooked only to pasteurization temperatures, so they must be refrigerated. Sterilized hams are usually available only in cans of under 3 pounds. Gelatin is often added in dry form to absorb the natural juices of the ham as it cooks.
- *Water-added ham.* Contains no more than 10% by weight of water

## CHEMIST'S CORNER 7-7

### Color Changes in Cooked Meats: Nitrites

Nitrite, a conjugate base of a weak acid named nitrous acid, provides color to processed meats by combining with the myoglobin pigment to produce nitrosyl-myoglobin. This resulting compound denatures during the cooking phase of the curing process to form a pink-colored compound called nitrosyl-hemochrome (Figure 7-17). One of the major problems of storing nitrite-cured meats is that continued exposure to oxygen and light oxidizes the iron from the ferrous ($^+2$) to ferric form ($^+3$), which results in a brown discoloration. Additional oxidation of the pigment-containing protein's porphyrin ring, instead of the iron, results in a very undesirable yellow or greenish color, making the meat unappealing to consumers (46).

**FIGURE 7-17** Color changes in cooked meats.

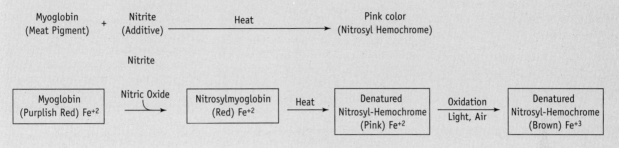

added. The added moisture contributes to a moist, juicy, and tender texture.

- **Imitation ham.** Ham that retains more than 10% moisture after curing.
- **Country ham.** Ham cured by the dry salt method and usually hickory smoked to develop a distinctive flavor.
- **Picnic ham.** Cured pork that comes from the front leg instead of back leg of the hog, and therefore cannot be labeled simply *ham*. This cut is less tender and higher in fat than regular ham.

**Bacon** Bacon is cured and smoked meat from the side of a hog. It should be balanced in its proportion of fat to lean. When cooked, bacon with too much lean will be less tender, whereas bacon with too high a proportion of fat will shrink too much.

**Sausage** Sausage originated in the Mediterranean (34). It is meat that has been finely chopped or ground and blended with various ingredients, seasonings, and spices. The seasonings usually include a curing salt, which is partly responsible for the distinctive flavor (60). It is then stuffed into casings or skins. Traditionally, the casings were made of the intestines of pigs or sheep, but now they are often manufactured from cellulose or collagen. Beef and pork, or a combination of the two, are the usual main ingredients. Other meats and meat combinations may be used, including veal, chicken, turkey, lamb, duck, rabbit, venison, and liver from any of several animals. Other ingredients that may be added include eggs, cream, oatmeal, breadcrumbs, potato flour, tripe, wine, and beer. Pork and/or beef fat are often added to boost the moisture content and enhance the texture. There are four major classifications of sausage:

- **Uncooked.** Made from ground, uncooked meat. Fresh pork sausage, bratwurst, mettwurst, and bockwurst are examples. New combinations of chicken, turkey, apple, and other lower-fat alternatives are available. Uncooked, fresh sausage has the highest moisture content, approximately 50–60% (60). Fresh sausage can be stored for several days, and requires cooking before ingestion.
- **Raw fermented.** Raw fermented sausage is kept at high temperatures, allowing the growth of bacteria. These bacteria produce lactic acid, which lowers the pH of the sausage (60). Examples of this type of sausage include merguez and Lebanon bologna, which is smoked. Raw fermented sausage must be cooked before eating.
- **Cooked.** Made from cured meat, which may be slightly smoked before being stuffed into the casings. Examples include hot dogs, bologna, and knockwurst. Chorizo, a spicy pork sausage native to Spain, may be sold fully cured and ready to eat or may require cooking at home.
- **Dry/semidry.** Made of cured meat that has been dried. Dry fermented sausage is thought to have originated in Italy in the early 18th century (60). Several decades later, the practice was adopted in Germany, where sausage became a major component of the local diet. Examples are pepperoni, salami, thuringer, and cervelat. Dried, cured sausage undergoes a ripening period in which the texture changes from a soft, pliable mass into a hard, sliceable, distinctly flavored sausage. The unique flavors of dry cured sausage result from the enzymatic breakdown of proteins, carbohydrates, and lipids to smaller compounds that exhibit intense aromas (30). Sausage is considered semidry when the weight decreases by 20%; these sausages have a moisture content of 35–50%. The pH of semidry sausage is around 5.0, and the water activity is above

0.86. Summer sausage is an example of semidry sausage. Dry sausage loses greater than 30% of its weight during the drying process, and generally has a moisture content below 35% (60). Italian Turista, salami, Spanish Salchichon, pepperoni, and French saucisson are types of dry sausage. Dry and semidry sausages can be fermented for various amount of time. A fermentation process of 7 days or less is considered rapid, while 3 weeks is considered regular and greater than 3 to 4 months is considered slow (60).

### Lower-Fat Processed Meats

Many processed meats contain 30 to 50% fat. Consumers have challenged processed meat product manufacturers by demanding foods that are lower in fat and cholesterol. Many processed meats are available in versions that are ≥95% fat-free.

Lower-fat processed meat products are produced by using leaner cuts of meat, adding more water, and/or including ingredients such as fiber, gums, modified starches, and whey protein concentrate (14). Water can be substituted for fat in processed meats as long as the total amount of fat and water does not exceed 40%, with a maximum fat content of 30% (11). Less fatty ingredients, including fat replacers (see Chapter 22 on fats and oils), may also take the place of more fatty ones.

The federal government used to define certain processed products by a minimum amount of fat, but these regulations have been changed in light of dietary recommendations. For example, cooked frankfurters had been required to contain about 30% fat, but a 1998 change in regulations lowered it to 20%. Sausages used to average 43% fat, but sausages are now available that do not exceed 15% fat (15).

### Mechanically Deboned Meat

The traces of meat that are left on the bones after butchering can be collected and sold as mechanically deboned meat. This is accomplished by grinding the remaining meat and bones together, and removing the bone by putting the mixture through a sieve. The resulting meat contains ground bone, bone marrow, and soft tissue and is most commonly used in further processed meat products. The presence of the bone increases the calcium and trace mineral content of the meat. Processed meat products containing up to, but no more than, 20% mechanically boned meat must include the designation "mechanically separated meat" on the food product's ingredient list.

### Restructured Meat

Restructured or fabricated meat is made from meat trimmings and/or lower-grade carcasses. It is similar to real meat in texture, flavor, and appearance, but is less expensive. The meat trimmings are broken down to particle size by flaking, shredding, grinding, or chopping, and are then bound together into uniform shapes and sizes. Some natural binding between the meat's proteins occurs, but binding is further accomplished by adding nonmeat ingredients such as egg albumen, gelatin, textured soy protein, and wheat or milk proteins (53). The uniformity in shape and weight of the types of products made possible with restructured meat makes it ideal for the fast-food industry and food service establishments (18).

# PREPARATION OF MEATS

Meat is usually the most expensive portion of a meal; therefore, its preparation is usually given extra consideration. It's important to observe the changes in the meat during heating, to look for signs of doneness, and to realize the differences between dry-heat and moist-heat preparations. Selecting a meat cut partially determines how the meat will be prepared. Some cuts are naturally tender, whereas others are tough, so preparation methods must vary accordingly. Tender cuts lend themselves to dry-heat methods such as roasting, broiling, grilling, and frying, whereas tougher cuts are better for long, slow, moist processes such as braising, stewing, or steaming.

Whether meat is prepared by dry-heat methods or by any of the various moist-heat methods, it should first be wiped with a paper towel to remove any surface moisture. Leaving water on the meat or washing it will result in a faded color and the loss of some water-soluble nutrients and flavor compounds. After it is wiped, the meat can be trimmed of any visible fat or connective tissue to reduce calories and increase tenderness. If it is a tougher cut, it can be tenderized according to the techniques discussed earlier.

For best results when preparing frozen meats, they should be thoroughly thawed in the refrigerator or microwave before cooking. Cuts prepared from the frozen state take longer to heat and are less energy and cost efficient. A frozen roast may take up to three times longer to prepare than a thawed roast. Frozen cuts are more difficult to heat evenly, and the center may remain frozen even though the outside looks perfectly done.

## Changes During Heating

### Tenderness and Juiciness

Cooking meats at the correct temperature for the right amount of time will maximize their tenderness, juiciness, and flavor. Although heat makes meat more palatable, exposing it to high temperatures for too long will toughen, shrink, and harden meat because such exposure shortens muscle fibers, denatures proteins, and causes the meat to dehydrate (2). Even with proper cooking, it is not unusual for a 4-ounce piece of meat to be cooked to 3 ounces. During heating, the collagen molecule begins to denature at 102°F (39°C), and collapses at 149°F (65°C), resulting in a considerable loss of volume and length in the meat (see Chemist's Corner 7-8). Another factor contributing to meat shrinkage is the freeing of some water as the meat's other proteins denature and lose their water-binding capacity. Tenderness starts to decrease as temperatures reach 104°F (40°C). Longer cooking at lower temperatures makes meat, especially the tougher cuts, more tender by breaking down the collagen, which often gelatinizes during cooling (48).

As has been mentioned, any fat in the meat melts as it is cooked, which increases tenderness, juiciness, and flavor. When meat is very lean, it may be desirable to add fat to it. Fat may be added

FIGURE 7-18 Sauces for beef.

## CHEMIST'S CORNER 7-8

### Effect of Temperature on Meat Components

Meat becomes tender when cooked due to breakdown of its protein, fat, and connective tissue with increasing temperatures (9). From a biochemical standpoint, specific components of meat are affected at different temperatures:

- 100° F: Proteins begin to unfold; the meat appears red, soft, and slippery
- 120° F: Proteins coagulate (clump together) and loose water; the meat appears very firm and pink
- 140° F: Connective tissue shrinks, more moisture is lost; the meat appears pinkish-brown and visibly loses juices
- 150° F: Connective tissue begins dissolving into a gelatin-like substance and proteins are densely packed; meat appears brown and shrunken, and starts to taste tough
- 170° F: Proteins are entirely coagulated and most of the moisture is lost; meat tastes hard and dry

by two older techniques known as **larding** and **barding**.

*Searing*  It was once thought that **searing** would help to keep the juices inside a piece of meat as it cooked. It is now known that roasts heated at low temperatures for the entire cooking time retain more juices than those that are seared. Searing still remains a valuable technique for increasing the flavor and color of meat, however, because it caramelizes the outside, sealing in the flavor. To sear a piece of meat, place the meat in a pan that is already extremely hot, and leave it in the pan long enough for it to form a rich, brown crust. Trying to move the meat too early in the searing process will destroy the crust formation (54).

*Blanching*  Another technique thought to lock in the juices is blanching. Meat is blanched by boiling it very briefly, but this method is no longer recommended, because water-soluble compounds such as vitamins, minerals, and flavor substances may be lost. In the end, proponents argue that neither

blanching nor searing makes any difference in moisture loss in meats exposed to prolonged heating (27).

## Flavor Changes

Natural compounds in meat yield that characteristic meat flavor, but other factors contribute to flavor as well, including protein coagulation, melting and breakdown of fats, organic acids, and nitrogen-containing compounds. The trace amount of carbohydrates in meat contributes to the special flavor of browned meat surfaces as these sugars react with proteins in the Maillard reaction, producing the desirable brown color. Storing meat for more than 2 days in the refrigerator or heating leftover meat can result in an unfavorable warmed-over flavor (WOF), which is best avoided by reheating the meat in a microwave oven (Chemist's Corner 7-9) (28).

## Flavor Enhancements

The flavor of baked or broiled meat can be enhanced by basting and seasoning. If the seasoning includes salt, however, some professional chefs recommend adding it only after the meat has been slightly browned, because salt draws out juices and retards browning. Meat is basted by brushing the meat drippings or fat-based marinade over its surface to help it retain moisture and flavor. Self-basting can be achieved by barding. Seasoning prior to heating may improve flavor if the seasoning becomes part of the crust. Marinating meat is a flavorful way to preseason it, whereas prepared sauces may be served with the meat (Figure 7-18). Sauces and their preparation are discussed further in Chapter 18. In addition to sauces, condiments can also

## CHEMIST'S CORNER 7-9

### Warmed-Over Meat Flavor

The warmed-over flavor in reheated meat is thought to be caused by the oxidation of the meat's unsaturated fatty acids, which results in various off-flavor substances (e.g., hexanal) (37). Warmed-over flavor is just one example of lipid oxidation, thought to be the major cause of quality deterioration in meats (19).

**Au Jus**—natural beef juices

**Béarnaise**—thick sauce of egg yolks, white wine, tarragon vinegar, herbs

**Béchamel**—seasoned white sauce

**Bercy Butter**—shallots cooked in white wine mixed with creamed butter and parsley

**Beurre Noir**—clarified butter with vinegar or lemon juice

**Bordelaise**—brown sauce with red wine, shallots or green onions, herbs, and lemon juice

**Brown (Sauce Espagnole)**—flavorful beef sauce used as baste for others

**Chasseur**—brown sauce with mushrooms, tomato sauce, tarragon

**Chili Salsa**—chopped tomato, onion, green chili pepper

**Choron**—béarnaise sauce and tomato

**Colbert**—béarnaise sauce and meat glaze

**Hollandaise**—thick sauce of egg yolks, melted butter, and lemon juice

**Madeira**—brown sauce and Madeira wine

**Maître d'Hôtel Sauce**—béchamel sauce with butter, lemon juice, parsley, and tarragon

**Marchand de Vin**—red wine, parsley, green onions, and lemon juice

**Meunière**—browned butter with lemon juice and parsley

**Mornay**—creamy cheese sauce

**Périgueux**—wine sauce with diced truffles

**Robert**—brown sauce with mustard, onion, tomato, and pickle

**Larding**  Inserting strips of bacon, salt pork, or other fat into slits in the meat with a large needle.

**Barding**  Tying thin sheets of fat or bacon over lean meat to keep the meat moist during roasting. The sheets of fat are often removed before serving.

**Searing**  Cooking that exposes a meat cut to very high initial temperatures; this is intended to seal the pores, increase flavor, and enhance color by browning.

be used to add flavor to meats. Those frequently served with meat include steak sauces, ketchup, seasoned butters, salsas, and chutneys and fruit sauces, such as mint sauce for lamb cuts.

# Determining Doneness

Several changes occur in meat during cooking, and a multitude of factors affect the cooking times of meats: the effects of carryover cooking; differences in the type, size, and cut of meat; the presence of bones, which conduct heat faster than flesh, or of fat, which acts as an insulator; the actual oven temperature; the temperature of the meat before heating; and variations in the degree of doneness preferred by the preparer. Various methods are used to determine doneness and sometimes more than one method is used. Those discussed below include internal temperature, time/weight charts, color changes, and touch.

## Internal Temperature

Using a meat thermometer is the most accurate method of determining doneness. There are several different styles of meat thermometers on the market; some are inserted into meats before heating and others, such as instant-read thermometers, can be inserted at any time. The thermometer should be inserted into the thickest portion of the meat and in such a way as not to touch any fat or bone. Meat thermometers should be thoroughly sanitized after each use. Table 7-7 gives the internal cooking temperatures indicating doneness for various meats. According to the USDA, the final internal temperatures for beef are as follows:

- *Rare:*       136°F–140°F (58°C–60°C)
- *Medium:*   160°F–167°F (71°C–75°C)
- *Well done:* 172°F–180°F (78°C–82°C)

**Carryover cooking** The phenomenon in which food continues to cook after it has been removed from the heat source as the heat is distributed more evenly from the outer to the inner portion of the food.

**TABLE 7-7    Internal Temperatures Recommended for Cooked Meat**

| Meat | Description | Color | Internal Temperature °F | Internal Temperature °C |
|------|-------------|-------|----|----|
| Beef | Rare | Rose red in center; pinkish toward outer portion, shading into a dark gray; brown crust; juice bright red | 140 | 60 |
| | Medium | Light pink; brown edge and crust; juice light pink | 160 | 70 |
| | Well-done | Brownish gray in center; dark crust | 170 | 77 |
| Veal | Well-done | Firm, not crumbly; juice clear, light pink | 165 | 74 |
| Lamb | Rare | Rose-red in center; pinkish toward outer portion; brown crust; juice bright red | 140 | 60 |
| | Medium | Light pink; juice light pink | 160 | 70 |
| | Well-done | Center brownish gray; texture firm but not crumbly; juice clear | 170 | 77 |
| Pork | | | | |
| Ham | | | | |
| Fully cooked or canned | Heated | Pink | 130–140 | 55–60 |
| Cook before eating | Medium | Pink | 140 | 60 |
| Smoked loin | Medium | Pink | 160 | 70 |
| Fresh rib, loin, picnic shoulder | Well-done | Center grayish white | 170 | 77 |

**Source:** USDA.

Most other meats are expected to reach an internal temperature of at least 140°F (60°C). In January 1993, following a highly publicized outbreak of *E. coli* in the Northwest, health departments across the United States increased the required preparation temperature for hamburgers served by eating establishments from 140°F (60°C) to 160°F (71°C).

When measuring internal temperature, it is important to adjust for **carryover cooking.** This can result in an average temperature increase of 10°F–15°F (6°C–8°C) for average-size roasts. Very large roasts can have as much as a 25°F (14°C) increase in temperature, whereas small cuts may rise only 5°F (3°C) in temperature. To adjust for this carryover cooking, most roasts should be removed from the oven when the internal temperature is 10°F–15°F (6°C–8°C) below the final desired degree of doneness. Meat cooked at a low temperature such as 200°F–250°F (93°C–121°C) will experience only minimal carryover cooking. Depending on their

size, roasts should be allowed to stand for 15 to 30 minutes in order to distribute the heat and juices.

## Time/Weight Charts

Time/weight charts, such as the one shown in Table 7-8, are useful in estimating roughly how long it will take to cook a piece of meat but are unreliable if used alone because of the many factors that can affect doneness. Instead, a combination of criteria is used to determine the doneness of meats.

These criteria include time/weight charts, along with color changes, internal temperature, and touch.

## Color Changes

Meat pigments change color as the meat is cooked. Doneness can be determined by observing the following colors in red meats:

- *Rare.* Strong red interior. Rare meat does not reach a final internal temperature considered microbiologically safe.

| TABLE 7-8 | Time/Weight Chart for Roasting Beef | | | | |
|---|---|---|---|---|---|
| Cut | Approximate Weight (pounds) | Oven Temperature (degrees F) | Approximate Cooking Time (minutes per pound) | | |
| | | | Rare | Medium | Well |
| Rib roast | 4 to 6 | 300 to 325 | 26 to 32 | 34 to 38 | 40 to 42 |
| | 6 to 8 | 300 to 325 | 23 to 25 | 27 to 30 | 32 to 35 |
| Rib eye roast | 4 to 6 | 350 | 18 to 20 | 20 to 22 | 22 to 24 |
| Boneless rump roast | 4 to 6 | 300 to 325 | — | 25 to 27 | 28 to 30 |
| Round tip roast | 3 1/2 to 4 | 300 to 325 | 30 to 35 | 35 to 38 | 38 to 40 |
| | 6 to 8 | 300 to 325 | 22 to 25 | 25 to 30 | 30 to 35 |
| Top round roast | 4 to 6 | 300 to 325 | 20 to 25 | 25 to 28 | 28 to 30 |
| Tenderloin roast | | | | | |
| Whole | 4 to 6 | 425 | 45 to 60 (total) | | |
| Half | 2 to 3 | 425 | 35 to 45 (total) | | |

- **Medium.** Rosy pink interior and not quite as juicy as a rare piece of meat.
- **Well done.** Brown interior. No traces of red or pink left. Moist, but no longer juicy.

Veal and pork are known as *white meats,* in part because they change from a pinkish to a whiter color as they are heated to the well-done stage. According to the USDA, pork should be heated at least to an end-point temperature of 160°F (71°C). Color may not be a good indicator for doneness in meat from older swine, which is often grayish-brown rather than pink (33). It is not recommended that color be used to judge the doneness of hamburger, either, because of the risk of *E. coli* 0157:H7 contamination.

## Touch

Doneness can be determined by the firmness of the meat. Some meat cuts such as steaks and chops can be judged for doneness based on their color and firmness. Pressing lightly on the center of the lean tissue can help to determine whether the meat is rare, medium, or well done (Figure 7-19). This technique takes a fair amount of experience to master and is most often used by professional chefs who frequently prepare steaks.

# Dry-Heat Preparation

Tender cuts are usually prepared by one of the dry-heating methods: roasting (baking), broiling, grilling, pan-broiling, and frying.

## Roasting

Roasting is the heating of moderate-to-large, tender cuts of meat in the dry, hot air of an oven. A roast will usually be at least 2½ inches thick and provide more than three servings. The meat is placed, fat side up (if it has any), on a rack in an open pan. The rack prevents the meat from sitting in its own juices, which would cause the meat to simmer rather than to roast. If a rack is not available, one can be made by lining up carrots and celery stalks lengthwise across the bottom of the pan. Figure 7-20 shows examples of cuts suitable for roasting.

Temperatures from 300°F–350°F (149°C–177°C) are recommended for roasting and should produce an evenly cooked, easy to carve, juicy, tender, flavorful roast with a greater yield than roasting at higher temperatures would have produced. Higher temperatures of 350°F–500°F (177°C–260°C) are recommended to produce roasts with deeply seared crusts in less time, but the higher oven temperatures cause greater shrinkage. In general, it usually takes 18 to 30 minutes of roasting time for every pound of meat. As previously mentioned, roasts should be removed from the oven slightly before their final desired temperature is reached and allowed to stand for 15 to 30 minutes in order for carryover cooking to occur. This will also make carving easier and result in a more evenly juicy roast.

## Broiling and Grilling

Smaller cuts of tender meat ranging from 1 to 3 inches in thickness can be broiled

**FIGURE 7-19** Touch as a test for doneness.

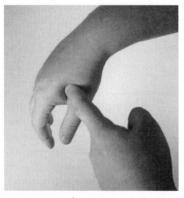

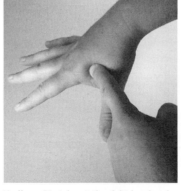

**Rare:** Shake, dangle, and relax right hand; pressing the area between thumb and index finger feels similar to rare steak—soft and yielding to slight pressure.

**Medium:** Stretch out the right hand and tense the fingers; the springy firmness is similar to the resistance felt in medium-cooked meats.

**Well done:** Harden the right hand into a tight ball; this hard and unyeilding feeling with all the springiness gone is how well-done meat feels.

**FIGURE 7-20**    Examples of tender roast cuts.

Ham roast

Leg of lamb

Pork center-cut loin

Veal roast

Digital Works

or grilled. High temperatures and short heating times keep the meat tender. Broiling and grilling times are based primarily on the meat's thickness and its distance from the heat (Table 7-9). Ovens, whether electric or gas, need at least 15 minutes to reach the desired temperature, whereas charcoal or wood fires need at least 25 minutes to burn down to the required heat. Beef retail cuts suitable for broiling include the following steaks in descending order of tenderness: filet mignon, strip loin, Delmonico, rib eye, top butt sirloin, chuck tender, and top round. A very light layer of oil on the meat will keep it from sticking to the grill, whereas using a marinade, spice rub, or adding sauces during basting will yield more flavor.

The goal in either broiling or grilling is to simultaneously heat the inside of the meat while achieving just the right degree of browning on the exterior. The thickness of the cut and the desired level of doneness dictate the intensity of the heat, which is controlled by altering the distance of the meat from the heat source, from 2 inches for cuts less than 1 inch thick, to up to 5 inches for thicker cuts. When broiling thicker steaks or those to be well done, the broiler rack in an electric oven should be lowered and the door left open to prevent steam from accumulating, thereby preventing the meat from browning. Gas broiler doors are left closed.

The oven, broiler, or grill should be preheated. Then the meat should be placed under the broiler or over the coals and heated until one side is brown. Tongs should be used to turn the meat, but if a fork is used, it is best inserted into the fat trim to avoid letting the juices escape. The second side is heated to the desired stage of doneness. When heating is complete, remove and serve immediately. One of the benefits of using a grill is that attractive, appetizing grill marks can be made by turning the meat over according to the pattern depicted in Figure 7-21.

### Pan-Broiling

Very thin cuts of meat, less than ½ inch, can be pan-broiled to achieve a tasty outside crust without overcooking the meat. In this method, heat is applied directly through the hot surface of a heavy pan or flat grill (Figure 7-22). Thin, tender cuts of beef steaks, lamb chops, and ground-beef patties are perfect for pan-broiling. Place the meat on the hot surface of the preheated pan with no added fat or oil. Any drippings should be drained during heating to prevent frying. The meat can be seasoned before, during, or after placing it on the pan.

### Frying

Sautéing, pan-frying, and deep-frying are suitable for tender, small pieces of meat that are low in fat or that have a breaded coating.

*Sautéing*    Sautéing is identical to pan-broiling except that a small amount of fat is heated to the sizzling point before the meat is added. Examples of sautéed meat dishes include liver and onions, veal Oscar, veal picatta, and veal cordon bleu. Liver should be salted after it is sautéed or else it will toughen and shrivel. Stir-frying is a type of sautéing that has become increasingly popular. For stir-frying, thin slices of meat are cooked in an oiled wok or other sloping-sided pan. The meat is stirred constantly over high heat for about 3 minutes to promote even heating. When the meat is done, it is moved to the side, and chopped vegetables are added to the pan. As soon as they are barely tender, they are mixed with the meat and any desired sauces or flavorings.

*Pan-Frying*    In pan-frying, more fat (but no more than up to ½ inch deep), lower heating temperatures, and longer cooking times are used than what is common in sautéing. Typically, pan-fried meat cuts are larger and include steaks (Figure 7-23), chops, and sliced pieces of liver. Meats are often seasoned and coated with flour or breading before pan-frying. The fat used in

### TABLE 7-9    Time/Weight Chart for Broiling Sirloin Steak

| Beef Cut | Approximate Thickness (inches) | Approximate Weight (pounds) | Distance from Heat (inches) | Approximate Cooking Time (total minutes) | | |
|---|---|---|---|---|---|---|
| | | | | Rare | Medium | Well |
| Sirloin steak | ¾ | 1¼ to 1¾ | 2 to 3 | 10 | 15 | — |
| | 1 | 1½ to 3 | 3 to 4 | 16 | 21 | — |
| | 1½ | 2¼ to 4 | 4 to 5 | 21 | 25 | — |

**FIGURE 7-21** Technique for making grill marks: Rotate clockwise a quarter of a turn.

**FIGURE 7-22** Pan-broiling.

1. Place beef in preheated frying pan.

2. Do not add oil or water. Do not cover.

3. Cook slowly (⁵⁄₈″ to 1″ cuts), turning occasionally. For cuts thicker than ¹⁄₂″ use medium to medium-low heat. For thinner cuts, use medium-high heat.

4. Pour off excess drippings as they accumulate.

5. Season if desired.

Digital Works

**FIGURE 7-23** Cuts of steak suitable for pan-frying (if less than ½″ thick).

a. Bottom round steak

b. Chicken steak

c. Flank steak

d. Shoulder steak

e. Eye round steak

f. Top round

g. Top sirloin

Digital Works

sautéing or in pan-frying should be vegetable oil or clarified butter. The low smoking temperatures of whole butter and margarine make them unsuitable for frying. An alternative to frying steaks and chops in oil is to use a nonstick pan or to sprinkle the pan with a thin layer of salt. The pan is heated until a drop of water hisses; the meat is then added, fried, and turned when the underside has reached the desired brownness.

*Deep-Frying*   Meat, with the exception of chicken-fried steak, is seldom deep-fried. When it is, the meat is usually cut into small pieces and dipped in seasoned flour or cornstarch, placed in a wire basket, submerged in oil preheated to 300°F–360°F (149°C–182°C), and heated until golden brown.

# Moist-Heat Preparation

Less tender cuts of meat, which tend to come from more heavily exercised muscles or older animals, are usually prepared by moist-heat methods such as braising, simmering/stewing, or steaming.

## Braising

Braising consists of simmering meat, in a covered pan, in a small amount of water or other liquid. It is ideal for less tender cuts such as beef chuck, round steak, and flank steak, because braising breaks down collagen and tenderizes the meat. Braising can transform a meat's texture from tough to fork-tender (59). Some smaller meat cuts such as round steaks, pork and veal chops, and organ meats are also good braisers. The most common braised meats are pot roasts, which are large cuts of meat cooked whole and served in slices covered with their own cooking liquid. Adding vegetables completes the meal and adds color. Chopped vegetables commonly added to pot roasts include potatoes, carrots, onions, celery, and tomatoes.

Although not necessary, browning the meat prior to adding the liquid improves the final color and flavor. Before browning, the meat should be dried with a paper towel, and it is sometimes dredged with seasoned flour. As with any browning, it is essential not to overcrowd the pan and to brown the meat in batches, if necessary.

After the liquid is added, the pan is covered and the liquid brought to a simmer; boiling must be guarded against because it will toughen the meat. The goal is to simmer the meat until it is tender. Doneness when braising is determined by fork tenderness. The flavor of the braising liquid can be enhanced by the addition of wine, soup stock, marinades, seasonings, or tomato products. Only enough liquid, no more than 1 inch, should be added to produce steam. If too much liquid is used, it can reduce the flavor by sheer dilution.

### Simmering or Stewing

Simmered or stewed meat is cooked completely submerged in liquid. The pan is covered, brought to simmering, not boiling, and cooked until the meat is tender. Fricassees are stews in which the meat is first browned in fat. Stews, unlike other simmered meats, are served in their own cooking liquid mixture, thickened or not, as desired, and usually contain vegetables added during the last hour of heating. Cured meats, such as corned beef or tongue and fresh beef brisket cuts, are commonly prepared by stewing. They are not browned first, and the cooking liquid, which has very little flavor, is usually discarded.

### Steaming

Steaming exposes food directly to moist heat. Meats can be steamed in a pressure cooker or in a tightly covered pan. They can also be wrapped in aluminum foil or placed in a plastic oven bag, which is then placed in a heated oven. Oven bags are heat-resistant nylon bags made to withstand oven temperatures in order to provide steam to foods that are being roasted. They are used to cook a variety of foods, but are most often used for cooking large cuts of meat such as turkey, ham, or beef roasts. Because the meat cannot be observed during heating in a pressure cooker, its doneness is determined by timing. Meats also heat very well in a crockery cooker, an electrical appliance that will gently steam meat to extreme tenderness with only a little added liquid. Depending on the size and toughness of the cut, this may take anywhere from 6 to 12 hours. The long heating time and relatively low

**FIGURE 7-24**    Carving across the grain.

grain direction

carving direction

PhotoDisc/Getty Images

temperature may pose food safety concerns, however (see Chapter 4).

### Microwaving

Microwave ovens are usually not the best option for cooking meats, except for thawing and reheating leftovers. They decrease juiciness, do not brown, and do not heat sufficiently to kill pathogens such as *Trichinella spiralis*. Microwaved meats do not taste the same as meats cooked by other time-tested methods, primarily because they do not get browned. Brown condiments such as Kitchen Bouquet, Worcestershire sauce, soy sauce, or steak or barbecue sauces can be used to add color to the meat or to cover it up, hiding the fact that the surface appears uncooked. Microwave browning skillets and grills are also available, but the flavor and texture problems remain the same. The power emissions from microwave ovens vary from brand to brand, so the manufacturer's instructions should be followed whenever a microwave is used for preparing meat or meat dishes.

## Carving

Meat should not be sliced in just any manner, because the way it is sliced affects its tenderness. The first step in slicing meat is to determine the direction in which the muscle fibers run, called the *grain*. This can be seen on the surface of the meat. It may be difficult to find the grain in larger cuts such as roasts, because they consist of parts of several different muscles, each with its own grain. When carving meats, it is important to cut across the grain to increase tenderness (Figure 7-24). Cutting across the grain shortens the muscle fibers into smaller segments, making the meat easier to chew.

## STORAGE OF MEATS

Meat contains high percentages of water and protein, both ideal for the growth of microorganisms. Consequently, meat should be stored in the refrigerator or freezer. Raw meat and poultry are stamped with "use by" dates on the packaging; they should be cooked or frozen by this date. After cooking, ground meat can be stored up to 2 days, and whole cuts of meat can be stored for 3 to 5 days.

# Refrigerated

Meats are best refrigerated at just above freezing (32°F/0°C), between 32°F and 36°F (0°C–2°C). They do not freeze until the temperature drops to below 28°F (–2°C). The best place to store meats in the refrigerator is in the coldest part. Many refrigerators have such an area or a compartment reserved for meat storage.

## Wrapping Meat

Most retail meats are packaged with plastic wrap and can be refrigerated in their original wrap for up to 2 days. After that time, the store wrapping should be removed and replaced by loosely wrapped plastic wrap, wax paper, or aluminum foil. Leaving the tight store wrapping on meat for more than 2 days creates moist surfaces, which promote bacterial growth and deterioration of the meat. Exceptions to this general storage guideline are hams and other processed meats that are high in salt. They should not be stored in aluminum foil because the salt's corrosive action on aluminum foil will cause discoloration of the meat. Cured meats are also high in fat, which quickly turns rancid when exposed to oxygen and light. For this reason, ham and other processed meats are best stored in the refrigerator in their original wrappings.

## Refrigeration Times

General guidelines suggest that fresh meat should not be stored in the refrigerator longer than 3 to 5 days, and that ground meats and variety meats should be cooked within 1 or 2 days (see back inside cover of this book) (51). Variety meats are more perishable than regular meat cuts and should be used within a day or two of purchase or frozen immediately. Cooked meat can be kept for about 3 to 4 days. If the meat needs to be kept longer than the recommended storage times, it should be frozen.

## Controlled-Atmosphere Packaging

One alternative to storing meats for long periods of time at refrigeration temperatures is a patented, controlled-atmosphere package (CAP) available only to meat wholesalers. It can extend the shelf life of fresh red meat from the current 2 days to up to 28 days. The process involves using a special package that allows the removal of oxygen and its replacement with a mixture of 70% nitrogen and 30% carbon dioxide (61).

# Frozen

Meats to be frozen should be wrapped tightly in aluminum foil, heavy plastic bags, or freezer paper and stored at or below 0°F (–18°C) (Figure 7-25). It is a good idea to first trim meat of bone and fat and to divide it up into individual servings before wrapping and freezing it. Most beef cuts can be kept frozen for 6 to 12 months, but ground beef should be frozen for no longer than about 3 months (see back inside cover of this book). The colder temperatures reached by commercial freezers for at least 20 days at 5°F (–15°C) can kill *T. spiralis*. If not frozen to this degree, pork should always be cooked to the recommended temperature of 160°F (71°C). Wrappers often hide the identity of their contents, so the packages of frozen foods should be labeled and dated. It is better to make more frequent purchases than to freeze meat for extended periods of time, which can reduce its quality.

The texture and flavor of thawed meats will be adversely affected if they are refrozen. Freezer burn, caused by loss of moisture from the frozen food's surface, can result if meat is stored longer than the recommended storage time or wrapped in materials that are not vapor proof or are punctured. The dehydration of freezer burn causes a discolored surface on the meat that becomes very dry, tough, and somewhat bitter in flavor when cooked.

**FIGURE 7-25** Wrapping meat for freezing (apothecary or drugstore method).

Wrapping Meat for the Freezer
(Apothecary or Drugstore Method)

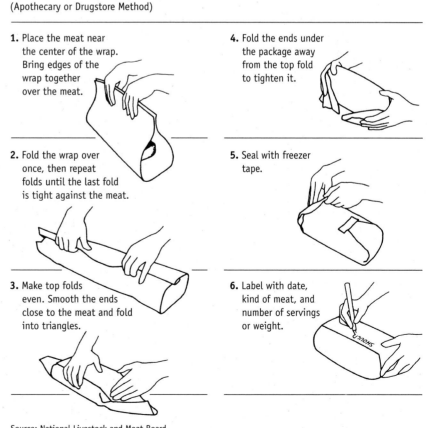

1. Place the meat near the center of the wrap. Bring edges of the wrap together over the meat.

2. Fold the wrap over once, then repeat folds until the last fold is tight against the meat.

3. Make top folds even. Smooth the ends close to the meat and fold into triangles.

4. Fold the ends under the package away from the top fold to tighten it.

5. Seal with freezer tape.

6. Label with date, kind of meat, and number of servings or weight.

Source: National Livestock and Meat Board.

# PICTORIAL SUMMARY / 7: Meat

Usually the most expensive item on a menu, meat serves as an important source of complete protein. In North America and Europe, the main sources of meat are herbivores, such as beef cattle, sheep, and swine.

## TYPES OF MEATS

**Beef.** Most beef is supplied by steers, male cattle that are castrated while young so that they will gain weight quickly. Heifers, females that have not borne a calf, are also used for meat.

**Veal.** Veal comes from male and female calves of beef (and dairy) cattle between the ages of 3 weeks and 3 months. These animals are fed a milk-based diet and have their movements restricted for a more flavorful and tender meat.

**Lamb.** Lamb comes from sheep less than 14 months old; the meat from older animals is sold as mutton.

**Pork.** Most pork comes from young swine of either gender. In the last 30 years, pork has been bred to be leaner and more tender.

## PURCHASING MEATS

**Meat inspection** is mandatory in the United States, but grading is voluntary. There are quality grades for beef, veal, lamb, and mutton. Factors considered in grading are color, grain, surface texture, and fat distribution. Yield grades are ranked from 1 (highest) to 5 (lowest), and indicate the amount of lean meat in proportion to fat, bone, and other inedible parts.

**Tenderness** in meats is due in part to natural influences such as the cut, marbling, animal age, heredity, diet, and slaughtering conditions. Meats can be artificially treated to make them more tender by aging, adding enzymes, salts, and acids, or subjecting them to mechanical or electrical treatments.

**Kosher meats** have met standards set by Jewish religious law.

**Variety meats** include the liver, sweetbreads (thymus), brain, kidney, heart, tongue, tripe (stomach lining), and oxtail of the animal.

**Processed meats** such as ham and sausage are preserved by curing, smoking, cooking, canning, or drying.

## COMPOSITION OF MEATS

Meats consist of muscle, connective tissue, adipose (fatty) tissue, and bone. In meat cuts, the fat deposited in the muscle is visible as white streaks called marbling.

In terms of nutrient composition, meat is primarily water, high-quality protein, fat, some minerals, and B vitamins. Meat is not a good source of carbohydrates, fiber, or vitamin C.

## PREPARATION OF MEATS

Meat should be sponged clean of any moisture with paper towels and trimmed of fat before being prepared. Doneness of meats can be determined by a combination of time/weight charts, color changes, internal temperature, and touch. Tender meats are best prepared by dry heat (roasting/baking, broiling, grilling, pan-broiling, and frying), whereas moist-heat methods (braising, simmering, stewing, and steaming) are best for tougher cuts. Common wholesale and retail cuts of meat are shown below:

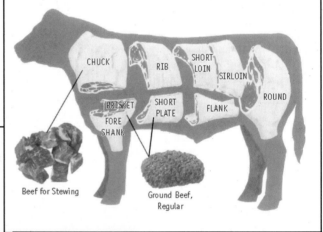

Beef for Stewing

Ground Beef, Regular

## STORAGE OF MEATS

All meats should be refrigerated or frozen according to recommended temperatures. They should be held in the refrigerator no longer than the suggested maximum times, usually 3 to 5 days, although ground and variety meats will last only 1 or 2 days. Most meats can be kept frozen for 6 to 12 months if properly wrapped to avoid freezer burn caused by moisture loss.

# CHAPTER REVIEW AND EXAM PREP

## Multiple Choice*

1. The most important influence on the tenderness of meat is:
   a. the animal's age
   b. diet
   c. cut (location on the animal's body)
   d. marbling

2. What is the most abundant protein in connective tissue?
   a. Cholesterol
   b. ATP
   c. Actin
   d. Collagen

3. Meats are good sources of the mineral _____, but poor sources of the mineral _____.
   a. iron, calcium
   b. chromium, calcium
   c. calcium, zinc
   d. zinc, iron

4. Which of the following retail cuts of beef would be classified as tender?
   a. Top loin steak
   b. Rump roast
   c. Brisket
   d. Chuck roast

5. For meat to be considered kosher, the animal must be:
   a. slaughtered in the presence of a rabbi
   b. slaughtered with a single stroke of a knife
   c. completely bled
   d. All of the above

6. What is the most flavorful, tender USDA grade of meat?
   a. Choice
   b. Select
   c. Prime
   d. Grade 1

7. What term is used to describe tying thin sheets of fat or bacon over lean meat to keep it moist during cooking?
   a. Larding
   b. Barding
   c. Searing
   d. Au jus

## Short Answer/Essay

1. Briefly describe the following components of meat: muscle tissue, connective tissue, adipose tissue, and bone.
2. Meat changes color during storage and preparation. Explain what is happening as meat turns from purplish red to bright red to brownish-red.
3. List the USDA quality grades for beef. How do these differ from the yield grade?
4. Discuss how the following factors affect meat tenderness: cut, age, heredity, diet, marbling, and slaughtering conditions.
5. What is rigor mortis? Describe the changes that occur in meat during aging.
6. List and briefly describe the various methods for artificially tenderizing meats.
7. Define these terms: *wholesale/primal cuts, IMPS, kosher meats, variety meats, processed meats, mechanically deboned meat,* and *restructured meat.*
8. Briefly describe four methods for determining the doneness of cooked meats.
9. Describe the general process of preparing meats by the following methods: roasting (include an explanation of carryover cooking), broiling, pan-broiling, braising, and stewing.
10. Discuss the special requirements for the storage of fresh meats, including temperature, packaging, and maximum storage time.

*See p. AK-1 for answers to multiple choice questions.

# REFERENCES

1. Abnet CC. Carcinogenic food contaminants. *Cancer Investigations* 25(3):189–96, 2007.

2. Ainwunmi I, LD Thompson, and CB Ramsey. Marbling, fat trim and doneness effects on sensory attributes, cooking loss and composition of cooked beef steaks. *Journal of Food Science* 58(2):242–244, 1993.

3. Alp S. [Bacterial resistance to antiseptics and disinfectants.] *Mikrobiyol Bulletin* 41(1):155–161, 2007.

4. Archer DL. Nitrite and the impact of advisory groups. *Food Technology* 55(3):26, 2001.

5. Armentrout J. Dry-aging beef pays off with big flavor. *Fine Cooking* 69:76, 2005.

6. Armentrout J. For juicier and tastier meat, try brining. *Fine Cooking* 67:74–76, 2004.

7. Bacus JN. Navigating the processed meats labeling maze. *Food Technology* 61(11):28–32, 2007.

8. Banasiak K. pH indicates quality of pork. *Food Technology* 60(12):12, 2006.

9. Barham P. What makes beef tender or tough? *Fine Cooking* 80:78–79, 2006.

10. Beerman DH. ASAS Centennial Paper: A century of pioneers and progress in meat science in the United States leads to new frontiers. *Journal of Animal Science* 87:1192–1198, 2009.

11. Beggs KLH, JA Bowers, and D Brown. Sensory and physical characteristics of reduced-fat turkey

frankfurters with modified corn starch and water. *Journal of Food Science* 62(6):1240–1244, 1997.

12. Berry BW. Low fat level effects on sensory, shear, cooking, and chemical properties of ground beef patties. *Journal of Food Science* 57(5):1205–1209, 1992.

13. Berry BW. Fat level, high temperature cooking, and degree of doneness affect sensory, chemical, and physical properties of beef patties. *Journal of Food Science* 59(1):10–14, 1994.

14. Berry BW. Sodium alginate plus modified tapioca starch improves properties of low-fat beef patties. *Journal of Food Science* 62(6):1245–1249, 1997.

15. Best D. Corporations invest in "back to basics" research. *Prepared Foods* 160(5):54–56, 1991.

16. Brewer MS, and FK McKeith. Consumer-rated quality characteristics as related to purchase intent of fresh pork. *Journal of Food Science* 64(1):171–174, 1999.

17. Bruce HL, SL Beilken, and P Leppard. Variation in flavor and textural descriptions of cooked steaks from bovine *m. Longissiumus thoracis et lumborum* from different production and aging regimes. *Journal of Food Science* 70(4):306–316, 2005.

18. Carter RA, et al. Fabricated beef product: Effect of handling conditions on microbiological, chemical, and sensory properties. *Journal of Food Science* 57(4):841–844, 1992.

19. Cheah PB, and DA Ledward. Catalytic mechanism of lipid oxidation following high pressure treatment in pork fat and meat. *Journal of Food Science* 62(6):1135–1138, 1997.

20. Chung KY and Johnson BJ. Application of cellular mechanisms to growth and development of food producing animals. *Journal of Animal Science* 86(14Suppl):E226–E235, 2008.

21. DeFreitas Z, et al. Carrageenan effects on salt-soluble meat proteins in model systems. *Journal of Food Science* 62(3):539–543, 1997.

22. Dorsa WJ. New and established carcass decontamination procedures commonly used in beef-processing industry. *Journal of Food Protection* 60(9):1146–1151, 1997.

23. Dowell P, and A Bailey. *Cook's Ingredients*. William Morrow, 1980.

24. Doyle MP. Dealing with antimicrobial resistance. *Food Technology* 60(8):22–29, 2006.

25. Eilert SJ, RW Mandigo, and SS Sumner. Phosphate and modified beef connective tissue effects on reduced-fat, high water-added frankfurters. *Journal of Food Science* 61(5):1106–1011, 1996.

26. Giese J. Developing low-fat meat products. *Food Technology* 46(4):100–108, 1992.

27. Gisslen W. *Professional Cooking*. Wiley, 1998.

28. Grun IU, J Ahn, AD Clarke, and CL Lorenzen. Reducing oxidation of meat. *Fine Cooking* 60(1):36–43, 2006.

29. Haard NF. Foods as cellular systems: Impact on quality and preservation. *Journal of Food Biochemistry* 19:191–238, 1995.

30. Hagen BF, JL Berdagué, AL Holck, H Næs, and H Blom. Bacterial proteinase reduces maturation time of dry fermented sausages. *Journal of Food Science* 61(5):1024–1029, 1996.

31. Hammerum AM and OE Heuer. Human health hazards from antimicrobial-resistant *Escherichia coli* of animal origin. *Clinical Infectious Disease* 48(7):916–921, 2009.

32. Harris JJ, et al. Evaluation of the tenderness of beef top sirloin steaks. *Journal of Food Science* 57(1):6–9, 1992.

33. Hauge MA, et al. Endpoint temperature, internal cooked color, and expressible juice color relationships in ground beef patties. *Journal of Food Science* 59(3):465–473, 1994.

34. Hui Y, et al. *Handbook of Food Products Manufacturing: Health, Meat, Milk, Poultry, Seafood, & Vegetables*. John Wiley and Sons, Hoboken, NJ, 2007.

35. Jhung MA, et al. Toxinotype *V Clostridium difficile* in humans and food animals. *Emerging Infectious Disease* 14(7):1039–1045, 2008.

36. Johnson BJ, and KY Chung. Alterations in the physiology of growth of cattle with growth-enhancing compounds. *Veterinary Clinics of North America Food and Animal Practices* 23(2):321–332, 2007.

37. Kerler J, and W Grosch. Odorants contributing to warmed-over flavor (WOF) of refrigerated cooked beef. *Journal of Food Science* 61(6):1271–1274, 1996.

38. Kim YH, JT Keeton, HS Yang, SB Smith, JE Sawyer, and JW Wavell. Color stability and biochemical characteristics of bovine muscles when enhanced with L- or D-potassium lactate in high-oxygen modified atmospheres. *Meat Science* 82(2):2234–2240, 2009.

39. Kinsman DM, AW Kotula, and BC Breidenstein. *Muscle Foods. Meat, Poultry, and Seafood Technology*. Chapman & Hall, 1994.

40. Kolok AS, and MK Sellin. The environmental impact of growth-promoting compounds employed by the United States beef cattle industry: History, current knowledge, and future directions. *Reviews of Environmental Contamination and Toxicology* 195:1–30, 2008.

41. Lonergan SM, et al. Porcine somatotropin (PPST) administration to growing pigs: Effects on adipose tissue composition and processed product characteristics. *Journal of Food Science* 57(2):312–317, 1992.

42. McWilliams M. *Foods: Experimental Perspectives*. Macmillan, 1997.

43. Milly PJ, RT Toledo, and S Ramakrishnan. Determination of minimum inhibitory concentrations of liquid smoke fractions. *Journal of Food Science* 70(1):M12–M17, 2005.

44. Ngapo TM, and C Gariepy. pH water capacity: Factors affecting the eating quality of pork. *Critical Review in Food and Nutrition Science* 48(7):599–633, 2008.

45. Nicolalde C, A Stetzer, EM Tucker, FK McKeith, and MS Brewer. Development of a model system to mimic beef bone discoloration. *Journal of Food Science* 70(9):575–580, 2005.

46. Pegg RB, and F Shahidi. Unraveling the chemical identity of meat pigments. *Critical Reviews in Food Science and Nutrition* 37(6):561–589, 1997.

47. Penfield MP, and AM Campbell. *Experimental Food Science*. Academic Press, 1990.

48. Pietrasik Z, JS Dhanda, RB Pegg, and PJ Shand. The effects of marination and cooking regimes on the

water-binding properties and tenderness of beef and bison top round roasts. *Journal of Food Sciences* 70(2):S102–106, 2005.

49. Pietrasik Z, and JAM Janz. Influence of freezing and thawing on the hydration characteristics, quality, and consumer acceptance of whole muscle beef injected with solutions of salt and phosphate. *Meat Science* 81(3):523–532, 2008.

50. Purviance J. Barbecuing tender ribs. *Fine Cooking* 93:46–49, 2008.

51. Q & A. How long is it safe to let cooked meat stay in the refrigerator? *Fine Cooking* 87:16, 2007.

52. Quinton RD, et al. Acceptability and composition of some acidified meat and vegetable stick products. *Journal of Food Science* 62(6):1250–1254, 1997.

53. Renerre M. Factors involved in the discoloration of beef meat. *International Journal of Food Science and Technology* 25:613–630, 1990.

54. Rosenfeld T. Sear, roast, and sauce. *Fine Cooking* 71:36, 2005.

55. Schlesinger C. Simple steps. *Fine Cooking* 64:42–46, 2004.

56. Serdaroglu M, K Abdraimov, and CA Onen. The effects of marinating with citric acid solutions and grapefruit juice on cooking and eating quality of turkey breast. *Journal of Muscle Foods* 18(2):162–172, 2007.

57. Spanier AM, Vercellotti JR, James C Jr. Correlation of sensory, instrumental and chemical attributes of beef as influenced by meat structure and oxygen exclusion. *Journal of Food Science* 57(1):10–15, 1992.

58. State of the food industry: Meat and poultry. *Food Engineering* 63(6):78, 1991.

59. Stevens M. Three ways to braise short ribs for the best flavor. *Fine Cooking* 77:42–48, 2006.

60. Toldra F and Reig M. Chapter 59: Sausages, from Hui Y, et al. *Handbook of Food Products Manufacturing: Health, Meat, Milk, Poultry, Seafood, and Vegetables.* John Wiley and Sons, 2007.

61. United States Food and Drug Administration, Economic Research Service. *U.S. Beef and Cattle Industry: Background Statistics and Information.* www.ers.usda.gov/news/BSECoverage.htm. Accessed April 27, 2009.

62. United States Food and Drug Administration, Food Safety and Inspection Services. *Fact Sheet: Meat packaging materials.* www.fsis.usda.gov/factsheets/Meat_Packaging_Materials/index.asp. Accessed April 24, 2009.

63. Vasconcelos JT, Sawyer JE, Tedeschi LO, McCollum FT, Greene LW. Effects of different growing diets on performance, carcass characteristics, insulin sensitivity, and accretion of intramuscular and subcutaneous adipose tissue of feedlot cattle. *Journal of Animal Science* 87(4):1540–1547, 2009.

64. Watch the water loss. *Food Manufacture* 66(4):39–43, 1991.

65. Wicklund SE, et al. Aging and enhancement effects on quality characteristics of beef strip steaks. *Journal of Food Science* 70(3):S242–S247, 2005.

66. Yin MC, and WS Cheng. Oxymyoglobin and lipid peroxidation in phophatidylcholine liposomes retarded by alpa-tocopherol and beta-carotene. *Journal of Food Science* 62(6):1095–1097, 1997.

# WEBSITES

The USDA's meat and poultry hot line can answer food safety questions:

**www.fsis.usda.gov/Food_Safety_Education/USDA_Meat_&_Poultry_Hotline/index.asp**

Questions about organic meat and other foods can be answered at:

**www.ams.usda.gov/AMSv1.0/Nop**

Meat and poultry labeling terms from the USDA can be found at:

**www.fsis.usda.gov/Fact_Sheets/Meat_&_Poultry_Labeling_Terms/index.asp**

This online video clip shows how pork bacon is processed:

**http://science.discovery.com/videos/how-its-made-bacon.html**

Several trade associations representing the interests of meat producers have websites.

American Association of Meat Processors:

**www.aamp.com**

American Meat Institute:

**www.meatami.com**

National Cattlemen's Beef Association:

**www.beef.org**

National Pork Producers Council:

**www.nppc.org**

American Sheep Industry Association:

**www.sheepusa.org/**

# 8 Poultry

The word *poultry* refers to all domesticated birds raised for their meat. Although chickens are the most popular poultry consumed, other species include turkeys, ducks, geese, guinea fowls, and pigeons (squabs). Game birds such as pheasant, wild duck, and quail are also consumed, but few of them reach the marketplace. Emus and ostriches, though not yet readily available in all parts of the country, are being bred for their lower-fat meat.

Despite the variety of poultry, chickens, raised by humans for over 4,000 years (11), remain the most common poultry consumed. Chickens are especially useful because both their meat

and their eggs are consumed. The popularity of chicken and turkey continues to increase at the expense of beef (18). In the last 40 years, production of broilers (young chickens) in the United States has increased from about 34 million to over 6 billion (3). Poultry is important to the diet, and the purpose of this chapter is to discuss poultry classification, composition, purchasing, preparation, and storage.

## CLASSIFICATION OF POULTRY

Ready-to-eat poultry is classified according to age and gender (Table 8-1). Classifications vary from species to species; chickens are classified as broilers, fryers, and so on, and turkeys as toms and hens.

In the past, there was a stewing hen classification in the chicken category, but such a designation is now rare. Younger poultry are usually preferred because they are more tender and have less fat than older birds.

## Chickens

Chickens sold on the market may be male or female, and differ in the age at which they are slaughtered and their weight. The younger chickens coming to market are classed as broilers/fryers, roasters, capons, and Cornish game hens.

### Broilers/Fryers

Broilers and/or fryers are chickens of either sex, slaughtered under 10 weeks of age (usually 7 weeks), and weighing approximately 3 to 5 pounds. They can be used not just for broiling and frying, as the names imply, but in any other way desired. At the market, these chickens will have soft skin, tender meat, and a flexible breastbone.

### Roasters

Roasters are older and therefore larger than broilers/fryers. These chickens are of either sex, are usually processed at 10 to 12 weeks of age, and weigh 6 to 8 pounds. The breastbone is less flexible than it is in broilers, having become calcified with age.

| TABLE 8-1 | Species and Classes of Poultry* | | |
|---|---|---|---|
| **Species** | **Class** | **Sex** | **Age** |
| Chicken | Cornish game hen | Either | 5–6 weeks |
| | Broiler or fryer | Either | Under 10 weeks |
| | Roaster | Either | Under 12 weeks |
| | Capon | Unsexed male | Under 4 months |
| | Hen, fowl, baking chicken, or stewing chicken | Female | Over 10 months |
| | Cock or rooster | Male | Over 10 months |
| Turkey | Fryer-roaster | Either | Under 12 weeks |
| | Young hen | Female | Under 6 months |
| | Young tom | Male | Under 6 months |
| | Yearling hen | Female | Under 15 months |
| | Yearling tom | Male | Under 15 months |
| | Mature or old | Either | Over 15 months |
| Duck | Duckling | Either | Under 8 weeks |
| | Roaster duckling | Either | Under 16 weeks |
| | Mature or old | Either | Over 6 months |
| Goose | Young | Either | |
| | Mature or old | Either | |
| Guinea | Young | Either | |
| | Mature or old | Either | |
| Pigeon | Squab | Either | |
| | Pigeon | Either | |

*The different species represent "kinds," while class is dependent on the bird's sex and age.

## Capons

Capons are neutered male chickens that usually reach the market under 4 months of age weighing 12 to 14 pounds. The tenderness and juiciness of the meat is comparable to that of broiler/fryers.

## Cornish Game Hens

Cornish game hens are bred by crossing a Cornish hen, a breed of chicken, with one of the other common breeds, such as White Plymouth Rock, New Hampshire, or Barred Plymouth Rock. The hens are slaughtered at 5 to 6 weeks, at which point they will weigh not more than 2 pounds. The meat is always very tender.

## Mature Chickens

Older adult chickens over 10 months of age, both female (hens, fowls, baking chickens, or stewing chickens) and male (cocks or roosters), have outlasted their breeding capabilities. Their meat is tougher, the skin coarser, and the breastbone less flexible. They are best used in stews, soups, and other slow-cooking dishes.

## Turkeys

The turkeys bred for their meat today look very different from the *Meleagris gallopavo silvestris* depicted in the familiar old paintings of pilgrims and Native Americans at the first Thanksgiving. Turkeys consumed today are actually descended from the *Meleagris gallopavo* domesticated by the Aztecs of Mexico. Right now, seven standard breeds of turkey exist, but only the broad-breasted white is of commercial significance.

Turkeys are classified as fryer-roasters, hens, and toms. Fryer-roasters are very young turkeys, under 12 weeks old, with a ready-to-cook weight of around 7 pounds. They are seldom found in the markets, however; young hens and toms are more often sold. A young hen will weigh less than a young tom of the same age. Young toms are usually processed at about 17½ weeks of age, while the hens are processed earlier, at 14½ weeks, when they weigh 26 and 14 pounds, respectively. The ready-to-cook weight varies from 8 to 15 pounds for a young hen and from 25 to 30 pounds for a young tom.

## Other Domestic Poultry

The flesh of ducks and geese is not as widely consumed as that of chickens or turkeys, and is considered a luxury food item by many people. Ducks are usually marketed when they are 7 to 8 weeks old and weigh 3 to 7 pounds in their ready-to-cook state. Geese are marketed at about 11 weeks of age and have a ready-to-cook weight of 6 to 12 pounds. Other birds such as guinea fowl, squab (young pigeon), quail, and pheasant are also sometimes consumed. Occasionally these birds may be served in restaurants as delicacies or special entrées. The immature version of these birds is preferred for consumption. For example, younger guinea fowl weighing 1¾ to 2½ pounds (live weight) are preferred over mature guinea fowl that are normally 1 pound heavier. Squab are processed just before they leave the nest, or at about 30 days of age.

# COMPOSITION OF POULTRY

The composition of poultry (muscle tissue, connective tissue, etc.) is similar to that of meat (see Chapter 7).

## Pigments

Turkeys and chickens have both white and dark meat, the lightness or darkness depending on the amount of myoglobin content in the muscle.

 **How & Why?**

*Why is the breast meat in chicken and turkey whiter than the thigh or drumstick?*

Higher amounts of the red-pigmented myoglobin are found in muscles that are used more frequently, such as those of the thighs and drumsticks (12, 21). In contrast, chicken and turkey breasts are more white because both these types of birds do almost no flying, and their meat (muscles) in these areas thus contains much less myoglobin. Wild birds such as ducks have darker breast meat because they actually use the muscles for flying.

# PURCHASING POULTRY

## Inspection

In 1968, the Wholesome Poultry Products Act made inspection of poultry shipped across state lines mandatory. It is also required that poultry sold within a state meet similar regulations, but these vary slightly from state to state. Poultry is inspected for wholesomeness before and after slaughter by a United States Department of Agriculture (USDA) inspector, who also ensures that the poultry is processed under sanitary conditions. Processing plants are encouraged to follow a Hazard Analysis Critical Control Point (HACCP) plan to minimize the risk of food-borne illness among consumers (6). Poultry that passes inspection is stamped with the USDA inspection mark.

## Grading

The grading of poultry is voluntary and is paid for by the producer. Three grades are used: A, B, and C. Grade A is the best and refers to a chicken that is full-fleshed and meets standards of appearance (Figure 8-1). The criteria used in grading are the conformation (the shape of the carcass), the fleshing (the amount of meat on the bird), the amount and distribution of fat, and freedom from blemishes such as pinfeathers, skin discoloration, broken bones, and skin cuts and tears (5). Poultry parts may also be graded USDA A, B, or C as well. In spite of the claims made by some chicken producers, skin color is not reflective of quality, but rather of the amount of xanthophyll and carotene plant pigments in the bird's diet.

The USDA grade shield shown in Figure 8-1 is used only when the poultry has been USDA graded. Because such grading is not mandatory, some poultry may be marketed under the proprietary grades established by individual packing houses, which may or may not match federal standards.

## Types and Styles of Poultry

Poultry comes to market in a number of different types and styles. *Type* refers to whether it is fresh, frozen, cooked, sliced, canned, or dehydrated. *Style* describes the degree to which it has been cleaned or processed, that is, live, dressed, ready-to-cook, or convenience categories. Live birds are rarely bought by the average consumer or restaurant. The other styles are far more prevalent.

- *Dressed.* Dressed birds are those that have had only the blood, feathers, and craw removed. The craw or crop is the pouch-like gullet of a bird where food is stored and softened.
- *Ready-to-cook.* Ready-to-cook poultry is **eviscerated**, free of blood, feathers, head, and feet; it is what is

**FIGURE 8-1**   USDA grades for poultry.

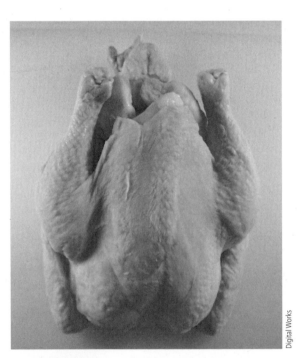

 Fully fleshed and meaty; uniform fat covering; well formed; good, clean appearance. This grade is most often seen at retail.

 Not quite as meaty as A; may have occasional cut or tear in skin; not as attractive as A.

Digital Works

typically found in the supermarket and in most food service facilities. In ready-to-cook poultry, the internal organs such as the heart, liver, neck, and gizzard (part of the bird's stomach) have been cleaned and had the fat removed, and are frequently put back inside the inner cavity, often in their own giblet bag.

- **Convenience.** For convenience, smaller pieces such as halves, breasts, drumsticks, thighs, and wings of both chicken and turkey are available.

Ground turkey and ground chicken products are also becoming increasingly popular, and are used in a variety of foods ranging from sandwich fillings to frozen entrées. Not all ground poultry products are created equal. Labels should be read carefully, because fat is sometimes added, which increases the total calorie and fat gram counts.

## Processed Poultry

Convenience is also available to consumers and food manufacturers in the form of processed poultry products. Processed chicken and turkey are commonly used in canned or dried soups, frozen dinners, potpies, sausages, hot dogs, burgers, and bologna. In addition, larger pieces of processed poultry meat minus the bone are sold as boneless turkey breast, roll, and ham. These meats are made from mechanically deboned poultry in which the bone fragments have been removed. The larger cuts are easy to carve and have a characteristic texture due to binders and other compounds that have been added (Chemist's Corner 8-1).

### Chicken Nuggets
What is in a chicken nugget? These are chicken pieces, either whole or composed of a paste of a finely minced combination of chicken meat and skin. Commercially, they are often made with a high proportion of chicken skin (100 calories/kcal per ounce) that provides a sticky consistency to hold the nugget together. Coated with batter or breadcrumbs, they are then normally deep-fried (commercially) or baked (home preparation). Both the skin content and frying contribute to making chicken nuggets a high-calorie food. Another possible ingredient in chicken nuggets sold at fast-food restaurants is monosodium glutamate (MSG).

## NUTRIENT CONTENT

The protein, carbohydrate, and vitamin content of poultry is somewhat similar to that of meats (see Chapter 7), with the exceptions listed below.

**Fat and Cholesterol.**   Contrary to the popular notion that poultry is always lower in fat and cholesterol, Figure 8-2 shows that, with the exception of a few higher-fat meat cuts, poultry is very similar to other meats in nutritive value.

In both chickens and turkeys, the dark meat is usually higher in fat, calories, and iron than white meat. It is only after removing the skin, about 100 calories (kcal) per ounce, that there is any significant difference in fat content between poultry and lean cuts of meat. Ducks and geese have a considerably higher fat content than chickens or turkeys, which increases their buoyancy in water. Emu and ostrich meat are lower in both calories and grams of fat (see Calorie Control).

Recently, chicken products with favorable fatty acid profiles have been developed through altering the oil content of chicken feed (15).

**Minerals.**   Unless specifically manufactured with less sodium, any processed poultry product (canned, dried, smoked, or self-basting) is higher in sodium than nonprocessed poultry. Processed poultry products are sometimes used as a substitute for the meat in foods such as hot dogs, bologna, and hamburgers, and lower-sodium varieties are available for these uses (9).

**FIGURE 8-2**   Comparing the calories and fat grams in poultry vs meat (3 oz).

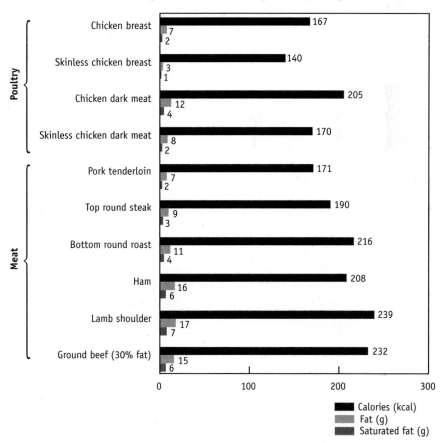

### Hormones and Antibiotics
The USDA does not allow the use of hormones in the raising of chickens. Antibiotics may be given to prevent disease

**Eviscerate**  To remove the entrails from the body cavity.

## CHEMIST'S CORNER 8-1

### Processing Poultry

The texture of processed poultry products is influenced by a variety of factors. First, the physical removal of meat from the bone mechanically causes a redistribution of the collagen fibers and myofibril proteins around the fat globules. This creates a more stable meat emulsion (17). Second, a brine mixture containing water, salt, and phosphates is added to improve flavor and cohesiveness. The phosphates in the mixture make the protein more absorbent to water by binding to the calcium and causing the protein fibers to relax. Gums such as carrageenan are then added to absorb water, creating a gel-like texture that prevents water loss during heating and makes slicing easier.

## CALORIE CONTROL
### Poultry

| Avoid fried poultry | baked | | fried |
|---|---|---|---|
| chicken (one breast with skin) | 187 | vs. | 364 calories (kcal) |

| Remove skin from poultry | no skin | | with skin |
|---|---|---|---|
| chicken (3 oz dark) | 170 | vs. | 205 calories (kcal) |
| turkey (3 oz) | 159 | vs. | 188 calories (kcal) |

Fried, with-skin poultry is almost double the calories of non-fried, no-skin poultry

**Daily Cap:** Limit meat servings to 5–6 ounces daily (two 3-oz servings; consider calories)
two 150-calorie servings = 300 calories (kcal) vs.
two 300-calorie servings = 600 calories (kcal)

**Portion Control:** One chicken wing with skin (baked) = 100 calories (kcal)
10 chicken wings = 1000 calories (kcal) & 70 grams of fat

**Calorie Crunchers:** Chicken nuggets (10 pieces) and duck with skin (3 oz) average 300 calories (kcal)

**Vegetarian Option:** 10 vegetarian nuggets (187 calories [kcal]) over 10 chicken nuggets (276 calories [kcal])

© 2010 Amy Brown

and increase feed efficiency. However, a "withdrawal" period is required prior to slaughter to ensure that there are no residues in the bird's system.

*Additives*   Additives are not allowed in fresh chicken. Processed chicken may contain additives; however, the chicken must be clearly labeled as containing them. Many of the additives used in poultry are also used in meats (see Chapter 7). Common additives used in poultry include salt, monosodium glutamate (MSG), and sodium erythorbate, which keeps processed poultry meat from changing color. Corn endosperm oil and *Tagetes erecta L.* (Aztec marigold flower petals that are dried and ground) are included in chicken feed to enhance the yellow color of chicken skin.

### Labeling

The United States Department of Agriculture regulates labeling of poultry products and a list of commonly used terms is available on the USDA website (www.fsis.usda.gov/FactSheets/Meat_&_Poultry_Labeling_Terms/index.asp). In 2008, a U.S. District Court judge ruled that a major meat producer could not advertise chicken products as "raised without antibiotics thought to lead to drug resistance in humans." This ruling came after competitors complained that this claim misled consumers, because all commercially raised chickens are antibiotic-free by the time they reach market (16).

### Standardized Poultry Buying

Similar to "The Meat Buyers Guide" for meat, another publication exists for the institutional purchasing of poultry, "The Poultry Buyers Guide." This guide details the different cuts of chicken, turkey, duck, goose, and game birds. "The Poultry Buyers Guide" allows standardization among poultry products so that there is some form of uniformity among companies selling and buying poultry meat.

## How Much to Buy

Ready-to-cook poultry contains a good deal of inedible bone and unwanted fat, which must be taken into consideration when deciding how much to buy. A good rule of thumb for most poultry is to buy ½ pound or slightly more per serving. The exceptions are ducks and geese, which have more fat to melt during cooking, resulting in less yield.

When purchasing a goose, plan on a bit over ½ pound per serving, and plan on 1 pound for ducks. Turkeys under 16 pounds, which have a higher bone-to-meat ratio, are best purchased at about 1 pound per person.

Common broiler-fryer chickens average 3½ pounds and yield four servings—two breasts, and two leg and thigh pieces. Chickens under 2½ pounds are not economical. Turkeys, especially full-grown toms weighing 18 pounds or more, provide the greatest yield per pound. One of the most economical ways to buy poultry is in its ready-to-cook whole state. Poultry purchased whole can be cut up following the steps illustrated in Figure 8-3.

# PREPARATION OF POULTRY

Throughout the world, chicken is the most widely eaten of all the types of poultry. In Mexico, cooked chicken is shredded to fill tacos, enchiladas, and tamales. The Chinese stir-fry freshly cut-up chicken with vegetables and soy sauce. Chicken Kiev is a Russian specialty consisting of boneless breasts that

**FIGURE 8-3**  Cutting up a chicken.

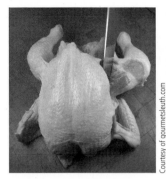

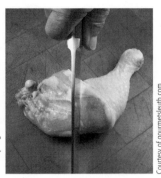

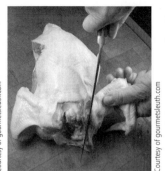

(1) Separate the leg from the breast. Pull the drumstick toward you. Use the tip of a knife to cut through the skin diagonally. Snap the thigh away from the backbone until the joint pops out of the back. Cut through the remaining skin.

(2) Optional: Cut through the leg joint to create the drumstick and thigh pieces.

(3) Dislocate the wings and then cut them away from the body.

(4) Whack through the ribs with a heavy chef's knife. Holding the chicken with the pointed end of the breast up, use a chopping motion to separate the whole breast from the back.

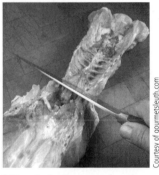

(5) Snap the backbone away from the breast. Hold the breast in one hand and push down on the backbone with the other. With this action, the wishbone is exposed. Cut along the wishbone to fully remove the back. Aim for the point where the wings join the breast, being careful to leave them attached to the breast. Save the back to use for stock.

(6) Cut the breast in half through the center of the cartilage. Cut off any pieces of wishbone and rib that remain attached to the breast.

(7) A whole chicken cut in 9 pieces.

are stuffed, rolled in a seasoned batter, and deep-fried. Paella, a Cuban favorite, is a combination of chicken with rice, tomatoes, sausage, and shellfish in one dish. In Africa, where peanuts are known as groundnuts, groundnut stew is made by simmering chicken with tomatoes and peanuts. In Japan, chicken may be marinated in a mixture of soy sauce, rice wine, and ginger before being grilled or steamed with cooked rice and egg. The resulting dish, called donburi, is very popular in that country. In India, chicken may be spiced and braised in a curry sauce or marinated in yogurt and spices before being roasted. The French are famous for *coq au vin,* or chicken braised in red wine,

and the Italians are known for roasting chicken with rosemary. Some chicken dishes commonly consumed in North America include fried chicken; chicken cordon bleu; chicken and dumplings; chicken à la king; chicken divan; and chicken pies, soups, and salads.

# Preparation Safety Tips

Raw poultry and meat should not be washed prior to preparation because this increases the danger of cross-contamination. Because most bacteria are on the surface of the meat or the inside cavity, washing may spread them

to counter surfaces, utensils, and read ready-to-eat foods (19).

## Thawing Frozen Poultry

Freezing will largely protect against bacterial growth while the poultry is frozen, but precautions should be taken during and after thawing, when any bacteria that are present may begin to grow. The refrigerator is the best place to thaw frozen birds, and its use requires planning ahead. It takes about a day for a 3½-pound chicken and 1 to 5 days for a turkey to defrost, depending on its weight (Table 8-2). When the cavity is sufficiently thawed, the package of internal organs should be removed, and the cavity rinsed. Thawing whole

| **TABLE 8-2**   Thawing a Turkey. The rule of thumb is about 24 hours of thawing for every 5 pounds of whole turkey | |
|---|---|
| Weight | Thawing Time in Refrigerator (40°F/4°C) |
| 8–12 lb | 1–2 days |
| 12–16 lb | 2–3 days |
| 16–20 lb | 3–4 days |
| 20–24 lb | 4–5 days |

poultry at room temperature, in the microwave oven, or under running cold water is not recommended.

### Stuffing

For food safety reasons, the USDA recommends that stuffing be prepared and cooked separately or, if not, at least checked with a meat thermometer to confirm that the internal temperature is at least 165°F (74°C). Prestuffed frozen poultry should never be thawed, but should be prepared, according to package directions, directly from the frozen state. The stuffing should be removed from leftover cooked poultry before the bird is refrigerated or frozen. Stuffing turkeys is not recommended as the stuffing may not reach 165°F (74°C).

### Brining

What is brining? It is soaking food in salty water (a brine). Applying this method to poultry prior to cooking increases its water content, resulting in a more juicy, flavorful meat. The poultry (whole or pieces) is placed in a large nonreactive pot, and covered with a brine solution. The simplest brine solution is water and salt, but sugar, herbs, and spices can also be added (8). Many formulations for brine solutions exist, but one example consists of water (1 gallon), salt (¾ cup), sugar (⅔ cup), soy sauce (¾ cup), and herbs (one teaspoon each of dried tarragon, thyme, and pepper). The water, salt, sugar, and soy sauce are boiled to dissolve the salt and sugar; the herbs are then added to the solution after it has been removed from the heat source. Once the brining solution is cooled, it is important that the chicken be completely submerged in it; do this by placing a heavy object on top of the pot's lid. Food safety is crucial, so the pot containing the submerged chicken in the brining

solution should be stored in the refrigerator. The brining process takes about 2 hours for chicken pieces and 4 hours for a whole chicken. The water and salt enter the muscle through diffusion and osmosis. Leaving the chicken for too long a time in the brining solution will cause the meat to become too mushy and salty.

## Changes during Preparation

Properly prepared poultry is tender and juicy, but overcooking causes the flesh to become dry, tough, and stringy. The skin of any poultry, which is primarily fat, can be removed before or after preparation, but if it is left on, it does contribute to flavor and juiciness. Fat that naturally melts off the bird during heating can be used to baste the poultry or to create sauces. Basting adds flavor and helps keep the meat tender and moist. Fat rises to the top of the drippings, so it may be easily removed before the drippings are used for gravy or sauce.

Reheated poultry, especially turkey, has a characteristic warmed-over flavor caused by the breakdown of fat (14). Microwave reheating results in less of this warmed-over flavor than reheating using conventional methods (4). The other changes that occur during preparation closely parallel those found in meats (see Chapter 7).

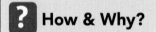

### How & Why?

*Why is there so much concern about food safety in poultry preparation?*

About one fourth of all chickens in the United States carry *Salmonella*, and about half carry *Campylobacter jejuni* (2). A national survey showed that although only about 4% of broilers tested positive for *Salmonella* before processing, the number rose to 36 % after the carcasses had been subjected to scalding, defeathering, evisceration, and chilling (7). For this reason, anything that comes in contact with raw poultry, including hands, cutting boards, sinks, utensils, dishes, and counters, should be cleaned and sanitized afterward.

## Determining Doneness

Poultry should always be heated until well done to enhance flavor and to minimize the risk of foodborne bacterial illnesses. Doneness may be determined by internal temperature, color changes, and/or touch and time/weight tables, each of which is discussed below.

### Internal Temperature

The best way to check poultry for doneness is to use a meat thermometer. It should be inserted into the thickest part of the breast, although it can also be inserted into the inner thigh. In either case it should not touch bone or fat. Poultry is sufficiently cooked when the internal temperature reaches a minimum of 165°F (74°C) for at least 15 seconds. The pop-up indicators that some poultry producers place in turkey breasts are not always reliable, so check for other signs of doneness. A thermometer placed in the center of any stuffing must reach a minimum temperature of 165°F (74°C) (Figure 8-4).

### Color Change

When the skin on oven-roasted chicken or turkey reaches a golden brown color, it is time to test for doneness. The juices coming out of the bird should have turned from pink to clear, and a bit of bone should be showing on the tip of the legs. When a turkey is roasted breast side up, the breast should be covered with metal foil or a bit of cooking oil to keep the breast from over-browning or burning. The foil should be removed 45 minutes to an hour before the end of heating to allow for final browning.

### Touch

When pressed firmly with one or two fingers, the well-done bird's flesh will feel firm, not soft. White meat may be firmer than dark meat, in part because certain proteins have a higher gel-forming ability in white muscle than they do in the dark muscles (3). Another way to tell whether or not the poultry is done through touch is to wiggle the drumstick—it should move easily in its joint.

### Time/Weight Charts

Time/weight charts appear on the packaging of all frozen and many fresh birds. It takes about 1½ hours in a 350°F

**FIGURE 8-4** Internal temperatures for a well-cooked turkey.

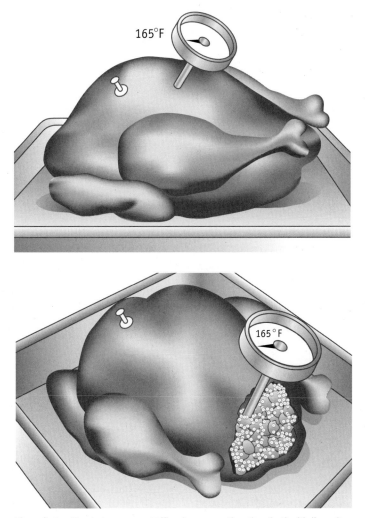

The safest option is to prepare stuffing in a pan rather than in the bird's cavity.

(177°C) oven to thoroughly cook a 3½-pound chicken. Preparation times for turkeys depend on their weight and are reduced for those roasted in one of the special oven bags (Table 8-3). Although there are time/weight charts for frozen turkeys, it is not recommended that they be cooked from the solidly frozen state, because they may not be heated through enough to destroy microorganisms.

# Dry-Heat Preparation

Some of the dry-heat methods of preparing poultry now discussed include roasting (baking), broiling or grilling, and several methods of frying.

## Roasting or Baking

Both whole or individual pieces of poultry can be roasted (baked). A heavy-duty roasting pan is used for the best result. Some people place the bird, breast up, directly on the pan, which promotes the loss of tasty juices to be collected later, or breast down, for more juicy breast meat. Others use a V-shaped rack or a flat rack to keep the bird elevated from the juices. The inside of the cavity of a whole bird is seasoned as desired, and the outside may be coated lightly with vegetable oil to prevent the skin from cracking and the meat from drying out. This process is further enhanced by basting the bird every 20 minutes with its own juices. The pan drippings (fat removed) also make an excellent natural sauce to pour over the chicken pieces, further adding to flavor. In fact, without these pan drippings, the meat may seem drier and less tasty. Margarine is not recommended for oiling the skin or for basting because of its low smoking temperature. Seasonings may be added as desired. Although surface seasonings do not add flavor to the flesh, the skin does become more flavorful when it has been browned to a certain crispness (Chemist's Corner 8-2). Salting the skin may dry it out and sometimes this is done purposely to create crispy skin (1). The outwardly salted bird is left uncovered in the refrigerator (4 hours to 2 days) to air dry. Sprinkling a slight amount of sugar on the skin will make it brown even more and perhaps cause darker spots. Salting and seasoning the inner cavity are optional, but the seasonings permeate from the inside out through the meat (via steaming) during baking, which results in a tastier meat.

The bird is then placed in an oven set at between 325°F and 350°F (163°C and 177°C) and baked for the allotted time (up to 1½ hours for a whole chicken):

- 20 to 25 minutes per pound for poultry up to 6 pounds
- 15 to 20 minutes per pound for poultry up to 15 pounds

**CHEMIST'S CORNER 8-2**

**Aroma of Roasting Chicken**

The classic aroma of roasting chicken comes from volatile compounds such as carbonyls and hydrogen sulfide (13).

**TABLE 8-3** Time/Weight Chart for Preparing Turkey at 325°F/163°C*

| Weight (pounds) | Cooked in Open Roasting Pan | | Cooked in Oven Cooking Bag | |
| --- | --- | --- | --- | --- |
| | Unstuffed (hours) | Stuffed** (hours) | Unstuffed (hours) | Stuffed** (hours) |
| 8–12 | 2¾–3 | 3–3½ | 1¾–2¼ | 2¼–2¾ |
| 12–16 | 3–4 | 3½–4¼ | 2¼–2¾ | 2¾–3¼ |
| 16–20 | 4–4½ | 4¼–4¾ | 2¾–3¼ | 3¼–3¾ |
| 20–24 | 4½–5 | 4¾–5¼ | 3¼–3¾ | 3¾–4¼ |

*These times are approximate and should always be used with a properly placed thermometer.

**The safest option is to prepare stuffing in a pan rather than in the bird's cavity.

**FIGURE 8-5**　Protecting stuffing from scorching.

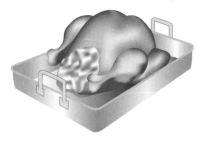

- 12 to 15 minutes per pound for poultry over 15 pounds

When birds are stuffed, cooking times must be increased by about 5 minutes per pound to make sure the stuffing is heated sufficiently to kill microorganisms all the way through. A small piece of aluminum foil placed over exposed stuffing in the final stages of baking will prevent it from scorching (Figure 8-5). As stated above, it is recommended that turkeys not be stuffed for food safety reasons.

Poultry may be **trussed** before roasting. This is usually done with turkeys because of their long preparation time. Figure 8-6 illustrates one method of trussing a bird. Wire clips, which frequently come with a turkey, will hold the legs in place without trussing. The wire clips should be temporarily removed when cleaning the bird prior to preparation. The wings can be tied up against the breast to prevent their edges from burning.

Birds to be roasted are placed, usually with the breast up, in a heavy-duty roasting pan on the lowest rack of the oven. The pan should have 2-inch sides; sides higher than 2 inches make basting difficult and prevent the lower portion of the bird from browning. Some cooks claim that the bird is juicier if placed breast down so that it can self-baste. However, eventually it must be turned over to brown the breast, and because this task is not easy, most people find that basting

**Truss** To tie the legs and wings against the body of the bird to prevent them from overcooking before the breast is done. It is also for presentation purposes.

**FIGURE 8-6**　Trussing poultry.

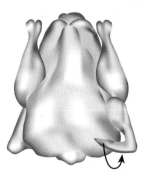

(1) Tuck the wings under the back to avoid overcooking.

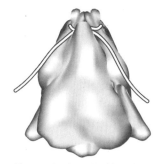

(2) Loop a butcher string (three times as long as the poultry) as shown.

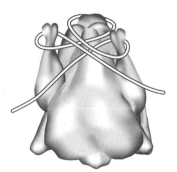

(3) Pull the ends of the string together and run them the breastbone.

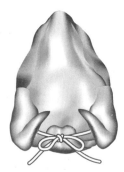

(4) Turn poultry over, tuck string under wings, and tie it over the neck flap.

a breast-up bird with accumulated pan juices and/or melted butter is quite satisfactory.

Ducks and geese, because of their high fat content, should be placed breast down after having had their skins thoroughly pricked to release excess fat during the cooking process. They are turned breast up about halfway through heating time. The skin is pricked again at least once during heating to facilitate fat drainage, and pan drippings are periodically removed during roasting. Duck and goose are sometimes preroasted for about 15 minutes and then prepared in the same manner as chicken. Cornish game hens are roasted the same way as broilers and fryers except that their cooking time is only about half an hour, unless they have been stuffed, in which case their baking time increases by 15 minutes.

**Basting**　Basting chickens and turkeys helps prevent drying of the skin and meat. This involves using a wide spoon or brush or a special tool called a baster to periodically cover the bird with liquid from the drippings, melted butter, or barbecue or other sauce. Any sauce

containing sugar (brown or white), such as barbecue sauce, will increase browning, possibly to an undesirable degree, and should be applied toward the end of the cooking process to avoid burning the sauce. The number and timing of bastings depend on the size of the bird and whether or not it has been covered early in the cooking with an oil-soaked cloth or other covering, but basting once every half hour is usually more than adequate. Basting helps the skin to brown, but to prevent overbrowning, tent the bird with aluminum foil two-thirds of the way through the cooking time.

Duck and goose do not need to be basted; they are so high in fat that they are self-basting. For the same reason, any stuffing for these fowl should be cooked separately, because it would become too soaked with fat if prepared in the cavity of the bird.

**Stuffing**　Stuffing refers to anything that is placed in the cavity of a bird during cooking. This is usually the familiar breadcrumb or cornbread stuffing; however, other foods such as vegetables and meats are sometimes stuffed in the bird's cavity. Dressing is distinguished from stuffing by being heated separately in a casserole or pan and served as a side dish.

The main ingredient of stuffing/dressing is cut cubes of day-old bread, packaged stuffing mixes, or rice. This starch-based foundation absorbs the juices released during cooking, which is why it is important that it be dried; otherwise, the dressing will be mushy. Bread cubes (¼ to ½ inch for turkey, smaller cubes for chicken) can be dried by spreading them out on a cookie sheet and baking them on low (275°F/135°C) for 15 minutes or leaving them out on the counter overnight. If grains such as rice are to be the main stuffing ingredient, they should be cooked and cooled before being combined with the other ingredients. Added to this bread or grain base is a **mirepoix** (meer-PWAH). Apricot or apple pieces, nuts, mushrooms, oysters, raisins, or other items may also be added to the base, according to personal preference. A liquid such as broth or water is then added to hold the mixture together, and eggs may be included to add cohesiveness. If the stuffing is going into the bird, only enough liquid to make the stuffing barely hold together should be added; if it is too moist, it will not be able to soak up juices.

All the ingredients should be lightly tossed together and then spooned into the poultry cavity. Stuffing should not be packed in, but should fill only three quarters of the cavity, because the stuffing will expand as it cooks. It is important to remember that stuffing a bird increases roasting time, and plan accordingly. The center of the stuffing needs to reach a final temperature of 165°F (74°C) in order to destroy microorganisms.

A stuffed bird should be allowed to stand for only a short time after being removed from the oven and before it is served. It should be refrigerated as soon as possible after that, and all the stuffing should be taken out of the bird's cavity before refrigerating. It cannot be stressed enough that stuffings, particularly those with eggs as an ingredient, are an ideal medium in which microorganisms can grow and flourish.

If stuffing is not used, then apple, potato, carrot, onion, or celery stalks may be placed in the cavity to absorb off-flavors. The fat and off-flavors absorbed

**FIGURE 8-7**   Carving roast chicken.

(1) Steady the chicken on a sanitary cutting board. To remove the leg, slice through the skin holding the leg to the breast.

(2) Push the leg down to partially dislodge the joint, cut through the meat between the leg and breast, then cut through the joint.

(3) To separate the breast meat, brace the chicken with a fork, slicing just inside the keel bone. Move the knife downward, pulling/cutting the breast section away from the rib cage.

Digital Works

by any such fruits or vegetables during cooking render them unappetizing to eat, and they are usually discarded.

*Carving*   Chicken is carved into the breast, leg, thigh, and wing pieces using the technique illustrated in Figure 8-7. Turkey should be allowed to stand for about 20 minutes after it is removed from the oven before carving. This allows the flesh to firm up and makes carving easier. Figure 8-8 demonstrates carving a turkey. Carve only what will be used immediately to avoid drying and cooling of the turkey meat pieces.

### Broiling or Grilling

Except when cooking a whole bird on a spit over hot coals, only cut-up poultry is used for broiling or grilling. It is frequently marinated or coated with butter and seasonings before being broiled or grilled. In the interest of food safety, marination must take place under refrigeration. A marinade must be fully cooked if it is to be served or used for basting. Failure to heat the marinade to a sufficient temperature to kill the bacteria that remain in it from the raw chicken may cause a foodborne illness. For the same reason, unless it is thoroughly washed in the interim, the plate used to carry the raw poultry to the grill should never be used to carry it back to the table after it has been cooked.

Vegetable sprays applied to the pan or grill help to prevent sticking. When an oven broiler is used, the poultry pieces are put skin side up on a rack in the broiler pan and placed approximately 6 inches below the heat source. The same procedure is used when

grilling over coals, except that the skin side goes down. The cooking time varies according to thickness, but in general, chicken takes 20 minutes per side. Turkey pieces are larger and so require longer cooking. Once the skin side is browned, use tongs to turn the poultry pieces over, because the piercing tines of a fork will allow juices to be lost. Sauces are best added during the last 15 minutes of preparation, because high heat readily burns sugar, which is the main ingredient of many barbecue sauces.

### Frying

Poultry pieces can be sautéed, pan-fried, deep-fried, or stir-fried.

*Sautéing*   Small poultry pieces are placed in a skillet or pan with a small amount of oil for quick preparation. Pieces must be turned to assure adequate doneness. Sautéing can also be used to brown larger poultry pieces prior to their being baked or braised to completion.

*Pan-Frying*   Pan-fried chicken pieces are usually breaded or floured before they are fried over high heat in approximately ¼ inch of fat. The breading adds texture and flavor and keeps moisture from being lost from the fried food; it

**Mirepoix**   A collection of lightly sautéed, chopped vegetables (a 2:1:1 ratio by weight of onions, celery, and carrots) flavored with spices and herbs (sage, thyme, marjoram, and chopped parsley are the most common).

**FIGURE 8-8**    Carving a turkey.

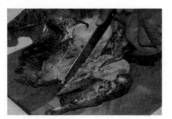

(1) Slice down between the thigh and the breast. Don't let the fork puncture the outside skin, causing juices to be lost.

(2) Push the thigh down (and bend under if necessary) to disjoint the bone; cut through the rest of the meat and any cartilage to separate it from the breast.

(3) Separate the thigh from the drumstick by cutting through the meat.

(4) Twist the thigh around the drumstick and cut through the joint.

(5) Slice the drumstick (and thigh; not shown).

(6) Remove the wing by first twisting and then cutting from the joint.

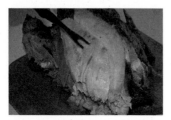

(7) Remove the skin back and slice the breast meat.

(8) Lift each slice using the knife and fork.

Courtesy of Amy C. Brown

also allows heat to be transmitted to the food without its absorbing as much fat. Fry with the skin side down first; when that side is brown, turn it over with tongs and brown the other side. Lower the heat and turn the pieces occasionally, for 30 to 45 minutes, or until done. If the poultry is placed in the oven following browning, the method of preparation is referred to as "oven fried," even though it is actually baked.

***Deep-Frying***    Deep-frying poultry pieces that have been breaded, floured, or battered involves submerging them completely in oil heated to between 325°F and 350°F (160°C and 180°C).

***Stir-Frying***    Stir-frying is lightly frying bite-size pieces of boned chicken

while stirring them frequently in a tiny amount of oil. Vegetables, also cut into small pieces, are usually added, along with soy sauce and/or other seasonings.

## Moist-Heat Preparation

Poultry can also be prepared by moist-heat preparations and some of those techniques now discussed include braising, stewing, poaching, and microwaving.

### Braising

Although braising, also called fricas-seeing, can be applied to any poultry, it is of particular value when it comes to preparing older, tougher birds. The

slow, moist heating tenderizes the meat and makes it easier to chew. The chicken or turkey is first cut into pieces and browned in a small amount of oil and/or butter; it may be floured or breaded first. Liquid is added, and the poultry is simmered in a tightly covered pan until tender. The initial browning is important because it helps create a rich flavor and holds in the juices. Desired seasonings are added with the liquid.

### Stewing

Any whole or cut-up fresh poultry can be covered in cold salted water and heated to the boiling point, at which point the heat is immediately lowered to a simmer. An average 3½ pound chicken usually takes about 2 to 2½ hours. The bones and skin may

# PROFESSIONAL PROFILE

### Food Safety Inspector and Food and Drug Branch Supervisor

Lance Wong majored in biology in college and soon after graduating decided to pursue a Master of Science degree in Public Health. His goal was to apply to law school with an emphasis in environmental health so that he could help protect the environment. Before applying to law school, Lance was checking the classifieds and saw a job as Food Safety Inspector at the Hawaii State Department of Health. That was 23 years ago. His law school plans long forgotten, Lance is now the Health Department's Food and Drug Branch Supervisor.

Lance says the Hawaii State Health Department has two divisions of Environmental Health: the Environmental Health Management division and the Environmental Health Services division. The Environmental Health Management division is concerned primarily with air and water quality, safe drinking water, solid and hazardous wastes, and wastewater disposal. It works closely with the Environmental Protection Agency at the federal level. The Environmental Health Services division is composed of Vector Control, Noise and Radiation, Sanitation, and the Food and Drug branches. The Sanitation branch has 25 food safety inspectors whose job it is to inspect restaurants, caterers, hotels, lunch wagons, and all other commercial food preparers. The Food and Drug branch, where Lance

**Mr. Lance Wong**

*Courtesy of Lance Wong*

now works (he has worked in both divisions), has 10 inspectors who inspect food manufacturers, bakeries, supermarkets, open markets (such as exist in Chinatown), pharmacies, and delis. His office works closely with the federal Food and Drug Administration in safeguarding the public's health by ensuring that foods, drugs, cosmetics, and medical devices are safe, effective, and properly labeled.

One of the best things he enjoys about his job is that "each day brings new challenges and learning experiences." He is responsible for supervising five inspectors. Each inspector is responsible for some 300 food establishments. "It's a people job," Lance says. "So much so, that all those psychology classes and Dale Carnegie seminars helped me to develop my people skills. You learn most of the work on the job, but science classes really gave me a good foundation."

He trains his staff on the importance of "talking to people, developing a positive rapport, and focusing on training and education rather than to be a heavy-handed enforcer." Some of the less positive aspects of the job Lance mentions include "irate managers (but only a small percentage); unsanitary conditions; and getting up close to rats, roaches, and flies." The primary goal, he says, is to "convey to the food establishments that food safety is their number one priority and that foodborne illness is bad for business."

---

or may not be removed from the pot, and dumplings, which are made from a dough mixture, can be placed gently on top of the simmering chicken 12 to 15 minutes before the end of preparation time.

## Poaching

Chicken pieces can be poached fairly quickly in a small amount of water. The chicken pieces, such as breasts, are placed in a frying pan and covered with 1⅓ cups water. The water is brought to a boil and then reduced to a simmer, and the chicken is cooked about 10 to 15 minutes or until tender.

## Microwaving

Microwave ovens do not always heat food deeply or evenly enough, and power levels vary from brand to brand, so it is suggested that stuffed poultry, particularly turkeys, be prepared in a conventional oven. The microwave manufacturer's instructions should be

followed for preparing all other poultry. This is equally true when it comes to thawing poultry or any other frozen food. Once thawed in the microwave oven, the poultry should be cooked immediately.

In general, microwave directions call for smaller pieces of poultry rather than whole fowl. If a recipe calls for chicken pieces, a microwave can be handy. The poultry pieces are arranged skin side up, with the thickest portions toward the outside of the dish and any loose flaps of skin tucked under. The dish is covered with wax paper or plastic wrap and cooked on high for about 8 minutes per pound, or according to the manufacturer's directions. Chicken breasts are heated on high for about 10 minutes or until well done. The pieces should be rotated at the 5-minute mark.

Flavor and appearance are enhanced if the pieces are initially covered with browning sauce, barbecue sauce, or some other topping.

Cooking is completed when the flesh is firm and fork tender, and the juices run clear instead of pink. However, using a thermometer to verify internal temperature is advised. The finished pieces should be left to stand about 5 minutes before serving. If they are to be used in a salad or other dish, it is best to chill them in the refrigerator for at least 2 hours. Two boned, skinned chicken breasts will yield 1 cup of cubed chicken meat.

# STORAGE OF POULTRY

Precautions should be taken in the handling of poultry because of the possibility of bacterial contamination. *Campylobacter* and *Salmonella* are two of the most common causes of foodborne illness.

Raw poultry is a major source of these two bacteria (see Chapter 4).

In 1993, the irradiation of poultry was approved for commercial use in the control of *Salmonella* following several studies demonstrating that irradiation reduces bacterial concentration (20). Irradiated poultry, however, is not sterile and should be handled using the same precautions used for any raw poultry. In 1992, the use of trisodium phosphate (TSP), a colorless, odorless, flavorless chemical mixture, also received approval for use by the poultry industry on poultry carcasses to further aid in reducing *Salmonella* contamination (see Chemist's Corner 8-3).

## Refrigerated

Fresh, ready-to-cook poultry can be kept safely in the refrigerator at 40°F (4°C) or below for up to 3 days (Chemist's Corner 8-4). It should be stored in the vapor-proof wrapping in which it is purchased because repackaging increases the risk of bacterial contamination. It is best kept in the bottom portion of the refrigerator to prevent its drippings from contaminating other foods. Chickens labeled "fresh" should not have been exposed to temperatures below 26°F (−3°C), the temperature at which chickens freeze.

## Frozen

Frozen whole poultry can be stored from 6 to 12 months at 0°F (−18°C),

> ## CHEMIST'S CORNER 8-3
>
> **How TSP kills bacteria**
>
> Trisodium phosphate (TSP) is approved for use as an antibacterial agent in pre- and post-slaughter chicken carcasses. TSP prevents bacterial growth by disrupting the cell membrane of the bacteria. When the cell membrane is broken, this allows contents of the bacteria to leak out, resulting in death of the bacterium (10).

> ## CHEMIST'S CORNER 8-4
>
> **Oxidation of Cooked Poultry**
>
> Dark meat has a higher myoglobin content than white meat does. Consequently, it is more easily oxidized because the iron in the myoglobin acts as a metal catalyst to speed up the reaction of the polyunsaturated fatty acids being oxidized. Oxidation of these polyunsaturated fatty acids found naturally in the meat results in disagreeable off-odors (14). As a result, chicken legs with their dark meat cannot be stored as long as chicken breasts.

whereas leftover cooked poultry can be frozen for up to 4 months. The meat will decline in moistness and eating quality if it is kept frozen beyond these recommended times. Breaded or fried poultry should never be thawed and refrozen.

# PICTORIAL SUMMARY / 8: Poultry

Humans domesticated chickens over 4,000 years ago. These days, the consumption of poultry, especially chicken and turkey, continues to increase in popularity.

## CLASSIFICATION OF POULTRY

Poultry, or domesticated birds raised for their meat, includes:

**Chicken**
- Broilers
- Fryers
- Roasters
- Capons
- Cornish Game Hens

**Turkey**
- Young Tom
- Young Hen

Duck

Goose

Pigeon

Guinea Fowl

Domesticated birds are classified according to age and weight, and the classifications vary from species to species. Chickens are sold as broilers and/or fryers, roasters, capons, and Cornish hens. The majority of the turkeys coming to market are young hens, hens, young toms, and toms.

## PURCHASING POULTRY

All poultry scheduled to be transported interstate must have the USDA stamp of approval. For birds sold intrastate, USDA inspection is voluntary. However, strict state inspection guidelines are enforced. These may vary slightly from state to state but are close to federal standards. USDA grade stamps indicate A, B, and C quality, with A being the best. Many processors use their own grading system and stamps.

The least expensive way to buy poultry is to purchase it as a ready-to-cook whole bird. The larger the bird, the more edible meat per pound. For chicken and turkey, approximately ½ pound of whole bird is needed for each serving.

Poultry is available for purchase in the following forms:
- Fresh
- Frozen
- Cooked
- Canned
- Dehydrated
- Live
- Dressed
- Ready-to-cook
- As convenience food

## COMPOSITION OF POULTRY

Nutritionally poultry, like meat, is a high-quality protein food. Contrary to the popular notion that poultry is always lower in fat and cholesterol, poultry is very similar to many other meats in nutritive value. Poultry does provide less fat if the skin is removed. The amount of myoglobin determines whether the flesh is white or dark.

## PREPARATION OF POULTRY

Poultry can be prepared in any number of ways:

**Dry-heat methods**
- Roasting
- Baking
- Broiling
- Grilling
- Frying

**Moist-heat methods**
- Braising
- Stewing
- Poaching

Regardless of the preparation method selected, poultry should always arrive on the table well-done as determined by the combined use of internal temperature, color changes, touch, and time/weight tables. Poultry is sufficiently cooked when internal temperature reaches 165°F (74°C) for 15 seconds.

Microwave ovens are not recommended for cooking poultry, except for smaller pieces. Thawing frozen poultry is best done in the refrigerator.

When handling fresh or frozen poultry, cleanliness and personal hygiene are of utmost importance in preventing foodborne illnesses.

**Cutting Up a Chicken**

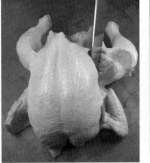

Courtesy of gourmetsleuth.com

## STORAGE OF POULTRY

Fresh poultry will keep in the refrigerator for up to 3 days, while frozen poultry will keep in the freezer for 6 to 12 months. All prepared foods should be refrigerated in covered containers and guarded against *Salmonella*.
- Store in the refrigerator a maximum of 3 days.
- Freeze for a maximum of 6 to 12 months.

# CHAPTER REVIEW AND EXAM PREP

## Multiple Choice*

1. What is the name for chickens that are older and larger than broilers/fryers?
   a. Capons
   b. Roasters
   c. Cornish game hens
   d. Roosters

2. Contrary to popular belief, the level of _____ in poultry is similar to that of other meats.
   a. carbohydrate
   b. white meat
   c. cholesterol
   d. fiber

3. How long should it take an 8- to 12-pound turkey to thaw in a 40°F refrigerator?
   a. 3–5 days
   b. 2–3 days
   c. 2–5 days
   d. 1–2 days

4. A mirepoix is defined as a 2:1:1 ratio of the following vegetables flavored with spices and herbs:
   a. carrots, onions, celery.
   b. potatoes, onions, carrots.
   c. onions, potatoes, carrots.
   d. onions, celery, carrots.

5. Poultry is sufficiently cooked when the internal temperature reaches what level?
   a. 125°F (52°C)
   b. 145°F (63°C)
   c. 165°F (74°C)
   d. 185°F (85°C)

*See p. AK-1 for answers to multiple choice questions.

6. A good rule of thumb is to purchase approximately _____ pound(s) of poultry (with the exceptions of geese, duck, or turkey) per person to be served.
   a. ¼
   b. ½
   c. 1
   d. 1½

7. What term is used to describe poultry that have had their entrails (inner organs) removed from their body cavity?
   a. Tressed
   b. Unstuffed
   c. Eviscerated
   d. Basted

## Short Answer/Essay

1. Describe how poultry is classified and then briefly discuss the various classifications of chicken.
2. What are the USDA grades for poultry and on what criteria are they based?
3. Purchasing poultry is often based on *type* and *style*. Briefly define these two terms.
4. Approximately how many pounds per serving would you purchase of chicken? Duck? Goose? Turkey?
5. Discuss the methods for determining the doneness of baked chicken or turkey.
6. Describe the basic steps involved in roasting or baking poultry.
7. Discuss the purpose and process of trussing, of basting, and of adding a mirepoix to stuffing.
8. How is poultry braised and what is the value of this preparation process?
9. What are the general recommendations for microwaving poultry?
10. What precautions should be taken when handling and storing poultry?

# REFERENCES

1. Anderson P. How to get crisp skin and juicy breast meat. *Fine Cooking* 70:40, 2005.
2. Djurdjevic N, SC Sheu, and YH Hsieh. Quantitative detection of poultry in cooked meat products. *Journal of Food Science* 70(9):586–593, 2005.
3. Ensminger ME. *Poultry Science*. Interstate Publishers, 1992.
4. Kerler J, and W Grosch. Odorants contributing to warmed-over flavor (WOF) of refrigerated cooked beef. *Journal of Food Science* 61(6):1271–1274, 1996.
5. Kinsman DM, AW Kotula, and BC Breidenstein. *Muscle Foods: Meat, Poultry, and Seafood Technology.* Chapman & Hall, 1994.
6. Kirkhorn SR. Food safety issues: A summary report of a panel session addressing pre- and post-harvest strategies to improve public health. *Journal of Agromedicine* 13(4):233–236, 2008.
7. Li Y, et al. *Salmonella typhimurium* attached to chicken skin reduced using electrical stimulation and inorganic salts. *Journal of Food Science* 59(1):23–24, 1994.
8. Masibay KY. Salt makes everything taste better. *Fine Cooking* 91:80, 2008.
9. Meullenet JF, et al. Textural properties of chicken frankfurters with added collagen fibers. *Journal of Food Science* 59(4):729–733, 1994.

10. Oyarzabal, O. Reduction of *Campylobacter* spp. by commercial antimicrobials applied during the processing of broiler chickens: A review from the United States perspective. *Journal of Food Protection,* 68:1752–1760, 2005.

11. Parkhurst CR, and GJ Mountney. *Poultry Meat and Egg Production.* Van Nostrand Reinhold, 1988.

12. Pegg RB, and F Shahidi. Unraveling the chemical identity of meat pigments. *Critical Reviews in Food Science and Nutrition* 37(6):561–589, 1997.

13. Penfield MP, and AM Campbell. *Experimental Food Science.* Academic Press, 1990.

14. Sárraga C, I Carreras, JA García Regueiro, MD Guàrdia, and L Guerrero. Effects of alpha-tocopheryl acetate and beta-carotene dietary supplementation on the antioxidant enzymes, TBARS and sensory attributes of turkey meat. *British Poultry Science* 47(6):700–707, 2006.

15. Schneiderová D, J Zelenka, and E Mrkvicová. Poultry meat production as a functional food with a voluntary n-6 and n-3 polyunsaturated fatty acids ratio. *Czech J. Anim. Sci.,* 52(7):203–213, 2007.

16. Shin A. Court orders Tyson to suspend ads for antibiotic-free chicken. *Washington Post*, page D01, May 2, 2008.

17. Tanaka MCY, and M Shimokomaki. Collagen types in mechanically deboned chicken meat. *Journal of Food Biochemistry* 20:215–225, 1996.

18. United States Department of Agriculture. Chicken consumption continues longrun rise. *Amber Waves.* April 2006. www.ers.usda.gov/ AmberWaves/April06/pdf/ ChickenFindingApril06.pdf.

19. United States Department of Agriculture. *Dietary Guidelines for Americans 2005.* Chapter 10, Food Safety. www.health.gov/ dietaryguidelines/dga2005/ document/html/chapter10.htm. Accessed 6/10/09.

20. United States Department of Agriculture. *Fact Sheets. Production and inspection: Irradiation resources.* www.fsis.usda.gov/Factsheets/ Irradiation_Resources/index.asp. Accessed 6/12/09.

21. United States Department of Agriculture. *Food Safety Information. The Color of Meat and Poultry.* www.fsis .usda.gov/PDF/Color_of_Meat_ and_Poultry.pdf. Accessed 5/5/09.

# WEBSITES

The USDA provides information on emus, ostrich, and rhea at this website:
**www.fsis.usda.gov/Fact_Sheets/ Ratites_Emu_Ostrich_Rhea/ index.asp**

The USDA's facts about poultry preparation can be found here:
**www.fsis.usda.gov/fact_sheets/ chicken_food_safety_focus/index.asp**

(Or, from **www.fsis.usda.gov**, click on "Fact Sheets" located in top horizontal bar. Click on "Poultry Preparation" located in right bar column. Scroll down to "Chicken Food Safety Focus.")

National Turkey Association:
**www.EatTurkey.com**

Find the calories in various chicken products and other foods, visit this website:
**http://caloriecount.about.com/ calories-chicken-ic0501**

# 9 Fish and Shellfish

Humans were eating fish, shellfish, and sea mammals long before they started cultivating plants or domesticating animals for food. Excavations of Stone Age sites have uncovered fishnets, spears, and fishing hooks made from the upper beaks of birds. Seafood is now the only major food source that is still hunted. Most other food sources are raised or grown.

At present, there are over 20,000 known species of edible fish, shellfish, and sea mammals. Of these, approximately 250 species are harvested commercially in the United States, of which millions of tons are being served up annually for the consumption of humans and domesticated animals. This chapter focuses on those species and examines their classification, composition, purchase, preparation, and storage.

## CLASSIFICATION OF FISH AND SHELLFISH

The staggering variety of creatures harvested from the water makes it difficult to classify them using only one set of criteria. As a result, several categories have arisen in order to distinguish them from each other: vertebrate or invertebrate, salt- or freshwater, and lean or fat.

Although these classifications are used to distinguish among different fish, a vertebrate could live in salt or fresh water, and be either lean or fat. The Food and Drug Administration (FDA) has attempted to standardize fish nomenclature by publishing the "Guide to Acceptable Market Names for Food Fish Sold in Interstate Commerce," and requiring that fish be named according to this publication (46). The FDA guide is the recommended way to classify fish and shellfish, but the three common methods mentioned above are now described: (1) vertebrate or invertebrate, (2) salt- or freshwater, and (3) lean or fat.

### Vertebrate or Invertebrate

This classification divides water animals according to the presence or absence of a backbone (Figure 9-1).

#### Vertebrate

The vertebrate category includes **finfish**, which obtain their oxygen from the

**Finfish** Fish that have fins and internal skeletons.

**FIGURE 9-1** Classification of fish.

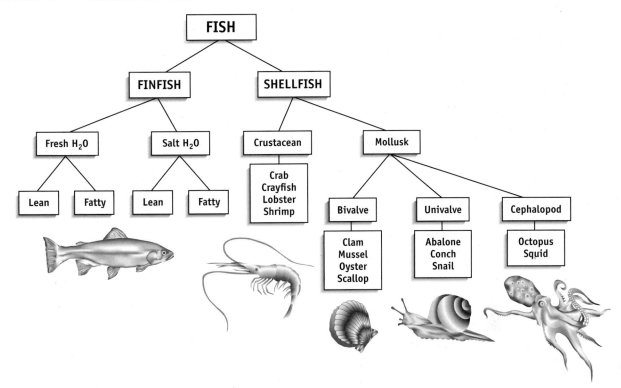

water through their gills, and sea mammals, all of which must get their oxygen from above the water's surface.

*Finfish* Finfish are found in the fresh water of rivers, lakes, and streams, and the salt water of oceans and seas. The most popular finfish in North America are tuna, cod, Alaska pollack, salmon, catfish, and flounder/sole.

*Sea Mammals* Sea mammals include dolphin, whale, and seal.

### Invertebrate

The invertebrate category includes shellfish, most of which have external skeletons or shells. The term *shellfish* is a commercial rather than a scientific classification, and includes the invertebrate **crustaceans** and **mollusks.** Examples of crustaceans are shrimp, crab, lobster, and crayfish. Mollusks include bivalves, univalves, and cephalopods. Bivalve creatures, including clams, oysters, mussels, and scallops, are contained within two hard shells that are hinged together. The univalves, such as conch and abalone, have only a single hard shell. Cephalopods, which include octopus and squid, have an almost rubbery

soft inner shell, which will be familiar to parakeet owners as a cuttlebone.

## Salt- or Freshwater

The majority of the fish eaten in the United States are taken from salty waters, but many also come from freshwater lakes, ponds, and streams. Saltwater fish often have a more distinct flavor than freshwater fish. Sole, however, is a very mild-flavored saltwater fish, and is one of several exceptions to the taste generalization (Table 9-1). Some saltwater fish other than sole are halibut, cod, flounder, haddock, mackerel, red snapper, salmon, shark, striped bass, swordfish, and tuna. Catfish, perch, pike, and trout are the most common freshwater varieties.

## Lean or Fat

Fish are sometimes identified by their fat content, but in this case, *fat* is a relative term. Fish are not very fatty compared to most other meats. A 3-ounce cooked portion of a lean fish (less than 5% fat) such as cod, pike, haddock, flounder, sole, whiting, red snapper, halibut, or bass contains less than 2.5 grams of fat

(Table 9-2). The same portion of fatty fish (more than 5% fat) yields 5 to 10+ grams of fat. Examples include salmon, mackerel, lake trout, tuna, butterfish, whitefish, and herring.

# COMPOSITION OF FISH

## Structure of Finfish

Regardless of their classification, fish are usually tender when they come to the table, and three structural factors contribute to this tenderness: collagen, amino acid content, and muscle structure.

---

**Crustacean** An invertebrate animal with a segmented body covered by an exoskeleton consisting of a hard upper shell and a soft under shell.

**Mollusk** An invertebrate animal with a soft, unsegmented body usually enclosed in a shell.

## Collagen

When compared with meat or poultry, fish muscle has lower amounts of collagen. The bodies of land animals average 15% connective tissue by weight, whereas fish are only 3% collagen (47).

## Amino Acid Content

Another reason fish is tender is that there is less of a certain amino acid (hydroxyproline) in the connective tissue. When fish is cooked, the collagen breaks down more easily at a lower temperature and converts to gelatin.

## Muscle Structure

Unlike mammals and birds, whose muscles are arranged in very long bundles of fibers, fish have muscles that are shorter (less than an inch in length) and are arranged into **myotomes,** which are separated by connective tissue called **myocommata** (Figure 9-2). This combination of structure and chemistry contributes to the characteristic flaking of prepared fish as the heat softens the collagen in the myocommata. The gelforming ability of the muscle proteins in fish can also contribute to a soft, tender, gel-like texture (18).

**FIGURE 9-2**   Fish muscle, unlike other meats, is arranged in layers of short fibers (myotomes) separated by very thin sheets (myocommata).

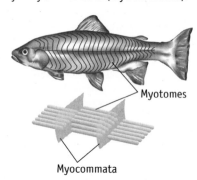

Myotomes

Myocommata

**Myotomes**   Layers of short fibers in fish muscle.

**Myocommata**   Large sheets of very thin connective tissue separating the myotomes.

**TABLE 9-1**   Common Fish and Shellfish Grouped by Flavor and Texture

| Texture | Flavor | | |
|---|---|---|---|
| | **Mild Flavor** | **Moderate Flavor** | **Full Flavor** |
| **Delicate** | Cod<br>Crab<br>Flounder<br>Haddock<br>Hake<br>Pollock<br>Scallops<br>Sole | Butterfish<br>Lake perch<br>Whitefish<br>Whiting | Mussels<br>Oysters |
| **Moderate** | Crayfish<br>Lobster<br>Pike (walleye)<br>Orange roughy<br>Shrimp<br>Tilapia | Mullet<br>Ocean perch<br>Shad<br>Smelt<br>Surimi products<br>Trout<br>Sea trout (weakfish)<br>Tuna (canned) | Bluefish<br>Mackerel<br>Salmon (canned)<br>Sardines (canned) |
| **Firm** | Grouper<br>Halibut<br>Monkfish<br>Sea bass<br>Snapper<br>Squid<br>Tautog (blackfish)<br>Tilefish<br>Wolffish | Catfish<br>Mahimahi<br>Octopus<br>Pompano<br>Shark<br>Sturgeon | Clams<br>Marlin<br>Salmon<br>Swordfish<br>Tuna |

**TABLE 9-2**   Lean vs. Fatty Fish: Fat Content of 3-Ounce Cooked Portions of Fish and Shellfish

**Lean Fish**

*Very Low Fat—Less Than 2.5 Grams Total Fat*

| | | | |
|---|---|---|---|
| Clams | Haddock | Pike (northern) | Red snapper |
| Cod | Halibut | Pike (walleye) | Snow crab |
| Cusk | Northern lobster | Pollock (Atlantic) | Sole |
| Blue crab | Mahimahi | Ocean pout | Squid |
| Dungeness crab | Monkfish | Orange roughy | Tuna (skipjack) |
| Flounder | Perch (freshwater) | Scallops | Tuna (yellowfin) |
| Grouper | Ocean perch | Shrimp | Whiting |

*Low Fat—More Than 2.5 Grams But Less Than 5 Grams Total Fat*

| | | | |
|---|---|---|---|
| Bass (freshwater) | Croaker | Salmon (pink) | Swordfish |
| Bluefish | Mullet | Shark | Rainbow trout |
| Blue mussels | Oysters (eastern) | Smelt | Sea trout |
| Catfish | Salmon (chum) | Striped bass | Wolffish (ocean catfish) |

**Fatty Fish**

*Moderate Fat—More Than 5 Grams But Less Than 10 Grams Total Fat*

| | | |
|---|---|---|
| Butterfish | Salmon (Atlantic) | Lake trout |
| Herring | Salmon (coho) | Tuna (bluefin) |
| Mackerel (Spanish) | Salmon (sockeye) | Whitefish |

*Higher Fat—More Than 10 Grams Total Fat*

| | |
|---|---|
| Mackerel (Atlantic) | Salmon (king) |

**Source:** U.S. Dept. of Fisheries.

# Pigments

When fish flesh is exposed to air during preparation, it will vary in color as a result of the presence of white, pink, or red pigments. The color of a fish's flesh depends on whether that fish relied predominantly on quick or slow movements to stay alive. Red or darker colored flesh, such as that seen in salmon, has a higher concentration of the "slow-twitch fibers" needed for long-distance swimming and endurance. White meat, like that of the sole, has more "fast-twitch fibers," which are designed for quick bursts of speed of brief duration between long periods on "cruise control." Some fish, such as tuna, are composed of both fast-twitch and slow-twitch fibers, giving them dark, light, and white meat. A higher fat content will also darken the color of the flesh, as seen in fatty fish such as mackerel and tuna.

The concentration of myoglobin contributes to the overall color of fish flesh. The more oxygen required by the muscle, the more myoglobin proteins are necessary, because they carry the oxygen. Unfortunately, a higher myoglobin concentration results in quicker rancidity because the iron in myoglobin accelerates the oxidation of fat found in the muscle (41).

# PURCHASING FISH AND SHELLFISH

Commonly purchased fish and shellfish and their uses are listed in Table 9-3. Retailers providing consumers with nutrition information must abide by the nutrition labeling values provided by the FDA for fish and shellfish (36). Fish processors may submit to inspection and grading on a voluntary basis.

## Inspection/Grading

Unlike inspection of meat and poultry, inspection of finfish is voluntary; inspection, when it occurs, is based on the wholesomeness of the fish and the sanitary conditions of the processing plant. The National Marine Fisheries Service of the U.S. Department of Commerce is responsible for fish inspections, which are paid for by the processor.

**TABLE 9-3   Types of Fish and Shellfish and Their Uses**

| Common Name(s) | Uses |
| --- | --- |
| American Pollock (Boston Bluefish) | Baked, broiled, pan-fried, steamed, or poached. |
| American Shad (Buck Roe; White Shad) | Baked, broiled, planked, stuffed, or sautéed. |
| Atlantic Croaker (Croaker; Hardhead) | Baked, broiled, poached, pan-fried, or oven-fried. |
| Blue Crab | Steamed (boiled), cakes, patties, deviled, stuffed, casseroles, salads, or appetizers. |
| Carp | Not a favorite of Americans; eaten by Europeans. |
| Catfish | Baked, broiled, grilled, barbecued, smoked, sautéed, or stuffed. |
| Cod (Codfish; Scrod) | Baked, broiled, poached, fried, or steamed; oven finish or deep-fry breaded portions and sticks. |
| Crayfish (Crawfish) | Like lobster; thick soup, crayfish bisque. |
| Dungeness Crab | Steamed, baked, broiled, simmered, casseroles, salads, appetizers, cocktails, and sauces. |
| Flounder (Blackback, Fluke; Summer Flounder, Winter Flounder) | Baked, broiled, poached, fried, steamed; oven finish or deep-fry breaded fillets, sticks, or portions. |
| Geoduck Clam | Steaks, fried, minced as dip or chowder; party snacks. |
| Haddock (Scrod) | Same as flounder. |
| Halibut (North Pacific Halibut) | Baked, broiled, poached, fried, or steamed. |
| Herring (Pacific Sea Herring) | Bait, oil, fertilizer; cooked or eaten as kippered herring. |
| Jonah Crab | Steamed, simmered, or broiled; casseroles, salads, appetizers, cocktails, and sauces. |
| King Crab | Used interchangeably with other crabmeat recipes; casseroles, salads, appetizers, cocktails, and sauces. |
| Lake Trout | Baked, broiled, poached, fried, steamed, or sautéed. |
| Lingcod | Broiled, sautéed, baked, poached, or deep fried. |
| Lobster (Spiny Lobster) | Baked, broiled, or simmered; variety of recipes for use. |
| Mackerel | Baked, broiled, fried, poached, or steamed. |
| Menhaden (Pogy; Fatback) | Seldom for human consumption. |
| Mullet (Black or Striped Mullet) | Deep-fried, oven-fried, baked, or broiled. |
| Ocean Perch (Redfish; Rockfish; Rosefish) | Baked, broiled, poached, or steamed fillets; oven fried or deep-fried, breaded, raw, or cooked portions. |
| Ocean Quahog Clam (Mahogany Quahog; Black Quahog) | Deep-fried; pan-fried patties; deviled clams; Manhattan clam chowder; clam cakes and rolls. |
| Oysters (Eastern or Atlantic Oyster; Pacific Oyster; Western Oyster) | Steamed, baked, sautéed, or used in variety of dishes. |
| Pacific Cod | Baked, broiled, poached, fried, or steamed; oven-finish breaded, cooked portions or sticks, deep-fried frozen breaded cuts. |
| Pompano (Cobblerfish; Butterfish; Pacific Oyster; Western Oyster) | Baked, broiled, pan-fried, or deep-fried. |
| Porgy (Scup) | Baked, pan-fried, or sautéed. |
| Rainbow Trout | Baked, broiled, pan-fried, poached, or steamed. |
| Red Crab | Broiled, baked, steamed, sautéed, or served cold; suitable in any recipe for crab. |
| Red Snapper | Broiled, baked, steamed, or boiled. |
| Rockfish | Baked, broiled, fried, and in chowders. |
| Sablefish (Black Cod) | Ready-to-eat; steamed; used in casseroles or salads. |
| Salmon | Baked, broiled, barbecued, fried, steamed, or poached; variety of recipes and dishes. |
| Sardines (Atlantic herring) | Ready-to-eat snack; convenience food. |
| Scallop | Boiled or sautéed, cocktails. |

*(continued)*

**TABLE 9-3** *(continued)*

| Common Name(s) | Uses |
|---|---|
| Sea Bass (Striped Bass) | Baked, broiled, pan-fried, oven-fried, or poached. |
| Shrimp (Northern Shrimp; North Pacific Shrimp; Southern Shrimp) | Simmered, baked, broiled, fried, or oven-finish; cocktail; hundreds of uses such as casseroles, salads, and sauces. |
| Smelt (Whiteball; Eulachon) | Broiled, fried, baked, or prepared in casserole. |
| Snow Crab (Tanner; Queen) | Used interchangeably with other crabmeat recipes. |
| Sole (Gray Sole; Witch Flounder) | Baked, broiled, fried, steamed, or deep-fried. |
| Spanish Mackerel | Baked, broiled, or smoked. |
| Squid (Inkfish; Bone Squid; Taw Taw; Calamari; Sea Arrow) | Fried or baked with a stuffing; salads; sauces; combination dishes. |
| Sturgeon | Specialty item. |
| Sunray Venus Clam | Chowder, fritters, patties, dips, and clam loaf. |
| Surf Clam | Steamed, fried, or broiled; in chowders, fritters, sauces, dips, or salads. |
| Swordfish | Baked, broiled, fried, poached, or steamed. |
| Tuna | As it comes from the can; variety of recipes. |
| Weakfish (Gray Sea Trout; Squeteagues) | Baked, broiled, sautéed, or pan-fried. |
| Whiting (Frost Fish; Hake; Silver Hake) | Baked, broiled, pan-fried, poached, or steamed; portions and sticks oven-finished. |
| Yellow Perch | Baked, broiled, or pan- or deep-fried. |

Only inspected finfish can be graded. Grading, too, is voluntary and paid for by the processor. Fish products are graded U.S. Grade A, U.S. Grade B, and substandard. Quality grades are based on appearance, texture, uniformity, good flavor, fresh odor, and an absence of defects. Breaded fish products are further evaluated in terms of their breading and bone-to-fish ratio. Labels should be read whenever possible to find out whether or not the fish product has been inspected and graded.

### ? How & Why?

*Why do salmon have that characteristic pink/orange hue?*

Sometimes a specific pigment adds a special hue. For example, a carotenoid pigment, astaxanthin, imparts a characteristic orange-pink color to certain salmon and trout that feed on insects and crustaceans containing this pigment.

## Shellfish Certification

The U.S. Department of Commerce also oversees the online publication of the Interstate Certified Shellfish Shippers List (ICSSL), which is a monthly PDF list of department-certified shippers of oysters, clams, mussels, and scallops (available at www.cfsan.fda.gov/~ear/shellfis.html). Only shellfish from these certified waters, which have been tested and found to be free of excessive levels of various microorganisms, can be sold for consumption. Wholesale containers of shellfish must then be labeled to include the harvester's name, address, and certification number, the date and location of harvest, and the type and quantity of shellfish. Shellfish that have been *shucked,* or removed from their shells, must also be tagged with a "sell by date" (for containers under 64 fluid ounces) or "date shucked" (over 64 fluid ounces). Food service operations are required to keep these tags for at least 90 days upon receipt. If shellfish are not properly tagged or if they are obtained from uncertified waters, the Department of Commerce may report the violation to the FDA, which is the regulatory agency with final jurisdiction over commerce in shellfish.

## Selection of Finfish

The criteria for selection of vertebrate and invertebrate seafood are very different and will now be described. Finfish can be purchased fresh or frozen, canned, cured, fabricated, or as fish roe.

### Fresh and Frozen Fish

Fish can be purchased fresh or frozen in several forms: whole, drawn, dressed, steaks, fillets, and sticks (Figure 9-3).

- **Whole fish.** The body is entirely intact.
- **Drawn fish.** Whole fish that have had their entrails (inner organs) removed.
- **Dressed fish.** The head, tail, fins, and scales have been removed in addition to the entrails.
- **Steaks.** Cut from dressed fish by slicing from the top fin to the bottom fin at a 90 degree angle at varying thicknesses. Steaks contain a portion of the backbone and other bones. Some steaks from large fish like tuna are commonly called fillets, even though they are actually steaks (23).
- **Fillets.** Made by slicing the fish lengthwise from front to back to avoid the bones. A fillet is one whole, boneless side of the fish (23).
- **Fish sticks.** Uniform portions cut from fillets or steaks. They can also be made from minced fish that is then shaped, breaded, and frozen.

*Variety of Finfish*   There are subtle differences in flavor even among different varieties of the same type of fish. For example, there are two types of salmon—Atlantic and Pacific. Among Pacific salmon there are five types—chinook (king), coho (silver), chum (keta), pink, and sockeye.

*Determining Freshness of Fish*
Sniffing for aroma may be the safest and easiest method of determining whether or not fish is fresh, but other criteria can be applied in addition to the "sniff test." When selecting whole fish, look for skin that is bright and shiny and eyes that bulge, are jet black, and have translucent corneas (the part surrounding the pupil). The fish should have a "fresh fish" aroma, tight scales, firm flesh, a stiff body, red gills, and a belly free of swelling or gas. The same criteria hold true for drawn fish with the exception of the potential gas-filled belly, which, of course, has been removed. At the market, look for fish that is stored on, but not directly in contact with, ice. The flesh should look shiny, moist, and plump (1). Avoid fish that looks spongy or gapes apart.

**FIGURE 9-3**  Forms of finfish available for purchase.

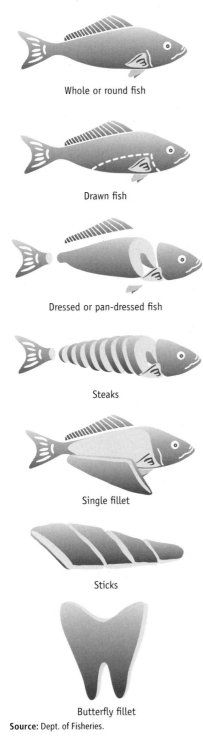

Whole or round fish

Drawn fish

Dressed or pan-dressed fish

Steaks

Single fillet

Sticks

Butterfly fillet

**Source:** Dept. of Fisheries.

**Rigor Mortis**  A stiff body is preferred when selecting a finfish because it is an indication that it is still in rigor mortis, which occurs after slaughter. Flesh that is allowed to go through rigor mortis (stiff to relaxed muscles) has a better texture and flavor. The water-holding capacity of the proteins is increased, which makes the flesh juicier than that of fish that have not undergone rigor mortis. For these reasons, it is better that handling, packing, processing, and freezing be avoided while fish are in the rigor state (5). It is also recommended that fish not be subjected to excessive stress prior to slaughter, if possible, because the resulting stronger rigor mortis is detrimental to texture (43).

Rigor mortis in fish can last anywhere from several hours to days, depending on the species, temperature, and condition of the fish when caught. Stiffness is delayed if caught fish are immediately placed on ice and kept chilled. Freshness is extended under these conditions because bacterial spoilage does not occur until after rigor mortis has passed. Freezing fish immediately after capture, rather than chilling them on ice and allowing rigor mortis to proceed until the muscles relax again, results in a tough-textured flesh. Cooking fish prior to rigor mortis also results in a tough texture.

**Phosphate Treatment of Fresh Finfish**  The meat of the fish should not be slimy, but this can be tricky to judge, because any slime present may have been produced by the fish having been soaked in a special phosphate-containing solution to prevent moisture loss. This solution increases the pH of the tissue, which denatures the proteins and makes them more capable of binding water. Fishermen frequently treat fish with this solution to cut down on the water loss, which might endanger their weight-based profits. Without this solution, fish that is refrigerated may lose up to 80% of its water-binding capacity within 5 days after harvest. The phosphate-containing solution restores the binding capacity of the muscle proteins and prevents the flesh from becoming dry and stringy. Treatment with phosphates also partially inhibits the oxidation of the natural fats in fish, which can result in "fishy" smells when the phosphates bind with the metal ions that promote oxidation.

**Signs of Decay in Fresh Finfish**  Other changes occur in a fish after death: the eyes flatten and become concave (although this may also be a result of the fish having been picked up by the eye sockets), the pupil turns gray or creamy brown, and the cornea becomes

**FIGURE 9-4**  A badly gaping fish fillet.

Digital Works

opaque and discolored. In addition, the bright red gills turn a paler brown, and as a result are sometimes removed.

When the gills turn brown and the eyes lose their bright look, the fish may be cut up as steaks, fillets, or fish sticks. Steaks and fillets should have a shiny, smooth surface that has no signs of curling at the edges. The pieces should be cut clean with no signs of blood, skin fragments, or loose bone, and they should be firm and free of **gaping** (Figure 9-4). Although gaping is a sign of aging, it may also be a result of rough handling, processing before rigor mortis is complete, the fish having been caught after spawning, or even genetics. Certain fish, such as bluefish and hake, are known to gape more easily (20). The physical reason behind gaping is the separation of the myotomes (14).

**Mercury Contamination**  Mercury occurs naturally in the environment, but it can also be released into the air through industrial pollution (19). Concerns about mercury contamination in fish date back to 1953 in Minamata, Japan, where certain manufacturing plants polluted the water with mercury (mercury can fall from the air into the water, where it becomes methylmercury). The fish there became contaminated and in turn caused health problems for about 120 people in the region who ate the fish. Infants that became sick had not eaten the fish, but their mothers had consumed the contaminated fish during pregnancy.

**Gaping**  The separation of fish flesh into flakes that occurs as the steak or fillet ages.

## How & Why?

### Why are some fish higher in mercury than others?

Mercury makes its way into water from either natural or industrial sources and is taken up into the gills of fish as they swim. It stays in the body for long periods of time, and small fish store this mercury in their flesh. Larger fish feed off of small fish. This means that larger fish are exposed to even higher levels of mercury because every time they eat a small fish they add mercury to their flesh. This is why larger, predatory fish and fish with long lives, such as swordfish, have the highest levels of mercury (45).

The mercury found in fish is different from the type of mercury used in thermometers or dental amalgam (45). Nearly all fish contain minute amounts of methylmercury that are not harmful to humans. However, mercury can accumulate to levels that are toxic to the developing central nervous system (brain and spinal cord), so its potential health risk is especially high to those who are in the early stages of brain development.

Because excess mercury may cause damage to the young, developing brain and nervous system (45), women of childbearing age, pregnant and/or nursing women, and children younger than 8 years of age are particularly vulnerable to its effects. The FDA issued the following recommendations in 2001 in an effort to prevent adverse effects on the brain development of unborn babies:

- Pregnant women, and those wishing to become pregnant, should not eat four types of fish—shark, swordfish, king mackerel, and tilefish—because they could contain dangerous levels of mercury (Figure 9-5).
- Pregnant women can safely eat up to 12 ounces a week of any other cooked fish—from canned tuna to shellfish to smaller ocean fish.
- Check local advisories about the safety of fish caught by family and friends in local lakes, rivers, and coastal areas.

Put into perspective, 12 ounces of cooked fish is equivalent to four 3-ounce servings (each the size of a deck of cards) or two

**FIGURE 9-5**    FDA fish intake recommendation for pregnant women, nursing mothers, and children under age 8.

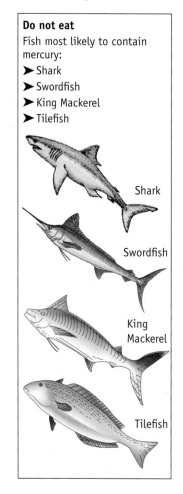

**Do not eat**

Fish most likely to contain mercury:
➤ Shark
➤ Swordfish
➤ King Mackerel
➤ Tilefish

Shark

Swordfish

King Mackerel

Tilefish

6-ounce servings. Fish with low, moderate, and high levels of mercury are listed in Table 9-4 (19).

While high levels of mercury are known to be harmful to developing fetuses, the effect of low levels of mercury—the amount we are exposed to when we eat fish—is uncertain. Reviews continue to be published on this subject (17). FDA testing revealed that the average amount of mercury in canned tuna products was well below the federal limit; however, some individual samples did test near or over the mercury limit.

Despite the uncertainty regarding the risk associated with mercury, the topic has received tremendous media coverage. A *Consumer Reports* article advised pregnant women to "avoid canned tuna entirely." This negative publicity resulted in reduced revenue for manufacturers of canned tuna products. In response, manufacturers are promoting the health benefits of tuna. Tuna, like many other fish, contain healthy types of fat, which reduce the risk of heart disease and other health problems. This is why it is commonly recommended that nonpregnant adults eat 2–3 servings of fish per week. Omega-3 fatty acids are present in high levels in some fish, and are now thought to be necessary in early brain development (7).

### How Much Fish to Buy

Part of selecting finfish is knowing how much to buy. A few general guidelines exist.

| **TABLE 9-4** Fish with Low, Moderate, and High Levels of Mercury | | |
|---|---|---|
| **Low levels of mercury** | **Moderate levels of mercury** | **High levels of mercury** |
| Salmon | Canned tuna (particularly albacore) | Pacific blue marlin |
| Shrimp | Cod (butterfish) | King mackerel |
| Scallops | Halibut | Tilefish |
| Catfish | Mahimahi | Shark |
| Tilapia | Grouper | Swordfish |
| Pollack | Striped marlin | |
| Clams | Orange roughy | |
| Sardines | Pollock | |
| Akule | | |
| Awa (milkfish) | | |
| Moi | | |
| Mullet | | |
| Opelu | | |
| Squid | | |
| Octopus | | |
| Fish sticks | | |
| Fish sandwiches | | |

Adapted from Dingeman R. Mercury content in fish raises concerns. *Honolulu Advertiser* July 27, 2003.

## NUTRIENT CONTENT

*Protein.* Fish is a high-protein food. In fact, fish is so high in protein—about 18 to 20%—that the food industry has devised the means to make protein concentrates by grinding whole fish, including the calcium-rich bones (if consumed), dehydrating it, and removing the fat to take away the fishy flavor. The resulting concentrate of between 70 and 80% pure protein is used as an additive in foods such as noodles to increase both their protein quality and their calcium content (39).

*Fat and Carbohydrates.* As a general rule, finfish are a low-fat food. Most fish are lower in fat than equivalent amounts of beef, pork, lamb, and even poultry. With the exceptions of mackerel, shark, herring, and eel, fish generally contain fewer than 160 calories (kcal) per 3-ounce cooked serving. Most of the calories in fish are derived from protein and fat, with few, if any, from carbohydrates. Shellfish contain carbohydrates in the form of glycogen, ranging in concentration from 1 to 3% by weight. The fat in fish is generally in low proportions, unless the fish has been fried. It should be noted here, however, that although fish and shellfish are relatively low in fat, squid and some crustaceans such as shrimp contain more than 100 milligrams of cholesterol per 100 grams (34).

*Functional Foods.* Even though fat does contribute to calories, the fat from fish is a good source of **omega-3 fatty acids** (Table 9-5), which are linked to several health benefits (25, 33, 48). The consumption of omega-3 fatty acids has been reported to be related to a decrease in the risk of heart disease (22, 26). It has also been suggested that they play a beneficial role in the alleviation of psoriasis and some inflammatory diseases, such as rheumatoid arthritis and lupus erythematosus (10, 25). High levels of omega-3 fatty acids are thought to be necessary for early brain development (7, 28).

*Vitamins and Minerals.* Fish is also a good source of the B vitamins—thiamin ($B_1$), riboflavin ($B_2$), niacin, $B_6$ (pyridoxine), and $B_{12}$—although small amounts of these water-soluble vitamins may be lost through decomposition, heating (cooking or canning), and/or extraction in water or salt solutions. The higher the fat content, the higher the levels of the fat-soluble vitamins A and D in the fish. Long before vitamin supplements became available, children were given (notoriously awful-tasting) cod liver oil as a dietary supplement of vitamin D to help protect them against rickets. Fish flesh is also a significant source of some minerals. Iodine is found primarily in salt-water fish. Sardines and salmon canned with the bones are good sources of calcium, and fish does contain some iron (27). For those watching their intake of sodium, dried or smoked fish have higher concentrations than the fresh forms.

### TABLE 9-5 Fish High in Omega-3 Fatty Acids (3-oz cooked portion)

**More Than 1.0 Gram**

Herring
Mackerel (Pacific, jack, Spanish)
Salmon (Atlantic, king, pink)
Tuna (bluefin)
Whitefish

**Between 0.5 and 1.0 Gram**

Bass (freshwater)
Bluefish
Mackerel (Atlantic)
Salmon (chum, coho, sockeye)
Smelt
Striped bass
Swordfish
Rainbow trout

About ½ pound of steaks makes an appropriate portion per person. Fillets or sticks require about ¼–⅓ pound per person. Purchases of ½–¾ pound for each person will be required when buying dressed fish, and ¾–1 pound per serving for whole or drawn fish.

***Processed Fish Products*** Fish mince, also called ground fish or fish hamburgers, can be prepared with mechanical deboners (Chemist's Corner 9-1) (31). This often results in a poor texture due to enzymes in the fish meat. Attempts to improve the texture of fish mince include the addition of cornstarch.

### Canned Fish

About half of all fish consumed in the United States is canned. Tuna accounts

> ## CHEMIST'S CORNER 9-1
>
> **Why Fish Mince Develops Bad Texture**
>
> Some fish species, including haddock, cod, and pollock, become tough when they are processed into fish mince. This is partly due to a chemical called trimethylamine oxide (TMAO) in the fish meat (50). TMAO is thought to help the fish regulate water and salt when they live in ocean water. After mincing and refrigeration, TMAO is broken down to trimethylamine (TMA) by bacteria, resulting in the "spoiled fish" odor. When fish are frozen instead of refrigerated, TMAO instead breaks down to dimethylamine (DMA) and formaldehyde. DMA does not cause a bad odor, but formaldehyde causes a biochemical reaction that creates links between protein molecules. These links cause the fish to bind water poorly, and the fish meat becomes dry and "cottony."

**Omega-3 fatty acids** A category of polyunsaturated fatty acids that includes eicosapentaenoic acid (EPA) and docosahexaenoic acid (DHA).

for approximately three fourths of canned fish consumption. Other commonly canned fish and shellfish include salmon, sardines, shrimp, clams, and crab (24). Canning alleviates the problem of the rapid perishability of fish.

*Tuna*   Six species of tuna are canned and sold in the United States: yellowfish, skipjack, bluefin, Oriental tuna, little tuna, and albacore. Canned tuna labeled "white" comes from albacore and is the most expensive. All other tuna is labeled "light meat tuna," although some of it can be quite dark. Canned tuna comes in three different styles: fancy or solid pack (a fillet or whole piece), chunk (large pieces), and flake (fine pieces or grated). Solid pack has the best appearance and is also the most expensive. Tuna may be canned in either water or oil, so buyers should examine labels for nutrition information. The total calories can vary drastically, depending on the canning medium. Each tablespoon of vegetable oil added to a can of tuna contains about 100 calories (kcal) and 15 grams of fat.

*Salmon*   Chinook (king) salmon is the most expensive of the canned salmons. In the less expensive ranges are sockeye (red salmon), coho (medium red), pink, and chum. Salmon is often packed with the bones, which increases the calcium content (if the bones are consumed).

*Sardines*   Sardines are always packed with their bones unless otherwise noted on the label. They come packed in tomato or mustard sauce or in oil. It is even possible to find them on the shelves packed in jalapeño sauce or plain water.

## Cured Fish

Fish may be cured by drying, salting, or smoking. Curing is one of the oldest ways of preserving fish. Although

### TABLE 9-6   Preserved Fish/Roe

| Type | Description |
| --- | --- |
| Anchovy | Tiny, very fatty fish with a powerful flavor, which are cured by having most of their fat content removed by pressure and fermentation. |
| Arbroath smokies | Haddocks or whiting that are brined. |
| Bismark herring | German herrings from the Baltic filleted and marinated in vinegar with onion rings for 2–3 days. |
| Bloaters | First developed at Yarmouth, England, in 1835, bloaters owe their special flavor to the activity of gut enzymes. They are dry-salted and smoked. |
| Block fillets | Haddock or whiting are brined, dyed to make them a bright yellow, then smoked. |
| Bombay duck | A well-known Indian condiment made from dried bummaloe. It is used as a condiment with curries. |
| Glasgow pales | So-called because after brining they are lightly smoked. |
| Katsuoboshi | A Japanese fried fish that can also be smoked. |
| Lutefisk | The reconstituted unsalted cod from Norway known as *stockfish* or *stockfisk*. |
| Kippers | Good-quality fresh herring, soaked in brine and then smoked. |
| Matjes herrings | Young Netherlands herring caught in the spring, before they become too fatty. |
| Migaki-nishin | Japanese dried fish fillets and abalone. |
| Roes and caviars | Roe is fish eggs. In the United States and many other countries, caviar is defined as the salted roe of any fish species. In Europe, caviar is defined as only the roe from sturgeons (beluga, osetra, and sevruga) originating in the Caspian Sea. |
| Rollmops | Herring fillets packed in spiced brine. |
| Sardine | Fish, cooked either by frying in peanut oil or steam cooking. |
| Smoked eels | Brined, dry-salted, and smoked. |
| Smoked salmon | Dry-salted with fine salt and smoked. Some curers add brown sugar, saltpeter, and rum. |
| Smoked sprats | The most famous are the *Kieler Sprotten*—brined sprats from Germany. |
| Smoked trout | Rainbow and brown trout are brined, speared on rods, smoked, then hot-smoked. |

distinctive tastes and prolonged keeping times are achieved using any of the curing techniques, curing can also harden the outer surfaces. Smoked salmon, smoked haddock (finnan haddie), pickled herring, and smoked herring, also known as kippered herring, are some of the more familiar forms of cured fish. Caviar, discussed below, also belongs in the cured category, because it is preserved by salting (Table 9-6).

*Anchovies*   Anchovies are tiny, bony fish that have been cured with salt. They come to the market either salt-packed or oil-packed and in cans as whole fish, fillets, or anchovy paste. Because of their strong flavor, anchovies are usually used as a garnish or in salad dressings and sauces rather than as a food in themselves. The salt-packed anchovies must first be rinsed, but their flavor tends to be superior to the oil-packed variety.

## Fabricated Fish

In an attempt to counter the twin problems of the expense and perishability of fish, several fish products, including fish sticks, fish cakes, nuggets, and simulated fillets, have been developed using fabricated fish. Fabricated fish products make use of the less popular species. The fish are mechanically deboned into a minced fish product, recovering 60 to 90% of the edible meat, and the flesh is then ground, seasoned, shaped, and breaded (31, 40). These products are commonly frozen for sale to the consumer.

The very high cost of genuine crabmeat has led to the introduction in this country of **surimi**, which has been used for centuries in Japan. Over 900 years ago, a Japanese fisherman discovered that fish would last much longer if it were minced, washed, mixed with salt and spices, ground into a paste, and then cooked (47). Today the deboned and minced fish is treated to produce

**Surimi** Japanese for "minced meat," a fabricated fish product usually made from Alaskan pollack, a deep-sea whitefish, which is skinned, deboned, minced, washed, strained, and shaped into pieces to resemble crab, shrimp, or scallops.

a pure-white product with a somewhat elastic, chewy texture that "sets" when the mixture forms a translucent, elastic, moist gel (18) (Chemist's Corner 9-2). Adding starch contributes to gelling. Adding starch or egg whites enhances binding, texture, and flavoring. Flavors such as salt and sugar are added directly. Red coloring is often added to impart the appearance of cooked crab legs.

Surimi at this stage cannot be consumed until it is cooked, and it is the method of cooking that determines the type of food produced (Figure 9-6). North Americans are most familiar with kamaboko, surimi that has been steamed and shaped into pieces resembling crab, shrimp, or scallops (32). Although the taste may be very similar to crab, the nutritional values are not. The resultant product usually has 75% less cholesterol than the original shellfish, but very little, if any, of the omega-3 fatty acids. It usually has more sodium because of the added salt.

### Food Additives in Fish
Preservatives are often added to fabricated fish products to maintain their shelf life. One such preservative is phosphate, often added because it increases water-binding capacity and contributes to a greater freezing stability (42).

### Caviar
Caviar, which has a mystique surrounding it as a food of the very rich, is really just fish eggs, also known as roe. The eggs of vertebrate fish, which are held together by a thin membranous sac, are only available from female fish

> ## CHEMIST'S CORNER 9-2
>
> **Surimi**
>
> Surimi's "springy" texture is derived in part from gums that are added to help form gels. Another influence is washing, which leaves behind the water-insoluble (myofibrillar) protein that gives surimi its elasticity and gel-forming capacity. Sugars such as sucrose and sorbitol are added as cryoprotectants to protect the myofibrillar proteins from denaturating during freezing (47). Starch granules make the surimi more compact by swelling with water around the protein matrix and filling in the interstitial spaces (51, 52). Improved gel strength is obtained by adding egg whites, which inhibit endogenous protease activity in fish flesh. Salt is then added to the surimi to solubilize its protein and produce a firm, elastic gel, and again later during freezing for stabilization (12).

during the spawning season and are highly perishable. The official definition of caviar varies according to the country in which it is sold. In the United States and many other countries, caviar is the clean, salted fish eggs of any fish species. The label is required to list the particular type of fish serving as the caviar source. In Europe, caviar is more narrowly defined by law as only the eggs of the Caspian Sea sturgeon.

The most expensive, largest-grained caviar comes from the beluga sturgeon. These fish can live for over 70 years and may grow to a length of 25 feet (21). Like chicken eggs, roe is very high in cholesterol—about 94 mg per tablespoon. It is also high in salt, but the best caviar is *malassol*, which in Russian means "little salt." To protect the taste of caviar, it is served with a bone or shell spoon, because metal imparts an off-flavor. It is sometimes served on a neutral-tasting bread that has been toasted on one side, with the caviar being gently placed on the untoasted side.

Freshwater roe is often breaded and fried, but the surrounding sac must be pierced first or it may explode during frying, causing severe burns. A major drawback to fresh fish roe is that it stays fresh for only a day or two at the most; it is usually preserved in a brine solution, which imparts a salty flavor, firms the roe, and extends its usable time. The roe sold in the unrefrigerated section of the supermarket has been pasteurized to extend its shelf life. Fish such as shad and herring from North Atlantic waters are popular roe sources, as are Pacific salmon and whitefish from the Great Lakes. Other roe sources include cod, carp, pike-perch, and gray mullet (Table 9-7) (21).

## Selection of Shellfish

The purchaser of shellfish is faced with several different forms from which to choose. The first decision is whether to buy them alive or processed.

**FIGURE 9-6** Surimi products.

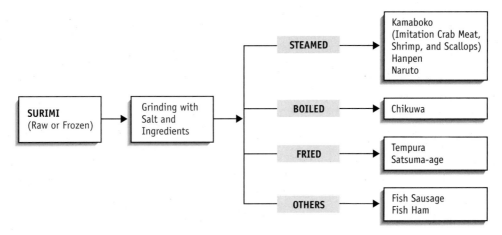

| **TABLE 9-7** | Sources of Caviar and Roe |
|---|---|
| **Sturgeon** | |
| Beluga Sturgeon | The beluga, the largest of the Caspian Sea sturgeon, produces the rarest and most expensive caviar. Beluga eggs are large and gray. |
| Osetra Sturgeon | Osetra caviar are more available than beluga. The medium-size eggs are gray-brown and have a nutty, meaty taste. |
| Sevruga Sturgeon | The smallest of the Caspian Sea sturgeon, the sevruga eggs are small and gray, and have a stronger, fishier taste than the other Caspian Sea caviars. Sevruga caviar is particularly popular in Russia and Europe. |
| **Other Roe-Producing Fish** | |
| Lumpfish | The lumpfish, caught off Iceland, produces small, colorful (black, red, or yellow) eggs that are popular as a garnish. |
| Salmon | The large, pinkish eggs of the salmon, from the Pacific Northwest, are also frequently used as a garnish. |
| Whitefish | The North American Whitefish produces small, golden eggs with a distinct crunch and mild flavor. |
| American Paddlefish | This denizen of the Mississippi River and its tributaries produces roe that looks like sevruga caviar. The small, gray eggs have a tangy flavor. |

## Purchasing Live Shellfish

Lobsters, crabs, oysters, and clams all may be purchased alive and in their shells. Shellfish are highly perishable, and to maintain their quality, must be kept alive until they are cooked, or, in the case of oysters, occasionally consumed raw.

### Selecting Live Mollusks

It is easy to tell whether or not crustaceans are alive, as they are normally active creatures; in contrast, determining the state of mollusks in their closed shells poses more of a puzzle. Tapping on the shell should cause it to close more tightly; the rule in most cases is that if the shell remains open, the mollusk is dead and should be discarded. The exceptions to this rule are mussels, which ordinarily gape, and longneck (steamer or soft-shell) clams, which normally have a gap in the shells where the "neck" protrudes.

Any shells that are broken, have a decaying odor, or float should be discarded.

### The "R-Month Rule"

An old rule of thumb held that shellfish should be eaten only during the months with the letter *r* in their names, because bacterial illnesses are more common in the warmer months of May through August. This is still a valid guideline, although modern methods of harvesting and storage provide a safer supply of shellfish year round.

## Purchasing Processed Shellfish

Shellfish can be sold raw within their shells, after removal of the shell, or after cooking. The shelf life of shellfish is longest when they are sold with their shells on, with minimal processing (15). They can be further processed by removal of the shell via steam, pressure, or flames.

The shells can then be removed from the meat by placing the shellfish in brine, where the shells sink and the meat floats.

Shellfish can also be bought cooked in the shell and chilled or frozen. Alternatively, the meat can be removed from the shell and sold fresh, chilled, frozen, canned, salted, smoked, or dried. Shellfish can be sold headless and in their shell, as in the case of shrimp or lobster tail, or they may be shucked, or removed from the shell, with a special knife.

### Shucking Shellfish

Shucked shrimp, scallops, oysters, and clams are often breaded and frozen. Shrimp may also be sold with the intestinal tract removed, a form known as "peeled and deveined." Shucking bivalves such as clams and oysters is a somewhat dangerous process (Figure 9-7). The hand holding the bivalve should be protected with a towel or a metal-mesh glove. The hinge is severed as the shells are pried apart, and the empty half of the shell is discarded; the muscle attachment to the other shell is spliced so the meat can be removed. An average worker can shuck almost 7 pounds in an hour, but automated shucking speeds up the process.

## Oysters

Oysters can be bought live in the shell, or shucked and then chilled, frozen, or canned. Live oysters should have tightly closed shells. Any gap between the shells means the oyster is dead and should be discarded. Select shucked oysters that are plump and full-bodied; about 1 cup is equal to one serving. If the oysters are in their shell, buy half a dozen per person. Three varieties of oyster are commonly available in the United States: eastern oysters from the

**FIGURE 9-7**   Steps to shucking oysters.

**1.** Hold oyster, flatter shell up, with a folded towel, and place the tip of the knife near the hinge at the pointed end.

**2.** Move the tip of the knife around to the front side. Pry and push the tip to bore into the shell until it pops open.

Atlantic coast, and Olympia (small) and Japanese (large) oysters from the Pacific coast. Oysters in the shell and well-refrigerated have a longer shelf life than other mollusks because their shells remain very tightly closed, whereas other shellfish have a tendency to gape, making them more susceptible to drying out and dying.

## Clams

Clams can be bought in the same forms as oysters, and, as with oysters, their shells should be closed tightly and there should be no decaying odor. About six to eight shelled clams are required per serving. A variety of clams can be purchased. Clams may be soft-shell or hard-shell. Soft-shell clams are not soft, but their shells are thin and brittle. Their neck (often called a foot) sometimes sticks out like a long, soft hose to siphon and release ocean water. Soft-shell clams are also known as longnecks or pissers because of this long tube. Soft-shell clams do not completely close so they are very susceptible to drying out and dying. A limp neck hanging out of the shell signals that the clam is dead and should be discarded (23). Soft-shell clams should be consumed within a day of purchase. They should never be eaten raw, which is why they are sometimes called steamers or fryers. Even if cooked, they are often more sandy or gritty than hard-shell clams because of their partially opened shell. As a result, they are often soaked in cold salted water in an attempt to eliminate the sand, and then served with the broth they were cooked in to rinse off any remaining grit.

**FIGURE 9-8**  Mussels are often steamed open and served in the half-shell.

The meat of hard-shell clams is less tender than that of soft-shell clams. East Coast hard-shell clams include cherrystones, which are the most common variety; littlenecks, which are the smallest and most tender; and chowders or quahogs (kwah-hahg), which are the largest. The larger the clam muscle is, the tougher the meat will be, so they are often chopped up and used in clam chowder or stews. West coast varieties include razor and pismo clams.

## Scallops

In North America, the only part of the scallop that is eaten is the creamy white or tan-colored abductor muscle responsible for opening and closing the shell to move it through the water. Scallops cannot close their shell tightly when taken from the water, so they are usually shucked and then sold fresh, frozen, or canned. This sweet-tasting mollusk varies in diameter from ½ to 2 inches, and about ¼ to ⅓ pound is an adequate portion for one person. The three main varieties of scallops available are bay scallops, which are small, sweet, and delicate; sea scallops, which are larger and not as delicate; and calico scallops, the least expensive and tiniest of all and the blandest in flavor. Calico scallops are often sold cooked or frozen.

## Mussels

The black or dark-blue colored shells of common mussels should be scrubbed free of barnacles, but the "beards" or black threads used to attach the shells to solid foundations in the ocean should not be pulled out until the mussels are ready to be cooked, because removing them kills the mussel (20). Mussels are

heated in their shells after being purchased live (Figure 9-8), or they are shucked and packed in brine. Extremely hollow or heavy-feeling mussels should be discarded, because they are either dead or filled with sand. Also available are the larger, green-lipped New Zealand mussels whose size makes them ideal for stuffing and baking.

## Abalone

Abalone is expensive because the supply is limited (20). These large mollusks are found mostly in the waters off California and northern Mexico. Unlike the other mollusks discussed, abalone have only one shell. Most of the animal consists of a massive, muscular foot. Only abalone with meat weighing at least ¼ pound may be legally harvested; some extremely large abalone yield as much as 3 pounds. The strict regulations governing the harvesting of wild abalone have led to farm-raised abalone, which are largely harvested in California and Hawaii (37).

## Lobsters

Lobsters are the largest of the crustaceans. They are mainly purchased as Northern (or Maine) lobster, or spiny or rock lobster varieties (Figure 9-9). Gourmet cooks prefer the female lobster for its finer flavor and because it contains "coral," or lobster roe, which is considered a delicacy. When cooked, the roe turns from dark green to red and is often used to color a sauce or served alone as a garnish. Another delicacy, found in both male and female lobsters, is the pale green liver, known as tomalley.

The majority of the meat from a lobster is in the tail, but there is also some in the claws of the Maine lobster. Lobsters are right-handed or left-handed, as

**FIGURE 9-9**  Northern lobster (*left*), spiny or rock lobster (*right*).

**FIGURE 9-10** Peeling and deveining shrimp.

  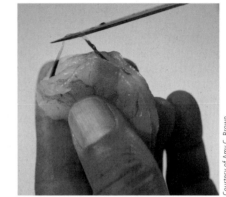

Courtesy of Amy C. Brown

indicated by which claw is the larger, and although the larger claw has more meat, that from the smaller claw is sweeter and more tender. Northern lobsters with one or both claws missing are sometimes sold as "culls." They are less expensive and are attractive to the buyer interested only in the tail meat. Unless they are canned or frozen, lobsters must remain alive until cooked, at which point their natural dark blue-gray or greenish color turns deep orange or red.

## Shrimp

The tail harbors most of the meat in shrimp. They are sold, headless, in either the raw shell-on (green shrimp), cooked shell-on, or cooked and peeled form. All three forms come both fresh and frozen, but the majority of shrimp are frozen.

### Peeling and Cleaning Shrimp

When shrimp are bought in their shells, they must first be peeled. Medium or large shrimp are then deveined, which involves removing the dark-colored "sand vein" that runs along the shrimp's back (Figure 9-10). The "sand vein" is usually left in small shrimp, where it is undetectable.

> **? How & Why?**
>
> **Why must the sand vein in large shrimp be removed?**
>
> The sand vein is actually the shrimp's intestines. In larger shrimp, it contributes a gritty, muddy taste if it is not removed.

After cleaning, the shrimp are dried by pressing them between paper towels to absorb as much moisture as possible. Before they are cooked, shrimp are somewhat grayish green, but they turn dark pink to borderline red when heated.

**How Much Shrimp to Buy** In general, the serving size averages ⅓ to ½ pound for headless, unpeeled shrimp, or ¼ to ⅓ pound for peeled and deveined shrimp. Shrimp are available in small, medium, large medium, large, and jumbo sizes, but these descriptions are not official nor are they used consistently by shrimp sellers (20). As a result, shrimp is purchased by "count per pound." The smaller the shrimp are, the higher the count per pound will be (Figure 9-11). For example, the number 51/60 on a shrimp package means that there are 51 to 60 individual shrimp in one pound. Seeing 21/25 on the label indicates that the shrimp in that designation are larger because there are 21 to 25 shrimp for the same weight of 1 pound. A U/15 or U/10A means "under 15" or "under 10" shrimp per pound. These would be the largest shrimp available. It is an incorrect but common practice to identify the largest jumbo-size shrimp as **prawns.** True prawns have lobster-like pincer claws and are otherwise different from shrimp. Another shrimp-related North American misnomer is the use of the word "**scampi**" for describing large broiled shrimp seasoned with butter and garlic.

**Canned Shrimp** Glass-like beads are sometimes found in canned shrimp, but they are completely harmless. They are formed during canning, specifically under the high heats of sterilization. Called struvite crystals, they consist of magnesium-ammonium phosphate compounds that form when the magnesium from seawater combines with the ammonia that is produced during heating of the shellfish's natural protein. Phosphate treatment prevents struvite crystal formation due to the phosphates binding with the magnesium. Struvite crystals can be crushed to a powder by a fingernail or dissolved by boiling for a few minutes in the weak acid of lemon juice or vinegar.

## Crab

Crabs can be sold live, raw, or cooked. Most of their meat is found in their claws and legs. The four top

**Prawn** A large crustacean that resembles shrimp but is biologically different. Large shrimp are often called by this name.

**Scampi** A crustacean found in Italy and not generally available in North America. The term is often used incorrectly to describe a popular shrimp dish.

**FIGURE 9-11** Shrimp size described by "count per pound."

Courtesy of Amy C. Brown

commercially harvested crabs are the blue crab from the Atlantic and Gulf coasts, stone crabs from Florida, Dungeness crabs from the Pacific coast, and, most expensive, king crabs from the northern Pacific waters. Soft-shelled blue crabs are considered a delicacy, particularly on the East Coast. These crabs are caught while molting, a process during which they shed their shell and have a soft exterior until the new surface is completely hardened. The process may take several days, during which time the crab is more vulnerable to predators, especially two-legged ones such as birds and humans.

***Canned Crab*** Canned crab may have a blue tint. This is caused by copper in the crab's blood combining with the ammonia in its flesh. Although the color may appear unappetizing, it is completely harmless.

### Crayfish
Referred to as either crayfish, crawdads, or crawfish, these small crustaceans average 4 ounces in weight. Crayfish are similar in appearance to lobsters but smaller, and their first pair of walking legs do not develop into huge, flesh-rich claws. Only their tails serve as a source of meat. They are found mainly in freshwater streams and ponds of the southeastern United States, especially Louisiana.

Crayfish are sold both head-on and tails only, fresh and frozen.

# PREPARATION OF FISH AND SHELLFISH

In the preparation of seafood, great care must be taken not to overcook it; cooking too long or at too high a temperature is the most common mistake when preparing fish or shellfish. It results in excessive flakiness, dryness, and flavor loss in fish and toughness in shellfish. For shellfish, however, the "well done rule" should always be observed, especially when microwaving, because of the

increased chance of their carrying foodborne illness. Shellfish are often simmered or steamed; the results with dry heat are harder to guarantee, because the meat may dry out and toughen. Nevertheless, with precautions such as breading taken against drying, shrimp, lobster tails, and half-shelled oysters and clams are fairly commonly baked broiled, or fried (see the color insert). Both temperature and time need to be carefully controlled in either dry- or moist-heat preparation of seafood. The following discussion refers to the preparation of finfish unless shellfish is specifically mentioned.

## Dry-Heat Preparation
Methods to prepare fish with the use of dry-heat include baking, broiling, grilling, and frying.

### Baking
Fish to be baked should be rinsed, patted dry with paper towels, and placed in a shallow pan. Season as desired, and place in a moderate oven (350°F–400°F/180°C–200°C). Baking time will vary depending on the shape and thickness of the fish, but a general rule of thumb is to bake up to 10 minutes per inch of thickness measured at the thickest diameter of the fish. Basting with butter or covering the fish with vegetables cuts down on moisture loss. Some prefer to prepare whole or drawn fish with the head and tail left on to help keep juices inside. Additional flavor can be added by filling the cavity with herbs and spices. Leaving the skin on whole or drawn fish seals in the moisture and flavor. Moisture loss may also be prevented by wrapping the fish in

---

✦ **CALORIE CONTROL**
**Fish and Shellfish**

Fish and shellfish are lower in calories and fat compared to an equal amount of meat (beef, pork, or lamb) or poultry. In fact, fish average 100–200 calories (kcal) for each 3 ounces. A few helpful hints are:

- *Water over oil.* Choose fish canned in water instead of oil.
- *Baked or broiled over breaded and fried.* The healthier option is to choose fish that has been baked, broiled, steamed, grilled, or cooked by any method other than breaded and fried.

© 2010 Amy Brown

**FIGURE 9-12**   Saucing a fish.

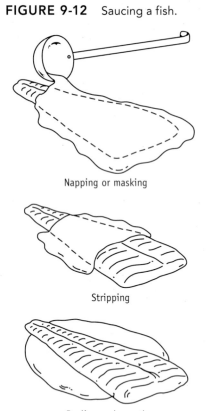

Napping or masking

Stripping

Pooling underneath

**Source:** Mizer et al., *Food Preparation for the Professional* (John Wiley & Sons, 1998).

foil, parchment paper, grape leaves, or leafy greens. These techniques of enclosing the fish technically result in a moist-heat cooking method because the fish steams in its own juices. The flavor, moisture, color, and texture of baked fish are often enhanced by the addition of sauces (Figure 9-12).

### Determining Doneness of Fresh Fish

Fish is done when it flakes easily with the gentle pressure of a fork without falling apart. The opaque look of fish that has been properly prepared is caused by denatured proteins that unwind and hook together with other proteins so that water can attach, resulting in a whitish hue. The presence of this "white" mesh results in a moist and tender flesh. Heating much beyond this stage tightens the protein bonds, shrinks the protein mesh, and squeezes out the water, resulting in tough, dry, unappetizing fish

**À la meunière** Fish seasoned, lightly floured, and sautéed in clarified butter or oil and served with a sauce made with butter and parsley.

flesh (16). Other signs of doneness include any bone being no longer pink and/or the flesh becoming firm, turning from translucent to opaque, and/or separating from the bone.

*Baking Shellfish*   Shellfish are often prepared by baking; examples include lobster thermidor, baked soft-shell clams, and oysters Rockefeller. In lobster thermidor, the lobster is split in half and baked. The meat is then extracted and mixed with a seasoned béchamel sauce before being put back in the lobster shell and baked again until golden brown and heated through. If soft-shell clams are to be baked in the oven, they are placed in a pan layered with rock salt and baked at 425°F (218°C) for about 15 minutes or until the shells open. Oysters Rockefeller is made by pouring a spinach mixture over half-shell oysters in a pan layered with rock salt. They are baked at 475°F (246°C) for about 10 minutes and then browned briefly under the broiler.

### Broiling

Dressed or filleted finfish or fish steaks are best broiled at 5 inches or less below the heat source. Lean fish should be coated with melted butter, margarine, or oil, but this step can be omitted with most fatty fish. Season the fish as desired, place it skin-side down on a pan that has been greased to avoid sticking, and broil it on one side until tender. Lobsters and large shrimp can also be broiled. Whole lobsters need to be killed and split before broiling, whereas lobster tails can be broiled whole.

### Grilling

Fish can be grilled on an outside grill or in the oven. Grilling is not recommended for delicate fish such as sole, because they may stick to the grill and fall apart easily. Fatty, firm-fleshed fish such as salmon, bluefish, and mackerel that have been drawn or cut into steaks are well suited for grilling. Also, larger shrimp may be put on skewers like kabobs and grilled. A fat coating such as oil or even mayonnaise can be applied to the fish to prevent it from sticking to the grill. The grill itself should be scraped of any residue and lightly oiled to prevent sticking. Steaks are seasoned as desired, and cooked on both sides if thick, but on only one side for

thin steaks or fillets. The fish should be about 4 inches from the heat source. When the fish flakes easily, serve it immediately. Drawn fish can be checked for doneness by slipping the tip of a paring knife into the back of the fish and pulling away. It is done if it clings briefly before giving way, but is overdone and dry if cooked to the "flakes easily" stage.

### Frying

Lean fish less than ½ inch thick, shrimp, and scallops will sauté nicely in a small amount of butter and/or oil. The fish is seasoned as desired and sautéed over medium heat until it is cooked about three quarters of the way through, at which time it is turned gently with a spatula and heated until the flesh flakes easily. Shellfish are best sautéed on high heat for a short time. Shrimp and scallops are ideal for this type of preparation. When done, scallops will be firm and look opaque, and shrimp will be opaque and pink.

*Sautéed Fish Variations*   Sautéed fish may be prepared **à la meunière** (a-lah-muhn-YAIR). The dish can be served amandine (with almonds), Florentine (with spinach), or à la belle (with mushrooms). A variation of this method that uses more fat is used to prepare trout and other small fish. They are seasoned, breaded or dipped in cornmeal or flour, and pan-fried until they are golden brown on both sides. Breading mild fish such as tilapia adds more body and flavor (35).

*Deep-Fried Fish*   Deep-frying is a popular method for preparing battered or breaded lean fish and shellfish (shrimp, scallops, clams, and oysters). Whole small fish, shellfish (which must first be shelled), fish fillets, or steaks are dipped in batter or seasoned breading mix before being deep-fried in oil until golden brown. The oil is heated to 350°F (180°C) for large fish and about 180°F (82°C) for small seafood such as fish strips, oysters, or clams (37). Fish is always fried alone because it imparts a fishy taste to the oil, which would be picked up by other foods fried in the same oil. Lean fish are preferred because unpleasant oily tastes often occur in fatty fish that are deep-fried.

# Moist-Heat Preparation

The moisture of fish is sometimes better retained through moist-heat preparation methods such as poaching, simmering, steaming, and microwaving.

## Poaching

Fish is a delicate food suitable for poaching. The lower water temperature of 160°F–180°F (71°C–82°C), which keeps bubbles small and clinging to the sides of the pan, protects the delicate flesh of fish. If a whole, drawn, or dressed fish is being poached, it can be wrapped in cheesecloth to hold it together. The liquid may be a **court bouillon** or a **fumet**. Although fatty fish such as salmon can be poached, white, lean fish such as cod, pike, haddock, flounder, sole, whiting, red snapper, halibut, and bass are best suited for this method. Sole fillets are thin enough to make paupiettes, or rolled fillets (Figure 9-13).

When poaching fish, the water should never be allowed to boil. Boiling causes flavor loss and toughens the fish, whereas low temperatures retain maximum flavor and moisture. A well-seasoned poaching liquid is also important. Seasonings and/or chopped vegetables such as tomatoes or shallots add flavor, texture, and color. The poaching liquid is often reduced and sometimes thickened for use as a sauce. The fish is placed in the middle of a baking or frying pan and cooking

**FIGURE 9-13** Rolled fish fillet (paupiette): Thin fillets (usually sole) are rolled with the skin side inside so the flesh cooks on the outside.

**Source:** Mizer et al., *Food Preparation for the Professional* (John Wiley & Son's, 1998).

liquid is added until it covers up to an eighth to a quarter of the fish's thickness. Some recipes call for covering the entire fish in liquid, but too much liquid may dilute delicate flavors. On the other hand, too little liquid will evaporate and cause the fish to dry out during cooking.

Fish fillets can be poached in an oven set at 350°F (180°C) or in a pan on top of a range set at poaching temperatures. The pan can be covered to trap more heat and moisture and to prevent volatile flavor compounds from escaping. This technique, when using only a small amount of liquid, is more akin to steaming, another moist-heat method.

## Simmering

Simmering uses slightly higher temperatures than poaching does—180°F (82°C) to just under boiling, where gentle bubbles rise but barely break the surface. This method is most often used to cook shrimp, even though the expression "boiled shrimp" is commonly used for the outcome of this process. Shrimp are often simmered and then chilled, shelled, and deveined for shrimp cocktail.

Lobster, crab, and crayfish may also be simmered. The live lobster, crab, or crayfish is killed by inserting it headfirst into boiling water that has been salted with 2 teaspoons per quart. Prior to placing crayfish in the water, the middle tail fin must be grabbed, twisted, and pulled to remove the stomach and intestinal vein. Lobsters will curl their tails when first dropped into the water, which may cause toughening. This is prevented by killing the lobster with the point of a sharp knife inserted directly between the head and the shell. A more expensive technique involves submerging the crustacean in a container of beer or wine, which inebriates it and causes it to relax. Once the shellfish is submerged, the water is brought back to a boil and then immediately reduced to a simmer. Heating time averages 5 minutes per pound for a lobster; a whole crayfish takes less than 7 minutes. When done, the crustacean is immediately removed from the water to prevent further cooking, drained well, and served at once with clarified butter and lemon (Chemist's Corner 9-3). Lobsters are often split in half at restaurants for the diners' convenience.

## Steaming

Fish can be steamed in the oven if it is tightly covered in a baking dish, aluminum foil, or parchment paper, or in a pan on top of the range. When fish is wrapped with parchment paper, along with seasonings and aromatic vegetables if desired, and cooked in the oven, this is known as cooking *en papillote*. When the fish is done, the parchment envelope puffs up, turns brown, and provides a dramatic presentation. Each person may then be served a portion still wrapped in its own paper package, making for a novel dining experience. Fish may also be cooked in foil envelopes, although these are generally removed before the fish is served at the table. Regardless of the way it is accomplished, steaming heats the fish in its own juices, which locks in the flavor and aroma.

***Steaming Shellfish*** Steaming can also be used to prepare lobster tails, clams, and mussels. Frozen lobster tails are thawed and "saddlebacked," which involves splitting the tail by

**Court bouillon** Seasoned stock containing white wine and/or vinegar.

**Fumet** A flavorful fish stock made with white wine.

**FIGURE 9-14**    How to saddleback a lobster tail.

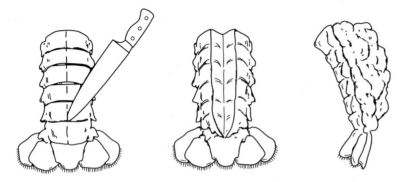

cutting through the hard top shell and pulling the meat out so it lies on top (Figure 9-14). The tail is then seasoned and steamed shell-down in a covered pan for a few minutes. Clams and mussels are steamed by placing them in a covered pot with a small amount of liquid on the bottom. Steaming clams or mussels just until the shells open does not kill microorganisms, so it is important to steam them for about 5 minutes or to a temperature of 145°F (63°C). Pressure steaming is not recommended because it tends to toughen both fish and shellfish.

***Clambakes Are Underground Steamings***    At a clambake, clams are actually steamed rather than baked. A hole a foot deep and 3 feet wide is dug into the sand and lined with smooth, round rocks. This serves as the base of a fire that will be kept going for 2 or 3 hours after the rocks and/or embers have been heated hot enough. The embers are raked over the rocks and removed, and soaked seaweed is placed over the rocks to a depth of about 6 inches. Chicken-wire mesh is laid over that to serve as a platform for a layer of hard-shell clams, which are then covered with sweet potatoes, followed by broiler chickens cut into quarters, partially husked corn, and then a layer of soft-shell clams. The whole pile is splashed with a bucket of seawater, covered with a wet tarp, and allowed to "bake," or rather steam, for about an hour. Doneness of the clams is tested by checking to see if their shells have opened. The chickens take longer and thus need to be tested for doneness separately.

### Microwaving

Almost any form of fish can be microwaved. If it is commercially frozen, the defrosting instructions on the package should be followed. In general, instructions call for arranging fish fillets or steaks or small fish in a single layer, with thicker portions toward the outside of a microwave-safe dish. Desired seasonings and dots of butter are added before covering with plastic wrap to trap the moisture. Poaching can also be done in the microwave oven.

### Raw Fish

*Sashimi* refers to raw fish that is consumed alone. *Sushi* refers to rice products. The term *sushi* is derived from the word "sour" and refers to the rice that is mixed with vinegar and molded into various shapes (rolls, rectangles, etc.). The pressed rice is often served with other ingredients—raw fish (sashimi), cooked fish or shellfish, eel, fish eggs, vegetables, and/or spices (8).

There are several different types of sushi:

- *Nigiri* is pressed, vinegared rice with something on top, usually raw or cooked fish.
- *Maki* is rice, plus fish or other ingredients, rolled in nori (seaweed) and cut into circles.
- *Uramaki* is similar, but the fish or vegetables are inside the nori and the rice is on the outside. These rolls are often coated in something, like sesame seeds or tobiko (flying fish roe).
- *Chirashi* is sushi with certain ingredients mixed directly into the rice.

***Ceviche***    Ceviche is raw fish that is prepared with an acidic marinade, lemon- or lime-juice based, that denatures the proteins and turns the flesh white. This type of preparation does not involve heating, and thus the fish should still be considered "raw" and treated accordingly.

***Raw Fish and Shellfish Warning***
The Centers for Disease Control and Prevention (CDC) warns about the hazards of eating raw fish or shellfish. Not only bacteria and viruses, but also parasites may pose a problem. Foodborne illnesses may result and in some cases lead to intestinal infections that are difficult to treat and cause further complications. Particularly vulnerable are pregnant or nursing women, the very young, the elderly, and anyone with a serious illness or compromised immune system. Raw fish have been known to harbor worms (anisakiasis parasites), and only heating to 145°F (63°C) for at least one minute or freezing the fish in a commercial freezer to 210°F (14°C) for 7 days ensures their destruction (see Chapter 4). Shellfish, especially mollusks, are particularly prone to carrying contaminants, because they are filter feeders whose usual habitat is in shallow waters, which are more likely to be subject to bacterial, viral, and chemical pollution. Consuming, or even shucking, raw oysters is a potential concern because they may carry *Vibrio vulnificus, V. cholera, V. parahaemolyticus,* Norwalk virus, or hepatitis A (29, 30).

# STORAGE OF FISH AND SHELLFISH

Fish can be purchased fresh, frozen, canned, or cured. Each style has its own storage requirements (see the back inside cover of this book), but it is important to stress once again that all fresh fish and shellfish are highly perishable and require that precautions be taken to ensure freshness. Although proper preparation helps to destroy microorganisms that occur naturally or are introduced during handling, fish, and especially shellfish, must be stored properly to reduce the risk of foodborne illness.

## Fresh Finfish

Fresh fish are best consumed within a day or two of purchase. Fish do not store well for longer periods because the flesh is much more perishable than animal tissue for several reasons. One of these is that all raw seafood carries some bacteria, which multiply rapidly above 40°F (4°C).

### Refrigerated

Fish should be stored in the coldest portion of the refrigerator. It should also be tightly wrapped to prevent odors from coming in contact with other foods (49). Fish bought wrapped in butcher paper should be rewrapped in plastic wrap and then in aluminum foil, but prepackaged fish and shellfish can be left in the original package in which they were purchased. Any exposure to oxygen increases perishability, because the high levels of polyunsaturated fatty acids in fish can be oxidized into compounds that affect odor and taste (3). Storing wrapped fish in a colander with ice will help to keep it cool (1). Another storage method is to wash and pat dry the fish, wrap it in wax paper, then place it in an airtight container (2).

### Spoilage Factors

Other factors that can contribute to spoilage are proteolytic enzymes, natural toxins, and contaminants. Proteolytic enzymes break down muscle proteins and provide amino acids for bacterial growth (4). Bacterial enzymes can also break down proteins to amino acids and elevate the levels of histamine, a toxin. Excessive consumption of histamine leads to a foodborne illness known as scombroid fish poisoning or scombrotoxism (discussed more fully in Chapter 4). Excessive histamine may accumulate in tuna, tuna-like fish, mahimahi, bluefish, and other species that usually have not been chilled immediately after being caught.

### Storing Caviar

Caviar is particularly sensitive to oxygen and cannot be left out in the air for more than 1 hour. Unopened caviar can be stored in the refrigerator for up to 3 months, but once opened, it should be consumed within 3 days.

## Fresh Shellfish

It is a good practice to eat fresh shellfish the day they are bought. If they must be kept, the storage requirements are varied and depend on the type of shellfish. Most fresh shellfish may be kept alive in cool, salty, wet environments, preferably in the refrigerator. Storing fresh shellfish on ice may kill them if they become submerged in fresh water from the melting ice. Live oysters, clams, and mussels should be well aerated in the refrigerator and not stored in plastic bags or in fresh water, where they will die. Any dead animals, indicated by an open shell or lack of response when tapped, should be discarded. Crabs, usually sold precooked, should be stored in the coldest part of the refrigerator and used within a day or two. Once cooked, all crustaceans must be refrigerated at temperatures below 40°F (4°C) and consumed within 2 days.

## Frozen

As mentioned in Chapter 28 on food preservation, the frozen-foods industry in North America began with fish because of Clarence Birdseye's accidental discovery while ice fishing that fish could be frozen prior to consumption. Freezing greatly extends the keeping time of fish that, depending on the type, can be stored in the freezer up to 9 months. It is absolutely necessary, in order to arrest microbial growth, to freeze fish if they are not cleaned (eviscerated) within 24 hours of being caught (6). Once cleaned, the general rule is that lean fish keep longer than fatty fish. Freezing lean fish often results in firmer fillets because of their low water-holding capacity; freezing fatty fish is limited by the deterioration of their fat content (lipid oxidation). Fish should be stored at 0°F (–18°C) or below and never refrozen once thawed. Prepackaged and frozen fish can stay in their original wrappers but should be kept airtight in order to prevent them from drying out.

Not all shellfish freeze as well as fresh fish. For example, freezing lobster (whole, cooked) results in tougher meat, off-flavors, and difficulty in removing the meat from the shell (9). On the other hand, cooked shrimp freezes fairly well.

### Thawing

Fish is best thawed by transferring it from the freezer to the refrigerator one day before preparation; once thawed, it should be cooked immediately. The exceptions are breaded frozen fish, or fish fillets or steaks weighing less than ½ pound; these should not be thawed before cooking because they will become mushy. Frozen, raw shellfish can also be prepared from the frozen state, whereas frozen precooked shellfish can be used as is after thawing.

Even though it is the most healthful and popular method of preserving fish, freezing tends to cause a reduction in quality, making fish dryer, tougher, and less springy, and possibly affecting the flavor (11) (Chemist's Corner 9-4).

## Canned and Cured

Canned fish can stay on the shelf for up to 12 months, but any dented, damaged, or bulging cans should be discarded. Unused fish from an opened can should be moved to a covered glass or plastic container and can be stored for up to a week in the refrigerator. Cured fish can be refrigerated, frozen, or canned. Chapter 28 discusses canning and curing in more detail.

> ### CHEMIST'S CORNER 9-4
>
> #### Effect of Freezing on Fish
>
> Freezing fish decreases its quality because the myofibrillar proteins are disrupted (denatured and/or aggregated) (11,12). The result is a subsequent loss in the muscle proteins' functional properties such as protein solubility, gel-forming ability, and water retention. These properties are important to quality. The tougher texture of frozen fish is thought to be due in part to the enzyme trimethylamine oxide (TMAO) demethylase, which breaks down TMAO to dimethylamine and formaldehyde. Formaldehyde is believed to be a cross-linking agent that is responsible for a tougher texture (12). Additives can counteract the negative effect of freezing on the quality of fish. Polymerized phosphates improve the texture of frozen fish by increasing water retention, reducing thaw drip, and decreasing cooking losses. Polyphosphates achieve this effect by increasing the binding of phosphates to meat proteins, breaking down actomyosin to actin and myosin, elevating pH, and improving ionic strength (13).

# PICTORIAL SUMMARY / 9: Fish and Shellfish

There are over 20,000 known species of edible fish, shellfish, and sea mammals. Of these, approximately 250 species are harvested commercially in the United States, with millions of tons annually being served up for the consumption of humans and domesticated animals.

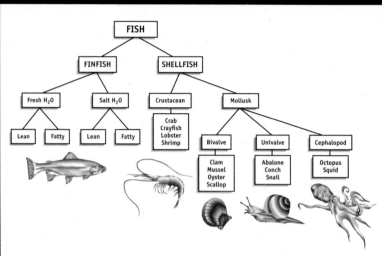

## CLASSIFICATION OF FISH AND SHELLFISH

Fish can be classified three different ways:
- Vertebrate or invertebrate: Vertebrate fish or finfish have fins and internal skeletons; invertebrate fish, or shellfish (mollusks and crustaceans), have external skeletons
- Saltwater or freshwater
- Lean or fatty

## COMPOSITION OF FISH

Fish and shellfish are more tender than other flesh foods, and nutritionally, 3 ounces of fish contain fewer calories than the same amount of beef, pork, lamb, or poultry. Fish is high in protein and relatively low in fat. Small amounts of carbohydrate may be present in fish in the form of glycogen. The fat in fish is polyunsaturated, and, depending on the fish, high in omega-3 fatty acids. Fish is also a good source of many B vitamins.

## PURCHASING FISH AND SHELLFISH

Inspection of fish is voluntary and is based on the wholesomeness of the fish and the processing plant. Only inspected fish products can be graded U.S. Grade A, U.S. Grade B, and substandard. Grades for shellfish such as shrimp and oysters are based on size. Fish can be purchased fresh or frozen in a variety of market forms, as well as canned, cured, and fabricated (surimi). Fish roe is also sold. Shellfish can be purchased alive, cooked in their shell, or shucked, to be refrigerated, frozen, or canned.

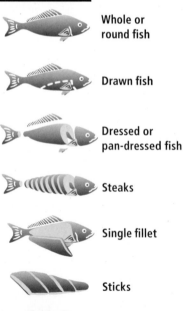

**Whole or round fish**

**Drawn fish**

**Dressed or pan-dressed fish**

**Steaks**

**Single fillet**

**Sticks**

**Butterfly fillet**

## PREPARATION OF FISH AND SHELLFISH

Overcooking is the most common mistake in the preparation of fish, resulting in excessive flakiness, dryness, and flavor loss. Dry heat is the most popular method of preparation, and includes baking, broiling, grilling, and frying. Moist-heat methods include poaching, simmering, steaming, and microwaving. Fish cook quickly and are done when the flesh turns from a translucent to an opaque color, is firm to the touch, separates from the bone (if present), and is moist and flakes easily at the segments without falling apart, and the bone is no longer pink.

## STORAGE OF FISH AND SHELLFISH

Fresh fish are best consumed within a day or two. If fish is purchased in butcher paper, it should be rewrapped with plastic wrap and aluminium foil. Pre-packaged fish and shellfish can stay in the original package. Fish should be frozen at 0°F (−18°C) or below and never refrozen once it is thawed. Breaded fish or fish fillets or steaks weighing less than $\frac{1}{2}$ a pound need not be thawed before heating. Most fresh shellfish must be kept alive prior to preparation. Canned fish can stay on the shelf for up to 12 months, but leftovers should be refrigerated in a glass or plastic container and used within 3 days.

# CHAPTER REVIEW AND EXAM PREP

## Multiple Choice*

1. A dressed fish is described as:
   a. whole fish with entrails removed.
   b. fish that has been sliced from top fin to bottom fin.
   c. fish body that is entirely intact.
   d. fish from which the head, tail, fins, scales, and entrails have been removed.

2. Which of the following fish are good sources of omega-3 fatty acids?
   a. Flounder and grouper
   b. Halibut and haddock
   c. Salmon and mackerel
   d. Puffer and cod

3. Identify the term used to describe the separation of fish flesh into flakes that occurs as the steak or fillet ages.
   a. Roping
   b. Gaping
   c. Stripping
   d. Slicing

4. What is the name of the process used to remove the "sand vein" or intestine from fresh shrimp?
   a. Deveining
   b. Sand veining
   c. Incising
   d. Shelling

5. Which of the following is a crustacean?
   a. Scallop
   b. Clam
   c. Oyster
   d. Crab

6. A food product consisting of compacted rice served with raw fish, cooked fish or shellfish, vegetables, and/or spices is called _____.
   a. sashimi
   b. sushi
   c. ceviche
   d. surumi

7. Canned fish can be stored on the shelf for _____ months.
   a. 3
   b. 6
   c. 12
   d. 24

*See p. AK-1 for answers to multiple choice questions.

## Short Answer/Essay

1. Describe the various ways in which fish and shellfish are categorized.
2. Describe the structural factors that make fish flesh so much more tender than beef or poultry.
3. Discuss the factors that affect the pigment of fish flesh.
4. Describe the nutrient content of fish, and explain why fish are said to have a greater nutritional value than other sources of protein.
5. Describe the qualities used as criteria in the inspection and grading of finfish.
6. Describe each of the different ways that vertebrate fish can be purchased: whole, drawn, dressed, steaks, fillets, and sticks.
7. What qualities should be considered when selecting vertebrate fish?
8. Define *caviar* and explain how this definition may vary in different countries. What is surimi, how is it prepared, and how is it used?
9. Define the following: *tomalley, prawns, scampi, struvite crystals,* and *fumet.*
10. Describe the basic methods of preparing vertebrate fish through baking, poaching, and steaming.

# REFERENCES

1. Armentrout J. Buying fish? Trust your nose. *Fine Cooking* 65:74, 2004.
2. Armentrout J. How to store fish. *Fine Cooking* 90:71, 2008.
3. Bandarra NM, et al. Seasonal changes in lipid composition of sardine (*Sardina pilchardus*). *Journal of Food Science* 62(1):40–42, 1997.
4. Benjakul S, et al. Physiochemical changes in Pacific whiting muscle proteins during iced storage. *Journal of Food Science* 62(4): 729–733, 1997.
5. Berg T, U Erikson, and TS Nordtvedt. Rigor mortis assessment of Atlantic salmon (*Salmo salar*) and effects of stress. *Journal of Food Science* 62(3):439–446, 1997.
6. Bett KL, and CP Dionigi. Detecting seafood off-flavors: Limitations of sensory evaluation. *Food Technology* 51(8):70–79, 1997.
7. Borresen T. Seafood for improved health and wellbeing. *Food Technology* 63(1): 88, 2009.
8. Bruno P. Sushi this way. *Hemispheres Magazine* February: 104, 2007.
9. Calder BL, et al. Quality of whole lobster (*Homarus americanus*) treated with sodium tripolyphosphate before

cooking and frozen storage. *Journal of Food Science* 70(9):C523–C528, 2006.

10. Calder PC. Session 3: Joint Nutrition Society and Irish Nutrition and Dietetic Institute Symposium on 'Nutrition and autoimmune disease' PUFA, inflammatory processes and rheumatoid arthritis. *Proceeding of the Nutrition Society* 67(4):409–418, 2008.

11. Careche M, and ECY Li-Chan. Structural changes in cod myosin after modification with formaldehyde or frozen storage. *Journal of Food Science* 62(4):717–723, 1997.

12. Chang CC, and JM Regenstein. Textural changes and functional properties of cod mince proteins as affected by kidney tissue and cryoprotectants. *Journal of Food Science* 62(2):299–304, 1997.

13. Chang CC, and JM Regenstein. Water uptake, protein solubility, and protein changes of cod mince stored on ice as affected by polyphosphates. *Journal of Food Science* 62(2):305–309, 1997.

14. Cheret R, et al. Effects of high pressure on texture and microstructure of sea bass (*Dicentrarchus labrax L.*) fillets. *Journal of Food Science* 70(8):E477–E483, 2005.

15. Clark JP. Navigating seafood processing. *Food Technology* 62(10):91–96, 2008.

16. Corriher SO. *CookWise*. Morrow, 1997.

17. Díez S. Human health effects of methylmercury exposure. *Reviews of Environmental Contamination and Toxicology* 198:111–132, 2009.

18. Dileep AO, et al. Effect of ice storage on the physiochemical and dynamic viscoelastic properties of ribbonfish (*Trichiurus spp.*) meat. *Journal of Food Science* 70(9):E537–E545, 2005.

19. Dingeman R. Mercury content in fish raises concerns. *Honolulu Advertiser* July 27, 2003.

20. Dore I. *Fish and Shellfish Quality Assessment*. Van Nostrand Reinhold, 1991.

21. Dowell P, and A Bailey. *Cook's Ingredients*. Reader's Digest, 1990.

22. Dubnov G, and EM Berry. Polyunsaturated fatty acids, insulin resistance, and atherosclerosis: is inflammation the connecting link? *Metabolic Syndrome and Related Disorders* 2(2):124–128, 2004.

23. Ehri A. Fish steaks vs. fillets. *Fine Cooking* 86:69, 2007.

24. Environmental Protection Agency. *Estimated per capita fish consumption in the United States: August 2002.* www.epa.gov/waterscience/fish/consumption_report.pdf. Accessed 06/30/09.

25. Fernandes G, A Bhattacharya, M Rahman, K Zaman K, and J Banu. Effects of n-3 fatty acids on autoimmunity and osteoporosis. *Frontiers in Bioscience* 13:4015–4020, 2008.

26. Galli C, and P Rise. Fish consumption, omega-3 fatty acids and cardiovascular disease. The science and the clinical trials. *Nutrition and Health* 20(1):11–20, 2009.

27. Gomez-Baauri JV, and JM Regenstein. Processing and frozen storage effects on the iron content of cod and mackerel. *Journal of Food Science* 57(6):1332–1336, 1992.

28. Greenberg JA, SJ Bell, and WV Ausdal. Omega-3 fatty acid supplementation during pregnancy. *Reviews in Obstetrics & Gynecology* 1(4):162–169, 2008.

29. Guillois-Bécel Y, et al. An oyster-associated hepatitis A outbreak in France in 2007. *Europe Surveillance* 14(10): pii:19144, 2009.

30. Huppatz C, et al. A norovirus outbreak associated with consumption of NSW oysters: Implications for quality assurance systems. *Communicable Disease Intelligence* 32(1):88–91, 2008.

31. Jahncke M, RC Baker, and JM Regenstein. Frozen storage of unwashed cod (*Gadus morhua*) frame mince with and without kidney tissue. *Journal of Food Science* 57(3):575–580, 1992.

32. Ma L, A Grove, and GV Barbosa-Canovas. Viscoelastic characterization of surimi gel: Effects of setting and starch. *Journal of Food Science* 61(6):881–883, 1996.

33. Nettleton JA. Concerning PUFA in fish. *Journal of the American Dietetic Association* 108(11):1830–1831, 2008.

34. Nettleton JA, and J Exler. Nutrients in wild and farmed fish and shellfish. *Journal of Food Science* 57(2):257–260, 1992.

35. Pendleton LG. Tilapia, fast flavorful. *Fine Cooking* 78:60–63, 2006.

36. Pennington JAT, and VL Wilkening. Final regulations for the nutrition labeling of raw fruits, vegetables, and fish. *Journal of the American Dietetic Association* 97:1299–1305, 1997.

37. Peterson J. *Fish and Shellfish*. Morrow, 1996.

38. Potter NN, and JH Hotchkiss. *Food Science*. Chapman & Hall, 1995.

39. Rakosky J. *Protein Additives in Food Service Preparation*. Van Nostrand Reinhold, 1989.

40. Regenstein JM. Total utilization of fish. *Food Technology* 58(3):28–30, 2004.

41. Rosell CM, and F Toldra. Effect of myoglobin on the muscle lipase system. *Journal of Food Biochemistry* 20:87–92, 1997.

42. Sathivel S. Chitosan and protein coatings affect yield, moisture loss, and lipid oxidation of pink salmon (*Oncorhynchus gorbuscha*) fillets during frozen storage. *Journal of Food Science* 70(8):E455–E459, 2005.

43. Sigholt T, et al. Handling stress and storage temperature affect meat quality of farmed-raised Atlantic salmon (*Salmo salar*). *Journal of Food Science* 62(4):898–905, 1997.

44. Srinivasan S, et al. Physiochemical changes in prawns (*Machrobrachium rosenbergii*) subjected to multiple freeze-thaw cycles. *Journal of Food Science* 62(1):123–127, 1997.

45. United States Food and Drug Administration. *Backgrounder for the 2004 FDA/EPA consumer advisory: What you need to know about mercury in fish and shellfish.* http://www.fda.gov/Food/FoodSafety/Product-SpecificInformation/Seafood/FoodbornePathogensContaminants/Methylmercury/ucm115662.htm

46. United States Food and Drug Administration. *Guidance for industry: the seafood list: FDA's guide to acceptable market names for seafood sold in interstate commerce.* Revised 2009. www.fda.gov/Food/GuidanceComplianceRegulatoryInformation/GuidanceDocuments/Seafood/ucm113260.htm. Accessed 7/7/09.

47. Vieira ER. *Elementary Food Science*. Chapman & Hall, 1996.

48. Weaver KL, P Ivester, JA Chilton, MD Wilson, P Pandey, and FH

Chilton. The content of favorable and unfavorable polyunsaturated fatty acids found in commonly eaten fish. *Journal of the American Dietetic Association* 108(17): 1178–1185, 2008.

49. Wempe JW, and PM Davidson. Bacteriological profile and shelf life of White Amur (*Ctenopharyngodon idella*). *Journal of Food Science* 57(1):66–68, 1992.

50. Wu TH, and PJ Bechtel. Ammonia, dimethylamine, trimethylamine, and trimethylamine oxide from raw and processed fish by-products. *Journal of Aquatic Food Product Technology* 17(1):27–38, 2008.

51. Yoon WB, JW Park, and BY Kim. Surimi-starch interactions based on mixture design and regression models. *Journal of Food Science* 62(3):555–560, 1997.

52. Yoon WB, JW Park, and BY Kim. Linear programming in blending various components of surimi seafood. *Journal of Food Science* 62(3):561–564, 1997.

# WEBSITES

Find the FDA's "Fish Encyclopedia" that provides a list of fish names (common and scientific) and photos at this site:

**www.fda.gov/Food/FoodSafety/Product-SpecificInformation/Seafood/RegulatoryFishEncyclopediaRFE/default.htm#mname**

Find more information on mercury via the search box found at the FDA's Center for Food Safety and Applied Nutrition (CFSAN) website on seafood information and resources:

**www.fda.gov/Food/default.htm**

Find more information on seafood and marine resources as the website of the National Marine Fisheries Service:

**www.nmfs.noaa.gov/**

The Interstate Shellfish Sanitation Conference (ISSC) promotes shellfish sanitation through the cooperation of state and federal control agencies, the shellfish industry, and the academic community.

**www.issc.org**

# 10 Milk

(porous bones) in later life, milk is a vital source of nutrition for millions of people. Although milk is rich in many nutrients, it is low in vitamins C and E, iron, complex carbohydrates, and fiber.

**FIGURE 10-1**  Dairy scene from ancient Egypt. Found in the tomb of Princess Kewitt.

**P**eople have been using milk as a food source for thousands of years. Records from ancient Babylon, Egypt, and India show evidence of cattle being raised for their milk (Figure 10-1). Milk is a unique beverage that provides complete protein, many of the B vitamins, vitamins A and D, and calcium. In fact, approximately 80% of the calcium ingested by Americans is derived from dairy products. Because a lack of dietary calcium causes poor bone development in children and is a risk factor for osteoporosis

This chapter focuses on cow's fluid milk—its composition and variations, the purchasing of milk products, its use in food preparation, and its safe storage. Cheese, butter, and frozen dairy products are covered in Chapters 11, 22, and 26, respectively. Although milk from other animals, such as goats, sheep, and camels, is a common part of the diet in some parts of the world, in this book, unless otherwise indicated, the word *milk* refers only to cow's milk.

# FUNCTIONS OF MILK IN FOODS

Numerous food products contain milk or some component of milk. The presence of milk in the diet is pervasive. Milk is itself a beverage, but there are also all sorts of drinks that use milk as a base—smoothies, milk shakes, yogurt drinks, eggnog, kefir, and more. Food products primarily made from milk include cheese, yogurt, sour cream, and whipped cream, to name just a few. Many foods rely on milk or ingredients derived from milk, including pizza, cheese soufflés, sandwiches, casseroles, quiches, sauces, processed meats, soups, dressings, infant formulas, coffee creams, food bars, sports nutrition products, breads, cereals, cakes, pies, puddings, cookies, ice cream, milk chocolate, caramels, frozen yogurt, and many other desserts.

Butter is made from milk and so all the foods incorporating this fat are dairy-based to some degree. The various functions and foods made with milk fat are discussed in Chapter 22 (see Table 22-1).

The food industry separates out the specific ingredients of milk as components in processed foods (Chemist's Corner 10-1) (10). The proteins are commonly added to many processed foods to improve their nutritive value. Certain proteins (caseinates) contribute to emulsifying and stabilizing, whereas others (whey proteins) assist with gelling. The milk sugar, lactose, aids with browning of baked goods, and is also important in the manufacture of confectionary and frozen desserts. Overall, milk contributes to processed foods by improving protein content (sports bars, chips, etc.), moisture, mixing ability (emulsification), foaming, texture, and flavor.

# COMPOSITION OF MILK

## Nutrients

The basic composition of milk remains the same regardless of its source. Milk is primarily water—87.4%. Figure 10-2 shows that the remaining 13% by weight consists of carbohydrate, fat, protein, and minerals (10). The high concentration of water gives milk a near-neutral pH of 6.6. Among domesticated cattle, the breed, stage of lactation, type of feed ingested, and season of the year all tend to slightly influence milk's content.

## Carbohydrate

Lactose, or milk sugar, is the primary carbohydrate found in milk—12 grams per 8-ounce cup. When bacteria in milk metabolize lactose, lactic acid is produced. The flavor of cheeses and fermented milk products such as yogurt and sour cream is, in part, derived from lactic acid. Lactose tends to be less soluble than sucrose, which may cause it to crystallize into lumps in nonfat dried milk and to produce a sandy texture in ice cream.

**Lactose Intolerance** This condition results in an inability to digest lactose, the form of sugar found in milk. Lactose-intolerant people lack sufficient quantities of lactase, the enzyme required to break the lactose disaccharide into its two monosaccharide units (41). It is a common food intolerance, affecting 30 to 50 million Americans, especially certain ethnic groups: it is present in 90% of Asian-Americans, 80% of Native Americans, 65% of African-Americans, and 50% of Hispanics (54). For people with lactose intolerance, fermented milk products are usually more easily digested than those that are not fermented.

**Medical food** A food to be taken under the supervision of a physician and intended for the dietary management of a disease/condition for which distinctive nutritional requirements are established by scientific evaluation.

**FIGURE 10-2**   Composition of Milk.

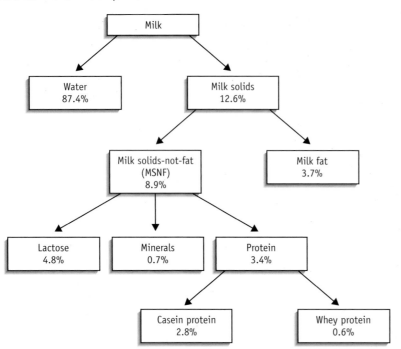

## Protein

The protein in milk is a complete protein; that is, it contains all the essential amino acids in adequate quantities to support growth and the maintenance of life. A cup of milk contains approximately 8 grams of protein. Two servings of milk or milk products a day provide almost half the protein recommended for a healthy adult woman, and one third that for a man.

***Casein and Whey***   The two predominant types of protein found in milk are **casein** and **whey** (Chemist's Corner 10-2) (10). Casein accounts for almost 80% of the protein in milk, whereas whey protein constitutes about 18%. Whey proteins consist primarily of lactalbumin and lactoglobulin (36). Whey is the liquid portion of milk that remains after cheese production (36). The nutritious whey protein can be isolated by putting the whey through an ultrafiltration process. These whey protein concentrates are used extensively by the food industry as emulsifiers and as foaming and gelling agents (42). Adding milk proteins to other foods generally improves their texture, mouthfeel, moisture retention, and flavor (52). Whey proteins are often added to foods to improve their protein profile, especially foods based on grains and beans that are low in lysine (40).

***Milk Allergy***   While food intolerance is the inability to digest or absorb certain foods, food allergy is an immune response to a molecule (usually a protein) found in food (54). Milk allergies are present in about 2 to 5% of children, but about 20% outgrow them by age 4, and 42% by age 8 (45).

## Fat

The fat in milk, called milk fat or butterfat, plays a major role in the flavor, mouthfeel, and stability of milk products (22). The creaminess of milk chocolate, for example, is due to its milk fat, which softens the characteristic brittleness of cocoa butter (8). Milk fat consists of triglycerides surrounded by phospholipid-protein membranes (lipoproteins), which allow them to be dispersed in the fluid portion of milk, which is primarily water. Milk fat contains substantial amounts of short-chain fatty acids—butyric, caprylic, caproic, and capric acids. The fatty acids in milk fat are approximately 66% saturated, 30% monounsaturated, and 4% polyunsaturated.

***Fat and Calorie Content of Milks***
An 8-ounce cup of fluid milk ranges from 86 to 150 calories (kcal) and 0 to 8 grams of fat. (See the fat and caloric content of various milk products in the Calorie Control feature.) Buttermilk, despite its name, contains only about

**Casein**   The primary protein (80%) found in milk; it can be precipitated (solidified out of solution) with acid or certain enzymes.

**Whey**   The liquid portion of milk, consisting primarily of 93% water, lactose, and whey proteins (primarily lactalbumin and lactoglobulin). It is the watery component removed from the curd in cheese manufacture.

2 grams of fat per cup, and fewer than half the calories of whole milk. This product is the fluid that remains when the fat from whole milk is removed to make butter—hence the name *buttermilk*. Other types of milk products vary greatly in their fat content per cup, from condensed milk, with about 27 grams, to fat-free (nonfat) milk, with less than half a gram.

*Cholesterol* Like other animal products, milk contains cholesterol—an average of 33 mg in a cup of whole milk, 18 mg in reduced-fat (2%) milk, and 4 mg in fat-free (nonfat) milk. The fat and cholesterol content of milk and other dairy foods such as cheese, butter, and ice cream drives some consumers to seek lower-fat alternatives to dairy products (See Calorie Control.) In fact, low-fat dairy products accounted for nearly 40% of new food products introduced to the market in the early 1990s (55).

## Vitamins

Milk contains vitamins A and D, riboflavin (B$_2$), and tryptophan, an amino acid important in the formation of the B vitamin niacin. It is low in vitamins C and E. Milk exposed to ultraviolet light loses riboflavin, so it is packaged in cardboard or opaque plastic containers to prevent the degradation of this vitamin by light (11).

### Vitamins A and D Fortification

Many milks are fortified with vitamins A and D. Vitamin D is found naturally in very few foods and was initially added to milk, a staple food, to reduce the incidence of rickets, a bone-softening condition in children that was at one time endemic in North America. Before the fortification of milk was widely practiced, many children grew up with severely bowed legs and other effects of vitamin D deficiency.

Because vitamins A and D are fat soluble, they are found in the milk fat of whole milk. For this reason, whole milk is not required to be fortified with either vitamin, although many milk manufacturers add both. However, vitamin A is diminished in reduced-fat (2% fat) and fat-free (nonfat) milks, dried whole milk, and evaporated skim milk, so vitamin A fortification is required. Fortification with vitamin D in reduced-fat and fat-free milks is optional, but 98% of milk processors add it anyway.

## ✳ CALORIE CONTROL

### Milk and Milk Products

- **Limit servings to 3 cups a day.** The daily recommendation for milk is two 1-cup servings, according to the Exchange Lists, or 3 cups, according to MyPyramid. Because a cup of milk averages 100 calories/kcal (ranges from 86 to 150 calories), 3–8 grams of fat, 12 grams of carbohydrate, and 8 grams of protein, it's important not to exceed the 3-cup daily limit.
- **Portion Control.** Remember that one large glass of milk may contain about two 8-ounce cups.
- **Choose lower fat options.** The recommended milk portions are for the low-fat (1% = 102 calories/kcal for each cup) or fat-free (nonfat or skim = 86 calories/kcal) milk options. Table 10-2 lists the calorie and fat contents of various dairy products and Table 10-3 provides a "choose more" and "choose less" list.
  - **Choose less often:** whole milk, chocolate milk, eggnog, evaporated milk, condensed milk, whole-milk yogurt, creams, and sour cream.
  - **Choose less often:** foods made with the above ingredients such as cream soups and ice cream.

© 2010 Amy Brown

### TABLE 10-2 Calorie (kcal) and Fat Content of Selected Milk Products*

| Milk Product | Nutrients/Cup | |
|---|---|---|
| | Calories (kcal) | Fat (g) |
| *Fluid Milk* | | |
| Fat-Free (Nonfat; Skim) | 86 | 0 |
| Low-Fat (1%) | 102 | 3 |
| Reduced-Fat (2%) | 121 | 5 |
| Whole | 150 | 8 |
| *Flavored Fluid Milk* | | |
| Chocolate | | |
| Low-Fat (1%) | 158 | 3 |
| Reduced-Fat (2%) | 179 | 5 |
| Whole | 209 | 9 |
| Eggnog | | |
| Reduced-Fat (2%) | 189 | 8 |
| Whole | 342 | 19 |
| *Canned Milk* | | |
| Whole Evaporated | 338 | 19 |
| Fat-Free (Nonfat) Evaporated | 199 | 1 |
| Sweetened Condensed | 982 | 27 |
| Sweetened Condensed (Fat-Free) | 632 | 0 |
| *Cultured Milk* | | |
| Buttermilk | 99 | 2 |
| Yogurt (Plain) | | |
| Fat-Free (Nonfat) | 137 | 0 |
| Reduced-Fat (2%) | 155 | 4 |
| Whole | 150 | 8 |
| Yogurt (Fruit Flavored) | | |
| Fat-Free (Nonfat) | 100 | 0 |
| Reduced-Fat (2%) | 231 | 3 |
| Whole | 250 | 6 |
| Sour Cream (1 tbs) | 31 | 3 |
| *Cream* | | |
| Half-and-Half | 315 | 28 |
| 1 tbs | 20 | 2 |
| Light Whipping | 698 | 74 |
| 1 tbs | 44 | 5 |
| Heavy Whipping | 821 | 88 |
| 1 tbs | 51 | 6 |
| Cream Substitute (1 tbs) | 20 | 2 |

*Milk averages 12 grams of carbohydrate and 8 grams of protein per cup.

**TABLE 10-3**    Dairy Products to "Choose More" or "Choose Less" to Reduce Dietary Fat

| Choose More | Choose Less |
|---|---|
| *Milk* | |
| Fat-Free (nonfat; skim) | Whole |
| Low-Fat (1%) | Evaporated |
| Reduced-Fat (2%) | Condensed |
| Fat-Free Dried | |
| Buttermilk | |
| Fat-Free Evaporated | |
| Fat-Free or Reduced-Fat Chocolate | |
| *Yogurt* | |
| Reduced-Fat (2%) | Whole-Milk |
| Low-Fat | Custard-Style |
| Fat-Free | |
| *Cream* | |
| Light Cream Cheese | Whipping Cream |
| Light Sour Cream | Half-and-Half |
| Mocha | Sour Cream |
| Nondairy Creamers | Sweet Cream |
| | Cream Cheese Spreads |
| | Cream Soups |
| | Creamy Dressings |
| *Cheeses* | |
| Lower-Fat Cheese (see Chapter 11) | Cheese over 6 g of fat/ounce |
| *Frozen Desserts* | |
| Sherbet | Ice Cream |
| Ice Milk | Frozen Whole-Milk Yogurt |
| Frozen Reduced-Fat (2%) Yogurt | |

**Source:** National Dairy Council.

Vitamin D fortification is required for evaporated whole and fat-free milks.

### Minerals

The major mineral in milk is calcium, with 1 cup of milk containing, on average, 300 mg. Two servings of milk a day provide a substantial portion of the 1,000 mg Dietary Reference Intake (DRI) for adults 19–50 years of age. Milk can also provide calcium in other forms, such as yogurt, pudding, ice cream, custards, hot chocolate, and cheese.

Other primary minerals found in milk and milk products include phosphorous, potassium, magnesium, sodium chloride, and sulfur. Although milk is rich in many minerals, it is low in iron.

## Color Compounds

Factors that contribute to the color of milk are fat, colloidally dispersed casein and calcium complexes, and water-soluble riboflavin ($B_2$). These compounds, by interfering with light transmission, contribute to milk's opaque, ivory color. The amount of carotene (a pigment found in some plants) in the cow's feed influences the color of its milk. Carotenoid pigments dissolved in the milk fat provide the yellowish tinge of butter and cream (see Chapter 22).

## Food Additives

The practice of adding vitamin D to milk began in the 1930s to reduce a public health problem: rickets, a bone disease in children (9). This practice, recommended by the American Medical Association's Council on Foods and Nutrition, nearly eliminated this disease in the United States. Vitamin A fortification was initiated in the 1940s because of the increasing popularity of reduced-fat and fat-free milk. Vitamin A is a fat-soluble vitamin, so it dissolves in the fatty portion of the milk. Fortifying these milks with vitamin A replaces that which is lost when the fat is removed.

Although not defined as a food additive, the hormone recombinant bovine growth hormone (rBGH) is given to approximately 5 to 30% of dairy cattle to make them produce about 10% more milk. Any health risk to humans from rBGH is considered unlikely as this protein hormone is digested in the stomach of the consumer. Steroid hormone use is not permitted in dairy cattle, as it is in cattle raised for their meat. Consumers do have the option of purchasing rBGH-free or certified organic milk and milk products (54).

### ? How & Why?

*Why does nonfat milk have a bluish hue?*

Removing any of the fat eliminates a proportional amount of carotenoid pigments and solids, resulting in the color changing from a yellowish white to the bluish hue seen in fat-free (nonfat) milk.

# PURCHASING MILK

## Grades

Milk is graded according to its bacterial count. The highest grade, Grade A, has the lowest count. The law requires that all Grade A milk and milk products crossing state lines be pasteurized. Although Grade A is the most common grade of milk sold, Grade B is also available. In addition, different grades exist for fat-free (nonfat) dry milk: U.S. Extra and U.S. Standard. Grading is voluntary and is paid for by the dairy industry. The USDA is responsible for grading; the U.S. Public Health Service recommends and enforces specific procedures for pasteurization (Grade A Milk Ordinance), laboratory tests, and sanitation at dairy farms and processing plants.

## Pasteurization

Milk is an excellent growth medium for microorganisms such as bacteria, yeast, and molds. In the early 1900s, it was frequently the vehicle for carrying such serious foodborne illnesses as

**TABLE 10-4**   Pasteurization Temperatures

| °F | °C | Time | Type of Pasteurization | Refrigeration Required |
|----|----|------|------------------------|------------------------|
| | | | Temperature* | |
| 145° | 63° | 30 minutes | Low-Temperature Longer-Time (LTLT) | Yes |
| 161° | 71.5° | 15 seconds | High-Temperature Short-Time (HTST) | Yes |
| 212° | 100° | 0.01 second | Higher-Heat Shorter-Time (HHST) | Yes |
| 280° | 138° | 2 seconds or more | Ultrapasteurization | Yes, but product has longer shelf life |
| 280°–302° | 138°–150° | 2–6 seconds | Ultrahigh-Temperature (UHT) | Not until opened |

*If the dairy ingredient has a fat content of 10% or more, or if it contains added sweeteners, the specified temperature shall be increased by 37°F/3°C.

typhoid, diphtheria, scarlet fever, and tuberculosis. Pasteurization, named after Louis Pasteur (1822–1895), its originator, was originally used to treat wine and beer, but soon came into use to treat milk as well, when it was found that heating milk for a short time to below its boiling point killed microorganisms. Pasteurization destroys 100% of pathogenic bacteria, yeasts, and molds and 95 to 99% of other, nonpathogenic bacteria. The process of pasteurization also inactivates many of the enzymes that cause the off-flavors of rancidity. Almost all milk sold commercially in North America is first pasteurized. In some states, where allowed by law, there is a small niche market for unpasteurized, or raw, milk.

To ensure that sufficient pasteurization has occurred, milk processors measure the activity of a specific enzyme found in milk, alkaline phosphatase. If this enzyme is no longer active, then the milk is safe for consumption. Pasteurization temperatures and times vary, but the ones most commonly used by milk processors are the first two listed in Table 10-4: 145°F (63°C) and 161°F (71.5°C). Even though pasteurized milk is no longer pathogenic, it will still spoil because the 1 to 5% nonpathogenic bacteria remaining convert lactose to lactic acid.

## Ultrapasteurization

A process called **ultrapasteurization** uses higher temperatures and shorter times (typically 280°F/138°C for at least 2 seconds) than regular pasteurization to extend the shelf life of refrigerated milk products (17). If this same treatment is combined with sterile packaging techniques, it is called **ultrahigh-temperature (UHT)**

processing. UHT processing destroys even more bacteria than standard pasteurization and increases the milk's shelf life. This milk is then packaged aseptically in sterile containers and sealed so that it can be stored unrefrigerated for up to 3 months (2). Once the aseptic seal is broken, the milk must be refrigerated. Originally, this preparation method was used on less-frequently purchased milk products such as whipping cream, half-and-half, and eggnog, but it is now used on a wider variety of products.

## Homogenization

Fat is less dense than water, causing it to float to the top of milk. This results in the thick layer of yellowish cream that rises to the top of unprocessed milk. **Homogenization** prevents this separation of water and fat known as creaming.

### Effect of Homogenization on Milk

Most milk in the United States is homogenized. This purely mechanical process has no effect on nutrient content; however, sensory changes do occur, resulting in a creamier texture, whiter color, and blander flavor. Homogenized milk also **coagulates** more easily, making puddings, white sauces, and cocoa more viscous. Its increased surface tension gives it a greater foaming capacity. Homogenized milk is also more prone to rancidity caused by oxygen being added to the double bonds of the unsaturated fatty acids. Pasteurizing milk before homogenization inhibits rancidity because the lipase enzymes responsible for breaking down fat are inactivated.

**? How & Why?**

*How is milk homogenized?*

The mechanical process of homogenization pumps the milk under a high pressure of 2,000 to 2,500 pounds per square inch through a machine that contains fine holes, which breaks up the fat globules. This decreases the fat globule size to less than 2 microns (Figure 10-3). The now very small droplets of milk fat are surrounded by a lipoprotein membrane, which prevents them from joining together and separating out. The liquid and fat components of the milk are now, in effect, homogenized.

**Ultrapasteurization** A process in which a milk product is heated at or above 280°F (138°C) for at least 2 seconds.

**Ultrahigh-temperature (UHT) milk** Milk that has been pasteurized using very high temperatures, is aseptically sealed, and is capable of being stored unrefrigerated for up to 3 months.

**Homogenization** A mechanical process that breaks up the fat globules in milk into much smaller globules that do not clump together and are permanently dispersed in a very fine emulsion.

**Coagulate** To clot or become semisolid. In milk, denatured proteins often separate from the liquid by coagulation.

**FIGURE 10-3** Homogenization. Fat globules are physically broken down into many more smaller globules to prevent the fat from grouping together and floating to the top of the liquid.

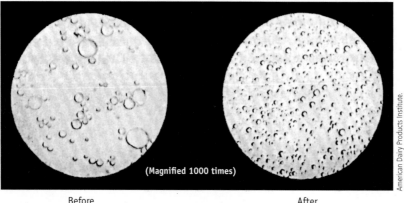

(Magnified 1000 times)

American Dairy Products Institute.

Before　　　　　　　　　　　　　　　　After

# TYPES OF MILK

About half the milk produced in the United States is sold as fluid milk and cream. Much of the rest comes to market as butter, cheese, and ice cream. The available market forms of milk include fluid milk—whole, reduced-fat (2%), low-fat (1%), fat-free (nonfat), UHT, chocolate, canned, and many others—dry milk, cream, and cultured milk products such as yogurt and buttermilk.

## Fresh Fluid Cow Milks

The three major types of milk considered "fresh fluid" cow milks are whole, reduced-fat and low-fat, and fat-free or nonfat.

### Whole Milk

To be classified as whole, milk must contain 3.25% milk fat and at least 8.25% **milk solids-not-fat (MSNF)** (Table 10-5). The milk is usually fortified with vitamins A and D, but this is optional (46). Figure 10-4 shows that milk is a nutritious beverage that delivers more nutrients for fewer calories than the same amount (one cup) of a soft drink.

**Milk solids-not-fat (MSNF)** Federal standard identifying the total solids, primarily proteins and lactose, found in milk, minus the fat.

### Reduced-Fat and Low-Fat Milk

These milks have had some of their fat removed so that milk fat levels are decreased to 2.0 and 1.0%, respectively, as noted on the carton. A minimum of 8.25% MSNF is necessary, but if it exceeds 10%, then the milk must be labeled "protein fortified" or "fortified with protein." The addition of milk solids improves the consistency, taste, and nutritive content of reduced- and low-fat milks. Vitamin A fortification is required, although the addition of vitamin D is optional.

Consumer interest in lower-fat products has resulted in a drastic downward trend in the consumption of whole milk (56). Between 1970 and 1990, reduced- and lower-fat milk sales increased by 300%, whereas sales of whole milk dropped by 50% (56). Overall, children in the United States are drinking less milk because they are drinking more of other beverages, such as soft drinks and fruit juices (32).

### Fat-Free or Nonfat Milk

*Nonfat* is synonymous with *fat-free*. Removing as much fat as technologically possible results in fat-free (nonfat) milk. The term *fat-free* replaced the older milk designation of *skim*. Fat-free milk should contain no more than 0.5% milk fat and a minimum of 8.25% MSNF. Vitamin A fortification is required, although the addition of vitamin D is optional. Less fat-free than reduced-fat milk is consumed in the United States. In fact, between

1988 and 1991, sales of reduced-fat milk increased 8 pounds per person, whereas sales of fat-free milk increased only 1 pound per person (56).

## Fresh Fluid Milks from Animals Other Than Cows

Not all milk comes from cows so other fresh fluid milks from other animals are now briefly addressed.

### Goat, Sheep, and Other Animal Milks

The history of the use of milk from various animals dates so far back that no one knows when the practice actually started. Cows provide almost all of the milk consumed in North America, but goats rank a close second in supplying milk to other regions such as Norway, Switzerland, the Mediterranean area, Latin America, and parts of Asia and Africa. Goat's milk is low in folate and vitamins D, C, and $B_{12}$. Some people in Spain, the Netherlands, Italy, and the Balkans also obtain milk from sheep. Camels provide milk for some people in the Middle East and central Asia, and reindeer are used for milk in the arctic region. The llama in South America sometimes serves as a source of milk, as does the water buffalo in parts of the Philippines, Asia, and India.

**TABLE 10-5**  Standards of Identity for Milk Products

| Category | Product | Pasteurization[a] / Ultrapasteurization[b] / UHT Processing[c] | Homogenization | Fat % (min./range) | MSNF % min. | Vitamin D | Vitamin A | Notes |
|---|---|---|---|---|---|---|---|---|
| FLUID | Milk | M a, b, c | Opt. | 3.25 | 8.25 | Opt. | Opt. | |
| FLUID | Reduced-Fat Milk | M a, b, c | Opt. | 0.5-2.0 | 8.25 | Opt. | M | |
| FLUID | Fat-Free Milk | M a, b, c | Opt. | <0.5 | 8.25 | Opt. | M | |
| CREAM | Half-and-Half | M a, b, c | Opt. | 10.5-18.0 | | | | |
| CREAM | Light Cream [Coffee Cream or Table Cream] | M a, b, c | Opt. | 18-30 | | | | |
| CREAM | Light Whipping Cream | M a, b, c | Opt. | 30-36 | | | | |
| CREAM | Heavy Whipping Cream | M a, b, c | Opt. | 36 | | | | |
| CAN | Evaporated Milk | M | M | 6.5 | 25.0 | M | Opt. | 60% H$_2$O removed |
| CAN | Condensed Milk | M a | Opt. | 7.5 | 25.5 | Opt. | | No vitamins added |
| CAN | Sweetened Condensed Milk | M | Opt. | 8.0 | 28 | | | 15% sugar added |
| DRY | Fat-Free Dry Milk | M | Opt. | Max 1.5 | | | | |
| DRY | Fat-Free Dry Milk Fortified with Vitamins A and D | M | Opt. | Max 1.5 | | M | M | |
| CULTURED | Sour Cream | M | Opt. | 18 | | | | |
| CULTURED | Acidified Sour Cream | M | Opt. | 18 | | | | |
| CULTURED | Sour Half-and-Half | M | Opt. | 10.5-18.0 | | | | |
| CULTURED | Acidified Sour Half-and-Half | M | Opt. | 10.5-18.0 | | | | |
| CULTURED | Yogurt | M a, c | Opt. | 3.25 | 8.25 | Opt. | Opt. | |
| CULTURED | Reduced-Fat Yogurt | M a, c | Opt. | 0.5-2.0 | 8.25 | Opt. | Opt. | |
| CULTURED | Fat-Free Yogurt | M a, c | Opt. | <0.5 | 8.25 | Opt. | Opt. | |

M = Mandatory
Opt. = Optional
a Pasteurization is mandatory but declaration of term *pasteurized* is optional.
b "Ultra-pasteurization" is an optional process—declaration of term *ultra-pasteurized* is mandatory on principal display panel, if applicable.
c UHT declaration is on label although not included in standards at this time.

Source: National Dairy Council.

# Flavored Fluid Milks

Flavored milk is created by adding a flavoring agent plus sugar to milk, thereby making the product more appetizing to customers. The first flavored milk was sold at the Wisconsin state fair in 1921. Since that time many different flavors have become available, including fruit, chocolate, vanilla, coffee, caramel, cappuccino, peanut butter, root beer, and vegetable (4). Because milk is 88% water, most flavoring agents are water based so they will dissolve and stay in solution within milk.

## Chocolate Milk

Chocolate flavoring comes from cocoa, natural or Dutch chocolate, chocolate liquor, or chocolate syrup. This flavor requires a stabilizer to prevent the cocoa from settling out of the milk (4). A common stabilizer is kappa II carrageenan (0.02–0.03%), which reacts with milk proteins to produce a gel, giving chocolate milk a thicker mouthfeel than regular milk. Other stabilizers include cellulose, guar gum, and xanthan gum, which can also add foaming characteristics to milk shake products. Guar or locust gum is used when a thickening quality is desired, such as in milk shakes.

Pasteurized milk containing 1.5% liquid chocolate, or 1% cocoa and 5% sugar, can be called chocolate milk. Chocolate milk is actually a suspension in which the continuous phase consists of milk fat and cocoa butter, whereas the dispersed phase includes the cocoa particles, fat-free (nonfat) milk solids, and sugar (23). Fortification with vitamins A and D is optional for chocolate whole milk, but fat-free chocolate milk must be fortified. Whole chocolate milk is high in calories, because the additional

**FIGURE 10-4** Selected nutrient content of milk (fat-free) vs. soft drink (1 cup, 8 fl oz, 245 g).

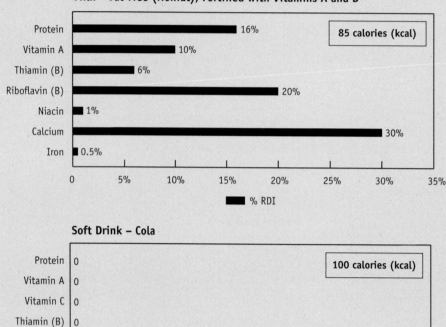

Milk – Fat-Free (Nonfat), Fortified with Vitamins A and D

85 calories (kcal)

| Nutrient | % RDI |
|---|---|
| Protein | 16% |
| Vitamin A | 10% |
| Thiamin (B) | 6% |
| Riboflavin (B) | 20% |
| Niacin | 1% |
| Calcium | 30% |
| Iron | 0.5% |

Soft Drink – Cola

100 calories (kcal)

| Nutrient | % RDI |
|---|---|
| Protein | 0 |
| Vitamin A | 0 |
| Vitamin C | 0 |
| Thiamin (B) | 0 |
| Riboflavin (B) | 0 |
| Niacin | 0 |
| Calcium | 0 |
| Iron | 0 |

sugar and cocoa contribute approximately 58 calories (kcal) per cup to its already higher calorie count, for a total of 209 calories (kcal) per cup. Chocolate milk made with reduced-fat milk contains 179 calories (kcal) and 5 grams of fat per cup, whereas the calories (kcal) and grams of fat in chocolate low-fat milk drop to 158 and 2.5, respectively.

### Eggnog

Packaged, commercially produced eggnog is manufactured to replicate a traditional rich holiday beverage made with eggs, cream or milk, nutmeg, and often spirits. It is sold predominantly during the Thanksgiving and Christmas

**Imitation milk** A product defined by the FDA as having the appearance, taste, and function of its original counterpart but as being nutritionally inferior.

holidays, and is defined as a pasteurized and homogenized mixture of milk, cream, milk solids, eggs, stabilizers, and spice. It contains 6 to 8% milk fat, about double that of whole milk, 9% MSNF, and 1% egg yolk solids. Other ingredients such as sugar, color and flavor additives, rum extract, nutmeg, and vanilla may be added. Although eggnog products vary widely and are available in low-fat versions, an average cup provides 342 calories (kcal) and 19 grams of fat, more than double that found in whole milk.

## Ultrahigh-Temperature Milk (UHT)

UHT milk is available in laminated aseptic cartons (43) that, when unopened, can be stored on shelves without refrigeration for up to 3 months. UHT milk has a "cooked" flavor at first, which tends to disappear with storage time. After 1 year,

UHT milk develops off-flavors described as "sweet," "flat," "musty," "rancid," and "chalky" (27). Chilling UHT milk before serving improves its taste. Once opened, UHT milk must be refrigerated and handled with the same care used for fresh milk. Although it is slightly more expensive than fresh milk and not widely distributed in the United States, UHT milk is ideal for boating, camping, hiking, and other situations in which refrigeration is not always available.

## Nutritionally Altered Fluid Milks

Some milks have their nutrient ingredients altered and a few examples include imitation, filled, low-sodium, and reduced-lactose milks.

### Imitation Milk

**Imitation milk** looks like milk but usually has little or no dairy content.

**FIGURE 10-5**   The REAL symbol denotes an item as being a real dairy product.

The Milk Board

Source: The Milk Board

Ingredients include water, corn syrup solids or sugar to replace the lactose, vegetable oils to replace the milk fat, protein from the sodium caseinate in soybeans, whey or milk solids-not-fat to substitute for the protein, and some stabilizers, emulsifiers, and flavoring agents.

Nutritionally, imitation milks are lactose free and so are useful for people with lactase deficiency, but sometimes sodium caseinate and whey are added, which makes these products inappropriate for people with milk protein allergies. The vegetable fat, like milk fat, may be high in saturated fatty acids from coconut or palm oils used in the manufacture of these products. The calcium and protein content is about half that found in regular milk. Although the nutrient content of milk is superior to that of imitation milk, the latter is less expensive. To emphasize their products' superiority over imitation milk, some dairy processors place a REAL seal shaped as a drop of milk on their products (Figure 10-5).

## Filled Milk

Filled milk is made by replacing all or part of the milk fat with a vegetable fat. Cholesterol levels drop to zero in filled milk, but if the fat substitute is a saturated vegetable oil such as coconut or palm oil, then there is a higher ratio of saturated to polyunsaturated fats. Both imitation and filled milks are regulated at the state level rather than at the federal level.

## Low-Sodium Milk

A cup of milk contains 120 mg of sodium, so people on sodium-restricted diets must watch their milk intake. Low-sodium milks containing only 6 mg per cup are available. These are produced by an ion exchange method that removes all but 5% of the original sodium.

## Reduced-Lactose Milk

Any pasteurized milk treated with the enzyme lactase will have most of its lactose converted to its two monosaccharides, glucose and galactose. This doubling of the sugar molecules results in a slightly sweeter flavor. The resulting milk is more easily digested by people who have some degree of lactose intolerance.

# Plant-Based "Milks"

Only reduced-lactose milk has the same high-quality nutrients as regular milk. For this reason, soy milk, rice milk, and almond milk should never be given to infants (birth to 1 year of age) or young children as a *sole* source of nutrition. These so-called milks are not true milks and will not nutritionally support the growth of an infant or very young child. Unsupplemented plant products do not contain the necessary nutrients required by a human infant and serious health problems could result. To reduce the risk of allergies, cow's milk is also not recommended for infants until they are 1 year of age. Only breastfeeding or infant formulas approved by the American Academy of Pediatrics are recommended during the infant's first year (25).

## Soy Milk

Soy milk is a milk-like product that is not from any mammal, but rather is made from soybeans that have been soaked, ground, and strained. Soy milk has been known for centuries in China and Japan and is believed to have been developed in China around 150 BC (37). Now it is consumed as a milk-like liquid, is used to make tofu (the cheese of soy milk), and is incorporated into some infant formulas. Some vegetarians use it as a substitute for cow's milk. As useful as it is, soy milk is lacking in certain nutrients. Only soy milks fortified with methionine (an essential amino acid), calcium, and vitamin $B_{12}$ should be substituted for cow's milk for growing children. On the other hand, it also lacks the carbohydrate lactose, which makes it ideal for people with lactose intolerance.

## Rice Milk

Another milk suitable for people with lactose intolerance is rice milk. It contains no lactose because it is made primarily from brown rice, filtered water, and a small amount of brown rice sweetener. Even though rice milk is made from rice, not all rice milks are gluten free, which would not be acceptable to people with celiac disease, an immune system condition that makes them allergic to gluten. Although rice contains no gluten, some rice milks are made using barley enzymes to convert the carbohydrates in brown rice to naturally occurring sugars. This is why rice milk is somewhat sweeter than soy milk. In addition, the flavor of either rice or soy milk can be enhanced by adding vanilla flavoring, making these milks even more acceptable to consumers than their plain counterparts.

## Almond Milk

Similar to soy and rice milk is almond milk, a milky drink made from ground almonds. The short shelf life of milk before the age of refrigeration meant that milk did not stay fresh very long. During medieval times, cooks turned to a milky liquid produced by grinding almonds or walnuts. Homemade versions can be made by combining ground almonds and water in a blender. Commercial almond milk products are available in plain, vanilla, or chocolate flavors.

## Other Nondairy Milks

- *Hemp milk* is a newer plant-based milk alternative. This nondairy milk with a slightly nutty flavor and a "putty" color is made by mixing water with the crushed seeds of the hemp plant.
- *Grain milk* may be made from oats, barley, wheat, rye, or flour mixed with water. It has a lower protein content and higher carbohydrate content than dairy milk.
- *Coconut milk* can be used as a cream substitute in some recipes (35). Coconut milk is not milk, but is rich, thick, and creamy enough to add texture to soups and desserts. It is most commonly used in Southeast Asian cuisine, particularly Thai. Coconut cream is the thickest portion, which separates to the surface in canned coconut milk. "Lite" coconut milk is canned coconut milk

Pat LaCroix/Stockfood Creative/Getty Images

with water added. Regardless of the fat content (which can be high), canned coconut milk (usually 5- to 16-ounce cans) needs to be shaken well to mix the "cream" with the water. After opening, coconut milk can be stored in a sealed container in the refrigerator for approximately 2–5 days.

- *Nuts,* such as hazelnuts and others, can also be ground and mixed with water to make nondairy milk.

## Canned Fluid Milks

Milk products are also available to consumers in canned form—whole, evaporated, and sweetened condensed milk.

### Whole Milk

Some ultrahigh-temperature milk is canned for export. Requirements regarding its content are similar to those for whole fluid milk sold in a carton.

### Evaporated Milk

Evaporated milk is produced by evaporating at least 60% of the water found in whole milk. By definition, evaporated milk contains at least 7.5% milk fat, contains 25.5% milk solids-not-fat by weight, and is fortified with vitamin

D. Evaporated milk provides 338 calories (kcal) and 19 grams of fat per cup. Stabilizers are often added to prevent separation of the fat during storage. Fat separation may also be prevented by turning the cans over every few weeks.

The evaporation process consists of initially exposing the milk for 10 to 20 minutes to a temperature of 203°F (95°C). This stabilizes the casein so that it will not coagulate during sterilization. In the next step of evaporation, the milk is heated to 121°F–131°F (50°C–55°C) at reduced atmospheric pressure, which allows the milk to boil without reaching the normal boiling temperature of 212.3°F (100°C). It is then homogenized, canned, and sterilized. The high temperatures of canning and the metal of the can may impart cooked and off-flavors to evaporated milk products.

Newer evaporation techniques produce evaporated milk by exposing it to ultrahigh temperatures for longer periods of time, placing it in sterilized cans, and aseptically sealing the cans. Compared to evaporated milk produced by the old method, UHT-evaporated milk is less viscous, less white, and has a different flavor.

Fat-free evaporated milk is produced from fat-free milk. By definition, it must contain less than 0.5% fat, at least 20% total solids, and vitamins A and D.

The lower fat content drops the calories (kcal) to 199 per cup and the fat to 0.5 gram. To reconstitute either regular or fat-free evaporated milk, combine equal volumes of the canned milk and water.

### Sweetened Condensed Milk

Sweetened condensed milk contains a high quantity of added sugar, which makes it ideal for preparation of desserts, especially pies and cheesecake. After whole milk has been evaporated by 50% by weight, 15% sugar in the form of dextrose or corn syrup is added. Sweetened condensed milk is defined as containing at least 28% total milk solids and about 8% milk fat. The extra sugar and highly concentrated nature of the nutrients in sweetened condensed milk make it very high in calories (982 kcal/cup) and fat (27 grams/cup). Unlike evaporated milk, sweetened condensed milk does not have to be sterilized, because its 40 to 45% by weight sugar concentration prevents microbial spoilage. The sugar content, either added or in the form of lactose, also contributes to the Maillard reaction. During heating the sugar combines with the protein in the milk to give it a light-brown color. This attribute makes sweetened condensed milk ideal for creating caramel-flavored or caramel-colored desserts such as pumpkin pie.

## Dry Milk

Fluid or canned milks are sometimes heavier to transport than milk purchased in dried form. The two main types of dry milk include nonfat dried milk and instant milk.

### Nonfat Dried Milk

Many milk products can be dried, but by far the most common product is nonfat dried milk (NFDM). NFDM can be made from whole, reduced-fat, low-fat, or fat-free milks, as well as buttermilk, but nonfat is usually selected because it contains the least amount of fat.

After the fat-free milk is dried, vitamins A and D may be added, although they are not required. Dried milks that originally contained some fat are required to be fortified with vitamin A and identified as "fortified with vitamins A and D." Overall, nonfat dried milk is nutritionally similar to fat-free milk, except that the levels of vitamins may be reduced by 20%. The absence of fat also gives it a long shelf life; nonfat dried milk can keep for about a year in a cool, dry storage area. The calcium and protein content of meals or beverages can be increased easily by adding nonfat dried milk to puddings, shakes, soups, casseroles, milk, and many other food combinations. Although it is inferior in taste to fluid milks, reconstituted nonfat dried milk is ideal for making batters and doughs for baked goods.

### Instant Milk

Instant milk is different from regular nonfat dried milk. It is manufactured by exposing nonfat dried milk to steam, after which it is redried. The double drying helps instant milk to achieve the ability to **agglomerate** and instantly dissolve in cold water. Generally, about ⅓ cup of instant nonfat dried milk will reconstitute into 1 cup of fluid milk, but manufacturers' instructions will vary. Allowing the reconstituted milk to refrigerate for several hours and adding a teaspoon of vanilla to a half gallon of reconstituted chilled nonfat dried milk make it taste more like regular milk. Also, more milk solids can be added by commercial food manufacturers to make it more flavorful and nutritious.

### ? How & Why?

*How is milk dried?*

Milk is dried by either spray, foam-spray, or roller drying processes. In spray drying, concentrated milk is sprayed into hot air, whereas foam-spray drying sends a jet of hot air into the concentrated milk. Roller drying consists of moving pasteurized fluid or condensed milk through two steam-heated rollers.

# Cultured Milk Products

Cultured or fermented milk products have been used for centuries, and have long been believed to have the potential to benefit human health (20). Some cultured milk products commonly consumed in North America are buttermilk, yogurt, acidophilus milk, kefir,

and sour cream. The one factor that all cultured milk products have in common is that they have had bacterial cultures added to them in order to ferment their lactose into lactic acid. The increased acid concentration causes the casein to precipitate out, which changes the milk into a more **curd**-like consistency. The type of bacterial culture inoculated into the milk largely determines the flavor of the resultant product. Bacteria also influence the quality of fermented dairy products by the amount and type of acids produced. Some protein is also broken down to provide nitrogen for bacterial growth, and this makes the curd softer and more digestible.

### Cultures Added to Milk

Microbial food cultures are composed of bacteria, fungi, and/or yeast (53). They include starter cultures, which are used in fermented foods such as yogurt, breads, and others. These microorganisms may be naturally present within the food, or added intentionally. The bacteria used to ferment dairy products in the United States are often "lactic acid producing bacterial cultures" or "propionic acid producing bacteria" (53). The function of these bacteria may be described as "culturing" or "action." In the United States, the only food for which specific strains of bacteria are mentioned in regulatory documents is yogurt—the U.S. regulatory documents specifically state that the culture must include *Lactobacillus bulgaricus* and *Streptococcus thermophilus* (53). In all other discussions of microorganisms in dairy products, there is no mention of specific strains or species. Thus, it is not possible to accurately document the composition of bacteria used in dairy products today. The U.S. government has granted "prior sanction" for the use of harmless lactic acid-producing bacteria in foods (53). Regulation of the bacteria in starter cultures used in meats and breads is similar to that of cultures used in dairy products.

When microbial food cultures are naturally present in a food, they are not considered an ingredient and thus are not required to be listed on the label (53). When cultures are added during the production process, the term "cultured" followed by the name of the substrate (i.e., "milk") must be listed as an ingredient.

### Safe Microorganisms for Foods

The U.S. Food and Drug Administration publishes a "Partial List of Microorganisms and Microbial-Derived Ingredients That are Used in Foods" on its website at www.fda.gov/Food/FoodIngredientsPackaging/ucm078956.htm. Similar lists are available from several other organizations:

- The European Food Safety Authority (EFSA) has published a list of organisms designated as Qualified Presumption of Safety (QPS) based on historical use and scientific research (18).
- The International Dairy Federation (IDF) publishes the Partial List of Microorganisms and Microbial-Derived Ingredients that are used in Foods (31).
- The IDF and the European Food and Food Cultures Association (EFFCA) have jointly published a list of microorganisms that are documented to have a history of safe use in the food supply (31).

### Buttermilk

Buttermilk contains little or no butterfat. Sweet natural buttermilk originally was the liquid left over after fresh cream had been chilled and churned to produce butter (57). Buttermilk is used in many baked products to add moisture and tanginess (3). Low- or full-fat buttermilk can be used, with the higher-fat products providing more richness. Natural buttermilk is often dried and used in baked products and ice cream because the phospholipids obtained from the fat-droplet membranes, which are broken down during churning, make it an excellent emulsifier. Now, most commercially available buttermilk is cultured by adding *Streptococcus lactis* bacteria to pasteurized nonfat, reduced-fat, or low-fat milk. The bacterial cultures in buttermilk convert the

**Agglomerate**  A process in which small particles gather into a mass or ball. In the case of milk, the protein particles regroup into larger, more porous particles.

**Curd**  The coagulated or thickened part of milk.

sugar (lactose) into lactic acid, which is the reason that buttermilk tastes more sour than regular milk. As the milk becomes more acidic, some of the proteins precipitate out because of its lower pH. This is *clabbering*, an older term used to describe the thickening of milk (57). The end result is that buttermilk is more sour, more bubbly, and thicker than regular milk. If a recipe calls for buttermilk, a substitute can be made by adding 1 tablespoon of white vinegar or lemon juice to 1 cup of milk and letting the mixture stand for 10 minutes.

Flavor may be enhanced by adding other bacteria, butterfat granules or flakes, natural sweeteners, citric acid (up to 0.15%), salt, and artificial flavors or colors. Fortification with vitamins A and D is optional. Cultured buttermilk must contain less than 0.5% milk fat and at least 8.25% milk solids-not-fat, and have an acidic pH of about 4.6. It is not mandatory that this pH be created by lactic-acid-generating bacteria, but when these bacteria are not used, the milk should be labeled "acidified buttermilk." A fat content between 0.5 and 2.0% changes the name to "cultured low-fat milk" or "acidified low-fat milk." A milk fat content over 3.25% is denoted by "cultured milk" or "acidified milk." Buttermilk has a longer shelf life than milk because its higher acid content inhibits the growth of spoilage-causing bacteria and its lower fat content makes it less likely to go rancid. Salt may be added to further inhibit bacterial growth.

## Yogurt

The term "yogurt" comes from the Turkish name "Yoghurmak," meaning, "to thicken" (44). This name reflects the thickening process that turns milk into yogurt. Yogurt is a fermented product created when bacteria are added to milk. These bacteria ferment the lactose in milk to lactic acid, a process that has been used in fermented dairy foods since prehistory (53). These bacteria influence the acidity, texture, and shelf life of the dairy products to which they are added. Traditionally, bacteria were naturally present and thus considered natural components. Now, these same bacteria may be added during the manufacturing process, but the end result is the same.

People in the Middle East have been eating yogurt for thousands of years. This smooth, semisolid, fermented dairy product can be made from whole, reduced-fat, or fat-free milks. Firm yogurt is known as the *set* style, whereas a more runny, semiliquid consistency is characteristic of *stirred* yogurt. Yogurt drinks are also gaining in popularity (39).

Yogurt is produced by mixing two types of bacteria, *Lactobacillus bulgaricus* and *Streptococcus thermophilous*, and adding them to pasteurized, homogenized milk to which some MSNF are also usually added. The whole mixture, including the bacteria, is held at a warm temperature (108°F–115°F/42°C–46°C) to allow fermentation to develop the desired consistency, flavor, and acidity. Although the milk used to make yogurt is usually slightly higher in lactose because of the addition of milk solids-not-fat, fermentation decreases the amount of lactose to 4% (49). During fermentation, the bacteria convert lactose to lactic acid, increasing the acidity (Chemist's Corner 10-3). In addition, folate (a B vitamin) levels increase as a natural by-product of bacterial growth.

*Active Culture Yogurts* Once the yogurt has reached the desired consistency, the fermentation and accompanying changes caused by the bacteria are discontinued. To accomplish this, bacterial growth can be inhibited by either chilling or heating the yogurt.

> **CHEMIST'S CORNER 10-3**
>
> **Flavor Compounds in Yogurt**
>
> The tartness of yogurt is partially derived from the bacterial conversion of lactose to lactic acid and acetaldehyde (26). A buttery flavor is also generated as bacteria break down the natural citric acid in milk to produce diacetyl.

If the yogurt is chilled, the culture remains alive, whereas the heated yogurt's cultures are destroyed. Yogurts containing live bacteria are labeled "with active yogurt cultures," "living yogurt cultures," or "contains active cultures." Only yogurt containing viable cultures is recommended for people with lactase deficiency or those taking antibiotics (33, 34). Consuming fermented milk products containing live cultures reportedly helps restore the normal intestinal bacteria eliminated by antibiotics, and has been associated with the treatment of diarrhea for centuries (7). Intestinal bacteria are beneficial because they help to produce some B vitamins and vitamin K.

*Other Ingredients Added to Yogurt* After fermentation, several ingredients, including gelatin and nonfat dried milk, may be added to yogurt to create a firmer texture, reduce the perception of acidity, or add color. The addition of sweeteners such as sugar, honey, fruit, fruit extracts, flavorings, and alternative sweeteners has increased the popularity of yogurt—over 85% of yogurt currently consumed in the United States is flavored. Yogurt with fruits blended throughout is called Swiss- or French-style yogurt, whereas yogurt with fruit on the bottom is known as sundae style.

*Official Yogurt Definition* To be called yogurt, the milk product must contain at least 8.25% MSNF and 0.5% acid. Fat content requirements for whole, reduced-fat, and fat-free yogurt are 3.25%, 0.5 to 3.0%, and less than 0.5%, respectively. There are no federal standards for frozen yogurt, an increasingly popular dessert choice for those watching their dietary fat and calorie intake.

*Calorie Content of Yogurts* The sweeteners and other flavorings in yogurt increase its caloric content. Plain whole yogurt at 150 calories (kcal) per cup jumps to 250 calories (kcal) when it is sweetened.

## Functional Foods—Probiotics

Yogurt is the most common vehicle for **probiotics** (e.g., *Lactobacillus* and *Bifidobacterium*) and **prebiotics** (21).

**Probiotics** Live microbial food ingredients (i.e., bacteria) that have a beneficial effect on human health.

**Prebiotics** Nondigestible food ingredients [generally fibers such as fructooligosaccharides (FOS) and inulin] that support the growth of probiotics.

The suggestion that probiotics have a positive impact on health by improving the intestine's microbial balance is attributed to the Nobel Prize-winning Russian scientist Elie Metchnikoff (1845–1916), who suggested that the long, healthy lives of Bulgarian peasants were due to their consumption of fermented milk products (48). Since then, it has been suggested that probiotics benefit health by relieving diarrhea and inflammation of the stomach and intestines; combating food allergies and certain cancers; and even by boosting immunity. Probiotics are theorized to exert their beneficial effects in several ways. They are known to create a healthy microbial balance in the intestines, aid digestion by producing helpful enzymes, prevent the attachment of harmful bacteria either directly as a barrier or indirectly through mucin (a mucoprotein) production, and stimulate immune function (48).

People in Europe and Japan have long recognized the possible benefits of probiotics and include them in a wide range of food products—yogurts, dairy-based beverages, breakfast drinks, health snacks or bars, luncheon meats, teas, puddings, and even candy (51). The Japanese probiotic drink Yakult™ is consumed by 24 million people daily (30). Probiotics are also sold in pill form.

Because of their newness on the market and the lack of scientific studies concerning their effectiveness, there may be some potential problems with probiotics. Questions exist about the ability of the microorganisms to stay alive, first during food processing and then during digestion (13). Another issue is dosage. What is the correct amount to consume and what is too much? It has been suggested that 1 to 10 billion live cells be consumed to ensure delivery of at least 100 million live cells per dose in the gastrointestinal tract (48).

The most important question to ask has been lost in the excitement of more new probiotic products entering the market: What is the effect of these extremely high microbial dosages on the natural intestinal tract environment? Is there a possibility that the artificial inflation of "good" bacteria, or unusual combinations of bacteria not seen before, might actually be "bad"?

Scientific research into the health benefits of probiotics has grown rapidly. The most well-recognized benefit of probiotics is the promotion of general digestive health and wellness (1). Other areas that have been studied in clinical trials include inflammatory bowel conditions, infant diarrhea, antibiotic-associated diarrheas (*C. dificile* colitis), infant colic, bacterial vaginosis, irritable bowel syndrome, oral health, lactose intolerance, immune support, and gut transit time.

The bacteria *S. salivarius,* also called BLIS K12, has recently been found to have protective properties in the oral cavity (1). This bacteria was first isolated from an adolescent boy who was resistant to common infections such as colds and coughs. The bacteria was then found to secrete a small protein, or peptide, that had anti-microbial properties. BLIS K12 was found to prevent the growth of harmful bacteria in the mouth. Periodontal disease, or infection of the gums in the mouth, is associated with other diseases, including cardiovascular disease. BLIS K12 has been shown to prevent ear infections (otitis media) and to be significantly more effective than oral antiseptics. Currently, it is sold as a powder, chewing gum, and chewable tablet.

While the final verdict concerning the health benefits of probiotics has not been determined, new fermented dairy products continue to be marketed for their so-called immune-enhancing benefits. Food corporations are inserting probiotics into yogurt and cultured drinks suggesting that they enhance immunity. Even dairy-based health "shots," small servings of probiotic-enhanced dairy products, are being marketed as a way to reduce cholesterol, improve digestion, or provide probiotics (44, 50). A large portion of the immune system encompasses the intestinal tract, and it is believed that probiotics may stimulate activity of immune cells (phagocytic cells, cytokines, and immunoglobins) and reduce gut inflammation (via cytokines) (38). Manufacturers have also been adding prebiotics such as inulin, oligofructose, and fructooligosaccharides (FOS) to probiotics because they act as food for microorganisms (38).

## Acidophilus Milk

Acidophilus milk is a cultured milk created with the assistance of *Lactobacillus acidophilus.* These bacteria break down lactose to glucose and galactose, resulting in twice as many sugar molecules. The resulting somewhat sweeter milk, usually packaged in cartons, is made by inoculating pasteurized milk with *L. acidophilus* culture and letting it incubate at 99°F (37°C) until a slight curd forms. A slightly acidic, sour taste also results, but this can be eliminated by mixing the bacteria directly into cold milk.

## Kefir

Kefir is a fermented milk product and probiotic that originated in Russia. It is sometimes referred to as "the champagne of milk" because of its bubbly, fizzy nature. Kefir is made by adding bacteria, *Lactobacillus caucasius,* and yeast, *Saccharomyces kefir* and *Torula kefir,* to milk (28). The milk is initially heated to 185°F (85°C) for half an hour and then cooled to 73°F (23°C), which allows the milk to ferment to a soft, foamy curd. The strong, tangy, sour taste comes from the formation of lactic acid. Kefir contains about 1% alcohol and a little carbon dioxide as a result of fermentation, and provides 250 calories (kcal) and 4.5 grams of fat per cup.

## Sour Cream

Cream can be soured by *Streptococcus lactis* bacteria or some other acidifying agent. Light cream or half-and-half is fermented at 72°F (22°C) until the acidity from lactic acid reaches 0.5%. A thicker sour cream is produced if MSNF, vegetable gums (carrageenan), or certain enzymes are added. To be labeled "sour cream," a minimum of 18% milk fat is required, although when sweeteners are added, the minimum milk fat content can be lowered to 14.4%. Lower-fat sour creams with half the fat content of standard sour cream, and sour cream substitutes, are also available.

# Creams and Substitutes

Creams come in a variety of different fat percentages and these can also be replaced by various cream substitutes.

## Cream

Cream is a collection of fat droplets that floats to the top of nonhomogenized whole milk. The heavier and thicker the cream is, the higher the fat content will be. Cooling the cream firms its fat globules and makes it even thicker. Creams vary in their milk fat content, ranging from a low of 18% to a high of 36%.

Cream manufacturers are not required to list the percentage of fat in the cream on the carton. The fat content of light or coffee cream is 18 to 30%, of light whipping cream, 30 to 36%, and of heavy cream or heavy whipping cream, not less than 36%. Combining cream with pasteurized or ultrapasteurized milk yields half-and-half, which is not true cream and contains only 10.5 to 18% fat.

Some whipping creams that are marketed contain added sugars and stabilizers to improve their taste and texture. Whipping cream is also sold in pressurized canisters, which provide the taste and texture of whipped cream in a convenient-to-use form. Another option is to purchase a whipped cream dispenser that comes with a nitrous oxide canister (Figure 10-6). The heavy cream will be whipped by the nitrous oxide charger and then can be held in the refrigerator for 2 weeks. Flavorings such as sugar (only confectioner's sugar or syrup, because regular sugar will clog the dispenser), coffee, cocoa, or liqueurs can be added.

**FIGURE 10-6**  Whipped cream dispenser.

Photo courtesy of iSi North America, Inc.

Photo courtesy of iSi North America, Inc.

### Cream Substitutes

Some of the whipped toppings in pressurized cans and tubs, as well as coffee creamers, dry mixes, imitation sour cream, and snack dips, are made from nondairy ingredients. These products came into being as low-cost substitutes that would last longer on the shelf. Nondairy coffee creamers can last over a year at room temperature. They often contain saturated fats, but many lighter substitutes have half the calories and one-third the fat of cream, and contain no cholesterol. One fat replacer can substitute for up to 100% of the fat in reduced-fat (2%) milk, sweet cream, sour cream, and butter. This product is made from milk protein that has been transformed into small, spherical globules; it has natural cream flavor and can be used in muffins, dressings, and low-fat soup (14). The nondairy whipped toppings in pressurized cans are made from water, vegetable oil, corn syrup solids, sodium caseinate or soy protein, emulsifiers, vegetable gums, coffee whiteners, and artificial flavors and colors.

# MILK PRODUCTS IN FOOD PREPARATION

## Flavor Changes

The bland, slightly sweet flavor of milk comes from its lactose, salts, sulfur compounds, and short-chain fatty acids. The percentage of fat determines the mouthfeel and body of a particular milk. Exposure to heat or sunlight, oxidation, the use of copper equipment or utensils, and the feed ingested by the source animal are just some of the other factors that can influence the flavor of milk (12). For example, an off-flavor develops when the amino acid methionine reacts with the sunlight-sensitive

riboflavin (vitamin $B_2$). The "cooked" flavor of heated milk develops in part because heating denatures whey proteins to release volatile sulfur compounds, which contribute to off-flavors (16). Dairy cattle that are allowed to eat wild onions, ragweed, French weed, beets, potatoes, cabbage, or turnips produce off-flavored milk (47).

## Coagulation and Precipitation

Some milk proteins coagulate or precipitate to form a solid clot, or curd, under certain conditions. These conditions include the application of heat and the addition of acid, enzymes, polyphenolic compounds, and salts.

### Heat

When milk is heated to near the boiling point, the whey proteins lactalbumin and lactoglobulin become insoluble, mesh with the milk's calcium phosphate, and precipitate, forming a film on the bottom and sides of the pan. This film can scorch easily. Scorching can be prevented by constant stirring, slow temperature increases, or use of a double boiler.

Casein will not coagulate with heat unless it is boiled for long periods of time. Canned evaporated milk, however, which contains higher concentrations of casein, may coagulate during the high heats of sterilization. This is prevented by warming the milk prior to sterilization.

*Why does a skin form on the surface of heated milk?*

This is caused by the evaporation of water, which is accompanied by an increased concentration of casein, fat, and mineral salts. This thin skin scorches easily; in addition, it can trap steam that is trying to escape and cause the milk to boil over. Several steps can be taken to avoid this problem, including using a lid, continual stirring during heating, floating a small pat of butter on top of the milk, or, in the case of hot chocolate, adding whipped cream or marshmallows.

## Acid

Adding acid to milk causes the casein (but not the whey proteins) in the milk to coagulate (Chemist's Corner 10-4). Casein precipitates when the normal 6.6 pH of fresh milk drops below 4.6. Sources of acids include acidic foods such as lemon and lime juices, tomato products, and certain fruits; or acid-producing bacteria in cultured milk products. Some foods must be carefully prepared because of the coagulating effect of acids on milk products. For example, extra caution is required when combining milk or cream with lemon-flavored tea, tomato soup, or coffee (which is acidic). The key to preventing the milk from coagulating is to add the acid to the milk base instead of the other way around. Avoiding high temperatures after milk has been mixed with acid also helps to prevent coagulation.

## Enzymes

Milk also coagulates and forms curds when it is combined with certain enzymes originating from animal, plant, or microbial sources. Enzymes used to coagulate milk include pepsin from the stomach of swine, proteases from fungal sources, and certain enzymes from fruits. The enzyme most commonly used to coagulate milk is **rennin,** which is used in the production of cheese and ice cream.

One of the major differences between coagulation caused by enzymes

---

### CHEMIST'S CORNER 10-4

#### Milk Coagulates as pH Drops

The calcium in milk exists in one of two forms: either combined with the casein protein, or as free calcium ions ($Ca^{++}$). At milk's near-neutral pH of 6.6, the casein combines with available calcium content, creating calcium caseinate. The casein complexes form a physical configuration known as a micelle, which is colloidally dispersed in the liquid portion of the milk (Figure 10-7) (5).

Normally, the casein proteins are negatively charged, causing them to repel each other in the fluid milk. Coagulation of these casein proteins occurs when the negative charges are neutralized by the hydrogen ions ($H^+$) from acid. When the pH drops to 4.6, casein becomes very insoluble and precipitates readily into a curd. This is milk's isoelectric point—the negative and positive charges on the molecules (i.e., casein) balance each other, and the overall charge is neutral. Milk products

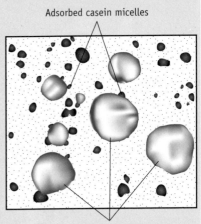

**FIGURE 10-7** The role of milk proteins and calcium in coagulation.

Adsorbed casein micelles

Homogenized milk fat globules

coagulated with acid are lower in calcium than those coagulated with an enzyme, because the acid releases calcium from the casein molecules, causing it to be lost in the whey:

$$\text{calcium phosphocaseinate} \xrightarrow{\text{acid}} \underset{\text{(gel)}}{\text{neutral casein}} + \underset{\text{(lost in whey)}}{Ca^{++}}$$

---

and that initiated with acid is that rennin-coagulated clots are rich in calcium and have a tough, rubbery texture, unlike those created by acid, which are less elastic and more fragile in consistency. Cottage cheese, which is normally coagulated by acid, contains less calcium per ounce (19 mg) than cheddar cheese (204 mg), which is usually made with rennin.

### Polyphenolic Compounds

Some fruits, vegetables, teas, and coffees contain polyphenolic compounds, which, when combined with milk, result in the precipitation of proteins. They also contribute to the curdling of cream or milk that sometimes occurs during preparation of scalloped potatoes or tomato or asparagus soup.

### Salts

When used in combination with milk, the salts in cured ham, some canned vegetables, and some seasonings can cause milk to curdle. Here again, to

prevent curdling, salt or salted foods should be added to the milk base instead of the other way around, and high temperatures must be avoided after the combination has been made.

## Whipped Milk Products

Liquid milk products such as cream, evaporated milk, and reconstituted nonfat dried milk can be made into a foam by whipping air into the liquid. During whipping, the protein in these milk products is mechanically stretched into thin layers that trap air bubbles, fat particles, and liquid (Figure 10-8).

**Rennin** An enzyme obtained from the inner lining of a calf's stomach and sold commercially as rennet.

**FIGURE 10-8**   Whipped foam structure. The numerous fat globules in whipped cream surround the air bubbles, giving solid reinforcement to the foam.

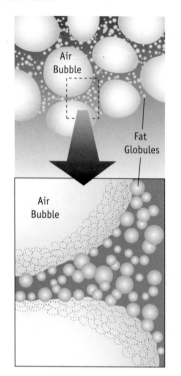

## Whipped Cream

Cream expands two to three times its volume when whipped. The stability of milk foams, especially whipped cream, is dependent on several factors: the fat content, the temperature of the cream, the age of the cream, the sugar content, the equipment used to whip the cream, and the length of whipping time.

**Fat Content**   The higher the fat content of the cream, the more stable the whipped cream will be. Solid fat particles create a more rigid foam. Heavy whipping cream beats more easily than lower-fat whipping creams, but becomes lumpy and buttery when overbeaten. An advantage to using heavy whipping cream (36 to 40% butterfat) is that its lower moisture content will prevent pastry crust from becoming soggy when it is filled with cream fillings. Whipping creams with a 30 to 36% fat content result in a softer, less stable foam (24).

Most whipping creams are sold unhomogenized to allow for easier aggregation of the fat globules. When the cream is homogenized, much of its protein surrounds the now smaller and more numerous fat globules instead of being available to envelop the air bubbles that are essential for foam formation. Vegetable gums and gelatin are sometimes added to improve the foaming ability of the commercial creams.

**Temperature**   Cooling cream increases its viscosity or firmness and its tendency to clump. For best results, refrigerate the cream, bowl, and beaters at 45°F (7°C) or less for at least 2 hours before whipping. Cream allowed to warm to room temperature or even to above 50°F (10°C) has more widely dispersed fat globules, which reduces the cream's ability to be whipped and creates a softer texture. The optimal temperatures for whipping various milk products are shown in Figure 10-9.

Most cream is pasteurized, but the heating process denatures an enzyme that helps fat globules to cluster, so nonpasteurized creams whip more readily and have a smoother texture (6). Although ultrapasteurized whipping cream takes even longer to beat to a peak-holding consistency, other ingredients can be added to improve

its whipping ability and dramatically extend its shelf life.

**Age**   The older that the cream is, the greater will be its viscosity and ability to foam. Whipping cream must be at least 1 day old in order to allow it to incorporate the air necessary for the optimal increase in volume.

**Sugar**   Sugar increases the stability of whipped cream, but it should be added gradually, toward the end of the whipping period. If added earlier, it will increase the whipping time and reduce overall volume and rigidity by delaying the clumping of fat. Sugar has the benefit, however, of lessening the likelihood of overbeating the cream. For the best stability, powdered sugar instead of granulated sugar should be used, because it dissolves more readily in the cold cream, and the cornstarch in the powdered sugar acts as a stabilizer.

**Whipping Time**   Physical agitation of the cream is necessary because it disrupts the phospholipid membranes surrounding the fat globules, preventing them from aggregating (29). Overbeating, however, is a common mistake. Overbeating for even a few seconds over the peak point turns whipped cream into butter and whey. The cream turns from glossy to

**FIGURE 10-9**   Optimal temperatures for whipping.

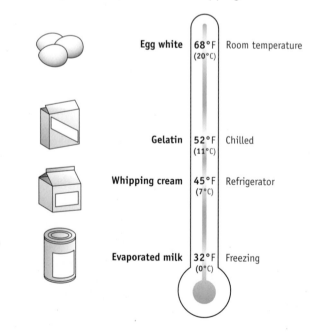

curdled looking. One way to salvage it is to add some more whipping cream and whisk it by hand gently into the overbeaten cream. To whip cream with an electric beater, it is best to beat it on medium high and then slow to a lower speed as soon as the cream starts to thicken. To check for sufficient whipping, the beating is stopped and the beaters are lifted to see if the cream is falling into glossy, large globs with soft peaks. The formation of stiff, yet moist, peaks signals the completion of the whipping process. The cream should be underbeaten slightly if ingredients such as sugar are to be whipped into the cream.

### Whipped Evaporated Milk

The high concentration of milk solids in evaporated milk makes it possible to whip it to three times its volume (Figure 10-10), but the flavor, texture, and stability are less acceptable than they are in whipped cream. The flavor of evaporated milk has a tendency to overpower other flavors; thus, it is best used with highly flavored foods. Evaporated milk foams are less stable than whipped cream foams, partly because of the former's lower viscosity and lower fat content. This can be compensated for to some degree by chilling the can of evaporated milk in the refrigerator for 12 hours or in the freezer until ice crystals form. Adding 1½ tablespoons of sugar per chilled cup of evaporated milk can further stabilize the protein and resulting foam. Lemon juice can also be used to stabilize the foam. The main advantage of whipped evaporated milk over whipped cream is its lower cost.

### Nonfat Dry Milk (NFDM)

Prepared nonfat dried milk powder can actually be whipped into a foam. This whipped milk product is very unstable, but it is much less expensive and lower in both calories and fat than whipped cream. It is prepared by dissolving equal parts of nonfat dried milk and cold water, chilling the mixture, and then beating until the mixture stands in soft peaks. Stability is increased by adding 1 tablespoon of lemon juice or 2 to 4 tablespoons of sugar during beating, which continues until the peaks bend over slightly on top.

# STORAGE OF MILK PRODUCTS

## Refrigerated

All fluid milk except unopened, aseptic packages of UHT pasteurized milk and certain canned milk products should be stored in the refrigerator. Milk should be removed only long enough to take what is to be used and then quickly returned to the refrigerator. Containers should be closed or covered to avoid rancidity, microbial contamination, and the absorption of odors from other foods in the refrigerator. Both oxidative and hydrolytic rancidity are potential problems in milk and milk products because of the substantial amounts of short-chain fatty acids. One should never drink milk directly from the container, because bacteria in the mouth can wash back into the product. Pouring unused milk products back into the original container is also not recommended, because microbial contamination could have occurred from exposure to the air or other sources. Storing milk in proper opaque containers will reduce its exposure to light, which can trigger oxidation, resulting in off-flavors and the loss of riboflavin ($B_2$). The shelf life of certain dairy products such as cottage cheese can be doubled with the addition of carbon dioxide, which disrupts microbial functions (15).

The following guidelines should be followed when storing milk products in the refrigerator:

- **Milk.** No more than 3 weeks.
- **Yogurt.** Best consumed within the first 10 days, but can last up to 3 to 6 weeks. If it separates, simply stir the liquid back into the curd before serving.
- **Buttermilk.** Best when used within 3 to 4 days after purchase, because it will continue to sour, but it can last up to 3 or 4 weeks.
- **Sour cream.** Unopened, up to 1 month, but is best when used within a few days.

## Dry Storage

Nonfat dry milk, ultrapasteurized milk, evaporated milk, and sweetened condensed milk are stored at or slightly below room temperature (72°F/22°C). Nonfat dry milk should not be exposed to moisture, because humidity will cause it to turn lumpy and become stale. Keeping containers tightly closed minimizes any contact with oxygen from the air. Nonfat dry milk stored in this manner will keep for about a year. Unopened cans of evaporated and sweetened condensed milks will keep up to a year in dry, ventilated areas, but double that if refrigerated. Both should be turned over every few weeks to prevent the solids from settling, thickening, and producing clots. Ultrapasteurized milk can be stored unopened at room temperature for up to 3 months. Once opened, all these milks must be treated like fresh milks and refrigerated. The sell by or pull date is the last day the item should be sold by the store, and this should always be checked before purchase.

**FIGURE 10-10**  Average volume yields for whipped milk products.

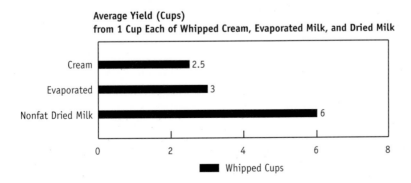

**Average Yield (Cups) from 1 Cup Each of Whipped Cream, Evaporated Milk, and Dried Milk**

# PICTORIAL SUMMARY / 10: Milk

About half the milk produced in the United States is sold as fluid milk and cream. Much of the rest comes to market as butter, cheese, and ice cream.

## COMPOSITION OF MILK

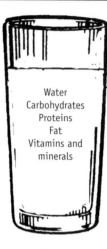

Water
Carbohydrates
Proteins
Fat
Vitamins and minerals

Almost 90% of milk is water.

Lactose, or milk sugar, is the primary carbohydrate found in milk.

The predominant proteins in milk are of high quality and consist of casein and whey proteins.

The fat in milk is known as milk fat or butterfat.

Milk is an excellent source of vitamins A and D (depending on fortification), riboflavin ($B_2$), and tryptophan, important in the formation of the B vitamin niacin. About 80% of the calcium ingested by Americans is derived from the dairy food group.

## PURCHASING MILK

USDA Grade A, the highest grade for milk, is based on the lowest bacterial count. Although grading milk is voluntary, pasteurization, the process of heating milk to kill harmful bacteria, is required for all milk. Homogenization prevents the separation of water and fat in milk.

## MILK PRODUCTS IN FOOD PREPARATION

Heat, the source animal's feed, oxidation, the use of copper equipment or utensils, and exposure to sunlight can all influence milk flavor.

The percentage of fat determines the mouthfeel and body of a particular milk.

Milk proteins coagulate to form a solid clot when exposed to heat, acid, enzymes, polyphenolic compounds, and salt.

Scorching milk can be prevented by constant stirring, slow increases in temperatures, and use of a double boiler.

Liquid milk products such as cream, evaporated milk, and reconstituted nonfat dry milk can be whipped into a foam.

## STORAGE OF MILK PRODUCTS

All fluid milk except unopened ultrahigh-temperature pasteurized milk and certain canned and dry milk products should be stored in the refrigerator. Containers should be closed or covered to avoid rancidity, microbial contamination, and the absorption of odors from other refrigerated foods. Oxidation caused by exposure to air can result in an off-flavor and losses of riboflavin ($B_2$). Nonfat dry milk, ultrapasteurized milk, evaporated milk, and sweetened condensed milk should be stored at or slightly below room temperature (72°F/22°C). Once opened, all these milks must be treated as fresh milks and refrigerated.

## TYPES OF MILK

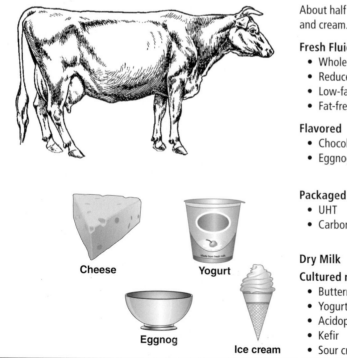

About half the milk produced in the United States is sold as fluid milk and cream.

**Fresh Fluid Milk**
- Whole
- Reduced-fat (2%)
- Low-fat (1%)
- Fat-free (nonfat)

**Flavored**
- Chocolate
- Eggnog

**Packaged**
- UHT
- Carbonated

**Dry Milk**

**Cultured milk products**
- Buttermilk
- Yogurt
- Acidophilus milk
- Kefir
- Sour cream

**Nutritionally Altered**
- Imitation
- Filled
- Low-sodium
- Reduced-lactose

**Plant-Based "Milk"**
- Soy, rice, almond
- Hemp, grain, coconut, and other nuts

**Canned**
- Whole
- Evaporated
- Sweetened condensed

**Dairy-Based Foods**
- Cheese
- Butter
- Ice cream

Soy milk

Butter

**Cheese**

**Yogurt**

**Eggnog**

**Ice cream**

Cream

# CHAPTER REVIEW AND EXAM PREP

## Multiple Choice*

1. The primary carbohydrate in milk is _____.
   a. sucrose
   b. glucose
   c. lactose
   d. fructose

2. The enzyme most commonly used to coagulate milk is _____, which is used in the production of cheese.
   a. lipase
   b. lactase
   c. rennin
   d. papain

3. What is the name of the heat treatment used to kill microorganisms in milk?
   a. Pasteurization
   b. Homogenization
   c. Filtration
   d. Evaporation

4. The curdling of milk or cream that sometimes occurs when preparing dishes such as scalloped potatoes or those containing tomato products or asparagus is due to the presence of _____ in these foods.
   a. phytochemicals
   b. sugar
   c. alkaline ingredients
   d. polyphenolic compounds

5. What are the two main types of protein found in milk?
   a. Rennin and albumin
   b. Whey and albumin
   c. Albumin and immunoglobulin
   d. Casein and whey

6. The ingredient in milk that contributes to a rich, creamy mouthfeel is _____.
   a. protein
   b. fat
   c. carbohydrate
   d. fiber

7. What are foods with live microbrial food ingredients that have a beneficial effect on human health called?
   a. Pasteurized
   b. Prebiotics
   c. Probiotics
   d. Medical foods

## Short Answer/Essay

1. Describe the nutrient composition of milk, including its primary carbohydrate, its predominant proteins, and the major vitamins and minerals it provides.
2. Describe the grading of milk and the process of pasteurization.
3. What is the difference between ultrapasteurization and ultrahigh-temperature processing?
4. Describe the purpose and process of homogenization.
5. Briefly describe the major characteristics of the following milk products: fluid milk, canned milk, dry milk, cultured milk products, and creams.
6. Define the following terms: *milk solids-not-fat (MSNF), agglomerate,* and *coagulate.*
7. List and explain the various ways that milk can be precipitated.
8. Discuss the effect of fat content, temperature, cream age, sugar, and whipping time on the stability of milk foams in whipped cream.
9. Exposing milk to light can trigger oxidation and the loss of which vitamin?
10. What are the recommended refrigerator storage times for fluid milk, yogurt, buttermilk, and sour cream?

*See p. AK-1 for answers to multiple choice questions.

# REFERENCES

1. Altaffer P and G Washington-Smith. Probiotic potential. *Nutraceuticals World* 12(2):34–35, 2009.

2. Amanter GF. Aseptic packaging of dairy products. *Food Technology* 37(4):138, 1983.

3. Anonymous. Can't find any buttermilk? *Fine Cooking* 88:83, 2007.

4. Aryana KJ. Flavored milks. In Hui Y, et al., *Handbook of Food Products Manufacturing: Health, Meat,* *Poultry, Seafood, and Vegetables.* John Wiley and Sons, 2007.

5. Aynie S, et al. Interactions between lipids and milk proteins in emulsion. *Journal of Food Science* 57(4):883, 1992.

6. Babcock CJ. *The Whipping Quality of Cream.* U.S. Dept. of Agriculture Bulletin No. 1075. Washington, D.C.: GPO, 1922.

7. Beausoleil M, et al. Effect of a fermented milk combining *Lactobacillus* *acidophilus* Cl1285 and *Lactobacillus casei* in the prevention of antibiotic-associated diarrhea: A randomized, double-blind, placebo-controlled trial. *Canadian Journal of Gastroenterology* 21(11):732–736, 2007.

8. Bolenz S, T Thiessenhusen, and R Schape. Influence of milk components on properties and consumer acceptance of milk chocolate. *European Food Research and Technology* 216:28–33, 2003.

9. Calvo MS, SJ Whiting, and CN Barton. Vitamin D fortification in the United States and Canada: Current status and data needs. *American Journal of Clinical Nutrition* 80(6):1710S–1716S, 2004.

10. Chandan R. *Dairy-Based Ingredients.* Eagen Press, 1997.

11. Choe E, R Huang, and DB Min. Chemical reactions and stability of riboflavin in foods. *Journal of Food Science* 70(1):R28–R36, 2005.

12. Christensen TC, and G Holmer. Lipid oxidation determination in butter and dairy spreads by HPLC. *Journal of Food Science* 61(3): 486–489, 1996.

13. Clemens R, and P Pressman. Probiotics and lessons learned from vitamin C. *Food Technology* 59(1):24, 2005.

14. Dairy ingredient replaces fat, adds flavor. *Food Engineering* 64(1):26, 1992.

15. Dairy Management, Inc. Increasing milk's competitiveness. *Dairy Research Focus* 2(4):1–2, 1996.

16. deMan JM. *Principles of Food Chemistry.* Van Nostrand Reinhold, 1999.

17. Ehri A. Ultrapasteurization: What does it mean? *Fine Cooking* 75:80, 2006.

18. EFSA. *The maintenance of the list of QSP microorganisms intentionally added to food or feed. European Food Safety Authority.* www.efsa .europa.eu/EFSA/efsa_locale-1178620753812_1211902221481 .htm

19. Euston SE, et al. Oil-in-water emulsions stabilized by sodium caseinate or whey protein isolate as influenced by glycerol monostearate. *Journal of Food Science* 61(5):916–920, 1996.

20. Fiander A, et al. Effects of lactic acid bacteria and fermented milks on eicosanoid production by intestinal epithelial cells. *Journal of Food Science* 70(2):M81–M86, 2005.

21. Floch MH, et al. Recommendations for probiotic use. *Journal of Clinical Gastroenterology* 40(3):275–278, 2006.

22. Francis LL, et al. Serving temperature effects on milk flavor, milk aftertaste, and volatile compound quantification in nonfat and whole milk. *Journal of Food Science* 70(7):S413–S418, 2005.

23. Full NA, et al. Physical and sensory properties of milk chocolate formulated with anhydrous milk fat fractions. *Journal of Food Science* 61(5):1068–1072, 1996.

24. Gardiner A, and S Wilson. Whipping egg whites and cream into stable, airy foams. *Fine Cooking* 66:74–75, 2005.

25. Gartner LM, et al. Academy of Pediatrics: Policy Statement. Breastfeeding and the Use of Human Milk. *Pediatrics* 115(2):496–506, 2005.

26. Hamann WT, and EH Marth. Survival of *Streptococcus thermophilus* and *Lactobacillus bulgaricus* in commercial and experimental yogurts. *Journal of Food Protection* 47:781, 1984.

27. Hansen AP. Effect of ultrahigh-temperature processing and storage on dairy food flavor. *Food Technology* 41(9):112, 1987.

28. Hertzler SR, and SM Clancy. Kefir improves lactose digestion and tolerance in adults with lactose maldigestion. *Journal of the American Dietetic Association* 103(5):582–587, 2003.

29. Hinrichs J, and HG Kessler. Fat content of milk and cream and effects on fat globule stability. *Journal of Food Science* 62(5):992–995, 1997.

30. Hollingsworth P. Yogurt reinvents itself. *Food Technology* 55(3):43–49, 2002.

31. International Dairy Federation. *Inventory of microorganisms with a documented history of use in foods,* Ch 2, No 377, 2002.

32. Johnson RK, C Frary, and MG Want. The nutritional consequences of flavored-milk consumption by school-aged children and adolescents in the United States. *Journal of the American Dietetic Association* 102(6):853–856, 2002.

33. Labell F. Yogurt cultures offer health benefits: Biotechnology to transform the yogurt of the future. *Food Processing* 50:130–138, 1989.

34. Laye I, D Karleskind, and CV Morr. Chemical, microbial, and sensory properties of plain nonfat yogurt. *Journal of Food Science* 58(5): 991–995, 1000, 1993.

35. Longbotham L. Instead of cream, try coconut milk. *Fine Cooking* 91:62, 2008.

36. Lucey JA. Milk protein gels. In Thompson A, M Boland, and H Singh, *Milk Proteins: from Expression to Food.* Academic Press, 2009.

37. Min S, Y Yu, and S St. Martin. Effect of soybean varieties and growing locations on the physical and chemical properties of soymilk and tofu. *Journal of Food Science* 70(1):C8–C12, 2005.

38. Ohr LM. Nutraceuticals and functional foods: Immunity boosters. *Food Technology* 59(1):65–70, 2005.

39. Oklahoma Cooperative Extension. *Fact Sheets. A Profile of Food.*

40. Onwulata C, and P Tomasula. Whey texturization: A way forward. *Food Technology* 58(7):50–54. 2004.

41. Ortolani C, and EA Pastorello. Food allergies and food intolerances. *Best Practices and Research in Clinical Gastroenterology* 20(3):467–483, 2006.

42. Otte J, et al. Protease-induced aggregation and gelation of whey proteins. *Journal of Food Science* 61(5): 911–915, 1996.

43. Petrus RR, EH Walter, JA Faria, and LF Abreu. Sensory stability of ultrahigh temperature milk in polyethylene bottle. *Journal of Food Science* 74:S53–S57, 2009.

44. Pszczola DE, and K Banasiak. Ingredients. *Food Technology* 60(5):45–92, 2006.

45. Reducing allergic reactions to milk. *Food Technology* 62(10):14, 2008.

46. Renken SA, and JJ Warthesen. Vitamin D stability in milk. *Journal of Food Science* 58(3):552–556, 1993.

47. Ronsivalli LJ, and ER Vieira. *Elementary Food Science.* Van Nostrand Reinhold, 1992.

48. Sanders ME. Probiotics: A publication of the Institute of Food Technologists' expert panel on food safety and nutrition. *Food Technology* 53(11):67–77, 1999.

49. Savaiano DA, and MD Levitt. Nutritional and therapeutic aspects of fermented dairy products. *Contemporary Nutrition* 9(6):1–2, 1984.

50. Sloan AE. Dairy's new dynamic. *Food Technology* 62(12):19, 2008.

51. Sloan AE. The top ten functional food trends. *Food Technology* 54(4):33–62, 2000.

52. Smith DM, and AJ Rose. Gel properties of whey protein concentrates as influenced by ionized calcium. *Journal of Food Science* 59(5):1115–1118, 1994.

53. Stevens HC, and L O'Brien Nabors. Microbial food cultures: A regulatory update. *Food Technology* 63(3):36–41, 2009.

54. Swientek B. Subtraction. *Food Technology* October 2008, pp. 45–52.

55. The 1992 supermarket sales manual. *Progressive Grocer* 71(7):80, 1992.

56. United States Department of Agriculture, Economic Research Service. *Livestock, dairy, and poultry outlook: Tables.* www.ers.usda.gov/Publications/ldp/LDPTables.htm. Accessed 6/09

57. Weil C. For tang and tenderness, bake with buttermilk. *Fine Cooking* 68:22, 2005.

## WEBSITES

The Dairy Industry's National Dairy Council website can be found here: **www.nationaldairycouncil.org**

This industry association site for consumers of dairy products has lots of information: **www.innovatewithdairy.com**

*Milk Matters* is a public health education campaign to promote calcium consumption: **www.nichd.nih.gov/milk**

The International Dairy Foods Association (IDFA) represents the nation's dairy manufacturing and marketing industries and their suppliers. IDFA is composed of three organizations: the Milk Industry Foundation (MIF), National Cheese Institute (NCI), and International Ice Cream Association (IICA): **www.idfa.org**

PhotoDisc/Getty Images

# 11 Cheese

Cheese is a preserved food made from the curd, or solid portion, of milk. Adding certain enzymes and/or acid to any type of milk causes the casein proteins and fat to coagulate and separate from the liquid portion, or whey (see Chapter 10). No one really knows when humans first started to consume cheese, but legend links its discovery to an anonymous shepherd who decided to carry milk in a bag made from a sheep's stomach. The sun warmed the bag, activating the natural rennin enzyme in the stomach lining and turning the liquid milk into a semi-solid clump. To the shepherd's surprise, the resulting mass was quite palatable. It probably did not take long after that for people to realize how useful this natural process would be in providing an edible food that could be transported.

Making cheese involves removing moisture from the curd to varying degrees after the whey is drained. The curd can then be treated in a variety of ways to produce over 2,000 varieties of cheese, some of which are shown in Figure 11-1. The whey also contains dissolved materials such as proteins, which can be processed to produce cheeses and other foods.

In the United States, the Food and Drug Administration says that cheese must be "a product made from curd obtained from the whole, partly fat-free/nonfat, or fat-free/nonfat milk of cows, or from milk of other animals, with or without added cream, by coagulating with rennin, lactic acid, or other suitable enzyme or acid, and with or without further treatment of the separated curd by heat or pressure, or by means of ripening ferments, special molds, or seasoning." This definition serves as the foundation for many different cheese varieties, but 15 varieties now account for most of the cheese consumed today. Appendix E lists the commonly used cheese varieties along with their shape, flavor, basic ingredient, ripening (aging) period, and mode of serving.

This chapter discusses the various ways to classify cheese, along with its nutrient composition, production, purchase, use, and storage.

## CLASSIFICATION OF CHEESES

There are many ways to classify cheeses. The differences among the numerous cheese varieties result from many variables, including milk source (cow, sheep, goat, camel, etc.); the fermentation and ripening conditions; the methods of pressing, sizing, and shaping; the type of bacteria utilized;

**FIGURE 11-1** Common cheeses.

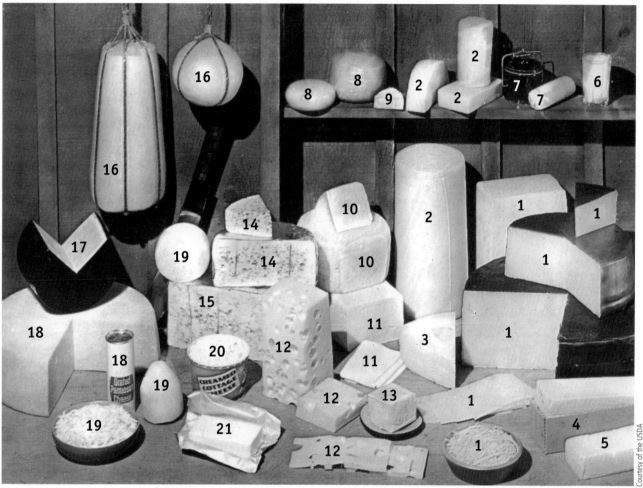

1. Cheddar
2. Colby
3. Monterey Jack
4. Pasteurized Process Cheese
5. Cheese Foods
6. Cheese Spreads
7. Cold-Pack Cheese Food or Club Cheese
8. Gouda and Edam
9. Camembert
10. Muenster
11. Brick
12. Swiss
13. Limburger
14. Blue
15. Gorgonzola
16. Provolone
17. Romano
18. Parmesan
19. Mozzarella and Scamorze
20. Cottage Cheese
21. Cream Cheese

Courtesy of the USDA

appearance; mode of packaging; and/ or even their place of origin (22). The most common methods of classifying cheeses, however, are by the processing method, the milk source, and moisture content. Both place of origin and moisture content are now briefly discussed.

## Place of Origin

Some types of cheeses are named for the location in which they originated. A few of the more common examples are listed in Table 11-1 and further elaborated on in Appendix E.

**TABLE 11-1** Location of Origin for Major Cheese Varieties

| England | Cheddar, Cheshire |
|---|---|
| France | Camembert, Brie |
| Italy | Parmesan, Gorgonzola |
| Netherlands | Edam, Gouda |
| Switzerland | Gruyere, Emmentaler |
| United States | Colby |

Adapted from (22).

## Moisture Content

The goal of making cheese is to remove water from the milk until a solid mass or curd remains. During this production process, the moisture content of cheese decreases, especially during aging. As cheeses age, they become drier and harder. Cheeses classified according to their moisture content are described as fresh, soft, semi-hard, hard, and very hard (34).

- *Fresh.* Fresh cheeses, also called country cheeses, are soft, whitish in color, and mild tasting. They are highly perishable, because their moisture content is over 80 percent, and they are not aged. They include cottage cheese and cream cheese, and ricotta, farmer's, pot, and feta cheeses.
- *Soft.* Soft cheeses, such as Brie, Camembert, and many Hispanic

cheeses, are aged for just a short time. Water content ranges from 50 to 75 percent.

- *Semi-hard.* Semi-hard cheeses contain 40 to 50 percent moisture. The best-known examples are Roquefort, blue, Muenster, brick, Gouda, Edam, Port du Salut, Gorgonzola, and Stilton.
- *Hard.* The moisture content of hard cheeses ranges from 30 to 40 percent. Cheddar and Swiss are examples of hard cheeses.
- *Very hard.* Parmesan and Romano are classified among the hardest cheeses. Very hard cheeses will not slice easily, but are easily grated or crumbled. They are aged the longest and have a water content of approximately 30 percent.

# CHEESE PRODUCTION

No two cheese varieties are produced by exactly the same method, which is why there are so many different types of cheese. Just think how many different types of cheddar cheese there are! However, the basic steps of making cheese—milk selection, coagulation, curd treatment (cutting, heating, salting, knitting, and pressing), curing, and ripening—are common to them all. The yield from 10 pounds of milk is approximately 1 pound of

cheese and 9 pounds of whey (31). On a smaller scale, that is equivalent to 1 cup (8 ounces) of milk providing about 1 ounce (specifically, 0.8 ounce) of cheese.

## Milk Selection

The first step in making cheese, and the one that has the greatest influence on its classification, is choosing the appropriate milk. Any mammal's milk can be made into cheese, but in the United States, pasteurized cow's milk is the most common source. However, interest in specialty cheeses made from milk other than cow's milk continues to grow (18). In Europe and the Middle East, a significant amount of cheese is made from either sheep's or goat's milk. Other animals whose milk may be used for making cheese include the camel (Iran and Afghanistan), reindeer (Lapland), horse (Mongolia), water buffalo (Philippines, India, and Italy), and the yak and zebu (China and Tibet).

The amount of fat found in cheese is determined by the type of milk from which it is made. Whole, reduced-fat (2 percent), or fat-free (nonfat) milk; buttermilk; cream; whey; milk solids-not-fat (MSNF); or any combination of these can serve as the basis for cheese making. Homogenized milk is usually selected for making soft cheese because homogenization makes the casein proteins coagulate more easily and the increased surface area of the broken-up fat results in a moister product. Homogenized

milk is not used to make hard cheeses, because the same trait that is desirable for softer cheeses creates a brittle texture during the long aging process (20).

Currently, the Food and Drug Administration (FDA) requires pasteurization for all fresh or soft-ripened cheeses, but allows the use of raw (unpasteurized) milk to make hard cheeses such as cheddar if they are aged at 35°F (4°C) or higher for at least 60 days (9, 33). Cheese aged for shorter periods than this are required to be pasteurized to reduce the growth of harmful bacteria such as *E. coli, Salmonella,* and *Listeria.* However, the U.S. National Academy of Science recommends that the law requiring pasteurization or a minimum 60-day aging should be replaced with standards that directly measure and limit the bacterial load allowable in cheese products (27).

## Coagulation

Cheese making starts with the coagulation of the casein protein in milk. The whey, which contains most of the water in milk, is removed as the curd forms. The protein, fat, and micronutrients from milk are hence concentrated within the curd. The type of coagulation method used determines many of the characteristics of the resulting cheese. The two main methods to aid coagulation utilize the action of enzymes and the action of acid (Figure 11-2).

**FIGURE 11-2**  Cheese is usually produced by coagulating (clotting) the milk with either an enzyme or acid.

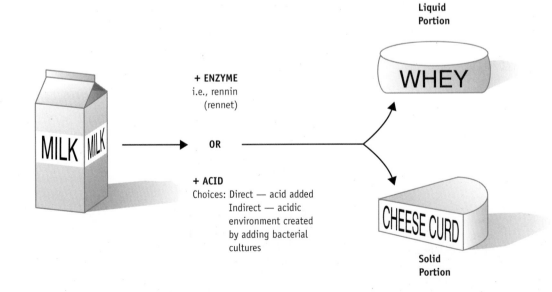

### Enzyme Coagulation

The enzyme most commonly used to coagulate milk in cheese making is rennin, obtained from milk-fed calves—specifically, their fourth stomach. Rennin, also called chymosin, is sold commercially as rennet. Other sources of available rennin include cows, pigs, plant sources, and genetically engineered bacteria (12) (Chemist's Corner 11-1). Alternatively, enzymes may be derived from bacterial starters (streptococci, lactobacilli), from certain molds, or from any of several other microorganisms. The different enzymes, bacteria, molds, and/or yeasts added during coagulation influence the flavor, texture, and color developed by the cheese during ripening (13, 39, 40).

The milk is usually heated in large vats at temperatures from 72°F–95°F (22°C–35°C) to provide an optimal environment for enzymes and bacterial activity, which contributes to the formation of curd (Chemistry Corner 11-2). Calcium chloride may be added to speed up coagulation and strengthen the curd's consistency. Coagulation with enzymes occurs in less than an hour, and creates a tough, rubbery curd. As the curd forms, the whey separates, but most of the milk's calcium remains in the curd.

### Acid Coagulation

There are two methods by which acid may be used to coagulate milk proteins and thus form cheese. The first method is simply to add acid directly to the milk. The second and more complex method is to inoculate the milk with a bacterial starter culture that acidifies the milk by converting lactose (milk sugar) to lactic acid (3).

Bacterial cultures have been used for centuries to produce fermented foods, and are carefully selected for their characteristic influences on a cheese's flavor and texture (4). It takes from 4 to 16 hours to coagulate milk with acid-forming bacteria. About one fourth to one half of the calcium in milk is lost in the whey during this process. The curd produced by acid has a soft and spongy texture. This texture is influenced by pH, becoming more solid and compact as the acidity increases (Figure 11-3). For instance, the pH of cream cheese determines its water-holding capacity (WHC) and therefore how much

> **CHEMIST'S CORNER 11-2**
>
> **Caseins and Coagulation**
>
> Milk contains phosphoproteins called caseins, of which there are four types (alpha-, beta-, kappa-, and gamma-caseins). These molecules form clusters with calcium and phosphate, forming small spherical structures called micelles. The addition of rennin to milk causes breakage of kappa-casein between two specific amino acid molecules. This breakage causes the spherical micelles to break apart and produce the coagulum that is used to make cheese (22).

**FIGURE 11-3** Cheese texture.

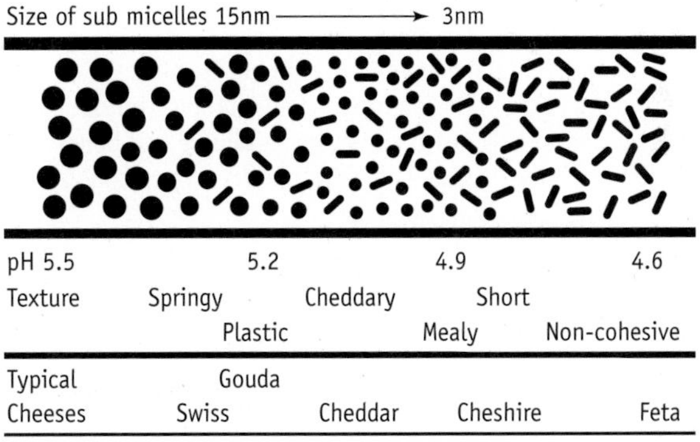
Size of sub micelles 15nm → 3nm
pH 5.5
5.2
4.9
4.6
Texture
Springy
Plastic
Cheddary
Mealy
Short
Non-cohesive
Typical Cheeses
Gouda
Swiss
Cheddar
Cheshire
Feta

> **CHEMIST'S CORNER 11-1**
>
> **Rennin's Role**
>
> Enzymes act on milk, causing it to clot or coagulate. The most common enzyme for this job is rennin, which coagulates milk by hydrolyzing casein. The rate of this reaction depends on the pH and ionic strength or salt/NaCl concentration of the medium. Commercial rennins in the United States include calf chymosin; bovine pepsin; porcine pepsin; microbial proteases (derived from bacteria such as *Cryophonectria parasitica, Mucor miehei* var. *lindt,* and *Mucor pusillus*); and chymosins generated from the fermentation of certain genetically engineered strains of bacteria (*Asperigillus niger* var. *Awamori, Escherichia coli,* and *Kluyveromyces lactis*) (19). All of these milk-clotting enzymes that cleave the polypeptide molecule are classified as aspartic proteases, because their active sites consist of aspartic acid residues.

> **NUTRIENT CONTENT**
>
> Cheese is a very nutrient-dense food. A pound of cheese is equivalent in protein and fat to approximately 1 gallon of milk. Cheeses with higher moisture levels are less concentrated in their nutrients than those containing less moisture. On average, 1 ounce of cheese (a 1¼-inch cube or a packaged slice of pasteurized process cheese) provides about 100 calories (kcal), most of which are derived from fat. Overall, the primary ingredients by weight are water, protein, and fat, and these nutrients vary in concentration according to the cheese.
>
> ***Fat.*** Fat is responsible for much of the satiety value, flavor, and texture of cheese. It is the fat content of milk, which is rather bland in taste, that is acted upon during ripening to result in a cheese with a very distinct flavor (17). Cheeses average 9 grams of fat per ounce (primarily saturated fat), which is equivalent to two pats of butter.
>
> *(continued)*

## NUTRIENT CONTENT

Cholesterol content averages 25 mg per ounce, but ranges from 0 mg in pot cheese to 40 mg per ounce in Gouda cheese. Dietary fat from cheese can be reduced by (1) using less cheese, or (2) using lower-fat forms.

*Lower-Fat Cheeses.*    The American Heart Association and the National Cholesterol Education Program define a low-fat cheese as one containing no more than 6 grams of fat per ounce (1). In addition, cholesterol should not exceed 20 mg, and sodium should be below 500 mg per serving (8). Only a few cheeses qualify as very low in fat, at less than 2 grams per ounce. These are Gammelost, a semi-soft blue cheese; sapsago, a pungently flavored, hard-textured, high-sodium cheese; and baker's, pot, or hoop cheese (28). The two most common fat-free (nonfat) milk cheeses available are ricotta and mozzarella. Replacing some of the fat in cheese with fat replacers can reduce the fat content to 3 grams per ounce. Another fat-reducing option is yogurt cheese, which contains only 1 gram of fat per ounce and is made by draining yogurt of its whey. Yogurt cheese is an excellent substitute for cheeses normally used to make cheesecake or other high-fat foods, because it lowers the fat content and takes on the other ingredients' flavors.

*Fat Influences Flavor.*    Fat is important to texture and flavor, so reducing the fat in cheese can result in a product with a rubbery texture and either no flavor or off-flavors (10). One reason that low-fat cheeses taste different is that their increased moisture content, resulting from decreased fat content, is more conducive to the growth of natural bacteria, which can produce compounds yielding flavors that are brothy, meaty, and bitter (23). Fat also masks certain off-flavors such as the bitter compounds produced from protein breakdown, which are detected when fat is reduced.

*Fat Influences Texture.*    Many reduced-fat cheeses behave differently when heated, which sometimes makes it difficult to substitute them in recipes calling for full-fat cheeses. Fat also contributes to the smooth, lubricated texture of full-fat cheese by breaking up the protein matrix. Removing this fat leaves more protein and increases its influence on texture (23).

*Protein.*    Cheese protein is complete and of high quality because it is of animal origin. One ounce of cheese contains approximately 7 grams of protein, equivalent to that in 1 cup (8 ounces) of milk or 1 ounce of meat. Cheese was being used as a meat substitute by monks in monasteries long before protein was recognized as an essential nutrient by food and nutrition scientists.

*Carbohydrate.*    Cheese contains very little carbohydrate because most of the lactose, the primary carbohydrate in milk, is drained off with the whey, and any remaining lactose is converted to lactic acid. Because the lactose is largely gone, some people with lactase deficiency (lactose intolerance) can consume cheese with little to no ill effects. As a result, ripened cheeses such as Swiss and cheddar are often suitable for lactase-deficient individuals (35), but process (processed) cheeses are usually not, because milk or whey is frequently added to them.

*Vitamins.*    Many of milk's vitamins are lost in cheese production. When whole milk is used to make cheese, the fat-soluble vitamins A and D remain in the curd, but many of milk's water-soluble vitamins are drained off in the whey. Only about one fourth of the riboflavin (vitamin $B_2$) and one sixth of the thiamin (vitamin $B_1$) remain in cheddar cheese, and much of the milk's original niacin, vitamin $B_6$, vitamin $B_{12}$, biotin, pantothenic acid, and folate drain off with the whey (30).

*Minerals.*    Cheese has a high concentration of minerals, specifically calcium, phosphorus, and zinc. There are approximately 200 mg of calcium in 1 ounce of cheese, compared to 300 mg in 1 cup of milk. Calcium content will vary depending on whether the cheese was coagulated with enzymes or acid: enzyme-coagulated cottage cheese contains twice the calcium as does acid-coagulated cheese. Cheese often has a high concentration of sodium, because salt is frequently added during manufacturing to aid in the process of ripening and to contribute to flavor. The amount of salt added to cheese varies widely, and low-sodium cheeses are available. Process cheese products tend to be especially high in sodium, averaging 400 mg per ounce.

moisture it retains (15). Acid-coagulated cheese is usually not aged because its high acidity inhibits the bacterial and mold growth that characterizes the aging process.

## Curd Treatment

The curd may be treated to remove more whey by cutting, heating, and salting. Further optional treatment includes knitting and/or pressing. Although a few chemical tests can be used to assess the progress of the curd through each of these treatments, it is often the experienced judgment of a cheese maker that determines when it is time for the next step (19).

- *Cutting.* Slicing the curd increases its surface area (Figure 11-4). Sometimes the curd is placed on strainers to remove even more whey.
- *Heating.* This encourages the evaporation of whey, which allows lactic acid to build up to create a firmer, more elastic texture. Heat also destroys certain undesirable microorganisms. After heating, the curds are washed with cold water to produce softer, higher-moisture cheeses.
- *Salting.* Salt controls the growth of bacteria and further dehydrates the curd. It also contributes to the flavor, texture, and appearance of cheese. The two major types of salting are dry salting and brine salting. Two factors influencing salting are pH and temperature—more salt is absorbed at a lower pH and higher temperature. Salted cheeses have a final salt concentration averaging 0.5–2 percent.
- *Knitting.* Some cheeses are *knitted*; that is, the curd is united or melted into a solid mass through the use of heat.
- *Pressing.* Pressing is another way to create a solid mass out of the curd, and the last step before ripening. Curds are physically pressed into compact masses by placing them in boxes or other containers under pressure.

## Curing and Ripening

Cheese becomes stronger in flavor as it ages. The aging process whereby cheese is converted from a bland, tough, rubbery, fresh curd into a unique cheese with its own mature flavor, aroma, and texture is called, often interchangeably,

# CALORIE CONTROL
## Choosing Cheeses

Being mindful of your cheese consumption is important if you're controlling calories because, in addition to some important nutrients, it delivers calories to the tune of:

1 ounce of cheese (size of whole thumb) = 100 calories (kcal) and 9 grams of fat

**How much is 1 ounce?** Not much, because 2 tablespoons equal 1 ounce, and it takes 3 teaspoons to equal 1 tablespoon. The popular slices of American cheese weigh slightly under an ounce. Caution is recommended, not only for just plain cheese, but also for all the foods made with cheese (pizza, calzones, cheese casseroles, etc.) and cheese dips.

**Fat-Free Cheeses:** Removing the fat from cheese reduces its calories by at least half (Table 11-2). Some of the lower-fat cheeses such as ricotta and cottage cheese are also lower in calories.

**Limit Portions:** When dishing up foods that include cheese, remember that a half cup of shredded cheese provides 228 calories (kcal). Doubling or tripling that amount does the same to the calories/kcal—456 and 684, respectively. One cheese stick is only 80 calories (kcal); however, 4–5 cheese sticks (round to 100 for easier counting) add up to 400–500 calories (kcal). Cheese can be a nutritious, delicious food, but it's best to limit portions as suggested by MyPyramid: 1½ ounces of fat-free natural cheese counts as one of the three daily milk servings recommended for adults. The Exchange Lists recognize 1 ounce of regular cheese or ¼ cup of ricotta or cottage cheese as a meat exchange.

**Suggested 100-Calorie Snacks:**

- One plain cheese stick (80 calories/kcal)
- Cheese and apple or crackers: An ounce of cheese sliced up and placed on apple slices (¼ medium [3"] apple = 25 calories/kcal) or 3 wheat crackers (10 calories/kcal each)

### TABLE 11-2   How Many Calories in Cheese?

**Cheese**

| 1-Ounce Servings | Serving[1] | Calories (kcal) | Fat (g) |
|---|---|---|---|
| • American, slice | Slice | 100 | 9 |
| • American, cheese spread | 1 oz | 82 | 6 |
| • Cheddar, regular | 1 oz | 114 | 9 |
| • Cheddar, fat-free | 1 oz | 40 | 0 |
| • Parmesan, grated | 1 oz | 122 | 8 |
| • Ricotta, whole | 1 oz | 50 | 4 |
| • Ricotta, fat free | 1 oz | 20 | 0 |
| • Swiss | 1 oz | 92 | 7 |

| ½-Cup Servings | Serving[1] | Calories | Fat |
|---|---|---|---|
| • Cheddar, shredded | ½ C | 228 | 14 |
| • Cheese stick | 1 stick | 80 | 4 |
| • Cottage cheese, 4% | ½ C | 120 | 5 |
| • Cottage cheese, 1% | ½ C[2] | 40 | 1 |
| • Ricotta, whole | ½ C[2] | 180 | 12 |
| • Ricotta, fat-free | ½ C[2] | 90 | 0 |

[1]Servings: oz = ounce; C = cup
[2]Serving Sizes = ¼ cup in the Exchange Lists

curing or ripening (29). Depending on the variety, cheeses are subjected to different temperatures (36°F–75°F/ 2°C–24°C) and humidities (higher for mold-ripened cheeses such as Roquefort and blue). Certain cheeses are treated in such a way as to develop a rind, which is simply the dried surface of the cheese. Ripening times range from 4 weeks to 2 years or longer.

Cheese flavor originates from a combination of over 300 different volatile and nonvolatile compounds that develop during curing and ripening (12). It is believed that some of these compounds originate from the milk, the activity of milk enzymes, and the starter bacteria. The skillful adjustment of curing techniques, along with the maintenance of the proper environment of temperature and humidity, creates the desired flavors, textures, and aromas of the multitude of cheeses available in our markets.

During ripening, several elements may be manipulated to affect the final product (Chemist's Corner 11-3).

## CHEMIST'S CORNER 11-3

### Cheese Ripening

There are three main steps in the ripening of cheese. The first is glycolysis, in which lactose is metabolized to form acetic and propionic acid, carbon dioxide, esters, and alcohols; this metabolism is initiated by the bacterial starter and other microbials present in milk (22). Next is lipolysis, the breakdown of fat molecules to free fatty acids, which are then further broken down to ketones, lactones, and esters by enzymes found in milk or added during manufacturing. The final step is proteolysis, the breakdown of proteins called caseins into peptides (small protein fragments) and amino acids. This is performed by enzymes found in the milk and by microorganisms from the starter and found naturally in milk.

**Curing** To expose cheese to controlled temperature and humidity during aging.

**Ripening** The chemical and physical changes that occur during the curing period.

**FIGURE 11-4**    Cutting cheese curd.

Courtesy of Karoun Dairies, Inc.

Added salt will draw out some of the remaining whey and inhibit bacterial growth, thereby slowing the ripening process. Bacteria and molds contribute to the development of flavors, aromas, and textures (Chemist's Corners 11-4 and 11-5). The choice of bacteria or mold, and whether they are internally or externally introduced during the cheese making process, will determine what kind of cheese is produced. Cheeses can be classified by the type of ripening (Figure 11-5).

The mold *Penicillium roqueforti* added to homogenized whole milk converts free fatty acids to smaller compounds, which impart the characteristic tangy flavor to blue cheese. The holes (eyes) in Swiss cheese are produced by gas-forming microorganisms that are active during the early part of ripening when the curd is pliable (Figure 11-6).

> ## CHEMIST'S CORNER 11-4
>
> **Bacteria and Cheese Flavor**
>
> Bacteria are responsible for some of the flavors found in cheese. Certain strains of bacteria produce lactic acid, which affects the degree of sharpness or mildness of cheese (26). Lactic acid bacteria are believed to impart flavor to cheese by reducing the oxidation–reduction potential, by protein hydrolysis, and by producing dicarbonyls. Certain dicarbonyls such as diacetyl, methylglyoxal, and glyoxal are thought to interact with amino acids via the Maillard reaction to produce flavor compounds (38). These same bacteria are present in yogurt, producing some of its flavor. Some of these "friendly" bacteria naturally produce other compounds that inhibit the growth of harmful bacteria. For instance, some strains produce a compound called bacteriocin that prevents the growth of the potentially infectious bacteria *Listeria* (26).

> ## CHEMIST'S CORNER 11-5
>
> **Enzymes and Cheese Flavor/Texture**
>
> One of the major biochemical processes responsible for cheese flavor and texture development during ripening is the breakdown of proteins via proteolysis (7). The breakdown of the casein structure into smaller peptides also reduces cheese rigidity and increases its flexibility (21). Proteolytic enzymes in cheese are derived from the milk, starter bacteria, coagulant, and other microorganisms (14). The enzymes break down casein to amino acids, which can then be further broken down to various substances, giving each cheese its unique flavor (36) (Figure 11-7). Proteolysis produces beneficial flavors in cheese, but also some bitter compounds (11). These various factors result in the wide variety of cheeses with different flavors.

Other processes that influence flavor during ripening include the hydrolysis of proteins to peptides (smaller protein molecules) and amino acids, the conversion of lactose to lactic acid, and the

**FIGURE 11-5**    Cheeses classified by their ripening methods.

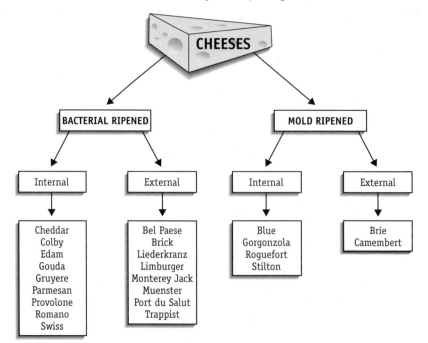

**FIGURE 11-6** The eyes in Swiss cheese are generated by carbon dioxide produced by the fermentation of lactate (from lactose) to proprionate and acetate.

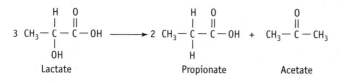

## Whey and Whey Products

Whey separated from its water content is low in fat and rich in nutrients. It contains the water-soluble whey proteins and most of the lactose, water-soluble vitamins, and minerals of the milk. Whey is highly perishable when fresh, so it is most often processed quickly into whey cheeses, dry whey, and modified whey products. In the past, the nutrient-rich whey was principally used to feed animals, but now the food industry is using pasteurized whey in a variety of products. Whey is found in baked goods, beverages, sauces, salad dressings, cheese products, canned fruits and vegetables, confections (caramels and other candies), dry mixes, frozen foods (fruits, vegetables, desserts), jams, jellies, meat, pasta, and milk products.

Two types of whey are incorporated into foods: sweet and acidic whey (5). Sweet whey results from coagulating milk with rennin, whereas acid whey originates from acid-coagulated milk. Besides having greater acidity, acid whey also contains higher mineral

breakdown of fatty acids into shorter, volatile fatty acids (14). Feta's unique flavor is partially derived from lipase, an added enzyme that breaks down milk fat (16). This enzyme, sold in powdered form, will deliver different flavors based on its source: calf lipases yield mild flavors while lamb lipases contribute sharper flavors.

Cheeses may exhibit different textures because of processing techniques during production, two of which are inoculation and kneading. The blue-veined cheeses have been inoculated with mold spores, whose growth within the cheese creates the blue veins. Mozzarella and provolone are ropy in texture because of kneading, which is the pulling and stretching of the curd after it has been knitted.

> **? How & Why?**
>
> **Why are cheddar cheeses labeled mild, medium, or sharp?**
>
> Like wine, the finished character of these cheeses is determined not only by the original ingredients, but by the maturation process. The flavor of cheddar cheese can range from mild to sharp depending on the duration of aging. Although the time cheddar is held for aging varies by the cheese company, the general guideline is that mild cheddar is aged for at least 60 days, medium for 3 to 6 months, sharp for a minimum of 9 months, and extra sharp for at least 15 months.

**FIGURE 11-7** How some cheese flavors develop.

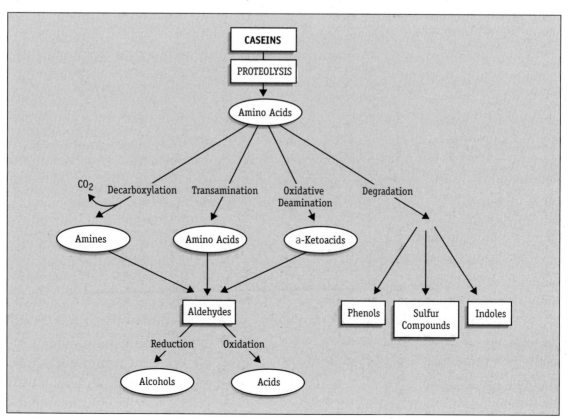

**FIGURE 11-8**   The manufacture of processed (process) cheese.

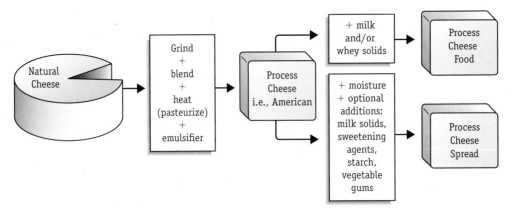

concentrations, because the acid releases the calcium from the casein molecule, causing it to be dispersed in the whey portion.

### Whey Cheeses

Scandinavian cheeses such as primost, mysost, and gjetost are very hard whey cheeses. They are made by evaporating the water until the whey is extremely concentrated. These cheeses tend to be sweet and lightly brown, a result of their lactose caramelizing in the process of water removal. Another type of whey cheese, ricotta, is produced by coagulating the whey with acid and high heat.

### Dry Whey

A large proportion of the dry whey produced in the United States is fed to livestock. The remainder is used as an ingredient in processed foods such as confections, soups, beverages, imitation cheeses, dessert toppings, nondairy coffee creamers, and sport supplements.

### Modified Whey Products

Condensed whey contains no more than 10 percent of its original water and is often used as an ingredient in process cheese food. Another product, sweet-type condensed whey, is used as

an ingredient for certain candies. Whey protein concentrate is whey concentrated to a minimum of 25 percent protein. Removing some lactose results in a partially delactosed whey. Also available is partially and/or totally demineralized whey, which has had some or all of its minerals removed.

## Process (Processed) Cheeses

**Process cheese** was patented in 1916 by James L. Kraft, who founded Kraft Foods (31). Process cheeses are all made from blended cheeses, but they differ based on their ingredients and manufacturing methods (Figure 11-8). Approximately one third of the cheese produced in the United States is used for the various kinds of pasteurized process cheeses. These include process cheese, cold-pack cheese, process cheese food, process cheese spread, and imitation cheese (30). These cheeses appeal to many consumers because of their uniform taste and creamy, meltable texture, longer shelf life, convenient packaging, and lower cost.

### Process Cheese

Process cheese is made by combining different varieties of natural cheese. Heating pasteurizes and stops further ripening of the cheese. Other ingredients added to produce a stable, homogeneous emulsion include emulsifying salts such as sodium citrate or sodium phosphate. During the emulsifying process, powdered milk, whey, cream or butter, and water may also be added.

The moisture content of the process cheese must not exceed 40 percent, and the fat content is similar to that of the natural cheese from which it is derived. When the mixture is partially cooled, it is formed into blocks; cut into slices, wedges, or other shapes; and packaged. American cheese made from blended cheddar cheese is a popular form of process cheese.

### Cold-Pack Cheese

Cold-pack or club cheeses were developed in 1918 by Hubert Fassbender, who found he could blend cheddar cheese, milk by-products, and spices to create a spreadable cheese for crackers and sandwiches. Because cold-pack cheeses are blended without being heated, they must be made from pasteurized milk products. In cold-pack cheese food, the original cheese may be combined with milk [whole, reduced-fat (2 percent), fat-free (nonfat), or buttermilk], MSNF, cream, or whey. It also may be sold in smoked form.

### Process Cheese Food

Process cheese food must be at least 51 percent natural cheese by weight, which is less than either process cheese or cold-pack cheese. The remaining ingredients may include milk, cream, oil, or whey (31). Process cheese food products, including Cheez Whiz™, Velveeta™, and Kraft Singles™, have a milder flavor and softer texture, and tend to melt more quickly.

### Process Cheese Spread

Process cheese spread is a softer, more spreadable product than process

**Process (processed) cheese**  A cheese made from blending one or more varieties of cheese, with or without heat, and mixing the result with other ingredients.

cheese food, but, like cheese food, must contain at least 51 percent natural cheese by weight (2). Ingredients such as sugar, dextrose, maltose, and corn syrup may also be added. The higher spreadability of process cheese spread, obtained by adding more liquid and an emulsifier and reducing the amount of milk fat, makes it ideal for use in sandwiches and on crackers.

## Imitation Cheese

Cheese analogues or imitation cheeses are cheese-like products in which the milk fat in natural cheese has been replaced with vegetable oil. These analogues are less expensive than natural cheese and are manufactured using a process similar to that used to make process cheese. Milk proteins such as calcium caseinate are mixed with a small amount of vegetable fat, water, salt, emulsifiers, and lactic acid before being heated to pasteurization temperatures for several minutes. The liquid is then poured into molds or formed into slices. The texture, flavor, and melting properties of imitation cheeses are similar to those of process cheese. Nutritionally, these analogues are lower in cholesterol and sodium, but equivalent in fat, although it is less saturated than that from natural cheese (32).

## Tofu and Other Non-Dairy Cheeses

The cheese made from soy milk is discussed in Chapter 13 under the section "Legumes" (25). Other types of non-dairy cheese include those made from rice or almond milk. However, some non-dairy cheeses do contain some milk and/or cream.

# Food Additives in Cheese

Cheese is a preserved food made from milk. As a result, several food additives are used in the manufacture of cheeses, especially process cheese. The additives in cheese will differ based on whether the cheese is unprocessed or process. Most unprocessed cheeses contain relatively few additives, including salt, enzymes, and annatto. The list of additives in process cheeses is much longer. (See Table 11-3 for a list of additives in both process and regular cheese.)

| TABLE 11-3 Selected Food Additives in Cheeses |
|---|

**Unprocessed Cheese—Typical Ingredients**

Milk, salt, enzymes, annatto, natamycin (other additives appearing less frequently are gums, powdered cellulose, potato starch, corn starch, calcium sulfate, or potassium sorbate).

**Process (Processed) Cheese—Typical Ingredients**

Milk, water, whey, milk fat, milk protein concentrate, whey protein concentrate, sodium phosphate, calcium phosphate, lactic acid, sorbic acid, sodium alginate, enzymes, apocarotenal, annatto, cheese culture (process cheese food or spreads that come packed in a jar or can often contain canola oil as one of the major ingredients and sugar as a sweetener).

| Food Additive in Cheese | Function |
|---|---|
| Annatto | A natural color often added to cheddar cheeses. |
| Apocarotenal | Natural orange carrot-like color. |
| Calcium phosphate | To improve texture, consistency, and spreadability of process (processed) cheese products. |
| Calcium sulfate | Acts as a desiccant (substance that absorbs water) to prevent caking in shredded cheese. Avoid if sensitive to sulfites. |
| Cheese culture | Bacteria added to produce cheese. |
| Enzymes | Natural enzymes used to produce cheese. |
| Gums (carageenan, guar gum, locust bean gum, and xanthan gum) | Thickening stabilizers derived from plants that provide a cohesive, rubbery texture. |
| Lactic acid (option—acetic acid, which is vinegar) | Used for pH control, flavor, and as a preservative because many bacteria are inhibited by low acidity. |
| Milk fat | Natural ingredient adding flavor and smooth mouthfeel. |
| Milk protein concentrate | Natural ingredient adding protein. |
| Natamycin | Mold inhibitor. |
| Potassium sorbate | Preservative that inhibits molds. |
| Sodium alginate | Extracted from seaweed; used to stabilize and thicken. |
| Sodium phosphate | Emulsifying, gelling, stabilizing, thickening. |
| Sorbic acid | Preservative that inhibits fungi. |
| Starch—potato starch, corn starch, cellulose | Plant products that prevent caking of shredded cheese. |
| Whey | Natural ingredient adding protein and lactose. |
| Whey protein concentrate | Natural ingredient adding protein. |

# PURCHASING CHEESE

## Grading

Not all cheeses are graded according to the United States Department of Agriculture (USDA)–defined U.S. Grades AA, A, B, and C (Figure 11-9), but those that are graded are evaluated based on their variety, flavor, texture, finish, color, and appearance. Exceptions to these criteria are Colby cheese, in which color is not considered, and

Swiss cheese, which is also graded for its salt level and eyes (holes) (6). Table 11-4 lists the federal standards for the maximum moisture and minimum milk fat content of various cheeses. U.S. grades have not been established for process cheese products. A Quality Approved inspection shield on the label means only that the cheese meets minimum quality standards and has been produced in a plant meeting USDA sanitary standards. An imitation cheese is defined as one that looks and tastes like the one it is intended to replace but is nutritionally inferior. Substitute

**FIGURE 11-9**   USDA grades of cheese.

**CHEDDAR CHEESE**

**U.S. Grade AA**
Fine highly pleasing cheddar flavor; smooth, compact texture; uniform color.

**U.S. Grade A**
Pleasing flavor; smooth compact texture but may have a few openings or curdy appearance; uniform color with possible white lines.

**U.S. Grade B**
Fairly pleasing flavor, but may contain certain undesirable flavors to a limited degree; texture may have several defects (mealy, coarse, crumbly, corky, etc.); color may possess several undesirable characteristics (dull, faded, salt spots, molded, etc.).

**U.S. Grade C**
May possess somewhat objectionable flavors and colors; texture may be loose; and color may be slightly bleached on the surface.

cheeses resemble the traditional product and meet the nutritional equivalency comparisons (30).

## Forms of Cheese

Cheese is sold in a variety of forms. The most common way to purchase cheese is in blocks that vary in shape, usually as rectangles, triangles, and circles. Food service establishments buy cheese in large 10+ pound blocks that may or may not be frozen depending on the type of cheese. Pre-shredded or sliced cheese makes preparation easier. Individually wrapped pre-portioned cheese sticks provide a high-protein/fat snack under 100 calories (kcal). This is one of the newest food products marketed under the "100-Calorie Cut-Off" trend (37), in which the food industry, to better serve calorie-conscious consumers, is packaging their products to yield individual 100-calorie (kcal) serving units. Process cheese, sold in jars or canisters, is also available, and it can be stored on the shelf before being opened. Less frequently used by consumers are the cheese powders used widely by the food industry for seasonings and flavorings. Finally, cheese can be melted into a sauce that can be used as a dip or poured over foods such as vegetables, baked potatoes, or nachos.

# FOOD PREPARATION WITH CHEESE

Cheese is most often used as an ingredient to add flavor, color, and texture in a variety of ways—on pizza; as a taco topping; and in cheese soufflés, sandwiches, casseroles, quiches, and sauces. Mozzarella, Parmesan, and ricotta cheeses are often found in Italian dishes; feta, in Greek dishes; Monterey Jack, in Tex-Mex dishes; and cheddar, in North American food dishes. Numerous dishes incorporate various cheeses; the most popular cheeses purchased in the United States for these purposes, in descending order, are cheddar, process, mozzarella, cream cheese, ricotta, Swiss, provolone, Muenster, Parmesan, Neufchâtel, and blue.

The two most important rules to follow when preparing foods with cheese are: (1) select the best cheese and (2) keep temperatures low and heating times short.

## Selecting a Cheese

The chemical composition of a cheese determines its functional properties and dictates how it will be used in food preparation. Some of these functional properties include shredability, meltability, oiling off, blistering, browning, and stretchability (6).

### Shredability

Not all cheeses shred uniformly, which may be important to food service operations interested in cost control. On average, 4 ounces of shredded cheese are equivalent to approximately 1 cup in volume. It may be possible to use less when using aged cheeses because of their strong flavor. Grating or chopping cheese increases the surface area and thereby increases the ease and speed with which it melts.

### Meltability

The higher the fat and moisture contents of a cheese are, the greater is its meltability (6). Aged or ripened cheeses such as cheddar and Swiss tend to melt and blend more easily when heated than the less-ripened cheeses. Process cheeses, which contain added water and emulsifiers, can be heated without the fat separating, blend more smoothly, and melt more easily than do unprocessed cheeses. Problems may result when using lower-fat cheeses because they separate more easily when exposed to high heat and their higher protein content makes them toughen as they are heated. For these reasons, they are not always good candidates for use in cooking.

**TABLE 11-4**  Federal Standards for Maximum Moisture and Minimum Milk Fat in Cheese and Cheese Products

| Cheese | Maximum Moisture % | Minimum Milk Fat % |
|---|---|---|
| *Soft, unripened* | | |
| Cottage (dry curd) | 80 | — |
| Cottage | 80 | 4 |
| Cottage (low-fat) | 82.5 | 0.5 |
| Cream | 55 | 33 |
| Neufchâtel | 65 | 20–33 |
| *Soft, ripened* | | |
| Brie | — | 27.7 ± 1.8 |
| Camembert | — | 24.3 ± 0.6 |
| Limburger | 50 | 25 |
| *Semisoft, ripened* | | |
| Asiago (fresh) | 45 | 27.5 |
| Blue | 46 | 27 |
| Brick | 44 | 28 |
| Gorgonzola | 42 | 29 |
| Monterey | 44 | 28 |
| Muenster | 46 | 27 |
| Roquefort | 45 | 27.5 |
| *Hard, ripened* | | |
| Cheddar | 39 | 30.5 |
| Colby | 40 | 30 |
| Edam | 45 | 22 |
| Gouda | 45 | 25.3 |
| Gruyere | 39 | 27.5 |
| Samsoe | 41 | 26.5 |
| Swiss | 41 | 25.4 |
| *Very hard, ripened* | | |
| Asiago (old) | 32 | 28.6 |
| Parmesan | 32 | 21.8 |
| Romano | 34 | 25.1 |
| *Pasta filata (plastic)* | | |
| Caciocavallo siciliano | 40 | 25.2 |
| Mozzarella | 52–60 | 18 |
| (low moisture, part skim) | 45–52 | 14 |
| Provolone | 45 | 24.8 |
| *Skim milk or low-fat* | | |
| Gammelost | 52 | — |
| Sapsago | 38 | — |
| *Pasteurized process* | | |
| Cheese | 43 | 27 |
| Pimento cheese | 41 | 29 |
| Cheese food | 44 | 23 |
| Cheese spread | 44–60 | 20 |
| *Cold-pack* | | |
| Cold-pack cheese | 42 | 27 |
| Cold-pack cheese food | 44 | 13 |

### Oiling Off

One drawback to using higher-fat cheeses in cooking is their greater tendency to *oil off*, which occurs when some free fat is released and glistens on the surface.

### Blistering

Heating cheese can sometimes cause blistering, or the formation of bubbles that may brown or pop. The number and size of the blisters that develop will depend on the cheese's age: large blisters tend to form when using excessively aged cheese, whereas numerous small blisters may be a sign that the cheese has not been aged very long.

### Browning

The browning of cheeses during heating, a result of the Maillard reaction, is desirable, but only up to a certain point. Too much browning occurs if there is an excess of sugars, amino acids, or lactose in the cheese.

### Stretchability

The functionality of a cheese also differs based on how well it stretches. The stretchability of a cheese depends on its concentration of calcium phosphate and its protein network structure (29). A tough, grainy texture results from the presence of too much calcium, whereas texture turns excessively soft when undergoing too much protein breakdown during aging (6).

> **? How & Why?**
>
> *Why is mozzarella cheese used on pizza?*
>
> Mozzarella cheese does not tend to oil off during cooking. As a result, mozzarella is the optimal cheese for pizza production because many people find a shiny sheen of oil on pizza unappetizing.

## Temperatures

Cooking temperatures for cheeses should be kept low, and heating times, short. High heat or prolonged cooking toughens cheese proteins and causes the fat to separate out, creating an oily, stringy, and inferior product. When using a microwave, it is best to use lower power settings—between 30 and 70 percent—for melting cheese. Cheeses used in sauces and other dishes should be added during the last stages of preparation to prevent separation. Adding a pinch of dry mustard to cheese sauces helps to bring out their flavor. One way to soften cream cheese without too much heat is to enclose it in an airtight zip-top plastic bag and briefly submerge it in hot water.

**FIGURE 11-10**   Cheese knifes shaped for function.

Andrew Twort/Alamy

Temperature is also important when serving cheese. Most cheeses (semi-hard and hard) reach their full flavor when taken out of the refrigerator and allowed to reach room temperature before serving. To prevent cheese from drying out, it is best either to cut cheese that has reached room temperature just before serving or to let people cut their own. Cream cheese, cottage cheese, and other unripened cheeses should, however, always be served chilled.

## Cutting Cheese

Cheese knifes come in many different shapes, each designed to be used with specific types of cheese (Figure 11-10) (42). A skeleton knife has large holes cut out of the middle of the blade, so that it can be used to cut soft cheeses without sticking to them. A steel string accomplishes the same purpose. Very hard cheeses are best cut with a strong triangular knife equipped with a short blade. The all-purpose cheese knife has a forked end that can be used to transfer individual cheese pieces to serving plates.

# STORAGE OF CHEESE

Cheese must be stored properly to prevent deterioration. Most cheeses should be refrigerated; some can even be frozen. Process cheese products can be stored in a cool, preferably dark, cupboard until ready for use, although refrigeration more effectively retains desirable qualities.

## Dry Storage

Many process cheese spreads, as well as dry, grated Parmesan cheese sold in cardboard containers, have long shelf storage times. Process cheese spreads may be safely stored on the shelf in jars at room temperature (70°F/21°C) for up to 4 months, but the quality is better retained if the product is refrigerated. Packaged dry, grated Parmesan cheese, one of the hardest cheeses, and other very hard cheeses can be kept at room temperature for up to 1 year. The exception is fresh Parmesan cheese, which has such high moisture content that it is sold in the refrigerator section.

## Refrigeration

Cheese should be stored in the refrigerator drawer, because this is where humidity and temperature are the highest. Most cheeses are best refrigerated in their original wrappers. Once opened, the cheese should be rewrapped as tightly as possible in its original wrapping or in aluminum foil, plastic wrap, or a sealable bag. This will prevent drying and absorption of odors from other foods. Because their odor may be picked up by other foods, strong-smelling cheeses should be double-wrapped. Firm cheese and blue cheeses are best stored wrapped in wax paper followed by plastic wrap, allowing the cheese to "breathe" and preventing contact with the plastic. Soft cheeses should be stored unwrapped in a container with holes poked in the lid. When eating cheese, take it out of the refrigerator at least 1 hour before serving so that it can soften and allow the full flavor to be experienced (42).

Properly wrapped cheeses are protected from the development of molds and their possible mycotoxins (24). Commercial efforts to reduce molds include coating the cheeses with a wax or resinous material and wrapping them in packaging film (39). In Europe, the cheese rind is sometimes coated with olive oil to protect it from bacterial contamination, and certain spices applied to the surface of some cheeses for flavor have been reported to have an antifungal effect (41).

Maximum storage time varies, because no two cheeses are alike. Ripened cheeses can be stored longer than unripened, softer cheeses. Fresh cheeses such as ricotta, cream cheese, and cottage cheese should be used within a week of their sell-by date. Process cheeses can be stored up to 4 months because they have been heat treated and contain mold inhibitors. Once opened, process cheeses should be refrigerated. Unopened Parmesan cheese stores the longest and can be kept up to a year, although the cheese may lose some of its flavor and further dry out. After being opened, Parmesan must be refrigerated. Dried-out cheeses can be salvaged by grating and storing them in the freezer for later use as toppings or in casseroles, sauces, or soups.

### ? How & Why?

*Why should moldy cheeses be discarded?*

Cheese that is kept too long can become moldy and/or dry. Most molds that develop on cheese are harmless, but because there are some molds that may produce toxins, all mold should be removed. The FDA Model Codes for Food Service recommend cutting 1 inch beyond the moldy area. Although mold-ripened cheeses such as Roquefort and blue have had special molds purposely added to them, they may develop a different-looking mold on the outside edges, and this should be removed. When soft, unripened cheeses develop molds, they should be discarded because of possible permeation throughout the cheese.

## Frozen

The water content of a cheese determines whether it can be successfully frozen. Most hard natural cheeses with a low water content can be frozen for up to 2 months and process cheeses for up to 5 months, but freezing is not recommended for soft cheeses having a high water content (30). The most suitable cheeses for freezing include brick, cheddar, Edam, Gouda, Gruyere, Parmesan, provolone, and Swiss. Freezing will change the texture and flavor to some degree, but the cheese should still be acceptable in quality (31).

# PROFESSIONAL PROFILE

**Judy Thompson MPH, RD— Registered Dietitian**

Courtesy of Judy Thompson

"When I was a sophomore at California State University at Long Beach," says Judy Thompson, "I didn't know what I wanted to do. I was getting the prerequisites out of the way when I enrolled in a health careers class and learned about registered dietitians [RDs]. I became interested in nutrition because nutrition was a very hands-on health strategy." As part of the course requirement, Judy shadowed a dietitian at the local veterans' hospital and found the work exciting.

Judy had already worked at a number of part-time jobs while in college; now she gained dietary experience by working as a hospital dietary aide (an entry position in a dietary department; see Chapter 30 on careers). She learned that variety in undergraduate work experience was as important as her dietitian courses: the path to becoming an RD involves life experience in addition to classroom and clinical experience. Dietary work experience, of course, increases a student's chances of obtaining a dietetic internship. (An internship in this areas is a 6- to 12-month program; however, the commitment is 2 years if combined with a master's degree.) These internships are available after graduation from an American Dietetic Association (ADA) approved program (some programs combine the degree with the internship).

After graduating from college, Judy traveled for a year, but went back to school after missing the classes and realizing that to become an RD, she would have to pursue her studies further. She applied for a master's degree program that incorporated the equivalent of an internship experience. Meanwhile, she was a dietary technician at a large hospital (500+ beds) where she was worked the tray line, collected menus, assisted with some of the supervisory food service needs, and interacted with patients.

Judy finished her master's degree in public health and then passed the Registered Dietetic Registration Examination to become a registered dietitian. Many dietitians (about 60 percent) obtain their master's degree in part because of increased job opportunities. Judy says she feels it is just as important to have one of the advanced professional certifications offered by the American Dietetic Association or American Diabetes Association (see Chapter 30). "I'm really, really glad that I have my Certified Diabetes Educator (CDE) qualification," she says, "because it expanded my role as a dietitian and also made me more employable."

Judy's employment track after becoming an RD started with a job in the Women, Infants, and Children (WIC) Program—a federally funded program to provide food vouchers and health and nutrition education for those at risk for certain health problems. She says she liked this job because it offered so much support to the practitioner (materials, resources, training) but felt it became a bit repetitive (similar clientele, similar nutrition issues). Next, she worked for a large medical clinic where doctors referred their patients to an RD. Collaborating with the patients' doctors regarding their health care was rewarding. Then, she moved on to work for a large hospital (500 beds) in the outpatient area. "I enjoy working with outpatients [those who have left the hospital but return for related treatment] because you are with them longer to help them problem-solve their lifestyle changes," Judy says. Inpatients are sometimes not ready for nutrition education until after they recover, and the hospital is not the best place to work with someone on making long-term lifestyle changes. However, she says she also enjoys inpatient work because it is "fast-paced and a good place to learn."

What drew Judy to work in a hospital? "It was so big that I was able to try different areas of focus such as cardiac, oncology [cancer ward], maternity, and diabetes. This variety made it interesting." Currently she is a team member in the hospital's outpatient diabetes education program where she not only teaches diabetic group classes and practices one-on-one patient counseling but also counsels patients on pre-diabetes, food allergies, weight gain, cancer risk reduction, and cholesterol or blood pressure reduction. "It's primarily patients with diabetes," she says, "because that's what's mostly covered at this time by insurance. Lack of reimbursement is a problem that the ADA continues to work on as an organization."

What does she wish she had known as a student? Judy says, "I wish I had been involved with a student dietetic association. This would have helped to give me more contact with and examples of real-life dietitians—to discover what they really did in the 'real world' out there."

Does she have any advice to give to students? "Talk to people who are in the field and find out what they do," she says. "There are a lot of different types of dietitians [hospital, community, long-term care, corporate, certified trainers; see Chapter 30 on careers]. Get involved in your dietetic association before and after graduation. Joining will show you how the ADA works at the national level. It will also give you a lot of peer support outside your job, and you'll learn how similar jobs are done in other places."

For best results, cheese should be frozen quickly, and this is best accomplished if it is in ½-pound pieces not more than 1 inch thick. Larger chunks will freeze more slowly, possibly resulting in crumbly cheese. It is best to freeze cheese in its original wrapper, but the next-best option is to use foil or plastic wrap designated for freezing. It should be wrapped tightly, with excess air being expelled, or else it can dry out.

Just as it is important to freeze cheese quickly, it is crucial to thaw it gradually. Gradual thawing is best achieved in the refrigerator over a period of a few days, after which the cheese should be used as soon as possible. Freezing certain cheeses may cause them to develop a dry and crumbly texture, but they may still serve as a shredded or cubed ingredient in a dish even if they would be undesirable in a sandwich or on crackers.

# PICTORIAL SUMMARY / 11: Cheese

Cheese is a preserved food made from the curd, or solid portion, of milk. One of the most nutrient-dense foods, cheese is most often used to add flavor, color, and texture to prepared foods.

## CLASSIFICATION OF CHEESES

Cheeses are most commonly classified by their moisture content (fresh, soft, semi-hard, hard, and very hard), processing method (process cheeses, whey products, and texture), and milk source (cow, sheep, goat, etc.). Cheese is primarily composed of water, fat, and protein. One ounce of cheese averages:

- **7 grams of protein.** High-quality protein equivalent to that found in 1 cup of milk or 1 ounce of meat.
- **100 calories (kcal).**
- **9 grams of fat.** A low-fat cheese is defined by the American Heart Association as one containing no more than 6 grams of fat per ounce.
- **0 grams of carbohydrates.**

The Art Archive/The Picture Desk Limited/Corbis

**Common cheeses**

|  |  |
|---|---|
| 1. Cheddar | 11. Brick |
| 2. Colby | 12. Swiss |
| 3. Monterey Jack | 13. Limburger |
| 4. Pasteurized process cheese | 14. Blue |
| 5. Cheese foods | 15. Gorgonzola |
| 6. Cheese spreads | 16. Provolone |
| 7. Cold-pack cheese food or club cheese | 17. Romano |
| 8. Gouda and Edam | 18. Parmesan |
| 9. Camembert | 19. Mozzarella and Scamorze |
| 10. Muenster | 20. Cottage cheese |
|  | 21. Cream cheese |

## CHEESE PRODUCTION

The basic steps of cheese manufacture include milk selection, coagulation (enzyme and/or acid), curd treatment (cutting, heating, salting, knitting, and pressing), and curing and ripening. Other products from cheese manufacture are whey products and process cheeses.

## PURCHASING CHEESE

A USDA Quality Approved inspection shield on the label means that the cheese meets minimum standards of flavor, texture, finish, color, and appearance. Of four USDA grades (AA, A, B, and C), Grade AA is the best. It is not mandatory, however, and U.S. grades have not been established for process cheese products. The most popular cheese in the United States is cheddar.

Courtesy of the USDA

## FOOD PREPARATION WITH CHEESE

Cheese can be consumed as is or as an ingredient in casseroles, pizza, soufflés, soups, salads, omelets, eggs, and other dishes. The cheese selected for preparing a particular food depends on its functional factors of shredability, meltability, oiling off, blistering, browning, and stretchability. Ripened cheeses heat better than soft cheeses. Temperatures should be kept low, and heating times for cheeses should be short.

## STORAGE OF CHEESE

Most cheeses are best refrigerated in their original wrappers. Freezing is not recommended for soft, high-water-content cheeses, but hard cheeses can be frozen for up to 2 months. Process cheese spreads packaged in jars are commonly stored on a cupboard shelf at room temperature for up to 4 months, but once opened, they must be refrigerated.

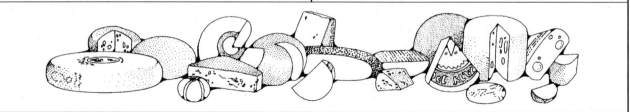

# CHAPTER REVIEW AND EXAM PREP

## Multiple Choice*

1. Based on their moisture content, how would Swiss and cheddar cheeses be classified?
   a. Fresh
   b. Soft
   c. Semi-hard
   d. Hard

2. Which of the following is the most common enzyme used to coagulate milk in cheese production?
   a. Glucagon
   b. Rennin
   c. Insulin
   d. Lipase

3. What is the name for the process by which cheese is exposed to controlled temperatures and humidity during aging?
   a. Curdling
   b. Boiling
   c. Curing
   d. Blistering

4. Enzymes are often used to coagulate milk, but milk can also be coagulated through the use of _____.
   a. lactose
   b. milk proteins
   c. bases
   d. acids

5. Which type of cheese is produced by a process of blending varieties of natural cheese, with or without heat, and mixing them with other ingredients?
   a. Colby
   b. Cheddar
   c. Process
   d. Feta

*See p. AK-1 for answers to multiple choice questions.

6. Cheese averages _____ calories (kcal) and _____ grams of fat per ounce.
   a. 50, 5
   b. 100, 9
   c. 150, 15
   d. 200, 20

7. The _____ content of cheese determines whether or not if can be frozen.
   a. carbohydrate
   b. fat
   c. protein
   d. water

## Short Answer/Essay

1. Describe some of the different ways in which cheese can be classified.
2. Approximately how many grams of fat per ounce are found in cheddar versus low-fat cheese?
3. What are the five steps of cheese production?
4. Describe the two methods of coagulating the casein protein in milk.
5. List and describe five methods of treating the curd.
6. Describe the purpose for and process of curing/ripening cheese.
7. How does Swiss cheese develop holes?
8. What are the two major proteins in whey? What major products are produced from whey?
9. What is process cheese? Briefly describe each of the following: cold-pack cheese, process cheese food, process cheese spread, and imitation cheese.
10. Cheeses differ in how they shred, melt, oil off, blister, brown, and stretch. Explain how each of the following functional characteristics affects selection of cheese used in food preparation: shredability, meltability, oiling off, blistering, browning, and stretching.

# REFERENCES

1. American Heart Association. *The Heart and Stroke Guide*. American Heart Association, 1999.

2. Blaesing D. Cheese spreads it on thick. *Prepared Foods* 166(5):70–71, 1997.

3. Branger EB, et al. Sensory characteristics of cottage cheese whey and grapefruit juice blends and changes during processing. *Journal of Food Science* 64(1):180–184, 1999.

4. Breidt F, and HP Fleming. Using lactic acid bacteria to improve the safety of minimally processed fruits and vegetables. *Food Technology* 51(9):44–48, 1997.

5. Burrington KJ. Food product development. Winning wheys. *Prepared Foods* 167(7):83–89, 1998.

6. Chandan R. *Dairy-Based Ingredients*. Eagen Press, 1997.

7. Cinbas T, and K Meral. Proteolysis and lipolysis in white cheeses manufactured by two different production methods. *International Journal of Food Science and Technology* 41(5):530–537, 2006.

8. Cowart VS, and P Gunby. Heart Association to endorse certain foods. *Journal of the American Medical Association* 260:1192, 1988.

9. Donnelly CW. Raw-milk cheeses can be produced safely. *Food Technology* 60(4):100, 2006.

10. Drake MA, et al. Sensory evaluation of reduced-fat cheeses. *Journal of Food Science* 60(5):898–901, 1995.

11. Fallico V, et al. Evaluation of bitterness in Ragusano cheese. *Journal of Dairy Science* 88(4):1288–1230, 2005.

12. Farkye NY. Contribution of milk-clotting enzymes and plasmin to cheese ripening. In EL Malin and MH Tunick, eds. *Chemistry of Structure-Function Relationships in Cheese.* Plenum, 1995.

13. Farkye NY, and CF Landkammer. Milk plasmin activity influence on cheddar cheese quality during ripening. *Journal of Food Science* 57(3):622, 1992.

14. Fox PF, TK Singh, and PLH McSweeney. Biogenesis of flavor compounds in cheese. In EL Malin and MH Tunick, eds. *Chemistry of Structure-Function Relationships in Cheese.* Plenum, 1995.

15. Gigante ML, M Almena-Almena, and PS Kindstedt. Effect of cheese pH and temperature on serum phase characteristics of cream cheese during storage. *Journal of Food Science* 62(1):53–56, 1997.

16. Gorder-Hinchy B. Make your own feta. *Fine Cooking* 100:74, 2009.

17. Holsinger VH. Nutritional aspects of reduced-fat cheese. In EL Malin and MH Tunick, eds. *Chemistry of Structure-Function Relationships in Cheese.* Plenum, 1995.

18. Johnson ME, and JA Lucey. Major technological advances and trends in cheese. *Journal of Dairy Science* 89(4):1174–1178, 2006.

19. Lee SJ, IJ Jeon, and LH Harbers. Near-infrared reflectance spectroscopy for rapid analysis of curds during cheddar cheese making. *Journal of Food Science* 62(1):53–56, 1997.

20. Malin EL, and EM Brown. Influence of casein peptide conformations on textural properties of cheese. In EL Malin and MH Tunick, eds. *Chemistry of Structure-Function Relationships in Cheese.* Plenum, 1995.

21. Malin EL, et al. Inhibition of proteolysis in mozzarella cheese prepared from homogenized milk. In EL Malin and MH Tunick, eds. *Chemistry of Structure-Function Relationships in Cheese.* Plenum, 1995.

22. *Manufacture and Use of Cheese Products.* New Zealand Institute of Chemistry. http://nzic.org.nz/ChemProcesses/dairy/3D.pdf. Accessed 5/09.

23. Misty VV. Improving the sensory characteristics of reduced-fat cheese. In EL Malin and MH Tunick, eds., *Chemistry of Structure-Function Relationships in Cheese.* Plenum, 1995.

24. Montagna MT, C Napoli, O De Giglio, R Iatta, and G Barbuti. Occurrence of aflatoxin m(1) in dairy products in southern Italy. *International Journal of Molecular Science* 9(12):2614–2621, 2008.

25. Murphy PA, et al. Soybean protein composition and tofu quality. *Food Technology* 51(3):86–88, 1997.

26. Nachay K. DNA data used in R&D. *Food Technology* 61(1):12, 2007.

27. National Academy of Sciences. *Scientific Criteria to Ensure Safe Food.* National Academies Press, Washington D.C., 2003.

28. Nonfat dairy products perform in cheesecake, pizza topping. *Food Engineering* 64(9):22, 1992.

29. Olsen NF. *Newer knowledge of cheese and other cheese products.* National Diary Council, 1992.

30. O'Mahony JA, PL McSweeney, and JA Lucey. A Model System for Studying the Effects of Colloidal Calcium Phosphate Concentration on the Rheological Properties of Cheddar Cheese. *Journal of Dairy Science* 89(3):892–904, 2006.

31. Pearl AM, C Cuttle, and BB Deskins. *Completely Cheese: The Cheeselover's Companion.* Jonathan David, 1978.

32. Potter NN, and JH Hotchkiss. *Food Science.* Chapman & Hall, 1995.

33. Pszczola DE. Say cheese with new ingredient developments. *Food Technology* 55(12):55–64, 2001.

34. Ruiz LP. Technology for healthier dairy foods. *Food Engineering* 61(12):57, 1989.

35. Scrimshaw NS, and EB Murray. The acceptability of milk and milk products in populations with a high prevalence of lactose intolerance. *American Journal of Clinical Nutrition* 48(Supp):1083–1147, 1988.

36. Sihufe GA, SE Zorrilla, and AC Rubiolo. Kinetics of proteolysis of beta-casein during ripening of fynbo cheese salted with NaCl or NaCl/KCl and ripened at different temperatures. *Journal of Food Science* 70(2):C151–C156, 2005.

37. Sloan AE. Top 10 functional food trends. *Food Technology* 60(4):22–40, 2006.

38. Steele JL. Contribution of lactic acid bacteria to cheese ripening. In EL Malin and MH Tunick, eds. *Chemistry of Structure-Function Relationships in Cheese.* Plenum, 1995.

39. Tarlov R. Discover the pleasures of a cheese course. *Fine Cooking* 25:56–59, 1998.

40. Trepanier G, et al. Accelerated maturation of cheddar cheese: Microbiology of cheeses supplemented with *Lactobacillus casei subsp. casei* L2A. *Journal of Food Science* 57(2):345–349, 1992.

41. Wendorff WL, and C Wee. Effect of smoke and spice oils on growth of molds on oil-coated cheeses. *Journal of Food Protection* 60(2):153–156, 1997.

42. Werlin L. Serving cheese. *Fine Cooking* 89:26, 2007.

# WEBSITES

Cheese information, including a cheese library, can be found at this website:

**www.cheeselibrary.com**

This website lists cheeses by names, countries, and kind of milk. It also lists vegetarian cheeses, cheeses in alphabetical order, cheese facts, and cheese recipes:

**www.cheese.com**

The American Cheese Society provides cheese standards, educational resources, and product promotion at this website:

**www.cheesesociety.org**

Watch a video how Swiss cheese curd is shaped, pressed, salted, and cured:

**http://science.discovery.com/videos/how-its-made-swiss-cheese-assembly.html**

# 12 Eggs

**B**ird eggs, long honored as symbols of fertility and life, have been part of our diet for thousands of years. All bird eggs are edible, highly nutritious, and neatly packaged in their own shells. An egg contains everything needed to sustain the life of a new chick.

The egg's nutritious contents make it susceptible to predation by other animals seeking food. Eggs were not always available or easy to find, so Americans solved this problem by breeding almost as many laying hens as there are people in the United States. Each of these laying hens produces about one egg per day.

Eggs are one of the most versatile of foods to prepare. In their most basic form, they can be cooked and eaten on their own—fried, scrambled, poached, and boiled. The unique physical and chemical properties of eggs also make them an invaluable ingredient in many prepared dishes. Just a few examples of how eggs are used in food preparation are listed below:

- Giving a foam structure to cakes and meringues
- Thickening custards and puddings
- Adding color to lemon meringue pie and eggnog
- Emulsifying mayonnaise and hollandaise sauce
- Leavening soufflés and popovers
- Binding ingredients in meatloaf and casseroles
- Coating foods prior to breading
- Glazing pastries and breads
- Clarifying liquids for soups

Understanding how these various roles in food preparation are fulfilled by eggs requires a general knowledge of them. Although all bird eggs are edible, this chapter is limited to those that are most commonly consumed—chicken eggs—and their composition, purchase, functions in foods, preparation, and storage.

# COMPOSITION OF EGGS

## Structure

The egg has five major components: the yolk, albumen (egg white), shell membranes, air cell, and shell (Figure 12-1). Each of these plays an important role in the egg's unique attributes that make it invaluable in food preparation.

## Yolk

The sunny yellow yolk situated in the center of the egg constitutes about a third (30 percent) of the egg's weight. Dense in

**FIGURE 12-1**    Composition of an egg.

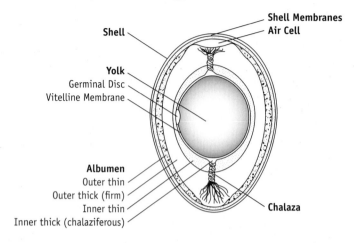

white determine differences in viscosity. Around the yolk is a layer of thick protein called albumin. The **chalazae** (ka-lay-zee) at the top and bottom of the egg anchor the egg yolk in the thick egg white surrounding it. They also secure the yolk to its **vitelline membrane** so it stays neatly centered in the middle of the egg.

## Shell Membranes

Between the egg white and the shell are two membranes, an inner and outer shell membrane. These press up against the shell and protect the egg against bacterial invasion.

nutrients, the yolk serves to nourish the chick. A white, pinhead-sized germinal disc sits on the surface of the yolk. This appears darker if the egg has been fertilized, but has no effect on the taste, functional properties, or nutritional value of the egg. Most eggs are screened to ensure they are not fertilized, but even if they do pass inspection undetected, fertilized eggs do not develop once refrigerated. The color of the yolk depends on the hen's diet. Pigments in the chicken feed, such as beta-carotene, cause colors ranging from pale yellow to deep red. Natural yellow-orange substances such as marigold petals are sometimes added to the chickens' feed to enhance the color of their egg yolks. Artificial color additives are not permitted.

## Albumen

The albumen, or egg white, accounts for almost three fifths (58 percent) of an egg's weight and is made up largely of water and protein (Chemist's Corner 12-1). Although it appears to be one mass, the egg white is actually constructed of layers differing in viscosity, alternating from thick to thin. The type and amount of proteins in various parts of the egg

**Chalaza (pl. chalazae)**  The ropy, twisted strands of albumen that anchor the yolk to the center of the thick egg white.

**Vitelline membrane**  The membrane surrounding the egg yolk and attached to the chalazae.

> ## CHEMIST'S CORNER 12-1

### Egg Proteins

Eggs, and especially the egg white, are composed of dozens of different proteins. Each of these proteins has its own characteristics and functions (Table 12-1). More than half of all the protein in eggs is ovalbumin. It plays an important role in egg preparation because it gels well and denatures easily when heated (5, 26). The next two proteins of any significance are ovotransferrin (conalbumin) and ovomucoid. Ovotransferrin helps to protect against bacterial contamination by forming a complex with iron, thereby making it unavailable to bacteria, which need iron for growth. Ovomucoid proteins, which are resistant to denaturation, inhibit the activity of trypsin, an enzyme that breaks down specific proteins. Lysozyme, an enzyme, fights certain bacteria by lysing (breaking chemical bonds by the addition of water) the cell wall of the bacterium. Ovomucin helps protect against microbial damage by reacting with viruses. Avidin in uncooked eggs binds to the B vitamin biotin and prevents it from being absorbed by the small intestine, which is why excessive raw egg white consumption is discouraged. However, once cooked, the avidin denatures and no longer inhibits the absorption of biotin. About 3 grams of protein are also found in the egg yolk in the form of lipoproteins—lipovitellin and lipovitellinin—which act as emulsifiers in food preparation.

| TABLE 12-1 | Protein Content of Egg White | |
|---|---|---|
| **Protein** | **Amount (%)\*** | **Properties** |
| Ovalbumin | 54 | Denatures easily |
| Conalbumin | 13 | Antimicrobial, complexes iron |
| Ovomucoid | 11 | Inhibits enzyme (trypsin) |
| Unidentified | 8 | Mainly globulins |
| Lysozyme | 3.5 | Antimicrobial |
| Ovomucin | 1.5 | Viscous, reacts with viruses |
| Flavoprotein | 0.8 | Binds riboflavin |
| Proteinase inhibitor | 0.1 | Inhibits enzyme (bacterial proteinase) |
| Avidin | 0.05 | Binds the B vitamin biotin (raw egg white only) |

\*Expressed as a percentage of the total protein content of the egg white.

## Air Cell

Between the two shell membranes at the larger end of the egg is a pocket of air known as the air cell. As a freshly laid egg cools, its contents contract, causing the inner shell membrane to separate from the outer shell membrane, forming the air cell.

## Shell

Nature's way of protecting the delicate internal contents of an egg is to surround it with a hard calcium carbonate shell (12 percent of an egg's weight). Eggshells are not solid but contain thousands of small pores, allowing an exchange of gases between the inner egg and the surrounding air. Shell color indicates the breed of the hen, but has no bearing on the nutrient content or taste of the egg. Brown eggs tend to cost more because they usually come from larger hens, which require more food and produce fewer eggs.

The shell is protected by the **cuticle** or **bloom.** Commercially sold eggs are washed, which removes this protective cuticle. Producers compensate for this loss by applying a thin coat of edible oil onto the shell (15).

# PURCHASING EGGS

In the United States, approximately 80 billion eggs are consumed each year (20). The majority (70 percent) are fresh eggs subject to United States Department of Agriculture (USDA) inspection and grading. The remaining eggs are processed (liquid, frozen, or dried eggs) for use in the food industry (food products) or food service (menu items), and their care is described more fully under storage. Egg substitutes and value-added eggs are discussed following an explanation of the inspection and grading of fresh eggs.

## Inspection

The Egg Products Inspection Act of 1970 requires that egg processing plants be inspected and that their eggs and egg products be wholesome, unadulterated, and truthfully labeled. This law is enforced by the USDA Poultry Division and applies to all eggs, whether imported or shipped intra- or interstate.

**FIGURE 12-2 Candling.** Eggs are automatically rolled against a light background during mass scanning, allowing checkers to remove those that have defects or that are cracked, soiled, or damaged.

Jupiterimages/Getty Images

### Eggs Failing USDA Inspection

Eggs that do not pass inspection, called restricted eggs, are not allowed to be sold whole to the consumer. Examples of restricted eggs are those with cracked shells, called *checks; leakers,* which have cracked shells as well as broken membranes; *dirties,* which have at least one fourth of their shell covered with dirt or stain; and *inedibles,* which have greenish egg whites or are fertilized, rotten, moldy, or bloody. Less than 1 percent of eggs have small blood spots, also known as *meat spots.* Although these are harmless, the eggs containing them are removed when found by electronic blood detectors used during grading. Inspection is intended to ensure that eggs with defects such as these are not sold to the consumer.

## Grading

Once eggs pass inspection, a producer can pay the USDA to have them graded for quality. The best-quality eggs are graded USDA Grade AA, followed by USDA Grade A. USDA Grade B, the lowest grade, is available to food service establishments and not sold directly to consumers. Grades AA and A eggs are the grades sold at supermarkets. Their firm, high albumens and yolks make them suitable for frying, coddling, poaching, and hard cooking. Grade B eggs have thinner whites and somewhat flattened yolks, making them better for scrambling or baking, or as ingredients in other food products.

### Grading Methods

Grades are determined by graders who incorporate three methods to judge the quality of eggs: **candling;** measuring Haugh (pronounced "how") units; and evaluating appearance, specifically that of the shell, white, yolk, and air cell. Commercial egg grades are assigned based on both interior and exterior quality. Interior quality is primarily determined by candling. The exterior quality is determined by the cleanliness of the shell, the shape of the egg, and the presence of shell irregularities such as pimples (calcium deposits) and weak shells.

*Candling* As the name implies, the original method of candling involved holding an egg up to the light of a candle to view its contents. Although it is still a good way to determine an egg's inner quality and to detect defects, eggs are now mechanically rotated over lights, many at a time, by rollers (Figure 12-2).

Candling is based on the principle that eggs start to deteriorate the minute they are laid, and these changes can be vaguely seen through an egg held

**Cuticle (bloom)** A waxy coating on an eggshell that seals the pores from bacterial contamination and moisture loss.

**Candling** A method of determining egg quality based on observing eggs against a light.

against a light. The whites become more thin and transparent as carbon dioxide departs through the shell; and the yolk membrane stretches as the yolk absorbs water from the white, which eventually will cause the yolk to break easily when the shell is cracked.

### ❓ How & Why?

*How does candling reveal whether an egg is fresh or aged?*

The yolk in a fresh, high-quality egg is suspended tightly by the chalazae, seen in candling only as a slight shadow. The yolks in older eggs, on the other hand, are surrounded by thinning, clearer egg whites and deteriorating chalazae. These older yolks lie closer to the shell because they are no longer suspended well by the chalazae; they are looser in consistency and cast a darker shadow. The egg's air cell, too, becomes wider or moves as the egg ages. Grade AA eggs must have an air cell depth smaller than ⅛ inch; Grade A eggs are limited to 3⁄16 inch; and Grade B eggs have no limit to the size of their air cell.

*Haugh Units*    The freshness of an egg can be detected by cracking it open onto a flat surface and looking at the height of its thick albumen. Fresh egg whites sit up tall and firm, whereas older ones tend to spread out. Professional graders cannot evaluate every egg for freshness, so an egg is randomly selected and measured using a special instrument called a micrometer (Figure 12-3). The Haugh unit, a numerical value reflecting an egg's freshness, is obtained by mathematically combining the thick albumen height with the egg's weight, and then using a formula or table to convert this number into a Haugh unit. As Haugh units decrease, so does egg quality, which is reflected in grading: Grade AA is given to eggs with a Haugh unit of 72 or higher, Grade A for a measure of 60 to 71, and Grade B for a measure of 31 to 59.

*Appearance*    Grading can also be based on the appearance of eggs broken on a flat surface. Graders evaluate the quality of an egg by observing the thickness of the albumen, the prominence of the chalazae, the roundness and firmness of the yolk, and the shape,

**FIGURE 12-3**    Measuring Haugh units to determine egg quality.

*Courtesy of USDA*

cleanliness, and texture of the shell. Figure 12-4 summarizes the characteristics of USDA Grades AA, A, and B eggs. The vast majority of eggs are graded on interior quality by candling. Grading of broken-out eggs is reserved for research purposes or random spot checks of candled eggs.

## Sizing

Sizing is not related to grading in any way. Eggs are sold in cartons by various sizes determined by a minimum

**FIGURE 12-4**    USDA grades for eggs.

A Grade AA egg will stand up tall. The yolk is firm and the area covered by the white is small. There is a large proportion of thick white to thin white.

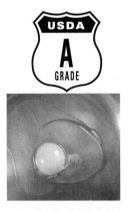

A Grade A egg covers a relatively small area. The yolk is round and upstanding. The thick white is large in proportion to the thin white and stands fairly well around the yolk.

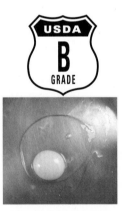

A Grade B egg spreads out more. The yolk is flattened and there is about as much (or more) thin white as thick white.

*Digital Works*

**FIGURE 12-5** Egg sizes determined by minimum weight (including shells) per dozen.

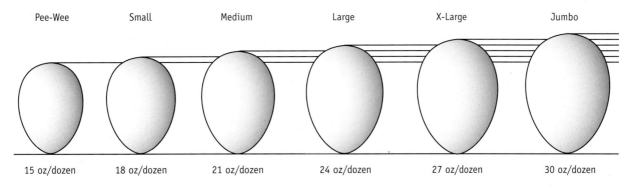

| Pee-Wee | Small | Medium | Large | X-Large | Jumbo |
|---------|-------|--------|-------|---------|-------|
| 15 oz/dozen | 18 oz/dozen | 21 oz/dozen | 24 oz/dozen | 27 oz/dozen | 30 oz/dozen |

## NUTRIENT CONTENT

While not originally considered a functional food, eggs are now promoted as a beneficial addition to our normal diet (12). Eggs are an excellent source of many nutrients—protein, fat, vitamins, minerals, and water—and possibly beneficial nonnutrients—choline, lutein, and zeaxanthin—although they contain very little carbohydrate or fiber. An average large egg provides about 75 calories (kcal), primarily from fat and protein.

*Protein.* One large egg contains approximately 7 grams of complete protein—4 grams from the white and 3 from the yolk. As a life-sustaining protein, the quality of protein in eggs is so high that it has become the standard by which researchers rate all other foods. As a result, the egg is often used as a **reference protein**.

*Fat.* One egg yolk contains approximately 5 grams of fat, predominantly in the form of triglycerides, phospholipids, and cholesterol. The fat is approximately 47 percent monounsaturated (2 grams), 37 percent saturated (slightly under 2 grams), and 16 percent polyunsaturated (slightly under 1 gram). Also available are eggs that are high in omega-3 (polyunsaturated) fatty acids, which are widely considered to be beneficial in the prevention of a variety of medical conditions, including heart disease and inflammatory disorders.

*Cholesterol.* Egg yolks are high in cholesterol, averaging 213 mg in a large egg (9). Dietary cholesterol has been reported by some researchers to increase blood cholesterol. As a result, the American Heart Association recommends no more than 100 mg of dietary cholesterol be consumed for every thousand calories (kcal), which generally limits egg consumption to no more than four egg yolks per week. Despite the dietary focus on lowering fat and cholesterol, eggs are not always easily replaced in recipes, and consumer surveys show that the amount of fat often determines whether or not a food is acceptable (3, 14).

Commercial food processors have also tried various methods to reduce the amount of cholesterol found in foods. One California poultry farmer developed a special chicken feed, using sea kelp, to reduce cholesterol content in large Grade AA eggs to only 125 mg each as compared with 213 mg in an average egg, but it also resulted in high iodine levels (7).

*Vitamins and Minerals.* Eggs are rich in certain vitamins and minerals. They are one of the few foods containing all the fat-soluble vitamins—A, D, E, and K—and naturally containing vitamin D. Certain water-soluble B vitamins, found primarily in the white, are also supplied by the egg in relatively high Reference Daily Intake percentages: 25 percent of vitamin $B_{12}$ and biotin, 15 percent each of riboflavin (vitamin $B_2$) and biotin, 12 percent of folate, and 11 percent of pantothenic acid.

In addition, eggs are a source of several minerals, especially selenium, iodine, zinc, and iron. Unfortunately, the iron in egg yolks is not very available because it binds to phosvitin, an egg protein that inhibits absorption. When eggs are overcooked, the iron in the yolk interacts with the small amount of sulfur found in the egg white to produce ferrous sulfide, which causes a characteristic strong off-odor.

weight for a dozen eggs in their shell (Figure 12-5). Unless otherwise designated, recipes are based on the use of large eggs. The contents of one large egg are equivalent to approximately ¼ cup. Table 12-2 shows that 4 to 6 eggs will fill 1 cup, depending on the size of the egg. Table 12-3 shows how eggs of other sizes are used when large eggs are not available.

## Egg Substitutes

Consumer demand for products lower in cholesterol has created a market for liquid egg substitutes, made by either omitting egg yolks, replacing egg yolks with vegetable oils, or removing some of the cholesterol in egg yolks (24). One drawback to egg substitutes is that they may be higher in sodium than regular eggs. On the positive side, egg substitutes are ultrapasteurized, packaged, and then refrigerated, preparing them for a 10-week shelf life (4). Pasteurization makes them ideal for use in uncooked foods such as eggnog, mayonnaise, and ice cream, in which raw eggs would pose a food safety risk. The ingredients and cholesterol contents of several brands of egg substitutes are listed in Table 12-4.

## Value-Added Eggs

There is a new type of egg on the market that takes into consideration the health of the consumer and that of the

**Reference protein** A standard against which to measure the quality of other proteins.

**TABLE 12-2**    Number of Eggs (Whole/Whites/Yolk) Equivalent to 1 Cup

| Egg Size | Whole | Whites | Yolks |
|----------|-------|--------|-------|
| Jumbo    | 4     | 5      | 11    |
| X-Large  | 4     | 6      | 12    |
| Large    | 5     | 7      | 14    |
| Medium   | 5     | 8      | 16    |
| Small    | 6     | 9      | 18    |

**TABLE 12-3**    Egg Equivalents: Number of Eggs Used to Replace Eggs of Other Sizes

Equivalent Number of Eggs

| Jumbo | X-Large | Large | Medium | Small |
|-------|---------|-------|--------|-------|
| 1     | 1       | 1     | 1      | 1     |
| 2     | 2       | 2     | 2      | 3     |
| 2     | 3       | 3     | 3      | 4     |
| 3     | 4       | 4     | 5      | 5     |
| 4     | 4       | 5     | 6      | 7     |
| 5     | 5       | 6     | 7      | 8     |
| 9     | 10      | 12    | 13     | 15    |
| 18    | 21      | 24    | 27     | 28    |
| 37    | 44      | 50    | 56     | 62    |

**TABLE 12-4**    Egg Substitutes Compared to Standard Egg

|   | Kcal | Fat (g) | Chol* (mg) | Protein (g) | Sodium (mg) |
|---|------|---------|------------|-------------|-------------|
| *Standard large egg (one)* | 75 | 5 | 215 | 6 | 65 |
| *Better'n Eggs™ (56 g)* 98% egg whites, water, natural flavors, sodium hexametaphosphate, guar gum, xanthan gum, color (includes beta-carotene). | 30 | 0 | 0 | 6 | 120 |
| *Egg Beaters™ (61 g)* 99% egg white, less than 1%: vegetable gums (xanthan gum, guar gum), color (includes beta-carotene), salt, onion powder, natural flavor, spices. | 30 | 0 | 0 | 6 | 115 |

*Cholesterol

**Sources:** www.betterneggs.com/products/nutritionView.cfm?prid=2, www.eggbeaters.com/products/original.jsp

laying hen. *Value-added* eggs have special attributes because of their nutrient content or the conditions under which the hens are raised. For instance, lower-cholesterol eggs are now on the shelf along with those that have higher omega-3 fatty acid levels or 300 percent more vitamin E, achieved by special feeding practices. In addition, value-added eggs include those that have been laid by hens that are free roaming, cage free, or fed natural grains without animal products—a plus for animal-friendly egg eaters (21). However, the buyer should be aware that these are marketing terms without official definitions.

# TYPES OF EGGS

All poultry produce eggs. Although chicken eggs are frequently used in the preparation of foods, many other eggs are available. There are other types of eggs that can be used and these include, but are not limited to, quail, duck, goose, turkey, and ostrich eggs (Figure 12-6). It's important to remember that these other egg varieties should always be well cooked because they are more apt to contain harmful bacteria.

# FUNCTIONS OF EGGS IN FOODS

Eggs are often combined with other ingredients. Their unique ability to flavor, color, emulsify or thicken, bind, foam, interfere, color, coat, leaven, clarify, and prevent crystallization makes them nearly indispensable in cooking (Figure 12-7). They are often used for these purposes in prepared foods such as snacks, entrees, and processed meats (22). Some of these unique qualities of eggs are now discussed.

## Emulsifying

Lecithin, found in egg yolks, is a natural emulsifying agent: one end of the molecule attracts water, while the other end is drawn to fat (1). Eggs thus help keep fat and water or other liquid compounds from separating, so they are often used to thicken and stabilize foods such as salad dressings, hollandaise and béarnaise sauces, mayonnaise, ice cream, cream puffs, and certain cakes.

## Binding

The high protein content of eggs makes them excellent binders. Fish, chicken, vegetables, and other foods are often dipped in beaten egg and then rolled in breading, batter, flour, or cereal. During cooking, heat coagulates the eggs' protein, which then acts as an adhesive, binding the other ingredients to the surfaces of the cooked material. Egg also binds together mixtures such as meatloaf, meatballs, lasagna, and manicotti. When the mixture cooks, the

**FIGURE 12-6** Types of eggs.

## CHICKEN EGGS

### Standard Eggs

Most eggs sold to consumers are white-shelled chicken eggs that have not been fertilized. Brown-shelled chicken eggs are also available. The type of hen breed determines the color of the shell. Shell color does not influence the egg's quality, flavor, or nutrient value.

### Fertile Eggs

Only fertilized eggs, usually identified by a red spot on top of the inner contents, can hatch into chicks (if incubated). Refrigeration stops the growth process. Fertilized eggs are not more nutritious than nonfertilized eggs. Fertilized eggs are actually removed from the eggs commonly sold to the consumer. Their disadvantage is that they are more costly to produce and do not store as well.

### Free-Range Eggs

The USDA has no standard definition for "free-range" eggs except that the hens must "have access" to outdoor areas. One reason that these eggs are more expensive is because eggs from penned hens are easy to harvest, whereas more labor is required to obtain eggs from free-range birds that, like all birds, instinctively hide their eggs.

### Organic Eggs

Commercial food for egg-laying hens cannot contain growth hormones. As a result, organic eggs are defined as those obtained from hens fed food ingredients that were grown without pesticides, herbicides, fungicides, or commercial fertilizers.

## NON-CHICKEN EGGS

### Duck Eggs

Because ducks need water, they are not as easily housed as chickens. As a result, harvesting of duck eggs does not lend itself well to mass production techniques. Their eggs are also richer due to more than triple the cholesterol (619 g) and double the fat (10 g) content compared to chicken eggs (212 mg cholesterol; 5 g fat). Historically, duck eggs are more commonly consumed in Asian than North American cuisine. Salted duck eggs and "1000 year old eggs" (a traditional Chinese delicacy of preserved eggs not more than 100 days old) are examples.

### Goose Eggs

Larger and more flavorful than either duck or chicken eggs, goose eggs are also much higher in cholesterol (1,227 mg) and fat (19 g).

### Ostrich Eggs

The largest of all eggs weighs about 3 pounds and is equivalent to about a dozen chicken eggs. Approximately 6 inches long and 5 inches wide, individual ostrich eggs are more a novelty than a common source of food for consumers.

### Quail Eggs

These dainty, delicate eggs are often hard-boiled. They are used most often as gourmet garnishes or salad ingredients.

### Turkey Eggs

Turkey eggs are bigger than chicken eggs but not as popular (primarily due to cost). The cost of turkey egg production is much higher because (1) the average chicken bred to produce eggs lays about an egg a day, whereas turkeys usually lay 1 egg every 3 days; (2) turkeys do not start laying eggs until they are about 10 weeks older than chickens are when they begin producing eggs; (3) chickens only weigh about 3 1/2 pounds and eat less than turkeys; and (4) turkeys weigh about 15 pounds and require more room.

**FIGURE 12-7** The function of eggs in food preparation.

| Food | Functions of Eggs |
| --- | --- |
| *Cereal, grains, and pasta* | |
| Noodles, pasta | Color, flavor, nutrition |
| Pancakes, crepes, waffles | Flavor, coagulation |
| *Desserts* | |
| Cakes, pastries | Foaming, coagulation, color |
| Candy | Inhibition of crystals |
| Custards, puddings | Coagulation, flavor |
| Doughnuts, croissants | Texture, flavor |
| Ice cream | Emulsification, texture |
| *Egg dishes* | |
| Meringues, soufflés | Foaming |
| Omelets, scrambled, poached | Coagulation, flavor |
| *Fish products and meat* | |
| Fish products (surimi) | Binding by coagulation |
| Meat (patties, sausages) | Binding by coagulation |
| *Salad dressings* | |
| Mayonnaise, salad dressing | Emulsification |

egg proteins firm and stabilize, providing structural strength. Too much egg, however, can cause foods to become overly firm and dry.

## Foaming

The capacity of egg whites to be beaten into a foam that increases to six or eight times its original volume is invaluable in food preparation (8). Egg-white foams are used to aerate and leaven a number of food products, such as puffy omelets, soufflés, angel food cake, sponge cake, and meringues.

The best eggs to use for an egg-white foam are fresh eggs, because they have thick egg whites, which contribute to a stable foam. Older eggs have thinner whites, which beat to a larger volume but are less stable and may collapse during heating. Consistency, regardless of an egg's age, is achieved in some food service establishments by the use of dried egg whites.

The formation and stability of egg white foams also depend upon the beating technique; the temperature; the type of bowl; the careful separation of yolks and whites; and whether or not sugar, fluid, salt, or acid have been added (13). The skillful control of all these factors yields the best possible egg-white foam.

### Beating Technique

Egg whites are best beaten with an electric mixer, but a wire whisk or a double-bladed rotary egg beater can also be used. Whichever device is used, the key is to whip the egg whites into very fine, delicate bubbles. Figure 12-8 shows the consistency of the foam at the various stages of foam formation—foamy, soft, stiff, and dry.

At first, beating speed should be moderately slow and even. The foamy stage consists of very large air bubbles, but as the foam thickens, smaller, finer bubbles begin to appear. When the egg whites are half whipped and beginning to hold their shape, the beating speed should be increased to medium or fast.

***Testing for Doneness*** Testing for doneness consists of stopping, lifting the beaters out of the foam, and seeing how the egg-white peaks form. Initially, the egg-white foam forms soft, shiny peaks that droop over without holding their shape, and the whipped foam slides around in the bowl. Beating should continue until the peaks are stiff and shiny but not dry. Whipping is complete at the stiff peak stage when the peak tips fall over slightly but keep their shape and the whipped foam sticks to the side of the bowl. Perfectly beaten egg whites will pass the inverted bowl test: they will cling to the inside of a bowl turned upside-down.

***Avoiding Overwhipping*** Excessive beating of egg whites occurs when the peaks stand tall, dry, and are no longer shiny. In addition, the protein films surrounding the air cells rupture, creating bubbles that are too large and unstable. It is important to beat egg

**FIGURE 12-8** Stages of foam formation.

(a) Foamy

(b) Soft peaks

(c) Stiff peaks

(d) Dry peaks

Digital Works

whites to the correct stage, because both over- and underbeaten foams will eventually collapse and separate into liquid pools at the bottom of the bowl.

## Temperature

The bowl, beater, and eggs should be at room temperature. The decreased surface tension of room-temperature egg whites allows them to whip more easily and to a larger volume than cold eggs. However, the stability of the foam deteriorates with continued exposure to room temperature. In addition, leaving eggs out for more than 30 minutes risks *Salmonella* growth. Although not highly recommended, eggs can be briefly submerged in a bowl of warm water before whipping.

## Bowl

Deep bowls with rounded bottoms sloping up into the sides are best, because they allow the egg whites to be picked up by the beater. Plastic bowls should be avoided, because their porous surface may harbor a thin film of grease from previous usage, which could interfere with foam formation (16). Even the smallest amount of fat will interfere significantly with an egg white's ability to foam.

> ### ? How & Why?
>
> **Why do beaten egg whites make a stable foam?**
>
> Vigorous beating or whisking of egg whites breaks the links between protein molecules, causing the protein molecule coil to unwind or become denatured. A foam structure is created when the unfolded proteins rearrange to construct films around the air cells (2). When the airy foam is heated, its air cells further expand, after which the egg proteins coagulate, solidifying the egg protein to create a firm, stable structure, higher in height than the same food made without egg-white foam.

## Separation of Eggs

Although fat contamination can come from plastic bowls or utensils, or from traces of cream, oil, butter, or other foods, the most common source is the egg yolk itself. Careful separation of the egg yolk and the white is imperative. Eggs can be separated using

**FIGURE 12-9**  Egg separator.

an egg separator, a small cup centered in a round frame. The central cup catches the yolk while slots around the frame let the white slip through to a container underneath (Figure 12-9). Another method is to break the egg into a clean funnel, which allows the white, but not the yolk, to run through. Egg yolk traces are sometimes removed by using the egg's shell, which has an affinity for picking up traces of yolk remaining in the egg white; however, this method of removal is not recommended because eggshells are a possible source of microbial contamination. For the same reason, eggs should not be separated by passing the yolk back and forth between the two shell halves.

## Sugar

Sugar stabilizes egg-white foam, but it also inhibits the mechanical coagulation of proteins necessary for foam formation. Therefore, it is best to add sugar near the end of the whipping time or volume may be compromised. One teaspoon of granulated or super-fine sugar per egg may be added only after soft white peaks have formed. Egg white sweetened with sugar is less likely to be overbeaten, and has a very fine texture with a smooth, satiny surface.

## Fluid

Adding fluid to egg whites increases the foam volume up to 40 percent, but decreases its stability.

## Salt

Salt decreases the stability and volume of egg-white foam, and for that reason is rarely added to egg whites.

## Acid

Normally, whole eggs are relatively neutral in pH (7.0 to 7.6), but egg whites by themselves tend to be alkaline (about 8.4 pH). In fact, alkalinity increases with age as the loss of carbon dioxide

increases the pH of the egg whites (up to 9.2) (16). Egg whites whip more easily into a stable foam when their pH is lowered slightly. Adding acid in the form of lemon juice or cream of tartar (the salts of tartaric acid) decreases the pH of egg whites, causing the egg proteins to become unstable and more likely to denature and whip into a foam. Adding too much acid, however, results in delayed foam formation and a much less stable foam. Proper acidity can be achieved by adding 1 teaspoon cream of tartar per cup of egg white or 1/8 teaspoon per egg white, or by adding 1/8 teaspoon of lemon juice or white distilled vinegar per egg white.

> ### ? How & Why?
>
> **Why were copper bowls once commonly used to whip egg-white foams?**
>
> It was long a common chef's practice to use copper bowls made specifically for beating egg whites. A unique reaction between the copper and the egg whites occurs: trace amounts of copper from the bowl combine with an egg-white protein, conalbumin, allowing the air bubbles in the foam to expand to a larger size as they are beaten. Although copper bowls did improve the whipping properties of eggs, they are no longer recommended because of toxicity risks associated with excess copper.

## Interfering

Eggs are often used in the preparation of frozen desserts such as ice cream, because they interfere with the formation of ice crystals. Similarly, in some candies, eggs are used to block the formation of large sugar crystals to create a smoother, more velvety texture.

## Clarifying

Egg whites are often used to **clarify** liquids. This is done by dissolving egg proteins, especially egg whites (albumen), in cold liquid, which is then heated. This causes the proteins to

> **Clarify**  To make or become clear or pure.

## CALORIE CONTROL

### Eggs—It's What You Serve with Them

- **Eggs Alone.** Eggs are relatively low in calories (kcal)—only 75 each for a large egg. That drops to 15 calories for the egg white if the yolk, with its 60 calories, is removed.
- **Portions Count.** Most people have at least two eggs at a time, and that adds up to 150 calories. A three-egg omelet would be 225 calories.
- **Breakfast Choices.** It's the foods consumed with eggs that really contribute to the total calorie count (Figure 12-10). Other options include vegetable omelets, fresh fruit, and/or a vegetable or fruit juice with breakfast.

**FIGURE 12-10**    Calorie content of three breakfast options.

### Breakfast #1 = 526 Calories

Danish (Large, 7 inches)

*Courtesy of Amy C. Brown*

### Breakfast #2 = 783 Calories

Two eggs, sausage, and hash browns

2 eggs = 180 calories (includes fat to fry)
2 sausage = 133 calories
1 cup hash browns = 470 calories

*Courtesy of Amy C. Brown*

### Breakfast #3 = 386 Calories

Two eggs, toast, butter, and orange juice

2 eggs = 180 calories (includes fat to fry)
1 slice whole-grain toast = 60 calories
1 pat butter = 36 calories
1 cup orange juice = 110 calories

*Courtesy of Amy C. Brown*

© 2010 Amy Brown

solidify, to attract other particles that may be clouding the liquid, and to rise with them to the surface for removal. This food preparation technique is used to make clear soups (see Chapter 15).

## Color

An egg's yolk contributes a golden brown color to yellow cakes, cookies, pastries, and even rolls, breads, and egg-containing noodles.

# PREPARATION OF EGGS

Eggs are extremely versatile and can be prepared alone or in combination with other foods (Figure 12-11). When eggs are the primary ingredient in a dish, it is best to use the highest quality (Grade AA or A) and those packed within 28 days of laying (20). Countless recipes that include eggs can be cooked by either dry- or moist-heat methods. Before these are discussed, a brief overview of the changes that can occur in prepared eggs is presented.

## Changes in Prepared Eggs

### Effects of Temperature and Time

The key to cooking eggs is to keep the temperature low and/or the cooking time short. Heating eggs at high temperatures and/or for long periods of time diminishes the eggs' texture, flavor, and color. Overheated proteins become tough and rubbery and shrink from dehydration, which is why overcooked scrambled eggs look curdled and feel dry and rubbery.

**FIGURE 12-11**    Food preparation with eggs.

| Eggs Prepared Alone: | |
| --- | --- |
| Fried | |
| Baked | |
| Simmered | |
| Coddled | |
| Poached | |
| Omelet | |
| Scrambled | |

| Foods Prepared with Eggs: | |
| --- | --- |
| Eggs benedict | Waffles |
| Meringue | Egg soup |
| Deviled eggs | Cream puffs |
| Pudding | Mayonnaise |
| Pancakes | Quiche |
| Egg salad | Fritatas |
| Mousse | Popovers |
| Eggnog | Cake |
| Soufflé | French toast |
| Crepe | Egg sandwich |
| Egg foo young | Cream pies |
| Custard | Salad dressing |

*Coagulation Temperatures*  Egg whites and yolks coagulate at different temperatures. Egg whites first start to coagulate at about 140°F (60°C) and become completely coagulated at 149°F–158°F (65°C–70°C). Slightly warmer temperatures of about 144°F–158°F (62°C–70°C) are needed for the egg yolks to start coagulating. This difference allows eggs to be cooked so their whites are firm but the yolks remain soft. An egg may be cooked at 142°F (61°C) for an hour and still have a soft yolk. Also, beaten eggs coagulate at a slightly higher temperature (about 156°F/69°C).

### Effects of Added Ingredients

Adding other ingredients to eggs changes their coagulation temperature. For example, incorporating milk into whole eggs in a custard dish increases the coagulation temperature to about 175°F (79°C). Sugar also increases coagulation temperature, whereas the addition of salt and/or acid lowers it. Eggs can curdle if too much of an acidic ingredient, such as tomato or vinegar, is added.

### Color Changes

Undesirable color changes may occur during egg preparation. Sometimes when eggs are overcooked or heated at too high a temperature, the sulfur in the egg white may combine with the iron in the yolk, the cooking water, or other iron sources to form ferrous sulfide, a green-colored compound with a strong odor and flavor. To eliminate the problem of "green yolk," it is best to use stainless steel equipment and low cooking temperatures, to avoid overcooking, to cool hard-cooked eggs quickly in cold water, and to serve them immediately. Another change in heated eggs, which is more difficult to prevent, is the slight browning that results from the Maillard reaction, in which egg proteins react with the few carbohydrates contained in the egg.

# Dry-Heat Preparation

Dry-heat preparation of eggs primarily involves frying and baking. Egg dishes that are commonly fried are fried eggs, scrambled eggs, and omelets. Baked egg dishes include shirred eggs, meringues (both soft and hard), and soufflés.

These dry-heat methods are now further discussed.

## Frying

A frying pan, a sauté pan (omelet pan), or even a griddle can be used to fry eggs. Cast iron pans work best for eggs if the pans are **primed** or **seasoned.** Priming is accomplished by rubbing a clean frying pan with a thin layer of vegetable oil and setting it on moderate heat, which is then briefly increased to high. Then it is removed from the heat and allowed to cool. Washing the frying pan with soap or cooking anything but eggs in it removes the primed surface. Nonstick pans do not need to be primed or seasoned. Frying is used to prepare fried and scrambled eggs and omelets.

*Fried Eggs*  For each fried egg, about 1 teaspoon or less of butter, margarine, or oil is added to a hot pan. Clarified butter can also be used; it will not burn like regular butter. To cut down on fat, a bit of fat may be spread on the pan's surface with a paper towel or waxed paper, or a vegetable spray may be applied to its surface before heating. Too little fat causes sticking, but excessive fat will result in greasy eggs. The fat should be hot enough to prevent the eggs from running, but not so hot that it toughens the egg proteins. The temperature is just right when a drop of water dropped into a hot pan sizzles instead of either rolling around or instantly vaporizing into the air.

Yolks are less likely to break open when the eggs are cracked if the eggs are allowed to warm very briefly in a bowl of hot water. Broken yolks can also be avoided by using fresh eggs and/or by first breaking the eggs into a bowl or other container rather than dropping them directly from the shell into a frying pan or griddle. Then, once the pan and the fat have been heated to the right stage, the eggs should be slid from the bowl, no more than two at a time, onto the pan or griddle. The heat should be lowered immediately to medium-high.

Coagulation is then allowed to occur according to the following "cook-to order" stages:

- *Sunny-side up.* The egg is cooked until the white is set and the yolk is still soft. The egg is not flipped.

Sunny-side up eggs may not be sufficiently cooked to eliminate bacteria, and thus some state health departments do not allow them to be served to the public. Covering the pan with a lid during cooking gives the yolk a rather opaque appearance, but eliminates any risk of an undercooked egg.

- *Over easy.* The eggs are flipped over when the whites are 75 percent set. Cooking continues until the whites are completely cooked but the yolks are still soft.
- *Over medium.* The same as over easy, except that the yolks are partially set.
- *Over hard.* The same as over easy, except that the yolks are completely set.

*Scrambled Eggs*  Scrambled eggs are beaten while raw until well blended and may be seasoned with salt and pepper or other seasonings. Liquid in the form of milk, cream, or water may be added to impart more body and/or flavor and a soft, creamy texture. The added liquid, a tablespoon or less for each egg, creates steam during cooking, which lifts the eggs and makes them lighter and fluffier. Too much liquid makes the eggs watery and forms small, tough, curd-like masses.

The beaten egg mixture is poured onto a heated surface, the heat is reduced, and the eggs are gently stirred as soon as they begin to coagulate. Too much stirring will break the egg into too many small pieces, so it is better to lift the cooked egg repeatedly with a spatula so the undercooked portions may slide underneath rather than literally to stir them. Scrambled eggs are finished cooking when they are set, yet still soft and moist. Like most egg dishes, they are best when served immediately. In restaurants or when cooking for large crowds, it is recommended that scrambled eggs be prepared in small batches, generally 3 quarts or less at a time (20).

*Omelets*  When eggs are beaten, cooked, and rolled into a cigar shape or folded into a flat half circle, the

**Prime (season)** To seal the pores of a pan's metal surface with a layer of heated-on oil.

resulting dish is called an omelet. Both plain (French or American-style) and puffy (fluffy) omelets can be prepared with or without fillings. Omelet preparation is considered so important by chefs that it is not unusual for a job applicant to be asked to chop an onion and make an omelet as part of the interview process.

Plain omelets consist of whole eggs, beaten, seasoned as desired, and poured into a prepared pan heated to medium-high. Once the mixture is in the pan, the heat is lowered to medium, and the mixture is not stirred. Uncooked portions are allowed to cook by lifting just the edges of the omelet with a spatula so the runny mixture flows underneath. When the top is firm, the omelet can be folded in half, rolled and folded over itself, or rolled and slid onto a dish (Figure 12-12). If fillings are to be added, they are placed on top of the omelet just before it is folded.

There are many opinions about how French and American omelets differ, but here are four basic differences:

- French omelets are never allowed to brown, whereas American omelets may have some color.
- French omelets never have texture lines, whereas an American omelet may have a few.
- Folding a French omelet occurs in threes (left over center, right over center) or complete rolling (like a rug), whereas American omelets are simply folded in half.
- The center of a French omelet is soft, whereas American omelet centers are fully cooked.

The fluffiness of puffy omelets is achieved by separating the yolks from the whites and whipping each portion separately. Seasonings and liquid, if added, are incorporated into the whipped yolks. The egg whites are whipped until stiff but not dry and then gently folded into the yolks. This mixture is poured into a preheated omelet pan or suitable frying pan with sloping sides, and the heat is reduced to medium. When the omelet is browned on the bottom, it is placed in a 350°F (177°C) oven for 5 to 10 minutes to allow additional rising and further coagulation of the surface proteins. The omelet is finished cooking when the top springs back from a gentle touch of the finger.

### Baking

Baking eggs and their ingredients leads to several different egg dishes: shirred eggs, meringues, and soufflés.

***Shirred Eggs*** Whole eggs that are baked and served in individual dishes are called shirred eggs. The egg is cracked, gently placed into a cup from which it can be rolled into a container coated with butter or margarine, and then baked in an oven at 350°F (177°C) until cooked to order.

**FIGURE 12-12** Omelet preparation.

(a)

(b)

(c)

(d)

(e)

(f)

(g)

Courtesy of the California Egg Board

(a) With wire whisk, beat eggs with salt and water just until well mixed, but not too frothy.

(b) Slowly heat a 9-inch skillet or omelet pan to medium heat. Temperature is correct when drops of cold water sizzle and roll off the pan's surface. Add butter and heat until it sizzles but is not yet brown. Gently pour the egg mixture into the heated pan.

(c) As egg mixture begins to set, use a spatula to push mixture back and allow unset egg to run onto pan's surface.

(d) Continue loosening and tilting until omelet is almost dry on top and golden-brown underneath.

(e) Add filling ingredients.

(f) Using spatula, fold omelet over itself.

(g) Tilt the pan and use the spatula to slide the omelet onto a plate.

*Meringue*   A meringue is an egg-white foam used in dessert dishes as a pie topping, a cake layer, or as frosting. It may also serve as a dessert on its own or be combined in other ways with dessert ingredients. Meringues are made by whipping egg white into foam and adding sugar, the amount of which determines whether the meringue is soft or hard.

Soft meringues are made with about 2 tablespoons of granulated (preferably superfine) sugar per egg white and are often used as pie toppings (e.g., lemon meringue pie). The sugar is gradually added to the egg whites—three will cover an average pie—and the mixture is whipped to the soft peak stage. The meringue is then spread immediately over the still-warm filling. A warm filling is necessary so the egg-white proteins can coagulate and bind to it. The whole pie with the meringue is then baked in the oven at between 325°F (163°C) and 350°F (177°C) for about 15 minutes. A temperature that is too low dries the meringue; a temperature that is too high shrinks it.

Some problems that can occur when preparing soft meringues are shrinking, **weeping,** and **beading.**

- *Shrinking.* To prevent the meringue from shrinking back and leaving an unsightly gap around the outside edges of the pie, it should be spread to slightly overlap the entire perimeter of the crust.
- *Weeping.* Also known as **syneresis,** weeping may be caused by underbeating the eggs, which leaves unbeaten whites on the bottom of the beating bowl, or by undercoagulation, created, for example, by placing meringue on a cold pie filling. A meringue can be protected from weeping by adding a teaspoon of cornstarch to the sugar before beating it into the egg whites.
- *Beading.* Undissolved sugar is the main cause of beading, but overcooking (overcoagulation) also contributes to this phenomenon. Beading can be avoided by using shorter cooking times and increasing the temperature up to 425°F (218°C).

Hard meringues are usually baked as cookies, but they can be formed into different shapes and used as decorations on puddings or other desserts. They are prepared with twice the amount of sugar used in soft meringues, about 4 tablespoons (¼ cup) per egg white. Confectioner's sugar is preferred over granulated sugar for use in hard meringues, because it is more evenly distributed through the beaten egg whites and lacks a gritty texture. Egg whites are beaten to the stiff stage, the sugar is beaten in, and the resultant meringue is shaped, placed on a parchment-covered baking sheet, and baked at the low temperature of 225°F (107°C) for about an hour or longer, depending on the size of the individual portions. When the meringue is delicately browned and the end product firm, the oven is turned off, the door left open, and the meringue left in the cooling oven for at least 5 minutes. Once the meringue is removed from the oven, the remainder of the cooling period should occur in a warm place free of drafts.

*Soufflés*   A soufflé is actually a modified omelet. The main ingredients of a soufflé are a thick base generally made from a **white sauce** or pastry cream, an egg-white foam, and flavoring ingredients. Initially, the egg yolks and whites are separated. A thick white sauce or pastry cream is prepared and combined with the egg yolks. Stiffly beaten egg whites are folded into the thick egg yolk mixture (Figure 12-13). For a main dish soufflé, flavoring ingredients such as diced or grated cheese, cooked meat, cooked seafood, and/or vegetables and seasonings are added to this mixture. Dessert soufflés will include sweet ingredients like sugar, chocolate, or fruit, but the process is the same.

Whichever the type of soufflé, the entire combination is gently poured into a lightly greased soufflé dish or other deep baking dish, placed in a larger pan of hot water, and baked in a moderate (350°F/177°C) oven for 50 to 60 minutes or until delicately browned and firm to the touch. Small, individual soufflés will take less time. The oven door should not be opened during baking until time to check for doneness, because it creates a draft that can cause the soufflé to fall. Doneness is determined by gently shaking the oven rack. If the center jiggles, even slightly, more baking time is required.

When combining beaten egg whites with other heavier mixtures, it is best to pour the heavier mixture onto the beaten egg whites. Then gradually, using a spoon or rubber spatula, combine the ingredients with a downward stroke into the bowl, across, up, and over the mixture. Come up through the center

**FIGURE 12-13**   Folding egg whites.

When combining beaten egg whites with other heavier mixtures, it is best to pour the heavier mixture onto the beaten egg whites. Then gradually, using a spoon or rubber spatula, combine the ingredients with a downward stroke into the bowl, across, up, and over the mixture. Come up through the center of the mixture about every three strokes and rotate the bowl during folding. Fold just until there are no streaks remaining in the mixture. Avoid stirring, which will force air out of the egg whites.

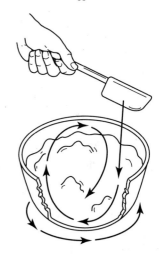

of the mixture about every three strokes and rotate the bowl during folding. Fold just until there are no streaks remaining in the mixture. Avoid stirring, which will force air out of the egg whites.

## Moist-Heat Preparation

Eggs can be prepared by moist heat using a variety of methods. Most common among these are "boiled" eggs, coddled eggs prepared in a cup, poached eggs, a variety of custards, and eggs that are prepared using the microwave. In all cases, eggs are cooked at simmering temperatures (see Chapter 5). Each of these methods and some of the egg dishes produced are now discussed in more detail.

**Weeping (syneresis)**   The escape of liquid to the bottom of a meringue or the formation of pores filled with liquid.

**Beading**   The formation of tiny syrup droplets on the surface of a baked meringue.

**White sauce**   A mixture of flour, milk, and usually fat.

### Hard or Soft "Boiled"

Although the term *hard-boiled eggs* is commonly used, eggs should actually be simmered and never boiled, because they will become tough and rubbery if so treated (6). The high heat of boiling also transforms the iron in the egg yolk into ferrous sulfide, causing the greenish-black color and unpleasant flavor found in the yolk of overly hard-cooked eggs (Chemist's Corner 12-2).

There are two methods for hard-cooking eggs: hot start and cold start. Each has advantages and disadvantages; each produces acceptable products.

*Hot-Start Method*  In the hot-start method, the water is heated to boiling and then the eggs are completely immersed in the boiling water. The heat is immediately reduced to simmer, and the eggs are cooked for 3 to 15 minutes, depending on the desired doneness:

- Soft        3 to 4 minutes
- Medium    5 to 7 minutes
- Hard       12 to 15 minutes

The cooked eggs are drained and then rinsed under cold running water to stop further cooking from residual heat. The extreme temperature change from hot to cold also helps loosen the egg's membrane from the shell, making it easier to peel. To further ease peeling, the first crack should be made at the air cell located at the larger end of the egg, and

**FIGURE 12-14**  Egg-cutting tools.

**Slicer**  A device that cuts a hard-cooked egg into neat slices with one swift stroke.

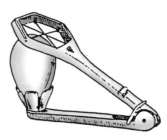

**Wedger**  Similar to a slicer, but it cuts the egg into 6 equal wedges rather than into slices.

then the egg rolled gently between the hands to break the shell all over. Peeling under cold running water also makes the job easier. Fresher eggs are harder to peel because the air cell is smaller and the membrane is tight against the cell wall. Although the larger air cell and higher pH of older eggs makes them easier to peel, they also tend to break more easily during heating.

The benefits of using the hot-start method are greater temperature control, eggs that are easier to peel, and a shorter total cooking time. A drawback is that lowering the eggs into boiling water may cause them to crack.

*Cold-Start Method*  In the cold-start method, the eggs are placed in a saucepan with enough cold water to cover them by at least an inch. The water is brought to a boil, immediately lowered to a simmer, and the eggs are then cooked to order:

- Soft        1 minute
- Medium    3 to 5 minutes
- Hard       10 minutes

Another way to prepare hard-cooked eggs from a cold start is to remove the pan from the heat as soon as the water boils, cover it tightly, and let it stand for 20 minutes. Cold-start eggs are less likely to crack during cooking.

The advantages to the cold-start method are that less attention to the process is required, the eggs are easier to add to the water, and they are less likely to break. On the other hand, starting eggs out in cold water may cause the egg white by the shell's surface to be more rubbery, and there is a greater chance of a greenish tint forming on the egg white.

Once cooked, eggs can be cut into slices or wedges using the equipment

shown in Figure 12-14. Dipping the knife in hot water before slicing keeps the hard-cooked eggs from falling apart. To tell a hard-cooked egg from a raw one, spin the egg on its side. A smoothly spinning egg is hard cooked, while one that wobbles out of balance is not.

> **? How & Why?**
>
> *Why is vinegar added to boiling water when cooking eggs?*
>
> Vinegar is often added to the water used to hard-boil eggs. There are two reasons for this. First, the acid in the vinegar dissolves the external surface of the shell, making it easier to peel away. Secondly, if the egg cracks while boiling, the vinegar will encourage the egg white to harden, preventing the entire egg from leaking out into the water.

> **? How & Why?**
>
> *Why do boiled eggs sometimes crack?*
>
> Eggs crack because the pressure created by fast-heating water pops the shell. Although simmering hard-cooked eggs makes them prone to cracking, this can be avoided by first warming eggs to room temperature in a bowl of warm water. Another way to reduce the chances of cracking is to push a sterilized pin or needle through the large end of the shell where the air cell is located. To reduce cracking still further, a spoon or other utensil may be used to place the eggs gently in the water.

## Coddling

Coddled eggs are prepared by breaking an egg into a small cup, called a coddler, made of porcelain or heat-proof glass with a screw-on top, and submerging the whole coddler in simmering water until the egg is cooked. The coddler should be buttered or greased before adding the raw egg. Cream or other flavorings such as ham or bacon are sometimes added before cooking. Once done, the egg is eaten directly out of the coddler.

## Poaching

Eggs are poached by being cracked and simmered in enough water to cover the egg by at least twice its depth. Fresh USDA Grade AA eggs are best to use for poaching, because the whites are firmer and less likely to spread out in the water and create *streamers,* floating strands of partially cooked egg whites. Salt (½ teaspoon per cup) and/or vinegar (1 teaspoon per cup) may be added to the water to speed coagulation and help to maintain a compact, oval shape of the egg. On the other hand, salt or vinegar will give the cooked egg a shinier, tougher, and, perhaps, more shriveled surface than one cooked in plain water. Poached eggs are cooked for 3 to 5 minutes, removed with a slotted spoon, drained, trimmed of any streamers, and served immediately. The well-poached egg should have a firm yolk and compact white. Poached eggs are commonly used for eggs Benedict, consisting of an English muffin layered with a slice of ham or Canadian bacon, followed by a poached egg, and topped with a dollop of hollandaise sauce.

## Custards

Custards are mixtures of milk and/or cream, sweeteners (sugar, honey), flavorings (vanilla, nutmeg, etc.), and eggs or egg yolks. Custards are thickened by the coagulation of egg proteins during cooking. These egg proteins denature when heated and recombine to form a network that sets or coagulates, at the right temperature, to form the solid gel of a custard. All custard dishes are very susceptible to microbial contamination and should be covered and refrigerated as soon as possible after preparation.

Custards are distinguished by whether they are sweet or savory, and by their preparation method: stirred or baked.

**Sweet and Savory Custards** Sweet custards are served as desserts in the form of puddings or as fillings for éclairs and pies. Savory (nonsweet) custards are used for dishes such as quiches. A popular quiche made with bacon and Swiss cheese is known as quiche Lorraine.

**Stirred Custard (Soft Custard or Custard Sauce)** The ingredients of this custard are stirred while being heated on the range over low heat or in a double boiler. The mixture retains a smooth, creamy, fluid consistency. Stirred custard is often eaten as a pudding; however, it may provide the base for many frozen desserts; be served as a sauce for cake, fruit, and other desserts; or be used to replace eggnog. The repeated stirring prevents the formation of a gel, so the custard mixture thickens instead of gels.

**Baked Custard** Baked custards are actually an example of dry-heat preparation. Both types of custards begin with the same ingredients, but are simply heated differently. Baked custard mixes are poured into ungreased custard cups that are placed in the oven, usually in a water bath (bain-marie), where they sit undisturbed and gel during baking. A water bath is made by filling a large, low-sided pan with 1 inch of hot water, into which the cups containing the custard mix are placed. The layer of water insulates the cups and prevents the outside of the custard from cooking to completion before the inside has had a chance to coagulate.

The internal temperature of custards should never be allowed to rise above the point of coagulation of the egg-liquid mixtures (185°F/85°C). Overheating causes the egg proteins to shrink, allowing liquid to be released from the egg and producing a product with a curdled, weepy, porous appearance. Another problem with baked custards is that they tend to have a runny texture, which makes them unsuitable for making solid pie fillings. As a result, custards to be used as pie fillings are often thickened with starch in the form of cornstarch or flour.

Custards should be baked at 350°F (177°C) until a knife inserted in the middle of the cup comes out clean—about 23 to 25 minutes for custard cups and 35 to 40 minutes for a casserole-size

dish. Just before it is completely done, the custard is removed from the oven and placed on a rack. Some additional cooking will inevitably occur during cooling, but can be minimized by using a cooling rack. Should the custard be overcooked, the cups can be set in ice water to stop further coagulation. Undercooking should likewise be avoided, because it will prevent the custard from setting properly.

## Microwaving

Eggs cook extremely rapidly in a microwave oven, so special caution should be taken to avoid overcooking. Manufacturer's instructions should be followed for microwave egg cooking. Whole eggs with intact shells should never be microwaved, because steam expanding within the shell can cause them to burst. The same principle applies to whole eggs out of the shell, because the vitelline membrane around the egg yolk traps steam and will burst if not punctured with a toothpick or the tip of a knife prior to going into the microwave.

**Fried** A browning dish is required to fry an egg in a microwave, and should be preheated on full power for 2 minutes, plus one additional minute for each egg being fried. About ½ teaspoon or less of fat per egg is melted in the dish before adding the cracked eggs from a bowl into the dish. The yolk membrane is punctured, and the dish is covered with plastic wrap, then microwaved on high for 45 seconds per egg or until the desired doneness is reached.

**Shirred** Shirred eggs are cooked in individual containers and are ideal for cooking in a microwave oven. The egg is placed in a custard dish, the yolk is punctured, the dish covered, and the egg is heated on medium for 45 to 60 seconds. It should be rotated a quarter turn at the half-minute mark.

**Scrambled** Before scrambling, 1 teaspoon of butter is melted in a 2-cup glass measure by setting the microwave very briefly on high. The beaten eggs are placed in the measuring cup and microwaved on high for 20 seconds. The egg mixture is then stirred, and the heating and stirring process is repeated one or two more times. Microwaving is completed when the eggs are just past the runny stage.

They should be allowed to stand one or two minutes if a firmer set is desired.

*Poached*    To poach an egg, ¼ cup of water, with a dash of vinegar and salt, is heated to a boil in a custard dish or 1-cup glass measure. The egg is dropped into the hot water, and the yolk is pierced with a toothpick. The dish is partially covered with plastic wrap and then heated at 50 percent power for about a minute, plus or minus 15 seconds. Allowing the cup to stand 2 to 3 minutes and gently shaking back and forth help to set the egg whites.

*Omelet*    Omelets can be prepared in the microwave by using a browning dish or 9-inch pie plate. Enough butter is added to slightly coat the bottom of the dish. It is melted on high and then spread evenly by tilting the container. The combined eggs, liquid, and seasonings are then poured into the container, covered with plastic wrap, and cooked on medium for 2 to 3½ minutes, or until the omelet is almost set. After removing the omelet from the oven, any fillers are added, and the omelet is folded over with a spatula. Puffy omelets are prepared in the same manner. An omelet cooked in a microwave will not brown unless a browning dish is used.

*Quiche*    A quiche dish or pie shell is filled with cooked vegetable and/or meat ingredients. Cream (preheated, unlike the cream in conventionally cooked quiches) is added to the beaten eggs, and the egg mixture is then poured over the vegetables and baked according to the manufacturer's guidelines.

# STORAGE OF EGGS

Eggs begin to deteriorate as soon as they are laid and lose quality very rapidly at room temperature. In fact, an egg will age more in 1 day at room temperature than in 1 week in the refrigerator. To ensure the freshness of whole or liquid eggs, they may be refrigerated, frozen, or dried. Proper storage is critical for eggs because they contain all the nutrients necessary to support the growth of a bird. Their nutrient content, especially water and protein, is ideal for bacterial growth and for this reason eggs should always be properly stored.

## Refrigerator

### Whole Eggs

An eggshell is not airtight. There are as many as 17,000 tiny pores over the shell's surface, so keeping eggs in the carton and refrigerated helps to retain their freshness. Several signs distinguish fresh eggs from those that have aged. Changes in proteins over time cause egg whites to thin. Fresh eggs also have more prominent, viscous chalazae on either side of the yolk than older eggs. In the process of aging, the vitelline membrane weakens and the yolk migrates or breaks (11). The size of an egg's air cell provides another indication of its age (Figure 12-15). The air cell gap between the membranes increases in size as the egg ages because moisture and carbon dioxide escape through the porous shell. Although the air cell size increases with age, the purpose of this space in every egg is to provide the chick with some air for its first breath, which it needs to break out of the shell.

Proper refrigeration of eggs helps to delay these changes and protects them from microbial growth, thus helping to maintain their quality. Many home refrigerators have built-in egg containers, but eggs retain their moisture better and keep longer if stored in the carton. It also helps prevent flavors and odors from being absorbed through their porous shells. Eggs should be stored in dry conditions, so damp cartons should not be used. Individual eggs should sit in the carton with their large ends up to prevent the air cell from moving toward the yolk (20). Washing eggs is not recommended, because this will remove the oil coating applied by the processor to prevent microbial growth and moisture loss.

### Shelf Life of Refrigerated Eggs

Refrigerated whole eggs should stay fresh for about 1 month even though the recommended storage time is 1 week. Separated egg yolks may be stored submerged in water in the refrigerator, but should be used within 2 days. Egg whites kept tightly covered in a glass container will last up to 4 days. When multiple cartons of eggs are purchased, the eggs should be used according to the "first in-first out" method—meaning that the oldest eggs are used first (20).

### Storage Eggs

Restaurants, food service institutions, and other food manufacturers must be especially careful about storing eggs, because they purchase such large quantities. **Storage eggs,** used by commercial food service establishments, are usually used within a month, but can be stored for up to 6 months. They are not available at the retail level. The coating of oil or plastic prevents microbial invasion and any loss of moisture or carbon dioxide.

### Pasteurized Eggs

The USDA regulations require that all liquid, frozen, or dried eggs be pasteurized or otherwise treated to protect against *Salmonella*. Commercial outlets frequently use refrigerated liquid eggs that have been pasteurized. Typical processed food products that may incorporate pasteurized liquid egg whites include baked goods, candies, and chilled or frozen desserts.

**Storage eggs** Eggs that are treated with a light coat of oil or plastic and stored in high humidity at low refrigerator temperatures very close to the egg's freezing point (29°F–32°F/–1.5°C–0°C).

**FIGURE 12-15**    The size of the air cell sac increases as eggs age.

After they are pasteurized, the advantages of liquid eggs over whole eggs or even frozen egg blends are convenience; consistent quality; microbial safety; and costs savings in terms of space, labor, and freezing (23). New to the scene are eggs pasteurized in their shells that are being marketed to the consumer (21).

## Frozen

Freezing a whole egg is not possible because the expanding liquids will cause it to crack. Food manufacturers solve this dilemma by breaking the eggs open at the processing plants where the contents are frozen whole (whites and yolk mixed together) or separated as whites or yolks. Prior to being frozen, the liquid whole eggs are usually pasteurized (140°F–143°F/ 52°C–55°C for 3½ minutes). Egg whites by themselves denature if pasteurized, so prior to this process they are often combined with a small amount of lactic acid and aluminum sulfate (Chemist's Corner 12-3). There are several drawbacks to using frozen pasteurized eggs: they are costly to freeze and keep frozen, they must be thawed, they are cumbersome to portion, and they have lowered functional quality (10).

Fortunately, separated egg whites are not adversely affected by freezing. Some commercially frozen egg whites have added stabilizers and whipping aids to improve their ability to form large, stable foams. For separated yolks, sugar, corn syrup, or salt is added (2 to 10 percent) to prevent them from becoming viscous and rubbery when thawed. Salt is used in frozen eggs only if they will be incorporated into sweet foods that will partially mask the salty taste. When freezing eggs at home, 1 tablespoon of sugar (corn syrup) or ½ teaspoon of salt

is added for every cup of blended eggs. Raw egg whites can be frozen with no special measures taken.

## Dried

Drying eggs is a simple process. Whole eggs or separated yolks are spray-dried to create a fine powder, which is mixed with anti-caking substances to prevent clumping. Egg whites are dried in different ways to form granule, flake, or milled textures. Dried eggs sometimes brown because of the Maillard reaction, but this can be prevented by removing glucose from the eggs before drying with the aid of an enzyme (glucose oxidase) or by yeast fermentation. Once dried, eggs can be stored in the refrigerator for up to a year, but they must be kept in tightly closed containers to prevent the clumping that can result from moisture accumulation.

Dried eggs, used extensively by food manufacturers, are particularly advantageous when storage and refrigeration space is limited. The major disadvantage of using dried eggs is that they lose many of the functional and sensory qualities of eggs, and are highly susceptible to bacterial contamination. Therefore, they should be used only when the end product will be thoroughly heated.

### Rehydrating Dried Eggs

Dried eggs are used in food preparation by adding them to water or by sifting them with dry ingredients. One egg can be reconstituted by sprinkling 2 tablespoons plus 1½ teaspoons of sifted egg powder over an equal amount of lukewarm water and beating until smooth. Combining ½ cup each of sifted egg powder and water produces the equivalent of three eggs. The mixtures should be used within 5 minutes. Table 12-5 lists the amount of frozen, refrigerated, or dried eggs needed to substitute for regular large eggs.

## Safety Tips

The chances of an egg being internally contaminated are relatively low, less than 1 in 10,000 commercial eggs (18). It is more common for contamination to

### CHEMIST'S CORNER 12-3

#### Egg Pasteurization

Aluminum sulfate is sometimes added to egg whites prior to pasteurization to stabilize conalbumin, an egg protein that is labile (unstable) to heat at pH 7.0. Most of the other egg proteins are stable to heat at this pH.

**TABLE 12-5   Substituting Frozen, Refrigerated, or Dried Eggs for Regular Large Eggs**

| Frozen or Refrigerated Liquid Eggs | Weights | Volume |
|---|---|---|
| *Whole Large Eggs:* | | |
| 1 | 1¾ oz | 3 tbs |
| 10 | 1 lb 1¾ oz | 2 C |
| 12 | 1 lb 5½ oz | 2½ C |
| 25 | 2 lbs 13 oz | 1 qt 1¼ C |
| 50 | 5 lbs 8 oz | 2 qt 2½ C |
| *Yolks:* | | |
| 10 | 7¼ oz | ¾ C |
| 12 | 8½ oz | ¾ C 2 tbs |
| 22 | 1 lb | 2 C less 2 tbs |
| *Whites:* | | |
| 10 | 11½ oz | 1¼ C 2 tbs |
| 12 | 14 oz | 1½ C 2 tbs |
| 14 | 1 lb | 2 C less 2 tbs |

| Dried Eggs | Weight/Volume | Water (to reconstitute dried eggs) |
|---|---|---|
| *Whole Large Eggs:* | | |
| 6 | 3 oz (1 C) | 1 C |
| 12 | 6 oz (2 C) | 2 C |
| 24 | 12 oz (1 qt) | 1 qt |
| 50 | 1 lb 9 oz (2 qt and ⅓ C) | 2 qt and ⅓ C |
| 100 | 3 lb 2 oz (2 qt and ⅔ C) | 1 gal and ⅔ C |
| 150 | 4 lb 11 oz (6 qt and 1 C) | 6 qt and 1 C |
| 200 | 6 lb 4 oz (2 gal and 1⅓ C) | 2 gal and 1⅓ C |

**Source:** American Egg Board.

occur during handling and preparation after the egg has been removed from its shell. Even so, eggs are an excellent breeding ground for microbes, and can become internally contaminated from a hen with a *Salmonella enteritidis* infection in her ovary or oviduct (17), or from absorbing bacteria through the pores. The latter can occur if the eggs are boiled and then cooled in the presence of infected water or an infected food handler. Externally, the eggs may also be exposed to *Salmonella enteritidis* by fecal contamination during egg laying. The Centers for Disease Control and Prevention implicated eggs as the source for 73 percent of *Salmonella enteritidis* outbreaks (25), and there is an increasing possibility that *Listeria monocytogenes,* which can grow at refrigerator temperatures and has already been observed on whole eggs, may also contribute to future outbreaks (19, 21). There are many precautions that can be taken to prevent foodborne illness from eggs:

### Purchase

- Only buy refrigerated eggs. Refrigerate promptly after purchase or during preparation.
- Open the carton to make sure eggs are clean and not cracked.

### Preparation

- Do not add raw egg to already scrambled eggs, a practice sometimes used in food service operations to increase the moisture content of dried scrambled eggs.
- Use an egg separator rather than passing the yolk back and forth between the two shell halves.
- Additionally, eggshells can be potentially hazardous if left in food products (20).
- Cook eggs until no visible liquid egg remains, especially when preparing French toast, scrambled eggs, poached eggs, and omelets.
- A knife inserted into baked egg dishes such as quiches, baked custards, and most casseroles should come out clean.
- Scrambled eggs should be held on cafeteria and buffet lines at appropriate temperatures.
- Be extra cautious when preparing lightly cooked egg dishes such as mousse, meringue, and other similar dishes, because they may not be sufficiently cooked to eliminate possible bacteria.
- The USDA recommends that consumers heat all egg dishes to 160°F (71°C). The FDA suggests 145°F (63°C) for at least 15 seconds as the final egg temperature to be reached by commercial food establishments.

### Consumption

- Raw eggs or undercooked eggs should never be consumed; this holds especially true for the very young, elderly, or immune compromised (people with conditions such as AIDS, cancer, etc.).
- The FDA recommends only pasteurized eggs for food items in which eggs are only lightly cooked or not at all, such as Caesar salad, uncooked hollandaise or béarnaise sauce, and homemade items such as mayonnaise, eggnog, and ice cream (16).

### Storage

- Always store eggs in the refrigerator, never at room temperature. Meringue-covered pies and other egg-containing foods should be refrigerated until served.
- Unopened liquid egg products can be stored for up to 7 days. Once opened, they should be used or discarded within 3 days.
- Dried egg mix should be used within 7 to 10 days of opening, and should be stored at a temperature of less than 50°F (10°C) (20).
- Cold egg dishes should be stored below 40°F (4°C); hot dishes should be kept above 140°F (60°C) for up to 1/2 hour, or 1 hour on a buffet according to the USDA (20).

The USDA provides more information about egg safety on its website: www.fsis.usda.gov/Fact_Sheets/Egg_Products_and_Food_Safety/index.asp and www.fsis.usda.gov/Fact_Sheets/Focus_On_Shell_Eggs/index.asp.

# PICTORIAL SUMMARY / 12: Eggs

As a life-sustaining protein, egg protein is so high in quality that it has become the standard (reference protein) by which researchers rate all other foods. The versatility of eggs, whether prepared alone or in combination with other foods, makes them nearly indispensable in cooking.

## COMPOSITION OF EGGS

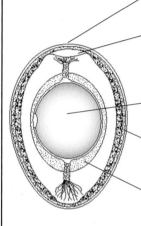

**Shell:** 12 percent of the egg's weight; porous calcium carbonate.

**Air cell:** located between the two shell membranes at larger end of the egg; becomes larger as egg ages.

**Egg yolk:** 30 percent of the egg's weight; dense in nutrients.

**Shell membranes:** inner and outer membranes protect the egg against bacterial invasion.

**Albumen** (egg white): 58 percent of the egg's weight; four layers, composed largely of water and protein.

**Nutrient Content:** Eggs are one of the very few foods containing all the fat-soluble vitamins (A, E, D, and K) and large amounts of certain water-soluble vitamins. Eggs contain all the essential amino acids, making them a good source of complete proteins. They contain very little carbohydrate or fiber.

A large egg contains:
- Calories: 75 kcal
- Protein: 7 grams
- Fat: 5 grams
- Cholesterol: 213 mg
- Vitamins: A, D, E, K, and several B vitamins
- Minerals: selenium, iodine, zinc, iron

## PURCHASING EGGS

Inspection of processing plants producing egg products is mandatory. Candling, Haugh units, and appearance are used to determine quality grading. Size, unrelated to grading, is determined by the weight of a dozen eggs.

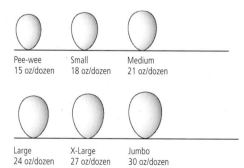

Pee-wee 15 oz/dozen    Small 18 oz/dozen    Medium 21 oz/dozen

Large 24 oz/dozen    X-Large 27 oz/dozen    Jumbo 30 oz/dozen

## TYPES OF EGGS

Chicken egg forms consist of standard (unfertilized white or brown), fertile, free-range, and organic. Non-chicken eggs include duck, goose, ostrich, quail, and turkey.

## FUNCTIONS OF EGGS IN FOODS

When combined with other ingredients, eggs have a unique ability to flavor, color, emulsify or thicken, bind, foam, interfere, and clarify.

- **Emulsifying:** Lecithin, found in egg yolks, keeps liquid compounds (like fat and water) from separating, thus thickening and stabilizing foods such as sauces and salad dressings.
- **Binding:** The high protein content in beaten eggs can act as an adhesive when cooked. Examples are egg used with breaded, fried foods and the addition of eggs to bind meatloaf.
- **Foaming:** Egg whites beaten into a foam increase their original volume six to eight times, and aerate and leaven food products such as soufflés and meringues. The best egg-white foam is made from fresh eggs at room temperature.
- **Interfering:** Because they interfere with the formation of crystals (ice, sugar), eggs are used to create a smooth, velvety texture in ice cream and candy.
- **Clarifying:** Egg whites are used to make clear soups.

## PREPARATION OF EGGS

To preserve the egg's texture, flavor, and color, it is best to keep the cooking temperature low and the heating time short. Eggs can be prepared in a great variety of ways, using either dry-heat or moist-heat methods.

**Dry-Heat Method**
- Fried, scrambled, omelets
- Baked (shirred, meringues, soufflés)

**Moist-Heat Method**
- Simmering
- Coddling
- Poaching
- Custards (stirred, baked)
- Microwaving

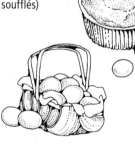

## STORAGE OF EGGS

Eggs begin to deteriorate as soon as they are laid, but can be preserved by refrigeration, freezing, and drying. Refrigerated eggs retain more moisture and keep longer when they are left in the carton.

The chances of an egg being internally contaminated are relatively low, but external bacterial contamination is possible and certain precautions can be taken to reduce this risk.

# CHAPTER REVIEW AND EXAM PREP

## Multiple Choice*

1. The white of an egg is also referred to as the
   _____.
   a. albumen
   b. beta-carotene
   c. yolk sac
   d. air cell

2. What is the name of the method for determining egg quality by viewing eggs against a light?
   a. Ripping
   b. Sizing
   c. Candling
   d. Emulsifying

3. In terms of food preparation, eggs are very functional for _____.
   a. emulsifying
   b. binding
   c. foaming
   d. All of the above

4. As eggs age, what happens to the size of the air cell?
   a. It increases.
   b. It decreases.
   c. It stays the same.
   d. It disappears.

5. What type of eggs should not be cooked in a microwave?
   a. Scrambled
   b. Poached
   c. Shirred
   d. Whole

6. What type of eggs should never be served to the very young, the elderly, or immune-compromised (HIV, cancer, etc.) individuals?
   a. Dried
   b. Pasteurized
   c. Liquid
   d. Raw

7. One large egg contains approximately
   a. 50 calories (kcal), 15 g of protein, 10 g of fat, and 150 mg of cholesterol.
   b. 75 calories (kcal), 5 g of protein, 5 g of fat, and 215 mg of cholesterol.
   c. 100 calories (kcal), 10 g of protein, 10 g of fat, and 250 mg of cholesterol.
   d. 150 calories (kcal), 15 g of protein, 15 g of fat, and 300 mg of cholesterol.

## Short Answer/Essay

1. List and describe the five components of an egg's structure.
2. Describe the protein, fat, cholesterol, vitamin, and mineral content of eggs.
3. List the USDA quality grades for eggs and describe the methods used to grade eggs.
4. Describe the following functions of eggs in food preparation: emulsifying, binding, interfering, and clarifying.
5. Explain how the following affect the quality of whipped egg whites: beating technique, temperature, bowl type, separation of eggs and whites, sugar, fluid, salt, and acid.
6. Describe the difference in rate of coagulation for egg whites and yolks, and explain how added ingredients change the coagulation temperature.
7. Discuss how undesirable color changes may occur during egg preparation and explain what can be done to prevent this.
8. Briefly describe the process of preparing the following types of egg dishes: over easy, shirred, soufflé, and poached.
9. What is a meringue? Describe the causes that may contribute to weeping and beading.
10. List at least six precautions that may be taken to prevent foodborne illness from eggs.

*See p. AK-1 for answers to multiple choice questions.

# REFERENCES

1. Anton M, and G Gandemer. Composition, solubility, and emulsifying properties of granules and plasma of egg yolk. *Journal of Food Science* 62(3):484–487, 1997.
2. Baniel A, A Fains, and Y Popineau. Foaming properties of egg albumen with a bubbling apparatus compared with whipping. *Journal of Food Science* 62(2):377–381, 1997.
3. Bringe NA, and J Cheng. Low-fat, low-cholesterol egg yolk in food applications. *Food Technology* 49(5):94–106, 1995.
4. Cyclodextrins remove cholesterol from eggs. *Food Engineering* 64(8):42, 1992.
5. deMan JM. *Principles of Food Chemistry*. Van Nostrand Reinhold, 1999.
6. Ehri A. How to boil an egg perfectly every time. *Fine Cooking* 83:71, 2006.
7. FDA orders egg company to halt sales. *Food Engineering* 63(3):36, 1991.
8. Gardiner A, and S Wilson. Whipping egg whites and whipped cream into stable, airy foams. *Fine Cooking* 66:74–75, 2005.

9. Gebhardt SE, and RH Matthews. *Nutritive Values of Foods*. Home and Garden Bulletin No. 72. U.S. Dept. of Agriculture, Washington, D.C.: GPO, 1994.

10. Giese J. Ultrapasteurized liquid whole eggs earn 1994 IFT Food Technology Industrial Achievement Award. *Food Technology* 48(9):94–96, 1994.

11. Jones DR, and MT Musgrove. Effects of extended storage on egg quality factors. *Poultry Science* 84(11):1774–1777, 2005.

12. Kovacs-Nolan J, M Phillips, and Y Mine. Advances in the value of eggs and egg components for human health. *Journal of Agriculture and Food Chemistry* 53(22):8421–8431, 2005.

13. Liang Y, and HG Kristinsson. Influence of pH-induced unfolding and refolding of egg albumen on its foaming properties. *Journal of Food Science* 70(3):C222–C230, 2005.

14. Light A, H Heymann, and DL Holt. Hedonic responses to dairy products: Effects of fat levels, label information, and risk perception. *Food Technology* 46(7):54–57, 1992.

15. McGee H. *On Food and Cooking: The Science and Lore of the Kitchen*. Macmillan, 1997.

16. Mermelstein NH. Pasteurization of shell eggs. *Food Technology* 55(12):72–73, 79, 2001.

17. Mollenhorst H, CJ van Woudenbergh, EG Bokkers, and IJ deBoer. Risk factors for *Salmonella enteritidis* infections in laying hens. *Poultry Science* 84(8):1308–1313, 2005.

18. Muriana PM, H Hou, and RK Singh. A flow-injection system for studying heat inactivation of *Listeria monocytogenes* and *Salmonella enteritidis* in liquid whole egg. *Journal of Food Protection* 59(2):121–126, 1996.

19. Palumbo MS, et al. Thermal resistance of *Listeria monocytogenes* and *Salmonella* ssp. in liquid egg white. *Journal of Food Protection* 59(11):1182–1186, 1996.

20. Powitz RW. Egg safety: Avoiding shell shock at retail. *Food Safety Magazine* 13(6):28–31, 2008.

21. Pszczola DE. Emerging ingredients: Still full of surprises. *Food Technology* 54(7):72–83, 2000.

22. Pszczola DE, and K Banasaiak. Ingredients. *Food Technology* 60(5):45–92, 2006.

23. Schuman JD, and BW Sheldon. Thermal resistance of *Salmonella* spp. and *Listeria monocytogenes* in liquid egg yolk and egg white. *Journal of Food Protection* 60(6):634–638, 1997.

24. Smith DM, et al. Cholesterol reduction in liquid egg yolk using betacyclodextrin. *Journal of Food Science* 60(4):691–694, 1995.

25. Tauxe RV. Emerging foodborne diseases: An evolving public health challenge. *Emerging Infectious Diseases* 3(4):425–434, 1997.

26. Yuno-Ohta N, et al. Gelation properties of ovalbumin as affected by fatty acid salts. *Journal of Food Science* 61(5):906–910, 1996.

## WEBSITES

The American Egg Board has an Incredible Egg website:
**www.aeb.org**

The Egg Nutrition Center has a website for egg education:
**www.enc-online.org**

The U.S. Poultry and Egg Association has a website focusing on research, education, and product promotion:
**www.poultryegg.org**

PhotoDisc/Getty Images

# 13 Vegetables and Legumes

Webster's dictionary refers to vegetables as "any plant" whose parts are used as food. In practice, a vegetable is the edible part of a plant (raw or cooked) accompanying the main course of a meal. Imagine a meal consisting of just meat, dairy products, and a starch. It is the vegetables (and fruits, discussed in the next chapter) that impart color and sometimes unique flavors and textures to meals. This chapter focuses on vegetables, specifically their classification, composition, purchase, preparation, and storage.

## CLASSIFICATION OF VEGETABLES

One method of classifying vegetables is to define them by the part of the plant from which they originated. For example, Figure 13-1 shows that vegetables may be derived from almost any part of a plant: roots (carrots, beets, turnips, and radishes); bulbs (onions and garlic); stems (celery and asparagus); leaves (spinach and lettuce); seeds (beans, corn, and peas); and even flowers (broccoli and cauliflower). In addition, there are foods that are routinely called vegetables and used as vegetables, but that are actually fruits. Botanically, fruits are the part of the plant that contains its seeds—specifically, the mature ovaries of plants. If it derives from a flower, then it is usually a fruit. The fruits most often seen masquerading as vegetables include tomatoes, squash, cucumbers, avocados, okra, eggplant, olives, water chestnuts, and peppers. Although many of the principles discussed in this chapter, such as composition and color, apply to both vegetables and fruits, the main focus here is on vegetables.

## COMPOSITION OF VEGETABLES

### Structure of Plant Cells

#### Cell Wall

Cells are the building blocks of both plant and animal organisms. One of the major differences between the two types of cells is that plants lack the skeletal structure that provides support in animal organisms. Instead, each plant cell gains its structural support by being surrounded by a sturdy wall (Figure 13-2). Contributing to the strength of these cell walls are several fibrous compounds that are indigestible by humans. Fiber

**FIGURE 13-1** Classification of vegetables.

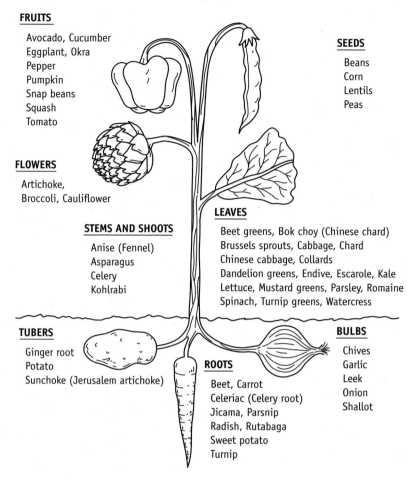

**FRUITS**
Avocado, Cucumber
Eggplant, Okra
Pepper
Pumpkin
Snap beans
Squash
Tomato

**SEEDS**
Beans
Corn
Lentils
Peas

**FLOWERS**
Artichoke,
Broccoli, Cauliflower

**STEMS AND SHOOTS**
Anise (Fennel)
Asparagus
Celery
Kohlrabi

**LEAVES**
Beet greens, Bok choy (Chinese chard)
Brussels sprouts, Cabbage, Chard
Chinese cabbage, Collards
Dandelion greens, Endive, Escarole, Kale
Lettuce, Mustard greens, Parsley, Romaine
Spinach, Turnip greens, Watercress

**TUBERS**
Ginger root
Potato
Sunchoke (Jerusalem artichoke)

**ROOTS**
Beet, Carrot
Celeriac (Celery root)
Jicama, Parsnip
Radish, Rutabaga
Sweet potato
Turnip

**BULBS**
Chives
Garlic
Leek
Onion
Shallot

**FIGURE 13-2** The structural differences between animal and plant cells.

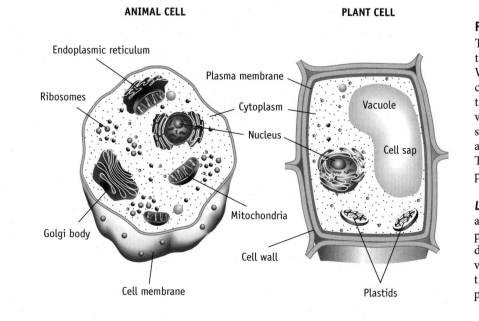

ANIMAL CELL

PLANT CELL

Endoplasmic reticulum

Plasma membrane

Ribosomes

Cytoplasm

Vacuole

Nucleus

Cell sap

Mitochondria

Golgi body

Cell wall

Cell membrane

Plastids

includes, but is not limited to, cellulose, pectic compounds, hemicellulose, lignin, and gums. Humans are unable to break down fiber because they lack the enzyme necessary to break down cellulose to the glucose molecules that the body can use.

Two indigestible fibers are pectic compounds and hemicellulose. These substances found within and between cell walls serve as a type of intra- and intercellular cement, giving firmness and elasticity to the tissues (53). The outer layer of the skin, peel, or rind has a higher proportion of cellulose and hemicellulose than the inner, thinner layers do. The surface cells of these protective layers secrete a waxy cutin, a water-impermeable coat that protects the plant.

Lignin is one of the few types of fiber found in foods that is a noncarbohydrate compound (it is made from polymers of phenolic alcohols). As vegetables mature, their lignin concentrations increase. This is why spinach stems and the inner cores of carrots, asparagus spears, and broccoli become tougher with age and do not soften when heated (50). Other fibrous compounds found in plants are gums, polysaccharides with a unique ability to absorb water and swell to several times their original volume. They are often added to processed foods such as ice cream, candies, and salad dressings to increase their viscosity. Examples of food gums derived from plants include algin, carob bean gum, carrageenan, gum arabic, gum guar, locust bean, gum tragacanth, and gum karaya.

## Parenchyma Cells

The most common type of cell in vegetables and fruits is the parenchyma cell. Within the jelly-like cytoplasm of these cells are the compounds responsible for the plant's starch content, color, water volume, and flavor. Several of these substances, such as starch and pigments, are stored in organelles called plastids. There are three types of plastids: leucoplasts, chloroplasts, and chromoplasts.

***Leucoplasts*** Leucoplasts store starch and some water. Starch stored by these parenchyma cells serves as the major digestible portion of the plant. Many vegetables and fruits are consumed for their starch content and the energy it provides.

### Chloroplasts and Chromoplasts

Certain plastids contain pigments responsible for the colors of many vegetables and fruits. The chloroplast plastids contain the chlorophyll that is essential for carbohydrate synthesis and provides the green color of plants. The orange-yellow colors of certain vegetables and fruits are derived from the chromoplast plastids that contain the carotene or xanthophyll pigments.

**Vacuoles**    Another function of the parenchyma cells is to store water and other compounds in sacs called vacuoles. The size of the cell is often related to the amount of liquid it holds, which determines the juiciness of the vegetable or fruit. The degree of juiciness ranges from very low in potatoes and bananas to very high in tomatoes and watermelon. The optimal water content provides **turgor** and crunchiness to leafy vegetables such as lettuce and spinach. Heat or humidity reduces turgor in plants. Other substances found in the cells' vacuoles besides liquid are the red-blue pigments known as anthocyanins (discussed below) and an array of flavor compounds such as saccharides, salts, and organic acids.

**Acids**    Organic acids found in the cell contribute to its pH and to the food's flavor and acidity. Most vegetables have a pH of about 5.0 to 5.6, with tomatoes slightly lower (pH 4.0 to 4.6) and corn, peas, and potatoes slightly higher (pH 6.1 to 6.3).

### Intercellular Air Spaces

Plant cells do not fit tightly next to each other, so intercellular spaces—the spaces *between* cells—fill with air, adding volume and crispness to vegetables and fruits. How close the cells are to each other influences the textural differences between fruits and vegetables in terms of their crispness. For instance, air spaces account for 20 to 25 percent of the volume of an apple, 15 percent of a peach, and only 1 percent of a potato. Without the air, both vegetables and fruits would be soft and flaccid.

**Turgor**    The rigid firmness of a plant cell resulting from being filled with water.

# Plant Pigments

Vegetables (and fruits) are clothed in all the colors of the rainbow, and so brighten meals that would otherwise look bland with only meat, dairy products, grains, and bread on the table. The selection of fruits and vegetables is often based upon the way they look, and color is an important attribute of a meal's appearance.

Plant pigments fall into three major groups: carotenoids, chlorophylls, and flavonoids (Figure 13-3). Carotenoids and chlorophylls are found in plastids and are fat soluble. Flavonoid pigments are water soluble, and have a tendency to be lost in cooking water.

### Carotenoids

Carotenoids (alpha-, beta-, and gamma-carotenes), along with lycopenes and xanthophylls, account for most of the yellow-orange and some of the red color of fruits and vegetables. Carotenes lend reddish-orange color to carrots and winter squashes. Lycopenes, which are a deeper red, provide the bright color of tomatoes (57). Light yellow xanthophyll pigments color pineapples.

Heat affects the color of vegetables, most likely because it modifies the pigments' chemical structure. Exposure to oxygen also causes oxidation of pigments and a resulting loss in color (4). Vegetables containing beta-carotene should not be overheated, because this pigment not only contributes to color but can also be converted to vitamin A; therefore, its destruction would be doubly undesirable (6) (Chemist's Corner 13-1).

### Chlorophyll

Chlorophyll, the pigment responsible for the green color of plants (Chemist's Corner 13-2), also makes possible the essential process of photosynthesis, in which leaves capture the sun's light energy to convert carbon dioxide and water to carbohydrates. It is not surprising that plants rich in chlorophyll include most of the leafy green vegetables

## CHEMIST'S CORNER 13-1

**Vitamin A Carotenoids**

The carotenoids contributing vitamin A activity include beta-carotene, alpha-carotene, and beta-cryptoxanthin (25). Beta-carotene can be split into two molecules with vitamin A activity (9).

**FIGURE 13-3**    Three classes of plant pigments.

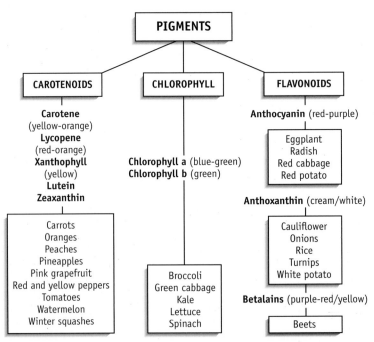

> **CHEMIST'S CORNER 13-2**

**Types of Chlorophyll**

There are two types of chlorophyll, and each produces a specific color in green plants. Chlorophyll a is blue-green, whereas the more common chlorophyll b is green. These chemical compounds differ in color because of their different structures—chlorophyll a has a methyl group ($-CH_2$) attached to one of the carbons on the chlorophyll molecule, whereas chlorophyll b has an aldehyde group ($-CHO$).

such as lettuce, spinach, broccoli, green cabbage, and kale.

In older plants or those picked and exposed to sunlight, chlorophyll is degraded, causing underlying pigments to show. This is why leaves may turn yellow in fresh parsley or broccoli florets left too long on the produce stand (61, 62). The process is similar to what happens in autumn, when the non-green colors, which have been in the leaves all along but masked by the darker, green chlorophyll, are allowed to show as the chlorophyll diminishes with the changing light and cooler temperatures.

### Blanching Enhances Green Color

Chlorophyll may also be hidden by the air between the cells in fresh vegetables. Sometimes the green color can be enhanced by blanching the vegetable, which causes the air to bubble away so that it no longer clouds the colors. The brighter green seen in frozen vegetables is caused by blanching them before freezing. Raw vegetable platters prepared by caterers are often more brilliant in color because the vegetables have been blanched and then dipped quickly in cold water to stop the cooking.

Chlorophyll, a fat-soluble pigment, is not as stable as carotenoid pigments. Color changes in green vegetables can be minimized by keeping heating times short, adding a small amount of sugar, and heating the food uncovered for the first few minutes to allow volatile organic acids to escape. Avoid heating green vegetables in a pan with the lid on because this causes the natural acids in the vegetables to concentrate and destroy the chlorophyll. A more alkaline cooking medium converts the chlorophyll pigment to a water-soluble form, which is what turns the cooking water green. After blanching, the vegetables are then "shocked" in cold water or ice water to stop carryover cooking.

? **How & Why?**

**Why do greens become duller in color when cooked?**

Although heating green vegetables causes them at first to turn a sharp, bright green, longer heating periods such as those required for canning cause them to turn a dull olive brown—the color often seen in canned spinach or peas (52). This heat-induced color change occurs when membranes that previously separated acids and pigments are disrupted, allowing acids to come in contact with chloroplasts. Hydrogens from the acids replace the magnesium in the chlorophyll, which results in an olive-brown compound known as pheophytin.

### Flavonoids

Flavonoid pigments include anthocyanins (red-blue), anthoxanthins (creamy to white), and betalains (purplish-red).

**Anthocyanins** Most of the red, purple, and blue colors seen in fruits and vegetables derive from anthocyanin. Although numerous fruits contain this pigment, it is found in only a few vegetables—red cabbage, eggplant, radish, and red potato (58). The color of the anthocyanins in these foods is affected by several factors, including changes in pH that may occur during simmering. Acidic tap water intensifies the red color of anthocyanins; alkaline water changes the reddish-blue hue first to an unappetizing blue and then to green. The latter process is sometimes observed in red cabbage and can be prevented by adding acid ingredients such as apple, lemon juice, or vinegar.

? **How & Why?**

**Why does red cabbage sometimes turn purplish-blue when cut with a knife?**

If a nonstainless metal knife is used to cut red cabbage, the anthocyanin pigment will react with the metal ions in iron, tin, or aluminum and create off-colors. Stainless steel and glass are consequently the best choices for the preparation and storage of foods rich in anthocyanins.

**Anthoxanthins** Anthoxanthins are actually a composite of compounds known as flavones, flavonols, and flavonones. They are the reason for the cream or white color of cauliflower, onions, white potatoes, and turnips. Further whitening can be achieved by adding acidic ingredients such as cream of tartar or vinegar. Anthoxanthins turn an undesirable yellow color in alkaline water, and can even change to blue-black or red-brown under excessive heating or in the presence of iron or copper.

**Betalains** Betalain pigments (red betacyanins and yellow betaxanthins) give beets their deep purplish-red color. It is important to leave beets unpeeled until after they are cooked in order to prevent their rich color from bleeding out into the water. For the same reason, 1 or 2 inches of stem should be left at the top during cooking. Acidic ingredients such as vinegar convert the purplish-red hue of beets to a brighter red. In an alkaline medium, the red color shifts to yellow.

## Plants as Functional Foods

Research now indicates that fruits and vegetables may carry a vast array of phytochemicals—nonnutritive compounds in plants that possess health-protective benefits (11, 47). Phytochemicals from foods, rather than manufactured supplements, are the preferred source. The **cruciferous** vegetables, as well as tomatoes, strawberries, pineapples, and green peppers, contain phytochemicals that appear to inhibit cancer in laboratory animals (Chemist's Corner 13-3).

Lycopene is the compound responsible for the red color of tomatoes and some fruits, such as watermelon. It is a powerful antioxidant, and a high intake of lycopene is linked to decreased risk of prostate and digestive tract cancers (2). Luteolin is a compound found in broccoli, celery, cabbage, spinach, green pepper, and cauliflower. In laboratory studies, luteolin has been shown to decrease the growth of cancer cells and have anti-inflammatory properties; however, many plant extracts have been reported to have similar properties. High dietary levels of soy are linked to decreased risk of breast and uterine cancers. Genistein, a compound in soybeans, is thought to be responsible for some of these effects. Epidemiologic evidence suggests that increased intake

**Cruciferous** A group of indole-containing vegetables named for their cross-shaped blossoms; they are reported to have a protective effect against cancer in laboratory animals. Examples include broccoli, brussels sprouts, cabbage, cauliflower, kale, mustard greens, rutabaga, kohlrabi, and turnips.

**Legumes** Members of the plant family *Leguminosae* that are characterized by growing in pods. Vegetable legumes include beans, peas, and lentils.

# NUTRIENT CONTENT

With few exceptions, fresh, unprocessed vegetables are naturally low in calories, cholesterol, sodium, and fat. Most vegetables are good sources of carbohydrate (including fiber), vitamins, and minerals—vitamin C, beta-carotene, certain B vitamins, calcium, and potassium. It has been suggested by researchers that this nutrient combination, in addition to other nonnutritive compounds found in a plant-based diet of vegetables, fruits, whole-grain foods, and **legumes** (especially soy products), contributes to a reduced risk for certain cancers (2).

*Cholesterol.* Plants contain no cholesterol. For example, peanuts are high in fat but do not contain even a trace of cholesterol. Only animals having a liver are capable of manufacturing cholesterol, so only animal products contain cholesterol. However, cholesterol may be added during processing of a plant-based food item.

*Fat.* Unprocessed vegetables contain little or no fat. The only vegetable foods high in fat are the processed vegetable oils derived from plant seeds—corn, peanut, rapeseed, cottonseed, safflower, and others.

*Carbohydrates.* Some vegetables are so high in complex carbohydrates that they can substitute for grain-based starches. Potatoes, legumes, and corn fall into that category. During the ripening process, vegetable sugars convert to starch for storage. Corn, carrots, and peas taste sweeter when harvested early because their sugars have less time to convert to starches. In fruits, the opposite occurs; starches convert to sugars, which explains why a ripe melon tastes sweeter than an unripe one.

*Fiber.* Many, but not all, vegetables are rich in dietary fiber. Fiber content varies a good deal among plant foods, as shown in Figure 13-4. For example, iceberg lettuce contains only 1 gram of fiber per cup; one would have to consume 20 to 30 cups of lettuce to obtain the recommended 20 to 30 grams of daily fiber, whereas only 1½ to 2 cups of kidney beans would have to be eaten to obtain a similar amount.

*Protein.* Vegetables lack certain essential amino acids and therefore are not a source of complete protein. The most complete sources of protein in the plant kingdom are

**FIGURE 13-4** Vegetables high in fiber (½ cup cooked unless otherwise noted).

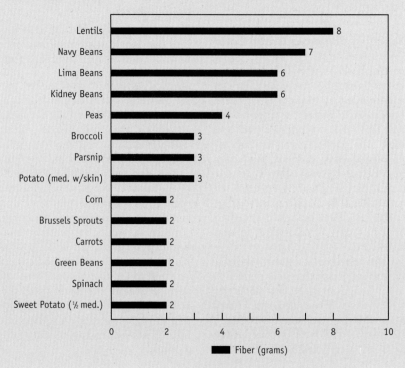

# NUTRIENT CONTENT (*continued*)

the legumes, which nevertheless tend to be low in the essential amino acid methionine (10). Most vegetables average only about 3 percent protein and are far from complete, so various combinations of legumes and cereals serve as a complete protein and act as a dietary staple for much of the world's population (12).

*Vitamins and Minerals.* Vegetables are usually higher in vitamins and minerals than fruits. Sprouted beans are high in vitamin C and riboflavin (vitamin B$_2$). Dark green, leafy vegetables are good sources of riboflavin, beta-carotene, vitamin C, and iron. Although 80 percent of the calcium in an average North American diet is derived from dairy sources, the next most important source is green vegetables such as broccoli. However, compounds called oxalates, found primarily in green vegetables, can bind to the calcium, zinc, or iron in these vegetables and may decrease their absorption in the body (5). Green vegetables are excellent sources of vitamin K and folate, a B vitamin, the need for which almost doubles during pregnancy.

Fruits and vegetables also contain electrolytes such as potassium and sodium, although the amount of sodium in fresh produce is of negligible significance in the human diet. Canning often increases the sodium content, but lower-sodium versions of many processed foods and beverages are now available on the market.

> ## CHEMIST'S CORNER 13-4
>
> ### Cancer and DNA Damage
>
> One of the ways that cells become cancerous is to sustain damage to their genetic material, or DNA. This damage can be caused by free radicals, which are highly reactive molecules that interact with and harm DNA. Certain metals, ultraviolet radiation, or chemicals such as cigarette smoke can form free radicals inside of cells. Antioxidants prevent the formation of free radicals by binding to these harmful chemicals or metals, and by activating proteins within cells that eliminate free radicals (23). By eliminating free radicals, antioxidants may help to prevent DNA damage, and possibly reduce the risk of certain cancers.

of dietary fiber, of which vegetables are an excellent source, decreases the risk of colon cancer (38). Vitamins including A, D, E and C have antioxidant and anti-inflammatory effects (29).

Studies on phytochemicals have prompted the National Cancer Institute to recommend an increased daily intake of fruits and vegetables, particularly cruciferous vegetables (40): three servings of vegetables and two servings of fruit each day.

## Additives

Fresh vegetables generally have few or no additives. Silicon dioxide may be used as an ink dilutent for marking fruits and vegetables. Some vegetables, such cucumbers, are coated in wax to prevent moisture loss. Canned vegetables may contain the preservative EDTA (ethylenediaminetetraacetic acid), as well as flavoring agents including salt and sucrose. Sodium bicarbonate (baking soda) is used to clean vegetables and to absorb

## CALORIE CONTROL
### Vegetables (1 cup cooked or 2 cups raw)

**Nonstarchy Vegetables: Average 50 (range: 25–100) Calories (kcal)**
Relatively low in calories. These vegetables make great snacks as they help you feel full (they are usually high in fiber and low in sugar), deliver lots of healthy phytochemicals, and keep calories and carbohydrates low (Table 13-1).

**Starchy Vegetables: Average 230 (range: 130–330) Calories (kcal)**
Provide three to four times the calories of nonstarchy vegetables due to their higher carbohydrate content (Table 13-1). On the other hand, for those who want to gain weight, starchy vegetables are an excellent calorie source that is low in fat (unless it's added).

**Recommended Intake:** The recommended daily intake of vegetables varies according to your caloric needs, the type of vegetable, and whether the vegetables are eaten cooked or raw. For instance, the United States Department of Agriculture (USDA) Food Guide/MyPyramid recommends:

- 2 cups/day for a 1,600-calorie (kcal) diet and up to 4 cups/day for a 3,000-calorie (kcal) diet
- Including dark green, orange/deep yellow, and starchy vegetables and legumes each week

The Exchange Lists recommend:

- 2–3 servngs of nonstarchy vegetables/day, with 1 serving equal to ½ cup of cooked vegetables or vegetable juice or 1 cup raw vegetables
- 6 servings/day of high-carbohydrate foods including starchy vegetables (based on a 1,300-calorie/kcal diet); serving sizes vary from ⅓–1 cup depending on the vegetable

**Meat Substitute (Dietary Guidelines):** 1 ounce of meat can be substituted with:

- ¼ cup cooked dry beans or tofu
- 1 tbsp peanut butter
- ½ ounce nuts or seeds

**TABLE 13-1**   Estimated Calories (kcal) in Nonstarchy and Starchy Vegetables (Based on 1 Cup Cooked or 2 Cups Raw; Calories/kcal Provided in Parentheses)

### Nonstarchy Vegetables

| 0–25 Calories (kcal) | 25–50 Calories (kcal) | 50–100 Calories (kcal) |
|---|---|---|
| Bamboo shoots (14) | Asparagus (43) | Artichoke and artichoke hearts (95) |
| Cucumber (16) | Broccoli (44) | Bean sprouts (65) |
| Eggplant (28) | Cabbage (23) | Beets (slices) (75) |
| Gourd (20) | Cauliflower (29) | Brussels sprouts (64) |
| Greens (collard, kale, mustard, | Celery (19) | Carrots (70) |
| turnip) (15) | Chayote, fruit or leaves (40) | Corn on the cob (7 inches) (77) |
| Lettuce (iceberg, shredded) (10) | Green beans (44) | Mixed vegetables (90) |
| Spinach (7) | Green onions (40) | Onions (90) |
| Salad greens (all types) (9) | Leeks (32) | Pea pods (67) |
| Tomato (25) | Mushrooms (42) | Peppers (green) (52) |
| Watercress (3) | Okra (33) | Rutabaga (66) |
| | Onion (55) | Vegetable juice (12 oz) (71) |
| | Pumpkin (49) | Water chestnuts (slices) (120) |
| | Radishes (25) | |
| | Seaweed, Kombu, Nori (28) | |
| | Squash (summer) (36) | |
| | Swiss chard (35) | |
| | Turnips (33) | |
| | Zucchini (29) | |

### Starchy Vegetables

| 100–200 Calories (kcal) | 200–300 Calories (kcal) | 300+ Calories (kcal) |
|---|---|---|
| Corn (132) | Beans, plain (236) | Baked beans, Boston (340) |
| Parsnips (125) | Cassava (221) | Hummus (398) |
| Peas (121) | Corn, creamed (213) | |
| Potato (1 med, 1 C) (160) | Lentils (230) | |
| Squash (winter) (115) | Plantain (1 med, 1 C) (218) | |
| Taro corm (192) | | |
| Yam or sweet potato (1 med, 1 C) (185) | | |

© 2010 Amy Brown

**Limit Servings of Starchy Vegetables.** Too much of a good thing, even if it's vegetables, can still contribute to an excess of calories. It would be difficult to get too many calories from greens; however, the calories from starchy vegetables can quickly add up.

1 cup dried beans, creamed corn, or cassava = about 230 calories (kcal)
2 cups                                      = about 460 calories (kcal)

**Limit Fried Food.** French fries are potatoes submerged in heated fat. Anything breaded and fried signals "loaded with fat and calories":

| Food | Serving Size | Calories (kcal) | Fat (grams) |
|---|---|---|---|
| French fries | Medium, 4 oz | **360** | 18 |
| Baked potato | 4 oz | 105 | 0 |
| | Medium | 160 | 0 |
| Onion rings (breaded, fried) | 4 rings | **230** | 12 |
| Onion | 1 cup | 55 | 0 |
| Zucchini (breaded, fried) | 4 oz | **277** | 16 |
| (stir fried, little fat) | 4 oz | 20 | 1 |

## Tips for Eating More Vegetables

- **Snacks.** Keep cut-up vegetables in a bowl of water in the refrigerator (or in an aerated bag) and include baby carrots, broccoli florettes, pepper (green, yellow, red) strips, and cauliflower. Use low-calorie dressings as a dip, but limit the amount to 1 tablespoon.
- **Vegetable soup.** Making homemade vegetable soup on the weekend yields a cup of soup for every day of the week.
- **Frozen vegetables.** Prepared as needed and/or cut into bite-size pieces and mixed in with sautéed onions, green peppers, and tomatoes.
- **Daily salad.** Increase greens by adding romaine lettuce and spinach, and increase fiber by topping the greens (usually low in fiber) with shredded carrots, cauliflower, beans, peas, broccoli, green pea pods, and corn.
- **Vegetable juice.** A 12-ounce (1½ cups) can of vegetable juice delivers only 70 calories (kcal) vs. the usual 150 calories (kcal) found in sodas.
- **Green and yellow.** Certain green vegetables are higher in folate (B-vitamin), and certain yellow-orange vegetables are rich in vitamin A.

© 2010 Amy Brown

excess acid in some canned products such as soups, and may also be used, along with sorbic acid, in pickled vegetables. BHT (butylated hydroxytoluene) and BHA (butylated hydroxyanisole) are preservatives and antioxidants that are used in dried potato flakes, cereals, and other processed foods.

# PURCHASING VEGETABLES
## Grading Vegetables

Because most fresh produce deteriorates too quickly to grade, most grading is at present voluntary; it is based on ripeness, color, shape, size, uniformity, and freedom from bruises and signs of decay. New labeling laws include USDA grading for the 20 most commonly consumed fresh vegetables (56). USDA grades are shown in Table 13-2.

### TABLE 13-2 USDA Grades for Vegetables and Fruits—Processed and Fresh

| Vegetables and Fruits | USDA Grade | What the Grade Means |
| --- | --- | --- |
| *Processed* | | |
| Canned and frozen. | U.S. Grade A (or Fancy) | This is the highest grade for canned and frozen vegetables. They are the most tender, succulent, and flavorful and have the best color for type. |
| | U.S. Grade B (Choice for fruits or Extra Standard for vegetables) | The majority of canned and frozen vegetables. They are not as tender or as well-colored as U.S. Grade A. |
| | U.S. Grade C (or Standard) | These canned and frozen vegetables are usually more mature, less tender, and not uniform in shape or color. They are an excellent buy when cost, but not appearance, is important. U.S. Grade C vegetables are frequently utilized for soups, stews, and casseroles. |
| | U.S. Grade D (or Substandard) | Lowest marketable quality. |
| Dried or dehydrated fruits. Fruit and vegetable juices, (canned and frozen). | U.S. Grade A | Very good flavor and color, and few defects. |
| Also includes preserves, jams, jellies, peanut butter, honey, ketchup, and tomato paste. | U.S. Grade B | Good flavor and color but not as uniform as A. |
| | U.S. Grade C | Less flavor than B, color not as bright, and more defects. |
| *Fresh* | | |
| The grade is more likely to be found without the shield. | U.S. Fancy | Premium quality; only a few fruits and vegetables are packed in this grade, so they are rarely seen by most people. U.S. Fancy is the highest grade used in judging the quality of fresh vegetables and is based on appearance and uniform shape. |
| | U.S. No. 1 | This is the most common grade for fresh vegetables and is given for those that appear fresh and tender, have good color, and are relatively free from bruises and decay. |
| | U.S. No. 2 | Intermediate quality between No. 1 and No. 3. |
| | U.S. No. 3 | Lowest marketable quality. |

# Selecting Vegetables

Vegetables come from living plants that grow in cycles with the passing seasons. Thus, the season of the year is the most important consideration when selecting vegetables (Figure 13-5). Figure 13 in the color insert shows some of the exotic vegetables that are available in many stores. Not all vegetables are in season at the same time, and they need to be selected accordingly to maximize quality. Choosing an out-of-season tomato that lacks color and flavor ultimately affects the quality of the resulting meal.

The amount to buy depends on the type of vegetable being purchased. Vegetables, especially leafy ones, tend to lose volume when trimmed and/or cooked, so consumers often purchase slightly more per person to make up for the anticipated waste. Specific tips on selection for each of the vegetables are discussed in alphabetical order (42).

### Artichoke

The artichoke is the immature flower head of a thistle. The edible "leaves" are actually petals (Figure 13-6). The furry center, known as the choke, must be removed before consumption. Occasionally baby artichokes are sold; they are tender throughout and may be eaten whole. Overmature artichokes are detected by leaves that separate and a woody texture. The edible portion of the artichoke includes the stem, the base (or heart), and the thicker bottom section of the leaves. Mature artichokes are usually trimmed as shown in Figure 13-7 prior to preparation. The most common variety is Green Globe. Selection varies according to season, with the

**FIGURE 13-5** Seasons for vegetables.

### Seasons for Vegetables

| Vegetable | January | February | March | April | May | June | July | August | September | October | November | December |
|---|---|---|---|---|---|---|---|---|---|---|---|---|
| Artichokes | | | ■ | ■ | ■ | ■ | | | | | | |
| Arugula | ■ | ■ | ■ | ■ | ■ | ■ | ■ | ■ | ■ | ■ | ■ | ■ |
| Asparagus | | | ■ | ■ | ■ | ■ | | | | | | |
| Avocados | ■ | ■ | ■ | ■ | ■ | ■ | ■ | ■ | ■ | | | ■ |
| Beans, shell | | | ■ | ■ | ■ | ■ | ■ | ■ | ■ | ■ | | |
| Beans, snap | | | | | ■ | ■ | ■ | ■ | ■ | ■ | ■ | |
| Beets, gold | | | | | | | ■ | ■ | ■ | ■ | ■ | ■ |
| Beets, red | | | | | | | ■ | ■ | ■ | ■ | ■ | ■ |
| Bok choy | | | | | | | | | ■ | ■ | ■ | |
| Broccoli | | | | | | ■ | ■ | ■ | ■ | ■ | ■ | ■ |
| Broccoli rabe | ■ | ■ | ■ | | | | ■ | ■ | ■ | ■ | ■ | ■ |
| Brussels sprouts | ■ | ■ | | | | | | | | ■ | ■ | ■ |
| Bulb fennel | ■ | ■ | ■ | ■ | | | | | ■ | ■ | | |
| Cabbages | ■ | ■ | ■ | ■ | ■ | ■ | ■ | ■ | ■ | ■ | ■ | ■ |
| Carrots | ■ | ■ | ■ | ■ | ■ | ■ | ■ | ■ | ■ | ■ | ■ | ■ |
| Cauliflower | | | | | | ■ | ■ | ■ | ■ | ■ | ■ | ■ |
| Celeriac | ■ | ■ | ■ | ■ | ■ | ■ | ■ | ■ | ■ | ■ | ■ | ■ |
| Chard | ■ | ■ | ■ | ■ | ■ | ■ | ■ | ■ | ■ | ■ | ■ | ■ |
| Chayote | ■ | ■ | ■ | | | | | | | | ■ | ■ |
| Chinese cabbage | | | | | | | | | | ■ | ■ | |
| Collards | | | | | | | | ■ | ■ | ■ | ■ | ■ |
| Corn | | | | | ■ | ■ | ■ | ■ | ■ | ■ | ■ | |
| Cucumbers | ■ | ■ | ■ | ■ | ■ | ■ | ■ | ■ | ■ | ■ | ■ | ■ |
| Eggplant, baby | | | | ■ | ■ | ■ | ■ | ■ | ■ | ■ | ■ | ■ |
| Eggplant, white | | | | | | | ■ | ■ | ■ | | | |
| Endive | | | | | | | | ■ | ■ | ■ | ■ | ■ |
| Escarole | ■ | ■ | ■ | ■ | ■ | ■ | ■ | ■ | ■ | ■ | ■ | ■ |
| Fiddleheads | | ■ | ■ | ■ | ■ | ■ | | | | | | |
| Frissée | ■ | ■ | ■ | ■ | ■ | ■ | ■ | ■ | ■ | ■ | ■ | ■ |
| Gourds | | | | | | | | | ■ | ■ | ■ | ■ |
| Jicama | ■ | ■ | ■ | ■ | ■ | ■ | ■ | ■ | ■ | ■ | ■ | ■ |
| Kale | | | | | | | | ■ | ■ | ■ | ■ | ■ |
| Kiwi | ■ | ■ | ■ | | | ■ | ■ | ■ | ■ | ■ | ■ | ■ |
| Kohlrabi | | | | | | ■ | ■ | ■ | ■ | ■ | ■ | |
| Leeks | | | | | | | | | | ■ | ■ | ■ |
| Leeks, baby | | | | | ■ | ■ | ■ | ■ | ■ | ■ | ■ | ■ |
| Lettuce, baby | | | ■ | ■ | ■ | ■ | ■ | ■ | ■ | ■ | ■ | |
| Lettuce, Bibb | ■ | ■ | ■ | ■ | ■ | ■ | ■ | ■ | ■ | ■ | ■ | ■ |
| Lettuce, iceberg | ■ | ■ | ■ | ■ | ■ | ■ | ■ | ■ | ■ | ■ | ■ | ■ |
| Lettuce, leaf | ■ | ■ | ■ | ■ | ■ | ■ | ■ | ■ | ■ | ■ | ■ | ■ |
| Mache | ■ | ■ | ■ | ■ | ■ | ■ | ■ | ■ | ■ | ■ | ■ | ■ |
| Mustard greens | | | | | | | ■ | ■ | ■ | ■ | ■ | |
| Okra | | | | | | | ■ | ■ | ■ | ■ | ■ | |
| Parsnips | | | | ■ | ■ | ■ | ■ | ■ | ■ | ■ | ■ | ■ |
| Peas, green | ■ | ■ | ■ | ■ | ■ | ■ | | | | | | |
| Peas, snow | ■ | ■ | ■ | ■ | ■ | ■ | | | | | | |
| Peppers, bell | | | | | ■ | ■ | ■ | ■ | ■ | ■ | ■ | ■ |
| Peppers, red | | | | | | | | ■ | ■ | ■ | ■ | |
| Potatoes, baby red | | | | | ■ | ■ | ■ | ■ | ■ | ■ | ■ | |

(continued)

**FIGURE 13-5** Seasons for vegetables. *(continued)*

| Seasons for Vegetables | | | | | | | | | | | | |
|---|---|---|---|---|---|---|---|---|---|---|---|---|
| Vegetable | January | February | March | April | May | June | July | August | September | October | November | December |
| Radicchio | ■ | ■ | ■ | ■ |  | ■ | ■ | ■ | ■ | ■ | ■ | ■ |
| Romaine | ■ | ■ | ■ | ■ | ■ | ■ | ■ | ■ | ■ | ■ | ■ | ■ |
| Rutabagas |  |  |  |  |  |  |  |  | ■ | ■ | ■ | ■ |
| Salsify | ■ | ■ | ■ | ■ |  |  |  |  |  | ■ | ■ | ■ |
| Scallions |  |  |  |  |  |  |  | ■ | ■ | ■ |  |  |
| Spinach | ■ | ■ | ■ | ■ | ■ | ■ | ■ | ■ | ■ | ■ | ■ | ■ |
| Squash, acorn |  |  |  |  |  |  |  |  | ■ | ■ | ■ | ■ |
| Squash, baby |  |  |  |  |  |  | ■ | ■ | ■ | ■ | ■ | ■ |
| Squash, butternut |  |  |  |  |  |  |  |  | ■ | ■ | ■ | ■ |
| Squash, cheese |  |  |  |  |  |  |  |  | ■ | ■ | ■ | ■ |
| Squash, crookneck |  |  |  |  |  |  | ■ | ■ | ■ | ■ | ■ |  |
| Squash, dumpling |  |  |  |  |  |  |  |  | ■ | ■ | ■ | ■ |
| Squash, nech |  |  |  |  |  |  |  |  | ■ | ■ | ■ | ■ |
| Squash, patty pan |  |  |  |  |  |  | ■ | ■ | ■ | ■ | ■ | ■ |
| Squash, spaghetti |  |  |  |  |  |  |  |  | ■ | ■ | ■ | ■ |
| Sunchokes | ■ | ■ | ■ | ■ | ■ | ■ |  |  |  |  | ■ | ■ |
| Tomatoes |  |  |  |  |  | ■ | ■ | ■ | ■ | ■ | ■ |  |
| Tomatoes, cherry |  |  |  |  |  |  | ■ | ■ | ■ | ■ | ■ |  |
| Turnips | ■ | ■ | ■ | ■ | ■ | ■ | ■ | ■ | ■ | ■ | ■ | ■ |
| Watercress | ■ | ■ | ■ | ■ | ■ | ■ | ■ | ■ | ■ | ■ | ■ | ■ |
| Zucchinis |  |  |  | ■ | ■ | ■ | ■ | ■ | ■ | ■ | ■ | ■ |

**FIGURE 13-6** The major edible portion of an artichoke is the fleshy ends of the "leaves" attached to the plant. The heart is also edible and considered a delicacy. It is reached by lifting out the cone and cutting out the choke (core).

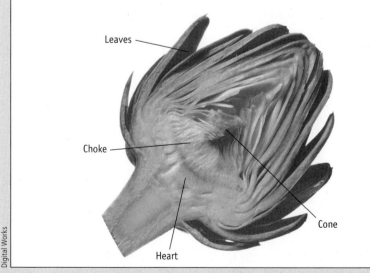

Leaves
Choke
Heart
Cone

Digital Works

**FIGURE 13-7** Preparing an artichoke.

Trim away both ends of the artichoke.

Scissors are used to cut away the barbs from the ends of leaves.

Digital Works

best-flavored artichokes available in the winter when the artichokes are heavy, compact, plump, and have bronze-tipped leaves (winter-kissed) from a light frost. Spring artichokes are globe-shaped, green, and have tight leaves. The shape becomes conical in the summer and fall, with the leaves becoming lighter in color and weight.

### Asparagus

Tenderness is the key consideration when selecting asparagus. Tight, compact buds and fresh, firm stalks that break with a crisp snap are desired. Opened buds and flattened or angular stalks are signs of tougher, less sweet asparagus. The size and thickness of the stalk or spear is not related to tenderness. The two main types of asparagus available in North America are the more popular dark green and the less often purchased white or light green.

### Beans (Green snap, green, wax, and yellow wax-podded beans)

Select sturdy, crisp pods with no sign of wrinkling, pitting, or bulging in the skin due to overmaturity. Green (green snap beans) and yellow (wax) are the two most common types of beans.

### Beets

Select beets that are smooth, firm, and free of any sign of dryness. Leaves should be fresh, tender, and clean.

### Broccoli

In fresh broccoli, the bud clusters should be firmly closed and the head dark green to purple in color. Stems should be light green and not pale, woody, hard, or dry. The signs of decay in broccoli are subtle, but avoid broccoli with florets that are starting to open or have an uneven, yellow-green color. Calabrese is the most common variety, but others include Emperor, Premium, Futura, Green Duke, Harvester, Atlantic, Topper 430, Gem, Medium Late 423, and Medium Late 145.

### Brussels Sprouts

Select those with compact leaves and clean stalk ends. Brussels sprouts with wilted, yellowish, puffy, or soft leaves are poorer in texture and flavor. Varieties such as Citadel, Lunet, Rampart, and Valiant differ in their bud-bearing characteristics.

### Cabbage

Look for cabbage with leaves firmly attached to the stem. The stem should be solid, hard, and heavy in relation to the head. The cabbage surface is usually smooth (except savoy cabbage), evenly colored, and free from any signs of dehydration. The most common variety is domestic cabbage, but others include Danish, the yellow-colored savoy, and red cabbage. Bok choy is Chinese mustard cabbage; it has large, sturdy, white stems and leaves, and a unique, mild flavor that enhances salads, stir-fry dishes, or soups.

### Carrots

Most carrots are sold by length and range from 2 to 8 inches long. The surfaces should be smooth, firm, and free from cracks, and the tops should be free from any sprouting green shoots, which would indicate an older carrot. The darker the orange color, the more beta-carotene the carrot contains. Pre-peeled carrots are also available. "Baby-cut" carrots are really just mature carrots cut and trimmed to represent young carrots (27).

### Cauliflower

Cauliflower, like broccoli, is a head of flowers. A fresh cauliflower head will not have started to bloom, spread, or look "ricey." The color ranges from white to creamy white. Lime-green is the color of broccoflower, a hybrid between broccoli and cauliflower. The jacket leaves at the cauliflower's base should be green and crisp. Popular varieties include Early Snowball, Super Snowball, Snowdrift, Pearl, Danish Giant, and Veitch Autumn Giant.

### Celery

The first thing to check for in celery is clean, smooth, tender-looking inner stalks. They should snap when broken and have a straight, solid cone in the center. Celery with signs of stringiness or roughness may be tough and should not be purchased. The leaves on top should be fresh, green, and free from signs of dehydration (wilting, color changes) or decay.

The green variety is the most popular, although Pascal, a lighter color variety, is growing in popularity because of its special flavor and lack of stringiness and toughness.

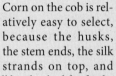

### Corn

Corn on the cob is relatively easy to select, because the husks, the stem ends, the silk strands on top, and the kernels can all be checked for freshness. Healthy husks and stems are a vibrant green and show no signs of discoloration or drying out. The stem ends are moist and not wilted or decayed. Many stores prefer that shoppers not rip open corn ears, because it causes the kernels to dry out and degrade, but when it is possible to do this, the husk should resist being pulled back and the kernels should be plump, even sized, and ripe-colored. Any indentations in the kernels or large or dark-yellow kernels indicate older, tougher, less flavorful corn. There are more than 200 varieties of corn, but the major type found commercially is Yellow Hybrid Sweet Corn.

### Cucumbers

The three types of cucumbers that are available include slicing or table, pickling, and greenhouse. Slicing cucumbers are found in most supermarkets and are best selected based on their firmness and fairly straight-sided oblong shape. The skin should be free from any soft spots or shriveling, and evenly colored, with no yellow evident. Typical supermarket varieties are heavily waxed to protect the fruit. Varieties include Ashley, Cherokee 7, Gemini, Hybrid Ashley, High Mark II, Long

Market, Marketer, Palomar, Poinsett, and Straight Eight. Hothouse or greenhouse cucumbers are elongated, seedless cucumbers with a mild flavor and no bitterness. European cucumbers are long and slender and usually come wrapped in plastic. Small- to medium-size cucumbers are used to manufacture pickles (see Chapter 28).

## Eggplant

Select the heaviest eggplant with at least a 3- to 6-inch diameter. Freshness in eggplants is judged by the condition of the skin. It should be smooth, firm, and glossy, with a deep purple, almost black color. Eggplants are also available in white and striped varieties, and, in these, any sign of yellow indicates aging. Flabbiness, softness, or dark bruise-like discolorations indicate poor quality. Chinese eggplants (also called Asian or Japanese eggplant) have a thinner skin and are long and thin, averaging 1 or 2 inches in diameter. American varieties include Black Beauty, Florida High Market, Florida Market, Myers Market, New Hampshire, and New York Purple.

## Exotic Vegetables

- *Adzuki bean.* Compared to other beans, adzuki beans have a slightly sweet flavor.
- *Amaranth.* A boiled leafy green vegetable prepared like spinach. Although not well known in the United States, amaranth is consumed extensively throughout Africa, Asia, the Caribbean, and Latin America. The greens are of considerable nutritional value (vitamins A and C, calcium, magnesium, and iron).
- *Chayote.* Pronounced "chy-o-tay," this native Mexican squash has a flesh that is crisp like a water chestnut (Color Figure 13). The chayote can be eaten raw in salads, or stuffed and baked.
- *Daikon.* This Japanese radish is gaining in popularity. Although frequently pickled in Japan, it serves as a good addition to green salads.
- *Jicama.* Pronounced "hé-cama," this legume grown in Mexico is

consumed fresh or is used like a radish in salads.

## Garlic

Roman gladiators ate garlic, believing it would make them stronger and capable of great feats, and even today many people believe it has curative powers against a host of ailments. Garlic is sold in whole heads. Select firm, well-filled cloves with skins that are not overly dry. Avoid those that are sprouting or have dark or rotted areas. Garlic can be peeled and minced by using a garlic press, or the clove can be crushed on a cutting board under the side of a chef's knife or heavy object such as the side of a can.

Garlic ranges in color from white to dark maroon, and varieties include Argentina White, California Early, California Late, Mexican Purple, and Mexican White. Elephant garlic, which is much larger, milder, and sweeter than regular garlic, is gaining in popularity, and is capable of reaching 6 inches in diameter and weighing up to 12 ounces.

## Ginger

This vegetable, often used as a spice in Asian, Indian, Moroccan, and Caribbean dishes, is selected by looking for a smooth, nonwrinkled skin with a fresh sheen and filled-out flesh. Choose ginger root that breaks cleanly with a snap and has the least number of knots.

## Greens

Greens are leafy vegetables that include collards, dandelion greens, kale, mustard greens, beet greens, endive, and Swiss chard. These excellent sources of vitamin A should be fresh, crisp, tender, and free from any drying, insect damage, or decay. Top varieties for each of the greens are as follows:

- *Collards.* Georgia, Louisiana Sweet, Morris Heading, and Vates
- *Dandelion.* Arlington Thick Leaf, Improved Thick Leaf, and Thick Leaf
- *Kale.* Dwarf Blue Curled Scotch, Dwarf Green Curled Scotch, and Tall Green Curled Scotch
- *Swiss Chard.* Burpee's Rhubarb Chard, Dark Green, White Ribbed, Fordhook Giant, Giant Lucullus, and Rhubarb Chard

## Leeks

This mild-flavored member of the onion family is traditionally used in soups and French dishes. Select leeks with tender leaves that are not too fibrous or coarse. Varieties include American Flag (London Flag), Blue Leaf, Carentan (Winter), and Musselburg.

## Lettuce

Lettuces are leafy vegetables that serve as the foundation for salads and as an ingredient in sandwiches or tacos (Table 13-3). Select lettuce with

| TABLE 13-3 | Major Varieties of Lettuce |
|---|---|
| **Variety** | **Characteristics** |
| Iceberg | Iceberg lettuce accounts for almost 10 percent of the entire produce market. Leaves are crisp and consistently mild in flavor, and have a high moisture content. |
| Boston or Bibb | Also known as butterhead, this lettuce has a very mild taste and pliable, soft leaves. Varieties include Bibb, Big Boston, May King, and White Boston. |
| Endive | Curly leaf ends distinguish this lettuce from the rest. The edges tend to be more bitter than the center. |
| Escarole | These crumpled-looking leaves, relatively fibrous and somewhat bitter tasting, are usually available in the Full Heart Batavian variety and are often served cooked or in soups. |
| Leaf lettuce | This lettuce has a crunchy texture not unlike iceberg lettuce, but it has more flavor. |
| Romaine or Cos | The darker green, sometimes reddish-tinged, flimsy leaves and stronger flavor differentiate this lettuce from iceberg. The pale, crisp cores, or hearts, of romaine are often sold for Caesar salad. |

fresh, crispy leaves and no signs of withering or dehydration. Iceberg lettuce can be easily cored by holding the head core-down and rapping it against the counter. This motion forces the core into the head, and it can then be easily twisted and removed. The lettuce head is then rinsed core-up so that the cold running water will spread through the leaves. All moisture should be removed after lettuce is washed to slow decay.

### Mushrooms

Mushrooms are actually a type of fungus that lacks chlorophyll. Select mushrooms with firm, smooth caps that are free of slime, mold, wilting, or bruises. The appearance of gills underneath the umbrella can indicate aging, although they are always visible in some varieties. If present, the gills should be clean, not too dark, and unbroken. Once purchased, mushrooms are often not washed, but instead stemmed and cleaned with a damp cloth or paper towel or scraped clean with a paring knife. They are very absorbent, so exposing them to water may ruin their flavor and texture, although many individuals subject them to a quick rinse just prior to preparation. There are quite a few varieties of mushrooms from which to select (Figure 13-8). "Wild" mushrooms should not be picked from fields, forests, or any other location, because even mushroom identification experts have been reported to make mistakes and get extremely ill.

### Okra

Okra consists of edible pods filled with seeds. The pods should be young and tender, and range from 1 to 3 inches in length. Fresh pods snap and puncture easily, whereas older ones are dry, fibrous, have a dull color, and contain hard seeds. Okra pods are usually green, but white and purple varieties are also available. Popular varieties of okra include Clemson Spineless, Dwarf Long Pod, French Market, and Perkins Spineless.

### Onions

The most common onion varieties are yellow, white, red, and green (Color Figure 8).

**FIGURE 13-8**   Varieties of cultivated mushrooms.

Shiitakes have a distinct, slightly smoky flavor that adapts well to other strong flavors; they're commonly used in Asian cooking.

Oyster mushrooms have a subtle flavor that's best paired with other simple ingredients.

Enoki mushrooms have a mild, pleasant flavor that isn't particularly distinctive.

Button mushrooms are the most common variety sold; they are somewhat nutty and creamy when raw and become earthy and rich when cooked.

Portabella mushrooms are simply cremini grown to gargantuan proportions; their flesh is denser and usually more fibrous.

Cremini are a darker variety of the standard button mushroom; they're firmer with a much more intense flavor.

Courtesy of the Mushroom Council

The yellow onion, which is the general all-purpose onion, should feel firm; the skin should be dry and papery, and the stem small. The skin should be free of any green sunburn marks, mold, or blackish decay spots. A typical fresh onion odor emits from well-selected onions. Onion flavor ranges from pungent to sweet. Vidalia onions are known for their distinctive taste, which is sweeter and milder than other onions because of their higher sugar and water content. This type of onion is preferred for onion rings, omelets, pizzas, marmalades, sautéing, or any other dish where a distinctive sweet onion flavor is desired. Varieties include Vidalia (small, sweet yellow), Texas 1015 (large yellow), White Boiler (small white), Medium White (large white), and Red. Only onions from specific counties in the state of Georgia may legally carry the Vidalia name (33).

Other types of onions and their relatives are as follows:

- **Red onions.** Commonly used raw in salads.
- **White or "boiling" onions.** Frequently added to stews.
- **Pearl onions.** Tiny onions that have a mild flavor and crisp texture. Pearl onions are often served with peas as a vegetable or as an ingredient in stews.
- **Green onions.** Also known as scallions, green onions are "onions to be" that have been harvested while the bulbs are still small and the leaves still growing. For scallions, tender, succulent, small to medium bulbs with springy leaves that hold their shape are preferred. Green onions are commonly used in salads, in Mexican dishes, for stir-frying, and alone as a garnish.
- **Shallots.** Related to the onion family, although their cloves more closely resemble garlic; they are a common ingredient in sauces.
- **Leeks.** A type of onion.

### Parsley
There are two kinds of parsley that differ quite a bit in taste—Italian (flat leaf) and Chinese (cilantro or coriander). Cilantro leaves are wider, droop more, and have a very distinctive fragrance.

### Parsnip
These long tapered vegetables look like bleached-out carrots, but are not eaten raw. They are excellent for soups.

### Peas
Green peas have a bulging pod, snow pea pods are flat, and sugar snap pod diameters range somewhere between green and snow peas. Unlike green peas, snow and sugar snap peas are consumed with their pods. All should be fresh, tender, and sweet. Avoid those that are in an advanced stage of maturity, which is signaled by grayish specks, yellow streaks, a darker color, dryness, and wrinkling.

### Peppers, Hot
Chili peppers come in a variety of colors, shapes, and degrees of "heat." Green, red, yellow, and black are the predominant colors (Figure 13-9). The skins of fresh chilies should always be bright, shiny, firm, and unwrinkled. Several of the more common types of chili peppers (Mexican varieties) include jalapeño, Anaheim, yellow chile, chile de arbol, poblano (ancho or pasilla), serrano, and habeñero, which is the hottest of all hot peppers.

### Peppers, Sweet
Sweet peppers are available in many colors—green, red, yellow, orange, purple, and brown. As peppers ripen, they become sweeter because of their increasing sugar content. Those with a bright, smooth, glossy color and thick, filled-out walls and a good solid weight are recommended, whereas those that are pitted, dull-looking, or shriveled should be avoided. Bell peppers with a more square or rectangular shape are easier to cut up. California Wonder is the most common variety of pepper, and others include Early Cal-Wonder, Burlington, Chinese Giant, Harris Early Giant, Neapolitan, and Yolo Wonder. Other sweet peppers include the round cherry, pimento, Holland, sweet Hungarian, Cubanelles, Italian, bull horn, and sweet banana wax.

### Potatoes
Numerous potato varieties are cultivated in South America, but in North America, three basic types are found in the market: russet/Idaho, white, and red (Color Figure 9). Russets are considered starchy potatoes and are preferred for baking; whites are best suited for roasting with meats or poultry; and reds, known as waxy potatoes, have the least starch and are good for simmering. New potatoes are immature, smaller versions of any mature potato variety. They have a delicate skin and waxy flesh that is ideal for simmering

**FIGURE 13-9** Selected varieties of hot peppers.

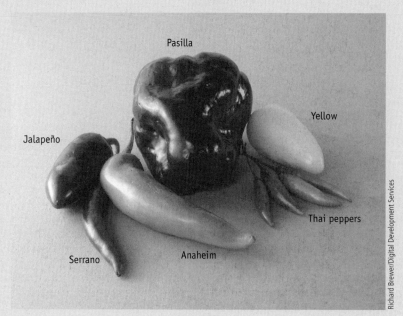

and for salads because they hold their shape. When selecting potatoes, look for those that are firm to the touch, and that have few eyes, good color, unshriveled skin, and a round or oblong shape (irregular shapes result in more waste during peeling). They should all be about the same size to ensure even cooking. Russets should have a net-like texture on their skins with no hint of green coloration, which indicates exposure to sunlight and possible presence of the toxin solanine. The varieties of each of the potatoes are as follows:

- *Russet.* Burbank, Centennial, and Norgold.
- *White.* Available in either long (White Rose) or round (Atlantic, Katahdin, Kennebec, Superior).
- *Red.* Norland, Red LaSoda, and Red Pontiac.
- *Others.* Varieties such as Yukon Gold, a medium-starchy all-purpose potato, and a wide array of small, fingerling potatoes are becoming increasingly available in the market.

### Radishes

Radishes are classified by shape: globular, oval, turnip-shaped, oblong, and long. The Red Globe type is the most popular and is selected based on a bright red color, smooth, firm skin, healthy roots, and crisp, white flesh. Avoid those that are cracked open, withered, spongy, and dry or rough looking. The leaves should be fresh and vibrant with no signs of decay. The daikon is a white, sweet, and elongated radish from Asia that is shaped like a carrot. Some markets also sell the large black radish, which is black on the outside and white on the inside.

### Rutabagas

Looking much like a turnip, rutabagas are more purple/beige in color, have a yellowish flesh, and have leaves that are thicker and smoother. The three most common varieties are American Purple Top, Laurentian, and the Thomson Strain of Laurentian. Like cucumbers, rutabagas are often sold heavily waxed.

### Spinach

Fresh, dark green spinach leaves with good turgor are preferred over those that are wilted, decaying, straggly, slimy, or have large, rough stalks. Spinach leaves may be wrinkled (savoy), flat, or semi-savoy, which is a combination of wrinkled and flat (sold for fresh and processed purposes). Young, tender leaves are somewhat lighter green and best used for salads, while the darker, more mature leaves are preferred for cooked preparations. Varieties include dark green Bloomsdale, Bloomsdale Long Standing, Old Dominion, and Virginia Savoy.

### Sprouts

The greenish-topped, white-stemmed alfalfa sprouts are selected based on very little breakage of the stems and freedom from wilting or slime. The sprout heads should be full, vibrantly green, and provide an overall nut-like flavor. Bean sprouts grown from mung beans are long, ivory-colored strands of uniform length and with a crisp, snappy texture. Signs of decay include darkened roots, sliminess, and a musty odor. Other sprouts, including radish, are also often available. The two dominant types of sprouts on the market are alfalfa and mung bean.

### Squash

There are two categories of squash: Summer or soft-skinned types and winter or hardshell types (Color Figure 10).

*Summer Squash* Summer squash include zucchini, yellow crookneck, yellow straightneck, chayote, cucuzza, scallopini, and sunburst. Look for small squash with a shiny skin and no bruises or scars. The smaller the squash is, the more tender its flesh, seeds, and skin. Pits in the skin are caused by chilling, whereas dull skin or discoloration may indicate old age.

*Winter Squash* It is difficult to determine the eating quality of winter squash because of their hard outer rind. Examples of winter squash include acorn, butternut, pumpkin, spaghetti, banana, turban, Hubbard, delicata, sweet dumpling, kabocha, and golden nugget. Select those that are dull and heavy for their size; as winter squash age, they become lighter in weight, indicating a dry and stringy texture. When selecting a pumpkin, look for a clean, uncracked surface with a bright orange color. Avoid those that show scarring, excess wrinkling, fading color, or soft spots on the rind.

### Sweet Potatoes

There are two types of sweet potatoes: soft-fleshed, which tend to have a more orange flesh and are soft, moist, and sweet when cooked; and firm-fleshed, which are lighter in color and dry and mealy when cooked. Select sweet potatoes that have large, uniform shapes with no signs of discoloration, wrinkling, or drying. If they are too tapered or pointed at the ends, they may be difficult to peel. Prominent sweet potato varieties include Beauregard, Jewel, Garnet, and Hernandez.

### Tomatoes

Tomatoes are often sold unripe because they are easily bruised during transportation (14). The color changes from green to bright red or reddish-orange as tomatoes ripen. Once ripe, they should give to gentle palm pressure; a watery consistency or slippery skin is a sign of being overly ripe. A sweet, delicate aroma often accompanies ripeness. Sometimes cutting the tomato open is the only way to determine if it is ripe for consumption.

There are several common types of tomatoes: mature green, which are best for slicing or garnishes; vine-pink/vine-ripe, suited for stuffed tomato entrees or salad wedges; Romas, which are elongated in shape, meaty, and flavorful, making them ideal for salads and sauces (see Figure 13-10); and cherry tomatoes, ranging from the size

**FIGURE 13-10** Selected varieties of tomatoes.

Beefsteak tomato

Roma tomato

Generic greenhouse tomato

Cherry tomato

Grape tomato

Pear tomato

Richard Brewer/Digital Development Services

of marbles up to ping-pong-ball size, which are excellent for salads, garnishes, or salad bars. These tomatoes soften at a slower rate than other tomatoes, allowing them to be harvested at a later date after they have had more time to ripen.

### Turnips

Soups and stews are the main dishes to which turnips are added. A firm, well-rounded turnip, free from shriveling or rough skin, is preferred. Common varieties include Amber, Golden Ball, Purple Top, Yellow Aberdeen, White Egg, and White Globe.

# LEGUMES

Beans, peas, and lentils serve as excellent sources of fiber, protein, iron, and complex carbohydrates (37). The single common identifying factor among all legumes is that they grow as seeds within a pod (Figure 13-11). Dried beans have served as a dietary staple since the Bronze Age.

Soybeans are unique in that, compared to other plant sources, they have relatively high protein and fat content, and can be utilized to make textured vegetable protein, soy milk (see Chapter 10), meat analogs, tofu, and fermented soybean foods (38).

## Textured Vegetable Protein

Textured vegetable protein (TVP) is derived mainly from soybeans, but other plants such as peanuts and cottonseed may be used in its composition. In TVP, the plant material has been altered into fibrous, porous granules that rehydrate rapidly. TVP is often used to extend ground meats in order to lower costs and/or reduce fat content (15). The USDA limits the use of TVP

in commercially prepared foods to no more than 30 percent of any particular product (49). One cup of TVP is usually rehydrated in a half cup of liquid before being added to hamburgers, hot dogs, chilies, tacos, spaghetti, lasagna, soups, meatloaf, pizza toppings, or other dishes. TVP burns easily, so it should be added only after the meat has been browned. It also retains moisture and contributes to texture by maintaining cohesiveness (15). TVP is available flavored or unflavored, and natural or caramel-colored.

### Meat Analogs

Meat analogs are imitations of meat products that are made by blending soy protein with other vegetable proteins, carbohydrates, fats, vitamins, minerals, colorings, and flavors. Although its use in the United States started in the 1960s, when the USDA allowed soy protein isolates to constitute 2 percent of cooked sausage, tofu dates back more than 1,000 years ago to China. Buddhist chefs created tofu as part of their vegetarian lifestyle (15). Today, there are meat analog products that can stand in for breakfast sausages, bacon, ham slices, and chicken or beef chunks. Even the classic hamburger and hot dog can be made with non-meat ingredients. These

products contain one third the fat of the meats they replace, and no cholesterol; however, depending on how they are made, the protein quality is relatively low, because soy protein lacks sufficient amounts of the essential amino acid methionine. They are also higher in sodium and more costly than the meats they simulate. However, when used with a mixed diet of grains and seeds, these meat analogs provide a balanced selection.

## Tofu

Tofu is the cheese made from soy milk. The protein in soy milk is precipitated out through the use of coagulants such as calcium sulfate dihydrate (increases volume), calcium chloride (increases firmness and calcium content), or magnesium chloride. After the soy milk coagulates and the liquid is removed, the curds are pressed, and these blocks of fresh tofu are sold in plastic containers filled with water, in aseptic packages with no water, or in bulk by selling several blocks together in buckets of fresh water. When refrigerated, tofu is relatively stable. Once opened, draining off the water prior to and during refrigeration and replacing it with fresh water increases its shelf life.

**FIGURE 13-11**   Dried beans.

Lima beans

Black beans

Blackeye beans

Garbanzo beans

Great Northern beans

Large Lima beans

Navy beans

Pink beans

Pinto beans

Red Kidney beans

Digital Works

## Fermented Soybean Foods

The fermentation of soybeans results in several different foods, including miso, natto, sufu, tempeh, and tamari.

- *Miso.* A fermented soybean paste, miso is used as a seasoning for soups, stews, dressings, dips, sauces, and gravies (19).
- *Natto.* Natto is a condiment made from fermented cooked soybeans held together by a sticky bacterial product.
- *Sufu.* Sufu is fermented tofu, also known as bean cake.
- *Tempeh.* A snack or meat alternative entrée, tempeh is made from fermented whole soybeans molded into a cake.
- *Tamari.* Tamari is naturally aged and fermented soy sauce. Its ingredients include soybeans, water, and salt, but not wheat. Other types of soy sauce (also known as *shoyu*) exist and those that contain wheat are also aged (about 6 months). A synthetic or nonbrewed soy sauce usually contains caramel color, hydrolyzed soy or vegetable protein, and corn syrup (31).

# PREPARATION OF VEGETABLES
## General Guidelines

Vegetables can be prepared by dry-heat methods (baking, roasting, sautéing, or deep-frying) or moist-heat methods (simmering, steaming, braising, or microwaving). Serving styles also vary and include plain, buttered, creamed, **au gratin,** glazed, **scalloped,** stuffed, or in soufflés, omelets, and cream soups. Regardless of the cooking method or serving style selected, some general principles governing the handling and preparation of vegetables should be followed:

- *Buying.* Purchase only the freshest possible vegetables in amounts that will be used within a few days.
- *Storage.* Store vegetables immediately at the appropriate temperature and do not leave them out of storage for any length of time unless they are being prepared. Leftovers should

**Au gratin** Food prepared with a browned or crusted top. A common technique is to cover the food with a bread crumb/sauce mixture and pass it under a broiler.

**Scalloped** Baked with milk sauce and bread crumbs.

Tofu has a bland flavor, and its texture varies in consistency and firmness. It is available in smoked, dried, seasoned, and sweetened varieties. The curd is often used in stir-fried dishes, sandwiches, soups, casseroles, and egg or meat dishes.

be refrigerated immediately and used within 3 days.

- *Washing.* All vegetables must be thoroughly washed (with a vegetable brush when appropriate) to remove soil, microorganisms, pesticides, and herbicides. Washing should be quick, because most vegetables absorb excess water when soaked. Many root vegetables, except beets and baked potatoes, are peeled of the outer layer that is normally washed.
- *Cooking liquid.* Vegetables should be cooked in as small an amount of liquid as possible; in many cases, leftover liquids may be saved for stock. Using a microwave minimizes the amount of water used.
- *Timing.* The cooking time should be as short as possible; most vegetables when heated too long will undergo undesirable changes in quality. Vegetables should be served promptly after cooking.

## ❓ How & Why?

### Why are some sweet potatoes called yams?

Contrary to popular belief, sweet potatoes and yams are not the same vegetable. True yams are large, yellow to white tubers sold mostly in Latino markets. The yam sold in North American markets is actually a sweet potato that was developed in the 1930s by a horticulturist who wanted to distinguish it from other sweet potatoes (42). For many years, the produce and grocery industries have used the term *yam* to describe sweet potatoes with a bright orange color and sweet, moist flesh. The USDA requires that these sweet potatoes described as *yams* must also carry a *sweet potato* label (45).

## Changes During Heating

When heated, vegetables undergo several changes in texture, flavor, odor, color, and nutrient retention. Understanding these phenomena can help to retain as much of their quality as possible during preparation.

## Texture

High temperatures gelatinize starch, decrease bulk by softening cellulose, and cause a reduction in turgor due to water loss. Although this is desirable when baking potatoes or cooking legumes, it is not recommended for most other vegetables; they should be heated until barely cooked. Acids or acidic foods, such as vinegar or tomatoes, should be added toward the end of the cooking time, because they make vegetables more resistant to softening and, by precipitating vegetable pectins, increase required heating time.

To compensate for turgor lost during processing, calcium salts are often added to pickles, canned vegetables, and fruits to make them firmer (30). The salts combine with pectic substances and become insoluble, firming the food's texture. Other calcium sources include molasses, hard water, and brown sugar. Sometimes the toughness of green beans and other vegetables actually results from preparing them in hard water. Adding alkaline ingredients such as baking soda has the opposite effect of breaking down cellulose and producing a very mushy texture.

## Flavor

Vegetables obtain their flavors from an assortment of volatile oils, organic acids, sulfur compounds, mineral salts, carbohydrates, and polyphenolic compounds. In general, to retain these flavor compounds, vegetables should be heated in as little water as possible and for as short a time as will do the job. There are exceptions; cooking for long periods with gentle heat yields milder flavors in onions. Garlic flavor also depends on whether it is raw or cooked. Raw garlic is very pungent, yet becomes mellow or sweet when cooked slowly, and even nutty and rich when cooked at length. Only low to medium-low heat should be used for cooking garlic because it cooks quickly and will become bitter when burned. Most of the substances that cause bitterness in vegetables such as cucumber and eggplant can be eliminated with peeling. Those that lurk under the skin can be drawn out prior to cooking by the technique of **degorging**. Droplets containing the bitter compounds are drawn out by osmotic pressure and can then be soaked up with a paper towel or rinsed off and patted dry.

## Odor

Food odors contribute to the perception of flavor, but some odors, such as that from cooked cabbage or onions, may be undesirable. These pungent odors are generated by sulfur compounds present in the *Cruciferae* family (the cruciferous vegetables) and the *Allium* genus (onions, garlic, shallots, leeks, chives) (54).

Garlic, onions, shallots, and leeks are odorless until they are cut or bruised, which allows an enzyme to contact a particular substrate to create a distinctive-smelling sulfur compound (Chemist's Corner 13-5) (36). The strong odors of cooked cabbage and onion can be reduced by shortening the heating time, adding a little vinegar to the cooking water, and/or by removing the lid occasionally during cooking to let volatile organic acids escape.

## Color

Both fat- and water-soluble pigments are affected by pH, heating, and the presence of metals (Table 13-4). Undesirable color changes can be prevented in several ways. Red cabbage, rich in anthocyanins, is prevented from turning blue if cooked with something acidic such as apples or a teaspoon of vinegar or lemon juice. Anthoxanthin-containing foods such as potatoes, rice, cauliflower, and onions are normally creamy white in color, but turn yellowish if exposed to hard water. This can be prevented by adding ½ teaspoon of cream of tartar, vinegar, or lemon juice to each gallon of water used. Adding baking soda to green vegetables makes them appear greener; however, this is not recommended due to the deleterious effect on B vitamins and texture.

Influences other than pigments on color include the Maillard reaction, the caramelization of sugars, and enzymatic browning (51). Enzymatic browning may be observed in cut-up potatoes that turn pinkish-brown when

**Degorge** To peel and slice vegetables, sprinkle them with salt, and allow them to stand at room temperature until droplets containing bitter substances form on the surface; the moisture is then removed.

## CHEMIST'S CORNER 13-5

### The Chemistry of Pungent Odors

Pungent odors from vegetables in the *Cruciferae* family and *Allium* genus are released by overheating. The heat triggers enzymes such as myrosinase that release an excess of hydrogen sulfide gas ($H_2S$) and convert sinigrin, a sulfur compound, to mustard oil (Figure 13-12). Garlic's sulfur compound, alliin, is odorless until the garlic is cut; cutting allows alliinase enzymes to come in contact with the alliin and convert it to thiosulfinates and volatile, odorous disulfides (60). Exposure to this vapor triggers tears during onion cutting. Two methods to prevent tearing include placing the onions in the refrigerator for an hour to slow down all chemical reactions or wearing a diving mask so the vapor cannot reach the eyes and nose. The thiosulfinates also act as antibiotics, which is probably why warriors rubbed cut onions on their wounds prior to the development of antibiotics. Leeks and scallions contain the same substances, but in lesser concentrations. Pyruvate (pyruvic acid) is also formed and can be measured as a direct equivalent to an onion's pungency (1). The flavor of cabbage is further affected by the sulphur compound (+)-S-methyl-L-cysteine sulfoxide as it converts to dimethyl sulfide.

**FIGURE 13-12** The chemistry of pungent odors. Heating vegetables in the *Cruciferae* family (such as cabbage) generates sulfur compounds.

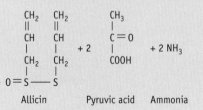

**Specific gravity** The density of a substance compared to another substance (usually water).

potatoes in water sprinkled with a little lemon or orange juice, or with cream of tartar added (1 teaspoon per quart) to increase acidity.

### Nutrient Retention

Careful preparation of vegetables conserves important nutrients. Because leaching is the greatest cause of mineral loss in vegetables, it is important to cook them using as little water as possible. In many instances, it is better to avoid immersing them in water altogether and instead revert to steaming, braising, baking, or microwaving. Other ways to minimize nutrient loss are to leave the skin on whenever possible, to cut vegetables into fewer, larger pieces rather than many smaller pieces, and to cook just to the point of doneness and no further.

Some nutrients may actually increase during food preparation, desirably or not. Frying vegetables increases the fat content of the finished product. Heating also increases the amount of protein available from legumes by destroying the enzymes known as protease inhibitors. Heat softens a food's fiber content and may even increase it as a percentage of weight after some natural water loss due to heating.

## Dry-Heat Preparation

### Baking

Some vegetables—especially potatoes, winter squash, onions, stuffed green peppers, and tomatoes—can be baked whole at approximately 350°F (177°C).

***Potatoes*** Potatoes are the most commonly baked vegetable and take approximately 1 hour, depending on their size and the oven temperature. Not all potatoes, however, lend themselves to baking. The starch content, density, and **specific gravity** of a potato determine whether it is a *boiler* or a *baker*. If a potato floats in a saline solution (½ cup salt in 5½ cups water), it is best for simmering. If it sinks to the bottom, it is best for baking. Russet and Idaho varieties, which are mealy potatoes, have a high starch content and are good for baking and mashing, because they yield a dry, light, fluffy texture. Cream of tartar is sometimes added to yield a whiter color (Chemist's Corner 13-6). When these

exposed to oxygen (22). Some potatoes turn dark blue-gray when oxygen reacts with their natural iron content. This discoloration can be prevented by blanching (43) or by soaking cut

**TABLE 13-4**  Pigment Colors Change in the Presence of Acid, Alkali, Heat, and Metals

| Pigment | Original Color | Acid | Alkali | Heat | Metal |
|---|---|---|---|---|---|
| *Carotenoid* | | | | | |
| | Orange-red | Lighter color | Brownish | Lighter orange | No effect. |
| | Yellow-orange | | | | |
| *Chlorophyll* | | | | | |
| Chlorophyll a | Blue-green | Gray-green | Bright green (chlorophyllin) | Initial bright green, then dull olive brown | Copper and zinc help to retain the green color of chlorophyll, but they are not used because of possible toxicity. |
| Chlorophyll b | Green | Olive brown (pheophytin) | Bright green (chlorophyllin) | Initial bright green, then dull olive brown | See above. |
| *Flavonoids* | | | | | |
| Anthocyanin | Red-purple | Red | Blue, purple, green | Dull reddish-brown | Copper, iron, aluminum, and tin change red-purple colors from green to slate blue. |
| Anthoxanthin | Cream/white | Whiter | Yellowish white | Little change; possible yellowing | Aluminum results in loss of white; yellow hue, whereas iron darkens the cream/white pigments. |
| Betalains | Purple-red/yellow | Red | Yellow | Darkens | Iron darkens, aluminum turns betalains a bright yellow. |

potatoes are whipped, they readily incorporate air and soak up added milk and butter. Medium-starch potatoes, such as white potatoes, are all-purpose and are suited for both baking and simmering.

Potatoes may be pierced with a fork before baking, or cooked for 20 minutes before piercing them to allow the steam to build up. Once they are in the oven, it is best to turn them every 20 minutes for even baking. Wrapping them in aluminum foil retains their moisture, resulting in a gummier potato, and produces a softer skin, which may or may not be desirable. Aluminum-foil-wrapped

potatoes are steamed, in addition to being baked.

Potatoes can be checked for doneness by squeezing them with an oven glove to see if they give in to pressure and feel soft under their skins. A fluffier potato results if the potato is massaged slightly before being slit open. Once opened, pushing both ends of the potato together plumps up the insides, making them still more fluffy and easier to empty.

### Roasting

Increasingly popular are roasted vegetables such as peppers, onions, and eggplant, which are generally sprinkled with oil and roasted at 375°F–425°F (191°C–218°C) until tender and well browned. Almost all vegetables can be roasted, and this technique creates a very flavorful vegetable as a result of its softer texture and sweeter taste after caramelization. A quick boil in salted water (parboiling) of potatoes prior to roasting will ensure that they will cook all the way through (55).

### Frying

Vegetables can be pan-fried, stir-fried, or deep-fried. The potato is the most commonly pan-fried of the vegetables.

***Stir-Frying***  Stir-frying combines a little oil with the vegetable's natural

moisture. For best results, tender, quick-cooking pieces such as mushrooms are cut into large, uniform slices, whereas less tender pieces such as carrots and celery are cut to expose the greatest amount of surface area to the heat. Although not necessary, a tight cover over the pan or wok will trap steam and reduce heating time, retaining maximum nutrients, color, and texture, although the vegetables will not brown as readily in the presence of trapped steam.

***Deep-Frying***  Deep-frying continues to be popular despite the associated increase in fat content. This method is used especially for french fries, onion rings, and breaded zucchini. High-starch russet and Idaho potatoes are preferred for french fries and potato chips (Chemist's Corner 13-7). Potatoes are selected for producing commercial potato chips based on the amount of sugars (specifically, reducing sugars) they contain: the less sugar, the more desirable the potato. These sugars contribute to excessive Maillard browning during frying and are considered a major quality defect (48). Other vegetables well suited for deep-frying are mushrooms, long beans, broccoli florets, okra, and eggplant.

> **CHEMIST'S CORNER 13-6**
>
> **Cream of Tartar and Potatoes**
>
> Oxidation of the potato's iron content from ferrous (+2) to ferric (+3) results in a dark pigment that makes the potato appear less white. Acid in the form of cream of tartar keeps mashed potatoes looking whiter by keeping the iron in the reduced ferrous state, preventing it from being oxidized.

> **CHEMIST'S CORNER 13-7**

**French Fry Phenomenon**

High-starch potatoes (russets, Idahoes, and Russet Burbanks) are used to prepare french fries that are crispy, golden on the outside, yet dry and fluffy on the inside. As a freshly cut fry is placed in hot frying oil, the outside starch granules swell and pull moisture from the interior of the potato. The heat seals the outside surface, preventing any more oil from being absorbed by the french fry (20).

## Moist-Heat Preparation

### Simmering

Vegetables should not be boiled, but instead simmered in as little water as possible to avoid nutrient loss and adverse effects on flavor, texture, and color. To simmer, vegetables (with the exception of potatoes) are added to lightly salted boiling water; when the water starts to return to a boil, the heat is immediately lowered to a simmer, and the vegetables are cooked until tender. Mushy textures result from overcooking vegetables or boiling them heavily so they bounce against each other. During simmering, the lid may be left off to allow volatile acids to escape, but leaving it on reduces the heating time. Approximate times for cooking vegetables vary from 5 to 30 minutes, with beets exceeding this general time frame by taking about 30 to 40 minutes for young beets and up to 2 hours for older beets.

Potatoes (peeled or unpeeled) are always started in cold salted water, brought to a boil, and lowered to a simmer. Waxy potatoes such as Round Reds have less starch than others and are more suited for simmering. Their higher moisture content and better ability to hold their shape make them appropriate for such dishes as potato salad, soups, casseroles, and scalloped potatoes.

Simmered vegetables may be served puréed or prepared as baby food. Young, tender vegetables are thoroughly scrubbed, cooked, and puréed. The puréed mixture is then heated to the boiling point, cooled, and served.

*Amount of Water*　There are two philosophies on the amount of water to use to cook vegetables properly: one is to use a minimal amount of water to reduce leaching, and the second is to use a large volume of simmering water to assure the vegetables are cooked as fast as possible. Some people feel that the speed of cooking in larger amounts of water far outweighs any nutrients lost in the cooking liquid; however, many professional chefs choose the first option.

### Steaming

Steaming sometimes takes a little longer than simmering (about 5 to 10 minutes more), but provides better retention of flavor, texture, and color. For steaming, a loose, shallow layer of vegetables is placed in a perforated basket inserted in a pan just above simmering water (Figure 13-13) and cooked, tightly sealed, until just tender (3). Pressure cookers are not recommended, because their high temperatures make vegetables mushy; legumes such as pinto beans are the exception.

*Foil-Wrapped Vegetables*　Yet another method to prepare vegetables using steaming is to wrap them in aluminum foil and heat them in the oven at 350°F–400°F (177°C–204°C) for about 20 to 30 minutes. Prior to placing them in the aluminum foil wrap, they can be tossed in oil with herbs (rosemary, thyme, sage); seasonings (garlic, salt; pepper upon serving); and a little fat (oil or butter). The vegetables heat in the flavored steam to become softer and tastier.

**FIGURE 13-13**　Equipment used to steam food.

Laura Murray

### Braising

Vegetables can be braised by first browning them in a sauté pan and then simmering them in their own juices (35). This method generates a tasty vegetable because of the caramelized flavor that develops during sautéing. A dash of sugar contributes to the browning and flavoring. Adding minced garlic or sliced onions supplies even more flavor. When caramelizing vegetables, know that there is a fine line between browning and burning. Adding a few tablespoons of water just as the vegetable, especially onion, is about to burn will prevent it from burning (49).

### Microwaving

One of the best ways to retain a vegetable's texture, color, and nutrient content is to cook it in a microwave oven because this method requires very little water and is fast enough to minimize loss of quality. Manufacturer's instructions should be followed for the specific vegetable being prepared, but general guidelines for microwaving vegetables include adding a minimal amount of water, covering with plastic wrap, and microwaving for about 3 to 10 minutes (rotate or rearrange in the middle of the time period) until fork tender. A more complete description of microwave preparation is described in Appendix A.

Frozen vegetables may be cooked in the microwave oven without adding water because water is already present. Canned vegetables are already cooked and need only to be reheated in the canning liquid and then drained. Home-canned vegetables low in acid concentration (above pH 4.6) such as green beans should not be microwaved, but rather boiled for 10 minutes to prevent possible foodborne illness (botulism).

## Preparing Legumes

Legumes are best prepared by simmering rather than boiling. One cup (½ pound) of dried beans usually yields 2 to 2½ cups of cooked beans; ½ cup of cooked dried beans or peas may be counted as a 1-ounce serving of meat or as one vegetable serving. Cooked legumes can be eaten plain or mixed and matched with other foods, as in red beans and rice, bean burritos, and tostadas.

There are three methods for preparing dried beans: overnight soak, short soak, and no soak. In the overnight soak method, beans are sorted and thoroughly rinsed and then immersed in water; those that rise to the top or are shriveled are discarded. The remaining beans are soaked in water amounting to three or four times their quantity; that is, 2 cups of dried beans require 6 to 8 cups of cold water. After soaking for approximately 10 hours, they are either drained and fresh water is added, or they are immediately simmered directly in their soaking liquid until tender. A lid is used but should be shifted slightly to the side to avoid a boil over. During simmering, water is added as needed to compensate for evaporation loss. Soaked beans require a long, slow cooking time to break down their hard-to-digest starches (Table 13-5).

In the short soak method, sorted and rinsed beans are brought to a full boil for 2 minutes, removed from the heat, and allowed to soak in the same *hot* water for 1 hour. They are then simmered as above in their soaking liquid.

Finally, beans can be prepared without soaking, but they take twice the amount of water and double the heating time; they also lose their skins more easily. The no soak method is used for lentils and split peas because their smaller size results in shorter cooking times.

When cooking legumes, it is important not to use hard water and not to add salt or acid in the form of tomato products or lemon juice until the beans are well cooked because these substances inhibit the softening of pectic compounds.

### Indigestible Carbohydrates

One pitfall of eating legumes is that flatulence often occurs. Gases (hydrogen, methane, and carbon dioxide) are produced by intestinal bacteria when they ferment the bean's indigestible carbohydrates, such as raffinose and stachyose (32). The problem can be minimized by soaking the beans, draining their water, rinsing them well before cooking, and then cooking them properly in fresh water, although this may result in a decrease in nutrients. Enzyme-containing commercial products such as Beano™ are now available on the market and may also be added to the food before consumption to assist in digestion.

# Preparing Sprouts

Any whole grain or legume can be sprouted (germinated) into fresh greens. The process of sprouting enhances flavor, nutrient content, palatability, and digestibility. Sprouting releases enzymes such as alpha-amylase that break down the starches into more readily digestible sugars (16).

The sprouting procedure starts with a jar equipped with a mesh top, which can be purchased at a health food store or made by topping a jar with cheesecloth and securing it with a rubber band. The seeds are rinsed and then placed in the jar with three times as much water as seed. An overnight soak in lukewarm water starts the majority of seeds on their way to sprouting, with small seeds needing only 4 hours and others, particularly seeds with very hard coats, requiring 2 days. Alfalfa and oat seeds do not need a presoak. Table 13-6 provides a list of seeds and preparation guidelines.

After soaking, the water is drained from the seeds and the jar is placed in a warm, dark place, positioned at an angle to let any water drain, and covered with a towel. Then the seeds must be rinsed at least twice daily, sometimes more often, with lukewarm water. Care must be taken to keep the seeds moist but not wet, because dried-out seeds do not sprout, whereas too much water causes spoilage. The rinsing process must be carried out gently and evenly so the seeds are not ripped from their tender sprouts.

Once the sprouts reach their full height, they are placed in direct sunlight to develop the chlorophyll, which will turn the leaves green within a day. Most sprouts are ready to eat within 3 to 6 days after presoaking. To make sprouts more pleasant to eat, the seed hulls can

| TABLE 13-5 Cooking Time for Beans (after a 10-hour soak) | |
|---|---|
| **Beans** | **Approx. Cooking Time** |
| Garbanzo (chickpeas) | 4 hours |
| Mung beans | 3 hours |
| Black beans | 1½ hours |
| Pinto beans | 2 hours |
| Soybeans | 2 hours |
| Black-eyed peas | 1 hour |
| Kidney beans | 1½ hours |
| Lentils, split peas (usually not soaked) | ¾ to 1 hour (half the time if soaked) |

| TABLE 13-6 Sprout Preparation | | | | | |
|---|---|---|---|---|---|
| **Seed** | **Quantity** | **Yield** | **Rinses** | **Time** | **Height** |
| Adzuki | ½ C | 2 C | 4 | 6 days | 1" |
| Alfalfa (salads) | 2 tbsp | 1 qt | 2 | 4 days | 1–2" |
| Beans (kidney, lima, fava, green, pinto) | 1 C | 4 C | 4 | 6 days | 2" |
| Chia | 2 tbsp | 3 C | 1 | 4 days | 1½" |
| Cress | 1 tbsp | 1½ C | 2 | 4 days | 1½" |
| Fenugreek (salad) | ¼ C | 4 C | 2 | 5 days | 3" |
| Garbanzo (chickpea) | 1 C | 3 C | 5 | 4 days | ½–1" |
| Guar beans (soak for 36 hours) | 1 C | 4 C | 4 | 5 days | 2–3" |
| Lentils | 1 C | 6 C | 2 | 4 days | 1" |
| Millet | 1 C | 2 C | 3 | 3 days | ¼" |
| Mung beans | 1 C | 4 C | 4 | 4 days | 2–3" |
| Peas | 1½ C | 4 C | 2 | 4 days | ½–1" |
| Radish | 1 tbsp | 2 C | 2 | 5 days | ½–1" |
| Red clover | 3 tbsp | 1 qt | 2 | 5 days | 1–2" |
| Soybeans | 1 C | 5 C | 8 | 5 days | ½–¾" |
| Sunflower | 1 C | 3 C | 2 | 2 days | ½" |
| Wheat | 1 C | 4 C | 3 | 4 days | ½" |

be removed by filling the sink with cold water, dropping in the sprouts, allowing the hulls to float to the surface, and skimming them off with a strainer.

Sprouts must be stored in the refrigerator and rinsed thoroughly before use. The FDA has warned that sprouts, especially alfalfa sprouts, have been responsible for several foodborne illnesses. As a precaution, some sprout preparers soak seeds in a very weak bleach solution (1 teaspoon per gallon of water) for 30 minutes before sprouting, followed by rinsing with clean water.

# STORAGE OF VEGETABLES

After vegetables are harvested, they are still viable and continue to respire by taking up oxygen and releasing carbon dioxide. This natural respiration contributes to the deterioration of their appearance, texture, flavor, and vitamin content. The faster a vegetable's respiration, the more quickly it deteriorates. Post-harvest **respiration rates** differ among vegetables, which explains why potatoes (yielding only about 8 mL of carbon dioxide per hour at 59°F/15°C) last longer than cabbage and green beans (yielding 32 and 250 mL of carbon dioxide per hour, respectively) (13).

Another factor contributing to produce spoilage is loss of moisture. Vegetables contain about 70 percent water just after being picked, but at that point the water is no longer replaced by the source plant's root or leaf system. Water loss, resulting in wilting and accelerated decomposition, can be avoided by maintaining a humid atmosphere, spraying with a fine mist, and/or covering the produce with an edible film or coating (Chemist's Corner 13-8). Optimal humidity for most vegetables and fruits is 85 to 95 percent (7). Ultimately, the key to properly storing vegetables is to slow their respiration and moisture loss, allowing them to stay fresh longer.

> ### CHEMIST'S CORNER 13-8
>
> #### Edible Films
>
> Some of the most common edible films are derived from natural substances such as zein (corn), chitosan (shellfish), casein (milk), and alginate (gum) (8). Other edible coatings include whey protein isolate (WPI), beeswax, or carnauba wax (44).

# Refrigerated

A cooler temperature is the most important factor in reducing respiration rates, and most fresh vegetables will last at least 3 days if refrigerated (21). Storage times for various vegetables are ultimately based on their water content. Vegetables with a high percentage of water, such as lettuce, tomatoes, and spinach, have shorter storage times than do vegetables with less water content, such as potatoes, carrots, and turnips. Because leaves draw moisture from the rest of the plant, removing the green tops of carrots, radishes, or beets increases their length of storage. Cooked beans will last up to 4 or 5 days in the refrigerator, and up to 6 months in the freezer.

## Special Storage Requirements

Some vegetables require special storage treatment. Certain fresh fruits such as avocados (in their skin), bananas, unripe tomatoes, and pears are best not refrigerated. Bean sprouts are best stored in a bowl of cold water in the refrigerator, and the water should be changed daily. Ginger root should be frozen or stored in an airtight container to trap its moisture. It will keep up to a week at room temperature and for a month in the refrigerator (59). Mushrooms will keep up to 5 days if refrigerated in a paper bag or basket, but will deteriorate rapidly if exposed to moisture or warm air. Plastic bags are not recommended for mushrooms, which lack the protective skins of vegetables and fruits, contain 90 percent water, and emit a water vapor that collects inside a plastic bag even if it is vented with small holes. The increased humidity causes brown and black blotches on white mushrooms. Eggplants deteriorate quickly in either warm or cold temperatures, so they should be kept in a cool location (in an aerated plastic bag), where they will normally keep for about a day or two. Asparagus lasts longest if treated like flowers and refrigerated with the stem ends in a jar filled with about 1 inch of water (26).

## Maintaining Moisture

An excellent way to retain moisture in vegetables is to store them in the refrigerator's crisper, which is designed for that purpose. If a crisper is not available, plastic bags with tiny holes are a good choice because the holes allow the food to breathe. Airtight plastic bags are not suitable for storing vegetables because they cause them to "stew," thus promoting spoilage. Paper towels wrapped around moist vegetables can help to avoid the spoilage caused by dehydration.

## Freezing

Fresh vegetables should not be frozen unless they are first blanched; their high water content causes undesirable texture changes when cell membranes burst upon freezing, and certain enzymes cause undesirable browning and deterioration (17, 18).

## Dry Storage

Proper storage does not automatically imply refrigeration. Tomatoes (unripe), eggplant, winter squash, tubers (potatoes), dried legumes, and most bulbs (onions) should never be stored in a refrigerator. Tomatoes are picked green while still firm in order for them to handle the rigors of transportation; however, once at their destination, their ripening can be accelerated by placing them in a paper bag. Keeping them in a bag protects tomatoes from direct sunlight, which softens rather than ripens them. Tomatoes placed in the refrigerator will never reach their optimal flavor and texture. In fact, flavor enzymes will be destroyed, and the tomatoes will lose their texture by becoming more mealy. They should be stored in the refrigerator only when fully ripe, to slow spoilage. Dried beans will keep up to a year stored in airtight containers in a dry place, but if kept too long and/or at high humidity (80 percent) and temperatures, dried beans will be difficult to cook (Chemist's Corner 13-9).

---

**Respiration rate** The rate of carbon dioxide produced from a given amount of produce over a certain unit of time.

> ## CHEMIST'S CORNER 13-9

### Legume Storage

It has been known since the 3rd century BC that certain changes occur in legume seeds during extended storage, especially at high temperatures and humidity, making them difficult to cook. This hard-to-cook phenomenon is a result of several hypothesized changes at the molecular level. The pectin-cation-phytate theory suggests that during storage, an intracellular enzyme, phytase, hydrolyzes phytin, resulting in the release of divalent cations. Once cooking begins, monovalent cations from the pectin located in the cell wall exchange with these divalent cations to form insoluble pectin. The walls, lined with this insoluble pectin, become extremely strong and make long-stored legumes difficult to cook. Another factor thought to contribute to this phenomenon is the interaction between the protein and starch. The proteins tend to over coagulate when heated, preventing starch from absorbing enough water to permit hydration and proper swelling (28).

### Storing Potatoes

Potatoes stored in the refrigerator undergo the conversion of their starch to glucose and develop an undesirable waxy consistency when cooked. French fries made from refrigerated potatoes turn an undesirable brown (20). Potatoes should not be exposed to sunlight, which will cause photosynthesis to take place, producing a bitter taste and, eventually, a potentially toxic compound called solanine, which can be seen as a slightly green tint on potato skins and sprouts. Eating large amounts of these green potatoes is not recommended because it may lead to poisoning (24). Potatoes will keep for a couple of weeks at room temperature stored in a basket or bag with holes, or longer in a cool root cellar (45°F–50°F/7°C–10°C). The exception is new potatoes, which are very perishable and should be used within a few days. Onions and potatoes should not be stored next to each other, because they shorten each other's shelf life.

## Controlled-Atmosphere Storage

Beyond simple refrigeration is a high-tech method of preservation called controlled-atmosphere storage, available to commercial food companies. This special method slows down the natural respiration of fresh vegetables by reducing the amount of oxygen and increasing the amount of carbon dioxide available to them while in storage (34). Such a contrived environment reduces a vegetable's respiration and/or metabolic rate to such a degree that lettuce can last up to 75 days. Other advanced storage methods include modified-atmosphere packaging (MAP), edible coatings, and plastic shrink- or stretch-wraps (41).

# PICTORIAL SUMMARY / 13: Vegetables and Legumes

Vegetables in raw or cooked form add color, flavor, and texture to meals as well as enhance a meal's overall nutritional value.

## CLASSIFICATION OF VEGETABLES

Vegetables may be derived from almost any part of the plant. Plant parts considered edible include the roots, bulbs, stems, leaves, seeds, and even flowers.

A few vegetables are actually fruits; that is, they are the part of the plant that contains its seeds.

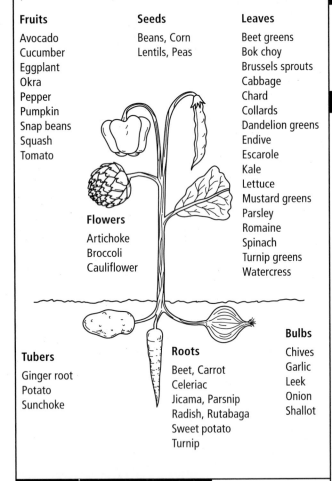

| Fruits | Seeds | Leaves |
|---|---|---|
| Avocado | Beans, Corn | Beet greens |
| Cucumber | Lentils, Peas | Bok choy |
| Eggplant | | Brussels sprouts |
| Okra | | Cabbage |
| Pepper | | Chard |
| Pumpkin | | Collards |
| Snap beans | | Dandelion greens |
| Squash | | Endive |
| Tomato | | Escarole |
| | | Kale |
| | | Lettuce |
| | | Mustard greens |
| | | Parsley |
| **Flowers** | | Romaine |
| Artichoke | | Spinach |
| Broccoli | | Turnip greens |
| Cauliflower | | Watercress |

**Bulbs**
Chives
Garlic
Leek
Onion
Shallot

**Tubers**
Ginger root
Potato
Sunchoke

**Roots**
Beet, Carrot
Celeriac
Jicama, Parsnip
Radish, Rutabaga
Sweet potato
Turnip

## COMPOSITION OF VEGETABLES

Most fresh, unprocessed vegetables are naturally low in calories, cholesterol, sodium, and fat. Vegetables are usually good sources of carbohydrates (especially fiber), certain vitamins/minerals, and nonnutritive compounds called phytochemicals, which possess health-protective benefits. Vegetables rich in fiber include:

- Kidney beans
- Navy beans
- Lima beans
- Lentils
- Peas
- Corn
- Parsnip
- Potato
- Broccoli
- Brussels sprouts
- Carrots
- Green beans
- Mung beans
- Spinach
- Sweet potato

## PURCHASING VEGETABLES

Season, ripeness, freshness, yield, and freedom from bruising or mold are considered when selecting vegetables. At present, the only USDA graded fresh vegetables are potatoes, carrots, and onions. Grading is voluntary and based on ripeness, color, shape, size, uniformity, and freedom from bruising and decay. Individual selection criteria vary according to the vegetable.

## LEGUMES

Beans, peas, and lentils grow as seeds within a pod. Soybeans are relatively high in protein and fat content, and can be utilized to make textured vegetable protein, soy milk, tofu, meat analogs, and fermented soybean foods—including miso, sufu, tempeh, and tamari.

## PREPARATION OF VEGETABLES

Because heat affects the vegetable's texture, flavor, color, and nutrient retention, limiting cooking time helps preserve both flavor and appearance. Acidity toughens vegetables, whereas alkalinity causes excessive softening. Heating vegetables in as little water as possible with the lid on is generally recommended, with a few exceptions, to retain flavor and nutrients. Vegetables may be prepared by either dry-heat (baking, roasting, frying) or moist-heat (simmering, steaming, braising, and microwaving) cooking methods. Regardless of the method selected, vegetables should be thoroughly cleaned before preparation to remove soil, bacteria, and pesticide residues.

## STORAGE OF VEGETABLES

Vegetables continue to respire after harvest, which contributes to the deterioration of their appearance, texture, flavor, and vitamin content. Refrigeration slows this process for most vegetables, except for tubers (potatoes), dried legumes, and most bulbs, which lend themselves to dry storage. Controlled atmosphere storage is a commercial process that extends vegetable shelf life by reducing oxygen and increasing the carbon dioxide in the surrounding air. Another commercial process that slows down respiration is coating vegetables with a thin, edible coating such as wax.

# CHAPTER REVIEW AND EXAM PREP

## Multiple Choice*

1. Which of the following would be classified as a root vegetable?
   a. Asparagus
   b. Broccoli
   c. Carrot
   d. Garlic

2. Carotenoids found in vegetables are susceptible to damage from which of the following factors?
   a. Light
   b. Heat
   c. Oxygen
   d. All of the above

3. What is the name of the pigment responsible for the green color of plants?
   a. Carotene
   b. Chlorophyll
   c. Lycopene
   d. Xanthophyll

4. Broccoli, cabbage, cauliflower, and kale belong to the family of _____ vegetables that may have protective effects against cancer.
   a. carotenoid
   b. phytochemical
   c. fibrous
   d. cruciferous

5. Tempeh is the name of a meat alternative made from whole _____ that have been fermented.
   a. pinto beans
   b. soybeans
   c. lentils
   d. navy beans

6. Legumes are a group of vegetables that include:
   a. tomatoes, eggplant, and zucchini.
   b. potatoes, peppers, and beets.
   c. brussels sprouts, cabbage, and cauliflower.
   d. beans, peas, and lentils.

7. Which vegetable should not be stored in the refrigerator?
   a. Potato
   b. Carrots
   c. Bean sprouts
   d. Squash

## Short Answer/Essay

1. List the different classifications for vegetables and provide examples for each group.
2. Explain how each of the following structural components of the plant cell relates to food preparation: cell wall, parenchyma cell, and intercellular air spaces.
3. What are the three major groups of plant pigments? Provide vegetable examples from each group.
4. Discuss how the plant pigments change color in the presence of heat and acid.
5. Describe the specific signs of ripeness for the following vegetables: asparagus, beans, broccoli, celery, garlic, mushrooms, and potatoes.
6. Define au gratin, scalloped, and degorging.
7. Explain why cabbage and onions have strong odors when cooked.
8. Discuss the various techniques for preventing undesirable color changes in certain vegetables.
9. Describe the three methods for preparing dried beans: overnight soak, short soak, and no soak.
10. Describe controlled-atmosphere storage and explain why it extends the shelf life of fresh vegetables.

*See p. AK-1 for answers to multiple choice questions.

# REFERENCES

1. Abayomi, LA, and LA Terry. Implications of spatial and temporal changes in concentration of pyruvate and glucose in onion (Allium cepa L.) bulbs during controlled atmosphere storage. *Journal of the Science of Food and Agriculture* 89(4):683–687, 2009.

2. Amin AR, O Kucuk, FR Khuri, and DM Shin. Perspectives for cancer prevention with natural compounds. *Journal of Clinical Oncology* 27(16):2712–2725, 2009.

3. Armentrout J. Vegetables, perfectly steamed and deliciously sauced. *Fine Cooking* 78:46, 2006.

4. Bao B, and KC Chang. Carrot juice color, carotenoids, and nonstarch polysaccharides as affected by processing conditions. *Journal of Food Science* 59(6):1155–1158, 1994.

5. Bhandari MR, and J Kawabata. Assessment of antinutritional factors and bioavailability of calcium and zinc in wild yam (*Dioscorea spp.*)

tubers of Nepal. *Food Chemistry* 85(2):281–287, 2004.

6. Birt DF. Phytochemicals and cancer prevention: From epidemiology to mechanism of action. *Journal of the American Dietetic Association* 106(1):20–21, 2006.

7. Brown, T, JEL Corry, and SJ James. Humidification of chilled fruit and vegetables on retail display using an ultrasonic fogging system with water/air ozonation. *International*

*Journal of Refrigeration* 27(8):862–868, 2004.

8. Buonocore GG, A Conte, and MA Del Nobile. Use of mathematical model to describe the barrier properties of edible films. *Journal of Food Science* 70(2):E142–E147, 2005.

9. Clinton SK. Lycopene: Chemistry, biology, and implications for human health and disease. *Nutrition Reviews* 56(2):35–51, 1998.

10. Coelho RG, and VC Sgarbieri. Nutritional evaluation of bean (*Phaseolus vulgaris*) protein: In vivo versus in vitro procedure. *Journal of Food Biochemistry* 18:297–309, 1995.

11. Craig WJ, AR Mangels, and American Dietetic Association. Position of the American Dietetic Association: Vegetarian diets. *Journal of the American Dietetic Association* 109(7):1266–1282, 2009.

12. De Lumen BO, DC Krenz, and MJ Revillesz. Molecular strategies to improve the protein quality of legumes. *Food Technology* 51(5):67–70, 1997.

13. Duck Soon An, EY Park, and DS Lee. Effect of hypobaric packaging on respiration and quality of strawberry and curled lettuce. *Postharvest Biology and Technology* 52(1):78–83, 2009.

14. Edan Y. Color and firmness classification of fresh market tomatoes. *Journal of Food Science* 63(4):793–796, 1997.

15. Egbert R, and C Borders. Achieving success with meat analogs. *Food Technology* 60(1):28–34, 2006.

16. Fahey JW. Underexploited African grain crops: A nutritional resource. *Nutrition Reviews* 56(9):282–285, 1998.

17. Fuchigami M, N Hyakumoto, and K Miyazaki. Programmed freezing affects texture, pectic composition and electron microscopic structures of carrots. *Journal of Food Science* 60(1):137–141, 1995.

18. Fuchigami M, K Miyazaki, and N Hyakumoto. Frozen carrot texture and pectic components as affected by low-temperature-blanching and quick freezing. *Journal of Food Science* 60(1):132–136, 1995.

19. Funaki J, and M Yano. Purification and characterization of a neutral protease that contributes to the unique flavor and texture of tofu-misozuke. *Journal of Food Biochemistry* 21:191–202, 1997.

20. Gardiner A, and S Wilson. The right potato for every recipe. *Fine Cooking* 70:76–77, 2005.

21. Haard NE. Foods as cellular systems: Impact on quality and preservation. *Journal of Food Biochemistry* 19:191–238, 1995.

22. Jaiswal AK, S Rupali, and G Gourav. Control of Maillard browning in potato chips by yeast fermentation. *Journal of Food Science & Technology* 40(5):546–549, 2003.

23. Kostyuk VA, and AI Potapovich. Mechanisms of the suppression of free radical overproduction by antioxidants. *Frontiers in Bioscience (Elite edition)* 1:179–188, 2009.

24. Langkilde S, et al. A 28-day repeat dose toxicity study of steroidal glycoalkaloids, α-solanine and α-chaconine in the Syrian Golden hamster. *Food and Chemical Toxicology* 47(6):1099–1108, 2009.

25. Lin SD, and AO Chen. Major carotenoids in juices of ponkan mandarin and liucheng orange. *Journal of Food Biochemistry* 18:273–283, 1995.

26. Lively R. Asparagus is at its best in spring. *Fine Cooking* 78:16, 2006.

27. Lively R. Sweet and crunchy carrots. *Fine Cooking* 70:24, 2005.

28. Lui K. Storage proteins and hard-to-cook phenomenon in legume seeds. *Food Technology* 51(5):59–61, 1997.

29. Mainardi T, S Kapoor, and L Bielory. Complementary and alternative medicine: Herbs, phytochemicals and vitamins and their immunologic effects. *Journal Allergy and Clinical Immunology* 123(2):283–294, 2009.

30. Manganaris, GA, M Vasilakakis, G Diamantidis, and I Mignani. Effect of post-harvest calcium treatments on the physicochemical properties of cell wall pectin in nectarine fruit during ripening after harvest or cold storage. *Journal of Horticultural Science & Biotechnology* 80(5):611–617, 2005.

31. Masibay KY. What is soy sauce? *Fine Cooking* 70:72, 2005.

32. Matella NJ, et al. Use of hydration, germination, and alpha-galactosidase treatments to reduce oligosaccharides in dry beans. *Journal of Food Science* 70(3):C203–C207, 2005.

33. McCue N. Vidalia onion: How sweet it is. *Prepared Foods* 165(10):67, 1996.

34. McKenzie MJ, LA Greer, JA Heyes, and PL Hurst. Sugar metabolism and compartmentation in asparagus and broccoli during controlled atmosphere storage. *Postharvest Biology and Technology* 32(1):45–56, 2004.

35. Middleton S. Quick-braising vegetables. *Fine Cooking* 71:41–45, 2005.

36. Miller RA, et al. Garlic effects on dough properties. *Journal of Food Science* 62(6):1198–1201, 1997.

37. Mitchell DC, FR Lawrence, TJ Hartman, and JM Curran. Consumption of dry beans, peas, and lentils could improve diet quality in the US population. *Journal of the American Dietetic Association* 109:909–913, 2009.

38. Murphy PA, et al. Soybean protein composition and tofu quality. *Food Technology* 51(3):86–88, 110, 1997.

39. Murthy NS, S Mukherjee, G Ray, and A Ray. Dietary factors and cancer chemoprevention: An overview of obesity-related malignancies. *Journal of Postgraduate Medicine* 55(1):45–54, 2009.

40. National Cancer Institute. *Lifestyle and Prevention.* http://prevention.cancer.gov/prevention-detection/lifestyle. Accessed 6/09.

41. Ngadi M, et al. Gas concentrations in modified atmosphere bulk vegetable packages as affected by package orientation and perforation location. *Journal of Food Science* 62(6):1150–1153, 1997.

42. Packer T. *1997 Produce Availability and Merchandising Guide.* Vance Publishing Group, 1999.

43. Pedreshchi F, P Moyano, K Kaack, and K Granby. Color changes and acrylamide formation in fried potato slices. *Food Research International* 38:1–9, 2005.

44. Perez-Gago MB, M Serra, M Alonso, M Mateos, and MA del Rio. Effect of whey protein- and hydroxyproply methylcellulose-based edible composite coatings on color change of fresh apples. *Postharvest Biology and Technology* 36(1):77–85, 2005.

45. Produce Marketing Association. *Produce Marketing Association Produce Manual,* 1997.

46. Raju J, and R Mehta. Cancer chemo-preventive and therapeutic effects of diosgenin, a food saponin. *Nutrition and Cancer* 61(1):27–35, 2009.

47. Riaz MN. Soybeans as functional foods. *Cereal Foods World* 44(2):88–92, 1999.

48. Rodriguez-Saona LE, RE Wrolstad, and C Pereira. Modeling the contribution of sugars, ascorbic acid, chlorogenic acid and amino acids to non-enzymatic browning of potato chips. *Journal of Food Science* 62(5):1001–1005, 1997.

49. Rosenfeld T. Caramelized onions. *Fine Cooking* 71:60–61, 2005.

50. Sanchez-Pineda-Infantas MT, G Cano-Munoz, and JR Hermida-Bun. Blanching, freezing and frozen storage influence texture of white asparagus. *Journal of Food Science* 59(2):821–823, 1994.

51. Sapers GM, et al. Enzymatic browning control in minimally processed mushrooms. *Journal of Food Science* 59(5):1042–1047, 1994.

52. Schwartz SJ, and TV Lorenzo. Chlorophylls in foods. *Critical Reviews in Food Science and Nutrition* 29(1):1–18, 1990.

53. Sila DN, C Smout, F Elliot, AV Loey, and M Hendrickx. Non-enzymatic depolymerization of carrot pectin: Toward a better understanding of carrot texture during thermal processing. *Journal of Food Science* 71(1):E1–E9, 2006.

54. Starkenmann C, B Le Calvé, Y Niclass, I Cayeux, S Beccucci, and M Troccaz. Olfactory perception of cysteine-S-conjugates from fruits and vegetables. *Journal of Agriculture and Food Chemistry* 56(20):9575–9580, 2008.

55. Steven M. Three steps to perfect oven fries. *Fine Cooking* 71:58–59, 2005.

56. Grading, certification and verification. *USDA Agricultural Marketing Service.* http://www.ams.usda.gov. Accessed 6/09.

57. Vertzoni MV, C Reppas, and HA Archontaki. Optimized determination of lycopene in canine plasma using reversed-phase high performance liquid chromatography. *Journal of Chromatography* 819(1):149–154, 2005.

58. Wang LS, and GD Stoner. Anthocyanins and their role in cancer prevention. *Cancer Letters* 269(2):281–290, 2008.

59. Wemischner R. Ginger: A fresh flavor that packs some heat. *Fine Cooking* 29:92–93, 1998.

60. Wolke RL. Clever kitchen tips and why they work. *Fine Cooking* 77:74, 2006.

61. Yamauchi N, and AE Watada. Pigment changes in parsley leaves during storage in controlled or ethylene containing atmosphere. *Journal of Food Science* 58(3):616–618, 1993.

62. Zhuang H, MM Barth, and DF Hildebrand. Packaging influenced total chlorophyll, soluble protein, fatty acid composition and liposygenase activity in broccoli florets. *Journal of Food Science* 59(6):1171–1174, 1994.

# WEBSITES

The "why and how" of blanching vegetables can be found at North Dakota State University's Extension Service (enter "freezing" in search box and click on "vegetables"):

**www.ag.ndsu.edu/extension/**

Oregon State University's Food Resource website lists numerous foods, including vegetables:

**http://food.oregonstate.edu**

*The Packer,* the weekly business newspaper of the produce industry, offers produce buyers *The Guide,* a publication to assist buyers purchasing produce:

**www.thepacker.com**

# 14 Fruits

Imagine bountiful cornucopias overflowing with **fruits**—nature's desserts available in almost every shape and color of the rainbow. Even in the times of the pharaohs, fruits were transported, sometimes with great difficulty, across continents and seas to eager consumers. Early explorers and migrating peoples carried fruits and their seeds to all parts of the world, so now many are grown in areas far from their original home. For centuries, dates were cultivated in North Africa, but now are also grown in California. Pineapples were once indigenous to South America. Lemons and limes originated in India; oranges, now a daily component in many North American breakfasts, came from southeastern Asia; and the kumquat was brought to North America from China and Japan. The kiwifruit from New Zealand is one of the newest fruits to have spread from its original home to North American markets. The Fruits and Veggies—More Matters™ campaign, sponsored by the National Cancer Institute and the Centers for Disease Control and Prevention, recommends that people consume a minimum of at least five servings of vegetables and fruits a day—three vegetables and two fruits (31).

This chapter looks at fruits—their classification, composition, purchase, preparation, and storage.

## CLASSIFICATION OF FRUITS

Fruits are the ripened ovaries and adjacent parts of a plant's flowers. They are classified according to the type of flower from which they develop: simple, aggregate, or multiple.

- *Simple fruits.* Develop from one flower and include **drupes, pomes,** and citrus fruits (oranges, grapefruits, lemons, limes, kumquats, and mandarins—tangerines and tangelos).
- *Aggregate fruits.* Develop from several ovaries in one flower. They include blackberries, raspberries, and strawberries.
- *Multiple fruits.* Develop from a cluster of several flowers. Pineapples and figs are two examples.

**Fruit** The edible part of a plant developed from a flower.

**Drupes** Fruits with seeds encased in a pit. Examples are apricots, cherries, peaches, and plums.

**Pomes** Fruits with seeds contained in a central core. Examples are apples and pears.

## Classification Exceptions

Sometimes it is difficult to tell the difference between a fruit and a vegetable. For example, is a tomato a fruit or a vegetable? The confusion over the tomato's classification even attracted the attention of the U.S. Supreme Court, which ruled it was a vegetable in 1893 (25). At the time, there was an import tax on vegetables but not fruits, so U.S. tomato growers changed the tomato from a fruit to a vegetable to protect themselves from foreign tomato growers trying to export to the United States. Despite the Supreme Court ruling, botanists beg to differ; botanically, the tomato grows from a plant's flower, so it is technically a fruit even though it is not sweet. Squash, okra, green beans, and cucumbers, too, are botanically fruits. Nuts are actually fruits, also, but they are seeds instead of fleshy fruits, so they are grouped separately.

Confusion of the opposite sort is stirred up by rhubarb, which is really a vegetable, but is usually treated as a fruit.

# COMPOSITION OF FRUITS

The cellular structure and pigments of fruits are similar to those of vegetables, described in Chapter 13. Organic acids, pectic substances, and phenolic compounds are also found in some vegetables, but have more relevance to fruits and are now discussed.

## Organic Acids

Natural sugars such as fructose, glucose, and sucrose are the major contributors to the sweetness of fruits, whereas the tart flavor component is partially due to organic acids located in the cell sap. Acidity varies with the maturity of the plant, usually decreasing as the fruit ripens. These organic acids found in fruits are either volatile or nonvolatile. Volatile acids vaporize during heating, whereas nonvolatile acids do not, but they can leach out when fruit is cooked in water.

The common organic acids in fruit include citric acid in citrus fruits and tomatoes; malic acid in apples, apricots, cherries, peaches, pears, and strawberries; tartaric acid in grapes; oxalic acid in rhubarb; and benzoic acid in cranberries. Cranberry juice also contains a unique blend of quinic and malic acid, with the ratio of the two so constant that it is used by juice processors to determine cranberry juice authenticity and to calculate the percentage of cranberry juice content in juice drinks (45). The oxalic acid in rhubarb can combine with calcium in the intestine to form calcium oxalate, an insoluble complex that cannot be absorbed. It may also combine with other minerals to form similar compounds.

### Acidity of Fruits

Acids cause most fruits to have a pH value below 5.0 (Table 14-1). The tartness of fruits is related in part to their acidic content. For instance, limes, lemons, and cranberries are very tart fruits, with the lowest pH values (around 2.0). The least

acidic fruits are more bland and sweet in flavor, and those with a pH above 4.5 most often serve as vegetables.

## Pectic Substances

Another group of compounds frequently found in fruit are pectic substances, of which there are three types: protopectin, pectin (pectinic acids), and pectic acid. *Pectin* is a general term describing this group of polysaccharides, which act as a cementing substance between cell walls and are partially responsible for the plant's firmness and structure (Chemist's Corner 14-1). Pectin is used commercially to

| pH | Foods |
|---|---|
| 2.2 | Lime juice |
| 2.3 | Lemon juice |
| 3.1 | Apples, boysenberries, grapefruit, prunes |
| 3.2 | Rhubarb |
| 3.3 | Apricots, blackberries |
| 3.4 | Strawberries |
| 3.5 | Orange juice, peaches |
| 3.6 | Raspberries, red sour cherries |
| 3.7 | Blueberries |
| 3.8 | Sweet cherries |
| 3.9 | Pears |
| 4.0 | Grapes |
| 4.2 | Tomatoes |
| 4.6 | Bananas, figs |
| 5.1 | Cucumbers |
| 5.2 | Squash |

**TABLE 14-1** Average pH Values for Selected Fruits

> ## CHEMIST'S CORNER 14-1
>
> ### The Chemical Structure of Pectin
>
> Depending on its source, pectin contains several different polysaccharides, but they all consist of methyl pectate, the compound responsible for gelling. Methyl pectate is a molecule that has a high water-holding capacity, with a structure of a long, thread-like chain of repeating galacturonic acid units obtained from the sugar galactose (Figure 14-1).
>
> **FIGURE 14-1** Methyl pectate.
>
>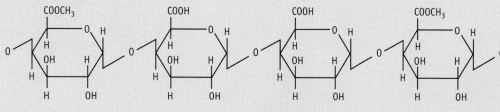
>
> Portion of a methyl pectate molecule

contribute to the gelling of fruit preserves. The string-like pectin molecules bond to each other under the right conditions to form a net-like solid structure that is able to trap water and form a gel. Pectin is found between the plant cells and within the cell walls, but not in the juice, so commercial sources of pectin include the pulp (pomace) remaining after apples are pressed for juice and the spongy **albedo** of citrus fruits.

## Use of Pectin by the Food Industry

In addition to its use in jams and jellies, pectin is used in several other foods as an emulsifier, stabilizer, thickener, and texturizer. Frozen foods benefit from pectin's ability to improve texture by controlling ice crystal size, preventing loss of syrup during thawing, and improving overall shape. Fruit pieces in yogurt are evenly distributed with the aid of pectin, and pectin imparts more body to diet soft drinks (79).

## Pectin Formation in Ripening Fruit

Prior to the formation of pectin, the pectic substance in immature fruit is protopectin, a large, insoluble molecule. As the fruit ripens, enzymes convert protopectin to the more water-soluble pectin (12). Enzymes play a key role in the softening of fruits, with the largest influence derived from those enzymes that break down pectin (39). Ripening mechanisms trigger the pectinase enzymes, which break down the pectic substances as the fruit ripens, and the degree of fruit softening is related to how many pectic substances were degraded (49). The stage of ripeness affects pectin concentration, and it is the pectin extracted from ripe fruits that is used to gel jams and jellies.

As fruit continues to ripen and becomes overripe, all the pectin gradually turns to pectic acid (Figure 14-2). Because neither protopectin nor pectic

**FIGURE 14-2**    Chemical breakdown of protopectin. Only fruits at the height of ripeness should be used for making fruit spreads without adding pectin.

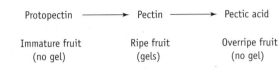

Protopectin ⟶ Pectin ⟶ Pectic acid

Immature fruit (no gel)    Ripe fruit (gels)    Overripe fruit (no gel)

acid can contribute to gelling, only fruit at the height of ripeness should be used for making fruit spreads without added pectin. Heating also converts pectin to pectic acids by hydrolyzing the chemical bonds holding the molecules together, causing the texture of the fruit to become soft and eventually mushy.

## Pectic Substances and Juice Cloudiness

When juice is extracted from fruits, pectic substances can sometimes cause it to cloud (48). Although this is desirable in orange juice, where it contributes to body, other juices, such as apple juice, are often more appealing to consumers if they are clear (9). One way to remove the cloudiness is through a clarification process in which enzymes such as pectinases are added to the juice to break down the pectin compounds responsible for juice cloudiness (20). Juice processors can also add enzymes to certain juices to increase juice extraction. Enzymes such as cellulase and hemicellulase break down the cellulose and hemicellulose in cell walls, releasing more juice (14).

# Phenolic Compounds

Another group of compounds found in fruits, phenolic compounds, is responsible for the browning and bruising that often occur in ripening fruit (35). These compounds, also known as tannins, are found predominantly in unripe fruits, giving them a bitter taste and leaving an astringent feeling in the mouth. Fruits containing phenolic compounds include apples, apricots, avocados, bananas, cherries, dates, grapes, nectarines, papayas, peaches, persimmons, pears, and strawberries (35).

## Phenolic Compounds and Enzymatic Browning

All of these foods turn brown from enzymatic browning, which, as you learned in Chapter 3, occurs in the presence of three substances: phenolic compounds, found within the cells; polyphenol oxidase enzymes (enzymes that oxidize phenolic compounds), also known as phenolase, catecholase, and tyrosinase; and oxygen, which enters the cells when the food is cut or bruised (Figure 14-3) (90). The polyphenol oxidase enzymes turn the color of the

**FIGURE 14-3**    Enzymatic browning in fruit is catalyzed (caused) by an enzyme (polyphenol oxidase).

**Albedo**  The white, inner rind of citrus fruits, which is rich in pectin and aromatic oils.

# NUTRIENT CONTENT

Nutritionally, fruits are high in carbohydrates, phytochemicals, and water, which composes 70 to 95 percent of fresh fruits. They tend to be low in calories, fat, and protein, as described in the Calorie Control feature.

*Fat.* The exceptions to the "fruits are low in fat" generalization are coconuts, avocados, and olives, as shown in Table 14-4. The fat in coconut is predominantly saturated, whereas that in avocado and olives is mostly monounsaturated. Fruits are cholesterol free because they are of plant origin and a liver is necessary to produce cholesterol.

*Carbohydrates.* Carbohydrates are the main source of calories in fruit (see Calorie Control). The carbohydrate in fruit starts out in the form of starch, but as ripening takes place, it is converted to sugars such as glucose, fructose, and sucrose. These sugars sweeten the fruit and increase its palatability. If the ripening process is inhibited, which occurs, for example, when bananas are refrigerated, the starch can not be hydrolyzed to sugar; but when fruits are allowed to ripen normally, their sugar content increases as they mature. Different types of fruits contain different amounts of carbohydrate: dates can run as high as 61 percent carbohydrate, whereas avocados contain only 1 percent carbohydrate. Canned or frozen fruits with added sugar have higher carbohydrate concentrations and thus greater caloric content. Food companies are now providing a wide assortment of processed fruit products with lower sugar concentrations.

*Vitamins and Fiber.* The United States Department of Agriculture (USDA) recommends that people "choose a variety of fruits and vegetables . . . by selecting from all five vegetable subgroups (dark green, orange, legumes, starchy vegetables, and other vegetables) several times a week." This helps to ensure an adequate intake of vitamin C, beta-carotene, and fiber, all of which have been reported to reduce the risk of cancer (82). Pectin and other soluble fibers may help lower high blood glucose and cholesterol levels in some people (13). Oranges are an excellent source of vitamin C, but fruits containing even more vitamin C than the orange include guava, kiwifruit, papaya, and strawberries, as shown in Figure 14-4. The vitamin C content is reduced if the fruit is exposed to air by bruising or cutting, or to alkali or copper through cooking or processing. Fruits high in beta-carotene (vitamin A) include mangoes, cantaloupe, apricots, persimmons, and plantains.

Not all fruits are high in fiber, but some good sources are raisins, pears, apples, raspberries, strawberries, prunes, and oranges. Removing the skin diminishes the fiber content; paring an apple, for example, removes 0.8 grams of fiber. Fruit juice has less fiber than the whole fruit, with most fruit juices averaging only about half a gram of fiber for each cup (47). In addition to being consumed for fiber, prunes and prune juice are often used as a laxative because they contain diphenylisatin, a compound that stimulates intestinal movement.

*Minerals.* Fruits are generally low in minerals, but there are some exceptions. Raisins and dried apricots are high in iron, and potassium can be obtained from avocados, apricots, bananas, dates, figs, oranges, plums, prunes, melons, and raisins.

*Phytochemicals.* Fruits are rich in dietary antioxidants that have purported health benefits—especially in possibly reducing the risk of heart disease and cancer (5).

These antioxidants protect against oxidative damage by being oxidized themselves, thereby protecting the structural and functional integrity of living material. The most abundant types of phytochemicals contained in fruits and vegetables are vitamin C; carotenoids (includes carotene, lycopene, xanthophylls, lutein, zeaxanthin); and phenolics (42). Fruits containing greater amounts of vitamin C than oranges include papaya, guava, cantaloupe, strawberries, honeydew melon, and kiwi. Good sources of carotenoids include persimmons, cantaloupe, mango, plantain, papaya, apricot, and guava. Phenolics can be found in highly colored fruits such as blueberries, blackberries, grapes, black currants, and elderberries, among others (68). Blueberries were consumed by World War I pilots who claimed the fruit improved their vision. Studies show that lutein and zeaxanthin may lower the risk of age-related macular degeneration—one of the leading causes of blindness in the elderly (78).

*(continued)*

phenolic compounds from clear to brown. These brown compounds, called melanins, although unappetizing, are safe to consume.

# Fruits as Functional Foods

Fruits and components of fruits are important contributors to the growth of the functional foods industry. Some, but not all, fruits are rich in certain types of antioxidants, fiber, vitamins, and/or minerals (88) (Chemist's Corner 14-2). Epidemiologic studies suggest diets rich in fruits and vegetables are associated with decreased rates of certain diseases. Furthermore, individual compounds in fruits have been isolated and studied to determine their potential beneficial effects (7).

## CHEMIST'S CORNER 14-2

### Berry Power

Polyphenols (phenolics) are chemical structures in berries that include flavonoids (anthocyanins, flavonols, and flavanols), tannins (proanthocyanidins, ellagitannins, and gallotannins), stilbenoids, and phenolic acids (73). Among berry phenolics, the anthocyanins (colorful pigments) have a wide range of bioactivities including antioxidant, anticancer, and anti-inflammatory properties.

Berries differ in their phenolic concentrations. For example, blueberries and cranberries contain predominantly proanthocyanidins, whereas blackberries, black raspberries, red raspberries, and strawberries contain predominantly ellagitannins. These differences also influence the berry's biological properties (73).

- *Cranberries.* Their bacterial antiadhesive properties are due to its oligomeric proanthocyanidins.
- *Blueberries.* The positive impact on neuronal function and behavior in aging animals may be due to their specific tannins.

## NUTRIENT CONTENT

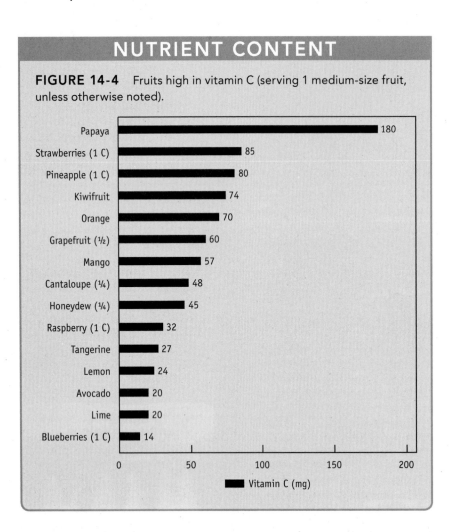

**FIGURE 14-4**   Fruits high in vitamin C (serving 1 medium-size fruit, unless otherwise noted).

- **Antioxidants.** Lycopene is a red compound found in watermelon and tomatoes that has antioxidant activity. Diets rich in lycopene are linked to a reduced risk of certain cancers, including prostate and colon cancer (7). Anthocyanins are another type of antioxidants that give berries their dark color. In laboratory tests, anthocyanins from berries reduce the growth of tumor cells and protect normal cells from damage (8). Vitamins A (beta-carotene), D, E, and C are powerful antioxidants; they also have anti-inflammatory effects.
- **Polyphenols.** High intakes of foods rich in compounds called polyphenols are linked to decreased risk of high blood pressure, heart disease, stroke, and dementia. High levels of polyphenols are found in pomegranate, grape seeds, and berries (33). These beneficial effects are thought to be due in part to the effect of polyphenols on blood vessels: the vessels relax and increase the delivery of blood to the brain and heart.
- **Fiber.** Certain fruits are a rich source of dietary fiber, which is linked to a decreased risk of colon cancer and high blood cholesterol (58).

---

### ✴ CALORIE CONTROL
### Fruits

#### Portion Control

- About 1½ to 2 cups of fruit a day are recommended (see www.mypyramid.gov)
- Approximately ½ a cup fits into a person's palm (Figure 14-5)
- Replace high-calorie/fat snacks with fresh fruit
- Enjoy fruits daily but do not overdo it, especially fruit juices (see Chapter 27)

**FIGURE 14-5**   Approximately 1/2 a cup.

Richard Brewer/Digital Development Services

#### How Many Calories in Fruits?

- Based on the Exchange Lists, each fruit serving—defined as 1 small (4 oz.) fresh fruit; ½ cup of fresh/canned fruit or unsweetened fruit juice; or 2 tablespoons dried fruit—provides about 60 calories (kcal), derived primarily from 15 grams of carbohydrate, with very little protein or fat (with the exceptions noted below).
- Rounding that number to 50 creates another method to control calories: counting in multiples of 50. This makes it easier to count total calories for each meal (Chapter 6).
- Tables 14-2 (whole fruit) and 14-3 (cups of fruit) show how much fresh fruit measured whole or by the cup delivers approximately 50 calories (kcal).
- Calorie values for fruits vary depending on how they are processed. These tables are for *fresh* fruit. Frozen fruits come close to these values *if* sugar has not been added. Canned fruit is much higher in calories because of the added syrup (which is "sugar"), unless it's a lower-sugar version (Figure 14-6).
- Read the Nutrition Facts label to determine the caloric content of processed foods.

*(continued)*

**TABLE 14-2** Whole Fresh Fruit Servings Equal to About 50 Calories (kcal)[1]

| ½ Fruit | 1 Fruit | 2 Fruits | Multiple Fruit |
|---|---|---|---|
| Apple (40 kcal) | Fig | Figs | Apricots (3 pieces) |
| Banana | Guava | Guava | Cherries (10) |
| Grapefruit | Kiwi | | Dates (2) |
| Mango | Nectarine (70 kcal) | | Grapes (15) |
| Papaya | Orange (70 kcal) | | Lychees (10) |
| Pear | Peach | | Raisins (2 Tbsp) |
| Persimmon (Japanese) | Persimmon | | |
| Pomegranate | Plum | | |
| | Tangerine | | |

[1] Same for frozen fruit with no added sugar. Canned fruit is higher in sugar, so read Nutrition Facts label for calories.

**TABLE 14-3** Cups of Fresh Fruit Equal to About 50 Calories (kcal)

| Under one cup fruit ½ Cup–¾ Cup | 1 Cup Fruit | Over one cup fruit 1¼ Cup to 1½ Cup |
|---|---|---|
| **½ Cup** | Blackberries | **1¼ Cup** |
| Applesauce (unsweetened) | Cantaloupe | Strawberries |
| Figs (dried) | Honeydew melon | Watermelon |
| Lychees | Poha berries | |
| Peaches (canned) | Promelo (pommalo) | |
| | Raspberries | |
| **¾ Cup** | | **1½ Cup** |
| Blackberries | | Ohelo berries |
| Blueberries | | Starfruit |
| Pineapple | | |

**FIGURE 14-6** Calorie content of canned fruit in water, light syrup, and heavy syrup.

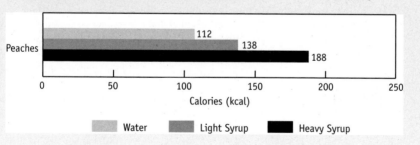

*Exceptions to the Rule:* A few fruits deliver more than the usual number of calories (Table 14-4). Both avocado and coconut derive some of these calories from their high fat content, while dried fruits that are very low in fat are a good source of calories for healthy adults (with no carbohydrate restrictions) wishing to gain weight.

**TABLE 14-4** Selected Fruits High in Calories

| Fruit | Serving Size | Calories (kcal) | Fat (grams) |
|---|---|---|---|
| Avocado | 1 med. | 300 | 30 |
| Coconut (dried, shredded) | ½ C | 250 | 21 |
| Dried fruit | ½ C | 200 | 0 |

© 2010 Amy Brown

# Food Additives in Fruits

Common food additives added to dried fruits to prevent browning include sulfites, sorbates, and benzoates ( 27). Some people are sensitive to sulfites, so the most frequent alternative used is ascorbic acid (vitamin C) (71). The FDA prohibits the use of sulfites on raw foods, with the exception of grapes. Foods containing sulfites must declare this on their labels. Sulfites can appear on labels under several names:

| | |
|---|---|
| Sulfur dioxide | Potassium bisulfite |
| Sodium sulfite | Sodium metabisulfite |
| Sodium bisulfite | Potassium metabisulfate |

Sorbic acid (sorbate) is an antifungal agent that aids in the preservation of various foods. Benzoic acid (benzoate) is a preservative often used in dried fruits or jams. Both substances are found naturally in cranberries, prunes, cinnamon, and cloves.

# PURCHASING FRUITS

Fruit consumption is on an upward trend. Not only does fruit look and taste good, but people feel that it is good for them (77). According to the USDA, the most frequently consumed fruits are bananas, apples, and oranges along with watermelons, cantaloupe, and grapes (62).

# Grading Fruit

Grading fresh produce is difficult, because the quality can change between the time it is graded and the time of purchase. Nevertheless, some fruit producers have fresh fruits graded by the USDA on a voluntary basis. USDA grading is based on size, shape, color, texture, appearance, ripeness, uniformity, and freedom from defects. Nutrient content between the grades does not differ to any great extent. The four grades for fresh fruit are U.S. Fancy, U.S. No. 1, U.S. No. 2, and U.S. No. 3 (81). U.S. Fancy grade fruit is the best and most expensive. Processed fruits (canned, frozen) are graded in a different manner, but the grading is still voluntary. USDA grades for canned or frozen fruits and vegetables are U.S. Grade A or Fancy, U.S. Grade B or Choice, and U.S. Grade C or Standard.

# Selecting Fruits

Fruits are selected for purchase based on the individual evaluation factors for each fruit, whether or not the fruit is graded, and the fruit's peak season (Figure 14-7). There are many varieties available, and the following sections provide selection tips for each fruit and explain how different varieties that exist for the same fruit differ (85).

### Apples

Apples are the second most commonly eaten fruit in the United States. Selection is based on whether the apple will be used for cooking or eaten raw (see Color Figure 14). Apples for cooking should be firm, be able to hold their shape, and remain flavorful after heating. The exception is applesauce, for which softer, fleshier apples such as a McIntosh are desired. The Red Delicious and Jonathan varieties are ideal for eating raw, but not for cooking, because they tend to collapse and lose much of their flavor when heated. The Rome Beauty and Newton Pippin are considered best for pies and sauces but are less acceptable for eating fresh. The remaining varieties can be used for either fresh eating or preparation. Choose apples with clean, smooth, and unbruised skins, with good color. Avoid apples that lack color; this indicates an immature fruit with less flavor. Avoid overmature apples, which will be mealy and soft under the skin.

There are over 1,000 varieties, but about a dozen constitute 90 percent of commercial apple production in the United States. The most common variety is the Red Delicious, followed by the Gala, Golden Delicious, Granny Smith, Fuji, McIntosh, Rome, Empire, York, Jonathan, and Idared (80).

### Apricots

Apricots originated in China thousands of years ago, but California is now the main supplier of apricots marketed in the United States. Apricot varieties available include Royal Blenheim, Castlebrite, Improved Flaming Gold, and Katy. The new varieties are easier to ship and maintain their freshness longer than older varieties. For best flavor, select soft, ripe apricots with as much golden orange color as possible. Although fresh apricots are available only during the summer, dried and canned apricots are sold year-round.

### Avocados

This Central American pear-shaped fruit with greenish-purplish skin was once considered exotic, but is now available throughout the year. The selection of avocados is based primarily on touch. If the skin gives slightly to gentle pressure, the fruit is ready for immediate use. Thick-skinned avocados can be tested for ripeness by inserting a toothpick in the stem end; if it moves in and out easily, the avocado is ripe. Most avocados are still hard when they arrive at the produce stand and will continue to ripen. Avoid avocados that are badly bruised or have soft, sunken spots.

California varieties such as the Hass, Fuerte, Bacon, and Zutano supply 85 percent of the U.S. market, with Florida providing the remaining 15 percent. The Hass, with its bumpy, dark skin and nutty flavor, is the most commonly sold variety. The others are larger and smoother skinned, but their lower fat content tends to make them less rich tasting.

### Bananas

Bananas are the world's chief tropical fruit. The United States is the world's leading banana importer, and Central

**FIGURE 14-7**   Peak seasons for various fruits.

| Seasons for Fruits | | | | | | | | | | | | |
| --- | --- | --- | --- | --- | --- | --- | --- | --- | --- | --- | --- | --- |
| Vegetable | January | February | March | April | May | June | July | August | September | October | November | December |
| Apples | ■ | | | | | | | | ■ | ■ | ■ | ■ |
| Apricots | | | | | ■ | ■ | ■ | ■ | ■ | | | |
| Avocados | ■ | ■ | ■ | ■ | ■ | ■ | ■ | ■ | ■ | ■ | ■ | ■ |
| Berries | | | | | | ■ | ■ | ■ | ■ | ■ | | |
| Blood oranges | ■ | ■ | ■ | | | | | | | | | |
| Boysenberries | | | | | | ■ | ■ | ■ | | | | |
| Cantaloupes | | | | | | ■ | ■ | ■ | ■ | | | |
| Cherries | | | | | | ■ | ■ | ■ | | | | |
| Figs | | | | | | | ■ | ■ | ■ | | | |
| Gooseberries | | | | | | | | ■ | | | | |
| Grapes | | | | | | ■ | ■ | ■ | ■ | ■ | ■ | |
| Kiwi | ■ | ■ | ■ | | | ■ | ■ | ■ | ■ | ■ | | |
| Kumquats | ■ | ■ | ■ | | | | | | | | | |
| Mangoes | ■ | ■ | ■ | ■ | ■ | ■ | ■ | ■ | | | | |
| Nectarines | | | | | | ■ | ■ | ■ | ■ | | | |
| Papayas | ■ | ■ | ■ | ■ | ■ | ■ | ■ | ■ | ■ | ■ | | |
| Peaches | | | | | | | ■ | ■ | ■ | ■ | | |
| Pears | ■ | ■ | | | | | | ■ | ■ | ■ | ■ | ■ |
| Pineapples | ■ | ■ | ■ | ■ | ■ | ■ | ■ | ■ | ■ | ■ | | ■ |
| Plums | | | | | | ■ | ■ | ■ | ■ | ■ | | |
| Raspberries | | | | | | ■ | ■ | ■ | ■ | ■ | | |
| Strawberries | | | | ■ | ■ | ■ | ■ | | | | | |
| Tangerines | ■ | ■ | ■ | | | | | | | | | |
| Watermelons | | | | | ■ | ■ | ■ | ■ | ■ | | | |

America is the major exporter. Bananas ripen best if the bunches are cut while still green. This practice also results in less bruising during handling, shipping, and marketing.

Plantains belong to the banana family, but they are so starchy that they are usually baked, boiled, or fried like potatoes, rather than eaten raw. They average 66 percent starch and only 17 percent sugar when ripe. When plantain skins are black, the fruit is usually ripe and ready to be consumed, but the amount of black on the plantain is a matter of personal preference. Saba and Mazano varieties are ripe when their skins are completely black.

The top two banana varieties are the bruise-resistant Gros Michel from Martinique, a long banana with a tapered tip, and the more bruise-prone Cavendish, which has a curved shape. Newcomer varieties include the Red Spanish or Red Cuban, Saba, and Manzano. Bananas contain less than 5 percent starch and at least 80 percent sugars (43). When choosing bananas for immediate consumption, pick those that are firm and greenish yellow to clear yellow in color.

## Berries

There are numerous varieties of succulent, delicate berries. They come into season in a succession that starts with the sweetest and ends with the least sweet berries; first come strawberries, then raspberries, blackberries, blueberries, gooseberries, red currants, and finally cranberries.

### *Strawberries*

Strawberries do not continue to ripen after they have been picked. For that reason they should ideally come to market when they are bright red and fully ripened. Strawberries are highly perishable because of their high respiration rate and susceptibility to fungal spoilage (74). If they are not to be eaten shortly after purchase, strawberries should be stored, unwashed, in the refrigerator. Rinsing them prior to refrigeration is not recommended because it dilutes the flavor and softens the texture. The short-lived strawberry season starts in April and lasts until sometime in July, but strawberries from Florida and other parts of the world, such as New Zealand and Mexico, have helped to extend the season. Common varieties include the Chandler, Douglas, Heidi, and Pajaro. Figure 14-8 shows techniques for removing the stem and core of a strawberry (known as *hulling*) prior to eating. Strawberries should be washed before hulling so they do not absorb water.

### *Raspberries*   Select raspberries that are plump, firm, and free from mold. Red raspberry varieties include the Willamette, Meeker, Heritage, and Sweetbriar. Munger raspberries are black.

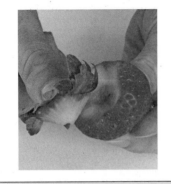

**FIGURE 14-8**   Hulling strawberries can be done with a paring knife (left) or a strawberry huller (right).

Richard Brewer/Digital Development Services

**FIGURE 14-9**   Deglet Noor dates.

Digital Works

*Blackberries*   Select blackberries that are plump and have an almost black color that covers the entire berry with no traces of green or red by the stem. Avoid berries with the little caps, which indicate immaturity. Native to North America and Europe, blackberry varieties include Boysen, Cherokee, Logan, Marion, and Olallie.

*Blueberries*   Select blueberries that are plump, smooth, and blue-black in color with a light grayish "bloom." Grown in the United States and Canada, blueberries come in Patriot, Bluetta, and Earliblue varieties.

*Gooseberries*   Similar in size and appearance to green seedless grapes, gooseberries are selected based on their smoothness and light green color. Most gooseberries, except for the Poorman variety, are so tart that they have to be cooked and sweetened to be eaten; they are primarily used to make pies and preserves.

*Red Currants*   These are also quite tart and generally cooked and sweetened or used primarily as a garnish. Red currants are best selected by picking those firmly attached to the stems.

*Cranberries*   North America is home to only three native fruits: the Concord grape, the blueberry, and the cranberry. The cranberry season starts in September and ends in November, which may explain why the cranberry is associated with the holidays that occur toward the end of the year. When selecting fresh cranberries, choose those that are firm, plump, and smooth-skinned. Four

varieties—Early Black, Howe, Searles, and McFarlin—supply most of the cranberries consumed.

### Cherries

Dark red or deep purple cherries are the tastiest. Choose sweet cherries that are plump, firm, and brightly colored with an almost mahogany-looking skin. For tart cherries, expect the skin to be lighter. The best choices are Bing, Lambert, Black Tartarian, Chapman, Borlat, Schmidt, and Republican. Tart or sour cherries such as the Montmorency tend to be lighter red in color and are more frequently used for pies. Royal Ann cherries are used to make maraschino and candied cherries.

### Dates

These long, brown, wrinkle-skinned berries from palm trees have been cultivated for centuries in the Middle East and North Africa, where they provided sustenance for many ancient desert nomads (84). Although dates appear to be dried, they are actually fresh. Most dates at the consumer level are sold packaged, ripe, and ready-to-eat. There are two basic classes of dates: soft and semi-dry, although a few are considered dry. Soft dates, such as the Medjool, Khadrawy, and Halawy varieties, contain more moisture, have a very soft, almost caramel-like flesh, and are popular with consumers as a snack. The semi-dry dates are important to food processors. Although there

are over 100 varieties, Deglet Noors constitute 95 percent of the market (Figure 14-9). The Zahidi date (semi-dry) has firmer flesh and is smaller than the Deglet Noor. Other varieties of dates include the Barhi, Thoory (the driest date), and Dayri.

### Figs

California is the largest producer in the United States of black and green figs. Originating in the Mediterranean basin, they have been cultivated longer than any other fruit in the world (86). The flowers on fig trees are never seen because they develop inside the fruit and produce the crunchy little seeds that give figs their unique texture. Black figs are selected for eating raw, and green figs are used primarily for cooking and canning. Dried, pressed figs have for centuries been served as a confection. When buying, avoid dry, hard figs or those with flat sides, splits, or mold. These sweet, small, pear-shaped fruits come in a number of varieties, Black Mission being the most common. Others include Brown Turkey, Brunswick, Celeste, and Kadota. Popular dried fig varieties include the Smyrna and Calimyrna.

### Grapefruit

Both pink and white/golden grapefruit are available. Grapefruit is at its best when the skin is shiny and smooth and both ends are flat. Avoid grapefruit with pointy ends, thick skins, or deep pores. The pink varieties, which are sweeter, include the popular Ruby Red and Star Ruby, followed by Burgundy Red, Foster Pink, and Marsh Pink. A popular white or golden grapefruit variety is the seedless Marsh.

## Grapes

Perhaps no other fruit is used quite as extensively and for as many different purposes as the grape. Grapes are consumed fresh or in the form of wine, or as raisins, juice, jams, or jellies. Most grapes in the United States are grown for the wine industry (72). Grapes are actually berries that grow in clusters, and they come in a wide variety of colors and shapes. Grapes do not get any riper or sweeter after being picked. Choose grapes that are firm, plump, and well-colored. Avoid those that are soft, wrinkled, moldy, or have bleached areas around the stem. Table grapes should be chilled for the best flavor and texture. Classification is based on color, and the most common commercial table grapes are the red Flame seedless and green Thompson seedless types. Other popular table grapes include Perlette, Lady Finger, Tokay, Cardinal, Catawba, Delaware, and Emperor. The fragile Concord grape is used primarily for making wine, juice, and jelly. Raisins are usually made from Muscat and Thompson seedless grapes.

## Guavas

Select guavas whose skins are yellow rather than green, and which yield to pressure and emit a fruity guava aroma. Slightly underripe guavas may be lightly cooked and sweetened. There are several varieties of guava, each with its own distinctive flavor, ranging from strawberry-like to pineapple-like to banana-like. The flesh may be white, red, or salmon in color. Seeds may be present or absent, depending on the type of guava.

## Kiwifruit

The New Zealanders changed the name of this fruit from *Chinese gooseberry* to *kiwifruit* as a salute to their native kiwi bird when they started marketing it to the United States. When ripe, kiwifruit yield to gentle pressure. The firmer the fruit is, the tarter the flavor will be. One kiwifruit provides more vitamin C than an orange. Kiwifruit can be peeled, sliced, and eaten, or cut in half and scooped out of its skin with a spoon, but it should not be cooked; heat destroys its color and texture. This fruit is commonly added to fruit salads, but kiwifruit cannot be used in the preparation of gelatin-based foods because an enzyme in kiwifruit, actinidin (a potent proteinase), prevents gelling (32). Grown in both New Zealand and California, kiwifruit are lemon-sized or smaller, oval, with fuzzy, brown exteriors and green interiors sprinkled with a ring of black seeds. They are available year-round.

## Kumquats

These small, orange-type fruits are native to China and are eaten whole without peeling. The Nagami and Meiwa are the two varieties commonly sold in the United States. They are most often found preserved in syrup in the specialty food section of the supermarket.

## Lemons

Size is not related to juiciness. Lemons with thin skins and those that are heavy for their size yield more juice than their thicker-skinned, lighter counterparts. Storing lemons in water will produce twice the juice when squeezed. Eureka and Hishon lemons are the two main commercial varieties. Other varieties include Avon, Bearss, Harney, and Villafranco. The Meyer lemon is a cross between a lemon and a mandarin and is prized for its floral, almost sweet flavor.

## Limes

When buying, select limes that are heavy for their size and have a thin, glossy skin. Sweet limes are available, but only the acidic varieties such as the Tahiti (Persian, Bearss, Idemore, and Pond) and Mexican types are grown in the United States. Key limes differ from the regular limes sold in the supermarket in that they are smaller—closer to cherry tomato size—and juicier, have a more potent lime flavor, and are usually more difficult to find and hence costlier.

## Mandarins

This category of orange-derived citrus fruit includes several familiar fruits, all of which are easily peeled. The most popular are tangerines, tangelos (a tangerine-grapefruit hybrid), and tangors (a tangerine-orange hybrid). Choose mandarins that are firm, full-colored, and covered with a thin, almost oily-feeling skin. Tangerine varieties include the Dancy and Robinson. A common tangelo variety is the Minneola followed by the Nova, Early K, Orlando, and Sampson. The most popular tangor variety is the Temple.

## Mangoes

The flavor of the golden-orange flesh of this tropical fruit has been compared to a cross between that of a peach and an apricot. Select mangoes that have very little green left and show some red, but are predominantly orange to yellow. The fruit should yield slightly to gentle pressure. Another indication of ripeness is a distinctive mango odor coming from the fruit, especially at the stem end. Over 1,000 varieties are available; small varieties include the Haden, Van Dyke, and Atalufo; medium varieties include the Francisque and Tommy Atkins; and the large varieties include the Keitt and Kent. The latter two varieties will stay green even when ripe, but most of the others will pass through a red phase and then turn yellow. Figure 14-10 illustrates how to cube a mango.

## Melons

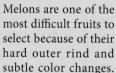

Melons are one of the most difficult fruits to select because of their hard outer rind and subtle color changes. There is a wide assortment of melons and each has its own selection nuances (see Color Figure 11). Melons fall into one of two categories: muskmelons, which include most melons, and gourds, such as the watermelon. Most varieties are sold slightly underripe and will have to ripen a few days at room temperature. When ripe, the stem end becomes slightly indented and yields to pressure. Another indication of ripeness

**FIGURE 14-10**　Cubing a mango.

© Stockbyte/RF/Getty Images

is a distinctive melon odor coming from the fruit, especially at the stem end. Overripe melons will have water-soaked spots, and a rattling sound will occur when the melon is shaken. Selection hints for the different melon varieties follow.

**Cantaloupe**　Look for a cantaloupe with a rough, rope-like netting over a golden-colored rind. Other signs of ripeness include cracks near the stem end, a cantaloupe aroma, and a springy response to firm pressure. Cantaloupe varieties include Top-score, Ambrosia, and Saticoy.

**Honeydew**　Ripe honeydews have a waxy-feeling, creamy-white rind, and honeydew aroma. For best flavor, choose one weighing at least 5 pounds.

**Persian**　These melons look like big cantaloupes, but the flesh is firmer, smoother in texture, and a deeper orange. Selection tips are similar to those for a cantaloupe except that more of the rind should show through the webbing.

**Watermelon**　The condition of the stem and the hollow sound when thumped are the two indicators most often used when selecting a watermelon. Do not try to evaluate a watermelon's ripeness by its green color, but rather look for a creamy or yellowish underside. A dull, smooth rind with a symmetrical shape are signs of a good

watermelon. Several varieties found in the market include Calsweet, Triplesweet, and Orchid Sweet (yellow flesh).

Figure 14-11 shows how to remove the seeds from a whole watermelon.

### Nectarines

Nectarines have been consumed for over 2,000 years and may soon exceed peaches in popularity. Many children prefer them because they have no fuzz. Select nectarines that yield to soft pressure and that have an orange-yellow background color between the red hues.

Almost the entire U.S. supply is grown in California, although some are imported from New Zealand and Chile. The favorite varieties include Firebrite, Red Diamond, Flavortop, Summer Grand, Flamekist, and Red Gold.

### Olives

Although olives are botanically considered a fruit, they are often treated as vegetables. The trees on which they grow can live 600 years or more and have one of the longest cultivation histories of any tree in the world (17). Unripe olives are green, and progressively darken the longer they are on the tree. By the time olives turn black naturally, the abundance of olive oil

they have produced makes them too ripe for consumption. The exception is the dried Greek olive, or salt-cured olive, which is treated with salt to remove the water and looks shriveled and oily. "Black" olives are actually picked green and processed in a curing solution that turns them black, removes their bitterness, and gives them their characteristic olive flavor. Green olives are fermented rather than subjected to the curing solution used to make black olives.

In North America, the three main types of olives are the California-style black ripe olive, the green, Spanish-style olive, and specialty olives such as Nicoise (neh-swahz), Kalamata (also spelled Calamata), Sicilian, and dried Greek. Most olives are sorted according to size and stored in cans or jars of brine. The variety of olive tree determines whether the olive is small, medium, large, extra large, jumbo, colossal, or super colossal (Figure 14-12).

### Oranges

Oranges, another popular fruit, are grown primarily in Florida and California. Even though the same varieties are grown in each state, the juiciness, color, and texture of their oranges differ because of variations in soil and climate. Thin-skinned, very juicy Florida oranges provide most commercial orange juice.

The sweetest, juiciest oranges are those picked at the peak of their season, and ideally this is when they come to market. Oranges that feel firm and heavy for their size are the best choice for juiciness. Different varieties are grown throughout the year, and some varieties, inconveniently for the grower, turn orange in the winter before they are ripe. Warmer weather then turns them back to green, so some growers in California expose them to ethylene gas—a gas naturally produced by ripening fruits—to increase the rate of ripening and return their color to orange. In Florida, dye is often used on oranges that do not respond to ethylene. Varieties include the Hamlin, Pineapple, Florida Valencia, and Pope Summer. California oranges are known for their fresh eating qualities and include the Navel and California Valencia

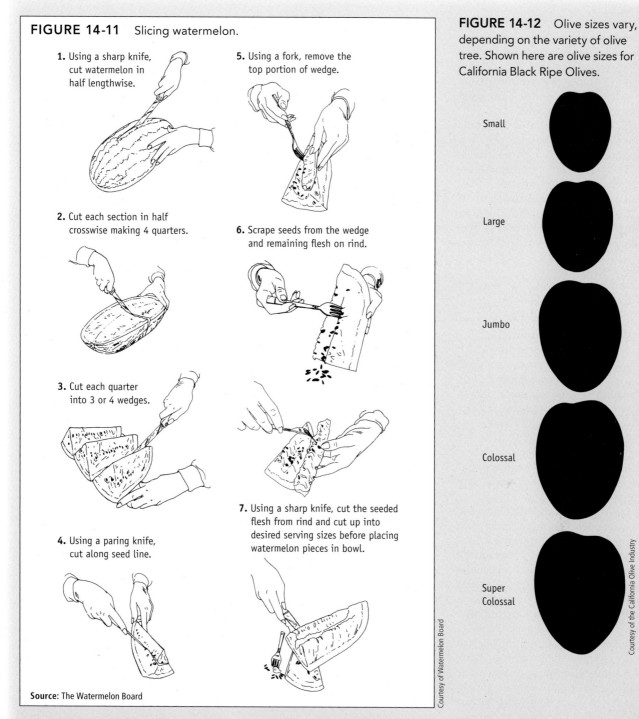

**FIGURE 14-11** Slicing watermelon.

1. Using a sharp knife, cut watermelon in half lengthwise.

2. Cut each section in half crosswise making 4 quarters.

3. Cut each quarter into 3 or 4 wedges.

4. Using a paring knife, cut along seed line.

5. Using a fork, remove the top portion of wedge.

6. Scrape seeds from the wedge and remaining flesh on rind.

7. Using a sharp knife, cut the seeded flesh from rind and cut up into desired serving sizes before placing watermelon pieces in bowl.

**Source:** The Watermelon Board

Courtesy of Watermelon Board

**FIGURE 14-12** Olive sizes vary, depending on the variety of olive tree. Shown here are olive sizes for California Black Ripe Olives.

Small

Large

Jumbo

Colossal

Super Colossal

Courtesy of the California Olive Industry

varieties. Because California Valencias are thin-skinned, they may also be used for juice. Sour oranges, including Seville and Chinotto, are sometimes sold in specialty markets. Blood oranges with their distinctive deep-red-colored flesh are a wintertime specialty.

### Papayas

Ripe papayas are yellow-orange, soft, and prone to damage, so they are shipped to the market in varying degrees of ripeness. The green papayas, which are one-quarter yellow, will ripen in 5 to 7 days; the half-green, half-yellow papayas will be ready in 2 to 4 days; and the papayas that are three quarters yellow-orange are at the ideal stage of ripeness. If they are already yellow-orange, they may be refrigerated and kept up to a week. In addition to the color factor, the skin of the papaya should be smooth, unblemished, and free of any dark or mushy spots. This tropical fruit comes in two major varieties, Hawaiian Kapoho and Sunrise.

### Peaches

Peaches are a very popular fruit. Choose peaches that are already ripe, preferably tree-ripened, and ready to eat. Peaches do not acquire additional sweetness after being picked, although they may become softer and juicier (69). Look for a yellowish or creamy background color. The amount of red is not always a good indication of ripeness, because some of the best peaches are pale-colored. Ripe peaches should yield to gentle pressure. Avoid those that are too firm, have any green background, or are bruised.

Hundreds of varieties are available, but the newer varieties of yellow freestones tend to dominate the market. They are larger, firmer, and more acidic than the older varieties. Clingstone peaches are still available fresh on the market, but most are used in commercial canning. The flesh of freestone peaches does not stick to the pit, whereas it does in the clingstone variety.

### Pears

Pears are generally from either the European or Asian type. Asian pears differ from European pears in that they vary in color from yellow to yellow-green and brownish-red, and they have a blander flavor and crispier texture that is closer to an apple. There are more than ten varieties of Asian pears, but the top three are Twentieth Century, Shinseiki, and New Kikusui. Skin color is a reliable indicator when selecting Asian pears, which, unlike European pears, are usually sold ripe. One way to test for ripeness is to press the neck near the stem; the fruit is ready if it gives off a sweet aroma. The graininess sometimes experienced when eating an unripe pear is from lignin deposits. European pears are juicy and come in four common varieties. Bartletts, the most common, arrive in late summer and are followed by three winter varieties: d'Anjou, Bosc, and Comice. European pears can be bought underripe and ripened at room temperatures. Pears bought ripe are best kept refrigerated in plastic bags. Pears are soft-skinned, so occasional surface blemishes should be expected.

### Persimmons

Although it is originally from Asia, this bright orange-red fruit is now supplied to the United States primarily from California. Commodore Perry introduced the fruit to North America in 1856 when he sent persimmon seeds from Japan to a friend in the United States (44). Fully ripe persimmons have slightly transparent skins revealing the seeds underneath, are a bright orange color, and are very soft; so soft, in fact, that they are rarely sold ripe on the market, but must be allowed to ripen after purchase or their high tannin content makes them extremely astringent. They become more sweet and juicy as they ripen, and are best consumed while still slightly crisp. Pointy Hachiya persimmons account for 90 percent of the U.S. market. Fuyu is the other main variety and looks more like a small, flat tomato.

### Pineapple

Rather than a single fruit, a pineapple is really a cluster of fruits of the Ananas tree. The pineapple is native to South America; it was discovered on the island of Guadeloupe by Columbus (52), and got its name from its similarity to a pinecone (30). It is the second most popular tropical fruit after the banana. Unlike bananas, pineapples do not ripen after being picked, and even though the outside color will change from green to yellow and the inside will get juicier, the sweetness does not increase. The best pineapples come fresh off the field after having been allowed to ripen to maturity. Most of the pineapples sold on the market, however, were cut while still green.

Color is not always a reliable indicator of a ripe pineapple. The best way to choose a pineapple from the market is by touch and smell. It should be plump, give slightly to pressure, and have a fresh, sweet, pineapple scent and be topped with fresh green leaves. Avoid fruit with soft spots, brown discoloration, a decaying base, or a fermenting odor. Sweetness often depends on variety. Gold is one of the sweeter pineapples available on the market. Figure 14-13 shows several techniques for cutting up a pineapple. Like those in kiwifruit, the enzymes in raw pineapple will prevent gelatin from gelling.

### Plums

Plums are available in two types: European plums (blue or purple, yellow, and green) and Japanese plums (yellow or red). Plums, like peaches, come in freestone and clingstone varieties; European plums are usually freestone. Some of the best-tasting varieties of Japanese plums include the Santa Rosa, Laroda, and Elephant Heart. European plums are firmer, milder in taste, and are often dried for prunes. Picked when still immature, plums continue to ripen and should be chosen when they give to gentle pressure and have reached full color. Pluots with their reddish, slightly orange color are a cross between a plum and an apricot.

### Pomegranates

These ruby-red, leather-skinned fruits are filled with pulp-laden seeds. The name came from the French word for "seeded apple." Choose the largest pomegranates available, because the larger the fruit, the more pulp will be surrounding each seed. Make sure the rind is fresh and not dried out. The bright red seeds may be eaten as is or they may be carefully juiced to extract the tart, bright red juice.

### Rhubarb

Rhubarb is actually a vegetable treated as a fruit, being sweet enough for pies once sugar is added. Only the stem of the plant is eaten; the leaves are so high in oxalic acid that they are toxic if eaten in large amounts. Field-grown rhubarb is larger, stringier, and tarter than greenhouse rhubarb. Select rhubarb stalks that are tender, juicy, pink to red in color, and free of fibrous development.

### Tangerines—see Mandarins.

### Tropical Fruits

As world transportation and marketing outlets have improved, more tropical fruits have found their way into the market (Color Figure 14). Increased consumer awareness of these fruits has led to higher demand for coconut, guava, mango, and papaya. Other tropical fruits gaining in popularity include the atemoya, breadfruit, carambola (star fruit), cherimoya, feijoas, granadilla, jaboticaba, loquat, lychee, mamey, mangosteen, passion fruit, papaw, plumcot, quince, sapodilla, sapote, and tamarind.

**FIGURE 14-13**   Cutting pineapple.

**PINEAPPLE CHUNKS**

**1.** Twist crown off pineapple.

**2.** Cut pineapple in quarters, lengthwise.

**3.** Trim each end, then cut away the center core strip and the fruit (as close to the shell as possible).

**4.** Cut the quartered fruit into bite size pieces.

**PINEAPPLE OUTRIGGER**

**1.** Cut pineapple in quarters, lengthwise, leaving crown attached to the shell.

**2.** Cut under the center core strip close to the shell with a thin, sharp knife. (Leave core strip connected at both ends of the shell.)

**3.** Cut crosswise through the loosened fruit, making wedge shaped chunks.

**4.** Slide wedges back into the shell. Spread the wedges as shown to create an outrigger (zigzag) effect.

**PINEAPPLE FRUIT SALAD BOAT**

Banana slices
Grapes
Apple wedges
Orange slices

Courtesy of Dole

**1.** Cut pineapple in half, lengthwise, leaving crown attached to the shell.

**2.** Remove fruit by cutting close to the shell with a thin, sharp knife or a curved fruit knife.

**3.** Cut away the center core strip, then cut the pineapple, and other fresh fruits into bite size pieces.

**4.** Mix all the pieces together in the pineapple shell. Sprinkle with coconut and serve.

## Superfruits

***Goji and Acai***   Goji berry (*Lycium barbarum*) and acai (*Euterpe oleracea*) are two fruits that have recently become popular. Both have been consumed for centuries in their native areas. Acai is a purple berry native to the region of South America near the Amazon River, where the river dwellers mix it with manioc flour (cassava) and consume it as a staple. Because it grows wild in the Amazon jungle and is highly perishable, it is exported as dried berries or in juice form for use in beverages, frozen yogurt or ice cream (often mixed with granola and banana), and desserts (11). Goji (Go-Gee) has been cultivated for 2000 years by the Chinese, who often dry this orange-red berry from the nightshade family (23).

The acai and goji berries are now being used outside of their native countries as dietary supplements and food additives for their purported health benefits. Manufacturers claim the antioxidants and other compounds in goji and acai can increase energy levels, protect the heart from chronic disease, and improve overall well being (73). As a result, these fruits have been categorized as *superfruits*, but that is only a marketing term with no official FDA or USDA definition.

Scientists have evaluated the claims made for goji and acai fruits, but the evidence is not conclusive. Laboratory tests confirm that the berries have high levels of antioxidants (21), and goji berries have a very large ORAC (Oxygen Radical Absorbance Capacity) score of 8,430 for each ounce (11) (Chemist's Corner 14-3). Clinical studies in humans have found high antioxidant levels in the blood of individuals consuming the juice or pulp (4, 40). A small clinical study reported subjective improvements in energy level and feelings of overall health with goji juice (3).

In experimental studies, extracts of the acai berries can inhibit the growth of cancer cells in vitro (within the test tube)

(26), have anti-inflammatory properties (40), and cause blood vessels to dilate in experimental animals (70). Currently, however, there is no clinical evidence to confirm the effect of goji or acai on long-term medical conditions such as heart disease, cancer, or immune function.

Although there is no real evidence yet (due to insufficient studies being conducted) to suggest they can drastically improve health, these fruits do appear to be safe when consumed in moderation. Consumers should be aware that acai products may also include guarana, which is high in caffeine and can cause its related side-effects.

*Blueberry*   Although blueberries are not a new fruit to North Americans. Little research exists on the effect of blueberries on health, but a few studies with rats suggest that a blueberry-enriched diet may protect against oxidative stress, reduce learning impairment, boost memory, and possibly partially protect the rodent brain against Alzheimer's disease (28, 34, 41).

*Cranberry*   Cranberry juice and cranberry extract have long been touted as useful in the prevention and treatment of urinary tract infections (bladder infections). Clinical studies suggest that the use of supplements or 100 percent juice may be beneficial in reducing the risk of infections, but not effective in treating infections once they start (36).

*Pomegrante*   Richer in antioxidants than blueberries or cranberries, pomegrante extract used in mice may reduce the risk for inflammatory arthritis symptoms (75).

When considering these findings about "superfruits," remember that one or even a few studies does not create a scientific consensus. Many of these studies were conducted on animals, so further research in humans must be conducted before benefits to human health are established.

> ## CHEMIST'S
> ## CORNER 14-3

### Measuring Antioxidant Power

Scientists back in 1993 developed an in vitro (in the test tube) method of determining the antioxidant capacity of various foods that they called ORAC (Oxygen Radical Absorbance Capacity) (11, 57). Fluorescence intensity is measured with and without the antioxidant to be tested and compared against the activity of a standard antioxidant (vitamin E). The test begins with the addition of a free-radical generator that produces peroxyl free radicals (such as an azo-initiator compound) to

a fluorescent molecule (such as beta-phicoerythrin or fluorescein). The mixture is heated, the free radicals damage the fluorescent molecules, and the loss of fluorescence is measured over time. The amount of reduced fluorescence curve is compared to standardized values created with a known antioxidant. Although Figure 14-14 shows that some fruits have high ORAC values, clinical research still needs to be conducted to determine if they are or are not relevant to what happens in the human body (57).

**FiGURE 14-14**   Oxygen Radical Absorbance Capacity (ORAC) values of selected fruits.

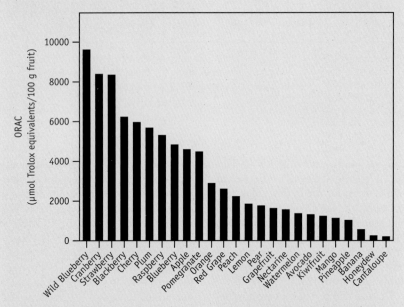

**Source:** Reprinted with permission from wolfe, K. L., et al, Cellular antioxidant activity of common fruits. Journal of Agricultural and Food Chemistry 56:8418–8426, 2008. Copyright 2008 American Chemical Society.

## Processed Fruits

Not so long ago, fresh fruit was available only in the summer, but refrigerated transportation now makes it possible to import fruits from other countries where growing seasons differ from those of North America. Canning, freezing, and drying are other ways of making fruit available all year. Home-freezing and home-drying are on the upswing. Conversely, home-canning knowledge is not as widespread as it used to be. Commercial canning remains responsible for providing the largest portion of processed fruits now reaching the market.

### Canned

Commercially, more fruits are canned than are frozen or dried. The most commonly canned fruits and fruit combinations are applesauce, peaches, pineapple, fruit cocktail, cranberries and cranberry products, and pears (87).

The advantages of canned fruits are their convenience, availability, and variety of forms: whole, half, sliced, chunks, crushed, sauce, or juice. Canned fruits have been cooked, which alters their taste and texture and depletes the water-soluble vitamin content if the cooking juices are not used.

***Canning Liquids***  According to the *Code of Federal Regulations,* fruits are canned in their own juice (no sugar added); light syrup (some sugar added, 10 to 14 percent density in fruit cocktail); or heavy syrup (heavier sugar concentration, 18 to 22 percent density in fruit cocktail) (83). The different packing liquids vary in their influence on the flavor and calorie content of the fruit (Figure 14-5).

***Grading of Canned Fruit***  Not all canned fruits are of the same quality. Voluntary USDA grading places the best fruit under U.S. Grade A or Fancy, with less perfect fruit placed under U.S. Grade B or Choice. Choice canned fruit is usually less expensive, and the fruit may be in smaller pieces than it would be as Grade A or Fancy, which makes it ideal for gelatin salads and fruit mixtures. U.S. Grade C or Standard covers fruits that have uneven pieces and some blemishes, making them best for jams, sauces, and other processed fruit products (81).

### Frozen

Some of the most commonly frozen fruits are cherries, strawberries, blueberries, sliced peaches, red and black raspberries, boysenberries, and loganberries. Fruits are simply frozen fresh with or without the addition of sugar or other sweeteners. Freezing a fruit retains its color and taste, but its texture decreases in quality, because the cell membranes have a tendency to rupture as the ice crystals expand during freezing (1). Fruits are softer and mushier than their fresh counterparts, but still can usually be used the same way as fresh fruit. Frozen fruit makes a quick phytochemical-rich dessert if microwaved for 20 to 30 seconds. When used for pies or other desserts, the fruit is usually partially defrosted in the refrigerator or in the microwave and all or some of the juice is drained away. The recipe can be altered for sugar content if the fruit has been sweetened.

***Signs of Refrozen Fruit***  Packages or bags that show signs of juice stains or have a heavy frost clinging to the outside should not be purchased, because these are signs that the fruit has been thawed and refrozen. Refrozen fruit is flaccid and less flavorful. A temperature of 0°F (−18°C) should be maintained at all times during storage and transport.

## Dried Fruits

Drying has been used as a method of preserving fruit for thousands of years. Commonly dried fruits, in descending order of popularity, include raisins, processed fruit snacks, prunes, and apricots (22). Dried apple, banana, and pear slices are also popular.

Drying makes these fruits available year-round. Dried fruits are often packed in small containers for easy transport in lunch bags or boxes, or by backpackers, campers, and seafarers. These dried fruits contain as much as 70 percent carbohydrate and, with the flavor and nutrients more concentrated, are an alternative to candy or other less nutrient-dense foods. Generally, dried fruit is a good buy if the cost does not exceed five times the price per pound for fresh fruit.

Most fruits are 85 percent water, but drying lowers that amount to below 30 percent. It takes about 5 pounds of grapes to produce 1 pound of raisins. The removal of water during drying increases resistance to microbial spoilage. Drying is also responsible for the loss of volatile substances in fruit and softening of the cellulose, resulting in a sweeter-tasting fruit.

### Fruit Leather

The beef jerky–like strips of fruit called *fruit leather* are made from a mixture of fruit juice concentrate. The fruit puree (a blenderized mix of fruit, water, and other ingredients) is dried on a flat surface in an oven, desiccator, or direct sunlight (38).

### Rehydrating Dried Fruit

Dried fruits can be rehydrated to some degree by adding ½ cup of water or other liquid, such as juice or liquor as called for in some recipes, for every cup of dried fruit and microwaving on medium high (70 percent) for 2 minutes.

## Fruit Juices

Fruit juices come to market in cans, bottles, and cartons. They arrive fresh, as frozen concentrates, or in powdered forms that often contain added sugar. Concentrated fruit juice has had about three fourths of its water removed. This may alter flavor somewhat, but the concentration in volume reduces shipping and handling costs.

Juices commonly sold in the supermarket include, but are not limited to, apple juice/cider, and orange, grape, cranberry, grapefruit, and prune juices (61). There has been an increase lately in the number of tropical juice mixes on the market. The labels on these mixes can be misleading, because the product can contain anywhere from 10 to 100 percent actual juice. The ratio of fruit juice to water determines the standard for fruit beverages under federal guidelines (Table 14-5). Other ingredients such as vitamins (vitamin C is most prevalent), preservatives, natural or synthetic flavorings, sweeteners, and colorings are frequently added to the juice.

Most fruits and vegetables contain very high proportions of water, which makes them easy to squeeze or pulverize into a juice. Two exceptions to this rule are bananas and avocados, which hold very little water. Orange juice sales lead the fruit juice market, followed by apple and grape juices (16). Although not a major threat to the market share of these traditional juices, tropical juices such as guava, mango, and kiwi are growing in popularity (54). As shown in Table 14-6, depending on the fruit or vegetable from which it is made, the juice may provide some vitamin A and/or C, making it more nutrient dense than a soft drink.

| TABLE 14-5  Fruit Juice Beverage Names Depend on the Percentage of Actual Juice | |
| --- | --- |
| Beverage Name | Fruit Juice (%) |
| Juice | Not less than 100 |
| Juice drink | Not less than 50 |
| Nectar | Not less than 30–40 |
| Ade | Not less than 25 |
| Drink | Not less than 10 |

**TABLE 14-6** Nutrient Comparison Between Various Fruits and Vegetable Beverages vs. Soda

| Beverage (8 oz/240 ml) | Calories | Vitamin A (RE)* | Vitamin C (mg) |
|---|---|---|---|
| *Soda* | *150* | *0* | *0* |
| *Juice* | | | |
| Apple, canned/bottled | 117 | 0 | 2 |
| Apricot nectar, canned | 141 | 332 | 2 |
| Carrot, canned | 95 | 6327 | 21 |
| Cranberry cocktail, bottled | 144 | 1 | 90 |
| Grape, canned/bottled | 155 | 2 | 0 |
| Grapefruit, fresh | 96 | 3 | 94 |
| Lemon, fresh | 60 | 5 | 112 |
| Lime, fresh | 66 | 3 | 72 |
| Orange, fresh | 111 | 50 | 124 |
| Papaya nectar, canned | 142 | 28 | 8 |
| Passion fruit (yellow) | 149 | 595 | 45 |
| Peach nectar, canned | 134 | 64 | 13 |
| Pineapple, canned | 157 | 0 | 26 |
| Prune, canned | 182 | 1 | 10 |
| Raspberry, bottled | 98 | 24 | 36 |
| Tangerine, canned | 124 | 104 | 55 |
| Tomato, canned | 42 | 136 | 45 |
| Vegetable (V8), canned | 46 | 283 | 67 |
| *RDI* | | | |
| Reference Daily Intake (RDI) | | 5000 | 60 |

*RE = Retinol equivalents—a measure of vitamin A activity based on an older estimation of how much vitamin A the body derives from carotenoids.

## Fruit/Vegetable Juice Processing

Extraction of juice is the first step in manufacturing fruit and vegetable juices. Other steps of juice production include clarification, deaeration (removing the air), pasteurization, determining concentration/additions, and packaging (bottle, can, or freezing) (63).

*Juice Extraction* Prior to juicing fruits or vegetables, most of them must be thoroughly washed; freed of any bruises, seeds, and mold; and often peeled and pitted. Citrus fruits are generally cut in half and squeezed, either mechanically or by hand. Although the pithy white parts of oranges and grapefruits containing bioflavonoids and vitamin C may remain, it is important to remove the skins in order to remove bitter-tasting and possibly toxic compounds. The skins of tropical fruits are also usually peeled, because they may have been grown in a country allowing the use of questionable pesticides. The skins of waxed produce and the leaves of carrots and rhubarb also must be removed. Cutting the fruits and vegetables into slices or chunks allows them to be placed in a juicer, where they can be processed into a beverage.

Different types of juice extractors result in varying amounts of pulp; some strain out most of the pulp, whereas others produce a thick, pulp-rich beverage. The most traditional commercial method of extracting juices is through the use of presses (16). Fresh fruit and vegetable juices taste best and are most nutritious when they are consumed immediately after preparation rather than stored for future use.

*Clarification* The natural pressing of plant material leaves a semifluid mass of cell wall material (cellulose and pectin). The small particles of suspended pulp remaining in pressed juice are called haze or cloud (15). Many consumers prefer crystal-clear fruit juice, so these particles are usually removed.

In tomato juice, the suspended pulp, which contributes to viscosity, is desired. If fresh tomato juice is left to stand, however, the natural enzyme pectin methyl esterase breaks down the pectin, resulting in a progressively thinner tomato juice over time. Tomato juice manufacturers can use a process called the hot-break process to inactivate the enzymes by quickly heating the product to 180°F (82°C). Allowing the enzyme activity to proceed and produce a tomato juice with a thinner consistency is called the cold-break process (63).

*Deaeration* Entrapped air in juices is removed by deaeration, a process that reduces undesirable changes resulting from the presence of oxygen. Juices are deaerated by spraying them into a vacuum deaerator. Deaeration improves shelf life, maintains flavor, and reduces the breakdown (oxidation) of vitamin C (63).

*Pasteurization* The high heats of pasteurization inactivate enzymes and destroy microorganisms that can cause foodborne illness. Some juices, especially fresh juices, are not pasteurized, which raises concerns when *Escherichia coli* outbreaks from nonpasteurized juices are reported (see Chapter 4). Pasteurizing juices diminishes their fresh-squeezed flavor, which can be partially replaced by adding back essence oils and folded peels from the original juice (46, 89). Alternatives to pasteurizing juices currently being studied include irradiation, hydrostatic pressure, ultrasound, high-intensity pulsed electrical fields, and oscillating magnetic fields (53) (see Chapter 28).

*Concentration/Additions* Blends of fruit flavors and juices are common and range from pure fruit juices to highly diluted, artificially flavored and colored drinks, which may contain little of the actual juice or its nutrients. The latter types of drinks may also be carbonated, and although they may be less thick and sweet than regular fruit juice, some are higher in sugar. Some juices are fortified with calcium to appeal to consumers interested in lowering their risk of osteoporosis (67).

The leading beverages to which fruit juice has been added include sparkling water, noncarbonated drinks, teas, and wine coolers (64). There are also "texturally modified" beverages that use real fruit pulp or pieces and other components to enhance mouthfeel (65).

## ? How & Why?

### How is the pulp removed from juice?

If a nonpulp texture is desired, fine filters or high-speed centrifuges are used to remove the pulp and associated haze. The latter separate contents based on density differences with the lighter substances separating to the top and the heavier ones settling to the bottom. Any minute pulp and colloidal particles still remaining can be removed by adding enzymes (Chemist's Corner 14-4). These enzymes break down pectic substances that settle to the bottom of the juice, followed by further filtering or centrifugation (9). "Liquification" enzymes are often added to clarify apple, cranberry, and grape juices (16).

## > CHEMIST'S CORNER 14-4

### Enzymes Reduce Haze

Cellulase and pectinase are added to juices to hydrolyze the cellulose and pectin, respectively (14). Amylase enzymes are added to eliminate starch haze caused by the presence of glucose polymers (15).

# PREPARATION OF FRUITS

## Enzymatic Browning

Certain kinds of fruit, when peeled or sliced, are susceptible to enzymatic browning (29). Inhibiting enzyme activity can be accomplished by denaturing enzymes, adding acid, lowering the storage temperature, and/or blocking exposure to oxygen through the use of coatings or antioxidants. Each of these methods of inhibiting enzymatic browning is now discussed.

### Denaturing Enzymes

Blanching foods by dipping them briefly in boiling water destroys the enzymes through denaturation. Fruits are not generally blanched, however, because heating already-ripe fruits makes them lose texture and flavor.

### Acid pH

Polyphenol oxidase enzyme activity is inhibited in the presence of acid (76). The optimal pH for this enzyme is 7.0, so the high acidic content of fruits such as oranges, lemons, and limes prevents enzymatic browning. Less acidic fruits such as peaches, apples, and bananas can be protected from enzymatic browning by coating them with lemon, lime, or orange juices, or with solutions of cream of tartar or the citric and acetic acids found in vinegar.

### Cold Temperatures

Cold temperatures reduce the rate of enzyme activity and thereby slow, but never completely inhibit, enzymatic browning (29).

### Coating Fruits with Sugar or Water

Cut fruits can be protected from exposure to oxygen by covering them with a light layer of sugar or syrup (55). Submerging fruits in water also blocks oxygen from contacting their surfaces.

### Antioxidants

Antioxidants such as ascorbic acid (vitamin C) and sulfur compounds prevent enzymatic browning by using up available oxygen (37). Ascorbic acid and sulfur solutions are available commercially for this purpose. Many dried fruits are treated in a similar manner. However, sulfur dioxide can cause allergic reactions in some people, so its use in salad bars to prevent browning was banned several years ago by the FDA (55). Pineapple juice, which is naturally high in sulfur compounds, can be helpful in inhibiting enzymatic browning when added to fruit salads or compotes.

## Changes During Heating

### Color

The pigments coloring fruit are the same as those discussed in Chapter 13 on vegetables: carotenoids, chlorophylls, and flavonoids (anthocyanins, anthoxanthins, and betalains). Fruit becomes more brilliant in color during ripening as chlorophyll breaks down and exposes the underlying pigments. Subsequent color changes in fruits are usually the result of a change in pH or a reaction with a metal salt and/or ethylene gas.

*pH*  Heating fruit may change the pH, which affects the color of some pigments. Sometimes a harmless blue discoloration is seen in baked products surrounding added ingredients such as cherries, cranberries, and walnuts; this is caused by the interaction of anthocyanins in these foods with the alkaline baking powder or soda in the flour mixture. To prevent this blue color, acid in the form of buttermilk or sour cream can be used to substitute for milk in these products, and walnuts can be roasted (about 10 minutes at 350°F/177°C), which removes the anthocyanin-containing skins (24).

*Metal Salts*  Canned fruits or vegetables contain acids, which often react with the tin lining to produce metal salts that change the color of pigments, particularly the red-blue anthocyanins. As a result, canned foods and juices containing anthocyanins are often stored in specially treated cans or glass to prevent food discoloration. Red, purple, or blue juices can also turn bluer when pineapple juice is added if iron has been introduced into the canning process from the pineapple processing equipment.

*Ethylene Gas*  One of the ways to facilitate the ripening of oranges and tomatoes to their optimal color is by exposing them to ethylene gas. This stimulates respiration, thereby accelerating the ripening process, but it may mean that the fruit deteriorates more quickly (59).

## ? How & Why?

### Why are some raisins brown and some "golden"?

The difference between light- and dark-colored raisins is that the former are often treated with sulfur dioxide, which prevents enzymatic browning (19).

# PROFESSIONAL PROFILE

**Janet Adams—Scientist for Frito-Lay, Incorporated**

Courtesy of Janet Adams, PhD

Janet Adams started out at Texas A&M University majoring in nutrition. During her junior year she was required to take a course in food science, a topic she had never even heard of before. That class sparked her interest in food science because it was not only about nutrition but also showed the many opportunities for employment in the food industry. Intrigued, she started talking to people in the department—faculty and food science students—about what job opportunities existed for her after graduation.

She also asked questions of her sister, who worked at Pizza Hut's corporate headquarters in Dallas, Texas. Through this contact, Janet was able to talk to someone in research and development (R&D) about career opportunities. She found out that Pizza Hut offered a 3-month internship, which she applied for and received.

Janet says, "Students from other universities can find out about internships by joining their local national food association branch [Institute of Food Technologists; www.IFT.org]; contacting IFT members; finding someone in the corporation to talk to [www.hoovers.com locates company information]; or summer work via internships." Some companies hire student lab technicians through temporary agencies, which is another way to learn about a company and obtain contacts. Networking is also important, she says. Students should attend food events and national seminars, and apply for food scholarships.

"Just getting experience, even if it's a part-time job, is key to entering the field," says Janet. "Another good source is the Grocery Manufacturers Association that provides scholarships, holds annual meetings, offers student memberships, and posts jobs, similar to IFT, on its website [www.gmaonline.org]."

After completing her Pizza Hut internship, Janet knew exactly what she wanted to do—work in the food corporation's R&D division. Formulation problem solving really intrigued her, and now she had her end goal and a new motivation to complete her undergraduate classes.

Just before her junior year ended, Janet switched her major from nutrition to food science, which she could do at Texas A&M University because it offers a major in both fields. (IFT lists universities offering food science degrees.) She started interviewing for jobs the summer before her December graduation, but her search revealed that she would need a master's degree to work where she wanted—in product development. Before enrolling in a master's degree program, she spent the summer working at a grocery store's bakery products manufacturing site. Her job was to write the HACCP (Hazard Analysis and Critical Control Point; see Chapter 4) system for the bakery.

Janet stayed at Texas A&M University for her Master of Science degree in Food Science, specializing in cereal science because of her interest in pizza dough, and eventually finished her research in flour tortillas. She was also active in the local IFT association as a student member and met two wonderful mentors (one worked at Frito-Lay, and the other was a sales representative for a food brokerage). She applied for various food corporation scholarships and asked her mentors questions about the industry and even which classes to take. They told her that she needed to set herself apart from other students by participating in internships.

Kellogg Company came to interview on campus, and Janet got a 6-month internship in its product development division. She worked on the bowl life of cereal and the effects of oven conditions on cereal texture.

Representatives for Frito-Lay also came to interview at the university, and Janet was hired upon graduating from her master's degree program. *Scientist* is her official title in the Product Development division there. Some of her work entails defining market trends, managing projects, suggesting what new products should be launched, writing specifications for new products, helping to develop new seasonings on potato chips and corn chips, and assisting recruiting teams that travel to various universities to discuss food science careers.

What does Janet wish she had known in college? She says she wishes that she had taken "certain electives in business—especially marketing, and also accounting and finance. Employees work cross-functionally in corporations all the time. For instance, every department has to save money, and it's also beneficial to understand the business language. Networking and joining clubs (student and professional) are also important. One of the most crucial things for students is to gain experiences in the work world. Try to obtain work in your field before you graduate. Don't be afraid to talk to anyone, and ask them questions."

What does Janet consider the biggest mistake that students make? "Students do not often understand what it is that they are getting into," she points out. "They have an idea about it, but they don't understand it conceptually. For example, there have been a few food science graduates that do not like developing new products; but they have a food science degree. So try to find out what the actual job entails before graduating with the degree. Also, have a backup plan such as combining a food science degree with a master's degree in business administration."

## Texture

Fruit is often served raw because the texture is usually more desirable than that of cooked fruit. Heat drastically softens fruit, sometimes to the point of mushiness (e.g., strawberries). During heating, several changes occur that contribute to this softness:

- The conversion of the fruit's protopectin to pectin
- The degradation of cellulose and hemicellulose
- The denaturation of the cell membrane proteins

This last change influences membrane function and the ability of the fruit to maintain its turgor.

*Osmosis*   During heating, the fruit's osmotic system of selective permeability is replaced with simple diffusion, contributing to the loss of shape in heated fruits. Normally, the cell walls of raw fruits have semipermeable membranes, which allow water, but not solutes such as sugars or minerals, to pass through the membrane. The water moves from a low solute concentration to a higher one, until the solute concentration is equal on both sides of the membrane. This is the principle behind the practice of spraying water on fruits displayed in supermarket stands to keep them crisp and turgid. The higher solute concentration in the cells draws the water into the fruit, causing cell enlargement and swelling, and keeping the fruits looking full and fresh. This is also the reason why lemons stored in water will produce about twice the juice as those that are not. Sprinkling sugar on fruit produces the opposite effect, causing fruit to become syrupy as water passes out from the cells to the surface where the solute concentration is higher (due to the sugar). This can backfire, however, if sugar is left on fruit too long, producing a soft, mushy texture.

Heating destroys the membrane's ability to prevent the loss or uptake of solutes through the cell wall. Solutes can then freely pass across the membrane until solute concentrations are equal on both sides.

Sugar may be added to fruit before it is cooked in water to replace the lost sweetness. This also helps the fruit to retain its shape and firmness. For the same reason, fruits are often heated in a syrupy solution of greater than 15 percent sugar. The sugar enters the cells as the water leaves. Sugar further inhibits loss of texture by interfering with the conversion of protopectin to pectin. Calcium salts and acids also help to firm fruit, which is why calcium is added to many canned fruits and vegetables, especially pears and tomatoes (10) (Chemist's Corner 14-5).

## ? How & Why?

### Why are canned fruits softer and shinier than fresh fruits?

Heating fruit in water causes the fruit's natural sugars to move out into the water and water to move into the fruit, resulting in the loss of shape. Fruits become more translucent and shiny when heated as water leaves the cells and the air between cells is expelled.

## > CHEMIST'S CORNER 14-5

### Calcium in Canned Fruits

Pectin often exists in nature as salts of calcium or magnesium. It is able to do so because of its abundance of carboxyl groups (–COOH). The bridges formed between the free carboxyl groups and the intracellular ions of calcium and magnesium strengthen the tissue of the plant and resist degradation during the heating of canning (60).

## Flavor

Sugars and acids contribute to the sweet and sour taste of fruits. The flavors are derived from the chemical configuration of compounds such as sugars, acids, phenolic and **aromatic compounds**, and **essential oils**. The flavor substances and volatile compounds that contribute to aroma can be lost during preparation, which is why fruits are either served raw or heated only for a minimal amount of time.

*Zest*   The peels of oranges, lemons, and limes serve as reservoirs for potent aromatic oils. The colorful outer layer of these citrus fruits, known as the *zest*, often contains more flavor than the fruit's juice (50). The zest is not the entire peel, but the outermost, bright-colored layer above the white pith (51). The zest can be removed one of three ways: grating with the smallest holes of a grater, using a zester tool, or by cutting with a sharp paring knife (Figure 14-15). Various forms of zest—flat strips, julienned slivers, or a fine grate—can be added to many dishes. The amounts of grated zest and unstrained juice obtained from selected citrus fruits are shown in Table 14-7. Large zest pieces can be strained out of a dish after they have contributed their flavors during cooking, or they can be left in the dish if blanched for two minutes to make them soft and palatable. Finely grated zest is often added to the batter of dessert dishes, where it is small enough to be evenly distributed.

# Dry-Heat Preparation

Fruits add variety to meals, and preparation techniques can make overripe fruits or otherwise unpalatable ones

**FIGURE 14-15**   Scraping the zest off citrus fruits.

**Grater:** The smallest holes on a grater can be used to collect zest, which is freed with the use of a pastry brush. Avoid scraping so hard that white pith ends up in the small pieces of zest.

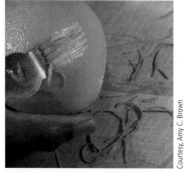

Courtesy, Amy C. Brown

**Five-hole zester:** The shallow blade of a zester does not go any deeper than the zest, making it easy to collect the flavorful and fragrant long strips of zest peelings.

**Aromatic compound**  A compound that has a chemical configuration of a hexagon.

**Essential oil**  An oily substance that is volatile (easily vaporized), with 100 times the flavoring power of the material from which it originated.

**TABLE 14-7** Amount of Zest and Juice Obtained from Citrus

| Citrus type | Grated zest, lightly packed | Juice, unstrained |
|---|---|---|
| 1 lime | 1–1½ tsp | about 3 tbsp |
| 1 lemon | ½–1 tbsp | ¼–⅓ C |
| 1 orange | about 1 tbsp | about ½ C |
| 1 grapefruit | about 1½ tbsp | about 1 C |

Source: © 1998 Molly Stevens, from Fine Cooking Magazine, 25:76, 1998. Reprinted by permission.

such as rhubarb and green apples appetizing. Before preparation, all fruits must be washed to remove pesticide residues, soil, and microorganisms. Because most fruit spoils faster if it is washed before refrigeration, it is best to rinse it just before use. Cooked fruit, with the exception of applesauce and fruit spreads (jams, jellies, etc.), should be served immediately after it has been prepared.

## Baking

Whole apples and some varieties of pears are good for baking. The apples may be left whole, or they can be cored, peeled, and placed in a baking dish. A sauce can be made from brown sugar, water, butter, cinnamon, and nutmeg boiled together and poured over the apples, which are then baked uncovered for approximately 1 hour at 350°F (177°C), during which they may be basted occasionally. Bananas, plums, peaches, and rhubarb can also be baked successfully. Fruit can be added to many baked goods to enhance their flavor, color, and texture. Cakes, pies, cookies, muffins, scones, and cobblers, as well as pancakes and waffles, all profit from partnerships with fruit.

## Broiling

A few fruits, such as grapefruit halves, pineapple halves, and bananas, can be broiled by placing them on a heavy-duty pan, which is then broiled until the fruit is golden brown. Often, a bit of melted butter mixed with sugar is drizzled on the fruit before broiling so that browning is assured.

### Frying/Sautéing

A small amount of fat in a frying pan can be used to sauté slices of apple, banana, cherry, and pineapple. Sometimes the fruit is sprinkled with white or brown sugar. Sautéing contributes to a slight increase in the fruit's caloric and fat content. Sautéed fruit may be served alongside roasted meats or as a dessert course.

## Moist-Heat Preparation

### Stewing/Poaching

Fresh fruits can be stewed, or, better yet, poached. For stewing, fruits are heated either without water (for fruit with a high water content), with very little water or juice, or with added sugar or syrup. Stewing fruits in a sugar syrup helps to maintain their shape.

For poaching, fresh fruit such as pears are gently lowered into boiling water or syrup, the heat is reduced to poaching temperatures, and the fruit is cooked until barely tender. This process does not take long, and the fruit should be removed from the heat as soon as cooking is complete. Very soft fruits are best drained immediately to prevent overcooking, whereas firmer fruits such as apples and pears are best left to cool in the poaching syrup in order to soak up more flavor and prevent wrinkling.

***Applesauce Preparation*** Applesauce is relatively simple to prepare from cored and chopped cooking apples. The peels may be removed for a smooth sauce, or left on to add a touch of color and increase the fiber of the applesauce. A pound of apples, about 1 to 1½ cups of water or cider, and a cinnamon stick (optional) is added to a saucepan. The pan is covered, and the mixture is brought to a boil and then reduced to a simmer and cooked for about 10 minutes, until tender. The pan is removed from the heat and the cinnamon stick is discarded. The apples are then pulverized with a potato masher or fork until smooth. Sugar may be added (1/2 cup per pound of apples). The applesauce should be served immediately or covered and refrigerated. Chunky applesauce is made by adding the sugar before heating and smashing the fruit only slightly.

### Preparing Dried Fruit

Dried fruit is usually soaked in water and then simmered in a covered pan, but some dried fruits, such as prunes, have a higher water content and do not need to be soaked before cooking. They may simply be covered with cold water and brought to a simmer. The natural sweetness of dried fruits usually makes it unnecessary to add sugar, but some fluid may be necessary to rehydrate the dried fruits. To add more flavor, some recipes call for juice or liquors instead of water.

## Fruit Spreads

Jam is a common type of fruit spread, but it is not the only one. The many different types of fruit spreads and their basic preparation are now discussed.

### Types of Fruit Spreads

The most commonly consumed fruit spreads are fruit preserves, jams, conserves, jellies, marmalades, and butters. Except for bananas and melons, most fruits lend themselves well to these methods of preservation. The type of fruit spread is determined by the amount of added sugar and the form of the fruit—as juice, or mashed, sieved, whole, halved, or in chunks:

- ***Preserves.*** Made from whole fruit, halves, or chunks.
- ***Jams.*** Made from ground or mashed whole cooked fruit.
- ***Conserves.*** Made from a mixture of fruits (usually citrus or at least with some citrus) to which nuts and raisins, but no sugar, are generally added.
- ***Jellies.*** Made from the juice of cooked fruit, with added sugar and pectin. For a clear jelly, the juice is first strained.
- ***Marmalades.*** Contain juice with thin slices of fruit and rind, especially citrus fruits.
- ***Butters.*** These thick and smooth fruit preserves are made from sieved, long-cooked fruit, are usually less sweet than jams and jellies, and do not keep as well as other fruit spread forms.

## Ingredients

Fruit spreads gel when pectin molecules bond to each other and form a mesh-like network. Pectin (1 percent), sugar (about 60 percent), and acid (pH 2.8 to 3.4) make the formation of the network possible.

*Pectin* Pectin makes fruit spreads gel (see the section "Pectic Substances" under "Composition of Fruits"). Although pectin is naturally found in fruits, concentrations vary according to the fruit, so more may need to be added when manufacturing certain fruit spreads. At least a 0.5 to 1 percent concentration of pectin needs to be present for fruit to gel, and if the fruit does not contain sufficient pectin, commercial pectin can be added in either powder or liquid form (Chemist's Corner 14-6). Special pectins are available for making fruit spreads lower in sugar.

*Sugar* Sugar is a natural preservative, and most fruit preserves are at least 50 percent granulated white sugar. For example, one recipe for strawberry jam calls for 5½ cups of strawberries (about 3 quarts), 8 cups of sugar, and one package of powdered pectin. Most microorganisms need water to survive, and when sugar osmotically pulls the water from their cell(s), they die. In addition to acting as a natural preservative, sugar helps to maintain the firmness of fruits, contributes to their flavor, and makes gelling possible—and adds calories. Most jams and jellies average 50 calories (kcal) per tablespoon, but this can be reduced by almost half [30 calories (kcal)/tablespoon] by using alternative sweeteners and a unique type of pectin that relies on calcium ions to create a net-like structure (56). Calcium ions contribute to the creation of cross-linking bridges, which help to gel fruit spreads, salad dressings, and diet jellies made without sugar.

### Lower-Calorie Fruit Spreads

Also available are calorie-free imitation fruit spreads. Another option for lower-sugar spreads is to purchase puréed baby fruits or to blend frozen fruits. Although all fruit spreads should be refrigerated after opening, it is doubly necessary to keep low- or no-sugar spreads refrigerated because there is little or no sugar to prevent microbial growth.

*Acid* Acid provides both flavor and gel formation. Hydrogen ions ($H^+$) from acid are needed to neutralize the negative charges that cause pectin molecules in water to repel each other. The pH range that allows pectin molecules to bind is between 2.8 and 3.4, with an optimal pH of 3.2. All fruit spreads should taste somewhat acidic. In fact, without acid, fruit spreads would be lacking in both flavor and gel strength. Some commercial pectins contain added acid. If the fruit mix is too sweet, 1 tablespoon of lemon juice, 1 tablespoon of vinegar, or 1/8 tablespoon of citric acid can be added for every 8-ounce cup of juice (2).

*Fruit* Because of their varying amounts of pectin and acid, some fruits are more suited for making fruit spreads than others (6). Table 14-8 shows which fruits contain sufficient pectin and acid for gelling. Adjustments for either low pectin or low acid can be made by combining one fruit with another, using commercial pectin, and/or adding lemon juice.

### Preparing Fruit Spreads

To prepare a good fruit spread, the correct balance of pectin, sugar, and acid must be attained. The fruit selected determines the amount of pectin and acid and the overall quality of the spread. Steps for preparing fruit spreads (see Figure 14-16) include:

1. Heat the fruit.
2. Add sugar and possibly pectin and acid.
3. Pour the mixture into sterile containers.
4. Seal the containers properly.

---

> ## CHEMIST'S CORNER 14-6

### Pectins and Methoxylation

Commercial pectins are sold according to their degree of methoxylation (DM). High-methoxyl pectins have a DM value of 50 to 80 percent, form gels in the presence of acid and sugar, and are most commonly used for the manufacture of fruit spreads. Low-methoxyl pectins have a 25 to 50 percent DM value and will form gels in the presence of divalent cations such as calcium (66). High-methoxyl pectin makes it possible for fruit spreads to gel through coordinate bonding with $Ca^{2+}$ ions or hydrogen bonding and hydrophobic interactions. In low-methoxyl pectin, gelation occurs because of ionic linkage via calcium bridges between two carboxyl groups belonging to two different chains in close proximity to one another (79).

---

| TABLE 14-8 | Fruit Pectin and Acid Content | |
| --- | --- | --- |
| **Adequate Pectin and Acid**[1] | **Low in Either Pectin or Acid** | **Low in Both Pectin and Acid** |
| • Apples (sour) | • Apples (ripe) | • Apricots |
| • Blackberries (sour) | • Blackberries (ripe) | • Blueberries |
| • Crabapples | • Cherries (sour) | • Figs |
| • Cranberries | • Chokecherries | • Grapes (Western Concord) |
| • Currants | • Elderberries | • Guavas |
| • Gooseberries | • Grapefruit | • Peaches |
| • Grapes (Eastern Concord) | • Grapes (California) | • Pears |
| • Lemons | • Loquats | • Plums (Italian) |
| • Loganberries | • Oranges | • Raspberries |
| • Plums (not Italian) | | • Strawberries |
| • Quinces | | |
| • Red currents | | |

[1] Enough natural pectin and acid to gel with added sugar.

**Source:** Andress EL, and JA Harrison. *Making jams and jellies. Jellied product ingredients.* www.uga.edu/nchfp/how/can_07/jellied_product_ingredients.html.

**FIGURE 14-16**    Preparing fruit spreads. Heat fruit, sugar, and possibly added pectin and/or acid sufficiently long enough (or it will not gel) before pouring into sterilized jars to set.

Digital Works

Ripe fruit, whole or cut up, is simmered in a small amount of water to liberate the pectin and juice. Approximately ¼ cup of water is used for every pound of fruit. Too much water will dilute the pectin and prevent gelation. If the fruit is extremely watery, simply crushing it may provide sufficient liquid. Boiling times vary from 10 to 15 minutes for soft fruit to about 20 minutes for hard, sliced fruit. Uncooked or undercooked fruit will not gel, because heat is necessary for the conversion of protopectin to pectin. Sugar and pectin and/or acid are added, after which the mixture is boiled to concentrate the ingredients. Too much sugar creates a gummy product, whereas too little prevents gelation. Some commercial pectins may be added to the fruit before the sugar, but there will be directions to that effect on the package.

One test to determine if the mixture is done is to dip a cool metal spoon in the boiling fruit, lift it up, and see if the mixture drops from the spoon in flat-sheet fashion. Heating the mixture beyond this stage will cause the pectin molecules to break down and so prevent the mixture from gelling; also, the color may get darker because of the caramelization of sugars. When done, the mixture is poured into sterile glass containers to within ¼ inch of the rim and sealed with two-part canning lids according to manufacturer's directions. Fading of the finished spread's color is prevented by storing the jars in a dark, cool place. Table 14-9 lists possible reasons for gelling problems.

# STORAGE OF FRUITS

Many types of fruits are picked and shipped to market in an unripe state because the hardship of transportation would damage delicate ripe fruits (18). Unripe fruit can be left at room temperature in a paper bag until ripe. Most European pears, for example, are marketed unripe, so it is best to store them outside the refrigerator in a loosely closed paper bag until they give to firm pressure. Commercially, fruit can be stored under controlled-atmosphere storage or with the aid of coatings or films as described in Chapters 28 and 13.

Fruits differ in their respiration rates, and thus in their rates of ripening. A few fruits, known as climacteric, will experience an increased phase of respiratory rate right before becoming fully ripe. These fruits continue to ripen after being harvested. Other fruits and most vegetables are nonclimacteric, and although they continue to respire at the same or even a lower rate after being harvested, they are best ripened fully before being harvested (Table 14-10).

## Storing Fresh Fruit

Ripe fruit with a high water content is best if consumed within 3 days of purchase. Grapes spoil quickly, so regardless of whether or not they are ripe, they

| TABLE 14-9    Gelling Problems and Their Causes | |
|---|---|
| **Problem** | **Possible Causes** |
| Sugar crystallization (mainly found in grape jelly) | Too much sugar <br> Not enough acid <br> Overcooking <br> Delay in sealing (may be prevented by permitting juice to stand overnight in cold place before making it into jelly) |
| Weeping | Cause not really known; may occur in jellies made from currants or cranberries high in acid |
| Cloudiness | Squeezing juice out of bag <br> Starch from apples used to make jelly (avoid pressure on jelly bag when straining) |
| Failure to gel | Improper balance between pectin, sugar, and acid <br> Lack of pectin or acid in fruit <br> Overcooking |
| Fermented jelly or mold formation | Jelly glasses not well sterilized <br> Jelly stored in warm, damp place <br> Jelly not completely sealed |

| TABLE 14-10    Selected Climacteric and Non-climacteric Fruits | |
|---|---|
| **Climacteric Fruits (continue to ripen after harvest)** | **Non-Climacteric Fruits (best ripened before harvest)** |
| Apple | Blueberry |
| Apricot | Cherry |
| Avocado | Citrus fruits |
| Banana | (grapefruit, |
| Breadfruit | lemon, orange) |
| Cantaloupe | Grapes |
| Guava | Melons |
| Peach | Olives |
| Pear | Pineapple |
| Plum | Strawberry |
| Tomato | |
| Tropical fruits (papaya, mango, passion fruit) | |

are stored, unwashed, in plastic bags in the refrigerator and washed just prior to consumption. Conversely, cooled apples will keep for weeks.

Once fruit is ripe, storage time may be increased by placing it in plastic bags punctured with air holes, and then in the refrigerator (see the back inside cover of this book). An exception is bananas, which are best stored at room temperature, because refrigeration interferes with their ripening process and causes their skin to blacken.

The following are storage requirements for specific fruits:

- **Cherries.** Cherries should be arranged in a single layer between paper towels, placed in a plastic bag, and refrigerated.
- **Dates.** Dates will keep for weeks at room temperature, up to 1 year in the refrigerator, and up to 5 years in the freezer.
- **Citrus fruit.** Citrus fruit can be stored at a cool room temperature (60°F–70° F/16°C–21°C) or in light refrigeration temperatures (45°F–48°F/7°C–9°C) for 2 to 3 weeks.
- **Pineapples.** Uncut ripe pineapples can be kept at room temperature for up 3 days.
- **Pomegranates.** Ripened pomegranates are kept at room temperature for short periods, but should be refrigerated if storage exceeds several days.
- **Guavas.** Once guavas ripen, they can be stored in the refrigerator for 1 to 2 days.

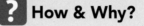

### ? How & Why?

**Why does one rotten apple spoil the barrel?**

An overripe apple stored with good apples will ruin the others by releasing ethylene gas, an agent that speeds up ripening. All bruised, dented, or otherwise damaged apples should therefore be removed before storing the other apples.

## Storing Canned Fruit

Canned fruits keep their quality longer if the cans are stored in a dry place with temperatures under 70°F (21°C). Bulging, dented, leaking, or rusted cans should always be discarded. Bulging may indicate the presence of the anaerobic bacteria that causes deadly botulism (see Chapter 4).

### Olives

The storage requirements for olives depend on whether they are packed in a can or jar. The unused portion from an open can of olives should be stored in the original brine, covered with breathable plastic wrap, and kept in the refrigerator for no more than 10 days; possibly harmful toxins develop when such olives are stored in sealed or airtight containers. The brine of olives in a jar has three times the salt concentration and a much lower pH than that found in canned olives, so it does not matter if the storage container is airtight or not.

# PICTORIAL SUMMARY / 14: Fruits

Throughout human history, fruits have been prized and their seeds transported wherever humans have settled, resulting in the great variety of fruits available today.

## CLASSIFICATION OF FRUITS

Fruits are classified according to the type of flowers from which they develop.

**Simple fruits** develop from one flower. They include:
- Drupes: Fruit with seeds encased in a pit, such as apricots, cherries, peaches, and plums.
- Pomes: Fruit with seeds contained in a central core, such as apples and pears.
- Citrus fruits: Oranges, grapefruits, lemons, limes, kumquats, and mandarins.

**Aggregate fruits** develop from several ovaries in one flower. They include blackberries, raspberries, and strawberries.

**Multiple fruits** develop from a cluster of several flowers. They include pineapples and figs.

## COMPOSITION OF FRUITS

The cellular structure and the pigments of fruits are similar to those in vegetables. The flavors of fruits are the result of combinations of sugars, acids, phenolic and aromatic compounds, and essential oils. The pH in fruits tends to be below pH 5.0. Pectin plays an important role in fruit ripening and the gelling of fruit spreads.

Nutritionally, fruits are low in calories, fat, and protein. The few fruits that are high in fat are coconut, avocado, and olives. Fruits are high in water and carbohydrates, the latter providing the majority of calories found in fruits. Some, but not all, fruits can be excellent sources of vitamin C, beta-carotene, iron, or potassium.

## PURCHASING FRUITS

Fruits are selected based on appearance, size, color, shape, uniformity, and freedom from defects. The four grades for fresh fruit are U.S. Fancy, U.S. No. 1, U.S. No. 2, and U.S. No. 3. Selecting fruits is based on grading factors as well as the variety, whether or not they are in season, and criteria that vary for each individual fruit.

Many fruits are available on the market in summer, but they can be provided throughout the year in various processed forms: canned,

frozen, dried, and juice. Refrigerated transport also makes fruits available year-round from countries with different growing seasons. Grades for canned or frozen fruits are U.S. Grade A or Fancy, followed by U.S. Grade B or Choice, and U.S. Grade C or Standard.

## PREPARATION OF FRUITS

Cooking alters the taste, texture, color, and shape of fruit, so fruit is often consumed in its raw state. Cooking fruit adds variety to meals, makes fruit like rhubarb and green apples palatable, and utilizes overripe fruits. Fruits should be prepared using a minimum of water, time, and heat.

Most fruit can be kept beyond the growing season by combining it with sugar in the following forms of fruit spreads:

**Preserves:** Made from whole fruit, halves, or chunks.

**Jams:** Ground or mashed whole cooked fruit.

**Conserves:** Made from a mixture of fruits (usually including citrus) to which nuts and raisins, but no sugar, are added.

**Jellies:** Made from the juice of cooked fruit, with added sugar and pectin.

**Marmalades:** Juice combined with thin slices of fruit and rind, especially citrus fruits.

**Butters:** Thick and smooth, made from sieved, long-cooked fruit; usually less sweet than jams and jellies, and more perishable than other fruit spreads.

## STORAGE OF FRUITS

Most ripe fruit should be stored in the refrigerator in plastic bags punctured with air holes, with the exception of bananas, which will turn brown if refrigerated. Unripe fruit is usually left at room temperature in a paper bag until ripe.

# CHAPTER REVIEW AND EXAM PREP

## Multiple Choice*

1. What is the term used to describe fruit with seeds contained in a central core, such as apples and pears?
   a. Drupes
   b. Frumps
   c. Seed cores
   d. Pomes

2. Identify the correct order for the chemical breakdown of protopectin as a fruit changes from immature to overripe.
   a. Protopectin to pectin to pectic acid
   b. Protopectin to pectic acid to pectin
   c. Protopectin to oxalic acid to pectin
   d. Protopectin to pectin to oxalic acid

3. The peels of which fruits are commonly used as zest to add flavor?
   a. Kiwifruit, watermelon
   b. Lemon, orange, lime
   c. Grapes, watermelon
   d. Banana, dates, mangoes

4. Jams are best described as fruit preserves made from _____.
   a. ground or mashed whole cooked fruit
   b. whole fruit, halves, or chunks
   c. the juice of cooked fruit with added sugar and pectin
   d. the juice of fruits with thin slices of fruit and rind

5. A fruit juice beverage that contains not less than 30 to 40 percent fruit juice would be classified as a(n) _____.
   a. juice drink
   b. nectar
   c. ade
   d. juice

*See p. AK-1 for answers to multiple choice questions.

6. Fruits are generally low in fat, with three exceptions that include
   a. avocados, coconut, and olives.
   b. figs, dates, and guava.
   c. olives, coconut, and pomegranates.
   d. dates, avocados, and dried fruit.

7. What fruit (medium size) has more vitamin C than a medium-sized orange?
   a. Papaya
   b. Kiwifruit
   c. Strawberries (1 cup)
   d. All of these choices

## Short Answer/Essay

1. What are the three classifications of fruits and how do they differ? Provide examples of each.
2. Discuss the three groups of pectic substances, how they contribute to the ripening of fruit, and how they are used by the food industry.
3. Certain fruits turn brown when cut open and exposed to air. Describe the process by which this occurs. What steps can be taken to inhibit browning?
4. Discuss the grading of fresh fruit and how it differs from the grading of canned or frozen fruit. What are the signs that frozen bags of fruit have been thawed and refrozen?
5. Describe the specific signs of ripeness for the following fruits: avocados, grapefruit, guavas, lemons, honeydew melon, and juice oranges.
6. List the names for the types of fruit juice beverages in descending order of their fruit juice concentration.
7. Describe the changes that occur when fruits are heated.
8. What is fruit zest and how is it used in food preparation?
9. Describe the differences between the following types of fruit spreads: preserves, jams, conserves, jellies, marmalades, and butters.
10. Explain the role that each of the following ingredients plays in the preparation of fruit spreads: pectin, sugar, acid, and fruit.

# REFERENCES

1. Alonso J, ME Tortosa, W Canet, and MT Rodrigues. Ultrastructural and change in pectin composition of sweet cherry from the application of prefreezing treatments. *Journal of Food Science* 70(9): E526–E530, 2005.

2. Altschul AM. *Low-Calorie Food Handbook.* Marcel Dekker, 1993.

3. Amagase H, and DM Nance. A randomized, double-blind, placebo-controlled, clinical study of the general effects of a standardized Lycium barbarum (Goji) Juice, GoChi. *Journal of Alternative and Complementary Medicine* 14(4):403–412, 2008.

4. Amagase H, B Sun, and C Borek. Lycium barbarum (goji) juice improves in vivo antioxidant biomarkers in serum of healthy adults. *Nutrition Research* 29(1):19–25, 2009.

5. Amin AR, O Kucuk, FR Khuri, and DM Shin. Perspectives for cancer prevention with natural compounds. *Journal of Clinical Oncology* 1;27(16):2712–2725, 2009.

6. Andress EL, and JA Harrison. *Making jams and jellies. Jellied product ingredients.* www.uga.edu/nchfp/how/can_07/jellied_product_ingredients.html. Accessed 7/9/09.

7. Ariga T. The antioxidative function, preventive action on disease and utilization of proanthocyanidins. *Biofactors* 21(1–4):197–201, 2004.

8. Bagchi D, CK Sen, M Bagchi, and M Atalay. Anti-angiogenic, antioxidant, and anti-carcinogenic properties of a novel anthocyanin-rich berry extract formula. *Biochemistry* 69(1):75–80, 2004.

9. Baker RA, and RD Cameron. Clouds of citrus juices and juice drinks. *Food Technology* 53(1):64–69, 1999.

10. Baker RA, and RG Hagenmaier. Reduction of fluid loss from grapefruit segments with wax microemulsion coatings. *Journal of Food Science* 62(4):789–792, 1997.

11. Bareuther CM. Superfruits. *Today's Diet and Nutrition* 3(5):64–68, 2007.

12. Batisse C, B Fils-Lycaon, and M Buret. Pectin changes in ripening cherry fruit. *Journal of Food Science* 59(2):389–393, 1994.

13. Bazzano LA. Effects of soluble dietary fiber on low-density lipoprotein cholesterol and coronary heart disease risk. *Current Atherosclerosis Report* 10(6):473, 2008.

14. Berne S, and CD O'Donnell. Filtration systems and enzymes: A tangled web. *Prepared Foods* 165(9): 95–96, 1996.

15. Beveridge T. Haze and cloud in apple juices. *Critical Reviews in Food Science and Nutrition* 37(1): 75–91, 1997.

16. Beveridge T. Juice extraction from apples and other fruits and vegetables. *Critical Reviews in Food Science and Nutrition* 37(5):449–469, 1997.

17. Boskou D. *History and Characteristics of the Olive Tree.* American Oil Chemists' Society, 1996.

18. Brownleader MD. Molecular aspects of cell wall modification during fruit ripening. *Critical Reviews in Food Science and Nutrition* 39(2): 149–164, 1999.

19. Cannellas J. Storage conditions affect quality of raisins. *Journal of Food Science* 58(4):805–809, 1993.

20. Chen CS, and MC Wu. Kinetic models for thermal inactivation of multiple pectinases in citrus juices. *Journal of Food Science* 63(5): 747–750, 1998.

21. Chin YW, HB Chai, WJ Keller, and AD Kinghorn. Lignans and other constituents of the fruits of Euterpe oleracea (Acai) with antioxidant and cytoprotective activities. *Journal*

*of Agriculture and Food Chemistry* 56(17):7759–7764, 2008.

22. Cohen J. Snack foods. *Progressive Grocer* 70(7):62, 1991.

23. Cooper CC. Functional foods. *Today's Dietitian* 10(10):42–44, 2008.

24. Corriher SO. *Cookwise: The Hows and Whys of Successful Cooking.* Morrow, 1997.

25. Cunningham E. Is a tomato a fruit and a vegetable? *Journal of the American Dietetic Association* 102(6): 817, 2002.

26. Del Pozo-Insfran D, SS Percival, and ST Talcott. Açai (Euterpe oleracea Mart.) polyphenolics in their glycoside and aglycone forms induce apoptosis of HL-60 leukemia cells. *Journal of Agriculture and Food Chemistry* 54(4):1222–1229, 2006.

27. Demirbuker D, I Arcan, F Tokatli, and A Yemenicioglu. Effects of hot rehydration in the presence of hydrogen peroxide on microbial quality, texture, color, and antioxidant activity of cold-stored intermediate-moisture sun-dried figs. *Journal of Food Science* 70(3): M153–M159, 2005.

28. Duffy KB, et al. A blueberry-enriched diet provides cellular protection against oxidative stress and reduces a kainate-induced learning impairment in rats. *Neurobiology of Aging* 29(11):1680–1689, 2007.

29. Fan X, BA Niemera, JP Mattheis, H Zhuang, and DW Olson. Quality of fresh-cut apple slices as affected by low-dose ionizing radiation and calcium ascorbate treatment. *Journal of Food Science* 70(2):S143–S148, 2005.

30. Flath RA. Pineapple. In *Tropical and Subtropical Fruits,* ed. P. Shaw, 157–183. Avi Publishing, 1980.

31. *Fruits and Veggies—More Matters.*™ www.fruitsandveggiesmatter.gov/. Accessed 6/09.

32. Funaki J, and M Yano. Inhibiting the activity of actinidin by oryzacystatin for the application of fresh kiwifruit to gelatin-based foods. *Journal of Food Biochemistry* 19:355–365, 1996.

33. Ghosh D, and A Scheepens. Vascular action of polyphenols. *Molecular Nutrition and Food Research* 53(3):322–331, 2009.

34. Goyarzu P, et al. Blueberry supplemented diet: Effects on object

recognition memory and nuclear factor-kappa B levels in aged rats. *Nutritional Neuroscience* 7(2): 75–83, 2004.

35. Groupy P. Enzymatic browning of model solutions and apple phenolic extracts by apple polyphenoloxidase. *Journal of Food Science* 60(3): 497–501, 1995.

36. Guay DR. Cranberry and urinary tract infections. *Drugs* 69(7): 775–807, 2009.

37. Gunes G, and CY Lee. Color of minimally processed potatoes as affected by modified atmosphere packaging and antibrowning agents. *Journal of Food Science* 62(3):572–575, 1997.

38. Huang X, and F Hsieh. Physical properties, sensory attributes, and consumer preference of pear fruit leather. *Journal of Food Science* 70(3):E177–E186, 2005.

39. Jackman RL, HJ Gibson, and DW Stanley. Tomato polygalacturonase extractability. *Journal of Food Biochemistry* 19:139–152, 1995.

40. Jensen GS, et al. In vitro and in vivo antioxidant and anti-inflammatory capacities of an antioxidant-rich fruit and berry juice blend. Results of a pilot and randomized, double-blinded, placebo-controlled, crossover study. *Journal of Agriculture and Food Chemistry* 56(18): 8326–8333, 2008.

41. Joseph JA, et al. Blueberry supplementation enhances signaling and prevents behavioral deficits in an Alzheimer disease model. *Nutritional Neuroscience* 6(3):153–162, 2003.

42. Kalt W. Effects of production and processing factors on major fruit and vegetable antioxidants. *Journal of Food Science* 70(1): R11–R19, 2004.

43. Ketiku AO. Chemical composition of unripe (green) and ripe plantain (*Musa paradisiaca*). *Journal of Food Science* 19:139–152, 1973.

44. Knight R. Origin and world importance of tropical and subtropical fruit crops. In *Tropical and Subtropical Fruits,* ed. P. Shaw, 1–120. Avi Publishing, 1980.

45. Kuzminski LN. Cranberry juice and urinary tract infections: Is there a beneficial relationship? *Nutrition Reviews* 54(11):S87–S90, 1996.

46. LaBell F. Fresh-squeezed flavorings. *Prepared Foods* 165(12):63, 1997.

47. Lanza E, and R Butrum. A critical review of food fiber analysis and data. *Journal of the American Dietetic Association* 86:732–743, 1986.

48. Lavons JA, RD Bennett, and SH Vannier. Physical/chemical nature of pectin associated with commercial orange juice cloud. *Journal of Food Science* 59(2):399–401, 1994.

49. Lee GH, et al. Ripening characteristics of light-irradiated tomatoes. *Journal of Food Science* 62(1):138–140, 1997.

50. Lively R. Sweet, juicy oranges taste better in winter. *Fine Cooking* 69:17, 2005.

51. Longbottom L. Lemon zest. *Fine Cooking* 78:20, 2006.

52. Magness JR. How fruit came to America. In *The World in Your Garden*. National Geographic Society, 1959.

53. McClements DJ. Ultrasonic characterization of foods and drinks: Principles, methods, and applications. *Critical Reviews in Food Science and Nutrition* 37(1):1–46, 1997.

54. McCue N. Beverages fixate on fruit flavors. *Prepared Foods* 165(7):59–60, 1996.

55. McEvily AJ, and R Iyengar. Inhibition of enzymatic browning in foods and beverages. *Critical Reviews in Food Science and Nutrition* 32(3):253–273, 1992.

56. McGee H. *On Food and Cooking: The Science and Lore of the Kitchen.* Macmillian, 1997.

57. Mermelstein NH. Determining antioxidant activity. *Food Technology* 62(11):63–66, 2007.

58. Murthy NS, S Mukherjee, G Ray, and A Ray. Dietary factors and cancer chemoprevention: An overview of obesity-related malignancies. *Journal of Postgraduate Medicine* 55(1):45–54, 2009.

59. O'Connor-Shaw RE. Shelf life of minimally processed honeydew, kiwifruit, papaya, pineapple and cantaloupe. *Journal of Food Science* 59(6):1202–1206, 1994.

60. Penfield MP, and AM Campbell. *Experimental Food Science.* Academic Press, 1990.

61. Petreyak RM. Beverages. *Progressive Grocer* 70(7):45–49, 1991.

62. Pollack SL. *Consumer Demand for Fruit and Vegetables: The U.S. Example.* www.ers.usda.gov/publications/wrs011/wrs011h.pdf. Accessed 12/04/09.

63. Potter NN, and JH Hotchkiss. *Food Science.* Chapman & Hall, 1995.

64. Proliferation of juice-added beverages continues. *Food Engineering* 62(11):31, 1990.

65. Pszczola DE. Drinks for everyone! *Food Technology* 49(9):30, 1995.

66. Raphaelides SN, A Ambatzidou, and D Petridis. Sugar composition effects on textural parameters of peach jam. *Journal of Food Science* 61(5):942–946, 1996.

67. Rediscovering calcium fortification. *Prepared Foods* 162(12):65, 1993.

68. Reyes-Carmona J, GG Yousef, RA Martinez-Peniche, and MA Lila. Antioxidant capacity of fruit extracts of blackberry *(Rubus sp.)* produced in different climatic regions. *Journal of Food Science* 70(7):S497–S503, 2005.

69. Robertson JA, et al. Relationship of quality characteristics of peaches (cv. Loring) to maturity. *Journal of Food Science* 57(6):1401, 1992.

70. Rocha AP, et al. Endothelium-dependent vasodilator effect of Euterpe oleracea Mart. (Açaí) extracts in mesenteric vascular bed of the rat. *Vascular Pharmacology* 46(2):97–104, 2007.

71. Rojas-Grau MA, A Sobrino-Lopez, MS Tapia, and O Martin-Belloso. Browning inhibition in fresh-cut 'Fuji' apple slices by natural anti-browning agents. *Journal of Food Science* 71(1):S59–S65, 2005.

72. Rommel A, R Wrolstad, and D Heatherbell. Blackberry juice and wine: Processing and storage effects of anthocyanin composition, color and appearance. *Journal of Food Science* 57(2):385–410, 1992.

73. Seeram NP. Berry fruits: Compositional elements, biochemical activities, and the impact of their intake on human health, performance, and disease. *Journal of Agriculture and Food Chemistry* 56(3):627–629, 2008a.

74. Shamaila M, WD Powrie, and BJ Skura. Sensory evaluation of strawberry fruit stored under modified atmosphere packaging (MAP) by quantitative descriptive analysis. *Journal of Food Science* 57(5):1168–1172, 1992.

75. Shukla M, K Gupta, Z Rasheed, KA Khan, and TM Haqqi. Consumption of hydrolyzable tannins-rich pomegranate extract suppresses inflammation and joint damage in rheumatoid arthritis. *Nutrition* 24(7–8):733–43, 2008.

76. Siddiq M, et al. Partial purification of polyphenol oxidase from plums (*Prunus domestica* L., cv. Stanley). *Journal of Food Biochemistry* 20:111–123, 1996.

77. Sloan A. Fruit frenzy. *Food Technology* 55(12):14, 2001.

78. Stahl W. Macular carotenoids: Lutein and zeaxanthin. *Developments in Ophthalmology* 38:70–88, 2005.

79. Thakur BR, RK Singh, and AK Handa. Chemistry and uses of pectin. *Critical Reviews in Food Science and Nutrition* 37(1):47–73, 1997.

80. United States Apple Association. *Core Facts.* www.usapple.org/consumers/applebits/core.cfm. Accessed 8/09/09.

81. United States Department of Agriculture, Agricultural Marketing Services. Grading, certification, and verification. http://www.ams.usda.gov/AMSv1.0/FreshFVGrading. Accessed 12/04/09.

82. United States Department of Agriculture. *My Pyramid*™. www.mypyramid.gov/index.html. Accessed 6/09.

83. United States Department of Health and Human Services, Food and Drug Administration. *Code of Federal Regulations* (21 CFR Part 145), 1997.

84. Vandercook CE, S Hasegawa, and VP Maier. Dates. In *Tropical and Subtropical Fruits,* ed. P. Shaw, 506–541. Avi Publishing, 1980.

85. Vance Publishing Group. *The Packer. 1997 Produce Availability and Merchandising Guide.* Vance Publishing Group, 1999.

86. Vinson JA. The functional food properties of figs. *Cereal Foods World* 44(2):82–87, 1999.

87. Wold M. Miscellaneous grocery. *Progressive Grocer* 70(7):70, 1991.

88. Wolfe KL, X Kang, X He, M Dong, Q Zhang, and RH Liu. Cellular antioxidant activity of common fruits. *Journal of Agricultural and Food Chemistry* 56:8418–8426, 2008.

89. Yulianti F, CA Reitmeier, BA Glatz, and TD Boylston. Sensory, flavor, and microbial analyses of raw, pasteurized, and irradiated apple ciders. *Journal of Food Science* 70(2):S153–S158, 2005.

90. Zhang X, and WH Flurkey. Phenoloxidases in Portabella mushrooms. *Journal of Food Science* 62(1):97–100, 1997.

# WEBSITES

The California Rare Fruit Growers, Inc., has compiled a list of fruit associations and organizations:
**www.crfg.org/related.org.html**

The North Carolina Cooperative Extension Office has produced several consumer publications on safe food handling, drying, jams and jellies, freezing, and more:
**www.ces.ncsu.edu/**
(Click on "Health and Nutrition" in left bar, scroll down to "Publications," and click on "Extension Publications for Health and Nutrition.")

Fruit Facts are a series of publications containing information on individual fruits, from avocado to white sapote:
**www.crfg.org/pubs/ff/index.html**

# 15 Soups, Salads, and Gelatins

The first impression of a meal is often a soup or salad. The quality of this introductory item often serves as an indication of what is to follow. Starting the meal with an excellent soup or salad is equivalent to putting your best foot forward. It is important to begin a meal with flavor and distinction as a prelude to the main course. The soup or salad serves as the curtain raiser, and can help introduce the meal in a colorful, entertaining manner.

The versatility of soups and salads allows them to be served not only before a meal, but also during or after, or even as a meal. Countless combinations are possible, and this probably explains why soups were some of the first meals ever prepared in a pot over an open fire.

## SOUPS

Soups are a liquid food that may or may not contain other solid ingredients such as meat, poultry, vegetables, and/or grain-based foods (bread, pasta, noodles, rice, barley, etc.). Stews differ from soups in that they are more solid than liquid in their ingredients and are a complete meal in themselves. Nevertheless, sometimes the difference between soups and stews is not always clear. In 16th-century France, specialized shops were set up among the ordinary trade shops to serve soup to the workers. A Parisian soup vendor had a Latin inscription above his door that read, "Come to me all of you whose stomachs cry out and I will restore you" (17). The restorative services of these soup shops eventually led to them being called restaurants, and our English word used for the evening meal, *supper,* is derived from *soup* (12). Pots were often used to prepare soups, which is why the word *pottage* is often associated with soups.

## Types of Soups

The enormous assortment of soups now available ranges from light soups that serve as appetizers to very heavy soups that can be offered as a main dish. Most soups are served hot, but there are exceptions, the most famous one being vichyssoise (vee-shee-swahz). Among the many accounts of its origin is the story of Louis Diat, chef of the Ritz-Carlton in New York, inventing this soup in 1917, and naming it after his native town in France (7). Another tale of its origins starts with King Louis XIV of France, who, suspicious that people were trying to poison him, ordered that all his food be sampled by an official taster. The king's taster felt the need for his own taster, who, in turn, also had a taster, and by the time this hot creamed leek and potato soup got to the king, it was cold, and so it has been eaten that way ever since (9).

**FIGURE 15-1**   Types of soups* (with selected examples).

| CLEAR AND THIN SOUPS | |
|---|---|
| Bouillon | Unstrained (unclear) (from simmered meat, poultry, vegetables, or grains) |
| Court bouillon | A special bouillon containing wine, vinegar, or lemon juice in which another ingredient has been pouched |
| Broth | Strained liquid (from simmered meat, poultry, vegetables, or grains) |
| Consommé | Cleared with egg white |

| THICKENED SOUPS | |
|---|---|
| Cream | • Added milk, cream, or sour cream<br>    Cream of asparagus<br>    Cream of cauliflower<br>    Cream of celery<br>    Cream of chicken<br>    Cream of corn<br>    Cream of mushroom<br>    Cream of tomato<br>    New England clam chowder |
| Purée | • Blenderized and strained ingredients |
| Bisque | • Combination cream and purée (usually crustaceans)<br>    Lobster bisque |
| Sauce | • Thickened with a velouté sauce (egg, butter, and cream) |
| Other Thickeners | • Thickened with the starch from potato, pasta, grain (rice), flour, cornstarch<br>    Egg drop soup (cornstarch)<br>    Vichyssoise (potato) |

| TRADITIONAL SOUPS |
|---|
| Avgolemono (Greek) |
| Legume soups—lentils, beans, peas |
| Minestrone (Italian) |
| Mulligatawny (India) |
| Pozole (Mexican) |

| COLD SOUPS |
|---|
| Borscht (Slavic) |
| Fanesca (Ecuador) |
| Gazpacho (Spanish) |
| Vichyssoise (French) |

| NOODLE SOUPS |
|---|
| Chicken noodle (American) |
| Pho (Vietnamese) |
| Ramen (Japanese) |
| Saimin (Hawaiian) |
| Thukpa (Tibetan) |
| Udon (Japanese) |

| FRUIT SOUPS |
|---|
| (served hot or cold) |
| "Frukt suppe" (Norwegian) |
| Sour soup (Vietnamese) |
| Winter melon (Chinese) |

| DESSERT SOUPS |
|---|
| Ginataan (Filipino) |
| Oshiruko (Japanese) |

*Many, but not all, soups are made from a liquid foundation of stock (white or brown from meat, poultry, fish, or vegetable) or broth (stock that has not been reduced).

**Stock**  The foundational thin liquid of many soups, produced when meat, poultry, seafood, and/or their bones, or vegetables are reduced (simmered) and strained.

**Broth**  Stock made from meat or meat/bone combinations and some water with little or no flavoring. Broths are seldom reduced (simmered until much of the water evaporates) and therefore not as strong-flavored as stocks.

**White stock**  The flavored liquid obtained by simmering the bones of beef, veal, chicken, or pork.

**Brown stock**  The stock resulting from browning bones and/or meat prior to simmering them.

Whatever the temperature, all soups are based on **stock** or **broth.** To this foundation are added other ingredients, lending each kind of soup its own name and unique characteristics. Stocks are very simple—primarily water with little or no flavor—and serve as foundations for other added ingredients. Building on this base, a clear, thin, or thickened soup can be made, then developed into any of the almost unlimited variations including, but not limited to, traditional, cold, noodle, fruit, and dessert soups (Figure 15-1). The different types of stocks are discussed below, followed by an examination of the two basic categories of soups: (1) clear or thin and (2) thickened, plus cream as an example of a thickened soup.

## Stocks

Cracked bones and water are often the main ingredients of meat stock. For this **white stock,** neck and knucklebones are preferred, because they contain more collagen (which converts to gelatin) and more flavorful extracts than any other bone in the animal (9). Although all bones are porous, splitting them open helps to release the gelatin, which imparts a rich thickness and body to the stock. When stock is made from meats, beef and chicken are the two most commonly used. Veal bones contribute the greatest amount of collagen, making the thickest, most gelatinous stocks, while beef bones contribute the richest, meatiest flavor. Many chefs use both. **Brown stock** has a deeper, caramelized flavor. Browning the bones and meat before adding water has the advantage of discouraging the stock from becoming cloudy. Heating the meat coagulates many of the proteins and traps minute particles that could otherwise cloud the stock. Regardless of whether the stock is brown or white, it is important to simmer rather than boil the stock; boiling would cause the particles floating to the top to churn back into the broth, turning it cloudy and less clean tasting (18). Although there is no standard formula for stocks, a general rule is that one pound of bones is required for every pint of water.

## Water Is the Main Ingredient

Pure, clean, cold water is the first ingredient of any stock. Spring or distilled water is preferred (although not always used due to cost), because tap water may carry the flavors of chlorine or other substances (11). Simmering the water and ingredients for half an hour up to several hours then generates flavor. Boiling should never be allowed because, in addition to causing clouding, it will toughen the meat and disintegrate any ingredients except bones. If additional water is required during simmering, hot water can be added to avoid cooling the entire stock.

## Flavoring Ingredients

A mirepoix (meer-pwah) made up of onions, celery, and carrots is often added to stocks. This standard mirepoix can be modified to produce a white mirepoix, in which leeks replace the carrots in order to create a near-colorless stock (5). Salt or other potent seasonings are not usually added until the last half hour, if at all, in order to prevent their becoming too concentrated as liquids evaporate during cooking. The exception is a **bouquet garni,** which can be added to dishes with a lot of liquid undergoing a long simmer. Standard ingredients of a bouquet garni include parsley, thyme, and a bay leaf, although other aromatics such as cloves, rosemary, sage, and garlic cloves can also be added (22). Cracked or ground pepper should never be added at the start of a long heating process because it will turn harsh and acrid in taste (17). Whole peppercorns may be used instead.

## Meat Stocks

Meat stock, based on cracked bones and sometimes raw meat, serves as the major ingredient for all meat soups and major meat sauces. Cooked meat will not yield the same flavor, although it is sometimes used. The more mature the meat, the better, because meat becomes more flavorful with age. In general, red meats contain more flavor than white (veal, pork). Some of the cuts most frequently used for meat stock include:

- **Beef.** Oxtail, chuck, shank, bottom round, and short ribs
- **Pork.** Hocks, ham bones, and Boston butt
- **Lamb.** Shank, leg, and shoulder

**Adding Meat and Bones** The first step in preparing soup stock made with meat and/or bones is to cube or grind the meat, and/or cut the bones into 3-inch sections. This increases the surface area and improves extraction.

To draw out even more flavor, some chefs first soak the meat or bones in a pot filled with cold water, using 2 cups of water to every cup of meat or pound of bones, or adding enough water to cover the packed ingredients by 1½ inches above them. After half an hour to an hour of soaking, heat to the boiling point and then reduce it to a bare simmer for up to 3 to 4 hours. The bones and meat heated in the water will release gelatinous particles as the stock simmers that will cloud and form a scum on top of the liquid. Many chefs skim this scum layer before adding any vegetables.

**Adding Vegetables** Vegetables are often not added until the last hour or half hour of preparation, both to prevent their becoming mushy with overcooking and to preserve the desired flavor, which would become too concentrated and bitter with the water evaporation that takes place during longer cooking. Sometimes, however, vegetables are placed in the stock at the beginning intentionally so they will get mushy and can be puréed to act as a thickener. The size that vegetables are cut also makes a difference—large chunks are best for long-simmering stocks, whereas small pieces are added near the end of heating time.

## Poultry Stocks

The more mature the poultry used to make stock, the more flavorful will be the liquid. Birds that have been free to roam and scavenge, called free-range birds, yield the most flavor of all (9). An entire cleaned chicken carcass or an assortment of meaty bones (ideally backs and necks) is placed in enough cold water to cover it by 1½ to 2 inches and brought to a simmer with the lid off. It is cooked at a simmer for about an hour (whole bird) to 4 hours (just bones) until the meat is tender and ready to be cut off in bite-size pieces. During this process, the frothy fat is repeatedly skimmed from the top as necessary. Once the meat is tender, the liquid is strained and returned to the pot before adding the chicken pieces, chopped vegetables, grain (such as rice or barley), and seasoning.

It is then covered and simmered for another hour. The vegetables often include, but are not limited to, onions, celery, parsley, leeks, carrots, parsnips, and turnips. Common seasonings are bay leaf, onion, garlic, pepper, sage, thyme, rosemary, or any other complementary herb. Chicken bouillon is sometimes added if the stock lacks flavor. Another method to add more flavor is to roast the chicken bones and vegetables before adding water.

## Fish Stocks

Most fish stocks use the backbones (called frames or racks), heads, and/or tails of lean white fish. Fatty fish are rarely used because of their strong oily flavors. The high gelatin concentration in the heads contributes to the flavor and body of the stock. Fish frames should be thoroughly washed and eviscerated, because remnants and organs, as well as gills and any attached skin, will give the stock an off-flavor (2). The eyes may also be removed. The bones are then combined with all the other ingredients in the liquid. Fish contain less gelatin than do meat and poultry, so the plain water usually used to start a soup stock is sometimes replaced by or fortified with chicken or vegetable stock, or bottled, but not canned, clam juice (11). The floating fish frames are occasionally pushed down into the liquid to release the flavorful compounds from the bones. Heating time should not exceed about half an hour, because more than that makes the stock too bitter or fishy tasting. For more flavor, chopped fennel leaves and the white portions of leeks

**Bouquet garni** A bundle of parsley, thyme, bay leaf, and whole black pepper rolled in a leek and tied together with twine.

are classic additions to fish stock, along with the standard combination of onions, carrots, and celery. The celery may be omitted for a milder flavor, and the carrots may be left out if a golden color is not desired in the stock. Sometimes the fish frames and vegetables are first heated in butter to extract more flavor and create a richer-tasting stock. Any added vegetables must be cut small because of the short cooking time.

### Shellfish Stocks
Shellfish stocks are usually made from shrimp, lobster, mussels, or clams, or from the shells of the first two. Leftover shellfish shells can be frozen until they are needed to make stock.

---

**?  How & Why?**

**Why is cold water used to start soups?**

The water should be cold to start with, because ingredients placed in cold water will transfer their flavor more efficiently to the liquid (8). If hot water is used, the stock will be less flavorful and less clear.

---

## Vegetable Stocks
Vegetable stocks have advantages over other kinds of stocks in that they are less expensive, less messy, and less time-consuming than their meat, poultry, or fish counterparts. Vegetables that lend themselves well to making stock include carrots, onions, leeks, shallots, garlic, celery, celeriac, parsnips, fennel, and tomatoes. Just as for poultry and beef, the more mature the vegetable, the more flavor in the extract. Vegetable stock takes only about half an hour to cook; any more time will turn the texture of the vegetables into mush and turn the flavor extract bitter. The standard vegetables and seasonings described under poultry stock are used when making vegetable stocks.

---

**Bouillon**  A broth made from meat and vegetables and then strained to remove any solid ingredients.

**Consommé**  A richly flavored soup stock that has been clarified and made transparent by the use of egg whites.

---

### Storage of Soup Stocks
Stock should never be placed in the refrigerator while it is still hot. The large volume of hot liquid can raise the internal temperature of the refrigerator to the point that the stock will not cool sufficiently within 2 hours, and may warm everything else in the refrigerator as well. A good way to cool the stock is to place the hot stockpot in a sink full of cold water and ice cubes until it becomes lukewarm, although the time spent in this procedure should not exceed 1 hour. After the stock is left uncovered for the first half hour and stirred occasionally to cool it, it should then be covered with a lid or an upside-down plate to prevent evaporation, which would cause the stock to become too concentrated. Refrigerated stock cools better in shallow pans. If covered, stock lasts up to 5 days, but is best if used within 2 days. Stock prepared for future use can be stored frozen for a couple of months (18). Any animal stock should be considered a potentially hazardous food when it comes to food safety and treated accordingly (see Chapter 4). Simmering stored stock for at least 10 minutes before use is a basic safety precaution.

## Clear and Thin Soups

Clear soups include bouillons and consommés. Thin soups are somewhat thicker than clear soups.

### Bouillon
**Bouillon** is the French word for broth. Bouillons are less gelatinous than stocks, and may be added to poultry and vegetable stocks to add flavor. Traditionally, this type of soup is called bouillon if it is based on beef, and court bouillon or fumet if it is prepared using fish. *Court* is the French word for "short," and it describes the preparation time of bouillon, which is much shorter than that for stocks.

### Consommé
A **consommé** is a perfectly clear beef bouillon. One raw egg white is added for every quart of stock, and the whole mixture is heated to boiling. The egg white will coagulate on the surface of the stock, forming what is referred to as a *raft*. As the stock simmers, the raft entraps loose particles. The resulting masses of unwanted material can be removed by straining. Because egg whites strip flavor from stock as well as

particles, some ground meat and finely chopped vegetables are generally mixed with the egg whites to add flavor back to the stock. Because *bouillon* is the French word for *broth*, and because it is so similar to consommé, the three terms are often used interchangeably (17).

### Thin Soups
Thin soups have a consistency between a clear liquid soup and a thickened soup. These soups are broth based, but typically contain a garnish of cooked meats, vegetables, and/or starch such as pasta, rice, or barley. Examples include tomato, onion, and noodle soups. Miso soup, originating in Japan, is an increasingly popular thin soup that often contains tofu blocks, seaweed, and green onions. There are many variations of soups with miso as the primary seasoning agent. Miso itself is made by mixing cooked beans (usually soybeans) with cultured rice or barley (called *koji*) and salt (Figure 15-2). This mixture is then fermented several weeks or as long as 3 years, depending on the variety of miso. Unpasteurized miso contains potentially beneficial microorganisms. Miso soup will vary in taste depending on the color of the miso soybean paste, which ranges from dark (reddish) to light (pale blond) (Figure 15-3). The darker misos are more bitter, earthy, and salty, whereas the lighter ones are more bland and less salty. In Japan, the larger grocery stores can offer up to 50 varieties of miso soybean paste.

## Thickened Soups

Any stock or broth can be thickened with cream, puréed vegetables, a sauce (velouté), or other thickeners such as added bread, noodles, grains (rice, barley, etc.), or plain starch (cornstarch). Adding starches to soups also reduces the perception of saltiness if the components of starch or any added gums bind to sodium (19).

### Examples of Thickened Soups
There are many varieties of thick soups. Chowders are usually fish-based soups thickened with vegetable pieces, such as potatoes, that have been cooked in milk. Gazpacho is an uncooked, thick, tomato-based soup with added vegetables and Latin American seasonings that is served chilled. Minestrone is an Italian soup that contains macaroni,

**FIGURE 15-2**    Miso for miso soup is traditionally made by mashing boiled beans (usually soybeans) underfoot like wine grapes, and mixing the mashed beans with cultured grain (called *koji*) and sea salt. This mixture is then fermented in wooden vats for several weeks or as long as 3 years, depending on the variety of miso. The miso maker in this photo wears special plastic foot coverings.

Courtesy of Sarah Chester, South River Miso Co

**FIGURE 15-3**    The flavor of miso soup depends on the type of miso soybean paste.

DAJ/RF/Getty Images

vegetables, and beans. Egg drop soup is a popular Chinese soup made by dropping a slightly beaten egg into simmering clear stock. In India, mulligatawny soup is popular; it consists of chicken, carrots, green pepper, and apple seasoned with curry, cloves, and mace. Borscht is a Russian/Polish soup that is based on beets and may or may not have vegetables and/or meats added to it (12). Thick soups also include any that are puréed or made from starchy vegetables that have been pulverized and sieved. Potatoes are a common thickener for a range of thickened soups.

***Cream Soups***    Cream soups are a type of thickened soup. These cream soups and **bisques** are made by adding cream and/or milk to a thickened, flavorful purée made from meats, poultry, fish, and/or vegetables. Common thickeners are white sauce (equal parts of flour + fat combined with liquid), potato starch, or puréed potatoes. Cream is usually the thickening agent in bisques. Cream of chicken is the most popular cream soup made from an animal source. The most common vegetable creamed soups include asparagus, broccoli, potato, spinach, tomato, celery, cauliflower, and corn.

***Preparation of Cream Soups***    In preparing creamed soups, flavorings are added to a thin white sauce, which is then combined with a liquid base that is either milk or white stock. Béchamel is a white sauce often used as a base for traditional cream soups because of its velvety, smooth, creamy consistency. Heavy cream and/or egg yolks may be added after the soup has been heated and seasoned, but these additions increase caloric and fat content. If only one vegetable is featured in the cream soup, it helps to add a few aromatic ones such as onion, garlic, leeks, and shallots to enhance the flavor. A garnish of parsley or a sprinkle of chives, paprika, or nutmeg makes a colorful addition to a bowl of cream soup.

***Lower-Fat Cream Soups***    Cream soups by definition contain fat, but substitutions can be made to create a lower-fat soup that will still retain

**Bisque**    Traditionally, a cream soup made from shellfish. Marketers sometimes label creamed vegetable soups as bisques.

most of the characteristics of a cream soup (Figure 15-4). The cream can be replaced by starch in the form of raw rice (½ cup) or potato (thinly sliced) added at the beginning of cooking (18). The starch gelatinizes, creating a thick, smooth consistency. Adding just a touch of cream rounds out the flavor of this starch-based cream soup.

### How to Avoid Curdling

One potential problem in making creamed soups is that the acid from ingredients such as tomatoes can cause milk proteins to curdle as the mixture heats. Curdling can be reduced by doing the following:

- Prepare a fat/flour mixture (see Chapter 17) with either the milk or the stock (but cold milk or cream should never be added to hot soup).
- Stir some of the hot soup into the cold dairy product to temper it before adding it to the warm ingredients.
- Do not allow the soup to come to a boil after adding any dairy product, particularly cheese.
- Add acid to milk rather than milk to acid.

### Safety Tips for Cream Soups

Because cream soups should not be boiled after the cream or milk has been incorporated, there is more of a chance for bacteria to grow. Therefore, these soups should be treated with the safety steps discussed in Chapter 4 to avoid foodborne illness.

# SALADS

The Romans are credited with inventing salads, because they brought garden delectables to the table, sprinkled them with salt and seasonings, and added vinegar and oil. Salt is such an important ingredient that the word *salad* is based on the Latin word for "salt." The practice of adding "salted" greens continued in the Middle Ages, when salads were a side dish that added greens to the massive amounts of meat served in castles.

Through the years, the concept of salads has evolved and branched out so much that almost any food imaginable can be made into a salad. Now salads are defined as a food item that extends

beyond the common green salad to which vegetables, fruit, cheese, and/or nuts may be added. Salads can be served as an appetizer, a main course, a side dish, or even dessert (Table 15-1). The most common types of salads are grouped by their main ingredient: leafy greens, vegetables, fruits, proteins, pastas/grains, gelatin, or some combination of these. Sometimes salads such as tuna, potato, rice, pasta, egg, chicken, and ham are categorized as "cooked," but in this text they are classified by their main ingredient. These will be discussed following the basics of creating salads, which are now available even at fast-food restaurants (21). Salads, dressings, and gelatin foods will be the remaining focus of this chapter.

## Salad Ingredients

The salad's base, also called an underliner, is the item that serves as the salad's first layer or foundation. It is usually lettuce or some other green, but the base might also be pasta, rice, cottage cheese, or gelatin. The predominant ingredient or main part of the salad on top of the

**FIGURE 15-4**   Making low-fat cream soup.

(1) Sauté aromatic vegetables (onions, garlic, shallots, or leeks) over medium heat. Adding herbs such as thyme, sage, or rosemary, improves flavor.

(2) Pour in the liquid (water, broth, milk, or a combination) and heat to a simmer. Use 5 to 6 cups of liquid for every 2 pounds of main vegetable to be added.

(3) Add a thickener in the form of rice (½ cup) or raw potato (sliced). Starchy ingredients thicken when they gelatinize, making the soup feel creamy.

(4) The time to add the main vegetable (in even, bite-size pieces) is when the time remaining to cook the thickener is equal to how long it will take to prepare the vegetable.

(5) When the thickener reaches its maximum viscosity, purée the soup in batches using a blender or food processor. Another option is to strain the soup to obtain a finer texture.

(6) Finishing touches include adding a little liquid to thin the purée if it is too thick, and/or adding a small amount of cream to give it a "cream" flavor. Serve hot or cold.

Digital Works

| TABLE 15-1 | Types of Salads |
| --- | --- |
| **Salad Type** | **Examples** |
| Appetizer | Dinner salad |
| | Shrimp cocktail |
| Main course | Chef's salad |
| | Chinese Chicken salad |
| | Fruit plate with yogurt or cottage cheese |
| | Taco Salad |
| Side dish | Coleslaw |
| | Pineapple-carrot salad |
| | Potato salad |
| | Three-bean salad |
| Dessert | Fruit cocktail |
| | Jello-fruit salad |
| | Melon and/or citrus wedges |
| | Waldorf salad |

base is called the body. Topping it all off is the dressing, which adds flavor and moistness. Tart dressings go best with greens or vegetables or protein-based salads, whereas sweetened dressings are usually reserved for fruit salads.

### Garnishes

Sometimes garnishes are added for eye appeal. A few of the more common garnishes are fruit wedges (lemon, orange), hard-cooked eggs (slices, halves, or wedges), tomatoes (cherry, slices, or wedges), mushrooms, pimentos, radishes, olives, pickles, watercress, mint, and parsley.

The flavor, texture, and color of salads can be made more interesting by adding toppings such as croutons, cheese, shredded vegetables (carrots, red cabbage), herbs (Figure 15-5), tender greens (Figure 15-6), flowers, seeds, and nuts.

**FIGURE 15-5**  Herbs add interest and flavor to salads.

Basil

Thyme

Mint

Sage

Savory

Rosemary

Oregano

Tarragon

Parsley

Digital Works

**FIGURE 15-6** Tender young lettuces and other greens add variety to salads.

Green Chard

Green Curly Kale

Frisee

Green Romaine

Arugula

Radicchio

Tango

Lollo Rosa

Green Oak Leaf

Red Leaf Lettuce

Red Chard

Red Mustard

Mizuna

Red Romaine

Spinach

Richard Brewer/Digital Development Services

## Toasting Nuts

Most nuts can be toasted by spreading them on a cookie sheet, placing them in the oven at 325°F–375°F (162°C–191°C), stirring every few minutes, and removing them after 5 or 10 minutes or when they have turned light brown (24). They should be allowed to cool before being added to the salad. Sometimes adding sugared nuts such as candied walnuts adds new flavor to a salad (16).

## Principles of Salad Preparation

There are several principles to keep in mind when constructing a salad. The ingredients must be of the finest quality because fresh dishes cannot hide inferior quality as cooked dishes sometimes can. The more assertively flavored a food, the smaller its role in the salad should be. Flavor variety is important to maintain a diner's interest; salads can include ingredients that are crunchy (croutons, toasted nuts, dried wonton pieces); sweet/juicy (mandarin slices, apples, pears, grapefruit, and other fruits); bitter (endive, frisée, kale); and salty/tangy (olives, capers, cheese) (20). Color and texture should also be considered when making salads. The ultimate goal is to assemble and arrange ingredients with a good balance of color, texture, and shape within the rim of the chilled salad plate (Table 15-2).

**TABLE 15-2**   Adding Color, Flavor, and Texture to Salad Greens

| Vegetables | Fruit | Protein Foods |
|---|---|---|
| Artichoke | Apple | Anchovies |
| Asparagus spears | Avocado | Cheese |
| Baby corn | Grapefruit | Chicken, turkey |
| Bell peppers | Olives | Egg slices |
| Broccoli | Orange | Ham, bacon |
| Carrots | Papaya | Nuts (sunflower, walnut) |
| Cauliflower | Pears | Salmon, sardines, shrimp, tuna |
| Celeriac (celery root) | Pimento | |
| Celery | Pineapple | |
| Cucumber | | |
| Hearts of Palm | | |
| Mushroom | | |
| Peas | | |
| Radishes | | |
| Red onion | | |
| Scallions | | |
| Tomato (cherry, sun-dried) | | |

---

## CALORIE CONTROL
### Fat and Fiber in Salads

Just because something is called a salad does not automatically mean it is low in fat or calories.

- *Limit Fatty Ingredients.* Many salads contain high-calorie/-fat ingredients such as meat, egg yolks, cheese, avocado, coconut, olives, nuts, and oil-based dressings. A chef's salad harbors large amounts of both calories and fat even before the dressing is added (Table 15-3).
- *Control Portions—Especially the Dressings.* Many salad dressings contain oil, which averages 120 calories (kcal) and 13 grams of fat per tablespoon. There is also a fair amount of sodium in many of the same ingredients and dressings.
- *Boost Fiber.* Salads are not necessarily good sources of fiber; iceberg lettuce contains only 1 gram per cup. Because the daily fiber recommendation is 25 grams for women and 38 grams for men, that adds up to 25–38 cups of lettuce a day, which is not practicable. So a salad's fiber content may be boosted by topping leafy greens with high-fiber vegetables such as broccoli, cauliflower, carrots, baby corn, cooked dried beans, and potato pieces with skins.
- *Limit Starchy Ingredients.* Starch adds approximately 100 calories (kcal) per slice of bread, or per ½ cup of grains (rice, bulgur, etc.), pasta (noodles, elbows, etc.), or starchy vegetables (see Chapter 13 for individual vegetables).

**TABLE 15-3**   Major Calorie Contributors to Chef's Salad

| Salad Ingredients | Calories (kcal) | Fat (g) |
|---|---|---|
| 1 oz beef | 75 | 6 |
| 2 oz turkey | 80 | 1 |
| 1 oz ham | 70 | 5 |
| 1 oz cheese | 114 | 9 |
| 1 hard-cooked egg | 74 | 5 |
| 3 ripe olives | 15 | 1 |
| Lettuce (2 C) | 18 | — |
| 3 tomato wedges | 19 | — |
| Total | 465 | 27 |

| Salad Dressings (2 tbsp) | Calories (kcal) | Fat (g) |
|---|---|---|
| Blue cheese | 154 | 16 |
| French | 134 | 12 |
| Italian | 137 | 16 |
| Russian | 151 | 16 |
| Thousand Island | 140 | 14 |

© 2010 Amy Brown

## Leafy Green Salads

In general, when people think of salads, leafy green ones are the most likely to come to mind; examples include garden, tossed spinach, mixed green, and Caesar salads. Greens range in flavor from extremely mild to tangy, bitter, and sharp. Iceberg lettuce is the most mild-flavored and the one most frequently used to fill out the body of salads. Its bland flavor almost necessitates that other, more distinctive-tasting, ingredients be added. Other salad greens include Romaine, Bibb, Boston leaf, and Ruby lettuces, and even spinach. If more assertive flavors are desired, escarole, Chinese (Napa) cabbage, kale, and either green or red cabbage make good choices. Flowers and herbs provide additional flavor and color, but should be added in limited amounts, with the exception of herb salad made only from leafy herbs (Color Figure 16).

### *Preparation of Green Salads*

Constructing a green salad starts with a thorough rinsing of the greens to remove any soil (Figure 15-7). Many greens are grown in the ground, and not adequately removing the soil leaves a gritty feel between bites of leaves. Any water left clinging to greens after washing will dilute the flavor, adversely affect the texture, and accelerate deterioration, so it should be removed either by draining greens in a colander, patting them between paper towels, or using a salad spinner. Next, the stems and/or cores must be removed. Greens should be refrigerated for 30 minutes to several hours prior to use in order to promote crispness. Once they are ready for the salad bowl, the greens are torn by hand into bite-size pieces. Hand-tearing the leaves is preferred over cutting with a knife because it reduces bruising and allows more salad dressing to be absorbed into the leaves. Commercial food service units often rely on quicker methods, however, and Figure 15-8 shows several techniques for cutting iceberg lettuce. Cabbage and carrots can also be prepared for salads (Figure 15-9).

## Vegetable Salads

Vegetable salads are those whose main ingredients are non-leafy vegetables. These are sometimes composed of

**FIGURE 15-7** Cleaning salad greens.

**Clean greens.** Briefly soaking and swirling the greens in a bowl of cool water dislodges grit. Lifting the leaves out with separated fingers is followed by replacing the water in the bowl and repeating the process until no grit remains.

**Dry greens with a salad spinner.** Water is removed by repeatedly spinning the leaves, draining the bowl, and tossing the leaves between spinnings.

a single vegetable, such as a tomato salad, or made up of a combination of vegetables. Examples of combination salads include cucumber and tomato; green bean, mushroom, and carrot; and coleslaw salads. Any vegetable is eligible, and those that are raw, cooked, canned, or marinated are all suitable candidates. Vegetables that are commonly used in salads include tomatoes, cucumbers, broccoli, cauliflower, beets, radishes, olives, green bell peppers, carrots, peas, mushrooms, onions, and celery. Tougher vegetables such as carrots and celery will have improved texture when shredded with a knife, grater, or food processor. Potato salad is best prepared with red or white boiling potatoes. They are less starchy and hold their shape when boiled better than the Russet or Idaho potatoes usually used for baking.

## Fruit Salads

Fruit salads are often served as a dessert or as a luncheon dish, and often combined with a higher-protein food such as cottage cheese or yogurt. Some of the fruits most frequently used for salads are apples, avocados, oranges, bananas, kiwi, pineapple, melons, grapefruit, pears, peaches, cherries, blueberries, strawberries, grapes, and raisins. Acidic fruits should be balanced with mild greens and a light sauce (17). Canned fruit should be thoroughly drained, and all fruit should be cut into pieces no bigger than bite-size (Figure 15-10). Some fruits undergo enzymatic browning when cut, which may be prevented by coating them lightly with citrus or pineapple juice, sugar, or an ascorbic acid (vitamin C) mixture. Fruit salads are often arranged, rather than tossed, in order to prevent their delicate flesh from bruising or damage.

## Protein Salads

Basing a salad on protein foods, such as cold meat, poultry, fish, eggs, or dairy products, helps to make it a main course. All except the dairy items must be used in their cooked form. Firm cheeses, such as cheddar or Swiss, are often cubed and added to salads. To quickly prepare chicken breasts specifically for salad, poaching works well (see Chapter 8). Four boned, skinned chicken breasts yield 2 cups of cubed chicken meat, which should be chilled by refrigerating at least two hours prior to assembly of the salad. Some chefs prefer roasted chicken for chicken salads.

## Pasta/Grain Salads

A bland background of pasta or grains serving as the body of a salad is a perfect invitation to more colorful and flavorful ingredients, especially if they are acidic or salty. The large number of pasta shapes (elbows, shells, noodles) and varieties of grains (rice, bulgur, wheat, etc.) allow the creation of an almost unlimited array of salads. Chapter 16 discusses the preparation of the grains and pastas themselves.

# Salad Dressings

There are as many different kinds of salad dressings as there are kinds of salads. The original dressing was probably

**FIGURE 15-8** Methods of preparing iceberg lettuce.

**Shredded**
*Manual*—Cut each head in half lengthwise. Then slice crosswise into ⅛–½" strips.
*Automatic slicer*—Crisp iceberg lettuce is the only lettuce suitable for shredding on slicers or slicing attachments. Use half heads; set the gauge for ⅛ to ½" strips.

**Rafts**
Cut each head crosswise into slices about 1" thick. A head should yield about 3 to 4 rafts.

**Chunks**
Slice head into rafts. Then cut each raft crosswise and lengthwise into 1½" square chunks.

**Large Leaves or Cups**
Begin with a full head, holding it core side down. Gently separate leaves and lift them away from the head. For smaller portions, cut heads in half, then lift off leaves.

**Wedges**
Small iceberg lettuce heads cut into quarters yield an average wedge. Wedges should weigh about 4 oz. each. Larger heads should be cut into six or eight wedges to yield similar size portions.

**Half Heads**
Cut a full head in half lengthwise. Then place cut side down on a board and make two or three parallel cuts partially through the half head. Present over a mound of shredded iceberg lettuce with the cuts spread open and filled.

**Full Heads**
Select a firm, springy head and core. Remove a slice from the core end, then pull out about a 3" heart from the center of the head and shred. Spread remaining leaves back from the center like a rose. Place shredded lettuce back into the center of the prepared head.

Digital Works

**FIGURE 15-9** Preparing various vegetables.

(a) Removing the cabbage core.

(b) Shredding cabbage.

(c) Grating carrots.

Digital Works

vinaigrette is varied by the types of oil and vinegar used.

*Oils* A vinaigrette may use a flavorful oil/vinegar duo such as extra virgin olive oil with red wine vinegar, walnut oil with a raspberry vinegar, or sesame oil with an oriental rice vinegar.

just plain salt. The word *salad* is derived from *herba salata,* Latin for salted greens (17). During Roman times, freshly picked greens were salted to enhance their taste, but now there are many different types of dressings, which

can be thought of as "sauces for salads" (Figure 15-11).

## Oil and Vinegar Dressings

**Vinaigrette** is the most basic of all dressings (10). The flavor of the

**Vinaigrette** A salad dressing consisting of only oil, vinegar, and seasoning.

**FIGURE 15-10** Preparing fruit for salad.

Trimming pineapples.

Cutting out pineapple "eyes."

Coring apples.

Peeling soft-fleshed fruits.

Preparing melon balls.

**FIGURE 15-11** Pouring dressing over salad.

Other salad dressings based on vegetable oil may rely on a neutral-flavored oil such as corn, cottonseed, soybean, or safflower. Winterized oils, often called salad oils, are those that have been treated so they do not cloud when refrigerated. Olive, peanut, walnut, hazelnut, and almond oils are sometimes chosen for their characteristic flavors. A nutty-flavored salad dressing can be made by adding toasted sesame oil to the dressing.

**Vinegars** Balancing the oil in a vinaigrette is some form of vinegar, of which the three most common types are wine, cider, and white/distilled (4). Vinegars differ in their base and ingredients, but they are similar in that they are all fermented from alcohol to acetic acid by the bacterium *Acetobacter* (14). The word *vinaigre* is French for "sour wine," and the higher the percentage of acetic acid (shown on the label in fine print) that the vinegar contains,

the more sour and stronger the vinegar will be. Vinegars were originally produced as a by-product of winemaking (3). Better-tasting vinegars are slowly fermented naturally, whereas those fermented with the aid of heat and chemicals often have harsh, metallic tastes (23). Some of the vinegars used in salad dressings include:

- **Wine.** Made from wines such as red, white, cabernet sauvignon, champagne, and sherry. These vary widely in terms of flavor and quality.
- **Cider.** A golden brown vinegar made from fermented apple juice; it has a strong flavor.
- **White/distilled.** A colorless, extremely acidic, almost flavor-free vinegar made from grain alcohol.
- **Herb.** Cider, white, or wine vinegar with herbs added; tarragon is a common herb-flavored vinegar.
- **Fruit.** Cider, white, or wine vinegar with fruits added; raspberry vinegar is increasingly popular.
- **Rice.** Made from rice wine, this vinegar has the lowest acidity, is clear to slightly yellow, and has a mild flavor.
- **Balsamic.** A rare, dark brown vinegar made from a very sweet grape concentrate; it has a slightly sweet taste. It is more costly than other vinegars because it is aged in wooden barrels for over 10 years. Most balsamic vinegar available in grocery stores is actually a red wine vinegar to which concentrated grape juice and perhaps caramelized sugar have been added.

The ratio of oil to vinegar varies among salad dressings. Flavor becomes more bland and less acidic as more oil is added, but shifts to a sharper, acidic taste when vinegar concentration increases. Sometimes lemon juice is added or replaces vinegar when a more lemony flavor is desired.

### Emulsified Dressings

Oil and vinegar normally separate if left to stand, and need to be mixed together before being applied to a salad. If the oil and vinegar mixture has been treated to create a permanent **emulsion**, however, the ingredients will not separate (see Chapter 22). One such emulsion is mayonnaise, which can be served as is in chicken or potato salad or in coleslaw, or as a base for many other salad dressings, including Chantilly, creamy Roquefort, Green Goddess, dill, Thousand Island, and Russian.

### Other Dressings

There are many other possibilities for salad dressing bases. Little or no oil is found in cooked dressings; these have a viscous consistency because of an added starch thickener. Some dressings are dairy based, with sour cream or yogurt as the main ingredient. A few dressings are made from fruit juice, and their lower calorie/fat content makes them a good alternative for health-conscious consumers.

### Adding Dressings to Salads

The best time to add dressing to a salad is just before serving, because the acidity of the dressing will start wilting the leafy green vegetables as soon as the dressing is added. It is best to chill the dressing for at least half an hour to improve its ability to coat the salad ingredients. Only enough to lightly coat the ingredients should be added; too much overwhelms the salad's flavor, while too little defeats the purpose of adding the dressing, leaving the flavor unenhanced. Salads are best tossed to evenly coat all the ingredients with a taste of dressing. Topping a salad with dressing without mixing is a less desirable way to serve any salad. Commercial salad dressings are often made with food additives that serve various functions (Chemist's Corner 15-1).

# GELATINS

Since gelatin salads rely on the formation of a gel, gelatins as a whole are discussed in this chapter, but many other foods include gelatin as an ingredient. What follows is a definition of gelatin, its nutritional value, and the preparation of gels that rely on numerous factors during gel formation.

## What is Gelatin?

Gelatin itself is a mixture of proteins extracted from the collagen found in the bones, hides, and connective tissue of animals. The principle commercial sources of gelatin are pork skins and the bones and hides of cattle (13). These sources of gelatin are sometimes avoided by certain vegetarians or people following kosher guidelines. Gelatin is available in either powder granules or sheets, the latter being used by commercial food manufacturers and institutions (1 teaspoon of gelatin powder is equivalent to 2 sheets). Standard amounts sold to consumers consist of uniform packets containing about 2½ teaspoons (7 grams) of powdered gelatin, which will usually gel about 2 cups of liquid. Gelatin is available in a colorless, almost transparent, odorless, and virtually tasteless state, or flavored, in combination with sugar, coloring, and flavoring agents.

## Is Gelatin Nutritious?

Even though gelatin is a protein from animal sources, it is low in tryptophan, an essential amino acid. Gelatin protein is therefore considered incomplete and of low **biological value.** A few gelatin manufacturers have added tryptophan to some of their products in an effort to remedy this problem, but gelatin's contribution to dietary protein is nevertheless limited by the small amounts—only 1 tablespoon (7 grams protein, 23 calories/kcal)—required to gel 2 cups of liquid. A serving size is about one fourth that amount, and so would contribute little in the way of nutrients.

## Preparation of a Gel

A gel is prepared by hydrating gelatin granules or sheets, dispersing them by heating and stirring the mixture, and then cooling it. Heating protein solutions decreases the protein's solubility and increases the interaction between dissolved proteins (1). Pieces of foods such as meats, vegetables, or fruits can be added during the cooling period when gelation or thickening takes place. The type of food incorporated into the gel determines whether the gel will be a salad, a dessert, or a main dish. A special type of gel that can be made

**Emulsion** A liquid dispersed in another liquid with which it is usually immiscible (incapable of being mixed).

**Biological value** The percentage of protein in food that can be utilized by an animal for growth and maintenance. High-quality, complete proteins are considered to have a high biological value.

with gelatin is an **aspic.** Sometimes aspics also have fruit or vegetable juices added, as in a tomato aspic.

## Phases of Gel Formation

The three phases of gel formation are hydration, dispersion, and gelation.

### Phase 1—Hydration

Gelatin can absorb five to ten times its weight when heated in water. Unflavored gelatin will clump unless it is first hydrated in cool or barely warm water before adding it to other ingredients. To separate the granules, the gelatin is sprinkled over the water, in the proportions of 1 tablespoon of gelatin to ¼ cup of water, and left to stand without stirring for 5 minutes. This initial hydration can be skipped when preparing flavored gelatin because it has been acidulated, or treated, so that it dissolves instantly when boiling water is added.

### Phase 2—Dispersion

After hydration, hot water or a flavorful liquid such as stock or juice is usually added to disperse the protein granules (Figure 15-12). Gelatin disperses at 100°F (38°C), and faster at higher temperatures. When adding hot liquid, it is important to stir, especially along the sides and bottom of the container where the granules may stick, to ensure sufficient dispersion. After the granules have been completely dispersed, cool liquid can be added. Cooling the remaining liquid not only speeds up the gelling process, but also prevents the loss of volatile substances that evaporate into the air when hot liquid is used.

### Phase 3—Gelation

Refrigeration is crucial to the gelling process. As the liquid containing the gelatin protein cools in the refrigerator, it will convert first to a **sol,** which has a liquid consistency. It is either left in the

**Aspic** A clear gel prepared from stock or fruit or vegetable juices.

**Sol** A colloidal dispersion of a solid dispersed in a liquid (see Chapter 3 for colloidal dispersions).

**FIGURE 15-12** Dispersion and gelation. Increasing the temperature by adding hot water causes the protein molecules to denature and unfold. Upon cooling, the protein molecules aggregate, forming a gel.

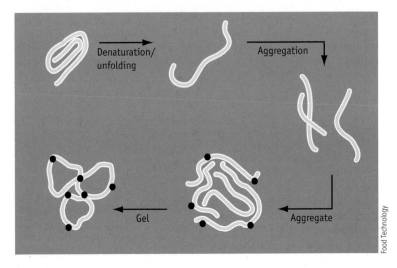

Denaturation/unfolding

Aggregation

Gel

Aggregate

Food Technology

mixing bowl, or poured into individual serving containers or into a larger mold that may be lightly coated with salad oil for easier unmolding. Molds vary in shape and can include any number of designs—a ring, a fish, a pineapple, or others.

Cooling the sol further (1 to 6 hours) will create a more solid-like gel. The result is a solid structure that can be used in gelatin-based molded salads (as well as desserts and plain-flavored gelatin). These can be topped with a sauce, a whipped topping, or mayonnaise, depending on the ingredients.

## Unmolding a Mold

Once the gel has cooled and set, it is ready to be served in its container(s) or removed from the mold(s) and placed on a serving dish. To unmold a gelatin, first slip the tip of a knife around the top edges of the mold to release some of the vacuum. Dip the mold in lukewarm water up to the upper edge of the container for a few seconds, but not long enough for the gel to start melting. Invert the mold over a plate, or, placing a plate over the mold, invert them together. A few shakes should then release the gel from the mold. The process should be repeated if the gel is still sticking to the container. A few drops of water or oil on the plate will allow the gel to slip to the center. Not all gels are firm after being unmolded; in that case,

they should be covered and put back in the refrigerator for further chilling before being served.

## Factors Influencing Gel Formation

Many things can affect gel formation, including the gelatin concentration, temperature, added ingredients, and whether or not the gelatin is whipped.

### Gelatin Concentration

Gelatin concentration should be approximately 1 to 2 percent, which is equal to a little under 1 tablespoon of gelatin per 2 cups of liquid (Chemist's Corner 15-2). Adding more gelatin allows the gel to set faster and be firmer, but it can also result in a tough, rubbery and almost bouncy texture.

>
> ### CHEMIST'S CORNER 15-2
>
> **Measuring Gelation Potential**
>
> The gelation potential of a particular gel is measured by the bloom test. Bloom is the amount of force required to compress a gel (6.67 percent concentration) a distance of 4 mL (6)—the higher the bloom, the smaller the amount of gelatin that is required for the gel to set (13).

Insufficient gelatin will result in a wobbly gel. The amount of gelatin required varies with each dish and its other ingredients. For example, food service organizations use about 2 ounces of unflavored gelatin for every gallon of liquid, but that same amount of water requires 24 ounces of gelatin if it is sweetened (5).

## Temperature

The faster the sol cools, the quicker it sets. The cooling process may be accelerated by adding ice cubes or crushed ice to the dispersed gelatin or by placing the container in a larger bowl filled with ice water. The container may also be placed in the freezer for a short time, but should not be allowed to freeze, because this damages the gelatin layers next to the pan. There are some disadvantages to quick cooling: fast-setting sols result in weak gels and are more likely to lose their structures when brought to serving temperature, and there may be more tough lumps formed during gelation.

## Sugar

Sugar delays gelling and weakens the gel's structure by competing with water for cross-linking sites. To compensate, more gelatin is added to mixtures containing sugar: approximately 2 tablespoons of gelatin for every half cup of sugar.

## Acid

A gel's optimal strength is found between pH 5 and 10. Acids weaken gels by hydrolyzing the gel's network of protein molecules. This is why it is important to avoid adding vinegar or acidic fruits and juices such as lemon or tomato juice to gels. These ingredients can decrease pH below 4, which can prevent gelation or result in a soft gel. This problem is avoided by doubling the concentration of gelatin used in such dishes.

## Enzymes

Proteolytic enzymes found in fresh pineapple (bromelin), papaya (papain), kiwi (actinidan), and figs (ficin) prevent and weaken gel formation by hydrolyzing the protein of the gel network. As a result, these foods are seldom added to gels. Heat denatures the enzymes in these fruits, so cooked or canned versions pose no threat to the gel's structure.

## Salts

Salts strengthen the structure of gels. Because natural mineral salts exist in hard water and milk, incorporating these two liquid sources into a gel will cause it to be firmer than usual. Adding calcium in the form of a salt has long been known to firm gels (15).

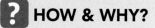

### ? HOW & WHY?

*How does gelatin contribute to gel formation?*

The long, thin protein strands of the gelatin give it its unique ability to gel. The gelatin protein molecules, each surrounded by water molecules, slowly start to form a network by creating cross-links (bonds or bridges) among themselves. The water remains trapped within the protein strands, creating an elastic solid (Figure 15-13).

## Added Solid Ingredients

The addition of solids, such as chopped fruits, vegetables, marshmallows, and nuts, also weakens gel formation. If any of these are acidic or contain proteolytic enzymes, the gel's structure is compromised even further, unless gelatin is added to compensate. In order to avoid lowering the gelatin concentration,

**FIGURE 15-13** Gelation phase. A gel forms when the protein molecules form a network, trapping water into a more solid mass.

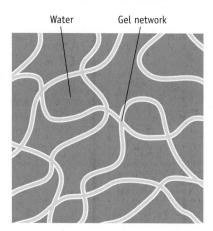

A gel forms when the protein molecules form a network, trapping water into a more solid mass.

either any added solids should be completely free of liquid, or the amount of other liquids added should be reduced to compensate for that contributed by the solids. It is best to use only a small amount of solid ingredients and to chop them up as finely as possible in order to decrease interference with the protein cross-links being formed. Solid ingredients should be added just before the sol has the consistency of beaten egg whites and is about to turn into a gel or else the solids will float to the top.

## Whipping

A whip or sponge is formed by combining a gel with beaten egg whites or whipped cream. Gels increase to two to three times their volume when whipped into a **foam** just before they are set. The final consistency should look like raw egg whites completely stirred into Jell-O. If the stirring occurs before the gel has had enough time to partially set, the foam will float to the top while the gel will set and sink to the bottom. Overgelling results in broken-up chunks of gel. Properly formed gels and foams can be used together to create a two-tone effect as seen in parfaits.

## Storage of Gelatin

Dried gelatin can be stored in a cool, dry place for many months. When stored too long, however, the gelatin becomes less soluble and loses its ability to hydrate; and without hydration, there is no gelation. Once prepared, gels are best served within 24 hours. Beyond this point, the gel continues to form bonds between the gelatin proteins, resulting in an increasingly firm, more compact gel as the spaces between the lattices in the gel network become smaller and smaller. Eventually some of the water will be squeezed out in a process known as **syneresis.** Most gels will lose their shapes when exposed to temperatures above 80°F (26.5°C) (13).

**Foam** A colloidal dispersion of a gas in a liquid.

**Syneresis** The oozing out of the liquid component of a gel.

# PICTORIAL SUMMARY / 15: Soups, Salads, and Gelatins

As the first act, a soup or a salad often sets the tone for the meal to come. But their versatility also allows them to take center stage as a complete meal in themselves.

## SOUPS

Soups are based on a stock, which is the flavorful liquid remaining from cooking bones, meat, and/or vegetables. Stock should be simmered, never boiled, and cooking time varies from half an hour to 12 hours. Hot stock should never be placed in the refrigerator without proper precooling.

The main types of stock are meat, poultry, fish, and vegetable.

**Meat/poultry stocks.** Simple stocks made from water and bones, cracked to release flavor, are called *white stocks*. If the bones are browned prior to simmering, it is a *brown stock*. If meat is added, it is usually raw. A bouquet garni or mirepoix of onions, celery, and carrots is often added to enhance flavor.

**Fish stocks.** Fatty fish are rarely used for stock because of their strong and oily flavor. Fish frames must be thoroughly cleaned and heating should not exceed about half an hour to preserve the delicate flavor. Another stock (chicken or vegetable) or bottled clam juice may be added to fortify the flavor of the base. Fennel and leek are classic vegetable additions, along with the standard onions, carrots, and celery.

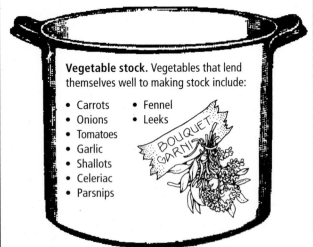

**Vegetable stock.** Vegetables that lend themselves well to making stock include:

- Carrots
- Onions
- Tomatoes
- Garlic
- Shallots
- Celeriac
- Parsnips
- Fennel
- Leeks

There are three different types of soups:
- **Clear and thin:** Removing the particles found in a stock results in a clear soup such as a bouillon or consommé. Thin soups have a consistency between a clear liquid soup and a thickened soup.
- **Thick:** Any stock can be thickened with starch or puréed vegetables to yield thick soups such as chowder, gazpacho, minestrone, egg drop, mulligatawny, and borscht.
- **Cream:** Adding cream and/or milk plus a thickening ingredient to a purée made from meats, poultry, fish, or vegetables results in a creamed soup, or bisque.

## SALADS

A great salad combines the freshest possible ingredients, with a good balance of flavor, color, texture, and shape.

**Types of salad.** The most common types of salad are grouped by their main ingredient: leafy green, vegetable, fruit, protein, pasta/grain, gelatin, or a combination of these. Leafy green salads are the most common, and examples include garden, spinach, mixed green, and Caesar salads.

**Salad dressings.** There are many types of salad dressings, but the most basic are dressings based on oil and vinegar. Emulsified dressings such as mayonnaise do not separate as regular oil and vinegar dressings do. Dressings may also be made with little or no oil, with dairy products such as sour cream or yogurt, or from fruit juice.

**Common garnishes:**

- Fruit wedges
- Hard-cooked eggs
- Tomatoes
- Mint
- Mushrooms
- Pimentos
- Radishes
- Parsley
- Olives
- Pickles
- Watercress

**Salad Toppings**

- Croutons
- Herbs
- Cheese
- Seeds, nuts
- Shredded vegetables
- Edible flowers

## GELATINS

**Gelatin dishes** are made from proteins extracted from the collagen found in animal skins, bones, and hides. Gelatin is prepared by hydrating the granules, dispersing the protein by heating and stirring the mixture, and cooling it until it sets. Factors influencing  these three stages of gel formation include gelatin concentration, temperature, acid, salt, sugar, enzymes, and added solid ingredients. Dried gelatin can be stored in a cool, dry area for months, but prepared gelatin salads should be served within 24 hours.

# CHAPTER REVIEW AND EXAM PREP

## Multiple Choice*

1. All soups are based on _____.
   a. water
   b. bouillon
   c. bouquet garni
   d. stock

2. What is the term used to describe a cream soup made from shellfish?
   a. Foam
   b. Bisque
   c. Sol
   d. Consommé

3. Thick soups are made from stocks that have usually been thickened with some type of _____.
   a. white stock
   b. brown stock
   c. bisque
   d. starch

4. Besides lettuce, what else may be used as a base for a salad?
   a. Pasta
   b. Rice
   c. Cottage cheese
   d. All of the above

5. Adding acidic fruits to a gelatin salad will decrease the pH and result in a salad that is too _____.
   a. soft
   b. rubbery
   c. firm
   d. tough

*See p. AK-1 for answers to multiple choice questions.

6. Gelatin concentration should approximate _____.
   a. 1–2%; 1 tablespoon of gelatin to 2 cups of water
   b. 4–5%; 4 tablespoons of gelatin to 2 cups of water
   c. 8–9%; 8 tablespoons of gelatin to 2 cups of water
   d. 10–11%; 10 tablespoons of gelatin to 2 cups of water

7. Fruits such as pineapple, papaya, kiwi, and figs are usually not added to gels because of their _____ content.
   a. acid
   b. sugar
   c. proteolytic enzyme
   d. salt

## Short Answer/Essay

1. All soups are based on stock. Describe the properties of a stock and list the four major types of stock. What are the basic steps involved in preparing a stock?
2. Explain the difference between white and brown stocks.
3. Define *mirepoix* and *bouquet garni*.
4. Soups differ in their consistency. Explain the differences among clear, thin, and thick soups. Define *bouillon* and *consommé*.
5. Provide several pointers for the preparation of each of the following types of salads: leafy green, vegetable, and fruit.
6. Discuss the recommended techniques for adding dressings to salads.
7. List and briefly discuss the three phases of gel formation.
8. Explain how the following elements influence gel formation: gelatin concentration, temperature, sugar, acid, enzymes, salts, and added solid ingredients.
9. What are the storage recommendations for dried gelatin and prepared gelatin?
10. Explain why water sometimes escapes out of aging gels. What is the name for this process?

# REFERENCES

1. Aguilera JM. Gelation of whey proteins. *Food Technology* 49(10):83–89, 1995.
2. Alford K. One stock makes three flavorful fish stews. *Fine Cooking* 18:57–60, 1997.
3. Baggs C. Vinegar adds SAS to any product. *Food Technology* 58(8):22, 2005.
4. Better Homes and Gardens. *Salads.* Better Homes and Gardens Books, 1992.
5. Bocuse P, and F Metz. *The New Professional Chef.* The Culinary Institute of America. Van Nostrand Reinhold, 2001.
6. Carr JM, K Sufferling, and J Poppe. Hydrocolloids and their use in the confectionary industry. *Food Technology* 49(7):41–44, 1995.
7. *The Chef's Companion: A Concise Dictionary of Culinary Terms.* Van Nostrand Reinhold, 1996.
8. Corriher SO. Make grease-free soups full of satisfying flavor. *Fine Cooking* 26:78–79, 1998.
9. DeGouy LP. *The Soup Book.* Dover Publications, 1974.
10. Deserio T. The four elements of vibrant vegetable salads. *Fine Cooking* 72:44–47, 2005.
11. Dragonwagon C. *Soup and Bread Book.* Workman Publishing Company, 1992.
12. Ensminger AH, et al. *Foods and Nutrition Encyclopedia.* CRC Press, 1994.
13. Gelatin Manufactures Institute of America. www.gelatin-gmia.com. Accessed 7/9/09.
14. Gullo M, L De Vero, and P Giudici. Succession of selected strains of *Acetobacter pasteurianus* and other

acetic acid bacteria in traditional balsamic vinegar. *Applied and Environmental Microbiology* 75:2585–2589, 2009.

15. Hongsprabhas P, and S Barbut. Protein and salt effects on $Ca^{2+}$ induced cold gelatin of whey protein isolate. *Journal of Food Science* 62(2):382–385, 1997.

16. Karmel E. Spinach salad gets a flavor makeover. *Fine Cooking* 70:54–55, 2005.

17. Kolpas N. *Soups.* Time-Life Books, 1993.

18. Peterson J. Simple steps to making versatile chicken stock. *Fine Cooking* 19:18–19, 1997.

19. Rosett TR, et al. Thickening agents' effects on sodium binding and other taste qualities of soup systems. *Journal of Food Science* 61(5):1099–1104, 1996.

20. Sinskey MH. Creating elegant winter salads. *Fine Cooking* 69:52–56, 2006.

21. Stein K. Healthful fast foods not part of healthful revenue. *Journal of the American Dietetic Association* 106(3):344–345, 2006.

22. Stevens M. Add flavor to soups and stews with a bouquet garni. *Fine Cooking* 17:74–76, 1998.

23. Stevens M. A cook's guide to vinegar. *Fine Cooking* 17:74–76, 1996.

24. Stevens M. Toast nuts to give them more flavor and crunch. *Fine Cooking* 30:74–75, 1998.

# WEBSITES

Learn the basic ingredients of the major stocks and soups at this website:
**www.chefdepot.net/sauces.htm**

Healthful salads at Medicinenet.com:
**www.medicinenet.com/script/main/art.asp?articlekey=60654**

Lots of information about gelatin from the Gelatin Manufacturers Institute of America:
**www.gelatin-gmia.com/html/qanda.html**

# 16 Cereal Grains and Pastas

The term "cereal" is derived from the Roman goddess of agriculture, Ceres (30), whose name was derived from the verb *creare*, which means "to create" (58). The word *cereal* refers to grains in general, not just those that may be poured out of a box at breakfast. Cereal grains are seeds from the grass family *Gramineae* (12). The edible portion of the plant is the caryopsis, also known as the grain, kernel, or berry (12). These seeds and their products may indeed be regarded now, as in ancient times, as "the staff of life." According to archaeological evidence, wheat and barley were being cultivated in the Fertile Crescent around 8000 BC, and rice was cultivated in Thailand as early as 4500 BC (22). The importance of grains to the well-being of humans is so great that breads are religious symbols throughout the world. In Christianity, many denominations symbolize the body of Christ with a form of bread, and during Judaism's Passover celebration, unleavened bread is served to symbolize the Jews' hasty exodus from Egypt, when there was no time to wait for bread to rise.

Grains are the world's major food crops, and there are numerous varieties (Figure 16-1). The most common grains are corn, rice, wheat, and barley, which together account for 95% of the world's production of grains. The remaining 5% consists of sorghum, millets, oats, and rye (Figure 16-2) (26). In the U.S. market, sales are highest for the whole grains of wheat, corn, rice, and oats (12).

Although breakfast cereals are a common form in which grains are consumed, they also find their way to the table as flour in baked goods and pastas, alone as vegetables (corn, hominy), as alcohol, and indirectly through the meat of animals that had consumed grains.

# COMPOSITION OF CEREAL GRAINS

## Structure

All grasses have individual kernels or grains, called caryopses, which are similar in structure. Each caryopsis

345

**FIGURE 16-1** Several of the many varieties of grain.

**Millet.** The seed of an annual, gluten-free grass that is widely eaten as a cereal in Africa and Asia. It is also used as a source of starch.

**Corn.** Indigenous to Mexico, corn is one of the most important cereals in the form of grain, meal, and flour. It is used to make corn bread and hominy, and is also an important source of starch and of cooking oil.

**Wheat.** Thought to have first been cultivated in the Nile region, it is the source of the highest-quality bread and baking flours. There are many different varieties: the durum wheat type is best known for making pasta.

**Oats.** Native to Central Europe, oats are used to make oatmeal and flour, and are often added to cakes and cookies.

**Barley.** Indigenous to the East, barley is used for making malt liquor, as a side dish similar to rice, and also in soups.

**Basmati rice.** Grown in the foothills of the Himalayas, this narrow long-grain rice is one of the finest. It should be soaked before cooking, and is the best rice to eat with Indian food.

Digital Works

has a protective outer **husk,** a **bran** covering, a starchy **endosperm,** and a **germ** (Figure 16-3).

## Husk

The husk, also called the chaff, protects the grain from frost, wind, rain, extreme temperatures, insects, and other potentially damaging environmental factors. Husks are not usually consumed but are sometimes processed into fiber supplements.

## Bran

Bran is about 14.5% of the grain by weight and is an excellent source of fiber and minerals. Just beneath the bran, there is a less fibrous coating called the aleurone layer, which contains protein, phosphorus, thiamin and other B vitamins, and some fat. Both the bran and the aleurone layer are removed from grains as they are processed into white flour.

## Endosperm

The endosperm makes up about 83% of the grain (12). The starch in the endosperm makes grains, especially

**Husk** The rough outer covering protecting the grain.

**Bran** The hard outer covering just under the husk that protects the grain's soft endosperm.

**Endosperm** The largest portion of the grain, containing all of the grain's starch.

**Germ** The smallest portion of the grain, and the embryo for a future plant.

**FIGURE 16-2** The most common cereal crops and their contribution to world grain production.

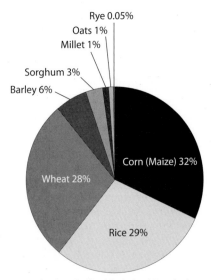

Rye 0.05%
Oats 1%
Millet 1%
Sorghum 3%
Barley 6%
Corn (Maize) 32%
Wheat 28%
Rice 29%

**Source for data:** Food and Agricultural Organization. FAOSTAT. http://faostat.fao.org/site/567/Desktop Default.aspx accessed 08/19/09.

**FIGURE 16-3** Structure of grains.

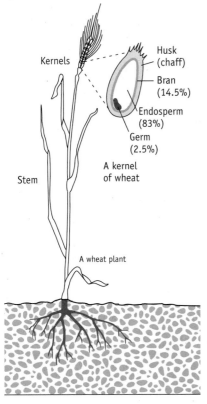

Kernels

Husk (chaff)

Bran (14.5%)

Endosperm (83%)

Germ (2.5%)

A kernel of wheat

Stem

A wheat plant

**Source:** Sizer and Whitney, *Nutrition Concepts and Controversies* (Wadsworth, 1997).

whole grains with their added fiber from the bran, excellent sources of complex carbohydrates (40). The endosperm serves as the basis for all flours, which are made by separating the endosperm from the husk, bran, and germ, and then milling it into fine powder (17). In whole-grain flours, the bran and germ are also milled into the flour—thus the name *whole grain*.

### Germ

The germ (or embryo) is found at the base of the kernel and accounts for only 2.5% of the grain's weight. Rich in fat, and with some incomplete protein, vitamins, and minerals, the germs are collected separately and sold as wheat germ, an excellent source of B vitamins and vitamin E. The fat content found in the germ makes it susceptible to spoilage, which is why wheat germ should be refrigerated. Also, because whole-grain flours contain the germ, they have a much shorter shelf life than pure white flours.

## Food Additives in Grain Products

So many grain products exist that it is not possible to categorize common food additives for this group of foods. However, the label on any package of processed foods containing grains will list the food additives, and many of these are described in Appendix F. The individual chapters on breads, cakes, pies, and pastries also contain information about food additives. A few food additives used in grain products are now addressed.

The stable fat-soluble vitamins (A, D, E, and K) can be added to the dough before baking. The nutrients not stable to heat (the B vitamins and vitamin C) can be applied by spraying them on the cereal after it has been heated. One of the reasons that cereals differ in their fortification levels is that certain cereals can handle the introduction of more nutrients than others without the taste being affected.

Butylated hydroxyanisole (BHA) and butylated hydroxytoluene (BHT) are often added to cereals to act as a preservative. Cereal grains would quickly stale without antioxidants that preserve product freshness. However, there is some concern about BHA and BHT, and so some consumers feel that other, less controversial antioxidants should be used (see Chapter 28 for further information).

# USES OF CEREAL GRAINS

Cereal grains are grown primarily for flour, pasta, and breakfast cereals. They are also used in the manufacture of alcoholic beverages and animal feeds.

## Flour

Flour is the fine powder obtained from crushing the endosperm of the grain. In the case of whole-grain flours, the bran and germ are also milled into the flour (44). Although any grain can be used to make flour, wheat flour is the predominant choice because it provides the protein structure that facilitates the rising of baked goods (see Chapter 17). Flour is used to make breads and an assortment of other baked products, such as biscuits, rolls, crackers, pretzels, chips, cookies, cakes, and pastries. It also plays an important role as a thickener (see Chapter 18).

## Pasta

Pasta is thought to have originated in China. It comes in a variety of shapes, and is sold in both dried and fresh forms.

## Breakfast Cereal

During the mid- to late 1800s, a vegetarian craze hit the United States. Among its advocates was the religious group called the Seventh-Day Adventists, who hired Dr. John Harvey Kellogg to be the director of a health sanitarium. Dr. Kellogg instituted many health-promoting innovations, among them the use of a vegetarian substitute, in the form of dry cereal, for the traditional breakfast of ham or sausage and eggs. Charles William Post, another Seventh-Day Adventist, visited the sanitarium, returned home with the dry breakfast cereal concept, and started a cereal company. At that point, Dr. Kellogg's brother, William Keith Kellogg, seeing the financial success experienced by Post, introduced Kellogg's Corn Flakes to the general public, and the dry cereal industry boom began.

The breakfast cereal market has grown since its early beginnings and is now categorized into ready-to-eat (RTE) cereals (cold cereals) and hot cereals. Over 75% of breakfast cereals are the RTE type made from wheat,

**Whole grains** Defined by the American Association of Cereal Chemists (AACC) as "Foods made from the entire grain seed, usually called the kernel, which consists of the bran, germ, and endosperm."

corn, or oats, and it is probable that neither of their originators could have foreseen the emergence of cereals with bits of chocolate or tiny colored marshmallows in the box. New cereal products are continually being introduced into the market by food companies. Even if they capture only 1% of the market, that translates into over $90 million (31).

**FIGURE 16-4**   Processing grains into cereal.

Extruded                    Flaked                    Granulated

Puffed                    Rolled                    Shredded

Digital Works

## ? How & Why?

*How do food companies create so many different shapes for their cereals?*

The varying shapes of ready-to-eat cereal available to consumers depend on whether they have been extruded (pressed through a die), flaked, granulated, puffed, rolled, or shredded (Figure 16-4). Extrusion cooking is popular for ready-to-eat cereals, after which they are shaped, toasted, flavored, colored, and sometimes enriched or fortified before being packaged and shipped to the consumer (18). Fortified cereals often provide a substantial percentage of a person's RDA for vitamins and minerals.

# Alcoholic Beverages

Various grains are used to make alcoholic beverages through the process of **fermentation**. For example, rice is used to make sake, a Japanese rice wine; and barley, rye, or corn may be used in brewing beer or distilling liquor.

# Animal Feeds

Grains are important in the manufacture of livestock and pet feed. Many dried dog foods list ground yellow corn as the first ingredient on their labels.

**Fermentation**  The conversion of carbohydrates to carbon dioxide and alcohol by yeast or bacteria.

## NUTRIENT CONTENT

A cup of cooked cereal, grain, or pasta contains about 160 calories (kcal), 30 grams of carbohydrate, 6 grams of protein, some vitamins and minerals, and a trace of fat. Whole-grain products provide additional fiber.

*Carbohydrate.*   Grains are an excellent source of complex carbohydrates that come in the form of breads, cereals, rice, pasta, and/or food products primarily made from these ingredients. MyPyramid recommends that at least half of the grain selections come from whole grains and/or their products, because of their health benefits (25).

*Protein.*   Although cereals supply almost half of the dietary protein worldwide (23), their protein is incomplete because grains are low in the essential amino acid lysine. They do, however, have adequate amounts of the amino acid methionine. For this reason they are often paired with legumes, which lack methionine but have sufficient levels of lysine, to achieve protein complementation.

*Fat.*   Cereals are very low in fat and contain no cholesterol. A slice of bread averages only 100 calories (kcal), but spreading 1 tablespoon of butter, margarine, mayonnaise, or peanut butter on the bread adds another 100 calories (kcal) and about 10 grams of fat.

*Vitamins and Minerals.*   Unfortunately, refined grains are also low in many vitamins and minerals, because the milling process that removes the husk, bran, and germ also removes nutrients. Certain vitamins are lost during milling: 77% of the thiamin ($B_1$), 80% of the riboflavin ($B_2$), 81% of the niacin, 72% of vitamin $B_6$, 50% of pantothenic acid, and 67% of folate. About 86% of vitamin E is also lost, but is not replaced by enrichment (49). Grains that are enriched have had certain nutrients added back to levels established by federal standards. The term "enriched baked products" refers to those that have had specific nutrients replaced: thiamin ($B_1$), riboflavin ($B_2$), niacin, folate, and iron.

A further difficulty in obtaining minerals from unleavened bread is that some minerals may not be absorbed as efficiently in the intestinal tract because they bind to phytate (phytic acid) (45, 59). This does not occur with leavened breads, because

## NUTRIENT CONTENT

yeast, commonly used to leaven bread during bread making, breaks down the phytate and prevents it from binding to minerals and hindering their absorption (54).

*Fiber.*   Consuming whole grains may also help meet the American Dietetic Association's recommendation that at least 25 (women) or 38 grams (men) of fiber be consumed each day. Whole-grain products are specifically a good source of soluble fiber, which has been shown to reduce high blood cholesterol and help stabilize high blood glucose (5); and of insoluble fiber, which may help to reduce the risk of colon cancer (5). Whole grains also contain phytochemicals, lignans, and phytoestrogens, compounds that are structurally similar to estrogenic substances and may have a protective effect against certain types of cancer (4). Following the wave of consumer interest in fiber, food companies have introduced high-fiber cereals that incorporate psyllium, a fiber that has been reported to help lower high blood cholesterol. Oat and rice bran have the same effect (35, 43). Fibers that are natural gums have also been added to foods such as breads, cookies, and crackers for their functional properties (48, 14). They make such products feel smoother in the mouth and cause them to hold their shape better in the package. Overall, the health-related benefits of fiber have resulted in many food companies touting their products as being high in fiber (46).

## ✳ CALORIE CONTROL
### Grain and Pasta Calories Quickly Add Up

- *Dietary Guidelines.* Grains are described in 1 ounce-equivalents (see conversion to serving sizes below) and approximately 6 ounces are recommended for a person consuming 2,000 calories (varies based on age and gender).
- 1 ounce of grains is equivalent to:
  - 1 slice of bread
  - ½ C cooked rice, pasta, or cereal
  - 1 C ready-to-eat cereal
  - 1 ounce of dry pasta or rice
  - 3 C popped popcorn
- The key to calorie control it so practice portion control, or else excess calories pile up quickly! It's important to include

whole-grain products because of their additional fiber, which over time may contribute to feeling fuller and more reasonable portion sizes.
- Watch the rice:
  - ⅓ cup — = 80 calories (kcal)
  - 1 cup (1 fist) — = 240 calories (kcal)
- Watch the pasta:
  - ½ cup (½ fist) — = 110 calories (kcal)
  - 1 cup (1 fist) — = 220 calories (kcal)
  - 4 cups (4 fists) — = 880 calories (kcal; excluding bread or sauce)

© 2010 Amy Brown

# TYPES OF CEREAL GRAINS

The three dominant grains grown within the last 5,000 to 8,000 years of human history appear to have been wheat, barley, and rice. The grains of primary importance in the world now in descending order are wheat, rice, corn, barley, millet, sorghum, oats, and rye. Any food made from a cereal grain is a grain product and examples of grains at www.MyPyramid.gov include bread, pasta, oatmeal, breakfast cereals, tortillas, and grits. Grain products are divided into two basic groups—whole grains and refined grains. The difference, as noted by MyPyramid.gov, is that whole grains contain the entire grain kernel—the bran, germ, and endosperm. Whole grains are composed of the intact, ground, cracked, or flaked kernel (27).

Examples at MyPyramid.gov include:

- whole-wheat flour or any other whole-grain flour
- bulgur (cracked wheat)
- oatmeal
- whole cornmeal
- brown rice
- less common whole grains: buckwheat, amaranth, millet, quinoa, sorghum, triticale

Refined grains have been milled, a process that removes the bran and germ. This is done to give grains a finer texture and improve their shelf life, but it also removes dietary fiber, iron, and many B vitamins. Some examples of refined grain products are:

- white flour
- degermed cornmeal
- white bread
- white rice

Most refined grains are *enriched,* meaning that certain B vitamins (thiamin, riboflavin, niacin folic acid), and iron are added back after processing. The removed fiber is not replaced. Some food products are made from mixtures of whole grains and refined grains. Whether a food is a whole- or refined-grain product can be quickly determined by reading the ingredients list.

Commercially, the food industry, previously dominated by refined grain products, is shifting toward whole-grain food products for consumers now interested in adding more fiber to their diets (50). Selected examples of foods higher in fiber include whole-wheat multigrain pizzas, 10-grain pancake mixes, whole-wheat bagels, instant quinoa cereal, and whole spelt tortillas.

# Wheat

More wheat is now produced throughout the world than any other grain. The country producing the most wheat is Russia, followed by the United States. About 75% of all of the harvested wheat is made into flour, and the remaining 25% is used for cereals, pasta products, animal feed, wheat germ, and wheat germ oil.

## Classification of Wheat

Wheat is divided into classes based on growing season, color, species, and/ or texture. There are 14 different species of wheat, and each of them has several different varieties. As a result, there are over 30,000 varieties throughout the world, but only three species—common, club, and durum—account for almost 90% of all the wheat grown in the world (Figure 16-5). Among the lesser-known wheats are spelt, emmer, and einkorn, referred to as ancient wheats because they were some of the earliest cultivated wheats (1).

*Growing Season*  The two major types of wheat defined by their growing season are winter wheat (hard) and spring wheat (soft). Hard and soft wheats also differ in their protein content, which makes wheat flour more suited than any other grain for a variety of different baking purposes. Hard wheats, of which durum is the hardest and highest in protein, are more suitable for bread and pasta production (24).

**FIGURE 16-5**    The three most common types of wheat.

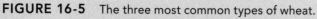

Common wheat    Club wheat    Durum wheat

Lois Frank

The majority of wheat grown in the United States is soft, or common, wheat. Soft wheat produces a flour that is lower in protein and is ideal for cakes, cookies, crackers, and pastries.

*Color*  Wheat may also be distinguished as being red or white, with various shades of yellows and ambers in between. Once it is milled, however, all flours appear white and lose many of their distinguishing color characteristics.

### Forms of Wheat

Flour is just one of the many possible forms of wheat. Other forms on the market include wheat berries, cracked wheat, rolled wheat, bulgur, farina, wheat germ, and wheat bran.

*Wheat Berries*  The simplest form of wheat is wheat berries, or groats, which are whole wheat kernels that have not been processed or milled. These take the longest time to cook compared to other wheat forms.

*Rolled Wheat*  Flattening wheat berries between rollers produces rolled wheat, a product similar to rolled oats, but different from the extruded wheat flakes used in breakfast cereals.

*Cracked Wheat*  This consists of wheat berries that are ground until they crack. It is available in coarse, medium, and fine grinds. Cracking wheat reduces the cooking time from 1 hour to about 15 minutes, resulting in a form of wheat that is more tender and less chewy than whole wheat berries.

*Bulgur*  Bulgur is wheat berries that have been ground even finer than cracked wheat. The berries have been partially steamed, dried, and then cracked to produce a more pronounced flavor. Bulgur is a common ingredient in tabouleh, a dish originally from Lebanon that is a salad of bulgur, vegetables, and herbs.

*Farina*  Farina (which may be familiar as the product known as Cream of Wheat™) is made by granulating the endosperm of the wheat into a fine consistency.

*Wheat Germ*  The germ of wheat is a good source of vitamin E (unless it has been defatted), some B vitamins, and fiber. Wheat germ contains polyunsaturated fat and, unless it has been defatted, it will become rancid if not refrigerated.

*Wheat Bran*  Wheat bran is a source of insoluble fiber—specifically, indigestible cellulose. The various fibers, along with their chemical compositions, are described in Chapter 3.

# Rice

Over half the world's population relies on rice as a staple food (9), so it's fortunate that rice has all eight amino acids in balanced proportions (43). In Asia, where 94% of the world's rice is produced, rice is so important that it is a symbol of life and fertility. This is why rice is sometimes thrown at the bride and groom at a wedding. Per capita yearly consumption in the Far East, where people may eat rice three times a day, averages 200 to 400 pounds, compared to only about 10 pounds in the United States (8). China and India lead the world in rice production.

There is some historical evidence that Alexander the Great may have brought rice to Greece after discovering it while invading China. The Moors carried rice to Spain, whence it was transported to the West Indies and South America by Spanish ships in the early 1600s. It is believed that rice was introduced into the southeastern United States during the mid-1600s when a ship from Madagascar was damaged by storms and its captain and crew took refuge in what is now South Carolina (22).

### Classification of Rice

Rice is classified according to its mode of cultivation, grain length, and texture. Over 90% of all rice is grown with its roots submerged in water, and is known as lowland, wet, or irrigated rice. It is not necessary for water to cover the base of the rice plant, but this method protects it from insects and weeds, which results in a better yield. Highland, hill, or dry rice is grown in areas of plentiful rain, where the hilly terrain prevents flooding. The length in relationship to the width of the grain determines whether rice is considered to be long-grain, medium-grain, or short-grain (Figure 16-6). Most long-grain rices, which are about four times longer than they are wide, cook to a drier, fluffier consistency, which allows the grains to separate. Medium- and short-grain rices have less amylose, which makes them stickier when cooked. The very starchy short-grain rice known as arborio gives Italian risotto its characteristically creamy texture. It is ideal for rice pudding (20). The sticky short-grain rice with an oval shape is also preferred by the Chinese and Japanese because of its easier handling with chopsticks and preparation of sushi (Figure 16-7). Dry, long-grain rice is very difficult to eat with chopsticks. Glutinous, or waxy, rice is especially sticky when cooked. The rice varieties consumed in North America are typically nonglutinous and include long-, medium-, and short-grain rice. The differences in stickiness among the rices is due in part to the

**FIGURE 16-6** Rice is classified by the length/width ratio of the grain.

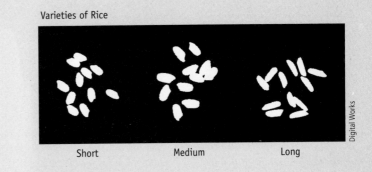

Varieties of Rice

Short      Medium      Long

Digital Works

**FIGURE 16-7** The sticky starch of short-grain rice makes it better suited than other rices for sushi and being picked up by chopsticks.

MIXA/Getty Images

fact that long-grain rices contain more amylose, whereas short-grain rices are higher in amylopectin.

### Forms of Rice

Rice comes in various forms, depending on its type and the way it is processed. These are white, converted, instant, brown, wild, glutinous, specialty rice, bran, and wild rice. Although there are more than 40,000 different varieties of rice worldwide, white rice is the most common (47).

**White Rice** White rice has been milled and polished to remove the husk, bran, and germ. By removing the bran layer and germ, all of the fiber and most of the B vitamins and iron are eliminated, although some of these are replaced in enriched grains. The exception is riboflavin ($B_2$), a B vitamin that is not normally added to rice because it turns the grains slightly yellow. The surface of white rice becomes smooth when it is placed in machines equipped with revolving bands of felt or leather, which rub off the grain's bran and some of the endosperm.

**Converted Rice** Converted or parboiled rice is long-grain rice that has been soaked, steamed under pressure, and dried before milling. It accounts for about 20% of all the rice sold in North America. Converted rice is commonly used in food service establishments because the grains stay firm and separate.

**Instant Rice** Instant rice is rice that has been cooked and then dehydrated, so it takes only a few minutes to prepare. The texture is inferior, however, and the grains have a tendency to split during cooking and become dry.

**Brown Rice** Brown rice has had only the hull removed, leaving the bran and germ intact. The presence of the outer covering of bran results in more fiber, but longer cooking times are required and the result is a tougher texture compared to white rice.

**Glutinous Rice** Also known as sweet rice, glutinous rice is slightly sweeter, stickier, and more translucent than regular white rice when heated. This rice is easily shaped and molded and is thus preferred for preparing rice dumplings, rice cakes, and sushi.

**Specialty Rice** A number of long-grain rice varieties known as specialty rices have nuttier tastes, separate easily, and are more expensive. These specialty rices are not enriched. Examples include basmati, jasmine, Texmati, Wehani, and wild pecan/popcorn rice.

**Rice Bran** Rice bran is just as effective in lowering high blood cholesterol as oat bran, but the bran from rice has a slightly different taste and a shorter shelf life than that from oats.

*Wild Rice*   Wild rice is not really rice, or even a grain, but rather a reed-like water plant (*Zizania aquatica*). It is harvested from areas in the Great Lakes region and in Canada where it grows wild, although increasingly it is deliberately cultivated. This food, historically consumed by Native Americans, has a nutty flavor and is often used to stuff game and poultry (21). It is sometimes mixed with dried fruits, mushrooms, or citrus. It may be mixed with brown or white rice for commercial purposes, and it contains twice the amount of protein and more B vitamins than white rice. Wild rice is cooked until the grains pop, revealing a creamy center. Cooking time varies from 40 minutes to more than 1 hour. Wild rice swells up to four times its size during cooking. Unlike traditional rice, wild rice does not absorb all of the water it is cooked in (21).

# Corn

Corn is native to the Americas, where fossilized corn pollen grains found near Mexico City have been estimated to be over 80,000 years old. Corn plays a very important role in the religious life of many Native American peoples, especially the Hopi in Arizona, among whom life from birth to death revolves around corn. Over 50% of the world's corn is grown in the United States. Outside the United States, corn is referred to as *maize* worldwide.

## Classification of Corn

Corn is classified according to its kernel type—dent, sweet, flint, popcorn, flour, and pod (Figure 16-8)—and by its color. Yellow and white predominate, but there is also red, pink, blue, and black corn as well as corn with bands or stripes.

*Dent Corn*   Dent corn, so named because each corn kernel has a small dent, accounts for over 95% of all corn grown in the United States. Although almost half is sold as livestock feed, the rest is stored as a buffer against the next year's crop, exported, or used by food manufacturers in the production of corn syrup, alcohol, starch, and canned cereal or other processed corn.

**FIGURE 16-8**   Corn is classified by its kernel type.

shalunishka/istockphoto.com

*Sweet Corn*   Most canned corn is derived from sweet corn, which can be either yellow or white. Sweet corn tastes best before the milky fluid in fresh corn kernels has had a chance to harden.

*Flint Corn*   Colonial settlers were taught to grow flint corn by the Native Americans. Grinding the extremely hard corn kernels makes a good quality cornmeal.

*Popcorn*   A special variety of corn with thick-walled kernels is used to make popcorn. During popping, the moisture inside the kernels heats into steam that builds up to 135 pounds of pressure per square inch, finally bursting the kernel open and exposing the puffy starch.

*Flour Corn*   This white or blue corn is common among Native Americans of the Southwest who have been using its flour as a staple for over 1,000 years. Blue corn chips are sold commercially on the same shelf as the standard potato, corn, and tortilla chips.

*Pod Corn*   Pod corn is a noncommercial corn in which each kernel is encapsulated in a separate husk, and the entire ear is wrapped in a large husk.

**? How & Why?**

**Why don't some popcorn kernels pop?**

Kernels that do not pop are believed to contain leaks that allow the steam to escape. Making sure the popcorn is stored in an airtight container, preferably refrigerated, reduces the number of UPKs—unpopped popcorn kernels.

### Forms of Corn

The bulk of corn grown in the United States is used for livestock feed. The remainder is used for human food, to make alcohol, and for its seed (6). Corn is consumed in its natural state on the cob, as kernels, or as processed foods, including hominy, hominy grits, cornmeal, cornstarch, corn syrup, and corn oil.

### Corn on the Cob and Kernel Corn

Corn on the cob is eaten directly off the cob. Corn kernels removed from the cob can be used as a vegetable in a wide variety of dishes. Fresh corn varieties include yellow and white corn, and a hybrid corn that contains both yellow and white kernels.

**Hominy**  The endosperm of this white corn is enlarged because it has been soaked in lye and dried (Figure 16-9). Hominy is enlarged kernels of hulled corn with the bran and germ removed. It is available in canned or frozen form and is particularly popular in the southeastern region of the United States.

**Hominy Grits**  Grits are coarsely ground grain. Hominy grits are obtained by grinding dried hominy into small uniform particles that are boiled as a breakfast dish in the southern United States.

**Cornmeal**  Meal is the coarsely ground version of any grain, so cornmeal is coarsely ground corn. The original corn can be whole or degerminated and either yellow or blue-white in color.

**FIGURE 16-9**  Hominy and hominy grits.

(a) Hominy is corn kernels that enlarge due to processing.

(b) Hominy grits are made by grinding hominy corn kernels.

Bread made from cornmeal has a shorter shelf life than wheat bread because cornmeal has a higher fat content than wheat flour.

### ? How & Why?

**How does cornmeal become a corn chip?**

Many commercial snacks, including corn chips and collets, are made from cornmeal. Fried corn chips are manufactured by the prolonged cooking of corn in water until it becomes soft. This resulting **masa** (50% water) is then extruded, cut into pieces, fried, seasoned, and salted. Collets are often cheese curls or balls of extruded cornmeal treated under high heat and pressure (Figure 16-10). The collets are coated with cheese seasonings. The general composition of a cheese coating consists of partially hydrogenated oil, cheddar cheese powder, acid whey powder, cheese flavor, and salt. The compositional differences in masa, seasonings, and shape result in different types of chips and collets. For example, nacho seasoning may include not only dried cheese powder (Romano, Parmesan, Cheddar, etc.) but also salt, sugar in the form of maltodextrin, tomato powder, monosodium glutamate, onion powder, citric acid, mustard, garlic powder, caramel powder, and colors. Colors used in the cheese coatings may be of vegetable origin (annatto, beet, beta-carotene, paprika, turmeric) or synthetic origin (FD&C-permitted colors) (15).

**FIGURE 16-10**  Collets are used to create cheese curl or ball cornmeal-based snacks.

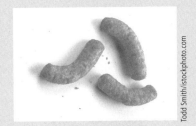

**Cornstarch**  Cornstarch is finely ground corn endosperm. It is used to thicken gravies, puddings, and sauces because of its high starch content (see Chapter 18).

**Corn Syrup**  Corn syrup is a viscous liquid consisting of fructose, glucose, and other sugars (maltose, dextrins) obtained by treating cornstarch with selected enzymes. Corn syrup may be light or dark; the dark is the stronger flavored of the two. High fructose corn syrup is a corn syrup that has been chemically altered to increase its sweetness and is discussed in Chapter 21.

**Corn Oil**  Corn oil is extracted from the germ of the corn kernel.

## Barley

Barley was one of the first grains cultivated by humans (34). It is a hardy plant that grows in a wide range of climates. The heavy "beards" growing on the grain made it a symbol of male potency for the ancient Chinese. Early Greek athletes believed that their sports ability was enhanced when they ate barley mush during training. Roman gladiators were called *hordearli*, meaning "eaters of barley." Spaniards introduced barley to South America in the mid-1500s, and from there it spread northward. Barley is now used primarily as a malt (see below); in cereals and soups; for livestock feed; in the manufacture of beers and whiskey; and in soups, salads, and stews (32).

### Forms of Barley

Used for thousands of years in the production of certain foods and beverages, barley is available in various forms: hulled, pot, pearled, flaked, barley grits, and malt.

**Hulled Barley**  Barley in its natural form is enclosed by a tough hull that can be removed leaving the bran intact.

**Masa**  A word that means "dough" in Mexico. It is made by cooking corn in water, after which it is ground into a pourable slurry. Masa is often used to make tortillas, tamales, and many commercial corn-based snacks.

*Pot Barley*   Pot barley has had the hull and some of the bran removed from the grain in the process called *pearling* (42).

*Pearled Barley*   In pearled barley, additional pearlings remove more of the bran, germ, and part of the endosperm, resulting in tiny grains with a polished pearl-like color. It is used to make barley flour, grits, and flakes, which can then be incorporated into breakfast cereals, breads, soups, cookies, and crackers. It is also cooked as is and added to soups, salads, and stews.

*Flaked Barley*   Flaked barley is rolled or pressed and used primarily as a hot cereal.

*Barley Grits*   Barley grits are barley grains that have been toasted and cracked into particles.

*Malt*   Malt is germinated (sprouted) barley that has been gently dried to stop the growth of germinating roots while leaving intact the enzymes that contribute to flavor and color. The enzymes are important for converting the starches to sugars such as maltose (Chemist's Corner 16-1). The barley grain is well suited for the production of the malt required in some alcoholic beverages, because the aleurone layer contains the enzymes needed to convert some of the endosperm starch to maltose. Also, the aleurone layer of barley is three to four cells thick, as compared to the single-cell layer in other cereal grains. About 4% of all barley is utilized in the production of malt, which can be purchased in dry or liquid form for use as food flavorings, as color additives in caramel, and as optional ingredients in baked products. Common foods to which malt is added include milk shakes, cereals, waffles, pancakes, ice cream, baked foods (cookies, crackers, pretzels, pizza, bagels), imitation coffees, and confections (32).

## Farro and Spelt

Farro is a rare grain with a nutty flavor, and is often confused with spelt due to their similar appearance (12). Spelt, actually a member of the wheat family, resembles wheat berries. It was first cultivated 9,000 years ago. It is a Middle Eastern cereal grass and is mentioned in the Old Testament, where Ezekiel 4:9

> **CHEMIST'S CORNER 16-1**
>
> **Malt Production**
>
> Malt is derived from maltose, the disaccharide naturally found in barley grain. Maltose is a part of the grain's starch, which first must be broken down to the smaller molecules of dextrins and maltose. Enzymes (alpha- and beta-amylase) found in the barley grain convert this starch to dextrin and maltose, resulting in a liquefied mass known as wort. The wort is evaporated under a vacuum to convert it into a syrup (80% solids), and the temperatures at which this occurs determine what type of malt extract is produced. These malt syrups can then be dried into "malts" that have the same color, flavor, and sweetness of their liquid counterparts without the enzyme activity (32).

says to take wheat, barley, beans, lentils, millet, and spelt, put them into a vessel, and make bread.

Farro should be soaked before cooking; it has a firm, chewy texture after cooking. Soaking is not required for spelt, which becomes soft and mushy with cooking (12).

## Millet

Millet grows wild in Africa and is thought to be one of the world's first cultivated grains (12). It is a staple in Asia and Africa, and is thought to have existed in China since 5500 BC.

One of the smallest among the cereal grains, millet contains insufficient amounts of gluten, the protein required for bread to rise properly (see Chapter 17), for use in leavened breads, but it is used in many countries throughout the world to make unleavened bread as well as beer. Millet is less common in the United States, where it is used primarily for birdseed, but it is grown widely in Russia, China, Japan, India, Africa, and southern Europe. The varieties of millet include common (also known as Indian, Proso, Hog, and Broomcorn), finger, foxtail, Japanese, little, and pearl or bulrush millet.

## Sorghum

Sorghum is a cereal grain of major importance in Africa and certain parts of Asia (56), Central America, Pakistan, and India. It is consumed in the form of food (usually porridge), alcoholic beverages, or livestock feed. Cushiony puffed sorghum grains can also be used as a biodegradable packing material.

## Oats

Oats are usually eaten as a hot or cold cereal or in breads, muffins, and cookies. A low protein content limits the use of oats for bread making. Approximately 85% of the oats grown in the United States are used for livestock feed.

### Forms of Oats

Oats are available in a variety of forms: groats, steel-cut, rolled, and bran (Figure 16-11).

*Oat Groats*   Oat groats are whole oats without the husks. They can be prepared like rice or used as an ingredient in other foods.

*Steel-Cut Oats*   Steel-cut oats are groats that have been cut lengthwise and then placed in cans or packages. They have a chewy texture, and are often imported from Scotland and Ireland, where they are more frequently consumed than they are in the United States.

*Rolled Oats*   Rolled oats have been heated and pressed flat with steel rollers. They come to market as regular old-fashioned, quick-cooking, or instant oatmeal. Instant oats have been steamed to pregelatinize the starch, allowing immediate hydration and making them, like all other pregelatinized cereals, the quickest to prepare. Rolled oats are often incorporated into granola, which is a mixture of toasted oats, nuts, dried fruits, and sweeteners. Muesli, the European version of granola, differs from it in that the oats have not been toasted and it usually contains less fat.

*Oat Bran*   This is the bran isolated from oats; it has been reported to have a lowering effect on high blood cholesterol.

**FIGURE 16-11**   Varieties of oats.

Steel-cut oats

Groats

Rolled oats

Quick oats

Instant oat flakes

Oat bran

The Quaker Oats Company

# Rye

Rye is second only to wheat as a grain used in bread making, but in the United States, rye is used primarily to produce rye crackers and whiskey. Compared to wheat, rye usually contains less protein (but still contains gluten) and starch, but more free sugars and dietary fibers (3). As a result, the loaf volumes of rye bread are half of that obtained for wheat products; however, rye breads are richer in flavor and aroma, and have a longer shelf life (Chemist's Corner 16-2) (57). Rye thrives in cold, wet weather and is widely grown and consumed in Scandinavia, Eastern Europe, and Russia. During the Middle Ages in parts of Europe, only the nobility were allowed to consume wheat bread, leaving the peasants to subsist on a type of rye bread. Some historians believe that ergot poisoning (Holy Fire or Saint Anthony's Fire), a foodborne illness caused by a fungus that grows on

grains, especially rye, was a contributing factor to the French Revolution: the peasants were consuming ergot-prone rye, whereas the ruling classes were eating wheat bread free of the taint of ergot

poisoning. Ergot poisoning can cause delusions, hallucinations, and erratic behavior, and it has been suggested that it may have been a factor in the Salem witch trials as well (22).

> **CHEMIST'S CORNER 16-2**

### Rye Proteins and Starches

The lower volume and sweeter taste of rye breads compared to wheat breads are due to the unique characteristics of rye's protein and carbohydrate content. Rye proteins are more water soluble than wheat proteins and exhibit the properties of both a viscous liquid and an elastic solid (55). Starches also play a key role in baked crumb structure, firmness during slicing, and chewiness. This contributes to the compact and

moist nature of rye bread. The starch granules found in rye bread are more easily disrupted and gelatinize at lower temperatures than wheat starch, which allows them to be degraded by alpha-amylase enzymes for a longer period of time during baking. Also contributing to the stickiness, moistness, and sweeter taste of rye bread is the ability of the pentosans (water-soluble, five-carbon carbohydrates such as D-xylose and L-arabinose) in rye to retain water (57).

## Other Grains

### Triticale

Triticale is a relatively new hybrid grain that was developed in 1875 by a Scottish botanist who crossed wheat (*Triticum*) with rye (*Secale*) to create a grain with the bread-making quality of wheat and the climate hardiness of rye. Triticale, one of the few fertile hybrids, has a higher protein content and more of the essential amino acid lysine than wheat. Lysine is the most common limiting amino acid in grains. The flour from triticale produces excellent loaf volume, but "sticky dough" problems prevent it from being widely used (42).

### Kamut

Kamut is an ancient Egyptian grain that is nutritionally superior to wheat in protein content, and has a rich, buttery texture (12).

### New Waves of Grain

Some new grains have entered the market, including amaranth, buckwheat, and quinoa.

Psuedograins or false grains are technically not grains because they come from broad leaf plants instead of grasses (12). Some of the more common examples include amaranth, quinoa, and buckwheat.

| **TABLE 16-1**   Gluten-Free Grains |
| --- |
| • Amaranth |
| • Buckwheat |
| • Corn |
| • Millet |
| • Quinoa |
| • Rice |
| • Sorghum |
| • Teff |

*Amaranth*   Amaranth is a high-protein, gluten-free grain that is becoming more available in the marketplace (Table 16-1) (41). It yields a flour that imparts moistness and a nutty flavor to baked products. The grain is so small that an expression in Africa states, "Some things are not worth an amaranth (seed)" (36). Amaranth was eaten by the Aztecs, who considered it a "superfood" due to its purported energy- and athletic performance-enhancing effects (12). It is one of the earliest known plants used as a food source. Amaranth is high in protein and fiber, and has double the amount of iron found in wheat. Due to its excellent nutritional profile, the National Academy of Sciences recommended that amaranth be promoted as an addition to the American diet (12). Amaranth can also be served like popcorn by popping it in a dry frying pan.

*Buckwheat*   Buckwheat is not related to wheat and is not, strictly speaking, a true grain. It is, in fact, the fruit of a leafy plant that is related to rhubarb. It is, however, generally categorized as a grain because of its use as a flour and cereal. The nut-like flavor of buckwheat flour makes it a popular addition to pancake flour and some breakfast cereals and crepes. The roasted, cracked, and granulated buckwheat is known as kasha, a dish made of buckwheat groats, popular in Eastern European countries.

*Quinoa*   Quinoa is from the seeds of the *Chenopodium* plant, and is technically a fruit, so it is gluten-free (12). A staple food of the Andean culture since 3000 BC and sometimes referred to as "Inca rice," quinoa is a very good source of plant protein (13%), higher in the amino acid lysine than wheat, and popular among vegetarians (13). The Incas of South America called quinoa the "mother of all grains" (12). The tiny, seed-like kernels must be rinsed before cooking to remove the naturally bitter coating of saponin on the surface, leaving a rather mild, pleasant-flavored grain that cooks very quickly. In some cultures quinoa was considered a sacred crop, and was a dietary staple served with potatoes and herbs.

# PREPARATION OF CEREAL GRAINS

Cereals in their natural form are nearly indigestible. The hard outer covering of the seeds of these grasses prevents their immediate consumption and can even break a tooth. This barrier is overcome by heating the grain in water, which softens the outer covering and makes the starchy endosperm digestible. Cooking also gelatinizes the starch, which improves the flavor and texture of cereals as the grains soften and expand. Gelatinization occurs when heated starch molecules absorb water and expand—a phenomenon seen when grains expand to two or three times their volume when cooked. The degree of expansion depends on the type of grain, and is partially caused by the escape of amylose and amylopectin from the starch granule. The desired results in prepared grains are most commonly achieved by moist-heat methods: boiling, simmering, microwaving, and baking in the presence of liquid.

## Moist-Heat Preparation: Boiling/Simmering

The type of grain dictates the amount of water to be added and the intensity or duration of heating. Table 16-2 lists recommended ratios of water to grain, cooking time, and yield for each of the different grain types. The two most important factors in grain preparation are the amount of water used and the exposure to heat.

### Pre-Prep

Whole grains are first rinsed thoroughly prior to cooking to remove any dirt or insect parts.

### Cooking the Grain

The most common preparation method for grains such as rice is the "absorption method." The grain is added to a measured amount of boiling water. Salt is added in the ratio of ¼ teaspoon per cup of uncooked grain to provide flavor. The pan is then covered, and the water is brought back to a boil. The heat is then immediately reduced, and the contents are allowed to simmer (covered) for the remainder of the preparation time. A steam cooker simplifies the process

## TABLE 16-2  Grain Heating Times

| Per ½ C Uncooked | Liquid (Cups) | Cooking Times | Yield (Cups) |
|---|---|---|---|
| **WHEAT** | | | |
| Whole berries | 1½ | 1 hour, 10 minutes | 1¾ |
| Cracked wheat | 1 | 15 minutes | 1 |
| Bulgur | 1 | 15 minutes | 1½ |
| Wheat flakes | 2 | 53 minutes | 1 |
| **RICE** | | | |
| Brown, long-grain | 1 | 25–30 minutes | 1½ |
| Brown, short-grain | 1 | 40 minutes | 1½ |
| Brown, instant | ½ | 5 minutes | 1¾ |
| Brown, quick | ¾ | 10 minutes | 1 |
| White, long-grain | 1 | 20 minutes | 1¾ |
| White, instant | ½ | 5 minutes (let stand 5 minutes) | 1 |
| Converted | 1½ | 31 minutes | 2 |
| Arborio | ¾ | 15 minutes | 1½ |
| Basmati, white | ½ | 15 minutes | 2 |
| Jasmine | 1 | 15–20 minutes | 1¾ |
| Texmati long-grain, brown | 1 (plus 2 tbs) | 40 minutes | 2 |
| Texmati long-grain, white | ¾ (plus 2 tbs) | 15 minutes | 1¾ |
| Wehani | 1 | 40 minutes | 1¼ |
| Wild | 2 | 50 minutes | 2 |
| Wild pecan | 1 | 20 minutes | 1½ |
| **BARLEY** | | | |
| Hulled | 2 | 1 hour, 40 minutes | 1¾ |
| Pearl | 1½ | 55 minutes | 2 |
| Quick | 1 | 10–12 minutes (let stand 5 minutes) | 1½ |
| Flakes | 1½ | 30 minutes | 1 |
| **CORN** | | | |
| Grits | ⅓ | 40 minutes (let stand 2–3 minutes) | ⅓ |
| **OATS** | | | |
| Groats | 1 | 6 minutes (let stand 45 minutes) | 1¼ |
| Steel-cut | 2 | 20 minutes | 1 |
| Old-fashioned rolled | 1 | 5 minutes | 1 |
| Rolled, quick | 1 | 1 minute (let stand 3–5 minutes) | 1 |
| Oat bran | 1 | 6 minutes | 1 |
| **RYE** | | | |
| Rye berries | 1½ | 1 hour, 55 minutes | 1½ |
| Rye flakes | 1½ | 1 hour, 5 minutes | 1¼ |
| **BUCKWHEAT** | | | |
| Groats, unroasted, whole | 1 | 15 minutes | 1¾ |
| Groats, roasted, whole | 1 | 13 minutes | 1½ |
| Kasha | 2½ | 12 minutes | 2 |

because the only step consists of adding the grain, unheated water, and seasoning to the container, which then regulates the temperature as needed to produce perfect rice. Regardless of which method is used, water is absorbed, after which the rice finishes cooking through the trapped steam (33).

*Grain consistency* If a pan is used, adding the grain to hot water results in a fluffier product; adding it to cold water yields a stickier grain. Stirring also affects stickiness, so the grain is initially stirred only as much as is necessary to disperse it and the salt evenly in the water. Stirring can be avoided by pouring the grain in a zigzag fashion over the entire surface of the boiling water for a more even distribution. Most grains should be stirred as little as possible while cooking. Stirring can cause a gummier texture because it makes the starch granules rupture prematurely. The exception to this rule is risotto, an Italian rice dish prepared by constant stirring.

### Determining Doneness

After the minimum amount of covered cooking time has passed, the grains are tested for doneness by tasting. The grain should be tender but have a slightly resistant core. Undercooked grains are difficult to chew and have a starchy, raw flavor. Overcooked grains may form a mushy, formless mash. Too much water contributes to stickiness, sogginess, and loss of nutrients, but insufficient water causes dry, toughened textures, and may even allow the grain to burn.

### Standing Time

Once cooked, the grain should stand for 10 to 15 minutes. This standing time allows steam to further separate the granules, creating a light, airy texture. To further this goal, after removing the saucepan from the heat, a fork can be used to fluff the grain by gently and quickly forming a pyramid with the grain in the pan. The fork handle is inserted into the pile in four places, moving it back and forth each time to create a ¼-inch tunnel for steam to escape from the pyramid. Placing a paper towel flat over the pan and then covering the pan with its lid allows the paper towel to absorb the rising steam. A modified version of this method is often used in the Middle East to ensure a light and fluffy grain.

### Sautéing and Baking

Although they are not as frequently used as the boiling method, there are two other procedures for preparing grains, both intended to add flavor to the finished dish. These are sautéing in oil and aromatics, referred to as the *pilaf method*, and baking. Grains such as rice and bulgur can be first sautéed in fat (1 tablespoon per cup of uncooked grain), after which boiling chicken broth or other stock, instead of water, is poured over the grain. It is then covered and simmered until done (33). Cereal grains can also be prepared in a casserole dish and baked in the oven if sufficient liquid is provided. The grain, usually rice, may already be cooked prior to being included with the rest of the ingredients that will be baked, but if not, boiling water is added to the dish. The casserole is usually covered to take advantage of the steam, which further aids in cooking its contents. Baking times vary according to the grain used, but average 20 to 30 minutes.

### Adding Seasonings

Regardless of the preparation method, grains are usually bland unless seasoning ingredients are added at the beginning of the cooking process. Chicken- or beef-flavored soup stock or bouillon can be used instead of water, giving the grain a slight chicken or beef flavor.

### Factors Influencing Grain Cooking

The form of grain, the presence of the bran or hull, the pH of the water, and the desired tenderness are factors that influence the amount of water to be used, the heat intensity, and the cooking time. Any reduction in particle size through cracking, rolling, cutting, or flaking decreases the heating time. For example, cracked wheat cooks in 15 minutes, whereas whole wheat berries may take over an hour. Removing the bran or hull also drastically cuts heating time. Brown rice takes about twice as long to cook as white rice, while barley with its hull intact takes

about 35 minutes longer than pearl barley. A more alkaline environment, such as the one created by using hard tap water, causes grains to cook at a faster rate and possibly to overcook because of increased breakdown of the cellulose.

### Hot Breakfast Cereals

Hot breakfast cereals account for 7% of all breakfast cereals. The most commonly consumed hot breakfast cereals are oatmeal, farina, hominy or corn grits, cream of rice, and bulgur. They are available in three forms: regular, quick-cooking, and instant. Most people like cereals prepared to a thick yet moist consistency, which can be varied by using more or less water. To prevent clumping, particularly when preparing the more finely granulated cereals such as farina and grits, it is best to sprinkle the cereal slowly over the boiling water while stirring so that each grain has a chance to be surrounded with hot liquid. Another way to prevent clumping is by mixing the cereal with cold water to make a slurry before adding it to the hot water. Once all the cereal is added, maintain a slow boil and stir occasionally until the grains are translucent. There are virtually no clumping problems with instant breakfast cereals, which simply require the addition of boiling water and a bit of stirring. One cup of uncooked cereal yields about four to six servings when cooked.

## Microwaving

Heating times are not significantly reduced with a microwave oven. Follow the manufacturer's directions for the amounts of required water, grain, and heating times. Add the grain to the boiling water, cover with a lid, heat according to the recommended time, stir at the halfway point, and let stand for 5 minutes before serving. Microwave preparation reduces cleanup because the same dish can be used for cooking and serving. The microwave oven is ideal for reheating because it yields a near "fresh cooked" flavor and texture, while reducing the tendency of grains to dry or overcook. It is also ideal for preparing instant hot cereals because single servings can be prepared directly in a microwave-safe serving bowl in a one-step process.

# STORAGE OF CEREAL GRAINS

## Dry

Dry grains, freed of their bran and germ, are best kept in airtight wrappings or containers in a cool, dry area free of rodents, insects, and other pests. Moisture is the biggest contributor to the deterioration of grains. The relative humidity in the environment determines the grain's moisture content, as grains take up moisture until equilibrium is reached with the atmosphere's water vapor. In practice, this means that a relative humidity of 70% or less is considered safe; microbial growth will occur above 75% relative humidity, resulting in extensive grain deterioration (23). Thus, once opened, grain packages should be tightly resealed or the grain placed in another airtight container that will protect it from air or animal invasion. Most grains, when stored properly, will keep for 6 to 12 months.

## Refrigerated

Whole grains should be refrigerated in airtight containers to retard rancidity and prevent mold growth, which can be caused by moisture. Usually, only whole or cooked grains are refrigerated.

Cooked grains will keep up to a week if they are tightly covered. The best way to reheat grains is in a microwave oven or in a covered saucepan on top of the range with about 2 tablespoons of water added for each cup of grain.

## Frozen

Cooked whole grains can be frozen for future use if they are tightly wrapped or placed in airtight containers. Uncooked grains should not be frozen, because freezing alters the protein structure in such a way that any baked products made from the grain will not rise as high.

# PASTAS

The Chinese developed noodles as early as 5000 BC, and Marco Polo, one of the first Europeans to reach the Orient, is credited with bringing them back from his travels and introducing pasta to Italy and the rest of Europe in the late 1200s. There is some evidence, however, of pasta having been made in Italy as early as 400 BC (22). Much later, in the late 1700s, pasta was introduced to North America by Thomas Jefferson after he visited Naples while serving as the American ambassador to France. Pasta now serves as a popular source of complex carbohydrates in many parts of the world.

Pasta, meaning "paste" or "dough" in Italian, is made predominantly from flour starch and water and describes thousands of extruded foods including macaroni, spaghetti, lasagna, vermicelli, and noodles (28).

These and other pastas are usually made from semolina, a flour derived from durum wheat, although other flours are sometimes used; flavorings and colorings can also be added (11). The highest-quality pastas are made from the higher-protein wheats. Durum flour's higher protein content makes it best suited to withstand the pressures of mechanical kneading and manipulation during commercial pasta production, as well as the heat during preparation (40). It is the protein in durum wheat flour that gives pasta its elasticity and helps it maintain its shape during cooking. Durum wheat is also higher in carotenoid pigments, which contribute to pasta's rich, golden color (19). The yellow color may also be derived from egg yolks, which are a common ingredient in pasta. Most pasta manufactured in North America is also enriched with several B vitamins and iron.

## Types of Pasta

The shapes by which pasta is identified are formed by placing the freshly made pasta dough in a cylinder and forcing it through holes in small discs (dies). Pasta dough is best extruded at 115°F (46°C), because temperatures higher than 140°F (60°C) will denature the protein and reduce the pasta's quality. Once pressed through the appropriate die, pasta is cut and then dried until the moisture level drops from 31% to 10 or 12% (51).

The type of disc used determines what kind of pasta is produced. Depending on the selected disc, pasta can be called spaghetti ("little strings"); linguine ("little tongues"); vermicelli ("little worms"); rigatoni ("grooved"); or fettuccine, capellini, cannelloni, tortellini, lasagna, ravioli, macaroni, noodles, wonton wrappers, and others (Figure 16-12). There are over 600 pasta shapes now in existence, but only about 150 are available in North America (38). "Long goods" such as spaghetti and linguini are most commonly consumed (41%), followed by "short goods" such as macaroni (31%) and noodles (15%), and specialty items such as lasagna and manicotti (13%) (29). An Italian website noted at the end of the chapter provides photos of many more different types of pasta.

## Pasta Nomenclature

Pasta, or "alimentary (nourishing) paste," is made by combining water with semolina flour and/or farina. Macaroni is a generic term for all types of dried pasta. Prior to pasta being dried in various shapes, optional ingredients may be added, such as vegetable purées, as in spinach pasta, and a variety of seasonings. Different types of pasta vary not only in shape, but also in their ingredients, which influences the following pasta nomenclature.

### Noodles

If eggs are added (at least 5.5% egg by weight), the pasta product is referred to as *noodles,* although eggless noodles are available on the market.

### Asian Noodles

Asian noodles are often made from flours other than the standard semolina flour or farina, and as a result, are often translucent or clear in appearance (16). This and the fact that they rarely contain eggs is why they are sometimes called imitation noodles in the West. Asian noodles may be made from rice, mung bean, taro, yam, corn, buckwheat, or potato flours. Examples of Asian noodles are rice, ramen, soba, and bean thread noodles (38). Ramen are instant Japanese noodles that have been previously dehydrated by frying, which makes them extremely porous and much more likely to absorb water than regular noodles. Adding water to ramen instantly rehydrates them, making them popular for use in soups and luncheon noodle meals.

### Whole Wheat

Pasta made from whole-wheat flour is slightly higher in nutrients and fiber than standard pasta, but it has a tougher texture and a stronger taste, and tends to disintegrate if cooked too long.

### Flavored

Vegetable purées made from spinach, tomatoes, or beets can be added to pasta to alter its color and flavor. A newer version of flavored pasta incorporates herbs and spices such as basil, garlic, parsley, and red pepper into the pasta dough.

### Fanciful

Unusual shapes such as dinosaurs and turtles have been developed to appeal to youthful consumers (29).

### High-Protein Pasta

Adding soy flour, wheat germ, or dairy products yields high-protein pasta products that contain 20 to 100% more protein than standard pasta (38).

### Fresh

Fresh pasta, found in the refrigerated section of the supermarket, has a higher moisture content, which gives it a softer consistency and shorter cooking time.

### Couscous

Couscous looks like a grain, but it is actually "Moroccan pasta" made from semolina that has been cooked, dried, and pulverized into small, rough particles the size of rice grains (7). Traditionally, lamb stew and many Middle Eastern dishes were served over a bed of couscous. Middle Eastern or Israeli couscous is larger than regular couscous and resembles small balls of pasta. It is usually toasted before preparation and retains a firm bite after cooking. In another part of the world, people in many African countries consume their couscous with milk for breakfast or as a starch with lunch or dinner (2).

**FIGURE 16-12** There is a rich array of pasta shapes and sizes.

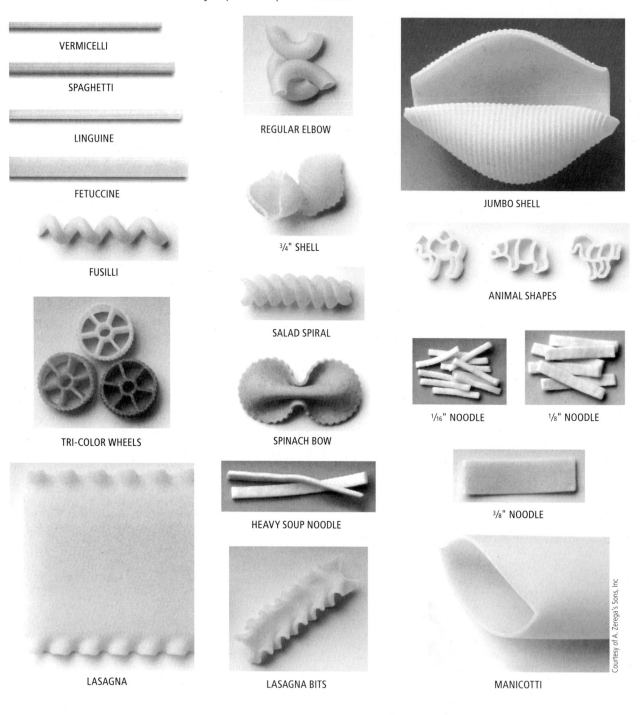

VERMICELLI

SPAGHETTI

LINGUINE

FETUCCINE

FUSILLI

TRI-COLOR WHEELS

REGULAR ELBOW

¾" SHELL

SALAD SPIRAL

SPINACH BOW

JUMBO SHELL

ANIMAL SHAPES

¹⁄₁₆" NOODLE

⅛" NOODLE

⅜" NOODLE

LASAGNA

HEAVY SOUP NOODLE

LASAGNA BITS

MANICOTTI

Courtesy of A. Zerega's Sons, Inc

# PREPARATION OF PASTA

## Moist-Heat Preparation

Pasta is easy to prepare by boiling or simmering (Figure 16-13). It is dropped into lightly salted boiling water, stirred to keep the strands or pieces from sticking together, and heated until it reaches the **al dente** (ahl-den-tay) stage. Properly prepared pasta is not excessively sticky, and once it is done, it is drained and ready to serve.

During heating, the majority of pastas expand to two or three times their original size. Flat noodles are an exception in that they increase in size only by half. The pasta-to-cooking-water ratio should ensure that the pasta is heated

**Al dente** Meaning "to the tooth" in Italian, it refers to pasta that is tender, yet firm enough to offer some resistance to the teeth.

**FIGURE 16-13** Preparing pasta.

1. In a large, heavy pot, bring 4 quarts of water for each pound of pasta to a full rolling boil. Add salt to taste.

2. For long pasta, add to boiling water in batches, pushing batch down as pasta softens; this avoids breaking. Short pasta can be added all at once.

3. Stir, carefully separating pasta pieces. Cover, return to boil, then immediately remove lid. Stir gently while cooking to prevent pieces from sticking together.

4. Cook pasta until tender but slightly firm (al dente). Taste to test doneness. Drain pasta, but do not rinse because starchy surface holds sauce better.

*Digital Works*

> ## CHEMIST'S CORNER 16-3
>
> ### Cloudy Pasta Water
>
> Cooking water becomes cloudy as pasta cooks because some of the pasta's starch leaks out into the water. The benefit of using this starch-enriched water to prepare the sauce is that it then "sticks" better to the pasta. The "gummy" nature of starch is responsible for the improved adhesion of the sauce (52).

in plenty of water: 4 quarts of water per pound of pasta is recommended. The water should remain at boiling temperature through the entire cooking period, and both undercooking and overcooking should be avoided. Undercooked pasta is identified by a white core of stiff, ungelatinized starch in the center, whereas overcooked pasta is mushy and limp in consistency and very bland (Chemist's Corner 16-3).

Water with an alkaline pH tends to cause stickiness (37). Oil may be added to the water to prevent the pasta pieces from sticking together, both during boiling and when the product is allowed to cool. Pasta may be rinsed with water after cooking to prevent sticking, but this may remove any B vitamins added during enrichment that have not already been lost in the cooking water.

Stickiness is influenced by the amount of water on the pasta's surface, the degree of force used to press the pasta through the disc, and most importantly, by the age of the pasta. Older pasta is not recommended because water is lost from the surface of pasta as it ages, causing a simultaneous loss in lubrication and increased stickiness (51).

For most regular pasta, cooking time is approximately 10 minutes, but fresh pasta and Asian noodles can take as little as 3 minutes. If cooked pasta needs to be held before serving, it is placed in a colander over a pot of hot, steaming water, or it can be cooled and later briefly dropped in simmering water to be reheated, then drained and served. Pasta should be slightly undercooked if it is to be stir-fried; baked; or added to soup, stews, or casseroles. Figure 16-14 shows the amount of pasta to use for a 1-cup serving.

**FIGURE 16-14** How much dried pasta to cook for a 1-cup serving.

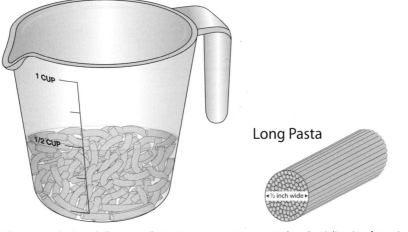

**Short pasta** (penne, shells, rigatoni): Use 2 oz. uncooked, or just over ½ cup.

**Long Pasta**

**Long pasta** (spaghetti, linguine, fettuccine): Use 2 oz. uncooked, or a bundh ½ inch in diameter.

## Microwaving

Like grains, pasta can be prepared in a microwave oven, but again, there is no significant savings in time compared to the conventional method. Follow the manufacturer's instructions for the recommended amounts of water and pasta and the required heating times. The pasta is stirred into very hot water, covered with a casserole lid or plastic wrap, and heated for the recommended amount of time. It should be stirred halfway through the cooking period, drained immediately when done, and served. Many new microwavable single-serve pasta entrees are being marketed.

# STORAGE OF PASTA

The storage of pasta depends on whether it is dried, fresh, or cooked. Dried pasta should be tightly wrapped and stored in a cool, dry place. Fresh pasta should be kept in the refrigerator until the "use by" date. It will be at its best for about a week and will keep in the freezer for about a month. Fresh pasta stored in modified-atmosphere packages may last up to 120 days, but there is an increased risk of microbial contamination because of the long storage time (24). The additional ingredients often found in fresh Asian noodles reduce their keeping time in the refrigerator to 2 days. Cooked pasta will keep for 2 to 3 days in the refrigerator and is easily reheated in the microwave oven under vented plastic wrap or by placing it in a pot of boiling water for half a minute.

---

## NUTRIENT CONTENT

Pasta is high in complex carbohydrates and naturally low in protein, fat, and cholesterol unless it is processed with eggs or fat. Naturally low-fat pasta may be beneficial to health, particularly for people who have diabetes or are interested in lowering their dietary fat and blood cholesterol levels (10, 53).Pasta, which used to have the reputation of being "fattening," only recently became popular when it was recognized as an excellent low-fat, low-calorie source of complex carbohydrates. Nevertheless, noodles, which are pastas made with eggs, have more fat and cholesterol than non-egg-containing pastas, and sauces made from butter, oil, cream, or meat will also add calories and fat. When pastas are stuffed with cheese or meat, as in tortellini and ravioli, the fat and calorie content soars even higher.

Traditional ramen, the Japanese noodle sold as an instant noodle or soup dish, is made from wheat noodles that have been deep-fried in highly saturated lard or palm oil and then dried, but lower-fat varieties are also available. Canned chow mein noodles have been fried in oil or fat and are high in calories—each cup contains about 240 calories (kcal) and 14 grams of fat (37).

# PICTORIAL SUMMARY / 16: Cereal Grains and Pastas

Cereal grains are seeds from the grass family. Serving as the world's major food crop, they have long been regarded as "the staff of life."

## COMPOSITION OF CEREAL GRAINS

**A kernel of wheat**

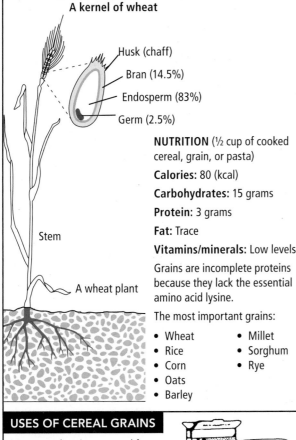

- Husk (chaff)
- Bran (14.5%)
- Endosperm (83%)
- Germ (2.5%)

Stem

A wheat plant

**NUTRITION** (½ cup of cooked cereal, grain, or pasta)

**Calories:** 80 (kcal)

**Carbohydrates:** 15 grams

**Protein:** 3 grams

**Fat:** Trace

**Vitamins/minerals:** Low levels

Grains are incomplete proteins because they lack the essential amino acid lysine.

The most important grains:

- Wheat
- Rice
- Corn
- Oats
- Barley
- Millet
- Sorghum
- Rye

## USES OF CEREAL GRAINS

Most cereal grains are used for
- Flour
- Pasta
- Breakfast cereals
- Alcoholic beverages
- Animal feeds

## TYPES OF CEREAL GRAINS

**Wheat:** Available as wheat berries, cracked and rolled wheat, bulgur, farina, wheat germ, and wheat bran.

**Rice:** Available as white, converted, instant, brown, wild, glutinous, specialty, or rice bran.

**Corn:** Can be classified according to its kernel type: dent, sweet, flint, popcorn, flour, and pod. Can be consumed as corn on the cob or kernel corn, hominy, corn or hominy grits, cornmeal, cornstarch, corn syrup, and corn oil.

**Barley:** Primarily used for the manufacture of beer and whiskey, and for livestock feed.

**Sorghum:** Major cereal grain in Africa and parts of Asia.

**Oats:** Available as groats, steel-cut, rolled, and bran.

**Rye:** Second only to wheat for bread making.

## PREPARATION OF CEREAL GRAINS

Heating dried cereals in water softens and gelatinizes their starch, creates an edible texture, and improves flavor. Gelatinization occurs when the starch molecules absorb water and expand. Although rice is the most commonly prepared grain in the world, all grains can be cooked in a similar manner, with just slight variations in the amount of water and length of heating. Cooking time decreases with any reduction in particle size due to cracking, rolling, cutting, or flaking.

## STORAGE OF CEREAL GRAINS

Grains are best kept in airtight containers in a cool, dry area free from rodents, insects, and other pests. Whole grains should be refrigerated to retard the rancidity that can occur due to their fat content. It is important to keep grains from becoming damp and subject to mold growth. Cooked whole grains can be refrigerated or frozen for future use.

## PASTAS

There are over 600 pasta shapes now in existence, although only about 150 are available in North America.

The higher protein content of durum wheat flour, also called semolina, makes it best suited for pasta production.

Nutritionally, pasta is an excellent source of complex carbohydrates. Although it is naturally low in fat, the fat and calorie count go up when it is served with sauces made from butter, oil, cream, and/or meat.

**Various pastas available:**

**Noodles:** Eggs are added (at least 5.5% egg by weight).

**Asian noodles:** These rarely contain egg and are often made from flours other than semolina.

**Whole wheat:** Made from whole-wheat flour.

**Flavored:** Vegetable purées are added for flavor and color.

**High-protein:** Added soy flour, wheat germ, or dairy products contain 20 to 100% more protein.

**Fresh:** Has a higher moisture content and faster cooking time.

**Couscous:** Made from semolina that has been cooked, dried, and pulverized. Popular in Africa and the Middle East.

## PREPARATION OF PASTA

Pasta is best prepared by boiling it in plenty of water and avoiding both under- and overcooking.

## STORAGE OF PASTA

Dried pasta should be stored in an airtight container and placed in a cool dry place.

# CHAPTER REVIEW AND EXAM PREP

## Multiple Choice*

1. Most of the vitamin E in grains is found in what portion of the kernel?
   a. Husk
   b. Germ
   c. Endosperm
   d. Bran

2. The endosperm of grains is primarily made up of:
   _____
   a. starch.
   b. protein.
   c. lipid.
   d. vitamins.

3. Which type of wheat is typically used to make the Lebanese dish called tabouleh?
   a. Rolled wheat
   b. Whole wheat berries
   c. Bulgur wheat
   d. Farina

4. Refined grains have four vitamins and one mineral added back to them after processing. What is the name for this process?
   a. Rolling
   b. Fortification
   c. Enrichment
   d. Regulating

5. After wheat, the second most common grain used in bread making is: _____
   a. corn (cornmeal)
   b. oats
   c. rye
   d. sorghum

6. Malt is made from germinated (sprouted) _____, so it serves as a source of gluten.
   a. wheat
   b. oat
   c. rye
   d. barley

7. What is the best wheat for producing cake flour?
   a. Soft
   b. Hard
   c. Red
   d. White

## Short Answer/Essay

1. Diagram the structure of a grain and briefly describe the four parts.
2. Describe the six different shapes that grains can be processed into during production of cereals.
3. Briefly define the following: *cracked wheat, bulgur, farina, converted rice, glutinous rice, dent corn, hominy, cornstarch, cornmeal, pearled barley, malt, millet, sorghum, rolled oats,* and *amaranth.*
4. How do the starch content and size of short-, medium-, and long-grain rice affect their consistency when they are prepared?
5. Briefly describe the basic steps for preparing grains through boiling/simmering.
6. What is the most significant contributor to the deterioration of grain during storage? Describe the recommended storage conditions for grains, including the maximum recommended storage time.
7. Pasta, meaning "paste" in Italian, is made from what two primary ingredients? Describe the following types of pasta, including how their ingredients or shapes differ from one another: macaroni, noodles, Asian noodles, flavored, fanciful, fresh, and couscous.
8. Describe how ramen is prepared. How would this influence nutrient content?
9. Briefly describe the moist-heat preparation process for pasta.
10. Define *al dente.*

*See p. AK-1 for answers to multiple choice questions.

# REFERENCES

1. Abdel-aal ESM, FW Sosulski, and P Huel. Origins, characteristics, and potentials of ancient wheats. *Cereal Foods World* 43(9):708–715, 1998.

2. Aboubacar A, and BR Hamaker. Physiochemical properties of flours that relate to sorghum couscous quality. *Cereal Chemistry* 76(2): 308–313, 1999.

3. Aman P, M Nilsson, and R Andersson. Positive health effects of rye. *Cereal Foods World* 42(8): 684–688, 1997.

4. Amin AR, O Kucuk, FR Khuri, and DM Shin. Perspectives for cancer prevention with natural compounds. *Journal of Clinical Oncology* 27(16):2712–2725, 2009.

5. Anderson JW, et al. Health benefits of dietary fiber. *Nutrition Review* 67(4):188–205, 2009.

6. Anonymous. *2009 World of Corn Report.* National Corn Growers Association www.ncga.com/files/pdf/2009WOC.pdf. Accessed 6/09.

7. Anonymous. From our test kitchen—couscous. *Fine Cooking* 63:74, 2004.

8. Anonymous. *Per capita rice consumption of selected countries.* University of Arkansas. www.uark.edu/ua/ricersch/pdfs/per_capita_rice_consumption_of_selected_countries.pdf. Accessed 6/6/09.

9. Barton FE, et al. Optimal geometries for the development of rice quality spectroscopic chemometric models. *Cereal Chemistry* 75(3):315–319, 1998.

10. Best D. Technology fights the fat factor. *Prepared Foods* 160(2):48–49, 1991.

11. Boyacioglu MH, and BL D'Appolonia. Durum wheat and bread products. *Cereal Foods World* 39(3):168–174, 1994.

12. Brannon CA. Ancient and alternative grains. *Today's Dietician* 9(5):1016, 2007.

13. Brinegar C. The seed storage proteins of quinoa. In S Damodaram, ed. *Food Proteins and Lipids,* 109–115. Plenum, 1997.

14. Butt MS, N Shahzadi, MK Sharif, and M Nasir. Critical review: Guar gum: A miracle therapy for hypercholesterolemia, hyperglycemia and obesity. *Food Science Nutrition* 47(4):389–396, 2007.

15. Chandan R. *Dairy-Based Ingredients.* Eagan Press, 1997.

16. Chansri R, C Puttanlek, V Rungsadthong, and D Uttapap. Characteristics of clear noodles prepared from edible canna starches. *Journal of Food Science* 70(5):S337–S342, 2005.

17. Clark JP. Ethanol growth inspires advances in cereal milling. *Food Technology* 60(9):7374, 2006.

18. Clark JP. Processing breakfast cereals to deliver nutrition. *Food Technology* 60(3):89–90, 2006.

19. Cole ME, DE Johnson, and MB Stone. Color of pregelatinized pasta as influenced by wheat type and selected additives. *Journal of Food Science* 56(2):488–493, 1991.

20. Dodge A. Making the creamiest rice pudding. *Fine Cooking* 43:73, 2001.

21. Dooley B, and L Watson. Wild rice—Unlock the flavor and texture with just the right amount of cooking. *Fine Cooking* 60:7173, 2003.

22. Ensminger AH, et al. *Foods and Nutrition Encyclopedia.* CRC Press, 1994.

23. Eskin NAM. *Biochemistry of Foods.* Academic Press, 1990.

24. Feillet P, et al. Past and future trends of academic research on pasta and durum wheat. *Cereal Foods World* 4(4):205–212, 1996.

25. Flight I, and P Clifton. Cereal grains and legumes in the prevention of coronary heart disease and stroke: A review of the literature. *European Journal of Clinical Nutrition* 60(10):1145–1159, 2006.

26. Food and Agricultural Organization. *FAOSTAT.* http://faostat.fao.org/site/567/DesktopDefault.aspx. Accessed 08/19/09.

27. Food and Drug Administration. *The scoop on whole grains.* www.fda.gov/ForConsumers/ConsumerUpdates/ucm151902.htm. Accessed 7/23/2009.

28. Gallagher E. Formulation and nutritional aspects of gluten-free cereal products and infant foods. In EK Arendt and F Dal Bello. *Gluten-Free Cereal Products and Beverages.* Academic Press, 2008.

29. Giese J. Pasta: New twists on an old product. *Food Technology* 46(2):118–126, 1992.

30. Gooding MJ. The wheat crop. In K Khan, and PR Shewry, *Wheat: Chemistry and Technology,* 4th ed. American Association of Cereal Chemists, 2009.

31. Grieder J. State of the industry report: Breakfast cereals in the U.S. *Cereal Foods World* 41(6):484–487, 1996.

32. Hickenbottom JW. Processing types, and uses of barley malt extracts and syrups. *Cereal Foods World* 41(10):788–790, 1996.

33. Iyer R. Basmati rice—The pilaf method. *Fine Cooking* 71:46–49, 2005.

34. Jadhav SJ, et al. Barley: Chemistry and value-added processing. *Critical Reviews in Food Science and Nutrition* 38(2):123–171, 1998.

35. Lee KW, et al. The effects of goami no. 2 rice, a natural fiber-rich rice, on body weight and lipid metabolism. *Obesity* 14(3):423–430, 2006.

36. Lehman JW. Case history of grain amaranth as an alternative crop. *Cereal Foods World* 41(5):399–411, 1996.

37. Malcolmson LJ, and RR Matsuo. Effects of cooking water composition on stickiness and cooking loss of spaghetti. *Cereal Chemistry* 70(3):272–275, 1993.

38. Margen S, et al., eds. *The Wellness Encyclopedia of Food and Nutrition.* Rebus, 1999.

39. Marquart L, and EA Cohen. Increasing whole grain consumption. *Food Technology* 59(12):24–31, 2006.

40. Mariani BM, MG D'egidio, and P Novaro. Durum wheat quality evaluation: Influence of genotype and environment. *Cereal Chemistry* 72(2):194–197, 1995.

41. Masibay KY. Taking control of gluten. *Fine Cooking* 92:8081, 2008.

42. Nelson KA, and SJ Eilertson. Cereal grain specialty products: Brief overview. *Cereal Foods World* 41(5):383–385, 1996.

43. Ohr LM. Go with the grain. *Food Technology* 60(9):6366, 2006.

44. Posner ES. Wheat flour milling. In K Khan and PR Shewry. *Wheat: Chemistry and Technology,* 4th ed., AACC International, 2009

45. Prasad AS, et al. Biochemical studies on dwarfism, hypogonadism and anemia. *Archives of Internal Medicine* 111:407–428, 1963.

46. Pszczola DE. Fiber gets a new image. *Food Technology* 60(2):43–54, 2006.

47. Pszczola DE. Rice: Not just for throwing. *Food Technology* 55(2):53–57, 2001.

48. Rideout TC, SV Harding, PJ Jones, and MZ Fan. Guar gum and similar soluble fibers in the regulation of cholesterol metabolism: Current understandings and future research priorities. *Vascular Health and Risk Management* 4(5):1023–1033, 2008.

49. Schroeder HA. Losses of vitamins and trace minerals resulting from processing and preserving foods. *American Journal of Clinical Nutrition* 24:562–573, 1971.

50. Sloan AE. Wholly grain! *Food Technology* 59(6):16, 2005.

51. Smewing J. Analyzing the texture of pasta for quality control. *Cereal Foods World* 42(1):8–12, 1997.

52. Stevens M. Question and answer. *Fine Cooking* 71:14, 2005.

53. Temelli F. Extraction and functional properties of barley beta-glucan as affected by temperature and pH.

*Journal of Cereal Science* 62(6): 1194–1197, 1997.

54. Turk M, NG Carlsson, and AS Sandberg. Reduction in the levels of phytate during wholemeal bread making: Effect of yeast and wheat phytases. *Journal of Cereal Science* 23:257–264, 1996.

55. Walker CE, and JL Hazelton. Dough rheological tests. *Cereal Foods World* 41(1):23–28, 1996.

56. Weaver CA, BR Hamaker, and JD Axtell. Discovery of grain sorghum germ plasm with high uncooked and cooked in vitro protein digestibilities. *Cereal Chemistry* 75(5):665–670, 1998.

57. Weipert D. Processing performance of rye as compared to wheat. *Cereal Foods World* 42(8):706–712, 1997.

58. Wrigley CW. Wheat: A unique grain for the world. In K Khan and PR Shewry, *Wheat: Chemistry and Technology*, 4th ed.

59. Zhou JR, and JW Erdman. Phytic acid in health and disease. *Critical Reviews in Food Science and Nutrition* 35(6):495–508, 1995.

# WEBSITES

The following is a website encyclopedia listing many foods (click on "grains" and "grain products"):

**www.foodsubs.com**

About.com's and Food-Info's galleries of common pasta shapes can be found at:

**http://italianfood.about.com/od/ pastabasics/ig/The-Pasta-Shapes-Gallery/ www.food-info.net/uk/products/pasta/ shapes.htm**

Take a look at the website for the American Association of Cereal Chemists (AACC), a nonprofit international organization specializing in the use of cereal grains in foods. It publishes the scientific journals *Cereal Chemistry* and *Cereal Foods World*:

**www.aaccnet.org**

PhotoDisc/Getty Images

# 17 Flours and Flour Mixtures

**F**lour is the fine powder derived from the endosperm portion of cereal seeds or other starchy foods. The most common source of flour is wheat, but any cereal grain can provide flour—oats, rye, barley, rice, corn, and others. Flour can also be made from non-cereal sources such as soybeans, potatoes, cattails, taro, arrowroot, and other starchy foods. The flours from these cereal and non-cereal foods were not available for over 99% of the 2 million years humans have been on earth because humans were primarily hunter-gatherers. It has only been within the last 1% of that time that the shift to domesticating plants and animals has occurred (58). About 10,000 years ago, humans are believed to have started eating a crude form of bread—a baked mixture of flour and water. In the time since humans first discovered that grains could be crushed for their starchy insides and mixed with other ingredients to provide more palatable nourishment, there has been a huge increase in the number and variety of baked goods. These range from basic staples, such as **yeast breads** and **quick breads,** to the crusty breads enjoyed by the French and the chewy bagels that have become so widely popular in recent years, to desserts such as cakes, cookies, and pastries. Flours from both wheat and non-wheat sources are also important for preparing many thickened gravies, sauces, puddings, and soups.

Baked products differ tremendously in their outward appearance and taste, but the foundation of them all is a flour mixture. The simplest flour mixture is one made from flour and water. Other ingredients that may be added include milk, fat, eggs, sugar, salt, flavoring, and leavening agents. Commercial manufacturers of baked products may also add certain additives. The ingredients of a flour mixture may be divided into two categories: (1) dry—flour leavening agents, sugar, and salt/flavoring or (2) liquid—water, milk, fat, and eggs. The types and proportions of these ingredients determine the structure, volume, taste, texture, appearance, and nutrient value of the finished baked product.

The principles of measuring the ingredients of a flour mixture and the descriptions of various mixing methods are discussed in Chapter 5. The purpose of this chapter is to review each of the

**Yeast bread** Bread made with yeast, which produces carbon dioxide gas through the process of fermentation, causing the bread to rise.

**Quick bread** Bread leavened with air, steam, and/or carbon dioxide from baking soda or baking powder.

367

basic ingredients in a flour mixture and its specific functions, along with preparation and storage guidelines for flour mixtures.

# FLOURS

Flours provide structure, texture, and flavor to baked products (Figure 17-1). Starch is one of the compounds in flour that strengthens the baked item through gelatinization and one of the factors that contributes to **crumb.** Crumb is partially created during baking by the number and size of air cells produced, the degree of starch gelatinization, and the amount of protein coagulation (49). A fine crumb is delicate with small, densely packed air bubbles, whereas a coarse crumb has large, often irregular air holes. A secondary function of the starch in flour is that it can be partially broken down by enzymes (amylases) into dextrin, malt, and glucose. These compounds add a slight sensation of sweetness, darken the crust color, and improve fermentation,

**FIGURE 17-1**    Flour from grains is the starting point of all baked products.

Baker Technology

making the mixture lighter in texture. The heat of baking, in the presence of moisture from vaporization, causes the dextrins to coat the crust with a shiny layer (49).

## Gluten

Other components in flour that play an important role in the structure of bakery items are the proteins that form **gluten.** These proteins, as well as those from eggs, contribute to the firming of the flour mixture, whereas sugar and fat act as tenderizing agents. The ability of a baked product to rise is directly related to its protein content (5). Because wheat flour has the highest concentration of the proteins that form gluten, it yields baked products with light, airy textures and is, therefore, most often preferred over other flours for baking.

### The Purpose of Gluten Formation

When flour is mixed with water, an elastic network forms when two types of proteins in flour, gliadin and glutenin, combine to yield the protein complex gluten (Figure 17-2). Gluten is both elastic and plastic. Its ability to expand with the inner pressure of gases such as air, steam, or carbon dioxide results from the combination of glutenin's elasticity and gliadin's fluidity and stickiness (1) (Figure 17-3). Bread dough rises as the gas resulting from the yeast or other leaveners, as well as the air bubbles entrapped in small pockets by **kneading,** expands and stretches the gluten strands upward

**FIGURE 17-2**    Gluten is formed from the combination of two wheat flour proteins: gliadin and glutenin.

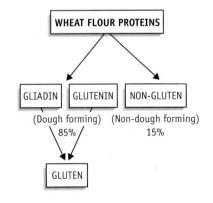

and outward. Then, during the temperature rises of baking, steam along with the expanding gases of carbon dioxide and ethanol cause the gluten to expand further. The baked product's structure sets when the heat from baking coagulates the proteins and gelatinizes the starch. If the oven door is opened frequently or if baking is stopped prematurely before these processes can occur and set the structure, the steam may be released from the gluten complex, causing the baked product to partially collapse.

In cakes and other pastries, gluten formation is not desired because it will result in a tough texture. Gluten formation is partially controlled in cake batters by using a low-viscosity batter with a high liquid content, rather than a high-viscosity dough (23).

### *Separating Gluten from Flour*

Gluten forms such a solid structure that it can be physically separated from the flour mixture by kneading a handful of dough under cold running water (62). This washes away the water-soluble proteins and frees the starch, which is lost down the drain, leaving a rubbery gluten mass. The gluten balls derived from cake, all-purpose, and bread flours are shown in Figure 17-4, and the larger, darker forms behind the gluten balls are the same items baked.

### Steps to Gluten Formation

The main purpose of combining the ingredients of a flour mixture prior to baking is to encourage the development of the gluten, which will contribute to the baked product's structural strength. The two major steps of gluten formation involve hydrating the flour mixture and kneading the dough.

### *Hydration*    Hydration of the flour proteins is the first step in gluten formation. Gliadin and glutenin form gluten, whereas the remainder of the flour proteins, consisting largely of albumin and globulin, become part of the dough or batter (34) (Chemist's Corner 17-1). The greater the protein content in the flour, the more water will be absorbed, in part because gliadin and glutenin absorb about twice their weight in water. The water helps to draw out the gluten-forming proteins from the crushed endosperm cells, and

---

**Crumb**  The texture of a baked product's interior.

**Gluten**  The protein portion of wheat flour with the elastic characteristics necessary for the structure of most baked products.

**Knead**  To work dough into an elastic mass by pushing, stretching, and folding it.

**FIGURE 17-3** Glutenin's elasticity (*left*), plus gliadin's fluidity and stickiness (*center*) combine to form gluten's consistency (*right*), which is ideal for preparing baked products.

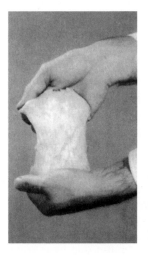

Baker's Digest

**FIGURE 17-4** The proportion of gluten derived from cake, all-purpose, and bread doughs. The ball in front is the gluten carefully extracted by running the dough through cold water to remove the starch. It was then baked to reveal the ball in back.

Cake

All-purpose

Bread

Baker's Digest

> ## CHEMIST'S CORNER 17-1

### Classifying Plant Proteins

Proteins are divided into three types: simple, conjugated, and derived (50). Plant proteins are simple proteins. All plant proteins are classified according to the Osborne system, which identifies proteins based on their solubility (ability to dissolve) in different solvents (Figure 17-5). The four major classification groups are albumin, globulins, prolamines, and glutelins. Gliadins are alcohol-soluble and belong to the prolamines group, while glutenins are alcohol-insoluble and classified as glutelins (50). The albumin fraction generally consists of enzymes. Common enzymes in flour include proteolytic enzymes; oxidative enzymes, which help bleach the flour; and phytase, which breaks down phytic acid (41). Prolamines and glutelins are the major storage proteins in most cereal grains (43). Prolamines have a high concentration of proline and lack the essential amino acid lysine (59). The main prolamine fractions for wheat and corn are gliadin and zein, respectively, whereas glutenin is from the glutelin group (19).

**FIGURE 17-5** The Osborne classification of plant proteins is based on the solubility of proteins in different solvents.

```
                    PLANT PROTEINS
          ┌──────────┬──────────┬──────────┐
          ▼          ▼          ▼          ▼
       ALBUMIN   GLOBULINS  PROLAMINES  GLUTELINS
       (water    (salt      i.e.,       i.e., glutenin
       soluble)  solution)  gliadin     (dilute acid
                            (aqueous     or alkali)
                            alcohol)
```

most doughs are 40% water by weight. Once hydrated, the two proteins start to form gluten's complex, intertwined network, which is filled with water in its inner spaces.

***Kneading*** Kneading is used extensively in bread making and briefly for biscuits and pastries. It alternately compresses and stretches the dough to increase gluten strength. Kneading also evenly distributes the yeasts to all their sources of food throughout the dough mass, redistributes the air bubbles, and warms the dough, increasing fermentation and carbon dioxide gas production. During kneading, the dough changes from a sticky mass to a smooth, stretchable consistency that is easily molded, yet springs back to light pressure (Figure 17-6). The "net" formed by the gluten stretches as the gases, air, steam, and carbon dioxide rise, but its elastic properties hold the general shape of the baked product.

**FIGURE 17-6**   Dough is developed by kneading.

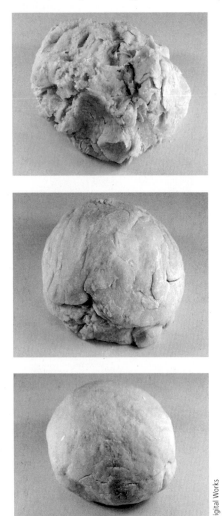

Digital Works

**FIGURE 17-7**   The role of lipids in gluten development. One theory is that they slide more easily past each other. Ultimately, this contributes to proper dough expansion.

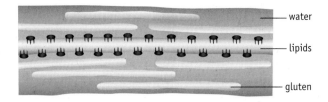

— water

— lipids

— gluten

### How & Why?

*How does kneading change the dough?*

At the molecular level, kneading realigns protein molecules so they run roughly in the same direction and are more likely to form the cross-links that make the dough stiffer (Figure 17-8). Sulfur-to-sulfur, or disulfide, bonds also help link the gluten proteins together. Kneading physically breaks the bonds between sulfur atoms, creates new ones, and allows the gluten molecule to stretch out in length (41).

### Dried Gluten

Dried gluten, sold as vital wheat gluten, is often used to increase the protein content of flours, cereals, meat analogs, pasta, and other bread products. It can also be used to manufacture simulated meat products for vegetarians. Adding vital wheat gluten to regular meat enhances its ability to be sliced and reduces cooking losses during processing. Increasing the gluten content of a flour will also contribute to improved texture and higher volumes during baking (14). Vital wheat gluten is added to some hamburger and hot dog buns to strengthen their structure so that they may hold heavy condiments and prevent the hot dog buns from breaking at their hinges (32).

### How & Why?

*How is gluten formed?*

Gluten is not present in dry, unprocessed flour. It is formed when the addition of water causes the proteins glutenin and gliadin to combine, forming a strong, stretchy protein (45). Fats such as oil and egg yolks coat individual gluten molecules, preventing them from forming large, interconnected proteins. Sugar prevents gluten production by binding to water, preventing it from interacting with glutenin and gliadin to form the gluten protein (35).

The lipids in flour also play a role in kneading, although their exact role is unknown. Lipids make up only about 1 to 1.5% of flour by weight, but without them, the dough will not rise. It has been suggested that lipids serve as an inner core surrounded by gluten and water (Figure 17-7). It has also been postulated that certain lipids link with flour proteins and gelatinized starch (35).

When kneading dough, it is important to avoid excess kneading. Too much kneading will break the gluten strands—the disulfide bonds between the gluten strands pull apart—resulting in a sticky, lumpy dough with little elasticity (11). Although it is difficult to overmanipulate a dough manually, it is fairly easy to do so when using a food processor, standing mixer, or food service dough developer.

**FIGURE 17-8**   Kneading realigns protein molecules. The physical pressure of kneading stretches out the gluten molecules cross-linked by sulfide bonds. Sheets of gluten (*far right*) give dough a smooth, fine texture.

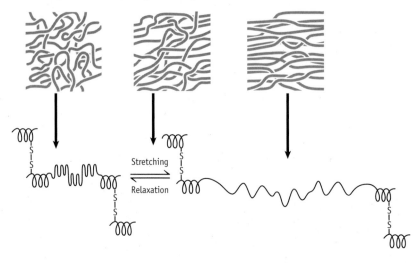

Stretching

Relaxation

# Cereal-Grain Allergies

Two types of allergic reactions can occur in response to exposure to cereals (50). Baker's asthma results from the inhalation of flour particles, and cereal allergies occur when an allergic individual eats certain types of cereals. In both types of allergies, the immune system produces a specific type of antibody, called IgE, that produces inflammation.

## Celiac Disease

Celiac disease is an immune disorder in which the body creates antibodies to gluten, a protein found in wheat, rye, barley, and some other grains (Chemist's Corner 17-2). The resulting inflammation damages the surface of the intestinal wall, eventually causing the villi (fingerlike projections important for absorption) to become "flattened" in appearance. Symptoms include abdominal bloating and pain, chronic diarrhea or soft stools, fatigue, weight loss, and, in infants, failure to thrive (growth in weight or height does not follow averages for their age). Celiac disease increases risk for thyroid

problems, eye inflammation, myocardiopathy (damage to the heart muscle), defective tooth enamel, intestinal cancer, and possible "brain fog" (50). Chronic malabsorption can result in malnutrition.

In the United States, the time between the symptoms' first appearance and diagnosis averages 10 years because celiac disease mimics many other diseases; but in Italy, where celiac disease is more common, the diagnosis takes less than a month. Celiac disease is genetically transmitted and tends to run in families of European descent (especially Irish), those with diabetes (8%), and those with Down syndrome (12%), and is also known to be triggered by surgery, pregnancy, viral infection, or severe stress.

Approximately 1% of the U.S. population—about 1 in 133 Americans—has celiac disease (12), with an average age at diagnosis of 40–50 years (8). Currently, the only treatment for celiac disease is to eliminate gluten sources from the diet.

## What Does "Gluten-Free" Mean?

Due to the increasing rates of celiac disease, sales of gluten-free products are expected to grow at an annual rate of 25% for the next several years (12). The Food and Drug Administration (FDA) proposed rules for the labeling of gluten-free products in January 2007. According to these rules, gluten-free is defined as a product containing "less than 20 ppm gluten." The term *gluten-free* also means that the product does not contain wheat, rye, barely, or

a cross between these grains; anything derived from these grains that has not been processed to remove gluten; or anything derived from these grains that has been processed for gluten removal but still contains ≥20 ppm gluten.

# Milling

Milling, or grinding, is the process in which the grain kernel is first freed of its bran and germ and then has its endosperm ground into a fine powder known as flour. Any grain can be milled, but wheat flour dominates the market. Other flours include rye, corn, buckwheat, and triticale flours, and flours made from legumes such as soybeans and peanuts.

Originally, a mortar and pestle were used to grind grain into whole-grain flour. This process did not remove any of the various kernel parts (bran, germ, aleurone, and endosperm) (56). The product produced by this technique is called whole meal and can be sifted to remove the bran. In Mesopotamia sometime around 800 BC, equipment was developed that used grinding stones moving in circles (Figure 17-9). Eventually mills, using flowing water for power, became capable of moving one circular stone over another stone that was held stationary.

## Five Steps of Milling

Milling today consists of five basic steps: breaking, purifying, reducing, sifting, and classifying.

**FIGURE 17-9** Different types of stone mills used for grinding grain.

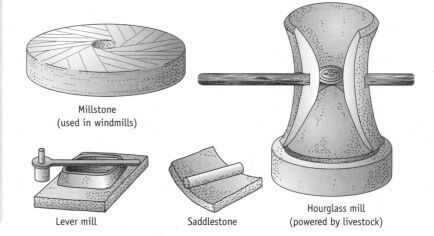

Millstone (used in windmills)

Lever mill

Saddlestone

Hourglass mill (powered by livestock)

**FIGURE 17-10** Machine milling. Break rollers are used to separate the grain's endosperm from the bran and germ.

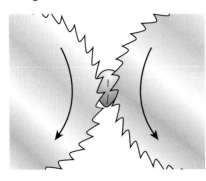

*Step 1—Breaking* In breaking, special machinery equipped with break rollers removes the bran and germ layers from the grain's endosperm (Figure 17-10). Each succeeding pair of steel break rollers is set closer together to remove more of the bran and germ. The result is called break flour. Break flour still has some bran attached to the endosperm layer.

*Step 2—Purifying* Next, the flour is moved through containers where blowing air currents remove any remaining bran. The endosperms, thus freed, or purified, from the whole grain and now free of bran, are known as middlings.

*Step 3—Reducing* At the reducing stage, 10 to 15 smooth-surfaced reduction rollers grind the middlings into flour. To prevent the starch granules from being damaged, the spaces between the rollers are progressively reduced.

*Step 4—Sifting* The flour is then sifted in **streams,** and this determines how the flour is classified.

---

**Stream** A division of milled flour based on particle size.

**Patent flour** Flour containing the finest streams produced during milling.

**Straight flour** Flour containing all the different types of streams produced during milling.

---

*Step 5—Classifying* Flour streams range from fine, or first break, to coarse, or clear, and the types of flour obtained from them range from **patent flour** to **straight flour.** Patent flour is divided into long, medium, or short patents depending on how much of the total endosperm was milled. Short patents come from the center of the endosperm, are high in starch, and are best for making pastry flour. Long patents contain more protein from the outer areas of the endosperm and are preferred for the production of bread flours. Clear flour, which is left over from patent flour, is further subdivided into fancy clear, first clear, and second clear. The coarser clear flours are primarily used for all-purpose flour. The quality and grade of the final flour are determined by which streams were combined in its production. The protein content varies according to the origin of the stream and the type of grain used (42). What follows is a discussion of flours from various wheat and non-wheat grains.

## Wet Milling

Wet milling is a technique used to produce pure starch and other by-products from corn. It differs from standard milling in that it does not involve any mechanical breaking or grinding (44). In this process, corn kernels are soaked in water for up to 36 hours. High-fructose corn syrup is produced via enzymatic isomerization of the glucose in corn syrup to fructose. Corn syrups can also be dried to produce corn syrup solids composed of glucose and maltodextrins. Another product than can be derived from corn syrup is crystallized fructose. The starch from corn can also be saccharified and fermented to form alcohol.

## Wheat Flour Classifications

The gluten-forming properties of wheat make it the most commonly milled grain and the major source of flour for bread making (30). The only other flour that comes close to wheat flour in utility for producing baked goods is rye, but even then it has to be mixed with about one fourth wheat flour.

Wheat flours are classified according to their protein or mineral content. United States and Great Britain flours are not standardized so they are distinguished by the percentage of their protein content. Germany and France use flour type numbers based on the flour's mineral or ash content.

### Percent Protein Content

Wheat flours are defined by their percent protein content (Table 17-1). Soft wheats have the least protein and highest starch content, making them ideal for the tender, fine crumb of crackers, cakes, and pastries. Hard wheats, with their higher protein content, yield high-protein flours preferred for making yeast bread (9). Durum wheat, with the most protein, is milled into semolina flour, which is used primarily for pasta.

### Mineral Content

Ash or mineral content is the portion of the food (or any organic material) that remains after it is burned at very high temperatures (incinerated). The mineral content of the endosperm, which is primarily starch, is much lower than the mineral content found in the grain's germ or bran. Flours containing more minerals derived from the outer part of the grain will have a higher numerical number. The amount of ash obtained from 100 grams of dry flour will vary in Germany from 405 (pastry flour; about 0.4% ash) to 1600 (whole wheat; 1.5% ash). French flour type numbers are based on the mineral content (in milligrams) per 10 grams of flour so Type 45 is standard pastry flour, and Type 150 is whole-wheat flour.

**TABLE 17-1    Percent Protein of Various Wheat Flours***

| Flour | Protein % |
|---|---|
| Gluten | 41 |
| Whole wheat (hard) | 14 |
| Durum wheat (hard) | 13 |
| Bread (hard) | 11 |
| All-purpose | 10 |
| Pastry (soft) | 9 |
| Cake (soft) | 8 |

*As protein increases, so does the "hardness" of the flour.

# Types of Wheat Flour

Despite a percent protein content classification, flours in the United States are identified by their common names. They vary based on the specific types of wheat used for their particular qualities, blending of different combinations, and the streams selected during milling. The result can be any of several different types of flours. Some of the more common are now discussed.

## Whole-Wheat Flour

Whole-wheat flour, also known as graham or entire-wheat flour, is made from the entire wheat kernel, including the bran, germ, and endosperm. One drawback to using whole-wheat flour alone in bread making is that the bran's coarse granules cut the gluten strands, decreasing the final volume of the bread. The flour is finely ground to reduce this effect and is often combined, usually half and half, with white flour for the same reason. A whole-wheat baked item with no added white flour will be dense and heavy. Whole-wheat flour contains fat from the germ, so it requires refrigeration to prevent rancidity.

### *Whole-Wheat Flour as a Functional Food*    According to the American Dietetic Association, increased intake of whole grains, including whole-wheat flour, is associated with a decreased risk of heart disease, certain types of cancer, stroke, diabetes, and some gastrointestinal diseases (4). The beneficial effects of whole grains are due in large part to their fiber content. High-fiber diets are associated with lower body weight, which in turn decreases the risk of diabetes and other diseases (53). The American Dietetic Association recommends a daily fiber intake of 25 grams and 38 grams for adult women and men, respectively. However, the average daily intake is estimated at only 15 g/day (53).

## White Flour

White flour is made from only the endosperm of wheat grain.

### *Bread Flour*    Most bread flour is a long-patent white flour made primarily from hard, winter wheat. The higher gluten content of this flour makes it ideal for making yeast breads and hard rolls that require elastic gluten for multiple rising periods.

### *Durum Flour (Semolina)*    Durum flour is made from hard winter durum wheat. It has the highest protein content (9–18%), making it best suited for the manufacture of pasta products. Disulfide, hydrogen, and hydrophobic bonds link the proteins in semolina flour, producing the viscoelastic properties of pasta (23). This cohesiveness is desirable in pasta because it prevents the dough from breaking apart during boiling.

### *All-Purpose Flour*    All-purpose flour, also known as family-type flour, contains less protein than bread flour does. The protein content of all-purpose flour averages about 11%, whereas bread flours that need strong gluten development have between 12 and 14% protein. Blending hard and soft wheat flours yields a flour that can be used for all purposes—breads, cakes, or pastry. Despite what appears as a uniform product, the actual protein content of all-purpose flour may differ from region to region and by brand. Regional preferences have influenced manufacturers to incorporate more protein in flour sold in the northern United States, where yeast breads are more popular, compared to the South, where biscuits requiring a softer flour are more routinely consumed.

### *Pastry Flour*    Also called cookie or cracker flour, cream-colored pastry flour is derived from soft wheat with short to medium patents. Its lower protein content of about 9% is preferred by commercial and professional bakers for preparing pastries, some cookies, sweet yeast doughs, biscuits, and muffins. Some people substitute a combination of all-purpose and cake flours for pastry flour.

### *Cake Flour*    The higher gluten content of regular flour would make cakes tough, so soft, extra-short patent wheat flours are used to make cake flour. It is pure white and has a very fine, silky, soft texture. Its lower protein content of only 8% and small particle size compared to all-purpose flour result in less gluten being formed, which gives cakes a fine grain, a delicate structure, and a velvety texture. Cake flour is treated with chlorine gas, which lowers the pH from about 6.0 to 5.0, bleaches the plant pigments, and improves baking quality (56). All-purpose flour can be used to make cakes, but they may have a slightly coarser texture and lower volume (63). Cake flour has the same amount of protein as pastry flour, but more starch.

### *Gluten Flour*    Gluten flour is made from wheat flour that has been milled in such a way as to retain the gluten. A small amount of this flour is used in combination with other flours (1 tablespoon for every 1 cup of other flour) to help heavy breads rise more readily. The extra protein binds with water and contributes to a moister bread, making it appear more fresh to the consumer. Vegetarians often use gluten flour to make seitan, a protein food with a springy texture and the ability to sponge up flavors from gravies or broths. Two factors to consider when using gluten are (1) all gluten products need to be refrigerated or frozen because they contain natural fats that can become rancid and (2) gluten causes digestive problems (abdominal bloating, pain, and diarrhea) for people with celiac disease.

### *Graham Flour*    Graham crackers were originally made with whole-wheat graham flour that was produced by combining a finely ground endosperm with a coarsely ground germ and bran. Most graham crackers today are imitations because they are made from regular wheat flour.

# Types of Non-Wheat Flour

Although most baked products are made primarily of flour from wheat, there are other, non-wheat flour sources. These include rice, rye, cornmeal, buckwheat, triticale, soybeans, potatoes, and even peanuts.

## Rice Flour

Naturally gluten-free rice flour is popular in Asian cultures, where it is used to make a variety of food products including rice noodles. It can also be used as a wheat substitute in baked goods (39).

## Chickpea Flour (gram flour or besan)

Common in India and Italy, this flour is produced from the chickpea legume.

### Rye Flour

The lower gluten potential of rye flour contributes to a very compact bread such as pumpernickel. In addition, rye flour contains a high concentration of water-soluble carbohydrates called pentosans, which give the flour a high water-binding capacity (60). As a result, the gases in rye bread do not expand very well (41). Wheat flours are often added to rye flour in a 1:4 ratio to create a more porous, lighter bread. The pentosans also are responsible for the characteristic stickiness of rye bread.

### Cornmeal Flour

Corn's chief protein is zein, which is incapable of mimicking the elastic or plastic properties of flours with higher gluten-forming potential (15). Cornbread and corn muffins do not rise to any great extent and tend to have a crumbly texture unless the cornmeal is combined with all-purpose flour. Cornmeal flour made from yellow, white, and even blue varieties of corn is available. Masa farina is a finely ground cornmeal flour made from corn that was presoaked in lime or lye. It is primarily used to make corn tortillas and tamales.

### Soy Flour

Soy flour is higher in protein than other flours because its source, the soybean, is actually a legume. It has more of the amino acid lysine than wheat, and is sometimes added to wheat flour to improve its protein profile (20). Soybean flour has a low gluten capacity, however, and so must be combined with wheat flour when it is used in baking. Up to 3% soy flour may be used in commercial white breads. Soy flours are generally added to many bakery products at levels of 3 to 6% (percent total formulation), although adding about 20% soy flour has been reported to result in a texture and color similar to those of whole-wheat bread (32).

### Buckwheat Flour

Buckwheat flour contains more starch and less protein than wheat flour and is used primarily in pancake and waffle mixes as well as in crepes and blinis. It is a popular wheat substitute in Europe, where it is used in breakfast cereals and breads (39). Buckwheat has a higher fiber content than regular wheat and better protein quality due to its high lysine content (38).

### Triticale Flour

This is a flour obtained from triticale, a hybrid of wheat and rye grains. Triticale flour is best used in a 1:3 ratio with white flour in the production of breads.

### Potato Flour

Cooked potatoes that have been dried and ground yield potato flour. The starch in potato flour increases loaf volume, which is why the liquid left over from cooked potatoes is sometimes used in homemade breads. Potato flour can be combined with other non-gluten flours to increase moisture content (39).

### Almond Meal Flour

Whole almonds can be ground to a fine powder to make almond flour. It is most often used to make cookies, cakes and desserts, and must be stored in the freezer (39).

### Amaranth

Amaranth is a grain that is high in protein and fiber, and has a nut-like flavor. In baked goods, it is usually combined with other types of flours (39).

### Coconut Flour

Coconut flour has a slightly sweet taste. It is high in fiber and contains more moisture than most non-gluten flours (39).

### Sorghum Flour

Sorghum has a stong flavor, and is used in small doses (about 1.8 cup per batch) in baked goods (39).

## Treated Flours

Wheat flours can be treated to improve their functional properties, resulting in aged, bleached, phosphated, self-rising, instant or agglomerated, or enriched flours. Many of these are used in the production of baked products.

### Aged Flour

Freshly milled flour is not white, and the resulting baked products are of poor quality. Historically, this problem was corrected by storing the flour for several months so it could age and become naturally bleached by the oxygen in the air. Natural aging is expensive, however, because of the required storage space, increased labor costs, and the risk of pest infestation.

### Bleached Flour

Unbleached flour has a yellow-brown color from the xanthophyll pigments found in the endosperm (56). These pigments range in color from yellow to red, and are commonly referred to as "carotenoid pigments." Because they are fat soluble, a lipophilic bleach is required to remove them from flour. The most commonly used agents are gases (including nitrogen oxides, chlorine dioxide, and chlorine) and benzoyl peroxide, which is also used for treating acne and whitening teeth.

Chlorine gas was used as early as 1912, and is still used to bleach all-purpose flour and in cake mixes. Nitrogen tetroxide is one of the oldest successful bleaching agents, and is also still in use. Benzoyl peroxide is the most commonly used agent in the United States and is a fine white powder added to flour at very low concentrations (parts per million).

Flours whitened by this method are labeled "bleached." The bleaching agents evaporate and do not leave residues or alter the nutrient value of the flour. Cake flour is always bleached with chlorine gas, which not only whitens the color, but also creates a very tender, fluffy baked product. All-purpose flours may or may not be bleached. Semolina flour is never bleached because its yellow hue contributes to the color of pasta (47).

Artificial bleaching not only oxidizes the flour's carotenoid pigments, but also improves the flour's functionality. The oxidation occurring during bleaching lightens the flour's color and increases the number of disulfide bonds between the protein chains, improving the strength and elasticity of the gluten (24). Oxidizing agents improve the volume, texture, and crumb structure of baked products (45).

### Phosphated Flour

Flour may be leavened by baking soda instead of baking powder if an acid, specifically monocalcium phosphate (no more than 0.75%), has been added. Flour thus phosphated also has the advantage of increased calcium content, with 68 to 165 mg of calcium versus 18 mg per sifted cup of unphosphated flour (18).

### Self-Rising Flour

Self-rising flour is all-purpose flour with the leavening agent and salt already added. The leavening agent in self-rising flour is a baking powder (baking soda combined with a salt, monocalcium phosphate), which also contributes a significant amount of dietary calcium. One cup of self-rising flour contains about 1½ teaspoons baking powder and ½ teaspoon salt. These amounts should be added to all-purpose flour when the recipe calls for self-rising flour.

### Instant or Agglomerated Flour

Instant flour mixes easily with water, readily gelatinizing without lumps, which makes it ideal for powdered soups, sauces, and gravies. This kind of flour may create a coarse texture in baked products, so it is not recommended for that use. Instant flour is created by passing flour through jets of steam, which wet it and allow it to stick together, or agglomerate, in very small particles. Heated chambers dry the uniformly sized particles so they flow freely and do not need to be sifted. When using it in recipes, 1 cup of all-purpose flour equals 1 cup minus 2 tablespoons of instant flour.

### Enriched Flour

Enriched flour is white flour to which the B vitamins thiamin ($B_1$), riboflavin ($B_2$), niacin, and folate as well as the mineral iron have been added in order to reach levels established by federal standards. Calcium is an optional enrichment nutrient. Fiber and other nutrients, such as vitamin E, that are lost with the removal of the bran and germ are not replaced (61).

# FLOUR MIXTURE INGREDIENTS

Flour mixture ingredients may include leavening agents, sugar, salt, liquid, fat, eggs, and, in some cases, commercial additives.

## Leavening Agents

The presence of a leavener causes the flour mixture to rise. Leaveners may be physical, biological, or chemical:

- **Physical leaveners.** Air and steam
- **Biological leaveners.** Yeast and bacteria
- **Chemical leaveners.** Baking powder and baking soda

Although flour mixtures can rise with the physical help of air and steam, most of the leavening is accomplished by carbon dioxide gas produced from either biological or chemical sources (25). The major biological sources of carbon dioxide are yeasts. Chemical leaveners, such as baking soda and baking powder, yield carbon dioxide when an alkali reacts with an acid in the presence of a liquid. In order to do its job, baking soda must have an acid ingredient added to the flour mixture, whereas baking powder has already had the acid incorporated. Acid ingredients are often added in the form of buttermilk (lactic acid), sour cream, chocolate, brown sugar (contains molasses), or molasses (aconitic acid).

The type of food is the primary determinant of what type of leavening agent will be used. Yeast breads are usually leavened with biological agents such as yeast. Quick breads, as well as cakes, cookies, and pastries, are leavened physically with air, steam, or carbon dioxide generated by baking soda and/or baking powder. Regardless of the source, leavening changes the baked product's volume, crumb, texture, and, ultimately, its flavor (Figure 17-11).

### Air and Steam

Air and steam are physical agents that help dough to rise.

**Air** Air is incorporated into almost all flour mixtures during mixing, during the creaming of fat and sugar, by sifting dry ingredients, or by using whipped egg whites (Figure 17-12).

**Steam** Water incorporated into flour mixtures produces steam when heated, expanding to 1,600 times its original volume. Steam, either from liquid or from other ingredients such as egg whites, is the primary leavening agent for piecrusts, pastry, cream puffs, and popovers.

### Yeast

The ability of yeasts (*Saccharomyces cerevisiae*), which are naturally found in air, water, and living organisms, to produce carbon dioxide through fermentation was probably discovered by accident. An Egyptian baker in ancient times is reputed to have set a flour-and-water dough aside in a warm place, where it became contaminated with yeasts from the air (18). The yeasts started to feed off the available sugar in the bread, producing carbon dioxide and water through the process of fermentation. This same scenario in the absence of oxygen results in the yeast producing ethyl alcohol, or ethanol (Figure 17-13).

**FIGURE 17-11** The role of leavening in bread volume and texture.

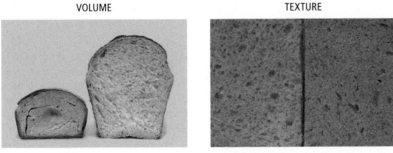

VOLUME — Unleavened · Leavened

TEXTURE — Well-leavened · Poorly leavened

Baker Technology

**Baking soda** A white chemical leavening powder consisting of sodium bicarbonate.

**Baking powder** A chemical leavener consisting of a mixture of baking soda, acid(s), and an inert filler such as cornstarch.

**FIGURE 17-12**    The role of air as a leavening agent. Shown here are three pound cakes made without baking powder and subjected to three different degrees of air incorporation.

**FIGURE 17-13**    Fermentation process.

$$\text{yeast} + 6\,O_2 + C_6H_{12}O_6 \longrightarrow 6\,CO_2\text{ (gas)} + 6H_2O$$
$$\text{yeast} + \text{oxygen} + \text{glucose} \longrightarrow \text{carbon dioxide} + \text{water}$$

$$\text{yeast (without oxygen)} + C_6H_{12}O_6 \longrightarrow 2\,C_2H_5OH + 2\,CO_2\text{ (gas)}$$
$$\text{yeast} + \text{glucose} \longrightarrow \text{ethanol} + \text{carbon dioxide}$$

Bread makers now add yeast directly to flour mixtures and control their growth through hydration, temperature, and salt concentration. During bread making, yeasts that have been alive but dormant become activated when they are hydrated in water at an optimal temperature, which varies according to yeast type. Once activated by the warm water, they are added to the flour mixture. After they are mixed in, the dough is kneaded and allowed to sit in a warm place as carbon dioxide generated by the growing yeast makes the dough rise. Yeasts multiply best at temperatures of 68°F–81°F (20°C–27°C), whereas fermentation is optimal at 81°F–100°F (27°C–38°C), specifically 95°F (35°C) (Figure 17-14). As the yeasts multiply and ferment in this warm environment, a small amount of sugar is sometimes added to serve as a food source for them. Adding too much sugar or salt, however, pulls water osmotically from the yeast cells and can literally dry them to death.

Yeasts are available in several forms, and are classified on the basis of their activity. Active yeasts include baker's yeast, brewer's yeast, and yeasts for alcoholic beverages. Inactive yeasts, such as dried brewer's yeast and primary-grown yeasts, are used primarily for their nutritional value and contribution to flavor (29).

In this chapter, the focus is primarily on baker's yeasts, which are sold as dry, fresh (compressed), or instant yeast.

### ❓ How & Why?

***If yeasts need sugar to ferment, how do they ferment and multiply in a flour mixture?***

The yeasts feed off saccharides derived from either the flour or sugar added to the dough. Flour contains about 1.5% sugar (dry weight basis), of which the glucose is quickly fermented. In addition, an enzyme in yeast (yeast invertase) hydrolyzes starch to glucose and fructose, whereas flour amylases break down some starch to maltose, which is then transported into the yeast cell where it is hydrolyzed to glucose. Sugar (sucrose) added to the flour mixture is also broken down by yeast invertase to glucose and fructose.

***Dry (Active) Yeast***    This is the most widely available type of yeast and is sold in small packets. This porous and free-flowing type of yeast can be stored at room temperature, but keeps longer when refrigerated or frozen. It becomes inactive if not used before the expiration

**FIGURE 17-14**    Yeast activity related to temperature.

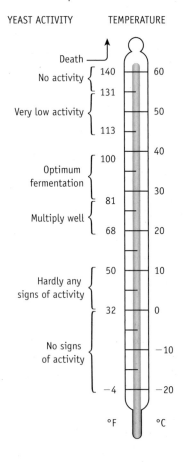

| YEAST ACTIVITY | TEMPERATURE |
| --- | --- |

date. Water warmed to 115°F (46°C) is ideal for rehydrating and activating the dry yeast. Once hydrated, it should not be exposed to temperatures below 100°F (38°C), which lower its activity and result in a sticky dough (Chemist's Corner 17-3), or above 140°F (60°C), which will kill it. In general, one package (¼ ounce or

## CHEMIST'S CORNER 17-3

### Cold Water and Yeast

Hydrating yeast with water that is too cool results in a sticky dough. Cold water slows cell membrane recovery in yeast and allows cell constituents to leach out. Sticky yeast dough can be caused by glutathione, which is naturally released from the yeast through their cell walls and into the dough when they are hydrated below 100°F/38°C. Glutathione is a natural reducing agent that disrupts the disulfide bonds between and within the protein molecules (see Figure 17-15). The proteins quickly unfold, resulting in a softer, stickier dough. This problem does not happen with fresh yeast, which disperses in cold water easily. Rather, it tends to happen with dry yeast, where optimal water temperature for cell restoration is 104°F/40°C.

**FIGURE 17-15** Glutathione released from yeast cells results in sticky dough.

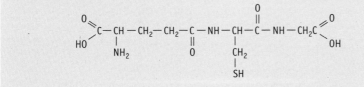

2 ¼ teaspoons) of active dry yeast is enough to leaven 4 to 6 cups of flour. It contains less moisture (8%) than compressed yeast and is therefore less susceptible to deterioration.

***Fresh Yeast*** Also called compressed or cake yeast, this type of yeast is sold as a semi-solid cake with about 70% moisture. It therefore has a short shelf life and develops mold easily, so refrigeration is required. Compressed yeast is not, consequently, a favorite of consumers, but it is preferred by professionals because it dissolves easily without excessively warm temperatures and produces more consistent results. Packs that are brownish (instead of cream-white), dried out, or have a bad odor should be discarded (Figure 17-16). Compressed yeast is

**FIGURE 17-16** The freshness of fresh yeast is tested with a fingernail scratch on the surface.

| Fresh yeast | Older yeast (dry storage) | Older yeast (airtight storage) |

reactivated by the addition of warm water (85°F/29°C).

### ? How & Why?

***How do brewer's, baker's, and nutritional yeast differ?***

Brewer's yeast is a brewery by-product that is used as a nutrient supplement. It is a source of B-complex vitamins, amino acids, chromium, calcium, phosphorus, potassium, and other nutrients. Brewer's yeast has a bitter taste and is made from the pulverized cells of *Saccharomyces cerevisiae* dried at high temperatures, so the yeast are killed and do not contribute to leavening.

Baker's yeast is used for leavening breads and other baked goods. It is dried at low temperatures so the yeast are not killed. It can be the same species as brewer's yeast (*S. cerevisiae*), or *Candida milleri* and *Lactobacillus sanfrancisco* for some sourdough breads.

Nutritional yeast is grown specifically for use as a supplement. Available in powder or flake form, nutritional yeast is grown on a molasses solution so it lacks the bitter taste of other yeasts—it actually has a sweet, nutty, cheesy flavor. It is the same species as brewer's yeast, dried at high temperatures so it is no longer alive. Nutritional yeast has a similar nutrient composition to brewer's yeast, but consuming more than 3 tablespoons per day may raise uric acid levels.

***Instant, Quick-Rising, or Fast-Acting Yeast*** Primarily for commercial bakers, this yeast strain reproduces more quickly than the traditional yeasts. Breads rise twice as fast with this yeast, eliminating the need for a second rising. However, there is less time for flavor to develop. Another drawback is that this type of yeast is extremely sensitive to temperature and moisture, and requires an activation temperature between 125°F and 130°F (52°C–54°C).

### Bacteria

Harmless bacteria that generate carbon dioxide are used as leavening agents in sourdough and salt-rising breads. These baked products depend on a **starter.** Two types of fermentation—one from bacteria and one from yeasts—contribute to the production of carbon dioxide. The bacteria also contribute a desirable, slightly sour flavor to certain baked products—a flavor that has become more and more popular in making the European-style rustic hearth breads. A sourdough starter must be kept alive and fed additional flour in order to keep on contributing to the leavening of breads.

### Baking Soda

Baking soda chemically yields carbon dioxide in the presence of moisture and an acid. Up to ¼ teaspoon of baking soda is required for each cup of flour to be leavened. Baking soda is not typically used by itself as a leavening agent because so much would be required that the flavor and appearance

**Starter** A culture of microorganisms, usually bacteria and/or yeasts, used in the production of certain foods such as sourdough bread, cheese, and alcoholic beverages.

> **CHEMIST'S CORNER 17-4**

**Acid and Baking Soda**

Organic acids give up protons to react with the sodium bicarbonate, eventually yielding carbon dioxide. Acids contribute to the crumb structure formation by providing, during mixing, small gas cell nuclei that serve as starting points for the carbon dioxide that will be produced (25).

**FIGURE 17-17**  Baking soda reactions.

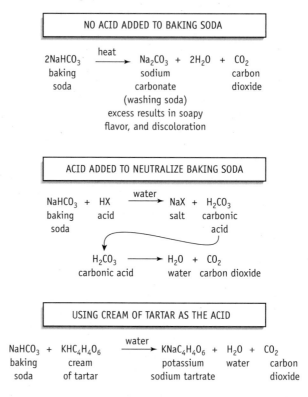

NO ACID ADDED TO BAKING SODA

$$2NaHCO_3 \xrightarrow{heat} Na_2CO_3 + 2H_2O + CO_2$$

baking soda → sodium carbonate (washing soda) excess results in soapy flavor, and discoloration + water + carbon dioxide

ACID ADDED TO NEUTRALIZE BAKING SODA

$$NaHCO_3 + HX \xrightarrow{water} NaX + H_2CO_3$$

baking soda + acid → salt + carbonic acid

$$H_2CO_3 \longrightarrow H_2O + CO_2$$

carbonic acid → water + carbon dioxide

USING CREAM OF TARTAR AS THE ACID

$$NaHCO_3 + KHC_4H_4O_6 \xrightarrow{water} KNaC_4H_4O_6 + H_2O + CO_2$$

baking soda + cream of tartar → potassium sodium tartrate + water + carbon dioxide

of the product would be adversely affected. Baking soda is only used when the flour mixture includes acid ingredients—lemon, vinegar, buttermilk, yogurt, molasses, brown sugar, cocoa, chocolate, citrus fruits, cream of tartar (¾ teaspoon per ¼ teaspoon baking soda), or sour milk (made by combining 1 cup milk with 1 tablespoon lemon juice or vinegar, or 1¾ teaspoons cream of tartar) (Chemist's Corner 17-4). The acid reacts with the baking soda and creates an intermediate compound, carbonic acid, that reacts to give off carbon dioxide and water (Figure 17-17). The immediate leavening effect caused by this reaction necessitates that any baked product prepared with baking soda be placed in the oven as soon as possible after it is mixed.

## Baking Powder

When baking powder is used, it is not necessary to add an acidic ingredient to the flour mixture in order to produce carbon dioxide because the acid has already been added. The inert filler in baking powder absorbs any excess moisture in the air, which would otherwise cake the powder and/or reduce its potency.

*Acid in Baking Powder*  Baking powder is a kind of enhanced baking soda that can be made by combining ¼ teaspoon of baking soda with ½ teaspoon of cream of tartar. In the past, cream of tartar was derived from the sediment that collected on the sides of wine casks, and was the most widely used acid in the manufacture of commercial baking powders (31). When a liquid is added to baking powder, the acid reacts with the alkaline baking soda to release carbon dioxide gas. The type of acid determines the speed with

which carbon dioxide is produced. Federal standards require that all baking powders must be formulated to be able to produce at least 12% carbon dioxide when water and heat are applied, yet many yield about 14% to allow for losses during storage (41). A nonlaboratory method to test the potency of baking powder can be conducted by mixing 1 teaspoon of baking powder with ⅓ cup of hot water to see if it causes bubbling.

*Types of Baking Powder*  The two main types of baking powder are fast, or single-acting, powder and slow, or double-acting, powder. Fast- or single-acting powder is available only to commercial bakers. A flour mixture

made with fast- or single-acting baking powder should be handled quickly and efficiently and placed in the oven as soon as possible, because it starts to produce carbon dioxide as soon as water is added. Any delay allows carbon dioxide to escape and decreases the ability of the mixture to rise. Approximately 1½ to 2 teaspoons of single-acting powder are required for every cup of flour.

Many commercial bakers use double-acting baking powder [sodium aluminum sulfate (SAS)-phosphate powder], which reacts twice: once when it is moistened and again during heating (Figure 17-18). Approximately 1 to 1½ teaspoons of double-acting baking powder are required for every cup of flour.

**FIGURE 17-18**  Baking powder reactions, specifically, double-acting SAS-phosphate.

*Step 1.*
$$2Na_2SO_4Al_2(SO_4)_3 + 6H_2O \xrightarrow{heat} Na_2SO_4 + 2Al(OH)_3 + H_2SO_4$$

SAS (sodium aluminum sulfate) + water → sodium sulfate + aluminum hydroxide + sulfuric acid

*Step 2.*
$$3H_2SO_4 + 6NaHCO_3 \longrightarrow 6H_2O + 3Na_2SO_4 + 6CO_2$$

sulfuric acid + baking soda (sodium bicarbonate) → water + sodium sulfate + carbon dioxide

Other types of baking powder may be used. Potassium bicarbonate is available for people on low-sodium diets. Commercial bakers use baking powders that are specifically designed for certain baked products. Ammonium bicarbonate, for example, is used for cookies that require very little water and have a high surface area; this allows the ammonia to completely evaporate, preventing a bitter flavor (25).

### Too Much/Too Little Leavening

Excess leavening results in a baked product that falls, and has a low volume and/or a coarse texture. Adding more than the required amount of SAS-phosphate powder can result in a bitter taste. Too much baking soda or pockets of baking soda created by inadequate mixing with the dry ingredients will cause the production of residues of sodium carbonate, resulting in a soapy flavor and discolored brown or yellow spots. This excess alkalinity can also affect the flavor of chocolate and turn its color slightly reddish, although the color change is desirable when preparing devil's food cake (Chemist's Corner 17-5).

Too little leavening results in a compact, heavy baked product. Unfortunately, this can also occur when the baking soda or powder has lost its potency through exposure to moisture or storage for over 6 months (31).

### Too Much/Too Little Flour

Too much flour results in a lower volume, an increased number of "tunnels," and a drier, tougher crumb. A baked product made with insufficient flour often has a course texture and a weak, possibly collapsible, structure.

## Sugar

Aside from contributing sweetness, sugar also influences the volume, moistness, tenderness, color, appearance, and caloric content of baked products.

### Functions of Sugar

The obvious sweetening role of sugar is apparent in baked products such as cakes, cookies, sweet breads, doughnuts, coffee cake, and sweet rolls. Many breakfast cereals contain sugar for sweetening power, but also to act as a protective coating (3). Other functions of sugar in flour mixtures include the following:

- Increases the volume of cakes and cookies by the incorporation of air into the fat during creaming (especially true for granulated sugars).
- Contributes to volume by providing food for the yeast—although too much sugar (over 12%) results in a proportional decrease in volume (Figure 17-19).
- Raises the temperature at which gelatinization and coagulation occur, which gives the gluten more time to stretch, thereby further increasing

**FIGURE 17-19** Sugar's influence on the volume of loaf bread.

Source: Schunemann, and Treu. *Baking: The Art and Science*. Baker Technology, 1988.

the volume of the baked product and contributing to a finer, more even texture (6).

- Increases moistness and tenderness, and also helps delay staling through the hygroscopic, or water-retaining, nature of sugar, thus improving the shelf life of baked products.
- Contributes to tenderness by competing with starch for the available water necessary for the hydration of flour proteins and eventual gluten development. The crust is initially very crisp, but becomes softer as sugar attracts moisture from the air or crumb (49).
- Helps to brown the outer crust of baked products through caramelization and the Maillard reaction (48).

### Types of Sugar

Different types of sugars have different weights per cup, which should be taken into consideration when making substitutions. As discussed in Chapter 21, alternatives to sugar are available and can sometimes serve as substitutes for sugar in baked products. Although no code of federal regulations exists specifying the amounts of honey to be used in baked products, the FDA states that honey breads and rolls should contain at least 8% honey (based on flour weight) and impart a noticeable flavor (2).

### Too Much/Too Little Sugar

Baked products made with too much sugar may fall, and may have a lower volume, a coarse grain, a gummy texture, and an excessively browned crust (Figure 17-20). Dough is considered sweet if it contains more than ½ cup of

**FIGURE 17-20** Excess sugar decreases cake volume and results in a gummy texture.

Photograph reproduced Courtesy of CHIPS Books, from *Baking: The Art and Science*, 3 ed,ISBN 0-9708584-8-5

sugar per 3½ cups of flour (1¼ ounces sugar per cup of flour). Too little sugar results in dryness, reduced browning, lower volume, and less tenderness.

## Salt/Flavoring

Small amounts of salt are added to flour mixtures for flavoring, for producing a firmer dough, for improving the volume, texture, and evenness of cell structure, and to prolong shelf life (51). Baked products made without salt tend to be bland, so flavor is one of the most important reasons for adding salt. It also plays a large role in firming the dough by adjusting the solubility and swelling capacity of the gluten (Figure 17-21) (21). Bakers often prefer flaky sea salt to granular table salt because the greater surface area of the more fragile flakes allows for greater distribution. Some people like the ease of adding the salt to the flour, whereas others dissolve it in the liquid for greater dispersion; however, the latter increases the risk of greater gluten formation and resulting toughness. Slightly less gluten is formed when salt is added to the flour mixture because fat may cover up some of the granules (13).

### Salt Controls Yeast Growth

In the production of yeast breads, salt helps control yeast growth. Without salt, fermentation would be too rapid and result in a sticky, difficult-to-handle dough (40). Too much salt, however, inhibits yeast activity, reducing the

**FIGURE 17-21**    Salt makes dough more elastic.

Courtesy of Amy C. Brown

Salt was added to the dough on the left making it much more pliable, which aids in gluten formation. No salt was added to the dough on the right.

**FIGURE 17-22**    Salt influences dough firmness.

No salt            Normal salt            Double salt

Baker Technology

amount of carbon dioxide gas produced and decreasing the volume of the loaf (Figure 17-22) (17). There are now available special formulations for salt-free breads for salt-sensitive consumers.

### Flavorings

Because the basic ingredients of all baked products are the same, all would taste very similar without variations in added flavorings. Flavor extracts, cocoa, melted baking chocolate, fruits, spices, nuts, and other flavorings, seemingly limited only by the baker's imagination, may be added to vary the taste experience.

### Too Much/Too Little Salt

Excess salt produces a firm dough with a low volume (because of partially inhibited fermentation), dense cells, and a too salty taste (49). Too little salt produces a flowing, sticky dough with a low volume, uneven cell structure, lack of color, and a bland taste (36).

## Liquid

Liquid in some form is required in flour mixtures to hydrate the flour and to gelatinize the starch. The water in the liquid also allows gluten to be formed (46), acts as a solvent for the dry ingredients, activates the yeast, provides steam for leavening, and allows baking powder or soda to react and produce carbon dioxide gas.

### Milk

It is not necessary to include milk in a flour mixture, but it is usually recommended over water, because it improves the overall quality of the baked product. In addition to contributing water, milk adds flavor and nutrients (complete

protein, B vitamins, and calcium), and contains certain compounds that help produce a velvety texture, a creamy white crumb, and a browner crust. Doughs made with milk are easier to shape, less sticky and heavy, and retain their shape better. They also tend to expand during fermentation without over- or underdevelopment, and retain gas better, which results in a higher volume. These positive attributes are the result of the presence of milk fat and a natural emulsifier, lecithin.

The lactose in milk participates in the Maillard reaction, resulting in a browner crust. Fresh fluid milk, buttermilk, nonfat dried milk, canned milks, yogurt, and sour cream can all be used at various times in baked goods, depending on the desired end product.

### Too Much/Too Little Liquid

Excess liquid may result in a very moist baked item with low volume. Too little liquid may produce a dry baked product that is low in volume and stales quickly.

## Fat

Fat incorporated into the flour mixture physically interferes with the development of gluten, creating a more tender crumb (54). In fact, the higher that the fat content is, the shorter the gluten strands, and the softer, more easily handled, and more pliable the dough will be. It is for this reason that fat is sometimes referred to as "shortening" (55). Fat lubricates the dough, preventing gluten formation through the creation of a continuous network of starch and protein (23). To maintain the correct consistency when fat is added to a flour mixture, the liquid is reduced by an amount equal to one fifth the amount of added fat. Although fat makes the dough softer and able to rise higher, adding too much of it (over 20% of the flour's weight) makes the baked product too "short" and results in a lower volume because of an insufficient gluten structure and tearing of the crumb (49).

## Functions of Fat

Fat performs many functions in baked goods. It acts as a tenderizer and adds volume, structure, flakiness, flavor, color, and a resistance to staling; it also plays a role in heat transfer (Figure 17-23).

*Fat Improves Volume* Fat increases the volume of the baked product as fat particles (crystals) melt during baking, making the batter more fluid and prone to expansion (10). The role of fat in improving loaf volume is also derived from its ability to stabilize large numbers of small air bubbles by adhering to their surfaces. This allows the bubbles to expand during baking without rupturing. Fat also contributes to volume because the creaming together of fats and sugars traps some air, which acts as a leavening agent during heating. The volume of baked breads increases by about 15 to 25% using 5% shortening (3 to 5% is generally used) (55).

*Fat Improves Strength, Crumb, and Flakiness* In the process of adding volume, fats provide strength to the baked product's structure. As a result, it is more resistant to shocks that otherwise might cause it to collapse during handling. The appearance of the crumb is also affected by fat concentration. Lower-fat baked products have a fine, velvety crumb, whereas those higher in fat lose this characteristic. The flakiness in piecrusts and pastries is entirely dependent on fat being incorporated into the flour (see Chapter 24).

*Fat Improves Flavor and Color* Fat is an important contributor to flavor. The moister crumb of many baked products and the smooth mouthfeel of many fillings used in cookies or pastries is heavily dependent on the presence of fats. The color is also influenced by the fat content and the type of fat used. Croissants owe much of their flavor and color to their large butter component.

*Fat Delays Staling* Another benefit of the inclusion of fats in flour mixtures is that they delay staling in the final baked product (55). Emulsifiers play a role here, also. The dry, hard, resistant, breakable crumb and crust of staled bakery items is not as evident in baked products made with fat and/or emulsifiers, because these ingredients interfere with the main mechanism of staling—recrystallization of starch molecules, specifically amylopectin. Starch gelatinizes in freshly baked products, but during cooling the randomly distributed amylose and amylopectin recrystallize. The amylose recrystallizes quickly, enabling the bread or other baked product to be sliced, whereas amylopectin takes several days and firms the crumb in the process (Figure 17-24).

## Types of Fat Used in Baked Goods

The fats most commonly used in baking are shortenings, unsalted butter, and margarine. Oil and lard are sometimes used. Shortenings have a much wider useful temperature range than either butter or margarine and have the added benefit of containing their own emulsifiers (about 3%) and 10 to 12% gas by volume (57). They are also less expensive. The emulsifiers in the shortening distribute the fat more finely throughout the batter, whereas the gas contributes to volume (Figure 17-25). Cakes made with hydrogenated vegetable shortenings rather than other fats have a higher volume and a finer, more even grain, but lack the flavor that butter provides. Emulsifiers by themselves increase volume, produce a more even, finer pore, and improve the shelf life of baked goods (37). Oil coats flour too thoroughly and prevents adequate gluten development and is, consequently, not often used in baked products, although in some cases oil is added to cake and quick bread recipes specifically for its tenderizing properties.

## Temperature of Fat

Except when it is to be used in making piecrusts and pastries, fat should be at room temperature for baking. A completely melted fat does not incorporate air well, and a cold fat does not disperse evenly with the other ingredients. It is important to distribute the fat

**FIGURE 17-24** How bread stales.

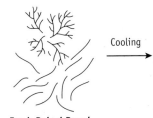

**Fresh Baked Bread**
The starch molecules, amylose (straight chain) and amylopectin (branched chain), found in bread taken right out of the oven are swollen and randomly positioned.

**Fresh Bread**
As the bread cools, the amylose molecules start to come together and crystallize, which changes the texture of the bread.

**Stale Bread**
Several days later, the amylopectin molecules come together and crystallize, further changing the texture of the bread. Reheating reverses this process to some degree.

**FIGURE 17-23** Fat influences the volume and texture of loaf bread.

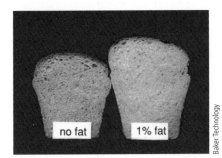

Baker Technology

**FIGURE 17-25** Emulsifier's influence on the pore pattern and volume of loaf bread.

No emulsifier                    Added emulsifier

*Courtesy of Baker's Digest*

throughout the batter so it can make its contribution to tenderness and volume.

### Lower-Fat Alternatives

The functional roles of fat in flour mixtures cannot be denied, but its elevation of the total fat gram and calorie counts has led to the development and use of several commercial flour blends that incorporate a fat substitute (7). Ingredients recommended for yielding a fat-like effect are various starches, gums, maltodextrins, beta-glucan, and both soluble and insoluble fibers (51). Although many baked goods, such as cakes, cookies, and muffins, are available in lower-fat or fat-free varieties, these do not necessarily have fewer calories. They may even have more calories because the fat has been replaced by a large quantity of sugar (26). Chapter 22 discusses fat alternatives in more detail, but the bottom line is that fat-free doughs perform poorly, and the technology to replicate the functional properties of fat with a substitute has not yet evolved. Even if achieved, transferring the fat-free technology from one product to another is not automatic, because what works for cakes may not necessarily be correct for tortillas.

### Too Much/Too Little Fat

Excess fat makes a batter too fluid, weakens its structure, and decreases the volume of the finished product. Too little fat makes a batter resistant to expansion during leavening and results in a tougher crumb.

## Eggs

Eggs are added to some flour mixtures to enhance their structural integrity, or for their contributions to leavening, color, flavor, and/or nutrient content (Figure 17-26). The delicate structure and fine crumb of many baked goods are fortified by the coagulation of egg proteins during baking. Air may be incorporated by adding beaten whole eggs. This reaches a different level of sophistication when egg whites are whipped into a foam and incorporated into cakes such as angel food cake, meringues, cookies, or other baked products. Eggs also contribute to leavening as their

**FIGURE 17-26** The influence of eggs on the volume and texture of loaf bread.

Without egg content        With whole egg

*Baker Technology*

liquid turns to steam when heated. Egg yolks act as emulsifiers and add flavor, nutrients, and color in the form of a yellower crumb and a browner crust. The emulsifiers, along with the fat content in an egg, delay staling and improve the shelf life of baked products. Adding eggs to a cake batter increases its nutritional content by supplying protein, the fat-soluble vitamins (A, D, E, and K), B vitamins, cholesterol, and fat. The appearance of many baked products may be enhanced with a shiny glaze made of egg white and sugar.

### Too Much/Too Little Egg

Excess egg causes a tough, rubbery texture in the baked product. Too little egg causes insufficient volume, and inferior structural strength, color, flavor, and nutrient content.

## Commercial Additives

The baking and milling industries often add commercial additives to the flour mixture to improve commercial production and the quality of the final baked product (52). Azodicarbonamide (ADA) and chlorine dioxide are added to freshly milled flour to induce maturation (22). Examples of other oxidants used in the flour industry include the bleaching agents benzoyl peroxide and acetone peroxide, and the dough improvers potassium bromade, potassium iodate, calcium bromate, calcium iodate, calcium peroxidate, and ascorbic acid. Malt, which is derived from germinated barley grains, is one of the oldest additives used by the baking industry. Malt contributes an enzyme that is a type of dough conditioner, but before the dough is conditioned, it is often aged. Both aging agents and dough conditioners are commercial additives that are now discussed in more detail.

### Aging Agents

Bakers noticed that bread products did not rise as well if freshly milled flour was used. Better baked products were obtained by using flour that had aged or yellowed for weeks. Because waiting for flour to slowly oxidize with the air (natural aging) was not always practical due to time and space constraints along with pest infestation, chemical maturing agents were added to the flour that would speed the aging process.

*Maturing Agents* Freshly milled wheat flour, called "green" flour, must mature before it is capable of being used to produce a strong dough (56). Maturing occurs over 1 to 2 months, during which time the flour is stored under ambient conditions. Flour can be chemically aged by the addition of the powdered maturing agents potassium bromate, ascorbic acid, and azodicarbonamide (49). For example, potassium bromate is known for providing improved gas retention, increased loaf volume, better dough-handling properties, and an improved crumb and texture (16).

### ? How & Why?

**Why does flour have to be aged?**

Aging increases the exposure to air (oxygen), which improves the strength of the resulting dough and results in faster mixing and reduced staling of the finished baked product.

### Dough Conditioners

Doughs are often aged with maturing agents, but also conditioned by adding dough conditioners to improve the effectiveness of the flour mixture. The reactions induced by these conditioners increase dough elasticity, improve machinability, and increase loaf volume. These additives, often mixed into dry flour, are referred to as dough conditioners or dough improvers, because their effects are seen only in the dough stage. Examples of dough conditioners include reducing agents, oxidizing agents, emulsifiers, and enzymes. The most commonly used dough conditioners are azodicarbonamide (ADA), potassium bromate, and ascorbic acid (56). Less commonly used are calcium bromate, potassium iodate, calcium iodate, and calcium peroxide.

### > CHEMIST'S CORNER 17-6

#### How Reducing Agents Work

The most common reductants are L-cysteine, the tripeptide glutathione (Y-L-glutamyl-L-cysteinyl-glycine), and bisulfite salts (56). L-cysteine is found in many natural proteins, glutathione is released from yeast cells, and bisulfite results from the reaction of sulfur dioxide with water and alkali. L-cysteine and bisulfite are the most economically viable reductants. Reductants react with disulfide bonds in dough, breaking them and reducing them to sulfhydryl groups. This results in a reduction in the molecular weight of the protein by decreasing polymerization, or the formation of large protein molecules.

*Reducing Agents* Reducing agents encourage gluten development and thereby shorten mixing time. They act opposite to oxidizing agents and actually disrupt the disulfide bonds, resulting in "disconnected" proteins that need less mixing (Chemist's Corner 17-6).

*Oxidizing Agents* Benzoyl peroxide, an oxidizing agent already mentioned in the discussion of bleached flour above, acts to whiten the flour and improve the functionality of the resulting dough. Oxidizers work by making more sulfur available for the disulfide bonds between proteins in gluten, which strengthens the dough. Oxidizers rebuild the bonds, preventing the dough from becoming too soft and sticky. Oxidation of lipids and lipid hydrolysis, resulting in increased free fatty acid content, occurs as flour matures.

*Emulsifiers* Common emulsifiers added include lecithin, monoglycerides, and diglycerides, which function to disperse fat ingredients more evenly through the flour mixture. Gas generated by yeast is more easily trapped, reducing **proofing** time and creating a more evenly textured baked product.

*Mix Time Reducers* These agents are added to reduce the time needed

when mixing dough. Proteases are the most common. They work by hydrolyzing gluten proteins to smaller sizes. This results in a smaller molecular weight of the proteins and a more elastic dough (56). These reactions occur early in the mixing process. However, proteases must be added carefully—addition of excess protease results in a weak dough. Salt causes dough to tighten, resulting in increased mixing time. By adding salt after the batter is well mixed, mixing time can be reduced by 10 to 20%. Lower pH decreases mixing time and may decrease the susceptibility of dough to overmixing. The addition of organic acids, found in sourdough breads, decreases mixing time by 50%. The effect of pH is reversible.

*Enzymes* Three types of supplemental enzymes are available to assist baked products: amylase, which converts starch to sugar; protease, which breaks down protein; and lipooxygenase, which bleaches the flour and strengthens the dough. Lack of natural amylase enzymes in the flour will result in baked products with a less than desirable volume, texture, and color (Chemist's Corner 17-7) (19). Formerly, amylase was obtained from malted wheat or barley flour, but now it is produced by fungal or bacterial fermentation. Loaf volume increases in breads made from doughs where amylase has been added. This occurs because amylase converts some of the flour starch to sugars for the yeast, which increases gas production. It also delays starch gelatinization during baking, resulting in more oven spring. Proteases have to be added carefully because too much will break down the gluten strands, but this helps soften strong flours with higher protein contents. Lipoxygenase enzyme increases gluten strength and makes a whiter loaf by releasing oxidizers that bleach natural flour pigments.

**Proof** To increase the volume of shaped dough through continued fermentation.

## CHEMIST'S CORNER 17-7

### Commercial Enzyme Additives

Enzyme supplementation lengthens shelf life and improves bread quality. Adding alpha-amylase to flours enhances the conversion of starch to maltose, which aids yeast fermentation and retards staling (33). The retrogradation thought responsible for staling is partially inhibited when alpha-amylase breaks down the size of the amylopectin molecule. Adding too much alpha-amylase, however, creates too many smaller molecules (dextrins) and results in a sticky dough. Proteases, which act on proteins, improve loaf uniformity and grain texture, and yield a softer crumb, when added to bread formulations. The breakdown of proteins catalyzed by the proteases makes the dough more elastic and easier to handle. Proteases are also thought to improve the crust color and flavor of baked products by hydrolyzing gluten, which releases more amino acids to participate in the Maillard reaction (27).

### Food Additives in Flours

The food additives listed on flour labels are different from the unlisted commercial additives added to flours to improve their functionality. For example, fat ingredients are often protected from rancidity by adding antioxidants such as butylated hydroxyanisole (BHA), butylated hydroxytoluene (BHT), vitamin E, and ascorbic acid (vitamin C). In addition to rancidity, breads and other baked products are naturally susceptible to molds. Adding mold inhibitors such

as calcium or sodium propionate extends the shelf life of breads.

# PREPARATION OF BAKED GOODS

## Doughs and Batters

Flour mixtures are either **doughs** or **batters,** depending on their flour-to-liquid ratio (Figure 17-27).

### Doughs

Doughs are classified by their moisture content into stiff or soft doughs.

***Stiff/Firm and Soft Doughs*** A stiff dough is created by adding only ⅛ cup of liquid per cup of flour, whereas a soft dough is made more pliable by adding ⅓ cup of liquid per cup of flour. The handling styles for these doughs also vary; some soft doughs, such as those prepared for yeast breads, rolls, and biscuits, are kneaded, whereas stiff doughs may be handled differently depending on whether they are used to make pasta, pie dough, or pastry. Cookie dough, generally a soft or stiff dough, is either shaped, rolled, or dropped.

***Refrigerated Doughs*** In addition to their ability to be handled, doughs

also differ from batters in that they can be refrigerated with little effect on their overall quality. This characteristic has been profitably exploited by food companies that sell convenient pop-open packages of dough for baked products such as rolls, cookies, croissants, and bread sticks.

### Batters

Batters, like doughs, are classified according to their moisture content, and may be either pour or drop batters (see Figure 17-27).

***Pour Batters*** Pour batters, which average from ⅔ to 1 cup of liquid per cup of flour, are liquid enough to be poured, piped (emptied from a piping bag), or spread (Figure 17-28). Common foods prepared from pour batters include pancakes, waffles, popovers, cookies, and cakes.

***Drop Batters*** Drop batters are too thick to be poured, so they are "dropped." Less liquid, ½ to ¾ cup per cup of flour, forms drop batters that are thick enough to be dropped or pushed onto a baking sheet. Examples of foods made from drop batters include muffins, quick tea breads, cream puffs, and some coffee cakes. In general, batters are best used immediately, although some refrigerator batters may be kept for a short period in the refrigerator.

**Dough** A flour mixture that is dry enough to be handled and kneaded.

**Batter** A flour mixture that contains more water than a dough and whose consistency ranges from pourable to sticky.

**FIGURE 17-27** Doughs and batters are flour mixtures that differ in their liquid content and general use.

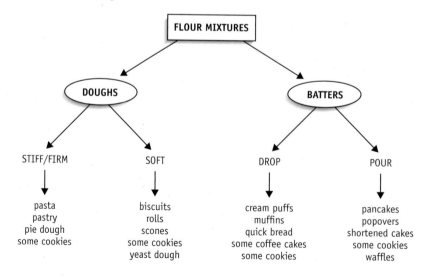

**FIGURE 17-28**  Batters can be spread, poured, or piped.

Spread                Poured                Piped

**FIGURE 17-29**  Flour compression test.

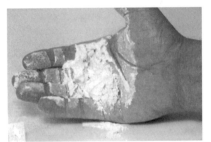

Flour is normal.

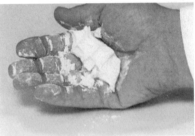

Flour is too moist and its baking quality is compromised.

## Changes During Heating

After the ingredients have been mixed and, if necessary, kneaded to develop gluten, the flour mixture is ready for baking, during which several changes take place. When heat is applied, the gases, such as steam, carbon dioxide, and air, almost immediately expand, creating a pressure that stretches the intricate, elastic network of gluten. The baked product rises in this fashion until the heat melts the fat; gelatinizes the starches; coagulates the flour, eggs, and/or milk proteins; and browns the outer surfaces by the caramelization of sugars, the dextrinization of starch, and/or the Maillard reaction. Once all this has been accomplished, the structure of the baked product is set and cooking is completed.

## High-Altitude Adjustments

When preparing baked products at high altitudes, several adjustments are necessary. The lower atmospheric pressure lowers boiling temperatures and raises the amount of water lost through evaporation. As a result, the leavening gases in baked products meet less resistance and tend to overexpand, and thereby collapse more easily. Recipes can be modified by using less leavening, fat, and sugar; adding a bit more flour and liquid; and baking at higher temperatures in order to set the batter before overexpanding can cause the cells to collapse.

# STORAGE OF FLOUR AND FLOUR MIXTURES

Flour should be stored in pest-proof containers and kept in a cool, dry place. White flour will keep in such conditions for about a year, whereas whole-grain flours, which still contain the fat-rich germ and can therefore turn rancid, should be refrigerated and can be held for only about 3 months (49).

## Dry Storage

Keeping the flour dry is important because moisture attracts insects. Even if it is kept completely dry, flour is still an attractive food source, and insects, beetles, and other pests can bore through paper, plastic bags, and even cardboard to get to it. Only metal, glass, or hard plastic, airtight containers keep pests out. In addition, some flours contain a chemical leavening agent that is triggered by moisture to release carbon dioxide. Figure 17-29 shows how the compression test can be used to determine if a flour is adequately dry.

## Cool Storage Temperatures Required

Flour should be kept cool to prevent the activation of its natural enzymes, which can cause it to deteriorate if it is stored too long. At first it is advantageous for flours to age, because this allows the amylase to break down starch into the smaller sugars (dextrin, malt, and glucose) that benefit the baked product (malt flour is often added to bypass this maturing stage). Storing the flour beyond its recommended storage time, however, results in a higher sugar content, which excessively browns the crust of white bread. Some flour also contains proteases, which break down its protein to the point at which the gluten-forming ability is compromised, but this is a problem only in low-gluten flours. White flour is not affected by its own proteases because it has enough initial protein. Another enzyme, lipase, breaks down any small fat component that may be present, resulting in rancid off-odors and flavors. In addition to rancidity, another drawback of fat deterioration is that the resulting baked product is usually lower in volume and exhibits large pores (49).

## Frozen

Kneaded flour mixtures can be frozen; after defrosting, they are ready to be shaped and baked. The dough may require being kneaded about 10 times after defrosting to improve its quality, because frozen dough loses some of its originally retained gases while in the freezer. Extended frozen storage can lead to a gradual loss of dough strength, which is why frozen doughs have a relatively short shelf life. Freezing damages the yeast cells and their resulting gassing power that results in longer proof times and smaller loaf volumes. It also reduces dough strength because ice crystals disrupt the gluten network. To overcome the problems of freezing dough, commercially prepared frozen doughs are made with more oxidants, emulsifiers, and gluten (28).

The storage guidelines for breads and other baked products made from flour mixtures, such as cakes, cookies, pies, and pastries, are discussed in their respective chapters.

# PICTORIAL SUMMARY / 17: Flours and Flour Mixtures

We owe the structure and texture of baked products largely to wheat flour, specifically to its starch and protein.

## FLOURS

- Starch strengthens mixture through gelatinization; adds texture to the mix
- Gluten proteins firm flour mixture when combined with water
- Dextrin, malt, and glucose in flour aid in fermentation, add sweetness, and help brown the crust

## TYPES OF FLOUR

**Wheat**
- Whole wheat
- White
- Bread
- Durum (semolina)
- All-purpose
- Pastry
- Cake
- Gluten
- Graham

**Non-wheat**
- Rice
- Chickpea
- Rye
- Cornmeal
- Soy
- Buckwheat
- Triticale
- Potato

**Treated**
- Aged
- Bleached
- Phosphated
- Self-rising
- Instant
- Enriched

## FLOUR MIXTURE INGREDIENTS

**Milk**
- Improves crumb and flavor
- Browns crust
- Adds nutrients

**Sugar**
- Sweetens
- Increases volume
- Adds moisture
- Improves color

**Eggs**
- Add structure
- Help leavening
- Improve color/flavor
- Add nutrients

**Liquid**
- Hydrates flour
- Gelatinizes starch
- Serves as a solvent for dry ingredients

**Leavening Agent**
- Increases volume
- Contributes to crumb, texture, flavor

**Salt**
- Adds flavor
- Firms dough
- Improves volume, texture, crumb
- Prolongs shelf life

**Fat**
- Tenderizes
- Increases volume
- Contributes structure, flakiness
- Adds flavor, color
- Increases resistance to staling

## TYPES OF FLOUR MIXTURES

**Doughs**

*Soft*
- Biscuits
- Yeast dough
- Some cookies
- Rolls
- Scones

*Stiff/Firm*
- Pasta
- Pie dough
- Pastry
- Some cookies

**Batters**

*Drop*
- Muffins
- Quick bread
- Some cookies
- Cream puffs
- Some coffeecakes

*Pour*
- Pancakes
- Popovers
- Shortened cakes

## PREPARATION OF BAKED GOODS

Baking a flour mixture expands the gases (steam, carbon dioxide, and air), which stretches the gluten network and causes the baked product to rise. During this time the fat melts; starches gelatinize; proteins coagulate (from the flour, eggs, and/or milk proteins); and outer surfaces brown by the caramelization of sugars, the dextrinization of starch, and/or the Maillard reaction. Heat ultimately sets the structure of the baked product.

## STORAGE OF FLOUR AND FLOUR MIXTURES

White flour can be kept up to a year if it is stored in a pest-proof container kept in a cool, dry place. Whole-grain flours, which contain the germ, need to be refrigerated and will keep up to 3 months. Any flour or dry flour mixture containing chemical leavening agents should be protected from moisture.

# CHAPTER REVIEW AND EXAM PREP

## Multiple Choice*

1. Gliadin and glutenin make up what protein complex in wheat flours?
   a. Albumin
   b. Gluten
   c. Quinoa
   d. Glucose

2. Too much kneading will result in which type of dough?
   a. Soft
   b. Uniform
   c. Inelastic
   d. Nonsticky

3. Durum or semolina flour is made from hard winter durum wheat and is typically used to make which type of product?
   a. Pasta
   b. Bread
   c. Cereal
   d. Baked

4. Using excessive leavening will produce what result in a baked product?
   a. High volume
   b. Low volume
   c. Smooth texture
   d. Moistness

5. Aside from sweetness, sugar influences which of the following characteristics of baked products?
   a. Volume
   b. Moistness
   c. Color
   d. All of the above

6. Stiff doughs are used primarily to make _____.
   a. pasta, pastry, and pie dough
   b. biscuits, rolls, and scones
   c. cream puffs, muffins, and quick bread
   d. pancakes, popovers, and shortened cakes

7. Which type of commercial agent encourages gluten development?
   a. Maturing
   b. Dough conditioning
   c. Reducing
   d. Oxidizing

*See p. AK-1 for answers to multiple choice questions.

## Short Answer/Essay

1. What is the essential difference between the preparation of yeast breads and that of quick breads?

2. What are the two wheat flour proteins that form gluten?

3. Describe the two major steps involved in gluten formation. How does gluten contribute to the formation of certain baked products?

4. What is the purpose of milling? Describe the various steps of this process.

5. Briefly describe how each of the following flours differs: whole wheat, white, bread, durum/semolina, all-purpose, pastry, and cake. Why are rye flour, cornmeal, and soy flour not used as extensively as wheat flour in the production of baked goods?

6. Wheat flours can be treated to improve their functional properties. Describe the following types of treated flours: aged, bleached, phosphated, self-rising, instant/agglomerated, and enriched.

7. Describe the purpose of a leavener. What are the three major types of leavening agents? What is a starter? Explain the difference between baking soda and baking powder. How do they behave differently as leaveners?

8. How do yeasts act as leaveners? List the three different types of yeast products available. What are the basic steps for incorporating yeast into a flour mixture? Discuss optimal temperatures for yeasts to multiply and ferment.

9. Discuss the problems that may occur when too little or much of the following ingredients is added to a flour mixture: flour, leavening, sugar, salt, liquid, fat, and eggs.

10. List several functions for each of the following ingredients of flour mixtures: flour, leavening, sugar, salt, liquid, milk, fat, eggs, and commercial additives.

11. How do the ingredients of and the preparation processes for doughs and batters differ?

# REFERENCES

1. Aamodt A, et al. Dough and hearth bread characteristics influenced by protein composition, protein content, DATEM, and their interactions. *Journal of Food Science* 70(3): C214–C221, 2005.

2. Addo K. Effects of honey type and level on the baking properties of frozen wheat flour doughs. *Cereal Foods World* 42(1):36–40, 1997.

3. Alexander RJ. Sweeteners used in cereal products. *Cereal Foods World* 42(10):835–836, 1997.

4. American Dietetic Association. *Get the whole grain picture.* www .eatright.org/cps/rde/xchg/ada/ hs.xsl/home_4028_ENU_HTML. htm. Accessed 7/27/2009.

5. Autio K, T Parkkonen, and M Fabritius. Observing structural differences in wheat and rye breads. *Cereal Foods World* 42(8):702–705, 1997.

6. Bean MM, and WT Yamazaki. Wheat starch gelatinization. 1. Sucrose: Microscopy and viscosity effects. *Cereal Chemistry* 55:936–944, 1978.

7. Blends reduce fat in bakery products. *Food Technology* 48(6):168–170, 1994.

8. Brannon CA. Ancient and alternative grains. *Today's Dietician* 9(5):10–16, 2007.

9. Carson GR, and NM Edwards. Criteria of wheat and flour quality. In K Khan and PR Shewry. *Wheat: Chemistry and Technology,* 4th ed. American Association of Cereal Chemists, 2009.

10. Castello P, et al. Effect of exogenous lipase on dough lipids during mixing of wheat flours. *Cereal Chemistry* 75(5):595–601, 1998.

11. Chen WZ, and RC Hoseney. Wheat flour compound that produces sticky dough: Isolation and identification. *Journal of Food Science* 60(3): 434–437, 1995.

12. Clemens R, and J Dubost. Catering to gluten-sensitive consumers. *Food Technology* 62(12):21, 2008.

13. Corriher SO. Salt makes bread doughs strong, but it can make pastries tough. *Fine Cooking* 25: 78–79, 1998.

14. Czuchajowska Z, and S Smolinski. Instrumental measurements of raw and cooked gluten texture. *Cereal Foods World* 42(7):526–532, 1997.

15. Dombrink-Kurtzman MA, and JA Bietz. Zein composition in hard and soft endosperm of maize. *Cereal Chemistry* 70(1):105–108, 1993.

16. Dupuis B. The chemistry and toxicology of potassium bromate. *Cereal Foods World* 42(3):171–183, 1997.

17. Eliasson AC, and K Larsson. *Cereals in Breadmaking.* Marcel Dekker, 1993.

18. Ensminger AH, et al. *Foods and Nutrition Encyclopedia.* CRC Press, 1994.

19. Eskin NAM. *Biochemistry of Foods.* Academic Press, 1990.

20. Faller JY, BP Klein, and FJ Faller. Characterization of corn-soy breakfast cereals by generalized procrustes analyses. *Cereal Chemistry* 75(6):904–908, 1998.

21. Fu BX, HD Sapirstein, and W Bushuk. Salt-induced disaggregation/solubilization of gliadin and glutenin proteins in water. *Journal of Cereal Science* 24:241–246, 1996.

22. Fustier P, and P Gelinas. Combining flour heating and chlorination to improve cake texture. *Cereal Chemistry* 75(4):568–570, 1998.

23. Gallagher E. Formulation and nutritional aspects of gluten-free cereal products and infant foods. In EK Arendt and F Dal Bello. *Gluten-free Cereal Products and Beverages,* 2008.

24. Gerrard JA, et al. Dough properties and crumb strength of white bread as affected by microbial transglutaminase. *Journal of Food Science* 63(3):472–475, 1998.

25. Heidolph BB. Designing chemical leavening systems. *Cereal Foods World* 41(3):118–126, 1996.

26. Hippleheuser AL, LA Landberg, and FL Turnak. A system approach to formulating a low-fat muffin. *Food Technology* 49(3):92–96, 1995.

27. James J, and BK Simpson. Application of enzymes in food processing. *Critical Reviews in Food Science and Nutrition* 36(5):437–463, 1996.

28. Koh BK, GC Lee, and ST Lim. Effect of amino acids and peptides on mixing and frozen dough properties of wheat flour. *Journal of Food Science* 70(6):S359–S364, 2005.

29. Labuda I, C Stegmann, and R Huang. Yeasts and their role in flavor formation. *Cereal Foods World* 42(10):797–799, 1997.

30. Leonard AL, F Cisneros, and JL Kokini. Use of the rubber elasticity theory to characterize the viscoelastic properties of wheat flour doughs. *Cereal Chemistry* 76(2):243–248, 1999.

31. Malgieris N. *Nick Malgieri's Perfect Pastry.* Macmillan, 1998.

32. Mannie E. Issues and answers in formulating with proteins. *Prepared Foods* 166(6):93–96, 1997.

33. Mannie E. Tapping the power of enzymes. *Prepared Foods* 167(5): 123–124, 1998.

34. Marcone MF, and RY Yada. A proposed mechanism for the cryoaggregation of the seed storage globulin and its polymerized form from *Triticum aestivum. Journal of Food Biochemistry* 18:147–163, 1995.

35. McCann TH, DM Small, IL Batey, CW Wrigley, and L Day. Protein-lipid interactions in gluten elucidated using acetic-acid fractionation. *Food Chemistry* 115:1–5, 112, 2009.

36. McWilliams M. *Foods: Experimental Perspectives.* Macmillan, 1997.

37. Mettler E, and W Seibel. Effects of emulsifiers and hydrocolloids on whole wheat bread quality: A response surface methodology study. *Cereal Chemistry* 70(4):373–377, 1993.

38. Mezaize S, S Chevallier, A Le Bail, and M De Lamballerie. Optimization of gluten-free formulations for French-style breads. *Food Engineering and Physical Properties* 74(3):E140–E146, 2009.

39. National Foundation for Celiac Awareness. *Getting Started—Celiac Disease and the Gluten Free Diet.* www.celiaccentral.org/About-Celiac-Disease/Diagnosis-Treatment/33/. Accessed 7/09.

40. Niman S. Using one of the oldest food ingredients—salt. *Cereal Foods World* 41(9):728–731, 1996.

41. Penfield MP, and AD Campbell. *Experimental Food Science.* Academic Press, 1990.

42. Petrofsky KE, and RC Hoseney. Rheological properties of dough made with starch and gluten from several cereal sources. *Cereal Chemistry* 72(1):53–58, 1995.

43. Phillips RD. Nutritional quality of cereal and legume storage proteins. *Food Technology* 51(5):62–66, 1997.

44. Posner ES. Wheat flour milling. In K Khan and PR Shewry. *Wheat: Chemistry and Technology,* 4th ed. AACC International, 2009.

45. Poulson C, and PB Hostrup. Purification and characterization of a hexose oxidase with excellent strengthening effects in bread. *Cereal Chemistry* 75(1):51–57, 1998.

46. Ruan RR, et al. Study of water in dough using nuclear magnetic resonance. *Cereal Chemistry* 76(2):231–235, 1999.

47. Samson MF, and MH Morel. Heat denaturation of durum wheat semolina beta-amylase: Effects of chemical factors and pasta processing conditions. *Journal of Food Science* 60(6):1313–1319, 1995.

48. Sapers GM. Browning of foods: Control by sulfites, antioxidants, and other means. *Food Technology* 47(10):75–83, 1993.

49. Schunemann C, and G Treu. *Baking: The Art and Science. A Practical Handbook for the Baking Industry.* Baker Tech, 1988.

50. Shewry PR, R D'Ovidio, Lafiandra D, Jenkins JA, Mills ENC and F Bekes. Wheat grain proteins. In K Khan and PR Shewry. *Wheat: Chemistry and Technology,* 4th ed. AACC International, 2009.

51. Shukla TP. Baking fat-free tortillas. *Cereal Foods World* 42(3):142–143, 1997.

52. Si JQ. Synergistic effect of enzymes for breadmaking. *Cereal Foods World* 42(10):802–807, 1997.

53. Slavin JL. Position of the American Dietetic Association: Health implications of dietary fiber. *Journal of the American Dietetic Association* 108(10):1716–1731, 2008.

54. Stauffer CE. *Fats and Oils: Practical Guides for the Food Industry.* Eagen Press, 1996.

55. Stauffer CE. Fats and oils in bakery products. *Cereal Foods World* 43(3):123–126, 1998.

56. Stauffer CE. *Functional Additives for Bakery Foods.* Van Nostrand Reinhold, 1995.

57. Thompson SW, and JE Gannon. Observations on the influence of texturation, occluded gas content, and emulsifier content on shortening performance in cake making. *Cereal Chemistry* 33:181–189, 1956.

58. Varughese G, WH Pfeiffer, and RJ Pena. Triticale: A successful alternative crop (part 1). *Cereal Foods World* 41(6):474–482, 1996.

59. Vasco-Mendez NL, and O Paredes-Lopez. Antigenic homology between amaranth glutelins and other storage proteins. *Journal of Food Biochemistry* 18:227–238, 1995.

60. Wang L, RA Miller, and RC Hoseney. Effects of (1-3)(1-4)-beta-D-glucans of wheat flour on breadmaking. *Cereal Chemistry* 75(1):629–633, 1998.

61. Wennermark B, and M Jagerstad. Breadmaking and storage of various wheat fractions affect vitamin E. *Journal of Food Science* 57(5):1205, 1992.

62. Wrigley CW. Wheat proteins. *Cereal Foods World* 39(2):109–110, 1994.

63. Yamazaki WT, and DH Donelson. The relationship between flour particle size and cake volume potential among Eastern soft wheats. *Cereal Chemistry* 49:649–653, 1972.

# WEBSITES

This website contains information on types of flour (for example, it explains what organic flour is), history of flour milling, and a flour milling chart:
**www.flour.com**

Want to know what flours to substitute for wheat flour? Type "flour substitutes" in the search box at this site:
**http://web.extension.uiuc.edu/state/index.html**

The Wheat Foods Council answers questions about wheat flours (type "wheat flours" in the search box):
**www.wheatfoods.org**

The National Institute of Diabetes and Digestive and Kidney Diseases offers

information on numerous digestive conditions. Click on "Celiac Disease" for information on gluten allergy:
**http://digestive.niddk.nih.gov/ddiseases/a-z.asp**

# 18 Starches and Sauces

**S**tarchy foods are the mainstays of diets throughout the world (12). In Ireland in the mid-1800s, the potato was so important to the diet that a widespread potato blight resulted in famine and starvation for many thousands of Irish people. In China, Japan, India, and other Asian countries, rice is the most important staple food, followed by a wide variety of noodles. Certain African countries rely on roots, tubers, and sorghum (a grain) as the main dietary starches.

As a complex carbohydrate, starch provides energy, and a well-balanced diet derives at least 55 to 65% of its calories from carbohydrates. Starches are made up of glucose molecules synthesized by plants through the process of photosynthesis, so they can be obtained only by consuming foods of plant origin. The glucose is converted to starch by the plant, and then utilized for energy or stored in the seeds, roots, stems, or tubers (enlarged underground stems). When foods from these sources are ingested by mammals, including humans, the digestion process converts the starch back to glucose. Starches provide 4 calories (kcal) of energy per gram.

The food industry makes widespread use of the capability of starches to thicken foods and to act as stabilizers, texturizers, water or fat binders, fat substitutes, and emulsification aids (24). Starches contribute to the texture, taste, and appearance of foods such as sauces; gravies; cream soups; Chinese dishes; salad dressings; and desserts, including cream pies, fruit pies, puddings, and tapioca. Cornstarch can be used to manufacture the corn syrups—including high-fructose corn syrup—that are discussed with other sweeteners in Chapter 21 (15). The preparations of cream soups and pie fillings are covered in Chapters 15 and 24, respectively, whereas this chapter covers the preparation of sauces with and without starch. First, let's take a look at starch sources, their structure, and their characteristics.

## STARCHES AS THICKENERS

### Sources of Starch

The word *starch* is derived from the Germanic root word meaning "stiff," and commercial starch lives up to the original meaning by acting as a thickening or gelling agent in food preparation. Some starches derived from plants can be considered food additives and are used in a wide variety of ways. Plants serve as the source of starch granules, which are the plant cell's unit for storing starch. Cereals such as wheat, rice, and corn are common sources of starch. Root starches include potatoes,

**FIGURE 18-1**   Starch granules differ in size based on their source.

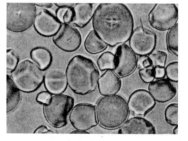

Corn

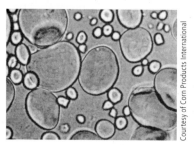

Potato

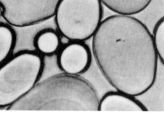

Tapioca

Wheat

*Courtesy of Corn Products International*

arrowroot, and cassava (tapioca) (28). Other sources of complex carbohydrates serving as dietary starch sources include dried beans, peas, and the sago palm. Starch granules differ in size and shape, depending on their botanical origin. Potato starch granules are the largest, while corn, tapioca, rice, and taro root starch granules are progressively smaller (Figure 18-1). Knowing the source of a starch is important when preparing foods because starches vary in flavor and viscosity.

## Cornstarch

The wet milling process is used to derive starch from corn, which is the major source (95%) of starch in the United States (2). The dried kernels are softened by soaking them in warm water containing sulfur dioxide. Once softened, the kernels are cracked, any extraneous material is removed, and the cracked kernels are ground and screened, or sifted down to yield starch and protein. The protein is removed and the starch is filtered, washed, dried, and packaged as cornstarch.

**Dextrose equivalent (DE)**
A measurement of dextrose concentration. A DE of 50 means the syrup contains 50% dextrose.

## Starch in Food Products

Starch serves several purposes in the food industry, including thickening agent, edible film, and sweetener source (dextrose and syrup) (Figure 18-2).

## Thickening Agent

Starch's main use in processed foods is as a thickening and/or gelling agent. Foods that are frequently thickened with starch include soups, sauces, pie fillings, gravies, chili, stews, cream-style corn, cream fillings, custards, fruit pie fillings, whipped toppings, and icings. Certain puddings, candies, gums, and salad dressings are also thickened with starch.

## Edible Films

Starch films are used as a protective coating for chewing gums, to bind foods such as meat products and pet foods, and to act as a base on the food for holding substances such as flavor oils in chocolates.

## Dextrose

Because starch consists of repeating units of glucose, it can be broken down into these individual units for use as a sweetener in the production of confections, wine, and some canned goods. The food industry refers to this glucose derived from starch as dextrose, and measures the degree of conversion from starch to glucose in **dextrose equivalents (DEs)** (31) (see Chapter 21).

**FIGURE 18-2**   Examples of foods containing starches.

*Bakery products*
   Cake mixes
   Cream pie fillings
   Cupcakes
   Frostings
   Glazes
   Fruit pie fillings
   Refrigerated cookie doughs

*Breadings*
   Breaded mushrooms
   Breaded zucchini
   Fried chicken
   Fried fish
   Onion rings

*Beverages*
   Carbonated sodas
   Fruit juices
   Fruit drink mixes

*Canned foods*
   Chili
   Chow mein
   Canned pastas
   Cream-style corn

*Condiments*

*Confections*
   Candies
   Fillings for chocolates
   Gum

*Dairy products*
   Cheese powders
   Cheese spreads
   Dips
   Sour creams
   Yogurts

*Dressings*
   Salad dressings
   Sandwich spreads

*Frozen foods*
   Ice cream
   Pasta
   Pot pies
   TV dinners

*Gravy and sauces*

*Meat products*
   Luncheon meats
   Sausage
   Turkey loafs

*Preserves*
   Jam
   Jelly

*Soups, stews*
   Chowders
   Dry soup mixes
   Soups
   Stews

*Toppings*
   Butterscotch
   Marshmallow

**Source:** Whistler and BeMiller, *Carbohydrate Chemistry for Food Scientists* (Eagen Press, 1997).

A hydrolysis product of starch is considered a starch hydrolysate (7). These starch hydrolysates have been used as sweeteners for over 3,000 years, starting most likely with candy made in China with the aid of maltose-bearing syrups derived from rice starch. The most common solid starch hydrolysates used now are dry maltodextrins (DE < 20) and corn syrup solids (DE = 24 to 48) (29).

### Starch Syrups

Over half of all the starch produced in the United States is eventually converted to syrups (12). Corn syrup made from corn starch is added to a large assortment of foods, including soft drinks, canned fruits, jams, jellies, preserves, frozen desserts, confections, frozen fruits, fountain syrups, and many others (2).

## Starch Structure

As discussed in Chapter 3, starch is a polysaccharide consisting of long chains of repeating units of glucose molecules linked together either in the form of amylose, which is made up of primarily linear molecules, or amylopectin, whose molecules are highly branched (35) (Chemist's Corner 18-1). The amount and proportions of amylose and amylopectin found in starches vary according to the starch's plant source (Table 18-1), but most starches contain about 75% amylopectin and 25% amylose (15). High-amylose starches are 40 to 70% amylose. All starches contain some amylopectin, but a few consist entirely of amylopectin, and these are known as *waxy* starches (6). It is the varying amylose content that causes texture differences in starch-containing

foods (27). Starches containing higher levels of amylose tend to gel, whereas starches containing higher levels of amylopectin are considered non-gelling, but are still somewhat gummy (33). Genetic engineering techniques can be used to produce any ratio of amylase or amylopectin required to achieve the starch's desired functionality (20).

**TABLE 18-1** Proportion of Amylose and Amylopectin in Various Starch Sources

| Starch | Amylose% | Amylopectin% |
|---|---|---|
| Potato | 21 | 79 |
| Tapioca | 17 | 83 |
| Corn | 28 | 72 |
| Waxy maize | 0 | 100 |
| Wheat | 28 | 72 |

### CHEMIST'S CORNER 18-1

**Amylose and Amylopectin**

The varying ratios of amylose to amylopectin cause the texture differences seen in different starches. Amylose consists of a linear molecule containing 500 to 2,000 glucose units linked through alpha-1,4 linkages. Amylopectin contains from $10^4$ to $10^5$ glucose units; they are highly branched and are bound with both alpha-1,4 and alpha-D-1,6 linkages (Figure 18-3) (23).

**FIGURE 18-3** Two types of starch.

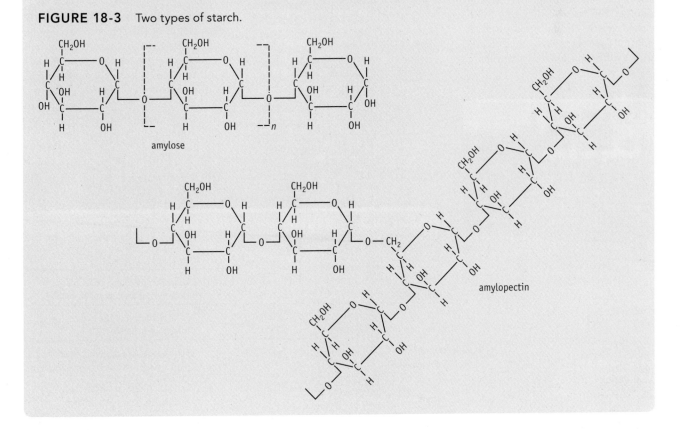

# STARCH CHARACTERISTICS

Starches have the capacity to undergo the processes of gelatinization, gel formation, retrogradation, and dextrinization. It is these capabilities that make them so valuable in food preparation (although some of these processes are more useful than others). The concentration of amylopectin and amylose in starch determines the degree to which these processes take place (Chemist's Corner 18-2). Starches can also be modified chemically or physically to better serve specific purposes.

## Gelatinization

**Gelatinization** occurs when starch granules are heated in a liquid (1). When the liquid is heated, the hydrogen bonds holding the starch together weaken, allowing water to penetrate the starch molecules, causing them to swell until their peak thickness is reached (14). The swelling of the starch granules increases their size many times over. The increased volume and gumminess associated with gelatinization radically changes the texture of many foods. Pasta, rice, oats, scalloped potatoes, and most sauces, soups, and puddings are very different in consistency before

> **CHEMIST'S CORNER 18-2**
>
> **Testing for Starch**
>
> Food scientists can determine if starch granules are either amylose or amylopectin by staining them under a microscope. If they turn blue, they contain more amylose. An amber-violet color indicates more amylopectin (21).

**Gelatinization** The increase in volume, viscosity, and translucency of starch granules when they are heated in a liquid.

and after cooking, and this is because of gelatinization (11).

### Factors Influencing Gelatinization

Gelatinization is dependent on several factors, including the amount of water; the temperature; timing; stirring; and the presence of acid, sugar, fat, and protein (Chemist's Corner 18-3). These factors need to work in synchrony in order for maximum gelatinization to occur.

*Water*  Sufficient water must be available for absorption by the starch. The amount of liquid needed for absorption depends on the concentrations of amylose and amylopectin in the starch (19). When preparing starchy foods such as grains or pasta, sufficient water is added to not only cover the food but

> **CHEMIST'S CORNER 18-3**
>
> **Controlling Gelatinization**
>
> Three substances that control gelatinization are sodium sulfate, sugar, and chlorinated flour. Sodium sulfate stabilizes starch granules by increasing their gelatinization temperature (15). The large negative charge and structure of sodium sulfate increase the starch's resistance to gelatinization. High sucrose concentrations also inhibit gelatinization, and may actually contribute to cakes falling if too much sugar is added to the batter. Chlorinated flour contains oxidized starch, which has a lower gelatinization temperature that is preferred for baking cakes.

also allow for evaporation and a two to three times expansion in volume.

*Temperature*  Starches do not dissolve in cold or room temperature liquids (28). In warmed liquids, the starch granules swell and burst, releasing more starch particles into the liquid. Table 18-2 shows that the temperature range within which gelatinization can occur varies according to the type of starch. Thickening usually begins at approximately 140°F (60°C). Some root starches, such as tapioca, have high concentrations of amylopectin, causing them to thicken at lower temperatures. Most starches gelatinize when heated above 133°F–167°F (56°C–75°C) (9). The temperature at which a particular starch gelatinizes generally falls within a narrow range, often referred to as its transition temperature (34). Larger starch granules, such as those of the potato, gelatinize at lower temperatures, while smaller granules, such as wheat, gelatinize at higher temperatures.

*Timing*  Heating beyond the gelatinization temperature decreases viscosity (Figure 18-4). Starch granules break apart when continued heating stresses the bonds holding them together.

*Stirring*  Stirring during the early formation of the starch paste, or the gelatinizing starch mixture, is required in order to ensure uniform consistency and to prevent lumps from forming. Continued or too vigorous stirring, however, causes the starch granules to rupture prematurely, resulting in a slippery starch paste with less viscosity.

*Acid*  Acids, such as lemon juice, wine, and vinegar, will weaken the

| **TABLE 18-2**   Critical Temperatures of Gelatinization | | |
|---|---|---|
| Starch Source | Critical Temperature, °F (°C) | Characteristics of Cooked Starch |
| Roots and tubers (potato and tapioca) | 133–158 (56–70) | Form viscous, long-bodied, relatively clear pastes; weak gel upon cooling. |
| Cereal grains (corn, sorghum, rice, and wheat) | 144–167 (62–75) | Form viscous, short-bodied pastes; set to opaque gel upon cooling. |
| Waxy hybrids (corn and sorghum) | 145–165 (63–74) | Form heavy-bodied, stringy, clear pastes; resistant to gelling upon cooling. |
| High-amylose hybrids (corn) | 212–320 (100–160) | Form short-bodied pastes; set to very rigid, opaque gel upon cooling. |

**FIGURE 18-4** Gelatinization. Starch granules heated in liquid initially swell, increasing viscosity. Heating beyond the maximum obtainable volume ruptures the starch granules, resulting in less viscosity.

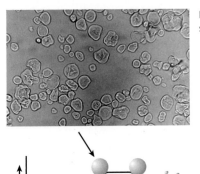

**Peak Viscosity**—Heating to 162°F (72°C) swells the cornstarch granules.

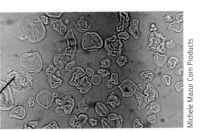

**Viscosity loss**—Continued heating to 194°F (90°C) causes some rupturing and viscosity loss.

Viscosity

Time/Temperature

**Uncooked cornstarch.**

Michele Mazur Corn Products

reduced granular swelling and starch–sugar and starch–water interactions.

***Fat/Protein*** Fat or protein delays gelatinization by coating the starch and preventing it from absorbing water.

# Gel Formation

Gelatinization must occur before the next step, gel formation, also called gelation (see Chapter 15) (Chemist's Corner 18-4). A fluid starch paste is a *sol*, whereas a semisolid paste is known as a *gel*. Not all starches will gel, but among those that do, the gel forms after the gelatinized sol has been cooled, usually to below 100°F (38°C) (1). Gel formation is dependent on the presence of a sufficient level of amylose molecules because amylose will gel and amylopectin will not. The linear amylose molecules form strong bonds, whereas the highly branched amylopectin molecules form bonds that are too weak to contribute to rigidity (37). The bonds that form between the amylose molecules create a three-dimensional network that traps water and increases the rigidity of the starch mass (3).

ability of starches to thicken. Specifically, a pH below 4.0 decreases a starch gel's viscosity. Any acidic fruit juices should be added to pie fillings or salad dressings after gelatinization has occurred. In order to maintain viscosity in canned high-acid foods, commercial food processors use chemically altered starches that are resistant to acid breakdown (see the section "Modified Starches" below).

***Sugar*** Sugar competes with starch for available water, delays the onset of gelatinization, and increases the required temperature (13). Adding too much sugar inhibits complete gelatinization and results in a thick, runny paste. Sugars differ in their ability to delay gelatinization, with the following sugars having the greatest impact (in order from least to greatest): fructose, glucose, lactose, and sucrose. Other factors contributing to the delayed gelatinization caused by sugars include

> **CHEMIST'S CORNER 18-4**
>
> **Gelatinization Under a Microscope**
>
> The process of gelatinization can be detected by looking at the starch granules with a microscope. Starch molecules are stored within the cell's starch granules, and the high degree of order within these granules is depicted by their birefringent nature, which enables them to refract light in two directions. In Figure 18-5, this birefringence is depicted by what look like dark crosses under a microscope providing polarized light. Once gelatinization occurs, this birefringence disappears, indicating a loss of crystallinity (26).
>
> **FIGURE 18-5** Cornstarch molecules.
>
>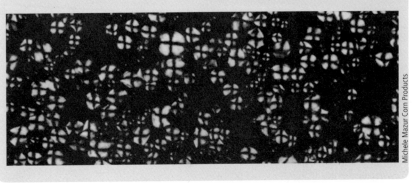
>
> Michele Mazur Corn Products

Regular cornstarch contains large amounts of amylose, which makes it a good gelling agent. Starches that contain high levels of amylose form more opaque gels than starches containing low levels, such as potato and tapioca, which result in more translucent gels. High-amylose starches have a much lower viscosity than other starches (15). Potato starches are commonly used in European recipes, and they are often used to thicken foods such as soups and sauces in which gel formation during cooling would be undesirable (28, 36). Arrowroot is composed of very small granules, creating very smooth sauces (28). It can be used in stir-fry sauces and fruit pie fillings. Waxy hybrids, a cross between corn and sorghum, contain very little amylose and do not gel (22). Commercially, regular cornstarch is most often used to thicken puddings and fruit juices, whereas waxy cornstarch is preferred in the manufacture of canned and frozen food products. Waxy maize starches are used in many frozen foods because other starches lose some of their thickening power when frozen.

## Retrogradation

As the gel cools, bonds continue to form between amylose molecules, and **retrogradation** occurs (17). This retrogradation is accelerated by freezing, so the starches used in frozen food products usually come from sources low in amylose, such as waxy corn or sorghum. The best way to prevent retrogradation is to use the gelled food as soon as possible.

## Dextrinization

Another process characteristic of starches is **dextrinization,** which results in an increase in sweetness. A side effect is that dextrinized starches

**Retrogradation** The seepage of water out of an aging gel because of the contraction of the gel (bonds tighten between the amylose molecules). Also known as *syneresis* or *weeping*.

**Dextrinization** The breakdown of starch molecules to smaller, sweeter-tasting dextrin molecules in the presence of dry heat.

lose much of their thickening power because they have been broken down into smaller units; thus, more flour is required to thicken gravy if the flour has been browned in the gravy-making process. The traditional darkened flour used in Louisiana gumbo has almost no thickening power left, although it adds tremendous flavor to the stew.

### ? How & Why?

*Why does toasted bread taste sweeter than untoasted bread?*

Toasting or browning breaks down amylose and amylopectin (Figure 18-6), and the resulting dextrins cause toast to taste noticeably sweeter than the original bread. Gravies and cooked commercial breakfast cereals also taste sweeter than their unprocessed ingredients because of this process.

## Resistant Starches

A small amount of starch, called resistant starch, is not digested in the small intestine and therefore does not contribute calories (18) (Chemist's Corner 18-5 and 18-6). Some researchers suggest

### > CHEMIST'S CORNER 18-5

**Types of Resistant Starch**

Resistant starches are divided into four categories: RS1, RS2, RS3, and RS4 (18). RS1 starches are physically located within the food matrix, and are thus inaccessible to enzymes in the digestive tract. RS1 starches are found in seeds, legumes, and unprocessed whole grains (25). RS2 starches are present naturally in a tightly packed radial pattern, and are dehydrated (18). This structure, like that of RS1 starches, makes RS2 starches difficult for digestive enzymes to access. RS2 starches include uncooked potato, green-banana flour, and high-amylose grains. RS3 starches are gelatinized and retrograded starches formed when crystalline structures are formed (25). These include cooked and cooled potato starch. RS4 starches are modified starches created by chemical treatment, such as distarch phosphate ester, starch ethers, and cross-bonded starches (18, 25). A fifth type of soluble polysaccharide is "resistant maltodextrin," which is not a true starch, but is formed from the rearrangement of starch molecules to an indigestible form (25).

**FIGURE 18-6**   Dextrinization. The formation of dextrins from amylopectin. Each G represents a glucose molecule.

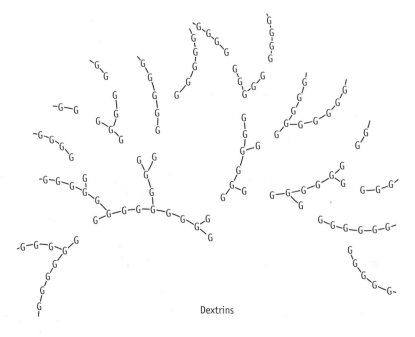

Dextrins

> **CHEMIST'S CORNER 18-6**

**Quantifying a Food's Resistant Starches**

There are four in vitro methods for quantifying resistant starches in food sources: (1) AOAC 2002.02/AACC Approved Method 32-40, (2) Englyst method, (3) AOAC 991.43, and (4) AOAC 2001.03 (25). In the United States, most resistant starches are analyzed as total dietary fiber using the AOAC Method 991.43. In this method, the food sample is hydrolyzed with alpha-amylase, protease, and amyloglucosidase, followed by 78% ethanol precipitation.

that it may be used like fiber in foods for weight loss purposes (32). Suggested benefits of resistant starch include improved glycemic (blood glucose level) management, colon health, weight management, and energy (18). The three types of resistant starch include physically inaccessible starch, resistant starch granules, and retrograded starch (5). Physically inaccessible starch includes those starch granules trapped in the food that are prevented from gelatinizing. Resistant starch granules are those that are indigestible because of their chemical configuration. Retrograded starch is formed during processing when the heating and subsequent cooling of a starch renders the molecules of amylose and amylopectin inaccessible to enzymatic hydrolysis (4).

## Modified Starches

Some starches are sold only to food service operations and food companies and are not encountered at the retail level. These starches have been altered to yield a wide variety of **modified starches,** extending their usefulness in food processing (8, 10). The modifications may affect the starch's gelatinization, heating times, freezing stability, cold-water solubility, or viscosity. Three types of modified starches are cross-linked starch, oxidized starch, and instant or pregelatinized starch.

### Cross-Linked Starch

One of the more commonly used modified starches, cross-linked starch has been treated chemically to link the starch molecules together with cross-bridges (Figure 18-7). Cross-linking makes a starch more heat resistant and less likely to lose viscosity when exposed to heat (16), making it ideal for use in cooked or canned foods such as pizza, spaghetti, cheese, barbecue sauces, pie fillings, bakery glazes, and puddings (21). Incorporating cross-linked waxy cornstarch into a pie filling significantly improves its appearance and texture when compared to one that utilizes unmodified cornstarch (Figure 18-8).

**FIGURE 18-7**   Cross-linking within the starch granule of a modified starch.

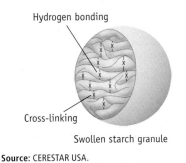

Hydrogen bonding

Cross-linking

Swollen starch granule

**Source:** CERESTAR USA.

**FIGURE 18-8**   Pies made with and without modified cornstarch.

Pie made with unmodified cornstarch exhibits syneresis.

Pie made with modified starch (cross-linked waxy cornstarch) exhibits good appearance.

Institute of Food Technology

### Oxidized Starch

Starches may also be modified by exposing them to chemical oxidizers. Oxidized starches become less viscous than cross-linked starches, but they are clearer and more appropriate for use as emulsion stabilizers and thickeners. The more powder-like consistency of oxidized starches also makes them ideal for dusting foods such as marshmallows and chewing gum (21).

### Instant or Pregelatinized Starches

These starches do not have to be heated in water to expand and gel (27). They have already been cooked and dried, so when cold water is added, they absorb it immediately and expand on the spot. Instant starches make it possible for the food industry to sell instant dry-mix puddings, gravies, and sauces. The dry-mix puddings made from instant pregelatinized starch, listed as modified cornstarch on the ingredient label, are easily prepared by adding cold milk and mixing briefly. Mixes that require heat for preparation contain plain "cornstarch."

## SAUCES

Because so many sauces contain starch, it is appropriate to discuss sauces in conjunction with starches. The purpose of a sauce is to enhance a food's flavor, texture, moisture, and appearance. The sauce itself should have a good consistency, a distinctive flavor, and a shine that is characteristic of that particular sauce. The number of different sauces, with wide variations in consistency, temperature, seasoning, and richness, makes categorizing them difficult. For purposes of our discussion, sauces will be distinguished by whether they are thickened or unthickened, and by their major unique ingredients. This section provides an overview of sauces along with their functions, discusses

**Modified starch**   A starch that has been chemically or physically modified to create unique functional characteristics.

their basic ingredients and preparation methods, and finishes with safe storage advice, because they often contain animal products (meat stock, milk/cream, and eggs).

# Functions of Sauces in Foods

Sauces perform a myriad of functions in foods, not limited to adding moistness, flavor, texture, body (especially to soups), and appearance through their rich color and shine. Fish, meats, and vegetables are much more appetizing and appealing when accompanied by sauces. Despite their positive benefits to the palate, sauces are often not prepared because of their delicate nature, their apparently mysterious preparation techniques, and the extra time and effort required to make them. However, a basic understanding of sauces will eliminate the mystery and open the opportunities for them to be incorporated more often in food preparation.

# Types of Sauces

The major sauces used in food preparation are thickened sauces, including cheese sauce, white sauce, and some gravies; and unthickened sauces, including other gravies; hollandaise; and butter, fruit, barbecue, tartar, and tomato sauces (Table 18-3). More frequently, condiments such as mayonnaise, relish, chutney, salsa, ketchup, and mustard serve the same purpose, although they are not generally referred to as sauces.

## TABLE 18-3    Thickened and Unthickened Sauces

| Thickened Sauces | Basic Ingredients | Uses |
|---|---|---|
| Cheese sauce | 1. White sauce with cheese added.<br>2. Melted cheese with cream or milk added. | Used for egg dishes, fondues, pasta dishes, and vegetables. |
| Custard sauce | Eggs, milk, flavoring. | Used in a variety of desserts. |
| Gravy | Meat stock or juices (extracted from meat during the cooking process) added to flour and possibly liquid. | Pan gravy is served with the meat from which it was derived or with mashed potatoes, vegetables, or breads. |
| White sauce | Butter, flour, milk, salt, and pepper. | Basic sauce that can be as thin or as thick as desired. Can have added ingredients. It can be used for fish, poultry, and vegetables. |
| Variations of white sauce | Variations of white sauce are used for caramel, butterscotch, lemon, honey, and other sauces. | These sauces are used for ice creams, steamed puddings, cottage pudding, soufflés, and bread puddings. |
| Barbecue sauce | Ketchup, onion, vinegar, Worcestershire sauce, and other seasonings. Green pepper optional. | For steaks, burgers, ribs, barbecued ribs, etc. |

| Unthickened Sauces | Basic Ingredients | Uses |
|---|---|---|
| Butter sauce | Butter, lemon juice, chopped parsley. | A sauce for fish dishes. |
| Chocolate sauce | Chocolate, water, and/or milk. | Used for cakes, ice cream, and parfaits. |
| Fruit sauces | Mostly mashed berries and sugar.<br>Optional: egg white, butter.<br>Added ingredients: sugar, water, cornstarch, etc.<br>(Adding cornstarch makes these thickened sauces)<br>    Apple<br>    Cranberry<br>    Currant jelly<br>    Orange with or without pineapple<br>    Orange-currant<br>    Raisin | Fruit sauces are usually not cooked. They are served over ice cream, cottage cake, cheesecake, or in parfaits, etc.<br><br>For game, goose, pork<br>For turkey<br>For game<br>For ham or tongue<br>For chicken, duck, ham, and lamb<br>For ham or tongue |
| Gravy | "Just the juices" extracted from meat during cooking (*au jus*). | Over meats, especially roast beef. |
| Hard sauce | Butter, powdered sugar, vanilla. | Creamed butter, sugar, and vanilla or other flavoring. This sauce can be used with any hot pudding, especially steamed plum pudding at Christmastime. |
| Hollandaise sauce | Butter, egg yolks, boiling water, seasoning, lemon juice. | An emulsion, similar to mayonnaise, that is beaten vigorously until smooth and creamy. This aristocrat of sauces is used mostly for fish, asparagus, broccoli, and cauliflower. |
| Tartar sauce | Mayonnaise, capers, olives, parsley, chopped pickles. | A favorite sauce for fish. |
| Tomato sauce | Tomatoes, onion, cloves, flour, fat, salt, and pepper. | Tomato sauces have a wide variety of uses with pasta and ketchup dishes and meats—especially hamburgers, hot dogs, and Mexican dishes. |

Whether thickened or unthickened, sauces can be grouped into mother sauces and small sauces.

## Mother Sauces

There are five groups of **mother sauces,** also known as grand, leading, or major sauces:

- Béchamel, or white, sauce
- Espagnole, or brown, sauce
- Hollandaise sauce
- Tomato sauce
- Velouté sauce

## Small Sauces

With the exception of the tomato and hollandaise sauces, the mother sauces are not usually used as sauces by themselves, but rather serve as the base for many other secondary or **small sauces.** Examples of small sauces include cheese, cream, curry, mushroom, and shrimp sauces (Table 18-4).

# Preparation of Thickened Sauces

Thickened sauces rely predominantly on starch gelatinization for their smooth texture. Proper preparation steps consist of selecting the right ingredients, combining them, and slowly heating the mixture to the point of gelatinization.

## Ingredients of Thickened Sauces

The three ingredients that serve as the foundation of a thickened sauce are a liquid, a thickening agent, and seasonings and/or flavorings.

*Liquid*   Any liquid can serve as the body of the sauce, but the most common are:

- White stock from chicken, veal, or fish (velouté)
- Brown stock from beef or veal (espagnole)
- Milk (béchamel)
- Clarified butter (hollandaise)
- Tomato juice or purée (tomato sauce)

*Thickening Agent*   A thickening agent, usually starch, may be added to the liquid to make it more viscous. Like gravies, sauces can be made with or without a thickening agent. Wheat flour is most frequently used as the thickening agent in North America and Europe, while cornstarch is preferred for Chinese food. The 10% by weight protein content of wheat starch creates a cloudier appearance and leaves a more floury taste than the purer starch content of cornstarch, which yields a glossy, translucent sheen. Arrowroot gives a result that is even clearer than cornstarch, but its cost limits its use. Pregelatinized or instant starches speed up the preparation process because they thicken immediately in cold water and do not have to be heated.

*Seasonings/Flavorings*   The most common seasonings and flavorings for starch-based sauces are salt, black pepper, white pepper, lemon juice, cayenne, herbs, and wine. Any acids in the form of vinegar, wine, tomato products, or lemon are usually added after gelatinization, because acid can break down the starch.

Many restaurants use **glazes** as flavorings for sauces. The word *glaze* is from the French *glace* for "glossy." This highly flavored concentrate, which congeals when refrigerated into a shiny, rubbery mass, may be obtained from meat (4), chicken, or fish stock. Because glazes take up much less shelf space than the stocks from which they originate and last for months when refrigerated, they are very convenient to use.

## Thickeners

One of the first steps in preparing a sauce is to add a starch thickener in the form of either a roux (roo), beurre manié (burr mahn-*yay*), or slurry.

*Roux*   Wheat flour is usually used in preparing a **roux,** rather than cornstarch, potato starch, or arrowroot. Hot liquid is gradually added to the cooked flour and butter, and this combination is cooked until it reaches the desired consistency, depending on what kind of sauce is being prepared.

There are three types of roux that serve as the foundation in making thickened sauces: white, blond, and brown. Variations in the heating times of the fat–flour combination cause the differences in the colors and flavors, with the most cooking producing the brown roux. As the roux cooks, it becomes darker and its starchy taste lessens, but its ability to thicken is also reduced as the starch molecules are broken down by heat (30). Thus, the darker the roux, the more of it will be needed to add to the liquid for thickening purposes.

| TABLE 18-4   Selected Small Sauces Derived from Mother Sauces | | |
| --- | --- | --- |
| **Mother Sauces** | **Base Ingredient** | **Small Sauces** |
| Béchamel (white) sauce | Milk | Cheddar cheese Cream Mornay Mustard |
| Espagnole (brown) sauce | Brown stock | Mushroom |
| Hollandaise sauce | Butter | Maltaise Mousseline |
| Tomato sauce | Tomato | Creole Portuguese Spanish |
| Velouté sauce | White stock Chicken stock Fish stock | Curry Mushroom Herb |

**Mother sauce**   A sauce that serves as the springboard from which other sauces are prepared.

**Small sauce**   A secondary sauce created when a flavor is added to a mother sauce.

**Glaze**   A flavoring obtained from soup stock that has been concentrated by evaporation until it attains a syrupy consistency with a highly concentrated flavor.

**Roux**   A thickener made by cooking equal parts of flour and fat.

**FIGURE 18-9**    Creating a beurre manié. Equal parts butter and flour are used to make a beurre manié (*left*). The blended beurre manié is whisked into the dish to be thickened (*right*).

Courtesy of Amy C. Brown

*Beurre Manié*    In a **beurre manié**, the butter and flour are not cooked (Figure 18-9). Also unlike a roux, which is added to a sauce at the beginning, a beurre manié is whisked in, bit by bit, to an already simmering sauce until it reaches its desired thickness just before serving. A beurre manié is not cooked, so it should only be used in small amounts to prevent the taste of the sauce from becoming starchy and unpleasant. A pot of stew (about 3 quarts) generally takes about 2 to 3 tablespoons of butter combined with an equal amount of flour. Once cooked, the stew needs to be set aside off the heat because extended simmering can also bring out the floury taste.

*Slurry*    The third type of thickener, **slurry**, is made by gradually mixing cold water, which will not cause the starch granules to expand, into either cornstarch or flour to make a fairly thin liquid. This slurry may then be mixed gradually into a simmering liquid sauce base. Under the simmering heat, the starch granules then expand and the sauce thickens. Typically, pan gravies are made with a slurry of flour and water. Slurry-thickened sauces are inferior to roux-thickened sauces, because a slurry can leave behind a starchy taste and the sauce that it has thickened will be less stable than other sauces.

## Preparing a Sauce from a Roux

The most common sauces that depend on the formation of a roux are white sauces and thickened gravies. A basic white sauce is a combination of butter, flour, milk, and seasonings. Whether it is a thin, medium, thick, or very thick white sauce depends on the amount of fat and flour added (Table 18-5). Veal, chicken, or fish stock can be substituted for the milk, and each will add its own distinctive flavor. Thickened gravy is easily created by adding a roux to the pan drippings of a roast.

### *Combining the Liquid and Roux*

Once the roux has been prepared by melting the fat, mixing in the flour, and gently heating the mixture until smooth and cooked to the desired degree of doneness (white, blond, brown), the next step in sauce making is to combine the liquid with the roux. The amount of liquid added to the roux depends on the desired thickness of the sauce. The temperature of the liquid also makes a difference. It is preferable to add room-temperature or warm, not hot, liquid into a moderately hot or warm roux. Neither the roux nor the liquid should be excessively hot. If the roux is cold, the liquid may be fairly hot, but cold liquid should never be added to a cold roux or the fat in the roux will solidify. Whatever the temperatures of the various elements, the key is to add the liquid gradually and to use a whisk to blend it into the roux.

### *Heating the Sauce*

The next step in sauce preparation is to heat the sauce just to the boiling point and then reduce it to a simmer until its maximum thickness is reached. The brief, initial boiling for 1 or 2 minutes removes the starch flavor. If the temperature is not reduced to a medium or low heat soon enough, the result will be a burned roux that is gritty and has an off-taste. Simmering the mixture for up to half an hour creates a velvety, smooth texture and further reduces the starchy taste of the flour.

### *Adding Seasonings/Flavorings*

The finishing step in thickened sauce preparation is the addition of seasonings and other flavorings by incorporating other ingredients, such as cream, egg yolks, and additional butter.

### *Cheese Sauce*

Adding cheese to a basic white sauce creates a cheese sauce. Several problems may occur in commercial cheese sauce production, including but not limited to poor texture, oiling off, syneresis, and oil streaking. Table 18-6 suggests several modifications to remedy these problems.

---

**Beurre manié** (pronounced burr mahn-YAY) A thickener that is a soft paste made from equal parts of soft butter and flour blended together.

**Slurry** A thickener made by combining starch and a cool liquid.

| **TABLE 18-5** | White Sauce Ingredient Proportions | | | | |
|---|---|---|---|---|---|
| Sauce | Fat | Flour | Liquid | Salt | Pepper |
| Thin | 1 tbs | 1 tbs | 1 C | ¼ tsp | dash |
| Medium | 2 tbs | 2 tbs | " | " | " |
| Thick | 3 tbs | 3 tbs | " | " | " |
| Very thick | 4 tbs | 4 tbs | " | " | " |

**TABLE 18-6**  Troubleshooting Problems in Commercial Cheese Sauce Production

| Symptom | Causes | Changes to Make |
|---|---|---|
| Poor texture | Viscosity too thick | • Decrease level of thickening agents and check thickening agent selections.<br>• Reduce heat exposure during processing.<br>• Minimize evaporation losses during processing. |
| | Viscosity too thin | • Increase level of thickening agent and check thickening agent selections.<br>• Increase heat exposure during processing. |
| | Lumpiness or gumminess | • Decrease level of thickening agents or whey protein concentrate.<br>• Increase blending time for uniform dispersion and hydration of thickening agents. |
| | Graininess | • Ensure uniform blending.<br>• Raise pH to avoid protein precipitation. |
| | Grittiness or sandiness | • Reduce level of salt in formulation.<br>• Reduce cooking to decrease extent of protein denaturation.<br>• Decrease lactose content to control crystallization. |
| Oiling off | Emulsion breakdown | • Increase levels of emulsifiers or emulsifying salts.<br>• Use less heat.<br>• Increase blending time to disperse ingredients more uniformly. |
| Syneresis, weeping | Colloidal system breakdown | • Increase stabilizer level and check stabilizer selection.<br>• Use more casein or cheese or less whey to bind water.<br>• Use less heat to decrease extent of protein denaturation.<br>• Ensure proper blending to hydrate stabilizer.<br>• Check mineral balance. |
| Oil streaking | Emulsion breakdown | • Increase levels of emulsifiers or emulsifying salts, use less heat, or increase blending action to disperse ingredients more uniformly. |
| Color streaking | Uneven color distribution | • Increase blending to disperse ingredients more uniformly. |
| | Poor color solubility | • Select color with solubility properties to match the system. |
| Skin formation | Air exposure and evaporation | • Stir and cover during usage and cover during storage. |
| Rancid flavor | Lipase activity | • Check incoming ingredients for off-flavors.<br>• Avoid mixing raw milk with pasteurized milk product. |
| Oxidized flavor | Oxidation | • Minimize exposure to light and maintain cool temperatures during product storage.<br>• Check incoming ingredients for off-flavors. |
| Bitter flavor | Protein breakdown to bitter peptides | • Use milder varieties or less aged cheeses.<br>• Check incoming ingredients for bitter notes.<br>• Check age of cheese sauce. |
| Improper flavor intensity | Flavor too strong | • Use mild-flavored or less aged cheeses.<br>• Decrease level of flavoring agents.<br>• Decrease amount of enzyme-modified cheeses in blend. |
| | Too little flavor | • Use strong-flavored or more aged cheese.<br>• Add more flavoring ingredients.<br>• Use enzyme-modified cheese in blend. |

## Preventing Lumps

The achievement of a smooth, lump-free sauce depends on the principle of separating the starch granules from one another to prevent them from being trapped in a ball surrounded by a film of gelatinized starch. The following are some of the ways to avoid lumps:

• The fat and flour in a roux should be blended until smooth before adding the liquid. If the flour is coated with fat in this way, it will not form lumps when it contacts the liquid.
• A small amount of sugar may be added to separate the granules, but care must be taken: too much sugar will cause the sauce to be irreversibly runny.
• A small amount of the starch (2 tablespoons) may be vigorously mixed with cold water in an enclosed jar before incorporating it into the rest of the liquid to be added to the roux.

# Preparation of Unthickened Sauces

Sauces prepared without a starch or any other thickening agent are considered to be unthickened. Briefly discussed below are examples of unthickened sauces such as some gravies and butter, fruit, tartar, barbecue, and tomato sauces. Some salad dressings also fall into this category of unthickened sauces; they are discussed in Chapter 15.

**FIGURE 18-10**   Preparing gravy through deglazing and reducing.

Brown ingredients slowly over heat.

Deglaze with water one cup at a time.

Further reduce to concentrate the sauce.

Add more water, deglaze, and reduce again.

**FIGURE 18-11**   Fat separator allows fat to rise to the top so remaining liquid can be poured out separately.

Digital Works

## Gravy

Gravy is made from the juices or drippings remaining in the pan after meat or poultry is cooked. The drippings can be served thickened and with added seasonings as described previously, or unthickened, which is commonly referred to as **au jus** (oh zhue), meaning "in its own juice." Roast beef is frequently served au jus. The basic method for making gravy from the pan drippings of roasted meat or poultry involves the five steps of degreasing, deglazing, reduction, straining, and seasoning (Figure 18-10).

**Au jus**   Served with its own natural juices; a term usually used in reference to roasts.

**Deglaze**   To add liquid to pan drippings and simmer/stir to dissolve and loosen cooked-on particles sticking to the bottom of the pan.

**Reduction**   The process in which a liquid is simmered or boiled until the volume is reduced through evaporation, leaving a thicker, more concentrated, flavorful mass; or the product of this process.

**Step 1—Degreasing**   Degreasing consists of separating the liquid from the fat. After cooking, the meat or poultry is removed from the pan so the liquids and residues on the bottom of the pan can be separated from the fat. There are several possible methods of degreasing the sauce:

- Refrigerating the mixture so the fat can rise to the surface, harden, and be removed.
- Tipping the pan so the fat can be skimmed or spooned off from the pan juices.
- Using special utensils that permit the fat and liquid to separate so that only the liquid is poured off (Figure 18-11).

**Step 2—Deglazing**   The next step in making gravy is to **deglaze** the pan. The liquid used may be water, but stock, wine, beer, milk, cream, tomato juice, or vegetable juice may also be suitable, depending on the kind of taste result that is preferred.

**Step 3—Reduction**   After deglazing comes **reduction,** which concentrates the volume and flavor of the contents in the pan.

**Step 4—Straining**   At this point, unthickened gravies are often strained through a cheesecloth, strainer, sieve, or china cap to remove large particles.

**Step 5—Seasoning**   Seasoning is the last step, because if it is done before reduction, the result might be an overly seasoned or salty sauce.

## Hollandaise Sauce

Egg yolks added to a base of butter are the main constituents of a hollandaise sauce. A small amount of water, salt, vinegar (or lemon juice), and cayenne are also incorporated. The key to this sauce is to heat the ingredients without curdling the egg yolks, and this is most easily accomplished by using indirect heat, usually in a double boiler. Hollandaise sauce is particularly susceptible to contamination because after the egg is added, it is not cooked to a temperature high enough (140°F/60°C) to kill bacteria.

## Barbecue Sauce

Barbecue sauces consist of tomato products and water, juice, or some other liquid as a base. Sugar, sautéed onions, vinegar, and seasonings or flavorings such as chili powder, pepper, salt, and dry mustard distinguish one barbecue sauce from another. All the ingredients are combined in a saucepan and simmered for at least 20 minutes.

## Butter Sauce

The simplest butter sauce is plain melted butter, but many variations are possible. These variations, such as clarified, brown, and black, are described more fully in Chapter 22.

### Tomato Sauce

As one of the mother sauces, tomato sauce is a popular base for many foods. Spaghetti sauce is one of the most common and familiar tomato sauces. Any of a number of ingredients are added—chopped onion; tomato purée; tomato paste; canned or fresh tomatoes; cooked ground beef or poultry; and Italian herbs such as basil, oregano, sage, and marjoram. Sometimes wine or stock (chicken or beef) is included as a liquid.

# STORAGE OF STARCHES AND SAUCES

The quality of dry starches deteriorates with improper storage. Like any other grain product, they should be kept in airtight containers and stored in a cool, dry place away from moisture, oxygen, light, and pests. Many foods made with starches contain eggs, milk, cream, or other dairy products, all of which make them prone to bacterial contamination and thus to foodborne illness. Sauces made with these ingredients should be heated to 165°F (74°C) and kept out of the temperature danger zone (40°F–140°F [4°C–60°C] for consumers; 41°F–135°F [5°C–57°C] for retailers). Thickened sauces should also be prepared, served, and stored with caution. These products should be stored in the refrigerator and never left to sit for long at room temperature.

### Fruit Sauce

Sauces based on sugar and fruit are relatively easy to prepare. Some fruit sauces are thickened by adding cornstarch or flour and briefly heating the concoction until the starch gelatinizes.

### Tartar Sauce

This sauce is prepared by adding chopped pickles, onions, capers, parsley, mustard, or shallots, or some combination of these, to a mayonnaise base. No cooking is required.

# PICTORIAL SUMMARY / 18: Starches and Sauces

As a complex carbohydrate, starch, in the form of potatoes, rice, noodles, and sorghum, is a mainstay of diets throughout the world. In food preparation, starch contributes to the texture, taste, and appearance of many foods such as sauces, gravies, cream soups, salad dressings, and desserts.

## STARCHES AS THICKENERS

Starches are derived from the seeds and roots of various plants. The size and shape of starch granules differ according to their source. Starches provide 4 calories (kcal) per gram and are a good source of energy. Starches are used in many food products as thickening agents, edible films, and sweeteners (dextrose and syrups).

## STARCH CHARACTERISTICS

When starch granules are heated in a liquid, they absorb the liquid, swell, and increase in viscosity and translucency —a process called gelatinization.

Factors affecting gelatinization include:

- **Water:** Should be sufficient for absorption by the starch.
- **Temperature:** Transition temperatures for various starches differ, but most are 133°F–167°F (56°C–75°C).
- **Timing:** Heating beyond the gelatinization temperature decreases viscosity.
- **Acidity:** A pH below 4.0 decreases the viscosity of a starch gel. Acidic fruit juices should be added after gelatinization.
- **Fat/Protein:** These delay gelatinization by coating the starch and preventing water absorption.

## SAUCES

Starches are used to thicken sauces and gravies. The basic ingredients of most sauces are liquid, optional thickening agent, and seasonings and/or flavorings.

The wide array of sauces makes it difficult to classify them, especially as they differ so much in consistency, temperature, seasoning, and richness.

Sauces used in food preparation:

**Thickened**
- Cheese
- White sauce
- Some gravies

**Unthickened**
- Hollandaise
- Butter
- Fruit
- Tartar
- Barbeque
- Tomato
- Some gravies

The five groups of mother sauces:

- Velouté
- Espagnole/brown
- Béchamel/white
- Hollandaise
- Tomato

}  These three serve as a base for secondary sauces.

## STORAGE OF STARCHES AND SAUCES

Dry starches are stored like any other grain product—in an airtight container in a cool, dry place away from moisture, oxygen, light, and pests. Foods and sauces made with starches are particularly prone to microbial contamination when they contain eggs, milk, cream, or any other dairy product, and therefore should be stored in the refrigerator.

PURE
**CORN STARCH**
FOOD SERVICE
NET WT. 18.5 oz

# CHAPTER REVIEW AND EXAM PREP

## Multiple Choice*

1. Amylose has a _____ structure, whereas amylopectin is _____ .
   a. branched, smooth
   b. linear, branched
   c. branched, linear
   d. linear, smooth

2. What is the word used to describe the breakdown of starch molecules to smaller, sweeter-tasting dextrin molecules in the presence of dry heat?
   a. Gelatinization
   b. Dextrinization
   c. Retrogradation
   d. Deglazing

3. Starches that have been altered to affect their heating times, viscosity, or gelatinization are referred to as:
   a. gelled starches.
   b. smooth starches.
   c. tough starches.
   d. modified starches.

4. A roux is a thickener made by cooking equal parts flour and _____ .
   a. fat
   b. water
   c. stock
   d. starch

5. What two ingredients are combined to make a slurry?
   a. flour, hot water
   b. flour, starch
   c. starch, cool liquid
   d. cool liquid, hot liquid

*See p. AK-1 for answers to multiple choice questions.

6. The increase in volume, viscosity, and translucency of starch granules heated in a liquid is referred to as _____ .
   a. gelatinization
   b. dextrinization
   c. retrogradation
   d. deglazing

7. The seepage of water out of an aging gel because of the contraction of the gel is called _____ .
   a. gelatinization
   b. dextrinization
   c. retrogradation
   d. deglazing

## Short Answer/Essay

1. Discuss the various ways in which starch is used in the food industry.
2. Briefly describe the basic structure of starch. What is unique about the structure of *waxy* starches?
3. Describe the various steps of the gelatinization process.
4. Explain how the following elements influence starch gelation: temperature, timing, stirring, acid, sugar, and fat/protein.
5. State the difference between a gel and a sol. Define *retrogradation*. What is the best way to prevent it from occurring?
6. Describe how the following starches differ in their gelling ability: starches with high levels of amylose (regular cornstarch), starches with low levels of amylose (potato and tapioca), and waxy hybrids.
7. Explain *dextrinization,* and describe the benefits and potential side effects of this process in food production.
8. Define *modified starch* and explain how it is used in the preparation of various food products.
9. Explain the relationship between a mother sauce and a small sauce. What is the difference between a roux, a beurre manié, and a slurry?
10. List the three ingredients that serve as the foundation of thickened sauces. Provide examples for each.

# REFERENCES

1. Aguilera JM, and E Rojas. Rheological, thermal and microstructural properties in whey protein cassava starch gels. *Journal of Food Science* 61(5):963–966, 1996.
2. Anonymous. *Tapping the treasure.* Corn Refiners Association. www.corn.org/tapping.pdf. Accessed 6/27/09.
3. Bora PS, CJ Brekke, and JR Powers. Heat induced gelation of pea (*Pisum sativum*) mixed globulins, vicilin

and legumin. *Journal of Food Science* 59(3):594–596, 1994.
4. Brown I. Complex carbohydrates and resistant starch. *Nutrition Reviews* 51(11):S115–S119, 1996.
5. Cairns P, et al. Physiochemical studies on resistant starch in vitro and in vivo. *Journal of Cereal Science* 23:265–275, 1996.
6. Carriere CJ. Evaluation of the entanglement molecular weights of maize starches from solution rheological

measurements. *Cereal Chemistry* 75(3):360–364, 1998.
7. Chronakis IS. On the molecular characteristics, compositional properties, and structural-functional mechanisms of maltodextrins: A review. *Critical Reviews in Food Science and Nutrition* 38(7):599–637, 1998.
8. Committee on Codex Specifications. *Food Chemicals Codex.* National Academy Press, 1993.

9. Ensminger AH, et al. *Foods and Nutrition Encyclopedia.* CRC Press, 1994.

10. Fadzlina ZA, AA Karim, and TT Teng. Physiochemical properties of carboxymethylated sago (*Metroxylon sagu*) starch. *Journal of Food Science* 70(9):C560–C567, 2005.

11. Galvez FCF, AVA Resurreccion, and GO Ware. Process variables, gelatinized starch and moisture effects on physical properties of mungbean noodles. *Journal of Food Science* 59(2):378–381, 1994.

12. Guzman-Maldonado H, and O Paredes-Lopez. Amylolytic enzymes and products derived from starch: A review. *Critical Reviews in Food Science and Nutrition* 35(5):373–403, 1995.

13. Hoover R, and N Senanayake. Effect of sugars on the thermal and retrogradation properties of oat starches. *Journal of Food Biochemistry* 20:65–83, 1996.

14. Hoover R, and T Vasanthan. Effect of heat-moisture treatment on the structure and physiochemical properties of cereal, legume and tuber starches. *Carbohydrate Research* 252:133–153, 1994.

15. Jane J. Starch: Structure and properties. In P Tomasik. *Chemical and Functional Properties of Food Saccharides.* CRC Press, 2004.

16. Kiribuchi-Otobe C, et al. Wheat mutant with waxy starch showing stable hot paste viscosity. *Cereal Chemistry* 75(5):671–672, 1998.

17. Klucinec JD, and DB Thompson. Amylose and amylopectin interact in retrogradation of dispersed high-amylose starches. *Cereal Chemistry* 76(2):282–291, 1999.

18. Leake LL. Analyzing resistant starch. *Food Technology* 60(7):67–69, 76, 2006.

19. Leslie RB, et al. Water diffusivity in starch-based systems. *Advances in Experimental Medicine and Biology* 302:365–390, 1991.

20. Liu Z, and JH Han. Film-forming characteristics of starches. *Journal of Food Science* 70(1):E31–E36, 2005.

21. Luallen TE. Starch as a functional ingredient. *Food Technology* 39(1):59–63, 1985.

22. Marrs WM. Blending starches, gums provides texture and processing benefits. *Prepared Foods* 166(10):63–66, 1997.

23. Mauro DJ. An update on starch. *Cereal Foods World* 41(10):776–780, 1996.

24. McCue N. Starches build flavor and functionality. *Prepared Foods* 166(2):73, 1997.

25. Mermelstein NH. Analyzing for resistant starch. *Food Technology* 63(4):80–84, 2009.

26. Penfield MP, and AM Campbell. *Experimental Food Science.* Academic Press, 1990.

27. Radosavljevic M, J Jane, and LA Johnson. Isolation of amaranth starch by diluted alkaline-protease treatment. *Cereal Chemistry* 75(2):212–216, 1998.

28. Rees N. Thickeners: A guide to the starches that make pie fillings, sauces, and gravies gel. *Fine Cooking* 81:30–31, 2006.

29. Schenck FW. Solid starch hydrolysates. *Cereal World Foods* 41(5):388–390, 1996.

30. Stevens M. Making and using a roux as a classic thickener. *Fine Cooking* 19:72, 1997.

31. Storz E, and KJ Steffens. Feasibility study for determination of the dextrose equivalent (DE) of starch hydrolysis products with near-infrared spectroscopy (NIRS). *Starch/Starke* 56:58–62, 2004.

32. Tharanathan RN. Starch: Value addition by modification. *Critical Reviews in Food Science and Nutrition* 45(5):371–384, 2005.

33. Thomas DJ, and WA Atwell. *Starches.* Eagan Press, 1999.

34. Vodovotz Y, and P Chinachoti. Thermal transitions in gelatinized wheat starch at different moisture contents by dynamic mechanical analysis. *Journal of Food Science* 61(5):932–937, 1996.

35. Wang LZ, and PJ White. Structure and properties of amylose, amylopectin, and intermediate materials of oat starches. *Cereal Chemistry* 71(3):263–268, 1994.

36. Wiesenborn DP, et al. Potato starch paste behavior as related to some physical/chemical properties. *Journal of Food Science* 59(3):644–648, 1994.

37. Yook C, UH Pek, and KH Park. Gelatinization and retrogradation characteristics of hydroxypropylated and cross-linked rices. *Journal of Food Science* 58(2):405–407, 1993.

# WEBSITES

The National Starch Food Innovation Company provides an excellent dictionary on food starch terms (click on "Resources" at the bottom of the home page for North America; then click on "Dictionary") as well as information regarding new innovations and market trends in starch products:
**www.foodinnovation.com**

# 19 Quick Breads

**A**rchaeological evidence suggests that the first bread eaten by human ancestors was a quick bread. It was probably made by mixing flour with water and baking the dough on hot stones. Quick breads are called *quick* because they are baked immediately after the ingredients have been mixed. There is no waiting, as in yeast breads,

for leavening (rising) to take place through the slow fermentation of yeast. Instead, quick breads are leavened during baking with air, steam, and/or carbon dioxide, which can be produced through the action of baking soda or baking powder (see Chapter 17). In short, quick breads are made quickly and without yeast, whereas yeast breads take hours because of the time required for the yeast to multiply before the cells are able to release enough gasses to raise the dough.

As seen in the previous chapter on flour mixtures, the basic bread ingredients are flour, liquid, salt, and leavening agent. All-purpose flour is most

commonly used in quick breads, but various other grain flours can be added to contribute to flavor, color, and texture. Occasionally, cake flour is used in the sweeter quick breads such as coffee cakes. Milk is the most frequently added liquid. Quick breads may also contain added fat, eggs, and sugar, making them sweeter than breads leavened with yeast. If fat is added, it is usually in the form of butter or margarine. The proportions of these various ingredients differ depending on the specific quick bread, but Table 19-1 lists some general guidelines.

This chapter focuses on the description and/or preparation of various quick

**TABLE 19-1** Quick Bread Ingredient Proportions

| Baking Quick Bread | Flour | Liquid* | Fat | Sugar | Eggs | Baking Powder | Salt |
|---|---|---|---|---|---|---|---|
| Biscuits | 1 C | ⅓ C | 2½ tbs | 0 | 0 | 2 tsp | ½ tsp |
| Cream puffs | 1 C | 1 C | ½ C | 0 | 4 | 0 | ¼ tsp |
| Muffins | 1 C | ½ C | 1–2 tbs | 1–2 tbs | ½ | 2 tsp | ½ tsp |
| Pancakes | 1 C | 1 C | 1 tbs | 1 tsp | 1 | 1–2 tsp | ¼–½ tsp |
| Popovers | 1 C | 1 C | 0–1 tbs | 0 | 2 | 0 | ½ tsp |
| Waffles | 1 C | ⅝ C | ¼ C | 1 tsp | 1 | 1–2 tsp | ¼ tsp |

*Liquid is milk, except for boiling water in cream puffs.

breads made from pour batters (pancakes, crepes, waffles, and popovers); drop batters (muffins, quick tea breads, coffee cakes, and dumplings); and doughs (unleavened—actually steam-leavened—breads such as tortillas, chapatis, crisp flat breads, and matzo; and biscuits, scones, and certain crackers that may be leavened with the action of baking powder or soda).

# PREPARATION OF QUICK BREADS

The two most important considerations when preparing quick breads are the consistency of the batter and the cooking temperature. In order to avoid undesirable gluten development, batters are mixed only until the dry ingredients are moistened. Thinner pour batters are used to make pancakes, crepes, waffles, and popovers. Thicker drop batters are used to produce quick breads such as muffins, Boston brown bread, corn bread, hushpuppies, quick tea bread, some coffee cakes, and dumplings. Even though quick tea bread and coffee cake batters are poured from the mixing bowl into baking pans, they are still called drop batters. Doughs contain more flour than drop batters do and are kneaded briefly. They serve as the basis for unleavened breads, biscuits, and scones.

## The Muffin Method

The muffin method is the basic method of preparing many quick breads. It consists of three steps:

1. Sift the dry ingredients together.
2. In a separate bowl, combine the moist ingredients.
3. Stir the dry and moist ingredients together with only a few strokes, until the dry ingredients are just moistened but still lumpy.

When kneading is called for, it is very brief, approximately 10 strokes. Overkneading creates too much gluten, which causes the finished bread to be dense and heavy. Unlike yeast breads, quick breads are baked immediately after the batter is mixed. For most quick breads the pans are greased, filled

two-thirds full, and baked at between 350°F and 450°F (177°C–232°C), depending on the type of bread. The bread is done when it is brown and passes the toothpick test. This consists of inserting a wooden toothpick vertically into the center, immediately withdrawing it, and checking to see if it is completely clean of batter. If there is no batter clinging to the toothpick, the bread is done.

# Additives Used in Quick Breads

Food additives are included in all kinds of baked goods, and quick breads are no exception. Calcium propionate, sodium propionate, sodium diacetate, and sorbic acid are preservatives that inhibit mold growth. Corn syrup provides the dual benefit of a sweetener and preservative. Sucrose, or table sugar, is another sweetener used for flavoring. Polysorbates act as emulsifiers to help ingredients mix well. Sodium citrate speeds color fixation of baked goods. Tartaric acid is a common additive that controls pH.

# VARIETIES OF QUICK BREADS

Quick breads can be categorized according to the flour mixture from which they originate: batter (pour or drop) or dough. Each of these flour mixtures differs in the amount of water that is added to the flour. The fluidity of the dough along with its ingredients and preparation method determine what type of quick bread is produced. Various quick breads are now described according to the classification of their flour mixtures.

## Pour Batters

Quick breads made from pour batters include pancakes, crepes, waffles, and popovers (Figure 19-1). These quick breads contain enough fluid so that their batters actually pour. Approximately two thirds to 1 cup of water is added for every cup of flour when making pour batters.

### Pancakes

Pancakes are made from a pour batter, the consistency of which is dependent upon the proportions of ingredients. The mixing technique and the griddle temperature are the key factors affecting pancake quality. The muffin method is used for mixing the liquid and dry ingredients. Too much stirring will result in dense, heavy pancakes, because stirring develops gluten and causes the carbon dioxide gas in the batter to escape.

To cook pancakes, a griddle or frying pan is heated and then tested by flicking a few drops of cold water onto the surface. If the water pops up and "dances" across the surface of the griddle, the temperature is just right; it is too cold if the water droplets stay on the surface and boil, and too hot if the water drops vanish instantly in a whisk of evaporation. The griddle should be lightly greased, although some griddles are specially coated so that adding a film of fat is unnecessary. For each standard-size pancake, one fourth cup of batter is gently poured onto the griddle. When bubbles start to appear over most of the pancake's surface, the underside should be a delicate brown, and it is ready to turn over. Any additions, such as fruit (e.g., blueberries) or nuts, are added before turning. The second side will usually not brown as evenly as the first, which always serves as the presentation side. For best results, turn the pancakes

**FIGURE 19-1** Quick breads made from pour batters—pancakes, crepes, waffles, and popovers.

The ratio of flour to liquid determines whether the mixture is a pour or drop batter or a dough.

**Pour batters** are quite thin.

Use for:
- Pancakes
- Crêpes
- Waffles
- Popovers

only once and do not press down on them with the spatula, as pressure will result in a too-flat, heavy product.

### Crepes

*Crêpe* is the French word for a thin pancake used to wrap other ingredients. The fillings in crepes may be quite sweet—syrups, creams, or fruit—in which case the crepes are classified as a dessert (crêpes suzette, crêpes Jacques, crêpes empire, and crêpes soufflé) or as a sweet-roll type breakfast item (5). Blintzes are sweet crepes that have been filled and then sautéed. Crepes can also be filled with nonsweet (savory) preparations such as chicken, meat, seafood, and vegetable combinations.

Lacking the baking soda and/or powder used in pancakes, crepes are much thinner. Crepe batters are best made in advance and allowed to sit in the refrigerator for several hours or overnight to allow the flour to absorb all the liquid. The batter is then thinned with milk. A special crepe pan, essentially a frying pan with about a 6-inch diameter, is often used, although it is not required. The pan is heated until hot, brushed lightly with melted butter, and filled with a thin layer of batter that is then spread quickly by tilting and rotating the pan. Care should be taken not to let batter slip up the sides. If the batter is too thick to pour, slightly more milk should be added; conversely, if large bubbles form as the crepe cooks, the batter is too thin and needs more flour. The crepe is turned with a flat spatula when it is golden-brown on the bottom, after which it is heated for another minute, and then gently slid from pan to plate. Crepes are usually stacked on top of each other to prevent them from drying out while others are made.

### Waffles

Waffles are made from a pour batter that contains more fat than a pancake batter. Folding beaten egg whites into waffle batter adds extra crispness and lightness. A waffle iron should be greased lightly with a vegetable spray. Most have an indicator that tells when the waffle iron has reached the proper temperature. Once that temperature is reached, the batter is poured into the middle of the waffle iron for even distribution. Pouring the batter from a pitcher, ladle, or measuring cup allows proper monitoring of the batter flow. When the waffle iron is two-thirds full, the cover is closed and the waffle is cooked for approximately 5 minutes, or until steam has stopped escaping from the waffle iron. If the lid offers any resistance, the waffle is not quite ready. Waffles are done when they are golden brown, crisp, and tender.

Waffle irons should never be submerged in water for cleaning. They are brushed free of crust and crumbs, and the outside is wiped clean with a damp cloth.

### Popovers

A popover is a puffy bread product that looks like an oversized, tall muffin, but with a consistency more like a warm roll. A popover has very thin, moist sides and a hollow center. The name of this quick bread comes from the behavior of the batter during baking: it expands to such a degree that it "pops over" the sides of the container. Popovers can be plain or flavored with herbs, spices, or cheese. They are made from one of the thinnest of all quick bread batters with a liquid-to-flour ratio of 1:1.

Gluten does not readily develop when there is so much liquid, so popovers rely on protein coagulation and starch gelatinization for structure. It is for this reason that at least two large eggs per cup of flour are added to a popover flour mixture (11). Lack of sufficient eggs in the mixture will result in a soggy, compact popover that will fail to expand properly. Egg proteins and starch both provide structural strength, while fat, either added as an ingredient or provided by the egg yolks, helps to make the popovers tender.

It is important not to add too much fat to the flour mixture, because it will weaken the popover structure, allowing steam to escape and hindering the formation of the cavity. Too much water will have the same effect. Another crucial factor in the formation of the cavity is the depth of the pans, which should be twice as deep as those used to make muffins. Although muffin pans or custard cups can be used, special popover pans are best. These pans enable the steam to collect in such a way that if the ingredients are prepared and baked properly, they literally pop up from the baking cup. The individual cups should be heavily greased and the oven and cups preheated while the batter is being made.

The muffin method is used to combine the ingredients; however, the batter is beaten until it is smooth and free of lumps. In the case of popovers, this additional beating is desirable because it gives the finished product its characteristic chewy texture. Overmixing is not a problem because of the large amount of water, which prevents excess gluten formation. When the batter is ready, the preheated cups are filled three-quarters full of batter and immediately placed in a hot (450°F/232°C) oven for about 15 minutes. The temperature is then lowered to 350°F (177°C) and the popovers are baked an additional 20 minutes, or until the proteins coagulate and the structure is properly set. The initial high temperature setting and the preheated muffin tins create leavening by converting the water into steam. Opening the oven or removing the popovers too soon can cause them to collapse. Once the popovers are baked and out of the oven, sogginess is prevented by making a small slit in the top to allow this steam to escape.

> ### ? How & Why?
>
> *Why do popover batters contain so much liquid?*
>
> The high liquid concentration of this pour batter allows the popovers to be leavened with steam, creating the characteristic large cavity in the center. The large amount of liquid in popover flour mixtures also allows gelatinization of the starch during baking, which is indicated in part by the soft, gel-like texture in the inside walls of a baked popover.

## Drop Batters

Drop batters do not contain as much water as pour batters—only about ½ to ¾ cup of water for every cup of flour. Examples of quick breads made from drop batters include muffins, quick tea breads, coffee cakes, and dumplings (Figure 19-2).

### Muffins

Muffins are one of the easiest quick breads to prepare. They usually contain flour, liquid, fat, egg, sugar, salt, a leavening agent, and a flavoring ingredient

**FIGURE 19-2** Quick breads made from drop batters—muffins, quick tea breads, coffee cakes, and dumplings.

**Drop batters** are thicker because they have more flour.

Use for:
- Muffins
- Quick tea breads
- Coffee cakes
- Dumplings

**FIGURE 19-4** Overstirring muffin batter results in tunnels.

Optimal stirring

Overstirred

Digital Works

that may also add texture. The general ratio of flour to liquid is usually 2:1. Muffins can be quite high in fat and sugar, with honey, molasses, brown sugar, or syrup sometimes taking the place of sugar. White flour can be partially replaced by other cereal products, such as cornmeal, oatmeal, or whole-wheat flour.

***Avoid Overmixing*** One of the keys to making a good muffin is to use the muffin method in mixing the ingredients, just barely moistening the dry ingredients. Some small lumps in the batter are desirable; a smooth batter means that overmixing has occurred, and too much gluten has been developed. Overmixing creates a muffin

with a smooth, peaked top and an interior that is tough and riddled with tunnels (Figure 19-4). The formation of tunnels occurs as the protein in the exterior portion of the muffin coagulates with the increased temperatures of baking, while the interior is still expanding. If there is too much gluten, it continues to stretch with the pressure of the expanding gases. The tunnels form when they are finally coagulated with the heat moving toward the center of the muffin. The likelihood of tunnels is reduced by using whole-grain flours or flours other than all-purpose. The bran from whole-grain flours cuts gluten and interferes with its development, whereas the lesser amount of protein in most other flours decreases their potential for gluten development (11).

***Avoid Undermixing*** Although both avoiding overmixing and using the appropriate flour help to avoid tunnels, undermixing leaves lumps that are too large, indicating insufficient gluten development and resulting

## NUTRIENT CONTENT

Quick breads consist primarily of carbohydrate, with varying amounts of fat depending on the type of quick bread (see Calorie Control). Quick breads are somewhat higher in both fat and sugar than yeast breads, although modifying the ingredients can produce quick breads with less fat and fewer calories.

***Healthful Recipe Modifications.*** The nutritional profile of certain quick breads can be altered to suit special needs by (8, 9):

Adding fiber.
- Substituting whole-wheat flour, bran, oatmeal, or wheat germ for about one fourth of the white flour to increase fiber content

Reducing sugar.
- Reducing sugar by 25% by using dried fruit purées

Reducing fat.
- Replacing hydrogenated vegetable shortening with nonhydrogenated margarine or vegetable oils to decrease the saturated fat content
- Adding starches to reduce fat content (in a few commercial cases)
- Substituting two egg whites for each egg
- Using flour blends made with a fat substitute (3) (for commercial bakers)

However, removing the fat from quick breads can cause several problems in the final product. For instance, low-fat tortillas may have poor dough-mixing properties due to a tough dough, chewy texture, cracking during storage, and decreased shelf life (12). Adding a carbohydrate-based fat replacer (hydrolyzed rice flour) that is hydrated with water mimics the texture and mouthfeel of fat while improving the dough's pliability and rollability and the freshness of the final tortilla (Figure 19-3).

**FIGURE 19-3** Low-fat tortillas.

Laura Murray

## ✦ CALORIE CONTROL

Low-carbohydrate diets cut back on the bread/cereal/pasta group because excessive intakes can contribute calories as well as approximately 15 grams of carbohydrate for each slice or cup. Although it's important to eat at least the minimum amounts of whole grains and breads for good health, it's equally critical to control this category for successful weight management.

- **Bread averages 80–200 calories (kcal) for each slice.** Count your recommended number of daily servings to stay within your calorie limit.
- **Limit the number of servings.** For instance, 3 pancakes (7") provide 513 calories (kcal), but 6 pancakes provide 1,026 calories.
- **Choose smaller servings.** For example:
  - Muffins: mini (1¼") = 68 calories (kcal), medium = 444 calories, extra-large = 660 calories
  - Taco shells: taco size = 50 calories (kcal), salad shell = 280 calories
  - Flour tortillas: 6" diameter = 94 calories (kcal), 12" diameter = 380 calories
  - Pancakes: 4" diameter = 84 calories (kcal), 7" diameter = 171 calories
- **Watch the butter, margarine, or preserves.** Calories also come from what is spread on breads. Fats provide about 100 calories (kcal) for each tablespoon, and preserves about half that amount (50 calories).
- **Each chip counts.** Chips (wheat, corn, etc.) are not calorie cheap. A few are fine because they only average 10 calories (kcal) each. However, they vary in size and thickness, so read the Nutrition Facts label. Most chips range from 5 to 40 calories (kcal) each. To find out the number of calories for each chip, divide the total calories per serving by the number of chips in each serving (i.e., 150 calories/kcal divided by 15 chips = 10 calories for each chip). The small snack bags average 150 calories (kcal) each. Never sit down to a big bag or bowl of chips because one big bag (11 ounces) yields 1,500 to 2,000 calories (kcal). If your calorie snack cap is 200, put 20 chips in the bowl or baggie to take to school, work, or elsewhere.
- **Don't dip, just skim.** Dips help you feel full because they are primarily fat. The exception is tomato-based salsa, which actually saves on calories. Skim the chip through a high-calorie dip for taste, rather than using the chip to scoop up a big dollop of dip.

| Dip | Serving | Calories (kcal) | Fat (g) | Carbohydrate (g) | Fiber (g) |
|---|---|---|---|---|---|
| Fat-based (e.g., Ranch) | 1 tablespoon | 50+ | 5 | 2 | 0 |
| Salsa | 1 tablespoon | 5 | 0 | 2 | 0 |

- **Fiber first.** Fiber sends signals of fullness to the brain and soluble fiber stabilizes blood glucose levels. Bread products are a major source of fiber in the diet. Choosing whole-grain products helps to increase daily dietary fiber intake. Crispbreads deliver more grams of fiber and are tasty when served with a small serving of fruit and/or cheese.

© 2010 Amy Brown

### TABLE 19-2   Quick Bread Problems and Their Causes

**Quick Bread Problems and Causes**

| Problems \ Causes | Excessive mixing | Oven too hot | Oven too cool | Too much flour | Baked too long |
|---|---|---|---|---|---|
| Peaked top | X | | | | |
| Smooth crust | X | | | | |
| Pale | X | | X | | |
| Burned | | X | | | |
| Tough, "elastic" | X | | | X | |
| Tunnels | X | | | | |
| Very compact | X | | | | |
| Too dry | | | X | X | X |

in a crumbly muffin that falls apart. Insufficient mixing also leaves the baking powder incompletely moistened, which results in a low-volume muffin. It is also important to beat the eggs separately before adding them to the liquid ingredients; insufficiently beaten eggs can concentrate on the outside of the muffin, where their coagulation results in an undesirable shiny, waxy appearance (11). A summary of these and other potential problems with muffins, along with their causes, is shown in Table 19-2.

**Added Ingredients**   Fruits or nuts may be added to the batter for flavor enhancement. To prevent their sinking to the bottom of the muffins, gently roll the fruit and nut pieces between lightly floured hands or thoroughly toss them in a bag containing some flour.

**Baking**   Once mixed, the batter and all its ingredients are ready for the muffin pan. Only the bottoms of the individual cups should be greased, because ungreased sides give the dough traction as it rises, thereby allowing it to rise higher. The muffin cups are filled two-thirds full, and the batter is baked at about 400°F (204°C) for about 20 to 25 minutes. The muffins will slide out of their individual cups more easily if the hot pan is first placed on a wet towel.

### Muffin Breads

Changes in the basic ingredients of a muffin recipe result in a variety of other quick breads, such as Boston brown bread, corn bread, hushpuppies, and a variety of tea breads.

**Boston Brown Bread**   Unlike most other breads, Boston brown bread is made with rye and graham flours, and is steamed instead of baked. The mixture is placed in containers (coffee cans or round cylindrical baking pans are an option), covered, and placed in a pan of simmering water. The containers are then covered and allowed to steam for 2 to 3 hours, depending on the size of the container. Following steaming, the covering is removed and the bread is placed in a moderate (350°F/177°C) oven for about 15 minutes to allow the top to dry out.

*Corn Bread* Corn bread flour mixtures are similar to a muffin batter, but include a combination of cornmeal (1¼ cup) and all-purpose flour (¾ cup). Sometimes small chunks of cheddar cheese, onions, chili peppers, corn kernels, and/or cooked meats such as ham or bacon are added. The muffin method is used to combine the ingredients, which can then be used to prepare corn bread, muffins, or sticks. To obtain the characteristic brown, crunchy crust and slightly gritty textured crumb, the pan should first be greased and heated to 425°F (218°C) before filling it with the batter. Baking times average 15 minutes for sticks and about 20 minutes for corn bread or muffins.

*Hushpuppies* Hushpuppies are a variation of corn bread and are made from a mixed, stone-ground cornmeal drop batter. They are shaped into small round or oblong balls and deep-fried in a skillet. This bread is usually made to accompany pan-fried fish, especially trout or catfish (4). The name is thought to have come from fishermen who threw them to their hungry, whining dogs to hush them up. Another possibility is that hushpuppies got their name from cowboys who used the bits of bread to quiet their dogs out on the range.

*Tea Breads* Tea breads are similar to muffins, but they are baked in a loaf pan. These are the sweetest of all quick breads and are frequently sliced and served, as the name implies, at teatime or as dessert. Those most commonly encountered are banana, zucchini, nut, carrot, cranberry, and blueberry breads. These breads are usually made from a drop batter that is placed in loaf pans and baked at 350°F (176°C) for about 1 hour. However, smaller containers can be used and the baking time reduced accordingly. Most tea breads keep well if they are wrapped tightly and stored in the refrigerator.

## Coffee Cakes

Coffee cakes usually contain nuts and/or raisins and are often topped with a brown sugar and butter mixture or other sweet topping. They may be mixed by the muffin method or by following the individual recipe instructions. Unlike tea breads, they are best served immediately after baking. Coffee cakes may also be made from yeast doughs; these are not considered to be quick breads.

## Dumplings

A dumpling is a small ball of flour (about an inch in diameter) combined with a few other ingredients. Dumplings are simmered briefly (5 to 20 minutes) in water, stock, or gravy and are commonly used as an ingredient in soups or stews. Plenty of liquid should be used to prevent them from becoming overcrowded in the pan or they will stick together and cook unevenly. Overcooking should also be avoided, because protein in the eggs usually used to bind dumplings together will toughen when heated too long.

# Doughs

Some quick breads are made from doughs that are briefly kneaded. The dough may be unleavened, producing a flat bread, or include leavening and be used to make biscuits or scones. As explained in Chapter 17, doughs can be either stiff/firm or soft depending on the amount of water that is added to each cup of flour (one eighth and one third cup of water, respectively). Doughs can be used to create quick breads, specifically unleavened (actually steam-leavened) breads such as tortillas, chapatis, crisp flat breads, and matzo; and those leavened with the assistance of either baking powder or soda—biscuits and scones (Figure 19-5).

## Unleavened Breads

Unleavened breads are the world's oldest breads and the easiest to prepare (1). Sometimes they are called *flat breads*, but that is an ambiguous term, because some breads leavened with yeast, such as pizza, pita, and English muffins, also appear flat (13). True unleavened breads are leavened by steam rather than some other agent. Flat breads can be steamed, oven- or skillet-baked, fried, grilled, or baked underneath the hot desert sand. Throughout the world, unleavened breads appear in different forms: tortillas and tacos in Central and South America; chapatis in India, Pakistan, and Iran; crisp flat breads in Sweden and other Scandinavian countries; and matzo in Israel. Flat breads represent a major form of wheat consumption in many Middle Eastern and North African countries (7).

*Tortillas* Mexico is the home of the tortilla, an unleavened, circular bread ranging in diameter from 4 to 14 inches. Tortillas may be prepared using either wheat flour or cornmeal.

- *Flour Tortillas.* In contrast to corn tortillas, the more pliable, softer flour tortillas are made from white flour, vegetable shortening or lard, water, and salt. After mixing, the dough is shaped into a circle, rolled or pressed into a flat, thin tortilla, and heated 30 to 60 seconds on each side on a hot griddle at 356°F–410°F (180°C–210°C) (10). Tortillas can be used to make a variety of Mexican dishes including burritos, enchiladas, and cheese quesadillas (15).

- *Corn Tortillas.* Corn tortillas have been a staple food in Mexico and Central America for many centuries. Today, sales of tortilla and tortilla chips continue to climb (2). The tougher-textured corn tortillas are prepared from masa, a cornmeal made from corn treated with lime, an alkaline solution. Commercially, tortillas are used to make nachos (deep-fried wedges of corn tortillas), tacos, enchiladas, burritos, and tostadas. Strips of tortillas can also be used to thicken traditional soups and stews.

**FIGURE 19-5** Quick breads made from doughs: biscuits; scones; and unleavened, actually steam-leavened, tortillas, chapatis, crisp flat breads, and matzo.

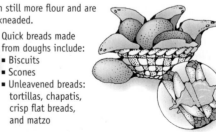

**Doughs** contain still more flour and are usually lightly kneaded.

Quick breads made from doughs include:
- Biscuits
- Scones
- Unleavened breads: tortillas, chapatis, crisp flat breads, and matzo

## ? How & Why?

### Why is lime added to corn when making cornmeal?

Whole corn has a hard outer husk, which is softened by boiling it in lime (calcium hydroxide) water, holding it for about 45 minutes, and steeping it at room temperature for an additional 12 to 18 hours (10). This cooked maize (nixtamal) is then cooled, washed, and ground into cornmeal. Water and some salt are then added to yield the final product called masa. The alkaline solution also increases the nutrient content of tortillas made from masa. It increases the mixture's calcium content and simultaneously releases niacin, a B vitamin, from the corn (14).

*Chapatis*  Common in India, Pakistan, and Iran, chapatis are one of the bread staples in these countries. They are prepared by mixing whole-wheat flour, water, clarified butter (known as *ghee* in Indian cuisine), and salt. This mixture is kneaded and rolled into very thin 6-inch circles before being heated on a hot griddle until the crust browns and starts to blister.

*Crisp Flat Breads*  Scandinavian countries are famous for their crisp flat breads made without any added leavening. The degree of flatness varies from paper thin for Norwegian *lefse* to the thicker crisp breads of Norway, Sweden, and Finland. Rye and wheat are the two most common flours used in their preparation, and the color varies from very light to brown.

*Matzo*  Because of their rapid exodus from Egypt in Biblical times, the Israelites did not have time to allow their bread to rise, so Passover (which celebrates the Exodus) is commemorated in part by consuming the unleavened bread known as matzo. Matzo meal is made from the crumbs of matzo and is used for breading and stuffing, for dumplings in Passover soup, in cakes, and as a binding agent.

### Biscuits

The word "biscuit" comes from the Latin term *bis cotus*, meaning "twice baked" (6). Although the technique is no longer practiced, people used to first cook biscuits in a hot oven and then dry them in a cool oven. Biscuits, which are relatively quick to prepare, rely on fat for shortening power, and on just the right amount of kneading to increase gluten formation. A biscuit dough contains flour, fat, milk, baking powder, and salt. Formerly, lard was usually the fat of choice, but butter may also be used, and margarine or hydrogenated vegetable shortening are common choices. The dough is much less sticky than muffin batter because the ratio of flour to milk is 3:1 (11).

During biscuit preparation, the dry ingredients are first mixed together; the fat is cut in with a pastry blender or two knives until particles the size of rice grains are formed; and then milk or another liquid is added to an indentation, or well, made in the center of the flour mixture. The flour is mixed only minimally in short dough, to minimize the development of a gluten network (6). Once the liquid is mixed in, the dough is then formed into a ball and placed on a lightly floured breadboard, where it is kneaded briefly, for about half a minute, until the stickiness disappears. The lack of liquid makes kneading necessary for gluten development. Overkneading causes toughening because of excess gluten formation, and allows carbon dioxide gas to escape, resulting in a compact, less tender biscuit.

Once kneaded, the dough is rolled out on a floured surface to the desired thickness: ¼ inch for plain biscuits, ½ inch or less for tea biscuits, and 1 inch or more for shortcake. It is then cut into round shapes using a sharp-edged cutter that has been very lightly dipped in flour. The dough rounds are placed on an ungreased cookie sheet 1½ inches apart for crisp sides. Inadequate spacing between the dough pieces will result in uneven or caved-in sides (Figure 19-6). Lightly brushing the tops with a thin coat of butter or margarine improves browning. Biscuits are baked at 425°F (218°C) for about 10 to 20 minutes or until the surface turns a golden brown color.

Buttermilk, whole-wheat flour, cheese, herbs, vegetables, and other ingredients can be added or substituted to produce variations of the basic biscuit. Buttermilk or any other acidic ingredient promotes the formation of biscuits with a white interior, whereas more alkaline ingredients, such as high levels of baking powder, contribute to a more creamy color.

### Scones

Scones contain eggs and milk or cream, which makes them richer than ordinary biscuits. Their flavor also differs from biscuits because they often contain pieces of dried fruits such as raisins, cranberries, blueberries, apples, apricots, currants, or cherries. They are baked at about the same temperature and for the same length of time as biscuits.

The eggs are beaten separately and combined with the cream. Fat is cut into the dry ingredients (flour, baking powder, sugar, and salt), a well is made, and the remaining liquid ingredients are poured into it. The dry and liquid ingredients are combined with a minimum number of strokes. The dough is placed on a floured board, patted

**FIGURE 19-6**  Baked products placed next to one another in the oven will tear on the sides that are touching.

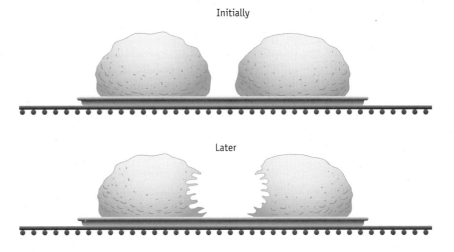

Initially

Later

# PROFESSIONAL PROFILE

Ever since Pamala Hayes was in grade school, she has loved to prepare food. Her mother was an excellent cook, and Pamala helped her every chance she got. A unique program in her Ohio high school allowed Pamala to work several hours a day in a hotel kitchen. She liked it so much that she started working in an Argentinean restaurant. On her high school's career day, she learned about the early admissions program to Johnson & Wales University in Rhode Island, from which she graduated with an Associate of Sciences degree in Culinary Arts. Part of her degree requirements was a 3-month internship that she completed at Disney World in Florida.

She liked her internship so much that she planned to work for Disney World, but instead she went to the World Center Marriott Resort in Orlando. Marriott International, Inc., is an international chain of more than 2,500 restaurants and hotels. Pamala started in the fine dining department there as a pantry/pastry chef and worked her way up through grill position, sauté position, and expeditor (grill and sauté coordinator, including vegetables, soups, and garnishes).

Marriage moved her to another town in Florida, where she got a job as a sauté/grill cook at a hotel restaurant. She found many of her ensuing jobs through word of mouth, the newspaper, or the Web (e.g., www.careerbuilder.com and www.monster.com). As a result, she worked as a sous chef for both Windham and Hilton hotels.

After her husband accepted a position at the University of Hawaii, Pamala had to contend with an unfamiliar job market, and a tight one at that. It had been relatively easy for her to find a job in Florida, a state in which restaurant workers are nonunionized, creating an environment in which there is generally a high level of turnover. In Hawaii, however, hotel industry culinary jobs are often unionized, and workers are less likely to leave their jobs. She

Pamala Hayes

also arrived during the slow season for tourism. Therefore, she accepted two part-time fry cook jobs—one at a Marriott and another at a private, free-standing restaurant.

As she worked her two part-time jobs, she kept looking for permanent work and finally found it at the Hilton Hawaiian Village Beach Resort and Spa as a tournant chef. This person fills in for the restaurant chef responsible for all the restaurants at a resort or large hotel. She then became the chef for one of the Hilton's many restaurants and is responsible for serving 700 people for breakfast, and about 150 people for each lunch and dinner. She works five 10-hour shifts a week, but those hours go up on holidays such as Mother's Day and Easter.

Pamala says that she prefers working at hotel restaurants because she likes the "structure, stability, consistency, and security. At free-standing restaurants, there was always the looming threat of disaster, and the work environment depended so much on the manager's personality."

When asked what she considers to be the biggest mistake that people make going into the restaurant business, Pamala says that people "assume that because they love to cook, they'll love to work in a kitchen. You have to love to *work*. And before they invest in a job, especially the time and expense of culinary school, they should work in a kitchen to see if they like it."

What does she think about working in a restaurant? "Restaurant work," says Pamala, "is standing on your feet, burns, knife nicks, cleaning up, and constant pressure to have everything done at a certain time. The best thing about it is the creativity and instant gratification. A scientist may have to wait 6 years to see any results, but in the restaurant business your results are right in front of you every day."

*Sodexho Inc., www.sodexhousa.com, branched off from Marriott Corporation, and is now the leading food service management corporation in North America. Sodexho manages the food services (cafeteria or restaurants) in more than 6,000 corporations, large hospitals, long-term care and retirement centers, schools, college campuses, and the military.

down to ¾-inch thickness, and cut into diamonds or other shapes. To produce a sheen, the tops are brushed with 2 tablespoons of the egg-cream mixture that has been set aside. If desired, the scones can then be sprinkled lightly with sugar.

## Crackers

Crackers are baked products low in both sugar and fat (6). These crisp breads may or may not be leavened (usually chemically leavened with baking soda or powder) (see leavening agents in Chapter 17). Holes are pricked into

cracker dough to prevent air pockets from forming and bursting during baking. Cracker dough develops a strong gluten network, so the protein content and quality of the dough are important in creating the right texture.

# PICTORIAL SUMMARY / 19: Quick Breads

The first bread eaten by our human ancestors was probably a quick bread, made by adding flour to water and baking the dough on hot stones. Quick breads are baked immediately after the ingredients have been mixed, and are either unleavened or leavened with air, steam, and/or baking soda or baking powder.

## PREPARATION OF QUICK BREADS

The basic ingredients of any bread are flour and water or other liquid, and possibly salt and a leavening agent such as baking soda or powder. Quick breads usually contain added fat, eggs, and sugar.

The two most important factors when preparing quick breads are:
- The consistency of the batter
- The cooking temperature

**The muffin method** is the basic method of preparing many quick breads. It consists of three steps:

1. Sift the dry ingredients together.

2. In a separate bowl, combine the moist ingredients.

3. Stir the dry and moist ingredients together with only a few strokes, until the dry ingredients are moistened but still lumpy.

### Quick Bread Problems and Causes

| Problems \ Causes | Excessive mixing | Oven too hot | Oven too cool | Too much flour | Baked too long |
|---|---|---|---|---|---|
| Peaked top | X | | | | |
| Smooth crust | X | | | | |
| Pale | X | | X | | |
| Burned | | X | | | |
| Tough, "elastic" | X | | | X | |
| Tunnels | X | | | | |
| Very compact | X | | | | |
| Too dry | | X | X | | X |

For most quick breads, kneading, if called for at all, is very brief, and the pan is greased, filled two-thirds full, and baked at between 350°F and 450°F (177°C–232°C), depending on the type of bread.

## VARIETIES OF QUICK BREADS

The ratio of flour to liquid determines whether the mixture is a pour or drop batter or a dough.

**Doughs** contain still more flour and are usually lightly kneaded. Quick breads made from doughs include:
- Biscuits
- Scones
- Unleavened breads: tortillas, chapatis, crisp flat breads, matzo, and certain crackers

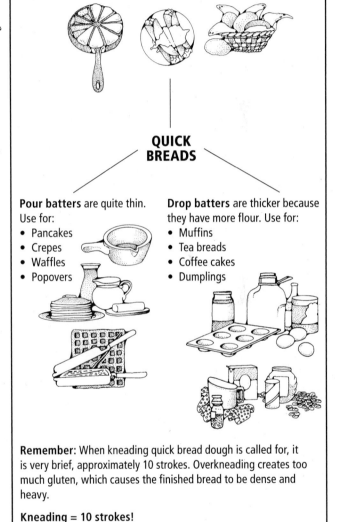

## QUICK BREADS

**Pour batters** are quite thin. Use for:
- Pancakes
- Crepes
- Waffles
- Popovers

**Drop batters** are thicker because they have more flour. Use for:
- Muffins
- Tea breads
- Coffee cakes
- Dumplings

**Remember:** When kneading quick bread dough is called for, it is very brief, approximately 10 strokes. Overkneading creates too much gluten, which causes the finished bread to be dense and heavy.

**Kneading = 10 strokes!**

# CHAPTER REVIEW AND EXAM PREP

## Multiple Choice*

1. The two most important factors when preparing quick breads are the _____ of the batter and the _____.
   a. thickness, flour
   b. fluidity, liquid
   c. consistency, temperature
   d. creaminess, fluidity

2. When using the "muffin method" to prepare quick breads, about how many strokes should you use when mixing the wet and dry ingredients together?
   a. One
   b. A few
   c. 5 to 10
   d. 20 to 25

3. What is the ratio of liquid to flour in popover batters?
   a. 1:3
   b. 2:1
   c. 2:3
   d. 1:1

4. What type of whole-wheat unleavened bread is commonly used in India, Pakistan, and Iran?
   a. Tortilla
   b. Ghee
   c. Matzo
   d. Chapati

5. To which of the following ingredients can the difference between scones and biscuits be attributed?
   a. Flour
   b. Eggs
   c. Baking powder
   d. Salt

6. A drop batter is used to prepare _____.
   a. pancakes
   b. coffee cakes
   c. crepes
   d. waffles

7. Biscuits, scones, and unleavened breads are made from a _____.
   a. dough
   b. pour batter
   c. drop batter
   d. thick batter

## Short Answer/Essay

1. Why are quick breads called *quick*? Give several examples of quick breads made from drop batters, pour batters, and doughs.
2. Which mixing method is used to make most quick breads? Explain the three basic steps.
3. Briefly describe how pancakes, crepes, and popovers are prepared.
4. Why do tunnels sometimes form when preparing muffins, and how can this be avoided?
5. How much of what other flour is added to corn bread flour to produce a baked product? What other ingredients may be added to corn bread flour mixtures?
6. What are tea breads? Provide several examples.
7. What are dumplings?
8. How are unleavened breads different from other quick breads? Provide four examples of unleavened breads.
9. Briefly describe the process by which biscuits are prepared.
10. How does the method for preparing scones differ from that for biscuits?

*See p. AK-1 for answers to multiple choice questions.

# REFERENCES

1. Alford J, and N Duguid. *Flatbreads and Flavors: A Baker's Atlas.* William Morrow, 2008.
2. Almeida-Dominguez HD, GG Ordonez-Duran, and NG Almeida. Influence of kernel damage on corn nutrient composition, dry matter losses, and processibility during alkaline cooking. *Cereal Chemistry* 75(1):124–128, 1998.
3. Blends reduce fat in bakery products. *Food Technology* 48(6):168–170, 1994.
4. Bocuse P, and F Metz. *The New Professional Chef.* The Culinary Institute of America. Van Nostrand Reinhold, 1996.
5. Friberg B. *The Professional Pastry Chef: Fundamentals of Baking and Pastry.* Van Nostrand Reinhold, 2002.
6. Gallagher E. Formulation and nutritional aspects of gluten-free cereal products and infant foods. In EA Arendt and F Dal Bello. *Gluten-free Cereal Products and Beverages.* Academic Press, 2009.
7. Gooding MJ. The wheat crop. In K Khan and PR Shewry. *Wheat: Chemistry and Technology,* 4th ed.

American Association of Cereal Chemists, 2009.

8. Hippleheuser AL, LA Landberg, and FL Turnak. A system approach to formulating a low-fat muffin. *Food Technology* 49(3):92–96, 1995.

9. Hogbin M, and L Fulton. Eating quality of biscuits and pastry prepared at reduced fat levels. *Journal of the American Dietetic Association* 92(8):993–995, 1992.

10. Martinez-Flores HE, et al. Tortillias from extruded masa as related to corn genotype and milling process. *Journal of Food Science* 63(8): 130–133, 1998.

11. McWilliams M. *Foods: Experimental Perspectives.* Macmillan, 2007.

12. Pszczola DE. Lookin' good: Improving the appearance of food products. *Food Technology* 51(11):39–44, 1997.

13. Qarooni J. Wheat characteristics for flat breads: Hard or soft, white or red? *Cereal Foods World* 41(5): 391–395, 1996.

14. Serna S, M Gomez, and L Rooney. Technology, chemistry, and nutritional value of alkaline-cooked corn products. In Y Pomeranz, ed., *Advances in Cereal Chemistry.* American Association of Cereal Chemists, 1990.

15. Werlin L. Making the best quesadillas. *Fine Cooking* 79:60–63, 2006.

# WEBSITES

View various interesting flat breads at this website:

**www.foodsubs.com** (click on "baked goods," and then on "flatbreads")

Want low-fat versions of flatbreads? Here they are at:

**www.fatfree.com** (click on "recipes by category," scroll down to "breads," and click on "quick breads")

Find step-by-step, simple instructions on how to prepare quick breads by Quaker Oats here:

**www.quakeroats.com** (hold the mouse over "Cooking & Recipes" and click on "Baking 101," then click on "Quick Breads")

# 20 Yeast Breads

When breads are leavened with the carbon dioxide gas produced by yeast, they are known as yeast breads. Baker's yeast, or *Saccharomyces cerevisiae,* is a one-celled fungus that multiplies rapidly at the right temperature and in the presence of a small amount of sugar and moisture. Yeast breads have been around since long before the birth of Christ, and no one knows exactly how it happened that people came to realize that naturally occurring, airborne yeast could be used as a leavening agent. The ancient Egyptians knew the technique and preferred the light airy texture of yeast-leavened bread, and the Romans adopted this procedure,

after which it became known throughout the Roman Empire. The basic principles for preparing yeast breads have not changed in the many intervening centuries. The subject of this chapter covers the preparation, types, and storage of yeast breads, including loaf breads (wheat, whole wheat, sourdough, and malt breads); specialty breads; rolls; pita bread; bagels; English muffins; pizza crust; pretzels and bread sticks; and raised doughnuts.

## PREPARATION OF YEAST BREADS

Yeast bread is prepared by mixing ingredients into a dense, pliable dough that is kneaded, allowed to rise by fermentation, and then cooked, typically by baking but sometimes by steaming or frying. Normally, the preparation of yeast bread is at least a 2½- to 3-hour

operation, which is one reason why many people choose to buy bread rather than to make it. Sourdough breads take even longer, requiring about 8 to 10 hours to fully develop their flavor. Regardless of the time involved, most yeast breads are prepared with the same basic ingredients.

## Ingredients

The fundamental ingredients of any yeast bread are flour, liquid, sugar, salt, and yeast; fat and/or eggs are optional (Table 20-1). Chapter 17, on flours and flour mixtures, explains the individual functions of these ingredients in more detail. Wheat is the most commonly used flour for making breads; rye is the next most common, followed by oat, barley, cornmeal, rice, and other flours. Because flours other than wheat have less gluten-forming ability, these flours must be supplemented by a certain amount of wheat flour in order to improve their baking quality (9). "White wheat" flour, a relatively

**TABLE 20-1  Yeast Bread Ingredients and Their Functions**

| Ingredient | Functions |
|---|---|
| Flour | Wheat flour is the most common flour used in yeast breads because it is the flour with the highest amount of gluten. Gluten is the protein that provides a firm, elastic structure to baked goods (32). All-purpose flour and bread flour (a specially formulated wheat flour that is high in gluten) can be used interchangeably in recipes. They are used either alone or combined with lower- or no-gluten flours, including whole wheat, rye, rice, or soy, and whole grains such as oats. |
| Leavening | Yeast is a living organism that is used primarily in bread baking. When yeast is mixed with liquid and some type of sugar and is kept at the proper temperature, it begins to ferment. The fermentation produces alcohol and carbon dioxide—the alcohol burns off during baking, while the carbon dioxide gas is trapped in the elastic network formed by the gluten, stretching and expanding the gluten, causing the dough to rise. Be sure to check the package for the yeast's expiration date both before purchasing and before using the yeast.<br>There are two major types of yeasts:<br>• *Active dry yeast* is dehydrated and can be purchased in ¼-ounce packets (envelopes) or in jars in the baking aisle of the grocery store. Available as (1) *regular active dry yeast* and (2) *fast* or *quick-rising active dry yeast* (leavens breads in one third to half the normal time of regular yeast).<br>• *Compressed, fresh yeast* is moist and *extremely* perishable. It is sold in 0.6-ounce and 2-ounce cakes in the refrigerated aisle at the grocery store. It should be stored in the refrigerator or freezer and then brought to room temperature before use. One ¼-ounce package yeast = 1 scant tablespoon dry yeast = one 0.6-ounce cake compressed fresh yeast. |
| Sugars/Other Sweeteners | Yeast breads need sugar to feed the yeast that grow and cause the bread to rise. As little as 1 or 2 teaspoons of sugar/sweetener give the yeast a boost and make the dough rise. Granulated sugar, brown sugar, honey, molasses, and maple syrup are the most common sweeteners added to bread flour mixtures. The added sugar also helps to bring out the flavors of the other ingredients in the bread. Sugar also caramelizes during baking, giving the yeast bread a nice brown crust. |
| Fat | The type of fat used impacts the taste and color of the baked bread. If fat is added, options include butter, margarine, shortening, and vegetable oil. Fat retards gluten development, so a bread loaf prepared with fat has a softer, smoother texture. The fat in bread also slows down moisture loss, so breads containing fat stay fresher longer than those without added fat. |
| Liquid | The liquid in yeast bread recipes dissolves and activates the yeast and moistens the flour so the gluten can develop. Liquid that is too hot will kill the yeast; liquid that is too cool will not activate the yeast. If the yeast is dissolved in the liquid, the temperature should be warm (105°F/41°C to 115°F/46°C for active dry yeast; 95°F/35°C for compressed, fresh yeast). If the dry yeast is combined with other dry ingredients first, then the liquid is added, the liquid should be very warm (120°F/49°C to 130°F/54°C).<br>Both water and milk are commonly used for preparing yeast bread. Milk makes dough rise more quickly because the yeast thrives on lactose—the natural sugar in milk. This natural sugar also aids in browning the baked bread. Milk also adds flavor and nutrients, and, like fat, helps keep bread fresher longer. |
| Oats | Oats are a very common ingredient in whole-grain breads. They add texture and a mild nutty taste to all types of yeast breads. |
| Eggs | Eggs enrich the taste, texture, and color of *some* yeast breads, adding strength to the elastic network formed by the gluten. They are most typically used in sweet yeast bread recipes. |
| Salt | Salt gives the yeast bread flavor. Although yeast bread can be made without salt, it is also added to inhibit yeast growth; otherwise, they would continue to multiply and produce too much carbon dioxide that would cause the bread to rise too much and possibly collapse. |

new addition to the market, retains the freshness, taste, and smoother mouthfeel of refined flour, while delivering the nutrient advantages of whole-grain flour—more minerals, B vitamins, and fiber (29, 32). Organic wheat flour is also available; however, some research indicates that conventionally cultivated wheat may be slightly better for bread making, whereas organic wheat flour may be better for cookie preparation (3). Some bread makers are experimenting with adding functional ingredients such as flax seeds (19), prebiotics, and antioxidants (34).

# Food Additives in Baked Products

Fresh bread baked from basic ingredients often does not have any additives other than those used to supplement the flour. Commercial bakers need to protect their baked products from mold, so they usually add calcium proprionate or a similar mold inhibitor such as sodium propionate, sodium diacetate, or sorbic acid. Corn syrup may be added as a sweetener and preservative. Polysorbates can be used as emulsifiers, while sodium citrate assists with color fixation in baked goods. Tartaric acid is used to control pH. Hydrocolloids are actually gums that may be of vegetable, animal, microbial, or synthetic origin. They are added to many products, including breads, to improve texture and viscoelastic quality, to slow starch retrogradation, to bind water, to replace fat, to improve the quality of stored products, and as a gluten replacement (6). Hydrocolloids improve mouth feel and acceptability, increase loaf volume, decrease firmness from staling, and improve the overall shelf-life of baked goods (Chemist's Corner 20-1). Other food additives that

## CHEMIST'S CORNER 20-1

### Hydrocolloid Food Additives

Hydrocolloid molecules contain numerous hydroxyl groups, making them hydrophilic (water loving) (6). The resulting ability of hydrocolloids to bind large amounts of water provides stability to products that are frozen and thawed. They can also increase the volume and moisture content of bread products. Specific types of hydrocolloids used in breads include carboxymethylcellulose (CMC) and guar gum in rye flour. Sodium alginate, kappa-carrageenan, xanthan gum, and hydroxypropylmethylcellulose (HPMC) improve dough stability during proofing. Certain hydrocolloids, including xanthan, alginate, guar gum, pectin, kappa-carrageenan, and HPMC positively affect the gelatinization process and pasting properties of wheat flour.

may be used in baked products include, but are not limited to, those listed in Table 20-2.

## Mixing Methods

The four best-known methods for mixing yeast breads are the straight dough, sponge, batter, and rapid mix methods. Regardless of the mixing method used, all ingredients should be brought to room temperature prior to mixing in order to obtain the desired dough consistency. It is the ingredients, their

| TABLE 20-2 | Selected Food Additives Used in Baked Products |
| --- | --- |
| **Food Additive** | **Description** |
| Ammonium sulfate | A popular yeast nutrient. |
| Ascorbic acid (or ascorbate) | A natural antioxidant known as vitamin C. |
| Azodicarbonimide (ADA) | An oxidant used in dough to give it strength, stability, and better flow and extensibility. |
| B-complex vitamins | A group of water-soluble vitamins such as $B_1$ (thiamin), $B_2$ (riboflavin), niacin, pantothenic acid, $B_6$ (pyridoxine), folate, and $B_{12}$ (cobalamin). |
| Bromate | An oxidant that is sometimes added to flour to enhance and further develop the gluten and make the mixing or kneading of the dough less difficult. The use of bromate has decreased throughout the baking industry. |
| Calcium peroxide | A peroxide that slowly releases oxygen. This action allows the dough to be softer and drier. More water can then be added to the dough and thus gives the dough better machinability. |
| Calcium propionate | A white powder used as a mold inhibitor in the baking industry. It is one of the preferential mold inhibitors in yeast-raised products because it has no effect on pH, which can affect yeast development. |
| Calcium sulfate | A compound of sulfate and calcium that is used to fortify breads and other baked products with calcium and also as yeast food. Calcium sulfate has minimal impact on yeast activity and is thus a preferential source of calcium in yeast-raised products. |
| Cultured whey | A natural mold inhibitor. It acts in a similar manner to calcium propionate. |
| Guar gum | Adding this natural gum to the dough increases yield, gives better/softer texture, improves handling, and increases shelf life. |
| Diacetyl tartaric acid esters of mono- and diglycerides (DATEM) | A product made from vegetable fat that increases the volume of yeast-raised baked goods. It improves dough stability and allows more tolerance in the handling and production of yeast-raised products. |
| Maltitol | A sugar alcohol or polyol. Used as a sweetener in baked goods. It has less caloric impact than traditional sweeteners, contributing slightly more than 2 calories per gram compared to 4 calories per gram in sugar. |
| Maltodextrin | Essentially a chain of molecules of the simple sugar glucose linked together. There are an average of seven glucose molecules linked together to form a maltodextrin molecule. |
| Modified wheat starch | Modification can add stability to the wheat starch, making baking easier, and producing a nicely textured finished product. |
| Mono- and diglycerides | Used as an emulsifier to help mix the oil and water components of the dough. Also used as a dough strengthener, it enhances texture of the bread and increases loaf volume. |
| Monocalcium phosphate | Used as a yeast nutrient, dough conditioner, and acid producer. Lowering the pH of the product inhibits mold growth. |
| Olestra | A fat substitute used to reduce calories. |
| Sodium bicarbonate | Baking soda. |
| Turmeric | A spice that is often used in curry. It is mainly used in baked goods as a colorant. Turmeric imparts a yellow color to the dough and finished baked product. |

**Source:** www.bread.com/glossary

## TABLE 20-3  Excessively Firm or Soft Dough Problems

| Excessively Firm | Excessively Soft |
|---|---|
| • Difficult to mix | • Tend to "flow" |
| • Difficult to weigh | • Sticky to work with |
| • Slower fermentation | • Dough pieces easily lose their shape |
| • Difficult to shape | • Finished product remains fairly flat |
| • Poor symmetry in baked goods | • Decreased volume |
| • Pale crust | • Crumb shows uneven, large cells |
| • Crumb shows dense cells | • Deep brown crust has a streaky texture |

amounts and types, and how much the dough is mixed that determine the dough's consistency. It is important for the dough to reach the desired degree of cohesion, because this influences the dough's handling characteristics and the final quality of the baked item (Table 20-3).

### Straight Dough Method

The straight dough method consists of placing all the ingredients into a bowl at the same time, where they are mixed. For automatic mixing, various dough attachments are available (Figure 20-1). Whether by hand or by machine, the dough is kneaded to develop the gluten and then allowed to rise once or twice before being shaped into a loaf or other form.

### Sponge Method

Combining the yeast with water and slightly over one third of the flour creates a foamy, bubbly mixture that looks like a sponge. This is allowed to ferment in a warm place for half an hour to an hour in order to become foamy and spongy, after which all the remaining ingredients (sugar, fat, and remaining flour) except salt are added to the mixture. Salt inhibits the yeast, so it is added last, after yeast activity is well under way (7). Then the dough is treated as a straight dough through kneading, bowl proofing, shaping, pan proofing, and baking.

### Batter Method

The batter method is the simplest of all the mixing techniques and requires no kneading after the ingredients have been mixed. It is a good method for the rapid production of the kinds of bread products required in food service operations. Once the ingredients are combined, they are beaten by hand, or by electric mixer or dough hook, to develop the gluten. The batter is ready when it no longer sticks to the sides of the bowl, but is still sticky itself. The batter method saves time and is often used for preparing rolls and hot dog and hamburger buns, although it may result in a bread with a more coarse and porous texture.

### Rapid Mix

The rapid mix method differs from the others in that it is used primarily with bread-making machines. Millions of North Americans own machines that make bread-making easier. Inside the bread machine is a nonstick pan with a kneading paddle, usually located on the bottom. The machine kneads the bread, allows it to rise, then bakes it in the same pan. For those who prefer more "hands-on" bread making, the dough mixture can be beaten by hand; alternatively, it can be mixed in the bread machine, then shaped by hand. Ready-to-use mixtures often come with these machines, but if using a recipe, the simplified process consists of placing warmed water, bread flour, yeast, and salt into the container, closing the lid, and pressing a button. Fresh bread can be completed within 2 to 4 hours.

## Kneading

Kneading develops the dough's gluten to its maximum potential (see Chapter 17). This step of bread making involves physically handling the dough until it achieves a smooth, soft, nonsticky surface and springs back when pressed gently. However, some traditional bread doughs do remain somewhat sticky after kneading. Kneading is not required if the batter method is used.

A kneading surface should be covered with a fine layer of flour, both to prevent sticking and to allow some flour to be kneaded into the dough. Kneading in too much flour should be avoided, however, because it will slow fermentation time, leaving the final product dry and streaked or heavy. Hands should be lightly floured prior to lifting or tipping the dough gently out of the bowl. The ball of dough is placed on the floured surface and the farthest edge of it is lifted up and folded toward the nearest edge, as shown in Figure 20-2. The fold is then pressed with the heels of both hands in a single rhythmic, rocking motion that pushes the dough fold, first down, and then away. Short presses that are neither too heavy nor too light are best. The mass is turned a quarter of a turn and the process is repeated. More flour may be necessary as kneading continues. To determine if more flour is required, hit the dough ball with an open hand, count to 10, and lift the hand off. If the hand sticks to the dough, it needs more flour (9). If the dough is too firm and inelastic, additional water may be required.

**FIGURE 20-1**  Types of dough mixers.

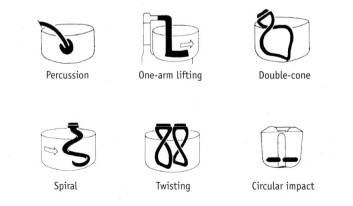

Percussion    One-arm lifting    Double-cone

Spiral    Twisting    Circular impact

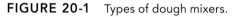

**Source:** Schumenann and Treu, *Baking: The Art and Science* (Baker Technology, 1988).

**FIGURE 20-2**    Kneading dough.

The dough's furthest edge is lifted up and folded toward the nearest edge.

The heels of both hands push firmly against the fold in a forward, down, and away motion.

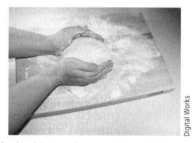

The dough is turned a quarter-turn and the process is repeated.

Observing an experienced bread baker helps in learning to determine the right amount of pressure and stretching and the length of kneading required to yield a dough ready for rising. In about 10 minutes, the dough usually signals it is completely kneaded by the development of a shinier surface that tends to "push back" when two fingers are gently pressed against the dough to test its gluten strength. Another way to determine if kneading is complete is to stretch some of the dough into a

**Proof**  To increase the volume of shaped dough through continued fermentation.

**FIGURE 20-3**   The way to test if dough has been sufficiently kneaded is to stretch it and see if the thin, translucent windowpane appears.

Courtesy of www.pinchmysalt.com

"gluten window" (Figure 20-3). Avoid excessively kneading the dough, because it results in a baked product with a coarse texture. An overkneaded dough will "snap back" when being rolled out due to a heavy gluten network that tightens the dough. Dough may similarly resist rolling out when it is too cold, but if temperature is the cause, the problem reverses itself when the dough is allowed to warm (31).

The effort and time involved in kneading can be eliminated by using a food processor or an electric mixer with a dough hook attachment (like those in Figure 20-1). It is important, however, to first learn to recognize the correct consistency for kneaded dough, because it is very easy to overknead a dough using a food processor or an electric mixer. On the other hand, it is very difficult to overwork a dough by hand.

# Proofing: Fermentation Causes the First Rising

After kneading, the dough's surface is greased by rolling its sides gently in a lightly greased bowl; this prevents it from drying out as it rises. It is then placed in the bowl, covered with a clean, moist dish or paper towel or plastic wrap, and allowed to rise or **proof** in a warm location undisturbed. Covering the bowl during proofing also helps to maintain humidity and prevent drying. For approximately the next hour, the bowl must be left in a preferably humid place that is slightly warmer than room temperature; approximately 85°F (30°C) is optimal. In these conditions, the dough can rise properly. Some traditional breads are left to rise more slowly in cooler (68°F/20°C) environments.

## Changes During Fermentation

As the yeast ferments, the dough will double in size (Figure 20-4) as carbon dioxide is produced by the yeast and as enzyme and pH changes take effect (Chemist's Corner 20-2).

Dough becomes more acidic (pH of 6.0 to 5.5–5.0) during fermentation because of the formation of carbonic acid, which is created when carbon dioxide combines with water, and because of the lactic and acetic acids produced by the yeast. The increased acidity may improve gluten's ability to hydrate with water. Acids also improve flavor, extend shelf life by inhibiting staling and mold growth, and reduce the stickiness of the dough. Alcohol is another by-product

**FIGURE 20-4**   First fermentation rise for dough. Shaped dough placed in a humid, warm, undisturbed place for rising (*left*); risen dough ready for punching (*right*).

Digital Works

### Enzymes in Yeast Dough

Enzymes naturally present in flour eventually influence the flavor, color, and texture of the baked bread. During fermentation, one of these enzymes, alpha-amylase, starts to degrade damaged starch granules found in flour to sugars. Among these sugars are dextrins that are then converted to maltose by another enzyme called beta-amylase (27). The most common commercial sources for the first enzyme, alpha-amylase, are fungal (*Aspergillus oryzae*), bacterial (*Bacillus subtilis*), and cereal (malted barley and wheat) (12). Maltogenic alpha-amylase is used in the baking industry because it has an anti-staling effect in baked products (35). Another important enzyme is the protease that breaks down the protein into peptides and amino acids. Proteases break down (hydrolyze) the protein bonds in the gluten proteins, making the dough more extendable. Sometimes professional bakers add enzymes to the flour to improve how the dough handles. However, caution must be used because excess protease enzymes will cause the bread to have decreased volume and an inferior texture.

produced during fermentation, but it evaporates during baking.

The amount of time it takes for the dough to rise will depend upon the type and concentration of yeast, the amount of available sugars, the temperature of its environment, salt concentration, and the mixing method: it may take from three quarters of an hour to 2 hours, or even overnight in the case of some European-style breads.

### Optimal Fermentation Temperatures

Yeasts are very sensitive to temperature extremes: they become activated at 68°F–100°F (20°C–38°C), slow down below 50°F (10°C), and die if exposed to temperatures at or above 140°F (60°C). Professional bakers use **proof boxes** to optimize fermentation conditions.

At home, the dough can be placed in a warm corner free of drafts, in a closed oven with a bowl of hot water on the shelf below, or in the oven after it has been heated for a minute or so, until the warmth is just beginning to be felt, and then turned off.

### Avoid Overfermentation

The first rise is completed when the dough has approximately doubled in size and two fingers pushed into the dough near the edge leave an indentation. As the dough rises during fermentation, the gluten stretches and becomes weaker; thus, rising should not be allowed to continue too long or the expanding dough will collapse. Allowing the dough to rise too high can also cause a coarse grain and a sour odor from the excess acid production. Overfermentation can also affect color; sugar must interact with flour proteins in order to create the desired browning of the crust through the Maillard reaction, and the greater sugar consumption by the yeast leaves too little available for browning. Similarly, because the Maillard reaction also contributes to the sweet, aromatic, and roasted flavors in baked products, overfermentation affects flavor (27).

## Punching Down— Second Rising

For most homemade and finer-textured breads, once the dough has risen to double its size, it is punched down and left to rise a second time (Figure 20-5).

**FIGURE 20-5** Punching down the dough allows excess gas to escape and redistributes the ingredients.

The dough can either be punched down while in the bowl, or placed once again on a lightly floured surface and gently pushed down in the center with a clenched fist, followed by about four kneading motions.

After punching, the dough is sometimes allowed to rise again to double its size, but no more, or the gluten will overstretch and cause the bread to fall. Some baked products are allowed to rise only once before shaping. When there is a second rising, it takes about half the time of the first. The completion of the second rising is signaled strictly by whether or not the dough has doubled its size and not by the finger indentation test.

### How & Why?

*Why does dough have to be punched down after the first rising?*

Punching down evens out temperatures; redistributes sugars, yeast, and gluten; breaks large air bubbles into smaller cells; and lets excess carbon dioxide gas escape. The bread would have large holes without the release of excess carbon dioxide.

## Shaping

After the bread has risen, it is ready for shaping. Breads can take on a wide variety of shapes. Figure 20-6 provides instructions for shaping rolls, braids, and other, more unusual bread shapes. The basic loaf of bread is shaped by first dividing the dough into the desired number of portions so the pan(s) will be at least half, but no more than two thirds, full of dough. The dough is shaped into an oblong, roughly rectangular, mound as long and as wide as the loaf pan. One third of each end is then folded under the mound. All the edges are pinched together to seal

**Proof box** A large, specially designed container that maintains optimal temperatures and humidity for the fermentation and rising of dough.

**FIGURE 20-6**    Various bread-shaping techniques.

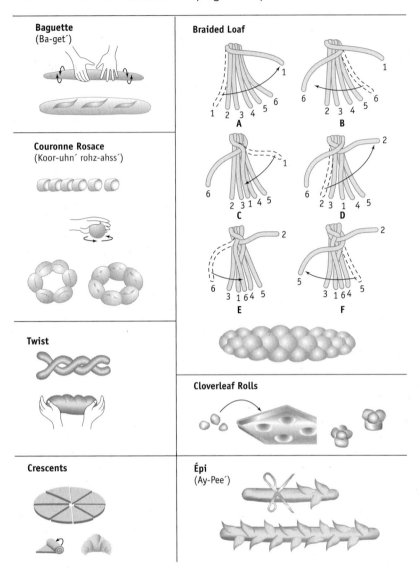

the seams, and the dough is then placed in the pan so the sealed edges are on the bottom and the dough is touching all four sides of the pan. The bottom and sides of the pan are sometimes greased so that the loaf can be easily removed, but some bakers leave the sides ungreased so that the bread will have more traction during rising.

## Selecting a Baking Pan

Yeast breads are usually baked in loaf pans (metal or glass), but can also be prepared with specialized loaf pans (French bread, baguette, etc.), cookie sheets, jelly roll pans, muffin tins, and even on stones (pizza stones, clay or quarry tiles). Table 20-4 lists the advantages and limitations of the various pan materials. Regardless of the pan used, the sides are usually well greased with solid shortening or a non-stick cooking spray so that the bread can rise without the dough sticking to the sides.

## Second Proofing (Optional)

A second proofing may occur in the pan or on a baking sheet, and it has an important effect on the quality of the finished bread. It is facilitated by placing the shaped dough in a warm, humid, and undisturbed environment. The pan is covered with a cloth, and the dough is usually allowed to double in size and take on the shape of the bread pan. However, the amount of rising allowed during proofing varies with the bread type. One of the purposes of proofing is to create a dough that is adequately aerated. If the dough expands beyond what is recommended, it leads to over-extension of the gluten, which causes the cell walls to break and collapse, the fermentation gas to escape, and, ultimately, a low volume in the finished product (Figure 20-7). Temperature is as important as timing; doughs that are too cool ferment too slowly, whereas those that are too hot produce breads that have a small volume, large cells, a pale crust, and a reduced shelf life. Humidity also plays an important role, as shown by the bread in Figure 20-8 that was proofed under conditions that were

| **TABLE 20-4**   The Effect of Different Pan Materials on Baking Yeast Breads | |
|---|---|
| Shiny aluminum | Provides even heating, but does not promote the development of bread's characteristic golden brown color and well-textured crust. |
| Dark heavy-weight steel | The choice of expert bread bakers. Promotes golden brown color and perfectly textured crust. |
| Dark nonstick | Keep bread from sticking, but they tend to brown bread more quickly, especially around the edges. Reduce the oven temperature called for in any recipe by 25°F/4°C. |
| Insulated | Consists of two thin sheets of aluminum with a layer of air between them. Breads may brown poorly, require longer baking times, and possibly stick. |
| Glass | Ovenproof glass baking dishes brown the bread well and you can see the browning of the crust on all sides. Reduce the oven temperature called for in any recipe by 25°F/4°C. |
| Baking stone (quarry or clay tiles, pizza stones) | Baking stones produce thick-crusted, country-style breads by replicating a brick or adobe hearth oven. |

**FIGURE 20-7**   Overproofing produces a low-volume bread due to collapsing cells and the escape of fermentation gases.

Optimal proofing

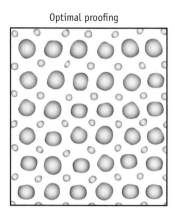

Overproofing

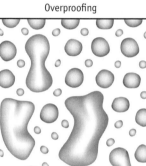

**FIGURE 20-8**   Bread proofed in conditions that were too dry; lack of humidity results in crust problems.

Crust faults due to drafty, final proofing

Uneven, rough crust

dough has been overfermented prior to baking. Underfermented dough, however, may not have sufficient oven spring. Temperature is also important for crust thickness, which is determined in part by the amount of vapor pressure (17).

### Changes During Baking

Baking changes the appearance, texture, flavor, and aroma of the dough as well as its structure. Once the dough is placed in the oven, the hot temperature kills the yeast, thereby stopping fermentation. It also inactivates enzymes, vaporizes the alcohol, and coagulates the protein that firms the dough.

Protein coagulates and starch granules start to swell and gelatinize at about 140°F (60°C), but because of the limited water content, gelatinization does not proceed as far as it might have. If the oven temperature is too high, the proteins will coagulate too soon, resulting in low volume; conversely, low oven temperatures will cause the structure to collapse.

If baked correctly, bread right out of the oven has an aroma all its own, but the wonderful smell is rapidly lost; this is believed to be the result of the evaporation into the air of volatile compounds (Chemist's Corner 20-3). Other components of flavor are developed in part by the breakdown of starch to dextrins and some caramelization, but primarily from the browning of the crust in the Maillard reaction. Substrates for the Maillard reaction in breads can be derived from any added milk, sugar, and/or egg.

The changes associated with baking require even heat exposure, so a minimum distance between baked items is necessary: at least 1 inch between rolls and about 3 inches between loaves (33).

too dry. While proofing is taking place (second proofing is usually about half the time of the first proofing), the oven may be preheated in anticipation of the next step—decorating and baking the bread.

## Decorating

In preparation for baking, the bread dough may be decorated with sesame, caraway, or poppy seeds by brushing the top of the loaf with a thin layer of egg white and sprinkling the seeds on top. Many types of breads are **scored** just before baking to allow them to rise evenly without tearing the crust. Another decorative touch, which also adds flavor, is to score the top of the bread with a sharp knife and pour a bit of melted butter into the slashes. Although scoring serves the dual purpose of decoration and allowing excess steam to escape, neither is strictly necessary. Milk brushed on the surface of the loaf

## Baking

A standard loaf of bread will bake in about 45 minutes. It is usually heated at 400°F (205°C) for the first 10 to 15 minutes, and then at 350°F (177°C) for the remaining 30 minutes. Temperatures and times will vary for different types of breads. The initial hot temperature contributes to **oven spring.** The initial increase in fermentation, enzyme activity, and softening of the ingredients also contribute to oven spring. If the initial temperature is not hot enough, the protein will not congeal properly and the yeast will continue to ferment, causing the dough to spill over the sides. This can also occur when the

before or during baking will give crusts a golden brown color due to the caramelization of the milk. The technique of brushing loaves with water or introducing steam into the oven will give loaves a crispier crust.

**Score**   To use a sharp knife or a special blade called a lame to create ¼- to ½-inch-deep slashes on the risen dough's top surface just prior to baking.

**Oven spring**   The quick expansion of dough during the first 10 minutes of baking, caused by expanding gases.

Baker Technology

### Bread Flavors and Aromas

Fermenting yeasts contribute to the unique flavor of baked products through the formation of volatile and nonvolatile products. Nonvolatile compounds are too large to become airborne and include amino acids and saccharides. These generate flavors via the Maillard reaction. Volatile substances contributing to the odor of fermenting dough include isobutanol, amylalcohol, isoamyl alcohol, phenylethylalcohol, esters, lactones, and organic acids (20). Many of the alcohols formed during fermentation, such as ethanol, n-propanol, isobutyl alcohol, and others, are lost into the air during baking (10). The odor of baked crust is derived in part from the formation of 2,5-dimethyl-4-hydroxy-3(2H)-furanone, which results from the thermal reaction between certain amino acids and sugars such as glucose, fructose, and rhamnose. Organic acids and carbonyl compounds also contribute to crust flavor. The unique flavor of French bread is thought to be due to its high level of acetic acid.

## Crumb Development

During baking, the dough converts from an elastic, undefined mass into a set structure with a defined bread **crumb** (15). The formation of a desirable crumb is dependent on gases produced during fermentation and proofing and air introduced during mixing and kneading.

Numerous small or medium cells result in a baked product with a fine and tender crumb, a large volume, and a longer shelf life. These cells multiply

**Crumb** The cell structure within the interior of a baked product that is revealed when it is sliced. Evaluation is based on cell size (called *open* if medium to large, or *closed* if small), cell shape, and cell thickness (thin walls occur in fine crumb, whereas thick walls predominate in a coarse crumb).

and enlarge during fermentation and proofing; during baking they will enlarge further, but will not multiply. Yeasts are not entirely responsible for cell number and size. Air incorporated into the flour mixture during mixing, punching, and kneading also contributes to the number of cells. This is why using sifted flour in baked products results in larger volumes than using unsifted flour (33).

? **How & Why?**

### How does crumb form in a loaf of bread?

The carbon dioxide produced during fermentation, proofing, and oven spring becomes many small bubbles entrapped in the gluten network (Figure 20-9). When the bread sets and the gas escapes, all that remains are the pores, or cells, forming the crumb of the baked product.

## Problems with Texture

An overfermented dough, with its large cells, gives the baked item a moth-eaten appearance and a coarse texture. Underfermented dough, in which the carbon dioxide was not properly distributed, results in a very dense loaf with thick cells, low volume, and a tough crust (Figure 20-10). Distribution of the carbon dioxide throughout the dough to create fine cells is accomplished through mixing, punching, kneading, and shaping. Bakery products in which the cells are either abnormally small or large have shorter shelf lives (33).

## Testing for Doneness

Bread can be tested for doneness by inverting the pan with one gloved hand, allowing the bread to drop into the other gloved hand so the bottom of the loaf is facing up, and tapping the bottom. If it rings hollow, it is done; if it does not, it is placed back in the pan and returned to the oven for an additional 5 to 10 minutes.

**FIGURE 20-9**   Fermentation gases produce the pores of the baked bread.

Laura Murray

**FIGURE 20-10**   Crumb quality depends on cell size.

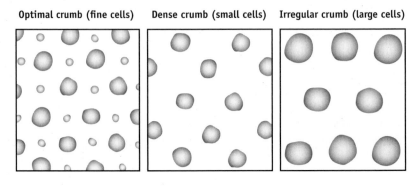

| Optimal crumb (fine cells) | Dense crumb (small cells) | Irregular crumb (large cells) |

**TABLE 20-5  Problems with Yeast Breads and Their Causes**

| Problems \ Causes | Improper mixing | Insufficient salt | Too much salt | Dough weight too much for pan | Dough weight too light for pan | Insufficient yeast | Dough proofed too much | Dough underproofed | Dough temperature too high | Dough temperature too low | Dough too stiff | Oven temperature too high | Proof box too hot | Green flour | Dough chilled | Too much sugar | Insufficient sugar | Dough too young | Dough too old | Improper molding | Insufficient shortening | Oven temperature too low | Over-baked | Dough too soft | Under-baked |
|---|---|---|---|---|---|---|---|---|---|---|---|---|---|---|---|---|---|---|---|---|---|---|---|---|---|
| Volume too low | X | | X | | X | X | | X | | X | X | | | | X | | X | X | | | | | | | |
| Volume too high | | X | | X | | X | | | | | | | | | | | X | | X | | | | | | |
| Crust too thick | | X | X | | | X | | | | | | | | | | | | | | | | | | | |
| Dry crumb | X | | | | | | | X | | | | | | | | | | | X | | | | | | |
| Moist crumb | | | | | | | | | | | | | | | | | | | | | | | | X | X |
| Streaky crumb | | | | | | | | | | | | | | | | | | | | X | | | | | |
| Gray crumb | | | | | | | X | X | | | | | X | | | | | | | | | | | | |
| Lack of shred | | | | | | X | | | | | | | | | | X | X | | | | | | | | |
| Coarse crumb | X | | | | X | X | | | X | | | | | | | X | X | X | | | | | | | |
| Poor texture, crumbly | | | | | | X | | | | X | | | | | | X | X | | | | X | | | | |
| Crust color too pale | | | | | | | X | | | X | | | | | X | | X | | | | | | | | |
| Crust color too dark | | | | | | | X | | X | | | | | X | | X | | | | | | | | | |
| Crust blisters | | | | | | | | | | | | | X | | | X | X | X | | | | | | | |
| Shelling of top crust | | | | | | X | | | X | | | | X | | X | | | | | | | | | | |
| Air pockets | | | | | | | X | X | | X | | | | | | | | | | | | | X | X | |
| Poor taste and flavor | | X | | | | | | | X | | | | | | | | | X | | | | | | | |
| Poor keeping qualities | | X | | | | X | | | X | | | | | | | X | | X | | X | X | | | | |

Another method consists of combining the hollow sound technique with an instant-read thermometer, which should read above 195°F (91°C) when inserted into the bread's interior. If the bread is not done but the top is already golden-brown, it is covered with aluminum foil or otherwise protected to prevent further browning. Once it is done, the baked bread is set out on wire racks for cooling.

A well-prepared bread product has certain characteristics: optimal volume, color, and flavor; a symmetrical shape with closed seams; a porous, pliable, firm, and even crumb; and a golden-brown, crispy crust. Major problems that may interfere with the attainment of these attributes, and their causes, are summarized in Table 20-5.

## Microwave Preparation

The use of a microwave for baking bread is not highly recommended; because no dry heat is available for dextrinization, the Maillard reaction, or caramelization, the crust will not brown. If a microwave oven is used, breads should be placed in a conventional oven during the last few minutes of baking to yield at least a partial crust. Individual microwave manufacturers' instructions should be followed.

## High-Altitude Adjustments

Baking at altitudes above 3,000 feet requires changes in recipe ingredient measurements, preparation time, and oven temperatures. Yeast breads rise faster at higher altitudes because there is less atmospheric pressure, so less leavening agent will be needed. Liquid also evaporates more quickly at higher altitudes, so more may be needed. Yeast bread should not be allowed to more than double in volume. Baking temperatures are increased slightly, by about 10°F–15°F (6°C–8°C), to help coagulate the proteins and prevent the gases from overexpanding the structure.

## NUTRIENT CONTENT

Breads are an excellent source of complex carbohydrates, providing about 15 grams of carbohydrate per 1-ounce slice, as well as 0–3 grams protein, 0–1 gram fat, and 80 calories (kcal). They may also, depending on the type of bread, contribute beneficial fiber to the diet (4). There may be up to 3 grams of fiber in a slice of whole-grain bread compared to only 1 gram of fiber in a slice of white bread.

*Fat.* Although many yeast breads contain very little if any fat (about 1 gram per slice), certain types such as croissants (12 grams of fat in a per 55-gram croissant), submarine rolls (4 grams of fat in a 135-gram slice), and hot dog/hamburger rolls (2 grams of fat in a 40-gram roll) contain more. Some baked products use fat substitutes to lower their fat content (30, 31).

*Specialty Breads.* The retail market now offers a wide variety of specialty breads, including those that are low in calories, low in fat, high in calcium, high in fiber, or a combination of these. Low-glycemic index breads (low-carbohydrate and/or high-fiber) are available to help people with diabetes maintain steady blood sugar levels (25, 26). The term low carb previously used for diet foods appears to be morphing in the bread category to cover a wide variety of baked products lower in digestible carbohydrates and higher in fiber that would benefit people's energy intake, blood sugar levels, and digestive health (29). Increasingly popular are the gluten-free specialty breads discussed in more detail under Specialty Breads.

---

### ✦ CALORIE CONTROL
### Yeast Breads

Many of the calorie control hints for quick breads discussed in the previous chapter apply to yeast breads.

- **Control Bread Portion Sizes.** A "standard" slice of bread delivers about 80 calories (kcal), but loaf sizes and slice thicknesses vary. Check the calorie information on the Nutrition Facts label of commercial breads.
- **Choose Whole-Grain Bread for Sandwiches.** Two slices of whole-wheat bread are lower in calories (160 calories/kcal) and higher in beneficial fiber than a croissant (231 calories/kcal) or a submarine roll (359 calories/kcal).
- **Thin vs. Thick Pizza Slices.** The caloric content of this popular bread-based food is partially based on the size and thickness

of the slice and its crust. Pizza slices vary considerably among restaurants, but generally a small, thin pizza slice provides 130 calories (kcal) and 4 grams of fat, whereas a large, thick pizza slice provides much more—400 calories (kcal) and 11 grams of fat. You can control pizza portion sizes by choosing thinner pizza, serving yourself smaller slices, and limiting the number of slices you eat. A meal of 3 large, thick slices adds up to about 1,200 calories (kcal) plus 33 grams of fat; a more reasonable portion of 3 small and thin or 1 large and thick pizza slice provides 400 calories (kcal) and 15 grams of fat.

© 2010 Amy Brown

---

# TYPES OF YEAST BREADS

The simplest yeast bread is made from flour, water, and yeast, but this basic formula has evolved into more complicated varieties that include loaf breads (white, whole-wheat, sourdough, and malt breads); specialty breads; rolls; pita bread; bagels; English muffins; pizza crust; pretzels and bread sticks; and raised doughnuts (Figure 20-11).

## Loaf Breads

### Wheat (White) Bread

Wheat, or white, bread is made with all-purpose flour; milk; water; and small amounts of sugar, salt, and yeast. Fat is an optional ingredient. Almost all of the white bread in the United States has been enriched.

### Whole-Wheat Bread

The bran in whole-wheat bread increases the fiber content from 1 to 3 grams per slice, but the sharp edges of the bran cut the strands of gluten in the dough, resulting in a shortened kneading time and a lower-volume loaf. If the bread is designated as "wheat bread," it has been made from a combination of whole-wheat and white flours, which produces a loaf with more volume and a lighter texture.

### Sourdough Bread

Sourdough bread has a unique tangy taste and chewy texture because it is made with a starter (sometimes called a

**FIGURE 20-11** Varieties of yeast breads.

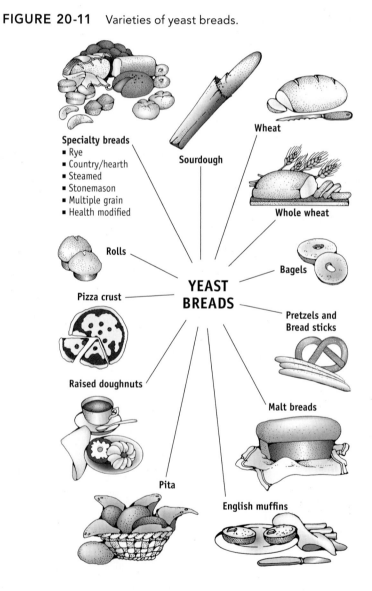

Specialty breads
- Rye
- Country/hearth
- Steamed
- Stonemason
- Multiple grain
- Health modified

Sourdough

Wheat

Whole wheat

Rolls

Bagels

Pizza crust

**YEAST BREADS**

Pretzels and Bread sticks

Raised doughnuts

Malt breads

Pita

English muffins

sponge), which consists of both yeast and lactic acid-producing bacteria. There are more than 50 species of lactic acid bacteria, including *Lactobacillus plantarum*, and more than 25 species of yeast (14). The lower pH of sourdough bread (4.0 to 4.8 compared to 5.1 to 5.4 pH for regular breads), which provides its characteristic texture and taste, results from the fermentation of the dough by these bacteria. Two cups of sourdough starter are equivalent to one small package of yeast. Sourdough yeast must be obtained to prepare a sourdough starter. Not ordinarily sold in the store, these unique sourdough yeasts can be purchased at various websites, gourmet cook stores, larger health food stores, cooking schools, or cooking clubs.

### ? How & Why?

#### Why is sourdough bread "sour"?

The *Lactobacillus plantarum* strain of bacteria that is widely used as a starter produces a lactic acid that results in a pleasantly sour taste (36). The lower pH also makes sourdough bread less sweet by inhibiting the amylase enzymes in the flour from breaking down starch into the sweeter-tasting maltose.

### Malt Breads

The malt in malt breads makes them sweeter, stickier, and heavier than other breads. Some researchers report that adding malted wheat to the dough improves bread quality (37). The added malt also contains enzymes that convert starch to sugars and digest some of the gluten. A weakened gluten structure can be avoided by using an inactive malt derivative and molasses, but the resultant bread product is then considered to be an imitation malt bread (13).

### Specialty Breads

Specialty breads differ from regular varieties either in their method of preparation or in their ingredients.

*Preparation Technique*   Some breads that are made in a different manner include:

- Wood-burning oven breads that are often referred to as country loaves or hearth breads
- Steamed bread that is processed with 15 to 20% steamed rye flour
- Stonemason bread containing grain that has been treated with moisture and then crushed or cut

Pumpernickel and rye-meal breads lack a crust because they are cooked under continuous steam, a process that breaks down the starch to sugar and imparts a sweet flavor and dark-colored crumb. Swedish rye crackers are baked in flat sheets. The unique flavor and greater mold resistance of rye bread is due to the release of lactic acid by bacteria during fermentation, which acidifies the dough (7). Rye breads also have a much harder crumb than wheat breads due to the smaller number of pores and greater concentration of large particles.

*Ingredient Variation*   The use of unique ingredients can also result in a specialty bread. Examples include breads made from flours other than wheat or rye, such as oat, barley, corn, rice, and others, to make three-, seven-, or nine-grain breads; breads prepared with extra dairy items such as butter, milk, buttermilk, whey, and yogurt; and those made with vegetable ingredients such as wheat germ, soy, bran, sesame, linseed, or spices (pepper or paprika). Note that garlic should not be added to bread dough formulations, because it causes low volume and decreases the dough strength to the point that it often breaks down during mixing. One theory behind this phenomenon is that allicin, a unique flavor compound in garlic, may interfere with the dough's disulfide bonds (23).

*Nutrient Modifications*   Specialty breads also include those that have had their nutrient value altered for people with health conditions that require dietary modifications (11). Graham bread is made without salt or yeast; other breads reveal their differences in their names—low-fat, low-calorie, low-sodium, and high-fiber. Gluten-free bread is produced for people with celiac disease, or gluten-sensitive enteropathy (2). The term *gluten-free* means without gliadin protein, and these products are increasing in popularity. A bread that is gluten free may still contain other proteins from flour and any dairy, eggs, or other protein-containing ingredient added to the flour mixture. Consumers avoiding gluten should be aware that products labeled "wheat free" may not be "gluten-free."

## Rolls

Rolls, including kaiser, submarine, sandwich, and others, contain the same ingredients used to make loaf bread. Some may contain additional fat, sugar, and/or eggs. The many available types of rolls vary not only in their ingredients, but in their shapes as well (Figure 20-12). Some professional bakers weigh the individual dough allotments for each kind of roll to ensure uniform proportions and appearance.

## Pita Bread

This circular Middle Eastern bread with a large hollow center is also known as *pocket bread*. It is prepared by flattening dough into thin circles about 9 inches in diameter and less than ¼-inch thick, and then baking them briefly in a very hot oven (500°F/260°C) until they form the characteristic two symmetrical halves (1). The moisture in the dough heats up so quickly under these high heats that it does not have time to escape through the pores. Instead, the dough balloons, and the bread bakes in that position, collapsing again upon removal from the oven. The entire process takes less than a minute. Another method is to cook the dough on a lightly greased griddle for 15 to 20 seconds, gently turn it over, continue heating for about 1 minute until bubbles appear, and then turn it back over until it balloons. Pita bread is handy in making sandwiches because it can be filled with a variety of foods, ranging from standard sandwich ingredients to vegetables and cooked meats. Almost any pita sandwich may be topped with lettuce, tomato, and onions.

## Bagels

Bagels are made by adding egg whites to the standard white bread formula. The dough is cut into 3-inch rounds with a ¾-inch center hole using a special bagel cutter (Figure 20-13). After the bagels rise, they are boiled in water, cooled, possibly brushed with an egg yolk and water mix, and then baked.

**FIGURE 20-12**    There are many varieties of both soft and hard rolls.

### Soft Rolls

Braided roll

Butter roll

Butter twist

Clover leaf roll

Double knot

Figure eight

Frankfurter roll

Hamburger roll

Maryland roll

Pan roll

Parker House roll

Poppyseed roll

Single knot

Snowflake roll

Spiral butter roll

Spiral roll

Square knot

Twin roll

Vienna bridge roll

Vienna roll

Whole-wheat roll

### Hard Rolls

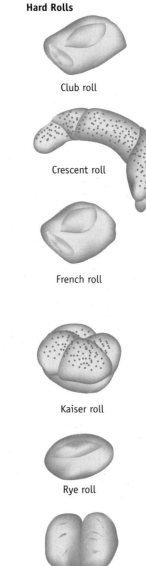
Club roll

Crescent roll

French roll

Kaiser roll

Rye roll

Water roll

**FIGURE 20-13**    A bagel cutter.

*Courtesy of King Arthur Flour "Baker's Catalogue" www.kingarthurflour.com*

### ? How & Why?

**Why are bagels boiled before baking?**

Boiling bagels creates a moist surface that does not brown as readily, so the crust remains lighter and less crunchy. The moisture also causes the starch to gelatinize into a thin, transparent coating that gives bagels their unique, soft, smooth crust texture.

## English Muffins

English muffins, unlike other breads, are baked on a greased griddle or pan. These yeast breads, usually about 4 inches in diameter when baked, are made by combining water or milk, sugar, salt, flour, yeast, and a little butter. After the first rising, they are shaped, with or without the use of muffin rings. If muffin rings are used, they are greased, dusted with cornmeal, and filled up to ½ inch with the dough. If rings are not available, the dough is placed on a board that has been lightly sprinkled with cornmeal, where it is pressed down to a thickness of ½ inch, cut into circles about 3 inches in diameter, and moved to a greased cookie sheet. English muffins undergo their second rising before being transferred by spatula to the greased griddle or to a pan where they will be heated until lightly browned.

## Pizza Crust

Pizza crust is made with yeast and hard wheat flour. It is allowed to rise only once, for 2 hours, and then baked in a very hot oven at 400°F (205°C) for about 25 minutes.

## Pretzels and Bread Sticks

Change the shape of the yeast dough used to make pizza crust or other baked products and bread pretzels or sticks can be formed. A small amount of dough is rolled into a "stick" that can then be baked into a bread stick or twisted into the shape of a pretzel. The difference between them and other baked products is that they are coated with butter and/or salt and are best consumed as soon as possible after baking.

## Raised Doughnuts

Raised doughnuts are made with bread flour and leavened with yeast. They are typically deep-fried in fat or baked in an oven. Cake doughnuts, on the other hand, are made with cake flour and leavened with baking powder (and are thus a type of quick bread).

# STORAGE OF YEAST BREADS

## Fresh

Freshly baked wheat bread is best when consumed within 1 or 2 days, whereas some rye breads can last an average of a week. When just out of the oven, bread can be kept warm by placing aluminum foil under the cloth in the breadbasket. If it will not be consumed right away, it should be completely cooled before being wrapped and stored in a dry, cool place at room temperature and away from sunlight (5).

> ### CHEMIST'S CORNER 20-4
>
> **The Chemistry of Staling**
>
> Some of the theories as to why staling occurs include possible aggregation of amylose molecules (16), recrystallization of the amylopectin (18), and/or a transfer of moisture from the gluten to the starch.

## Staling

Bread gradually loses freshness once it is baked. In fact, staling begins as soon as the bread leaves the oven (Chemist's Corner 20-4) (6). This unavoidable part of bread preparation is thought to be responsible for the loss of an estimated 3 to 5% of all baked breads sold in the United States (35). Staling occurs as the crust toughens, the crumb become less elastic and more firm, and the soluble starches lose moisture and flavor (18, 21). These undesirable changes are thought to be due primarily to retrogradation (crystallization) of the starch molecules released during gelatinization (8). Moisture levels also play an important role in the staling of bread as water moves from the center of the loaf toward the crust (28).

### Preventing Staling

Staling is best prevented by keeping the bread away from air. Several techniques include wrapping breads in plastic or paper bags, adding moisture retainers such as fat or sugar (nonfat French bread stales quickly), and/or freezing. The staling caused by retrogradation is reversible when the bread is warmed, but returns upon cooling. Reheating the bread in an oven at 125°F–145°F (52°C–63°C) for a few minutes recreates many of the characteristics of fresh bread, especially if a damp cloth or paper towel is placed over the bread during reheating. This method is not recommended for microwave ovens, which cause the bread to become tougher, rubbery, and more difficult to chew (24).

### Anti-Staling Additives

Commercial bakers add mono- and diglycerides or fat to bread doughs to help prevent staling (22), and sodium or calcium propionate to retard mold and the bacteria that cause **rope.** These bacteria, which live in the soil, can contaminate grains and the flour made from them. The spores survive baking, germinate, and feed off the carbohydrates they obtain by decomposing the bread's starch. The end result is a bread that appears fine on the outside, but is internally mushy or stringy to the point that it can be pulled into "ropes." Breads affected with rope smell like overripe melons. This condition may mistakenly be attributed to incomplete baking, but consuming rope-contaminated bread can cause vomiting and diarrhea. Rope contamination is more apt to occur in the summer months (33).

## Refrigerated

Bread should be refrigerated immediately in the warm temperatures and moist humidities of tropical regions. In less humid areas, refrigerating bread is not recommended because it speeds staling.

## Frozen

Freezing is one of the best ways to maintain some of the texture and flavor of freshly baked bread. If crusty breads are not going to be consumed within 1 day, or soft breads within 2 days, freezing is the optimum storage method (5). Most breads can be frozen for 2 or 3 months. Before freezing, "cater wrap" the bread in two layers of plastic wrap (5). Alternatively, the bread should be wrapped in heavy-duty aluminum foil and dated. Frozen bread should be removed from the freezer and thawed at room temperature in the wrapper. Thawing in a hot oven results in a soggy, flavorless loaf. Home-baked bread can get back its freshly baked flavor if the top portion of the foil covering the thawed bread is opened and it is placed in a preheated 250°F–300°F (121°C–149°C) oven for about 10 minutes. The foil will keep the loaf warm during slicing and serving.

Unbaked bread dough can be frozen for up to 2 weeks by first shaping the dough and then wrapping it in freezer paper or foil. It should thaw and rise to double its height before being baked. It can also be placed overnight in the refrigerator and then allowed to rise for 2 hours.

> **Rope** The sticky, moist texture of breads resulting from contamination by *Bacillus mesentericus* bacteria.

# PICTORIAL SUMMARY / 20: Yeast Breads

The basic principles of preparing yeast breads have not changed since they were baked by ancient Egyptians, who knew the techniques and preferred the light, airy texture of yeast-leavened bread.

## PREPARATION OF YEAST BREADS

As with any bread, the basic ingredients of yeast breads are flour, water (or other liquid), and a leavening agent—in this case yeast. Sugar, salt, and fat and/or eggs are often added.

Normally, the preparation of yeast bread is at least a 2½- to 3-hour operation, and involves the following steps:

**Mixing:** Ingredients are usually combined using one of these four common mixing techniques:
- *Straight dough method.* All ingredients are placed into the bowl at once and mixed.
- *Sponge method.* Yeast is combined with water and slightly over one third of the flour. The "sponge" is then allowed to rest in a warm place for a half hour to an hour and then other ingredients are added.
- *Batter method.* Ingredients are combined and beaten by hand or with an electric mixer. No kneading is required.
- *Rapid mix method.* This method is used primarily with bread-making machines. All ingredients are added simultaneously.

**Kneading:** The dough is handled until it is smooth, has a soft, nonsticky surface, and springs back when pressed gently.

**Rising:** As the yeast ferments, the dough doubles in size.

**Punching down:** Most breads need to be punched down after they double in size, and allowed to rise a second time.

**Shaping:** Breads can take a wide variety of shapes. Usually the pan is at least half but no more than two-thirds full of dough.

**Proofing:** The dough rises in the pan or sheet one last time.

**Decorating:** Dough may be sprinkled with seeds or brushed with egg white. Many breads are scored (cut across the top) to allow them to rise evenly without tearing the crust.

**Baking:** A standard loaf of bread will bake in about 45 minutes. It is usually heated at 400°F (200°C) for the first 10 to 15 minutes, then baked at 350°F (177°C) for the time remaining. The use of a microwave is not recommended because the crust will not brown. Baking time must be adjusted at high altitude.

## TYPES OF YEAST BREADS

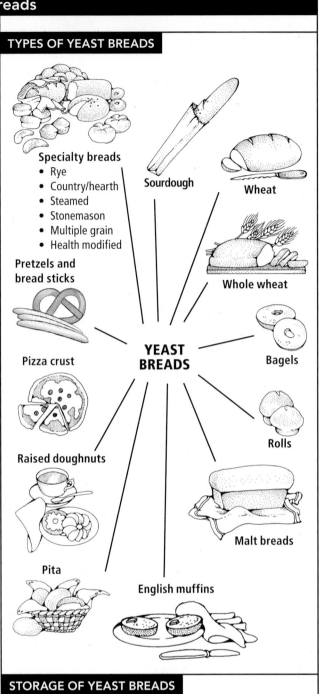

**Specialty breads**
- Rye
- Country/hearth
- Steamed
- Stonemason
- Multiple grain
- Health modified

**Pretzels and bread sticks**

**Pizza crust**

**Raised doughnuts**

**Pita**

**Sourdough**

**Wheat**

**Whole wheat**

**Bagels**

**Rolls**

**Malt breads**

**English muffins**

**YEAST BREADS**

## STORAGE OF YEAST BREADS

Fresh bread should be properly wrapped and stored up to 2 to 3 days in a dry place. If not used within a few days, bread should be frozen to prevent contamination and spoilage. In most cases, room temperatures are best for storing bread for short periods of time.

# CHAPTER REVIEW AND EXAM PREP

## Multiple Choice*

1. Breads leavened with carbon dioxide produced by baker's yeast are known as _____.
   a. unleavened breads
   b. crisp breads
   c. yeast breads
   d. baker's breads

2. What is the term used to describe the quick expansion of dough caused by expanding gases during the first 10 minutes of baking?
   a. Quick rise
   b. Gas expanse
   c. Gas rise
   d. Oven spring

3. Whole-wheat bread is a great source of _____, whereas white bread is a poor source.
   a. carbohydrates
   b. fiber
   c. fat
   d. yeast

4. When baking at altitudes above 3,000 feet, liquids need to be _____ and temperatures need to be _____.
   a. increased, increased
   b. increased, decreased
   c. decreased, increased
   d. decreased, decreased

5. Before it is baked, bagel dough undergoes what cooking process?
   a. Toasting
   b. Microwaving
   c. Boiling
   d. Braising

6. The method of mixing ingredients and beating them together by hand or a mixer without kneading is known as the _____ method.
   a. straight dough
   b. sponge
   c. batter
   d. rapid mix

7. Scoring a bread refers to:
   a. shaping the dough.
   b. cutting across the top to prevent the crust from tearing.
   c. punching down on the dough to release the gasses.
   d. rating the bread's baked flavor.

## Short Answer/Essay

1. What are the fundamental ingredients of yeast breads?
2. Briefly explain the following methods for mixing yeast breads: straight dough, sponge, batter, and rapid mix.
3. What is the purpose of kneading, and how can one determine when kneading is complete?
4. What are the recommended conditions for the first rising of the dough? Approximately how much time should be allowed for the process? What are proof boxes?
5. What is the purpose for punching down the dough and sometimes allowing it to rise a second time?
6. Define the following terms: *score, oven spring, proofing,* and *rope.*
7. Describe the general baking conditions for a standard loaf of bread, including proper temperature and time frame. Describe the changes that occur during the baking of yeast bread.
8. Why is the use of a microwave not highly recommended for yeast bread preparation?
9. Briefly describe how the ingredients and/or preparation for the following yeast breads differ from those of a standard white loaf bread: sourdough bread, malt bread, English muffins, pizza crust, and pumpernickel bread.
10. Baked products start to stale the minute they leave the oven. Describe the process of staling and discuss the steps that can be taken to prevent or delay it.
11. List the possible causes for the following problems associated with yeast bread preparation: low volume, poor flavor, and crust color too dark.

*See p. AK-1 for answers to multiple choice questions.

# REFERENCES

1. Alford J, and N. Duguid. *Flatbreads and Flavors. A Baker's Atlas.* William Morrow, 2008.
2. Alvarez-Jubete L, EK Arendt, and E Gallagher. Nutritive value and chemical composition of pseudocereals as gluten-free ingredients. *International Journal of Food Science and Nutrition* 22:1–18, 2009.
3. Amarjeet K, WS Singh, K Kamaljit, and DS Kler. Quality parameters of wheat grown under organic and chemical systems of farming. *Advances in Food Sciences* 28(1):14–17, 2006.
4. Anderson JW, et al. Health benefits of dietary fiber. *Nutrition Review* 67(4):188–205, 2009.
5. Anonymous. Q&A. *Fine Cooking* 60:20, 2003.
6. Arendt EK, A Morrissey, MM Moore, and F Dal Bello. Gluten-free

breads. In EK Arendt and F Dal Bello F. *Gluten-free Cereal Products and Beverages.* Elsevier, 2008.

7. Autio K, T Parkkonen, and M Fabritius. Observing structural differences in wheat and rye breads. *Cereal Foods World* 42(8):702–705, 1997.

8. Cauvain S. Stalling staling of bakery products. *Prepared Foods* 168(3):69–75, 1999.

9. Clayton B. *Bernard Clayton's New Complete Book of Breads.* Simon and Schuster, 2006.

10. deMan JM. *Principles of Food Chemistry.* Van Nostrand Reinhold, 1999.

11. Ensminger AH, et al. *Foods and Nutrition Encyclopedia.* CRC Press, 1994.

12. Eskin NAM. *Biochemistry of Foods.* CRC Press, 1990.

13. Gobbetti M, et al. Free D- and L- amino acid evolution during sourdough fermentation and baking. *Journal of Food Science* 59(4):881–884, 1994.

14. Gobbetti M, M De Angelis, R Di Cagno, and C Guiseppe Rizzello. Sourdough/lactic acid bacteria. In EK Arendt and F Dal Bello F. *Gluten-free Cereal Products and Beverages.* Academic Press, 2008.

15. Hayman D, RC Hoseney, and JM Faubion. Bread crumb grain development during baking. *Cereal Chemistry* 75(4):577–580, 1998.

16. Huang JJ, and PJ White. Monoglyceride interaction in a model system. *Cereal Chemistry* 70(1):42–47, 1993.

17. Jefferson DR, AA Lacey, and PA Sadd. Understanding crust formation during baking. *Journal of Food Engineering* 75(4):515–521, 2006.

18. Keetals CJAM, et al. Structure and mechanics of starch bread. *Journal of Cereal Science* 24:15–26, 1996.

19. Kristensen M, et al. Whole flaxseeds but not sunflower seeds in rye bread reduce apparent digestibility of fat in healthy volunteers. *European Journal of Clinical Nutrition* 62(8):961–967, 2008.

20. Labuda I, C Stegmann, and R Huang. Yeasts and their role in flavor formation. *Cereal Foods World* 42(10):797–799, 1997.

21. Lahtinen S, et al. Factors affecting cake firmness and cake moisture content as evaluated by response surface methodology. *Cereal Chemistry* 75(4):547–550, 1998.

22. Mettler E, and W Seibel. Effects of emulsifiers and hydrocolloids on whole wheat bread quality: A response surface methodology study. *Cereal Chemistry* 70(4):373–377, 1993.

23. Meuser F, JM Brummer, and W Seibel. Bread varieties in Central Europe. *Cereal Foods World* 39(4):222–230, 1994.

24. Miller RA, and RC Hoseney. Method to measure microwave-induced toughness of bread. *Journal of Food Science* 62(6):1202–1204, 1997.

25. Nordlee JJA, and SL Taylor. Immunological analysis of food allergens and other food proteins. *Food Technology* 49(2):129–132, 1995.

26. Ostman Em, AH Frid, LC Groop, and IE Bjorck. A dietary exchange of common bread for tailored bread of low glycemic index and rich in dietary fiber improved insulin economy in young women with impaired glucose tolerance. *European Journal of Clinical Nutrition* 60(3):334–341, 2006.

27. Partnership yields new Maillard flavor systems for microwave foods. *Food Engineering* 65(3):36–37, 1993.

28. Potus J, A Piffait, and R Drapron. Influence of dough-making conditions on the concentration of individual sugars and their utilization during fermentation. *Cereal Chemistry* 71(5):505–508, 1995.

29. Primo-Martin C, et al. The role of the gluten network in the crispness of bread crust. *Journal of Cereal Sciences* 43(3):342–352, 2006.

30. Pszczola DE. Ingredients for bread meet challenging 'kneads.' *Food Technology* 59(1):55–63, 2005.

31. Reduced-fat bakery foods: Meeting the taste challenge. *Prepared Foods* 162(8):79–80, 1993.

32. Rogers DE, et al. Stability and nutrient contribution of beta-carotene added to selected bakery products. *Cereal Chemistry* 70(5):558–561, 1993.

33. Schunemann C, and G Treu. *Baking: The Art and Science. A Practical Handbook for the Baking Industry.* Baker Tech, 2007.

34. Seidel C, et al. Influence of prebiotics and antioxidants in bread on the immune system, antioxidative status and antioxidative capacity in male smokers and non-smokers. *British Journal of Nutrition* 97(2):349–356, 2007.

35. Shalani D, and N Lakshmi Devi. Development and acceptability of breads incorporated with functional ingredients. *Journal of Food Science and Technology* 42(6):539–540, 2005.

36. Si JQ. Synergistic effect of enzymes for breadmaking. *Cereal Foods World* 42(10):802–807, 1997.

37. Wehrle K, and EK Arendt. Rheological changes in wheat sourdough during controlled and spontaneous fermentation. *Cereal Chemistry* 75(6):882–886, 1998.

# WEBSITES

Descriptions of various doughs and breads can be found here:

**www.foodsubs.com** (click on "Baked Goods," then "Breads," and then "Doughs")

You can find video demonstrations on various aspects of food preparation, including how to knead dough at these sites:

**www.foodnetwork.com** (Search for "How to Knead dough")

**www.quakeroatmeal.com** (click on "Cooking & Recipes" then "Baking 101," and then "Yeast Breads")

PhotoDisc/Getty Images

# 21 Sweeteners

**FIGURE 21-1**  Sugar cane and sugar beets are the primary sources of sucrose (table sugar).

Teubner/StockFood Creative/Getty Images

Wally Eberhart/Visuals Unlimited/Getty Images

Just as bears are drawn to honey, people seek numerous methods to extract sweeteners from the natural world. Sugar cane and sugar beets deliver the most sugar to satisfy the human taste buds (Figure 21-1). The history of people's interest in these plants dates back to Alexander the Great, who was enticed by the "honey-bearing rods" (sugar cane) he encountered near the river Indus. Christopher Columbus brought sugar cane to the West Indies, where it has been a major cash crop ever since. Jesuit missionaries followed by introducing sugar cane into Louisiana in 1751, and in the mid-1800s sugar beets were planted in the United States.

Although sugar is the most widely used sweetener in food preparation, other types of sweeteners may be used, including syrups, sugar alcohols, and **nonnutritive sweeteners.** And sweetness is only one of the many ways in which sugar functions in foods (Table 21-1). In baked goods, sugar

**Nonnutritive sweeteners**  Food additives requiring FDA approval that provide sweetness with no or insignificant amounts of energy (calories/kcal). Also known as alternative sweeteners, sugar substitutes, sugar replacers, and macronutrient substitutes.

435

**TABLE 21-1**   Functions of Sugars in Foods

| Functions in Foods: | Baked Goods (Yeast) | Beverages | Cakes/ Cookies | Cereals | Confections | Dairy | Frozen | Jams/ Jellies | Processed Foods |
|---|---|---|---|---|---|---|---|---|---|
| Aeration | X | | X | | | | | | |
| Boiling Point (Increase) | | | | | X | | | | |
| Browning (Maillard Reaction) | X | | X | X | | | | | |
| Bulking/Bodying Agent | | X | X | | X | X | X | X | |
| Caramelization | X | | X | | X | | | | |
| Crystallization | | | X | | X | | | | |
| Fermentation | X | | | | | | | | X |
| Flavor Enhancer | | X | X | X | X | X | X | X | X |
| Foam Formation | | X | | | X | | X | | |
| Freezing Point (Decrease) | | | | | | | X | | |
| Glaze Formation | X | X | X | | | | | | |
| Moisture Retention | X | | X | | X | | | | |
| Preservative | | | | | X | X | | X | X |
| Shelf Life (Extender) | | X | X | | | | | | X |
| Solubility | | X | | | X | | | X | X |
| Sweetener | X | X | X | X | X | X | X | X | X |
| Tenderizer | X | X | X | X | X | | X | X | |

produces a finer texture, enhances flavor, generates browning of the crust, promotes fermentation of yeast breads, and extends shelf life by virtue of its ability to retain moisture. Sugar also gives body to soft drinks—so much so that sugar-free drinks need additional ingredients to replace the viscosity provided by sugar so they will not seem flat (36). Sugar offsets the acidic, bitter, or salty tastes of certain foods, such as tomato sauces, chocolate, and sodium-processed meats (ham, bacon, etc.).

Adding sugars to increase food palatability remains a common practice and a major use of sweeteners within the food industry. According to the Dietary Guidelines for Americans, added sugars are defined as sugars or syrups added to foods during processing or preparation. This includes a wide variety of sweeteners such as white sugar, brown sugar, raw sugar, corn syrup, corn-syrup solids, high-fructose corn syrup, malt syrup, maple syrup, pancake syrup, and fructose sweetener (53).

Sugar's many important functions are discussed later in this chapter, following a look at the different types of sweeteners (natural and nonnutritive) and their nutrient contributions.

# NATURAL SWEETENERS

As people searched for new ways to sweeten food, they discovered that plants produce abundant natural sugars through the process of photosynthesis (Figure 21-2). Natural sugars come from many different plants, but most of the sweeteners used in foods are extracted from sugar cane, sugar beets, maple trees, and corn. Lactose in milk is the only sweetener of animal origin, and it is not very sweet. Fruit sugars are derived from specific fruits, while honey is harvested from bees (who produce it from flower nectars).

Sweeteners can be placed into three major groups based on their different chemical structures, which influence their functions in foods and beverages:

- Sugars
- Syrups
- Sugar alcohols

## Sugars

Sugar is what usually comes to mind when the word *sweetener* is used. Once it has been extracted from its plant source,

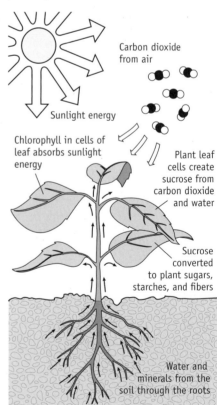

**FIGURE 21-2**   The process of photosynthesis.

Carbon dioxide from air

Sunlight energy

Chlorophyll in cells of leaf absorbs sunlight energy

Plant leaf cells create sucrose from carbon dioxide and water

Sucrose converted to plant sugars, starches, and fibers

Water and minerals from the soil through the roots

sugar becomes a refined carbohydrate, providing a pleasing flavor and packing 4 calories (kcal) per gram. Sugar's ability to sweeten makes it the number-one food additive, by weight, in the United States. Processed foods such as regular soft drinks, sugars and candy, cake, cookies and pies, fruit drinks (fruitades and fruit punch), dairy desserts and milk products (ice cream, sweetened yogurt or milk), and sweetened grain products (coffee cake, doughnuts, waffles, etc.) all contribute to sugar consumption (53). Sugar is also found in less obvious foods, such as most commercially produced catsup (29% sugar) and fruit canned in heavy syrup (18% sugar). Many other foods that contain relatively large amounts of sugar, such as nondairy creamers (up to 65% sugar), do not taste particularly sweet.

The many kinds of sugars differ in their individual characteristics and functions in food. The sugars discussed now include sucrose, glucose, fructose, lactose, and maltose.

## Sucrose

Sucrose, or table sugar, is derived from either sugar cane or sugar beets. Sugar cane has been used as a source of sugar for centuries. It first grew in India, where it is still a major crop. Sugar beets are a more recent addition, a German scientist having first extracted sugar from this plant in 1794. Today, sugar cane provides about 60% of the sugar consumed in the United States; white sugar beets, a relative of the red beet commonly eaten as a vegetable, provide the rest. The sugar obtained from sugar cane and beets is extracted through a number of different refining and purifying steps after harvesting, including crushing, extraction, evaporation, and separation. Once processed, there is no difference between the sugar from these two sources.

### Obtaining Table Sugar from Plants

Harvested sugar cane is washed and then machine-shredded. Heavy steel rollers crush and squeeze the sweet juice from the cane. Water is sprayed on the cut and pulverized sugar cane, allowing more juice to seep out and be collected. The process of extracting sucrose from sugar beets differs in that they are washed, sliced, and then soaked in vats of hot water to remove the sugar. The extracted juices of both cane and beets are then heated and concentrated in evaporation tanks to create a thick syrup known as molasses. Vacuum equipment lowers the boiling point of the syrup so that it may be concentrated without burning. Large sugar crystals form as the solution becomes saturated. The crystals are then separated from the molasses liquid by a centrifuge. Centrifuges spin solutions at very high speeds to separate particles and/or liquids according to density.

### Types of Sucrose

Once refined, the sugar is further processed to produce the commercial products known as raw, turbinado, white, powdered, fruit, Baker's Special, sanding, liquid, and brown sugars (Table 21-2).

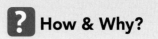 **How & Why?**

*Why does brown sugar have a tendency to dry out and become hard?*

Drying and hardening occur because even though the molasses film over the sugar keeps brown sugar soft and pliable at first, it hardens as soon as the water from the molasses evaporates. Airtight containers, preferably sealed plastic bags, are best for storing brown sugar. Hard sugar can be softened by draping a moist paper towel over the sugar within its container for about 12 hours (sugar draws the moisture to itself). Brown sugar can also be temporarily softened by briefly heating it in an oven at 250°F–300°F (121°C–149°C) or in a microwave. Alternatively, a piece of bread may be placed in the sugar container for several hours so the brown sugar absorbs moisture from the bread (56). Storing brown sugar in a plastic bag in the refrigerator is another way to increase humidity and delay dehydration.

| TABLE 21-2 | Different Forms of Available Sucrose |
|---|---|
| Raw sugar | The sugar that is extracted from sugar cane juice (not beets) without any further refining. As a result, it contains natural contaminants such as soil, insect parts, yeast, molds, waxes, and lint. The FDA banned its sale to the public. |
| Turbinado sugar | A coarse, crunchy, amber-colored raw sugar that has been centrifuged and purified with steam. Sometimes labeled "raw sugar," although it is not truly raw. Demerara sugar is the English version of turbinado sugar that differs in its larger crystal size. |
| White sugar | Made by further refining raw sugar by repeatedly washing and filtering until the rinse liquid is a clear, colorless syrup. The syrup is then boiled until it crystallizes. Crystals are separated by size into "fine," or table sugar, and "superfine" or "ultrafine," which are used by the food industry for cake baking, dry mixes, candy coatings, and mixed drinks. |
| Powdered sugar | Made by pulverizing white granulated sugar. Also called confectioners' or icing sugar, it is frequently combined with an anti-caking substance such as cornstarch, silica gel, or tricalcium phosphate to keep the powder soft and pourable. |
| Invert sugar | A powdered sugar of the finest grain that is used to prepare glossy icings and frostings without any trace of grain or grittiness. |
| Fruit sugar | Very finely granulated sucrose. Its uniform crystal size allows it to remain evenly disbursed in a mix. Used in dry mixes such as gelatins, puddings, and drink bases. |
| Baker's Special | Even more finely granulated than fruit sugar. Used primarily by the baking industry in cookies, cakes, and doughnuts. |
| Sanding sugar | Large-granule sugar often used to decorate the tops of baked goods because it does not melt during baking and it sparkles attractively. |
| Liquid sugar | A solution containing a highly purified sugar that is used in canned foods, beverages, confections, baked goods, frozen foods, and ice cream. |
| Brown sugar | Made by adding molasses syrup to white sugar. The amount and type of molasses determine the grade, with higher-grade brown sugars having a darker color and stronger flavor. These are used when strong flavors are desired, as for baked beans, minced meat, plum pudding, and gingerbread cookies. |
| Muscovado sugar | A softer, finer brown sugar also known as Barbados sugar that imparts a butterscotch flavor to baked goods. |

## Glucose

Glucose, also known as dextrose, is the basic building block of most carbohydrates and is the major sugar found in the blood. Chief plant sources of this monosaccharide are fruits, vegetables, honey, and corn syrup. Glucose can be obtained from starch by commercially treating it with heat and acids or with enzymes. Food companies often use glucose, which is half as sweet as sucrose, in candies, beverages, baked goods, canned fruit, and fermented beverages. The baking industry relies on glucose to enhance crust color, texture, and crumb; as a component in dry mixes for baking; and to temper the sweetness of sucrose (49).

## Fructose

Fructose, also called levulose or fruit sugar, is found naturally in fruits and honey, and is the sweetest of all granulated sugars. It is not related to the fine-textured form of fruit sugar used in dry mixes. This sugar is rarely used in food preparation, because it causes excessive stickiness in candies, overbrowning in baked products, and lower freezing temperatures in ice cream. It is used primarily in the manufacture of pharmaceutical products. It is added to foods and beverages in the form of high fructose corn syrup (HFCS) that is between 42% and 55% fructose (16).

Fructose is sweeter than sucrose, but it still provides the same 4 calories (kcal) per gram found in any other carbohydrate. Diabetics should be aware that while the blood sugar may increase more slowly with fructose than with glucose, it will still be raised.

## Lactose

The sweetness of milk comes from its lactose content. Lactose, a disaccharide, is the least sweet of all sugars and is extracted from whey (a by-product of cheese production) for commercial use in baked products, where it aids in browning. It cannot be fermented, however, so it is rarely used in preparation of yeast bread products or alcoholic beverages that depend on fermentation. Lactose is also used in pharmaceutical products as a filler in pills because its gritty, hard crystals flow freely and are easily compacted for tablet pressing.

## NUTRIENT CONTENT

White granulated sugar is 99.9% pure carbohydrate, so table sugar or any food made with it is a source of refined carbohydrate.

One health problem currently associated with excess sugar consumption is dental caries (cavities). The relationship between sugar and obesity—another widespread problem—is a subject of debate among scientists and the Sugar Association. Obesity has reached epidemic proportions in the United States, and people going on weight-reducing diets often lose weight by cutting back on sweets. The Dietary Guidelines' recommendation is to choose foods/beverages with little added sugars, while the World Health Organization has advised that sugar intake should be less than 10% of calories (39). The American Heart Association recommends no more than 100 to 150 calories (kcal) per day from sugar in women and men, respectively (4). In the United States, regular soft drinks are the main dietary source of simple sugars (22).

One way to reduce refined sugar intake is to reduce the use of table sugar and consumption of processed foods made with sugar. Food and beverage labels can list sugar as sucrose, glucose (dextrose), lactose, maltose, brown sugar, honey, molasses, and corn syrup. The words "no sugar added" on a label do not necessarily mean there is no sugar in the product; it may be present naturally.

## Maltose

Maltose, also called malt sugar, lends certain milk shakes and candies their characteristic malt taste. It is used primarily as a flavoring and coloring agent in the manufacture of beer. During the malting process, barley and other grains are treated to convert the grain's starch to maltose (see Chapter 27).

# Syrups

Syrups are sugary solutions that vary widely in viscosity, carbohydrate concentration, flavor, and price (Chemist's Corner 21-1). The more common ones include corn syrup, high-fructose corn syrup, honey, molasses, maple syrup, and invert sugar. The most frequently purchased syrups in the supermarket are maple-flavored syrups, which are primarily a blend of corn syrup, cane sugar syrup, and honey (50).

## Corn Syrup

A by-product of cornstarch production, corn syrups are viscous liquids containing 75% sugar and 25% water. Corn syrups, rather than granulated sugar, are often used in soft drinks and processed foods to reduce manufacturing cost. Dried corn syrups or corn syrup solids are used in dry mixes for beverages, sauces, and instant breakfast drinks.

***How Corn Syrup Is Made***    Corn syrup is made commercially by adding a weak acid solution to cornstarch, then

## > CHEMIST'S CORNER 21-1

**Measuring Sugar Solutions**

Various instruments are used to measure the amount of sugar solids in a syrup or liquid. The concentration of sugar solutions can be measured using a hydrometer, refractometer, or flowmeter. The hydrometer relies on the Archimedes principle by comparing the weight of a plummet in juice with its weight in water. Refractometers measure how much light is bent by the sugars in the liquid; the more concentrated the liquid, the more light is refracted. A flowmeter judges density by how quickly a liquid flows. The concentration measured by these instruments is converted into a degrees brix value, which is the food industry's standard of identifying the sugar concentration in syrups or liquids (15).

boiling, filtering, and evaporating the mixture until the right sugar concentration is achieved. Adding different enzymes to cornstarch can convert the starch into glucose, maltose, or small polysaccharides known as dextrins (Figure 21-3). Beta-amylase enzymes yield more maltose, whereas glucoamylase enzymes result in a higher-glucose syrup, which is sweeter and less viscous. Manufacturers can modify the

**FIGURE 21-3**  Corn syrup, obtained from hydrolysis of cornstarch, is a solution of sugars dissolved in water. Smaller sugars taste sweeter, whereas longer ones contribute to viscosity.

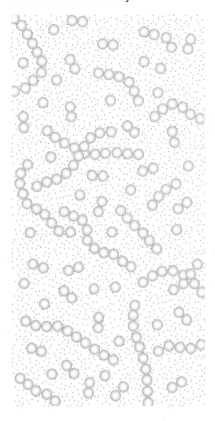

characteristics of a corn syrup to their specific requirements by controlling the type of saccharides formed.

*Dextrose Equivalents*   The sugar that comes from corn is most often called dextrose, rather than glucose. Thus, corn syrups are rated according to their dextrose content, which is measured in **dextrose equivalents (DEs)** (Chemist's Corner 21-2). A complete conversion of starch to glucose has a DE value of 100. **High-conversion corn syrups** are less viscous, have greater sweetening power, are able to promote fermentation, and contribute to browning. The less viscous nature of high-DE syrups is due to smaller molecules such as glucose, fructose, and maltose. In comparison, lower-conversion syrups containing a higher concentration of larger molecules are more viscous, cannot be fermented, and are less sweet.

> ## CHEMIST'S CORNER 21-2
>
> ### Dextrose Equivalents
>
> The dextrose equivalent (DE) value is a ratio of the number of reducing sugars over the total number of glucose molecules.
>
> $$DE = \frac{\text{Number of reducing sugars}}{\text{Total number of glucose molecules}} \times 100$$
>
> | DE | Conversion |
> | --- | --- |
> | 28–37 | Low |
> | 38–47 | Regular |
> | 48–57 | Intermediate |
> | 58–67 | High |
> | >68 | Extra high |
> | 80–100 | Crude corn sugars |

### High-Fructose Corn Syrup

High-fructose corn syrups (HFCSs) were first invented by Japanese scientists in the 1970s (Chemist's Corner 21-3). Their intense sweetness contributed to their widespread use in food products: baby foods, bakery products, canned fruits, carbonated beverages, confections, dry bakery mixes, fountain syrups/toppings, frozen fruits, fruit juice drinks, frozen desserts, jams, jellies, preserves, meat products, pickles, condiments, and table syrups. HFCS replaced sucrose in sodas/pops to the point that the beverage industry alone uses 90% of all HFCSs (13). Sweetened soda/pop drinks account for one third of all added sugars in the diet (22).

HFCS is approximately 40% fructose and 50% glucose. The greater sweetening power of HFCS means that less is

> ## CHEMIST'S CORNER 21-3
>
> ### HFCS Production
>
> HFCS is made by first treating cornstarch with alpha-amylase, producing shorter sugar chains called polysaccharides. Glucoamylase is then added to break the polysaccharides into glucose molecules. Finally, the insoluble glucose isomerase enzyme is added; this chemical reaction causes partial enzymatic conversion of glucose to fructose (24).

needed, so it is much cheaper to use than sugar. Its clarity and colorlessness also contribute to its industrial popularity.

> ## 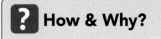 How & Why?
>
> #### Is HFCS natural?
>
> In recent years, there has been much debate in the food industry over HFCS and the use of the term "natural." The Food and Drug Administration (FDA) does not define the term "natural," but it does have a longstanding policy that "natural" products do not contain any artificial or synthetic substance that would not normally be expected in the food (24). This includes artificial flavors or colors. Because HFCS is altered in the manufacturing process, the FDA has stated that HFCS does not qualify as "natural."

*The HFCS and Obesity Controversy*
The significant increases in HFCS consumption over recent decades have prompted some to suggest that HFCS is linked to the concomitant rise in obesity and its related health problems during recent years. The American Medical Association recently performed a review of scientific literature, and reported that there was insufficient evidence that HFCS was any more dangerous than other sweeteners (6). Similarly, other authors have concluded that the increases in obesity, diabetes, and related diseases are more closely linked to the overall increase in calorie and sugar intakes, rather than to HFCS specifically (17). Caloric intake from soft drinks and other carbohydrate-based beverages is a major source of excess calories that contribute to obesity.

### Honey

The world initially depended primarily on honey as a sweetener. Bees collect the thin, watery nectar of flowers and,

> **Dextrose equivalent (DE)** A measurement of dextrose concentration. A DE of 50 means the syrup contains 50% dextrose.
>
> **High-conversion corn syrups** Corn syrups with a dextrose equivalent over 58.

during the flight back to their hive, convert it through enzymatic action into fructose and glucose molecules. The bees deposit the nectar in honeycombs, where most of the water evaporates to create a thick, sweetened syrup, which is further flavored with enzymes added by the bees. It takes 2 million flowers to produce enough nectar to make 1 pound of honey, and the average worker bee makes only ½ of a teaspoon of honey in its entire lifetime.

Honey can range in color from clear to dark brown, and can be fluid, viscous, or solid (48). The taste, color, flavor, and crystallization of honey are all influenced by the plant from which it is derived. Honey can be analyzed to determine its pollen content when its origin is in question. Honey varieties with fewer crystal nuclei will crystallize at a slower rate, resulting in larger crystals that settle at the bottom of the container. Examples of this type of honey are buckwheat and honeydew (48). Crystallized honey can be warmed to liquefy the mixture, but the honey does not usually recrystallize uniformly.

**Sugars in Honey**   Honey typically consists primarily of sugars other than sucrose: fructose (40%), glucose (35%), sucrose (2%), and traces of other carbohydrates (48). Honey also contains maltose (1.5–4%), along with various other sugars present in less than 1% concentrations (including isomaltose, turanose, trehalose, erlose, maltotriose, melecitose, and raffinose). To protect the consumer from honey that has been extended by the addition of sucrose, the FDA limits commercial honey to no more than 8% sucrose. Over 180 substances, including beeswax, minerals, and water (18%), are found in honey. Examples of enzymes found in honey include alpha- and beta-amylase, glucose oxidase, catalase, phosphatase, and maltase (48). *Clostridium botulinum* spores are also often present in honey and pose a hazard to children under 1 year of age. Honey is therefore not recommended for infants, whose systems are not yet able to handle the spores as those of older children and adults can.

**Removing Honey from the Comb**   Honey was originally sold in the comb, but now it is generally extracted by processors by cutting the comb on one side and releasing the honey in a centrifuge.

Honey can also be collected by crushing the combs and straining out the thick fluid. The extracted honey is heated to 140°F (60°C) for 30 minutes to destroy most microorganisms, then filtered and packaged in airtight containers. Some small producers sell unfiltered honey as well as comb honey.

**Whipped, Creamed, Granular, and Infused Honeys**   Prior to packaging, honey may be processed into various forms. Whipped or creamed honey has had some of the fructose removed, resulting in a thicker consistency. Dried, granular honey is used in baked products, confections, and dry mixes. Infused honey has been heated with a flavor or herb to give it a unique taste (Table 21-3).

**Storing Honey**   Regardless of the form that the honey is in, its naturally high sugar content prevents the growth of bacteria; therefore, honey can remain shelved for years without spoiling. Stored for long periods of time, however, it can harden as its sugar precipitates into crystals. If this occurs, it can be softened by warming the jar in hot water for an hour, or setting the opened jar in the microwave on the low defrost setting.

**Substituting Honey for Sugar in Recipes**   Honey can be substituted for table sugar in recipes if a few guidelines are followed (12):

- In baked products, no more than half the granulated sugar should be replaced with honey.
- Use 1 part honey for every 1¼ parts sugar.
- Reduce the liquid in the recipe by ¼ cup because honey is largely water.
- Add ½ teaspoon baking soda for every cup of honey to reduce the acidity and weight of honey.

Honey has a more pronounced flavor than sugar, and this will affect the final flavor of the product. It also has a tendency to increase the browning of baked products (10). Adding ⅛ of a teaspoon of baking soda allows even browning; reducing oven temperatures by 25 degrees helps prevent overbrowning. The stickiness of measuring honey can be minimized by coating the inside of a measuring cup with water or a very thin layer of vegetable oil before measuring.

### ? How & Why?

**Why are there different flavors and colors of honey?**

The flavor and color of the honey depend on the type of flower visited by the bees. There are over 300 varieties of honey, the most popular being alfalfa and clover. Honeys are blended by the bees as they collect nectar from an assortment of flowers. Normally, honey is a golden amber, but the darker the color, the stronger the flavor. For example, Australian eucalyptus honey has a reddish-brown color and a strong, tangy flavor. Acacia honey, in contrast, is almost clear and has a very delicate flavor and aroma.

### Molasses

Molasses is the thick, yellow to dark-brown liquid by-product of the juice of sugar cane or beets. The liquid is repeatedly boiled, but for the end product to be called molasses, it must contain no more than 75% water and 5% mineral ash. Most of the sugar in molasses is sucrose, which renders the product darker with each boiling. The syrup's ultimate color determines its grade. Blackstrap molasses, the most concentrated in syrup and minerals, is the darkest in

---

**TABLE 21-3**   Infused Honey Recipe

Add 1 cup of honey to a saucepan with a flavoring (mint, ginger, lime, rosemary, cinnamon stick, orange, etc.).

Heat it to boiling and immediately reduce to a simmer for 5 minutes, then let stand for 10 minutes (24 hours for cinnamon stick).

Strain the infused honey while it is still warm.

Cooled infused honeys are good in hot or cold teas, or on overcooked carrots, French toast, seasoned pork, glazed chicken, cornbread, and numerous other foods.

color and most bitter, and is used primarily for industrial purposes and cattle feed, although it is available for home consumption. The thick gumminess of molasses is due to its content of fiber (hemicellulose and pectin), waxes, proteins, and dextran (57). Most commercial grades of molasses are actually blends of different types of molasses.

### Foods Made with Molasses

Molasses is used both in food preparation and in the making of rum. Its main use is in baking, where it enhances the flavor of breads, cakes, and cookies. A few other foods that incorporate molasses are baked beans, glazes for hams and sweet potatoes, cookies, and candies such as toffees and caramels. Fermenting molasses yields rum, an alcoholic beverage that is distilled and generally aged for 5 to 7 years. Quicker aging periods of 1 to 4 years are used for rapidly fermented light rums.

## Maple Syrup

Native Americans were the first to collect the sap from maple trees and boil it into a smooth, tasty syrup. Long ago, sap was harvested by drilling holes into a maple tree, inserting a spout, and catching the fluid in a bucket positioned under the spout. Newer methods eliminate the buckets, instead utilizing a network of plastic pipelines attached to the trees. The pipeline carries the sap, a clear, almost tasteless, watery liquid, directly to the sugarhouse, where it is boiled down. Sap is collected in the late winter and early spring during the few weeks when the days are relatively warm, but the nights are still cold. Vermont, Maine, New Hampshire, northern New York, and parts of Canada, where the dramatic rise and fall in spring temperatures trigger the flowing of the sap, all have the ideal environment for maple syrup harvesting.

### Maple Syrup Colors   The flavor and color of maple syrup develop during the boiling of the initially colorless sap. Government standards specify that maple syrup must contain at least 65.5% sugar among its other ingredients, such as acids and salts. Maple syrup is graded and sold by color and ranges from light amber, or Fancy, to the darkest color, known as Commercial. The darker the color of

the maple syrup, the more pronounced its flavor will be. The lightest colored syrups have the most delicate flavors.

### "Real" vs. Blended Maple Syrups

Because it takes about 40 gallons of sap to produce 1 gallon of maple syrup, most "maple syrup" sold today is blended with corn syrup and/or cane sugar syrup. Many companies add artificial maple flavorings to foods, but real maple syrup has a unique flavor and smoothness not duplicated by substitutes. Pure or blended maple syrup is commonly poured over pancakes, waffles, and French toast or added as an ingredient in maple butter, cream, and candy.

### Maple Sugar   Maple sugar is a product of maple syrup. It is made by further boiling the syrup until most of the water evaporates and the sugar **crystallizes** out of the syrup. About 8 pounds of maple sugar are produced from 1 gallon of maple syrup.

## Natural Evaporated Cane Juice

This product is made from sugar cane grown on the Hawaiian Islands and produced from extracts in the first pressing of sugar cane (46). Natural evaporated cane juice retains some of the natural molasses, giving it a unique flavor. Two forms are available: *Natural White* has a hint of molasses flavor, and *Premium Maui Gold*™ has a rich, robust flavor.

## Invert Sugar

**Invert sugar** is available only in clear, liquid form and is sweeter than granulated sugar. This type of sugar resists crystallization and is commonly used by professional confectioners to achieve a smooth, melt-in-the-mouth texture in candies.

### How Invert Sugar Is Made   Invert sugar is made commercially by dissolving sucrose in water, heating the solution, and adding either an acid such as cream of tartar or an invertase enzyme such as sucrase, which hydrolyzes the sucrose into two equal portions of glucose and fructose. This process is called inversion (Chemist's Corner 21-4). The use of cream of tartar or sucrase inhibits crystallization. In addition, the acidity of cream of tartar (tartaric acid) has the added benefit of preventing the natural decomposition

> ### CHEMIST'S CORNER 21-4
>
> **Invert Sugar Production**
>
> Invert sugar is produced by adding an acid to a liquid solution of sucrose, causing the sucrose to break down to d-glucose and d-fructose. It is called "invert" sugar because the optical rotation of the molecules reverses (or inverts) from +66 degrees to −22 degrees when sucrose is broken down to fructose and glucose (54).

of monosaccharides into bitter, brown-colored substances, which occurs when they are exposed to hard water or any other alkaline medium. The amount of cream of tartar added depends upon the percentage of invert sugar concentration desired.

### Invert Sugar in Foods   The confectionary industry uses invert sugar to develop the soft, fluid center of certain chocolates. See Chapter 25 for more information on how invert sugar is utilized in preparing confectioneries.

# Sugar Alcohols

Foods containing sugar alcohols can be labeled "sugar free" because they are low in calories (16). Sugar alcohols are not carbohydrates, but the alcohol counterparts of specific carbohydrates. Although they are found naturally in fruits and vegetables, they are also synthesized by hydrogenating certain sugars (Chemist's Corner 21-5). Sugar alcohols are also referred to as polyols, polyalcohols, and polyhydric alcohols (54). Polyols approved for use in the United States include erythritol, isomalt, lactitol, maltitol, mannitol, polyglycitols (hydrogenated starch

> **Crystallization** The precipitation of crystals from a solution into a solid, geometric network.
>
> **Invert sugar** An equal mixture of glucose and fructose, created by hydrolyzing sucrose.

## CHEMIST'S CORNER 21-5

### Sugar Alcohol Production

Sugar alcohols are polyols, which is short for polyhydric alcohols (44). Listed below are several examples of sugars and their alcohol counterparts. These sugar alcohol compounds differ from their monosaccharides by a slight arrangement of their atoms; the carbohydrate's hydroxyl group (–OH) is replaced with an aldehyde or ketone group ($C_5O$).

| Sugar | Alcohol Counterpart | Calories (kcal)/Gram |
|---|---|---|
| Sucrose | Sorbitol | 2.6 |
| Mannose | Mannitol | 1.6 |
| Maltose | Maltitol | 2.1 |
| Xylose | Xylitol | 2.4 |

Polyols occur naturally in certain foods, but for commercial use they must be prepared in a laboratory. Hydrogenation of fructose, derived from sucrose or starch, produces mannitol (54). Mannose, glucose, or fructose is hydrogenated (hydrogen is added) to yield mannitol; glucose is hydrogenated to generate sorbitol. Sorbitol is the reduced form of dextrose, and is the most widely available polyol (56). Hydrogenation is achieved using the catalyst Raney nickel (54). Xylitol is a pentose derived from xylose, which comes from xylans, gummy polysaccharides found in plant parts such as wood pulp, sugar cane bagasse, wheat straw, oat hulls, corncobs, and birch wood (56). Xylitol is formed by hydrogenation of D-xylose in these products (54). Erythritol is produced from glucose through fermentation. Maltitol is a 1,4-linked molecule of glucose and sorbitol produced by hydrogenation of high-maltose syrups derived from enzymatic hydrolysis of starch (54). Lactitol is produced by reduction of the glucose in lactose (54). Other examples of sugar alcohols include the hydrogenated starch hydrolysates, which are a mixture of hydrogenated oligo- and polysaccharides (32).

hydrolysates), sorbitol, and xylitol (37). Isomalt is derived from sucrose, and has been used in Europe since the 1980s (45). Maltitol appears to be the closest equivalent to sucrose in solubility and functionality (43). Although most sugar alcohols are primarily sold as ingredients to food manufacturers, isomalt is available to professional chefs (32). Sugar alcohols are 40% to 100% as sweet as sucrose but are resistant to

digestion by intestinal enzymes, so they contribute few calories and are digested and absorbed more slowly (19). Sugar alcohols also have low hydroscopicity: erythritol has the lowest, followed by mannitol, isomalt, lactitol, maltitol, sorbitol, and xylitol (54).

### Sugar Alcohols in Foods

The sugar alcohols' ability to contribute sweetness, combined with their tendency to be slowly absorbed, makes them useful ingredients in various dietetic foods. In addition, sugar alcohols produce a negative heat of solution when they are dissolved (54). This means that they produce a characteristic cooling sensation when they dissolve in the mouth, making them useful for such products as sugarless gums, dietetic candies, sugar-free cough drops, throat lozenges, breath mints, and tablet coatings. Not all sugar alcohols are equal in their cooling effects: xylitol has a strong cooling effect, while maltitol and isomalt do not.

Sugar alcohols do provide some calories, but fewer than sucrose, which contributes 4 calories (kcal) per gram. According to the FDA, their caloric values are as follows (19, 37):

| | |
|---|---|
| 0.2 kcal/g | erythrol |
| 1.6 kcal/g | mannitol |
| 2.0 kcal/g | isomalt and lactitol |
| 2.4 kcal/g | xylitol |
| 2.6 kcal/g | sorbitol |
| 3.0 kcal/g | polyglycitols and hydrogenated starch hydrolysates |

The low caloric content and cooling sensation of sorbitol, mannitol, and xylitol make them attractive to food manufacturers as sweetening agents, especially in dietetic candies (as mentioned above). Other products utilizing sugar alcohols to reduce sugar content include jellies, jams, chocolate, hard-boiled candies, chewing gums, and ice cream (54).

Sugar alcohols have other advantages as well; for example, they are cariostatic, or cavity preventing, because they cannot be digested by the bacteria responsible for dental caries (cavities). They also function as **humectants** and emulsifiers, extending the shelf life of processed foods. Sorbitol, the most widely used sugar alcohol, is frequently used as a humectant to maintain moistness in marshmallows and shredded coconut.

### Problems with Sugar Alcohols

One drawback of sugar alcohols in dietetic foods is that they are more slowly absorbed from the small intestine than other sugars, which can lead to diarrhea, abdominal pain, and gas (16). For this reason, consumption of food products containing over 30 grams of sorbitol is not recommended, and only limited quantities of xylitol are allowed in special dietary foods (18).

# NONNUTRITIVE SWEETENERS

Nonnutritive sweeteners provide minimal to no energy (47) and must be approved by the Food and Drug Administration (FDA) (37). The FDA also defines **Acceptable Daily Intake (ADI)** values for each of the five nonnutritive sweeteners currently in use.

**Humectant** A substance that attracts water to itself. If added to food, it increases the water-holding capacity of the food and helps to prevent it from drying out by lowering the water activity.

**Acceptable Daily Intake (ADI)** The amount of a food additive that can be safely ingested daily over a person's lifetime.

The five nonnutritive sweeteners currently approved by the FDA for use in the United States are saccharin, aspartame, acesulfame-K, sucralose, and neotame (Table 21-5) (31, 37). They are also known as *intense sweeteners*—defined as substantially sweeter than sucrose (by weight ranging from 30 to several thousand times as sweet as sucrose). Although aspartame provides the same 4 calories (kcal) per gram as sucrose, so little of the sweetener is used that its caloric contribution is negligible. One other nonnutritive sweetener, alitame, has been developed and is awaiting approval by the FDA (16). In addition, manufacturers are trying to regain approval for cyclamates, which had been approved but were later banned in the United States as a potential carcinogen. Neohespiride, which is

1,500 times sweeter than sucrose and has a licorice-like flavor, is approved as a food additive but not as a sweetener in the United States.

Nonnutritive sweeteners have a history of controversy regarding their safety in the human diet, yet they enjoy continued popularity among diabetics, people watching their weight, and individuals trying to prevent tooth decay (55). The food industry attempts to satisfy the market by providing a wide variety of foods containing one or more of the FDA-approved nonnutritive sweeteners. In descending order, the most common foods sold to consumers that contain nonnutritive sweeteners are diet soft drinks, tabletop sweeteners, pudding, gelatin, yogurt, frozen desserts, powdered drinks, cakes, cookies, jams, jellies, and candy (8).

One drawback of nonnutritive sweeteners is that they lack the important functional characteristics of sugar: bulking, binding, texturing, and fermenting. However, certain compounds can be added to foods to compensate for the lost characteristic of bulking. These include cellulose (3), maltodextrin (also used for its binding property), the sugar alcohols, and polydextrose. Polydextrose provides a texture similar to sugar, with only 1 calorie (kcal) per gram, and is currently approved for use in frozen dairy desserts, baked goods and mixes, confections and frostings, hard and soft candy, chewing gum, gelatins, puddings and fillings, and salad dressings (18).

## Saccharin

Saccharin was discovered as a sweetener in 1879 by Constantin Fahlberg (7). The researcher noticed that his dinner roll tasted strangely sweet and traced it back to a saccharin substance he had accidentally spilled on his hands while working in his university research lab (35). Saccharin is known by the brand name Sweet'N Low™ (47). It can be used in a variety of products, including baked or processed foods. Saccharin's major drawback, at least for some people, is its bitter aftertaste, which can be masked only partially by blending it with other sweeteners (18).

### Concerns About Saccharin

The controversy over saccharin's safety peaked when researchers in a Canadian

study reported an increased incidence of bladder cancer in rats fed very high amounts of saccharin (5% to 7.5% of the diet)—the human equivalent of drinking at least 800 diet sodas a day (25). Responding to that study, the FDA proposed a ban on saccharin in 1977. When letters of protest poured into Congress, a congressional moratorium was placed on the FDA ban, along with the requirement that all saccharin-containing products carry a public health warning (34). In 2000, President Clinton signed legislation relieving manufacturers of the requirement to include a warning label on products containing saccharin (26).

The Office of Technology Assessment, a research arm of Congress that attempts to review scientific matters objectively, concluded that saccharin is a potential cause of cancer in humans, although it is among the weakest carcinogens ever detected.

## Aspartame

Like saccharin, aspartame was discovered by accident. In 1965, James Schlatter was doing research on ulcer drugs at G.D. Searle Co. when he licked his finger to pick up a piece of weighing paper and noticed that the finger tasted sweet. He realized that the sweetness came from an earlier spill in the laboratory (47). Schlatter discovered spartame which is a synthetic combination of two amino acids, aspartic acid and phenylalanine. These amino acids provide 4 calories (kcal) per gram, but this energy value is insignificant because so little is needed to produce intense sweetness. FDA-approved in 1981, aspartame is now sold as NutraSweet™, Equal™, and Equal Granular™. Equal Granular combines aspartame and maltodextrin, a nonsweet bulking agent derived from cornstarch, which provides 4 calories (kcal) per gram (5). Equal is a blend of dextrose, maltodextrin, and aspartame. Soft/pop drinks account for over 70% of the aspartame consumed (16).

**No-observed-effect level (NOEL)** The level or dose at which an additive is fed to laboratory animals without any negative side effects.

## TABLE 21-4  Approved Nonnutritive Sweeteners

**Approved Alternative Sweeteners**

| Sweetener | Chemical Structure | Sweetness (sucrose = 1) | Taste Characteristics | Uses | ADI* (mg/kg of body weight) OR equivalent |
|---|---|---|---|---|---|
| Saccharin | | 200–700 | Slow onset, persistent aftertaste, bitter at high concentrations | Used in soft drinks, assorted foods, and tabletop sweeteners | 5 |
| | | | | | 8 packets** |
| Aspartame (Nutrasweet™) | | 180 | Clean, similar to sucrose, no bitter aftertaste | Approved for use in tabletop sweeteners, dry beverage mixes, chewing gum, beverages, confections, fruit spreads, toppings, and fillings | 50 |
| | | | | | 15 diet soda cans |
| Acesulfame-K (Sunette™) | | 130–200 | Rapid onset, persistent, side-tastes at high concentrations | Approved for use in tabletop sweeteners, dry beverage mixes, and chewing gum | 15 |
| | | | | | 25 diet soda cans |
| Sucralose (Splenda™) | | 600 | Can withstand high temperatures without losing flavor | Approved for use in soft drinks, baked goods, chewing gums, and tabletop sweeteners | 5 |
| | | | | | 5 diet soda cans |
| Neotame | | 8000 | Clean, sweet, sugar-like taste; enhances flavors of other ingredients | Under review | 18 |

*Acceptable Daily Intake (ADI) values for aspartame and acesulfame-K are FDA. Values for saccharin and sucralose are United Nations Joint FAO/WHO Expert Committee on Food Additives (JECFA).

**One Sweet'N Low™ packet contains 30mg; one soft drink averages 125 mg.

### Aspartame Side Effects

As with saccharin, several research studies have questioned the safety of aspartame, and there does appear to be a small subgroup in the population that is sensitive to one or more of its breakdown products (aspartic acid, phenylalanine, and methanol) (55). Common complaints among this subpopulation include headaches, dizziness, mood changes, and nausea. In addition, research suggests a possible increased risk some types of cancer in rats (55). Although these side effects are controversial, there is no question that aspartame should not be consumed by individuals with phenylketonuria (PKU), a rare genetic disease afflicting one out of every 15,000 infants. For this reason, food products containing aspartame as an additive must carry the following warning: "Phenylketonurics: Contains Phenylalanine."

Although aspartame does not have the bitter aftertaste of saccharin, it loses it's sweetness when exposed to heat or acids. It can be used in baked goods if encapsulated in a hydrogenated fat coating that melts at the end of baking.

## Acesulfame-K

Acesulfame-K was discovered in 1967 and, like the other artificial sweeteners, was stumbled upon by accident. Sold as Sunette® and Sweet One (47), acesulfame-K was FDA approved in 1988 for use in tabletop sweeteners, dry beverage mixes, and chewing gum (51). It is stable to heating and cooling, but has a bitter aftertaste.

## Sucralose

Sucralose, marketed as Splenda™ in the United States and Canada (47), was discovered in 1976, and approved by the FDA in 1998 (28). It is known as E number E955 within the European Union. A student (Shashikant Phadnis) discovered this nonnutritive sweetener when working in a laboratory in London at King's College, where he mistook the word *testing* for *tasting*.

Splenda's website states, in part, that "Splenda . . . is not sugar, it is a non-caloric alternative to sugar. . . . Splenda Brand Sweetener is manufactured through a patented multi-step process that starts with cane sugar and selectively replaces 3 hydrogen–oxygen groups on the sugar molecule with 3 chlorine atoms. Chlorine is present naturally in many of the foods and beverages that we eat and drink every day ranging from lettuce, mushrooms, and table salt. In the case of sucralose, its addition converts sucrose to sucralose."

This nonnutritive sweetener is stable at high temperatures, making it suitable for use in baked products. Sucralose is used in over 100 products, such as carbonated soft drinks, cakes, muffins, juices, and gums, and as a tabletop sweetener (11, 43). At one point, Splenda became so popular that demand was starting to surpass supply. Although the final scientific conclusions regarding the safety of any nonnutritive sweetener continue to be controversial (27), Splenda currently appears to be less questionable than aspartame.

## Neotame

In 2002, the FDA approved the newest nonnutritive sweetener on the scene, neotame (33). It is 8,000 times sweeter than sugar and calorie free. Like aspartame, neotame is made from the amino acids aspartic acid and phenylalanine. However, neotame is not significantly metabolized to phenylalanine, so no warning label is required for people with phenylketonuria (42). This tabletop sweetener has the potential to replace both sugar and high-fructose corn syrup (41). Only about 6 milligrams of neotame are needed to sweeten a 12-ounce can of soda.

## Stevia: Dietary Supplement and GRAS Additive

Currently, products sweetened with stevia are being rapidly introduced into the U.S. market (46). The leaves of the South American plant *Stevia rebaudiana* contain components that are 300 times sweeter than sucrose (21). However, a sometimes bitter, astringent aftertaste may follow depending on the

extracted and/or modified substance. Confusion results when the term *stevia* (an herb from the plant's leaves) is used interchangeably with the names of its extracted and/or modified components (see Chemist's Corner 21-6 for clarification).

Though the infusion of stevia products into the U.S. market is a recent development, the leaves have been used in Paraguay, Venezuela, Brazil, and Colombia as a sweetener for more

than 1,500 years (45, 47). Guarani native inhabitants of Paraguay used stevia to sweeten a bitter drink called mate. Sweeteners made with stevia were actually introduced to the Japanese market in the early 1970s (45, 48).

Although certain country agencies approve stevia as a sweetener, others, including the European Commission, do not. Stevia was banned in the United States until 1995, when the 1994 Dietary Supplement Health and Education Act led the FDA to permit stevia as a dietary supplement, but not as a food additive. This is why the product can be purchased at health food stores or in the nutritional supplement section of stores instead of next to other sweeteners.

Large U.S. corporations began to develop and use their own stevia derivatives in their soft drinks, and the FDA recently granted Generally Recognized as Safe (GRAS) status to one of stevia's extracted components, rebaudioside A (46). This compound is 200 times sweeter than sugar, and has a clean taste useful in many foods and beverages. Truvia™ is a calorie-free sweetener made from rebiana, a 97% pure extract of rebaudioside A that is manufactured by Cargill (www.cargill.com). Truvia is available as a tabletop sweetener that contains erythritol and other flavoring agents. Rebiana is made by drying stevia leaves, then steeping them in fresh water. The sweet components of the leaves are then purified from the water. This extract has a purer taste and lacks the licorice flavor found in cruder stevia mixtures.

### Safety of Stevia

Stevia has been found to be mutagenic (causes mutations due to disrupted DNA) in bacteria (30) and possibly rats (40), but the "weight of the evidence" from the majority of studies suggests that it is safe, and some government agencies consider it safe due to its longstanding use in humans (47). Others wait for additional research, especially in the areas of male reproduction and interference with carbohydrate absorption (20).

## Pending Nonnutritive Sweeteners

Cyclamates were discovered in 1937 by a university graduate student, Michael Sevda. They are 30 times sweeter than sucrose and stable to heat. In 1970, cyclamates were banned for use in the United States because studies indicated they might be carcinogenic. They are, however, permitted for use in low-calorie foods in more than 50 countries, and in Canada for tabletop sweeteners and pharmaceuticals (38). In 1980, the FDA rejected a petition for cyclamate approval, but the nonnutritive sweetener continues to be under review.

Another nonnutritive sweetener pending approval by the FDA is alitame, a peptide formed from the two amino acids aspartic acid and alanine (38, 55).

## Other Sweeteners

Around the world, the search continues for a sweetening substance without the caloric content of sugar. Several sweeteners from a variety of sources are being investigated or are awaiting approval. Chemical structures vary tremendously and include peptides, amino acids, carbohydrates, inorganic salts, and synthetic compounds. Several proteins with sweet tastes, including thaumatin, talin, brazzein, monellin, mabinlin, and pentadin, have also been identified (47).

- *Glycyrrhizin.* An extract from the licorice root, it is 50 to 100 times sweeter than sucrose and is used only in confections.
- *Dihydrochalcones.* Obtained from citrus peel, these compounds are several hundred times sweeter than sucrose, with a slow taste onset and a lingering aftertaste.
- *L-sugars.* The chemical mirror image of natural sugars, these are not metabolized by body enzymes and are noncarcinogenic.
- *Neohesperidine.* This licorice-flavored substance can enhance the taste and mouthfeel of beverages and is actually listed as a flavor ingredient, but not a sweetener.
- *Thaumatin.* An extract of the fruit of a West African plant, *Thaumatococcus danielli* (47), it is one of the sweetest substances known—1,600 times sweeter than sucrose. It has a licorice-like taste when used in high concentrations, is very stable to heat and acid, and is used in chewing gums (52).
- *Tagatose.* This natural substance is found in dairy products (1.5 calories

per gram). It offers the bulk of sucrose and is almost as sweet. Unlike other sweeteners, it is considered Generally Recognized as Safe (GRAS) and may be used in the U.S. food supply (37).

- *Trehalose.* Also on the GRAS list is trehalose, which is naturally found in honey, mushrooms, and lobster but is commercially made from starch. It benefits frozen foods by protecting their cell structure so their texture is better (37).
- *Isomaltulose.* Used as a sugar substitute in Japan since 1985, this alternative sweetener provides 4 calories (kcal) per gram and is about as sweet as table sugar (37). Also known as Palatinose, NRGylose, or Xtend Isomaltulose, this disaccharide is manufactured enzymatically from sucrose via bacterial fermentation.
- *Brazzein.* Csweet™ uses this protein, derived from the West African fruit *Pentadipandra brazzeana Baillon*, to provide sweetness (45–47). The taste of brazzein is similar to table sugar, with no significant aftertastes. Because it is a protein, diabetics do not include it when counting carbohydrates.
- *Mabinlin.* This relatively new sweetener is a protein molecule found in the seeds of the *Capparis masaikai* plant (47).

# FUNCTIONS OF SUGARS IN FOODS

Sugars serve many functions in foods beyond merely providing sweetness (Table 21-1). Even when it comes to sweetening ability, however, various sugars differ due to their unique chemical arrangements. These structural differences also influence how each dissolves, crystallizes, browns, melts, absorbs water, contributes to texture, ferments, and preserves food.

## Sweetness

Sugars are not equal in their ability to sweeten bland foods or minimize sour and bitter tastes, as shown by Table 21-5, which compares the relative sweetness

**TABLE 21-5**    The Relative Sweetness of Sweeteners Compared to Sucrose

| Sweetness | Sweetener (Sucrose = 1) |
|---|---|
| ***Less Sweet than Sucrose*** | |
| Lactose | 0.4 |
| Maltose | 0.5 |
| Sorbitol | 0.6 |
| Galactose | 0.6 |
| Glucose | 0.7 |
| Mannitol | 0.7 |
| Xylose | 0.7 |
| Invert Sugar | 0.7 |
| Xylitol | 0.8 |
| | |
| ***Sweeter than Sucrose*** | |
| Invert Sugar Syrup | 1.6 |
| Fructose | 1.7 |
| Cyclamate | 30 |
| Glycyrrihizin | 50 |
| Acesulfame-K | 130 |
| Aspartame | 180 |
| Stevioside | 300 |
| Saccharin | 500 |
| Sucralose | 600 |
| Thaumatin | 1,600 |
| Alitame | 2,000 |
| Talin | 2,500 |
| Monelin | 3,000 |
| Neotame | 8,000 |

of sugars, sugar alcohols, and nonnutritive sweeteners to sucrose, which is scored as 1. Even the type of sweetness differs among sweeteners, as observed when comparing the tastes of table sugar (sucrose), honey, molasses, and corn syrup. Temperature also influences sweetness; cold foods and drinks usually taste sweeter than their hot counterparts. Other variables are the pH, other food ingredients, and the taster's sensitivity to sweetness (14). The dominant determinants of sweetness, however, are the type of sugar used and its concentration.

## Solubility

Syrups owe their existence to sugar's ability to dissolve in water. Solubility, which varies from sugar to sugar, is determined by measuring how many grams of sugar will dissolve in 100 milliliters of water. Fructose is the most soluble, followed by sucrose, glucose (dextrose), maltose, and finally lactose (2). The solubility of a sweetener influences the perceived mouthfeel and texture of a food or beverage. For example, the least soluble sugar, lactose, is

**FIGURE 21-5**  Effect of temperature on the solubility of sucrose.

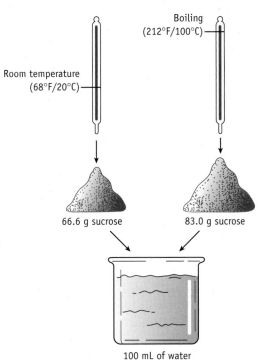

as stirring. Agitation should also be avoided during the cooling process, as it may result in crystal formation. Another way to control crystallization is by being selective in the type of sugar used. Sugars with low solubility, such as lactose, have a greater tendency to crystallize, whereas fructose, with its high solubility, does not. Invert sugar and corn syrup also resist crystallization, which is why they are often used in confectionary production.

## Browning Reactions

Browning reactions involving sugars (**reducing sugars**) and proteins (usually from milk) are due to the Maillard reaction (see Chapter 3). Food products relying on browning from the Maillard reaction include microwaved baked products that have incorporated fructose or dextrose sugars. These reducing sugars should not be added to powdered beverage mixes because they may cause browning during storage. Instead, sucrose is added to such products.

## Caramelization

**Caramelization** results from heating sugars. Sucrose heated in a dry pan will start to melt into a clear, viscous mass when heated to about 320°F (160°C). If heating continues to 338°F (170°C), the melted sugar mass will become smooth and glossy and start to caramelize (1). Sugars differ in the temperatures at which they melt. Fructose, for example, caramelizes at a slightly lower temperature, about 230°F (110°C).

Caramelization is the result of chemical reactions that break the sugar

responsible for the gritty texture of ice cream that has partially thawed and been refrozen. Increasing the temperature of a sugar solution allows more sugar to dissolve, resulting in a supersaturated solution when it cools (Figure 21-5). In turn, increasing the sugar concentration raises the boiling point of water (Table 21-6). (See Chapter 3 for more discussion of saturation and solutions.)

## Crystallization

Crystallization is a vital process in the manufacture of candy. The development or inhibition of crystal formation

determines the finished product's quality (23). (See Chapter 25 for more on candy.) The goal in preparing noncrystalline candies is to prevent the sugar in solution from precipitating out in the form of crystals, causing an undesirable grainy texture. The formation of one crystal can start a domino effect, triggering the entire mixture to crystallize. Small foreign particles, changes in temperature, or nicks or cracks on the container's surface may also serve as the starting point for crystal formation. In baked goods, the high oven temperature causes the surface of the dough to dehydrate, inducing sugar crystallization (29). This results in the crunchy, sweet coating of brownies, pound cake, cookies, and muffins.

### Preventing Crystallization

To prevent crystallization when heating sugar solutions, the sides of the pan can be kept cleared of any particles with a brush dipped in water, or the pan can be covered to generate steam, which will have the same effect. Once the condensed steam has done its job, the lid should be removed to allow water to evaporate. Heating is then continued with no further agitation such

| **TABLE 21-6** Adding Sucrose to a Solution Increases Its Boiling Temperature* | | | |
|---|---|---|---|
| | | Boiling Point | |
| % Sucrose | % Water | °F | °C |
| 0 | 100 | 212 | 100 |
| 40 | 60 | 215 | 101 |
| 60 | 40 | 217 | 103 |
| 80 | 20 | 234 | 112 |
| 90 | 10 | 253 | 123 |
| 99.6 | 0.4 | 340 | 170 |

*The boiling point corresponding to each sugar concentration differs for different sugars.

**Source:** Food Technology.

**Reducing sugars** Sugars such as glucose, fructose, maltose, and others that have a reactive aldehyde or ketone group. Sucrose is not a reducing sugar.

**Caramelization** A process in which dry sugar, or a sugar solution with most of its water evaporated, is heated until it melts into a clear, viscous liquid and, as heating continues, turns into a smooth, brown mixture.

molecules into smaller pieces, which develop a darker color and complex flavor (29). Caramelized sugars are less sweet but more flavorful than the original sugar and may even be slightly bitter. The darker the caramel, the less sweet it is. The food industry relies on caramelized sugars to give a distinct flavor and color to food products such as puddings (flan), frostings, ice cream toppings, and dessert sauces. Candies made using the principles of caramelization include, as the name suggests, caramels, as well as peanut brittle.

# Moisture Absorption (Hygroscopicity)

The **hygroscopic** nature of sugars is responsible for their influence on a food's moistness and texture. The degree to which sugars draw moisture from the air differs depending on the sugar, with fructose having the most moisture absorption capability. Foods made with high-fructose sugars such as honey and molasses are noted for retaining their moisture. In fact, cookies and other baked goods made with honey stay moist even to the point of losing their desired texture. These and other baked products, when made with sucrose, will retain freshness longer. The moisture-absorbing property of sugar makes it imperative that baking mixes be stored in airtight containers; otherwise, moisture drawn to the mixture will lower the quality of the baked product. Unfortunately, sugar itself gets lumpy if it absorbs too much moisture from the air, so additives are incorporated into commercial baking mixes to prevent lumping or caking of ingredients.

# Texture

The texture of many processed or prepared foods relies on sweeteners, especially sucrose. Without sugar, soft drinks feel flat in the mouth, so bulking agents are often added. Sometimes other carbohydrates such as inulin (a polysaccharide) are added to increase viscosity and add a creamy, fat-like consistency to a liquid. Inulin occurs naturally in plants and is used primarily as a texturizer to give body to beverages, improve the texture of low-fat ice creams, make creamier sauces, and help aerate nonfat icings (44). Sugars attract water, causing baked goods to retain a soft, moist texture and prevent drying (29). Moist cakes can be dusted with confectioners sugar just before serving to further attract moisture and create a sticky, sweet coating. Sugar also acts to prevent overdevelopment of protein and starch structures in baked goods, which would make the dough tough (29). In meringues, sugar forms a supportive coating around the air bubbles within the batter, which prevents collapse of the bubbles. This keeps the meringues light and fluffy.

# Fermentation

Many alcoholic beverages and quite a few other foods around the world rely on the ability of carbohydrates to be fermented. Fermentation plays a role in producing beers, wines, cheeses, yogurts, and certain breads. The following conditions are desirable for fermentation:

- The presence of a yeast, mold, or bacterial culture. Even natural yeasts in the air can cause foods to ferment, and in some cases spoil, by

metabolizing their available sugars, a circumstance that probably led to the discovery of yeast-bread baking.
- A food source, usually a carbohydrate, for the microorganisms. Yeast bread rises because of the carbon dioxide gases produced by the yeasts feeding off the carbohydrates in flour. Any sugar except lactose can be fermented to carbon dioxide and alcohol by yeast organisms.
- The correct temperature to help the microorganism to grow.
- Conducive salt and acid concentrations.

# Preservation

High concentrations of sugar can act as a preservative by inhibiting the growth of microorganisms. Sugar was used to preserve jams, jellies, and other fruit spreads long before either canning or freezing methods were developed. The osmotic pressure created by the high concentration of sugar dehydrates the bacteria or yeast cells to the point of inactivation or death.

# Leavening

Sugars promote the leavening of baked goods (29). When sugar is added to a bread batter, its particles break up the dough, creating pockets of air. During baking, these air pockets expand, causing the batter to rise.

# Other Uses

Sugars can also impart other characteristics, such as crust formation, coating, creaming, and surface cracking, to food products.

**Hygroscopic** Having the ability to attract and retain moisture.

# PICTORIAL SUMMARY / 21: Sweeteners

From earliest times, the taste of sweetness has attracted people and has led them to search for new ways to extract it from the world around them. Today sugar is the number-one food additive, by weight, in the United States.

## NATURAL SWEETENERS

As people searched for new ways to sweeten foods, they discovered that the only sweetener of animal origin was lactose, found in milk. Plants, however, produce abundant natural sugars through photosynthesis. Once extracted from its source, sugar becomes a refined carbohydrate, with 4 calories (kcal) per gram.

Other sweeteners besides sugar are used in food preparation.

Sweeteners can be grouped into four categories:

### 1. GRANULATED SUGARS

- **Sucrose** (table sugar) is derived from either sugar cane or sugar beets. Once processed, the resulting sugars are the same, but are then refined into these commercial products:

White sugar    Powdered sugar    Liquid sugar
Raw sugar    Fruit sugar    Brown sugar
Turbinado sugar    Baker's Special
Sanding sugar

- **Glucose** (dextrose) is the basis of most carbohydrates and is the major sugar in our blood. Sources are fruits, vegetables, honey, and corn syrup. It is one half as sweet as sucrose.
- **Fructose** is found naturally in fruits and honey, and is the sweetest of all the granulated sugars.
- **Lactose**, found in milk, is the least sweet of all sugars.
- **Maltose**, also called malt sugar, provides the characteristic "malt" flavor to milkshakes, but is mostly used to make beer.

## 2. SYRUPS
(A by-product of cornstarch)
- Corn syrup
- High-fructose corn syrup
- Honey
- Molasses
- Maple syrup
- Invert sugar

## 3. SUGAR ALCOHOLS
(Found naturally in fruits and vegetables, or synthesized from certain sugars)
- Sorbitol
- Mannitol
- Xylitol
- Maltitol
- Isomalt
- Lactitol
- Erythritol

## NONNUTRITIVE SWEETENERS

FDA-Approved Nonnutritive Sweeteners
- Saccharin
- Acesulfame-K
- Neotame
- Aspartame
- Sucralose

Dietary Supplement/GRAS Food Additive
- Stevia

## FUNCTIONS OF SUGARS IN FOODS

Sugars have many more functions in foods besides merely providing sweetness, and various sugars differ in their sweetening ability. The unique chemical arrangements of the various sugars influence how they are used in food preparation:

**Sweetness:** The type of sugar and its concentration determine sweetness. Fructose is the sweetest.

**Solubility:** Solubility is determined by measuring how many grams of sugar will dissolve in 100 mL of water. Fructose is the most soluble.

**Crystallization:** Sugars with low crystallization, such as lactose, have a greater tendency to crystallize, while fructose, with its high solubility, does not invert. Invert sugar and corn syrup also resist crystallization, which is why they are often used in confectionary production.

**Browning reactions:** Two major types of browning involving sugars are the Maillard reaction, dependent on protein (amino acids) and sugar (reducing); and caramelization, dependent on dry heat.

**Moisture absorption (hygroscopicity):** The hygroscopic nature of sugars influences the moistness and texture of food to which they are added. Fructose has the best ability to absorb moisture from the air and impart it to food.

**Texture:** Many foods rely on sucrose for body and texture.

**Fermentation:** Sugars (except lactose) serve as fuel for yeast during fermentation, especially in certain baked products and alcoholic beverages.

**Preservation:** High concentrations of sugar act as a preservative.

**Leavening:** Sugar promotes leavening of baked goods.

In baked goods, sugar produces a finer texture, enhances flavor, generates browning of the crust, promotes fermentation of yeast breads, and extends shelf life by virtue of its ability to retain moisture. Sugar gives body to soft drinks and helps offset the bitter, acidic, or salty taste of certain foods, such as tomato sauces, chocolate, and sodium-processed meats (ham, bacon, etc.).

# CHAPTER REVIEW AND EXAM PREP

## Multiple Choice*

1. Raw sugar is made from _____.
   a. beets
   b. sugar cane
   c. sugar beets
   d. molasses

2. Which of the following is the sweetest of all granulated sugars?
   a. Fructose
   b. Powdered sugar
   c. Turbinado sugar
   d. White sugar

3. Children under one year of age should not be fed honey because it may contain _____.
   a. *C. botulinum* spores
   b. *E. coli*
   c. *Salmonella*
   d. *G. lamblia*

4. The nonnutritive sweetener aspartame is made from which two amino acids?
   a. Glutamic acid and glutamine
   b. Glycine and phenylalanine
   c. Lysine and aspartic acid
   d. Aspartic acid and phenylalanine

5. What is the name of the process by which dry sugar is melted to form a brown, viscous mixture?
   a. Fermentation
   b. Caramelization
   c. Maillard reaction
   d. Crystallization

6. An important function of sugar is solubility, which is defined as:
   a. the degree to which a sugar draws moisture from the air.
   b. a heated sugar solution that turns brown.
   c. the amount of sugar that dissolves in 100 mL of water.
   d. precipitation out of the solution in the form of crystals.

7. High concentrations of sugar in jams act as a preservative due to _____.
   a. fermentation
   b. hygroscopicity and osmotic pressure
   c. caramelization
   d. crystallization

## Short Answer/Essay

1. Discuss the individual functions of sucrose, glucose, fructose, lactose, and maltose in food preparation.
2. Discuss the differences among the following sugars: raw, turbinado, white, powdered, fruit, Baker's Special, sanding, liquid, and brown.
3. Discuss the differences among the following syrups: corn syrup, high-fructose corn syrup, honey, molasses, maple syrup, and invert sugar.
4. What is a dextrose equivalent (DE), and how is the DE value of a substance determined?
5. How are sugar alcohols different from other sweeteners? How are they utilized in food products?
6. What nonnutritive sweeteners are currently available in the marketplace? Discuss their basic chemical structures and their sweetness relative to sucrose.
7. Define *solubility* and list the sugars according to their solubility rates, from lowest to highest.
8. Describe the ways to prevent crystallization when heating sugar solutions.
9. Describe the two types of browning reactions involving sugars and how these processes are utilized by the food industry.
10. Define *hygroscopicity* and describe the role this characteristic of sugars plays in food preparation.

\* See p. AK-1 for answers to multiple choice questions.

# REFERENCES

1. Ahmed Z, H Banu, F Akhter, KM Faruquzzaman, and S Haque. Concept on sugar—a review. *Online Journal of Biological Sciences* 1(9):883–893, 2001.

2. Alexander RJ. Sweeteners: Nutritive. *American Association of Cereal Chemists*, 1998.

3. Ang JF, and GA Crosby. Formulated reduced-calorie foods with powdered cellulose. *Food Technology* 59(3): 35–38, 2005.

4. Anonymous. *Association recommends reduced intake of added sugars.* American Heart Association. http://americanheart.mediaroom.com/index.php?s=43&item=800. Accessed 11/15/09.

5. Anonymous. *Ingredients. Equal sweetener.* www.equal.com/products/ingredient.html. Accessed 11/15/09.

6. Anonymous. *Report 3 of the Council on Science and Public Health (A-08):* *The Health Effects of High Fructose Syrup.* American Medical Association. www.ama-assn.org/ama/no-index/about-ama/18641.shtml. Accessed 7/09.

7. Bakal A. A satisfyingly sweet overview. *Prepared Foods* 166(3):47–49, 1997.

8. Best D, and L Nelson. Low-calorie foods and sweeteners. *Prepared Foods* 162(7):47–57, 1993.

9. Brusick DJ. A critical review of the genetic toxicity of steviol and steviol glycosides. *Food and Chemical Toxicology* 46:S83–S91, 2008.

10. Cardetti M. Functional role of honey in savory snacks. *Cereal Foods World* 42(9):746–748, 1997.

11. Chapello WJ. The use of sucralose in baked goods and mixes. *Cereal Foods World* 167(5):133–135, 1998.

12. Charlton J. Discover honey's many flavors. *Fine Cooking* 22:74, 1997.

13. Coulston AM, and RK Johnson. Sugar and sugars: Myths and realities. *Journal of the American Dietetic Association* 102(3):351–353, 2002.

14. Deliza R, HJH MacFie, and D Hedderley. Information affects consumer assessment of sweet and bitter solutions. *Journal of Food Science* 61:1080–1084, 1996.

15. Demetrakakes P. Sweet success: The latest options for on-line measurement of sugar solids. *Food Processing* 57(9):83–86, 1996.

16. Duffy VB, and M Sigman-Grant. Position of the American Dietetic Association: Use of nutritive and nonnutritive sweeteners. *Journal of the American Dietetic Association* 104:255–275, 2004.

17. Fulgoni V. High-fructose corn syrup: Everything you wanted to know, but were afraid to ask. *American Journal of Clinical Nutrition* 88(6):1715S, 2008.

18. Giese JH. Alternative sweeteners and bulking agents: An overview of their properties, function, and regulatory status. *Food Technology* 47(1):114–126, 1993.

19. Grabitske HA, and JL Slavin. Low-digestible carbohydrates in practice. *Journal of the American Dietetic Association* 108(10):1677–1681, 2008.

20. Grieger L. Managing diabetes with sugar alternatives: The sweet 'n' lowdown. *Today's Dietician* 10(9):20–22, 2008.

21. Guens JC. Stevioside. *Phytochemistry* 64(5):913–921, 2003.

22. Harrington S. The role of sugar-sweetened beverage consumption in adolescent obesity: A review of the literature. *Journal of Scholarly Nursing* 24(1):3–12, 2008.

23. Hartel RW, and AV Shastry. Sugar crystallization in food products. *Critical Reviews in Food Science and Nutrition* 1(1):49–112, 1991.

24. Heller L. HCS is not 'natural', says FDA. *Decision News Media,* April 2, 2008.

25. Kalkhoff RK. Symposium on sweeteners. Policy statement: Saccharin. *Diabetes Care* 1(4):209–210, 1978.

26. Karottki SL. Saccharin warning removed. *Food Technology* 55(2):10, 2001.

27. Kille JW, et al. Sucralose: Lack of effects on sperm glycolysis and reproduction in the rat. *Food and Chemical Toxicology* 38(2):S19–S29, 2000.

28. LaBell F. Sucralose sweetener approved. *Prepared Foods* 167(5):135, 1998.

29. Masibay KY. Sugar: More than just sweet. *Fine Cooking* 96:100, 2009.

30. Matsui M, et al. Evaluation of the genotoxicity of stevioside and steviol suing six in vitro and one ivovi mutagenicity assays. *Mutagenesis* 11:573–579, 1996.

31. Mattes RD, and BM Popkin. Non-nutritive sweetener consumption in humans: Effects on appetite and food intake and their putative mechanisms. *American Journal of Clinical Nutrition* 89(1):1–14, 2009.

32. McNutt K, and A Sentko. Sugar replacers: A growing group of sweeteners in the United States. *Nutrition Today* 31(6):255–261, 1996.

33. Mermelstein NH. Neotame is approved as nonnutritive sweetener. *Food Technology* 56(8):22, 2002.

34. Mermelstein NH. Washington news. Advisory panel says that saccharin should remain listed as carcinogen. *Food Technology* 51(12):24, 1997.

35. Mitchell ML, and RL Pearson. In LO Nabors and RC Gelardi, eds., *Alternative Sweeteners.* Marcel Dekker, 1991.

36. Myer S, and WE Riha. Optimizing sweetener blends for low-calorie beverages. *Food Technology* 56(7):42–45, 2002.

37. Nabors LO. Regulatory status of alternative sweeteners. *Food Technology* 61(5):24–32, 2007.

38. Nabors LO. Sweet choices: Sugar replacements for foods and beverages. *Food Technology* 56(7):28–45, 2002.

39. Nishida C, and F Martinez Nocito. FAO/WHO scientific update on carbohydrates in human nutrition: Introduction. *European Journal of Clinical Nutrition* 61(Suppl 1):S1–S4, 2007.

40. Nunes APM, et al. Analyis of genotoxic potential of stevioside by comet assay. *Food Chemistry and Toxicology* 45:662–666, 2007.

41. Ohr LM. A sampling of sweeteners. *Prepared Foods* 167(3):57–63, 1998.

42. Prakash I, et al. Neotame: The next-generation sweetener. *Food Technology* 56(7):36–40, 2002.

43. Pszczola DE. Ingredients: Synergizing sweetness. *Food Technology* 60(3):69–79, 2006.

44. Pszczola DE. Products and technologies: Ingredients. Sweet beginnings to a new year. *Food Technology* 53(1):70–76, 1999.

45. Pszczola DE. Sweeteners for the 21st century. *Food Technology* 62(11):49–57, 2009.

46. Pszczola DE, and K Nachay. IFT plants a rich crop in Anaheim. *Food Technology* 63(5):47–56, 2009.

47. Rhandir R, and K Shetty. Chapter 14: Biotechnology of Nonnutritive Sweeteners. In K Shetty, G Paliyath, AL Pometto, and LE Levin. *Functional Foods and Biotechnology.* CRC Taylor and Francis Group, 2007.

48. Rybak-Chielewska H. Chapter 6: Honey. In P Tomasik, ed., *Chemical and Functional Properties of Food Saccharides.* CRC Press, 2004.

49. Shallenberger, RS. Sweetness theory and its application in the food industry. *Food Technology* 52(7):72–74, 1998.

50. Spreads and syrups. *Progressive Grocer* 70(7):78, 1991.

51. Sweetener. *Food Engineering* 62(2):24, 1990.

52. Thaumatin: The sweetest substance known to man has a wide range of food applications. *Food Technology* 50:74–75, 1996.

53. Thompson FE, TS McNeel, EC Dowling, D Midthune, M Morrissette, and CA Zeruto. Interrelationships of added sugars intake, socioeconomic status, and race/ethnicity in adults in the United States: National Health Interview Survey, 2005. *Journal of the American Dietetic Association* 109(8):1376–83, 2009.

54. Wang YJ. Chapter 3: Saccharides: Modifications and Applications. In P Tomasik, ed., *Chemical and Functional Properties of Food Saccharides.* CRC Press, 2004.

55. Whitehouse CR, J Boullata, and LA McCauley. The potential toxicity of artificial sweeteners. *American Association of Occupational Health Nurses Journal* 56(6):251–259, 2008.

56. Wolke RL. How and why. *Fine Cooking* p75, 2006.

57. Xu Y, S Barringer, and V Alvarez. Cause and prevention of cane molasses gelling. *Journal of Food Science* 70(8):C461–C464, 2005.

# WEBSITES

Click on "Sugars" at the Oregon State University's well-linked website on foods:

**http://food.oregonstate.edu**

Visit the Sugar Association's website:

**www.sugar.org**

The Calorie Control Council, an association of low-calorie/reduced-fat food manufacturers, has a website:

**www.caloriecontrol.org**

Information on high-fructose corn syrup can be found here:

**www1.lsbu.ac.uk/biology/enztech/hfcs.html**

# 22 Fats and Oils

Fats and oils (referred to collectively in this chapter as *fats*) are very important in the diet because they contribute to the flavor, color, texture, and mouthfeel of numerous foods (35). The first dietary fat that was used by humans probably came from animal carcasses. Humans quickly learned that melted fat improved the flavor of meat and other foods. The domestication of animals provided another major source of fat—butter. The heavy cream-colored liquid rising to the top of milk

was soon separated, churned, and used in a variety of ways. As populations grew, a more abundant source of fat was utilized by extracting the oils from plant seeds. The type of plant seed determined what kind of vegetable oil was produced: for example, safflower, sunflower, sesame, cottonseed, olive, corn, and others. People also learned that many kinds of nuts—almonds, walnuts, peanuts, and macadamia nuts, to name a few—provided a variety of oils. A few fruits high in fat such as coconut and avocado were also used to generate oils. The Egyptians derived fats from several plant sources, including almonds, olives, safflower seeds, and sesame seeds.

As time passed, technology allowed the development of new fats from classic sources. Food shortages experienced during wartime triggered the demand for a butter replacement, and margarine was created by the thickening of vegetable oil, a process first introduced in the 1860s. The continued thickening of vegetable oils past the "soft margarine

stage" resulted in the thick, white paste of vegetable shortenings used for frying.

Soon fats were being used not only for preparing foods, but also as an ingredient. Food manufacturers added fats to their food products to ease production and handling and improve storage stability (4). Today, fats are often added to various processed food products because of their unique shortening powers, melting points, plasticities, and solubilities. A wide variety of food products incorporate fat as an ingredient, such as breads; dairy foods; numerous processed foods such as potato and tortilla chips; and desserts, for example, cakes, icings, cookies, pies, pastries, and some frozen desserts.

Fats are not only used in the preparation and formulation of many foods, but are also added directly to finished foods; for example, butter or margarine is spread on breads or vegetables; mayonnaise is mixed into salads (potato, tuna, and chicken); and salad dressings are served on greens. As a result, so many foods contain fats and oils through both preparation and formulation that it is sometimes difficult to find foods without fat. The excess of fat in many diets and the related health risks associated with too many calories and grams of fat in the diet have led to the development of fat replacers.

This chapter focuses on the above topics and covers the various functions of fat in foods, different types of fat, fat replacers, the use of fat in food preparation, and storage requirements. This and the preceding chapter on sweeteners come before the chapters concerning various desserts because so many of these desserts incorporate both sweeteners and fats. Chapter 3 covers the chemical composition and classification of fats.

# FUNCTIONS OF FATS IN FOOD

Nothing else can mimic the unique properties that fats impart to foods (Table 22-1). As a result, it is difficult to achieve the functional characteristics that fats have in foods through substitution of other ingredients. The unique chemical configuration of the fat molecule (specifically the length and saturation of the three fatty acids on the glycerol molecule of a triglyceride) is key to the uses of fats in foods. It contributes to the functions of fat in heat transfer, shortening power, and emulsions, and influences the fat's melting point, plasticity, solubility, flavor, texture, appearance, satiety, and nutrient content, which will be discussed in greater detail.

## Heat Transfer

A major function of fats is their ability to act as a medium for heat transfer. Numerous meals use fat to transfer heat

**Shortening**　A fat that tenderizes, or shortens, the texture of baked products by impeding gluten development, making them softer and easier to chew.

**TABLE 22-1**　Functions of Fats in Foods

| Function | Examples |
| --- | --- |
| Heat transfer | Sautéing, pan-frying, deep-frying |
| Shortening power | Biscuits, pastries, cakes, cookies |
| Emulsions | Mayonnaise, salad dressings, sauces, gravies, puddings, cream soups |
| Varying melting points | Candies |
| Plasticity | Confections, icings, pastries, other baked goods |
| Solubility | Fats do not dissolve in water, yielding unique food flavors/textures and foods such as salad dressings |
| Flavor/mouthfeel | Flavor (butter, bacon, fried foods), lubricity, thickness, cooling |
| Textures | Creaminess, flakiness, tenderness, elasticity, cutability, viscosity |
| Appearance | Sheen, oiliness, color |
| Satiety | Fats contribute to "feeling full" |
| Nutrients | Provide energy: 9 calories (kcal)/g; carry fat-soluble vitamins |

to foods without burning them—butter in the frying pan, oil in the deep-fryer, and peanut oil in the wok. The amount of fat used can range from the minimal quantities used in sautéing, to the moderate levels used in pan-frying, to enough to completely submerge a food, as in deep-frying.

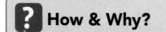

## ? How & Why?

### How is food heated in deep-frying?

In deep-frying, food is quickly cooked in several stages involving moisture transfer, fat transfer, crust formation, and interior cooking (52). As soon as the food is submerged in fat, the water on the food's surface vaporizes into the surrounding oil, which draws the moisture within the food toward its surface (Figure 22-1). A layer of steam forms around the food, protecting it from the high temperatures of frying and preventing it from becoming saturated with oil, although some amount of oil is transferred into the food through the pores from which the water escaped. In the next stage the crust browns, in part because of the Maillard reaction, and becomes somewhat larger and more porous from the water being driven out of the food by the frying heat. Most of the oil that has been absorbed remains in the crust and outer layer of the fried food. Finally, the inner core of the food cooks through heat penetration rather than by direct contact with the heated fat.

## Shortening Power

The **shortening** power of certain fats makes them essential in the preparation of pastries, piecrusts, biscuits, and cakes. The more highly saturated fats tend to have a greater shortening power. Mixed into a flour mixture, fat separates the flour's starch and protein and, when heated, melts into the dough. This creates air spaces that give the finished baked product its characteristic delicate texture. A fine grain is created from certain cake and cookie batters with the use of shortenings that gently encase the numerous air bubbles, serving as a

**FIGURE 22-1**　In fried food, oil is absorbed and water leaves as steam, contributing to a crisp, moist surface.

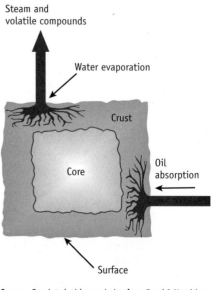

**Source:** Reprinted with permission from *Food & Nutrition Encyclopedia.*

starting point for the air to expand and increase overall volume. Baked goods become more tender, up to a point, as fat concentration increases. The exact role of fat in the formation of various bread products and their lower-fat alternatives is discussed in more detail in Chapters 19, 20, and 23.

## Emulsions

All foods contain some amount of liquid, and if fats or oils are present, then the combination results in some type of **emulsion** (49) (Chapter 3). There are two types of lipid emulsions:

1. Oil-in-water (Figure 22-2), in which oil droplets are dispersed throughout the water
2. Water-in-oil, in which water droplets are dispersed throughout the oil

Most food emulsions are of the first type—oil-in-water. Examples of such natural emulsions include milk, cream, and egg yolks. Emulsions can be quite viscous and thick, or more liquid and less stable. Examples of prepared foods that are emulsions (oil-in-water) include mayonnaise, salad dressings, cheese sauces, gravies, puddings, and cream soups. The less common water-in-oil emulsion, in which the smaller amount of water is dispersed in the fat, is found in foods such as butter and margarine.

### Emulsifiers

There are three parts to an emulsion:

- The dispersed or discontinuous phase, which is usually oil.
- The dispersion or continuous phase, which is most likely water-based.
- An emulsifier, which is a stabilizing compound that helps keep one phase dispersed in the other.

The two phases of emulsions are kept apart by surface tension, and the boundary

**FIGURE 22-3**   Mechanism of emulsifiers in oil-in-water and water-in-oil emulsions.

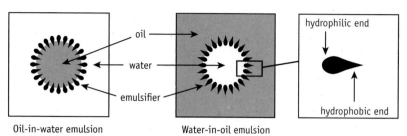

Oil-in-water emulsion          Water-in-oil emulsion

**FIGURE 22-4**   Emulsifier added to cake batter.

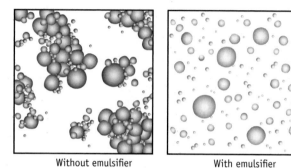

Without emulsifier          With emulsifier

between them is called the *interface*. The emulsifier migrates to this interface and acts as a **surfactant,** lowering the surface tension between the dispersed and continuous phases so that the two phases mix more readily (15).

An emulsifying agent can act as a bridge between oil and water because it is a two-part molecule, one portion being *hydrophilic* ("water-loving"), whereas the other is *hydrophobic* ("water-fearing") (Figure 22-3). It has been suggested that this balance of water-loving and water-fearing (or lipid-loving) portions allows emulsifiers to act like a "zipper" in drawing the water and oil phases together (7).

Emulsifiers not only work with liquids, but also with gaseous phases. Figure 22-4 shows how an emulsifier added to cake batter disperses the air bubbles, resulting in a cake with a finer crumb.

Mono- and diglycerides are the most frequently used emulsifiers in the food industry (Chemist's Corner 22-1). They are added to foods to increase or improve emulsion stability, dough strength, volume, texture, and tolerance of ingredients to processing (6). Other emulsifiers are phospholipids (lecithin from egg yolks), milk proteins, soy proteins, gelatin, gluten, vegetable gums such as carrageenan, and starches. Ground paprika, dried mustard, and other finely ground herbs or spices, which are often included in salad dressings, also act as emulsifiers. Emulsifiers synthesized by the food industry include polysorbate 60 and propylene glycol monoesters.

### Stability of Emulsions

Emulsions can be temporary, semi-permanent, or permanent and differ

**FIGURE 22-2**   Oil-in-water emulsion; the oil is dispersed in water.

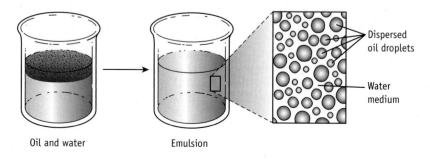

Oil and water          Emulsion          Dispersed oil droplets / Water medium

**Emulsion**  A liquid dispersed in another liquid with which it is usually immiscible (incapable of being mixed).

**Surfactant**  Surface-active agent that reduces a liquid's surface tension to increase its wetting and blending ability.

in their degree of viscosity and stability. Stability is defined by the degree to which the liquids stay in emulsion regardless of gravity, agitation, long storage times, extreme temperatures, surface drying, or added salt.

*Temporary Emulsions* Temporary emulsions are the least viscous and stable; they separate on standing when left alone. Such temporary emulsions include oil and vinegar salad dressings, in which the oil rises to the top of the vinegar. These emulsions must be shaken each time they are used in order to reform the emulsion.

*Semipermanent Emulsions* The tendency of the emulsion to separate is decreased by adding stabilizers. They create a viscosity similar to soft yogurt. Examples of semipermanent emulsions include commercial French and Italian salad dressings.

*Permanent Emulsions* Permanent emulsions are very viscous and stable, to the point that they do not separate. Mayonnaise is a permanent emulsion

**Polymorphism** The capability of solid fats to change into several crystalline forms, each with its own melting point, crystal structure, and solubility.

in which the major ingredient is vegetable oil (dispersed phase). The added egg yolk in mayonnaise contains lecithin, which acts as an emulsifier to keep the oil dispersed in the liquid, usually vinegar or lemon juice (continuous phase). The high oil content of mayonnaise contributes about 100 calories (kcal) and 11 grams of fat in every tablespoon. However, lower-fat alternatives are available.

## Melting Point

Not all fats melt at the same temperature. Each food fat has a unique melting temperature range that depends on the mixture of fatty acids contained in its triglycerides. These fatty acids are single molecules that each have a distinct melting point. Ultimately, a fat's melting point is determined by the following four characteristics of its predominant fatty acids:

- Degree of saturation
- Length
- Cis-trans configuration
- Crystalline structure

### Degree of Saturation

Most plant oils contain more polyunsaturated fatty acids than saturated fatty acids, which causes them to be liquid at room temperature. Animal fats tend to have more saturated fatty acids, causing them to be solid at room temperature.

### Length of the Fatty Acids

The length of the fatty acids can alter these general rules, as even saturated fats with shorter carbon chains can have lower melting points than those with longer ones. Butyric acid and stearic acid are saturated fatty acids found in butter. However, butyric acid has only 4 carbons and thus melts before stearic acid, which is 18 carbons long. Coconut oil is a saturated oil containing short fatty acids, which causes it to remain solid at room temperature. However, it will quickly liquefy if the bottle is held in a person's warm hand.

### Cis-Trans Configuration

Another significant structural difference that affects melting point is whether the fatty acid has more cis or trans double bonds (Chapter 3). A fatty acid with a trans configuration has a

higher melting point than an identical fatty acid with a cis form at the double bond. For example, oleic acid, an 18-carbon fatty acid with one double bond in the cis form, has a melting point of 57°F (14°C), whereas the same fatty acid in the trans form is called eladidic acid and has a melting point of 111°F (44°C). Hydrogenation, a commercial process that adds hydrogen to the double bonds of the unsaturated fatty acids, changes the cis form to a trans form.

### Crystalline Structure

The fourth influence on the melting point of fats is its crystalline structure, the arrangement of the fatty acids on the triglyceride molecule. How they are packed, or crystallized, in the solid phase of the fat determines at what point the fat will melt. This principle is very important to chocolate manufacturers: the larger the fat's crystals, the higher the melting point will be, which allows chocolate to be held in the hand without melting.

*Types of Crystals Affect Food Quality* Most fats exhibit **polymorphism,** which is the ability to exist in more than one crystalline form. Fat crystals are classified as alpha ($\alpha$), beta prime ($\beta'$), or beta ($\beta$).

The melting point of fats rises as the crystal sizes increase from alpha to beta prime, and eventually to beta (Figure 22-5). The rate of cooling dictates the type of crystals formed. Rapid cooling results in unstable alpha crystals with a waxy, transparent consistency. Alpha crystals are extremely fine and very unstable, melting readily and recrystallizing into the larger, more stable beta prime form. These beta prime crystals can be obtained by agitating the fat during cooling, which should be conducted at an intermediate rate. Beta prime crystals are best for food preparation, because they yield fine-textured baked goods and smooth-surfaced hydrogenated vegetable shortenings (37). Extremely slow cooling or long storage times form the most stable, or beta, crystals, which have an opaque look but produce a sandy, brittle texture (20).

## Plasticity

The plasticity of fat is its ability to hold its shape but still be molded or shaped under light pressure. Plasticity

**FIGURE 22-5**  Fat forms. The crystalline form of fat (alpha, beta prime, or beta) influences its melting point and texture.

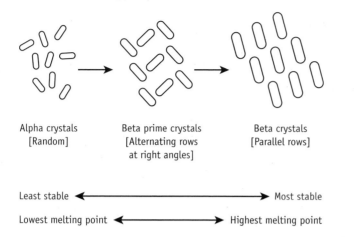

Alpha crystals
[Random]

Beta prime crystals
[Alternating rows
at right angles]

Beta crystals
[Parallel rows]

Least stable ⟷ Most stable

Lowest melting point ⟷ Highest melting point

determines a fat's spreadability. It is an important characteristic in the preparation of confections, icings, pastries, and other baked products. Although most fats look solid at room temperature, they are actually composed of liquid oil with a network of solid fat crystals holding it in place. This combination allows the fat to be molded into various shapes. Chilled butter has very little plasticity as compared to hydrogenated vegetable oil, or shortening. The more unsaturated a fat is, the more plastic it will be. Temperature also influences plasticity. For example, hard fats such as butter become soft and more spreadable when warmed.

## Solubility

Fats are generally insoluble in water. That is why oil floats above the vinegar in a salad dressing. Fats are actually defined as fats because they do not dissolve in water, but will dissolve (become soluble) in organic compounds such as benzene, chloroform, and ether.

## Flavor

The taste of fried foods such as breaded poultry or fish, french fries, potato chips, and doughnuts is one of the most obvious contributions of fat to flavor in foods. The flavor developed in certain foods by fats is very difficult to duplicate. For example, fats give butter, bacon, and olive oil their own distinctive tastes. Fats not only contribute their own flavor to foods, but also absorb fat-soluble flavor compounds from other foods. Sautéing garlic, onions, and herbs in oil releases their flavorful and aromatic compounds, while also lending them a smooth, rich mouthfeel.

## Texture

Fats also contribute texture to foods. Consider how fat gives textures to flaky pastries, smooth chocolates, half-melted ice cream, whipped cream topping, and crispy fried foods. The texture of baked products would not be the same without fat's positive influence on tenderness, volume, structure, and freshness (see Chapter 17).

The higher the fat content in ice cream, the smoother and creamier is the mouthfeel. The tenderizing effect of fats makes foods easier to chew and causes foods to feel more moist in the mouth (60). The lubricating action of fat moistens certain foods such as crackers and chips in which saliva would not be enough. These dry foods are processed in the mouth much more easily if they are coated with an oil or served with a high-fat dip or spread.

## Appearance

Foods are made more appealing by pigments located in a food's natural fats. Milk would be chalky white or bluish if not for its natural fat-based pigments giving it a more appealing color. The soft yellow hue of butter was found to be so important to consumers that attempts were made to duplicate it in margarine. Fat also coats food with a sheen of delicate oil that improves the appeal of chicken, pastries, chocolate, and many other foods.

## Satiety or Feeling Full

Fats induce a sense of fullness, or satiety. Foods and beverages containing fat help to delay the onset of hunger pangs by two methods: (1) fats take longer to digest than carbohydrates and proteins, and (2) fats delay the emptying of the stomach contents, which makes a person feel full longer.

# TYPES OF FATS

Through the years, fat's desirability in foods and multiple roles in food preparation have led to many different types of fats being obtained from both animal and plant sources. The most abundant sources of fats and oils in the diet are those of animal origin such as meats, poultry, and dairy products. Plants also contribute to fat in the diet, and those with the highest fat levels include nuts, seeds, avocados, olives, and coconut. The different types of fats such as butter, margarine, shortenings, oils, lard, and cocoa butter, as well as fat replacers, are now discussed.

## Butter

If you leave unhomogenized milk alone it will separate into its watery and fatty portions. The cream that floats to the top is used to make butter. In short, butter is made from the cream of milk. A full 10 cups (2½ quarts) of milk are used to generate one stick (¼ pound) of butter. The United States Department of Agriculture (USDA) defines butter as 80% milk fat, no more than 16% water, and 4% milk solids. Salt and coloring additives such as extract of annatto seed or carotene may or may not be added.

Butter is responsible for the desirable mouthfeel of baked goods and the flakiness and tenderness of pastries, and increases shelf life by decreasing moisture loss and slowing staling of starches (10). Other properties of butter include aeration of dough, emulsification, and flavor extension.

**FIGURE 22-6** Butter production.

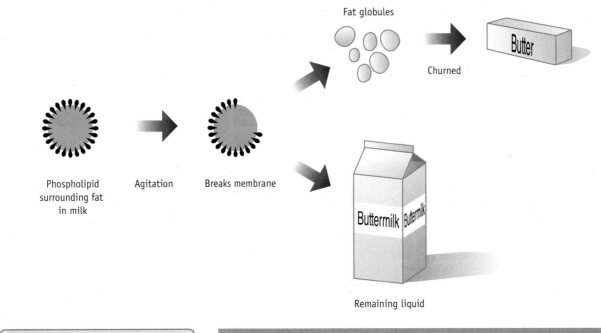

Phospholipid surrounding fat in milk → Agitation → Breaks membrane → Fat globules → Churned → Butter

Buttermilk — Remaining liquid

---

## ? How & Why?

### *How is butter made?*

Dramatic changes occur when the cream of milk is converted to butter: milk, an oil-in-water emulsion, reverses to a water-in-oil emulsion, butter. For this to occur, the membranes (phospholipid) around the fat globules have to be mechanically broken down to release the milk fat so it can lump together (Figure 22-6). Methods of doing this include stirring cooled cream or, in commercial operations, centrifuging the cream to expel the water. The fatty portion of the cream separates out as a soft, yellowish solid; these are granules of butter the size of corn kernels. Liquid drained from the process is collected and sold as buttermilk, a tangy tasting, opaque, reduced-fat milk by-product. The butter granules are washed and then churned at slower speed until they are mixed into a smooth, homogenous paste. Any remaining water is drained, and salt is sometimes added at this point for flavor and to act as a preservative.

In commercial dairies, the process of making butter begins with a cream that is concentrated up to 80% fat, and then further concentrated to 98%. It is first pasteurized to destroy pathogenic bacteria, cooled, and then recombined with the milk solids and water.

---

## NUTRIENT CONTENT

***Excess Dietary Fat and Health.*** Pure fats contain no carbohydrates, proteins, vitamins, or minerals—just fat. They also contain very little water. In contrast, foods such as butter contain about 16% water and small amounts of protein and other nutrients. Diet margarines can contain up to 50% water, which causes them to spatter when placed in a hot pan.

The important functions and attributes of fat have led to its incorporation into many foods, perhaps more so than needed. It is generally agreed by researchers that the average American diet contains too much total fat, even though the percent of total calories from fat has been decreasing. The health concern comes from fat's probable relationship to obesity (18). Most American adults are overweight, and a growing number are obese, which usually reflects an imbalance between energy intake and energy expenditure (8, 42).

***Recommended Fat Intake.*** Recommendations exist for both the amount and type of fat. The Dietary Guidelines for Americans 2005 recommend that adults keep dietary fat intake between 20% and 35% of calories with no more than 10% derived from saturated fat (65). However, the American Heart Association is more specific when it suggests that "overweight" Americans should consume no more than 30% of total calories from fat, and that all adults limit saturated fat to less than 7% of calories (22). Both recommend that adults consume less than 300 mg a day of dietary cholesterol, and The National Institute of Health's National Cholesterol Education Program focuses more on blood lipid values when it suggests keeping total blood cholesterol below 200 mg/dL (43).

A Food Marketing Institute annual trends survey found that almost three fourths of the people surveyed who reduced fat in their diets did it primarily "to improve their family's health," whereas only one fourth did so for "weight control" (57). Unfortunately, "America got fatter on fat-free" because the focus on fat has not always been accompanied by a clear message that calories, even low-fat ones, still contribute to weight gain (56).

***Trans-Fatty Acids.*** Despite increasing consumer awareness about the need to reduce trans-fatty acids, their knowledge about food sources of different fats has remained low (21). Small amounts of trans-fatty acids are found naturally in some foods (beef, butter, etc.), but the largest dietary sources are derived through the

*(continued)*

## TABLE 22-2 Types of Butter

| | |
|---|---|
| **Sweet cream** | Cultured by being acidified by lactic bacteria, this butter is made with or without added salt. |
| **Whipped** | A popular variant in texture is whipped butter, which is lighter in weight than regular butter and easier to spread because it has been aerated with air or nitrogen gas. The lower density results in half the calories (kcal) (50 vs. 100 per tablespoon). However, more of the whipped butter may be needed to achieve the same flavor as imparted by sweet cream butter. |
| **Compound (flavored or composed butter)** | A softened butter mixed with one or more flavors, such as garlic, lemon, honey, wine, herbs, or nuts. |
| **Powdered** | One of the newer forms of butter, powdered butter is made by removing the fats in natural butter and chemically treating, drying, and combining it with various ingredients such as whey solids, maltodextrins, guar gum, corn syrup solids, and color additives. It can be reconstituted with hot water and used in food service establishments to add flavor to vegetables, sauces, soups, and baked products. It is lower in calories and cholesterol than natural butter. |
| **Clarified** | High temperatures will burn regular butter. It must be specially treated before it can be used in certain types of food preparation involving high heats, because its milk solids burn very easily. The milk solids are removed by melting butter over low heat and allowing it to cool until the heavier, cloudy milk solids have settled out. The clear liquid portion is then gently poured off. Butter so treated, called **clarified butter** (*ghee* in Indian cuisine), is almost 100% fat. Ghee is a major fat used in Indian cooking. The smoke point of clarified butter is much higher than that of regular butter. |
| **Brown or black** | The tendency of butter's milk solids to burn can also have a positive effect. It allows the nuttier-flavored brown butter, or beurre noisette, and black butter, beurre noir, to be made by heating butter over low heat until it turns the desired color. |

## NUTRIENT CONTENT

process of hydrogenation. Hydrogenation of vegetable oils is performed to increase stability, solidity, and shelf-life. An estimated 80% of dietary trans-fatty acids come from partially hydrogenated oil and/or the foods made with this ingredient—vegetable shortenings, frying oils, margarines, baked products, confectionary products, deep-fried products, and many processed snack foods (39). These trans-fatty acids appear to have adverse effects on blood lipids and are associated with elevated risks of coronary heart disease (39). In 2006, the Food and Drug Administration (FDA) required that trans-fatty acids be listed on the label of all foods containing them (39, 46). The food industry responded with novel ways of reducing or replacing partially hydrogenated fats in numerous food products (62). The American Heart Association recommends that trans-fatty acid intake be limited to less than 1% of total daily energy, or about 2 grams for a 2,000-calorie (kcal) diet (22).

*Omega-3 Fatty Acids.* On a positive note, **omega-3 fatty acids** from fish or algae are increasingly found in many food products, such as breads, pasta, and yogurt (46). These fatty acids may benefit health by helping to prevent heart disease, hypertension, and thrombosis (coagulation of the blood in the heart or veins), and acting against autoimmune diseases such as lupus and arthritis (9). They are also thought to be possibly beneficial for brain health (reduced risk of Alzheimer's disease, improved memory and mood); inflammatory conditions (arthritis, psoriasis); and cancer (9, 47).

*Avoiding a Fat Deficiency.* Completely eliminating all sources of dietary fat is not recommended. A total lack of fat can cause a deficiency of essential fatty acids (linoleic acid and linolenic acid), resulting in failure-to-thrive (stunted growth) in children and eczema (red, itchy, scaly skin) in both children and adults. Dietary fat should never be restricted in healthy children under two years of age (17), and at least 3 to 5 grams of essential fatty acids (especially linoleic acid) equivalent to 15 to 25 grams of total fat are recommended daily for adults. Keeping a certain amount of fat in the diet is also important because it serves as a carrier for the fat-soluble vitamins: A, D, E, and K (58). Particular types of fat such as the omega-3 fatty acids are thought to benefit health, as discussed above and in Chapter 9.

Once it is formed, commercial butter is divided into blocks that are individually wrapped. Waxed paper is usually used to prevent odor absorption from other foods. Butter bought at the market has usually been cut into quarter-pound segments, and rewrapped. It is sold in half-pound or 1-pound packages. Grading is voluntary on the part of the processor. Characteristics used to evaluate the quality of butter include texture, flavor, color, and salt content (Figure 22-7).

### Types of Butter

Butter can be purchased in several forms that range in taste, consistency, and color (Table 22-2).

**Clarified butter** Butter that will not burn because its milk solids and water have been removed.

**Omega-3 fatty acids** Polyunsaturated fatty acids in which the first double bond is three carbons from the methyl ($CH_3$) end; examples are eicosapentaenoic (EPA) and docosahexaenoic acid (DHA).

## ✴ CALORIE CONTROL
### Sensible Fat Choices

**Major Sources of Dietary Fat.** Fat delivers more than twice as many calories per gram as carbohydrate or protein, so it's important to know dietary sources when planning a calorie-controlled diet. North Americans typically get most of their fat from:

- Meats
- Dairy products (whole milk, butter, cheese, cream, ice cream, and whipped cream)
- Commercial fats and oils

### Steps to Lower Fat Calories:

- **Monitor Portion Sizes.** A small amount, the size of your thumb, delivers approximately 100 calories (kcal) and 10 grams of fat. That's any of the following foods:
  - *1 tablespoon butter, margarine, or mayonnaise.* Fat is filling and delicious, but use it sparingly.
  - *1 tablespoon vegetable oil or salad dressing.* This amount poured over low-calorie greens adds 126 calories (kcal) and 14 grams of fat. Low-calorie dressings are an option, or limit regular dressings to one tablespoon.

- *1 ounce nuts.* Ever sit down to a bowl of nuts? Nuts are a healthy snack, but a reasonable portion should be no more than the size of your thumb, unless you are trying to gain weight.
- *1 ounce cheese.* Most people like cheese, but it takes a cup of milk and all its calories to make an ounce.
- **Watch for Invisible Fats.** Some fats—such as the fat in poultry skin, steak trim, vegetable oil, butter, margarine, shortening, lard, and tallow—are clearly visible. Fat becomes invisible, however, when it acts as an ingredient in candies, cakes, sauces, snack foods, and other food products or is concealed as the marbling within high-fat meats.
- **Plants High in Fat.** A few plant foods are high in fat and these include nuts, seeds, avocados, coconuts, and olives.
- **Portion Control for Nuts.** Nuts average 10 calories each, depending on the type, so about 10 nuts deliver 100 calories (kcal). A tablespoon provides about 50 calories (kcal) and ¼ cup can yield approximately 200. See Table 22-3 for more precise measurements.
- **Portion Control for Peanut Butter and Other Nut Butters.** Spread these thin, because 1 tablespoon yields almost 100 calories (kcal) and 10 grams of fat.

| **TABLE 22-3** Calories (kcal) in Nuts: Per Single Nut, Tablespoon, and ¼ Cup | | | |
|---|---|---|---|
| **Nut** | **Each** | **Tablespoon** | **¼ Cup** |
| Almond | 8 | 52 | 206 |
| Brazil | 31 | 55 | 218 |
| Cashew | 9 | 50 | 197 |
| Macadamia | 19 | 60 | 241 |
| Peanuts | 6 | 53 | 214 |
| Pecans (half) | 14 | 44 | 176 |
| Pistachio | 3 | 44 | 176 |
| Pumpkin Seeds | 2 | 18 | 71 |
| Sunflower Seeds | negligible | 54 | 207 |
| Walnuts (half) | 13 | 48 | 163 |

© 2010 Amy Brown

**FIGURE 22-7** USDA grades for butter.

## Margarine

During the Napoleonic Wars, the short supply and rationing of butter led Napoleon III to organize a contest to find a suitable butter replacement. And so it was that, in 1869, a French pharmacist and chemist, Hippolyte Mege Mouries, won the contest by developing oleomargarine. During World War II, when a law prevented the coloring of food products, margarine was introduced to U.S. consumers in a form unappetizingly lard-like and flat white in color. Eventually the law was repealed, and yellow margarine is now a staple in the North American market.

### Composition of Margarine

Standard stick margarine must contain at least 80% fat, about 16% water, and 4% milk solids, which is very similar to butter's general composition. Contrary to popular belief, regular margarine contains as many calories as butter. However, the fat sources differ and lower-fat versions are available. Margarine may be made from soybean, corn, safflower, canola, or other partially hydrogenated vegetable oils. Some soft margarines are made without the hydrogenation of oils. In addition, margarines usually contain the following:

- Cultured skim milk
- Emulsifiers such as lecithin
- Mono- and diglycerides
- Preservatives such as sodium benzoate, potassium sorbate, calcium disodium EDTA, isopropyl citrate, and citric acid
- Vitamins A and D
- Flavorings, usually diacetyl
- Food colorings, usually annatto and/or carotene

Diacetyl is added to margarine for flavoring because it is largely responsible, in addition to short-chain fatty acids, for butter's characteristic flavor (Figure 22-8).

### Types of Margarine

Some types of margarines include whipped margarine; light blends of margarine and butter containing about 60% fat; diet or reduced-calorie margarine, which has a higher water content; and imitation margarine, identified as "vegetable oil spread." Imitation margarines generally average half the fat of regular margarines. Plant stanol/sterol-fortified margarines are also available to consumers.

## Shortenings

Shortenings are plant oils that have been hydrogenated to make them more solid and pliable. Soybean oil is the major source of hydrogenated shortening and serves as a common frying oil. In the manufacture of shortenings, the soybean oil is hydrogenated until it reaches a solid consistency and then whipped or pumped with air to improve plasticity and give

it a white color. Many shortenings are also **superglycerinated,** making them ideal for baking applications needing solid fat, especially for flaky pastries and cakes containing more sugar than flour (60).

## Oils

Vegetable oils are derived from a variety of seeds, fruits, and nuts (12). The most common vegetable oils used in food preparation come primarily from soybeans, rapeseed (canola oil), sunflower seed, corn, cottonseed, and safflower seeds. Fruit oil sources include the avocado, coconut, palm kernel, palm, and olive. A few examples of nut sources include almond, peanut, and walnut. Oils differ dramatically in their taste, color, and texture, depending not only on their source, but also on their method of extraction.

### Extracting and Refining Oils

Oils are obtained from plant sources through the processes of extraction and possibly refining.

*Extraction*   The first step in producing any oil is to extract it from its original food source—seeds, fruit, or nuts. Oils are removed from their plant sources by one of the following three extraction methods:

- *Cold-pressing.* Mechanically pressing the seeds against a press, called cold pressing. Cold-pressed oils are sometimes sold as "specialty" oils and are usually not refined. Examples of unrefined, cold-pressed oils, such as peanut and olive oils, have the full flavor of the plants from which they were pressed.
- *Expeller-pressing.* Squeezing the seeds at very high pressures, which may generate some heat. These oils

still retain most of their flavor, color, aroma, and nutrients.
- *Chemical solvents.* Chemically removing the oil from the seeds with solvents (13). Most inexpensive commercial brands of oil are extracted using chemical solvents.

*Refining*   After an oil is extracted, it is either left unrefined or purified (refined) to produce a neutral, clean-flavored oil (Table 22-4). Natural impurities such as water, resins, gums, color compounds, soil, and free fatty acids exist in extracted oils. If these substances are not removed, they adversely affect the oil's flavor, color, clarity, smoke point, and shelf life.

**Unrefined Oils**   Some oils are sold without being refined. The aroma from the oil's original source often lingers, so these unrefined oils are used for salad dressings and as a result are frequently called salad oils. Sometimes their intense flavors dominate the food, which may or may not be desired. They cannot be exposed to high temperatures because of the risk of smoking (their smoke point is 320°F [160°C] or less), so they are limited to sautéing or low-heat baking uses. They are usually slightly higher in nutritional value than refined oils, but their free fatty acids can detract from the oil's flavor and make them prone to rancidity.

**Refined Oils**   Refining produces a neutral, low-aroma, bland-flavored oil. Many refined oils lack any distinguishing characteristics, a factor desired by chefs who do not want the oil influencing the flavor of the food being prepared (Table 22-3). These oils have higher smoke points, making them more suitable for frying. As a result of these two advantages, most commercially produced oils are extracted by the use of heat and solvents and then

**FIGURE 22-8**   Diacetyl and other short-chain fatty acids contribute to a "buttery" flavor in butter.

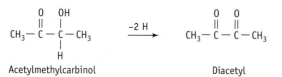

Acetylmethylcarbinol                                   Diacetyl

Diacetyl formation

**Superglycerinated** Describes a shortening that has had mono- and diglycerides added for increased plasticity.

| TABLE 22-4 | Selecting a Refined Cooking Oil | | |
|---|---|---|---|
| **Name** | **Description/Uses for Refined Oils** | **Fat Type*** | **Smoke Point** |
| **Almond** | Nut oils are best used in cold dishes; heat destroys their delicate flavor. | Mono | 495°F (257°C) |
| **Avocado** | This rather unusual light, slightly nutty tasting oil is considered primarily to be a novelty. | Mono | 520°F (271°C) |
| **Canola** | Of all oils, canola is second highest in monounsaturated fatty acid content. (Olive oil has more, but its flavor is not as mild.) Refined canola oil's mild flavor and relatively high smoke point make it a good all-purpose oil. | Mono | 400°F (204°C) |
| **Corn** | Made from the germ of the corn kernel, this oil's abundance makes it one of the most common vegetable oils in the United States. Many chefs like its mild, almost buttery flavor that becomes slightly toasty when used in pan-frying. It is high in polyunsaturated fat and used to make margarine, salad dressings, and mayonnaise. | Poly | 440°F (227°C) |
| **Grape seed** | This light, medium-yellow, aromatic oil is a by-product of wine making. It is used in salads and some cooking and in the manufacture of margarine. | Poly | 400°F (204°C) |
| **Olive** | Olive oil is a monounsaturated oil extracted from tree-ripened olives. Olive oils range from light amber to green in color and bland to extremely strong in flavor. Olive oil is graded according to its degree of acidity and the process used to extract the oil (Table 22-6). | Mono | Unrefined 320°F (160°C); Extra Virgin 406°F (208°C); Virgin 420°F (216°C); Extra Light 468°F (242°C) |
| **Peanut** | Peanut oil is made from pressed, steam-cooked peanuts. Peanut oil has a bland flavor and is good for cooking because it doesn't absorb or transfer flavors. | Mono | 450°F (232°C) |
| **Safflower** | Sunflower oil is an excellent all-purpose oil made from the seeds of safflowers. Safflower oil is a favorite for salads because it doesn't solidify when chilled and has a pale yellow color and a bland flavor. The refined version has a high resistance to rancidity. | Mono | 450°F (232°C) |
| **Sesame** | There are two types of sesame oil made from pressed sesame seeds: light (made with untoasted sesame seeds) and dark (made with toasted sesame seeds). Light sesame oil has a nutty flavor and is especially good for frying. Dark sesame oil (Asian) has a stronger flavor and should only be used in small quantities for flavoring foods—stir-fries, baking, sauces, and spreads. It has a rich, smoky, sesame aroma; nutty taste; dark brown color; thick consistency; and cloudy appearance. It is used a great deal in Chinese and Indian cooking. Just a few drops add a flavor that enhances many foods. | Poly | 410°F (210°C) |
| **Soybean** | This oil is an all-purpose oil because it has very little flavor. Highly refined soybean oil is reasonably priced, very mild, and versatile, accounting for over 80% of all oil used in commercial food production in the United States. Almost any product that lists vegetable oil as an ingredient probably contains refined soybean oil. | Poly | 492°F (256°C) |
| **Sunflower** | Sunflower oil is an all-purpose oil made from sunflower seeds. A pale, bland-tasting oil very similar to safflower oil, but not as widely available. | Mono | 450°F (232°C) |
| **Vegetable** | An inexpensive and all-purpose blend of oils made from plant sources such as vegetables, nuts, and seeds. Read the label to see the predominant oil(s) used in vegetable oil; most are made from soybeans. | Varies | Varies |

*Type refers here to the *predominant* fatty acid in the oil. All vegetable oils contain a combination of saturated, monounsaturated, and polyunsaturated fatty acids. The proportional differences determine how vegetable oils differ in their cooking properties and health effects.

**Source:** Adapted from http://missvickie.com/howto/spices/oils.html#Refined%20Cooking%20Oils and Fine Cooking (April/May):78, 2001.

refined. Refining, which results in oil that is 99.5% pure, consists of five steps: degumming, neutralizing, washing and drying, bleaching, and deodorizing (Figure 22-9). Once refined, these oils can be used as medium-heat cooking oils (225°F–350°F/107°C–177°C), high-heat cooking oils (350°F–450°F /177°C–232°C), and deep-frying oils (450°F+/232°C+).

## Types of Oils

Many different types of oils are available for food preparation purposes. Their individual characteristics determine their applications, and vary depending upon whether they are refined (Table 22-4) or unrefined (Table 22-5). The first factor to consider when selecting an oil is its flavor or lack thereof (34). Frying oils are mild flavored, bland, and stable to heat. Oils ideal for frying include soybean, corn, and safflower. Cottonseed oil, however, is the leading choice in food service operations for frying potato chips and for producing baked goods and snacks. This oil is preferred because of its low risk of developing and imparting off-flavors, and its relatively low price.

Another bland-flavored oil is canola, so named because it was developed in Canada; "canola" is a contraction of "Canadian oil, low acid." This oil is derived from the light, clear oil of rapeseed; it has a bland flavor and high monounsaturated fatty acid content. Rapeseeds originally contained high levels of erucic acid and glucosinolates, which, in large amounts, were found to cause cancer in laboratory animals. However, new genetic varieties contain minimal amounts of these substances and the FDA has allowed the sale of canola oil.

Strong-flavored oils such as peanut oil or olive oil vary widely in quality and character. Because their flavors are distinctive, these oils must be used carefully

**FIGURE 22-9** Steps to refining edible vegetable oils.

---

**Degumming**
When combined with water, certain impurities in oil form gums. These gums are removed by adding hot water to the oil and spinning it at high speeds to separate the oil from the gums.

**Neutralizing**
Free fatty acids are removed by adding an alkaline medium to convert the fatty acid to an insoluble soap, which settles to the bottom of the neutralizing tank. Newer methods use a centrifuge to separate the major layers according to specific gravity.

**Washing/Drying**
Traces of soap created by the neutralizing process are removed by washing the oil with water. The water is drained, and the oil is dried under a vacuum.

**Bleaching**
Colored matter in the oil is removed by adding absorbent materials, such as fuller's earth or activated carbon. The absorbed colored matter is then filtered out.

**Deodorizing**
Passing steam through the heated oil removes volatile compounds that cause off-odors—aldehydes, free fatty acids, hydrocarbons, ketones, and peroxides.

---

in foods. Peanut and sesame seed oils are more costly than many others, but their unique flavors make them the oils most commonly used in Chinese stir-fry dishes. Refined peanut oil is less expensive and is very heat stable, making it ideal for high-heat sautéing and frying. Peanut oil's flavor is preferred by some snack food manufacturers for their products (20).

*Olive Oil* Olive oil, which is considered a specialty oil, is more expensive than most other vegetable oils. Despite its higher price, olive oil consumption has increased among health-conscious consumers because of its high monounsaturated fatty acid content (78%) (66). A qualified health claim regarding reducing the risk for coronary heart disease has been approved by the FDA for olive oil (28). Part of this claim states, "Limited and not conclusive scientific evidence suggests that eating about 2 tablespoons (23 grams) of olive oil daily may reduce the risk of coronary heart disease due to the monounsaturated fat in olive oil."

When it comes to quality, the FDA has no grades for olive oils, but following Italian law, olive oils are classified according to acidity: the lower the acid content, the better the grade (Table 22-5). Price does not always reflect quality; olive oils are best judged by whether or not they have a clear, deep color (usually, but not always, green) and a distinct olive aroma and flavor. Some of the compounds contributing to the flavor and aroma of Extra Virgin olive oil are volatile and are lost when heated, so it is best used in cold salad dressings and as a final flavoring added to a dish. Milder olive oils are preferred for sautéing. Unrefined olive oils are also popular in Italian dishes and salad dressings for their full flavor.

*Tropical Oils* Longer shelf lives are obtained in food products using tropical oils such as coconut, palm, and palm kernel oils. Their higher saturated fat content has made them popular in the past with the food industry. Common foods made with tropical oils include cereals, candy, baked items, chocolate coatings for ice cream bars, pressurized whipped toppings, and dog and cat food. Because saturated fats contain no double bonds, they do not break down as easily. Therefore, they do not become rancid as quickly as unsaturated fats when subjected to oxygen, heat, and light. Tropical oils received negative publicity, however, when the consumption of saturated fats was linked to an increased risk of heart disease (29). As a result, many food manufacturers and even some fast-food enterprises have switched from tropical oils to vegetable oils for frying. Tropical oils are still used for some confectionaries, such as chocolate coatings for ice cream bars, because they become firm but melt quickly in the heat of the mouth.

*Oils as Functional Foods* One of the newest modified oils on the market is Enova™ (or Econa™) oil, which is now the best-selling vegetable oil in Japan (3). Its Japanese manufacturer claims that compared to other cooking oils, less of the Enova oil is stored as body fat after being consumed (36), and that the oil lowers blood lipids (69). Enova oil is made from a combination of soy and canola oils in which many of the triglycerides (3 fatty acids on a glycerol molecule) have been converted to diglycerides (70% are 2 fatty acids on a glycerol molecule).

---

| **TABLE 22-5** Olive Oils* Vary According to Acidity | |
|---|---|
| **Extra Virgin** | The highest-quality and best-tasting olive oil. This olive oil is the least acidic (no more than 0.8% acidity) and produced solely from mechanical presses (no chemicals or heat are used). The olives are ground to a paste before the oil is cold-extracted through either a centrifuge or a hydraulic press (24). The result is a high-quality oil with a strong olive flavor and a greenish tint from the presence of chlorophyll pigments. |
| **Virgin** | Also produced by cold pressing (not more than 2% acidity). |
| **Olive Oil or Pure Olive Oil** | This is a blend of Virgin olive oil with refined olive oil. This combination creates a lower acidity than refined olive oil, but also less intense flavor and color than either Virgin or Extra Virgin olive oil. About 70% of all olive oils sold in the United States are either Olive Oil or Pure Olive Oil (29). |
| **Light or Extra-Light** | Refers to color, fragrance, and taste, but not fat or calorie content. This refined oil is lighter in color than corn or safflower oil, and almost as mild. |
| **Olive-Pomace** | This oil may not be called olive oil. It is produced less expensively by extracting the oil from olives through both cold pressing and the use of solvents. The resulting oil is refined and then blended with Virgin olive oil to improve its taste, odor, and color. |

*Olive oils deteriorate with time and are best in the first year after bottling, but still fine to use through the next year. Check the olive oil's "sell by" date.

It tastes and looks like regular vegetable oil, but, according to the manufacturer, the shape of the diglycerides prevents their reformation into a triglyceride after digestion. Normally, fats consist primarily of triglycerides that are broken down into 3 fatty acids and a glycerol molecule in the digestive tract. These components are re-synthesized back into triglycerides in the cells of the small intestine. Enova's manufacturer states that its diglycerides hinder this process because "the middle position of the glycerol backbone is vacant and the body's enzymes for resynthesis are predominantly specific for the 2-monoglycerides." Although no adverse effects are claimed by the manufacturer, there have been no long-term studies of human consumption (2).

Other oils are marketed as functional foods due to their favorable fatty acid profile or other purported benefits (45). Flaxseed oil is rich in alpha-linolenic acid, which has been shown to decrease blood pressure. Rice bran oil is hypoallergenic and rich in antioxidants such as tocopherols, tocotrienols, gamma oryzanol, phytosterols, pholyphenols, and squalene. Olive oil and canola oil are valued for their high omega-3 fatty acid content (51). Other functional oils include borage oil for its gamma-linolenic acid content, sunflower oil for its high vitamin E level, and fish oil for its high content of omega-3 fatty acids (45).

## Winterized Oils

Some vegetable oils, when stored in the refrigerator, do not stay completely liquid. The cooler temperatures may result in cloudiness from the crystallization of certain fatty acids that have a higher melting point than their neighboring fatty acids. This cloudiness may be eliminated by **winterizing** the oil.

**Winterizing** A commercial process that removes from vegetable oils the fatty acids that have a tendency to crystallize and make the oils appear cloudy.

**Hydrogenation** A commercial process in which hydrogen atoms are added to the double bonds in monounsaturated or polyunsaturated fatty acids to make them more saturated.

Commercial salad dressings and so-called salad oils are usually made with winterized oil (27). Unwinterized vegetable oils that have crystallized in the refrigerator are perfectly edible and will revert to their clear character if allowed to come to room temperature.

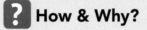

### How & Why?

*Why does refrigerated oil look cloudy?*

The cloudiness that occurs in some refrigerated oils is caused by the solidification of fatty acids. The cloudiness disappears once they reach room temperature again.

## Hydrogenated Oils

**Hydrogenation** makes fats and oils more solid, allows them to be heated to higher temperatures without smoking, and increases their shelf life or that of the foods coated with them (Chemist's Corner 22-2). Through this process, vegetable oils may be converted to spreadable hydrogenated shortenings or margarines. Too much hydrogenation, however, will cause the product to become brittle and hard. In addition to affecting plasticity, hydrogenation contributes to making piecrusts flaky and puddings creamy.

### Side Effects of Hydrogenation

One of the side effects of hydrogenation is that more trans than cis configurations are created at the double bonds between carbon atoms. The benefits of a higher concentration of trans-fatty acids are a rise in the fat's melting point, increased solidity, and lengthened shelf life. Among other health risks, trans-fatty acids have been reported to increase the risk for heart disease (41), and the Dietary Guidelines recommend keeping intakes as low as possible. This heart-health concern led the FDA to mandate that food products list the amount of trans fat per serving on Nutrition Facts labels beginning in 2006 (44).

## Lard/Tallow/Suet

Lard, which is the fat from swine, was the major shortening in use in the early 1900s. Tallow is also an animal fat, but it is derived from beef cattle or sheep. Suet is the solid fat found around the kidneys and loin of beef and sheep. These animal sources of fat are primarily saturated fat. They cannot be used for their shortening power in food preparation without first being rendered (melted down); for commercial use, the rendered fat is then deodorized. Antioxidants are often added to lard to increase shelf life. Lard produces poor textures in cakes and icings; therefore, it is used primarily in pastry piecrusts, commercial frying, and regional cooking.

### CHEMIST'S CORNER 22-2

#### Hydrogenation

The process of hydrogenation is facilitated with the aid of a metal catalyst (nickel, copper, platinum, or palladium) and the presence of pressure and heat (Figure 22-10). The catalysts are removed after the process is completed. The degree of hydrogenation, or the number of hydrogen atoms added, determines the firmness of the final product.

**FIGURE 22-10**  Hydrogenation: Hydrogen atoms can be added to double bonds in unsaturated fatty aids in the presence of a catalyst, pressure, and heat.

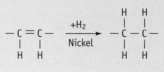

### Interesterification

Lard was largely replaced by the shortenings that appeared on the market in the 1950s, but lard usage increased again in the 1960s when the process of **interesterification** was introduced. Certain fats, such as cocoa butter substitutes and lard, have unacceptable textures until they are modified by interesterification. Lard is naturally too grainy and soft at room temperatures, but becomes extremely hard when refrigerated. Interesterification creates a smoother-textured lard with a slightly higher melting point, which allows the lard to retain its shape at room temperature. Another food application dependent on interesterification is emulsifiers, which are incorporated into numerous processed foods to improve the functionality of a fat.

## Cocoa Butter

Cocoa butter originates from the seeds of the *Theobroma cacao* tree. Its melting point is just below body temperature, making it perfect for "melt-in-the-mouth" chocolate confections and candies. Cocoa butter is also used in the manufacture of nonfood products such as soaps and cosmetics.

# FAT REPLACERS

One of the biggest challenges facing the food industry is reducing the fat content in foods (40). However, reducing fat in foods can cause problems. Fat reduction affects important functional properties such as flavor, appearance, texture, mouthfeel, handling, preparation, and storage stability (63). In addition, food engineers become exasperated when consumers demand low-fat foods but simultaneously value taste above nutrition when it comes to selecting food products (19).

The number of fat replacers and their use continue to increase as the food industry tries to meet the consumer demand for better-tasting, low-fat food items (5). Dietary fat consistently ranks at the top of the list of consumer nutrition concerns. Some selected processed foods using fat replacers include dairy products such as cheese, sour cream, butter, and margarine; meat products such as sausages and hamburgers; frozen desserts, including ice cream and yogurt; baked goods such as cakes, biscuits, and muffins; and frostings, sauces, and gravies (68). The most commonly used fat-modified products are milk, chips or snack foods, salad dressings, mayonnaise, and sauces (30).

## Types of Fat Replacers

There is no official classification of fat replacers or any standard method of naming them. The term *fat substitute* is often used interchangeably with *fat replacer*. However, *replacer* is a more general term describing any ingredient used to replace fat, which can include substitutes, mimetics (imitators), analogs, and extenders (31).

- *Fat substitutes.* Physically resemble fats, are often lipid-based, and usually replace the fat in foods on a weight-to-weight basis to duplicate the functional properties of fat (50).
- *Fat mimetics.* Water-soluble, often protein- or carbohydrate-based ingredients that mimic the mouthfeel of fat, recapitulating its creaminess and smoothness (50). They improve the texture of low-fat foods, especially cheeses (1). They do not replace fat by weight, as do the fat-soluble substitutes and extenders (64).
- *Fat analogs.* Have the characteristics of fat but with fewer calories because of their altered digestibility.
- *Fat extenders.* Optimize fat functionality so that less fat is required.

## Composition of Fat Replacers

Fat replacers are made from a variety of ingredients such as synthetic fats, microparticulated proteins, starch, fiber (cellulose, gums, etc.), and even dried fruit puree; research continues on others. Synthetic fat replacers approved by the FDA are Simplesse™ and olestra. Other synthetic fats awaiting approval are esterified propoxylated glycerols (EPG) and trialkoxytricarballate (TATCA).

Fat replacers, regardless of the name, are commonly grouped according to whether their chemical structure is carbohydrate, protein, or fat based (Table 22-6).

## Carbohydrate-Based Fat Replacers

Most of the fat replacers used by the food industry are based on carbohydrates (32, 50). Fibers, gums, pectin, cellulose, and starches bind with water, swell, and impart some of the texture, mouthfeel, and opacity of fat (27).

## Protein-Based Fat Replacers

Milk (whey) or egg proteins usually serve as the source for protein-based fat replacers, but for many years isolated soy protein has been used in foods, particularly ground meat products. The USDA allows isolated soy protein to be added at certain percentages to ground meat, poultry products, cooked sausages, and cured pork (27). Protein-based fat replacers are often used in meats and other food products that have to be refrigerated or frozen.

### Simplesse

An example of a protein-based fat replacer is Simplesse, which provides only one seventh the number of calories found in fat (38, 61). It is made from the whey of milk or from egg-white proteins that have been reduced to tiny particles through a process called microparticulation (Figure 22-11) (50). These protein droplets break down when heated so Simplesse cannot be used in baking or frying (11, 55).

## Lipid-Based Fat Replacers

Chemically modifying the molecular structure of fats can result in fat-based fat replacers that have fewer calories than fat (33). Often the chemical changes

> **Interesterification** A commercial process that rearranges fatty acids on the glycerol molecule in order to produce fat with a smoother consistency.

**TABLE 22-6**   Fat Replacers Classified as Carbohydrate-, Protein-, or Fat-Based

| Class | Trade Names | Composition | Functional Properties |
|---|---|---|---|
| **Carbohydrate-Based** | | | |
| Cellulose | Avicel® cellulose gel, Methocel™, Solka-Floc®, Just Fiber | Cellulose ground to microparticles | Water retention, texturizer, thickener, mouthfeel, stabilizer |
| Dextrins | Amylum, N-Oil® | Sources include tapioca | Gelling, thickening, stabilizing, texturizing |
| Fiber   • Grain | Opta™, Oat Fiber, Snowite, Ultracel™, Z-Trim® | Sources include oat, soybean, pea, and rice hulls or corn or wheat bran | Moisturizer, mouthfeel |
|   • Fruit | Prune paste, dried plum paste, Lighter Bake, WonderSlim, fruit powder | Sources include fruits (prunes and plums) | Gelling, thickening, stabilizing, texturizing |
| Gums (xanthan, guar, locust bean, carrageenan, alginates) | Kelcogel®, Keltrol®, Slendid™ | Hydrophilic colloids or hydrocolloids | Water retention, texturizer, thickener, mouthfeel, stabilizer |
| Maltodextrins | CrystaLean®, Lycadex®, Maltrin®, Paselli® D-LITE, Paselli EXCEL, Paselli SA2, Star-Dri® | Sources include corn, potato, wheat, tapioca | Gelling, thickening, stabilizing, texturizing |
| Polydextrose | Litesse®, Sta-Lite™ | Water-soluble polymer of dextrose containing minor amounts of sorbitol and citric acid | Moisture retention, bulking agent, texturizer |
| Starch (modified food starch) | Amalean® I & II, Fairnex™ VA15 & VA20, Instant Stellar™, N-Lite, OptaGrade®, Perfectamyl™ AC, AX-1, & AX-2, PURE-GEL, STA-SLIM™ | Sources include potato, corn, oat, rice, wheat, tapioca starches | Gelling, thickening, stabilizing, texturizing |
| **Protein-Based** | | | |
| Microparticulated protein | Simplesse® | Whey, milk, or egg protein | Mouthfeel |
| Modified whey protein concentrate | Dairy-Lo® | Whey protein | Mouthfeel |
| Other | K-Blazer, ULTRA-BAKE®, ULTRA-FREEZE™, Lita® | Egg white, milk, and corn protein | Mouthfeel |
| **Fat-Based** | | | |
| Emulsifiers | Dur-Lo®, EC™-25 | Vegetable oil mono- and diglycerides | Mouthfeel |
| Salatrim | Benefat | Short- and long-chain acid triglyceride molecules | Mouthfeel |
| Lipid analogs | Olean | Sucrose and edible fats and oils | Mouthfeel |
| Fat extender | Veri-Lo | Oil-in-water emulsion | Mouthfeel |

**Source:** Adapted from *Fat Replacers: Food Ingredients for Healthy Eating,* Calorie Control Council, www.caloriecontrol.org/fatreprint.html, 2002; and The American Dietetic Association's Position Statement on Fat Replacers, *Journal of the American Dietetic Association* 98(4):463–468, 1998.

involve inhibiting absorption or shortening the length of the fat's fatty acid. Short- and medium-chain fatty acids provide fewer calories than larger ones. This is why butyric acid (4 carbon atoms) yields fewer calories than palmitic acid (16 carbons) (25).

Fat extenders can reduce the amount of fat in mayonnaise and salad dressings by as much as 70% (33, 54). Calories are reduced by diluting the fat with an "extender" such as water in the form of an oil-in-water emulsion.

## Olestra

Another fat-based fat replacer on the market is olestra used in snack foods (crackers, potato and tortilla chips), fried and baked goods, and dairy products. Procter & Gamble researchers were trying to locate an easily digested fat for premature infants, but instead found a substance not broken down by the body at all (67). Previously known as sucrose polyester, olestra gained FDA approval in 1996 and is marketed under the brand name Olean™. It is

**FIGURE 22-11** Microparticulation. The process of reducing the original size of a substance to numerous tiny particles.

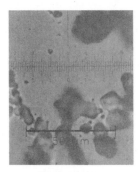 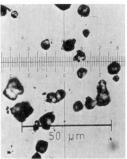 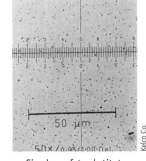

Powdered sugar
(10–30 microns)

Nonfat chocolate
(5–10 microns)

Simplesse fat substitute
(0.1–3.0 microns)

made from sugar and vegetable oil in a process in which the 3-carbon glycerol molecule in the oil is replaced by a large sucrose with 6 to 8 fatty acids attached. This molecule is so large that it moves through the digestive tract before enzymes have time to digest the fatty acids (20).

Olestra is stable during heating, but it reduces the body's absorption of vitamins A and E (38), so the FDA requires that fat-soluble vitamins be added to products made with olestra. Side effects of too much olestra consumption may include diarrhea, cramps, and gas.

# FOOD PREPARATION WITH FATS

Very few fat replacers can function in heat transfer during food preparation the way fat can. Fats allow the transfer of heat during frying, sautéing, stir-frying, pan-frying, and deep-frying (see Chapter 5). This section focuses on how to take care of the fat used in frying, and describes food preparation techniques that reduce the amount of fat transferred to the food.

## Frying Care

Foods fry better if the preparer knows which fats are best for frying, understands optimal frying temperatures, uses recommended equipment, and maintains the fat's frying optimal ability with optimal care.

## Can Any Fat Be Used for Frying?

Not every fat is suited for the very high temperatures of deep-frying, which average 350°F–450°F (177°C–232°C). The high temperatures of deep-frying allow foods to be heated more quickly than if they were boiled. The fats commonly used for frying must be 100% fat, and include vegetable oils (except for olive or sesame oil) and hydrogenated shortenings (without additives such as emulsifiers). The vegetable oils most frequently used include cottonseed, corn, canola, peanut, and safflower.

Many vegetable oils are also chosen for frying because they have little flavor of their own and will not overpower the flavor of very lightly seasoned or bland foods. Conversely, some foods call for butter as a sautéing fat to enhance flavor, but the heat must be carefully controlled because the water and milk solids in butter cause it to spatter and burn more easily. Margarine is not recommended for frying because, in addition to containing water, it has a low smoke point. The water will splatter, and foods fried in fats with low smoke points develop unpleasant flavors.

## Smoke Point

Select fats with high **smoke point**—above 420°F (216°C) or higher—for frying. This temperature is much higher than the boiling point of water (212°F/100°C), and even higher than frying temperatures that range from 350°F–450°F (177°C–232°C). Because fat boils at a much higher temperature than water, fat that starts to bubble is

very dangerous because it has reached its boiling point. It should be immediately removed from the heat source. It is also important to select a fat with a smoke point above the frying temperature or else it will break down. It may overheat and decompose into glycerol and its individual fatty acids. The glycerol is further broken down (hydrolyzed) to a steel-blue smoke called acrolein. Acrolein's sharp, offensive odor warns people of its presence. The smoke is not only extremely irritating, but even harmful to the mucous membranes of the mouth and nasal passages.

Table 22-7 lists the smoke points of various fats. Selecting fats with smoke points above 420°F (216°C) for commercial frying automatically excludes olive oil, lard, and vegetable shortenings. Hydrogenated shortenings with added mono- and diglycerides are not recommended for frying. Their fatty acids are easily removed from the glycerol molecule, which is then free to form acrolein.

## Flash Point and Fire Point

A more serious problem than smoking occurs when an oil is heated to its **flash point,** or to about 600°F (316°C). Increasing the heat even higher to 700°F (371°C) will result in the fat reaching its

| TABLE 22-7 Smoke Points of Selected Frying Fats and Oils | |
|---|---|
| Fat/Oil | Smoke Point |
| Vegetable shortenings + emulsifier | 356°F –370°F (180°C –188°C) |
| Lard | 361°F –401°F (183°C –205°C) |
| Vegetable oils | 441°F –450°F (227°C –232°C) |
| Most olive, virgin oils | 391°F (199°C) |
| Corn oil | 440°F (227°C) |
| Soybean oil | 492°F (256°C) |

**Smoke point** The temperature at which fat or oil begins to smoke.

**Flash point** The temperature at which tiny wisps of fire streak to the surface of a heated substance (such as oil).

**fire point.** If this occurs, water should not be used to put out the fire. Fire extinguishers with a *C* designation should be used to extinguish a fire caused by fat. If an extinguisher is not available, it may be possible to smother the fat-fueled fire with a pan lid or large amounts of baking soda.

## Controlling the Temperature of Frying Fats

It is difficult to detect overheating visually. Compounding this problem, heated oil is always hotter than it appears. These two problems can cause overheating, which contributes to the rapid deterioration of fat through **polymerization**. Any egg yolks used in the coating of battered foods also contribute to the darkening effect on the fat. Further, the increased viscosity of overheated fat results in higher fat absorption rates in the fried foods, making them greasy (48). One way to control the temperature of cooking fats and prevent excess absorption is to use thermostatically controlled deep-fryers, but it is recommended that these thermostats be checked for accuracy routinely.

### Avoid Too-Low Temperatures

Although it is important not to overheat frying fats, it is equally important not to let temperatures drop too low because this may lead to excessive fat absorption, resulting in soggy, greasy fried food. Temperatures quickly drop when large quantities of frozen food are added to hot oil. To combat this problem and help stabilize the temperature, the food should be added in batches so that the oil has sufficient time to reheat to the correct temperature. It is also important that the food pieces in a batch be the same size, so they finish cooking at the same time.

**Fire point** The temperature at which a heated substance (such as oil) bursts into flames and burns for at least 5 seconds.

**Polymerization** A process in which free fatty acids link together, especially when overheated, resulting in a gummy, dark residue and an oil that is more viscous and prone to foaming.

### Perfect Browning of Fried Foods

If temperatures are correctly controlled, the result will be a food that has a crisp, golden crust surrounding a tender, perfectly cooked center. The key in deep-frying is to ensure that the food's inside is sufficiently cooked without overdoing its outside. Fried foods cook on the principle that frying temperatures convert the food's water to steam, which then escapes, keeping the food cool and preventing it from burning and/or absorbing fat. Eventually, however, the amount of steam decreases, allowing the outside to brown. Foods left too long in the fryer after all the steam has escaped will have burned crusts and excess fat absorption.

High-moisture foods such as french fries need to be cooked at lower temperatures or the outside will turn crispy before the inside has had a chance to cook. Steam trapped by the hard crust will cause the food to become limp as it cools. Conversely, low-moisture foods need higher temperatures so they will cook quickly, leaving oil no time to enter the food. Other determinants in temperature selection to obtain the best crust color are the amount of food, the length of time it is submerged in oil, the temperature of the food, the oil quality, and the food's shape and size, porosity, and type of coating (53).

### Optimal Frying Temperatures

The optimal frying temperature is 375°F (191°C), with higher temperatures (375°F–390°F/191°C–199°C) required for smaller pieces of food, and lower temperatures (350°F–365°F/177°C–185°C) for larger pieces of food (52).

### Recommended Equipment

Frying temperature is not the only factor influencing the fried food product's quality. It is important to use stainless-steel equipment; iron, and especially copper or copper alloys such as brass, may increase rancidity. Hoods or exhaust systems above the fryer should be cleaned frequently so that accumulated particles do not drip back down into the fat. Deep, narrow containers are recommended for deep-frying, because shallow, wide pans increase the surface area, lowering the smoke point through greater exposure to air. The fryer should be filled no more than one half to three fourths full of oil. As fat

is absorbed by the foods, it should be replaced with fresh fat. However, fresh fat should never be added to fat that is rancid, foaming, or dark, because it will not overcome these defects and will deteriorate very quickly.

### Optimal Frying Conditions

The fats in a fryer go through stages that influence the quality of the fried product. At the new and break-in stages, foods absorb too little oil; just the right amount is absorbed at the fresh and optimal levels that follow; and then too much soaks in at the degraded and runaway phases (59). Many professional chefs claim that foods fry best in oil that has been used at least once. Desirable browned crusts occur when oils pick up proteins and carbohydrates from the foods that have been fried in them. Eventually, however, the browning becomes too dark, and the fat must be replaced. Also, as the fat deteriorates, the surface tension of the frying oil decreases, making foods more likely to soak up the fat. Repeated use of a frying fat will also lower its smoke point, because each heating hydrolyzes some of the triglycerides into smaller molecules.

### Avoid Water

Foods should be as free of surface moisture as possible before being submerged in the heated fat. Water causes spattering of hot oil, which can cause burns; it requires more energy to maintain temperatures; it may result in longer frying times; and it causes the fat to break down chemically, reducing its frying life.

### Avoid Food Particles

Inevitably, particles of food or breading break off or fall through the basket and build up in a deep-fryer over time. These food particles should be filtered out daily (or every 8 hours of use), or they will darken the oil's color, lower its smoke point, and reduce its keeping time. On the other hand, excessive filtering introduces oxygen into the oil, resulting in rancidity, gum development, and foaming, the latter observed as a persistent layer of bubbles on the surface.

### Cool the Frying Fat

A frying fat should theoretically stay fresh for several months if it is cooled immediately after use and stored in an airtight container in a dark, cool place. Refrigeration will also increase its shelf life. Large commercial fryers contain too much fat to be cooled completely and then efficiently reheated, so they are turned down to approximately 225°F (107°C). Decreasing the temperature during downtime prevents the fat's breakdown and extends its usefulness.

### When to Discard the Used Frying Fat

There is no easy method for determining when oil that has been used repeatedly should be discarded. Generally, the oil becomes darker and more viscous, smokes easily, begins to have a rancid odor, and starts to impart off-flavors to the foods fried in it. The first indication that an oil needs to be replaced is usually that its color and the color of the food fried in it start to darken. This transformation takes place right before the flavor and odor of the oil start to deteriorate. An experienced person can tell by looking at it if the oil needs changing, but food service establishments may purchase a commercial kit that allows anyone to determine an oil's freshness by checking its color against the kit's standard. A further indication that the oil is too old is that the food fried in it is greasier than normal because of increased oil absorption. Other factors to consider in the decision to discard oil include the type of oil used, the type of foods being fried, the number of times the oil has been used, the presence of many particles, excessive foaming or smoking, and the quality of the foods cooked in the oil (52). Guidelines for preserving frying oils are summarized in Figure 22-12; Table 22-8 provides a list of problems that may arise and what causes them.

# Lower-Fat Preparation Techniques

Dietary fat consumption may be reduced by following MyPyramid (added fats/fats in high-fat animal products fall into the "discretionary calories" group); following a meal pattern that is lower in fat, especially the saturated and trans types; relying on lower-fat or nonfat cooking methods; and reducing the fat in recipes.

## Fats Preferred for Health

Once overall fat intake is reduced following these guidelines, the next step is to modify the types of fat that are ingested. Monounsaturated fats are preferred over polyunsaturated, which in turn are recommended over saturated fats. Table 22-9 shows that, compared to other cooking oils and fats, canola oil contains one of the highest levels of monounsaturated fatty acids (58%). In the same category, olive, avocado, almond, and apricot oils tend to impart more flavor but are more expensive. Safflower oil scores highest in the category of polyunsaturated oils. Saturated fats such as coconut, palm, and palm kernel oils and butter should be avoided according to certain dietary guides. Butter is often chosen, however, for its unique flavor or by those concerned about the trans-fatty acids found in margarines and other partially hydrogenated fats. Although butter and margarine contain approximately the same number of calories and grams of fat, the fat in butter is primarily saturated, whereas that from margarine is more unsaturated. Lard, the saturated fat from swine, is best replaced by vegetable shortening, but even the latter is partially saturated.

## Reducing Fat by Healthy Methods

Dietary fat intake may also be lowered by selecting a cooking method that does not rely on fat. All of the moist- and dry-heat cooking methods, with the exception of frying, lend themselves to fat-free preparation of foods. Even frying, specifically sautéing and stir-frying, is acceptable if the right type of fat is chosen and a minimal amount of it is used. Pan-frying and deep-frying are the only two methods for which it is essentially impossible to lower the amount of fat used.

## Modifying Recipes to Reduce Fat

Another way to reduce fat in food preparation is to focus on the recipes. The following foods are the main contributors

**FIGURE 22-12**　Preserving frying oils.

### Equipment
- Stainless steel
- Deep, narrow containers for deep frying
- Fill no more than one-half to three-fourths full
- Clean commercial fryers at least once a week
- Keep hood/exhaust system above fryer clean

### Heating
- Most frying occurs between 350–450°F (177–232°C)
- Avoid high temperatures (undercooked inside, overly crispy/brown outside)
- Avoid low temperatures (soggy fried food)
- Heat no longer than necessary
- Allow sufficient time for oil to reheat between batches
- Avoid large quantities of frozen foods
- Do not overheat to flashpoint, about 600°F (316°C)
- Check thermostat for accuracy

### During Frying
- Use oils with smoke points above 420°F (216°C)
- Avoid exposure to oxygen/air, light, salt (season after frying), certain metals (iron, copper, nickel)
- Only completely dry food should be submerged
- Filter particles of batter/flour and/or food
- Limit egg yolks used in the batter or flour (darkening effect on the fat)
- Monitor freshness of frying oils by checking their color against a standard
- Add fresh oil daily to commercial fryers in order to replace entire amount every three to five days

### Storage
- Store unopened containers in cool, dry place
- Store opened containers airtight in refrigerator
- Commercial fryers: reduce temperatures to about 225°F (107°C)
- Discard dark, gummy, or repeatedly used oil

**TABLE 22-8**    Problems in Deep-Frying Oils and Their Causes

| Problem | Possible Cause |
|---|---|
| Fat darkens excessively | • Overheating of fat<br>• Defective equipment<br>• Inadequate filtering<br>• Inadequate cleaning of equipment<br>• Use of inferior fat<br>• Foreign material entering the fat |
| Fat smokes excessively | • Overheating of fat<br>• Defective equipment<br>• Inadequate filtering<br>• Use of inferior fat<br>• Poor ventilation<br>• Inadequate cleaning of equipment |
| Life apparently gone from fat: fat won't brown the food; fat won't hold the heat | • Too low a frying temperature (including faulty thermostat)<br>• Lack of proper kettle recovery<br>• Not cooking long enough<br>• Excessive foam development<br>• Overloading kettle<br>• Improper preparation of food |
| Fat foams excessively and prematurely | • Overheating of fat or failure to reduce heat when not in use<br>• Insufficient filtering<br>• Failure to clean and rinse equipment adequately<br>• Brass or copper being used in kettle<br>• Use of inferior fat<br>• Salt or food particles getting into the fat |
| Food greasy; too much absorption | • Frying at too low a temperature<br>• Slow temperature recovery due to poor equipment<br>• Overloading kettle<br>• Frying fat has broken down<br>• Improper preparation of food<br>• Keeping food in fat after it is done, or insufficient draining after removal |
| Fat has "objectionable" odor or flavor | • Excessive crumb or foreign material in kettle<br>• Use of poor-quality food<br>• Detergent film due to insufficient rinsing<br>• Use of inferior or broken-down fat<br>• Excessive fat absorption<br>• Holding cooked food too long |

**Source:** CONCISE ENCYCLOPEDIA OF FOODS & NUTRITION. EBOOK Copyright 1995 by TAYLOR & FRANCIS GROUP LLC - BOOKS. Reproduced with permission of TAYLOR & FRANCIS GROUP LLC - BOOKS in the format extranet posting and in the format Presentation via Copyright Clearance Center.

to fat in recipes: meats, dairy products (including whole milk, cheese, cream, ice cream, whipped cream, and butter), commercial fats and oils, avocado, coconut, olives, nuts, and seeds.

Processed foods such as cakes, cookies, pies, snacks, and others that are made with these ingredients are also high in fat. Many recipes could simply have their fat content reduced or another ingredient substituted without affecting overall quality (Table 22-10). Sometimes the fat can be removed altogether. Following MyPyramid will automatically eliminate recipes that are too high in fat and that cannot be adequately modified. A good rule of thumb is that any recipe exceeding 20 grams of fat is probably too high in dietary fat for people consuming three meals a day. Other ways to reduce the amount or modify the type of fat in the diet include the following:

- Fruit preserves and honey can replace butter on breads.
- Mustard, ketchup, or low-fat salad dressing or mayonnaise may substitute for regular mayonnaise in sandwiches or salads.

- Purees of fruits such as plums, dates, apples, and figs may replace some, but not all, of the fat in recipes for baked products.
- Crumb crusts can replace standard piecrusts.
- Double-crust pies can be converted to one-crust pies, automatically cutting fat by close to 50%.
- A nonfat condiment such as salsa, relish, or chutney can replace some of the butter or sour cream toppings on baked potatoes.

# STORAGE OF FATS

Storage of fat depends on its type. Fats such as butter and margarine are best stored in the refrigerator. Butter will keep for months in the freezer, but margarines do not freeze as well because their emulsions may separate under such conditions. Shortenings and most oils are usually stored at room temperature and should be kept tightly covered in a dark spot on the cupboard shelf; however, they are best refrigerated because they will keep longer. Olive oil has a shorter shelf life than most vegetable oils and should be refrigerated fairly soon after opening. Monounsaturated fats such as olive oil usually keep for about 1 year, and unrefined polyunsaturated fats for about half a year.

## Rancidity

Rancidity is the chemical deterioration of fats, which occurs when the triglyceride molecule and/or the fatty acids attached to the glycerol molecule are broken down into smaller units that yield off-flavors and rancid odors. The longer a fat is stored, the greater the possibility of its becoming rancid. Fats and oils used in cooking tend to become rancid because they are exposed to oxygen, heat, and light. For this reason, they should be checked frequently for rancidity. Rancid fats should be discarded because they will adversely affect flavor if used to make cakes, cookies, or other baked goods. Rancid fat will also ruin the flavor of sautéed or fried foods and cause problems during heating because of its lower smoke point.

**TABLE 22-9** Comparison of Dietary Fats*

| DIETARY FAT | Cholesterol mg/tbs | % Monounsaturated Fat | % Polyunsaturated Fat | % Saturated Fat |
|---|---|---|---|---|
| **Highest in monounsaturated** | | | | |
| Olive oil | 0 | 77 | 9 | 14 |
| Canola oil | 0 | 58 | 36 | 6 |
| Peanut oil | 0 | 49 | 38 | 13 |
| **Highest in polyunsaturated** | | | | |
| Safflower oil | 0 | 12 | 79 | 9 |
| Sunflower oil | 0 | 20 | 69 | 11 |
| Corn oil | 0 | 25 | 62 | 13 |
| Soybean oil | 0 | 24 | 61 | 15 |
| Cottonseed oil | 0 | 19 | 54 | 27 |
| **Hydrogenated** | | | | |
| Margarine | 0 | 48 | 34 | 18 |
| Vegetable shortening (Crisco) | 0 | 43 | 31 | 26 |
| **Highest in saturated** | | | | |
| Coconut oil | 0 | 6 | 2 | 92 |
| Butter (fat) | 33 | 30 | 4 | 66 |
| Palm oil | 0 | 39 | 10 | 51 |
| Lard | 12 | 47 | 12 | 41 |

*Fatty acid content normalized to 100 percent.

**TABLE 22-10** Replacing Fatty Ingredients in Recipes

| Higher-Fat Item | | Replaced by Lower-Fat Item | |
|---|---|---|---|
| 1 oz (1 sq) | Baking chocolate | 3 tbs | Powdered cocoa + 1 tbs margarine |
| 1 C | Butter | 1 C | Margarine (lowers saturated fat) |
| 1 oz | Cheese | 1 oz | Lower-fat cheese |
| 1 C | Cream (heavy) | 1 C | Evaporated skim milk |
| | | 2/3 C | Nonfat milk + 1/3 C vegetable oil |
| 1 C | Cream cheese | 1 C | Reduced-fat cottage cheese + 4 tbs margarine + salt to taste + milk for blending |
| 1 | Egg (large) | 1 | Egg white + 1 tsp vegetable oil |
| | | 2 | Egg whites |
| | | 1/4 C | Egg substitute |
| 1 C | Fat | 1/3 C | Applesauce + 2/3 C fat |
| 1 C | Milk (whole) | 1 C | Nonfat or reduced fat milk |
| 1 tbs | Salad dressing | 1 tbs | Low-calorie salad dressing |
| 1/2 C | Shortening | 1/3 C | Vegetable oil |
| 1 C | Sour cream | 1 C | Plain yogurt |
| | | 1 C | Reduced-fat cottage cheese (blended) |

## Types of Rancidity

There are two basic types of rancidity: hydrolytic rancidity, which occurs when *water* breaks larger compounds into smaller ones; and oxidative rancidity, in which the double bond of an unsaturated fatty acid reacts chemically with *oxygen* to result in two or more shorter molecules.

*Hydrolytic Rancidity* Fats become rancid when exposed to water, usually the water found frozen on food to be fried. The addition of water hydrolyzes the bonds in the triglyceride, causing it to break down into smaller compounds. Catalyzing this reaction are lipase enzymes and heat. This hydrolytic rancidity has implications for deep-frying. Placing cold, wet food in heated frying oil introduces water, making the oil prone to hydrolytic rancidity. Conversely, fats that have not been heated are more prone to hydrolytic rancidity because the lipase enzymes have not yet been destroyed by heat. Butter left out at room temperature, which is ideal for the lipase enzyme, quickly decomposes; therefore, butter is often refrigerated or frozen. Butter also contains water, which is the reason it has a tendency to go rancid. Butter's volatile short-chain fatty acids, such as butyric and caproic acids, create a rancid odor and off-flavor when released into the air. The long-chain fatty acids are also freed, but they are not volatile and therefore do not contribute to the odor of rancid butter.

*Oxidative Rancidity* Fats can also become rancid when they are exposed to the oxygen in air. The higher the degree of unsaturation, the more likely it is that the fat will be subject to oxidative rancidity. This is why saturated and hydrogenated fats were popular in the past with some food manufacturers and food service establishments.

Unlike hydrolytic rancidity, the rancidity due to oxygen occurs in a series of steps (Chemist's Corner 22-3). The initiation period is slow and is triggered by light; high temperatures; table salt; food particles in the frying oil; and certain metals such as iron, copper, and nickel. This initial stage is followed by a quicker, irreversible, and self-perpetuating chain reaction. Oxygen atoms attach to the carbons

> CHEMIST'S
CORNER 22-3

### Oxidative Rancidity

The three stages of oxidative rancidity, also known as oxidation, involve initiation, propagation, and termination (Figure 22-13) (26).

- **Initiation.** Initiation occurs when the loosely held hydrogen atom at the double bond of an unsaturated fatty acid is lost, forming a free radical (R •), which is very reactive.
- **Propagation.** In the propagation stage, an oxygen combines with the free radical, forming a compound (peroxide-free radicals) that can remove another hydrogen near a double bond and yield another free radical. This reaction can be repeated several thousand times until most double bonds on the fatty acids have been removed. Further breakdown of the hydroperoxide molecules into smaller units such as acids, alcohols, aldehydes, and ketones results in rancid off-odors and flavors.
- **Termination.** The chain reaction is terminated when all the free

**FIGURE 22-13**   Stages of oxidative rancidity.

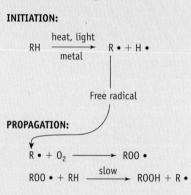

radicals have reacted with other free radicals or antioxidants and/or there are no more hydrogens at the unsaturated fatty acids' double bonds to react with oxygen.

> CHEMIST'S
CORNER 22-4

### Measuring Antioxidant Activity and Oxidation

Antioxidant activity and oxidation can each be measured by several different methods. Measures of antioxidant activity include oxygen radical absorbance capacity (ORAC), total radical-trapping antioxidant parameter (TRAP), and cellular antioxidant capacity (CAP), the latter being the most biologically relevant (14). The most popular techniques for measuring oxidation are the thiobarbituric acid (TBA) test and the peroxide value (PV) (50). TBA is a molecule that, when added to the substance to be tested, reacts with molanyl aldehyde produced from fatty acids that have three or more unsaturated double bonds. PV is the milliequivalents of iodine released from the reaction of lipid hydroperoxides with hydrogen iodide.

next to the double bond of the fatty acid, creating very reactive and unstable molecules called free radicals. These free radicals contribute to the further breakdown of fats into smaller compounds, resulting in unpleasant odors and off-flavors. Once this process starts, it is difficult to stop because the free radicals generated by the reaction create more free radicals, and this domino effect continues until all the double bonds have been used in the process. Antioxidants, found naturally in the fat or commercially added, inhibit oxidative rancidity and extend shelf life (Chemist's Corner 22-4).

*Flavor Reversion*   Food manufacturers must also deal with **flavor reversion,** a type of characteristic flavor change that occurs even before actual rancidity begins. The odor and flavor of oils, particularly those having high linolenic acid levels, can be altered by light and heat, which convert the fatty acids to volatile compounds, causing off-odors. Only a small amount of oxygen needs to be present to oxidize linolenic acid. The odors and flavors produced by flavor reversion depend on the type of oil. Soybean oil initially becomes "beany" and then "fishy," but the two most commonly used oils, cottonseed and corn, are very resistant to flavor reversion.

### Preventing Rancidity

Rancid products have reduced shelf lives and must be discarded. In the past, cereal manufacturers incorporated predominantly saturated fatty acids such as coconut and palm oils

into their products to reduce the risk of rancidity. More recently, public concern over saturated fat and its relationship to blood cholesterol levels contributed to increasing use of unsaturated oils and new ways of deterring rancidity.

*Avoid Oxygen and Heat*   One method of inhibiting rancidity is to pack food items high in unsaturated fatty acids, such as potato or tortilla chips, in vacuum packs or nitrogen to prevent contact with oxygen.

Several protective measures can also be taken to prevent rancidity of the oils and fats themselves. Vegetable oil bottles should be recapped immediately after use to minimize exposure to oxygen. Storing a bottle of oil on the shelf near the range, where heat is constantly being generated, is not recommended. The bottles are best kept in cool, dry places away from air, light, high temperatures, and exposure to metals such as iron and copper. In warmer climates, they fare better in the refrigerator.

*Antioxidants*   The USDA's Code of Federal Regulations defines antioxidants as substances used to preserve food by retarding deterioration,

**Flavor reversion**   The breakdown (oxidation) of an essential fatty acid, linolenic acid, found in certain vegetable oils, leading to an undesirable flavor change prior to the start of actual rancidity.

> **CHEMIST'S CORNER 22-5**

**Antioxidants in Action**

Antioxidants prevent oxidation of fats either by (1) being oxidized themselves, (2) donating their hydrogen to a fatty acid as a reducing agent, and/or (3) sequestering metals such as a chelating agent. Substances that assist with the second mechanism of action by transferring hydrogen are called reducing agents. Examples include vitamin C, vitamin E (tocopherols), beta-carotene, flavonoids, erythorbic acid, ascorbyl palmitate, and sulfites.

The third method involving chelating agents works when they "attach" to metal ions such as copper or iron and prevent them from catalyzing oxidation. The word *chelate*, from *chela* or claw, refers to the claw-like manner in which the agent binds to the metal through coordinate covalent bonds. Examples of chelating agents, also known as metal sequesterers, include ethylenediaminetetraacetate (EDTA), citric acid, and phosphates (26).

**FIGURE 22-14** Commonly used commercial antioxidants.

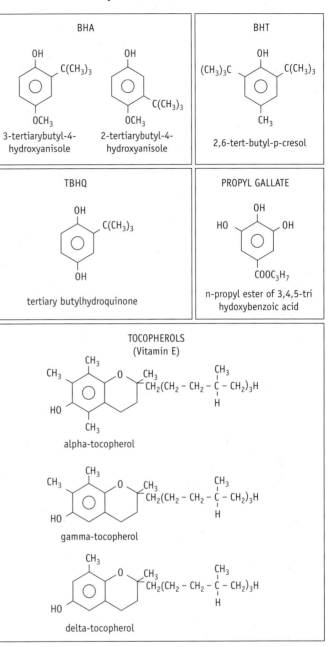

rancidity, or discoloration due to oxidation. Antioxidants, natural and commercial, are added to foods containing large amounts of unsaturated fats in order to prevent rancidity (Chemist's Corner 22-5). Foods to which antioxidants are commonly added include dry cereals, crackers, nuts, chips, and flour mixes.

Naturally occurring antioxidants include vitamins E and C, lecithin, flavonoids, and gum guaiac. Many vegetable oils naturally contain vitamin E. Commercial antioxidants permitted by the FDA include butylated hydroxyanisole (BHA), butylated hydroxytoluene (BHT), propyl gallate, and tertiary butyl hydroquinone (TBHQ) (Figure 22-14) (23).

# PICTORIAL SUMMARY / 22: Fats and Oils

Fats and oils are important components in well-prepared and good-tasting food. Their unique properties not only contribute to taste, texture, and nutrition, but also greatly influence food preparation.

## FUNCTIONS OF FATS IN FOOD

Properties of fat that affect food preparation:

**Heat transfer:** Fats act as a medium to transfer heat and prevent food from burning.

**Shortening power:** Fats are essential in the preparation of pastries, piecrusts, biscuits, and cakes. The more highly saturated the fat is, the greater the shortening power.

**Emulsions:** All foods contain some liquid, and if fats or oils are present, the combination is an emulsion, such as mayonnaise.

**Melting point:** Fats have a range of melting points. Most plant oils are liquid at room temperature, whereas animal fats are solid.

**Plasticity:** The plasticity of fat is its ability to hold its shape but still be molded and spread; this influences its use in icings and the like.

**Solubility:** Fats are generally insoluble in water.

**Flavor/Satiety:** Fats contribute their own flavor as well as release the flavors and aromas of other foods when cooked with them. Fats also provide a creamy texture to foods, and because they take longer to digest than carbohydrates and proteins, they induce a sense of fullness or satiety.

**Nutrition:** Fats and oils contribute essential fatty acids and calories to the diet. According to the current dietary goals 20% to 35% of caloric intake in adults should be derived from fat; however, children under 2 years of age should not be restricted in their fat intake.

## TYPES OF FATS

**Fats are derived from both plant and animal sources.**

**Margarines** are vegetable oil spreads.

**Shortenings** are hydrogenated oils.

**Vegetable oils** are derived from plants, primarily the seeds of soybeans, corn, cottonseed, rapeseed, sunflower, and safflower.

**Cocoa butter,** which is used in the manufacture of chocolate candies, is made from the seeds of the cacao tree.

**Butter** is made by churning the cream from milk.

**Lard** is derived from the fat of swine, and **tallow** from beef or sheep fat.

## FAT REPLACERS

Fat replacers are increasingly used in the food industry, and include synthetic fats (olestra), proteins, starch, fiber, and even dried fruit purée.

## FOOD PREPARATION WITH FATS

In food preparation, fats and oils are primarily used for frying (sautéing, pan-frying, deep-frying) and as shortening agents in pastries and other baked goods.

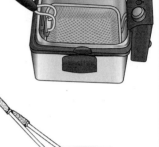

Vegetable oils with high smoke points are used for deep-frying, and their care consists of several steps that must be taken to ensure their preservation. A frying temperature that is too high or too low can lead to excessive fat absorption by the food.

Fat content can be reduced during food preparation by carefully regulating the amount of fat and by making substitutions.

## STORAGE OF FATS

Fats are best stored in the refrigerator, and butter will keep for months in the freezer. Oils, except olive oil and cold-pressed oils such as walnut oil, can be stored tightly covered for long periods

of time at room temperature. The longer a fat is stored, the greater the possibility of its becoming rancid (oxidative or hydrolytic) and producing undesirable off-odors and unpleasant flavors.

# CHAPTER REVIEW AND EXAM PREP

## Multiple Choice*

1. Emulsifying agents act as a bridge between which two substances?
   a. Water and water
   b. Water and oil
   c. Oil and oil
   d. Water and vinegar

2. Increasing the chain length of a saturated fatty acid will _____ its melting point.
   a. increase
   b. decrease
   c. not influence
   d. saturate

3. Shortenings are solid at room temperature because they have gone through what type of process?
   a. Fermentation
   b. Clarification
   c. Emulsification
   d. Hydrogenation

4. The fat replacer olestra is composed of sucrose and 6 to 8 _____ .
   a. amino acids
   b. fatty acids
   c. phospholipids
   d. triglycerides

5. Which type of rancidity results from the exposure of fat to oxygen?
   a. Regular rancidity
   b. Oxidative rancidity
   c. Hydrolytic rancidity
   d. Flavor rancidity

6. Every vegetable oil has a smoke point and most start at temperatures:
   a. within the temperature danger zone of 40°F–140°F (4°C–60°C).
   b. below boiling at 212°F (100°C).
   c. at frying temperatures of 350°F–450°F (177°C–232°C).
   d. above 420°F (216°C).

7. Frying oil needs to be replaced when:
   a. it starts to show excessive foaming or smoking.
   b. too much fat is being absorbed by the fried foods.
   c. it starts to show a darker, more viscous quality.
   d. All of these qualities

## Short Answer/Essay

1. List and briefly describe some of the functions of fats in food preparation.
2. What types of foods rely on the shortening power of fats?
3. Define *emulsion*. Discuss the three parts of an emulsion, and the different types of emulsions. What is the difference between an oil-in-water emulsion and a water-in-oil emulsion?
4. Discuss the four factors that determine a fatty acid's melting point and affect solidity at room temperature.
5. What is the importance of a fat's plasticity in food preparation?
6. Define the following: *clarified butter (ghee), smoke point, fire point, winterizing, hydrogenation, interesterification,* and *flavor reversion.*
7. Many different types of fat are available for food preparation. Describe the source, production, and general use of the following fats: butter, vegetable oil, shortening, margarine, and lard.
8. Describe the differences among the terms *fat replacer, fat substitute, mimetics,* and *extenders.* Describe the structural differences among carbohydrate-, protein-, and lipid-based fat replacers.
9. Discuss the factors to consider when selecting a frying fat. Describe the optimal frying temperatures and optimal frying conditions.
10. Define *rancidity*. What is the difference between hydrolytic and oxidative rancidity?

*See p. AK-1 for answers to multiple choice questions.

# REFERENCES

1. Akoh CC. Scientific status summary: Fat replacers. *Food Technology* 52(3): 47–52, 1998.

2. Balogh A, O Csuka, I Teplan, and G Keri. Phosphatidylcholine could be the source of 1,2-DAG which activates protein kinase C in EGF-stimulated colon carcinoma cells (HT29). *Cell Signal* 7(8):793–801, 1995.

3. Banasiak K. Industry news. *Food Technology* 58(8):10, 2004.

4. Best D. The challenges of fat substitution. *Prepared Foods* 160(6):72–76, 1992.

5. Best D. Technology fights the fat factor. *Food Development* 160(2):48–49, 1991.

6. Boyle E. Monoglycerides in food systems: Current and future uses. *Food Technology* 51(8):52–59, 1997.

7. Boyle E, and JB German. Monoglycerides in membrane systems. *Critical Reviews in Food Science and Nutrition* 36(8):785–805, 1996.

8. Centers for Disease Control and Prevention. *US obesity trends.* www.cdc.gov/obesity/data/trends.html. Accessed 11/20/09.

9. Chan EJ, and L Cho. What can we expect from omega-3 fatty acids? *Cleveland Clinic Journal of Medicine* 76(4):245–251, 2009.

10. Chandan R. *Dairy Based Ingredients.* Eagan Press, 1997.

11. Clark JP. Diet trends affect processing. *Food Technology* 59(3):59–60, 2005.

12. Clark JP. Fats and oils processors adapt to changing needs. *Food Technology* 59(5):74–76, 2005.

13. Clark JP. Processing: Extraction—a widely used, complex process. *Food Technology* 58(7):78–80, 2004.

14. Clemens R, and P Pressman. Oxidative stress: Defense or disease. *Food Technology* 61(11):17, 2007.

15. Colbert LB. Lecithins tailored to your emulsification needs. *Cereal Foods World* 43(9):686–688, 1998.

16. Coupland JN, et al. Solubilization kinetics of triacyl glycerol and hydrocarbon emulsion droplets in a micellar solution. *Journal of Food Science* 61(6):1114–1117, 1996.

17. Craig WJ, AR Mangels, and American Dietetic Association. Position of the American Dietetic Association: Vegetarian diets. *Journal of the American Dietetic Association* 109(7):1266–1282, 2009.

18. Drewnowski A. The real contribution of added sugars and fats to obesity. *Epidemiology Review* 29:160–171, 2007.

19. Drewnowski A. Why do we like fat? *Journal of the American Dietetic Association* 97(7):S58–S62, 1997.

20. Dziezak JD. Fats, oils, and fat substitutes. *Food Technology* 43(7):66–74, 1989.

21. Echel RH, et al. Americans' awareness, knowledge, and behaviors regarding fats: 2006–2007. *Journal of the American Dietetic Association* 109(2):288–296, 2009.

22. *Fat.* American Heart Association. www.americanheart.org/presenter. jhtml?identifier=4582. Accessed 11/20/09.

23. Fats and fattening: Fooling the body. *Prepared Foods* 166(7):51–52, 1997.

24. Firenze C. What's the difference between extra-virgin, virgin, and other olive oils? *Fine Cooking* 78:20, 2006.

25. Garcia DJ. Omega-3 long-chain PUFA nutraceuticals. *Food Technology* 52(6):44–49, 1998.

26. Giese J. Antioxidants: Tools for preventing lipid oxidation. *Food Technology* 50(11):73–78, 1996.

27. Giese J. Fats, oils, and fat replacers: Fats and oils play vital functional and sensory roles in food products. *Food Technology* 50(4):78–84, 1996.

28. Giese J. News & analysis. FDA allows health claim for olive oil. *Food Technology* 58(12):22, 2004.

29. Good chemistry: How form shapes function in food oils. *Journal of the American Dietetic Association* 91(7):777–778, 1991.

30. Harrison GG. Reducing dietary fat: Putting theory into practice. Conference summary. *Journal of the American Dietetic Association* 97(7):S93–S96, 1997.

31. Jonnalagadda SS, and JM Jones. Position of the American Dietetic Association: Fat replacers. *Journal of the American Dietetic Association* 105(2):266–275, 2005.

32. Katz F. Functional fruit: Fruits as fat replacers providing texture, water holding, mouthfeel. *Food Processing* 57(7):74–75, 1997.

33. Kosmark R. Salatrim: Properties and applications. *Food Technology* 50(4):98–101, 1996.

34. LaBell F. Cottonseed oil for frying and baking. *Prepared Foods* 166(2):75, 1997.

35. Lui X, et al. High hydrostatic pressure affects flavor-binding properties of whey protein concentrate. *Journal of Food Science* 70(9):C581–C585, 2005.

36. Maki KC, et al. Consumption of diacylglycerol oil as part of a reduced-energy diet enhances loss of body weight and fat in comparison with consumption of a triacylglycerol control oil. *American Journal of Clinical Nutrition* 76(6):1230–1236, 2002.

37. McWilliams M. *Foods: Experimental Perspectives.* Prentice Hall, 2007.

38. Mela DJ. Nutritional implications of fat substitutes. *Journal of the American Dietetic Association* 92(4):472–476, 1992.

39. Mermelstein NH. Analyzing for trans fat. *Food Technology* 63(8):71–74, 2009.

40. Morris CE. Balancing act: Engineering flavors for low-fat foods. *Food Engineering* 64(8):77–78, 1992.

41. Mozaffarian D, A Aro, and WC Willett. Health effects of trans-fatty acids: Experimental and observational evidence. *European Journal of Clinical Nutrition* 63(Suppl 2):S5–S21, 2009.

42. National Center for Health Statistics. *Prevalence of overweight, obesity and extreme obesity among adults: United States, trends 1960–62 through 2005–2006.* www.cdc.gov/nchs/data/hestat/overweight/overweight_adult. htm. Accessed 11/20/09.

43. *National Cholesterol Education Program.* www.nhlbi.nih.gov/about/ncep/index.htm. Accessed 11/27/09.

44. Niederdeppe J, and DL Frosch. News coverage and sales of products with trans fat: Effects before and after changes in federal labeling policy. *American Journal of Preventative Medicine* 36(5):395–401, 2009.

45. Ohr LM. Following functional oils. *Food Technology* 63(2):65–68, 2009.

46. Ohr LM. Nutraceuticals: Functional fats. *Food Technology* 60(3):81–84, 2006.

47. Ohr LM. Nutraceuticals & functional foods: Functional fatty acids. *Food Technology* 59(4):63–68, 2005.

48. Orthoefer FT, S Gurkin, and K Liu. Dynamics of frying. In EG Perkins and MD Erickson, eds., *Deep Frying: Chemistry, Nutrition and Practical Applications.* AOCS Press, 1996.

49. Osborn-Barnes HT, and CC Akoh. Applications of structured lipids in emulsions. In CC Akoh, ed., *Handbook of Functional Lipids.* Taylor and Francis, 2006.

50. Owusu-Apenten R. *Introduction to Food Chemistry.* CRC Press, 2005.

51. Palmer S. The fairest fats of them all (and those to avoid). *Today's Dietician* 10(10):36–40, 2008.

52. Paul S, and GS Mittal. Regulating the use of degraded oil/fat in deep-fat/oil food frying. *Critical Reviews in Food Science and Nutrition* 37(7):635–662, 1997.

53. Pinthus EJ, P Weinberg, and IS Saguy. Oil uptake in deep-fat frying as affected by porosity. *Journal of Food Science* 60(4):767–769, 1995.

54. Pszczola DE. Functional ingredients enhance value of low-fat foods and microwaved foods. *Food Technology* 46(4):116, 1992.

55. Reducing acid taste in low-fat dressings. *Food Engineering* 64(5):46, 1992.

56. Schwartz NE, and ST Borra. What do consumers really think about dietary fat? *Journal of the American Dietetic Association* 97(Suppl): S73–S75, 1997.

57. Sloan AE. Consumer project trends: Fats and oils slip and slide. *Food Technology* 51(1):30, 1997.

58. St. Angelo AJ. Lipid oxidation in foods. *Critical Reviews in Food Science and Nutrition* 36(3):175–224, 1996.

59. Stauffer CE. *Fats and Oils: Practical Guides for the Food Industry.* Eagen, 1996.

60. Stauffer CE. Fats and oils in bakery products. *Cereal Foods World* 43(3):121–126, 1998.

61. Stern JS, and MG Hermann-Zaidins. Fat replacements: A new strategy for dietary change. *Journal of the American Dietetic Association* 92(1):91–93, 1992.

62. Tarrago-Trani MT, et al. New and existing oils and fats used in products with reduced trans-fatty acid content. *Journal of the American Dietetic Journal* 106(6):867–880, 2006.

63. Thomas DJ, and WA Atwell. *Starches.* Egan Press, 1999.

64. Turn of the lites. *Food Engineering* 42(9):41, 1991.

65. United States Department of Agriculture. *Dietary Guidelines for Americans 2005.* www.health.gov/DietaryGuidelines/. Accessed 11/27/09.

66. Visioli F, and G Claudio. The effect of minor constituents of olive oil on cardiovascular disease: New findings. *Nutrition Reviews* 56(5):142–147, 1998.

67. Williams J, and FH Mattson. Discovered fat substitute known as olestra. *San Diego Union Tribune,* June 1, 1997.

68. Yackel WC, and C Cox. Application of starch-based fat replacers. *Food Technology* 46(6):146, 1992.

69. Yamamoto K, et al. Long-term ingestion of dietary diacylglycerol lowers serum triacylglycerol in type II diabetic patients with hypertriglyceridemia. *Journal of Nutrition* 131:3204–3207, 2001.

# WEBSITES

To find the fat grams (or any other nutrient) of most foods at this site, click on "What's in Food" in the left column, then click on "Look up calories or nutrients in a food" on the right. Calories are listed as "energy (kcal)":

**www.nutrition.gov**

The Calorie Control Council (lobby group) has a glossary of fat replacers:

**www.caloriecontrol.org/articles-and-video/feature-articles/glossary-of-fat-replacers**

Here is the website for Olean fat replacer:

**www.olean.com**

The American Oil Chemists' Society (a science-based organization that students can join for free) has a useful website:

**www.aocs.org**

# 23 Cakes and Cookies

A basic flour mixture serves as the foundation for quick and yeast breads, but it is also the basis for their sweeter cousins, cakes and cookies. Cakes, which are believed to have originated in Egypt as round, flat, unleavened breads, are essentially sweetened breads, whereas cookies can be considered to be little cakes. Ingredients can be combined in several different ways and styles, creating confections ranging from simple sugar cookies to elegantly decorated, many-tiered wedding cakes. In the not-too-distant past, such items were created "from scratch," but now the vast majority of cakes and cookies are made from packaged mixes that come ready to be combined with liquid ingredients. To make the process even easier, there are cookies in the supermarket's refrigerator section that are sold ready to bake. This chapter discusses the different types, the nutrient content, and the preparation and storage of cakes and cookies, whose ingredients, along with their specific functions, were discussed in Chapter 17 (on flours and flour mixtures).

## TYPES OF CAKES

As shown in Figure 23-1, cakes are classified according to whether or not they contain fat. The majority of cakes are either shortened or unshortened; chiffon cakes make up a third category (Figure 23-2). The variety of different cakes that fall into these three basic categories is endless, but some of the more common cakes are briefly described below.

## Shortened Cakes

**Shortened cakes,** also called butter or conventional cakes, are usually leavened with baking powder or baking soda, although steam generated from the liquid ingredients and air incorporated during the mixing process also contribute to leavening. Cakes were not always the delicate creations they are today because at one time they were leavened with yeast. The refined cake grain that is available today therefore was not always possible; the heavy ingredients, such as sugar and fat, were not able to rise easily without the leavening agents baking soda and baking powder. Baking soda did not exist before the 1840s, and baking powder appeared in the 1860s. Examples of shortened cakes leavened with baking soda or powder include the standard yellow, plain white, chocolate (devil's

**Shortened cake** A cake made with fat.

**FIGURE 23-1**  General classification of cakes.

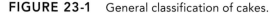

CAKES

Shortened
(butter or conventional cakes)
Leavened with baking powder or soda

Chocolate    Pound    White
Fruit    Spice    Yellow

Unshortened
(sponge or foam cakes)
Leavened with beaten egg whites

Angel
Sponge

Chiffon
(hybrid of shortened and unshortened)
Leavened with both baking powder or
soda and beaten egg whites

Chocolate chiffon
Lemon chiffon

**FIGURE 23-2**  Types of cakes: Pound cake, chiffon cake, angel food cake.

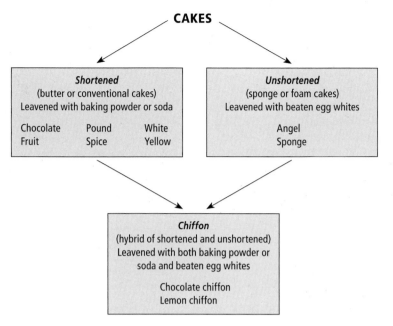

Chiffon cake

Pound cake

Angel food cake

Richard Brewer/Digital Development Services

food), spice, and fruitcakes. Pound cake is a compact, shortened cake leavened only by air and steam. In contrast to other shortened cakes, pound cake contains equal amounts of fat and sugar in addition to cake flour (or, less commonly, all-purpose flour), large amounts of egg, and flavoring. Grinding the sugar with a food processor makes the pound cake even lighter and smoother.

***Bundt Cake***  This dessert is prepared in a Bundt™ pan that did not really sell very well until 1966, when a Bundt cake won second place in a baking contest sponsored by Pillsbury.

***Butter Cake***  This is the standard cake commonly used at birthdays, weddings, and graduations. Numerous flavors exist, but the most common are white, yellow, and chocolate.

***Carrot Cake***  This sweet spice cake more closely resembles a quick bread (see Chapter 19 on quick breads). Carrot cakes are very moist, dense cakes, prepared either layered, in a loaf pan, or as cup cakes. Basic ingredients include grated carrots, nuts, and cinnamon. Other added ingredients contributing to the cake's moistness and density include oil, applesauce (or apples), raisins, or well-drained crushed pineapple. Carrot cakes are served plain or topped with cream cheese icing.

***Cheesecake***  Cheesecake fillings are made from cheese—cream cheese, cottage cheese, or ricotta cheese—mixed with eggs, sugar, and flavorings. The crust can be made from graham cracker crumbs, wafer crumbs, gingersnaps, finely ground nuts, or pastry. The ingredients are baked in a springform pan or cheesecake pan and may include a top layer of sour cream. Cheesecake styles vary in texture from light and airy to heavy and dense according to their ingredients, which vary among regions. In North America, it is common to have cream cheese as the cheese of choice. Adding heavy cream and egg yolks creates a richer, denser cheesecake, whereas pot or farmer's cheese yields a more tangy, softer type of cheesecake. Neufchatel cheese and gelatin are preferred in France, whereas emulsified cornstarch and eggs create a unique smooth-style cheesecake often sold in Japan.

***Ciambellone***  This ring-shaped cake is lightly sweetened and flavored with dried fruit and lemon zest.

***Coffee Cake***  Often served with coffee, these sweet, rich cakes and breads are usually served at breakfast, brunch, or afternoon tea. Coffee cakes may contain fruits, nuts, spices, chocolate, fruit, jam, streusel, or cream cheese (see Chapter 19 on quick breads).

***Cup Cake***  A small, individual cake baked in a paper-lined, cup-shaped mold (usually a muffin pan) that can be made from many different cake batters.

***Devil's Food Cake***  A rich, chocolate layered cake with a "reddish" hue.

***Fruitcake***  This traditional British Christmas cake is full of fruit (candied and dried), nuts, and spices. Fruitcake is laced with alcohol (usually brandy) and sometimes covered with marzipan and royal icing.

*German Chocolate Cake*   The "German" portion of the name originates from the Dallas homemaker who used Baker's German Sweet chocolate (sweeter than baking chocolate) to create this chocolate-buttermilk cake. Its typical coconut-pecan frosting also differentiates it from other chocolate cakes.

*Ice Cream*   Ice cream serves as the center of a butter cake or the "cake" is pure ice cream shaped like a cake.

*Mooncake*   These round-shaped Chinese confections are traditionally consumed each fall during the Harvest Moon Festival. The larger, brighter moon during this time allowed harvesting to continue into the night. About the size of a cup cake, these very dense, rich cakes are cut into small wedges and consumed with Chinese tea.

*Muffin*   Muffins are miniature versions of short breads (see Chapter 19, Quick Breads).

*Pound Cake*   The name is derived from the original British recipe that utilized one pound each of butter, sugar, flour, and eggs. Although no longer the case, the cake still retains much of its original richness and density.

*Upside-Down Cake*   After baking, this cake is literally turned upside-down. Chopped or whole fruits such as pineapple or cherries are placed on the bottom of the pan on top of a layer of brown sugar before the batter is poured in. Thus, when the cake is inverted after baking, the bottom layer becomes a decorative, moist topping.

**Unshortened cake**   A cake made without added fat.

**Chiffon cake**   A cake made by combining the characteristics found in both shortened and unshortened cakes.

## Unshortened Cakes

**Unshortened cakes** are also known as sponge or foam cakes and include angel food, sponge, meringue, dacquoise, génoise, and roulade. The term *sponge* in food preparation is frequently used to denote foods made with beaten egg whites. They contain very little, if any, fat—just flour, sugar, and egg whites. The light, delicate structures of angel food and sponge cakes rely on steam and air from foamed, or beaten, eggs as the major leavening agent. Angel food cakes are made with beaten egg whites, whereas sponge cakes are made with whole eggs, which contribute to the latter's rich, yellow color. They can be eaten plain or topped/filled with frostings, whipped cream, preserves, fruit, nuts, chocolate, or other flavorings.

*Angel Food Cake*   Whipped egg whites and the lack of fat contribute to the very light, airy texture and taste of this cake. A special angel food cake pan with a tube in the middle creates the hole in the middle of the cake and prevents it from falling on itself.

*Boston Cream Pie*   This "pie" is really a cake. The "cream" is really vanilla custard spread between two layers of sponge cake. A chocolate glaze is poured over the top. It's called "Boston" cream pie because it was first created by a Boston hotel chef around 1855.

*Dacquoise (da-kwoz)*   This French cake, named after a town called "Dax," consists of alternating layers of meringue and buttercream. Ground nuts are usually added to the meringue, and sometimes whipped cream may replace the buttercream.

*Génoise (zg-eh-nwoz)*   Génoise is an Italian cake named after Italy's city of Genoa. The use of whole eggs and melted butter makes it different from other sponge cakes. It may also be sliced into layers separated by chocolate, fruit, pastry cream, or whipped cream.

*Meringue (muh-rang)*   Meringue layers form the basis of this cake, which is usually topped with soft spreads such as marzipan and whipped cream.

*Petit Four*   These small multi-layer cakes, averaging only 1-inch square, are usually elaborately decorated with icing.

*Roulade (roo-lahd)*   This is a rolled sponge cake filled with various ingredients that can include preserves, frostings, nuts, or other flavorings. A roulade is created by placing the cake batter in a sheet pan (the type often used for preparing cookies). The thin baked cake, which is only about ½ inch thick, is rolled while still warm to set its log-like shape, unrolled when cooled, spread with fillings, rerolled, and then either left plain on its outer surface or covered with frosting or powdered sugar.

*Tiramisu (teer-a-me-sue)*   This Italian dessert takes its name from the Italian words "tira" and "su," which together mean "pull up"—probably due to its caffeine-laden ingredients, coffee and cocoa. Tiramisu basically consists of layers of sponge cake or ladyfinger cookies soaked in coffee and alcohol (wine, brandy, or liqueur), spread with the unique Italian ingredients of mascarpone cheese and/or zabaglione, and then topped with sprinkles of finely grated chocolate. Mascarpone (mas-kar-pone-eh) is a soft Italian cheese that is very high in fat because cream is added to the cow's milk before it is made into cheese. Zabaglione (za-bal-yone-ee) is a custard or sauce made by whipping egg yolks, sugar, and wine (usually Marsala) or liqueur over a boiler.

## Chiffon Cakes

**Chiffon cakes** are a hybrid of shortened and unshortened cakes that were not developed until the 1920s. Fat, usually from vegetable oil and egg yolks, is combined with foamed egg whites, cake flour, and leavening agents. This yields a cake with a texture that is light and airy, yet richer and denser than angel food. A tube pan and extra baking powder are required to lift the cake during baking. Common examples of these variations of angel food cakes include lemon and chocolate chiffon cakes.

# PROFESSIONAL PROFILE

Willie Byrd had no idea what he wanted to major when he started attending Alabama Agriculture and Mechanical University in Normal, Alabama. He liked the food science courses he read about in the course catalog. He had always been curious about the chemical composition of foods and wondered why different foods have different tastes.

"Food chemistry," he says, "allows you to understand the 'why.'" Willie went on to graduate with a major in food science and a minor in chemistry. Wanting to work in private industry, he thought a master's degree would prepare him for better future employment and salary opportunities. He says, "Even though a lot of it has to do with experience, sometimes degrees help if two job candidates have the same experience, but one holds a higher degree." As a result, Willie continued his education and obtained a Master of Science degree in Food Science and Technology.

Despite his long-term goal to work in the food industry, Willie's first job out of college was teaching at the university's department of food science for 2 years as an instructor. While attending an Institute of Food Technologists' convention in 1986, he met a Kellogg Company representative who invited him to Battle Creek, Michigan, for an interview. The company paid all expenses. Two weeks later Willie was offered a job along with a substantial salary increase, benefits, and promotion opportunities.

"I had no intent of leaving teaching immediately," he says, "but the difference in financial rewards in education compared to those offered by corporate America were too great. I also thought the experience would be valuable for teaching as well as for working in private industry. My plan was to work for the Kellogg Company for 3 years and then return to academia, but that was 20 years ago."

"My first job," Willie continues, "was as a research chemist developing new laboratory assays for measuring various food components. We used near infrared light [NIR] to measure proteins, fats, glycerol, and sugars. I spent 2 years in that position before being offered a position as a food technologist in a new plant. Some of my work responsibilities included ensuring that food formulas were correct and that equipment used in production was set to the right parameters (right oven temperature; right humidity to puff cereals)."

Willie held that job for a year before returning to corporate headquarters as a manager in product evaluation and stability. He managed eight people as the Product Evaluation and Stability Manager—four in the product evaluation department and four in the stability department. He says, "Stability testing is necessary for shelf-life determination to guarantee freshness. For example, we make sure packaging material is properly tested to ensure desired shelf life at various temperatures. In the product evaluation department, sensory panels evaluate the texture, flavor, and appearance of foods."

His next move in the corporation was as Quality Technologist in new product innovation, a position that entailed writing specific standards for new products—shape, size, flavor, sugar concentration, and texture (crispy, glassy, gummy, or dry). These standards

**Willie Byrd, Director, Manufacturing Audit Programs, Corporate Quality, Kellogg Company**

Courtesy of Willie Byrd/Kellogg Corporation

are taken to manufacturing plants, where employees are trained in how to produce and evaluate the new product. Shortly thereafter, he switched to being Senior Quality Technologist of existing brands. At the time, he worked with three manufacturing plants on day-to-day issues. After this position, he became the Quality Manager of two cereal plants. A few of his duties included ensuring sanitation and food safety, implementing HACCP programs, training people in good manufacturing practices (GMPs), and working with government officials (FDA, USDA, OSHA, EPA). Willie then left the cereal division of quality control and moved to the snacks division.

Currently, Willie's job title is Associate Director of Snacks Quality Programs & Global Performance. He ensures that the right quality programs are in place; his global performance role is to work with the quality managers in Kellogg Company facilities around the world. "I like quality control," he says, "because I enjoy producing a product that I know the consumers are going to enjoy. Quality control is not negotiable. This is a company with sales near $13 billion, and I have to stand up for quality."

When asked what he wished he had known in college that he knows now, Willie says that "the university prepared me well for the job market, but once you have a job, the question is, how do you keep it? No one sits down and tells you these things. Ask your bosses, 'What are the unwritten rules?' There are a lot of politics in a business corporation. Politics is something you have to witness, experience, and understand. It's a critical thing. It's not always the knowledge. Perception is very important.

"For instance, if you are a relatively quiet person, people will perceive you as not wanting to be engaged, not being a strong team player, or being an independent individual, and a person who does not want to stand up in the front firing line. You need to speak up and be heard, interact well with your peers, be a team player, and always give more than 100%. Just doing your job is not always enough, and it is not always about what you know. If you know what to do, do it. If you do not know, ask.

"Also," he added, "a lot of people fail to understand that simply going to your boss and asking, 'Is there something that I should have done differently or is there something I can do for you?' is perfectly fine. Superiors will see that you are willing to learn and grow."

When asked what else he could have done in college and what college students might do to improve their opportunities, Willie says, "I wish I had challenged myself to ask more questions and learn about what's happening outside of the university in terms of job opportunities and careers. It's important to seek out a mentor at the university and in corporate business (your field of discipline). It's a mistake not to seek out an experienced person in the organization who can guide you. You come in not knowing, thinking you just have do your job every day, and sometimes your boss just does not tell you some of the important things that you should know."

# PREPARATION OF CAKES

As discussed earlier, the ingredients used to make shortened and unshortened cakes differ; different mixing methods also result in different cakes. Some of the most common ways of combining the ingredients of a shortened cake are the conventional, conventional sponge, pastry-blend, single-stage (quick-mix), and muffin mix methods. These mixing methods apply to other foods besides cakes, and are discussed in Chapter 5.

Overall, the flour mixtures that produce cakes and cookies are very similar to those used to make breads. The significant differences are that they are sweeter and often have added flavorings not typically used in breads.

## Ingredients

Cakes have a higher proportion of sugar, milk, and fat to flour than do breads, and the flour used is usually cake flour (Chemist's Corner 23-1). Both flour and eggs contain the proteins that contribute strength and structure to cakes (9). Fat and sugar have the opposite effect, softening the cake's structure by providing moisture and tenderness. Too much flour and too many eggs may make the cake tough and/or dry, whereas too much fat and sugar may weaken the cake to the point that it does not set. Ultimately, the goal is to create cakes that have the strength to hold together, but are still tender and moist.

### Flour

Cake flour establishes a crumb structure in cakes (Chemist's Corner 23-2). The flour's starch gelatinizes, and its proteins form gluten to provide a structural network (12). The structural strengthening effect of cake flour and egg is balanced by the tenderizing effect of the sugar and fat ingredients.

### Sugar

Sugar's multiple functions in cake preparation include (1) sweetening, (2) increasing volume, (3) browning the crust, and (4) increasing shelf life. The higher volume seen in cakes made with sugar is due to sugar's ability to delay gelatinization. With sugar, the cake has more time to rise during baking before the starch gelatinizes and sets the cake's structure.

For many years, the weight of the sugar in cake mixtures could not exceed that of flour because higher proportions of sugar would interfere too much with the gelatinization of starch and the hydration of proteins, causing the cake to collapse. Now, high-sugar (high-ratio) cake mixes with a sugar-to-flour ratio ranging from 1.25:1 to 1.40:1 are common as a result of improvements in cake flour and shortenings. The extra sugar results in cakes with greater moisture content, which also improves their shelf lives.

### Three Formulas for "High-Ratio" Cakes

There are three basic formulas for preparing the sweeter "high-ratio" cakes that contain more sugar than flour (9). Following these ingredient proportions will ensure a high-ratio cake that is not too dry or too moist.

1. The sugar should weigh the same as or slightly more than the flour. It is the *weight* and not the volume that counts. Remember, 1 cup of sugar (7 ounces by weight) weighs more than 1 cup of flour (4½ ounces by weight).
2. Eggs should weigh almost as much as or slightly more than the fat. Because 1 large egg weighs about 1¾ ounces, a recipe using 4 ounces of butter would call for 2 large eggs (3½ ounces).
3. The liquid ingredients (including eggs) should weigh the same as or more than the sugar.

### Fats

Fats such as butter and shortening also contribute to (1) tenderness, (2) volume, (3) moistness, and (4) flavor. These attributes are best achieved by fats other than vegetable oil, which does not entrap air during creaming. The purpose of creaming is to beat tiny air bubbles into the fat, so vegetable oils are generally not used (except for tea breads like carrot cake and commercial cake mixes) because they completely engulf and eliminate air bubbles, resulting in a decreased volume and harsh crumb (10). Air bubbles are not as easily incorporated into oil as they are into butter, so oil-shortened cakes rely on chemical leaveners like baking soda or physical leavening from whipping air into the batter, especially the egg whites. However, some people add olive oil, which contains natural emulsifiers, to cakes shortened with more solid fats to make them more tender and moist (20). Oil coats the flour proteins, preventing them from adhering to water; this reduces gluten formation and leaves more moisture in the batter. The key is not to add too much oil; otherwise, the cake becomes too heavy and compact.

> ### CHEMIST'S CORNER 23-1
>
> #### Cake Batters
>
> Most cake batters are considered oil (fat) in water emulsions. A cake batter consists of two phases: a continuous, aqueous phase, holding the dissolved solutes of sugar, salt, and leavening salts; and the dispersed particles too large to go into solution—the colloidal proteins and the suspended starch granules, fat globules, and gas cells (19). The aqueous phase allows the sugar to dissolve while suspending the flour particles. Air bubbles are usually held in the solid fat rather than the aqueous phase, but as the cake batter heats during baking, the air bubbles transfer from the fat to the aqueous/foam phase where they expand, contributing to volume. The high water content in cake formulas creates a low-viscosity batter that minimizes the formation of gluten, which contributes to the tenderness of cakes (12).

> ### CHEMIST'S CORNER 23-2
>
> #### Chlorinated Cake Flour
>
> Cake flour is often chlorinated to break the bonds (hydrogen and peptide) within and between flour proteins. This is believed to improve dispersion of ingredients, increase swelling of the starch granules, and improve baking quality (19).

## Eggs

Eggs are added to help strengthen the structure, as well as to increase leavening, to act as emulsifiers, and to add color and flavor.

## Milk

Milk is usually the main liquid in cake preparation (the aqueous phase described in Chemist's Corner 23-1). It hydrates the dry ingredients, dissolves the sugar and salt, provides steam for leavening, and allows baking soda or powder to react and produce carbon dioxide gas.

## Leavening Agent

Both cakes and cookies are leavened with gas produced by either baking soda, baking powder, air, and/or steam. The amount of chemical leavening agent used is dependent on how much flour is used. For every cup of flour, high-ratio cakes use 1 teaspoon of baking powder or ¼ teaspoon of baking soda (9).

## Additional Ingredients

Salt is an important ingredient because it is a flavor enhancer. Also, flavoring agents such as vanilla, chocolate, spices, fruits, and nuts are commonly incorporated into the basic flour mixture.

**Food Additives in Cakes** Cakes made from scratch do not contain additives other than standard ingredients—sugar, salt, and possibly chemical leaveners. Other additives are introduced when certain commercialized cake mixes, cakes, or toppings are purchased. The assortment of available food additives is relatively large, but some of the more common ones are listed in Table 23-1. Surfactants are often added to commercial cake mixes, and these are explained in Chemist's Corner 23-3.

| TABLE 23-1 Selected Food Additives in Cakes | |
|---|---|
| **Food Additive** | **Function in Food** |
| Annatto | Natural colorant with orange hue |
| Artificial color and flavorings | FDA-approved substances providing color and flavor |
| Calcium propionate | Preservative that inhibits mold growth |
| Carob bean gum | Stabilizer and thickener |
| Cellulose gum | Stabilizer and thickener |
| Citric acid | Preservative, antioxidant, pH control agent, sequestrant |
| Glycerin | Humectant (draws moisture to itself to decrease staling) |
| Guar gum | Stabilizer and thickener that improves texture |
| High-fructose corn syrup | Sweetener |
| Lactose | Sweetener (disaccharide derived from milk) |
| Maltodextrin | A carbohydrate produced from cornstarch that enhances stability and flavor |
| Modified cornstarch | Drying agent, formulation aid, processing aid, surface-finishing agent |
| Monocalcium phosphate | Leavener |
| Mono- and diglycerides | Emulsifiers |
| Potassium chloride | Salt substitute |
| Potassium sorbate | Preservative |
| Sodium caseinate | Milk proteins that provide emulsification and dispersion |
| Sodium propionate | Preserves freshness by inhibiting molds |
| Sorbitan monostearate | Emulsifier, stabilizer, and thickener |
| Xanthan gum | Stabilizer and thickener that improves texture |

## Other Factors

In addition to ingredients and mixing methods, four other factors to consider when baking cakes are:

- The type of pans to use and their treatment
- Timing
- Temperature
- Testing for doneness

These factors vary depending on whether the cake is shortened or unshortened.

## Preparing Shortened Cakes

Shortened cakes are the most commonly prepared cakes, especially for birthday and wedding celebrations. They can be made from scratch or purchased as a boxed mix in the supermarket.

Conventional round cake pans are typically used, but large rectangular pans create large surfaces that can be covered with a limitless number of icing decorations or messages.

### Type and Treatment of Pans

Pan characteristics affect cake quality, so it is important to select the best

## NUTRIENT CONTENT

Cakes consist of flour mixtures, extra sugar, and sometimes fat, so they are often high in carbohydrates, fat, and calories. The flour contributes some protein, but the total protein content is more dependent on other ingredients such as milk, eggs, and occasionally, nuts.

Any nutrient modification of cake mixtures usually focuses on the fat content. Fat is not always easy to replace, but fine-tuning the ingredients can reduce fat content somewhat (6). Fat's function as a moistener can be partially fulfilled by substituting yogurt, nonfat sour cream, or applesauce. The flavor lost by the removal of the fat can be replaced, in part, by adding more vanilla or another flavor extract. When baking from scratch, the fat in chocolate items can be reduced by using unsweetened cocoa powder, which contains only 6 grams of fat per ounce compared to the 15 grams found in the same amount of unsweetened chocolate. Food companies have more flexibility than the consumer in reducing the fat content of cakes, because the thicker, richer consistency often provided by fat can be partially replaced by using the commercial fat substitutes available to the food industry, but not, as yet, to the consumer (18, 24). There are now several commercial cake mixes on the market that are 94% fat-free by weight.

### CALORIE CONTROL
#### Cakes

Cakes average 225 calories (kcal) and 10 grams of fat per slice. However, this will vary depending on the type, with angel food cake being the lowest in fat and calories, and fruitcake being the highest, as shown in Table 23-2.

- **Control Portion Sizes:** To limit the number of calories in your dessert, serve reasonable slice sizes (and eat only 1 slice). The calorie counts for each slice in Table 23-2 are usually based on one-eighth of an 8-inch round pan. If the serving size is one-fourth of the cake, it provides double the calories found in a standard-sized slice. Half a cake provides about 900 calories (kcal).
- **Choose Plain Instead of Frosted:** Leaving off the frosting reduces the sugar, fat, and calorie content of cake desserts (chocolate frosting, for example, provides about 82 calories/kcal for each tablespoon).

**TABLE 23-2**   Calories in Cakes

| <200 Calories (kcal) | 200–300 Calories (kcal) | 300+ Calories (kcal) |
|---|---|---|
| Angel food (73) | Coffee cake (220) | Spice (327) |
| Pound (116) | Boston cream (231) | Pineapple upside down (367) |
| Cupcake, iced (131) | Chocolate, iced (234) | Fruitcake, 1 × 3 inch (417)[1] |
| Fruitcake, 1 × 1 inch (140)[1] | Yellow (245) | |
| Sponge (187) | Coconut (250) | |
| | Banana (250) | |
| | Devil's food (250) | |
| | Cheesecake (256) | |
| | Gingerbread (263) | |
| | Carrot, iced (280) | |
| | German Chocolate (280) | |

[1] Notice the serving size of fruitcake determines its calorie content.

© 2010 Amy Brown

pan for the job and to prepare it properly. Dull, rough-surfaced pans are best for baking cakes because they absorb heat more readily, resulting in the cake baking more quickly and having a larger volume, a finer grain, and a more velvety texture. Crumb formation is partially dependent on the degree of rising that occurs when the cake batter is first placed in the oven, and rapid heat absorption plays a role in this. On the other hand, shiny surfaces reflect heat, which causes the cake to take longer to bake, resulting in a coarser grain and lower volume. The weight, or gauge, of the pan's metal also affects quality: the heavier the pan, the better. When using glass pans for baking cakes, baking temperatures should be lowered by 25°F (14°C), because glass pans lead to shrunken corners from overcooking and cause the exterior crust to brown readily—a condition desirable for breads, but less so for cakes.

***Pan Preparation***   The pans are prepared prior to mixing the batter. The bottom is greased, but the sides generally are not; the ungreased sides provide traction, allowing the rising mixture to reach its full volume. Dusting the pan's greased surface with cake flour, or cocoa if it is a chocolate cake, makes later removal of the cake from the pan easier. After dusting, if the pan is turned upside down and tapped to remove excess cake flour, it will prevent the cake from having a mottled bottom. Waxed or parchment paper may also be placed in the bottom before greasing to allow for easier cake removal.

### Temperature/Timing

Prompt transfer of the batter into pans and placement in a properly heated oven is another important factor in cake quality. If a batter is allowed to stand too long after pouring, carbon dioxide and air will escape. This reduces the volume and increases the coarseness of the cake's cells. Immediately after mixing, the pan should be filled with cake batter between half and two-thirds full and placed immediately in a preheated oven to ensure proper leavening. In order to avoid uneven baking and burning from hot spots, pans should not be allowed to touch each other or the sides of the oven, nor should they be placed directly above or below each other (Figure 23-3).

Most shortened cakes are baked at 325°F–350°F (163°C–177°C). In general, layered cakes (8-, 9-, or 10-inch round pans) bake in about half an hour, whereas thicker loaf cakes take three quarters of an hour to an hour to bake. Cupcakes normally take only about

**FIGURE 23-3**   Pan placement. Adequate air circulation is achieved by placing the pans on the middle rack and at least 1 inch from each other and the oven sides.

Digital Works

**FIGURE 23-4**   Testing a cake for doneness. The cake is done baking if a light touch to the center leaves no imprint or if a wooden toothpick inserted in the center comes out clean.

Digital Works

**FIGURE 23-5**   Cakes should be cooled on a rack. This allows them to set and prevents sogginess.

Digital Works

20 minutes because of the small quantity of batter in each cup.

## Changes During Baking

Heat plays several roles during cake baking. It increases volume by expanding air, steam, and carbon dioxide. It sets the structure by coagulating protein and gelatinizing starch. The heat flows from the edges toward the center of the pan, so cakes become rounded on the top as their interiors continue to rise after the outside portions of the cake have started to set. In addition, heat browns the crust via the Maillard reaction and the caramelization of sugars. If the oven temperature is too low, leavening gas is lost from the batter before there is a chance for coagulation of the proteins and gelatinization of the starch. As a result, the cake produced will have low volume, thickened cells, and possibly an indentation in the center. Excessively high temperatures create a crust before the cake can rise, resulting in a hump formed as the interior of the cake continues to rise (19).

## Testing for Doneness

When cakes are nearing doneness, they start to "wrinkle" at the pan edges. They should be removed from the oven before a gap forms between the cake and the pan. One method to test for doneness in shortened cakes is to insert a cake tester or toothpick in the center of the cake. If it comes out clean, with no batter clinging to it, the cake is done. If a moister cake is desired, it is best to remove it from the oven while a few crumbs are still sticking to the toothpick. Another way to test for

doneness is to touch the top of the cake lightly with a finger. If it springs back, the cake is done (Figure 23-4). Testing for doneness should be reserved until the last few minutes of heating time, and done as infrequently as possible to keep from making unnecessary holes with the tester and to avoid drafts from the open oven door. Every time the door is opened, the oven temperature drops and baking time is prolonged. The drafts may even cause the cake to fall.

## Cooling

Once the cake is done, it should be removed gently from the oven and allowed to cool on a rack for 5 to 10 minutes. The warmer interior of the cake needs a chance to become firm, so any drastic, sudden movement or lack of adequate cooling jeopardizes its structure. The rack allows even air circulation under the cake; this prevents condensation and sogginess (Figure 23-5). Shortened cakes are usually removed from the pan before they are completely cooled, usually about 10 to 15 minutes after being taken out of the oven. Once the cake has cooled, a spatula or knife can be inserted and moved around the edges of the pan before inverting the cake onto a cake plate. Cakes should have a fine crumb, a tender texture, optimal volume, and a lightly browned, delicate crust. Common cake faults and their contributing causes are listed in Table 23-3.

## High-Altitude Adjustments

As in bread preparation, ingredients must be modified for cake preparation at altitudes higher than 3,000 feet (see

the back inside cover of this book). The lower atmospheric pressure at higher elevations reduces the need for baking powder or soda. Also, water evaporates more quickly at higher altitudes, and thus the concentration of sugar increases. Structural strength can be improved by adding 1 to 2 tablespoons of cake flour, increasing the amount of water, and reducing baking powder and sugar quantities (Table 23-4). Increasing the baking temperature 10°F–15°F (6°C–8°C) increases the rate at which the cake sets by speeding the coagulation of the protein and the gelatinization of the starch. Cake mix boxes provide altitude adjustment instructions.

## Microwave Preparation

It is possible to prepare cakes in the microwave oven, but the process has not yet been perfected. All cakes are better prepared in a conventional oven. Still, microwavable cake mixes are available and some of these contain special ingredients, such as xanthan gum, to increase moisture retention (25). Microwaved cakes are usually cooked in about 10 minutes and rise higher than conventional cakes. They lack the characteristic browning and crust formation expected from conventional cakes. The lack of crust is often masked with frosting; however, it is more difficult to hide the soft, uneven tops that may occur without crust formation.

Packaged microwave cake mixes usually include a microwave-safe baking pan. If not, round pans are preferred over square or rectangular ones because the corners of cakes baked in the latter are prone to burning. Two or more pans filled with cake mix can be cooked at the same time as long as the air can

**TABLE 23-3**   Cake Problems and Their Causes

| Problems \ Causes | Over mixing | Under mixing | Batter too firm | Too much flour | Too much air | Too much sugar | Too little sugar | Too much fat | Too little fat | Too much egg | Too much liquid | Too little liquid | Too little salt | Over baking | Under baking | Too high temperature | Too low temperature | Too little or old leavening | Too much leavening | Uneven oven heat | Too much flour | Oven door opened too early |
|---|---|---|---|---|---|---|---|---|---|---|---|---|---|---|---|---|---|---|---|---|---|---|
| Decreased volume | X | X | | | | | X | | | X | | | | | | X | X | | | | | |
| Increased volume | | | | | X | | | | | | | | | | | | | | | | | |
| Too brown | | | | | | X | | | | | | | | X | | X | | | | | | |
| Too pale | X | | | | | | X | | | | | | | | X | | X | | | | | X |
| Falls in center | | X | | | | X | | X | | | X | | | | X | | X | X | X | | | X |
| Uneven shape | | | | | | | | | | | | | | | | | | | | X | | |
| Peaked | X | | X | | | | | X | | | | | | | X | | X | | | X | | |
| Cracks on top | X | | X | | | | | | | | | | | | X | | | X | | | | X |
| Tunnels | X | | X | | | | | X | X | | | | | | X | | | X | X | | | |
| Tough | X | | | X | | X | | X | | | | | X | | | | X | | | | X | |
| Dry/crumbly | | X | | X | | X | | X | | | X | | | X | X | | | X | | | X | |
| Soggy | | X | | | | X | X | | | X | | | | | X | | | X | | | | |
| Heavy | X | | | | | X | X | | | X | | | | | X | | | | X | | | |
| Shrinks | X | | | | | | | | | | | | | | | X | X | | | | | |
| Flat taste | | | | | | | | | | | | | X | | | | | | | | | |

---

**TABLE 23-4**   Altitude Adjustments for Shortened Cakes

| Ingredient Adjusted | | 3,000 Feet | 5,000 Feet | 7,000 Feet |
|---|---|---|---|---|
| Baking powder | For each teaspoon, decrease | ⅛ tsp | ⅛–¼ tsp | ¼ tsp |
| Sugar | For each cup, decrease | 0–1 tbs | 0–2 tbs | 1–3 tbs |
| Liquid | For each cup, add | 1–2 tbs | 2–4 tbs | 3–4 tbs |

circulate; otherwise, they are prepared one at a time or baked in one 16-cup fluted pan. Some cake mixes on the market can be baked in either conventional or microwave ovens. As always, follow the manufacturer's instructions.

# Preparing Unshortened and Chiffon Cakes

Unshortened cakes rely on foam formation for their structure, making their preparation slightly more involved than that for conventional, shortened cakes. The preparation of angel food, sponge, and chiffon cakes (hybrids of shortened and unshortened cakes) are discussed below.

## Angel Food Cake

The three basic ingredients of angel food cake are egg whites, sugar, and cake flour. Small amounts of cream of tartar, salt, and flavoring are also added. The tender texture of an angel food cake is partially the result of the cake flour; the majority of leavening is caused by steam produced by the evaporation of liquid from egg whites (7). All angel food cake ingredients should be at room temperature; however, for food safety reasons, if eggs are too cold, they may be dipped briefly in warm water and used immediately. A stable egg-white foam, for which the eggs must be at room temperature, is necessary to form the cake's basic structure. Sugar contributes further to the stability of the foam through the incorporation of air into the mixture as it is added, which allows the formation of small air bubbles. It has a tenderizing effect on the cake by interfering with gluten development, and it raises the coagulation temperature of the egg proteins and the gelatinization temperature of starch. Too much sugar, however, will cause the cake to collapse.

*Mixing Technique*   Proper mixing technique is essential when making angel food cake. The sugar is added after the egg whites have begun to foam. It should be added very gradually or it will pull water from the egg whites, creating a syrupy foam and producing a low-volume cake. Salt and flavoring are also added at this time. The cake flour is then sifted gradually over the egg-white foam to prevent its weight from collapsing the air cells. The ingredients must be thoroughly blended, while avoiding overmanipulation, which would reduce tenderness and volume (Figure 23-6).

*Temperature/Timing*   The prepared batter is poured quickly into an ungreased tube pan; a spatula is run through the batter, sealing it to the sides of the pan; and then the pan is placed in the lower third of a preheated moderate (350°F/177°C) oven. An oven that is too cool will cause a low-volume cake because the sugar will absorb liquid from the egg whites, turn syrupy, weep out of the batter, and disrupt the air cells. An oven that is too hot will set the cake's exterior before the cake has been fully expanded and baked through, resulting in a low-volume, dense cake (7).

**FIGURE 23-6**   Under- and over-mixing angel food cake.

Undermanipulation: uneven grain, course texture, low volume

Optimal folding of ingredients: fine grain, light spongy texture, high volume

Overmanipulation: compact grain, low volume

During baking, the light batter rises and relies on the additional support of the central tube (Figure 23-7). As the cake bakes, the proteins coagulate, stabilizing the air cells; water evaporates from the fluid mixture to create a more rigid structure; starch gelatinizes, further contributing to structure; and browning on the surface occurs because of the Maillard reaction (5). Baking time is approximately 45 minutes.

When the cake is done, it is inverted in its pan and allowed to stand for about 1½ hours in this position to stretch and strengthen its structure. For serving, it is best to cut or separate angel food cake into pieces by using the special divider made for that purpose (it looks like a large comb), or by using a sawing motion with a serrated knife (Figure 23-8). Conventional knives will crush the delicate cake.

The few ingredients and minimal number of steps needed for the preparation of angel food cake may make it

**FIGURE 23-7**   The pan for angel food cake.

**FIGURE 23-8**   Cutting angel food cake. Unlike other cakes, the delicate structure of angel food cakes requires that they be cut with a comb-like utensil or serrated knife.

seem simple, but the delicacy of the egg-white foam can make it a tricky procedure. Chapter 12 on eggs describes how to prepare an egg-white foam and explains the many factors that influence its formation.

---

**❓ How & Why?**

**Why is cream of tartar added to angel food cakes?**

Cream of tartar is an acid that strengthens the egg-white foam by denaturing the protein, thus stabilizing the air cell structure. The acidic environment created by the cream of tartar helps increase tenderness and contributes to the white color of angel food cake by whitening the flour's natural yellowish, anthoxan-thin pigments.

---

## Sponge Cake

Sponge cake is similar to angel food cake, except that there are two foams—an egg-white foam and an egg-yolk foam—and lemon juice often replaces the cream of tartar as the acid ingredient. (The preparation of egg-white foams is discussed in Chapter 12 on eggs.) There are three possible methods for preparing sponge cake.

*Method 1*   In the first method, the egg whites and yolks are first separated and each is beaten separately. Sugar, and possibly vanilla extract, are beaten into the whipped yolks. The cake flour is then gently folded in, followed by the egg-white foam.

*Method 2*   Another method, called the syrup or meringue method, creates a finer-textured end product. A syrup is made by cooking two parts sugar with one part water and boiling the mixture until it has reached the softball stage (238°F–240°F/109°C–116°C) (see Chapter 25). Egg whites are beaten with cream of tartar until they are stiff but have not yet formed dry peaks. Then, as the egg-white foam is beaten constantly, the hot syrup is poured into it in a fine stream. When this meringue is completed, beaten egg yolks are combined with lemon juice and folded into the mixture, and, finally, cake flour is sifted over and lightly folded into the other ingredients.

*Method 3*   In the third method, whole eggs are beaten until foamy and pale yellow in color, then a small amount of cream of tartar or lemon juice is added. The mixture is beaten until stiff before the sugar is added in 2-tablespoon increments. A presifted flour-and-salt combination is then sifted into the egg mixture and folded into the batter.

## Chiffon Cake

Chiffon cakes are more tender than either angel food or sponge cakes because of added vegetable oil. A chiffon cake is prepared by folding whipped egg whites into a mixture of cake flour, sugar, beaten egg yolks, and oil. The batter is poured into an ungreased tube pan and baked in a preheated oven set at 325°F (163°C). When baking is completed, the pan is inverted over a cooling rack and allowed to cool for at least 20 minutes before being turned out onto a cake plate.

### Type and Treatment of the Pans

When preparing unshortened cakes, pans are left ungreased in order to provide traction so that the batter can achieve optimal volume; this also prevents angel food cake from falling out of its pan while it is cooling upside down. Tube pans provide structure, help set the delicate cake by allowing heat to reach a greater surface area, and allow for easier cake removal.

### Temperature/Timing

Unshortened cakes are baked in a moderate 350°F–375°F (177°C–191°C) oven for approximately three-quarters of an hour to an hour. Excessively high temperatures toughen the cakes' delicate structure, coagulate the top before the air and steam have had sufficient time to accomplish leavening, and may burn the crust.

### Testing for Doneness

The unshortened cake is cooked when the surface is lightly brown and springs back when touched. Using a toothpick to test for doneness does not work for unshortened cakes.

# FROSTINGS/ ICINGS

Frostings or icings, either uncooked or cooked, are spread on baked products such as cakes, cupcakes, and pastries. In addition to improving the appearance, flavor, and texture of baked dessert goods, frostings retain moisture and increase shelf life. Some of the more popular types of frostings are flat (also known as simple), decorating, buttercream, and whipped cream. More sugary, syrup-like toppings known as ganache and fondant (see Chapter 25 on candies) can serve as both frostings and fillings. In this chapter we refer to these sugary spreads as *frostings,* which is the common practice in the United States; in other countries, *icings* is the more popular term because confectioners'/powdered sugars are known as icing sugars.

## Flat Frostings

Uncooked flat or ornamental frostings are the simplest to prepare. Flat frostings are made by beating powdered sugar (10X) with water, milk, or cream, and a flavoring. Powdered sugar is used rather than granulated because it dissolves instantly in unheated liquids. Unfortunately, the presence of cornstarch (about 3%) in powdered sugar gives uncooked frostings a slight raw taste. Also, unless a fat is added, these frostings have a tendency to become dry and to crack. Flat frostings are most frequently used on coffee cakes and sweet rolls.

## Decorating Frosting

Decorating frosting, also called royal or ornamental frosting, is traditionally used for decorating wedding cakes. It is similar to flat frosting, but beaten egg whites are added to impart a firmer structure to the icing when it has dried. The hard surface is flat, smooth, and glossy. Decorating frosting is ideal for the delicate work of creating lettering, borders, flowers, and any other structures added to the cake's surface. Although this frosting is useful for its decorative powers, the flavor is not only lacking, but also sometimes has a "pasty" consistency.

## Cooked Frosting

Cooked frostings are often made with granulated sugar that is dissolved in heated water; additional ingredients may include egg whites, cream of tartar, or corn syrup. If confectioners' sugar is used, heating allows the formation of a sugar syrup that will serve as the frosting's foundation. A common cooked frosting, either granulated or confectioners' sugar, is buttercream frosting. This frosting is used primarily as a filling and icing for pastries and cakes, and frequently to create rose or leaf decorations.

As the name implies, buttercream frostings have butter (preferably sweet, unsalted) added to make them creamier, easier to spread, more flavorful, and less likely to crack. They are made by combining whipped butter and/or shortening with a boiled sugar syrup and flavorings (vanilla, chocolate, raspberry, lemon, orange, coffee, liqueur such as Grand Marnier, etc.). An additional layer of flavor can be added by spreading the cake with a mixture of jam thinned with liqueur or adding toppings (22).

Modifications of the basic sugar-based buttercream frostings include those in which soft whipped butter is beaten or folded into a meringue or beaten egg yolks. Pastry chefs use three types of buttercream icings for their delicate pastries—Swiss, Italian, or French (4).

- *Swiss buttercream.* The easiest buttercream to prepare. Egg whites and sugar are mixed together over heat (a bowl over a water bath). After the sugar melts, the egg whites are whipped into stiff peaks before adding the butter and flavorings.
- *Italian buttercream.* The lightest and sweetest of all buttercreams. A hot sugar syrup is whipped into the already whipped egg whites.
- *French buttercream.* The rich, heavy consistency of this buttercream is obtained by whipping the hot sugar syrup into whipped egg yolks (or whole eggs).

## Whipped Cream Frosting

Whipped cream can be used as a frosting or even a filling. Cocoa or fruit purees can be added to color and flavor the whipped cream.

## Ganache

*Ganache* is the French word for the rich, smooth, and shiny chocolate-cream mixture used as both a frosting and a filling for certain cakes and pastries (Figure 23-9). Prepared from a heated mixture of chopped chocolate and heavy cream, ganache is poured over baked items, where it eventually hardens into a chocolate-based covering. A buttercream consistency can be obtained by whipping soft butter into the basic ingredients, resulting in ganache beurre.

FIGURE 23-9 Ganache poured over a cake creates a chocolate-based topping.

Scott & Zoe/FoodPix/Getty Images

FIGURE 23-10 How to create the swirly marbled look.

Dave King/StockFood Creative/Getty Images

## Garnishes

Numerous garnishes, from figurines to candles, can be added to cakes, and are only limited by the imagination. Figure 23-10 shows how dragging a skewer (or toothpick, chopstick, table knife, or fork) can create a swirling marble look from two contrasting liquids placed next to each other (2).

## STORAGE OF CAKES

Cakes stale fairly quickly. Staling can be prevented to some degree by keeping them covered. Placing half an apple in the cake box also seems to extend the cake's shelf life. Frosting the cake as soon as it cools is another method to slow down moisture loss. The amount and type of sweetener used in the preparation of a cake affect its ability to be stored. As mentioned before, high-sugar-content cake mixes have longer shelf lives than lower-sugar-content cakes. Substituting honey for part of the sugar in cake contributes to even more moisture retention; it contains fructose, which is extremely hygroscopic, or water-loving. However, the fructose and glucose in honey may result in excess browning due to the Maillard reaction.

Freezing is another method to deter staling. Cakes that have been wrapped airtight can be frozen unfrosted, or with one of the types of frostings that freezes well. Cakes with fruit should not be frozen, because defrosting will cause the fillings to run, resulting in soggy cakes. Frozen cakes keep up to 3 months if frosted, and up to 6 months without frosting.

## TYPES OF COOKIES

The word *cookies* refers to "small cakes." In fact, cookies contain many of the same ingredients as cakes except that the proportion of water is low, whereas those of sugar and fat are high. The word *cookie* is derived from the Dutch word "koekje" or "koekie." In the United Kingdom, cookies are called biscuits; in Spain, they are called galletas; in Germany, they are called kels; and in Italy, they are called biscotti. The joyofbaking.com website says that every country has its favorite cookie: in North America, it's the chocolate chip; in the United Kingdom, it's shortbread; in France, it's sablés and macaroons; and in Italy, it's biscotti.

Cookies and small cakes were popularized by the English ritual of afternoon tea, a practice that did not exist before the 19th century. Until then, lunch was eaten quite early in the day, whereas dinner was served at 8 or 9 p.m. Then, about 1830, Anna, the Duchess of Bedford, asked for tea and light refreshments (now cookies and small cakes) in her room one afternoon. She started inviting friends and soon elegant tea parties were fashionable and serving afternoon tea became a common custom.

There are hundreds of different cookie recipes, and this enormous variety is possible because a wide range of flavoring agents may be added. Chocolate chips, nuts, coconut, fruit, marshmallows, peanut butter, and many other ingredients find their way into cookies. The seemingly endless assortment of cookies (see Figure 23-11) makes it difficult to categorize them, and not all fit neatly into one classification. In general, however, the fluidity of the batter or dough determines which

**FIGURE 23-11**   A selection of cookies.

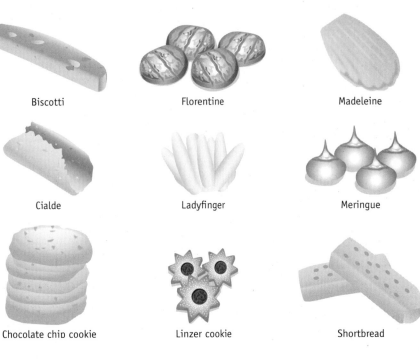

Biscotti

Florentine

Madeleine

Cialde

Ladyfinger

Meringue

Chocolate chip cookie

Linzer cookie

Shortbread

of the following six categories—bar, dropped, pressed, molded, rolled, or icebox/refrigerator—cookies fall into.

## Bar Cookies

The most fluid of cookie batters is used to make bar cookies such as brownies. The batter is baked in a pan instead of on a baking sheet and cut into individual pieces, or bars, with a knife or spatula.

**Brownies**   A rich, chocolate cake named after its color. Brownies may or may not contain nuts.

**Lemon Bars**   Lemon bars are bar cookies that have a sweet, rich, tart, lemony custard top on a shortbread base.

## Dropped Cookies

Dropped cookie batter is literally dropped from a spoon or portion control scoop onto a baking sheet. The batter contains just enough flour so the cookie will not spread out like a pancake when it is dropped onto the baking sheet. Figure 23-12 shows the desired spread of cookies during baking. Both spread and surface cracking are used to determine the baking quality of cookies (19). Chocolate chip, oatmeal-raisin, and meringue cookies are examples of dropped cookies.

**_Chocolate Chip Cookies_**   One of the most popular cookies in North America. According to Nestlé's website, this famous cookie was created in the early 1930s by Ruth Wakefield, owner of the Toll House Inn in Massachusetts. The generally accepted story is that she created incredible desserts that drew people to the inn. She was making a batch of "Butter Drop Do cookies" (a favorite Colonial recipe) with a Nestlé Semi-Sweet Chocolate bar cut up into small pieces, but they never melted into the batter. Ruth's recipe was published in newspapers, and the cookie's popularity grew. She reached an agreement with Nestlé, Inc. that allowed them to print the Toll House Cookie recipe on the wrapper of their Semi-Sweet Chocolate Bar in exchange for being supplied with all of the chocolate she could use to make her cookies. After 1939, Nestlé introduced tiny pieces of chocolate in convenient, ready-to-use packages.

**_Cialde (chee-al-day)_**   Italian, anise-flavored cookies. Cialde can be eaten plain or stuffed with fillings (usually fruit).

**FIGURE 23-12**   Desired spread of a dropped cookie.

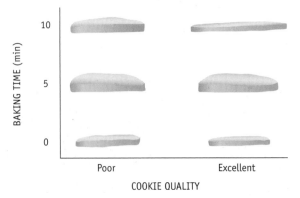

***Florentines***   Round, wafer-thin cookies made from nuts and candied fruit that is coated with a sugar syrup. This chewy cookie is coated with chocolate on one side.

***Fortune Cookies***   Invented by an Asian living in California soon after 1900, fortune cookies are plain, thin, crisp wafer cookies. While still warm, the cookie is folded around a small strip of paper, which has a "fortune" printed on it. Cooling makes the cookie turn crisp and hard, so the only way to retrieve the "fortune" is to break the cookie open.

***Madeleines***   Special pans with shell-shaped indentations are required to produce these often buttery-lemon flavored cookies that also come in other flavors.

***Meringues (muh-rangs)***   These raindrop-shaped cookies have a delicate structure based on sugared, whipped egg whites (see Chapter 12 on eggs).

***Wafer***   Wafers are sweet, crisp, very thin and flat, dry cookies that may have an indented diagram or picture. They are consumed as is or served with ice cream.

## Pressed Cookies

The flour mixture for pressed cookies is viscous enough to be stuffed into a pastry bag or cookie press and forced out through cookie dies. Examples of pressed cookies include tea cookies, ladyfingers (used to make tiramisu), and coconut macaroons.

***Ladyfingers***   These light, sweet sponge cake cookies look like big, flat fingers due to their whipped egg white content. Ladyfingers are created by piping génoise sponge cake batter into flat strips. These airy cookies can be eaten alone, but are most commonly consumed as part of tiramisu. The different styles of ladyfingers demonstrate how many variations of the same cookie may be available. For making tiramisu, prepackaged cookie-style ladyfingers from Italy are recommended over those from North America, which

are usually sold fresh-baked in supermarkets and are soft and cake-like. The Italian cookie-style ladyfingers will hold their shape when soaked in espresso, whereas those that are fresh-baked will be too soggy to hold their structure in the tiramisu dessert (13).

***Macaroons***   As a close cousin to the meringue, macaroons are mound-shaped cookies often made from shredded coconut, sweetened condensed milk, and almond or vanilla extract. A typical macaroon is lightly browned and crisp on the outside, but soft and chewy on the inside.

***Russian Tea Cookie (Mexican Wedding Cakes)***   These round, mounded cookies are based on a flour–butter–nut dough that is baked and then rolled in powdered sugar.

## Molded Cookies

This dough is heavy enough to be formed or molded into balls, bars, or other shapes before being placed on the baking sheet. Molded and other cookies sometimes have powdered sugar sprinkled on top, which can be applied evenly by using a mesh tea strainer. Peanut butter cookies are an example of molded cookies.

***Almond Cookies***   Chinese cookies are light and delicate with an aroma of almonds derived from almond extract.

***Biscotti***   These Italian cookies are first baked as a loaf and then cut into strips and baked again. The double baking creates a hard texture, making them excellent for serving with coffee or cappuccino. They are often flavored with almonds, anise, or chocolate. Traditionally, biscotti cookies were dipped in a dessert wine after dinner, and because they could be stored for long periods of time, they provided nourishment for seamen, explorers, and soldiers.

***Corico***   Hispanic markets carry these lightly sweetened cookies made with cornmeal.

***Peanut Butter Cookies***   An unknown physician invented peanut butter for a chewing-challenged patient in

1890 by grinding peanuts through a meat grinder. However, George Bayle, Jr. mechanized the process and began selling peanut butter to the public. Peanut butter cookies were credited to George Washington Carver (1864–1943), who came up with more than 100 ways to use peanuts.

## Rolled Cookies

Rolled cookies are made from a slightly heavier dough than molded cookies. The dough is rolled out on a lightly floured board and cut into the desired shape. Using too much flour on the cutting board leads to hard-textured cookies. Powdered sugar may be used in place of flour on the cutting board, but obviously it will make the cookies slightly sweeter. Any leftover dough from the initial cuttings of cookies can be salvaged to make more cookies by rerolling it into a ball, spreading it out, and cutting it again without concern about additional gluten development. The most common type of rolled cookie is the sugar cookie. Shortbread cookies are also rolled.

***Butter Cookies***   Thin, crisp cookies made with butter, flour, and sugar. Butter cookies come in various shapes— circles, ovals, and squares.

***Cannoli Shell***   Italians sometimes stuff these cookies with a ricotta cheese-based sweet filling.

***Gingerbread***   Ginger is the major spice used to make gingerbread cookies, cake, and houses. Cookies are often shaped into gingerbread men. The amount of ginger added to each of these desserts determines their spiciness and whether their color is light or dark.

***Linzer Cookies***   These cookies are uniquely identified by variously shaped hole(s) through which preserves can be viewed. Linzer cookies are white on top because of a fine dusting of confectioners' sugar.

***Mandelbrot***   Although similar to almond-flavored biscotti, these cookies are smaller and softer. They are typically sold in Jewish markets.

*Sablés* "Sablé" is a French word meaning "sand," and these cookies are so named because of their sandy, crumby texture. Fluted edges characteristically distinguish sablés from other short cookies.

*Shortbread* Scotland is home to shortbread, a centuries-old baked product that preceded butter cookies. These rich, dense, crumbly cookies have a melt-in-the-mouth texture because they are made with a 3-2-1 proportion of flour, butter, and sugar. Sometimes ground rice or corn flour is added to further define the delicate texture that crumbles easily in the mouth. "Short" is an old term for "crumbly texture"—hence the name "shortbread." The fat used in shortbread recipes interferes with gluten development and results in "short" strands of gluten that create crumbly cookies. ("Short" is a term that is now more commonly attributed to "shortening," or the fat that contributes to this type of texture, even though authentic shortbread cookies are made with real butter.)

*Sugar* Cookies derived from a sugar-based dough (plus flour, butter, eggs, baking powder, vanilla, and salt) that is rolled and then cut into various designs by cookie cutters. The baking time determines if they are soft and chewy or crisp. Sugar sprinkles often decorate the top.

## Icebox/Refrigerator Cookies

The same kind of dough used for rolled cookies can be formed into a cylinder, wrapped, and placed in the refrigerator to harden. The chilled dough is then sliced into thin discs for baking. The commercially prepared cookie doughs sold in the refrigerator section of the supermarket are of the icebox/refrigerator type.

## Cookies as Functional Foods

While cookies are not generally marketed for their health benefits, some manufacturers have begun to add healthful ingredients in order to market them as functional foods. For example, red palm olein has been added to cookies to increase their vitamin E content (1). Minerals such as magnesium and calcium or multivitamin mixtures can be added to cookies (3). Probiotics, fiber, and omega-3 fatty acids are also used as additives to improve the nutritional profile of baked sweets. Beta-glucan, a natural dietary fiber, can be used to replace shortening in batter in order to reduce fat content (14).

# PREPARATION OF COOKIES
## Ingredients and Mixing Methods

The type of cookie to be prepared determines the mixing method, but for most types the conventional cake method is used. The degree of gluten development is not as important for cookies as it is for cakes, which need a soft, delicate crumb, so all-purpose flour, rather than cake flour, is usually used for cookies. Cake flour is preferable if a puffy, soft cookie is desired, as noted in Table 23-5 and Chemist's Corner 23-4. Because lower-protein flours such as cake flour absorb less water than those that are higher in protein, more water is available for steam

generation, resulting in a more puffed cookie. Also, the chlorination of cake flour tends to inhibit the spread of cookie dough, so flat cookies are less likely to result.

Once the ingredients are chosen based on whether a flat or puffy cookie is desired, they are usually mixed together just until moistened, because development of gluten is not necessary. The cookies are then ready for baking unless chilling is required. Overmixing will cause the cookies to be hard and because of the addition of too much air, which facilitates the formation of a protein foam.

| TABLE 23-5 Change the Ingredients, Change Cookie Character ||
| --- | --- |
| **Ingredient** | **Result** |
| *For puffy, soft, pale cookies:* | |
| Cake flour (low-protein, acidic) | More steam and puff; less browning |
| Shortening (high melting point) | Less spread |
| All brown sugar (hygroscopic, acidic) | Soft and moist; less spread when used with egg |
| Egg | Moisture for puff; less spread with acidic ingredients |
| *For thin, crisp cookies:* | |
| All-purpose flour (high-protein) | Browning |
| Butter (protein) | More spread; browning |
| Baking soda (alkaline) | Browning |
| Corn syrup (glucose) | Browning; crisp |
| White sugar (sucrose) | Crisp |
| No egg | No puff; more spread |

**Source:** Copyright © Shirley O. Corriher. From an article originally published in *Fine Cooking Magazine* 24:74–75, 1998.

**TABLE 23-6   Selected Food Additives in Cookies**

| Additive | Function |
|---|---|
| Calcium lactate | A salt produced when lactic acid reacts with calcium carbonate. Used in foods as baking powder. |
| Carnauba wax | A wax to add cohesion and shine to coatings. |
| Cornstarch | Anticaking agent, drying agent, formulation aid, processing aid, surface-finishing agent. |
| Malic acid | Flavoring. A colorless substance that occurs naturally in foods, e.g., unripe fruit. |
| Potassium sorbate | Preservative. This salt of sorbic acid is used to inhibit molds and yeasts in many foods. |
| Soybean lecithin | Emulsifier. Assists the mixing of lipid-based ingredients with water-soluble ingredients. |
| Whey | A dairy protein by-product used as a binder or extender to provide a uniform texture. |

## Food Additives in Cookies

Commercial cookies often contain food additives added to maintain their quality and consumer appeal. The type of food additive varies according to the product and the manufacturer, but all must be listed on the label. Like those used in cakes, cookie additives can vary widely, but a few are used more frequently than others (Table 23-6.) Note that a few commercial cookie manufacturers do not use additives.

# Baking Cookies

## Type and Treatment of the Pan

Cookie baking sheets are preferred for all except bar cookies. Their low or nonexistent sides allow hot air to circulate and bake the cookies evenly. A shiny top surface and a dull bottom allow even browning. Pans are usually greased for dropped, bar, or rolled cookies, but not for pressed, molded, or icebox/refrigerator cookies. Cookies should be placed far enough apart on the baking sheet to prevent them from touching or flowing together during baking. To prevent spreading, cookie dough should be placed on cool cooking sheets or on a sheet of aluminum foil or parchment. Cookies bake better if the pan is placed in the middle or top rack in the oven with at least 2 inches between the pan and the oven wall. Better air circulation can be achieved with unrimmed cookie sheets that have one or two raised edges. Burned cookie bottoms may be prevented by inserting one pan

into another, leaving a pocket of air between them (17).

## Temperature/Timing

Usually, temperatures ranging between 325°F and 375°F (163°C and 191°C) are used in the baking of cookies. Exceptions include meringue or sponge cookies, which bake at about 225°F/107°C. Higher temperatures help prevent the dough from spreading and facilitate browning, but if the temperature is too high, excessive drying and browning will result. Baking times average between 10 and 30 minutes, depending on the type of cookie.

## Testing for Doneness

Cookies are done when the browning is complete and the centers are

cooked. The easiest and most tasty way to determine doneness is to split a sample cookie open and do a taste test. Once done, cookies should be removed immediately from the pan and placed on a cooling rack. They should have a crisp or chewy texture, a uniform shape, even browning, and good flavor. The flexibility of a cookie often observed after it is taken out of an oven soon disappears, in part due to the sucrose crystallizing out of solution (19). Cookies burned on the bottom can be salvaged by rubbing them on the coarse side of a grater. Further excess browning can be prevented by double-panning, or using two cookies sheets of the same size, one placed on top of the other. Poor characteristics in prepared cookies and their causes are explained in Table 23-7.

## High-Altitude Adjustments

When baking at higher altitudes, a slightly higher temperature may be needed. A decrease in the quantity of baking powder and sugar or more flour may also be necessary.

## Microwave Preparation

Heating time for cookies is so short that using a microwave oven is often impractical and may be detrimental to cookie quality. There are many prepared cookie mixes that come with microwave instructions, however.

**TABLE 23-7   Cookie Problems and Their Causes**

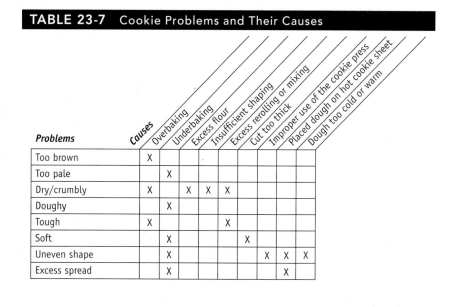

| Problems | Overbaking | Underbaking | Excess flour | Insufficient shaping | Excess rerolling or mixing | Cut too thick | Improper use of the cookie press | Placed dough on hot cookie sheet | Dough too cold or warm |
|---|---|---|---|---|---|---|---|---|---|
| Too brown | X | | | | | | | | |
| Too pale | | X | | | | | | | |
| Dry/crumbly | X | | X | X | X | | | | |
| Doughy | | X | | | | | | | |
| Tough | X | | | | X | | | | |
| Soft | | X | | | | X | | | |
| Uneven shape | | X | | | | | X | X | X |
| Excess spread | | X | | | | | | X | |

## NUTRIENT CONTENT

The high sugar, fat, egg, and flavoring concentrations contribute to the higher calorie and fat content of cookies. Standard cream cookie filling may contain up to 40% fat, though glycerin-based fat-free fillings are available (16).

The carbohydrate (largely sugar) content is of concern to people with diabetes or hypoglycemia, but sugar-free cookies are now available for such people (21). Research continues in the development of a cookie dough containing no added simple sugars (26). One problem in reducing the amount of sucrose in cookies is that it diminishes the bulking properties contributed by sugar. Commercial cookie manufacturers get around this by using bulking agents such as maltodextrins, sugar alcohols, polydextrose, cellulose, and insoluble fiber compounds (8).

## ✳ CALORIE CONTROL

Cookies average 50 calories (kcal) each even though the food industry has marketed many new products that are lower in fat or even fat free (23). Often, when the fat is removed, more sugar is added.

- Regular cookies (2 inches wide) provide 50 calories (kcal) each
- Jumbo cookies (size of 10 regular cookies) provide 500 calories (kcal) each

### TABLE 23-8   Calories in Cookies (average size)

| <50 Calories (kcal) | 50–100 Calories (kcal) | 100+ Calories (kcal) |
|---|---|---|
| Animal cracker (13) | Chocolate sandwich (50) | Biscotti, almond (120) |
| Butter (23) | Chocolate chip (53) | Oatmeal raisin (200) |
| Wafer (26) | Fig bar (56–110) | Brownie, 4" square (224) |
| Ginger snap (29) | Peanut butter (71) | |
| Fortune (30) | Ladyfinger (4×2×1) (87) | |
| Graham cracker (30) | Macaroon, coconut (97) | |
| | Brownie, 2" square (112) | |

© 2010 Amy Brown

# STORAGE OF COOKIES

Airtight containers are best for maintaining cookie freshness. As soon as they are cool, cookies are transferred to a flat dish or plate and covered with plastic wrap or metal foil. They may also be arranged in layers in a covered cookie jar or plastic zipper bag. Some commercially packaged cookies can keep for months. Most cookies (except those with fresh fruit fillings) are ideal for freezing because their relatively low moisture content means fewer ice crystals will form (15).

# PICTORIAL SUMMARY / 23: Cakes and Cookies

The simplest cookies and the most elaborate wedding cakes are really the same basic flour mixture as quick and yeast breads—but made with a higher proportion of sugar, fluid, and fat. And what a difference those additions can make!

## TYPES OF CAKES

**Shortened cakes:** Also called butter or conventional cakes, these are made with fat. Examples are white, yellow, chocolate, and spice cakes.

**Unshortened cakes:** Also called sponge or foam cakes, these are made without fat, and rely on whipped egg-white foam for structure. Angel food and sponge cakes are two examples.

**Chiffon cakes:** These combine the characteristics of both shortened and unshortened cakes. Examples include lemon and chocolate chiffon cakes.

## PREPARATION OF CAKES

Compared to breads, cakes have a higher proportion of sugar, milk, and fat to flour. Structural strength is provided by flour and eggs, with the latter also contributing to color, flavor, and leavening, and acting as emulsifiers. Fats and sugar add tenderness and volume, whereas sugar also contributes to sweetening and browning of the crust. Cakes are leavened by baking soda or powder, air, and/or steam, but a true pound cake is leavened only by air and steam. Flavorings provide variety in cakes.

Nutritionally, an average cake serving yields about 10 grams of fat and 225 calories (kcal). Per slice, angel food cake is the lowest in fat and calories [0.1 gram/143 calories (kcal)], whereas fruitcake (1 × 3 inches) is the highest in calories [15 grams/417 calories (kcal)].

Four considerations in baking cakes are the type and treatment of the pans, timing, temperature, and testing for doneness. Commonly used mixing methods include conventional, conventional sponge, pastry-blend, quick-mix, and muffin mix. Regardless of the type of mixing method used, too much or little stirring can cause problems, and special measures are taken at altitudes higher than 3,000 feet.

## FROSTINGS/ICINGS

Either cooked or uncooked, frostings are spread on baked dessert products to improve their appearance, flavor, texture, and shelf life. The most common types of prepared frostings are:

- Flat (simple)
- Decorating (royal or ornamental)
- Buttercream (Swiss, Italian, French)
- Whipped cream
- Ganache and fondant

## STORAGE OF CAKES

Cakes do not retain their freshness for very long and should be protected from any exposure to air. Iced cakes can be frozen and will keep their quality for up to 3 months, whereas unfrosted cakes can be kept frozen for up to 6 months.

## TYPES OF COOKIES

In general, the fluidity of the batter or dough determines the type of cookie: bar, dropped, pressed, molded, rolled, and icebox/refrigerator cookies. All contain flour, sugar, salt, fat, and liquid. Cookies are usually higher in sugar and fat than cakes, although low-fat and nonfat cookies are available.

## PREPARATION OF COOKIES

The type of cookie determines the mixing method. All-purpose flour is usually used instead of cake flour because the degree of gluten development does not dramatically affect cookies. With the exception of bar cookies, which are baked in a pan, baking sheets are the preferred pan for baking cookies.

## STORAGE OF COOKIES

As with cakes, cookies' exposure to air should be minimized. Airtight containers or wrappings are crucial to maintaining the freshness of home-baked cookies. Most cookies freeze well due to their low moisture content.

# CHAPTER REVIEW AND EXAM PREP

## Multiple Choice*

1. What is the name for a cake made with fat?
   a. Chocolate
   b. Shortened
   c. Unshortened
   d. Flat

2. Which category of cakes combines qualities of both shortened and unshortened cakes?
   a. Leavened
   b. Creamed
   c. Angel food
   d. Chiffon

3. The cream of tartar in angel food cakes serves which of the following purposes?
   a. Leavening agent, strengthens egg-white foam, and acts as a preservative
   b. Strengthens egg-white foam, increases tenderness, and whitens color
   c. Increases tenderness, whitens color, and improves volume
   d. Leavening agent, increases tenderness, and firms egg-white foam

4. Which type of cooking vessel is used for baking bar cookies?
   a. Baking sheet
   b. Baking pan
   c. Skillet
   d. Sauce pan

5. Cake flour is not recommended for making most cookies because it is _____ in protein than hard wheat flour, which reduces cookie spread.
   a. lower
   b. higher
   c. richer
   d. softer

6. Pastry chefs use three types of _____ frosting to fill their delicate pastries.
   a. flat
   b. decorating
   c. buttercream
   d. ganache

*See p. AK-1 for answers to multiple-choice questions.

7. Cakes with a higher _____ content tend to retain their moisture and have longer shelf lives.
   a. butter
   b. sugar
   c. protein
   d. flourd

## Short Answer/Essay

1. Describe the differences among shortened, unshortened, and chiffon cakes.
2. Which ingredients are used in a higher proportion in cakes than they are in breads? What characteristics does each of the ingredients, including cake flour, contribute to the prepared cake product?
3. What types of pans are best for cake preparation (i.e., smooth or rough, dull or shiny, metal or glass, light or heavy gauge)? How should the pans be prepared prior to pouring in the batter?
4. Describe the baking process for cakes, including the filling of the pan, the proper baking temperature and time, and how to test for doneness.
5. What adjustments must be made when preparing cakes at high altitude (over 3,000 feet)?
6. Which ingredients contribute to the tender texture of angel food cake? How do the ingredients and preparation methods for sponge cake differ from those for angel food cake?
7. How does the treatment of pans used to prepare unshortened and shortened cakes differ?
8. List several methods that may be used to deter the staling of cakes and cookies.
9. How does the proportion of ingredients in cookies and cakes differ, and how does this affect starch gelatinization and gluten formation?
10. List the possible causes of the following problems associated with cake preparation: low volume, soggy crumb, and tunnel formation.

# REFERENCES

1. Al-Saqer JM, et al. Developing functional foods using red palm olein. IV. Tocopherols and tocotrienols. *Food Chemistry* 85(4):579–583, 2004.

2. Anonymous. Garnish like a pro. *Fine Cooking* 91:73, 2008.

3. Anonymous. *Market brief. The functional food and nutraceuticals market in Spain.* Agriculture and Agri-Food Canada. www.ats.agr.gc.ca/eur/4275-eng.htm. Accessed 7/09.

4. Armentrout J. From our test kitchen. Buttercream 101. *Fine Cooking* 78:12, 2006.

5. Arunepanlop B, et al. Partial replacement of egg white proteins with whey proteins in angel food cakes. *Journal of Food Science* 61(5):1085–1093, 1996.

6. Baggett N. Chocolate quartet: Classic cakes for an enlightened repertoire. *Eating Well* 4(1):67–73, 1993.

7. Braker F. When it comes to angel food cakes. *Fine Cooking* 22:43–45, 1997.

8. Bullock LM, et al. Replacement of simple sugars in cookie dough. *Food Technology* 46(1):82–86, 1992.

9. Corriher S. For great cakes, get the ratios right. *Fine Cooking* 42:78, 2001.

10. Corriher SO. Mixing matters. when to cream, cut in, whisk, fold, or stir. *Fine Cooking* 61:18, 2004.

11. Dodge AJ. Many cake recipes call for beating eggs into the batter one at a time. *Fine Cooking* 76:20, 2006.

12. Gallagher E. Formulation and nutritional aspects of gluten-free cereal products and infant foods. In EK Arendt and FD Bello. *Gluten-Free Cereal Products and Beverages.* Academic Press, 2008.

13. Giannatempo L. Tiramisù, the ultimate "pick-me-up." *Fine Cooking* 77:63–64, 2006.

14. Kalinga D, and VJ Mishra. Rheological and physical properties of low fat cakes produced by addition of cereal beta-glucan concentrates. *Journal of Food Processing and Preservation* 33(3):384–400, 2009.

15. Klivans E. Freezing cookies. *Fine Cooking* 32:13, 1999.

16. LaBell F. A less-filling filling. *Prepared Foods* 162(12):67, 1993.

17. Middleton S. Pros pick the best baking sheets. *Fine Cooking* 26:55–57, 1998.

18. Mun S, YL Kim, CG Kang, KH Park, JK Shim, and YR Kim. Development of reduced-fat mayonnaise using 4alphaGTase-modified rice starch and xanthan gum. *International Journal of Biological Macromolecules* 1;44(5):400–407, 2009.

19. Penfield MP, and AM Campbell. *Experimental Food Science.* Academic Press, 1990.

20. Revsin L. Old-fashioned cakes with a subtle twist. *Fine Cooking* 43:59–63, 2001.

21. Sangronis E, and M Sancio. Development and characterization of rice bran cookies. *Acta Cientifica Venezolana* 41(3):199–202, 1990.

22. Seeley KE. A piece of cake. *Fine Cooking* 78:64–69, 2006.

23. Sigman-Grant M. Can you have your low-fat cake and eat it too? The role of fat-modified products. *Journal of the American Dietetic Association* 97(7):S76–S81, 1997.

24. Smith P. "Lite" cakes: A matter of formula manipulation. *Bakers Digest* 13:31, 1984.

25. Xanthan gum eliminates problems in microwave cakes. *Food Engineering* 61(9):54, 1989.

26. Zoulias EI, V Oreopoulou, and E Kounalaki. Effect of fat and sugar replacement on cookie properties. *Journal of the Science of Food and Agriculture* 82(14):1637–1644, 2002.

# WEBSITES

View various cakes and cookies at this website:

**www.foodsubs.com** (click on "baked goods," then "cakes" and "cookies")

A professional baking association website providing lots of information under the "resources" tab:

**www.bakingbusiness.com**

Helpful baking tips and information can be found at this website:

**www.joyofbaking.com/cakes.html**

# 24 Pastries and Pies

Pastry is essentially a type of bread product that is characteristically flaky, tender, crisp, and lightly browned. This delicate combination is not always easy to achieve. Unlike many other food preparations, pastries can be intimidating because they are made with precisely measured ingredients, in a time/temperature-sensitive manner, and with an artistic touch. These labor-intensive pie shells and desserts are the true test of a food preparer's skill.

## TYPES OF PASTRY

So many variations of pastries exist that there is no one standard way to classify them (Figure 24-1). The relatively compact texture of piecrusts, the delicate flakes of an apple turnover, strudel that is almost chewy, and the numerous fine sheets of baklava represent just a few examples of pastry varieties. The major differences among pastries result from the mixing method in which the fat is introduced to the flour, the type of fat (butter, lard, shortening, oil, etc.), the major leavening source, the addition of sugar or salt, and the baking technique.

## Nonlaminated and Laminated Pastries

This chapter will focus on the two main types of pastry, which are defined based on whether the fat ingredient is (1) cut into the flour mixture (nonlaminated), or (2) repeatedly folded into the dough through **lamination.**

## Plain Pastry (Nonlaminated)

**Plain pastry** is also known as pie or shortcrust pastry, and, as the name implies, it is used for piecrusts, tarts, tartlets, and galettes. This type of pastry is more compact and crumbly, rather than light and flaky. The chilled fat cut into smaller pieces is cut into or rubbed into the flour and a minimum of cold water is added to bring the dough together. The dough is then chilled to relax the gluten before it is rolled out with a minimum of strokes into a circular shape to fit into a pie pan. The resulting pies are defined by their fillings and can include, but are not limited to, those listed in Table 24-1.

**Lamination** The arrangement of alternating layers of fat and flour in rolled pastry dough. During baking, the fat melts and leaves empty spaces for steam to lift the layers of flour, resulting in a flaky pastry.

**Plain pastry** Pastry made for producing piecrusts, quiches, and main-dish pies.

498

**FIGURE 24-1** An assortment of puff pastries.

Richard Brewer/Digital Development Services

| TABLE 24-1 | Types of Pies | | | | |
|---|---|---|---|---|
| **Dessert** | **Tart*** | **Tartlet*** | **Galette** | **Main Dish** |
| Pies are baked in sloping pie pans. **Diameter:** 8 or 9 inches | European cousins of the dessert pies. Tarts are open-faced pies baked in a short-sided, fluted-edged pan with a removable bottom. **Large tarts:** 9 or 10 inches **Small tarts:** 3½–4 inches | Single-serving versions of tarts: **Miniature tartlets:** 2–2½ inches **Miniature barquettes:** 3 inches | Free-form pies made without pans. The dough is rolled out like a large cookie; the fruit fills the center, but a 2-inch border is left that is then pleated in folds over the fruit, which shows through the center. | Deep dishes are lined with pie dough to encapsulate hearty ingredients. The pie is served to accompany a meal or as an entire meal. |

|  Possible Fillings | |
|---|---|
| **Dessert / Tart / Tartlet / Galette** | **Main Dish** |
| Fruit<br>• Apple<br>• Apricot<br>• Blueberry<br>• Cherry<br>• Key lime<br>• Lemon<br>• Peach<br>• Plum<br>• Raspberry<br>• Rhubarb<br>• Strawberry          Vegetable<br>  • Pumpkin<br>  • Sweet potato<br>Cream<br>  • Banana<br>  • Chocolate<br>  • Coconut<br>Custard<br>Chiffon<br>Ice cream<br>Meringue<br>Pecan | Fish<br>Mincemeat<br>Potpies<br>Potato<br>Quiches<br>Shepard's |

*Tarts and tartlets differ from pies in that they are freestanding on the plate.

### Hot-Water Crust Pastry (Nonlaminated)

This pie pastry serves as the base for potpies filled with meat, beans, and/or vegetables. The less flaky consistency of this pastry results because lard is melted in boiling water and poured into the flour to create a mixture that is then kneaded into a dough. The dough is then molded around the container that will be filled with meats and other ingredients before being baked.

### Short or Sweet Dough Pastry (Nonlaminated)

More sugar goes into this pastry dough to create a sweeter flavor and more crumbly crust. Tartlets and some crisp cookies (ginger snaps) are made with short or sweet dough.

## Brioche Pastry (Nonlaminated)

Brioche (bree-osh) pastries are made from a sweet yeast dough that is not laminated. It is commonly used to make airy buns and loafs.

# Choux Pastry (Nonlaminated)

The dough used to make choux pastries is called pâte à choux (paht ah shoo). Another name for the dough is choux paste because it is more like a thick paste than a dough. The most common pastries made from this dough include cream puffs, profiteroles (little cream puffs), and éclairs. A distinctive method is used to prepare this dough: 14 whole eggs are beaten into 1 pound of flour mixture that is then cooked until it forms a smooth paste. The thick fluid paste cannot be folded so it is poured or piped unto baking sheets that are then placed in a hot oven. Unlike puff pastry that uses fat to "puff" up in volume, pâte à choux pastry relies on its high water content from the eggs to steam open the pastry into a round or oblong shape (steam leavened). The steam opens up the mounds of paste, creating a cream puff look that is similar to a cabbage head; in fact, pâte à choux is French for "cabbage paste" (8). The hollow (cleaned out) center is filled with a whipped cream or other types of filling and sometimes topped with powdered sugar or chocolate **glaze.**

# Puff Pastry (Laminated)

**Puff pastry** and other similar flaky types of pastry also differ from the pie variety in that they are very light, airy, and flaky. The dough contains no yeast or chemical leaveners, but it can increase up to eight times its original size. This is equivalent to a one fourth inch of pastry expanding to 2 inches. This increase in size is due to leavening resulting from the moisture and air expanding in the heat of a hot oven. The unique flakiness is also achieved through a laminated dough obtained by folding fat into the

dough. Lamination is a series of folding, rolling, and turning manipulations that can create up to as many as 1,000 alternating layers of fat and a non-yeast dough (1). The necessity of folding many layers together, even when there are considerably fewer than 1,000, makes puff pastry very time-consuming to produce. As a result, puff pastry can be purchased ready-made in the freezer section of some stores. Despite the time involved, puff pastry remains popular and is primarily found in desserts such as patty shells, tarts, and cream horns.

Patty shells are small, cup-shaped shells usually made from puff pastry that can be filled with creamed dishes of meat, poultry, fish, or vegetables. Their versatility allows them to be served as hors d'ouevres, entrees, or desserts. If they are going to hold desserts such as ice cream, puddings, glazed fruit, and other sweet fillings, they are usually made from a sweet dough that will hold its shape better. Bakeries often sell fresh patty shells, while supermarkets sell the frozen, unbaked versions.

## Quick (Blitz) Pastry (Laminated)

Quick or blitz pastry (blitz is German for "quick attack") is a type of puff pastry that is quicker and easier to prepare than the regular form. Although faster to make than the time-consuming puff pastry, it does not rise as high. Quick pastry combines the mixing technique of plain pastry with the rolling and folding technique of puff pastry. Examples of quick pastries are Napoleons and tart shells. Cream-filled pastries such as Napoleons are often referred to as French pastries. The word strudel refers to a Central European pastry believed to have originated in Austria. Strudel, a type of puff pastry, has an elastic nature that is derived from the flour's high gluten content, minimal fat content, and lack of sugar.

## Phyllo (Filo) Pastry (Laminated)

The word phyllo (which can also be spelled filo or fillo; pronounced fee-low) means "leaf" in Greek and describes the characteristic tissue-thin sheets of dough that are made primarily with flour, water, and a small amount of oil. The numerous sheets are brushed with butter between layers that can then be rolled or folded into a variety of shapes. The Greeks and people from Middle Eastern cultures learned about preparing phyllo pastries from the Turks, who were in their region

during the 400-year rule of the Ottoman Empire. As a result, similar phyllo pastries that are called by various names exist within the region and beyond. The type of filling dictates what each of these phyllo pastries is called:

- *Baklava:* Honey and nuts
- *Patatopita:* Potatoes
- *Spanakopita:* Spinach and feta cheese (Ispanakborek in Turkey)
- *Tiropita:* Cheese (Peynir borek in Turkey; Burekas in Israel; Gibanica in Serbia)

## Croissant Pastry (Laminated)

This puff pastry obtains its flaky consistency from a laminated yeast dough. The pastry's classic crescent shape originates from a Turkish legend.

## Danish Pastry (Laminated)

The centers of this type of yeast-leavened puff pastry often contain sweet fillings such as jam, custard, or cream cheese, and the pastry is usually covered with sugary toppings. Examples of Danish pastries include cheese pockets, envelopes, pinwheels, and turnovers.

# PREPARATION OF PASTRY

Pastries are the most delicate of all baked products, and their success depends on the right proportions of ingredients and the correct preparation technique. It takes skill to correctly distribute the fat and develop the gluten to the point of creating a crust that is flaky, tender, and crisp. Each type of pastry, especially the various types of puff pastries, has its own ingredient requirements and unique instructions for mixing, rolling, filling, and baking, but the general guidelines are discussed below.

## Ingredients of Pastry

Most pastry flour mixtures contain at least four ingredients: flour, fat, liquid, and salt. Eggs and sugar, the latter sometimes added for its flavor and browning properties, are optional. Except for croissant, Danish, and brioche doughs, which make use of yeast, the leavening agents for most pastry doughs are steam and air. The type and quantity of each of these ingredients determine the final quality of the pastry product. Table 24-2 compares

---

**Glaze**   A sugar-coated icing poured over pies or pastries that hardens to provide flavor and structure. The word *glaze* is used both for pastries and soup stocks, but they have different meanings.

**Puff pastry**   A delicate pastry that puffs up in size during baking because of many alternating layers of fat and flour.

**TABLE 24-2** Pastry Dough Formulas Compared to Bread Dough (Based on 1 Pound of Flour)

| | Basic Bread Dough | Nonlaminated Doughs | | | | Laminated Doughs | | | |
|---|---|---|---|---|---|---|---|---|---|
| | | Pie Dough | Short Dough | Brioche Dough | Pâte à Choux | Quick Puff Pastry | Puff Pastry | Croissant Dough | Danish Dough |
| Flour | 1 lb (455 g) | 1 lb (455 g) | 1 lb (455 g) | 1 lb (455 g) | 1 lb (455 g) | 1 lb (455 g) | 1 lb (455 g) | 1 lb (455 g) | 1 lb (455 g) |
| Butter (Margarine in Danish) | 2 oz (55 g) | 10 oz (285 g) | 14 oz (400 g) | 5 oz (140 g) | 6½ oz (185 g) | 1 lb (455 g) | 1 lb (455 g) | 11½ oz (325 g) | 17 oz (485 g) |
| Yeast | 1 oz (30 g) | 0 | 0 | 1 oz (37 g) | 0 | 0 | 0 | 1¼ oz (30 g) | 1½ oz (40 g) |
| Sugar | 1 oz (30 g) | 0 | 6 oz (170 g) | 2½ oz (70 g) | 0 | 0 | 0 | 7 tsp (35 g) | 1½ oz (40 g) |
| Eggs | 0 | 0 | 1 | 4 | 14 | 0 | 0 | 0 | 2 |
| Water | 0 | 3 oz (90 mL) | 0 | 0 | 0 | 3⅓ C (800 mL) | 3 oz (90 mL) | 6½ oz (195 mL) | 0 |
| Milk | 8 oz (240 mL) | 0 | 0 | 1½ oz (45 mL) | 0 | 0 | 0 | 9 oz (270 mL) | 6½ oz (195 mL) |
| Salt | 2¼ tsp (11.25 g) | 2½ tsp (12.5 g) | 0 | 2 tsp (10 g) | 1¼ tsp (6.25 g) | 2½ tsp (12.5 g) | 2½ tsp (12.5 g) | 2½ tsp (12.5 g) | ⅔ tsp (3 g) |
| Other | 0 | 3 oz (85 g) lard | 1 tsp (5 mL) vanilla extract | 0 | 0 | 0 | 1½ tsp (7.5 mL) lemon juice | Few drops lemon juice | ⅔ tsp cardamom |
| Mixing/ Production Method | Dissolve the yeast in warm milk. Form the dough by combining the flour, sugar, and salt. Mix in the butter. The dough is kneaded, proofed, formed, and possibly proofed again. | Combine flour and salt. Add butter and lard into flour mixture by cutting them into marble-sized pieces. Add water and mix just until the dough holds together. | Mix for a few minutes at low speed: butter (soft), sugar, egg, and vanilla. Mix in flour, but until just incorporated. | Dissolve the yeast in warm milk. Add the sugar, eggs, and salt. Add the flour. Add the butter. The dough is kneaded before being proofed in the refrigerator, shaped, and proofed again. | Boil a mixture of the water, butter, and salt. Stir in the flour and cook the roux for a few minutes. Cool slightly before adding the eggs. | Combine flour and salt. Cut butter into large pieces and add to flour mixture without thorough mixing. Add water just until dough holds together. | Form butter block by mixing together most of the butter, ⅓ of the flour, and lemon juice into a block. Form dough by mixing the remaining ingredients. | Form butter block by mixing together butter, some flour, and lemon juice. Form the dough by dissolving the yeast in milk and kneading in the remaining ingredients with the dough hook. | Form margarine block. Form the dough by dissolving the yeast in the cold milk and eggs and adding the remaining ingredients. |

the ingredients of these pastry doughs against each other and against bread dough. Pie and pastry doughs contain more fat than other baked products. In fact, they include 50% as much fat as they do flour, by weight, and may contain as much as an equal amount of fat and flour. Compare this to bread doughs, which usually contain only about 12% fat. Although the basic ingredients of pastries do not vary much, a wide assortment may be prepared, partially because the flour mixtures can be handled so many different ways.

Among all the doughs listed in Table 24-2, short dough is most commonly used in the preparation of pastries (8). The term *short* describes the tenderness of the resulting baked product. Shortbread cookies are very tender, and give way very easily when eaten. The tenderness of shortbread cookies is achieved by using a lot of sugar and shortening to interfere with gluten development.

Regardless of the dough, the ingredients in a flour mixture are either tenderizers or tougheners (Table 24-3). Each

of these ingredients is now discussed in more detail with regard to their ability to tenderize or toughen pastry.

**TABLE 24-3** Tenderizers and Tougheners in Flour Mixtures

| Tenderizers | Tougheners |
|---|---|
| Acid | Egg white or whole egg |
| Egg yolk | Flour |
| Fat | Milk |
| Sugar | Water |

## Flour

Pastry quality depends heavily on the type of flour used, its amount, and how it is handled. Pastry flour and the unbleached all-purpose variety are popular. There is less protein in pastry flour, and professional bakers prefer it, but the easy availability of all-purpose flour makes it more popular with the general public. Cake flour, which is even lower in protein than pastry flour, is sometimes used. If all-purpose flour is used, however, the gluten formation increases, which may result in a tougher pastry, so additional fat is sometimes necessary to increase tenderness.

Too much of any type of flour will toughen pastry, because one of the major influences on tenderness is the concentration and distribution of gluten. Excess gluten formation toughens pastry. The flakiness of pastry is attributed in part to its limited gluten formation achieved by using a very small amount of chilled water, coating the flour with fat, and chilling the dough briefly before handling. Other techniques that can contribute to tenderness include adding acid in the form of unflavored vinegar, lemon juice, yogurt, or sour cream to break apart the long gluten strands; handling the dough minimally (the warmth of the hands can melt the fat); and using as little water as possible because it facilitates gluten formation (4, 16).

## Fat

The proportion of fat is probably the most important determinant of quality in pastry, especially in creating flakiness. Large amounts of fat are required to produce a flaky crust (Figure 24-2), with the proportion varying depending on the type of pastry. The specific factors influencing flakiness and tenderness are discussed below.

*Flakiness*   Fat—the size of its particles, its firmness, and how evenly it is spread—is the major contributor to flakiness. In the baking industry, flake consistency is described as long flake (very flaky), short flake (medium flaky), or mealy crust (no flake). Flake size is controlled by fat particle size. When fat is cut in small, cold pieces that are incorporated into the dough without being creamed or absorbed by the flour, the fat will then melt during baking, and leave empty spaces where steam may collect to leaven and lift the layers of dough. These "pockets" vacated

**FIGURE 24-2**   The role of fat in producing flaky pastry.

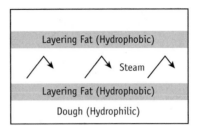

- Water in the dough turns to steam upon baking

- The layering fat creates an impervious layer

- The steam stays inside each dough layer, forcing it to expand because of the pressure it develops underneath each impervious fat layer

- Fat melts and is absorbed by the dough, improving the eating properties of the pastry

**FIGURE 24-3**   Pastry flakiness.

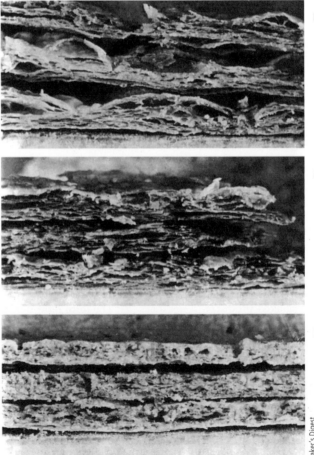

Flakiest

Moderately flaky

Least flaky

Baker's Digest

by fat create the characteristic flakiness of pastry (Figure 24-3). The more pockets there are in the pastry (sometimes created by many layers), the more flaky the pastry will be.

To maximize flakiness, keep everything cold—fat, flour, water, countertop (a cold marble pastry board is an ideal surface), and hands (use fingertips and minimize contact). Sometimes the dough is placed in the refrigerator between rollings. The process starts with cold fat that is cut into chilled flour to form a fat/flour mixture. Keeping the fat cold increases the flakiness of pastry in two ways: (1) less fat is absorbed by the flour, and (2) more pea-size balls of fat are dispersed and surrounded with flour to become pockets of air during baking (Figure 24-4). The fat is added to the flour before any liquid to protect the flour and its proteins from water, which would increase the formation of gluten (4).

The type of fat used in making the pastry also affects flakiness (see Chapter 22 for descriptions of the various fats). Firmer, plastic, 100% fats such as

**FIGURE 24-4**  How fat is added influences pastry flakiness.

Tender pastry results when fat (such as oil) mixes thoroughly with the flour.

Flaky pastry is achieved when cold fat (such as butter) is cut into pea-size balls and lightly combined with the flour.

hydrogenated shortenings and lard produce the flakiest pastries, although lard is being used less often because of its tendency to go rancid and its high saturated fat content. Shortening is softer and more pliable than cold butter, so it coats the flour more easily and can be rolled out even when refrigerated. Butter and margarine may be used in pastry making, but their water content causes increased gluten formation. Butters vary in water content, with the best high-quality butters having the lowest water content, making them the number-one choice for the finest pastries. Because of butter's desirable flavor, it is often selected to prepare pastries; unlike dough containing margarine, however, when a butter-containing dough is refrigerated, it tends to harden and requires more effort to roll out (5). The food industry has combined the positive aspects of both butter and hydrogenated shortenings by producing a butter-flavored hydrogenated shortening. Oil is the least desirable fat for making flaky pastry because it coats each flour particle, resulting in an extremely tender but mealy (grainy) texture (12).

***Tenderness***  Tenderness differs from flakiness, and in fact, factors that contribute to one may detract from the other (Table 24-4). Tenderness is described as the ease with which pastry gives way to the tooth. Fat is a major ingredient in pastry production because it increases tenderness through the inhibition of gluten development. This occurs when fat coats the flour. To achieve maximum tenderness in pastry, the fat must be thoroughly combined with the flour. For most pastries, including piecrusts, the ideal is a combination of tenderness and flakiness achieved by having some fat absorbed by the flour and leaving some

fat in pea-size pieces to melt and let off steam during baking. Tough, dry, or flat-flavored pastry may result from excess gluten formation caused by not cutting enough fat into the flour; by using more flour than necessary, especially during rolling; by adding too much water; or by excessive manipulation of the dough. It is also important to work quickly: the greater the amount of time allowed between adding water to the dough and baking the pastry, the less tender the pastry will be, because the gluten will have more opportunity to be hydrated by the water (13).

On the other hand, too little gluten development results in a pastry that is too tender and crumbly. This occurs when fat is cut into pieces that are too small; when oil, rather than shortening, is used as the fat; when too little water is added; or when the dough is undermanipulated. It may also happen when conditions are too warm during handling, causing fat to melt and coat the flour, inhibiting gluten development. When using butter, the goal is to work

quickly and to put the bowl back in the freezer for about 5 minutes whenever the butter starts to soften (4).

***Fat Increases Calories***  With the delicious tenderness and flakiness provided by fat comes a high fat-gram and calorie count. Pies, for example, average about 14 grams of fat and 300 calories (kcal) per serving (an eighth of a 9-inch pie). Among pies, some of the lowest and highest in terms of fat and calories (kcal) are strawberry pie [9 fat grams/228 calories (kcal)] and pecan pie [22 fat grams/541 calories (kcal)].

### Liquid

The liquid (usually water or milk) component of pastry dough is important for leavening, hydration, and the crispiness of the crust. Pastry is leavened by steam, so liquid is necessary for that purpose. Liquid is also needed to hydrate the proteins so gluten can develop, and to dissolve the salt. The pastry structure is set primarily by the coagulation of flour proteins during baking. As little liquid as possible should be added when making pastry, because too much water will cause shrinkage and a tougher crust from excess gluten development (13). The effect on gluten formation means that liquids tend to make pie and pastries more tough; however, too little water results in a crumbly crust.

More liquid may be needed at higher altitudes where water evaporates more quickly. Vinegar or lemon juice may be added to the cold liquid, on the principle that acidic ingredients contribute to more tender pies and pastries because acid helps to inhibit gluten formation (5).

**TABLE 24-4**  Making a Tender, Flaky Pastry

| To Make *Tender* Pastry | Why |
| --- | --- |
| Blend soft fat into the flour before adding any liquid. | Fat coats the proteins and prevents them from forming gluten. |
| Instead of water, use an ingredient that is part fat, such as sour cream, cream, or egg yolks. | Gluten can't form without water, and the additional fat contributes to tenderness. |
| Add acid to the dough in the form of lemon juice, vinegar, or sour cream. | Acid breaks long gluten strands. |

| To Make *Flaky* Pastry | Why |
| --- | --- |
| Keep the fat cold and in large pieces (pea-size). | Large, cold pieces will remain firm in the oven long enough to create flakes. |
| Flatten large pieces of cold fat. | Chunky pieces will make holes in the crust rather than act as spacers. |

**Source:** Corriher S. The secrets of tender, flaky piecrust. *Fine Cooking* 17:78–79, 1996.

## CALORIE CONTROL
### Pastries and Pies

Pastries contain more fat than cakes, so they are usually higher in calories as well, ranging from 143 calories (kcal) per slice for coconut cream pie to 541 per slice for pecan pie that is topped with nuts (see Table 24-5).

- **Control the Number of Servings.** Avoid seconds. If 1 slice of pie (⅛ of 8")

provides 300 calories (kcal; without whipped cream or ice cream), then 2 slices provide 600 calories (kcal). Be aware that just half a pie (4 slices) holds 1,200 calories (kcal).

- **Control Portion Sizes.** Cut a normal pie slice in half to reduce the calories (kcal) to 150.

| TABLE 24-5  Calories and Fat in Pies[1] and Pastries (calories/fat grams) | | |
|---|---|---|
| **<200 Calories** | **200–300 Calories** | **300+ Calories** |
| Fruit turnover (83/4) | Toaster pastry (204/5) | Lemon meringue pie (302/10) |
| Coconut cream (143/8) | Cream puff, 3 1/2 × 2" (238/17) | Cherry pie (304/13) |
| Fruit strudel (195/8) | Chocolate mousse pie (247/15) | Banana cream pie (308/16) |
| | Apple fritters, 2 (260/11) | Chocolate cream pie (343/22) |
| | Eclair, 5 × 2 × 3/4" (262/16) | Pecan pie (541/22) |
| | Coconut custard pie (270/14) | |
| | Blueberry pie (271/12) | |
| | Apple pie (277/13) | |

[1]⅛ of 8 inch pie unless otherwise specified

© 2010 Amy Brown

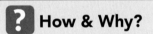

## How & Why?

### *Why is pastry so crisp?*

A crisp crust is created, in part, by water evaporation. In addition, the very little water that is used in pastry preparation inhibits starch gelatinization, which would make the pastry less crisp (15). Other factors influencing crispness are baking time and temperature; thickness of the rolled dough; whether the rolled dough is baked immediately or refrigerated or frozen first; and the moisture content of pie fillings.

## Eggs

Egg yolks are high in fat, so they tend to tenderize pastries. On the other hand, whole eggs or their whites will make pastries tougher by acting as binders or by adding more liquid, so they should be used in moderation. Eggs also add color, flavor, and richness to pastry dough. Sometimes the egg yolk alone is used because the water and protein content of the white contributes to toughness. Egg yolk can also be mixed with a minute amount of water and then

brushed over the pastry to produce a golden crust.

## Salt

In pastry, the only function of salt is to add flavor. It can be omitted, but the crust will lack flavor. Sometimes sugar is added for the same reason, and it has the additional benefit of browning the crust.

## Sugar

Sugar sweetens the pie or pastry flour mixture, but it is not always added. If added, it acts as a tenderizer by competing with the flour for water, which prevents gluten formation.

## Flavorings

Several different flavorings give pies and pastries their unique tastes that are sometimes difficult to duplicate. Some examples include alcohol, essences, extracts, and flavored oils. The use of vanilla extract is well known, but extracts of lemon, orange, pineapple, and other fruits also exist. Flavored oils are also available for some of these same fruits and others, such as coconut-flavored oil for making coconut cream pie.

*Alcohol*   Many foods can have their flavors improved through the addition

of wine, liqueurs, and spirits (see Chapter 27 on beverages). Liqueurs are especially flavorful because these are alcohols sweetened with sugar. A common example of an alcohol-based flavoring is amaretto, an almond-flavored liqueur used in the manufacture of macaroons.

*Essences*   A few drops of these aromatic flavors distilled from fruits, spices, and other plants are enough to change a food's flavor profile.

*Extracts*   Extracts are basically essences that have been diluted in alcohol because the latter acts as a preservative. Common extracts include almond, peppermint, and vanilla. Using some of these extracts to flavor alcohol results in schnapps, a general term used to describe a flavored alcoholic beverage. Schnapps can then be used to flavor foods or drinks.

*Flavored Oils*   Essences diluted in oils include those obtained from various fruits (apple, cherry, grapefruit, lemon, lime, orange, tangerine, etc.). They are more potent in their flavor than extracts that are alcohol-based. Nuts can also have their flavors removed by cold-pressing. However, heat will destroy the flavor of nut oils, so it is best to use them in cold dishes or after heating is complete.

## Thickeners (for Fillings)

Pies get their dense or viscous quality through thickeners. These thickeners are often starch-based, but non-starch-based thickeners include egg yolks, cream, gums or other fibers, and gelatin. Starch-based thickeners are discussed in Chapter 18 on starches and sauces because they are also used to thicken soups, stews, sauces, puddings, and gravies. Some of the common starches used to thicken pies—arrowroot, cornstarch, ClearJel®, and tapioca—are now explained in further detail.

*Arrowroot*   The low boiling point of this high-gloss starch makes it undesirable for fruit fillings because it will thicken before the fruit boils.

*Cornstarch*   This common low-gloss thickener made from corn is used to thicken fruit fillings as long as they are not acidic. Low-pH ingredients interfere with the ability of cornstarch to gel. As a result, lemon, lime, or similar acidic foods are not added to pie filling until after it has completed cooking. Another drawback to using cornstarch is that it might leave a "starchy-tasting" residue.

***ClearJel®*** Commercial bakers frequently use this modified cornstarch to thicken fruit pie fillings. This chemically altered cornstarch is superior to regular cornstarch because it resists breakdown from acidic ingredients such as citrus fruits, is stable to high temperatures, and reduces the tendency of pie fillings to "weep" during storage.

***Tapioca*** Two types of high-gloss tapioca are available to thicken pies and pastry ingredients—quick-cooking (pearl) and powder. The quick-cooking tapioca, which can also be ground into a powder, is usually used in two-crust pies because it does not always completely dissolve and looks like little "eyes." Tapioca powder is a flavorless thickener that holds fruit fillings together.

***Adding Thickeners*** The above starches thicken only when heated. The process begins by slowly stirring cold liquid into the fine white starch powder (about 2 tablespoons for a pie) until both are thoroughly mixed into a paste. The paste is combined with sugar and then either (1) combined with the fruit, where it will be heated in the pie or pastry during baking, or (2) whisked into a hot liquid and heated to gelatinize the starch. Heating in hot liquid should not exceed 1 minute or else the starch granules will rupture. As a result, the starch will lose its thickening power and may even become thin again. It may also lump, as it would if the hot liquid were added to the starch.

# Mixing

One of the main differences among pastry doughs is how the fat is introduced into the flour mixture. Plain or pie pastry relies on the fat of pea-size balls being worked into the flour, whereas puff and other types of flaky pastry depend on the process of lamination, in which numerous folds and turns create layers of fat within the dough. Potpie flour mixtures are based on melted fat being worked into the flour.

## Mixing Plain (Pie) Pastry

The classic pastry method is used to mix the ingredients of plain pastry. Flour and salt are first sifted together and then chilled. Cold fat is cut into the chilled flour and salt mixture using a pastry blender or forks, or by crisscrossing two knives together until the particles of the mixture are reduced to the size of peas (Figure 24-5). Large chunks of butter should not remain in the mix (2). This can also be achieved in a food processor. The loose mixture is then bound together using as little water and manipulation as possible. Cold or ice water is sprinkled evenly, one tablespoon at a time, over the flour. A spray bottle of chilled water achieves the same results. After each addition, the mixture is tossed lightly with a fork or pastry blender until the flour is just moistened (Figure 24-6). Only just enough water is used to moisten all the flour and make the particles hold together. Overmixing or adding water beyond the "just moistened" stage results in a tough pastry. When the dough no longer clings to the side of the bowl, it is pressed lightly into a flat disc, wrapped in waxed paper or plastic wrap, and refrigerated for about 15 minutes (overnight is best, but time-consuming) to chill the fat. It is then ready for rolling. It is best to avoid overworking the dough during mixing or rolling because it elongates the gluten strands, creating a chewy pastry instead of one that crumbles lightly.

## Mixing Puff Pastry

There are two separate mixings in the preparation of puff pastry. The first mixing is for the fat component, or butter block, consisting of fat, flour, salt, and perhaps an acid. These ingredients are mixed together by hand and shaped into a ¼-inch thick, 12-inch-square block. The second mixing is for the dough, made from flour, salt, water, and often a little fat. Puff pastry usually contains both cake and all-purpose flours because its dough is manipulated more than that of plain pastry, and the inclusion of cake flour lessens the likelihood that excessive gluten will form. The flours are sifted together and chunks of butter, if used, are cut into the flour with the fingertips until the mixture forms particles the size of coarse crumbs. An electric mixer might be used as an alternative to the hands. The mixture is formed into a mound with a well in the center into which the water and salt are poured. The flour or flour–fat combination is integrated into the water with both hands, and more water is added, if needed, to make a dough that is sticky but manageable. Both the butter block and the dough are refrigerated for 30 minutes. Refrigeration keeps the fat firm and prevents it from being absorbed by the flour. After refrigeration, the dough and the butter block are ready to be folded and rolled together.

**FIGURE 24-5** Fat cut into pieces of the correct size.

Fat is too lumpy.

Fat is just right.

Fat is too fine.

**FIGURE 24-6**    Mixing pastry ingredients.

1. The flour and salt are mixed together with a pastry blender and the fat is cut into the size of tiny peas.

2. Water is sprinkled into the flour, a tablespoon at a time. A fork is used to mix it with the flour until all the flour is moistened.

3. The dough is mixed thoroughly until the sides of the bowl are clean, indicating a correct amount of gluten development.

***Cream Puffs and Éclairs***    A variant of this method is used for pâte à choux pastry dough. This flour mixture, used to prepare cream puffs and éclairs, may be referred to as a paste rather than a dough. It has the highest proportion of liquid of any of the pastry doughs, which contributes to the very light weight of the resulting pastries. The large cell formation that occurs, leaving a hollow middle, is produced by steam leavening during baking.

The liquid, usually water, is combined with the fat and salt in a saucepan and heated to boiling (Figure 24-7). As the mixture boils, the flour is added (often all at once) and constantly stirred to keep lumps from forming. The paste is then cooked and stirred for about 3 minutes until it is sufficiently dried out to pull away from the sides of the pan. The paste is then transferred to a bowl, where it must be allowed to cool only slightly before eggs are added, one or two at a time. When the dough is glossy and soft enough, it is placed in a pastry bag, which is used to extrude it onto the baking sheet into small rounds for cream puffs or long, narrow shapes for éclairs. During baking, pâte à choux expands in such a way as to leave a hollow center. This will be filled with whipped cream for cream puffs or cream filling for éclairs. The top is covered with a chocolate syrup. Specific problems in preparing pâte à choux pastry and their causes are listed in Table 24-6.

**FIGURE 24-7**    Preparing a pâte à choux.

1. Pâte à choux in progress

2. Ready to come off the heat

3. Add eggs one at a time

4. Mixing the dough

5. Pipe dough onto sheet pan

**TABLE 24-6**  Choux Pastry Problems and Their Causes

| Problem | Cause |
|---|---|
| Volume of the pastries too small. | Insufficient roux formation; starch insufficiently gelatinized; batter too firm, not enough eggs added. |
| Contours of the pastries are diluted; they expand too much laterally. | Batter too soft, eggs added too quickly, or too many eggs added. |
| Small volume, thick crust, dense cells. | Lack of steam during baking. |

# Rolling

Both plain and puff pastry dough must be rolled, with minimum hand contact, in order to spread the fat and gluten in fine sheets layered on top of each other in the lamination process. Overmanipulating the dough by rolling it too much, too hard, or too often will decrease the flakiness, tenderness, and crispness of the pastry.

## Chilling the Dough

The first step in rolling pastry dough is to let the dough chill in the refrigerator for a set amount of time (minutes, hours, or days) to make it easier to handle and to keep the fat from melting into the flour. Properly wrapped dough can be refrigerated up to 4 days or be frozen for up to 6 months (16). Chilling the dough in the refrigerator allows the flour more time to rehydrate and gives the gluten strands an opportunity to relax so that during baking they can expand at the same rate as the gases. Gluten is like a rubber band in that it wants to snap back after being rolled or stretched out, so refrigeration relaxes the dough and makes it easier to manipulate. Once it is taken out of the refrigerator, the cooled dough is allowed to sit at room temperature until it is malleable enough to roll.

## Rolling Surface

A cold surface is best for rolling out the dough, which is why marble rolling boards are often used. Any other surface can be cooled by placing an ice water–filled roasting pan on it for a few minutes and then drying it well (16). Dough tends to stick if the surface or fat ingredients are not cold enough. Too much liquid also causes a sticky dough. Some of the surface moisture on a dough is eliminated by sprinkling the rolling surface lightly with flour. Another option is to rub a minimal amount of flour into a pastry cloth covering a cutting board. The rolling pin can also be covered in cloth and/or lightly floured to prevent sticking, but the amount of added flour should be kept to a minimum.

## General Guidelines

Next, the dough is placed on the rolling surface and flattened slightly on top with a hand or the rolling pin to provide a starting point. Short strokes are made with the rolling pin from the center outward in a circle until the dough is flattened to ⅛ inch thick. Less pressure is used at the end of each roll to avoid excessive thinning of the edges. Lifting the edge of the dough intermittently and dusting the surface with flour as needed will help prevent sticking. It is best to roll as lightly and as little as possible and to avoid rolling repeatedly in one area to obtain the desired thickness. This will toughen and shrink the pastry by creating too much gluten.

Rolling the pastry too thin will make it too weak to hold fillings and may cause it to become too brown or to burn during baking. Pastry dough rolled too thick and used as a top crust may form a raised dome when baked.

## Rolling Plain Pastry

For plain pastry, the dough is rolled out in a circle 1 to 2 inches larger than the bottom of the pan. The wider diameter allows sufficient dough to cover the sides of the pie pan. Roll the dough one crust at a time. Once the dough is rolled, the simplest way to transfer it to the pan is to fold it in half and then in quarters to form a wedge shape (Figure 24-8). Rolling the dough on wax paper is an option that makes it easier to pick up, because only the edges of the wax paper need be lifted to accomplish the folding. The wedge is placed in the pie pan with the point in the center and the outside edge on the pan rim and then unfolded so it covers the bottom and sides of the pan. It should be large enough to cover the pan without stretching. Stretching dough to fill a pie pan may cause it to shrink back during baking. The dough is then pressed gently to eliminate any air bubbles, and squeezed together to patch up any tears or holes. Bottom crusts to be baked empty and filled later must be pricked with a fork so air can escape. Bottom crusts are not pricked if the filling is to be cooked in them, because pricking allows the filling to leak out.

The top crust is placed over the filling using the wedge procedure, or by lifting the wax paper and dough together, gently turning it over onto the filling, and slowly peeling off the wax paper. Top crusts are usually pricked with a fork, slashed with a thin knife, or opened with picturesque designs to create vents that allow the steam to escape during baking (Figure 24-9). This prevents the contents from boiling over. A decorative open latticework crust can also be made by cutting strips of rolled dough and arranging them

**FIGURE 24-8**  Transferring a rolled piecrust to a pie pan.

**Method 1:** Fold the rolled-out circular pie dough in half and in half again for easy transfer; unfold in pie plate.

**Method 2:** Roll the pie dough onto the rolling pin, place on edge of pie plate, and gently unroll.

Digital Works

**FIGURE 24-9** There are many artistic ways of venting a pie's top crust; this allows steam to escape so the contents don't boil over.

Direct cuts

Curled lattice

**FIGURE 24-10** Decorative piecrust edges.

Free form cuts

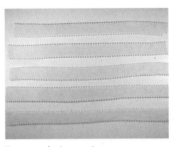

True woven lattice, step 1

Plain fork

Free form cuts

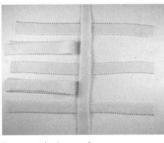

True woven lattice, step 2

Herringbone

Free form cuts

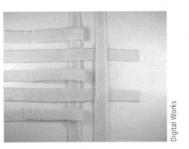

True woven lattice, step 3

Digital Works

Flutes

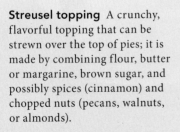

Free form cuts

Scallops

Rope

Digital Works

**Streusel topping** A crunchy, flavorful topping that can be strewn over the top of pies; it is made by combining flour, butter or margarine, brown sugar, and possibly spices (cinnamon) and chopped nuts (pecans, walnuts, or almonds).

on top of the filling. Another option, especially with one-crust apple pies, is to sprinkle the filling with a **streusel topping.** Adding a crimped edge gives pies a finished look and helps to seal top and bottom crusts together in a two-crust pie. The crust edges can be pinched together with fingers, a utensil handle, or the prongs of a fork to make various designs (Figure 24-10).

## How & Why?

### How do you avoid a soggy bottom crust?

To avoid a soggy bottom crust, it is important to fully thicken the filling before adding it into the pie shell. Another method is to prebake the bottom crust on the lower shelf of an oven set at 450°F (232°C) for at least 3 minutes, and then move it to the center shelf at 350°F (177°C) where it is baked for the remaining three quarters of an hour or until the filling is ready. The baked pastry is then chilled before lightly coating the bottom crust with cold fat and adding the filling. Another method to prevent a soggy crust is to prepare a **mealy** crust, which is more resistant to fluid absorption than a flaky crust. Using vegetable oil, completely melting the fat, or cutting up the fat to resemble cornmeal results in a mealy pastry.

**Alternative Piecrusts**  Flour is not the only ingredient that can be used to create piecrusts. Other options, not strictly considered pastry, include crushed graham crackers, cereal flakes, granola, and cookies. Crisp cookies make a crust with the best texture, and a food processor can easily pulverize the cookies (7). All can be made into a cohesive mass by mixing them with a little fat and sugar. This type of crumb crust is then shaped by laying plastic wrap over the cookie crumbs (to prevent sticking to hands) before pressing them against the sides of the pie pan. It is then baked or left unbaked, depending on what the individual recipe indicates.

## Rolling Puff Pastry

Production of puff pastry, which is higher in volume and fat than other pastries, relies on first folding the fat block and dough together before rolling. Repeated folding creates numerous layers of alternating fat and dough. If the fat does not stay as a separate layer, the pastry will have fewer layers and a gummy texture (8). These numerous folds contribute to the puffing up that occurs when the steam generated during baking forces the layers apart.

### Laminating Puff Pastry Dough

To begin, the chilled dough is rolled into a ¼-inch-thick rectangle. The chilled fat, which is in a block two-thirds the size of the dough rectangle, is placed on top of the dough. Then the dough and the butter are folded together (Figure 24-11). The package of dough is then turned 90 degrees, so the fold is running vertically, and rolled out to its original size. The fat must be at the right temperature; if it is too warm, it will flow into the dough, and if it is too cold, it will damage the dough (17). If butter is used, it can be partially softened by pounding it with a rolling pin to make it malleable while still cool (11). The folded, layered dough is then chilled, folded, and rolled again. The process is repeated two or more times before it is ready to be cut. Figure 24-12 shows the results of insufficient or excessive folding of puff pastry dough. Cutting must be done with a sharp knife or the layers will press together, drastically decreasing volume during baking.

The several extra layers of fat in puff pastry dough make its resulting baked products extremely tender, rich, flaky, delicious, and, of course, high in calories.

### Frozen Rolled Phyllo Dough

Phyllo dough that has already been rolled is available in the frozen-food section at supermarkets. It is sold as thin sheets, approximately 12 inches wide and 16 to 20 inches long, that are rolled and packaged into long, thin cardboard boxes. Defrosting the phyllo dough slowly results in the best pastry texture. The dough should be allowed to defrost in the refrigerator for about 12 hours and for an additional hour at room temperature so that the sheets separate more easily. The phyllo dough, which contains no eggs and is extremely thin, should be protected from drying out by covering it with a layer of plastic wrap and then a clean, moist towel (14). The phyllo dough can be brushed with melted butter to keep it supple, then cut and rolled or shaped into the desired pastry (Figure 24-13). Brushing each phyllo dough sheet with a thin layer of fat keeps it supple; too much, however, will weaken the dough, and too little will result in a thick, heavy pastry (18). Tearing of the delicate sheets is not uncommon during handling; this can be remedied by covering the rip with another sheet of phyllo dough. Phyllo dough that is ripping excessively may have already been accidentally defrosted at the supermarket or have dried out with age. Neither problem can be detected by looking at the frozen phyllo dough, so it is best to purchase it at a store with a high turnover rate, such as a market specializing in Middle Eastern foods.

# Fillings

The list of possible pie and pastry fillings runs the gamut from fruits to nuts. Main-dish pie options include quiches with egg, vegetable, and meat fillings. Then there are desserts such as fruit pies, cream pies, ice cream pies, chiffon pies, custard pies, and meringue pies. There are one-crust versions of pies known as tarts (Figure 24-14), and smaller versions, called tartlets, that are produced as individual servings. It is obvious that there is a pie to fit almost any taste at almost any meal.

## Fruit Fillings

Successful fruit pie fillings depend on the proper combination of fruit, fruit juice, sweetener, and starch thickener. Gelatin is sometimes used to glaze fruit pies or tarts that are not to be baked. Fruit for pies can be fresh, frozen, cooked, canned, or even dehydrated if it is soaked and simmered before being used. Sugar is usually added, but too much will draw water out of the fruit, causing it to become shriveled and tough. The usual thickening agent is cornstarch or tapioca, but modified starches such as waxy maize create a clear gel and are best for fruit pies. Flour is not recommended because it has a tendency to cloud the filling.

There are three methods of preparing fruit pies:

- The old-fashioned method
- The cooked-juice method
- The cooked-fruit method

## Old-Fashioned Method  In the old-fashioned method, the sweetener, spices, and starch thickener are mixed

**Mealy**  A pastry with a grainy or less flaky texture, created by coating all of the flour with fat.

**FIGURE 24-11**　Rolling puff pastry.

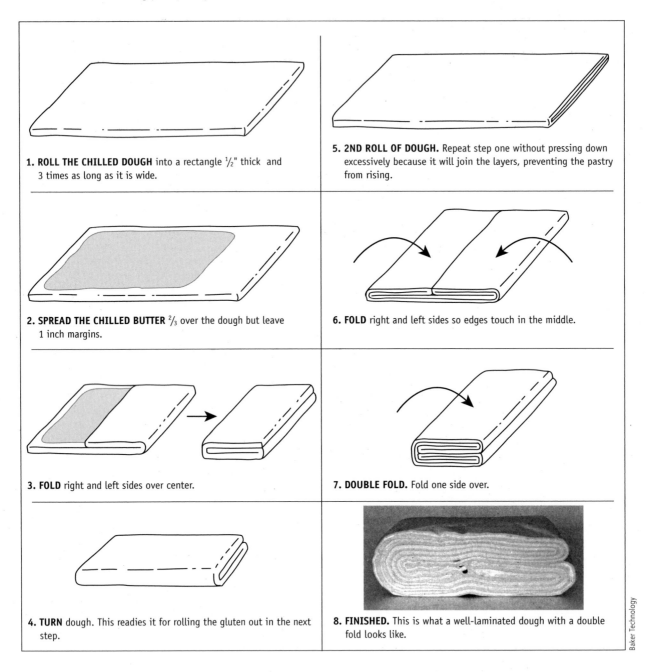

1. **ROLL THE CHILLED DOUGH** into a rectangle ½" thick and 3 times as long as it is wide.

2. **SPREAD THE CHILLED BUTTER** ⅔ over the dough but leave 1 inch margins.

3. **FOLD** right and left sides over center.

4. **TURN** dough. This readies it for rolling the gluten out in the next step.

5. **2ND ROLL OF DOUGH.** Repeat step one without pressing down excessively because it will join the layers, preventing the pastry from rising.

6. **FOLD** right and left sides so edges touch in the middle.

7. **DOUBLE FOLD.** Fold one side over.

8. **FINISHED.** This is what a well-laminated dough with a double fold looks like.

Baker Technology

**FIGURE 24-12**　Puff pastry that was insufficiently or excessively folded.

Insufficient folding

Optimall folding

Exessive folding

Baker Technology

together and then combined with the fruit. This filling is then placed in the unbaked pie shell. Thin slices of butter are sometimes strewn across the top of the filling before it is covered with a top piecrust and baked.

***Cooked-Juice Method***　The cooked-juice method allows the uncooked fruit to retain more of its shape, flavor, and texture. Most frozen

**FIGURE 24-13** Handling phyllo dough.

1. Cover phyllo dough so it won't dry out. Work with one sheet at a time; cover the rest with plastic and a moistened kitchen towel.

2. To keep the dough supple, brush the dough with melted butter. Begin by buttering the edges, which dry out faster, and then work toward the center.

3. Use a sharp knife or scissors to cut the dough to size. Buttering and layering the sheets before cutting is faster and keeps the phyllo more pliable.

4. Repair tears by pasting on a piece of phyllo from an extra sheet with melted butter.

5. Versatile phyllo dough can be made into many shapes and will hold sweet or savory fillings. Shown here: folded triangle and a molded cup.

Digital Works

**FIGURE 24-14** Tarts are European versions of pies.

SharonDay/istockphoto.com

or canned fruit pies and berry pies can be prepared by this method. The juice is drained from the fruit and water is added if needed. The liquid is brought to a boil, and a mixture of starch dissolved in cold water is stirred into the boiling liquid, followed by the sweetener and spices. When the mixture is thick and clear, it is cooled slightly and poured over fruit that has been previously arranged in a prebaked crust. The pie is then refrigerated.

***Cooked-Fruit Method*** If not enough juice is available from the selected fruit, the other option is the cooked-fruit method. This method is the same as the cooked-juice method except the fruit is added to the thickened juice and the mixture brought back to a boil. The results are poured directly into a prebaked pie shell. No further baking is required unless it is a two-crust pie.

### Cream Fillings

Cream pie fillings are made by heating a mixture of milk, sugar, flavoring, cornstarch, and often egg until the starch gelatinizes, then pouring this mixture into a one-crust prebaked pie shell. (The principles of starch gelatinization are covered in Chapter 18.) Whole, low-fat, or nonfat milk, light or heavy cream, or evaporated milk may be used. Cream pies such as chocolate, banana, or coconut are often topped with a whipped topping or meringue.

### Custard Fillings

Custard pies are one-crust pies made with a milk and egg filling; examples include pumpkin and coconut pies. Cheesecake, in spite of its name, is really a custard pie, because it relies on the coagulation of eggs for structure. The custard pie mixture is thickened by the eggs; an average custard pie contains 2 eggs per 1 cup of milk. During baking, the egg proteins coagulate and set the filling. Because the egg-based custard filling coagulates much more quickly than the crust can bake, the two are often baked separately, with the cooled filling being placed in the cooled shell right before serving. Another method is to start the custard pie on the bottom rack of a hot 425°F–450°F (218°C–232°C) oven for the first 10 minutes. Then, after the crust sets, to ensure that it will be crisp, the oven temperature is lowered to 325°F–350°F (163°C–177°C), where it remains for the rest of the time, usually about 25 to 40 minutes.

### Chiffon Pies

The use of gelatin in their preparation distinguishes chiffon pies from other types. The foam structure of chiffon pies imparts a light, fluffy texture. Egg whites or whipped cream is often an ingredient in these delicate pies. Only the shell is baked, so nondairy whipped toppings or rehydrated dried egg-white powder whipped into a foam can serve as substitutes for raw whipped egg whites to eliminate the risk of *Salmonella* contamination sometimes posed by raw eggs.

### Meringue Pies

Meringue pies are simply cream or chiffon pies covered with meringue (see Chapter 12). The meringue should be spread to cover all parts of the crust edge so it will not shrink back during baking. The meringue is swirled on top and quickly browned in the oven. For serving, an easy way to cut through the meringue topping is to butter or oil the knife blade.

### Pastry Fillings

Possible fillings for pastries include custards, creams, Bavarians, mousses, jams, fresh or cooked fruits, chocolate, and even cheese, cooked poultry, fish, or ham. The preparations of these fillings are discussed in the respective chapters devoted to their food type. Cones and pastry bags are frequently used to fill pastries; they are usually made from canvas, paper, or some other material (Figure 24-15).

## Toppings

Pie fillings can be crowned by a wide variety of toppings that contribute additional textures and flavors. A two-crust pie has another crust to cover its filling, whereas a single-crust pie can vary in its topping from plain (nothing added) to whipped cream, meringue, glaze, or a crumb topping. The simplest decoration is sprinkled sugar (large crystals such as turbinado sugar) over a pie crust lightly brushed with a beaten egg white (10). The more complicated toppings, whipped cream and meringues, are discussed in Chapter 10 (on milk) and Chapter 12 (on eggs), respectively. Glazes and crumb toppings are briefly discussed below.

### Glazes

An icing unique to pastries can be prepared by combining confectioners' sugar, heavy cream, corn syrup, and a flavoring (vanilla, orange, chocolate, etc.) (6). The glaze needs to be thick enough to hold its structure upon cooling, but sufficiently liquid to drape over the pastry. If it is too thick, cream can be added a tablespoon at a time until the right consistency is reached.

### Crumb Toppings

A crunchy, crumbly texture is added to the top of fruit pies (especially apple) by topping them with a crumb topping. The texture and the size of the crumb can be controlled by changing the proportion of the ingredients—flour, butter, sugar, and salt (8). Crumb topping is essentially a pie dough with added sugar that is mixed, collected together into

**FIGURE 24-15**    Parchment cones and pastry bags.

Making a pivot point

Rolling into funnel shape

Sealing the filled cone

Creating small opening

Filling pastry bag

Hand position for piping

Digital Works

a clump, and refrigerated. The difference is that instead of rolling, it is broken apart into the desired crumb sizes and sprinkled over the pie filling to act as a loose top crust. Texture is affected by the ingredients: dusting with confectioners' sugar or adding melted butter makes it look more like coffee cake; brown sugar increases the crunch and caramelized look (9).

In addition to being used as a topping for pies, crumb toppings are important for the preparation of fruit cobblers. These are dishes in which fruits coated with sugar water are then mixed and sprinkled with crumb toppings before being baked in a deep dish. The flour mixture in and around the fruit gives cobblers their baked bread-like characteristics. Cobblers are often served topped off with whipped cream.

## Baking

Whether using a conventional oven, convection oven, or microwave oven, the type of pan, the temperature, the time, and the method of testing for doneness are important to the finished pastry product.

### Pans

Either Pyrex™ glass pans or pans with dull finishes are best for pies because they help absorb the heat. Shiny metal pans, which deflect heat, and thick metal pans, which take too long to heat, are not recommended (16). Piecrusts are usually baked in 8- or 9-inch pans with sloping sides and an edge designed to catch escaping juices. The size of the pan determines the number of pie slices:

- 8-inch pan  = 4 slices of pie
- 9-inch pan  = 6 slices of pie
- 10-inch pan = 8 slices of pie

Puff pastries are baked on cookie sheets or specially prepared pans. Pies and puff pastries should be baked in the center of the oven, making sure the pan or cookie sheet does not touch the sides. Custard-based pies, however, should be placed on the bottom rack so the crust will bake quickly and not have time to get soggy. An extra crispy, flaky bottom crust can be obtained

**FIGURE 24-16** Creating a foil cover.

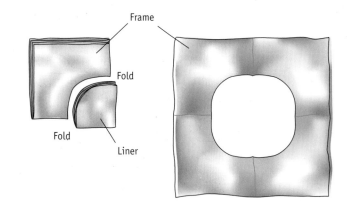

Unfolded Foil Frame

by placing a cookie sheet on the rack prior to heating the oven. The direct contact between the bottom of the pie pan and the cookie sheet delivers more direct heat.

### Temperature/Timing

The oven should be preheated and hot at approximately 425°F–450°F (218°C –232°C) for baking pies and puff pastries. Slower baking would contribute to shrinking and result in hard, flat pastries. The high heat helps to generate steam, melt the fat, and set the gluten. Adjustments to temperature are sometimes made, depending on the type of filling.

Pie shells that are **blind baked** tend to bubble, lose their shape, and brown unevenly. These faults can be prevented by pricking the pie dough with a fork and/or weighting it down with beans, which are removed a few minutes before the baking is completed. Aluminum foil placed around the edge of the piecrust can prevent it from overbrowning or burning (Figure 24-16).

Fillings should not be added until immediately before baking. It is important not to prick the bottom crust of pies to be baked already filled, because the juices will run through. The bottom crust can be moisture-proofed by several methods: brushing beaten egg glaze or fruit preserves on the pie dough and then placing it in the oven for a short prebaking; placing a very thin layer of melted butter or flour on the bottom crust; or making sure the

filling is very hot as it is poured into the crust.

Several decorative touches can be added just before baking. For example, the top of a two-crust pie can be brushed lightly with water and sprinkled lightly but evenly with granulated sugar or coarse decorating sugar. For a shiny or glazed look, the top can be brushed with an egg wash (milk and beaten egg yolk combined with an equal amount of water). Be aware, however, that although glazes make pastry crusts appealing, they also make them tough. A crunchy crust is achieved by sprinkling about 2 tablespoons of sugar (white, brown, sanding, turbinado) on the egg wash (3).

## Testing for Doneness

A lightly, delicately browned crust signals that baking is complete. Adding a thin layer of butter on the crust helps it to brown. Once done, all pies should be cooled in the pan on a wire rack to prevent moisture condensation, which results in a soggy lower crust. Table 24-7 lists several reasons why the baked pastry's appearance, tenderness, texture, or flavor may not be up to par.

**Blind bake** To bake an unfilled piecrust.

**TABLE 24-7**  Pastry Problems and Their Causes

| Problems / Causes | Over-mixing/handling | Under-mixing | Too much fat | Too little fat | Too much flour (in mix or on board) | Too much liquid | Too little liquid | Over-baking | Under-baking | Rolled too thick | Rolled too thin | Oven temperature too low | Filling too moist | Pastry stretched too tight | Not pricked with fork | Fat cut too finely | Dough stored too long in refrigerator |
|---|---|---|---|---|---|---|---|---|---|---|---|---|---|---|---|---|---|
| Tough | X | | | X | X | X | | | | | | | | | | | |
| Too tender (falls apart) | | X | X | | | | X | | | | | | | | | | |
| Crumbly | | X | X | | | | X | | | | | | | | | | |
| Doughy | | | | | | X | | | X | | | | | | | | |
| Dry/mealy | | | | | | | X | | | | | | | | | X | |
| Too brown/burned | | | | | | | | X | | | X | | | | | | |
| Pale/dull color | | | | X | X | X | | | X | X | | X | | | | | |
| Soggy bottom crust | | | | | | | | | | | | X | X | | X | | |
| Shrinks | X | | | | | | | | | X | X | | | X | | | X |
| Blistered shell | | | | | | | | | | | | X | | X | X | | |

# STORAGE OF PASTRY

Pastries are best consumed when freshly baked, but most keep longer when refrigerated than they would if left at room temperature, and some can even be frozen. Pastry doughs freeze up to 6 months. Unbaked pies will last about 4 months in the freezer, whereas baked berry pies can be frozen for 6 to 8 months.

Not all pies can be frozen, especially if they contain egg and milk products, which may separate out. Custard and cream pies are frozen commercially, but freezing is not recommended for home recipes. As with all foods containing milk and/or eggs, custard, cream, and meringue, these kinds of pies should be kept refrigerated to avoid the growth of bacteria and resulting foodborne illness.

# PICTORIAL SUMMARY / 24: Pies and Pastries

Pastry is essentially a bread that is flaky, tender, crisp, and lightly browned. But this delicate combination is not always easy to achieve, and piecrusts and pastries are the true test of a food preparer's skill.

## TYPES OF PASTRY

**Nonlaminated**
- Plain pastry (pie or shortcrust)— piecrusts, tarts, tartlets, galettes
- Hot-water crust—main-dish pies
- Short or sweet
- Brioche
- Choux

**Laminated**
- Puff
- Quick (blitz) puff
- Phyllo (filo)
- Croissant
- Danish

## PREPARATION OF PASTRY

The main ingredients in pastry are flour, fat, liquid, and salt (egg is optional). Either pastry or all-purpose flour can be used. Fat is probably the most important ingredient, because it contributes the most to a pastry's flakiness and tenderness as well as its taste. It also, however, contributes to making most pastries high in fat and calories (kcal). Liquid is needed to develop gluten and leaven the pastry with steam, and salt is needed for flavor.

Plain and puff pastries are the most delicate of all baked products, and their preparation requires skill to correctly distribute the fat and develop gluten to the point of creating a crust that is flaky, tender, and crisp. Mixing; rolling; type of filling used (fruit, cream, custard, chiffon, or quiche ingredients); decorations (sugary, shiny, glazed, or frosted surface); and baking all contribute to the success of the finished product.

## STORAGE OF PASTRY

When held for later consumption, pastry and pastry products will keep longer when refrigerated than they will at room tempera-  ture. Pastries vary in how successfully they may be frozen. Custard, cream, and meringue-topped pies should not be frozen, but need to be refrigerated to avoid bacterial growth and thus prevent food-borne illness. Frozen pastry crusts should go as quickly as possible from the freezer into a preheated oven.

# CHAPTER REVIEW AND EXAM PREP

## Multiple Choice*

1. Which word is used to describe a pastry with a grainy or less flaky texture, created by coating all of the flour with fat?
   a. Tender
   b. Moist
   c. Tough
   d. Mealy

2. The proportion of what ingredient is probably the most important determinant of quality in pastry, especially in regard to flakiness?
   a. Flour
   b. Sugar
   c. Liquid
   d. Fat

3. The liquid component of pastry dough is important for which of the following?
   a. Leavening
   b. Hydration
   c. Crispness of the crust
   d. All of the above

4. Which two types of pastry must be rolled, with a minimum of hand contact, in order to spread the fat and gluten in fine sheets layered on top of each other?
   a. Plain and thick
   b. Plain and puff
   c. Puff and thin
   d. Thin and laminated

5. If you are baking a pie that has six slices, what size pie pan should you use?
   a. 9-inch
   b. 6-inch
   c. 10-inch

*See p. AK-1 for answers to multiple choice questions.

6. Cream puffs and eclairs are made from a _____ pastry dough.
   a. phyllo
   b. puff
   c. choux
   d. brioche

7. Which starch-based thickener should not be used for fruit pies because it will thicken before the fruit boils?
   a. Arrowroot
   b. Cornstarch
   c. ClearJel®
   d. Tapioca

## Short Answer/Essay

1. Describe the characteristics of good-quality pastry.
2. What is the difference between plain and puff pastry? Give examples of each.
3. How do pastry flour mixtures differ from those used to prepare bread doughs? Explain the role of each of the following ingredients used in pastry preparation: flour, fat, liquid, eggs, and salt.
4. Explain the difference between flakiness and tenderness and how each is achieved during pastry preparation.
5. Fat plays a critical role in pastry production. Explain how fat contributes to a flaky pastry through the process of lamination.
6. Flour is not the only ingredient that can be used to make piecrusts. What are some of the other options?
7. Briefly define the following terms: *mealy crust, streusel topping,* and *blind bake.*
8. List the basic ingredients of the following types of pie fillings and explain how the fillings are prepared: fruit, cream, custard, and chiffon.
9. What is the best temperature for baking pastries? Why is this temperature recommended?
10. State the possible causes of the following problems associated with pastry preparation: low volume, insufficient flaking, fat flows out.

# REFERENCES

1. Amendola J, and Rees N. *The Baker's Manual.* Van Nostrand Reinhold, 2002.
2. Bruce E. Foolproof all-butter pie pastry. *Cook's Illustrated* 76:20–21, 2005.
3. Chang J. How to make a rustic fruit tart. *Fine Cooking* 73:50–54, 2005.
4. Corriher S. The secrets of tender, flaky piecrust. *Fine Cooking* 17:78–79, 1996.
5. Dodge AJ. Baking classic American pies. *Fine Cooking* 29:73–77, 1998.
6. Dodge AJ. Pure vanilla flavor. *Fine Cooking* 71:67–71, 2005.
7. Dodge AJ. Stylish tarts from a quick crust. *Fine Cooking* 68:50, 2005.
8. Friberg B. *The Professional Pastry Chef.* Van Nostrand Reinhold, 2002.
9. Kalen W. Apple desserts with easy crumb toppings. *Fine Cooking* 67:62, 2004.
10. Kreitler AE. Clever crust. *Fine Cooking* 95:80, 2008.
11. Malgieri N. *Nick Malgieri's Perfect Pastry.* Macmillan, 1998.
12. Masibay KY. The whys of pies. *Fine Cooking* 95:86, 2008.
13. McWilliams M. *Foods: Experimental Perspectives.* Macmillan, 2007.
14. Mushet C. Sweet, crisp baklava. *Fine Cooking* 92:63, 2008.

15. Penfield MP, and AM Campbell. *Experimental Food Science.* Academic Press, 1990.

16. Purdy SG. *As Easy as Pie.* Macmillan, 1990.

17. Schunemann C, and G Treu. *Baking: The Art and Science. A Practical Handbook for the Baking Industry.* Baker Tech, 2007.

18. Stevens M. Baking flaky pastries with phyllo dough. *Fine Cooking* 20:18–19, 1997.

# WEBSITES

*Fine Cooking* magazine offers lots of tips, information, and even videos:
**www.finecooking.com/videos/**

The following is a resource website for information on baking pies:
**www.bettycrocker.com** (click on "How-To," then "Baking Basics," then "Pies")

This website has lots of information on baking tarts:
**www.joyofbaking.com/PieAndTart Recipes.html**

# 25 Candy

The words *love* and *sweets* have been interwoven in everyday language through such terms of endearment as *sweetheart, honey,* and *sugar.* There are few who disapprove of love, or dislike sweets such as candy. This is especially true in the United States, which produces more candy than any other country in the world. Despite the high production of confections in the United States, people from Europe actually consume the most candy (34). However, Americans are not far behind

and approximately half of their intake is in the form of chocolate with the rest from other types of sugar confectioneries (12).

In pursuit of the gratification of the sweet tooth, candy making has developed into a many-faceted art. No one really knows how candies first got their start, but evidence points to the ancient Arabs, who used refined sugar to make medicinal lozenges. One of the factors supporting the role of Arabs as the first makers of confections is that "candy" is derived from *qand*, the Arabic word for sugar (2). Sugar (specifically sucrose), along with its close relative, corn syrup, are the foundational ingredients of almost all candies and essential to the process of confectionery production. Sugar is mixed with liquid and other ingredients into a solution that is then heated to concentrate its sweet taste. The type and concentration of ingredients, the temperature to which the mixture is heated, the degree to which it is then cooled, and any stirring or manipulation that follows determine the type of candy produced.

This chapter discusses the classification of candies and their basic preparation methods.

## CLASSIFICATION OF CANDIES

There are thousands of different candies, but they can be classified according to their ingredients and/or preparation method (Table 25-1). Two major classifications in the confectionery industry are (1) syrup phase or fat phase, and (2) crystalline or noncrystalline (amorphous).

### Syrup Phase or Fat Phase

One way of categorizing candies according to ingredients is to divide them into candies termed *syrup phase* (or *sugar phase*) and *fat phase*. Most candies are syrup phase, meaning they are made from a **simple syrup**

**Simple syrup** A basic mixture of boiled sugar and water.

**TABLE 25-1   Types of Candy**

| Class | Comments |
|---|---|
| Candied or crystallized fruits | Certain fruits (cherries, citrus peels, plums, etc.) are prepared and dropped into a syrup at 234°F (112°C). They are simmered until clear, spread on a screen, and dried until no longer sticky. |
| Caramels | Caramels contain sugar, corn syrup, milk, cream, and butter; and are cooked to stiff-ball stage of 246°F (119°C). After being poured into a buttered pan and allowed to cool, they are turned out and cut into squares. |
| Chewing gums | Chewing gum is not eaten; it is chewed. Its sugar content allows it to be classed as a candy. |
| Chocolates | Chocolate liquor is the fundamental ingredient of most chocolate products. Other ingredients can include cocoa butter, sugar, milk products, and flavorings. |
| Fondants | For fondants, the syrup is cooked to the soft-ball stage (238°F/114°C), poured into large platters, cooled until lukewarm, then stirred and kneaded until smooth. |
| Fudges | Fudge is made by gently boiling sugar and corn syrup with milk to 238°F (114°C), adding butter, cooling, then beating until it holds its shape. It is spread in a buttered pan and, when hard, cut into squares. |
| Glacé fruits and nuts | Similar to candied fruits, but the fruit is not cooked. The fruit or nut is dipped into the syrup, which has been cooked to 300°F (149°C), and dried on a rack. |
| Hard candies | Hard candies are the simplest form of candy—they are made mainly from sugar and syrup and are usually boiled to 300°F (149°C). They come in all shapes, sizes, colors, and flavors. |
| Jellies | The main ingredients for jellies (jelly beans, gum drops, etc.) are sugar, corn syrup, and a gelling agent such as gelatin, natural gums, pectin, or starch. |
| Licorice | Licorice sticks are made with flour, molasses, sugar, and corn syrup, and flavored with licorice extract. |
| Marshmallows | Marshmallows are made by whipping a combination of sugar, corn syrup, gelatin, and/or egg whites. This makes a light, fluffy-textured candy that can be served plain or as a filling for Easter eggs. |
| Marzipan | Egg whites are beaten and mixed with almond paste plus sugar. It is not cooked. After standing 24 hours, it can be shaped, stuffed, or dipped. |
| Nougats | The syrup is cooked to 246°F–250°F (119°C–121°C) and then poured over the well-beaten egg whites, or gelatin, or both. Fat is added and it is beaten until cool. |
| Peanut brittle | The syrup is cooked to 300°F (149°C). It is then poured over the nuts spread out on a pan, allowed to cool, and then broken into pieces. |
| Popcorn balls | The sugar, corn syrup, and water are cooked to the medium-crack stage (280°F/138°C), flavoring is added, and the mixture is poured over the popcorn and stirred. Then, with buttered hands, the mixture is shaped into balls. |
| Spun sugar | The syrup is boiled to 310°F (154°C). Then the pan is put into cold water to stop cooking, and back into a warm pan before it crystallizes. Using a wooden spoon, the threads of syrup are wrapped across the greased handles of two wooden spoons that have been anchored into a drawer. |
| Taffy | Similar to a caramel mixture except more concentrated and pulled to incorporate air. |
| Toffee | Toffee is hard caramel. |

mixture. Examples of some syrup-phase candies are hard candy, fondants, marshmallows, nougats, jelly beans, gums, caramels, and fudges. These candies are basically sugar with added flavorings. When chocolate or nut pastes such as peanut butter are used, the candy is considered fat phase. A combination of both fat and syrup phases is found in candies such as chocolate-covered candy bars.

# Crystalline or Non-crystalline (Amorphous)

Another way to classify confectioneries is based on the method of preparation, which determines whether the candy will be crystalline or noncrystalline in nature. **Crystalline** candies, in which the sugar is present in the form of crystals, are soft, smooth, and creamy; examples are chocolate, creams, fudge, fondant, nougats, marshmallows, pralines, lozenges, and divinity. **Noncrystalline,** or

**Crystalline candy** Candies formed from sugar solutions yielding many fine, small crystals.

**Noncrystalline (amorphous) candy** Candies formed from sugar solutions that did not crystallize.

amorphous (without form), candies are those in which the sugar is present in an uncrystallized form. Examples include caramel; toffee; taffy; hard candy; brittles; and gummy candies such as jelly beans, gummy bears, fruit slices, and gumdrops. The difference in texture between the two types depends on how the candy's ingredients are combined and/or manipulated.

# PREPARATION OF CANDY

When it comes to preparation, candies are temperamental. Confectionery production is highly sensitive to timing, temperature, and the skill of the preparer, making it an art successfully executed only with practice and patience. The obscure aura surrounding candy making is further blurred by the fact that there are even more recipes than the number of candies available. For example, there are over 1,000 different formulas for making marshmallows alone (3). This chapter describes the preparation of candies based on whether they are crystalline, noncrystalline, or chocolate. A brief summary of confectionery preparation follows in order to put the various types of candies in perspective.

## Steps to Confectionery Preparation

Controlling crystallization is one of the most important aspects of producing confections (9). The preparation of many, but not all, confections can be generally summarized by four basic steps that revolve around controlling the formation of sugar crystals (29):

*Step 1:* Creating a syrup solution
*Step 2:* Concentrating the contents of this mixture via heating and evaporation
*Step 3:* Cooling
*Step 4:* Beating

**Nuclei**  Small aggregates of molecules serving as the starting point of crystal formation.

These steps and how they differ in the preparation of crystalline and noncrystalline candies are now discussed.

The preparation of many crystalline and noncrystalline candies starts out the same—a syrup or sugar solution is heated to melt the sugar, evaporate the liquid, and concentrate the sugar. The formation of sugar crystals from this syrup solution is the basis of crystalline candies, whereas the goal in preparing noncrystalline candies is to inhibit their formation. This is achieved in part by the way their sugar solutions are cooled and beaten. Rapid cooling and beating results in crystalline candies, whereas slow cooling without agitation forms noncrystalline candies (1).

## Crystalline Candies

Two types of crystalline candies exist based on the size of their crystals: (1) candies with large crystals such as rock candy, and (2) cream candies with crystals too small for detection by the tongue (fudge and fondant). Crystals are a compilation of loosely packed sugar molecules organized around **nuclei.** The size of the sugar crystals is determined by the rate or speed of nuclei formation. If the nuclei appear slowly in the syrup solution, there is more time for the sugar molecules to aggregate around the nuclei and form large crystals. The goal in preparing crystalline candies is to develop numerous, very fine nuclei in the syrup solution, which will serve as the foundation of the sugar crystals. Confectioners generate small nuclei by (1) controlling the form and content of sugar, (2) controlling the temperature, and (3) stirring correctly. As the solution cools, the sugar hardens out, or crystallizes, creating a candy.

### Candies Start with a Syrup Solution

Sugar is dissolved in water until no more sugar can be added (this is known as a saturated solution, having about 1 pound of sugar for every cup of water). Sucrose is the most common sugar used, but sometimes another sugar such as glucose, invert sugar (a mixture of glucose and fructose), or corn syrup is added to make sucrose more soluble and less likely to form large crystals. However, this functional benefit could also cause a problem because too many added

monosaccharides may make the syrup so runny that it never crystallizes. Invert sugar may be purchased commercially or made by adding an acid such as cream of tartar to sucrose (see Chapter 21). When added to chewy candies, invert sugar's hygroscopic nature prevents the candies from drying out. The benefit of adding corn syrup is that, like invert sugar, it contributes to chewiness but also adds viscosity, slows the dissolving rate of candies in the mouth, and strengthens the structure of sugar crystals so they are less likely to be affected by temperature or mechanical shock (30).

In the confectionery business, glucose is referred to as dextrose, and fructose as levulose (30). Both of these sugars crystallize more slowly than sucrose and, when combined, are sweeter than sucrose. Other foods, such as chocolate, fat, milk, cream, and eggs, are often added and help to interfere with large crystal formation.

### Heating the Syrup

Temperature is important in confectionery production because it influences crystallization at all stages of heating and cooling. The next step in creating confections is to heat the syrup mixture. The purpose is to increase the amount of sugar that can be added to the solution so that it becomes supersaturated (see Chapter 21 on sweeteners for more on saturated and supersaturated solutions). As the heat evaporates the water, the syrup solution becomes more concentrated. As the concentration increases, so does the boiling point. Temperatures exceed the normal boiling point of pure water and leave behind the solids, primarily sugars, which then form into a hard mass.

Controlling the degree of sugar concentration determines what kind of candy will be produced. However, it's difficult to measure changing concentrations accurately, so boiling point is used as an indirect measure of concentration (9). Table 25-2 shows the final temperatures of syrups for each of the different types of candies. For example, a syrup heated to a high temperature will result in a harder candy (peanut brittle) than one heated to a lower temperature (jelly or fudge). Lower temperatures evaporate less water, and the more water a candy mixture retains, the softer its consistency. Even the weather is a factor

**TABLE 25-2**   Candy Syrup Temperatures and Doneness Tests

| Candy | Doneness Test | Final Temperature of Syrup Begins At* | | Description of Test† |
| | | °F | °C | |
| --- | --- | --- | --- | --- |
| Jelly | | 220 | 105 | Syrup runs off spoon in drops that merge to form a sheet. |
| Syrup | Thread | 230 | 110 | Syrup spirals a 2-inch thread when dropped from spoon. |
| Fondant Fudge Panocha | Soft ball | 234 | 112 | Syrup forms a soft ball that flattens out between fingers. |
| Caramels Nougat | Firm ball | 244 | 118 | Syrup forms a firm ball that retains its shape when held between fingers. |
| Divinity Marshmallows Popcorn balls | Hard ball | 250 | 121 | Syrup forms a ball that is hard enough to hold its shape, yet plastic enough to roll out. |
| Butterscotch Hard candies Toffees | Soft crack | 270 | 132 | Syrup separates into threads that are hard and brittle under water, but become soft and sticky when removed from the water. |
| Brittle Glacé Some hard candies | Hard crack | 300 | 149 | Syrup separates into threads that are hard and brittle, but do not stick to fingers. |
| Caramel | Light brown liquid | 338 | 170 | No cold water test. The sugar liquefies and becomes light brown. |

*For each increase of 500 feet in elevation, cook the syrup to a temperature 1°F lower than temperature called for at sea level. If readings are taken in Celsius (centigrade), for each 900 feet of elevation, cook the syrup to a temperature 1°C lower than called for at sea level.

†Remove pot from heat while testing to avoid overcooking. Drop 1 drop from an eyedropper of syrup into fresh cold water, which is replaced with each test.

when making the syrup solution because high-humidity days are not conducive to making confections.

**? How & Why?**

*Why is the weather considered by confectionery makers when preparing candy?*

Simple syrups can absorb moisture from the air during a humid day, causing a dilution of the syrup mixture and making the candy softer than desired. As a result, cool, clear days when there is little humidity in the air are preferred for candy making (1).

*Over- and Underheating*   One problem sometimes encountered in confectionery cooking is overheating the solution, which results in a too-hard, sometimes excessively brittle candy. Overheating can also cause color and flavor changes. On the other hand, too cold a temperature results in a too-soft, runny consistency.

There are two ways to determine the correct final temperature during candy making: (1) using a candy thermometer, and (2) the cold-water test.

*Thermometer Test*   Candy thermometers have been specifically designed for the high heats that syrup mixtures reach, and have a clip on the side to hold them inside the pan. Home-use candy thermometers have a range of about 100°F–320°F (38°C–160°C). Commercial models are longer to fit larger pans and have a broader temperature range of 60°F–360°F (16°C–182°C) (31). A candy thermometer's bulb should be completely immersed in the mixture without touching the bottom of the pan. The top of the mercury line should be read by bringing the eyes level with the thermometer, not the thermometer to eye level.

Before each use, and especially at high altitudes, it is important to check the candy thermometer for accuracy. This is accomplished by determining the boiling point of water on the thermometer, which should be 212°F (100°C) at sea level, and 1°F lower for each 500-foot increase in elevation above sea level (1°C lower for each 900 feet in elevation). If the boiling point of the thermometer is below the standard 212°F (100°C), then the number of degrees below this point is subtracted from the recipe's cooking temperature. Once the thermometer has been used to make candy, it helps to immerse it immediately in very hot water to make it easier to clean and to prevent breakage (31).

*Cold Water Test*   The older and still viable method of testing the temperature of candy mixtures is the cold water test, which measures the syrup's consistency. This test is carried out by placing a very small amount of the candy mixture, about 1 drop from an eyedropper or ¼ teaspoon, into a cup of very cold (not ice) water, after which the drop is observed for softness or firmness and possibly double-checked by pinching it between two fingers. Drops of the syrup, when cooled at successive stages during the cooking process, will form first a soft, then a firm, and finally a hard ball. Because the mixture can move from one stage to the next fairly rapidly, the saucepan should be removed from the heat source before any testing is done. Experience is the best teacher in learning the precise look and feel of the candy mixture at different stages.

**Avoiding Agitation during Heating**

Stirring, or agitation, is another factor governing crystal formation. It is important to avoid vigorous boiling or stirring when heating the solution to its final temperature. Syrup splashing onto the sides of the pan can prematurely **seed** the solution and initiate crystal formation. One sugar crystal falling

**Seed**  To create nuclei or starting points from which additional crystals can form.

into the mixture from the side can start a chain reaction, resulting in a large sugar mass. This can be prevented by covering the boiling syrup solution with a lid, creating steam that returns the syrup on the sides of the pan back into the solution before it can form crystals. The pan must then be uncovered later in cooking to allow the water to evaporate and the temperature to rise. To prevent crystals from forming during this final cooking stage, the sides of the pan may be wiped often with a wet pastry brush or a damp paper towel or cheesecloth. If this is not done, large crystals, instead of the desired small ones, will likely form.

## Cooling and Beating

Although agitating the mixture can result in some premature crystal formation during heating, the crystallization desired for candy making begins during the cooling of a supersaturated solution. After the correct temperature is reached, the solution is cooled immediately, without any additional movement, not even to stir in flavoring or move the thermometer or spoon in the mixture. As the mixture cools, it becomes supersaturated, which allows for the formation of nuclei. Immediate cooling also prevents further evaporation of the water.

*Cooling Is Crucial*  Cooling at the correct rate is crucial for candy production. Sugar molecules move rapidly in a hot solution, but drastically slow down as the temperature drops. Small crystals are less likely to aggregate in a hot solution where molecules are rapidly moving. At the same time, the small crystals that do occur grow large because the greater frequency of contact among sugar molecules encourages their growth. The syrup mixtures for crystalline candies such as fudge and fondant are quickly cooled to slow molecular movement, and stirred to form small crystals by fostering aggregation of the molecules. Figure 25-1 shows that starting off with a greater number of small nuclei results in smaller crystals.

*Stirring During Cooling*  Stirring the mixture after it cools promotes the formation of numerous small crystals that contribute to a smoother consistency in the candy, but this is done only after the desired cooler

**FIGURE 25-1**  Correct cooling is critical to the formation of many small crystals.

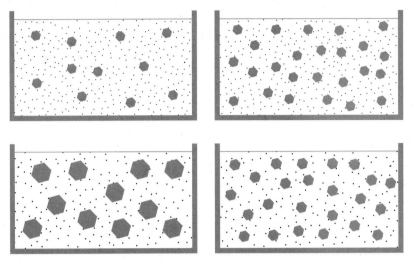

Smoother candies result if sugar crystals are kept small. This can be achieved by starting off with more (*top right*) rather than less (*top left*) seed crystals so there are more sites for the remaining sugar to crystallize on. A greater number of starting points results in smaller crystals (*bottom right*).

temperature has been reached. Then the crystallization is initiated by beating the mixture rapidly until its shiny, glossy appearance turns dull. If it is cooled too long before beating, the development of the finer crystals necessary to produce a smooth, crystalline candy will be inhibited.

## Types of Crystalline Candies

In the preparation of crystalline candies, small crystals form when many nuclei quickly appear. This allows less time for the sugar molecules to collect around the nuclei, and divides the available sugar among a larger number of nuclei, meaning less sugar aggregates around each nucleus. The smooth, creamy texture of fondant, fudge, and divinity depends on the formation of numerous, small sugar crystals (Figure 25-2). A discussion of these crystalline candies and their preparation follows.

*Fondant*  Fondant is a crystalline candy made from a sugar syrup that has been crystallized into a creamy, white paste. It serves as the soft, rich filling of many chocolates, mints, and candy bars (Chemist's Corner 25-1).

## NUTRIENT CONTENT

Candy contains primarily carbohydrate—from refined sugar—and possibly fat—from butter, cocoa butter, or other fats. See Calorie Control for their contribution to dietary calories.

The Food and Drug Administration (FDA) does not allow the use of alternative sweeteners in confectioneries, so reducing calories in candies is not easy, even though consumer demand for such products continues to increase (3, 18). Since the 1960s, however, certain gums and breath mints have been made with alternative sweeteners (38), and the FDA policy may soon be amended or revoked in response to numerous requests (27).

Despite this hurdle, one option for reducing calories in confectioneries is to use polydextrose, a low-calorie bulking agent that is allowed in hard and soft candies and in frostings (35). Another option is the sugar alcohols sorbitol and xylitol, sometimes found in mints and chewing gums (14). These sugar alcohols used by the confectionery industry are not carcinogenic, unlike some other alternative sweeteners (35).

**FIGURE 25-2**   Sugar crystal size changes during candy preparation.

 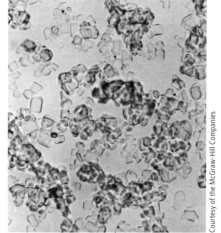

Courtesy of the McGraw-Hill Companies

## CHEMIST'S CORNER 25-1

### Liquid Fondant

The fondant of cream-filled centers is often prepared with an invertase enzyme that slowly hydrolyzes the sucrose to invert sugar. The higher solubility of this invert sugar compared to sucrose makes it melt under the chocolate layer, turning the center into a creamy liquid (30).

Fondant is also used as an icing to glaze and decorate pastries, sealing baked products from air and the loss of moisture (15). Professional bakers buy it ready-made or in powder form, but large amounts can also be prepared by hand.

Fondant preparation starts with sugar, preferably glucose, which is combined with half as much water and a little cream of tartar or corn syrup. The mixture is heated in a covered saucepan to a rapid boil. The solution is then cooked to the soft-ball stage, or 234°F–240°F (112°C–115°C). After reaching the final temperature, the mixture is poured onto a flat marble slab and cooled to about 120°F (49°C). The fondant should be poured, with no scraping or shaking, over a wide surface for even cooling. When the proper temperature is reached, a spatula or wooden spoon is used to mix the candy until it becomes white and creamy. The color and consistency come from the formation of fine, uniform crystals (24). At this point, ½ teaspoon of vanilla or other flavor extracts or colorings may be added. The mixture is then kneaded until smooth. The end product should be stored overnight or even several days in a tightly covered jar until it becomes soft enough to use (15). It can keep up to several weeks on the shelf if kept in an airtight container in a cool, dry location away from high humidity, and up to several months once it is surrounded by chocolate.

**Fudge**   Fudge is the most popular of all crystalline candies. It is made by adding chocolate, cream, milk, and butter to a simple syrup. Table 25-3 lists the functions of dairy ingredients in the manufacture of fudge and other candies. After the ingredients have been heated to the soft-ball stage (234°F–240°F/112°C–115°C), the mixture must be cooled to about 120°F (49°C). The fudge is then beaten until the gloss dulls and the texture turns creamy smooth. Beating fudge too long hardens it, but this can be corrected with recooking (Figure 25-3). After cooling and beating, the fudge should be poured at once into a greased pan. Covering it will prevent its drying, as will waiting until serving time to cut it into squares. Panocha or penuche is similar to fudge except it is made with brown, instead of white, sugar and contains no chocolate or cocoa.

**Divinity**   Divinity is prepared like fondant and fudge up to the point of beating. When the mixture has reached 250°F–266°F (121°C–130°C) (the hard-ball stage), it is poured slowly over stiffly beaten egg whites, with continuous beating. The beating continues until the candy holds its shape when dropped from a spoon. At this point, it can be poured into a buttered pan or dropped by spoonfuls onto a buttered surface.

| TABLE 25-3 | Functions of Dairy Ingredients in Confections |
|---|---|
| **Dairy Ingredient** | **Function in Confections** |
| Milk fat | • Gives rich flavor, texture, and mouthfeel. |
| | • Provides lecithin and monoglycerides, which aid in emulsification. |
| | • Contains precursors of key flavor compounds formed by cooking. |
| | • Acts as a flavor carrier. |
| | • Provides a moisture barrier. |
| | • Inhibits bloom defect. |
| | • Oxidation and/or hydrolysis produces distinctive desirable flavor compounds. [However, under certain conditions of processing and storage, off-flavors (rancid, fishy, stale) may develop.] |
| Milk proteins | • Aid in emulsification and miscibility of various ingredients. |
| | • Assist in whipping and foam formation. |
| | • Absorb moisture due to water-binding capacity. |
| | • With heat, can yield protein fiber network that gives rigidity, shape, texture, and chewability. Furthermore, impede movement of water molecules, leading to retardation of sugar crystallization. |
| | • Contribute to flavor and color development. |
| | • Improve nutritional profile. |
| Lactose | • Imparts chewiness and graininess. |
| | • Causes sandiness defects under certain conditions, e.g., at high lactose concentrations. |
| | • Acts as a flavor carrier. |
| | • Polymerizes to form a flexible and extensible matrix (textural effect). |
| | • Caramelizes to generate color and flavor compounds in caramels and toffees. |
| | • Provides reducing sugar moiety that can participate in Maillard browning, which affects color and flavor. |
| | • Acts as a precursor of pyrazines (nutty flavor), maltol and isomaltol (caramel flavor), furfurol and furfuraldehyde (cooked cereal). |
| | • Contributes to dark color and bitter taste defects under certain conditions of overcooking. |

**FIGURE 25-3**  Beating fudge.

Adequately beaten fudge is smooth and the gloss has just dulled.

Overbeaten fudge becomes hard and loses its gloss.

# Noncrystalline Candies

Noncrystalline candies are easier to make because the goal is to ensure that the sugar does not crystallize. Two major methods are used to inhibit crystallization: creating very concentrated sugar solutions and/or adding large amounts of **interfering agents** to block the sugar molecules from clustering together. The high viscosity of the concentrated syrup inhibits crystallization by impeding molecular movement.

**Interfering agent** A substance added to the sugar syrup to prevent the formation of large crystals, resulting in a candy with a waxy, chewy texture.

### Concentrating the Sugar Solution

Using high temperatures to evaporate much of the water results in a very concentrated syrup, and the degree of evaporation determines the percentage of moisture in noncrystalline candies. Sugar-syrup-based candies containing 2% moisture or less include hard candies, sour balls, butterscotch, and nut brittles. A higher moisture content of 8% to 15% is found in caramel and taffy, and an even higher level of 15% to 22% moisture is contained in marshmallows, gumdrops, and jelly beans (30).

### Interfering Agents

Interfering agents may inhibit crystallization in several ways:

• By inhibiting nuclei formation
• By physically coating the crystals, which prevents their growth
• By decreasing water activity, which leaves less water available for the sugar to dissolve in

The two main interfering agents used in confectionery production are corn syrup and cream of tartar, but certain other ingredients added in large amounts can also contribute to preventing crystallization. Corn syrup inhibits crystallization because its glucose chains form a complex that impedes the movement of the sugar and water molecules. The inhibiting action of cream of tartar results from the formation of invert sugar from sucrose; the invert sugar then acts similarly to corn syrup by getting in the way of the sucrose molecules. Commercial invert sugar achieves the same purpose. Both interfering compounds provide more time for cooling and/or beating, which promote small crystal formation (1).

### Types of Noncrystalline Candies

Noncrystalline candies vary in their degree of sugar solution concentration

**CHEMIST'S CORNER 25-2**

**Citric Acid in Candies**

Citric acid is added to some candies for flavor. The only problem is that the acid, being very hygroscopic (tending to draw moisture from the air), turns the candy somewhat liquid so that it sticks to the wrapper. The confectionery industry has bypassed this side effect of citric acid by encapsulating candies in a film or coating (28). However, the coating's effectiveness does not last as long as the hard candy, which is why very old candy left out in a bowl turns "sticky soft."

**FIGURE 25-4**   Spreading peanut brittle to cool.

Thick gloves protect against the hot candy that is stretched as thin as possible.

and the types of interfering substances added (Chemist's Corner 25-2). The preparation of a few of these noncrystalline candies is now discussed in more detail.

*Hard and Brittle Candy*   All but 1% of the moisture from simple syrup is removed when making hard candies (7). Flavorings and colorings are then added, with centers or fillings being optional. Brittle candies such as peanut brittle and toffee have baking soda added. To make brittle, the syrup is heated to the high temperatures of caramelization and then spread out on a hard surface to cool (Figure 25-4). The bubbles in these candies come from the carbon dioxide gas produced by the baking soda.

*Caramels*   To inhibit the formation of crystals when making caramel, large amounts of interfering substances such as fat, cocoa butter, concentrated milk products, and/or corn syrup are added to the sugar syrup. The amount of fat added to caramels is high—about ¼ cup of fat for every cup of sugar. The result is a candy with a waxy, chewy texture. The dairy products contribute protein, calcium, and flavor. The milk also influences the characteristic color of caramel candies. Ironically, the tan color is not from caramelization, but from the Maillard reaction, which occurs as sugars interact with the milk proteins (19). The color starts to develop when the temperature reaches

about 325°F (163°C); however, if it is allowed to become too dark, a bitter taste results (24).

*Taffy*   Taffy is very similar to caramel except that it is made from a more concentrated solution, which makes it firmer. The other difference is that, once solidified, the syrup mass is pulled to aerate the mixture by incorporating air bubbles. This pulling transforms the mass into a candy that is lighter, chewier, and paler than caramel. Different flavorings can be added to create varied types of taffy.

*Aerated Candies*   Corn syrup serves as the foundation for aerated candies such as marshmallows, jelly beans, and gumdrops. Candy can be aerated physically (pulling), chemically (adding sodium bicarbonate), or by the addition of foams for structure; marshmallows, for example, incorporate an egg-white foam (1). Gummy textures are formed by adding gelling agents such as starch, gelatin, pectin, and gums to the sugar. Gelatin can be used in confectionery production to stabilize marshmallows and other aerated products, to add elasticity to gums and jelly beans, and to soften and bind water in caramels and other chewable sweets (32).

# CHOCOLATE

Chocolate is the chief ingredient for many different types of confections. It is consumed in the form of chocolate bars; as a syrup topping; or as a coating (or center) for candy bars, candies, ice cream, cookies, nuts, and fruit (6). The origin of chocolate confections starts with the tropical cocoa or cacao tree (Figure 25-5). Warm, moist climates near the equator provide the best environment for cacao tree growth. Botanists believe the tree originated in the Amazon-Orinoco river basin in South America; the crop is now cultivated primarily in West Africa and Brazil.

The tree's cacao beans played an important role in the traditions, religion, and legends of the Aztecs. Historians suggest that it was the Aztecs who first came up with chocolate beverages, and who believed that the cocoa tree had a divine origin, leading to its botanical name, *Theobroma*, meaning "food of the gods." The Aztec's term for their drink, "bitter water" or "cocoa water," was *xocolatl*, which the Spanish later converted to "chocolate." Hernando Cortez brought the cocoa beans from Central America back to Spain in 1758, and from there they quickly spread to other parts of Europe. The popularity of chocolate has led to a hypothesis that for some people its addictive-like quality and resulting "chocolate binges," especially as they relate to depression, may parallel other addictions (36).

**FIGURE 25-5** Cacao tree with pods (fruit).

Courtesy of the Chocolate Manufacturer's Association

# Chocolate Production

The steps of producing chocolate include removing the chocolate liquor from the cocoa beans, developing flavor and consistency through the process of conching, tempering the chocolate to increase its resilience to temperature changes, and finally molding the chocolate into the finished product (see Figure 25-6 for an overview).

## Obtaining Chocolate Liquor from Cocoa Beans

The first step in chocolate production is the manufacture of chocolate liquor, the basic ingredient for all chocolate products. After being picked, cocoa beans are scooped out of the pods in which they are packed and heaped into large piles, covered with banana leaves, and allowed to ferment. The fermentation alters the seed coat, making it easy to remove, and modifies the beans' flavor and color (21). The beans are then dried to a 7% moisture level to give them good keeping quality before being shipped to a chocolate manufacturer.

Upon arriving at the manufacturing plant, the beans are blended into various combinations to obtain specific flavors and colors, which are further developed by roasting. The hull and the germ of the cocoa beans are then removed, and what remains is called the nibs. These nibs, containing 54% cocoa butter, are ground very fine under heat to yield chocolate liquor (Figure 25-6). Grinding is very important to the final chocolate's mouthfeel (9). The cocoa butter in this chocolate liquor has a melting point just below body temperature, and is responsible for chocolate's melt-in-the-mouth appeal and brittle snap at room temperature (15, 30).

## Conching

The next step in manufacturing good-quality chocolate is conching, during which the chocolate's characteristic flavor and consistency are developed. Warmed chocolate (usually between 70°F and 160°F/21°C and 71°C) is kneaded (conched) and aerated by machines to increase its smoothness, viscosity, and flavor (30). Several ingredients may be added during this time. These include flavorings, additional cocoa butter, and lecithin, which increase viscosity and cause the chocolate to develop a velvety

**FIGURE 25-6** The process used to manufacture chocolate.

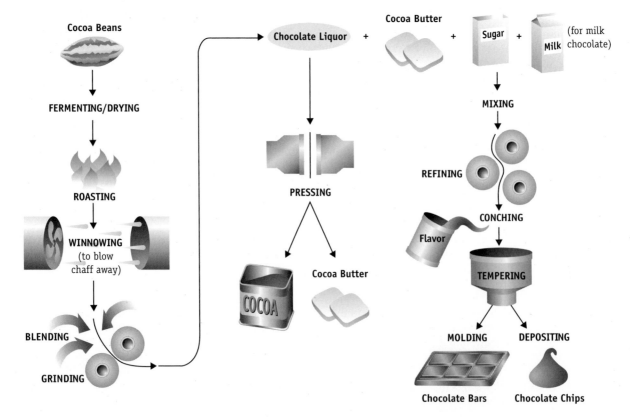

smooth coating. Conching complete, the chocolate may be poured into blocks and cooled before being packaged; at this point, it can be stored without any major problems for up to a year (10).

### Tempering

Chocolate may also undergo **tempering** before it is formed and packaged. The three basic steps in tempering chocolate are melting, cooling, and rewarming (22). Proper tempering is what gives high-quality chocolate that "snap" when bitten into or broken in half (Chemist's Corner 25-3). Tempered chocolate is often used to **enrobe** candies and other foods such as ice cream bars, fruits like raisins and strawberries, cookies, nuts, and doughnuts (37). The chocolate manufacturing industry often automates the enrobing process by pouring a sheet of molten chocolate over products placed on a moving belt (39). Chocolate can also be used to enrobe soft, fat centers such as truffles, nut pastes, and peanut butter (11). Untempered or improperly tempered chocolates may exhibit gray surfaces, or bloom, on the surface of hardened chocolate. Bloom is considered a defect in chocolate manufacturing not only because of its discoloration, but also because it affects flavor (20).

Fat bloom can be distinguished from sugar bloom by its greasy feeling. Bloom and other problems of chocolate production along with several solutions to them are listed in Table 25-5.

***Factors Affecting Tempering*** The correct chocolate, the temperature, and the timing are crucial factors in tempering. A certain amount of fat is needed to produce a smooth, glossy coating, so using the right chocolate is imperative. Baking chocolate and cocoa, for example, are too low in fat for dipping candies and will not respond to tempering. The weather also affects chocolate consistency; too much humidity can interfere with the solidification of the chocolate. Clear, cool days not exceeding 70°F (21°C) are preferred (31). A minimum of 1½ pounds of chocolate is required for tempering. Using any less than this amount will yield results that are too unpredictable under the rapid temperature changes that will be necessary to temper the chocolate (24). When used for coating candies, 1½ pounds of chocolate will cover about 90 ¾-inch centers (31). Prior to heating, the chocolate should be chopped or grated to allow for faster, more uniform melting. Chopping is easier if the chocolate is at room temperature and a clean, dry cutting board is used.

The three most common tempering methods are tabliering, seeding, and the cold-water method.

***Tabliering Method*** Melting the chocolate pieces is the first step in the tabliering technique. It is never done over direct heat; a double boiler is used in order to keep temperatures from escalating too high. A bowl placed over a saucepan of simmering water might also be used. The bowl should not touch the water and should be large enough to prevent steam escaping from the saucepan from condensing and coming in contact with the chocolate. In either case, the chocolate is stirred constantly to promote melting and prevent burning. The spoon or spatula should be wooden or plastic because these materials are poor conductors of heat. A metal utensil will conduct heat away from surrounding areas of the chocolate. This cools the chocolate and makes the mixture more saturated.

For the same reason, as well as to prevent lumping, cold liquids in the form of milk, cream, or liqueur should never be directly added. On the other hand, melting the chocolate at too high a temperature or too quickly can scorch it or cause the cocoa butter to separate out, forming undesirable lumps.

---

### > CHEMIST'S CORNER 25-3

#### Tempering and Chocolate Melting Point

Cocoa butter exhibits polymorphism (many shapes), meaning that it can take on several crystal forms, of which only one is stable (9). Tempering causes the fatty acids in cocoa butter to change into the stable beta form. Table 25-4 shows that fatty acids in the stable beta form have the highest melting point. Slow cooling is recommended over rapid cooling to avoid the conversion to the unstable alpha phase. However, rapid cooling can be remedied by slow heating, which causes the crystals to convert from alpha to beta prime, and finally to the more stable beta form (37).

#### TABLE 25-4
Polymorphism and Melting Points of Cocoa Butter

| Fatty Acid Crystal Form Phase | Melting Point °F | °C |
|---|---|---|
| α | 73 | 23.3 |
| β' | 82 | 27.5 |
| β | 93 | 33.8 |

---

### ? How & Why?

**Why does a grayish film sometimes form on chocolate?**

The grayish bloom, which usually occurs after the chocolate has been stored or allowed to get warm, is caused by one of two things: (1) fat crystallizing onto the surface or (2) moisture in the air reacting with the sugar on the surface (23, 30). Fat bloom may be caused by improper crystallization; incorrect temperatures in heating or cooling; insufficient stirring; or inadequate conditions where the chocolate is prepared, such as less than optimal room temperatures, humidity, or drafts. It also may occur when the soft fat centers or added fats of candies are incompatible with cocoa. Even if the chocolate has been tempered well, it may bloom if stored for too long a time (9). Sugar bloom occurs when chocolate is stored under damp conditions (sugar in the chocolate absorbs water) and then exposed to a drier environment (moisture on the surface evaporates, causing sugar to recrystallize on the surface) (1).

---

**Tempering** The process of heating and cooling chocolate to specific temperatures, making it more resistant to melting and resulting in a smooth, glossy, hard finish.

**Enrobe** To coat food with melted chocolate that hardens to form a solid casing.

## TABLE 25-5   Confectionery Problems, Causes, and Solutions

| Problem | Causes | Solution |
|---|---|---|
| Soapy, rancid flavor | Lipase activity | Check incoming ingredients for off-flavors. |
| Burned, medicinal, fishy, or chemical taste | Oxidation of oils or fats during product storage | Minimize exposure to light and maintain cool temperatures. |
| | | Check incoming ingredients for off-flavors. |
| Scorched or burned taste | Overheated sugar and milk solids | Control temperatures and times during heating and cooling. |
| Lack of brown color and flavor notes | Insufficient Maillard browning | Increase amount of milk products. Use a slower cooking process. |
| | Too little caramelization | Increase sugar content. |
| | | Cook at higher temperatures and/or for longer times. |
| Poor caramel texture | Too thick or poor elasticity | Decrease milk protein level. |
| | Too thin | Increase milk protein level to increase viscosity. |
| Gritty or grainy texture | Change in sugar crystal type | Adjust balance or levels of sugar types for proper crystallization. |
| | Large sugar crystals formed | Reduce lactose content. |
| | | Check seeding and agitation during heating and cooling. |
| | Sugar crystals formed on product surface | Optimize storage conditions (humidity and surface temperature) to prevent migration of fat and water to surface. |
| Bloom (white, dusty appearance) | Fats not in proper phase | Check melting temperatures and cooling temperatures and times. |
| | | Check amount and type of seed materials to ensure proper tempering. |
| | | Increase tempering time. |
| | Incompatible fats | Use fats with melting points that are appropriate for the processing conditions. |
| | | Avoid lauric fats in coatings. |
| | | Include milk fat in formula. |
| | Improper storage temperatures | Store at recommended temperatures (80°F–86°F/27°C–30°C). |

Source: Reprinted from Dairy-Based Ingredients by Ramesh Chandan, Eagan Press, St. Paul, MN, USA. Copyright © 1997, Elsevier.

When the temperature reaches about 115°F–120°F (46°C–49°C), it is ideally kept there for about 30 minutes with continued stirring. The temperature should not be allowed to exceed 125°F (52°C). The chocolate is then cooled by removing it from the heat and placing one third of it on a marble slab (Figure 25-7). It is scraped back and forth until it cools to just under 80°F (26°C), at which point it will crystallize and thicken, signaling that it is to be added back to the other two thirds. Then it is again warmed slowly to 85°F–90°F (29°C–32°C), where it reaches a consistency that can be worked. The entire process should be repeated if the temperature is allowed to rise above 91°F (33°C) (24).

**Seeding Method**   Slight variations of the tabliering method have resulted in two other tempering techniques, the seeding and cold-water methods. In seeding, already tempered chocolate is melted in the double boiler, but then removed from the heat source, at which time more unmelted tempered chocolate pieces are added. The chocolate is cooled to 80°F (26°C) and stirred for 2 minutes at this temperature before being warmed again to 85°F–90°F (29°C–32°C) (15).

**Cold-Water Method**   The cold-water method consists of cooling the heated chocolate by placing the bowl of chocolate in a larger bowl of ice water instead of pouring it onto a slab.

## FIGURE 25-7   Further cooling and manipulation of chocolate.

Working chocolate on marble helps to keep it cool.

Shaping chocolate in a plastic-lined frame.

The chocolate must be stirred as the bowl is lowered into the ice water, because it starts to set immediately. As in the other methods, the chocolate is then rewarmed to 85°F–90°F (29°C–32°C) so that it can later be cooled and shaped.

### Tempering for Chocolate Dipping
A slight variation is made in the tempering process if the chocolate is to be used for dipping fruits, nuts, or other items (Figure 25-8). After it reaches its maximum temperature, it is cooled to about 83°F–85°F (28°C–29°C), the ideal temperature for dipping. Items must be dipped quickly, because the temperature range in which the chocolate stays at the right consistency is extremely narrow. It is kept from getting too cool by leaving it in the double boiler on reduced heat. After the items have been dipped, they are held up for a moment in the air, allowing the excess to drip off; then they are placed on wax paper in such a way that the hardening chocolate does not form an unsightly line or an undesirably large base. A broad base is caused by using too much chocolate or having the temperature too high.

### Commercial Chocolate Coating
The food industry uses four major steps to coat nuts, raisins, and other

**FIGURE 25-8**   Tempering chocolate for dipping.

Chocolate chopped for uniform melting.

Checking chocolate temperature.

Dipping item to be coated in chocolate.

**FIGURE 25-9**   Coating nuts with chocolate.

Hazelnuts with semi-rough finish

Almonds with semi-rough finish

Raisins with wax finish

Richard Brewer/Digital Development Services

small foods (Figure 25-9). Chocolate is unique in that its coating solidifies upon cooling rather than being absorbed by the food (8). Roasted nuts sprayed with chocolate yield a semi-rough surface. Their surfaces become smoother as the coated nuts tumble in the mixer. The glossy, shiny exterior of chocolate-covered nuts is often achieved by adding a natural gum to seal the chocolate, a shellac solution, and a final layer of edible wax.

### Coatings

The inconvenience of tempering and the risk of bloom can be entirely avoided by using **nontempered coatings.** Known as *coating chocolate, chocolate icing, confectionery coating,* or *nontemp chocolate*, these products are not true chocolates. The cocoa butter in nontemp chocolates has been replaced with other fats, such as hydrogenated palm kernel, lecithin, soy, or cottonseed oils.

Nontempered coatings are available in white, milk, and dark chocolate flavors, and are easier to use than tempered chocolate, but they fall short of real chocolate in flavor and texture. They are accordingly reserved for use as glazes or decorations rather than as fillings or major ingredients (15). Their major benefit is that items can be dipped in them without the preparer having to follow the detailed, time- and temperature-dependent steps of tempering.

## Types of Chocolate Products

Most chocolate products start with chocolate liquor. Chocolate manufacturers create different chocolates based on how much cocoa butter and or other ingredients are added, and how the mixture is manipulated. Chocolate confections are made by combining cocoa butter, cocoa

powder, sweeteners, milk (sweetened condensed milk, whole milk powder, or evaporated milk), and other ingredients such as nuts (9). Chocolate standards of identity have been established to define various chocolates, and they often contain more cocoa butter than cocoa powder.

### Baking Chocolate

Baking chocolate is chocolate liquor that has cooled and solidified into cakes. It is also sold as bitter, unsweetened baking chocolate, but in any form it must have a chocolate liquor content of at least 35%. Other possible ingredients include sugar, cocoa butter, lecithin, and flavoring (24). Lecithin is often used in confections because it can replace some of the cocoa butter in chocolate bars (32).

### Cocoa

Removing most of the cocoa butter from chocolate liquor results in cocoa. This reduces the fat content to 10% to 24% in cocoa, compared to 38% in eating chocolate and 50% in unsweetened baking chocolate (26). The lower fat proportion gives cocoa a very concentrated chocolate flavor. Cocoa is made by pressing much of the cocoa butter out of the heated liquor as it hardens into cakes. The cakes are then ground into cocoa powder that comes in a wide variety of colors, particle sizes, and flavors (9). Two main types of cocoa are available: natural cocoa, which

**Nontempered coating** A coating resembling chocolate that is not subject to bloom because it is made with fats other than cocoa butter.

is slightly acidic, and the less bitter cocoa, which may be called alkalized, Dutch-processed, or Dutch cocoa. A Dutchman started one of the first hard chocolate businesses by inventing the process of *Dutching*, which consists of treating crushed cocoa beans or chocolate liquor with an alkali (usually potassium carbonate or sodium carbonate) to raise the pH. The degree of alkalization of the cocoa determines its color, which ranges from deep reddish brown to charcoal black. As the color darkens, the chocolate's flavor becomes more bitter and astringent. Europeans prefer the darker, reddish-colored Dutch cocoa for their recipes, whereas North Americans predominantly use natural cocoa (26).

## Semisweet or Sweet Chocolate

Granulated sugar and extra cocoa butter are added to the chocolate liquor to produce the sweeter, smoother taste of semisweet or sweet chocolate. Semisweet chocolate must contain at least 15% chocolate liquor, whereas equal parts of chocolate liquor and sugar are present in sweet chocolate (24).

## Milk Chocolate

Milk chocolate candy bars were first produced in 1875 when a Swiss manufacturer added condensed milk to chocolate liquor and other ingredients. Condensed milk had just been invented by another Swiss, Henri Nestlé. A little over 25 years later, in 1903, Milton Hershey formed the first company to mass-produce milk chocolate in the United States, which later became the largest chocolate manufacturing plant in the world. The name of the Pennsylvania town where the factory was built was eventually changed to Hershey (15). Today, more than 60% of all chocolate sold in the United States is milk chocolate (6).

Milk chocolate has a lighter color and sweeter, milder flavor than other chocolates. It is made like semisweet and sweet chocolate except that it contains less chocolate liquor (10% minimum), more granulated sugar (40% to 55%), cocoa butter, and dried whole-milk solids (12% minimum) (19). The milk is added in powder form to prevent the chocolate's consistency from becoming too liquid. At the molecular level, milk chocolate is a suspension in which the continuous phase is the cocoa butter and milk fat, whereas the dispersed phase consists of the cocoa, sugar, and nonfat milk solids (16).

## Imitation Chocolate

Some or all of the cocoa fat is replaced with vegetable fat in imitation chocolates. These less costly versions of chocolate, which are also less likely to melt during the warmer summer months, are sometimes used to coat candies, ice cream bars, and crackers.

## White Chocolate

Technically, white or ivory chocolate is not chocolate, nor is it recognized as such by the FDA, because it contains no chocolate liquor or cocoa (17). Its basic ingredients are sugar, cocoa butter, milk, natural flavor, lecithin, and vanillin. If the cocoa butter is removed from white chocolate, the product is referred to as *white coating* or *confectionery coating*.

---

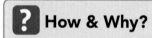

**? How & Why?**

### Why do chocolates taste different?

The various types of chocolate just discussed, along with the ratio of added ingredients, affect flavor, but so do the origin of the cocoa bean and differences in roasting and alkalization techniques, which result in variations of cocoa powder colors, aromas, and flavors (4). The amount of cocoa not only affects flavor—which becomes stronger as more cocoa powder is added—but actually determines how chocolate is defined (5):

- Unsweetened chocolate = 100% cocoa
- Bittersweet chocolate = 35% or more cocoa
- Semi-sweet chocolate = 15–35% cocoa

---

## Functional Chocolate

A new trend in chocolates is to maximize their health benefits (25). Although the high fat and sugar content of chocolates prohibits them from being labeled as *functional*, some manufacturers are adding ingredients known to benefit health (see Chapter 1). For example, fiber and omega-3 fatty acids or flavonols (antioxidants) have been added to boost the benefits of chocolate (40). Dark chocolates are extolled for their higher antioxidant concentrations compared to milk chocolate (13). Even low-carbohydrate chocolates made with less sugar are now available (32).

# STORAGE OF CANDY

Storage requirements vary depending on the candy. Those lowest in water content, such as hard candies and brittle, keep indefinitely if properly wrapped, because they do not support the growth of microorganisms. Extended storage and subsequent exposure to moisture in the air, however, can cause their surfaces to become gummy. Candies such as fudge and fondant, which have a higher moisture content, get softer and smoother in texture if left in an airtight container, but after about a day, a graininess develops as the sugar crystals become larger. Ingredients other than sugar, such as fat or milk products, are subject to rancidity, which results in off-flavors and odors; however, this degradation can be delayed by refrigeration or freezing. Ingredients that improve shelf life include humectants, which act to hold moisture. Common humectants incorporated into candies include glycerin (glycerol), sorbitol, pectins, and gums (30).

## Shelf Life of Chocolate

Chocolates can stay on the shelf for over a year as long as they are not subjected to wide fluctuations in temperature and humidity. The exceptions are milk and white chocolates, which have a slightly shorter shelf life of about 8 months to 1 year due to their milk content.

In order to maximize its shelf life, chocolate should be properly wrapped and stored in a cool, dark place, the ideal temperature being 65°F (18°C) with the humidity at 50% (24). Even slight melting may result in fat bloom as the cocoa butter crystals form a grayish or whitish film on the chocolate's surface. Sugar bloom, which is rougher in appearance and texture, can result under conditions of high humidity.

# PICTORIAL SUMMARY / 25: Candy

The United States produces more candy than any other country in the world. Americans eat an astonishing 24 pounds of confectioneries per person per year, with a little over half of this in the form of chocolate.

## CLASSIFICATION OF CANDIES

There are thousands of different candies, but they can be classified according to their ingredients and/or preparation method.

### Classification by ingredients:

*Syrup phase (or sugar phase):* Most candies fall into this category; they are made from a simple syrup, which is a mixture of boiled sugar and water. Composed of sugar and flavorings, syrup-phase candies include:

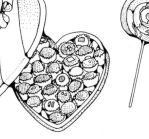

- Hard candy
- Nougats
- Caramels
- Marshmallows
- Jelly beans
- Gums
- Fudges
- Fondants

*Fat phase:* When chocolate and/or peanut butter are added, the candy is considered fat phase.

*Combination of both fat- and syrup-phases* can be found, for example, in a chocolate-covered candy bar.

### Classification by preparation method:

*Crystalline candies*, which are soft, smooth, and creamy, are formed from sugar solutions yielding many fine, small crystals. They include fudge, fondant, and divinity.

*Noncrystalline candies* are formed from sugar solutions that did not crystallize, and are amorphous, or without form. They include caramel, toffee, hard candy, and gummy candies.

## PREPARATION OF CANDY

The art of confectionery production, highly sensitive to timing, temperature, and skill, requires patience and practice.

The preparation of many, but not all, confections can be generally summarized by four basic steps:

1. Creating a syrup solution

2. Heating this mixture to concentrate the contents through evaporation

⎤ These steps are basically the same for crystalline and noncrystalline candies.

3. Cooling

4. Beating

*Crystalline candies:* Goal is to form sugar crystals, so rapid cooling and beating are used.

*Noncrystalline candies:* Goal is to avoid the formation of sugar crystals, so cool slowly, without agitation.

## CHOCOLATE

**Chocolate** from the tropical cacao tree is the main ingredient of many different types of candies. Cocoa beans are ground very fine and heated to yield chocolate liquor, the basic ingredient for all chocolate products—cocoa, baking chocolate, milk chocolate, semisweet chocolate, and sweet chocolate. Other ingredients can include cocoa butter, sugar, milk products, and flavorings.

## STORAGE OF CANDY

Candies lowest in water content (hard candies/brittle) keep the longest if properly wrapped; however, extended storage times or exposure to moisture can cause their surfaces to become gummy. Fat or milk added to candies makes them prone to rancidity, but refrigeration or freezing delays this process. Humectants added to candies improve their shelf life.

Chocolates can stay on the shelf for over a year, except for milk and white chocolates, which last about 8 months to 1 year. All chocolate should be properly wrapped and stored in a cool (65°F/18°C, humidity 50%) dark place. Unsightly grayish or whitish film observed on a chocolate's surface is bloom resulting from either fat or sugar crystallizing on the surface.

PhotoDisc/Getty Images

# CHAPTER REVIEW AND EXAM PREP

## Multiple Choice*

1. Fudge is classified as what type of candy?
   a. Noncrystalline
   b. Crystalline
   c. Divinity
   d. Hard

2. The size of the sugar crystals is determined by the rate or speed of _____ formation.
   a. nuclei
   b. fat crystal
   c. syrup
   d. invertase

3. Identify the correct order of steps, from beginning to end, for preparing most types of candy.
   a. Cooling, beating, heating, creating a syrup solution
   b. Creating a syrup solution, heating, cooling, beating
   c. Beating, heating, cooling, creating a syrup solution
   d. Creating a syrup solution, heating, beating, cooling

4. Which ingredient is used to ensure that the sugar does not crystallize when preparing noncrystalline candies?
   a. Emulsifiers
   b. Sugars
   c. Interfering agents
   d. Seeds

5. What is the name of the process by which warmed chocolate is kneaded and aerated by machines for several days?
   a. Blooming
   b. Enrobing
   c. Tempering
   d. Conching

6. The process of heating and cooling chocolate to make it shiny and more resistant to melting is called _____ .
   a. conching
   b. seeding
   c. tempering
   d. blooming

7. White chocolate is technically not chocolate because it does not contain _____ .
   a. invert sugar
   b. chocolate liquor or cocoa
   c. cocoa butter
   d. conched chocolate

## Short Answer/Essay

1. Many (but not all) candies can be prepared by using which four basic steps?
2. How do the principles of preparation differ for crystalline and noncrystalline candies? Give examples of each type of candy.
3. Why is glucose, invert sugar, or corn syrup added to sucrose when making crystalline candies? What happens if too much is added?
4. Describe the influences of the following factors on crystalline candy formation: overheating, underheating, and weather.
5. Explain the proper use of a candy thermometer, including how to check the thermometer for accuracy, how to immerse it in the mixture, and how to read the thermometer.
6. How do beating and cooling the syrup solution affect the formation of crystals in candy?
7. List several examples of interfering agents used in the preparation of noncrystalline candies. By what three mechanisms do they inhibit crystallization?
8. Explain how the following chocolate products differ according to the amount of cocoa and other ingredients they contain: baking chocolate, cocoa, milk chocolate, imitation chocolate, and white chocolate.
9. Discuss the processes of conching and tempering in the production of chocolate, and describe the desired results they produce.
10. What are the possible causes of the following problems associated with candy preparation: rancid flavor, lack of brown color, and gritty texture?

*See p. AK-1 for answers to multiple choice questions.

# REFERENCES

1. Alexander RJ. *Sweeteners: Nutritive*. Eagan Press, 1998.
2. Alikonis JJ. *Candy Technology*. Avi Publishing, 1979.
3. Altschul AM. *Low-Calorie Food Handbook*. Marcel Dekker, 1993.
4. Anonymous. Cocoa origin affects end product. *Food Technology* 63(5):47–91, 2009.
5. Anonymous. Tasting dark chocolate. *Fine Cooking* 83:53, 2007.
6. Chandan R. *Dairy-Based Ingredients*. Eagan Press, 1997.
7. Chin M, and D Frick. Formulating a color delivery system for hard candy. *Food Technology* 49(7):56–61, 1995.
8. Clark JP. Coating is critical to many foods. *Food Technology* 58(8): 82–83, 2004.

9. Clark JP. Crystallization is key in confectionery processes. *Food Technology* 58(12):94–96, 2005.

10. Clark JP. Lessons from chocolate processing. *Food Technology* 61(12):89–91, 2007.

11. Choi YJ, KL McCarthy, and MJ McCarthy. Oil migration in a chocolate confectionery system evaluated by magnetic resonance imaging. *Journal of Food Science* 70(5):E312–E317, 2004.

12. *Confectionery*: 2003. U.S. Census Bureau. www.census.gov/industry/1/ma311d03.pdf. Accessed 7/09.

13. Corti R, AJ Flammer, NK Hollenberg, and TF Lüscher. Cocoa and cardiovascular health. *Circulation* 19(10):1433–1441, 2009.

14. Deshpande A, and AR Jadad. The impact of polyol-containing chewing gums on dental caries: A systematic review of original randomized controlled trials and observational studies. *Journal of the American Dental Association* 139(12):1602–1614, 2008.

15. Friberg B. *The Professional Pastry Chef*. Van Nostrand Reinhold, 1996.

16. Full NA, et al. Physical and sensory properties of milk chocolate formulated with anhydrous milk fat fractions. *Journal of Food Science* 61(5):1068–1072, 1996.

17. Giannatempo L. Shopping for white chocolate. *Fine Cooking* 91:78, 2008.

18. Izzo M, C Stahl, and M Tuazon. Using celluose gel and carrageenan to lower fat and calories in confections. *Food Technology* 49(7):45–50, 1995.

19. Jeffrey MS. Key functional properties of sucrose in chocolate and sugar confectionery. *Food Technology* 47(1):141–144, 1993.

20. Kinta Y, and T Hatta. Composition and structure of fat bloom in untempered chocolate. *Journal of Food Science* 70(7):S450–S452, 2004.

21. LaBell F. Cocoa: A bean of many flavors. *Prepared Foods* 166(8):81, 1997.

22. Loisel C, et al. Fat bloom and chocolate structure studied by mercury porosimetry. *Journal of Food Science* 62(4):781–788, 1997.

23. Loisel C, et al. Tempering of chocolate in a scraped surface heat. *Journal of Food Science* 62(4):773–779, 1997.

24. Malgieris N. *Nick Malgieri's Perfect Pastry*. Macmillan, 1989.

25. McShea A, E Ramiro-Puig, SB Munro, G Casadesus, M Castell, and MA Smith. Clinical benefit and preservation of flavonols in dark chocolate manufacturing. *Nutrition Reviews* 66(11):630–641, 2008.

26. Medrich A. Rediscovering cocoa. *Fine Cooking* 17:68–73, 1996.

27. Nabors LO. Intense sweeteners: Acesulfame K, alitame, aspartame, saccharin, sucralose. *Manufacturing Confectioner* 70(11):65, 1990.

28. Ohr LM. A means to various ends. *Prepared Foods* 166(11):67, 1997.

29. Penfield MP, and AM Campbell. *Experimental Food Science*. Academic Press, 1990.

30. Potter NN, and JH Hotchkiss. *Food Science*. Chapman & Hall, 1995.

31. Pritchard A. *Anita Pritchard's Complete Candy Book*. Harmony Books, 1978.

32. Pszczola DE. Ingredient developments for confections. *Food Technology* 51(9):70, 1997.

33. Pszczola DE. Ingredients. Confection: A sweet acronym. *Food Technology* 58(10):50–65, 2004.

34. Seligson FH, Krummel DA, and JL Apgar. Patterns of chocolate consumption. *American Journal of Clinical Nutrition* 60(suppl):1060S–1064S, 1994.

35. Shinsato E. Confectionery ingredient update. *Cereal Foods World* 41(5):372–375, 1996.

36. Smit HJ, EA Gaffan, and PJ Rogers. Methylxanthines are the psycho-pharmacologically active constituents of chocolate. *Psychopharmacology (Berl)* 176(3–4):412–419, 2004.

37. Stauffer CE. *Fats and Oils: Practical Guides for the Food Industry*. Eagan Press, 1996.

38. Vink W. Applications in confectionery: Tableted confections. *Manufacturing Confectioner* 70(11):77, 1990.

39. Wichchukit S, MJ McCarthy, and KI McCarthy. Flow behavior of milk chocolate melt and the application to coating flow. *Journal of Food Science* 70(3):E165–E171, 2005.

40. Wright R. Sweet opportunities in functional confectionery. *Nutraceuticals World* 12(3):60–67, 2009.

# WEBSITES

This is a website on everything candy by the National Confectioners Association:

**www.candyusa.org**

Tour the world's largest chocolate factory via the Web:

**www.hersheys.com/discover/chocolate.asp**

This is the official site for the United States chocolate industry. It includes links to the Chocolate Manufacturer's Association, the American Chocolate Institute, and the World Cocoa Foundation:

**www.worldcocoafoundation.org**

# 26 Frozen Desserts

Frozen desserts have remained popular through the ages. One of the earliest known frozen desserts was made with winter snow (from the mountains of what is now Turkey) mixed with fruit and drizzled with molasses. In AD 62, the Roman emperor Nero sent slaves to the mountains to retrieve ice, which was then flavored with nectar, fruit pulp, and honey to be enjoyed by him and his court. Marco Polo brought the

**Stabilizer** A compound such as vegetable gum that attracts water and interferes with ice crystal formation, resulting in a smoother consistency in frozen desserts.

formula for water ices to Europe from Asia, where they had been enjoyed for at least a thousand years. George Washington and Thomas Jefferson were both reputed to have made ice cream in their homes, and Dolly Madison is known to have served ice cream at the White house in 1812 (21). By 1840, ice cream had moved from palaces and mansions to the streets of America's largest cities. The ice cream scoop was invented 66 years later, and by 1921 ice cream was served to immigrants arriving on Ellis Island as part of their first American meal. Now American consumption of ice cream and other frozen desserts is higher than ever.

## TYPES OF FROZEN DESSERTS

Ice cream and other commercially frozen desserts, from simple water ices to elaborate ice cream cakes, are probably the most commonly consumed desserts in North America. Newer frozen desserts or modifications of old ones

appear on the market regularly in the form of pies, mousses, cakes, parfaits, pudding sticks, frozen yogurt, popsicles, and new flavors of ice cream.

What makes one frozen dessert different from another? It is the ingredients, especially the type and proportion of fat (milk fat) and milk solids-not-fat (MSNF)—proteins and lactose. In the United States, the government regulates the names of frozen desserts based on the types and quantities of ingredients that go into them. Another factor that differentiates desserts is the way in which these and other ingredients are combined. Common ingredients other than milk fat and MSNF include sugar, **stabilizers** (gums), emulsifiers, water, air, and flavorings. Ice cream, imitation ice cream, sherbet, sorbets, water ices, frozen yogurt, and still-frozen desserts, their ingredients, and their preparation are the subjects of this chapter.

### Ice Cream

Americans consume more ice cream than any other nation in the world, undeterred by the fact that it is the frozen

**FIGURE 26-1**   Ice cream is the most popular of the frozen desserts—and also the highest in fat.

Don Farrall/PhotoDisc/Getty Images

dessert with the highest fat content (Figure 26-1). The many variations of ice cream and some of their compositions are described in Table 26-1. Ice cream is a food prepared by simultaneously stirring and freezing a pasteurized mix of dairy (milk, cream, butterfat, MSNF, etc.) and nondairy ingredients (sweeteners, stabilizers, emulsifiers, possibly eggs, colors, and flavors) (Chemist's Corner 26-1).

## Contents of Ice Cream

Legal definitions for ice cream differ among countries, but in the United States, by law, regular ice cream must contain a certain amount of fat to be called *ice cream*—at least 10% milk fat (by weight). Some super-premium brands of ice cream have up to 20% milk fat, making them very rich and creamy. The legal definition prevents some manufacturers from using less cream to reduce their costs, while still claiming that their product is *ice cream*. Another ingredient, other than milk fat, that must be included in ice cream is milk MSNF, which varies in amount depending on the milk fat percentage (18). Less MSNF content is required for ice creams made with bulky flavoring ingredients such as chocolate or fruit (14). The amount of MSNF is important because it influences ice cream's flavor, **body,** texture, dispensing qualities, shelf life, and nutritive value, specifically in terms of protein, B vitamins, vitamin A, and calcium (22).

Vegetable fats are often used in other countries, but it has to be milk fat in the United States. Milk fat in frozen desserts comes primarily from milk products such as cream, butter (unsalted), sweetened condensed milk, or whole milk. The egg yolk that is used in some ice cream may be an additional source of fat and also

| TABLE 26-1 | Types of Ice Cream |
|---|---|
| **Dessert** | **Description** |
| Bombe | A decorative ice cream, in two or more different flavors, layered in a mold. |
| Frozen custard | Ice cream that contains 1.4% egg-yolk solids. Also known as French ice cream. |
| Ice cream | Frozen mixture of dairy ingredients containing at least 10% milk fat and weighing at least 4.5 pounds to the gallon. Marketing terms include regular ice cream, economy ice cream (lower price), premium ice cream (higher fat content), and super-premium ice cream (high fat content—at least 12% milk fat—and the best ingredients). |
| Ice cream cone | Born at the 1904 St. Louis World's Fair when a vendor selling ice cream ran out of dishes and created a "cup" from waffles being sold in a nearby stand. |
| Light ice cream | Contains 50% less total fat or 33% fewer calories than regular ice cream. |
| Low-fat ice cream | Contains no more than 3 grams of total fat per serving. |
| Neapolitan | Alternating lengthwise layers of 2 to 4 ice cream flavors. |
| Nonfat ice cream | Contains no more than 0.5 grams of total fat per serving. |
| Parfait | Alternating layers of ice cream and fruit or syrup in a tall, slender glass. |
| Reduced-fat ice cream | Contains at least 25% less total fat than regular ice cream. |
| Scoop ice cream | The two main types of scoop ice cream are uncooked and cooked. Both contain cream, sugar, and flavoring, but cooked ice cream adds at least 1.4% egg yolk solids. Thickening agents are sometimes substituted for eggs. |
| Sundae | Ice cream topped with syrup, nuts, whipped cream, and a cherry. There are many stories about their origin; one is that when ice cream became illegal to sell on Sunday (blue laws) in the United States, vendors got around this law by adding chocolate, nuts, and whipped cream so that it was not called *ice cream*. |

**Body** The consistency of frozen desserts as measured by their firmness, richness, viscosity, and resistance to melting.

**FIGURE 26-2**   Percent milk fat in low-fat, premium, and super-premium ice creams.

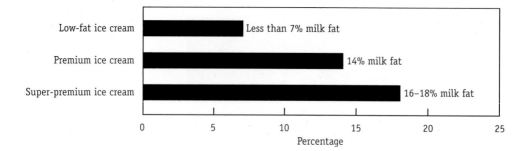

provides cholesterol, as do cream and butterfat.

The percent milk fat differentiates low-fat (formerly called ice milk), premium, and super-premium ice creams from one another (Figure 26-2). Expensive ice cream brands labeled "premium" and "super premium" are denser, smoother, richer, and higher in calories because they use heavier cream than that used in standard ice cream. Federal and state regulations determine the minimum levels of fat and MSNF for many frozen desserts (12, 19). Because milk fat is the most expensive ingredient, the more that is used, the higher the frozen dessert's cost. Other ingredients usually found in ice cream include eggs, sugar, stabilizers (gums, gelatin), flavoring, and coloring. When real cocoa or other forms of chocolate are added, the fat content rises even further.

### Low-Fat Ice Cream

Low-fat ice cream, previously called ice milk, usually contains more sugar than milk (by weight) and less fat. Low-fat ice cream was first introduced during the Great Depression as a lower-cost (and not very popular, according to some reports) alternative to ice cream (14). After regular-fat ice cream, lower-fat ice creams are the second most popular product, accounting for approximately one fourth of the U.S. market (24). At present, about one third of the nonfat milk sold commercially in the United States is used in the manufacture of ice cream and related products.

At one time, the milk fat content of any frozen dairy product sold as ice cream had to be at least 10% of its weight. Aiming to encourage lower-fat eating patterns, the FDA modified its restrictions and introduced the definitions for commercial low-fat ice cream

based on a half cup serving (3, 23). These definitions are based on comparison of the referenced product with either an average of the leading regular brands or the company's own brand:

- Reduced-fat ice cream: at least 25% less fat than the original product
- Light-fat ice cream: 50% less fat per serving or 33% fewer calories
- Low-fat ice cream: less than 3 grams of fat per serving
- Fat-free (nonfat) ice cream: less than 0.5 grams fat per serving

### Ice Cream as a Functional Food

The increased attention toward healthier lifestyles has inspired product development in the ice cream arena. Some manufacturers are offering ice creams with vegetable flavors—cucumber, carrot, pumpkin, and beet. Blueberries, known for their possible health links to the heart, eyes, and brain (memory), are becoming a popular flavor in frozen desserts. Green tea–flavored ice cream is growing in popularity in Japan. In addition, ice creams have been fortified with probiotics, plant phytosterols, and iodine (in Russia) (26). Other novel ice cream additives include natural fruits and fruit extracts (1). Manufacturers are also offering low-carbohydrate, lactose-free, sugar-free, and other nutrient-modified ice creams to accommodate dietary needs or habits (25).

## Imitation Ice Cream

Replacing the milk fat and MSNF in ice cream with other ingredients results in imitation ice cream. Mellorine is a frozen dessert similar to ice cream except that the milk fat has been replaced with other fats, usually vegetable fat. Another type of imitation ice cream

contains soybean products such as tofu, soy protein, or soy isolate, and milk fat is usually replaced by corn oil. Parevine is a frozen dessert that is free of any milk fat or MSNF, and meets the kosher requirements set by Jewish food laws (17). An example would be a frozen dessert made with tofu. Frozen desserts made for diabetics usually replace the sucrose with sorbitol.

## Gelato

Gelato has been called the "Italian word for ice cream" (2). Just as there are many ways to make ice cream, so there are numerous methods of producing gelato. As a result of these differing preparation methods, gelato tends to be creamier in the North of Italy, but icier in the South of the country. Compared to ice cream, gelato contains less cream and more milk, which makes it denser because milk incorporates less air than cream. It is also much more smooth and creamy than ice cream because it is churned more slowly and served at higher temperatures.

## Frozen Yogurt

Frozen yogurt sales crept along for two decades after the product was first introduced, but now frozen yogurt is not only available in most large shopping malls (Figure 26-3) but also accounts for about 4% to 5% of U.S. frozen dessert purchases (24). The increasing demand for frozen yogurt is due in part to its taste, texture, lower fat content, "good-for-you" image, and effective marketing strategies (14, 18). Most frozen yogurt contains no live yogurt cultures, although some manufacturers ensure that live and active cultures are present in their frozen yogurt.

**FIGURE 26-3**  Frozen yogurt is a popular dessert.

## Sherbet

Sherbets (often incorrectly pronounced *sherbert* in American English) fall into a category of frozen desserts that is separate from ice creams (Table 26-2). The base of these frozen desserts is iced, sweetened fruit juice or purée, so their fat content is lower than that of low-fat ice cream. The law states that they must contain less than 2% milk fat. Sherbets are often made with egg whites and/or gelatin to give them a creamy consistency. To compensate for the body lost because there is less fat, more sugar is added to the basic milk and fruit juice mixture. This makes sherbet's caloric content similar to, and sometimes higher than, that of ice cream (Figure 26-4), even though fat provides more than twice as many calories as sugar (4 calories/kcal per gram of sugar vs. 9 calories/kcal per gram of fat).

## Sorbet

Sorbets differ from sherbets in that they are made without fat, eggs, gelatin, or dairy products (milk solids-not-fat) and as such have a harder consistency than either ice cream or sherbet. They consist of puréed fruit or fruit flavoring and a sugar syrup made of equal amounts of sugar and water simmered together. Countless flavors are possible and include any of the fruits or combinations thereof.

## Water Ices

Water ices (glacés), made from a base of sweetened water and fruit juice, lack both fat and MSNF. They may contain gelatin, vegetable gums, egg whites, flavorings, and/or colorings. Popsicles are water ices that evolved from frozen lemonade on a stick (16). Granites from Italy are flavored ices that contain fruit and are quickly frozen to promote the formation of the large ice crystals that give them a rough texture. They are generally stirred a few times during freezing to further promote large crystalline formation. The word *granites* is often used interchangeably with *granitas,* which are technically frozen fruit slushes, shaved ice, or beverages and not frozen desserts. The word *granita* comes from the Italian word meaning "grainy" (20). Though made from a similar mixture, they are different in that they are frozen solid. Another type of ice dessert is frappé, which is crushed and served as a slush after it is frozen.

## Still-Frozen Desserts

A still-frozen dessert is one that is not stirred during freezing. Examples include mousses, bombes, and parfaits. The light, airy, smooth, velvety texture of these desserts comes from the incorporation of whipped egg whites or whipped cream into the mixture. The foam structure acts as a physical barrier, preventing large ice crystals from forming.

# PREPARATION OF FROZEN DESSERTS

The preparation of frozen desserts can be very time consuming. Coupled with the convenience and wide availability of commercial frozen desserts, this results in food service establishments and individuals not routinely preparing ice cream or other frozen desserts. Some frozen desserts, however, do lend themselves to preparation on certain occasions, and this section discusses that option.

## Factors Affecting Quality

The structure of frozen desserts depends on the crystallization of water from a sugar mixture (19). In many frozen desserts, these crystals are made by either churning a mixture while it is in the process of freezing, or placing it in a mold where it is allowed to freeze (6). In preparing frozen desserts, the three general factors crucial to their quality are flavor, texture, and body.

| TABLE 26-2 | Types of Non–Ice Cream Frozen Desserts |
| --- | --- |
| **Dessert** | **Description** |
| Frozen yogurt | Frozen dessert made from a cultured dairy product with added sweeteners and flavors. |
| Gelato | Frozen dessert with intense flavor and colors. It contains sugar, milk, cream, egg yolks, and flavoring. |
| Mellorine | A product made without milk fat. Other fats from either animal or vegetable sources can be used as long as fat content is not less than 6%. |
| Mousse | A French term meaning "foam" or "froth" and used to describe an airy, rich dish that can be hot, cold, or frozen. The dish can be a savory main meal (meat, fish, shellfish, cheese, or vegetables) or a dessert that is cold or frozen. The fluffy texture is from either whipped cream or egg whites, and its structure is often strengthened by adding gelatin. |
| Sherbet | Frozen dessert containing less than 2% milk fat and often more sugar than ice cream. |
| Sorbet | Similar to sherbet, but it contains no dairy, fat, egg, or gelatin ingredients. |
| Still-frozen desserts | Frozen desserts not stirred during freezing—mousses, bombes, and parfaits. |
| Water ice (glacé) | Frozen dessert made from sweetened water and fruit juice. |

## CALORIE CONTROL
## Frozen Desserts

The frozen desserts listed in Figure 26-4 average approximately 300 calories (kcal) for each cup, but range widely from 200 to 540 calories (kcal). Because the suggested calorie caps for an entire meal for women and men average 400 and 600 calories (kcal), respectively, it's easy to see how daily frozen desserts can add on the pounds. It's alright to enjoy desserts as long as they are within a person's daily calorie limit.

- Practice portion control. Satisfy the sweet tooth with a ½-cup serving.
- Avoid multiple servings. If you splurge, stick to one cup or large scoop (300 calories/kcal), not two (600 calories/kcal) or three (900 calories/kcal).
- Choose lower-calorie toppings. Opt for fruit and sprinkles over nuts, whipped cream, or sauces (caramel, chocolate) that deliver 50–100

calories for each ounce. Chocolate and caramel syrups, though fat free, provide 54 and 50 calories (kcal) per tablespoon, respectively; whipped cream, 25 calories (kcal) and 3 grams fat per tablespoon.
- Try popsicles or ice cream bars that average about 100–150 calories (kcal) each (read Nutrition Facts label).
- Go light on premium ice creams.
- Soft serve appears "light" but is actually higher in calories than regular ice cream.
- Sherbet may be lower in fat than ice cream, but it has a higher sugar content so the calorie counts are nearly the same.
- Gelato delivers 420 calories (kcal) and 24 grams of fat for each cup.

Frozen yogurt is not calorie free. Brands differ, so read the Nutrition Facts label.

**FIGURE 26-4**　Calories and fat in frozen desserts (per cup).

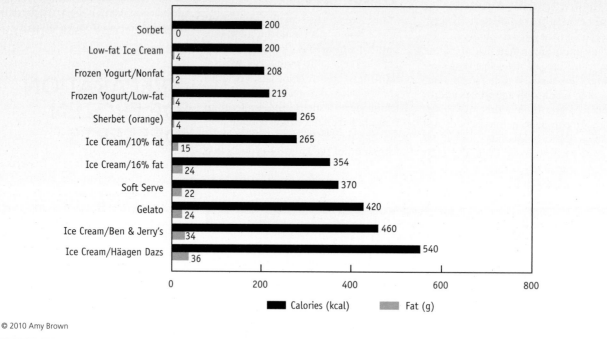

Calories (kcal) and fat (g)/cup, vanilla flavor

| Dessert | Calories (kcal) | Fat (g) |
|---|---|---|
| Sorbet | 200 | 0 |
| Low-fat Ice Cream | 200 | 4 |
| Frozen Yogurt/Nonfat | 208 | 2 |
| Frozen Yogurt/Low-fat | 219 | 4 |
| Sherbet (orange) | 265 | 4 |
| Ice Cream/10% fat | 265 | 15 |
| Ice Cream/16% fat | 354 | 24 |
| Soft Serve | 370 | 22 |
| Gelato | 420 | 24 |
| Ice Cream/Ben & Jerry's | 460 | 34 |
| Ice Cream/Häagen Dazs | 540 | 36 |

■ Calories (kcal)　■ Fat (g)

© 2010 Amy Brown

## Flavor

When making frozen desserts, it is important to remember that cold temperatures mute flavors, and therefore the mixture must be boldly flavored before freezing. The number of possible flavors that may be added to the foundation mixture of frozen desserts is almost limitless. When it comes to ice cream, however, vanilla is the most popular, and chocolate ranks second (25). Other flavor choices include, but are certainly not limited to, butter pecan, strawberry, Neapolitan, chocolate chip, cookies and cream, cherry, peach, pineapple, orange, raspberry, and coffee. Flavor may be further enhanced by the addition of nuts, candies, cookies, and other ingredients.

## Texture

A smooth texture is preferred in most frozen desserts, with the exception of frozen ices and granites. All are crystalline products whose textures are dependent on the formation of ice crystals. The size of the ice crystals determines the smoothness of the frozen dessert. The smaller and more evenly distributed these crystals are, the smoother the texture will be. Larger crystals yield a coarser texture. How is ice crystal size determined? It is directly related to the number of nuclei. The greater the number of nuclei present, the smaller the crystal size, and the

## PROFESSIONAL PROFILE

Drs. Kristen and Jonathan Gray are food scientists, married to each other, and working for different food companies in Chicago, Illinois. Kristen's food science specialty is in the area of scientific and regulatory affairs. She says that this area is "growing very quickly and it provides a way to work in the food industry without working directly on the bench or in a manufacturing facility." Jonathan says his job as a flavor chemist is bench related and entails a lot of "studying ingredient interactions, evaluating new ingredients, and patenting new ingredients or technologies. Other types of chemists working in the food industry," he adds, "are flavor chemists or analytical chemists who often focus on quality control or nutrition labeling requirements."

Like her husband, Kristen began studying chemistry in college (University of Illinois), but switched to microbiology. She enrolled in graduate school to study food science at Purdue University. The school introduced her to two summer internships. The first was at the United States Department of Agriculture (USDA), where she worked in its Eastern Regional Research Laboratory in Pennsylvania. The second internship was at the Food and Drug Administration (FDA), where she worked on risk assessment projects. Her husband, whom she met at Purdue University, did his undergraduate work at the University of Arkansas and completed an internship at Procter & Gamble studying food ingredient interactions.

After leaving graduate school, Jonathan went to work for Nestlé R & D, and she did a postdoc in industry. She says, "A postdoc allows corporations to hire you under contract without opening a new full-time position, but I learned a lot in the Food Safety Group [Hygenic Design Group]. We were looking at vending style coffee machines and testing the equipment for food safety."

"Postdoc positions are listed as any other job," she says. "However, it's best to contact industry directly because they do not need

**Kristen M. Gray, PhD**
**Technical Regulatory**
**Specialist—**
**Wm. Wrigley Jr. Company**

**Jonathan Gray, PhD**
**Food Chemist—Kraft Foods**

to open an official position for a postdoc. All you need to do is send in a résumé and cover letter requesting a postdoctoral position."

Dr. Kristen Gray's next job in the food industry, after some time in the cosmetic industry, was in Scientific & Regulatory Affairs with the William Wrigley Jr. Company, where she switched from bench work to deskwork. "Bench work is not for everybody, and it wasn't for me anymore. My husband still loves bench work." She prefers work that checks for regulatory compliance, handles consumer complaints, and ensures product safety. She says, "Really, I do very different things from day to day. For example, I do formula review (final product), ingredient safety, and ingredient review (raw materials). I work with everybody—legal department, product developers trying to figure out what's acceptable or not, some technical consumer questions, and I deal with a lot of big issues such as allergens and food sensitivities because there are so many different regulations nationally and globally."

"Label compliance is another important area," she adds, "and the rules to list ingredients and display food labels differ so much internationally, too. A lot of this job is keeping up with regulatory changes."

What is the one thing that Kristen and Jonathan learned that would help students? "How important an internship is—it is really, really important. Internships set you apart from the other food science candidates looking for corporate positions."

What did they think is the biggest mistake that students make? "Students not taking enough initiative or responsibility for their own careers. Also, really learning the food science information in college because you are going to need to know it. The information learned in the core food science classes is applied in the everyday work of a food scientist."

smoother the consistency (4). Other factors determining texture are the frozen dessert's content of milk solids-not-fat, fat, sugar, and other substances. About 50% of the total solids in ice cream and many other frozen desserts consist of dairy products (3).

Sugar lowers the freezing temperature of water, so lower-fat ice cream, sherbet, and water ices, which all contain more sugar and less fat, feel harder and colder on the tongue than ice cream.

Monosaccharides lower the freezing point even more than sucrose; consequently, corn syrup and dextrose (glucose) are commonly added to commercial frozen desserts (19). The freezing point is lowered with added sugar so that the ice cream mixture does not freeze completely. Otherwise, ice cream could not be scooped. The semi-softness of ice cream is maintained because as the water freezes, the other ingredients become more

concentrated, preventing the frozen dessert from freezing into a solid block of ice.

Other factors contributing to texture are air cells, emulsifiers, and the type of treatment. Ice cream is a foam of air cells, each surrounded by a layer of fat coated with emulsified protein films, and a network of ice crystals (Figure 26-5). The millions of air cells found in every bite of ice cream make the frozen mixture light and airy.

**FIGURE 26-5**    The structure of ice cream. Ice cream is a foam of air bubbles trapped in frozen liquid. The liquid contains the dissolved sugar and milk solids.

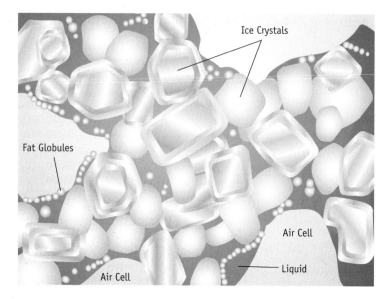

locust bean, tragacanth, xanthan, and gelatin (30). These food additives (see Appendix F and Chapter 29) are primarily of natural origin, although a few are chemically modified. The food industry uses gums because they contribute to body, resistance to melting, viscosity, and reduced ice crystal formation during storage (15). Stabilizers are common ingredients in frozen desserts because they produce a smooth body and texture. They work by attaching to any water that is freed from the melting ice crystals during an increase in temperature such as the opening of a freezer door. The water bound to gums cannot attach to existing ice crystals and make them larger, which would adversely affect quality (29). The use of stabilizers also reduces the amount of cream needed to maintain body, which reduces production cost. Using less cream also results in fewer calories and this appeals to dieters.

*Ice Cream Shrinks as It Ages*    The volume of ice cream shrinks as it ages due to the collapsing films around the air cells (Figure 26-6). Another cause of shrinkage is mechanical compaction caused by a dipper being pressed against ice cream during scooping (21).

## How & Why?

### Why are MSNF sometimes added to frozen desserts?

One reason MSNF or fat is added to frozen desserts is to help produce a smooth texture (12). The addition of MSNF results in smaller ice crystals, tinier air cells, and thinner air cell walls. Fat assists in separating water molecules and coating the frozen crystals to create the sensation of smoothness (21). Use of excess MSNF, however, produces a sandy texture from the precipitation of lactose, the least soluble of sugars (18).

### Body

Commercial ice cream has more body than homemade ice cream because of the added stabilizers (no more than 0.5%). These include vegetable gums such as sodium carboxymethylcellulose (CMC), carrageenan, guar gum, agar, acacia, alginate, furcelleran, karaya,

Emulsifiers such as mono- and diglycerides, egg yolk, and/or lecithin are often added to create a smoother texture by stabilizing the dispersion of air in the foam. Rapid freezing and correct storage temperatures also contribute to the smooth texture of frozen desserts. Commercial ice creams are aged (held at refrigerator temperatures for a given period of time). Glycerol monostearate (GMS) may be added to destabilize the emulsion by displacing protein from the fat globules, allowing them to partially aggregate. This results in a smooth texture and more resistance to melting (8).

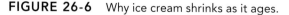

## NUTRIENT CONTENT

Fat content is often the factor that differentiates frozen desserts from one another. In order to meet the demand for lower-fat products, food companies have provided an army of reduced-fat frozen dessert products that have challenged ice cream's historical share of the market. The new products use skim milk, fat substitutes, emulsifiers, and stabilizers. Fat and sugar are difficult to replace in frozen desserts, which rely on these ingredients to provide smooth mouthfeel and body. The functions and replacement of sugar and fat are discussed in Chapters 21 and 22, respectively.

**FIGURE 26-6**    Why ice cream shrinks as it ages.

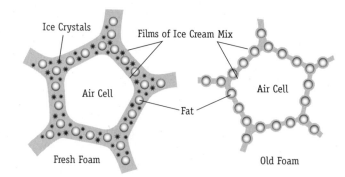

***Heat Shock***   One of the biggest problems for ice cream manufacturers is **heat shock.** After ice cream is produced, its ice crystals continue to grow until they eventually become detectable within the mouth and/or ice crystals grow around the edge of the ice cream (26). Quality changes from smooth and creamy to rough and slightly off-flavored. Ice crystals cannot be detected in new ice cream because they are numerous and small. During storage they undergo recrystallization, a process that effects changes in their number, size, and shape. Each time the temperature increases, the smaller ice crystals disappear, whereas the larger ones keep growing. The number one way to prevent recrystallization is to avoid heat shock. Other production steps to reduce heat shock are to freeze the ice cream quickly and avoid temperature fluctuations during storage and transport. Stabilizers are often added to ice cream to combat heat shock by entrapping melted water so that it cannot contribute to the creation of larger ice crystals.

***Overrun***   The purpose for pumping air into the original frozen dessert mixture to expand it and create **overrun** is twofold. The first reason is that it improves body by making it softer and creamier. The numerous small air cells created by overrun prevent the ice cream from being too hard, dense, or cold. This is one of the reasons why homemade ice cream has a harder consistency than commercial ice cream. Commercial ice cream usually contains an overrun of 70% to 100% (21). A 100% overrun means that the volume of air is equal to the volume of the mix before it was churned. Too much overrun damages body by creating a foamy texture, but too little results in heavy, compact ice cream with a coarse texture. The finest ice creams have between 3% and 15% overrun.

The second reason is that it is economically advantageous to maximize product volume; however, excessive overruns are prevented by federal standards that protect customers by setting the minimum weight for ice cream at 4.5 pounds per gallon (the exception is reduced-fat ice cream at 4 pounds per gallon) (9). Ice cream with lower overrun and higher fat content is classified as premium ice cream, whereas very low overruns and high fat content—at least 12% milk fat (16% to 18% preferred)—and the best ingredients constitute super-premium ice cream. Overruns greater than 150% are possible if more stabilizers and emulsifiers are added, but the product is then classified as a "frozen dairy dessert" instead of ice cream.

# Mixing and Freezing

## Ice Cream

Commercial mixes are used by the food industry to produce many different kinds of ice cream (Figure 26-7). Cream is the main ingredient; this provides a smooth mouthfeel. In addition, sugar and flavorings such as vanilla bean, chocolate, strawberry, mint, pistachio, candy, cookie pieces, brownie chunks, and caramel or chocolate sauce are added.

### Cooked vs. Uncooked Ice Cream

A basic custard serves at the base for cooked ice creams. Custards are created by stirring egg yolks into milk, sugar, and cream over low heat. More specifically, the cream is heated to scalding, and the eggs and sugar are beaten into a smooth paste and then added to the cream. The mixture is then heated until it thickens while being stirred constantly with a whisk or wooden spoon. The mixture should be heated to at least 180°F (82°C) to protect against the growth of any *Salmonella* that may be present. The mixture is removed from the heat before the flavoring agents are added. The addition of flavorings may be delayed until after the ice cream has been churned (see below) and is still soft to prevent the breakage of ingredients such as nuts and chocolate chunks. The ice cream mixture is cooled for several hours or overnight in the refrigerator before it is processed into a frozen product.

Uncooked ice cream can bypass the cooking and refrigeration stage. Care should be taken to ensure that only pasteurized products are used in either cooked or uncooked ice cream to reduce the risk of foodborne illness. The consistency of uncooked ice cream is closer to that of sherbet, which has a grainier consistency.

**FIGURE 26-7** Making a frozen dessert from a commercial mix.

A complete mix comes all ready to just add water and blend. It can be combined with hundreds of flavors to produce an endless variety of ice cream desserts.

Reconstituted mix is transferred to the freezer equipment.

Ready to serve.

**Heat shock** Repeated cycles of temperature fluctuations from cold to warm and back that cause larger ice crystal growth, reducing frozen dessert quality.

**Overrun** The volume over and above the volume of the original frozen dessert mix, caused by the incorporation of air during freezing.

**TABLE 26-3**   Ice Cream Flavorings. The number of flavorings added to ice creams continues to grow.

| Flavoring | Examples |
|---|---|
| Fruit | strawberry, peach, mango, kiwi, lemon, orange, dried fruit (raisins), jam, etc. |
| Nuts | walnut, peanut, almond, pistachio, pecan, etc. |
| Chocolate | chips, shavings, chunks |
| Beverages | coffee, tea, rum |
| Candy/Cookies | marshmallow, candy bar pieces, gum, cookie dough |
| Spices | ginger, cinnamon, mint, lavender |

**Source:** Adapted from Lebowitz D. Scooped. One easy method, countless incredible ice creams. *Fine Cooking* 99:46–53, 2009.

This can be compensated for to some degree by using thick fruit pulp (10). Food safety guidelines dictate that uncooked ice cream, commercial or homemade, should never be prepared with raw eggs because of the risk of *Salmonella* foodborne illness.

**Flavorings**   Whether it is cooked or uncooked, the number and combination of flavorings added to ice cream continues to grow (Table 26-3). If fruit is used for flavoring, it should first be combined with a sugar syrup or liqueur to eliminate the unpleasant effect of solidly frozen fruit chunks on the texture. The acidity of fruit flavors is balanced by sugar; thus, double the normal amount of sugar is usually added to frozen desserts such as fruit sherbets (27).

**Heating and Aging**   Commercial preparation of ice cream consists of combining the liquid ingredients in a mixing vat where they are heated to 104°F (43°C), warm enough to dissolve the added sugar and other dry ingredients. The ice cream mix is pasteurized and homogenized; the latter improves the overall texture and body of ice cream. Commercial mixes are then aged from 3 to 24 hours at 40°F (4.4°C) in vats. During this stage, several beneficial changes take place: The fat solidifies, and there is a swelling of milk proteins, gelatin, and other stabilizers, which increases the viscosity of the mix. Smoother texture, improved body, and resistance to melting also result from the aging process (21).

**Churning and Freezing**   After being aged, ice cream mixes are ready to be frozen. During freezing, the ingredients are churned for several reasons. Churning promotes the formation of numerous small nuclei necessary for a smooth, velvety texture. A second reason to whip the mixture is to encourage the homogenized (smaller and separated) fat globules to partially disrupt their membranes and rejoin with similar disrupted fat globules to form a larger mass that is detected by the consumer as "creamier" (Figure 26-8). It also aids in the incorporation of air to increase volume.

**FIGURE 26-8**   Restructuring of fat during ice cream churning.

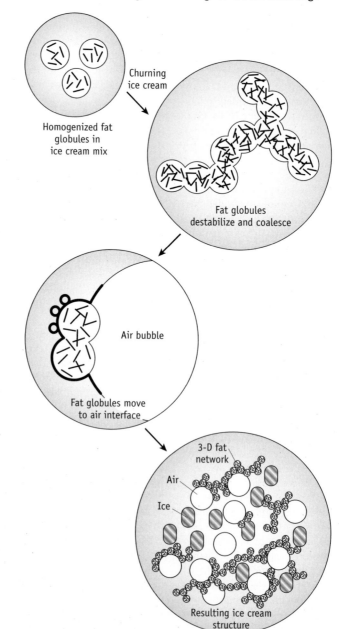

**FIGURE 26-9**   Ice cream freezer.

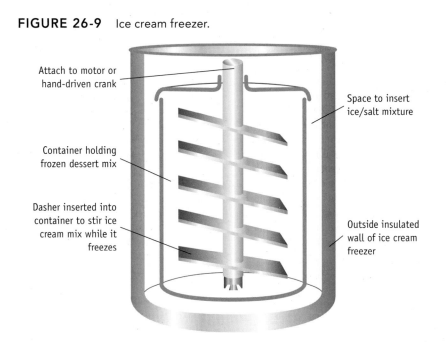

Attach to motor or hand-driven crank

Space to insert ice/salt mixture

Container holding frozen dessert mix

Dasher inserted into container to stir ice cream mix while it freezes

Outside insulated wall of ice cream freezer

The two major noncommercial methods for mixing ice cream ingredients are the electric mixer method (ice cream machine) and the old-fashioned hand-cranking method. In both methods, the ice cream freezer canister is filled only two-thirds full to allow for the natural expansion that occurs with freezing (water expands when frozen) and the incorporation of air during churning. This container, equipped with a dasher for rotating the mixture, is traditionally surrounded with a combination of rock salt and crushed ice (1:4–6 ratio, with the ratio depending on the dessert) to promote freezing to temperatures below 32°F (0°C) during churning (Figure 26-9; Chemist's Corner 26-2). The dasher should be turning before the mixture is added. Some electric models use a canister with liquid inside that is frozen instead of ice and salt.

It is best to fill the space around the canister half full with crushed ice and then add the salt to prevent it from aggregating at the bottom. The remaining salt and ice can be mixed together and then added, or layered alternately on top of the ice around the canister. Crushed rather than cubed ice is used because of its greater surface area, and rock salt is preferred over table salt, which tends to cake, is more expensive, and dissolves too quickly. To maintain the cooler temperatures, it is important not to discard the liquid that results from the salt melting the ice. The cooler temperature of the outside container necessitates the constant churning of the ice cream mixture in order to prevent large ice crystals from forming on the perimeter. The sharp edge of the revolving dasher moves ice crystals forming on the container's wall toward the center to promote uniform crystal size formation. It also helps to begin the churning process with everything cold by freezing the canister overnight and refrigerating the ice cream mixture until it's completely chilled (7).

> ## CHEMIST'S CORNER 26-2
>
> **Salt's Effect on Freezing**
>
> The freezing point of water decreases 6.7°F (3.72°C) for each gram molecular weight (58 grams) of sodium chloride (NaCl) added to 1 liter of water. Every gram of melting ice requires 80 calories (kcals) of energy, which is absorbed by the ice–salt mixture around the ice cream canister (17).

> ## ? How & Why?
>
> **Why is rock salt added to ice cream machines?**
>
> Salting the ice surrounding a churning ice cream canister speeds the melting of the ice, which absorbs energy (heat) away from the contents of the canister. It keeps the canister and its contents cooler. The goal in adding an ice and salt mixture is to remove sufficient heat from the ingredients in the canister so that they start to freeze. The melting ice signals this absorption of heat.

**Speed of Churning/Freezing**   The number of revolutions per minute of the mixer is low at first, but is increased near the end to facilitate the formation of small crystals. This is based on the principle that slow freezing results in large crystals, whereas quick freezing creates tiny ice crystals (4). The speed of the dasher also determines how much air is incorporated into the mixture. A faster dasher introduces more air, creating a lighter ice cream. Slowing down the dasher traps less air, yielding a denser ice cream (13). The ice cream is finished churning in about 20 minutes, when the dasher becomes difficult to rotate.

**Storing and Hardening**   The dasher is then carefully removed and the ice cream is packed down and put in the household or institution freezer for about 4 to 6 hours before serving. Ice cream may be served immediately after preparation, but its texture will be somewhat soft. If the ice cream is too grainy, a common fault, it may be because of one of several preparation errors: filling the container more than two-thirds full; using too much salt, which freezes the ingredients too rapidly and does not allow enough time for the formation of small crystals; and/or churning too rapidly. Freezing will also have an impact on the ice cream's texture. The initial churning freezes about 33% to 67% of the water, whereas the second freezing during storage, called the hardening phase, freezes an additional 23% to 57% of the water (19).

### Frozen Yogurt

An ice cream freezer is used to prepare frozen yogurt. The process is similar to that for making ice cream, except that the ingredients are yogurt based. Emulsifiers may also be added.

### Sherbet

The procedure for preparing sherbet is again similar to preparing ice cream, but the ingredients are syrup based rather than milk based. No egg yolks are used. The sugar concentration is key to determining the final flavor of the sherbet. This can be fine-tuned either by tasting or by using an instrument that measures sugar concentration (Chemist's Corner 26-3).

### Sorbet

Sorbets are relatively easy to prepare. A purée of either fresh or frozen fruits is combined with a sugar syrup and then frozen in an ice cream maker. Fruits that are beyond ripe, but not rotten, are best because they yield the most intense flavors. A sweet fruit should be the primary choice and may be balanced by one that is more tart, such as strawberries or plums (5). Herbs such as rosemary, thyme, or basil may also be added as a secondary flavor (10). If the fruit is less than ripe, it can be poached in the sugar syrup to soften the flesh and make it easier to purée. Poaching is always recommended for pineapple and kiwis, no matter how ripe, in order to destroy certain enzymes that will interfere with freezing.

*Sorbet Syrup Preparation* The syrup of equal parts sugar and water is simmered prior to adding the fruit purée. It is best if the syrup is still warm when the purée is added in order to bring out the most flavors from the fruit. Chilling the mixture before churning speeds up the freezing process, but the mixture should first be tasted for any necessary flavor adjustments, because once frozen, the flavors are less intense. In its liquid state, the sorbet mixture should have the desired flavor but be too potent to drink. Tart citrus flavors such as lemon, grapefruit, or lime can be added if the mixture is too sweet.

*Churning Sorbets* The final stage of sorbet preparation consists of placing the mixture in an ice cream freezer that incorporates air into the fruit syrup, creating a light, airy texture. After it thickens, the sorbet should be transferred to an airtight container and frozen for at least a couple of hours before consumption, although it may be consumed immediately. Sorbet made this way will keep for up to 2 weeks (5). The use of stabilizers such as gelatin, pectin, agar, or gum tragacanth improves the shelf life of sorbets by reducing the tendency of their frozen liquids to separate from the ice crystals (10).

### Water Ices

Water ices are made by combining one part sugar with four parts liquid and flavoring. This flavored syrup is then placed in a mold, covered with aluminum foil, and frozen. Some frozen ices, such as granites, are stirred periodically during freezing to promote large ice crystal formation. Too much sugar will prevent the mixture from freezing, while too little results in an unappetizing hard-frozen mass.

### Still-Frozen Desserts

Still-frozen desserts are frozen without stirring, usually in molds. The mixtures for these vary tremendously in their ingredients and preparation. Their light, airy texture and high volume are derived from whipped cream or egg-white foams that are folded into the other ingredients. The smoothness of still-frozen desserts is often the result of emulsifying agents having been added to prevent the formation of large crystals during freezing. Such emulsifiers include eggs, gelatin, corn syrup, and/ or cornstarch. It is best to remove still-frozen desserts from the freezer a half hour prior to serving, but care should be taken with regard to food safety because many contain eggs or cream.

## Food Additives in Frozen Desserts

Food manufacturers of frozen desserts rely on several additives to maintain the quality, appearance, and flavor of their products. In contrast, the initial quality of homemade ice cream lasts only about 3 days. To combat the quality deterioration, commercial frozen dessert manufacturers incorporate a variety of food additives into their frozen desserts—sweeteners, stabilizers, emulsifiers, colorants, and flavorings (Table 26-4).

# STORAGE OF FROZEN DESSERTS

Ice cream is best stored at temperatures of 0°F (−18°C) or below for a month or two. A thin, plastic film is sometimes used inside the carton to cover commercial ice cream. If this is lacking, a sheet of wax paper can be pressed against the ice cream before resealing the carton, or the entire carton can be placed in an airtight plastic bag. This prevents absorption of other food odors in the freezer and exposure to moisture buildup, which can promote the formation of large crystals. Ice crystals do not remain the same size as when they formed, but slowly enlarge, so the consistency of frozen desserts rarely stays the same throughout their storage life.

## Texture Changes

The gradual change in texture often observed in ice cream, lower-fat ice cream, and sherbet results from transportation from the store, repeated removal of the frozen dessert from the freezer, and/ or freezer temperature fluctuations.

### CHEMIST'S CORNER 26-3

**Measuring Sugar Concentration**

A saccharometer, also known as a syrup-density meter, hydrometer, or Baumé hydrometer, measures the sugar concentration of a liquid. The calibration of the hollow glass tube ranges from 0° to 50° and is read as "degrees of Baumé." The recommended range for sherbets, sorbets, and ices is about 16° to 20° Baumé (10).

**TABLE 26-4** Common Food Additives Used in Ice Cream

| Food Additive | Function | Food Use and Comments |
|---|---|---|
| Agar | Thickener | A seaweed derivative. |
| Alginate | Stabilizer | Seaweed derivatives that are used as stabilizers retain moisture, make foods creamier, and extend shelf life. |
| Carrageenan | Emulsifier, stabilizer, thickener | Type of seaweed (Irish moss) that stabilizes foods by making them gel. |
| Cellulose | Emulsifier, stabilizer, thickener | Component of all plants. Inert bulking agent in foods. |
| Citric acid | Preservative, antioxidant, pH control agent | Widely distributed in nature in both plants and animals. |
| Guar gum | Stabilizer, thickener, texturizer | Extracted from the seeds of the guar gum plant (legume grown in India). |
| Maltodextrin | Flavor enhancer, stabilizer | A carbohydrate produced from cornstarch that enhances stability and flavor. |
| Mono- & diglycerides | Emulsifiers | Derived from lipids. |
| Xanthan gum | Stabilizer | Usually produced from cornstarch. |

**FIGURE 26-10** Ice cream quality is best when first purchased because numerous small crystals are present (*left*). Any increase in temperature causes the smallest crystals to melt (*center*). The extra water is taken up by the remaining crystals, making them larger and the ice cream more grainy (*right*).

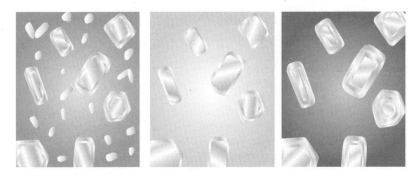

During any one of these occasions, some of the ice crystals melt into water, which attaches to neighboring crystals and refreezes. Consequently, the crystals get larger and larger, creating a coarse texture (28). Over time, the number of ice crystals decreases at the same time that the average ice crystal size increases. Thawing frozen desserts, even partially, and then refreezing them has a negative effect on their quality (Figure 26-10). It is best, therefore, to keep them at freezing temperatures and protect them as much as possible from changes in temperature.

## Scooping Frozen Desserts

In spite of the hazard to its quality, slightly warmer ice cream is preferred during serving because it is easier to scoop and more flavorful. A half gallon can be easily sliced or scooped when left in the refrigerator for about 15 minutes. Scooping downward into the container to gather a compact ball of ice cream or sherbet results in fewer servings per tub than does skimming the dipper over the surface. Dipping the dipper in cool water before scooping also helps to keep the ice cream from sticking to the dipper.

# PICTORIAL SUMMARY / 26: Frozen Desserts

From AD 62, when Roman emperor Nero sent runners to the mountains to fetch ice to be flavored with nectar, fruit pulp, and honey for him and his court, frozen desserts have maintained their popularity. Today, commercially frozen treats are probably the most commonly consumed desserts in North America.

## TYPES OF FROZEN DESSERTS

Federal and state regulations determine the minimum levels of fat and milk solids-not-fat (MSNF) for many frozen desserts.

What makes one frozen dessert different from another? The ingredients, proportions, and the way they are prepared.

**Ice cream:** Ice cream is prepared by simultaneously stirring and freezing a pasteurized mix of dairy (milk, cream, butterfat, etc.) and nondairy ingredients (sweeteners, stabilizers, emulsifiers, possibly egg, colors, and flavors). By law regular ice cream must contain at least 10% milk fat and 20% MSNF, with premium ice creams containing at least 14% milk fat.

**Lower-fat ice creams** are compared to a referenced product (either an average of the leading brands, or the company's own brand):

- Reduced-fat ice cream: at least 25% less fat than the original product
- Light-fat ice cream: 50% less fat per serving or 33% fewer calories than original product
- Low-fat ice cream: less than 3 grams of fat per half-cup serving
- Fat-free (nonfat) ice cream: less than 0.5 grams fat per serving

**Imitation ice cream:** Milk fat is replaced with other ingredients such as vegetable fat or corn oil. Other imitation ice cream products contain soy products.

**Gelato:** The "Italian" version of ice cream made with less cream and more churning to yield a very smooth, creamy, intensely flavorful dessert.

**Frozen yogurt:** Full-fat frozen yogurt has almost as many calories (kcal) as regular ice cream (about 280 kcal per 8 oz.) but nonfat yogurt averages only 180 kcal per 8 oz., plus zero grams of fat.

**Sherbet:** Sherbets must contain less than 2% milk fat, and their creamy consistency often comes from egg whites and/or gelatin. More sugar is added so their caloric value is actually very similar to regular ice cream's.

**Sorbet:** Sorbets consist of puréed fruit or fruit flavoring and a sugar syrup. Sorbets have a harder consistency than ice cream or sherbet.

**Water ices (glacés):** Water ices, made from a base of sweetened water and fruit juice, lack both milk fat and MSNF. Examples: popsicles, granites, and frappes.

**Still-frozen desserts:** These desserts are not stirred while freezing. Examples include mousses, bombes, and parfaits. The use of whipped egg whites or whipped cream results in an airy, velvety texture, without large ice crystals.

**Ice cream and other frozen desserts continue to be popular.**

## PREPARATION OF FROZEN DESSERTS

The first step in preparing a frozen dessert is to combine the ingredients, which may or may not be cooked. Frozen dessert mixtures are then either churned using an ice cream freezer (electric or hand-cranked) until they are frozen (ice cream, sherbet, sorbet, frozen yogurt) or placed in a mold (sorbet, water ices, still-frozen desserts) where they are allowed to freeze undisturbed.

Flavor, texture, and body all contribute to the quality of frozen desserts. The texture of frozen desserts is based on their crystalline nature—the smaller and more evenly distributed these crystals are, the smoother the texture. Larger crystals yield a coarser texture. Other factors contributing to texture are fat, MSNF, air cells, emulsifiers, and processing. Body is defined as the product's combined firmness, richness, and viscosity.

Among the different flavors of ice cream . . .

. . . **vanilla** is the most popular, followed by **chocolate**.

## STORAGE OF FROZEN DESSERTS

Ice cream is best stored at temperatures of 0°F (−18°C) or below for 1 to 2 months, and is best if protected as much as possible from fluctuations in temperature.

# CHAPTER REVIEW AND EXAM PREP

## Multiple Choice*

1. By law, regular ice cream must contain what percentage of milk fat?
   a. 10%
   b. 20%
   c. 30%
   d. 40%

2. What characteristic(s) of ice cream is/are influenced by the milk solids-not-fat (MSNF) content?
   a. Flavor
   b. Body
   c. Texture
   d. All of the above

3. The freezing point for water _____ when salt (sodium chloride) is added.
   a. stays the same
   b. increases
   c. decreases
   d. varies

4. Sherbet is different from ice cream in that it is _____ based, not milk based.
   a. syrup
   b. yogurt
   c. sorbet
   d. ice

5. Why does commercial ice cream have more body than homemade ice cream?
   a. Larger batches are made.
   b. High-tech equipment is used.
   c. Secret recipes are used.
   d. Stabilizers are used.

6. Commercial ice creams often have _____ , measured in percent of increased volume from having air pumped into the mixture, which makes them softer, creamier, and more economical.
   a. overrun
   b. churning
   c. emulsifiers added
   d. milk-solids-not-fat added

7. A frozen dairy dessert containing no fat is called _____ .
   a. gelato
   b. sherbet
   c. mellorine
   d. sorbet

*See p. AK-1 for answers to multiple choice questions.

## Short Answer/Essay

1. Describe how the basic ingredients of the following frozen desserts differ: ice cream, low-fat ice cream, imitation ice cream, frozen yogurt, sherbet, sorbet, water ices, and still-frozen desserts.
2. Smooth texture is important to the majority of frozen desserts. Describe the factors that affect texture in frozen dessert preparation.
3. Explain how commercial ice cream manufacturers add body to their products.
4. Describe the basic steps for preparing ice cream.
5. How does the preparation of still-frozen desserts differ from that of most other desserts that are frozen?
6. Why (for what purpose) is commercial ice cream sometimes covered with a thin, plastic film for storage?
7. Describe how the thawing of frozen desserts reduces their quality.
8. What is the best technique for scooping ice cream with a dipper? Explain why.
9. Explain why salt is added to the crushed ice to facilitate the freezing of homemade ice cream.
10. What are the recommended storage temperatures and times for ice cream?

# REFERENCES

1. Anonymous. Frutarom pushes concept of functional ice cream. *Confectionery News.* June 6, 2005. www.confectionerynews.com/Processing-Packaging/Frutarom-pushes-concept-of-functional-ice-cream. Accessed 7/09.
2. Anonymous. You say gelato, I say ice cream. *Fine Cooking* 87:71, 2007.
3. Chandan R. *Dairy-Based Ingredients.* Eagen Press, 1997.
4. deMan JM. *Principles of Food Chemistry.* Van Nostrand Reinhold, 1990.
5. Deville D. Cool sorbets, intensely flavored. *Fine Cooking* 16:67–71, 1996.
6. Donovan MD. *The New Professional Chef. The Culinary Institute of America.* Van Nostrand Reinhold, 1995.
7. Driscoll M. Review: The scoop on ice cream makers. *Fine Cooking* 73:26, 2005.
8. Euston SE, et al. Oil-in-water emulsions stabilized by sodium caseinate or whey isolate as influenced by glycerol monostearate. *Journal of Food Science* 61(5):916–920, 1996.
9. Food and Drug Administration. Federal Register Advance Notice of Proposed Rulemaking—70 FR 56409—September 27, 2005: Frozen Desserts; Petition to Revoke Standards for Goat's Milk Ice Cream and

Mellorine and to Amend Standards for Ice Cream and Frozen Custard, Sherbet, and Water Ices; Petition to Amend Standards for Parmesan and Reggiano Cheese. www.fda.gov/food/labelingnutrition/foodlabeling guidanceregulatoryinformation/regulationsfederalregisterdocuments/ucm076969.htm. Accessed 12/07/09.

10. Friberg B. *The Professional Pastry Chef.* Van Nostrand Reinhold, 2002.

11. Goff HD. Partial coalescence and structure formation in dairy emulsions. In S Damodaram, ed., *Food Proteins and Lipids.* Plenum, 1997.

12. Guinard JY, et al. Sugar and effects on sensory properties of ice cream. *Journal of Food Science* 62(5):1087–1094, 1997.

13. Jay S. What to look for in an ice cream machine. *Journal of Food Science* 31:50–53, 1999.

14. Low-fat frozen desserts: Better for you than ice cream? *Consumer Reports* 57(8):483, 1992.

15. Martin DR, et al. Diffusion of aqueous sugar solutions as affected by locust bean gum studied by NMR.

*Journal of Food Science* 64(1):46–69, 1999.

16. McGee H. *On Food and Cooking: The Science and Lore of the Kitchen.* Macmillan, 1997.

17. McWilliams M. *Foods: Experimental Perspectives.* Macmillan, 2007.

18. O'Donnell CD. Dairy derivatives that deliver. *Prepared Foods* 166(6):129, 1997.

19. Penfield MP, and AM Campbell. *Experimental Food Science.* Academic Press, 1990.

20. Plue N. Refreshing granitas. *Fine Cooking* 59:46, 2003.

21. Potter NN, and JH Hotchkiss. *Food Science.* Chapman & Hall, 1999.

22. Pupillo M. *From the Cow to the Cone: How Ice Cream Is Made.* International Dairy Foods Association. www.idfa.org/facts/icmonth/page6.cfm. Accessed 7/09.

23. Pupillo M. *Ice cream labeling: What does it all mean?* International Dairy Foods Association. www.idfa.org/facts/icmonth/page5.cfm. Accessed 7/09.

24. Pupillo M. *Just the facts: Ice cream sales and trends.* International Dairy Foods Association. www.idfa.org/facts/icmonth/page2.cfm. Accessed 7/09.

25. Pupillo M. *What's Hot in Ice Cream?* International Dairy Foods Association. www.idfa.org/facts/icmonth/page3.cfm. Accessed 7/09.

26. Pszczola DE. Ingredients. Innovative chills ahead for frozen desserts. *Food Technology* 59(3):40–51, 2005.

27. Smith D. Sugar in dairy products. In N Pennington and C Baker, eds., *Sugar: A User's Guide to Sucrose.* Van Nostrand Reinhold, 1990.

28. Sutton RL, D Cooke, and A Russell. Recrystallization sugar/stabilizer solutions as affected by molecular structure. *Journal of Food Science* 62(6):1145–1149, 1997.

29. Sutton RL, and J Wilcox. Recrystallization in model ice cream solutions as affected by stabilizer concentration. *Journal of Food Science* 63(1):9–11, 1998.

30. Whistler RL, and JN BeMiller. *Carbohydrate Chemistry for Food Scientists.* Eagen Press, 1997.

# WEBSITES

You can find history, science, and recipes for ice cream at:

**http://results.about.com/ice_cream/**

Prepare various frozen desserts using this website:

**www.bartleby.com/87/0026.html**

You can find nutrition information on various frozen desserts from Nutrition Action at:

**www.cspinet.org/nah/6_98des.htm**

PhotoDisc/Getty Images

# 27 Beverages

All living organisms require water for survival. This essential nutrient can be absorbed by some organisms through cell membranes, cell walls, or the skin, and in higher organisms is consumed in the form of foods and/ or directly by drinking. Human bodies are approximately 65% to 70% water, so fluids must be consumed daily to maintain that supply. Helping to meet this need are foods such as fruits, vegetables, milk, meat, and eggs, which all contain approximately 70% water.

Beverages, however, contribute a greater share to the human requirement for water, and they do this in sometimes mundane, sometimes delicious ways. Most beverages used for obtaining fluids fall into one of the following seven categories: water, carbonated beverages, functional beverages, fruit/ vegetable beverages (see Chapter 14), aromatic beverages (coffee and tea), dairy beverages (see Chapter 10), and alcoholic beverages (27). This chapter discusses these beverages (except for fruit and vegetable beverages, which are discussed in Chapter 14), specifically their composition, processing, preparation, and storage.

## WATER

Water is so vital to human life that its source determined where early humans lived—near streams, rivers, lakes, and springs. A person can survive up to 40 days without food, but only about 7 days without fresh water or water-based liquids. While the majority of earth is covered with water, only 3% of it is suitable for drinking; the rest is saltwater or unusable (82). About three fourths of freshwater used in the United States comes from surface water sources, and the remaining one fourth comes from ground water (28).

Bottled water is now the second most popular beverage in the United States (23). Soft drinks are number one, but they are also 90% water (or 98% for diet soft drinks). Water and other beverages continue to increase in popularity at the expense of carbonated drinks, and more than 24 million Americans now drink bottled water (3). Water contains no calories or vitamins. Its predominant mineral content, usually either calcium or sodium, varies according to region and determines whether the water is hard or soft (see Chapter 3). In some locations, the water is so high in calcium that it serves as a significant source of daily dietary calcium (59). The quality of bottled water is regulated by the Food and Drug Administration (FDA), whereas tap water quality is the responsibility of the Environmental Protection Agency (EPA) (2). The wide number of variations on plain water available on the market is now discussed in more detail (see Chapter 3 on food composition for the background information on water's chemical composition, content in foods, specific heat, freezing point, melting point, boiling point, altitude influences, soft versus hard water, and functions of water in food—heat transfer, universal solvent, chemical reactions, and food preservation).

## Types of Water

There are now so many different types of water available that bottled water is not just a trend, it is a permanent change in the beverage industry (44)—so much so that both Coca-Cola and PepsiCo have entered the bottled water market.

Bottled water can be made from various types of water—mineral, deionized, distilled, and sparkling.

### Mineral Water

Unless it has been distilled or deionized, all water is **mineral water** because it naturally contains dissolved mineral salts. Natural springs tend to carry higher concentrations of minerals, such as sodium chloride, sodium bicarbonate, sodium carbonate, salts of calcium and magnesium, and sometimes iron or hydrogen sulfide, and these give mineral water its characteristic taste. Bottled water labeled "natural" has an unaltered mineral content. Examples of bottled mineral water from springs, some of which are flat and some sparkling (i.e., carbonated), include Perrier (Figure 27-1), Evian, Vichy, Contrexeville, and Apollinaris. Bottled water may be labeled **spring water, well water,** or **artesian water.**

### Deionized Water

Deionized water is **purified water** that has had all of its mineral content removed. The deionization process removes all the mineral salts (via ion exchange), leaving only pure water. Use of deionized water can make beverages taste less flavorful.

### Distilled Water

**Distilled water** is also purified water, but it is formed by converting water into steam and then condensing it in another, cooler container. The purpose of distillation is to separate a liquid from other particles or liquids. The steam leaves behind the dissolved particles and/or other liquids with higher boiling temperatures. The steam funnels upward into another chamber, where it is collected as it condenses back into a liquid. This process not only removes dissolved minerals but also destroys pathogens. Sometimes normal concentrations of minerals are added back into the water (27).

Distilled water is preferred for making ice cubes because they will be clear rather than cloudy, as they are when made from tap water. The cloudiness is caused by dissolved air bubbles, which are removed by the process of distillation. The same result can be obtained by boiling tap water for several minutes. Whether the water is distilled or boiled, the resulting ice cubes last longer because they are denser. Ice cubes can also pick up off-flavors from being held in the freezer too long with other foods. Rinsing the ice cubes before putting them in beverages helps to eliminate any possible freezer odors (41).

---

**Mineral water**  Water from natural springs that has a strong taste or odor due to small amounts of salts of calcium, magnesium, and sodium (sodium bicarbonate, sodium carbonate, sodium chloride), and sometimes iron or hydrogen sulfide.

**Spring water**  Water that, according to the FDA requirements, flows from its source without being pumped and contains at least 250 parts per million of dissolved solids.

**Well water**  Water pumped from an aquifer, an underground source of water.

**Artesian water**  Water that has surfaced on its own from an aquifer, rather than being pumped.

**Purified water**  Water that has undergone deionization, distillation, reverse osmosis, or any other method that removes minerals, chemicals, and flavor.

**Distilled water**  Water that has been purified through distillation to remove minerals, pathogens, and other substances.

**FIGURE 27-1**    Carbonated mineral spring water.

Silberkorn/istockphoto

### Sparkling Water

Carbon dioxide gas contributes the fizz and bubbles to sparkling water, which is available naturally from underground sources (17). Tap water that has been commercially filtered and carbonated is called seltzer. Club soda (soda water or plain soda) is somewhat different from carbonated water in that, in addition to having been filtered and carbonated, it has had mineral salts such as bicarbonates, citrates, and phosphates of sodium added. Sometimes club sodas, flavored club sodas, and seltzers serve as a source of sodium (13).

## Contaminants in Water

The U.S. Environmental Protection Agency (EPA) is responsible for monitoring water sources and setting safe levels of potentially dangerous compounds. Regulated contaminants include infectious organisms, disinfectants, inorganic and organic chemicals, and radionuclides. A list of these EPA regulated contaminants can be found at www.epa.gov/safewater/contaminants/index.html.

# CARBONATED BEVERAGES

Carbonated soft drinks, sodas, or "pops" are the most widely consumed type of beverage in North America, even more so than tap water (83). In the past, cola drinks were the most commonly purchased soft drinks, followed by lemon-lime, orange, ginger ale, root beer, and grape drinks (80). Trade journals, such as *Beverage Digest* for the beverage industry, provide the latest data on what and how much is selling (76). Beverages in this publication are categorized as carbonated, noncarbonated, and bottled water. In recent years, carbonated soft drink consumption has been decreasing (78), due in part to concerns about the high-fructose corn syrup and artificial sweeteners contained in sodas (3). What are people actually buying when they purchase soft drinks? About 90% to 98% is sparkling water, 10% is sugar, and the remainder consists of additional flavors, colors, acids, and preservatives (Chemist's Corner 27-1).

> ### CHEMIST'S CORNER 27-1
> #### The Water in Soft Drinks
>
> The water used in soft drinks is relatively chemically pure compared to that obtained from municipal drinking water. Tap water cannot be used in soda production because the alkalinity is too high, causing it to neutralize the acids; minerals such as iron and manganese interfere with coloring and flavoring agents; and residual chlorine is detrimental to flavor. Bottling plants treat water to be used for sodas by deionization; chemically precipitating the minerals; using activated charcoal to remove undesirable flavors, odors, and chlorine; deaeration to remove oxygen; and a final filtration to remove any remaining compounds passing the carbon filter (64).

## Early Soft Drinks

The major ingredient in soft drinks, sparkling water, was created in 1772 by the chemist Joseph Priestley, who added carbon dioxide, a colorless, nontoxic gas, to water. One of the early methods of creating carbon dioxide consisted of acidifying baking soda (sodium bicarbonate), which is where the term *soda* originates. The word *soft* was originally used to distinguish these beverages from so-called "hard" alcoholic drinks (27). The carbonated bubbles also increase the pressure in the can so that, when opened, it sounds like a "pop," another name given to soft drinks. In addition to providing sparkling bubbles, carbon dioxide provides a tingling mouthfeel, an effervescent appearance, and a preservative effect to the beverage.

In the 1800s, another chemist, Benjamin Silliman, began selling carbonated soda water to the public (27). Flavored soda water became popular after 1830, and in the late 19th century a druggist invented a flavored soda by adding an extract from the African kola nut. Coca-Cola was invented in 1886 by a pharmacist, John Pemberton, as a hangover remedy (55). The list of Coca-Cola ingredients has been guarded for over 100 years, but one researcher claims the original Coca-Cola recipe included citrate, caffeine, extract of vanilla, seven flavoring oils, fluid extract of coca (cocaine), citric acid, lime juice, sugar, water, caramel, and alcohol (70). The use of cocaine became illegal in the 1930s. Another pharmacist who created a soft drink was Charles Hires, who added a unique extract to sparkling water and introduced root beer.

## Soft Drink Processing

Numerous soft drink companies started with closely guarded flavored syrup formulas. Water is added to the syrup, resulting in a mixture that is then carbonated by adding carbon dioxide gas. Today, soft drinks are carbonated by placing a syrup/water mixture in a pressurized carbon dioxide container known as a carbo-cooler. After interacting with the carbon dioxide and cooling off, the newly carbonated syrup water is pumped to a filler, which then allocates a set amount of the soft drink into a sterile container (38). Added sweeteners, usually in the form of corn syrup, sucrose, or monosaccharides, contribute flavor and increase density, viscosity, mouthfeel, and calorie content. Sucrose was originally used to sweeten soft drinks, but was gradually replaced by the less expensive corn syrups. The fructose in corn syrup is sweeter than sucrose, so less is required.

Various acids also contribute to the flavor of soft drinks: citric acid from citrus fruits contributes tartness, usually to fruit drinks; phosphoric acid, usually used in cola drinks and root beers, is flat and sour; and malic acid from apples is used for its long-lasting flavor (37, 64). Preservatives added to soft drinks usually consist of benzoic and sorbic acids and their calcium, potassium, and sodium salts (38).

## Soft Drink Health Concerns

Recent news stories have brought to light the role that soft drinks may play in the obesity epidemic. The American Heart Association stated, "Soft drinks and other sugar-sweetened beverages are the primary source of added sugars in Americans' diets. On the basis of the 2005 US Dietary Guidelines, intake of added sugars greatly exceeds discretionary calorie allowances" (48). Beverages are a significant source of dietary calories. In fact, about half of the caloric intake among Americans comes from high-fructose corn syrup–sweetened beverages (63). As a result, some people have suggested the use of a tax on certain foods (soft drinks, fast foods, and snacks) as a preventive measure (i.e., to curb consumption) (71). Portion sizes are also a concern; the first popular soft drink was originally sold in a 6.5-ounce bottle that has now grown to 12 ounces (1½ cups). Avoiding regular soft drinks may be necessary for weight management; one study found that reduction in liquid calorie intake had a greater effect on weight loss than reduction in solid calorie intake (20).

Some researchers have suggested that the exposure of tooth enamel to the low pH of some soft drinks poses the risk of erosion; in addition, prolonged contact with the sugar in these drinks is a risk factor for dental caries (21). These and other issues regarding the health risks of soft drinks remain open to debate.

## Diet Soft Drinks

Diet soft drinks use alternative sweeteners such as aspartame, saccharin, acesulfame-K, or sucralose. The result is an increase in water content from 90% to 98%, and a decrease in sugar to zero, which has a detrimental effect on mouthfeel (37). Bulking agents such as carboxymethyl cellulose or pectin are often added to counter this side effect (64). Diet soft drinks were introduced in the 1950s and have steadily grown in popularity, with diet colas now accounting for at least one third of the soft drink market (15, 58).

## Food Additives in Soft Drinks

Additives in drinks commonly make up only 2% to 3% of the beverage weight, but can have a large influence on taste. They are often used to add definition and character, to balance the acid or sugar content, to improve nutrition, and to act as a preservative (37).

Only a few substances are added to soft drinks other than the sweetener or alternative sweeteners (see Chapter 21 on sweeteners), and these include acids, preservatives, colors, and flavors (Table 27-1).

# FUNCTIONAL BEVERAGES

Functional beverages are the latest breakthrough in the manufacture and sale of beverages (9, 10, 26, 51, 86–88). The term "functional" refers to health benefits from foods or beverages exceeding the two standard contributions of foods—energy and essential nutrients. An older term, "new age beverage," was coined in the mid-1990s to represent all beverages that did not fit into the traditional categorization of drinks (45). The newer term "functional beverages" took over because many of these beverages serve a function other than simple hydration.

## Types of Functional Beverages

It is difficult to categorize these drinks, so the terminology remains fluid as new beverages emerge, often combining attributes of existing categories or creating new ones (67). A few, but not all, of these functional beverage categories are now discussed: sports and performance, enhanced water, ready-to-drink tea, enhanced fruit, soy, enhanced dairy, herbal, nutraceutical, natural, energy, and smart beverages.

### Sports and Performance Beverages

Sports drinks have long been mainstream, but recently have grown in popularity to the point of sometimes replacing soda/pop beverages. They were initially designed for and targeted toward sport and exercise enthusiasts to replace the lost fluid, electrolytes, and carbohydrates of intense exercise (Figure 27-2) (3). Physical performance is often enhanced by these sport, isotonic, or recovery beverages that prevent dehydration, replace important electrolytes, and provide extra energy in the form of carbohydrates. Carbohydrates are available in many forms, but usually include sucrose, glucose, and maltodextrin. Adequate hydration, according to the American College of Sports Medicine, is achieved by most people if they drink about 17 ounces (500 mL) of fluid approximately 2 hours before exercise and at regular intervals during physical exertion (24).

| TABLE 27-1   Selected Food Additives in Soft Drinks | |
| --- | --- |
| **Acids** | |
| Citric acid | Natural acid obtained from citrus fruits, strawberries, raspberries, and black currants. |
| Malic acid | Natural acid obtained from apples, cherries, peaches, and plums. |
| Phosphoric acid | A stronger acid usually used in cola drinks. Although caustic at high concentrations, much lower levels are used for food use. Phosphorus is present in most animal and plant cells. Cola drinks contain about 13–19 mg/100 mL, compared to 200–800 mg/100 g in cheese, eggs, meat, fish, bread, and nuts. |
| **Preservatives** | |
| Benzoic acid | Inhibits molds and yeasts. |
| Sorbic acid | Better at lower pH levels than benzoic acid. |
| **Colors** | |
| Natural | Derived from plant products—fruits, vegetables, herbs, nuts, spices, and oils. |
| Artificial | Synthetically manufactured. |
| Caramel | Made from sugar to provide the characteristic brown color of cola drinks. |
| **Flavors** | |
| Natural | Derived from plant products—fruits, vegetables, herbs, nuts, spices, and oils. |
| Artificial | Synthetically manufactured. |

**FIGURE 27-2**   Functional drinks are gaining in popularity over soft drinks.

However, physical performance lasting under 1 hour does not benefit from the carbohydrates in isotonic drinks. Another benefit of sport or isotonic drinks is that they provide the electrolytes potassium and sodium in the form of monopotassium phosphate, potassium chloride, sodium chloride, and sodium citrate. They typically contain 14–15 grams of carbohydrates and at least 70 mg of sodium per 8 ounce serving. Specialized sports drinks designed for endurance events have an even higher sodium content—about 200 mg per 8 ounces (3).

### Enhanced Water
Fitness or enhanced waters are one of the newest functional products on the market. These beverages are primarily water with a new flavor, process, or ingredient added primarily for increasing the water's palatability. Although most of these products are lightly flavored waters with fewer calories than sports/performance drinks, a growing market, especially in Japan, exists for desalinated deep seawater, which sells for 3 to 5 dollars a bottle. Proponents claim that this seawater, pumped from the depths of the ocean by pipes submerged over 3,000 feet under the surface, is free of modern impurities and may benefit health.

### Ready-to-Drink Teas
Ready-to-drink teas arrive already brewed and packaged for immediate use. New tea lines continue to be produced, offering consumers a variety of flavors, decaffeinated options, and herbal blends marketed as having a medicinal value.

### Enhanced Fruit Drinks
Functional fruit juices offer consumers low-sugar options, fruit concentrates, and mild carbonation. Although numerous new functional drinks are rapidly appearing on the market, one of the oldest functional beverages is cranberry juice, recommended to reduce or treat urinary tract infections (40). Normally, the bladder is sterile, but bacteria such as *Escherichia coli* can cause an infection that can enter by the urinary tract, pass the bladder, and end up in the kidneys, where it can cause serious damage if not treated. The unique blend of organic acids in cranberry juice appears to have an inhibitory effect on potentially harmful bacteria.

### Soy Beverages
Soy functional beverages offer consumers the health benefits of soy. Derived from soybean protein, soymilk offers a vegetarian option for adults (but not for infants, who need breast milk or infant formula).

### Enhanced Dairy Drinks
Yogurts and smoothies are now being enhanced with added fiber, probiotics, or a variety of dietary supplements. Claims made for the benefits of these products range from aiding digestion (probiotics) to increasing regularity (fiber). An example is lassi, a traditional yogurt-based drink from India made by blending yogurt, water, ice, salt, pepper, and Indian spices. The sweeter version is mixed with sugar instead of spices (78).

### Herbal Beverages
Adding an herb to a drink results in an herbal beverage. In the United States, beverages containing popular herbal supplements are now sold (Table 27-2) (12). Herbal additives include hydroxyl citric acid, which is claimed to improve brain function (26). Additional supplements may include vitamins, minerals, bee pollen, chromium, carnitine, picolinate, and selenium, to name a few (6, 19).

**TABLE 27-2   Selected Herbs and Supplements in Functional (Nutraceutical) Beverages**

| Herb | Popular Claim in the Press |
|---|---|
| Echinacea | "Boosts immune system" (conflicting evidence) |
| Ginkgo biloba | "Sharpens the mind" (some evidence) |
| Ginseng | "Boosts energy" (conflicting evidence) |
| Guarana | "Stimulant" (due to caffeine) |
| Kava | "Alleviates stress" (evidence of liver toxicity if consumed in excess) |
| Saint John's wort | "Herbal antidepressant" (mild depression evidence) |

| *Supplement* | |
|---|---|
| Carnitine | "Muscle growth" (minor evidence of minimal effect) |
| Glucuronolactone | "Fight fatigue and improve well-being" (no evidence); precursor of taurine |
| Inositol | "Part of B-complex" |
| Taurine | "Amino acid" used in combination with carnitine for muscle growth (no evidence) |

## Nutraceutical Beverages

Functional beverages delivering health benefits have also been described as *nutraceutical* beverages and are often marketed as having beneficial health effects. Forty years ago, Japanese merchants were already selling nutraceutical beverages, elixirs aimed at countering stress, fatigue, and the effects of alcohol (42). Ingredients include green tea, antioxidants, soy, fiber, probiotics, phytochemicals, vitamins, minerals, and even oxygen (66, 75). Despite any concerns about overblown health claims, these functional beverages continue to appear on the market. The shift appears to be moving from sugary to more functional drinks (79).

*Nutraceutical Trends* The latest focuses in the nutraceutical beverage line include joint health, cancer (antioxidants), heart health, brain health, women's health, and weight management (61). Ingredients such as glucosamine and chondroitin sulfate and methylsufonylmethane (MSM) are being added to drinks sold to benefit joints (62). Lycopene, the red pigment from tomatoes, is being added as an antioxidant. Pomegranate and mangosteen juices are claimed to be high in fruit-based antioxidants. Heart-friendly ingredients include omega-3 fatty acids and fruit phytochemicals. Blueberries and a few other botanicals are suggested to aid memory, alertness, and cognitive function. Women's health nutrients include, but are not limited to, calcium, folic acid, soy isoflavones, and docosahexaenoic acid. The plant compound inulin serves as the foundation of some nutraceutical drinks. This carbohydrate-based fiber promotes the growth of beneficial bacteria in the large intestine, which is thought to reduce the risk of colon cancer (50). Weight-loss products are now focusing on fiber, resistant starches, chromium, and oat beta-glucan (61). The low-carbohydrate product trend that peaked in 2004 is being replaced by interest in drinks with a low glycemic index (79).

## Natural Beverages

Leading the trend toward more natural foods are many of the major beverage companies, which are replacing the high-fructose corn syrup in their products with cane sugar (85). In addition, many have also switched from previously used alternative sweeteners to stevia-based sweeteners that are perceived by the public as more natural because the leaves come from the stevia plant (see Chapter 21).

## Energy Beverages

Energy drinks are high in calories (usually from carbohydrates) and added stimulants such as caffeine. They may also include other ingredients such as water-soluble B vitamins, herbs, amino acids, and/or other flavorings. Some energy drinks have been banned in some European countries because of their high caffeine content. In the United States, energy drinks are the fasting growing portion of the beverage market, and the most popular among adults age 18–24 (79). The number of new energy drinks on the market increased from 6 in 2000 to over 200 in 2006 (26).

## Smart Beverages

Certain compounds claimed to boost brain power are added to beverages sold as "smart" drinks—despite the fact that there is very little to no research supporting such claims. Substances said to stimulate mental activities that are added to smart drinks include amino acids such as choline, L-cysteine, taurine, and phenylalanine (45).

# COFFEE

Coffee is one of the two most common aromatic beverages, the other being tea (Figure 27-3). An *aromatic beverage* is defined as one generating a pleasant odor from plants or spices. Coffee consists primarily of water with some additional compounds extracted from coffee beans during brewing. Scandinavians

---

## ✳ CALORIE CONTROL
### Beverages

Though water is calorie free, many other ingredients added to water are not.

- **Drink Water.** Plain water may not be as tasty as sugared and/or flavored beverages, but it meets the primary need of the body—hydration.
- **Choose Lower-Calorie Flavored Beverages.**
  - Bouillon (low-sodium) (0 calories/kcal)
  - Club soda (0 calories/kcal)
  - Coffee (0 calories/kcal without added sugar, milk, or cream)
  - Diet soda (0 calories/kcal)
  - Enhanced waters (0 to 100 calories/kcal each; read Nutrition Facts label)
  - Sugar-free drinks (liquids or mixes) (0 calories/kcal)
  - Tea (unsweetened) (0 calories/kcal)
  - Tomato or vegetable juices (70 calories/kcal for 12 ounces)
  - Tonic water (0 calories/kcal)
- **150 Calories (kcal) for Each 12-ounce Beverage.** A general rule of thumb is that the average 12-ounce beverage (soda can size) delivers approximately 150 calories (kcal). This adds up quickly: 300 calories (kcal) for two cans, or 450 calories (kcal) for three cans.
- **Beverages Can Contribute Excess Sugar.** Sugar or high-fructose corn syrup is often added to soft/soda drinks. In fact, the equivalent of 10 teaspoons of sugar is added to a *single* 12-ounce regular beverage, which exceeds the 2005 Dietary Guideline recommendation of a maximum of 8 teaspoons of sugar a day based on a 2,000-calorie (kcal) daily intake (84).
- **Fruit Juice Is Nutritious, But Sugary Too.** Surprisingly, fruit juices pack a lot of calories, making high juice intake a possible hidden source of excess calories. Only 8 ounces (1 cup), which is less volume than a soda, delivers 100 to 150 calories (kcal).
  - Limit fruit juice to 2 cups daily
  - Dilute fruit juice (1 cup fruit juice plus 1 cup water; add sweetener if needed)
  - Replace fruit juice with actual fruit
- **Alcoholic Beverages Also Deliver Calories.**

**FIGURE 27-3**   Coffee follows tea as one of the most common aromatic beverages.

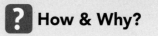
Henrik Sorensen/The Image Bank/Getty Images

The beans also darken from green to brown during roasting. Unroasted green coffee beans would have little, if any, flavor and aroma.

Roasting time determines whether a coffee is classified as a light, medium, or dark roast. The darkest roasts, in ascending order, are French (New Orleans), Spanish (Cuban), and Italian (espresso) roasts. Europeans prefer darker roasts, whereas Americans gravitate toward medium-roast coffees.

> ### ? How & Why?
>
> *Why do roasted coffee beans have an oily coating?*
>
> The oily surface of roasted coffee beans results from the breakdown (hydrolysis) of fats (fatty acids released from triglycerides) that occurs during the heat of roasting.

consume more coffee overall than any other nation in the world, with Americans second in consumption (69).

Coffee was probably discovered in Ethiopia around the 3rd century AD, when an Arabian goatherd named Kaldi noticed that his goats became particularly frolicsome after eating certain berries. The berries, of course, turned out to be from the coffee plant. Although many varieties of this plant are grown, only two, *Coffea arabica* and *Coffea robusta*, constitute 99% of the commercial crop. A third variety is *Coffea liberica* (27). Brazil and Colombia are the world's top coffee-producing countries, but other areas with tropical rain forest climates conducive to coffee production include Indonesia, Mexico, Central America, and several African countries. Knowing where a coffee is grown is important to coffee producers, because the flavor and quality of coffee vary according to the region where the particular coffee plant is grown.

## Coffee Processing

Coffee is made from beans that are picked, partially dried, processed to remove the hull, and either roasted and ground immediately, or exported to a processing plant for roasting and grinding. A small percentage arrives on the market as whole beans. Most imported beans are shipped green, because they can be stored in this state with little loss of quality. Coffee beans can also be processed to remove the caffeine, and liquid coffee extract can be used to produce instant coffee.

### Removing the Hull

Coffee plants, which are evergreen, generate numerous white flowers that yield a cherry-like fruit containing two seeds, the coffee "beans," enclosed in a tough skin. Coffee beans are usually handpicked and then subjected to one of two treatments to remove their hulls and free their greenish-blue berries. Countries with a limited water supply, such as those in the Middle East, historically have used the dry method where they sun-dry the beans. In the wet method, the berries are washed and then hulled by machine and partially fermented. Coffee processed by the wet method is more expensive, but experts agree that it has a better flavor than coffee processed by the dry method. Whichever method is used, the beans are then graded for size, type, and quality. It takes about 2,000 hulled coffee beans to produce 1 pound of coffee.

### Roasting

Roasting develops the beans' flavor, aroma, and appearance (35). The beans' physical and chemical makeup changes during the process. Physically, the beans dehydrate, becoming lighter and more porous as their surface area is increased.

### Grinding

Once roasted, the beans are ground to create more surface area from which the hot water can extract the compounds that contribute to flavor, aroma, and appearance. The type of grind—fine, drip, or regular—depends on the equipment that will be used to brew the coffee. Coffee beans ground too fine for their equipment deliver a bitter cup of coffee, whereas coarse grinds produce a weak, flavorless coffee. Generally, fine grind is used for vacuum and pressure pots, drip grind for drip pots, and regular for percolators. Particle size does not differ very much among these three main grinds, although fine has a greater number of smaller particles, whereas regular has the largest particles.

### Decaffeination

Currently, more than 20% of the coffee consumed is decaffeinated. Caffeine is usually removed by soaking green coffee beans in steam or water and extracting the caffeine from the water with a solvent such as methylene chloride or ethyl acetate. When used in large amounts, methylene chloride has been shown to cause cancer in laboratory animals, so the FDA limits its concentration to 10 parts per million. Ethyl acetate occurs naturally in fruits and vegetables. Some manufacturers choose nonchemical decaffeination methods, such as filtering the soaking water with activated charcoal. Such coffee is labeled as "water-processed" decaffeinated coffee. Regardless of which method is used, not all the caffeine is removed. Most decaffeinated coffees are about 97% caffeine free. Scientists are working to produce caffeine-free coffee beans through genetic engineering (4, 8).

### Instant Coffee

There are two major methods for manufacturing instant coffee: spray drying and freeze-drying. In the first method, a strong extract of coffee is sprayed through a jet of hot air, evaporating the water and leaving dried coffee particles. Freeze-drying is accomplished by freezing coffee concentrate into a solid mass and breaking the mass into small particles, which are then heat-dried in a vacuum.

# Composition of Coffee

Coffee's aroma, taste, and stimulating qualities are derived from substances extracted from the beans when hot water is poured over them. These substances are almost completely depleted from the ground coffee in this way, so the grounds should never be reused. There are more than 18 classes of flavor compounds in ground coffee, but to simplify the discussion, only the volatile compounds, bitter substances, and methylxanthines will be reviewed.

### Volatile Compounds

The roasting of green coffee beans creates many volatile substances, which give coffee its characteristic fragrance (72). Not all of the compounds responsible for the aroma of coffee have been deciphered, but over 600 volatile compounds have been identified so far (53). Heat vaporizes these substances into the air, where they signal to the human nose that the coffee is ready.

**Polyphenol** An organic compound with two or more phenols—carbon atoms structured into an aromatic ring with one or more hydroxyl (–OH) groups.

**Methylxanthine** A compound that stimulates the central nervous system.

---

**? How & Why?**

*Why does coffee lose its flavor as it cools or stands?*

The instability of some of the volatile compounds causes the flavor of roasted coffee to deteriorate quickly as it cools. In addition, when coffee is kept hot too long, evaporation causes the loss of many of these compounds and their accompanying flavors.

---

### Bitter Substances

The bitter, slightly sour taste of coffee is partially due to its organic acids. Chlorogenic acid, which constitutes about 4% of roasted coffee beans, contributes to over half of the acid found in a cup of coffee. Two other components contributing to the bitter taste of coffee are the **polyphenol** compounds and caffeine. Coffee should never be boiled, because temperatures at or above boiling increase the solubility of these compounds, resulting in coffee that has moved from pleasantly bitter to unpalatable.

### Methylxanthines

Coffee contains two substances—caffeine and theobromine—that belong to a group of compounds called **methylxanthines** (Figure 27-4). Methylxanthines can have either positive or negative effects, depending upon the individual. Possible side effects of caffeine, other than alertness and enhanced exercise performance, include temporarily increased heartbeat, metabolism, and stomach acid; sleep disturbance; and dilation and/or constriction of certain blood vessels. It also has diuretic effects, causing increased urination and increased loss of calcium through the

**FIGURE 27-4** Methylxanthines: Compounds found in certain foods and beverages that stimulate the central nervous system.

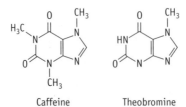

Caffeine          Theobromine

---

urine. In addition, withdrawal symptoms, including headache, fatigue, moodiness, depression, and anxiety, often appear after abrupt cessation of habitual caffeine use (73). Caffeine is also present in over 60 other plants, including tea leaves, cocoa beans, and the kola nut, but coffee and tea contain the highest concentration of caffeine. Theobromine is slightly less stimulating than caffeine; it is also found in cocoa and chocolates.

The amount of caffeine in a cup of coffee depends on which brewing method is used. Brewed drip coffees have the most caffeine, about 132 to 180 mg per cup, whereas decaffeinated coffees have the least, about 3 mg per cup. In comparison, tea contains about 40 mg of caffeine per cup. Caffeine also sells soft drinks, as evidenced by many of the top-selling soft drinks containing caffeine. Added to this list of caffeinated drinks are the popular "energy shots" sold at check-out counters. Countering the caffeine-loaded beverages are caffeine-free sodas, decaffeinated coffee and tea, herbal tea, and caffeine-free chocolate and cocoa substitutes.

# Types of Coffee

So many different types of coffee beverages exist that they can become confusing. Coffees vary depending upon the beans from which they are produced, how the bean is processed (ground, instant, decaffeinated, or espresso), or what other ingredients are added (specialty coffees, imitation, café au lait, flavored, or iced). Each of these different types of coffee is now further described.

### Types of Beans

*Coffea arabica* produces the finest coffee, whereas *Coffea robusta* yields a strong but inferior quality coffee. However, almost all coffee sold in the United States is blended and known by a brand name rather than by its variety.

**Blended Coffees**    Blending different varieties of beans allows the best characteristics of each to be fully expressed. Coffee beans found in nature rarely possess the taste, aroma, body, and consistency of blended varieties, because the beans used for blends are chosen so that the best qualities of each type complement those of the others (27). Sometimes a vegetable called chicory is added to blends to reduce their cost. This root is dried, roasted, and ground to look like coffee, but its use produces a heavier, more bitter, and darker cup of coffee.

Some desired qualities and the coffees that contribute these qualities to a blend include:

- Richness/body (Sumatran, Java)
- Sweetness (Venezuelan, Haitian)
- Extra sweetness (Mature Java, Mysore)
- Flavor/aroma (Sumatran, Colombian, Celebes)
- Brightness/acidity (Costa Rican, Central American coffees)

**Unblended Coffees**   Lately the demand for unblended coffees such as Mocha, Java, and Kona has increased. Many specialty coffee roasters are also offering popular "estate" coffees, or coffees grown in specific locations, resulting in certain characteristic flavors.

### Types of Processing
Coffee quality changes depending on how the coffee beans are processed—ground, instant, decaffeinated, or espresso.

- *Instant coffee.* This coffee is convenient, is less expensive, and constitutes about one fifth of all coffee sold. Its quality, unless it is decaffeinated, is inferior to that of freshly brewed coffee, which limits its use in restaurants.
- *Decaffeinated coffees.* The quality of these coffees varies, ranging from a detectable difference from regular coffee to no discernable difference.

- *Espresso coffee.* The word *espresso* is Italian for "pressed out," describing how high-pressure steam is used to force water through the ground coffee. This "express" pressing only takes 25 to 30 seconds (36). Espresso is usually served in small coffee cups because it is very dark and strong. This is partly because the beans are roasted until they are almost black, and twice the normal amount of ground coffee is used (Table 27-3).

### Types of Ingredients
Coffee can be combined with other ingredients to yield many different and flavorful products:

- *Specialty coffees.* Blends of coffee flavored with chocolate, sugar, almonds, and even liquor. They are higher in price but preferred by certain coffee connoisseurs.
- *Café Brulot.* A classic Creole strong coffee made with brandy, citrus zest, and spices.
- *Irish coffee.* A strong black coffee with added Irish whiskey, a little sugar, and regular or whipped cream.
- *Imitation coffee.* A cheaper version of coffee made from roasted and ground grains. Molasses is often added to cover the grainy taste and aroma.
- *Café au lait, caffè latte, or café con leche.* Coffee made with milk (usually a double-strength coffee mixed with equal parts of hot milk). Thai coffee is unique in that it utilizes condensed milk, making it much heavier and sweeter. It is also poured over ice to chill the beverage.

- *Flavored coffees.* Coffees containing flavored syrups made from sugar, water, natural flavor, and a preservative.
- *Iced coffee.* A popular beverage in Japan and increasingly common in North America.

## Preparation of Coffee
Many coffee-drinking consumers judge an establishment by its coffee. The key to making a good cup is to extract the compounds that contribute to good flavor and aroma, while simultaneously limiting the extraction of bitter substances. Three distinct layers form a cup of coffee—the bottom first layer called the shot, the middle section called the body, and the crema portion on top. Much of the flavor is derived from the oils that float to the top crema area. The process of preparing coffee focuses on various elements that go into an enjoyable cup of coffee. These include the type of coffee selected (see above), freshness, water-to-coffee ratio, water, temperature, brewing time, brewing equipment, and holding time. Each of these factors influences the coffee beverage's final quality, and all are now explained in more detail.

### Coffee Freshness
Coffee flavor and aroma begin to deteriorate as soon as the coffee beans are roasted. Deterioration and staling become even more rapid once the coffee is ground, so it is usually packaged in vacuum-packed cans: the fresher the bean, the better the coffee. Also for this reason, many people buy whole beans and grind them on a daily basis for ultimate freshness. It is best to grind the coffee beans before each brewing. Opened containers of ground beans should be discarded after 1 week. Coffee, whether ground or in bean form, is often stored in the freezer to maintain its freshness. Unopened, sealed packages of whole coffee beans store fairly well in a cool, dry place, but the freezer is best.

### Water-to-Coffee Ratio
The normal ratio of coffee to water ranges from 1 to 3 tablespoons of coffee (1 tbs = weak, 2 tbs = medium, 3 tbs = strong) for each 6-ounce cup, or about

| TABLE 27-3 | Types of Espresso Coffee |
|---|---|
| Espresso | A very dark, very strong coffee. Espresso is usually served in small cups. It can also be used as the foundation for the espresso beverages listed below. |
| Caffè mocha | Espresso and hot chocolate (mocha) mixed together in a tall glass, possibly topped with whipped cream and sprinkled with cocoa powder. |
| Caffè latte | Primarily steamed milk with espresso poured into it (whereas cappuccino consists of espresso with foamed milk on top). |
| Cappuccino | Some steamed milk is added to espresso that is then topped with the creamy foam created from steamed milk. May be dusted with sweetened cocoa powder, cinnamon, or nutmeg. |
| Mocha | Espresso combined with hot chocolate and foamy steamed milk. |

1 pound of coffee for every 1¾ to 2 gallons of water. Measurements for coffee pots are based on an average 6-ounce coffee cup. Personal preferences and geographical locations dictate the most appropriate ratio.

## Water Type

Coffee is 98% water, so the type of water selected to prepare it makes a difference. Fresh, cold, tap water is usually best for making coffee. Bottled water is preferred if the tap water has off-flavors or if a coffeemaker that may trap mineral buildup is used. Hot water has a tendency to produce a flat, stale cup of coffee. Deionized or distilled water tends to produce a sour cup of coffee. Water that is naturally soft is best, although chemically softened water is not recommended, because it filters more slowly and the sodium ions may combine with the fatty acids of the beans to form soaps (81).

## Water Temperature

The water to be used for coffee should not be boiled because the dissolved oxygen will escape, resulting in flat-tasting coffee.

Boiling also causes a loss of volatile compounds responsible for flavor; thus the aroma of boiling coffee smells good, but the coffee itself lacks flavor. It is best to warm the water to just below boiling temperatures (190°F–200°F/88°C–93°C) in order to extract just the right amounts of substances from the coffee grind. At these temperatures, about 20% of the weight of the grounds is extracted, which makes the best cup of coffee. Lower temperatures do not extract sufficient flavor, whereas higher temperatures extract too many bitter compounds.

## Brewing Time

Brewing time usually averages about 5 minutes, depending on the type of grind. A fine grind requires less time for extraction than a coarse grind. The fineness or coarseness of grind selected is often determined by the brewing equipment used (Table 27-4).

## Brewing Equipment

Different types of coffee-making equipment are shown in Figure 27-5. Materials used to make coffee equipment should be stainless steel, glass, earthenware, or enamelware; uncoated metals leave an undesirable aftertaste. After use, the equipment must be washed to remove the natural oils that accumulate and may become rancid. Thorough rinsing removes any soap residue. Mineral deposits in the drip spout can be prevented by pouring through it a combination of 1 cup of hot water and 1 cup of white vinegar (once every 3 months). The basic styles of brewing equipment rely on drip, vacuum, percolator, filter, and steeping mechanisms.

| **TABLE 27-4**   Brewing Time Increases with Larger Grind Sizes | | |
|---|---|---|
| Grind | Brewing Time (Minutes) | Usual Equipment |
| Very fine | 1–2 | Espresso, filter |
| Pulverized | 2–3 | Turkish coffeepot |
| Fine | 2–4 | Vacuum |
| Drip | 4–6 | Drip pot |
| Regular | 6–8 | Percolator/steeped |
| Coarse | 6–8 | Percolators with aluminum filters |

**FIGURE 27-5**   Coffee brewing equipment.

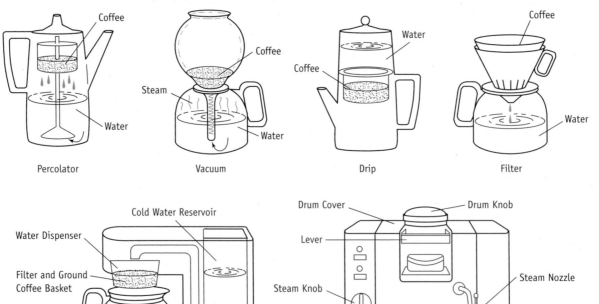

Percolator          Vacuum          Drip          Filter

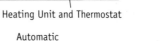

Automatic

Espresso

*Drip Coffeemaker*  Automatic or nonautomatic drip coffeemakers are probably the most popular devices for making coffee. They consist of a brewing basket with a perforated bottom, a filter that goes in the basket, and a coffee pot to catch the dripping coffee that results from the heated water passing through the coffee grounds.

*Vacuum Coffeemaker*  The vacuum coffeemaker looks something like an hourglass. The cold water goes in the lower portion, and fine coffee grounds go in the top. The seal between the two is tight, creating a closed system that allows steam to build up in the bottom. The pressure from the steam forces the water through the funnel to the top, where it comes in contact with the coffee grounds. The heat is then reduced to allow the water to stay on top for 1 to 4 minutes. Removing the pot from the heat source cools the water and creates a vacuum that pulls the coffee down. Two potential problems with vacuum coffeemakers are that the coffee may boil, and it may not be hot enough when served.

*Percolator*  In a percolator, a brewing basket is placed on the top of a tube whose base rests on the bottom of the pot. A tight lid with a glass dome is placed on the pot. Heating forces the water up the tube into the dome and down over the coffee. Electric percolators are considered best, because both the heat and timing are automatically regulated. Stove-top percolators must be timed carefully, and coffee quality is harder to control. Overpercolating by either method results in flavor loss. When the coffee is ready to serve, the grounds should be removed to prevent them from absorbing the aroma of the coffee.

*Filter*  The filter method is relatively easy, consisting only of two steps. A coffee filter is placed in a filter holder, which is then set on top of an empty coffee pot or coffee cup. Coffee grounds are placed in the filter, and enough hot water is poured through to match the number of cups desired.

*Steeping Method*  The steeping method requires only a pot and a lid. It is a straightforward method ideal for use on a picnic or camping trip. Regular ground coffee is heated in water to just below the boiling point and held for several minutes. The mixture is then filtered to remove the larger grounds, while the remaining fine particles sink to the bottom. Large quantities, known as *camp coffee,* can be steeped by wrapping and tying up the grounds in cheesecloth.

## Holding Time

For the best flavor, piping hot coffee should sit for 3 to 5 minutes before being sipped in order to allow the flavor to mellow. Coffee can be held at 185°F–195°F (85°C–88°C), but the longer it is held, the more the flavor deteriorates. Coffee should never be held for over an hour or the loss of flavor will be substantial. Higher temperatures will also negatively affect the quality of the coffee. Reheating is not recommended because it causes coffee to lose much of its flavor and aroma due to volatile compounds being lost, destroyed, or altered.

## Storage of Coffee

Whole, roasted coffee beans should be transferred from their paper bag to a tightly sealed glass or metal container and stored in the refrigerator, freezer, or other cool, dry place. Properly stored, they will retain their freshness for up to 3 weeks. Whole coffee beans can be frozen for several months, and there is no need to thaw the beans before grinding. For the freshest taste and aroma, they should be ground just before brewing. Coffee beans contain sufficient fat to make them good candidates for rancidity with improper or over-long storage; this will also cause them to lose volatile essential oils. Most ground coffee on the market is vacuum-packaged, but once opened, it starts to lose its freshness. For maximum keeping time, ground coffee should be protected against any exposure to air or moisture by being stored in an airtight container in the refrigerator or freezer.

## TEA

Despite the popularity of coffee, its consumption has been steadily decreasing over the last several years as consumption of various kinds of teas and other alternative beverages has increased (18). Tea remains one of the most popular beverages, with about one third of Americans drinking tea every day (43). In fact, next to water, tea is the most consumed beverage in the world (47).

Tea's origins can be traced to ancient China. Legend has it that around 2737 BC, Chinese emperor Shen Nung was boiling water when some wild tea leaves fell into the pot. The aroma was so pleasant that he tasted the brew and declared it good (47). The plant from which the leaves fell was an evergreen shrub, the *Camellia sinensis.* It is believed to have been first grown in India. In the 16th century, traders from Europe carried tea leaves and the knowledge of how to make the brew home from the Far East, and by the 18th century it had become the national drink of England. A tea tax that increased the cost of tea imported by American colonists is said to have triggered the Boston Tea Party of 1773 and the start of the American Revolutionary War (27).

Tea comes from many regions throughout the world. The United States imports most of its tea from Sri Lanka (Ceylon) and India, but other main tea-producing countries include China, Indonesia, and Japan. As with coffee, the country of origin and its climate influence the flavor characteristics of each tea variety, but the method of processing is even more important.

## Tea Processing

Only the smallest and most tender leaves are picked (Figure 27-6). About 1,000 leaves are needed to make 1 pound of manufactured tea. Processing consists of withering, rolling, oxidizing, and firing. Withering of tea leaves is accomplished by spreading them out in thin layers to expose them to warm air for 6 to 18 hours to reduce their moisture content from 55% to 65%. Rolling disrupts cell structure within the tea leaves, allowing oxidation to occur. Oxidizing, also called *fermenting,* is the process whereby the natural enzymes in the leaves, released by rolling, cause the leaves to become darker. Firing dries the leaves by passing them through a hot dryer for 20 minutes, inactivating the enzymes and decreasing their leaf moisture (64).

**FIGURE 27-6** Harvesting tea leaves.

travelphotographer/istockphoto.com

## Types of Tea

The three major categories of tea that result from different processing techniques are black (fully fermented), green (unfermented), and oolong (partially fermented) (47) (Figure 27-7). The majority, about 77%, of commercially produced tea is black tea, the predominant choice in North America, Great Britain, Europe, and some Asian countries. People in China, Japan, Korea, and India prefer green tea, which accounts for 21% of the produced tea. Oolong tea, which is consumed primarily in southeastern China, accounts for the remaining 2% of tea production (1). These teas may be further distinguished by added flavorings. In addition, white tea and various herbal teas are available on the market.

### Black Tea

The leaves are allowed to wither and dry under hot blown air (80°F–200°F/ 27°C–93°C) before being rolled to break the membranes between the cells. Enzymes, naturally occurring in the leaves, then oxidize the polyphenolic compounds in the cells and create changes in color, taste, and aroma.

### Green Tea

For green tea, the leaves are not oxidized, but rather heated and steamed before being rolled. The heat prevents "fermentation" by inactivating the enzymes (22). The lack of oxidation results in the "green" tea that is lighter (greener) than black tea. Matcha is a particular

**FIGURE 27-7** Three major types of tea based on processing technique.

**BLACK TEA (77%)**

Leaves rolled and exposed to air. Enzyme oxidizes polyphenols, darkening the leaves. Catechins: 3%–10%

**GREEN TEA (21%)**

Leaves steamed or heated. This destroys enzymes that oxidize polyphenols. Leaves dried. Catechins: 30%–42%

**OOLONG TEA (2%)**

Leaves repeatedly rolled and dried. This oxidizes polyphenols. Catechins: 8%–20%

type of green tea created by grinding the entire leaf and dissolving it in water, as opposed to steeping the leaves in a bag or strainer (47). The debate over whether green tea is beneficial against cancer continues (16). In 2006, the FDA reviewed 105 articles and other publications and determined that there was insufficient evidence to date to allow a claim that green tea reduces the risk of heart disease.

### Oolong Tea

Oolong tea is processed to be somewhere between black and green tea. It is only partially oxidized and is often perfumed and flavored with jasmine flowers.

### White Teas

These teas are produced from the tip buds of a special tea plant produced only in a province of China (29).

### Flavored Teas

These are teas flavored with natural oils, spices, and dried flowers or fruits. Almond, orange peel, cinnamon, and lemon peel are a few such teas.

### Herbal Teas

Tea beverages derived from plants other than *Camelia sinensis* may be called *herbal teas* (29). In Europe, herbal teas are referred to as "herbal tisane" (49). Technically these are not teas, because they do not contain tea leaves; they take their name from the fact that they are brewed in a similar fashion to tea. A wide variety of leaves, flowers, barks, and/or fruits of plants other than the tea plant are used to make herbal teas, which are sometimes called *infusions*. Examples of herbal teas include chamomile, ginseng, spearmint, rose hip, and raspberry. Health and therapeutic benefits are often claimed for these and other herbal teas, but research is still inconclusive. On the other hand, it has been reported that sassafras tea contains the carcinogen safrole; that comfrey in large amounts causes liver damage; that lobelia can cause vomiting; and that woodruff, tonka beans, and melilot are anticoagulants that can result in bleeding (71, 77).

### Specialty Teas

This term may be used to refer to teas that are exotic and sometimes flavored with added ingredients (49). The term "premium" is often used interchangeably with "specialty" in the U.S. tea market.

## Grades of Tea

The grade of a tea is based primarily on leaf size. The leaves are separated by screens with different sizes of holes. Quality decreases as the size of the leaf increases. The largest leaves are generally packaged as loose tea and are named orange pekoe (peck-oh), pekoe, and pekoe souchong. Smaller or broken leaves are usually used in tea bags and called broken orange pekoe, broken orange pekoe fannings, and fannings (27). Tea bags became popular when an American merchant sent tea leaf samples wrapped in silk bags to customers, who took advantage of this to brew individual cups of tea. The leaves in tea bags are usually blended to yield a characteristic flavor associated with a given brand.

## Composition of Tea

Over 300 compounds have been found in black tea, and about 30 in green tea. Some of the major constituents, other than water, in a cup of tea are polyphenolic compounds (such as **catechins** and flavonols), methylxanthines, and volatile compounds. Catechins tend to be found in higher concentrations in young tea leaves than in older leaves (25). The concentration of polyphenols also varies depending on whether the tea is black, green, or oolong. Polyphenolic compounds and/or their oxidative products are largely responsible for the flavor and strong astringency of tea. Green tea has the greatest amount, which gives this tea its metallic taste. Phenolic compounds are believed to have possible antioxidant and anticarcinogenic properties (5). On the other hand, polyphenols bind iron and prevent some of it from being absorbed by the intestine if tea is consumed with a meal containing iron (57). Methylxanthines, such as caffeine and theobromine, are found in tea, but in smaller quantities than are found in brewed coffee. Green tea is an excellent source of fluoride, and a moderate source of folate, a B vitamin. Other compounds in tea, theaflavins, help decrease the bitterness of the caffeine, whereas **tannins** contribute to its astringency and

characteristic reddish-brown color. Very few nutrients, other than water, fluoride, and folate (B vitamin), are found in tea.

## Health Benefits of Tea

Tea, particularly green tea, has grown in popularity due to its purported health benefits. Its cancer-preventing effects are supported by in vitro, animal, and epidemiologic studies (14, 52, 89, 47).

## Preparation of Tea

The goal in making tea, as in coffee making, is to extract the compounds responsible for good flavor and aroma, while limiting the extraction or development of bitter substances. Fresh tea may be consumed hot or iced, but the brewing process is the same for both.

### Brewing Tea

Tea is easy to prepare. Distilled water is best because minerals and chlorination are not present to interfere with the tea's flavor. Water is heated to just below the boiling point (212°F/100°C), poured over a tea bag or 1 teaspoon of loose black tea per 6-ounce cup, and allowed to steep for 3 to 5 minutes. Then the tea or bag is removed, and the beverage is served hot or with ice (38). In any case, it is important to avoid boiling the water or steeping the leaves too long, both of which will extract more of the bitter compounds, increase the degree of astringency, and elevate caffeine concentration. The caffeine content of tea can be reduced by 50% to 75% if the tea bags are soaked in cold water overnight. If the opposite is preferred, stronger teas can be prepared by increasing the amount of tea leaves rather than the brewing time, because the longer the tea steeps, the more bitter it will become.

**Catechins** Flavonoid pigments that are a subgroup of the flavonol pigments.

**Tannins** Polymers of various flavonoid compounds, of which some of the larger ones yield reddish and brown pigments.

The addition of milk decreases the bitterness of tea because milk proteins bind the astringent tannic acids (41). Using glass, pottery, china, or stainless steel teapots will keep metallic tastes to a minimum. The teapot should always be covered to prevent the escape of volatile compounds.

### Iced Tea

The story goes that a vendor selling hot tea on a blistering, humid day at the 1904 World's Fair in St. Louis was not having much luck until he dropped a couple of ice cubes into the brew. Iced tea now accounts for more than half the tea consumed in the United States. It starts out hot, prepared with double the amount of tea leaves used to brew tea to be consumed hot, and it can be poured over ice cubes immediately or allowed to cool before the ice is added.

Sun tea is a popular means of creating iced tea. This is made by placing 8 to 10 regular-size tea bags in a 1-gallon jar of cold water and setting it out in the sun for 2 to 4 hours. The bags are then removed and the brew refrigerated. Overnight tea can be made by placing 6 regular-size tea bags in 1 quart of cold water, covering the brew, and placing it in the refrigerator for at least 8 hours.

Fruit-flavored and herbal teas also make good iced teas. Flavored ice cubes made of citrus, cranberry, or other fruit juice add a special taste to iced teas. Sugar and lemon are often added to iced tea; mint, less often. Lemon combats the astringency and lightens the tea by bleaching the tannins. It also prevents the tea from clouding by creating an acidic environment, which inhibits the precipitation of compounds formed by caffeine and the aflavins. "Ready-to-drink" iced teas were introduced in 1991 and are available unsweetened or sweetened, in green and white tea varieties, and flavored.

### Instant Tea

Instant iced tea can be made by mixing soluble tea powders with water and ice. Instant tea powder is manufactured by dehydrating a strong concentration of tea and sometimes mixing it with sugar and other flavorings.

### Microwaving

A single serving of tea can be made in the microwave oven by placing one tea bag in a 6-ounce microwave-safe cup filled with cold water, microwaving on high for 30 seconds without boiling, and letting it stand for 30 seconds before removing the tea bag. Iced tea can be made the same way using a 12-ounce glass and ½ cup of cold water. It is allowed to cool a few minutes before adding ice.

## Storage of Tea

Proper storage is important in maintaining the quality of tea. It keeps longer than coffee, but when it is stored improperly, it is susceptible to oxidation as well as to the loss of volatile compounds. Tea should be stored in airtight containers at temperatures below 85°F (30°C).

# DAIRY BEVERAGES

Milk-based beverages contain more nutrients, in the form of protein, calcium, and B vitamins, than most other beverages. The choices are many, beginning with whole milk and moving to reduced-fat (2%), low-fat (1%), fat-free (nonfat), and reconstituted nonfat dried milk (NFDM). Any of these can be used to make hot chocolate, milk shakes, and chocolate or other flavored milk drinks. Some drinks formulated to enhance dietary intake, known as *nutritional beverages,* may be made with a milk base. Fermented or cultured dairy beverages such as kefir, acidophilus milk, and buttermilk are also available. These and a variety of other dairy products are explained more fully in Chapters 10 (on milk) and 26 (on frozen desserts). Smoothies are a unique way of blending dairy and fruit flavors, are often perceived as healthy, and are becoming increasingly available (68).

## Cocoa Beverages

Cocoa powder serves as the basis for either cold or hot cocoa beverages (see Chapter 25 on candy for more information on chocolate). United States Patent number 5264228 states that a commercial dry cocoa beverage mix is "prepared by mixing cocoa powder, non-fat milk solids, maltodextrin, an emulsifier and an artificial sweetener to make a mixture which then is ground to disperse the emulsifier and to reduce lumps. The ground mixture then is agglomerated and after a resting period of up to 120 seconds, the agglomerates are dried." Adding hot water will create a hot chocolate. Regular sweetener can also be used; North American formulas tend to be sweeter than those formulated in Europe.

Noncommercial cocoa beverages can be prepared by either the syrup or paste method. In the syrup method, a syrup is made by briefly boiling either cocoa or chocolate in sugared water. Hot milk is then added into the syrup base. The paste method relies on boiling the same ingredients with some cornstarch before adding the milk. The cornstarch gel creates a more full-bodied hot chocolate beverage. Beverages made by the paste method also have less of a tendency to settle out than do syrup-based cocoa beverages.

# ALCOHOLIC BEVERAGES

Alcoholic beverages fall into three major categories: beer, wine, and spirits. Figure 27-8 shows that, although serving sizes for these drinks differ in volume, each contains the same amount of alcohol—1 ounce. The alcohol in drinks is in the form of ethanol ($C_2H_5OH$), which is formed by fermentation of the pyruvate found in plants with a high carbohydrate content (64). When a person drinks any type of alcoholic beverage, the ethanol is absorbed into the bloodstream, where it reaches its maximal concentration (blood alcohol concentration, or BAC) in about 45 to 75 minutes after drinking.

The most common plants used in alcohol production include barley, wheat, corn, and grapes. The differences among alcoholic beverages stem from how they are manufactured. Beers are fermented by the action of yeast on grain, primarily barley, whereas wines are produced by fermentation the sugars derived from fruit, primarily grapes. Hard alcohols contain more alcohol and are called *spirits* because they contain (via distillation) the "spirit or soul" of fermented grain or fruit mixtures. Over 50 years ago, apple cider was considered an alcoholic beverage, but the term,

**FIGURE 27-8**  All these drinks contain the same amount of alcohol—1 ounce.

One 1.5 oz
glass of liquor

One 5 oz
glass of wine

One 12 oz
wine cooler

One 12 oz
beer

for which there is no legal definition, is now used to describe fresh-pressed apple juice with a darker color and less clarity than regular apple juice because of its suspended solids (74). However, the term *hard cider* is often used to describe an alcoholic beverage.

## Calorie (kcal) Content

Regardless of their source, alcoholic beverages have little to offer in the way of nutrients. Alcohol does provide calories (7 calories/kcal per gram), however, and lower-calorie alcoholic drinks are appearing on the market as weight-conscious consumers become a growing segment of the population. Mild vodkas and whiskeys, with half the alcohol, are available in Europe (60). In North America, it is more common to find "lite" beers and wine coolers. The term *lite* refers to lower-calorie beer. The word *light* is used to refer to beer that is lighter in color than the dark beers, so some confusion is possible between the two terms. Normally, beer averages about 150 calories (kcal) per 12 ounces, whereas lite beers contain about 100 calories (kcal). A martini made with vodka or gin will average 150 calories (kcal), whereas a gimlet can run as high as 350. Several dealcoholized beverages that have fewer calories than their original counterparts are available (7).

## Beer

*Baere* is the German word for "barley" and the root of the word *beer*. Barley is the principal grain used in the production of beer, although other grains can serve as a source of carbohydrates, which are eventually fermented by yeast to alcohol. Different types of beer result from brewers using different grains and processing techniques.

### Malt Production

Regardless of the grain, the yeast cannot utilize the starch unless it is first converted to simple sugars or malt. Malt is an extract of the grain's sugars (maltose, maltotriose, and glucose) and the compound giving beer its characteristic color and flavor: the darker the malt, the darker the beer.

The natural process of germination activates starch-breaking enzymes that supply sugars, such as malt, to the growing seedlings. In order to produce malt, beer manufacturers must first allow the barley to germinate by soaking the grain in water, a production step known as *steeping* (46). Germination is then halted by drying the grain, a process called *kilning*, which results in malt. Timing is important; the germination process must be stopped after the enzymes such as alpha- and beta-amylases are released, but before they have converted the starch naturally present in barley to sugar. Steeping generally occurs for several days, followed by 5 days of kilning at about 60°F (16°C) (9). After germination, the malt is crushed to separate the kernels from the husk.

The barley malt, called *grist,* is made further available to the yeasts by the next step, called *mashing,* in which malted barley is mixed with water and heated to gelatinize its starches.

Mashing converts the nonfermentable starches into simpler sugars, mostly in the form of maltose (9). As a result, the liquid fraction resulting from mashing, called *wort*, is very high in sugars capable of being fermented by yeast (64). The separation of this liquid fraction is called *lautering* (9).

### Brewing Beer

The next stage in beer production, *brewing*, is such a critical step that the entire process of manufacturing beer is often referred to as brewing. Extracting plant materials by pouring hot water over them describes the brewing of coffee, tea, and beer. Beer brewing further consists of adding **hops** to the wort and boiling the mixture. Brewing also concentrates and sanitizes the wort, inactivates enzymes, precipitates proteins contributing to haze, and caramelizes the sugars.

The hops added during brewing are from the hemp family of plants, *Cannabaceae*. Different hops can be used to achieve different flavors, aromas, and bitterness in the brew (9). Hops also have the effect of inhibiting the growth of certain types of bacteria that cause beer to spoil. After hops are added, the mixture is then transferred to a *whirlpool* where it is cooled and clarified (9).

---

**? How & Why?**

*Why are hops important to beer brewing?*

Hops contribute several substances to beer flavor: essential oils (responsible for aroma and flavor), bitter resins, and the tannins that contribute to color and astringency. The bitterness of hops is lost during storage when the resins are broken down (oxidized and polymerized) (47).

---

### Adding Yeast

After cooling, this mixture is ready to be inoculated with *Saccharomyces carlsbergensis* yeast. This ferments for approximately 9 days to produce a

---

**Hops**  The dried fruit of the *Humulus lupulus* plant, which grows in the Pacific Northwest of the United States.

beverage containing about 4.6% alcohol by volume (46), though the alcoholic content of beer can range from 2% to 8% (27). Yeast converts the sugar in the brew to alcohol, and produces carbon dioxide (9). Yeast is also important in determining the dryness or sweetness of beers. The degree to which sugar is converted to alcohol is termed *attenuation*, and high attenuation yields dry (nonsweet) beers, whereas low attenuation produces sweet beers. To stop the production of alcohol from yeast, the temperature is quickly lowered to 32°F (0°C), and the beer is filtered to remove the yeast and other suspended particles.

### Lagering

The next step, called *lagering,* involves storing the beer in tanks for several weeks to months, allowing the development of flavor and body, and the settling of particles. The particles, traces of degraded proteins and tannins, will cause an unsightly haze in chilled beer, so they are filtered out to give the beer a clear appearance (56). Enzymes may be added to further degrade the proteins, or earths or clays may be utilized to adsorb these materials (64).

### Draft Beer and Pasteurized Beer

The yeast in a bottle of beer brought back to room temperature will continue to ferment up to a certain concentration of alcohol, creating undesirable pressure within the sealed bottle. To avoid this, most beer is either cold filtered (draft) or pasteurized to completely halt fermentation. Filtering removes the yeast, whereas pasteurization kills it. Some of the off-flavors in canned and bottled beer result from the heat of pasteurization. Draft beer is filtered, packaged in kegs, and refrigerated, resulting in a better flavor than that of pasteurized beer. Beer that has been filtered and then refrigerated is identified as having undergone cold pasteurization.

### Types of Beer

The two main categories of beer, ales and lagers, are now briefly explained (9).

*Ales*    Ales are produced with yeasts that form clumps, or flocculate, at the top of the fermentation tank, and prefer temperatures around 60°F–70°F (16°C–21°C). Ales are aged for several weeks at about 50°F (10°C) and are characteristically complex and flavorful (30). They are usually served at room temperature.

*Lagers*    These beers have a lighter, cleaner taste than ales and are usually served chilled. Lagers are produced with yeasts that flocculate at the bottom of the tank and prefer cooler temperatures (approximately 50°F/10°C). They are aged for weeks or months at temperatures of 32°F–45°F (0°C–7°C).

### Home Brewing

In addition to differing in whether they have been cold filtered (draft) or pasteurized, beers differ in where they have been produced—at home, or by a small or large commercial brewer. Home brewing has become popular, and brew pubs and small brewing companies are enjoying widespread patronage (18). These small brewers produce much of the regional and specialty beer: ales, pale and bitter, light and dark, porter, bock, stout, and flavored beers.

### Beer Deterioration

Beer flavor deteriorates the minute it is packaged. Exposing the beer to oxygen contributes to oxidation and the production of substances (i.e., carbonyl compounds) that taste "pasty, papery, or like cardboard" (11).

### Drinking and Storing Beer

Beer is best served chilled, but not too cold. Light lagers should be served at 38°F–52°F (3°C–1°C), light ales at 54°F–58°F (12°C–14°C), and rich ales and lagers at 57°F–65°F (14°C–18°C) (30). While beer can be consumed straight from the bottle, pouring it into a glass releases aromatics that enhance the flavor. The "correct" pouring technique is to hold the glass at a 45 degree angle and pour the beer in evenly and slowly, while slowly straightening the glass to its upright position. This should result in about one inch of foam on top. Beer is best stored in a dry, dark location free of temperature fluctuations. The best storage temperature is 55°F–60°F (13°C–16°C).

## Wine

Wines are made from the fermented juice of fruits, usually grapes (*Vitus vinifera*). The grapes used in winemaking are carefully cultivated (9). Compared to table grapes, wine grapes are smaller, have thinner skin, and have a higher sugar content. Wine grapes are harvested when their sugar content is highest, because yeast will convert the sugar to alcohol. In contrast, table or eating grapes are usually harvested when their sugar content is at its lowest.

After the yeast ferments the wine, it is usually racked and aged. These three major steps of wine production are now briefly explained.

### Fermentation

Wine production involves the growth of grapes, their harvesting, the extraction of their juices, and finally fermentation (Figure 27-9). The sugar in grapes contributes to yeast fermentation, whereas their acidic concentration (pH = 3) discourages the growth of most other microorganisms. Yeasts produce many of the flavors found in wine, so selecting a particular wine yeast strain is very important. Yeasts produce certain compounds (higher alcohols, acids, and esters) that all contribute to flavor quality (39). Fermentation results in the conversion of grape juice to ethyl alcohol (ethanol) (9).

The sweetness and/or dryness of a wine does not always correspond to the alcohol content resulting from fermentation. A dry wine containing 14% alcohol can be made sweeter by adding juice or sugar. Sweet wines are not always lower in alcohol because distilled spirits may have been added. In general, wines average 11% alcohol.

> ### ? How & Why?
>
> *Why are wines described as sweet or dry?*
>
> The degree to which sugars are fermented to alcohol determines whether the wine is sweet or dry. Sweet wines contain higher concentrations of unfermented sugars than do dry wines; almost all of the carbohydrate in dry wines has been converted to alcohol.

### Racking

After fermentation, the wines undergo *racking*, during which they are allowed to stand to settle out the yeast cells and finely suspended material. The wine is

**FIGURE 27-9**  Grapes arrive at the winery in gondolas from the vineyard, ready to be crushed into juice for fermentation. Gondolas are usually pulled by a tractor through the vineyard, where grapes are picked either by hand or by mechanical harvester.

The Wine Institute

**FIGURE 27-10**  Fifty-gallon oak barrels age wine and impart flavor and complexity. Wine may be aged in oak barrels for an average of 1 to 3 years. Many fine wines will also benefit from further aging once in the bottle.

alvarez/istockphoto.com

then drawn out, leaving behind the sediment or *lees.* Racking was classically done in large 50- to 1,000-gallon barrels, but is now often achieved with large tanks.

### Aging

The next step is aging, in which the wine is stored in casks (often made of oak) or tanks for several months or years to allow the remaining traces of sugar to ferment and the further development of flavor (Figure 27-10). Most of the wines with a final alcohol concentration under 17% are filtered (cold pasteurized), or heat pasteurized just before bottling (Chemist's Corner 27-2). The exception is sparkling wine, in which the carbon dioxide content acts as a partial preservative. Sulfur dioxide ($SO_2$) is sometimes added to wines to inhibit microbial growth and the enzymatic browning that occurs when the phenolic

> ### CHEMIST'S CORNER 27-2
>
> #### Filtering Wines
>
> Tartrates naturally found in grape juice tend to crystallize as salts of tartaric acid. If these salts are not removed from wine by ion exchange treatments or a combination of chilling and filtering, the salts will slowly appear as glass-like crystals in the bottom of stored wine bottles (64).

compounds found in grapes are oxidized by phenolase enzymes (64).

#### Evaluating Wines

Appearance or clarity, mouthfeel, taste, and aroma are characteristics used to judge various wines. In terms of appearance, wine may be clouded by bacteria or trace metals. The wine's astringency and body (viscosity) contribute to the mouthfeel. Body is the weight of the wine in the mouth—light, medium, or heavy bodied: the more alcohol, the fuller the body (34). Another way to gauge alcohol content is to swirl the wine in a glass and watch how the liquid streams back down the sides. Thicker rivulets, also called *legs* or *tears,* often indicate a higher alcohol level (34). The major taste difference in wines is their balance between sourness and sweetness. The longer the taste remains in the mouth after being swallowed, the longer the finish and the better the wine will be. Aroma, also known as bouquet or nose, comes from over 200 volatile compounds, among them **congeners,** which contribute to aroma as well as to hangovers (55). Aromatic wines such as vermouth are flavored with the flowers, leaves, roots, or bark of plants. The aromas and tastes of wine are more easily detected if the wine is aerated with oxygen by swirling it in the glass. To detect any sediment or cloudy, brownish hues indicating spoilage, a wineglass is held up to the light. A dry, crumbly cork means that the wine was improperly stored (34).

> **Congener**  Alcohol by-product such as methanol or wood alcohol.

## How & Why?

**What are Congeners?**
Congeners (amines, amides, acetones, histamine, polyphenols, and methanol) are produced during alcoholic fermentation or added during production. Alcoholic beverages high in congeners include red wine, cognac, tequila, and whiskey (65).

## Selecting a Wine

Choosing a wine can be confusing because over 10,000 different strains of *Vitis vinifera* exist and the names of vineyards or regions on the labels of wines are unfamiliar to most consumers. If a particular growing year results in an especially good wine, it is known as a **vintage** year.

*Types of Wines*  Table 27-5 lists some of the common types of wines. Some of these wines are named after their grape of origin—such as Cabernet Sauvignon, Concord, Muscatel, Pinot, Riesling, and Zinfandel. Others are named for their region of origin (more common with French wines)—Bordeaux, Burgundy, Chablis, Madeira, Moselle, Port, Rhine, and Sauterne. Wines may also be classified based on certain characteristics, such as sweetness or dryness, carbonation, or color. As a result, wines are sometimes categorized as appetizer wines, sweet dessert wines, sparkling wines, and red or white table wines. Champagne, named after a region in France, is a wine that has been carbonated—either naturally in the bottle, a method known as the champagne method, or mechanically by adding carbon dioxide to the wine. Only sparkling wines made in the Champagne region of France can be called Champagne (31). Champagne flutes are designed so that their shapes help the bubbles to last longer.

### TABLE 27-5  Wines

| Product | Description | Uses |
|---|---|---|
| Apple wine (hard cider) | Apple cider or juice that has been allowed to ferment. | Served cold or hot as a beverage. Distilled to make apple brandies such as applejack (an American product) or Calvados (from Normandy, France). |
| Aromatic wine | A fortified wine flavored with one or more aromatic plant parts such as bark, flowers, leaves, roots, etc. Vermouth is one of the best-known types of aromatic wines. | An aperitif (drink served before a meal to stimulate the appetite) that is best when poured over ice. Mixer for cocktails and similar drinks. |
| Bordeaux | A wine produced in the Bordeaux region of France. | During meals or with the dessert. Good when chilled slightly and served in elongated Bordeaux glasses. |
| Brandy | Distillate from a wine (hence, the characteristics of each product stem from those of the original wine, the type of distillation, and the aging process). | After-dinner drink. In desserts and other dishes. |
| Burgundy | A wine produced in the Burgundy region of France. May be red, white, or sparkling. | During meals or with the dessert. Good when chilled slightly. |
| Cabernet | Wine made from the Cabernet Sauvignon grape, which was brought from Bordeaux, France, to California. | With or after a meal. Pairs well with steak, but not spicy foods. |
| Chablis | An excellent dry white wine (with a green-gold tint) from the French town of Chablis. However, the name is sometimes applied to similar dry, white wines made elsewhere. | The best wine for serving with oysters. Also good with fish, hors d'oeuvre, seafood, and shellfish. |
| Champagne | A sparkling wine that is made by allowing wine from Pinot grapes to undergo a second fermentation after a small amount of sugar has been added to the bottle. In France, the name Champagne is limited to the sparkling wines produced in the province of Champagne. | An aperitif that is served chilled. However, it may also be served at any time during any meal. A tulip-shaped glass helps to retain the bubbles. |
| Chianti | Red wine from the Tuscany region of Italy that is often sold in a round-bottom flask placed in a straw basket. However, the best wine comes in tall bottles that can be binned for aging. | With meals, particularly when Italian meat or pasta dishes are served. |
| Claret | A dry, red Bordeaux wine made from Cabernet Sauvignon grapes. | With or after a meal. |
| Cognac | Brandy that is double-distilled from wine made in the Charente district of France. | After-dinner drink. In desserts and other dishes. |
| Dry wines | A wine that is not sweet or sweetened. (In other words, all or most of the natural sugar content has been converted to alcohol.) | With or after a meal. |
| Fortified wines | Wines that have had their natural alcohol content increased by the addition of a brandy. | With dessert or after dinner. Should be served in small, narrow glasses. |
| Honey wine (mead) | An ancient type of wine that was made from fermented honey flavored with herbs. | With meals. |

**TABLE 27-5** Wines (continued)

| Product | Description | Uses |
|---|---|---|
| Light wines | A wine that has a low alcohol content. | With meals. |
| Madeira | One of the wines made on the island of Madeira. Madeira wines are the longest-lived (they keep for many years without deterioration) of any of the wines. | Depending upon the type of wine, it may be served at various parts of the meal. |
| May wine | A light, white Rhine wine that is flavored with the herb woodruff. | Served chilled in a punch bowl with pieces of fresh fruit floating on top. |
| Moselle wines | Light wines (the alcohol content is usually about 10% or less) made in the valley of the Moselle River in Germany, which lies to the west of the Rhine. | With lunch or dinner. |
| Mulled wines | Heated, sweetened, spiced wine served in a cup. | Served during the winter holidays. |
| Muscatel | A sweet, fortified wine made from Muscat grapes. | Served with dessert. |
| Perry (pear wine) | Light wine made from pear juice. | With meals. |
| Pinot | Wine made from Pinot grapes. | Starting material for making champagne. Served with meals. |
| Port | The type of fortified wine that originated in the town of Oporto in Portugal. | With dessert or after dinner. |
| Pulque | Fermented juice of the agave plant that grows in Mexico and in the southwestern United States. A common drink in Mexico. | Used to make a distilled liquor, or used shortly after its preparation because it does not keep well. |
| Red wines | Wines produced from dark-colored grapes that are fermented together with their skins (which contain most of the color pigments). | Served at meals featuring beef or lamb dishes, other stews flavored with wine. (In the latter cases, wine is served after the meal.) |
| Resinated (Greek wines) | Greek wines that contain a resin that imparts a pine-like flavor. | Best served with mild-flavored main dishes made from fish, pork, or poultry. |
| Rice wine (sake) | A Japanese wine made from fermented white rice. Although made from grain and sometimes referred to as beer, sake's alcohol content is similar to wine's. | With meals at Japanese restaurants. May be served hot. |
| Riesling | White wine made from the Riesling grape, which is considered to be the finest wine grape grown in Germany. | With meals. |
| Rhine wines | Wines from grapes grown in the Rhine River Valley of Germany. (The wines range from dry and light to rich and sweet.) | Depends upon the characteristics of the wine. |
| Rosé wines | Rose-colored wines produced by fermenting dark-colored grapes. The best rosé wines are made from Grenache grapes. | With cold foods and light meals, or when either a red or white wine may be used. |
| Sauternes | Wines made in the Sauterne district of Bordeaux, France, from grapes withered somewhat by a *Botrytis* mold that is also called "noble rot." | Should be served cold at the end of a meal, preferably one at which no other wine has been served. Serve in small, narrow glasses. |
| Sherry | A fortified wine made by a process similar to the one developed in Jerez de la Frontera, Spain. (Sherries range from pale-colored dry wines, to rich, sweet ones.) | Depends upon the characteristics of the particular wine. |
| Sparkling wines | Wines that are bubbly with carbon dioxide gas by virtue of having undergone a second fermentation initiated by the addition of a small amount of sugar. | Accompaniments to any part of a meal. |
| Sweet wines | Fortified wines that contain considerable amounts of unfermented sugars. (The addition of extra alcohol prevents the fermentation of the sugars that are present.) | Served as dessert. Should be served in a small, narrow glass. |
| Table wines | Unfortified wines of low to moderate alcoholic content. (They usually contain 14% or less alcohol.) | Served with meals. |
| Tokay | A rich white dessert wine made in Hungary that comes in dry and sweet varieties. | At meals or with desserts, depending upon whether the dry or sweet variety is served. |
| Vermouth | A fortified wine that is flavored with a variety of aromatic herbs and comes in dry and sweet varieties. | Preparation of martinis or other cocktails. Sweet Italian vermouth is often served on ice as an aperitif. |
| White wines | Made by fermenting grapes separated from their skins in order to keep the content of colored pigments low. | Served at meals featuring fish, pork, poultry, seafood, shellfish, or other bland-flavored items. |
| Zinfandel | A red wine made from Zinfandel grapes grown in California. | At meals featuring beef or lamb dishes other than stews that contain wines. |

**Why are wines decanted?**

A wine decanter can be used to separate the sediment from the wine and improve its flavor. Decanting is simply pouring the wine into a glass container, such as a decanter or carafe (32). Decanting allows young wines to aerate, which improves their flavor and increases the temperature for optimum drinking. Older wines are decanted to remove any sediment that develops over time. Decant wines by letting the unopened bottle stand for 12 hours (30 minutes for young wines), removing the cork as gently as possible without disturbing the sediment, wiping the top and inside neck of the bottle with a cloth to remove any dust or mold resulting from the aging process, and pouring the wine slowly in one continuous pour to prevent the sediment from mixing with the wine (32).

### The Colors of Wine

Wines may be white, red, or pink. White wines can be made from lightly colored (white or green) grapes; or they can be made from red grapes by one of two methods (Chemist's Corner 27-3). The first method is to remove the skins, pulp, and seeds from the red grapes prior to pressing, and the second is to gently press the red grapes and collect the juice early enough so that the pigments from the skin have not had time to release. Red wines are produced by leaving the skins, pulp, and seeds in the juice. Pink wines are made by adding a small amount of red wine to white wine (64). The traditional guideline has been that red wine goes with red meat, white wine goes with white meat, and a dessert wine is served after dinner/supper,

**White Wine from Red Grapes**

If white wines are made with red grapes, any remaining pigments can be removed with ion exchange, anthocyanase enzymes, or activated charcoal treatments (64).

but in reality, the decision should be made entirely on the basis of which wine one prefers (33).

### Food Additives in Wines

Sulfites are usually added to red wines (not white) to maintain their red or pink colors. A certain subset of the population is sensitive to sulfites. They can avoid these additives by reading the beverage's ingredient label (54).

## Spirits

As mentioned previously, distilled beverages are often called spirits because they embody the *spirit* of the fermented mixture. They are also often referred to as *hard* because they contain more alcohol than beer or wine. The boiling point of alcohol (173°F/78°C) is almost 40 degrees lower than that of water, so alcohol can be easily vaporized, cooled, condensed, and collected. Without this distillation, the alcoholic concentration of most alcoholic drinks would not exceed 15%, because that is the highest alcohol concentration that yeast can tolerate.

Spirits, or liquors, include whisky, rum, vodka, brandy, and gin, among others (9). Liquors have no added sugar and are at least 35% alcohol by volume. Liqueurs, on the other hand, are sweetened and flavored with various compounds, including fruits, herbs, spices, or nuts.

Liquor is created via distillation of a mash that begins with boiling fermented fruit or grains (9). Alcohol separates from the water as steam, which is collected in a cooling tube and then recooled so it returns to a liquid state. Alcohol that is repeatedly distilled can result in 95% pure ethanol.

While wine and beer continue to age and develop flavor after bottling, liquor does not. It is best to store liquor in a cool, dry location.

### Proof

The amount of alcohol in spirits is called its **proof.** An 80-proof liquor contains 40% alcohol by volume. Most liquors are 80 proof or 100 proof (40% to 50% alcohol, respectively), but some concentrated forms of whiskey and rum are 150 proof (75% alcohol), and grain alcohol can be up to 190 proof (95% alcohol). Liquor has a much higher proof than beer or wine. Beer is generally 8 to 12 proof, corresponding to an alcohol content of 4% to 6%. Wine is 14 to 28 proof, or 7% to 14% alcohol (9).

### Common Spirits

Table 27-6 lists some of the common distilled spirits. Gin and most whiskeys are derived from fermented grains. Vodka is made from either grains or potatoes. Liqueurs, which are sweet, syrupy, flavored liquors, are produced by steeping herbs or fruits in strong spirits before distillation. Rum is made from products of sugar cane. Tequila, a less common liquor, comes from the fermented sap of the Mexican-grown mescal plant. Spirits can also be produced by distilling the alcohol in wines. This process results in such beverages as brandy, which is distilled wine; cognac, which is double distilled brandy; and fortified wines, to which brandy is added to double the alcohol content.

**Vintage** The year in which a wine was bottled; especially, an exceptionally fine wine from a year with a good crop.

**Proof** Alcoholic strength indicated by a number that is twice the percent by volume of alcohol present.

| TABLE 27-6 | Spirits |
|---|---|
| **Type of Liquor** | **Production** |
| Brandy | Brandies are distilled from either a wine or a fermented mash of fruit. They are often aged for 2 or more years to mellow the harshly flavored constituents common to distilled liquors. |
| Gin | A fermented grain mixture is distilled to yield a strongly alcoholic mixture, which is then either redistilled or mixed with flavoring derived from juniper berries or other botanical substances. |
| Liqueur or cordial | A distilled liquor is either mixed or redistilled with one or more flavoring materials such as fruits, fruit peels, herbs, spices, flowers, cocoa, coffee, or roots. Liqueurs may contain from 2½% (minimum) to 35% added sugar. |
| Rum | Rum is distilled from a fermented sugarcane product. The stronger-flavored rums are usually aged for 3 or more years. Sometimes the distillation residue is added to subsequent rums to produce a strongly flavored product. |
| Vodka | Produced from starchy materials such as potatoes and/or grains (the starch is converted to sugar by enzymes). The distillation process used for vodka yields a product high in alcohol and low in congeners. The distillate is usually run through charcoal to remove any unwanted components. |
| Whiskey:<br>  All types | Produced from malted and unmalted grains. The enzymes in the former convert the starch in the latter to sugar. Distillation of the fermented grain mash is conducted so as to yield a product rich in congeners. The newly distilled liquor is usually aged in wooden barrels. Various whiskeys are often blended. |
|   Bourbon | Made from either malted or unmalted grain, mainly corn. The distillation mixture is a sour mash containing about ¼ old mash (previously fermented) and ¾ new mash. Aging is done in charred barrels. |
|   Canadian | Produced in Canada from malted barley and unmalted corn, rye, and wheat. Distilled as other whiskeys, then aged for 3 or more years. Usually, Canadian whiskeys with different characteristics are blended. |
|   Irish | Malted barley and unmalted barley, corn, oats, rye, and wheat. Product of either the Republic of Ireland or Northern Ireland. The fermented mash, made from barley and other grains, is distilled in 3 stages. (Hence, it is "triple distilled.") Irish whiskeys are aged for at least 4 years. Many are blended. |
|   Rye | Produced from rye and grain and distilled to produce a high content of congeners, then aged in charred barrels. |
|   Scotch | Produced in Scotland from mainly malted barley. The fermented mash is distilled in 2 stages. (Hence, it is "double distilled.") Newly distilled Scotch whiskey is aged from 3 to 4 years in barrels previously used for whiskey or wine, then blended with other whiskeys. |

# PICTORIAL SUMMARY / 27: Beverages

Most beverages fall into one of the following seven categories: water, carbonated, fruit/vegetable, functional, aromatic (coffee and tea), dairy, and alcoholic.

## WATER

Water contains zero calories or vitamins, and its mineral content varies according to region. Water is an essential nutrient required for the very existence of life. Plain water, the simplest of all beverages, from surface or underground sources, is either hard or soft. Other types of water include mineral, deionized, distilled, and sparkling waters.

## CARBONATED BEVERAGES

Adding sweetener, flavors, colors, acids, and preservatives to sparkling water creates a limitless number of carbonated drinks, known as soft drinks, sodas, or pops. Using alternative sweeteners increases the water content from 90% to 98%, and decreases the normal 10% sugar level to zero.

## FRUIT AND VEGETABLE BEVERAGES

Fruit and/or vegetable juices vary widely in their nutrient content. These drinks range from pure juices extracted from fruits and vegetables to highly diluted, artificially flavored and colored drinks (see Chapter 14 on fruits).

## FUNCTIONAL BEVERAGES

Functional beverages represent all beverages that do not fit into the traditional categorization of drinks, and include beverages identified as sports and performance drinks (low-carbohydrate, high-electrolytes), enhanced water (flavored), ready-to-drink teas (lower calories), enhanced fruit drinks (low-sugar), soymilk, enhanced dairy (probiotics or fiber) drinks, herbal drinks (containing herbs), nutraceuticals (health benefit), natural beverages, energy drinks (high-carbohydrate or -caffeine), and smart beverages (alleged brain stimulators).

## COFFEE

Coffee is made from beans that are processed by hull removal, roasting, and grinding. The goal of making a good cup of coffee is to extract enough of the compounds contributing to good flavor and aroma, but to limit the extraction of bitter substances.

Factors influencing the quality of coffee:

- Types of coffee
- Coffee quality
- Coffee freshness
- Water-to-coffee ratio
- Water type
- Water temperature
- Brewing time
- Brewing equipment

Henrik Sorensen/The Image Bank/
Getty Images

Once roasted, coffee beans will stay fresh for 2 to 3 weeks, whereas the freshness of ground coffee lasts only a few days.

## TEA

Tea is one of the most popular drinks in the world and can be served hot or iced. The three basic categories of tea—black, green, and oolong—depend on the type of processing, which consists of withering, rolling, fermenting, and firing. Tea grade is based primarily on leaf size. Tea keeps best in airtight containers placed in a dry atmosphere.

## DAIRY BEVERAGES

Milk-based beverages are high in protein, calcium, and B vitamins. Examples of dairy beverages include fluid whole, reduced-fat (2%), low-fat (1%), fat-free (nonfat), and dried milks. These milks can be incorporated into other beverages such as hot chocolate, milk shakes, and chocolate or other flavored milk drinks.

## ALCOHOLIC BEVERAGES

Alcoholic beverages fall into three major categories: beer, wine, and spirits (hard liquor).

Beers and wines are fermented with the action of yeast on barley and grapes, respectively.

Spirits contain the distilled alcohol derived from fermented grain or fruit mixtures.

BEER　　WINE　　　SPIRITS

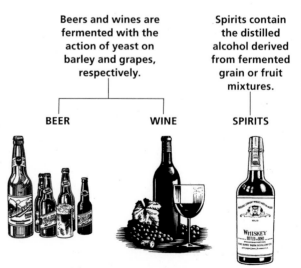

# CHAPTER REVIEW AND EXAM PREP

## Multiple Choice*

1. Deionized water is basically purified water without
   _____.
   a. color
   b. minerals
   c. carbonation
   d. steam

2. The fizz or bubbles in sparkling water are attributed to which chemical compound?
   a. Oxygen bromide
   b. Carbon carbonate
   c. Carbon dioxide
   d. Galactose

3. What factor determines whether a coffee is classified as a light, medium, or dark roast?
   a. Grinding level
   b. Roasting time
   c. Roasting temperature
   d. Decaffeination

4. What chemical compounds found in green tea act as antioxidants and may have anticarcinogenic properties?
   a. Phenolic compounds
   b. Phospholipids
   c. Pyroles
   d. Phosphates

5. What is the principal grain used in beer production?
   a. Wheat
   b. Hops
   c. Malt
   d. Barley

*See p. AK-1 for answers to multiple choice questions.

6. The substance(s) _____ is/are added to some nutraceutical beverages for joint health.
   a. glucosamine and chondroitin sulfate
   b. omega-3 fatty acids
   c. chromium
   d. mangotseen juice

7. Sulfites are often added to red wines to maintain their
   _____.
   a. flavor
   b. yeast fermentation
   c. color
   d. alcohol content

## Short Answer/Essay

1. Describe how the following types of water differ: mineral, deionized, distilled, and sparkling.
2. Describe the general ingredients in soft drinks and their function.
3. Define the following types of functional beverages: *sport, herbal, nutraceutical, energy, smart,* and *fun.*
4. What is the purpose of roasting and how does it affect the quality of coffee?
5. After roasting, the coffee beans are ground. Why are coffee beans ground, and how does the size of the bean influence the type of equipment that will be used?
6. What compounds are responsible for the bitter, slightly sour taste of coffee? What preparation mistake causes a higher concentration of these compounds to be processed?
7. Describe the differences among the following types of coffee beverages: espresso, caffè mocha, caffè latte, and café au lait.
8. How should roasted coffee beans be stored so that their freshness is maintained?
9. Explain the difference between black, green, and oolong teas.
10. Describe how each of the following types of alcoholic beverages is produced: beer, wine, and spirits.

# REFERENCES

1. Agarwal R, and H Mukhtar. Cancer chemoprevention by polyphenols in green tea and artichoke. In *Dietary Phytochemicals in Cancer and Prevention Treatment.* American Institute for Cancer Research, 35–50. Plenum, 1996.

2. Albrecht JA, and DN Nero. Position of the American Dietetic Association: Food and water safety. *Journal of the American Dietetic Association* 109(8): 1449–1460, 2009.

3. Anonymous. Adults pass on soda: Choose water, energy drinks and sports drinks. *ADA Times* 6(3):10, 2009.

4. Ashihara H, H Sano, and A Crozier. Caffeine and related purine alkaloids: Ciosynthesis, catabolism, function and genetic engineering. *Phytochemistry* 69(4):841–856, 2008.

5. Athar M, et al. Resveratrol: A review of preclinical studies for human cancer prevention. *Toxicology and*

*Applied Pharmacology* 224(3): 274–283, 2007.

6. Banner RJ. A new age for herbal drinks. *Food Processing* 58(7):35–36, 1997.

7. Bellamy G. Drinkable desserts: Hot-selling no alcohol drinks range from bottled beverages to calorie-conscious cocktails. *Restaurant Hospitality* 77(4):112–113, 1993.

8. Black R. *GM decaf coffee plant created.* www.news.bbc.co.uk/2/hi/science/ nature/3002112.stm. Accessed 12/09/09.

9. Brannon CA. Alcohol: Functional food or addictive drug? *Today's Dietician* 10(12):8–13, 2008.

10. Brannon CA. Phoods and bepherages: The phuture of phunctionality. *Today's Dietician* 9(9):12–18, 2007.

11. Brody AL. Packaging. Micro-oxygen packaging retains quality in beer. *Food Technology* 59(4):69–71, 2005.

12. Brown D. Consumers dive into fortified drinks. *Journal of the American Dietetic Association* 102(11):1602–1604, 2002.

13. Bubbles fizz while flavor holds appeal. *Food Engineering* 64(12):28, 1992.

14. Butt MS, and MT Sultan. Green tea: Nature's defense against malignancies. *Critical Reviews in Food Science Nutrition* 49(5):463–473, 2009.

15. Byrne M. Sweetners: A matter of taste. *Food Engineering International* 24(1):49–52, 1999.

16. Cabrera C, R Artacho, and R Gimenez. Beneficial effects of green tea—A review. *Journal of the American College of Nutrition* 25(2):79–99, 2006.

17. Carbonated waters. *Consumer Reports* 57(9):569–571, 1992.

18. Charlet K. Hot drinks cool down beverage momentum. *Prepared Foods* 166(5):57–58, 1997.

19. Charlet K. Trickle down theory. *Prepared Foods* 168(4):67–68, 1999.

20. Chen L, et al. Reduction in consumption of sugar-sweetened beverages is associated with weight loss: The PREMIER trial. *American Journal of Clinical Nutrition* 89(5):1299–1306, 2009.

21. Cheng R, H Yang, MY Shao, T Hu, and XD Zhou. Dental erosion and severe tooth decay related to soft drinks: A case report and literature review. *Journal of Zhejiang University* 10(5):395–399, 2009.

22. Cho HY, SJ Chung, HS Kim, and KO Kim. Effect of sensory characteristics and non-sensory factors on consumer liking of various canned tea products. *Journal of Food Science* 70(8):S532–S538, 2005.

23. Clemens R, and P Pressman. Hyperhydration—Can too much fluid hurt? *Food Technology* 59(4):16, 2005.

24. Convertino VA, et al. American College of Sport Medicine position stand: Exercise and fluid replacement. *Medicine and Science in Sports and Exercise* 28(1):I–VII, 1996.

25. Dreosti IE. Bioactive ingredients: Anitoxidants and polyphenols in tea. *Nutrition Reviews* 54(11):S51–S56, 1996.

26. Eckert M, and P Riker. Overcoming challenges in functional beverages. *Food Technology* 61(3):20–26, 2007.

27. Ensminger AH, et al. *Foods and Nutrition Encyclopedia.* CRC Press, 1994.

28. *Factoids: Drinking water and ground water statistics for* 2008. United States Environmental Protection Agency. www.epa.gov/safewater/databases/pdfs/data_factoids_2008.pdf. Accessed 8/09.

29. Friedman M, et al. Distribution of catechins, theaflavins, caffeine, and theobromine in 77 teas consumed in the United States. *Journal of Food Science* 70(9):C550–C557, 2005.

30. Gaiser T. Beer 101. *Fine Cooking* 87:20–21, 2007.

31. Gaiser T. Beyond Champagne. *Fine Cooking* 76:22–23, 2006a.

32. Gaiser T. Simple decanting instructions. *Fine Cooking* 92:26–27, 2008.

33. Gaiser T. Wine and cheese. *Fine Cooking* 69:32, 2005.

34. Gaiser T. Wine lingo demystified. *Fine Cooking* 84:20–21, 2007.

35. Geiger R, R Perren, R Kuenzli, and F Escher. Carbon dioxide evolution and moisture evaporation during roasting of coffee beans. *Journal of Food Science* 70(2):E124–E130, 2005.

36. Geise J. Developments in beverage additives. *Food Technology* 49(9):63–72, 1995.

37. Giannatempo L. Espresso: It's not just strong coffee. *Fine Cooking* 77:70, 2006.

38. Giese JH. Hitting the spot: Beverages and beverage technology. *Food Technology* 46(7):70–80, 1992.

39. Gil JV, et al. Aroma compounds in wine as influenced by apiculate yeasts. *Journal of Food Science* 61(6):1247–1249, 1996.

40. Guay DR. Cranberry and urinary tract infections. *Drugs* 69(7):775–807, 2009.

41. Hillman H. *Kitchen Science.* Houghton Mifflin, 1989.

42. Hollingsworth P. Functional beverage juggernaut faces tighter regulation. *Food Technology* 54(11):50–54, 2000.

43. Hollingsworth P. It's tea time. *Food Technology* 56(7):16, 2002.

44. Hollingsworth P. Profits pouring from bottled water. *Food Technology* 56(5):18, 2002.

45. Hollingsworth P. Redefining New Age. *Food Technology* 51(8):44–51, 1997.

46. Hoseney RC. An overview of malting and brewing. *Cereal Foods World* 39(9):675–679, 1994.

47. Ilka J. Can green tea every day keep cancer away? *Today's Dietician* 11(3):70–71, 2009.

48. Johnson RK, et al. Dietary sugars intake and cardiovascular health. A scientific statement from the American Heart Association. *Circulation* 120(11):1011–1020, 2009.

49. Keating B. U.S. tea industry: Heating up and getting hotter. *Nutraceuticals World* 7(9):46–49, 2004.

50. Kelly G. Inulin-type prebiotics: A review. (Part 2). *Alternative Medicine Review* 14(1):36–55, 2009.

51. Lal GG. Getting specific with functional beverages. *Food Technology* 61(12):24–31, 2007.

52. Larsson SC, and A Wolk. Tea consumption and ovarian cancer risk in a population-based cohort. *Archives of Internal Medicine* 165(22):2683–2686, 2005.

53. Lee TA, R Kempthorne, and JK Hardy. Compositional changes in brewed coffee as a function of brewing time. *Journal of Food Science* 57(6):1417–1419, 1992.

54. Lin SC, and G Georgiou. A biocatalyst for the removal of sulfite from alcoholic beverages. *Biotechnology Bioengineering* 89(1):123–127, 2005.

55. McGee H. *On Food and Cooking: The Science and Lore of the Kitchen.* Macmillan, 1997.

56. McMurrough I, et al. Haze formation shelf-life prediction for lager beer. *Food Technology* 53(1):58–62, 1999.

57. Mennen LI, R Walker, C Bennetau-Pelissero, and A Scalbert. Risks and safety of polyphenol consumption.

*American Journal of Clinical Nutrition* 81(1 Suppl):326S–329S, 2005.

58. Modified oatrim finds use in healthy beverages. *Food Engineering* 65(1):22, 1993.

59. Morr S, E Cuartas, B Alwattar, and JM Lane. How much calcium is in your drinking water? A survey of calcium concentrations in bottled and tap water and their significance for medical treatment and drug administration. *HSS Journal* 2(2): 130–135, 2006.

60. New generation of beverages will have lower alcohol levels. *Food Engineering* 61(9):60, 1989.

61. Ohr LM. Nutraceuticals. A functional foods sneak peek. *Food Technology* 60(5):95–96, 2006.

62. Ohr LM. Nutraceuticals. Joint health. *Food Technology* 60(1): 57–59, 2006.

63. Popkin MB, et al. A new proposed guidance system for beverage consumption in the United States. *American Journal of Clinical Nutrition* 83(3):529–542, 2006.

64. Potter NN, and JH Hotchkiss. *Food Science.* Chapman & Hall, 1995.

65. Prat G, A Adan, and M Sánchez-Turet. Alcohol hangover: A critical review of explanatory factors. *Human Psychopharmacology* 24(4):259–267, 2009.

66. Pszczola DE. How ingredients help solve beverage problems. *Food Technology* 55(10):61–73, 2001.

67. Pszczola DE. Lookin' good: Improving the appearance of food products. *Food Technology* 51(11):39–44, 1997.

68. Pszczola DE. Sipping into the beverage mainstream. *Food Technology* 53(11):78–88, 1999.

69. Regular or decaf? Coffee consumption and serum lipoproteins. *Nutrition Reviews* 50(6):175–178, 1992.

70. Rice M. Author claims he found secret recipe for Coke. *Arizona Daily Sun* April 25, 1993.

71. *Roehr B.* "Soda tax" could help tackle obesity, says US director of public health. *British Medical Journal* 339:b3176, 2009.

72. Sakano T, et al. Improvement of coffee aroma by removal of pungent volatiles using a-type zeolite. *Journal of Food Science* 61(2):473–476, 1996.

73. Satel S. Is caffeine addictive?—A review of the literature. *American Journal of Drug and Alcohol Abuse* 32(4):493–502, 2006.

74. Semanchek JJ, and DA Golden. Survival of Escherichia coli 0157:H7 during fermentation of apple cider. *Journal of Food Protection* 59(12):1256–1259, 1996.

75. Shah N. Functional foods from probiotics and prebiotics. *Food Technology* 55(11):46–53, 2001.

76. Sicher J. *Beverage Digest.* www .beverage-digest.com/about.html. Accessed 12/12/09.

77. Siegal RK. Herbal intoxication: Psychoactive effects from herbal cigarettes, tea, and capsules. *Journal of the American Medical Association* 236:437–476, 1976.

78. Sloan AE. New directions in drinks. *Food Technology* 62(5):19, 2008.

79. Sloan AE. A state-of-the-industry report. Healthy vending and other emerging trends. *Food Technology* 59(2):26–35, 2005.

80. Soft drinks and mixes. *Progressive Grocer* 70(7):49, 1991.

81. Spiller GA. The coffee plant and its processing. In *The Methylxanthine Beverages and Foods: Chemistry, Consumption and Health Effects.* Alan Liss, 1984.

82. Tarver T. Ensuring water quality. *Food Technology* 62(1):38–42, 2008.

83. *Trended per capita beverage consumption.* Nestle Waters. www.nestle -watersna.com/Menu/AboutUs/ Performance/Trended+Per+Capita+ Beverage+Consumption.htm. Accessed 8/09.

84. United States Department of Agriculture. 2005 *Dietary Guidelines for Americans.* www.health.gov/ DietaryGuidelines/. Accessed 8/29/09.

85. Wright R. Functional beverages: Thriving or surviving? *Nutraceuticals World* 12(6):28–37, 2009.

86. Wright R. Nutraceutical beverage showcase: Energy, sports, and wellness drinks. *Nutraceuticals World* (79):30–41, 2004.

87. Wright R. Nutraceutical beverage update. *Nutraceuticals World* 9(9):42–60, 2006.

88. Wright R. Nutraceuticals coast in the beverage market. *Nutraceuticals World* 11(7):38–52, 2008.

89. Zhang M, AH Lee, CW Binns, and X Xie. Green tea consumption enhances survival of epithelial ovarian cancer. *International Journal of Cancer* 112(3):465–469, 2004.

# WEBSITES

The Cook's Thesaurus provides detailed information on beer, wines, and liquors:
**www.foodsubs.com**
(click on "liquids" and then "alcohol")

Oregon State University offers links to drinks:
**http://food.oregonstate.edu/be.html**

Find the latest top 10 soft drinks at this website:
**www.beverage-digest.com**

The website of the American Beverage Association provides a wealth of information on non-alcoholic beverages:
**www.ameribev.org/**

*Beverage World* magazine publishes articles of interest for the beverage industry:
**www.beverageworld.com/**

This website offers videos showing how wine and rum are made:
**http://science.discovery.com/videos/ how-its-made-food-and-drink/**

anthonysp/istockphoto.com

# 28 Food Preservation

The World Health Organization estimates that about 20% of all food is lost to food spoilage. Spoilage contributes to the average North American family discarding about one fourth of the food it purchases (86).

**Yeast** A fungus (a plant that lacks chlorophyll) that is able to ferment sugars and that is used for producing food products such as bread and alcohol.

574

Some of the biological, chemical, and physical changes that lead to spoilage are preventable. Preservation methods that make this possible are discussed after a brief introduction on why food spoils.

## FOOD SPOILAGE

The foods we eat are all derived from living matter, making them subject to the natural process of decomposition. Food not only decomposes but also is lost or spoiled by being consumed by creatures other than humans—rats, mice, flies, and microorganisms. Food spoilage is obvious and detectable due to changes in appearance, taste, texture, and/or odor that all affect eating quality (4). In contrast, food contamination, which can also occur with food spoilage, is often undetectable (see Chapter 4 on food safety).

Not all foods spoil at the same rate. Foods are classified as perishable, semi-perishable, or nonperishable, depending on their susceptibility to spoilage. The foods most perishable are those with large concentrations of protein

and/or water, which accelerate the microbial and chemical processes of decomposition. For example, fish, seafood, meat, eggs, and dairy products, which all have a high protein and water content, are very perishable. Watery fruits and vegetables such as tomatoes, peaches, berries, and leafy vegetables are also highly perishable. Semi-perishable foods contain less water and include those, such as potatoes, carrots, beets, turnips, onions, and apples, that are edible for several months if stored under the proper conditions. Processed nuts, cereals, dried tea leaves, pastas, and dried beans and peas are nonperishable because they contain very little water and will keep for many months with little loss of quality.

The biological, chemical, and physical changes in food that contribute to spoilage are now briefly discussed.

## Biological Changes

The prime biological factors involved in food spoilage are microorganisms such as bacteria, **yeasts,** and molds.

Just like people, these tiny organisms need food to survive, so food is a natural target. The most common foods spoiled by bacteria include meat, eggs, milk, and opened canned goods. Foodborne illnesses caused by bacteria are discussed in Chapter 4.

Yeasts prefer high-sugar foods, such as fruits, vegetables, and fruit preserves, and can cause unwanted **fermentation** of fruits and fruit juices in the presence of the proper amounts of oxygen, moisture, and acidity (pH). The naturally occurring yeasts found in the air normally pose no threat. The moisture in foods, however, can encourage their growth to a point that becomes unacceptable. Any method that keeps the moisture content low will be successful in their control.

Molds and their toxins are discussed in Chapter 4. These microorganisms, like yeasts, prefer high-sugar foods, but are particularly drawn to cheese and bread. The appearance of their bloom on foods indicates that spoilage has begun. Molds are easily spread through the air, are very resistant to drying, and can be difficult to control by the means used for bacteria and yeasts. Commercial food enterprises employ vacuum pumps to remove oxygen from containers, because molds cannot grow in its absence.

The many weapons in the arsenal used to fight spoilage resulting from microbial action include boiling, refrigeration, drying, and curing with high concentrations of sugar or salt. These methods, which destroy or inhibit the growth of microorganisms, are discussed later in this chapter under the section "Food Preservation Methods."

## Chemical Changes

Chemical reactions or changes also contribute to food deterioration. Enzymes play a significant role in catalyzing these reactions and can be categorized by the substance on which they act (substrate) or their mode of action (80). Proteases, also called proteolytic enzymes, split proteins into smaller compounds (Chemist's Corner 28-1). Fish has many more active proteases than meat, which is one of the reasons fish deteriorates so quickly. Lobsters are also prone to proteolytic breakdown, which occurs the minute they expire. Unless lobsters

---

**CHEMIST'S CORNER 28-1**

### Proteinases

Proteases can be further classified into proteinases. These enzymes split proteins into smaller compounds, such as peptones and proteoses, which are then broken down to polypeptides and peptides. The latter two are then split by peptidases into amino acids. The various protein enzymes break protein down into a liquefied mass because the amino acids are water soluble (70).

---

are kept alive to the very last second, the proteases cause the lower abdomen to partially liquefy, with the result that the tail meat becomes crumbly when cooked (70).

Lipids are broken down by enzymes called lipases, which degrade the triglycerides of fat into glycerol and fatty acids. Further degradation leads to rancidity, or off-odors and off-tastes.

Enzymes that decompose carbohydrates are carbohydrases, each named after the particular sugar on which it acts. Another group of enzymes serves to oxidize compounds and these include ascorbic acid oxidase, peroxidase, tyrosinase, and polyphenolase. The latter two enzymes are involved in enzymatic browning, which leads to unappetizing brown discoloration in some fruits and vegetables. Hydrolysis may also contribute to the deterioration of foods. See Chapter 3 for a more detailed discussion of these chemical reactions.

## Physical Changes

Physical changes, unlike chemical changes, do not result in the formation of new compounds. Common physical changes occurring in foods as they spoil are evaporation, drip loss, and separation. Water evaporates out of improperly stored foods, creating an unattractive, dried-out appearance and possible undesirable flavor changes. Water can also be lost (syneresis) out of foods such as gelatins, yogurt, and sour cream as they age. Separation of water and oil occurs in such foods as nonhomogenized milk, mayonnaise, salad dressings, and high-moisture

---

cheeses when they are stored too long or are frozen and later thawed.

# FOOD PRESERVATION METHODS

For over 5,000 years, humans have been preserving foods by drying, salting, and fermentation. Ironically, the demands of war have triggered the monumental developments in food preservation techniques. Napoleon's need for a safe and portable food supply for his armies in the late 1700s and early 1800s led to the discovery of canning. Nicolas Appert is thought to have invented the process (27). World War II led to the development of **dehydrated** foods such as instant potatoes and eggs. The American Red Cross provided irradiated milk in the food packages given to prisoners of war (68). The Vietnam War spurred the refinement of the process of **freeze-drying,** which allowed for the development of complete, lightweight foods that could be carried into the field easily and transformed into ready-to-eat meals by adding water.

Because of newer preservation techniques and advances in refrigeration and transportation, people now enjoy a wide variety of foods, including out-of-season and exotic foods from all over the globe.

Chemical protection prevents changes in food composition typically due to exposure to gases, moisture, or light (51). Glass and metal provide an absolute barrier to these agents, while plastic is more permeable and only provides partial protection (52, 82). Biological protection is the prevention of contact with microorganisms, rodents,

---

**Fermentation** The conversion of carbohydrates to carbon dioxide and alcohol by yeast or bacteria.

**Dehydrate** To remove at least 95% of the water from foods through exposure to high temperatures.

**Freeze-dry** To remove water from food when it is in a frozen state, usually under a vacuum.

insects, and other animals. Physical protection focuses on preventing mechanical damage (51).

The various methods of drying, curing, fermentation, pickling, and canning are now discussed followed by separate sections on cold, hot, and other preservation methods. Nutrient retention during preservation is the last topic discussed.

# Drying

Drying is the food preservation process that consists of removing the food's water, which effectively inhibits the growth of microorganisms and simultaneously makes the food lighter to transport. Bacteria and molds need approximately a 15% moisture level to survive, whereas yeast needs at least a 20% moisture content. As mentioned in Chapter 3, the water activity of food ($a_w$) is the measurement of water content in food (25). The $a_w$ required for preservation varies by food, but is generally a value below 0.6. Once dried, the food can be eaten as is, or, unlike microorganisms, be rehydrated (have water added), which changes its size, color, flavor, and texture (25).

## Sun-Drying

Many early cultures subsisted throughout the year on naturally dry foods such as nuts, grains, and dried legumes. The discovery that fruits, vegetables, and meats could be dried in the sun was a natural extension of this practice (86). The sun provided the heat for evaporation for many centuries, and continues to do so in various countries around the world.

## Commercial Drying

Because sun-drying takes a long time and exposes foods to the weather and to the action of insects, most foods are now dried by various commercial processes, although raisins are still sun dried (Figure 28-1).

The most important types of commercial drying are conventional

**FIGURE 28-1** Grapes drying in the sun to yield raisins.

George D. Lepp/Documentary/Corbis

(heat), vacuum (pulls the water out), osmotic (water drawn out by osmosis), and freeze-drying (ice crystals vaporize).

***Conventional Drying*** Conventional drying uses heat to evaporate the water. In one method, the food is spread on a slatted floor or on shelves within kilns or drying rooms. A blower then passes hot air from a heater over and through the food. In tunnel drying, food is placed on trays or "cars," which are moved through a tunnel of carefully controlled hot air. Liquids can be dried by either spray drying or drum drying. In the former, a fine spray of the liquid is dried very quickly in midair. Spray drying is used to produce such foods as nonfat dried milk and some types of instant coffee. Drum drying occurs when liquid is poured over the very hot surface of a drum dryer, an apparatus resembling a large barrel. The dried food can then be peeled off like tissue paper, ground into flakes, and packaged. Some mashed-potato flakes and quick-cooking hot cereals are dried in this way.

***Vacuum Drying*** Vacuum drying dehydrates foods to very low moisture levels (1% to 3%) through the use of a vacuum. Milk, tomato paste, orange juice, and coffee are often concentrated by vacuum drying. The food is placed in a chamber, and the surrounding pressure is reduced below atmospheric pressure, which lowers water's boiling point. The water is then more easily boiled off at this lower boiling temperature.

***Osmotic Drying*** Osmotic drying is not often used commercially. It relies on the reuse of a strong syrup with a high sugar concentration that osmotically draws water from the object being dried. Ocean Spray's Craisins are cranberries dried through osmotic drying (18).

***Freeze-Drying*** Freeze-drying consists of first freezing the food and then placing it in a vacuum, where the ice **sublimates** to a vapor (Figure 28-2). This process of sublimation is the most effective method for drying foods, because it does not subject the food to high heat, which alters a food's flavor, color, and structure. Freeze-dried products yield the highest quality, store indefinitely, and can be reconstituted easily. However, the process is more costly than conventional drying.

**Sublimation** The process in which a solid changes directly to a vapor without passing through the liquid phase.

**FIGURE 28-2**   Typical freeze-drying process.

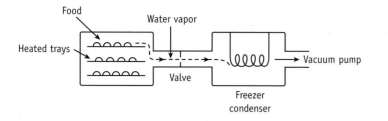

**FIGURE 28-3**   Selected food products produced by fermentation.

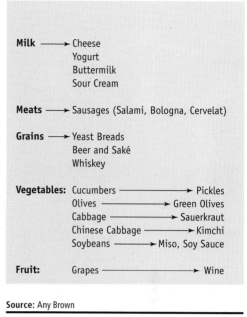

**Source:** Any Brown

*Pretreatments*   Some fruits are pretreated prior to drying. Plums have their skins *checked* before being dried into prunes by dipping them in lye or very hot water. This process cracks the skin, thereby improving skin texture and shortening the drying time by exposing more surface area to drying. Another pretreatment consists of blanching certain vegetables, such as potatoes or carrots, prior to drying to prevent enzymatic browning. The heat of this brief boiling disrupts the enzymes responsible for browning and inhibits the discoloration. Some fruits, such as apricots and peaches, are dipped in a sulfite solution or exposed to sulfur dioxide gas.

> ### ❓ How & Why?
>
> **Why is sulfur dioxide often listed as an ingredient in dried fruits?**
>
> To preserve their natural color and prevent spoilage, some foods, such as apricots and peaches, are dipped prior to drying in a sulfite solution or exposed to sulfur dioxide gas. Sulfur protects against enzymatic browning and the loss of vitamins A and C. However, it destroys thiamin (vitamin B₁), and in certain sensitive individuals it may cause headaches or allergic reactions, and even coma and death.

## Curing

High concentrations of salt bind to the water in food, making it unavailable to microorganisms. The earliest recorded use of salt as a preservative dates back to 3000 BC, when salt was used to **cure** fish. One of the earliest methods of preservation was to rub the surface of meats and fish with salt. Native Americans used salt to preserve some of their foods; the Hopi people traveled long distances to the salt mines within the Grand Canyon to obtain salt for this purpose.

Corned beef is a cured meat. The word *corn* refers to the Latin word for "grain," which in this case means grains of salt (72). Today, the most commonly cured meats include ham, sausages, hot dogs, bacon, and bologna.

## Smoking Cured Meats

Cured meats sometimes undergo the optional treatment of smoking for added flavor. Smoking is one of the oldest methods of preserving meat and fish (57). Meats are placed in smokers where they are exposed to the smoke of burning wood. The type of wood selected (sawdust, mesquite, hickory, oak, and various combinations of these woods) determines the resulting flavor of the meat (86). There are some health concerns with smoked foods, however, as they have been linked to cancer in laboratory animals and possibly humans (57, 66).

## Fermentation

For thousands of years, fermentation has been used both for the production and preservation of various foods (Figure 28-3). In the third century BC, laborers building the Great Wall of China were fed fermented vegetables as part of their rations. Throughout Asia, vegetables are still commonly fermented. In North America, foods most often preserved by fermentation are cucumbers, olives, and cabbage.

## Pickling

Pickling preserves food by acidification of food products through the addition of acid and/or fermentation (20). Vinegar is often used because the acidity of the vinegar keeps many microorganisms in check. In the Middle East, vinegar was used as early as 1000 BC to preserve such foods as fruits, onions, and walnuts. The food was simply covered with vinegar, boiled, and sealed in a container. It was then allowed to stand for at least 3 weeks to give the vinegar enough time to penetrate all parts of the food.

The acid created by fermentation of natural sugars in most pickled foods is lactic acid (20). Salt is often added to draw liquid out of the vegetable being pickled and to inhibit microbial growth. *Clostridium botulinum* is one important food contaminant inhibited by acidification because it cannot grow below a pH of 4.6.

Most people associate pickles with those made from cucumbers, but pickled foods include beets, cauliflower pieces, green tomatoes, green beans,

> **Cure**   To preserve food through the use of salt and drying. Sugar, spices, or nitrates may also be added.

**FIGURE 28-4**  Cucumbers are converted into pickles through the pickling preservation process.

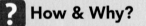
Courtesy of Presto®

and pickling (see above), in which acid is added in the form of vinegar, resulting in quick pickles. During the process of fermentation, a 10% salt solution serves as the liquid in which the cucumbers are submerged and allowed to ferment for several weeks. In this fermentation period, bacteria normally found on vegetables break down the sugar in the cucumbers. The salt penetrates the cucumbers, and the brine concentration is increased to 15%, except in the processing of dill pickles. Once fermented, pickles (whole or sliced) are placed in warm water; packed in glass jars; covered with a combination of vinegar, sugar, spices, and garlic; and pasteurized. It is this canning process, rather than fermentation, that preserves the pickles.

chilies, bell peppers, and sliced Jerusalem artichokes (Figure 28-4). These foods are preserved in vinegar and salt, with spices often added to enhance their flavor. For example, pickling changes the texture, color, and flavor of cucumbers to create pickles (20). Although pickling is supposed to preserve foods, those that need to be stored long-term are canned for safety reasons. They are also sometimes pasteurized to reduce microbial growth and preserve acidification (20).

The FDA regulates acidified foods, which requires electronic filing of the preservation process (20).

### ? How & Why?

*How are pickles made?*

Cucumbers can be manufactured into sweet, sour, dill, kosher dill, and other pickles by one of two methods: the longer process of fermentation, which yields brined pickles (with acid produced from the bacteria);

**Edible coating**  A thin layer of edible material, such as natural wax, oil, or petroleum-based wax, that serves as a barrier to gas and moisture.

## Edible Coatings on Foods

A unique food preservation method is to surround the food with an **edible coating.** The purpose of edible coatings is fourfold (29, 46):

1. To increase shelf life by acting as a barrier to moisture, oxygen, carbon dioxide, volatile aromas, and other compounds whose loss would lead to deterioration (81).
2. To impart improved handling characteristics, such as the ability to bend more easily without breaking.
3. To improve appearance through increased gloss and color.
4. To serve as a vehicle for added ingredients such as flavors, antioxidants, antimicrobials, etc.

### Composition of Edible Coatings

Edible coatings can be produced from carbohydrate, protein, or lipid materials. The most common edible coatings are lipid-based (beeswax, candelilla wax, carnuba wax, rice bran wax); oils (paraffin oil, mineral oil, vegetable oils); and petroleum-based waxes (paraffin, polyethylene wax) (9, 81). Carbohydrate coatings include starches, cellulose, seaweed extracts (carrageenan, alginates), pectinates, and chitosan. Previously, sugar was

used to coat foods to prevent decay and moisture loss (e.g., sugar-coated nuts). Protein films can be made from foods through the use of gelatin, collagen, whey protein, corn zein, soy protein, and wheat gluten (9).

### Commonly Coated Foods

Edible coatings are commonly used on vegetables and fruits such as cucumbers, tomatoes, peppers, eggplants, pumpkins, summer squash, apples, bananas, guavas, mangoes, papaya, melons, nectarines, and citrus fruits. They are also used to coat candy; cheese; nuts; dried fruit (prevents stickiness, especially in raisins and dates); eggs in their shell (as a moisture and bacterial barrier); and processed meats (especially sausages), poultry, and fish (46).

## Canning

Canning is a two-step process. First the food is prepared by being packed into containers, which are then sealed. Then the containers are *canned,* or heated to ensure that all microorganisms are destroyed.

As previously mentioned, during the Napoleonic wars, Napoleon was having difficulty feeding his troops. He offered a prize to the person who could discover a new method of preserving food. The winner was Nicolas Appert, who invented the canning process in the late 1790s. Food was placed in glass jars or canisters (cans), boiled, and then sealed shut. The jars were then boiled a final time, creating a vacuum. Many people believed that lack of oxygen in the cans preserved the foods. Almost a hundred years later, Louis Pasteur discovered that the real reason canning was successful was because the high boiling temperatures destroyed harmful bacteria. The heat processing also destroyed the enzymes responsible for the deterioration of foods, thereby protecting canned food from both harmful microorganisms and natural spoilage (87).

### Preparing Food for Canning

Fruits, vegetables, and meats are the most frequently canned foods. All are prepared prior to canning by either *hot* or *cold* packing. Hot-pack canned foods are heated to at least 170°F (77°C) in syrup, juice, or water prior to being poured into **sterilized** jars. This initial heating drives out much of the air, allows the food to be packed tightly in the jar, gives a translucent appearance to fruit, and reduces the heating time. In the cold-pack method, food is placed directly into the sterilized jars. The jars are then filled with boiling liquid. In both cases, some space is left at the top of the jar (the amount of space depending on the product and canning method). Air bubbles are removed by gently sliding a rubber spatula or knife between the jar's side and the food, and a small amount of air, known as the headspace, is left at the top of the jar (86). This headspace is required for pulling a vacuum and ensuring a seal. Jars with nicks or cracks should not be used for canning because they are not guaranteed to provide a reliable seal.

## Two Methods of Canning

The two major techniques for canning foods are boiling-water processing and pressure canning. The method chosen depends primarily on the pH of the food. A potentially deadly bacterium, *Clostridium botulinum,* cannot survive in an acidic environment, so boiling temperatures are sufficient to process most fruits and tomatoes whose pH falls below 4.6. For varieties of tomatoes that are less acidic, lemon juice is sometimes added to the tomatoes to avoid the possibility of food poisoning. For foods above pH 4.6, the United States Department of Agriculture (USDA) states, "To prevent the risk of botulism, low-acid and tomato foods [not canned according to USDA recommendations] should be boiled 10 minutes at altitudes below 1,000 feet. For altitudes at and above 1,000 feet, add 1 additional minute per each 1,000 feet." Another way to heat foods with pHs above 4.6 (meats and vegetables) is to use a special piece of equipment known as a pressure canner that allows higher temperatures to be reached.

The majority of *C. botulinum* food poisoning cases in the United States are traced back to errors in home canning. In an effort to discourage home canning, the USDA no longer provides information booklets showing how to can foods at home.

# COLD PRESERVATION
## Refrigeration

Refrigeration slows down the biological, chemical, and physical reactions that shorten the shelf life of food. About one half of the foods consumed in the United States is refrigerated or frozen, compared to about one third that is canned (42). Prior to the 1900s, refrigeration consisted of iceboxes filled with ice cut from lakes. In the 1850s, beer brewers began using mechanical refrigeration. Later, in the 1880s, meat packers started using refrigerated railroad cars to ship carcasses rather than live animals. However, home refrigeration was not widespread in the United States until the 1940s.

### Refrigerating Food

Not all foods have to be refrigerated, but all perishable foods should be refrigerated as soon as possible, preferably during transport, to prevent bacteria from multiplying. Placing produce in the crisper section of the refrigerator limits the amount of oxygen available to the foods, which slows their metabolism and prolongs their life and quality. Vegetables and fruits often continue to ripen and exchange gases with the environment after harvesting, even though they are cut off from their roots or leaves. When the fruit or vegetable's stored nutrient supply is finally depleted, the cells slowly die, and spoilage occurs as enzymes that break down, soften, and brown the tissue are released.

Plastic bags used to store vegetables or fruits should have holes in them in order to let in a small amount of oxygen to prevent tissue death, but not enough to speed up deterioration. Trapping these gases by using a closed bag on the counter speeds ripening and eventual deterioration.

### Refrigeration Temperatures and Times

For safety purposes, refrigerators should be kept between just above freezing to no more than 40°F (4°C). A study reported that one fifth of household refrigerators surveyed were set at or above 50°F (10°C), whereas the majority were set at about 45°F (7°C) (47).

Refrigerator storage times vary according to the food (see back inside cover of this book). Hot foods can be cooled down by placing them in shallow containers and setting the containers in cold water. Stirring during cooling distributes the heat and speeds the cooling process (86).

## Freezing

Serendipity played a part in the discovery that foods could be frozen commercially. Clarence Birdseye was ice fishing when he noticed that a fish he pulled out of the water froze in midair. He experimented and found that fish frozen immediately after being caught tasted fresh when thawed and prepared weeks later. In 1930, Birdseye started the frozen food industry (70).

In the past, people had used nature's snow and ice to preserve foods in the winter of colder climates. In fact, Sir Francis Bacon is reputed to have died of pneumonia contracted while he was trying to preserve chickens by stuffing their cavities with snow (86).

It is now known that freezing foods at 0°F (−18°C) or below is the least damaging to the food's original flavor, nutrient content, and texture compared to most other preservation methods. Freezing makes water unavailable to microorganisms. In addition, chemical and physical reactions leading to deterioration are slowed by freezing. Some oxygen is still present, however, allowing these reactions to continue, with the result that frozen foods have a shorter shelf life than canned foods. Shelf life is extended further for most supermarket products that are flash frozen in a very cold industrial freezer (38). The faster the food freezes, the less will be the damage that occurs to the food.

### What Foods Can Be Frozen?

The food's composition determines whether, and for how long, it can be stored in the freezer. Certain fruits, vegetables, and liquid dairy products do

**Sterilization** The elimination of all microorganisms through extended boiling/heating to temperatures much higher than boiling or through the use of certain chemicals.

not freeze well. The plant tissues of un-blanched fruits and vegetables are irreversibly damaged during freezing. Dairy products may have their original distribution of fat and water permanently altered to the point of becoming unacceptable in quality. The higher a food's fat content, the shorter its life span in the freezer, because fat can become **rancid** even when frozen. Conversely, foods containing very little water, such as dried fruits and beans, grains, nuts, and coffee, can have their useful life span doubled if they are frozen (Chemist's Corner 28-2).

## How Long Can Food Be Frozen?

Specific maximum freezer storage times for various foods are listed on this book's back inside cover, but in general, foods can be kept frozen for 2 to 12 months. The "first in, first out" rule should be followed: foods should be stored so that those most recently bought or frozen are the farthest back in the freezer, thereby moving food into a more easily accessible position as the freezer time increases.

## Four Problems with Frozen Food

Despite its effectiveness as a food preservation method, freezing has several potential disadvantages. These include freezer burn, cell rupturing, fluid loss, and recrystallization.

***Freezer Burn***   **Freezer burn** occurs when an opaque surface partially covers food that has been frozen too long or improperly (73). It alters the texture, color, and flavor of foods and is due to a loss of moisture that occurs when air spaces are left in a package, or the package is torn, or moisture-proof paper is not used. Moisture loss occurs through sublimation, or the transition from ice directly to vapor. Once the food is thawed, the texture of freezer-burned food is spongy, often reducing its quality to the point that it may be discarded (75). Frost-free freezers and the use of individually frozen products increase the risk of freezer burn (73).

The best way to prevent freezer burn is to wrap foods properly in airtight, vapor-resistant material, tape the package tightly, date it, and use the food before the optimal storage time has passed. An effective method for meats is to triple-wrap them, first with plastic wrap, then aluminum foil, and finally freezer paper secured with freezer tape. Masking tape should not be used, because the adhesive is not made to withstand freezing temperatures. The practice of keeping an up-to-date inventory on the freezer door showing the date on which each food was frozen avoids the costly waste of freezer-burned food. For commercially frozen foods, the use of vacuum packaging has almost eliminated the problem of freezer burn.

***Cell Rupturing***   Water expands when it freezes, and the resulting ice crystals pierce the food's cell walls, rupturing them and causing the food to take on an inferior texture. For this reason, some foods with a very high water content, such as lettuce, milk, tomatoes, and cottage cheese, should not be frozen. One way to minimize this problem is with rapid freezing, which results in smaller ice crystals, less cell rupturing, and a higher quality (6). Rapid freezing is accomplished commercially by the use of liquid nitrogen (−320°F/−196°C), which freezes foods in a few minutes rather than the 6 or more hours needed in a home freezer. Several other commercial methods of freezing that overcome the cell-rupturing problem include the following (40, 86):

- ***Air-blast freezing.*** Frigid air is blown on foods as they pass on a conveyor belt.
- ***Plate or contact freezing.*** Foods are placed between two plates of metal while being cooled by refrigerants.
- ***Immersion freezing.*** Foods are submerged in a low-temperature brine.
- ***Cryogenic freezing.*** Incorporates very low temperatures (−140°F/−60° C) with liquid nitrogen, liquid carbon dioxide, or their vapors.

***Fluid Loss***   Most frozen meats lose fluid when thawed. This is known as *drip,* and although it is red, it is not blood, but is actually water being lost from the cells. If too much fluid is lost, the meat's texture will be dryer when cooked than the fresh product would have been.

***Recrystallization***   Another problem with freezing is recrystallization, which occurs with longer storage times and temperature fluctuations caused by opening and closing the freezer door. Numerous small ice crystals melt, then combine on refreezing to form larger crystals, thus affecting the texture and quality of foods such as ice cream. During a lengthy power failure, the freezer door should be kept shut, and dry ice (solidified carbon dioxide) should be placed in the freezer and/or refrigerator as soon as possible. The addition of 25 pounds of dry ice will keep a 10-cubic-foot freezer below freezing for 2 to 3 days. A much smaller amount is required for the refrigerator and refrigerator freezer. Once the power returns, any thawed foods can be refrozen as long as the temperature has stayed below 40°F (4°C).

---

> ## CHEMIST'S CORNER 28-2

### Glass Transition Temperature

The shelf life of frozen products is extended if their water movement is limited. Deterioration results when water migrates during storage, resulting in large ice crystal formation. One way to avoid this is to raise the food's glass transition temperature ($T_g$), the temperature at which the food is said to be in a *glassy state*. This glassy stage is reached when the food's pure water freezes and leaves behind a very viscous concentration of solutes that trap the remaining water. The trapped water cannot crystallize. Food components that significantly raise $T_g$ are higher-molecular weight compounds such as protein, polydextrose, and gums. $T_g$ is not increased by low-molecular weight compounds such as sucrose, sorbitol, lactitol, and certain maltodextrins (61).

---

**Rancidity**  The breakdown of the polyunsaturated fatty acids in fats that results in disagreeable odors and flavors.

**Freezer burn**  White or grayish patches on frozen food caused by water evaporating into the package's air spaces.

# HEAT PRESERVATION

Heat in its various forms can be used as a preservation method, because many of the microorganisms responsible for food spoilage or foodborne illnesses are susceptible to heat. Heating methods include boiling, pasteurization, and ohmic heating.

## Boiling

The simplest heat preservation method, which has been used for centuries, is boiling. Ten minutes of boiling renders most foods free from microorganisms, but to remain that way, they must not be allowed to touch any unsterile object or be exposed to the air. Boiling temperatures can be exceeded by using pressure canners, autoclaves, and other instruments.

## Pasteurization

Another method of preserving liquids and extending their shelf life through heat is **pasteurization,** which destroys nonspore-forming pathogenic microorganisms. Many organisms that cause spoilage are destroyed as well. Milk is the most common beverage treated by pasteurization (discussed in more detail in Chapter 10), but fruit juices and other beverages can also be pasteurized. Nutrient losses during pasteurization are small, with flavor changes being the most significant alteration (3). The USDA has approved the use of steam pasteurization for meat, in which whole beef carcasses are sprayed with a blast of steam, killing microorganisms on the surface (15). Steam-pasteurized meat is not considered kosher (see Chapter 1) because there is no rabbinical supervision and the necessary temperatures required for kosher acceptance are not reached, as such temperatures would cook the edges of the carcass meat.

## High-Temperature Pasteurization

Using temperatures higher than those used for conventional pasteurization is described as high-temperature, short-time (HTST), and ultrahigh temperature

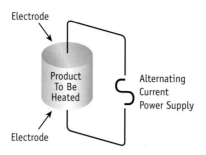

**FIGURE 28-5**   Ohmic heating.

Electrode

Product To Be Heated

Alternating Current Power Supply

Electrode

(UHT) (see Chapter 10 on milk). HTST temperatures and times start at 161°F (72°C) for 15 seconds, whereas UHT temperatures are above 280°F (138°C). Shorter times mean that there is minimal degradation of the products (55).

## Ohmic Heating

One of the latest preservation techniques through the use of heat is **ohmic heating.** Other names for the same process include resistance heating, joule heating, or electroheating (Figure 28-5) (72). Not all foods can be subjected to this type of preservation as they need to be in a semi-suspended state, such as a sauce (91). The benefit of ohmic heating is that it reduces the need for excess heat, resulting in a higher-quality food (44). Foods currently preserved with ohmic heating include liquid eggs, orange juice, and other fruit juices (19).

# OTHER PRESERVATION METHODS

## Irradiation (Cold Pasteurization)

Irradiation is being promoted as today's pasteurization (65). Although **irradiation,** also known as cold pasteurization, has been used for a variety of applications, it is a relatively recent food preservation method. Food irradiation was in practice as early in 1895 (65). It was approved in 1963 by the FDA for the control of insects in wheat and wheat flour (65). The Army Medical Department started to study the safety of irradiated foods in 1955, and the process was adopted in the early 1970s for

foods consumed by astronauts (26). In 2002, the concern with school-related foodborne outbreaks led the USDA to approve the use of irradiated meat for the school lunch program, but this practice was not adopted until later (62). In 2008, the FDA allowed irradiation of fresh and bagged spinach (65). Electron beam sterilizers have been used for the last 15 years. Listed below are several other examples of how irradiation is used:

- Sterilization of medical equipment (instruments, surgical gloves, alcohol wipes, sutures, etc.)
- Sterilization of consumer products (adhesive bandages, contact lens cleaning solutions, cosmetics, etc.)
- Foods for immune-compromised hospital patients (e.g., AIDS, cancer, or transplant patients)
- Spices and seasonings used in products such as sausage and certain baked goods

### Irradiated Foods

Irradiation of foods has expanded rapidly since it was first approved as shown in Table 28-1. Not all foods can be irradiated as it may affect the texture of certain plant foods (Chemist's Corner 28-3).

### The Irradiation Process

Common types of irradiation in the environment include visible light, radiofrequency, infrared light, ultraviolet light, and microwaves (33). Gamma rays, x-rays, and electron beams are stronger forms of irradiation called "ionizing" radiation because they produce ions, or electrically charged particles.

**Pasteurization**  A food preservation process that heats liquids to 161°F (72°C) for 15 seconds, or 143°F (62°C) for 30 minutes, in order to kill bacteria, yeasts, and molds.

**Ohmic heating**  A food preservation process in which an electrical current is passed through food, generating enough heat to destroy microorganisms.

**Irradiation**  A food preservation process in which foods are treated with low doses of gamma rays, X-rays, or electrons.

**TABLE 28-1** Foods Approved for Irradiation in the United States (33, 65)

- Wheat flour
- White potatoes
- Certain fruits and vegetables
- Seeds for sprouting
- Herbs and spices
- Meat
- Shellfish
- Poultry
- Fresh shell eggs
- Dry enzyme preparations
- Dry spices or seasonings

> **CHEMIST'S CORNER 28-3**
>
> **Irradiation and Softening of Food**
>
> Irradiation induces a loss of firmness, or softening, in some types of fruits and vegetables. This occurs because irradiation causes depolymerization of cell wall polysaccharides, cellulose, and pectin in the produce (33). Also contributing are irradiation's effects on the activity of the enzymes pectinmethylesterase and polygalacturonase, found in the cell wall.

Food is irradiated by passing it through an irradiator, an enclosed chamber where it is exposed to an ionizing energy source. A number of ionizing energy sources may be used (gamma rays, X-rays, or electrons), but gamma rays are the most common (Chemist's Corner 28-4). Irradiation causes free radicals, or highly reactive charged molecules, to be created from water (33). These free radicals act to disrupt DNA and other components of microorganisms and prevent them from multiplying (33, 65). Thus, irradiation works by breaking down the chemical bonds within the DNA and other molecules in the cells. The treated food does not become radioactive itself, any more than

> **CHEMIST'S CORNER 28-4**
>
> **Irradiation Dosages**
>
> Gamma radiation uses the radioactive form of Cobalt ($^{60}$Co), and possibly Cesium ($^{137}$Cs) to produce gamma rays (65). The amount of radiation energy absorbed from these gamma rays is expressed in kilograys (kGy). For example, one kilogray equals 1,000 grays, and 1 gray equals 1 joule per kilogram. Irradiation dosages from low (under 1 kGy) to medium (1 to 10 kGy) kill insects, larvae, and pathogenic bacteria in wheat and wheat flour; inhibit sprouting in potatoes; and slow ripening and spoilage in fruits. Higher dosages (10 to 50 kGy) are enough to sterilize foods for use by astronauts or immune-compromised hospital patients (63). The Environmental Protection Agency limits the amount of radiation to 10 kGy per food. The FDA sets limits specific to different types of food: 1 kGy for fruit, 3 kGy for poultry, and 30 kGy for spices (65).

does a briefcase passing through an airport security check. The energy of the radiation used is not high enough to induce the changes in atomic nuclei that result in radioactivity (65). It also does not create significant radioactive waste. Cobalt-60, one of the radioactive metals used to produce gamma rays, decays by 50% in 5 years, and can then be recharged and reused. However, Cesium-137 requires 30 years to decay by 50% (65).

## Effects of Irradiation on Foods

Treating foods with gamma radiation renders them less susceptible to deterioration and destroys many, but not all, microorganisms. For example, although it destroys bacteria, yeasts, and molds, the dosage used on foods for pasteurization is too low to eradicate *C. botulinum* spores, and it may not eliminate smaller viruses (42). Irradiation reduces spoilage, lengthens shelf life, sterilizes or kills insects, and can protect the public from many bacterial foodborne diseases (Figure 28-6) (41). Meat can keep three to four times longer in the refrigerator when irradiated (2), and irradiation does not change the aroma, taste, or texture of ground beef (78). Research has demonstrated that irradiated fresh fruit retained desirable quality for up to 16 days, compared to nonirradiated fruit, which spoiled within 6 days.

Not all produce can be irradiated successfully, however. Many fruits and vegetables become mushy and discolored, and actually spoil faster with irradiation (90). In fact, only about 10 different fruits experience an extended shelf life with irradiation. Irradiation causes certain fruits and vegetables to lose firmness; this effect can be reduced in tomatoes and apples by dipping them in calcium solution prior to irradiation and by storing them in modified-atmosphere packaging (33). Irradiation is not thought to affect the trans-fatty acid content of foods (34). Irradiation can reduce vitamin C levels by converting ascorbic acid to dehydroascorbic acid, but this effect is insignificant (33).

The FDA investigated the possibility that furan, a carcinogen, was produced during food irradiation (33). Studies

**FIGURE 28-6** Irradiation inhibits mold formation in strawberries refrigerated for 17 days.

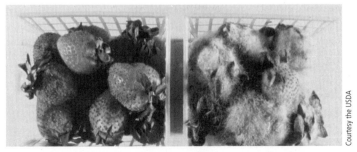

Irradiated (1 kGy)                    Nonirradiated

Courtesy the USDA

showed that 5 kGy irradiation did not induce detectable furan production in most fresh fruits and vegetables.

### Irradiation Pros and Cons

Despite official approval from many organizations (FDA, USDA, World Health Organization, United States Army, American Medical Association, American Dietetic Association, Institute of Food Technologists, and others) (83), food irradiation has aroused considerable controversy and is the subject of extensive studies (48, 60, 68, 85). Many companies are reluctant to use irradiation because of consumer fears or lack of information (50). Consumer concerns include possible nutrient loss and the environmental hazards arising from the use of radioactive materials in irradiation facilities. There is also the question of free radicals produced by irradiation; however, the FDA states that these are generated by other heating methods. Proponents argue that irradiation eliminates the need for some chemical fumigants and preservatives, reduces spoilage losses, and increases food safety by destroying harmful microorganisms, especially *E. coli* and *Salmonella* (48).

Consumer acceptance of irradiation may actually be higher than is indicated by media coverage. In fact, consumer concern about food irradiation has decreased (49). Survey participants of one study declared more concern for bacteria, pesticide residues, food additives, animal drug residues, and growth hormones (56). The major consideration for consumers is that irradiated food be clearly labeled so that they can make the choice for themselves. The FDA requires that irradiated food be labeled with the radura symbol or include the words *treated with radiation, treated with irradiation,* or the word *irradiated* as part of the product name (Figure 28-7) (77).

**FIGURE 28-7** The green radura symbol identifies irradiated food.

**FIGURE 28-8** The effect of pulsed light on *Staphylococcus aureus*.

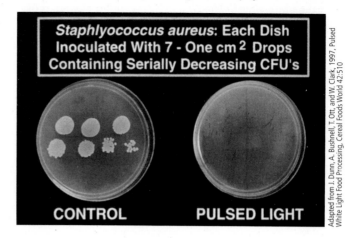

In 2007, the FDA proposed a change to irradiation labeling, recommending that only foods in which irradiation caused a "material change" in the food should be required to include the radura logo and the statements concerning irradiation (33). Material changes include any nutritional, functional, or organoleptic modifications to the food.

## Pulsed Light

Another type of preservation technique, which uses the visible spectrum of radiation and is currently undergoing approval by the FDA, is called pulsed light (PureBright) (39). This method works by exposing the food to intense, very brief flashes of light, which disrupt the cell membranes of bacterial cells, but not those of the surrounding food (Figure 28-8) (23, 32). Pulsed light also kills fungi, spores, viruses, protozoa, and cysts (31).

## High-Pressure Processing

High-pressure processing is sometimes referred to as high-hydrostatic pressure processing (HHP), ultra-high pressure processing (UHP) (5), or **pascalization**—after Blaise Pascal, a 17th-century French scientist who described how contained fluids are affected by pressure. Although discovered in 1899, treating foods with high pressure is a relatively new method of preservation, and one still under development. The pressure that pascalized

foods are exposed to is tremendous. A normal car tire holds about 30 pounds of pressure per square inch, whereas the pressure applied to foods in pascalization is at least 50,000 pounds per square inch, applied for 15 minutes. High pressure may preserve foods by inactivating the cells of the bacteria or yeast. HPP is currently used for the processing of meats, juice, fruit, and shellfish and the list continues to grow (22, 54, 58, 76).

> **? How & Why?**
>
> *How does pascalization work?*
>
> Hydrostatic pressure deactivates various microorganisms and numerous enzymes responsible for food deterioration (63). The high pressure of pascalization kills many bacteria, yeasts, and molds (14). However, bacterial spores remain resistant and must be treated with acid to block their ability to germinate. Thus, naturally acidic foods such as yogurt, tomato products, and sliced fruit are best suited for pascalization. The process also denatures proteins and disrupts noncovalent bonds without interfering with the food's overall structure.

**Pascalization** A food preservation process utilizing ultrahigh pressures to inhibit the chemical processes of food deterioration.

**FIGURE 28-9**   Lightning or high-energy ultraviolet rays trigger ozone formation in the atmosphere. They do so by rupturing oxygen molecules, which produce fragments that then combine with other oxygen molecules to form ozone ($O_3$).

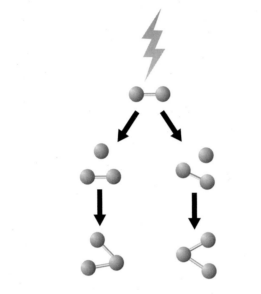

Source: Food Technology.

## Ozonation

The fresh, clean, invigorating smell in the air after a thunderstorm is due to ozone formed by the action of lightning (Figure 28-9). Ozone (an oxidizing agent) is commercially produced by exposing oxygen to an electrical current (37). The oxygen molecules ($O_2$) are broken apart, creating numerous oxygen fragments that reunite with other oxygen molecules to yield ozone ($O_3$) (88).

It has been known for decades that ozone is an effective disinfectant and sanitizer for many food products (69, 89). Ozone effectively kills viruses, molds, and bacteria (88) and controls the growth of microbial organisms, including *E. coli* and *Salmonella,* in packaged foods (45). Its antimicrobial effects are due to its ability to disrupt the cell walls of bacteria and viruses (88).

The FDA has allowed ozone to be used in the treatment of bottled drinking water, in meat-aging coolers where frozen beef carcasses are stored (37), and, since 1997, on food. In Europe, especially Germany and France, ozone is the principal means of sanitizing the public water supply, rather than chlorine, which is used in the United States (37).

## Aseptic and Modified Atmosphere Packaging

Aseptic and modified-atmosphere packaging (MAP) have been used for many years on food products (8, 64). Foods commonly packaged in these ways include processed meats and cheese, lunch kits, prepared poultry, ground turkey, pastas, sauces, snack foods, juices, and liquid dietary supplements (Figure 28-10).

### Aseptically Packaged Food

Food that is *aseptically packaged* is sterilized, packed, and sealed in a sterilized container under sterile conditions (53).

**FIGURE 28-10**   Aseptic and modified atmosphere packaging.

Digital Works

Common foods that are aseptically processed include baby foods, cheese sauces, puddings, and tomato paste (21).

### Modified-Atmosphere Packaging (MAP)

Aseptically treated foods are sometimes stored within modified-atmosphere packaging (MAP) (43, 74). This type of packaging changes the air's composition around the food to prolong its shelf life; specifically, it works by reducing oxygen and increasing carbon dioxide (67). Fruits and vegetables are often packaged with MAP because it reduces their respiration (16). Cured cheese is commonly packaged in modified-atmosphere packaging in which the carbon dioxide has been flushed out (12). Most meat and poultry is packaged using modified-atmosphere packaging (9), and applications for bakery products continue to grow (17).

# NUTRIENT RETENTION

Foods and beverages may retain their nutrients longer than most people realize (30). The macronutrients—carbohydrates, fats, and proteins—are relatively unaffected by processing. The same is true for fat-soluble vitamins and minerals; however, both are susceptible to liquid losses that occur through boiling or meat juice drippings. The nutrients most sensitive to loss are the water-soluble vitamins. These include vitamin C (ascorbic acid) and the B vitamins (especially folate). Boiling or cooking food in liquids may reduce these nutrients, but approximately 75% or more of the nutrients are usually retained.

It is difficult to discuss nutrient retention, because so few nutrient retention data exist. One of the reasons for the limited information is that measuring vitamins and minerals in the laboratory is costly and time-consuming. As a result, available data usually concern vitamin C, and usually center on only one preservation technique. Even the gold standard of nutrient retention data, the USDA Table of Nutrient Retention Factors (2007), states, "Nutrient data are frequently lacking for cooked foods." This source lists nutrient retention for only some selected foods under a few standard preparation techniques (frying, boiling, grilling, etc.). The resource available on the Internet does not include nutrient loss data for any of the preservation processes discussed in this chapter. The USDA Table of Nutrient Retention Factors showing the percentage of nutrients retained in foods prepared by different methods may be viewed at www.ars.usda.gov/Services/docs.htm?docid=9448.

The lack of research notwithstanding, it is known that even unprocessed fresh-cut produce lose vitamin C over time. For example, researchers reported the following vitamin C losses in fresh-cut produce after 6 days (36):

- 5% in mangos, strawberries, and watermelon
- 10% in pineapple
- 12% in kiwi
- 25% in cantaloupe

About 10% to 20% of the ascorbic acid was reported to be lost in fresh-cut potatoes stored under various types of modified-atmosphere packaging (MAP) (84). A few limited studies concerning vitamin C retention in foods that have been canned, irradiated, pasteurized, or high-pressure processed are now discussed.

Canning will preserve foods longer than most other methods of food preservation, with the exception of drying. Although canned foods will keep approximately 1 year in a cool, dry place, the heat treatment used in canning may adversely affect the food's texture, color, and flavor (79). It may also cause the loss of some heat-sensitive nutrients.

Minerals and macronutrients are not affected by canning, but vitamins are susceptible to loss. Thiamin $(B_1)$ and vitamin C may experience 50% to 80% losses during canning. Most other vitamin losses range from 10% to 30% (86).

Researchers at a USDA research center reported that the vitamin C content in fresh-cut cantaloupe was not affected by irradiation treatment (35). Pasteurization, however, reduced vitamin C levels in pineapple juice by approximately 37% (1). Storing the unpasteurized juice at room temperature reduced vitamin C content by 60%. Unlike pasteurization, which utilizes a heat treatment, high-pressure processing is done without heat. As a result there is a slightly better retention of vitamin C using the high-pressure preservation process than pasteurization for orange juice (71).

# PICTORIAL SUMMARY / 28: Food Preservation

The World Health Organization estimates that about 20% of the world's food supply is lost to spoilage. The average North American family discards about one fourth of the food it purchases because of spoilage. Understanding the biological, chemical, and physical changes that contribute to food spoilage can help prevent food waste.

## FOOD SPOILAGE

Foods are organic in nature and are subject to spoilage. Food deterioration decreases both the quality and safety of food due to a combination of changes:

**Biological:** Microorganisms are responsible for some of the biological reactions in foods that contribute to their deterioration.

**Chemical:** Chemical reactions that contribute to spoilage often involve enzymes.

**Physical:** Physical reactions include evaporation, water drip, and separation.

Some of these changes are preventable, some can be alleviated, and others will occur to some degree despite current preservation methods.

## FOOD PRESERVATION METHODS

Keeping spoilage to a minimum is the objective of both food manufacturers and consumers. Food preservation methods include:

- Drying
- Curing
- Fermentation
- Pickling
- Canning
- Edible coatings
- Aseptic and modified-atmosphere packaging
- Heat preservation
- Cold preservation
- Irradiation
- Pulsed light
- Ozone
- Pascalization

## COLD PRESERVATION

One half of the foods consumed in the United States are refrigerated or frozen. Refrigeration slows down the biological, chemical, and physical reactions that shorten the shelf life of food. For safety purposes, refrigerators should be kept between just above freezing and no more than 40°F (4°C), and foods should be frozen at 0°F (−18°C) or below, for the least damage to the food's original flavor, nutrient content, and texture.

## HEAT PRESERVATION

Heat in its various forms can be used as a preservation method, because many of the microorganisms responsible for food spoilage or foodborne illnesses are susceptible to heat. Heating methods include boiling, pasteurization, and ohmic heating.

## OTHER PRESERVATION METHODS

**Irradiation:** Food is passed through an enclosed chamber, where it is exposed to an ionizing energy source, such as gamma rays, X-rays, or electrons, which renders them less susceptible to deterioration and destroys many, but not all, microorganisms. This method is controversial and under study.

**Pulsed light:** Currently undergoing FDA approval, this method exposes food to intense, very brief flashes of light, disrupting bacterial cell membranes but not the surrounding food.

**Pascalization:** A relatively new method, pascalization subjects food to very high pressure, a process that kills many bacteria, yeasts, and molds, although bacterial spores remain resistant. Acidic foods are best suited for this method, because the acid will prevent the spores from germinating.

**Aseptic and modified-atmosphere packaging:** Aseptically packaged food is sanitized, packed under sanitized conditions, and stored on the shelves at room temperature. Modified-atmosphere packaging changes the air composition around the food, decreasing oxygen and increasing carbon dioxide.

Digital Works

## NUTRIENT RETENTION

Food and beverages retain many of their nutrients. Length of storage may have a greater impact on nutrient loss than the type of processing. Carbohydrates, fats, protein, fat-soluble vitamins, and minerals are relatively unaffected by processing. The nutrients most sensitive to loss are the water soluble-vitamins—vitamin C and the B vitamins (especially folate).

# CHAPTER REVIEW AND EXAM PREP

## Multiple Choice*

1. In fermentation, yeast or bacteria convert which macronutrient to carbon dioxide and alcohol?
   a. Fat
   b. Water
   c. Protein
   d. Carbohydrates

2. Which form of food preservation uses vinegar to increase acidity and keep microorganisms in check?
   a. Pickling
   b. Fermentation
   c. Freeze-drying
   d. Vacuum drying

3. The ideal temperature for refrigeration is between 32°F/0°C and _____.
   a. 40°F/4°C
   b. 45°F/7°C
   c. 55°F/13°C
   d. 60°F/16°C

4. Which heating preservation method passes an electrical current through the food?
   a. Steaming
   b. Ohmic heating
   c. Boiling
   d. Pasteurization

5. Irradiation is effective for eliminating bacteria, yeasts, and molds, but it may not eradicate which specific microorganism?
   a. *E. coli*
   b. *Salmonella*
   c. *C. botulinum* spores
   d. Insects

*See p. AK-1 for answers to multiple choice questions.

6. Modified-atmosphere packaging (MAP) operates by reducing _____ and increasing _____ in the air around the food.
   a. acid; alkalinity
   b. oxygen; carbon dioxide
   c. oxygen; ozone
   d. humidity; carbon dioxide

7. White or grayish patches on frozen food are due to _____.
   a. rancidity
   b. humidity
   c. freezer burn
   d. irradiation

## Short Answer/Essay

1. Give examples of foods that are perishable, semi-perishable, and nonperishable.
2. Describe the biological, chemical, and physical changes that can occur in foods.
3. Briefly explain the following food preservation methods: curing, fermentation, and pickling.
4. Briefly define the following terms: *dehydrated, freeze-drying,* and *sublimates.*
5. Explain the difference between boiling-water processing and pressure canning.
6. What is the recommended refrigeration temperature?
7. Describe the four potential disadvantages of freezing foods: freezer burn, cell rupturing, fluid loss, and recrystallization.
8. Discuss four ways in which heating is used to preserve foods.
9. Define *irradiation.* Describe the pros and cons of this process.
10. Describe the processes of aseptic and modified-atmosphere packaging (MAP).

# REFERENCES

1. Achinewhu SC, and AD Hart. Effect of processing and storage on the ascorbic acid (vitamin C) content of some pineapple varieties grown in the Rivers State of Nigeria. *Plant Foods for Human Nutrition* 46(4):335–337, 1994.

2. Alur MD, SP Chawla, and PM Nair. Microbiology method to identify irradiated meat. *Journal of Food Science* 57(3):593–595, 1992.

3. Argaiz A, Pérez-Vega O, and López-Malo A. Sensory detection of cooked flavor development during pasteurization of a guava beverage using R-index. *Journal of Food Science* 70(2):S149–S152, 2005.

4. Ashie INA, and BK Simpson. Effects of hydrostatic pressure on alpha-macroglobulin and selected proteases. *Journal of Food Biochemistry* 18:377–391, 1995.

5. Balasubramaniam VM. Preserving foods through high-pressure processing. *Food Technology* 62(11):33–38, 2008.

6. Berne S. New techniques yield a tastier freeze. *Prepared Foods* 163(7):109–115, 1994.

7. Brody AL. Case-ready fresh red meat: Is it here or not? *Food Technology* 56(1):77, 2002.

8. Brody AL. Exhibits capture innovation in packaging. *Food Technology* 59(9):88–92, 2005.

9. Brody AL. Packaging. Edible packaging. *Food Technology* 59(2):65–66, 2005.

10. Brody AL. Packaging for nonthermally and minimally processed foods. *Food Technology* 59:75–77, 2005.

11. Brody AL. RFID moves ahead. *Food Technology* 60(9):76–79, 2006.

12. Brody AL. Say "Cheese"—and package it, please! *Food Technology* 55(7):76, 2001.

13. Brody AL. What's fresh about fresh cut. *Food Technology* 59(11):74–77, 2005.

14. Cano MP, A Hernandez, and B De Ancos. High pressure and temperature effects on enzyme inactivation in strawberry and orange products. *Journal of Food Science* 62(1):85–88, 1997.

15. Castillo A, et al. Decontamination of beef carcass surface tissue by steam vacuuming alone and combined with hot water and lactic acid sprays. *Journal of Food Protection* 62(2):146–151, 1999.

16. Charles F, J Sanchez, and N Gontard. Modeling of active modified atmosphere packaging of endives exposed to several postharvest temperatures. *Journal of Food Protection* 70(8):E443–E449, 2005.

17. Clark JP. Drying of foods. *Food Technology* 60(12):90–91, 2006.

18. Clark JP. Drying still actively researched. *Food Technology* 56(9):76, 2002.

19. Clark JP. Equipment catching up with theory. *Food Technology* 59(9):84–86, 2005.

20. Clark JP. New issues with acidified foods. *Food Technology* 63(2):76–80, 2009.

21. Clark JP. Processing. Aseptic processing: New and old. *Food Technology* 58(11):80–82, 2004.

22. Clark JP. Processing. High–pressure processing research continues. *Food Technology* 60(2):63–65, 2006.

23. Clark JP. Processing. Pulsed electric field processing. *Food Technology* 60(1):66–67, 2006.

24. Clark JP. Understanding beverage pressurization. *Food Technology* 62(6):114–118, 2008.

25. Clark JP. Understanding drying. *Food Technology* 97–100, 2007.

26. Cliver DO, et al. Position of the American Dietetic Association: Food irradiation. *Journal of the American Dietetic Association* 96(1):69–72, 1996.

27. Cowell ND. More light on the dawn of canning. *Food Technology* 61(5):40–44, 2007.

28. Culora J. The safety of beverages in plastic bottles. *Food Safety Magazine* 15(2):52–55, 2009.

29. Debeaufort F, JA Quezada-Gallo, and A Voilley. Edible films and coatings: Tomorrow's packages: A review. *Critical Reviews in Food Science* 38(4):299–313, 1998.

30. Dewanto V, et al. Thermal processing enhances the nutritional value of tomatoes by increasing total antioxidant activity. *Journal of Agriculture and Food Chemistry* 50(10):3010–3014, 2002.

31. Dunn J, et al. Pulsed white light food processing. *Cereal Foods World* 42(7):510–515, 1997.

32. Elmnasser N, S Guillou, F Leroi, N Orange, A Bakhrouf, and M Federighi. Pulsed-light system as a novel food decontamination technology: A review. *Canadian Journal of Microbiology* 53(7): 813–821, 2007.

33. Fan X. Irradiation of fresh fruits and vegetables. *Food Technology* 62(3):36–43, 2008.

34. Fan X, and SE Kays. Formation of trans fatty acids in ground beef and frankfurters due to irradiation. *Journal of Food Science* 74(2):C79–C84, 2009.

35. Fan X, et al. Combination of hot-water surface pasteurization of whole fruit and low-dose gamma irradiation of fresh-cut cantaloupe. *Journal of Food Protection* 69(4):912–919, 2006.

36. Gil MI, E Aguayo, and AA Kader. Quality changes and nutrient retention in fresh-cut versus whole fruits during storage. *Journal of Agriculture and Food Chemistry* 54(12):4284–4296, 2006.

37. Graham DM. Use of ozone for food processing. *Food Technology* 51(6):72–75, 1997.

38. Harris LJ. Food science. Freezing and thawing 101. *Fine Cooking* 67:84, 2004.

39. Hoover DG. Minimally processed fruits and vegetables: Reducing microbial load by nonthermal physical treatments. *Food Technology* 51(6):66–71, 1997.

40. Hung YC, and NK Kim. Fundamental aspects of freeze-cracking. *Food Technology* 50(12):59–61, 1996.

41. Johnson J, and M Marcotte. Irradiation control of insect pests of dried fruits and walnuts. *Food Technology* 53(6):46–51, 1999.

42. Katz F. Battling bacteria: New ways for meat, poultry processors to attack contamination. *Food Technology* 57(6):83–84, 1996.

43. Katz F. Is it time for changes in meat packaging and handling? *Food Technology* 51(6):99–100, 1997.

44. Kim HJ, et al. Validation of ohmic heating for quality enhancement of food products. *Food Technology* 50(5):253–262, 1996.

45. Klockow PA, and KM Keener. Safety and quality assessment of packaged spinach treated with a novel ozone-generation system. *Food Science and Technology* 42(6):1047–1053, 2009.

46. Krotchta JM, and C De Mulder-Johnston. Scientific status summary. Edible and biodegradable polymer films: Challenges and opportunities. *Food Technology* 51(2):61–72, 1997.

47. Labuza TP, and AE Sloan. *Food for Thought.* Avi Publishing, 1977.

48. Loaharanu P. Cost/benefit aspects of food irradiation: A summary of the report and recommendations of the working group of the FAO/IAEA/WHO International Symposium. *Food Technology* 48(1):104–108, 1994.

49. Lusk JL, JA Fox, and CL McIlvain. Consumer acceptance of irradiated meat. *Food Technology* 53(3):56–59, 1999.

50. Marcotte M. Irradiated strawberries enter the United States market. *Food Technology* 46(5):80, 1992.

51. Marsh K, and M Bugusu. Food packaging—Roles, materials, and environmental issues. *Journal of Food Science* 72(3):R39–R55, 2007.

52. Marsh K, and B Bugusu. Food packaging and its environmental impact. *Food Technology* 6(4):46–50, 2007.

53. Mermelstein NH. Aseptic bulk storage and transportation. *Food Technology* 54(4):107–110, 2000.

54. Mermelstein NH. High-pressure processing reaches the United States market. *Food Technology* 51(6):95–96, 1997.

55. Mermelstein NH. High-temperature, short-time processing. *Food Technology* 55(6):65–70, 2001.

56. Mermelstein NH. Washington news. Irradiation of shell eggs approved. *Food Technology* 54(8):30, 2000.

57. Muratore G, and F Licciardello. Effect of vacuum and modified-atmosphere packaging on the shelf life of liquid smoked swordfish (*Xiphias gladius*) slices. *Journal of Food Science* 70(5):C359–C363, 2005.

58. Myer RS, et al. High-pressure sterilization of foods. *Food Technology* 54(11):67–72, 2000.

59. Nacha K. Analyzing nanotechnology. *Food Technology* 61(1):34–36, 2007.

60. O'Bryan CA, PG Crandall, SC Ricke, and DG Olson. Impact of irradiation on the safety and quality of poultry and meat products: A review. *Critical Reviews in Food Science and Nutrition* 48(5):442–457, 2008.

61. O'Donnell CD. Stabilizing the big freeze. *Prepared Foods* 165(1):45–46, 1996.

62. Olsen DG. Food irradiation future still bright. *Food Technology* 58(7):112, 2004.

63. Olsen DG. Irradiation of food. *Food Technology* 52(1):56–62, 1998.

64. Palaniappan S, and CE Sizer. Aseptic process validated for foods containing particulates. *Food Technology* 51(8):60–68, 1997.

65. Palmer S. Irradiation: What it is, what it does, and how it affects the food supply. *Today's Dietitian* 11(1):32, 2009.

66. Phukan RK, et al. Dietary habits and stomach cancer in Mizoram, India. *Journal of Gastroenterology* 41(5):418–424, 2006.

67. Poubol J, and H Izumi. Shelf life and microbial quality of fresh-cut mango cubes stored in high $CO_2$ atmospheres. *Journal of Food Science* 70(1):M69–M74, 2005.

68. Pszczola DE. How *Food Technology* covered irradiation over the years. *Food Technology* 51(2):49, 1997.

69. Richardson SD, et al. Chemical byproducts of chlorine and alternative disinfectants. *Food Technology* 52(4):58–61, 1998.

70. Ronsivalli LJ, and ER Vieira. *Elementary Food Science*. Van Nostrand Reinhold, 1992.

71. Sanchez-Moreno C, et al. Impact of high pressure and pulsed electric fields on bioactive compounds and antioxidant activity of orange juice in comparison with traditional thermal processing. *Journal of Agriculture and Food Chemistry* 53(11):4403–4409, 2005.

72. Sastry SK, and Q Li. Modeling the ohmic heating of foods. *Food Technology* 50(5):246–249, 1996.

73. Schmidt SJ, and JW Lee. How does the freezer burn our food? *Journal of Food Science Education* 8(2):45–52, 2009. www3.interscience.wiley.com/cgi-bin/fulltext/122278975/HTMLSTART. Accessed 8/09.

74. Sebranek JG, MC Hunt, DP Cornforth, and MS Brewer. Perspective. Carbon monoxide packaging of fresh meat. *Food Technology* 60(5):184, 2006.

75. Simatos D, and G Blond. DSC studies and stability of frozen foods. *Advances in Experimental Medicine and Biology* 302:139–155, 1991.

76. Sizer CE, VM Balasubramaniam, and E Ting. Validating high-pressure processes for low-acid foods. *Food Technology* 56(2):36–42, 2002.

77. Smith JS, and S Pillai. Irradiation and food safety. *Food Technology* 58(11):48–54, 2004.

78. Sommers CH. Food irradiation is already here. *Food Technology* 58(11):22, 2004.

79. Stanley DW, et al. Low temperature blanching effects on chemistry, firmness, and structure of canned green beans and carrots. *Journal of Food Science* 60(2):327–333, 1995.

80. Sucan MK. Identifying and preventing off-flavors. *Food Technology* 58(11):36–40, 2004.

81. Talens P, and JM Krochta. Plasticing effects of beeswax and carnauba wax on tensile and water vapor permeability properties and whey protein films. *Journal of Food Science* 70(3):E239–E243, 2005.

82. Tarver T. Novel ideas in food packaging. *Food Technology* 62(10):54–59, 2008.

83. Thayer DW, et al. Radiation pasteurization of food. *Council for Agricultural Science and Technology* 7:1–10, 1996.

84. Tudela JA, et al. L-galactono-gamma-lactone dehydrogenase and vitamin C content in fresh-cut potatoes stored under controlled atmospheres. *Journal of Agriculture and Food Chemistry* 51(15):4296–4302, 2003.

85. United States Department of Health and Human Services. Food and Drug Administration. Irradiation in the production, processing, and handling of food; Final rule (21 CFR Part 179). *Federal Register* 62(232):64107–64121, 1997.

86. VanGarde SJ, and M Woodburn. *Food Preservation and Safety: Principles and Practice*. Iowa State University Press, 1994.

87. Van Loey A, et al. Potential *Bacillus subtilis* alpha-amylase-based time-temperature integrators to evaluate pasteurization processes. *Journal of Food Protection* 59(3):261–267, 1996.

88. Wolke RL. Ozone in the kitchen. *Fine Cooking* 90:76–77, 2008.

89. Xu L. Use of ozone to improve the safety of fresh fruits and vegetables. *Food Technology* 53(10):58–62, 1999.

90. Yu L, CA Reitmeier, and MH Love. Strawberry texture and pectin content as affected by electron beam irradiation. *Journal of Food Science* 61(4):844–846, 1996.

91. Zoltai P, and P Swearingen. Product development considerations for ohmic processing. *Food Technology* 50(5):263–268, 1996.

# WEBSITES

The Food Processing Suppliers Association has its own website:

**www.fpsa.org/**

*Prepared Foods* and *Food Processing* are trade publications for food manufacturers; here are their websites:

**www.preparedfoods.com** and
**www.foodprocessing.com**

Here is a virtual link to the food processing industry:

**www.foodonline.com**

The Centers of Disease Control & Prevention answers questions on food irradiation:

**www.cdc.gov/ncidod/dbmd/
diseaseinfo/foodirradiation
.htm#whichprevent**

wsmahar/istockphoto.com

# 29 Government Food Regulations

Food laws and regulations date back to ancient Greece and Rome. One of the first recorded food laws made it illegal to dilute wine with water (19). Later, during the Middle Ages, trade guilds policed themselves to ensure that teas and spices were pure. This protected them against people taking unfair trade advantage by selling inferior, cheaper products because some traders did try to cheat. Yet problems continued, even in the New World. In 1630, a colonist of the Massachusetts Bay Colony was whipped for selling "a water of no worth" as a cure for scurvy (40).

In the United States, the public's demand for laws to protect consumers against contaminated foods helps to make this country's food supply the safest in the world (32). The government has responded at all levels—federal, state, and local—with laws and regulations to protect the foods and beverages consumed by the public (18). Companies can be temporarily shut down until they comply with the regulations and may in rare instances be put out of business. Despite these threats, most companies voluntarily comply with regulations in order to maintain consumer confidence in their products (40).

This chapter provides an overview of the United States government's food regulations. The various federal (not state) agencies involved include the Food and Drug Administration, the United States Department of Agriculture, the Environmental Protection Agency, the Centers for Disease Control and Prevention, and other regulatory agencies (Figure 29-1). Also discussed are the two major international agencies that develop global food standards— the Food and Agriculture Organization and the World Health Organization of the United Nations (10).

## FEDERAL FOOD LAWS

There was a time in the United States and other countries when laws related to the food supply did not exist. Serious problems stimulated the formation of new laws and agencies were given the responsibility of recommending and or enforcing these regulations. A brief review of the major laws passed in the United States is now discussed.

**FIGURE 29-1** Major United States federal agencies involved in food regulation.

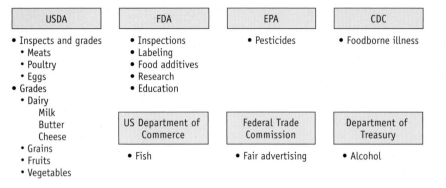

| USDA | FDA | EPA | CDC |
|---|---|---|---|
| • Inspects and grades<br>  • Meats<br>  • Poultry<br>  • Eggs<br>• Grades<br>  • Dairy<br>    Milk<br>    Butter<br>    Cheese<br>  • Grains<br>  • Fruits<br>  • Vegetables | • Inspections<br>• Labeling<br>• Food additives<br>• Research<br>• Education | • Pesticides | • Foodborne illness |

| US Department of Commerce | Federal Trade Commission | Department of Treasury |
|---|---|---|
| • Fish | • Fair advertising | • Alcohol |

ingested a poisonous "Elixir of Sulfanilamide" (which contained a highly toxic chemical analog of antifreeze) (19). Several federal agencies were established to ensure consumer food safety. Unfortunately, the result was a fragmented and overlapping system of regulations that to some extent still exists today (23).

# Numerous Government Agencies

Numerous laws regulating food were introduced, passed, and enforced over the years (the updated list of FDA Milestones can be viewed at www.fda.gov/AboutFDA/WhatWeDo/History/Milestones/default.htm) (Figure 29-2). However, there was no systematic application from a single government agency (17). To assist in correcting this problem, the United States congress passed The Safe Food Act of 1997. The purpose

# Food and Drug Act (1906)

The federal Food and Drug Act of 1906, sometimes referred to as the Pure Food Law, was the beginning of the government's involvement in food regulation. This followed a series of nationally publicized blunders by manufacturers, including reports following the 1898 Spanish-American War that many soldiers had become ill after being fed spoiled canned meat and contaminated flour. Around the same time, it was found that some dairies were using formaldehyde as a milk preservative. One notorious incident during the early 1900s involved store owners who substituted poisonous mushrooms for edible varieties.

Upton Sinclair's novel *The Jungle*, published in 1905 and very widely read, exposed the filthy conditions in Chicago's meatpacking plants. Animals that had died of disease were sometimes processed into meat products, and it was not uncommon to find foods contaminated with filth, including the hairs, urine, and feces of rats and mice; insects; maggots; larvae; parasitic worms; and excrement from humans and animals. Sausage meat often contained ground rats, other vermin, and even human fingers.

Sinclair's book added to the public groundswell of demand for the government to take action. The passage of the 1906 Federal Food and Drug Act followed, but the list of injurious foods and manufacturing processes continued to grow, resulting eventually in the need for modifications to the original bill.

# Food, Drug, and Cosmetic Act (1938)

The food laws were rewritten, and in 1938 the Food, Drug, and Cosmetic Act was passed. This was brought on, in part, by the deaths of 100 people who

**FIGURE 29-2** Time line of selected congressional acts and amendments legislating the U.S. government's role in regulating the food supply.

| 1906 | Pure Food and Drug Act (FDA) |
|---|---|
| 1906 | Federal Meat Inspection Act (USDA) |
| 1914 | Federal Trade Commision Act (FTC) |
| 1935 | Federal Alcohol Administrative Act (U.S. Dept. of the Treasury) |
| 1938 | Federal Food, Drug, and Cosmetic Act (FDA) |
| 1944 | Public Health Services Act (USDA) |
| 1946 | Agricultural Marketing Act (USDA) |
| 1947 | Federal Insecticide, Fungicide, and Rodenticide Act (EPA) |
| 1954 | Pesticide Residue Amendment to 1938 (EPA) |
| 1956 | Fish and Wildlife Act (U.S. Dept. of Commerce) |
| 1957 | Poultry Products Inspection Act (USDA) |
| 1958 | Food Additives Amendment to 1938 |
| 1960 | Color Additives Amendment to 1938 |
| 1966 | Fair Packaging and Labeling Act (FDA) |
| 1967 | Wholesome Meat Act (USDA) |
| 1970 | Egg Products Inspection Act (USDA) |
| 1972 | Federal Environmental Pesticide Control Act (EPA) |
| 1974 | Safe Drinking Water Act (EPA) |
| 1977 | Saccharin Study and Labeling Act |
| 1990 | Nutrition Labeling and Education Act (FDA) |
| 1996 | Food Quality Protection Act |
| 1996 | Pathogen Reduction Hazard Analysis and Critical Control Points (HACCP) Systems Final Rule |
| 1997 | Pathogen Reduction Act |
| 1997 | Safe Food Act |
| 2002 | Country of Origin Labeling (or COOL) Requirements |
| 2002 | Bioterrorism Preparedness Act |
| 2004 | Project BioShield Act |
| 2004 | Food Allergy Labeling and Consumer Protection Act |

of this legislation was to create a single independent agency to oversee food safety (8). As of this date, however, four federal agencies are primarily responsible for ensuring the safety of the food supply: the Food and Drug Administration (FDA), the U.S. Department of Agriculture (USDA), the Environmental Protection Agency (EPA), and the Centers for Disease Control and Prevention (CDC). Other agencies involved in monitoring food safety include the Department of Commerce, the Federal Trade Commission, the Public Health Service, the Consumer Product Safety Commission, and the Department of the Treasury (17). These latter agencies have jurisdiction over food that travels across state lines.

# FOOD AND DRUG ADMINISTRATION

The oldest federal consumer protection agency in the United States is the Food and Drug Administration (FDA), which is now located in the Department of Health and Human Services. The FDA's job is to enforce the Food, Drug, and Cosmetic Act of 1938. This act ensures that the food, drugs, and cosmetics purchased by consumers are safe, wholesome, and produced under sanitary conditions. As a result, the FDA has jurisdiction over the production of all foods that pass across state lines. The major exceptions are the meat, poultry, and egg industries, which are regulated by the USDA. The main duties of the FDA are to inspect facilities and manufacturing processes, set standards, oversee food labeling, and regulate **food additives.**

## Research/Education

Two other main functions of the FDA that are not as well known are to (1) conduct research and (2) educate the public about food and nutrition. One of the reasons that these two duties are part of the FDA's mission is that the FDA's first chief was a university professor. In 1883, Dr. Harvey W. Wiley was a chemistry professor who left Purdue University to become Chief of the Bureau of Chemistry (later named FDA). During the early FDA days when Dr. Wiley was in charge, there were no laws about how companies could use chemical preservatives in

food and beverages (19). To remedy this situation, Dr. Wiley tried to introduce new legislation. However, this required hard data, so in an attempt to generate support for his food safety laws, Dr. Wiley recruited scientists. They studied the safety of foods in the marketplace and took the findings to the public. The heavy emphasis on research from a scientific perspective continues to be an important function of the FDA. The results of this research are reported in two major FDA publications: (1) *FDA Consumer* and (2) *FDA Drug Bulletin.* The information summarized in these journals often covers products that come before the FDA for approval.

# The Code of Federal Regulations

Dr. Wiley, known as the "crusading chemist," and his staff worked to have many food laws passed at a national level. The specific regulations enforced by the FDA can be found in the Code of Federal Regulations (CFR), Title 21 (a new edition is published annually every April 1) (15). Any changes that are made during the year can be viewed in the Federal Register online at www.fda.gov/ AboutFDA/WhatWeDo/Laws/default. htm. That publication covers food, drugs, and cosmetics; the publication focusing only on food is the FDA's **Food Code.** The Food Code is used as a reference for food safety guidelines in food establishments to promote safe practices and uniformity across the country (2). These provisions are compatible with the Hazard Analysis and Critical Control Point (HACCP) system (see Chapter 4). They are used by local, state, and federal regulators to update their own food safety rules. In 2005, the Food Code was updated to comply with the Food Allergen Labeling and Consumer Protection Act of 2004 and to update definitions of potentially hazardous foods due to microbial contamination (2). The Food Code can be viewed at www.fda.gov/ Food/FoodSafety/RetailFoodProtection/ FoodCode/FoodCode2005/default.htm.

# FDA Inspections

One of the ways that the FDA ensures compliance with its laws is to conduct inspections. These unannounced onsite

inspections of food manufacturers and processors are possible because the Food, Drug, and Cosmetic Act prohibits the sale of adulterated products. Such foods are defined as "defective, unsafe, contaminated with filth, or produced under unsanitary conditions."

Inspection is the chief tool for determining whether or not food manufacturers meet these standards. The FDA, based in Rockville, Maryland, has about 1,700 inspectors to cover over 90,000 FDA-regulated businesses (39). These inspectors are located in regional, district, and local offices spread throughout 157 cities across the country. Although they cannot visit every site, they do inspect more than 20,000 facilities a year to determine whether products are being prepared, processed, and packaged under sanitary conditions.

Companies comply with FDA laws through the use of Good Manufacturing Practice (GMP) Regulations. These regulations set the requirements for quality controls such as sanitation inspection. They are subject to constant scrutiny and modification (19).

## FDA Enforcement of Its Laws

Prior to the 1906 passage of the Pure Food Law, judges had no authority to stop rampant food adulteration. For example, a company could sell "fruit jam" made of water, sugar, grass seed, and artificial color. This kind of unethical practice not only cheated the consumer but also gave the offending manufacturer a higher profit margin than a legitimate competitor. The FDA now has the legal power to enforce the sanitation standards set by law. Now, if a shipment is found to be unfit for consumers, it is either detained in order to determine whether it should be destroyed,

**Food additive** A substance added intentionally or unintentionally to food that becomes part of the food and affects its character.

**Food Code** An FDA publication updated every two years that shows food service organizations how to prevent foodborne illness while preparing food. Available at www.fda.gov/Food/FoodSafety/ RetailFoodProtection/FoodCode/ default.htm.

or conditionally released back to the manufacturer until it is brought into compliance with the law. Another tool of the FDA is **product recall**. Once a product is recalled, the manufacturer has three alternatives:

- It can allow the FDA to dispose of the food product.
- It can contest the government's charges in court.
- It can request permission of the court to bring the product into compliance under the law.

Over 3,000 products are withdrawn from the marketplace every year, either by recall or court-ordered seizure. There are three levels of FDA recall: Class I—dangerous or defective products that could cause serious health problems or death; Class II—products that cause a temporary health problem or pose a slight threat of a serious nature; and Class III—products that are unlikely to cause any adverse health reaction, but violate FDA labeling or manufacturing regulations. The two most common reasons for the FDA to recall foods or cosmetics are mislabeling and microbial contamination (47). A list of current products being recalled is available at www.recalls.gov/food.html.

### Allowable Contaminants

Eliminating all contaminants in foods is an unattainable goal. The FDA responded to this reality by setting "acceptable levels of filth" in certain foods. For example, it is extremely

---

**TABLE 29-1**   What the FDA Allows in Selected Foods

The list below represents what the FDA considers an acceptable level of contaminants in the foods listed.*

| Food | Acceptable Levels of Filth |
|---|---|
| Chocolate | Fewer than one rodent hair/100 g |
| Coffee beans | Less than 10% insect-infested |
| Fig paste | Fewer than 13 insect heads/100 g |
| Fish (fresh frozen) | Less than 5% "definite odor of decomposition" |
| Mushrooms (canned) | Fewer than 20 maggots of any size/100 g (drained) |
| Peanut butter | Fewer than 30 insect fragments/100 g |
| Popcorn | Less than one rodent pellet/sample or one rodent hair/2 samples |
| Spinach | Fewer than 50 aphids, thrips, or mites/100 g |
| Tomato paste | Fewer than 30 fly eggs/100 g |

*Adapted from data provided by the FDA at www.fda.gov/Food/GuidanceComplianceRegulatoryInformation/ GuidanceDocuments/Sanitation/ucm056174.htm

---

difficult, if not impossible, to ensure that all flour is free of insect parts, or that spinach does not contain a single aphid. As a result, a 1956 decision was passed stating that insect and worm parts fall under the legal principle of *de minimis non curat lex*, Latin for "the law does not concern itself with trifles" (40). Some of these official "trifles" are listed in Table 29-1.

## FDA Standards

In addition to addressing "filth" and ensuring the correct ingredients are used in foods, the FDA became responsible for making sure that canned goods were filled to their capacity. Some manufacturers were filling their cans half full with food and using water to make up the difference. To combat these unfair practices, the FDA established three "content" standards: (1) **Standards of Identity**, (2) **Standards of Minimum Quality**, and (3) **Standards of Fill**.

### Standards of Identity

Standards of Identity were established for common food products purchased by consumers in order to ensure quality and consistency, and to curb unfair business practices. A Standard of Identity defines a food's composition by the specific ingredients it must contain. When a food does not adhere to this Standard of Identity, it cannot be labeled by its common name. For example, raisin bread cannot be called "raisin bread" unless the raisins equal 50% of the flour's weight. Fruit jam cannot be sold as "fruit jam" unless it consists of at least 45% fruit (46). Beef stew must have at least 25% beef. Some of these foods for which a Standard of

Identity exists include ketchup, cheese, ice cream, frankfurters, bread, mayonnaise, salad dressings, soups, and jams (Figure 29-3) (35). If a company wishes to sell any food that falls into a Standard of Identity, it must include the list of ingredients identified as that food by the FDA. Nevertheless, even with a Standard of Identity designation, a food must still list the ingredients on the food's label.

**FIGURE 29-3**   FDA Standards of Identity.

Food manufacturers have to list the ingredients in standardized foods, which include the following product categories.

- Milk and cream
- Cheeses and related products
- Frozen desserts
- Bakery products
- Cereal flours and related products
- Macaroni and noodle products
- Canned fruits
- Canned fruit juices
- Fruit butters, jellies, and jams or preserves
- Fruit pies
- Canned vegetables
- Vegetable juices
- Frozen vegetables
- Eggs and egg products
- Fish and shellfish
- Cacao products (cocoa, chocolate)
- Tree nut and peanut products
- Margarine
- Sweeteners and table syrups
- Food dressings and flavorings (mayonnaise, salad dressing, vanilla flavoring)

---

**Product recall**   Civil court action to seize or confiscate a product that is defective, unsafe, filthy, or produced under unsanitary conditions.

**Standards of Identity** Requirements for the type and amount of ingredients a food should contain in order to be labeled as that food.

**Standards of Minimum Quality** Minimum quality requirements for tenderness, color, and freedom from defects in canned fruits and vegetables.

**Standards of Fill**   The amount of raw product that must be put into a container before liquid (brine or syrup) is added.

## Standards of Minimum Quality

These standards ensure that foods must meet a minimum level of quality. For instance, string beans cannot be too "stringy," peas cannot be too hard, and cream-style corn cannot be too watery. The Standards of Minimum Quality are mandatory and are not to be confused with the voluntary grades of A, B, C (or Fancy, Choice, Standard) set by the USDA for canned fruits and vegetables. Canned foods that do not meet the Standards of Minimum Quality are labeled "Below standard in quality; good food—not high grade."

## Standards of Fill

Until a Standard of Fill regulation was put into effect, manufacturers did not have to fill a can to the top with its ingredients. Cans were often only partially packed, with the remainder of the volume being filled with fluid. Consumers are now safe from this type of manufacturing fraud. The Standards of Fill require manufacturers to have a minimum weight for solid food. The food is measured after the liquid is drained from the can. Cans must be filled to their maximum capacity (usually 90%) with solid ingredients.

# Food Labeling

Interest in food labeling began in the 1960s. Consumers began to be concerned about processed foods and started demanding to know exactly what was in the foods they were ingesting. As a response to this public concern, the government developed new labeling regulations. It became a legal requirement (through the Fair Packaging and Labeling Act) to include the following information on the food label:

- The ingredients of packaged (canned, bottled, boxed, and wrapped) foods listed in descending order by weight. Food additives, colors, and chemical preservatives are also required to be listed on the label.
- The name and form (crushed, sliced, whole) of the product.
- The net amount of the food or beverage by weight, measure, or count. In addition to net weight, two other types of weight measurement include drained weight and solid content (Figure 29-4).

**FIGURE 29-4** Drained weight and solid content as weight measurements.

Drained weight. Net drained wt. 9.25 oz.    Solid content. Wt. of pineapple 14 oz.

Digital Works

**FIGURE 29-5** Mandatory nutrient labels assist consumers in making healthy choices.

New title signals that the label contains the newly required information.

The list of nutrients covers those most important to the health of today's consumers, most of whom need to worry about getting too much of certain items (fat, for example), rather than too few vitamins or minerals, as in the past.

Calories from fat are shown on the label to help consumers meet dietary guidelines that recommend people get no more than 30 percent of their calories from fat.

% Daily Value shows how a food fits into the overall daily diet.

## Nutrition Facts

Serving Size 1/2 cup (114g)
Serving Per Container 4

| Amount Per Serving | |
|---|---|
| **Calories** 90   Calories from Fat 30 | |
| | **% Daily Value*** |
| **Total Fat** 3g | 5% |
| Saturated Fat 0g | 0% |
| Trans Fat 0g | 0% |
| **Cholesterol** 0g | 0% |
| **Sodium** 300mg | 13% |
| **Total Carbohydrate** 13g | 4% |
| Dietary Fiber 3g | 12% |
| Sugars 3g | |
| **Protein** 3g | |

| Vitamin A 80% | • | Vitamin C 60% |
|---|---|---|
| Calcium 4% | • | Iron 4% |

*Percent Daily Values are based on a 2,000 calorie diet. Your daily values may be higher or lower depending on your calorie needs:

| | | Calories: | 2,000 | 2,500 |
|---|---|---|---|---|
| Total Fat | Less than | | 65 g | 80 g |
| Sat Fat | Less than | | 20 g | 25 g |
| Cholesterol | Less than | | 300 mg | 300 mg |
| Sodium | Less than | | 2,400 mg | 2,400 mg |
| Total Carbohydrate | | | 300 g | 375 g |
| Dietary Fiber | | | 25 g | 30 g |

Calories per gram:
Fat 9  •  Carbohydrates 4  •  Protein 4

The label will now tell the number of calories per gram of fat, carbohydrates, and protein.

Daily values are new. Some are maximums, as with fat (65 grams or less); others are minimums, as with carbohydrates (300 grams or more). The daily values on the label are based on a daily diet of 2,000 and 2,500 calories. Individuals should adjust the values to fit their own calorie intake.

- The name and address of the manufacturer, packer, or distributor.
- The nutrient content depicted as Nutrition Facts (Figure 29-5).

The FDA is responsible for ensuring that consumers are informed about a packaged product's ingredients and nutrient content. The exceptions are meat and poultry products, which are regulated by the USDA (27).

## Nutrition Facts Label

Food labeling was not always mandatory. It became required for most foods in 1994 when the nutrient label on foods changed to Nutrition Facts (Figure 29-6). The Canadian nutrition label is also called Nutrition Facts, but Canada requires two labels—one in English and one in French (5). Food labels provide consumers with pertinent information in making reasonable food choices. The various sections of the food label known as Nutrition Facts are now explained (42).

### Serving Size and Serving Number

The first items on the food label are serving size and the number of servings per container. These should be checked carefully because serving sizes were previously decided by manufacturers. They sometimes reported nutrient contents for foods high in calories or fat based on unreasonably small serving sizes to disguise their true contents. For example, a candy bar containing 220 calories (kcal) would contain only 90 calories (kcal) per serving if the manufacturer divided it into 2½ servings. New regulations require that a serving size be defined as the "amount of food customarily eaten at one time."

Nevertheless, in some instances, they may still be smaller (35).

### Calories (kcal)

The calories (kcal) are listed below the serving information on the Nutrition Facts label. Included are the calories (kcal) from fat.

### Nutrients

The five major nutrients listed on Nutrition Facts include (1) fat grams (total, saturated, and trans), (2) cholesterol (milligrams), (3) sodium, (4) total carbohydrate (dietary fiber and sugar), and (5) protein. The "% Daily Value" is provided for these nutrients as well as those for vitamins A and C, calcium (Ca), and iron (Fe). Canadian Nutrition Facts use different daily values for certain vitamins and minerals, so it is difficult to use a single label for a product sold in both countries (5).

### Daily Values

Daily Values are nutrient standards derived from the Daily Reference Values (DRV) and Reference Daily Intakes (RDI). Daily Reference Values refer to fat, saturated fat, trans fat, cholesterol, carbohydrates, fiber, sodium, and potassium. Reference Daily Intakes cover other nutrients (protein, vitamins/minerals). Daily Values are based on a daily diet of 2,000 or 2,500 calories (kcal), and are mandatory for 10 food components while optional for 22 others (26).

### Other Nutrition Facts Information

The Nutrition Facts label also includes specific names of flavorings, colorings, and spices. FDA-certified food color additives such as FD&C Yellow No. 5 or 6, Red No. 3 or 40, Blue No. 1 or 2, and Green No. 3 must be declared on all foods except butter, cheese, and ice cream (35). Beverages claiming to contain fruit juice(s) are now required to list the percentage of real fruit juice (22).

At the bottom of the Nutrition Facts portion of the label, some manufacturers may also give the American Diabetes Association's exchange calculation based on their "Exchange Lists for Meal Planning" for diabetics.

### Food Labeling Exemptions

Although not all foods require labeling, about 90% of all processed foods carry nutrition information such as that shown on the cans of tuna and cola in Figure 29-6. Some foods do

**FIGURE 29-6** Nutrition Facts labels for tuna and cola.

| Nutrition Facts | Amount/serving | % DV* | Amount/serving | % DV* |
|---|---|---|---|---|
| Serv. Size ⅓ cup (56g) | **Total Fat** 1g | 2% | **Total Carb.** 0g | 0% |
| Servings about 3 | Sat. Fat 0g | 0% | Dietary Fiber 0g | 0% |
| **Calories** 80 | Trans Fat 0g | 0% | | |
| Fat Cal. 10 | **Cholest.** 10g | 3% | Sugars 0g | |
| *Percent Daily Values (DV) are based on a 2,000 calorie diet. | **Sodium** 200mg | 8% | **Protein** 17g | |

Vitamin A 0% • Vitamin C 0%
Calcium 0% • Iron 6%

## Nutrition Facts

Serving Size 1 can (360mL)

**Amount Per Serving**

**Calories** 140

| | % Daily Value* |
|---|---|
| **Total Fat** 0g | 0% |
| Sat. Fat 0g | 0% |
| Trans Fat 0g | 0% |
| **Sodium** 300mg | 1% |
| **Cholesterol** 0g | 0% |
| **Total Carbohydrate** 13g | 12% |
| Dietary Fiber 0g | 0% |
| Sugars 36g | |
| **Protein** 0g | 0% |

*Percent Daily Values are based on a 2,000 calorie diet.

not require a nutrition label and these include raw foods (fresh fruits, vegetables, and fish), game, restaurant food, deli and bakery items, foods containing insignificant amounts of nutrients (coffee, tea, spices), bulk food, infant formula, medical foods, donated foods, and foods in packages weighing less than half an ounce, such as gum or breath mints (22).

## Food Allergens

Each year, 200 people die from allergic reactions to food ingredients. People with food allergies now have more protection from the Food Allergen Labeling and Consumer Protection Act (FALCPA) that was passed in 2004. Prior to that date, manufacturers did not have to list food allergens on their labels. The act now requires the labeling of any food containing a major allergen, defined by the FALCPA as a protein from one of the eight foods listed in Table 29-2. These allergens account for 90% of all food allergies (9).

*Gluten Free* The gluten-free labeling law was mandated by the FALCPA. According to Codex and FDA regulations, the "gluten free" label can only be used on foods with less than 20 ppm gluten (21). However, private organization also provide labels with more strict standards of 3 to 10 ppm.

## FDA Allowed Claims on Labels

Prior to the 1990s, claims that linked a nutrient or food to a disease, in either a positive or a negative way, were not permitted. Now, under the current

| TABLE 29-2 Eight Major Allergens Identified by the Food Allergen Labeling and Consumer Protection Act |
| --- |
| Peanuts |
| Soybeans |
| Cow's milk |
| Eggs |
| Fish |
| Crustacean shellfish |
| Tree nuts |
| Wheat |

regulations, the FDA allows three types of claims on food or dietary supplement labels: **health claims**, **nutrient content claims,** and **structure/function claims** (www.fda.gov/Food/LabelingNutrition/LabelClaims/ucm111447.htm).

Health claims are marketing factors that can impact consumer interest in a product (20). The FDA categorizes health claims into unqualified or qualified health claims. The FDA Center for Food Safety and Applied Nutrition (CFSAN) website mentioned at the end of this chapter has more information on health claims.

 **How & Why?**

***What is the difference between "unqualified" and "qualified" health claims?***

Both health claims characterize a relationship supported by scientific evidence between a specific food or one of its components and a disease or health-related condition. **Unqualified health claims** (authorized health claims) must be supported by significant scientific agreement among qualified experts that the claim is supported by the totality of publicly available scientific evidence (44). In contrast, **qualified health claims** are supported by scientific evidence, but do not meet the significant scientific agreement standard. As a result, they must be accompanied by a disclaimer or other qualifying language to accurately communicate the level of scientific evidence supporting the claim. Both types of claims may be used on conventional foods or dietary supplements (44).

### Qualified Health Claims

The use of qualified health claims began after an attorney, Jonathan Emord, filed a lawsuit against the FDA in the 1990s (3). Qualified health claims are those that are based on "emerging evidence" of a product's ability to reduce the risk of a given disease (20). A health claim by definition has two essential components: (1) a substance (whether a food, food component, or dietary ingredient) and (2) a disease or health-related condition. The FDA permits the use of one of its qualified health claims to

be placed on the label if (1) language already approved by the FDA is used, (2) the weight of the evidence supports it, and (3) the company petitioned the FDA to use the language and received written notification. Sixteen qualified health claims have been approved, dealing with cancer, cardiovascular disease, cognitive function, and birth defects (3). About 24 or 25 claims have been filed but not yet approved (20). The list of qualified health claims authorized by the FDA continues to grow, but Table 29-3 lists those approved at the time of this textbook's publication date. An updated list can be found at www.fda.gov/Food/LabelingNutrition/LabelClaims/QualifiedHealthClaims/default.htm. In the past, other health claims were reviewed and tentatively found not to have sufficient basis to be authorized by

**Health claims** Describes a relationship between a food, food component, or dietary supplement ingredient and reducing risk of a disease or disease-related condition.

**Nutrient content claims** Food label descriptions communicating the amount of a nutrient or dietary substance contained in a food or beverage.

**Structure/function claims** Statements identifying relationships between nutrients or dietary ingredients and a body function.

**Unqualified health claim** An FDA term describing a relationship between a food, food component, or dietary supplement ingredient and reduced risk of a disease or health-related condition. Significant scientific agreement supports these authorized claims.

**Qualified health claim** An FDA term describing a relationship between a food, food component, or dietary supplement ingredient and reduced risk of a disease or health-related condition. Although the "weight of the evidence" qualifies the statement as a health claim, these claims are not held to the standard of significant scientific agreement required for unqualified health claims.

**TABLE 29-3 Qualified Health Claims Allowed by the FDA**

- Cancer Risk
  - Tomatoes and/or tomato sauce (prostate, ovarian, gastric, and pancreatic cancers)
  - Calcium (colon/rectal cancer or recurrent colon/rectal polyps)
  - Green tea
  - Selenium
  - Antioxidant vitamins

- Cardiovascular disease risk
  - Nuts
  - Walnuts
  - Omega-3 fatty acids
  - B vitamins
  - Monounsaturated fatty acids from olive oil
  - Unsaturated fatty acids from canola oil
  - Corn oil

- Cognitive Function
  - Phosphatidylserine

- Diabetes
  - Chromium picolinate

- Hypertension (and pregnancy-induced hypertension and preeclampsia)
  - Calcium

- Neural tube birth defects
  - 0.8 mg folic acid

Source: Adapted from the FDA's website at www.fda.gov/Food/LabelingNutrition/LabelClaims/QualifiedHealthClaims/ucm073992.htm.

the FDA for use in food labeling (26). Some of these include the link between antioxidant vitamins and cancer, zinc and the immune systems of elderly people, and green tea and heart disease (37).

The FDA ranks qualified health claims by the strength of the underlying scientific evidence (3). Claims are rated as A, B, C, or D, with A being the strongest. The original language of the 1993 Nutrition Labeling and Education Act allowed only unqualified food health claims, which required "significant scientific agreement" (SSA); the FDA was later forced by a lawsuit to provide guidance for the use of qualified health claims (3). Qualified health claims are generally approved within 270 days of filing (20). When a claim is approved, the FDA posts a letter to its website. Unqualified health claims may take twice as long (540 days) to receive approval.

Approval of a claim follows the following basic steps: (1) definition of the substance/disease relationship, (2) identification of relevant studies, (3) classification of studies, (4) rating of the studies, (5) rating the strength of the evidence, and (6) reporting a rank (20).

Strict requirements now govern how and when the FDA-approved health claims may be used. The food must contain enough of the nutrient to contribute at least 10% of the Daily Value and must not contain any nutrient or food substance that increases the risk of a disease or health condition. For example, whole milk cannot bear the calcium/osteoporosis claim because of its high saturated fat content. Excess saturated fat has been linked to heart disease. On the other hand, low-fat and nonfat milk do qualify for the calcium/osteoporosis claim (11).

### Nutrient Content Claims

Nutrient content claims describe the level of a nutrient or dietary substance in a product by using terms such as *free, high,* and *low* (Table 29-4). They may also compare the level of a nutrient in a food to that of another food by using terms such as *more, reduced,* and *lite.*

**Drug** A product able to treat, prevent, cure, mitigate, or diagnose a disease or disease symptom.

### Structure/Function Claims

Structure/function claims describe the role of a nutrient or dietary ingredient intended to affect normal structure or function in humans. For example:

"calcium builds strong bones"
"fiber maintains bowel regularity"
"antioxidants maintain cell integrity"

The manufacturer is responsible for ensuring the accuracy and truthfulness of these claims. Although not pre-approved by the FDA, they must be truthful and not misleading. If a dietary supplement (rather than food) label includes such a claim, it must have a "disclaimer" on the label. Disclaimers state that the FDA has not evaluated the claim and that the product is not intended to "diagnose, treat, cure or prevent any disease" (only a drug can legally make such a claim).

***Dietary Supplements*** Under the 1994 Dietary Supplement Health and Education Act (DSHEA, pronounced de-shay), dietary supplements are treated as foods (50). Health claims can be made for dietary supplements as long as they are not disease claims, which are only allowable for **drugs.** Manufacturers of dietary supplements that make structure/function claims on labels must submit a notification to the FDA no later than 30 days after marketing the dietary supplement. Further information can be found in the "Structure/Function Claims Small Entity Compliance Guide" at www.fda.gov/Food/GuidanceComplianceRegulatoryInformation/GuidanceDocuments/DietarySupplements/ucm103340.htm.

## Food Additives

Food additives are substances added to food that become part of the food and affect its character. The FDA is responsible for regulating more than 3,000 food additives that are used in the food industry. Some of the more common additives and their functions in foods are listed in Appendix F. The four main purposes of these food additives are discussed in Chapter 3 on food composition, and common examples have been provided throughout the text. The purpose of this section on food additives is to describe the FDA's role in their approval and safety.

## TABLE 29-4 · Nutrient Content Claims Allowed for Food Labeling

| Description | Amount/Serving |
|---|---|
| Free* | No nutrient or in very small amounts—calories (kcal) <5, fat <½ g, saturated fat <½ g, cholesterol <2 mg, sugar <½ g, and sodium <5 mg |
| Low | <40 calories (kcal)<br><3 g fat<br><1 g saturated fat<br><20 mg cholesterol<br><140 mg sodium (very low sodium <35 mg) |
| Extra Lean | <5 g fat<br><2 g saturated fat<br><95 mg cholesterol |
| Lean | <10 g fat<br><4 g saturated fat<br><95 mg cholesterol |
| Light/Lite | ⅓ fewer calories<br>Half the fat<br>Can also describe color or texture |
| Reduced or Less | 25% less than regular product |
| More | 10% or more of Daily Value compared to reference food |
| High | 20% or more of Daily Value for the nutrient |
| Good Source | 10% to 19% or more of Daily Value for the nutrient |
| More Than | 10% or more of a Daily Value than comparison food |
| Fewer Than | 25% less than comparison food |
| Healthy | <3 g fat<br><1 g saturated fat<br><480 mg sodium<br><60 mg cholesterol |

*Instead of "free," manufacturers may also use "no," "zero," "without," "trivial source of," "negligible source of," and "dietarily insignificant source of."

### ? How & Why?

**How do additives obtain FDA approval?**

Additives are never given permanent approval, but are subject to review to determine whether the approval should continue, be modified, or be withdrawn. The process of approving a new additive may take years and cost millions of dollars. An additive is first tested on laboratory animals to determine whether it affects life span, cancer rates, incidence of birth defects, allergies, or other health problems. A hundredfold margin of safety is applied; that is, if 100 mg of the substance is the minimum level that causes no harmful effect in laboratory animals, then manufacturers may use no more than 1 mg in foods given to humans (30). This safety margin sets the **Acceptable Daily Intake (ADI)** for human consumption (49). Only food additives on the **Generally Recognized as Safe (GRAS) list**, plus those more recently approved by the FDA, are allowed in the U.S. food supply (19).

## Safety of Food Additives

The United States government defines *safety* as "reasonable certainty . . . that the substance is not harmful under the intended conditions of use" (40). Two amendments to the Food, Drug, and Cosmetic Act of 1938—the Food Additives Amendment (1958) and the Color Additives Amendment (1960)—state that the FDA must approve the safety of all food additives. Prior to the 1958 Food Additives Amendment, the burden was on the FDA to prove that an additive was harmful before a court order could be issued banning its use. After the amendment was passed, the tables were turned and it became the manufacturer's responsibility to convince the FDA that the additive was safe through extensive testing on laboratory animals.

## Exempt from Definition of Food Additive

The law excludes substances that are on the GRAS list or have prior sanction from being defined as a food additive.

*GRAS List*   Substances on the GRAS list are not officially considered food additives, so they are exempt from the legal requirement of proving their safety (36). However, GRAS substances are continually reevaluated, and as new methods for testing are developed, some substances are removed from the GRAS list (34). This list was originally formed when the FDA appointed a group of scientific researchers to compile a list of substances considered safe. The purpose of the GRAS List was to avoid the time and expense of testing the safety of every substance.

*Prior Sanction*   These substances were used in foods prior to the passage of the 1958 Food Additives Amendment. As a result, they fall into the category of previous approval or prior sanction.

## The Delaney Clause—Does the Additive Cause Cancer?

The **Delaney Clause** (1958) tightened the regulations on food additives, but

**Acceptable Daily Intake (ADI)** The amount of food additive that can be safely ingested daily over a person's lifetime.

**Generally Recognized as Safe (GRAS) list** A list of compounds exempt from the food additive definition because they are generally recognized as safe based on "a reasonable certainty of no harm from a product under the intended conditions of use."

**Delaney Clause** A clause added to the Food, Drug, and Cosmetic Act of 1938 stipulating that "no additive shall be deemed to be safe if it is found to induce cancer when ingested by man or animal."

has subsequently been reinterpreted and modified as the techniques for identifying carcinogenic compounds have improved (37). The Department of Health and Human Services has estimated that food additives are linked to only 1% of all cancers. The FDA has banned all food additives that were found to be carcinogenic in laboratory animals, with the exception of saccharin and a few food colorings. Other additives whose safety has been questioned include food colorings, nitrites (see Chapter 7 on meats), pickling additives, BHA/BHT, and pesticides.

*Saccharin*   High doses of saccharin have been linked to bladder cancer in rats, but the Office of Technology and Assessment concluded that saccharin is one of the weakest food carcinogens detected (33).

*Food Colorings*   There are two categories of FDA approved food additives—certifiable and exempt (43). Certifiable color additives include those that are man-made and derived from petroleum and coal. Certifiable additives are tested by the FDA for purity; batches that are certified are then issued a certification lot number. Certified color additive have prefixes, such as FD&C, D&C, or Ext D&C. Approximately 30 colors are approved for use in foods, half of which are synthetic and classified as certified (45). Exempt color additives are those derived from plant, animal, or mineral sources, such as caramel color and grape color extract (43). Several of the food colors have tested positive as carcinogens (33). The FDA banned the use of certain dyes in foods, but there have been temporary lifts on the bans, allowing foods containing these colors to remain on the market (7).

*Pickling Salt and Vinegar*   In China and Japan, it is a common practice to preserve foods by pickling in salt and vinegar or another acid.

Researchers speculate that the high incidence of stomach cancer in both countries is directly linked to the high intake of pickled vegetables and fish (6).

*BHA and BHT*   Butylated hydroxyanisole (BHA) and butylated hydroxytoluene (BHT) act as antioxidants to help prolong storage life. They are often found on the ingredient lists of cereals, potato chips, chewing gum, and baked products (13). Scientific evidence shows that the compounds both stimulate and inhibit cancer in animals. However, the amount used in food is extremely small and does not reach the cancer-producing levels used with rats.

*Pesticides—Unintentional Additives*   Pesticides are among the more than 12,000 substances that unintentionally become a part of the food chain (33). Until recently, pesticides were classified as food additives, but the Food Quality Protection Act of 1996 modified the Delaney Clause by redefining *food additive* to exclude pesticide residues found on foods. The FDA regulates the use of various pesticides, and constantly monitors pesticide residue levels to ensure that the risk of food contamination remains low (14). Nevertheless, accidental poisonings do occur, and 80% of American shoppers perceive pesticide residues as a major health concern (48).

Allowable levels for pesticides vary from country to country. In 1987, a review of the FDA's pesticide program by the U.S. General Accounting Office and the Congressional Committee on Energy and Commerce reported that there was inadequate monitoring of pesticide residues on imported foods (47). This resulted in passage of the Pesticide Monitoring Improvements Act of 1988, which mandates pesticide monitoring, foreign pesticide usage information, and pesticide analytical methods. An FDA study in 1991 revealed that only about 3% of the foods sampled contained unacceptable levels of pesticides (41). However, according to the General Accounting Office, the FDA's pesticide tests do not screen for several pesticides that most seriously affect health. The FDA lists the pesticides they screen for online (12).

**? How & Why?**

*Do food additives cause cancer?*

Over the last 50 years, cancer death rates, with the exception of lung cancer, have remained relatively the same or slightly decreased. If food additives influenced cancer death rates, an increase in the general population dying of cancer would most likely have been documented. It has been estimated that if there is a risk, food additives would be responsible for less than 1% of all cancers.

### Genetically Modified Organisms (GMOs)

The definition and labeling requirements of GMOs are discussed in Chapter 1 on food selection.

## The Bioterrorism Preparedness Act

Through the Bioterrorism Preparedness Act of 2004 and other legislation, the U.S. government made it mandatory for all food facilities to register with the FDA, provide prior notice of food imports, and maintain necessary records to assist the FDA in determining whether food poses a threat of serious adverse health consequences or death. The regulation covers all food manufacturers, ingredient suppliers, packaging suppliers, and food distributors. Foreign facilities providing food for American consumption are also required to register. Electronic registration via the FDA's website at www.fda.gov/Food/GuidanceComplianceRegulatory Information/RegistrationofFoodFacilities/default.htm is encouraged. Failing to register is a criminal act (5).

## U.S. DEPARTMENT OF AGRICULTURE

The U.S. Department of Agriculture (USDA) was originally established for three reasons:

- To increase the income of farmers by developing methods for improving productivity and by generating new markets for farm products
- To reduce hunger and malnutrition
- To inspect and **grade** farm products

**Grading** The voluntary process in which foods are evaluated for yield (a 1 to 5 grading for meats only) and quality (Prime, Choice, AA, A, Fancy, etc.).

The USDA and its responsibilities have expanded tremendously. This chapter, however, focuses only on the USDA's inspection and grading services. A major difference between inspection and grading is that inspection is mandatory and grading is voluntary. Another big difference is that the USDA both inspects *and* grades meat, poultry, and eggs, but only grades dairy products, grains, fruits, and vegetables. These services are conducted through the USDA's Food Safety and Inspection Service (FSIS). Some overlap exists, in that the FDA is responsible for food products containing relatively small amounts of meat (3%) and poultry (2%) (40).

## USDA Inspections

The terrible conditions of the early meat-packing plants in 1906 led Congress to pass numerous acts of food-related legislation. This allowed the federal government to inspect plants producing meats, poultry products, and eggs. The goal was to ensure that both the processes and the products were sanitary. As a result, products containing foods that cross state lines must be inspected at the federal level. Products that pass the USDA inspection are then given the blue inspection stamp (Figure 29-7).

State inspection regulations for intrastate commerce (foods that never leave the state, i.e., cross the state line and enter the federal jurisdiction) must equal or exceed USDA standards. Each state established its own guidelines, and this is why the rules (especially with regard to health inspections) sometimes differ between states and even the federal government.

The special legislation affecting USDA inspections includes the following:

- *Federal Meat Inspection Act (1906).* The USDA is responsible for inspecting meat to ensure it is safe, wholesome, and accurately labeled.

- *Agricultural Marketing Act (1946).* The USDA's Agricultural Marketing Service administers the inspection and grading of raw and processed foods such as cereal, dairy, fresh fruits and vegetables, poultry, eggs, fish, and shellfish.
- *Fish and Wildlife Act (1956).* Governs the inspection of fish and shellfish.
- *Poultry Products Inspection Act (1957).* Requires inspection of all poultry.
- *Wholesome Meat Act (1967).* Meats that do not travel interstate must meet state inspections equal to federal guidelines. This act also includes provisions for inspection of foreign processing plants.
- *Egg Products Inspection Act (1970).* Mandatory inspection of plants producing egg products.
- Pathogen Reduction Hazard Analysis and Critical Control Points (HACCP) Systems Final Rule (1996). Meat and poultry plants must implement mandatory HACCP programs.
- *Pathogen Reduction Act (1997).* A zero tolerance is set for *E. coli* in foods.
- *Farm Security and Rural Investment Act (2002).* The 2002 Farm Bill requires **Country of Origin labeling.** The product is exempt if it is an ingredient in another food such as salmon in sushi, nuts in a candy bar, or tenderloin in beef Wellington. Hot dogs and luncheon meats are also exempt.

For 90 years inspectors relied on sight, touch, and smell to detect spoiled meat (Figure 29-8). Their tests for microorganisms took hours or even days. Recent laws governing meat and poultry safety have radically changed how inspections are conducted (25). To ensure that their plants and equipment are clean, the FSIS (under the HACCP Systems Final Rule) requires that meat and poultry plants incorporate Sanitation Standard

Operating Procedures (SSOPs) (4). Food manufacturers are now required to develop and implement HACCP programs to identify critical points where microbial contamination could occur, install controls to prevent or reduce those hazards, and document the process (24). Quick microbiological tests that detect *E. coli, Salmonella,* and other bacteria are encouraged. Each meat and poultry manufacturing plant must also adopt HACCP programs to eliminate hazards at every point in the production process. Similar HACCP programs have been recommended for the egg and fruit/vegetable juice industries (28).

## USDA Grading

Only food that has passed inspection may be graded for quality. Unlike inspection, however, USDA grading is voluntary. For those food companies that wish to pay for USDA grading, it is available for meat, poultry, eggs, dairy products, fresh fruits and vegetables, and some fish, shellfish, and cereal. There are two types of grading, quantitative and qualitative.

### Quantity Grades or *Yields*

Quantitative grading is reserved for meats. It describes the *yield* or the ratio of the lean or muscle tissue to fat, bone, and refuse on the animal's carcass. The yield grade ranges from 1 to 5, with 1 representing the highest yield and lowest waste (Figure 29-9). This type of grade is not commonly seen at the consumer level.

### Quality Grades

Consumers are more familiar with quality grades. These are based on a food's appearance, texture, flavor, and other factors, depending on the particular food. Figure 29-10 shows the various quality grades and their characteristics for different types of foods.

**FIGURE 29-7** USDA inspection stamps.

**Country of origin labeling** The required identification of the country of origin on the label for fresh red meat (beef, pork, lamb, veal), marinated products (marinated meats), seafood, produce, and peanuts.

**FIGURE 29-8** USDA inspectors look for spoiled meat. The Federal Meat Inspection Act of 1906 required that all interstate beef be inspected. However, some believe that the old "sniff and poke" methods need to include more routine and rapid tests detecting pathogenic bacteria.

Vito Palmisano/Stone/Getty Images

**FIGURE 29-9** USDA Grade Yield stamp.

(The differences between individual grades for meat, poultry, fish, dairy, vegetables, and fruits are discussed in their respective chapters.) Figure 29-11 shows how different grades of fruit may be used for various food dishes. For example, high-quality U.S. Grade A strawberries would be used to top strawberry shortcake, whereas lower-quality U.S.

Grade B strawberries would be reserved for strawberry pie and even lower U.S. Grade C strawberries would be used for jam.

## Irradiated Foods

Most foods that have been irradiated are labeled with a radura or proper wording to inform consumers. See Chapter 28 on food preservation for the specific requirements.

## Organic Foods

The definition and labeling requirements of organic foods can be seen in Chapter 1 on food selection.

## Kosher/Halal Foods

More and more foods are being sold under kosher and/or halal designations. See Chapter 1 on food selection for information on the influence of religion on food choices, along with the definitions of what is kosher and halal.

## Country of Origin Labeling (COOL)

The USDA approved a country of origin labeling (COOL) law (21). This law applies to beef, lamb, pork, chicken, fish, shellfish, peanuts, fresh and frozen fruits and vegetables, pecans, macadamia nuts, and ginseng. Exempt foods include those served in foodservice

**FIGURE 29-10** USDA quality grade stamps for specific food types.

USDA CHOICE
(Meat)

USDA A GRADE
(Poultry)

USDA A GRADE
(Eggs)

U.S. GRADE NO. 1
(Fresh fruits & vegetables)

USDA AA GRADE
(Butter)

USDA U.S. EXTRA GRADE
(Instant nonfat dry milk)

**FIGURE 29-11**   The use of different fruit grades in food preparation.

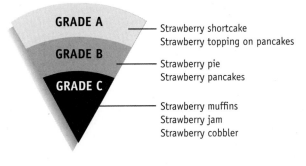

GRADE A ——— Strawberry shortcake
Strawberry topping on pancakes

GRADE B ——— Strawberry pie
Strawberry pancakes

GRADE C ——— Strawberry muffins
Strawberry jam
Strawberry cobbler

establishments (such as restaurants) or sold by companies with less than $230,000 in annual sales, exporters, and butcher shops and fish markets (21). COOL labeling was originally set to go in effect in September 2004, but was stalled by lobbyists (29) until March 16, 2009, when it finally went into effect. COOL is not intended to provide any information concerning food safety, but only provides consumers with data on the country from which food originated (29).

COOL labeling categories include the following (21):

- **U.S. only origin.** For animals born, raised, and slaughtered in the United States.
- **Foreign only origin.** For products produced entirely outside the United States.
- **Mixed origin.** This category must be labeled with the country of birth, raising, and slaughter.

Approved COOL labels include statements such as "Product of USA," or may appear as a check box (1).

# ENVIRONMENTAL PROTECTION AGENCY

The Environmental Protection Agency's (EPA) goal is to protect human health and the environment. In terms of food, it determines the safety of new pesticides and sets tolerance levels for pesticide residues in foods. The major law empowering the EPA with the authority to regulate the use of certain pesticides

is the 1947 Federal Insecticide, Fungicide, and Rodenticide Act (Figure 29-1). This legislation is an exception to the Delaney Clause, allowing certain pesticides that are known carcinogens to be used on raw agricultural commodities as long as the benefits from their use outweigh the risks from their presence (49). It is illegal for foods to bear traces of pesticides in excess of the safe tolerance levels set by the EPA. Fresh fruits, vegetables, and grains, especially those from foreign countries, are continually inspected by the EPA for pesticide residues.

# CENTERS FOR DISEASE CONTROL AND PREVENTION

The Centers for Disease Control and Prevention (CDC) has many functions for protecting the health of Americans and in relationship to food that focuses on preventing and controlling foodborne diseases. This federal agency dates back to World War II. Initially, the Public Health Service established the Office of Malaria Control to prevent malaria in army camps. Today, one of the CDC's responsibilities is to track outbreaks of foodborne illnesses across the country, determine their cause, and prevent their recurrence. State health departments report foodborne illness outbreaks (and infectious disease data) to the CDC. The federal agency then uses this information to conduct surveillance and keep records on a national scale (38).

# OTHER REGULATORY AGENCIES

## U.S. Department of Commerce

It may seem odd, but the U.S. Department of Commerce (USDC) is responsible for the inspection of fish and fish products. Specifically, the National Marine Fisheries Service within the USDC oversees these voluntary inspections and gradings. Fish processing plants pay for the service. Fresh fish are not graded because deterioration begins the minute fish are caught. However, grading is available for frozen fish products such as fish sticks, fillets, steaks, and breaded/precooked fish portions. Shellfish require more caution, so oysters, clams, mussels, and whole scallops are sold to consumers only if harvested from certified waters. The Interstate Certificate Shellfish Shipper List, published monthly by the FDA, describes these certified waters. The updated version of the shellfish list is available online at https://info1.cfsan.fda.gov/shellfish/sh/shellfis.cfm.

## Federal Trade Commission

The major purpose of the Federal Trade Commission (FTC) is to prevent fraud, deception, and unfair business practices. In terms of food and dietary supplements, the FTC protects consumers from false or deceptive advertising. The FTC enforces the sections of the Food, Drug, and Cosmetic Act that prohibit the sale of products that are misrepresented by advertising or by misleading words, designs, or pictures on the label (19). "Cease and desist" orders are issued by the FTC against a company or individual it believes to be engaging in unlawful trade practices.

## Department of the Treasury

Alcoholic beverages fall under the jurisdiction of two agencies within the Department of the Treasury, the Internal Revenue Service (IRS) and the Bureau of Alcohol, Tobacco, and Firearms. The IRS enforces federal laws regulating alcoholic beverages because alcohol is subject to federal tax. The Bureau of Alcohol, Tobacco, and Firearms is partially responsible for enforcing the laws that regulate the production, distribution, and labeling of alcoholic beverages. A third federal agency involved in alcohol regulation is the FDA, which is responsible for wines with less than 7% alcohol.

## State Agencies

Products made and sold exclusively within a state often have to meet state food safety regulations. These are frequently administered at the federal level. However, each state has slightly different laws unique to its specific locality, needs, and problems.

# INTERNATIONAL AGENCIES

Global food standards are established by two major international agencies that are part of the United Nations—the Food and Agriculture Organization and the World Health Organization (10).

Under joint effort, they created the **Codex Alimentarius Commission** in 1962 to provide global guidelines concerning food hygiene, food additives, residues of pesticides and veterinary drugs, contaminants (environmental and industrial), and food labeling (31).

## The Food and Agriculture Organization

The Food and Agricultural Organization (FAO) was established in 1944 in response to the nutrition awareness created by World War II. Headquartered in Rome, Italy, the FAO's main objectives are to:

- Eliminate hunger
- Improve nutrition by enhancing food production and distribution through research and technical assistance and educational programs
- Maintain statistics on world food production

## The World Health Organization

The main objective of the World Health Organization, headquartered in Geneva, Switzerland, is to develop international cooperation for better health conditions. It aims to achieve this objective through the following goals:

- Conduct research on nutrition, cancer, vaccines, and nuclear hazards
- Expand government health programs by opening clinics, dispatching experts, and training medical personnel
- Organize educational campaigns to control diseases

## European Regulation

The European Parliament established regulations for nutrition and health claims (16). They were updated to provide definitions of nutritional claims, which address nutritional properties; health claims, which address the relationship between food and health; and risk-reduction claims, which include claims that a food significantly reduces the risk of a certain disease.

**Codex Alimentarius Commission**
The international organization that develops international food standards, codes of practice, and other guidelines to protect consumers' health.

# PICTORIAL SUMMARY / 29: Government Food Regulations

Ever since it became illegal to dilute wine with water in ancient Greece and Rome, food laws and regulations have been instituted to protect the consumer. Thanks to public demand for legislation against contaminated foods, our country's food supply is among the safest in the world.

## FEDERAL FOOD LAWS

From the Food and Drug Act of 1906 (the Pure Food Law) to the Bioterrorism Preparedness Act of 2002, the U.S. government has been involved in regulating the country's food supply. One of the most important of these laws is the Food, Drug, and Cosmetic Act, which became law in 1938.

## FOOD AND DRUG ADMINISTRATION

 The Food and Drug Administration (FDA) is a branch of the Department of Health and Human Services.

The FDA:

- Enforces the Food, Drug, and Cosmetic Act of 1938, which ensures that those products are safe, wholesome, and produced under sanitary conditions.
- Regulates the production of all food involved in interstate marketing with the exception of the USDA-regulated meat, poultry, and egg industries. This includes the following:
  - Inspecting facilities and manufacturing processes
  - Setting standards
  - Labeling
  - Regulating food additives
- Conducts research and educates the public about food and nutrition.

## U.S. DEPARTMENT OF AGRICULTURE

 The U.S. Department of Agriculture (USDA) was first established to help farmers improve productivity and find new markets, reduce hunger, and inspect and grade farm products. Since then, its role has expanded tremendously.

The USDA inspects and grades:

- Meat
- Poultry
- Eggs
- Dairy products
- Grains
- Fruits and vegetables

USDA Choice (Meat)   USDA A Grade (Poultry)   USDA A Grade (Eggs)

U.S. Grade No. 1 (Fresh fruits and vegetables)   USDA AA Grade (Butter)   USDA U.S. Extra Grade (Instant nonfat dry milk)

## ENVIRONMENTAL PROTECTION AGENCY

 The Environmental Protection Agency (EPA) determines the safety of new pesticides and sets tolerance levels for pesticide residues in foods. Fresh fruits, vegetables, and grains, especially those from foreign countries, are continually inspected by the EPA for pesticide residues.

## CENTERS FOR DISEASE CONTROL AND PREVENTION

The Centers for Disease Control and Prevention (CDC) has many functions related to preventing and controlling diseases, and one of them is to track outbreaks of foodborne illnesses across the country, determine their cause, and prevent their recurrence. State health departments report their data on such outbreaks to the CDC, which uses this information to conduct surveillance and to maintain records on a national scale.

## OTHER REGULATORY AGENCIES

- **U.S. Department of Commerce (USDC):** The National Marine Fisheries Service within the USDC oversees the voluntary inspections and grading of fish and fish products, which are paid for by fish-processing plants.
- **Federal Trade Commission (FTC):** The FTC maintains free and fair competition in the economy, and one method of achieving this goal is to protect consumers from false or misleading advertising.
- **Department of the Treasury:** The Bureau of Alcohol, Tobacco, and Firearms within the Department of the Treasury is responsible for enforcing the laws that regulate the production, distribution, and labeling of alcoholic beverages.
- **State Agencies:** Products made within a state often have to meet safety regulations unique to the state as well as federal regulations.

## INTERNATIONAL AGENCIES

- The Food and Agricultural Organization (FAO)
- The World Health Organization (WHO)

## CHAPTER REVIEW AND EXAM PREP

### Multiple Choice*

1. Which of the three standards set by the FDA regarding content and quality specify the amount of raw product that must be put into a container before liquid is added?
   a. Standards of Identity
   b. Standards of Fill
   c. Standards of Minimum Quality
   d. Standards of Recall

2. In order for the term *fat free* to appear on a food label, the product must contain less than how many grams of fat?
   a. 1.2 g
   b. 3.4 g
   c. 1 g
   d. 11.2 g

3. On the Nutrition Facts label, the % Daily Value is based on what number of calories?
   a. 1,500
   b. 2,000
   c. 2,500
   d. 3,000

4. The amount of a food additive that can be safely ingested daily over a person's lifetime is referred to as the
   _____.
   a. GRAS list
   b. Delaney Clause
   c. Acceptable Daily Intake
   d. safe quantity

5. One of the main jobs of the Centers for Disease Control and Prevention is to _____.
   a. inspect and grade farm products
   b. reduce hunger and malnutrition
   c. monitor food labeling
   d. track outbreaks of foodborne illnesses

*See p. AK-1 for answers to multiple choice questions.

6. The FDA defines a health claim as
   a. a food able to improve a specific health condition.
   b. a food, food component, or dietary supplement ingredient that may reduce disease risk.
   c. not well enough established to meet the significant scientific agreement standard.
   d. supported by "significant scientific agreement."

7. "No additive shall be deemed safe if it is found to induce cancer when ingested by man or animal" is known as
   a. Good Manufacturing Practice (GMP) Regulations.
   b. the Food Code.
   c. COOL.
   d. the Delaney Clause.

### Short Answer/Essay

1. List the major government agencies responsible for food regulation.
2. What are the major food regulation responsibilities of the FDA?
3. Discuss the process of product recall.
4. Define each of the following: *Standards of Identity, Standards of Minimum Quality,* and *Standards of Fill.*
5. The Fair Packaging and Labeling Act requires what information to be placed on food labels? What are the exceptions?
6. Describe the labeling requirements of COOL.
7. Define the following: *ADI, GRAS list,* and *Delaney Clause.*
8. Which government agency is responsible for grading? What foods are most commonly graded? Discuss the two types of grading.
9. Pesticide regulation is under which government agency's control?
10. The U.S. Department of Commerce is responsible for the inspection of what type of food?

## REFERENCES

1. Anonymous. Country of Origin Labeling (COOL). *Retail Food and Beverage* 6(3):14–15, 2008.
2. Banasiak K. News. Updated Food Code includes allergens. *Food Technology* 59(11):13, 2005.
3. Barlas S. Qualified health claims update. *Nutraceuticals World* 9(1):20–21, 2006.
4. Berne S. Simplifying sanitation. *Prepared Foods* 166(3):80–84, 1997.

5. Bodor A. U.S. regulatory update. *The Manufacturer Confectioner* 83(8):45–52, 2003.
6. Brenner H, D Rothenbacher, and V Arndt. Epidemiology of stomach cancer. *Methods in Molecular Biology* 472:467–477, 2009.
7. Collins TFX, et al. Teratogenic potential of FD&C red No. 3 when given in drinking water. *Food Chemistry and Toxicology* 31(3):161–167, 1993.

8. Durbin RJ. The Safe Food Act of 1997. *Food Technology* 52(1):112, 1998.
9. Duxbury D. Laboratory. Testing for food allergens. *Food Technology* 60(1):62–63, 2006.
10. Ebert AG. IFT's Codex Alimentarius activities. *Food Technology* 56(1):22, 2002.
11. Farley D. Look for "legit" health claims on foods. *Focus on Food*

Labeling: *An FDA Special Report* (May):21–28, 1993.

12. Food and Drug Administration. *Residue Monitoring Reports.* www.fda.gov/Food/FoodSafety/FoodContaminantsAdulteration/Pesticides/ResidueMonitoring Reports/default.htm. Accessed 12/12/09.

13. Giese J. Antioxidants: Tools for preventing lipid oxidation. *Food Technology* 50(11):73–78, 1996.

14. Giese J. Pesticide residue analysis in foods. *Food Technology* 56(9):93, 2002.

15. Government Publishing Office. *Code of Federal Regulations (CFR).* Title 7 (Part 52): [Section 1]84–89. www.gpoaccess.gov/cfr/index.html. Accessed 9/9/09.

16. Gruenwald J. Regulation renovation. *Nutraceuticals World* 10(1):26–27, 2007.

17. Hutt PB. A guide to the FDA modernization Act of 1997. *Food Technology* 53(5):54–60, 1998.

18. Institute of Medicine/National Research Council. *Ensuring safe food: From production to consumption.* Commission to ensure safe food from production to consumption. Institute of Medicine & National Research Council. National Academy Press, 1998.

19. Janssen WF. *The U.S. Food and Drug Law: How It Came, How It Works.* DHHS pub. (FDA) 79–1054. Washington, D.C.: GPO, 1979.

20. Kuhn ME. Making sense of health claim regulations. *Food Technology* 62(6):70–75, 2008.

21. Kupper C. Too much labeling information? New labeling regulations for 2008. *Today's Dietician* 10(5):8–13, 2008.

22. Kurtzweil P. Good reading for good eating. *Focus on Food Labeling: An FDA Special Report* (May):7–13, 1993.

23. Looney JW, PG Crandall, and AK Poole. The matrix of food safety regulations. *Food Technology* 55:60–76, 2001.

24. McCue N. Getting a handle on HACCP. *Prepared Foods* 166(6):109–116, 1997.

25. McNutt K. Common sense advice to food safety educators. *Nutrition Today* 32(3):128, 1997.

26. Mermelstein NH. A new era in food labeling. *Food Technology* 47(2):81–96, 1993.

27. Mermelstein NH. Nutrition labeling. Regulatory update. *Food Technology* 48(7):62–71, 1994.

28. Morris CE. HACCP update. *Food Engineering* 69(7/8):51–56, 1997.

29. Nachay K. Commodities get cool. *Food Technology* 6(3):42–44, 2009.

30. Potter ME. Risk assessment terms and definitions. *Journal of Food Protection* (Supp):6–9, 1996.

31. Randall AW. Scientific basis for setting food standards through Codex Alimentarius. *Food, Nutrition and Agriculture* 31:7–11, 2002.

32. Rodricks JV. Safety assessment of new food ingredients. *Food Technology* 50(3):114–117, 1996.

33. Sandler RS. Diet and cancer: Food additives, coffee, and alcohol. *Nutrition and Cancer* 4(4):273–279, 1983.

34. Scheuplein RJ. Perspectives on toxicological risk; An example: Food-borne carcinogenic risk. *Critical Reviews in Food Science and Nutrition* 32(2):105–121, 1992.

35. Segal M. Ingredient labeling: What's in a food? *Focus on Food Labeling: An FDA Special Report* May (Suppl):21–28, 1993.

36. Smith RL, et al. GRAS flavoring substances. *Food Technology* 55(12):34–55, 2001.

37. Stauffer JE. Food quality protection act. *Cereal Foods World* 43(6):444–445, 1998.

38. Tauxe RV. Emerging foodborne diseases: An evolving public health challenge. *Emerging Infectious Diseases* 3(4):425–434, 1997.

39. Taylor MR. Reforming food safety: A model for the future. *Food Technology* 56(5):190–194, 2002.

40. Thonney PF, and CA Bisogni. Scientific status summary. Government regulation of food safety: Interaction of scientific and societal forces. *Food Technology* 46(1):73–80, 1992.

41. U.S. Food and Drug Administration. FDA's pesticide residue program: Residues in foods—1990. *Journal of the Association of Official Analytical Chemists* 74:121Z–141A, 1990.

42. U.S. Food and Drug Administration. *The Food Label.* http://vm.cfsan.fda.gov/~dms/fdnewlab.html. Accessed 8/09.

43. U.S. Food and Drug Administration. *How safe are color additives? FDA Consumer Health Information.* December 10, 2007.

44. U.S. Food and Drug Administration. *Questions and Answers: Qualified Health Claims in Food Labeling— Draft Report on Effects of Strength of Science Disclaimers on the Communication Impacts of Health Claims.* September 28, 2005. www.fda.gov/Food/LabelingNutrition/LabelClaims/QualifiedHealthClaims/ucm109470.htm. Accessed 9/9/09.

45. U.S. Food and Drug Administration. *Summary of Color Additives for Use in United States in Foods, Drugs, Cosmetics, and Medical Devices.* www.fda.gov/ForIndustry/ColorAdditives/ColorAdditive Inventories/ucm115641.htm#table1A. Accessed 8/09.

46. U.S. Food and Drug Administration. *What guidance does FDA have for manufacturers of Fruit Jams (Preserves), Jellies, Fruit Butters, and Marmalades?* http://vm.cfsan.fda.gov/~dms/qa-ind5c.html. Accessed 8/09.

47. Venugopal R, et al. Recalls of foods and cosmetics by the U.S. Food and Drug Administration. *Food Protection* 59:876–880, 1996.

48. Wessel JR, and NJ Yess. Pesticide residues in foods imported into the United States. *Reviews of Environmental Contamination and Toxicology* 120:83–104, 1991.

49. Winter CK, and EJ Francis. Scientific status summary: Assessing, managing, and communicating chemical food risks. *Food Technology* 51(5):85–92, 1997.

50. Yetley EA. Implementing DSHEA— Progress made, but much remains to be done. *Food Technology* 53(7):24, 1999.

# WEBSITES

Department of Health and Human
Services (DHHS)
**www.hhs.gov**

Food and Drug Administration (FDA)
**www.fda.gov**

The Food and Drug Administration's
food topics page
**www.fda.gov/Food/default.htm**

FDA Information on Qualified
Health Claims
**www.fda.gov/Food/LabelingNutrition/
LabelClaims/QualifiedHealthClaims/
ucm073992.htm**

Centers for Disease Control and
Prevention (CDC)
**www.cdc.gov**

National Institutes of Health (NIH)
**www.nih.gov**

Department of Agriculture (USDA)
**www.usda.gov**

Marketing and Regulatory Programs
Agricultural Marketing Service (AMS)
**www.ams.usda.gov**

Animal and Plant Health Inspection
Service (APHIS)
**www.aphis.usda.gov**

Grain Inspection, Packers, and
Stockyards Administration (GIPSA)
**www.gipsa.usda.gov**

Research, Education, and Economics
**www.ree.usda.gov/**

Agricultural Research Service (ARS)
**www.ars.usda.gov**

National Institute of Food and Agriculture (NIFA); formerly the Cooperative
State Research, Education, and Extension Service (CSREES)
**www.csrees.usda.gov/**

Economic Research Service
**www.ers.usda.gov**

National Agricultural Statistics Service
**www.nass.usda.gov/**

Food Safety and Inspection Service (FSIS)
**www.fsis.usda.gov**

Environmental Protection Agency (EPA)
**www.epa.gov**

Office of Prevention, Pesticides, and
Toxic Substances
**www.epa.gov/oppts/**

Office of Pesticide Programs
**www.epa.gov/pesticides**

Department of Commerce
**www.commerce.gov**

National Oceanic and Atmospheric
Administration
**www.noaa.gov**

National Marine Fisheries Service
**www.nmfs.noaa.gov**

Federal Trade Commission (FTC)
**www.ftc.gov**

Department of the Treasury
**www.ustreas.gov**

Bureau of Alcohol, Tobacco, Firearms,
and Explosives
**www.atf.gov**

United States Customs & Border
Protection Service
**www.cbp.gov/**

# 30 Careers in Food and Nutrition

"**W**hat are you going to do when
you graduate?" is a common
question asked of college students.
Many students select a major based on
their career goal or their interest in a
particular class. The majority change
their major at least once before graduating. Deciding on a career is not only
challenging but stressful as well because this decision often determines a
person's next major direction in life.

The purpose of this chapter is to assist students in exploring various career
opportunities in food and nutrition. An
overview of various jobs in the food and

nutrition field along with their respective curriculum requirements is now
provided.

## THREE MAJOR FOOD AND NUTRITION AREAS

There are three major branches in the
food and nutrition field: nutrition science and dietetics; food science; and food
service (Figure 30-1). Each discipline has
several different professional associations
(Table 30-1). Each association has an annual convention at which members network and learn the latest developments
in their field; sponsor affiliates at the state
level, each with its own yearly conven-
tions and other meetings throughout
the year; publish a journal; and host a
website. Dr. Lucia Kaiser, a nutrition professor at the University of California at
Davis, advises students to "go to professional society meetings and find out what
the professionals are doing."

Depending on the size of the academic institution, these three areas of
food and nutrition are represented by
one major or by separate majors under
entirely different departments. Not all
universities offer degrees in all three
academic areas, so sometimes a student must transfer to pursue a specific
career choice. When making a decision on which university or college to
attend, students should consider location, price, program emphasis, faculty,
job placement, and financial aid. These
three possible majors are now further
explained to assist students in their
search for food- and nutrition-related
careers.

## NUTRITION SCIENCE AND DIETETICS

Nutrition science focuses on what happens to food in the body. This science
looks at how nutrients break down, get

**FIGURE 30-1**   The three main categories in the field of food and nutrition.

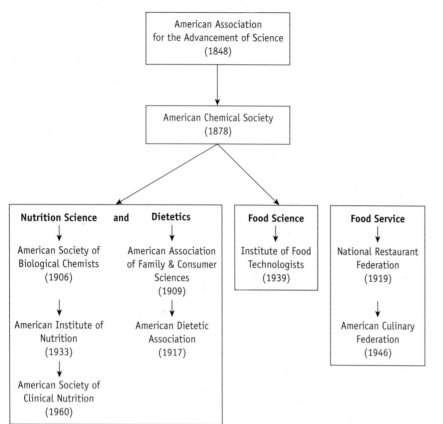

absorbed, enter cells, react within a living system (biochemistry), and affect health through deficiencies and excesses, and at how **medical nutrition therapy** can be used to treat certain diet-related health conditions. Students pursuing a major in nutrition often have two options: (1) nutrition science and (2) dietetics. A strong scientific foundation is required for both career options.

## Nutrition Science

The nutrition science emphasis prepares students for graduate school or professional school in the areas of nutrition, food science, medicine, dentistry, veterinary, physical therapy, pharmacy, physician's assistant, chiropractor, naturopathy, and other health-related professions (Table 30-2). See the section in this chapter on graduate school for more information. Although not part of Table 30-2, additional careers include medical writer (www.amwa.org) and medical illustrator (www.ami.org).

### Nutritionist

The word *nutritionist* is a generic term that has no professional definition. Even a nutrition undergraduate degree

| TABLE 30-1 | Selected Professional Associations in the Food and Nutrition Fields | | |
|---|---|---|---|
| **Field** | **Professional Association*** | **Journal** or Books** | **Website**** |
| Nutrition Science and Dietetics | American Dietetic Association | *Journal of the American Dietetic Association**** | www.eatright.org |
| | American Society of Clinical Nutrition (ASCN) | *Journal of Nutrition* | www.ascn.org |
| | | *American Journal of Clinical Nutrition* | www.nutrition.org |
| | American Board of Nutrition | (no journal) | www.acbn.org/ |
| | American College of Nutrition | *Journal of the American College of Nutrition* | www.amcollnutr.org |
| Food Science | Institute of Food Technologists | *Food Technology* | www.ift.org |
| | | *Journal of Food Science* | |
| | American Chemical Society | (see website—too numerous to list) | www.acs.org |
| | American Association of Cereal Chemists | *Cereal Foods World* | www.aaccnet.org |
| | American Oil Chemists' Society | *Journal of the American Oil Chemists' Society* | www.aocs.org |
| | | *Journal of Surfactants and Detergents* | |
| | | *Lipids* | |
| Food Service | American Culinary Federation | (books on website) | www.ciachef.org |
| | National Restaurant Federation | (books on website) | www.restaurant.org |
| | Foodservice Professionals Network | (books on website) | www.fspn.org |
| | International Association of Culinary Professionals | (books on website) | www.iacp.com |
| | Dietary Managers Association | *Dietary Manager Magazine* | www.DMAonline.org |
| | National Society for Healthcare Foodservice Management | *Innovative Magazine* | www.HFM.org |
| | American Society for Hospital Food Service Administrators | (books on website) | www.ashfsa.org |

*Student memberships are available.

**Job opportunities are often posted in back of the journal or on the website.

***The journal has a section called "New in Review" that summarizes the major journal articles of interest to food and nutrition professionals; employment services or agencies that locate jobs; and courses that help you prepare for the RD exam.

**TABLE 30-2** American Medical Association's List of Allied Health Careers*

| Accrediting Agency | Occupations | Website |
| --- | --- | --- |
| Accrediting Council for Occupational Therapy Education (ACOTE) | Occupational Therapist<br>Occupational Therapy Assistant | www.aota.org/nonmembers/area13/ |
| Accreditation Review Commission on Education for the Physician Assistant (ARC-PA) | Physician Assistant | www.aapa.org/prov.html |
| American Association of Colleges of Pharmacy | Pharmacist<br>Pharmacist Technician | www.aacp.org |
| American Art Therapy Association (AATA) | Art Therapist | www.arttherapy.org |
| American Board of Genetic Counseling (ABGC) | Genetic Counselor | www.faseb.org/genetics/abgc/abgcmenu.htm |
| American Orthoptic Council (AOC) | Orthoptist | www.othoptics.org/page11.html |
| American Physical Therapy Association (APTA) | Physical Therapist<br>Physical Therapist Assistant | www.apta.org |
| Commission on Accreditation of Allied Health Education Programs (CAAHEP) | Anesthesiologist Assistant<br>Athletic Trainer<br>Cardiovascular Technologist<br>Cytotechnologist<br>Diagnostic Medical Sonographer<br>Electroneurodiagnostic Technologist<br>Emergency Medical Technician-Paramedic<br>Health Information Administrator<br>Health Information Technician<br>Kinesiotherapist<br>Medical Assistant<br>Medical Illustrator<br>Ophthalmic Medical Technician/Technologist<br>Orthotist/Prosthetist<br>Perfusionist<br>Respiratory Therapist (Advanced)<br>Respiratory Therapist (Entry-Level)<br>Specialist in Blood Bank Technology<br>Surgical Technologist | www.caahep.org/ |
| Commission on Accreditation for Dietetics Education (CADE) of the American Dietetic Association | Dietetic Technician<br>Dietitian/Nutritionist | www.eatright.org/cade/ |
| Commission on Dental Accreditation (CDA) of the American Dental Association | Dental Assistant<br>Dental Hygienist<br>Dental Laboratory Technician | www.ada.org/prof/ed/accred/index.html |
| Commission of Opticianry Accreditation | Ophthalmic Dispensing Optician<br>Ophthalmic Laboratory Technician | www.nao.org/coa/htm |
| Council for Accreditation of Counseling and Related Educational Programs (CACREP) | Community Counselor<br>Marriage and Family Counselor/Therapist<br>Mental Health Counselor<br>Student Affairs Practitioner<br>School Counselor | www.counseling.org/CACREP/main.htm |
| American Speech-Language-Hearing Association, and Audiology Foundation of America | Audiologist<br>Speech-language Pathologist | http://stats.bls.gov/oco/ocos085.htm<br>www.asha.org<br>www.audfound.org |
| Council on Rehabilation Education | Rehabilitation Counselor | www.core-rehab.org/ |
| Joint Review Committee on Education in Radiologic Technology (JRCERT) | Radiation Therapist<br>Radiographer | www.jrcert/prg/ |
| Joint Review Committee on Educational Programs in Nuclear Medicine Technology (JRCNMT) | Nuclear Medicine Technologist | www.nucmednet.com/joint.htm |

*(continued)*

| **TABLE 30-2**   American Medical Association's List of Allied Health Careers\* *(continued)* | | |
| --- | --- | --- |
| **Accrediting Agency** | **Occupations** | **Website** |
| National Accrediting Agency for Clinical Laboratory Sciences (NAACLS) | Clinical Laboratory Technician/Med Lab Tech-Associate Degree | www.naacls.org/ |
| | Clinical Laboratory Technician/Med lab Tech-Certificate | |
| | Clinical Laboratory Scientist/Medical Technologist | |
| | Histologic Technician/Technologist | |
| | Pathologists' Assistant | |
| National Association for Schools of Music (NASM) | Music Therapist | www.arts-accredit.org/nasm/nasm.htm |

**Source:** American Medical Association. All rights reserved.
- Contact individual schools to determine prerequisites that differ among health careers.
- Full list available at www.ama-assn.org [click on "Education and Careers," then "Careers in Health Care," then "Health Care Careers Directory"].
- Salary ranges for the various health professions are found by clicking on "Health Care Income Ranges."

by itself does not qualify a person to counsel patients about their diets. As a result, individuals with nutrition degrees from colleges or universities usually continue their education (either through a dietetic internship or graduate school) to become qualified as registered **dietitians** or nutrition professors/ nutrition researchers/extension agents, respectively.

Currently in the United States, only registered dietitians are reimbursed by insurance companies through the referring physician's insurance code (7). Reimbursement practices are slowly changing for registered dietitians, who can now be reimbursed directly for some medical conditions covered by Medicare (the federal health insurance program for patients who are 65 and older). However, some people pursue the nutritionist title and seek direct payment instead of third party reimbursement (getting paid through an insurance company). Food and nutrition professionals can also obtain professional liability insurance (3).

The website www.quackwatch.org/ 04ConsumerEducation/nutritionist.html provides a more detailed explanation of the various routes of becoming certified as a nutritionist. Professionals generally accept the nutrition certification routes shown in Figure 30-2. According to the Quackwatch website, certain other nutrition "certifications" should be viewed skeptically as "Questionable Credentials."

## Dietetics

Three major career options are available in dietetics: registered dietitian (RD; title initials vary in Canada); Dietetic Technician, Registered (DTR); and Dietary Clerk (or Dietary Aide). Not every nutrition department has a dietetics program and not every community college has a dietetic technician program accredited by the Commission on the Accreditation of Dietetics Education (CADE) of the American Dietetic Association (ADA), but those that do can be found at www.eatright.org/ cade (click on "Accredited Education

**Medical nutrition therapy** The provision of nutrition care in clinical settings to manage nutrition-related diseases such as heart disease, diabetes, cancer, renal disease, liver failure, and others. Registered dietitians, as part of the health care team and often referred by a physician, contribute to the patient's care by providing this dietary therapy.

**Dietitian (registered dietitian or RD)** A health professional who counsels people about their medical nutrition therapy (diabetic, low-cholesterol, low-sodium, etc.). Registration requirements consist of completing an approved 4-year college degree, exam, internship, and ongoing continuing education. Categories exist for generalist, specialty, and advanced practices.

**FIGURE 30-2**   Nutrition certification routes.

**American Society of Clinical Nutrition (ASCN) (www.ascn.org)**

Clinical arm of the American Society for Nutritional Sciences (ASNS) (www.nutrition.org). Previously called the American Institute of Nutrition. Open to individuals who have published meritorious nutrition research.

**American Board of Nutrition**

People with graduate degrees (PhD, MD) can obtain certification in clinical nutrition (MDs only) or human nutritional sciences (MDs and PhDs). This board certification requires passage of a comprehensive examination.

**American College of Nutrition (www.amcollnutr.org)**

Certified Nutrition Specialist (CNS) credential to those with an accredited master's or doctoral degree who also have clinical experience.

Programs" in left column to see Coordinated and Didactic Programs to become a registered dietitian and Dietetic Technician Programs to become a dietetic technician). Valuable information for students about the dietetics field is available at www.eatright.org/students.

## Registered Dietitian (RD)

Dietitians (dieticians) counsel people about diet. They help prevent and treat diet-related illnesses by promoting healthy eating habits, recommending dietary modifications through medical nutrition therapy, and supervising the preparation and serving of meals at hospitals and other institutions. They are encouraged to base client care decisions on the *Nutrition Care Manual* (www.nutritioncaremanual.org) and Evidence-Based Nutrition Practice Guidelines, which are clinical practice recommendations based on the latest research summaries (www.adaevidencelibrary.com). Both of these resources are available to ADA members and the *Nutrition Care Manual* is sometimes available through libraries that maintain a subscription. It is also important for dietitians to approach client care using the **nutrition care process and model** (NCPM), which is a systematic problem-solving method (5), and to document the process within the medical record, because "If it's not documented, it didn't happen" (4).

## Academic Requirements

The Commission on Dietetic Registration (CDR) defines a registered dietitian (RD) as an individual who meets the following requirements:

1. Completed a degree from a CADE-approved 4-year undergraduate program or its equivalent. A baccalaureate degree is recommended, but students may hold a different undergraduate degree or be enrolled in a graduate program. As long as they fulfill all the ADA course requirements, they can receive a **verification form** by the dietetic program director (each program's approved courses vary slightly).
2. Been accepted to and completed a supervised practice program [internship; or this practical experience can be part of step 1 if the undergraduate degree is in the form of

**FIGURE 30-3**   Registered dietitian assisting patient with dietary choices.

Tom McCarthy / PhotoEdit

a Coordinated Program (CP) that includes the internship as part of the educational experience].
3. Passed the Registration Examination for Dietitians. (The five major exam areas as of 2007 include nutrition care process, foodservice systems, management, food and nutrition sciences, and counseling/communication.)
4. Pays the annual registration maintenance fee.
5. Stays updated through a Professional Developmental Portfolio (PDP) by accruing 75 hours of approved continuing professional education every 5-year reporting period.

Without completing the above dietetics track, few individuals will find work in the area of dietary counseling. A professor stated, "I even know PhDs who could not find community work in nutrition because they lacked a dietetics background." She added that it is "better to get the right coursework in graduate school, even if you do not plan to do an internship right away." A dietetics verification form opens more career doors, even for those who eventually want to teach at the college level.

### Supervised Practice Programs

**Internships** provide supervised experience in a clinical setting that averages a year in length, but can range from 6 months to 2 years. The longer or part-time programs often combine the experience with a master's degree.

About 45% of all dietitians hold a master's degree, and 4% go on to complete their doctorates (15). Internships are competitive, so students need a strong grade point average (GPA) and three letters of recommendation. It also helps to have nutrition or food-related work experience and extracurricular activities (student nutrition club, health fairs, nonprofit organizations, etc.). The currently approved list of internships can be found at found at www.eatright.org/cade (click on "Accredited Education Programs"). Students should be aware that the military also offers internships through the U.S. Army and U.S. Air Force (see Military Careers under "Websites"). The U.S. Navy hires registered dietitians,

**Nutrition care process and model** A standardized model intended to guide registered dietitians and dietetic technicians, registered in providing high-quality nutrition care.

**Verification form** Documentation provided by the dietetic program director to an individual who has completed the undergraduate ADA course requirements approved by the ADA for that particular academic institution. This form is required for (1) acceptance into an ADA internship and (2) taking the ADA examination.

who enter the service as officers. All three branches require a minimum of 3 years of service, and if continued, the military offers excellent retirement packages (12).

***Professional Developmental Portfolio (PDP)*** Once a person becomes a registered dietitian, professional skills and registration are maintained through the accrual of Continuing Professional Education Units (CPEU). These are obtained by attending seminars, workshops, national ADA meetings, academic course work, or other preapproved avenues (see www .eatright.org) that are required to maintain registration. Otherwise,

re-registration requires that the professional exam be retaken. To maintain registration, a registered dietitian pays an annual fee to the ADA and maintains a Professional Development Portfolio to accumulate 75 hours of continuing education units every 5 years (1). The website www.cdrnet.org/ pdrcenter/portfolioTOC.htm lists the five steps of the Professional Development Portfolio process used to record CPEUs.

A less formal way of staying up to date without CPEU credit is through the ADA's "Daily News," a daily e-mail newsletter that lists the top food and nutrition news stories. Another avenue is to read RDsWeighIn, which is a blog

written by ADA media spokespeople and available at www.rdsweighin. typepad.com (10).

## Types of Dietitians
The three types of dietitians include generalist, specialist, and advanced practitioner (14, 19).

A general practitioner is a dietitian who practices in one of the following major areas: clinical, administrative, community health, research, education, food and nutrition management, or consultation and business (private practice) (Table 30-3). Figure 30-4 shows the percentage of each type of dietitian working in the field according to a survey by the ADA.

---

**TABLE 30-3** Types of Dietitian Positions and Their Major Duties

| Type of Dietitian | Employer Example | General Duties |
|---|---|---|
| Clinical | The majority (66%) of dietitians work as clinical dietitians in hospitals, clinics, health maintenance organizations (HMOs), nursing homes, and home care.<br><br>Federal dietitian jobs can be found at www.usajobs.gov. The U.S. Army, Air Force, and Navy employ registered dietitians (see Military Careers under "Websites"). | Responsible for meeting the nutritional needs of patients by providing regular diets or medical nutrition therapy (MNT) prescribed to patients by their physicians. Assess patients' conditions and nutritional needs, plan a nutrition therapy program, implement the plan, and evaluate the results (16). |
| Administrative or Chief Dietitian | Same as above. | Responsible for the management of the nutrition department and/or food service department of hospitals, schools, nursing homes, and prisons. Duties include supervising clinical and food service dietitians; formulating department policies; training; supervising kitchen personnel; coordinating menus; purchasing food, equipment, and supplies; and ensuring food safety and sanitation of the department. |
| Community (Public Health or International Health) Dietitian | Jobs are primarily in government organizations such as state or county health departments, cooperative extension agencies, the federal government, or international organizations such as WHO (www.who.int/employment/en/) and FAO (www.fao.org/VA/Employ.htm). Federal dietitian jobs are listed at www.usajobs.gov. The American Overseas Dietetic Association is at www.eatrightoverseas.org. | Work in disease prevention and health promotion for the community. Projects include WIC (Women, Infants, and Children), cholesterol screening programs, hypertension awareness campaigns, Meals on Wheels, and elderly feeding programs. National nutrition programs can be found at http://fnic.nal.usda.gov (click on "Nutrition Assistance Programs"). |
| Corporate Dietitian | Jobs exist in health care and food corporations, grocery chains, or specialized commodity boards (American Beef Council, Dairy and Nutrition Council, American Egg Board, etc.). Corporations can also hire dietitians as sales representatives for their food products, pharmaceuticals, and dietary supplements. | Public relations, marketing and sales of food products, and nutrition consulting. |
| Private Practice or Consulting Dietitian | Self-employed dietitians work independently and may contract with hospitals, clinics, nursing homes, physician's offices, HMOs, day-care centers, schools, restaurants, sport teams,* and employee fitness programs. | Clinical nutrition or food service management consulting. It helps to have hospital experience. An increasingly popular role for the dietitian is as a "health coach" (11). |
| Research or Teaching Dietitian | Dietitians who have an advanced degree (doctorate) work for universities or medical schools (usually in the capacity of a professor), corporations (food or pharmaceutical), or the government (USDA, FDA, National Institutes of Health, etc.). | Research and/or teaching. |

*Many students are interested in being a sports nutritionist, but few full-time positions exist. Consultant dietitians are usually hired to provide nutrition guidance to teams and athletes, and not all teams hire dietitians.

**FIGURE 30-4**   The percentage of registered dietitians employed in various areas of dietetic practice.

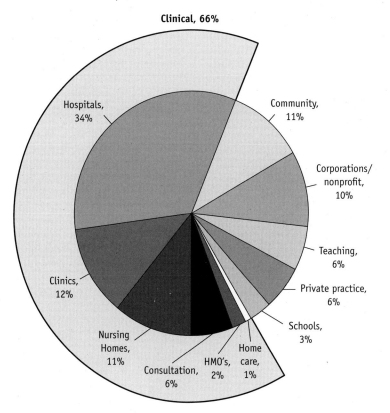

Clinical, 66%

Hospitals, 34%

Community, 11%

Corporations/ nonprofit, 10%

Teaching, 6%

Private practice, 6%

Schools, 3%

Clinics, 12%

Nursing Homes, 11%

Home care, 1%

HMO's, 2%

Consultation, 6%

(AADE) at www.diabeteseductor.org provides tools for "diabetes educators to help patients change their behavior and accomplish their diabetes self-management goals." The desired outcome of diabetes education is measurable behavior change, specifically eating a healthy diet, being active, self-monitoring, taking medication, problem solving, reducing risks, and coping in a healthy manner. The AADE is an accrediting body for diabetes education to provide "physician offices, clinics, pharmacies and community centers with a way to ensure that they are offering their patients comprehensive, effective diabetes self-management education."

> ### ? How & Why?
>
> ### *How does a dietitian work for the U.S. Public Health Service?*
>
> The U.S. Public Health Service (PHS) Commissioned Corps led by the U.S. Surgeon General is one of seven uniformed services of the United States (i.e., Army, Navy, Air Force, Marines, Coast Guard, and National Oceanic and Atmospheric Administration). Its mission is public health, not defense. RDs work in clinical or research settings; in testing; in inspection of food production facilities for the FDA and USDA; conducting research at the NIH and CDC; coordinating nutrition programs for the Indian Health Services, Health Resources and Services Administration, Bureau of Prisons, and the Centers for Medicare and Medicaid Services; developing national health and nutrition programs and dietary guidelines; and responding to public health emergencies (9). More information can be found at www.usphs.gov.

A dietitian may specialize in one or more particular areas of practice, and obtain supplemental certifications. Examples of these specialties include the following:

- **Board-Certified Specialization.** Board-certified specialties can be obtained through the ADA (www.cdrnet.org/certifications/CDRcertification.htm) in the following areas: Gerontological Nutrition (CSN), Oncology Nutrition (CSO), Pediatric Nutrition (CSP), Renal Nutrition (CSR), and Sports Dietetics (CSSD).
- **Certified Diabetes Educator (CDE).** Registered dietitians can become specialized in counseling clients with diabetes by contacting the National Certification Board for Diabetes Educators (NCBDE) at www.ncbde.org (www.cdecb.ca in Canada).
- **Certified Nutrition Support Dietitian (CNSD).** This specialization offered by the American Society of Parenteral and Enteral Nutrition (ASPEN) (www.nutritioncare.org) is for parenteral (feeding patients

through a tube that enters the vein) and enteral (nutrition through a tube that enters the gut) feedings.
- **Dietetic Practice Groups (DPG).** A list of ADA Dietetic Practice Groups (DPGs) with their contact information is found at www.eatright.org.
- **Certificate of Training.** A certificate of training (which may or may not contribute to a specialty) can be obtained by participating in workshops on adult (www.cdrnet.org/wtmgmt/CertificateOfTraining.htm) and child (www.cdrnet.org/wtmgmt/childhood.htm) weight management that are occasionally offered by the ADA (not CDR).

Advanced practitioners are dietitians who have acquired an expert knowledge base in a more complex practice that often includes a master's or doctoral degree. Although the CDR does not offer any advanced-level certifications, an available option is to become Board Certified in Advanced Diabetes Management (BC-ADM). The American Association of Diabetes Educators

### Licensure

The ADA website warns that "the majority of states have enacted laws which regulate the practice of dietetics. Should you plan to practice dietetics in these states it is important that you contact a state regulatory agency prior to practicing dietetics. In many states it is a VIOLATION of state law to practice dietetics (consult patients) without a license." The agency contact information for each individual state is available at www.cdrnet.org.

### Dietetic Technician, Registered (DTR)

Dietetic technicians, registered (DTRs) assist registered dietitians (17). The official route to becoming a DTR involves completing an ADA–approved program (www.eatright.org). Three pathways are listed on the ADA website for becoming a dietetic technician (2):

- Pathway I: A 2-year associate degree from a CADE-accredited Dietetic Technician Program. Many programs are offered through community colleges that offer a 2-year associate's degree.
- Pathway II: A 4-year bachelor's degree (CADE-accredited Didactic Program in Dietetics) and completion of a supervised practice (accredited for Dietetic Technicians).
- Pathway III: A 4-year bachelor's degree (CADE-accredited Didactic Program in Dietetics) without the supervised practice (effective 2009).

However, despite the degree recommendation, some dietary technicians advance from entry-level dietary clerk positions without obtaining an official degree. Next, the individual passes the CDR Registration Examination for Dietetic Technicians and pays the annual registration maintenance fee. Fifty hours of CPEUs are required every 5 years to maintain registration.

### Dietary Clerk or Dietary Aide

Dietary clerk or dietary aid jobs are entry-level positions, usually in a hospital dietary department. Students can often fill these entry-level positions to increase their experience and improve their chances of obtaining a dietetic internship. To locate these types of jobs, students can search the local newspaper or hospital human resources boards on the Internet, or contact the administrative dietitian directly. Briefly explain that you are a student majoring in nutrition and that you are seeking experience in the form of either shadowing one or more dietitians for a day, volunteering, working part time, or working full time during the summer. Sometimes these starter jobs lead to full-time positions. Ashorkor Tetteh, a person contemplating a career change, said, "Work experience has taught me that what a job entails on a day-to-day

basis can turn out to be very different from what is initially imagined." To find a registered dietitian to shadow, contact him/her directly at local hospitals; use the websites www.aadenet.org for certified dietetic educators and www .nutritioncare.org for a certified nutrition support dietitian; or attend local or state ADA meetings (ADA state dietetic associations can be found at www .eatright.org).

# FOOD SCIENCE

Food science (food technology) focuses on food outside of the body. Students study food using the scientific disciplines of chemistry, biology, microbiology, mathematics, physics, and engineering. Specifically, food science determines the physical and chemical constituents of food and how these substances change during product development, production, processing, preserving, packaging, distribution, preparation, and storage.

## Food Scientist

The majority of food scientists (food technologists) work in the food industry for corporations in their processing, ingredient, and/or manufacturing plants (Figure 30-5). Others work for the government (Department of Agriculture, Food and Drug Administration), universities or colleges (professors or extension agents), or international agencies such as the Food and Agricultural

**FIGURE 30-5**    Food scientist working at a food corporation.

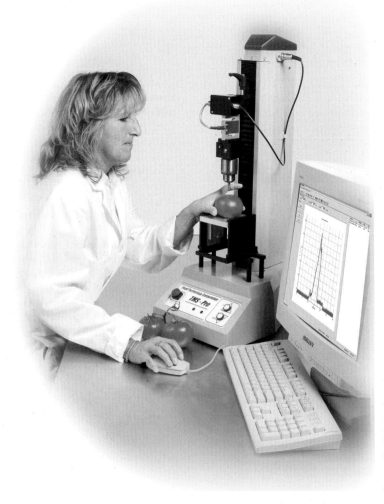

Courtesy of Food Technology Corporation

Organization (FAO) and World Health Organization (WHO). Related occupations include, but are not limited to, food chemists, dairy scientists, and microbiologists.

If food science is your major, consider that major food companies are regionally located, so food scientists often relocate to the food corporation's headquarters, laboratory, or plant. Before graduating, Dave Gaidos, Product Manager for General Mills, selected his job by searching in the library to determine where the food companies were located.

### Academic Requirements

To become food scientists, students graduate with a bachelor's degree in either food science, food technology, or food engineering from a program listed with the Institute of Food Technologists (IFT); check the IFT website (www.ift.org) for a list of these programs. Students with undergraduate degrees in nutrition, related areas, or even unrelated areas may also apply if they have completed the necessary prerequisite courses. When selecting either an undergraduate or graduate school to attend, be aware that not all food science or nutrition programs offer the IFT-recommended courses; it is important to visit the IFT website for a list of qualified schools. Students pursuing graduate degrees often do so for research, teaching, or management positions. The IFT also lists available jobs in the food industry that are sometimes higher in pay than those in other food and nutrition areas (13).

### Internship or Co-Op Experience

To improve a food science student's opportunity for a job after graduation, Dave Gaidos and others working for food corporations say the best advice they can give to students is to complete an **internship** or **co-op** (Table 30-4). Students often find these opportunities through their food science department, recruiters visiting the campus (larger food science departments), or food corporations. A survey by the National Association of Colleges and Employers discovered that corporations offer jobs to about half of the students who complete an internship or co-op. Fellowships and scholarships are also offered by

**TABLE 30-4   Selected Corporations Offering Summer Food Science Internships***

Campbell Soup Company

ConAgra

Dairy One

Frito-Lay

General Mills

Gorton's

Häagen-Dazs Company

H.J. Heinz Company

The International Food Network

Kellogg Company

Kraft Foods/Kraft Kitchens

M&M/Mars Incorporated

Nabisco

National Starch & Chemical Company

Nestle

Pepperidge Farm

PepsiCo Incorporated

Pepsi-Cola Company

Pfizer Incorporated

Pillsbury Technology Center

Taco-Bell

Unilever United States Incorporated

Warner-Lambert

International Internships available at: The International Association for the Exchange of Students for Technical Experience (IAESTA; www.iaeste.org)

*Seek out other internships by contacting the corporation.

corporations and some food associations (see www.allfoodbusiness.com/national_assoc.php for a partial list of food associations). Lisa Pannell, Food Scientist in Product Development for General Mills, recommends an international internship. As a student, she received an internship in a Holland cheese factory by applying through the International Association for the Exchange of Students for Technical Experience (www.IAESTE.org).

### Contacting Companies

Contact information for various corporations is available at www.hoovers.com, www.thomasregister.com, or, more specifically, the Thomas Food and Beverage Market Place (www.tfir.com), which lists food/beverage-related companies. Two other related sites are the Food and Beverage Export Directory and Search Engine (www.foodcontact.com) and the Food and Consumer Products Manufacturers of Canada (www.fcpmc.com).

### Types of Food Scientists

Once hired, food scientists often specialize in one or more of the positions listed in Table 30-5. A few tips from food scientists follow. Barbara Garter, the Senior Product Development Scientist at Kellogg Company, advises students not to "try and prove themselves to be all knowing, but rather develop skills in networking, collaborating, and teamwork to be really successful in the real world." For instance, she said, the important knowledge about food ingredients and how they function in foods was "learned on the job" and not in school. Food ingredient suppliers, she said, "were critical in helping me gain knowledge. Knowing who to call with technical questions is a very useful skill." Carrie Yonts, Corporate Analytical Chemist at the J.M. Smucker Company, stated that it is important for food chemists to have experience running equipment at food companies because "wet [liquid or bench] chemistry is used in college laboratories." She added, "Students should force themselves to become familiar with running the equipment and troubleshooting problems. Instrumentation knowledge is both practical and looks great on a résumé."

**Internships** Corporate internships are temporary job positions (usually for 3 months during the summer or for 6 months, which includes a summer and a semester) in industry, government, or academia. Dietetic internships are supervised practice experiences that average 12 months (6 to 24 months).

**Co-op** Work-study program with a corporation that is often customized for the student.

**TABLE 30-5** Selected Food Scientist Positions in Food Companies

| Food Scientist Position | General Duties |
| --- | --- |
| Research and Development (R&D) | Formulation of new food and beverage products, technologies, or manufacturing methods. |
| Food Engineers | Focus on the process of producing new or existing food and beverages (process design, energy efficiency, product manufacturability, heat profiles, equations, mass balances, and process optimization). |
| Quality Control (QC)/Assurance | Food and beverages must meet quality standards at each processing step. Sampling schemes, analytical methods, and statistical process/quality control. |
| Food Microbiology/Food Safety | Brand-new product lines need to be tested for food safety at every step of production—new ingredients, sanitation on the line, bacterial testing, and making sure plants are compliant to GMPs. HACCP for meat and poultry. |
| Sensory Evaluation Specialist | Consumer testing determines if the consumer likes the product. See Chapter 2 on Food Selection under subjective and objective food evaluation. |
| Flavor Chemist | Food or beverage flavor needs to be appealing to consumers and maintain its flavor through processing and storage. |
| Lab Technician | Works in the lab with duties such as batching syrups, doing finished product checks (BRIX, sulfuric acid content, etc.), micro testing of the water, etc. |
| Plant Manager | Managing all aspects of plant operations and supervising employees. |
| Ingredient Manager | Selecting ingredient vendors, making sure all vendors are following GMPs, and ensuring the quality and food safety of ingredients. |
| Marketing and Sales | Ensuring that the food or beverage product is sold to pay for the corporate structure. |
| President/Chief Executive Officer/General Manger | Executive management leading the food or beverage corporation. |
| University professor | Focusing on teaching, research (grant writing and publishing), and service (committees, community presentations, etc.). |
| University extension specialist (land grant universities) | Communicating and assisting the public with research reported by university researchers. |

# Food Science Technician

Food science technicians assist food scientists in food development, quality control, laboratory research, food safety, and other aspects of food production. To become a food science technician, students obtain a 2-year degree from a community college or technical school listed on the IFT website.

# FOOD SERVICE

A wide selection of careers is available in the food service field, which deals with two basic areas: (1) management and (2) food preparation. As a result, the curriculum is either business based for the management side (restaurant management, lodging management, travel and tourism) or culinary based for the food preparation aspect of food service, which is now further explained.

## Academic Preparation

Culinary institutes, cooking schools, and apprenticeships are three ways to obtain professional culinary training (Figure 30-6). Programs range from 2 months to 4 years in length depending on whether a 4-year bachelors degree, 2-year associates degree, or less-than-1-year diploma or certificate is awarded. Important factors to consider when choosing an educational option are accreditation, cost, student-to-faculty ratio, and placement rates. Location is also important as 4- and 2-year degrees are often offered by private institutions, community colleges, or hotel/restaurant management schools within a university. Many private schools teaching food preparation exist. The American Culinary Federation (ACF) is considered the primary certification institution for culinary arts professionals and establishes 14 levels of culinary certification including those for becoming a master chef (6). Its website at www.acfchefs.org lists accredited schools. Other listings of culinary/cooking schools can be found at www.allculinaryschools.com/featured/bpArts/, www.fspronet.com/schools.htm, or www.shawguides.com. Students should weigh the sometimes high tuition costs against the probability of obtaining an executive chef position. Lower-level food service positions do not always pay as well.

- **Culinary Institutes.** These are the prestigious, very expensive educational routes toward becoming a certified culinary professional. A degree specialization exists for baking and pastry.
- **Cooking Schools.** Smaller versions of culinary institutes providing specialized training via a certificate or diploma. Many mainstream cooking schools and programs focus on French cuisine, but specializations for international cuisines can include Italian, Chinese, Japanese, Thai, and others.
- **Apprenticeships.** On-the-job-training is available through numerous internships or apprenticeships for students wishing to specialize in food service. The largest employers are food service management companies. They offer internships throughout the year in finance and accounting, culinary arts, front office, and banquets/catering. The ACF lists over 80 apprenticeships in schools by state at www.acfchefs.org.

**FIGURE 30-6**   Culinary jobs provide careers for those interested in food preparation.

Tim Pannell/Corbis Yellow/Corbis

## Types of Food Service Culinary Positions

### Chef

Chapter 6 on meal management describes the different types of administrative (executive, production manager, or sous chef) and cook (sauce, fish, vegetable, soup, roast, broiler, pantry, pastry, and relief) positions in a food service organization.

### Food Service Manager

A business-based degree from a business or hotel and travel school assists graduates in preparing for jobs as food service managers in several venues—restaurants, hotels, catering, schools, universities, and the travel industry (air, train, ship).

### Sales and Marketing

One of the jobs available to graduates is as a sales representative for food equipment and food products. Ray Kroc, founder of McDonald's, started as a sales representative selling cups and then milk shake mixers to restaurants.

### Dietary Manager

Dietary managers work for institutions that serve food, specifically hospitals, nursing homes, other health care facilities, correctional institutions, and schools. Students can become dietary managers by enrolling in certain community college programs that vary in length from 12 to 18 months. The Dietary Managers Association (DMA) (www.dmaonline.org) offers certificates to students meeting the requirements to become Certified Dietary Managers (CDM). Lists of DMA-approved programs (schools) are available at their respective websites.

After finishing the program, students must pass an examination and apply for a certification through the Certifying Board of Dietary Managers. To maintain certification, graduates obtain 45 hours of CPEUs every 3 years.

## Food Service Certifications

Several certifications are available in the culinary field, including (18):

- Certified Culinary Professional (CCP) (www.iacp.com)
- Foodservice Management Professional (FMP) (www.nraef.org)
- Certified School Foodservice and Nutrition Specialist (SFNS) (www.schoolnutrition.org)
- Certified Culinarian (CC) (www.acfchefs.org)
- The Research Chefs Association (RCA) (www.culinology.com) includes chefs, food scientists, salespeople, and other food professionals interested in the development of new products and services (8). RCA has approximately 10 Culinology programs within various universities and colleges and offers two certifications: Certified Research Chef (CRC) and Certified Culinary Scientist (CCS).

## GRADUATE SCHOOL

A professor from one of the top nutrition graduate schools said, "Many students have little vision of why they are in graduate school, but they score extremely high on their **GRE**s and have high undergraduate GPAs." It's important to ask yourself what exactly your career goal is and to plan accordingly. Also, if you are planning to do research in graduate school, it's a good idea to research the different graduate schools. Graduate schools listed by discipline can be found in *Peterson's Guide to Graduate Schools* or similar publications. These books (and related websites) list graduate schools by their field, location, cost, number of students accepted, number of graduates, and other pertinent information helpful to applying students. Both the ADA and IFT also list graduate schools in their respective websites. You can also e-mail a specific department chairperson to locate the chair of the department's graduate program and the student president of the graduate club. International students can find contacts by getting in touch with the international club on campus.

## Prerequisites

You do not always need an undergraduate degree in dietetics, nutrition science, or food science to be accepted into a graduate school program. However, certain courses are usually required for admittance to individual graduate programs, and it is up to the individual to determine if those have been met. Once in graduate school, many

**GRE** The **G**raduate **R**ecord **E**xamination® general test measures verbal reasoning, quantitative reasoning, critical thinking, and analytical writing skills.

nutrition majors take undergraduate dietetic classes to qualify for the verification form so that they may apply for an internship and become a registered dietitian. Sometimes 400-level courses (senior level) apply to a graduate degree that usually consists of 600- to 700-level courses. It is also possible for students to pursue a master's degree while they take dietetic courses.

## Academic Requirements

In general, a master's degree in nutrition takes an average of 2 years to complete, and a doctorate takes 3 to 5 years (Figure 30-7). Research conducted by master's students is called a thesis and a dissertation for doctoral candidates. The length of the doctoral degree depends on the university, the

**FIGURE 30-7** General steps toward obtaining a master's or doctoral degree in food or nutrition sciences.

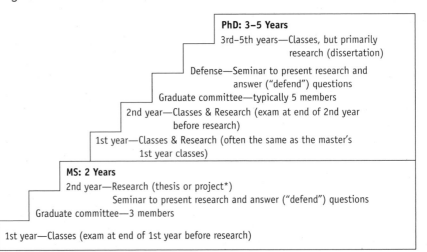

**PhD: 3–5 Years**
3rd–5th years—Classes, but primarily research (dissertation)
Defense—Seminar to present research and answer ("defend") questions
Graduate committee—typically 5 members
2nd year—Classes & Research (exam at end of 2nd year before research)
1st year—Classes & Research (often the same as the master's 1st year classes)

**MS: 2 Years**
2nd year—Research (thesis or project*)
Seminar to present research and answer ("defend") questions
Graduate committee—3 members
1st year—Classes (exam at end of 1st year before research)

**Graduate School**

*Projects are expanded term papers allowing students to enroll in more courses. They are not recommended for graduate students wishing to pursue careers in research or teaching.

## PROFESSIONAL PROFILE

**Dr. Liz Applegate, Senior Lecturer/Sports Nutritionist/Entrepreneur**

As an undergraduate, Liz Applegate was both a biochemistry major at the University of California at Davis (UC Davis) and a professional triathlete. She had planned to go on to medical school, but got married and remained at UC Davis. "I honestly randomly decided to get a doctorate in nutrition," she says. "I had my first child the same year that I received my PhD."

Liz had no idea what she was going to do upon graduation, but she was racing well and had sponsors. "I was a professional triathlete," she says, "and people started asking me questions. *Runner's World* contacted me 20 years ago to write stories, and I've been writing a monthly column for them ever since. So that's how it happened, but I loved it."

Meanwhile, UC Davis needed someone to fill in for a professor who was retiring. "My teaching career naturally combined with my working with athletes and it just all evolved. Next," she says, "I was also doing a lot of TV and radio work as a nutrition expert, and then I started getting requests to write books.

"I was very fortunate," Liz continues, "and I have had many wonderful opportunities." The university then formalized what she was doing into a joint position—teaching in the Department of Nutrition and holding the position of Director of Sports Nutrition in the Department of Intercollegiate Athletics. She was also the sports nutritionist for the Sacramento Kings and the Oakland Raiders sports teams. Liz feels that she was contacted to hold the nutritionist positions because of her public appearances and writings, and the well-respected reputation of UC Davis. Although she is often asked to endorse products, Liz says, "I don't do that, but I certainly can discuss the positive benefits of healthy eating and performance."

Liz now enjoys teaching and putting science in simple terms, a skill her students certainly appreciate. She currently teaches 600 students a quarter at UC Davis and has received the Excellence in Teaching Award. Her biggest advice to students regarding education is to get a very sound undergraduate degree. "If you want to be a sports nutritionist," she says, "get a master's degree with the registered dietitian qualification or a doctoral degree, because then you will have a university affiliation."

Liz warns students, "The mistake I made while I was in college was not respecting English classes as valuable tools of communication." She focused primarily on science courses. Giving a speech was difficult for Liz back then. "I could not bear the fact of standing in front of somebody," she says. "Writing was also challenging, but a professor forced me to go to the library, read up on a subject, and write about it until I improved.

"If I were a student again," Liz continues, "I would go back and take more English and journalism courses. Students should definitely take advantage of their school time, especially in any courses dealing with communication—oral or written." Liz should know; she is now frequently the keynote speaker at scientific, industry, and athletic meetings; is the author of six nutrition-related books (including a textbook) and over 300 articles for national magazines; and has been a guest on more than 200 international, national, and local radio and television shows including *Good Morning America*, CNN, and ESPN.

individual's undergraduate degree (it does not have to be food or nutrition), and whether or not the student stayed in the same school for his/her degrees (BS, MS, PhD). Students apply to both the graduate school of a university and their desired department through the chair of the graduate program. Many schools, but not all, require that the student seek out and be accepted by a major professor (with or without funding) prior to acceptance into the department. If accepted into a department, graduate students may apply for teaching or research assistantships (TA or RA, respectively). If available, these jobs are 20-hour-a-week positions assisting faculty in either teaching or research. They are recommended for students interested in becoming professors and/or those needing financial assistance. If the school enrollment deadline is missed or the student is not sure what major to select, he/she can enroll as an "unclassified" graduate student in some, but not all, universities.

## Examination Requirement

Graduate and professional schools often require that individuals interested in applying take the Graduate Record Examination (GRE) (www.ets.org). The biggest mistake that undergraduate students make is not planning far enough ahead so they can take the exam in time for the results to be submitted before the application deadline for their chosen post-baccalaureate program. Professional schools all have their own individual exam requirements, such as the Medical College Admission Test (MCAT), for example. Foreign students will need to submit their **Test of English as a Foreign Language (TOEFL)** scores (www.ets.org).

## Graduate Degree Jobs

### Researcher

Many graduate students think they will be researchers upon graduation because that is their main focus in graduate school. The larger food companies and certain government agencies do hire food and nutrition researchers (usually PhDs); however, most university research is

**FIGURE 30-8**   Professors usually obtain doctoral degrees to teach at colleges and universities.

Comstock Select/Corbis

primarily conducted by professors. Research positions at universities that are not affiliated with a teaching position are often supported by "soft" money. This means that grants are obtained to keep the position open. Grant funding success rates often range from 15% to 30%. Funding agencies often provide the percent of grant applications funded.

Summer internship programs in biomedical research (clinical research) exist for students at the National Institutes of Health, the number-one federal government agency funding health-related research (see www.training.nih.gov/student). Students interested in becoming researchers may want to determine which researcher they would like to work under before applying to graduate schools. This advice also holds true for students who wish to become professors; however, many other factors are important when choosing a graduate school.

### Professor

To become a professor, many (but not all) individuals complete post-doctorates to gain more experience before applying for teaching and/or research positions at colleges and universities. Most 4-year universities and colleges hire individuals with doctoral degrees for **tenure track** positions. University instructor or part-time positions and professor positions at community colleges sometimes accept those with master's degrees, although the trend is toward doctoral degrees. Professors

focus on teaching, research, and service for their advancement (Figure 30-8). In terms of research, they must be skilled at grant writing, procuring funding, and conducting and publishing research. The number of papers published in peer-reviewed journals (top-tiered journals are preferred), student-evaluation scores, and community service (speeches, committee work, etc.) are also important when a person is reviewed for advancement.

It is highly recommended that the new faculty member focus on a particular research subject (preferably, but not necessarily, their graduate research topic) to create a series of publications that builds their expertise and support

**TOEFL**   The Test of English as a Foreign Language™ measures the ability of non-native speakers to understand English as it is spoken, written, and heard in colleges and universities.

**Tenure track**   The process in which new faculty members are hired as assistant professors and are on probation for approximately 5 years. If their yearly contract is renewed for 5 years in a row based on satisfactory performance, they may apply for tenure (permanent hire) and promotion to associate professor. The next and final step is promotion to full professor.

for future funding and tenure. New faculty members are encouraged to apply for tenure-track positions and to find a mentor. Established researchers may allow them to join in on research projects in order to add their names to grants and research articles (as a coinvestigator and coauthor). The mentor could also teach the new faculty member to understand university politics, a learned skill whose importance should never be underestimated.

---

### ? How & Why?

**How do I find research articles for my papers?**

Research articles can be found online through two major indexing websites in the field of food and nutrition: (1) Medline with free access to abstracts at www.pubmed.gov and (2) Food Science and Technology Abstracts® (FSTA), which is the world's largest database of information on food science, food technology, and nutrition and is available through subscription, usually paid for by certain university libraries. Both undergraduate and graduate students should be familiar with both resources for writing their research papers.

---

## Professional Schools

An undergraduate degree including the proper prerequisites can be used to apply for professional schools—medical, dental, veterinary, physical therapy, pharmacy, psychotherapy, and others. Students must complete the recommended prerequisites for these schools; these are often best obtained from an advisor specializing in this field. Premedical and other professional student clubs where students can obtain knowledge and guidance usually exist on campuses.

Naturopathic schools are also available, but students should verify that the school is accredited (www.aanmc.org/the-schools.php), as there are many "schools" that are not. Note that only certain states license naturopathic practitioners; the American Naturopathic Medical Association (www.anma.org) provides more information.

## Pharmaceutical and Nutraceutical Industries

Individuals with food and nutrition degrees can also work for pharmaceutical or nutraceutical companies that produce and sell drugs, food products, or dietary supplements (vitamins, minerals, herbs, etc.). Students can potentially create connections with people in these industries through:

- Internships
- Networking at industry conferences
  - Natural Products Association (formerly NNFA) via www.naturalproductsassoc.org
  - Expo East & West (naturalproductexpo.com); three national conferences occur each year—the largest is in the West (spring); there is also one in the East (fall) and one held internationally
- Reading top trade journals
  - *Nutraceuticals World* (www.nutraceuticalsworld.com)
  - *Nutrition Industry Executive*
  - *Nutrition Today*
- Being active on industry trade websites
  - www.Newhope.com
  - www.NPICenter.com
  - www.Crnusa.org
  - www.Navigator.com

# PICTORIAL SUMMARY / 30: Careers in Food and Nutrition

## THREE MAJOR FOOD AND NUTRITION AREAS

There are three major categories for food/nutrition careers: nutrition science and dietetics, food science, and food service.

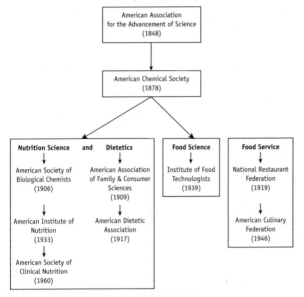

American Association for the Advancement of Science (1848)

American Chemical Society (1878)

**Nutrition Science** and **Dietetics**

American Society of Biological Chemists (1906) → American Association of Family & Consumer Sciences (1909)

American Institute of Nutrition (1933) → American Dietetic Association (1917)

American Society of Clinical Nutrition (1960)

**Food Science**
Institute of Food Technologists (1939)

**Food Service**
National Restaurant Federation (1919)

American Culinary Federation (1946)

## NUTRITION SCIENCE AND DIETETICS

Nutrition science focuses on what happens to food in the body. This science looks at how nutrients break down, get absorbed, enter cells, react within a living system (biochemistry), and affect health through deficiencies and excesses, and how medical nutrition therapy can be used to treat certain diet-related health conditions.

### Jobs in Nutrition Science and Dietetics

- **Graduate or Professional School.** The nutrition science emphasis prepares students for graduate school or professional school in the areas of nutrition, food science, medicine, dentistry, veterinary, physical therapy, pharmacy, physician's assistant, chiropractor, naturopath, and other health-related professions.
- **Nutritionist.** The word *nutritionist* is a generic term that has no professional definition.
- **Registered Dietitians (dieticians).** Dietitians counsel people about diet; they reduce dietary risk factors for various diseases and treat illnesses by recommending dietary modifications through medical nutrition therapy. Types of dietitians include clinical, administrative, community health, research, education, and private practice.

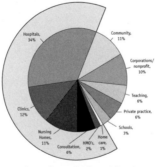

- **Dietetic Technician, Registered (DTR)** assists registered dietitians.
- **Dietary Clerk or Aide** jobs are entry-level positions usually located in a hospital dietary department.

## FOOD SCIENCE

Food science (food technology) deals with food outside of the body. Specifically, food science determines the physical and chemical constituents of food and how these substances change during product development, production, processing, preservation, packaging, distribution, preparation, and storage.

### Jobs in Food Science

Jobs in food science include, but are not limited to, flavor chemist, lab technician, plant manager, ingredient manager, marketing and sales, upper management (president, CEO, general manager), and university professor or extension specialist.

- **Food Scientist.** The majority of food scientists (food technologists) work in the food industry for corporations in their processing, ingredient, and/or manufacturing plants. Others work for the government (USDA, FDA), universities or colleges, or international agencies (FAO, WHO).
- **Food Science Technicians** assist food scientists in food development, quality control, laboratory research, food safety, and other aspects of food production.

## FOOD SERVICE

A wide selection of careers is available in the food service field, which deals with two basic areas: (1) management (restaurant management, lodging management, travel and tourism) and (2) culinary-based for the food preparation aspect of food service. Culinary institutes, cooking schools, and apprenticeships are three ways to obtain professional culinary training.

### Jobs in Food Service

- **Chef (administrative and various food service stations)**
- **Food Service Manager**
- **Dietary Manager**
- **Sales and Marketing**

## GRADUATE SCHOOL

In general, a master's degree in nutrition takes an average of 2 years to complete, whereas a doctorate usually takes 3 to 5 years.

### Jobs after Graduate School

- **Researcher.** Larger food companies and certain government agencies hire food and nutrition researchers (usually PhDs); however, most university research is primarily conducted by professors.
- **Professor.** Professors focus on teaching, research, and service for their advancement.
- **Professional Schools.** Students may also pursue professional degrees—medical, dental, veterinary, physical therapy, etc.
- **Pharmaceutical and Nutraceutical Industry jobs.** A wide range of positions exists in these two major fields that can sometimes be filled with graduates of food and nutrition degrees.

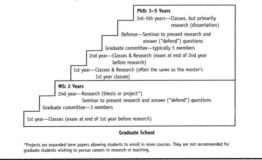

**PhD: 3–5 Years**
3rd–5th years—Classes, but primarily research (dissertation)
Defense—Seminar to present research and answer ("defend") questions
Graduate committee—typically 5 members
2nd year—Classes & Research (exam at end of 2nd year before research)
1st year—Classes & Research (often the same as the master's 1st year classes)

**MS: 2 Years**
2nd year—Research (thesis or project*)
Seminar to present research and answer ("defend") questions
Graduate committee—3 members
1st year—Classes (exam at end of 1st year before research)

**Graduate School**

*Projects are expanded term papers allowing students to enroll in more courses. They are not recommended for graduate students wishing to pursue careers in research or teaching.

# CHAPTER REVIEW AND EXAM PREP

## Multiple Choice*

1. The majority of registered dietitians are employed within a _____ setting.
   a. school
   b. private practice
   c. community
   d. clinical

2. One of the major professional associations established for food scientists is the _____.
   a. American Dietetic Association
   b. Institute of Food Technologists
   c. American Culinary Federation
   d. Foodservice Professionals Network

3. The job of ensuring that food and beverages meet quality standards at each processing step is the responsibility of the _____.
   a. plant manager
   b. quality control department
   c. sensory evaluation specialist
   d. lab technician

4. A person who specializes in counseling people about diabetes is called a _____.
   a. registered dietitian
   b. certified diabetes educator
   c. food counselor
   d. community nutrition specialist

5. The research paper completed by a doctoral candidate is called a _____.
   a. dissertation
   b. thesis
   c. research project
   d. defense

6. One of the major professional associations established for food service positions is the _____.
   a. American Dietetic Association
   b. Institute of Food Technologists
   c. American Culinary Federation
   d. Foodservice Professionals Network

7. Culinary training is available through
   a. the American Dietetic Association.
   b. certifications.
   c. culinary institutes, cooking schools, and apprenticeships.
   d. the Institute of Food Technologists.

## Short Answer/Essay

1. Define the three major career areas in food and nutrition.
2. How would a student pursue a career in dietetics?
3. How would a student pursue a career path in food science?
4. What are the general steps for pursuing a graduate degree?
5. List some of the jobs available to food science graduates.
6. List and briefly define the different types of dietitians.
7. List and briefly define the different types of culinary certification.
8. Explain some of the specializations available to dietitians.
9. What is a nutritionist?
10. How does the academic preparation of a dietetic technician differ from that of a registered dietitian?

*See p. AK-1 for answers to multiple choice questions.

# REFERENCES

1. Aasse S. Make the most of your professional development portfolio. *Journal of the American Dietetic Association* 109(7):1152–1154, 2009.

2. American Dietetic Association. *Registration eligibility requirements for dietetic technicians.* www.cdrnet.org/certifications/rddtr/pathwaysdtr.htm. Accessed 09/09/09.

3. Anonymous. Professional liability insurance FAQ. *Journal of the American Dietetic Association* 109(6):981, 2009.

4. Brummitt PS. The importance of careful documentation. *Today's Dietitian* 10(1):22–24, 2008.

5. Bueche J, et al. Nutrition Care Process and Model Part I. The 2008 update. *Journal of the American Dietetic Association* 108(7): 1113–1117, 2008.

6. Chipps RL. Chef and RD partnerships: Recipe for success. *Today's Dietitian* (January): 30–31, 2008.

7. Hager MH. Clinical privileging for registered dietitians: A regulatory perspective. *Journal of the American Dietetic Association* 107(4):558–560, 2007.

8. Hartel RW, and CP Klawitter. *Careers in Food Science: From Undergraduate to Professional.* Springer Science and Business Media, New York, 2008.

9. Huy J, and CJ Pauli. Opportunities with US Public Health Service. *Journal of the American Dietetic Association* 109(5):809, 2009.

10. Lipscomb R. ADA member benefits update. *Journal of the American Dietetic Association* 109(4):596–599, 2009.

11. Lipscomb R. Health coaching: A new opportunity for dietetics professionals. *Journal of the American Dietetic Association* 106(6):801–804, 2006.

12. Mathieu J. RDs in the military. *Journal of the American Dietetic Association* 108(12):1984–1987, 2008.

13. Mermelstein NH. 2007 IFT membership employment and salary survey. *Food Technology* 68(2):42–60, 2008.

14. Puckett RP, et al. American Dietetic Association Standards of Practice and Standards of Professional Performance for registered dietitians (generalist and advanced) in management of food and nutrition systems. *Journal of the American Dietetic Association* 109(3):540–543, 2009.

15. Rogers D. Dietetic salaries on the rise. *Journal of the American Dietetic Association* 106:296–305, 2006a.

16. Rogers D, and JA Fish. Distinguishing entry-level RD and DTR practice: Results from the 2005 Commission on Dietetic Registration Entry-Level Dietetics Practice audit. *Journal of the American Dietetic Association* 106(6):965–972, 2006b.

17. Rogers D, and JA Fish. Entry-level dietetics practice today: Results from the 2005 Commission on Dietetic Registration Entry-Level Dietetics Practice audit. *Journal of the American Dietetic Association* 106(6):957–964, 2006c.

18. Shaddix CK, and JA Bell-Wilson. Finding your niche: Certification options for the RD. *Today's Dietitian* 9(3):42–46, 2007.

19. Steinmuller PL, et al. American Dietetic Association Standards of Practice and Standards of Professional Performance for registered dietitians (generalist, specialty, advanced) in sports dietetics. *Journal of the American Dietetic Association* 109(3):544–552, 2009.

# WEBSITES

Colleges and Universities

This is a virtual library of 45,000+ college catalogs (cover-to-cover) for 2-year, 4-year, graduate, professional, and international schools. Included at this nonprofit website is a list of available majors, virtual tours, website links, and campus maps.

**www.collegesource.org**

This website, targeted toward the food industry audience, provides a career center, university links (those offering food degrees plus a full list of U.S. universities), organizations (associations, government, institute), resources (publications, industry links), a news center to stay updated, and directories on food ingredients, equipment, services, and companies.

**www.foodinfonet.com**

Other useful websites include:
Guide to colleges
**www.petersons.com**

Virtual campus tours
**www.campustours.com**

Financial aid
**www.finaid.org and www.fastweb.com**

Culinary Scholarships
**www.iacp.com**
**www.nraef.org/scholarships**
**www.hsmai.org**
**www.aiwf.org**
**www.starchefs.com**
**www.jamesbeard.org**

Job search engines
**www.adacareerlink.org or www.ift.org**
**www.careerbuilder.com**
**www.monster.com**
**www.hotjobs.com**

Salaries
**www.salary.com**
**www.payscale.com**

Jobs in Dietetics
**www.eatright.org**
(Enter "ADACareerLink" in the search box. Available to ADA members, but student memberships are also available.)

Jobs in Food Science
**www.ift.org**

(Click on "Employment" in left bar; under "Job Seekers," click on "View all Jobs")

U.S. Federal Jobs
**www.usajobs.gov**

U.S. Department of Labor
**www.bls.gov**

U.S. Public Health Service internship opportunities through Commissioned Officer Student Training and Extern Program (COSTEP). Find paid experiences at CDC, FDA, USDA, Indian Health Service, NIH, EPA, and other federal offices.

**www.usphs.gov/student/**

Military Careers available at:
**www.todaysmilitary.com/careers/job-listings/dietitians**
**www.goarmy.com/amedd/m_spec/corps_careers.jsp**
**www.navy.com/careers/healthcare/medicalservicecorps/clinicalcareproviders/dietetics/**
**www.airforce.com/careers/job.php?catg_id=3&sub_id=4&af_job_id=220**

# Appendixes

# Appendix A: Food Preparation Equipment

Appendix A provides a brief overview of food preparation equipment—primary equipment, auxiliary equipment, and utensils.

# PRIMARY EQUIPMENT

Primary equipment consists of ranges, ovens (conventional, convection, and microwave), refrigerators, and usually dishwashers.

## Ranges

Ranges can have open or flat top surfaces with electrical or gas burners (Figure A-1).

## Ovens

The conventional oven is located below the range, but it can also be a separate unit (Figure A-2).

**FIGURE A-1**   Ranges.

Open top

Flat top

Primarily used for baking and roasting, it is also used for braising, poaching, and simmering.

Ovens rely on hot air for heating food, primarily by convection, but conduction and radiation can also occur.

Baked foods rely on freely moving currents for the transfer of heat, so it is important to ensure that baking pans are placed on the racks in such a way as to allow the efficient flow of air currents. Figure A-2 shows some of the types of

**FIGURE A-2**   Four types of ovens.

Conventional oven                Double-deck convection oven

Convection oven

Stack (or deck) oven (typically installed one on top of another)

The Vulcan-Hart Co.

ovens that are available to food service establishments. They include:

- **Stack or deck oven.** Each component of the stack has a separate thermostat.
- **Convection oven.** Hot air is circulated by a fan, baking contents more quickly.
- **Revolving or carousel oven.** Trays rotate like a Ferris wheel, ensuring an even temperature.
- **Impingement oven.** Hot-air jets heat food more quickly (5).
- **Infrared oven.** Heat is generated by a very hot infrared bulb.
- Brick-lined or hearth oven.
- **Pizza oven.** Reaches very hot temperatures.
- Microwave oven.

## Refrigerators/Freezers

The proper refrigeration and freezing of foods is one of the most important factors in preventing foodborne illness. Refrigerator temperatures should be maintained at or below 40°F (4°C), and freezers or freezer compartments at 0°F (−18°C). Household refrigerators are classified by the location of the freezer—above, below, or beside the refrigerator section (Figure A-3). Food service establishments usually have a walk-in refrigerator and freezer, which may range in size from a small closet to a large room. Other types of freezers/refrigerators include reach-ins, roll-ins, and pass-throughs (Figure A-4).

**FIGURE A-3**   Refrigerator-freezers.

One-door, freezer compartment inside

Freezer below food compartment

Freezer above food compartment

Side-by-side refrigerator/freezer

**FIGURE A-4**   Food service refrigerators and freezers.

Reach-in                    Walk-in

Roll-in

Courtesy of Victory Refrigeration, part of AFE

## Dishwashers

The two categories of dishwashers are household dishwashers and commercial dishwashers.

The three basic types of household dishwashers are:

*Built-in dishwasher.*—Integrated under the counter to match the cabinets.

*Portable dishwasher.*—Used as soon as the hoses are attached to the kitchen faucet; one hose drains into the sink.

*Convertible dishwasher.*—Used as a portable or installed permanently as a built-in.

Commercial or food service dishwashers are so large that they often require a separate room.

## Equipment Standards and Safety

The National Sanitation Foundation (NSF) seal of approval assures buyers of food service equipment that certain standards of sanitation and safety have been met in its design and production. This nonprofit organization is interested in the promotion of public health and has established minimum standards of construction for food service equipment (3). Information about equipment or approved manufacturers can be obtained by writing to NSF Testing Laboratory, Inc., PO Box 130140, Ann Arbor, MI 48113 (or e-mailing info@nsf.org). Another private organization overseeing the safety of electrical equipment is the Under-writers Laboratory (UL), which ensures that an electrical appliance, cord, or plug has passed certain tests for electrical shock, fire, and other related injuries (6).

# AUXILIARY EQUIPMENT

In addition to the primary equipment in a kitchen, auxiliary equipment includes fryers, broilers, steamers, grills, cutting equipment, mixers, and coffee/tea makers.

## Griddles

Griddles supplement range units. Their larger, flat, smooth surfaces are ideal for preparing eggs, hamburgers, pancakes, French toast, and hash browns (Figure A-5). Food service griddles contain a drip cup to collect draining fat. Preparation is easier when grill surfaces are primed by smearing them with oil followed by a brief heating. To maintain the primed surface, griddles are never washed with soap and water, but scraped clean, wiped with a grease mop, and then polished with a soft cloth.

## Tilting Skillets

Found only in large food service operations, the tilting skillet, brazier, or fry pan can be used to make anything from chili to poached eggs. The wide range of temperature settings stretches from low braising to high frying heats. As a result, it can be used as a fry pan, brazier, griddle, stockpot, steamer, or steam table. The entire skillet can be tilted to pour out liquid-based contents (Figure A-6).

**FIGURE A-5**   Griddle.

**FIGURE A-6**   Tilting skillet.

## Broilers and Grills

The difference between broiling and grilling is the heat source: the broiler's heat is above the food whereas the grill's is below the food. Temperature control is achieved by moving the grid up or down. Heat for these may be provided by wood, charcoal, electricity, or gas.

## Steamers

Two basic types of steamers are used in food service organizations: cabinet or compartment steamers and steam-jacketed kettles (Figure A-7).

*Cabinet steamers.*—Stacked one above the other with the door of each sealed tight with clamps.
*Steam-jacketed kettles.*—Used more for fluid-type foods such as soups and stews. Range in size from 1 quart to 200 gallons. The steam is not generated inside the kettle, but is circulated between the double-layered metal plates of the kettle's outer shell. A handle is used to tilt the entire steam-jacketed kettle to pour out the food.

Steam heats foods by moist heat. Most steamers in a food service establishment steam under pressure, which is measured by a gauge in pounds per square inch (psi). Pressure steamers allow food to heat to temperatures higher than boiling, which decreases cooking time. Vegetables can be cooked at pressures of 10 to 15 psi, reaching a temperature of 250°F (120°C). Lower pressures and temperatures (5 to 10 psi, 225°F/105°C) are used to cook meats, which would fall apart at the higher pressures required for vegetables. Cooked vegetables and even

**FIGURE A-7**   Food service steamers.

The Groen Co./Unified Brands

Steam-jacketed kettle

Cabinet steamer

some fish maintain their texture, color, taste, and nutrients best when they are properly steamed or microwaved. Rice, pasta, poultry, eggs, fish, and shellfish may also be steamed, but the flavor of meats and poultry will usually be diminished by the process.

Safety is particularly important with steamers. They should never be run without water, and they should be periodically checked to ensure that safety valves are working. They should never be opened until the pressure has gone down, and then should always be opened away from the face.

## Deep Fryers

Breaded fish and vegetables, fried chicken, and french fries are some of the foods commonly prepared in deep fryers. Frying is similar to boiling, except that in frying the liquid is fat, which can reach higher temperatures than water. Food is loosely placed in a wire basket, which is then submerged in heated oil. When the food floats to the top of the oil, it can be considered cooked. The basket is then removed and set aside so that the oil can drain from the food. The fryers themselves may be small enough to be portable or so large that they are floor mounted (Figure A-8). Most deep fryers have automatic heat controls.

## Woks

This large bowl-shaped pan is central to Chinese cooking. It comes equipped with (1) a metal ring to fit over a range burner (Figure A-9), or (2) self-contained with an electrical cord. The most time-consuming step in using a wok is cutting the

**FIGURE A-8**   Deep fryer.

© Frymaster, LLC, 8700 Line Ave, Shreveport, LA 71106.

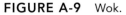
**FIGURE A-9**   Wok.

foods into many small, uniform pieces. The actual cooking of the foods is a quick process, lasting approximately 5 to 10 minutes. It starts with high heat under the wok, which has been lightly coated with oil (usually sesame or peanut oil). The foods that take the longest to cook are added first. The food is stirred rapidly for a few minutes, for even cooking, and then the heat is lowered and the pan covered so the steam thus generated can complete the process.

## Crockery

Crockery, or electric slow cookers, have been popular for some 40 years and are particularly good for moist-heat cooking of meat and legumes. Crockery cooking is long and slow, with controlled heat that needs little or no supervision. A meal can be started in the morning that will be ready to eat by dinnertime. Because there is some evidence that crockery may not keep food sufficiently hot for the entire duration of cooking, its use has lately been discouraged by some food experts because of the risk of foodborne illness.

## Cutting Equipment

Meat slicers, food choppers, and grinders are common pieces of equipment in food service establishments (Figure A-10).

### Meat Slicer

Carelessness in the use of a meat slicer causes more food service accidents than any other kind of equipment. The following safety tips apply: The machine should always be unplugged when not in use. After the slicer is plugged in, the blade control is adjusted for the desired slicing thickness, and the blade guard positioned. The food, usually boneless meats, but possibly cheese, vegetables, fruits, and even bread, is then placed on the carriage and held there firmly with the guard before the switch is turned on. The carriage is moved back and forth by its handle in a smooth motion.

The equipment should be thoroughly sanitized after use and between different types of foods, especially with raw meats. The cord must be removed from the socket and the blade control set at zero before cleaning. Metal utensils should never be used to scrape food from the blade because they may nick the slicer. Manufacturer's instructions should be followed in removing the various parts and subjecting them to the sanitizing solution and to rinsing and drying. The blade guard should be replaced immediately to prevent any risk of cuts. The use of protective gloves through the whole cleaning process is highly recommended.

**FIGURE A-10**   Cutting equipment.

Slicer

Food chopper

Grinder

2007 Hobart Corporation

### Food Chopper or Cutter

Another potentially dangerous piece of food service equipment is the food chopper or cutter. The key to preventing injuries here is to turn the machine off, allow the knife blades to come to a rest, and flip the safety catch on before removing the food with a bowl scraper. The hands should never go into the bowl. The guard can be raised to remove any remaining food. Meats with bones or gristle should not be processed with food choppers because they will damage the gears and knives.

# Mixers

Mixers are convenient for controlling the rate at which ingredients are combined. They are used to prepare whipped cream, beaten egg whites, and mashed potatoes. In the food service industry, models range in size from tabletop to floor size.

Attachments vary from a paddle for general mixing, to whips for cream or eggs, to dough arms for kneading yeast dough (Figure A-11). Additional attachments may be added, including a shredding, grating, or slicing attachment and a grinder for meats and other foods. Some home mixers have similar attachments.

**FIGURE A-11**   Tabletop mixer and three typical attachments: (a) wire whip—incorporates air, (b) flat beater—general mixing, and (c) dough hook—mixing heavy doughs.

a

b   c

Photo courtesy of KitchenAid® Home Appliances

For safety's sake, attachments must be securely in place before the machine is turned on, and it is best to disconnect the power entirely before removing them. Spoons or hands in the bowl during mixing are not recommended, but rubber scrapers can be used occasionally to scrape down the sides of the mixing bowl.

### Blenders and Food Processors

Blenders and food processors allow further refinements to mixing food (Figure A-12). In blenders, the blades or mixing component is on the bottom. They are used for everything from making milk shakes to blending the vegetables used in making gazpacho, a Spanish cold soup. Food processors are more versatile and allow cutting, chopping, grinding, slicing, and shredding foods, and even kneading dough. They come with specialized blades for accomplishing all these tasks, and many even come with a juicing attachment.

**FIGURE A-12**   Blenders and food processors.

Blender          Food processor

Photos courtesy of KitchenAid® Home Appliances

**FIGURE A-13**  Coffee makers and tea dispensers.

Electric urn

Iced tea dispenser

Courtesy of CecilWare Corporation

Coffee brewer and dispenser

Drip coffee maker

**FIGURE A-14**  Common pots and pans.

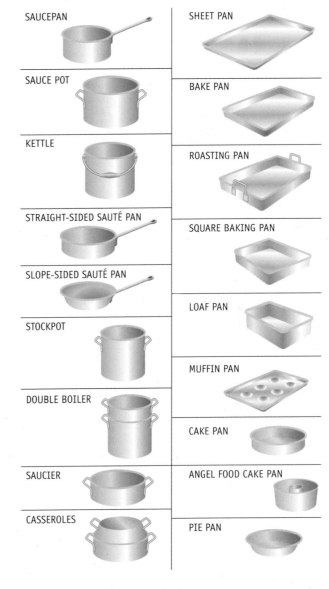

SAUCEPAN

SAUCE POT

KETTLE

STRAIGHT-SIDED SAUTÉ PAN

SLOPE-SIDED SAUTÉ PAN

STOCKPOT

DOUBLE BOILER

SAUCIER

CASSEROLES

SHEET PAN

BAKE PAN

ROASTING PAN

SQUARE BAKING PAN

LOAF PAN

MUFFIN PAN

CAKE PAN

ANGEL FOOD CAKE PAN

PIE PAN

## Coffee Makers

Many homes and food service venues use automatic coffee makers daily (Figure A-13). Food service operations serve coffee from an electric urn or automatic coffee brewer. Electric urns are connected to a hot water source and automatically shut off after the coffee is finished brewing. The hot water running through a coffee brewer stops after the decanter or pot is full.

## Pots and Pans

Pots and pans are distinguished from one another by their size, shape, and handle (Figure A-14). Another defining feature of a pot, pan, or kettle is their capacity defined by (1) the number of quarts, or (2) inches (baking pans and skillets are described in inches; e.g., 8- or 9-inch cake pan) (6).

Pots have two handles and are used when preparing large quantities. Saucepans and frying pans have a single long handle and less capacity. Saucepans are usually straight-sided, whereas frying pans can be either straight- or slope-sided. The flattest pans are those used for baking and roasting. The sturdiest, and thickest, pans are used for roasting. A double boiler consists of a bottom pan in which water is heated, and a top pan containing a food item that must be kept below the boiling point. These are used for preparing certain sauces and to keep food hot without burning.

## Pot and Pan Materials

Pots and pans are made from a variety of materials (Figure A-15):

- Aluminum, copper, and stainless steel
- Nonstick coatings
- Cast iron
- Glass and glass/ceramic combinations

## Pan Shapes

Cakes come in a variety shapes because the pans can be round, square, or oblong. There are special pans for making sponge and angel food cakes; the angel food cake pan usually has a tubular segment in the middle that separates from the sides for easy removal of the cake. Spring-form pans allow one to "spring" open the sides for easy removal of the cake. Cookie sheets, sometimes referred to as baking sheets, have no sides (except one or two that are raised for handling), allowing the hot air to flow evenly over the cookies. Heavy-duty sheet pans with four sides can also be used for preparing cookies and myriad other foods such as

## FIGURE A-15    Materials that make the pot.

**Stainless steel:** Poor heat conductor, which is why they are often bottom-coated with copper.

**Nonstick coatings:** Tolerate high heats and eliminate sticking that can occur with stainless steel.

**Aluminum:** Excellent heat conductor and lightweight, but it reacts with foods that are acidic, alkaline, or sulfurous.

**Anodized aluminum:** Surface is electrochemically sealed to make it nonreactive.

**Cast iron:** Superb at retaining heat, but slow to heat or cool; needs to be completely dried and primed.

**Enameled cast iron:** Benefits of cast iron without the maintenance problems; however, enamel coating may chip with abrasion and wear.

Digital Works

biscuits, bread, pizza, breadcrumbs, roasted nuts, and even some meats. Full-sheet pans are used in restaurants for bulk baking, whereas half-sheets (half the size of full sheets) are reserved for home use (4). Some of the half-sheets purchased at supermarkets may warp at temperatures over 300°F (149°C) but not pans made of heavy-duty aluminum or steel. The aluminum pans tend to be more popular because their lighter color reflects heat, which helps to prevent overbrowning and baking (4). Also, they do not rust. The darker the pan, as seen in steel pans or those coated with a nonstick surface, the darker the cookies.

### Pan Colors

One advantage of the darker sheet pans is that they absorb heat, resulting in a crisper crust for pizza and fruit pies. Regardless of the color, professional bakers use kitchen parchment on baking sheets to prevent sticking, to move items around with ease, and to protect against burning (4).

### Best Heat Conductors

Aluminum, copper, and combinations of copper and stainless steel are the best conductors of heat (6). Aluminum accounts for more than half the cookware sold in the United States, but it is very lightweight and prone to denting. Aluminum may also react chemically with many foods, particularly those high in acid, and it is not recommended for storing foods. Copper is an excellent heat conductor, but it is costly and requires special care. A further disadvantage is that excessive copper may dissolve into the food being prepared, causing nausea and vomiting. Therefore, copper pans are usually lined with stainless steel or tin. Stainless steel is known for its durability and easy cleaning; however, it is a poor conductor of heat and tends to generate hot spots, which may scorch the food. To keep this from happening, the bottoms of many stainless steel pans are coated with copper or aluminum.

### Nonstick Pans

Nonstick pans reduce the amount of fat needed to prevent sticking, but their surfaces are easily scratched, so plastic, rubber, or wooden utensils are recommended.

### Cast Iron Pans

Cast iron pots and pans are heavy, heat slowly, rust easily, and are difficult to clean. They do, however, retain high temperatures for longer periods of time, heat evenly, and add extra iron to the diet. Acidic foods such as tomato sauces tend to absorb more iron: 5 mg of iron are absorbed for every 3 ounces of spaghetti sauce cooked in a cast iron pan.

Cast iron pots and pans may be cleaned in one of two ways. The first involves a preliminary priming or conditioning of the pan with a very thin coat of vegetable oil, after which it is heated and cooled. A primed pan is cleaned by scraping and wiping away food particles after each use. Reconditioning may be done whenever necessary. The second method is to wash the pan with soap and water, heat it to dry, and then coat it with a minute amount of oil. This second method is more likely to remove any traces of rancid fat, which can impart an off-flavor to any food subsequently prepared in the pan.

### Glass Pans

Heat-proof glass, such as Pyrex, and glass/ceramic combinations, such as Corningware, break more easily than metal-based pots and pans, but have the advantage of not reacting with foods. Most casserole pans, which are usually oval or oblong with low sides, are made of such materials. Baking temperatures should be reduced by 25°F (4°C) when using tempered glass. The newer versions of glass/ceramic materials can be moved from the range or oven to the refrigerator or freezer, and later be taken from the cold and placed directly into the oven or microwave. Glass pots and pans are not allowed in food service operations, however, because of possible breakage and liability problems.

# Utensils

Utensils are vital items needed for cutting, stirring, turning, measuring, and serving food. The utensils covered in this section include knives and utensils used in preparation, measuring, and serving.

## Knife Utensils

Knives are to the chef as brushes are to the artist. Some people consider them to be the most important tools in food preparation. Knowing the different kinds of knives, their particular tasks, and how to care for them is crucial to the preparation of foods. The food to be cut determines what type of knife should be used. The basic knife starter set consists of chef's, paring, slicing, boning, and utility knives. The first three types of knives often complete the set of many home kitchens. Common knives used in food preparation are shown in Figure A-16.

***Chef's Knife***   The chef's, or French, knife is one of the largest and serves as an all-purpose knife for cutting meats and for mincing, dicing, and chopping a variety of foods. Chef's knives are commonly available in blade lengths of 6, 8, and 12 inches, and the side of their blades can be used to crush garlic cloves, ginger slices, and peppercorns (2).

**FIGURE A-16**   Common knives used in food preparation.

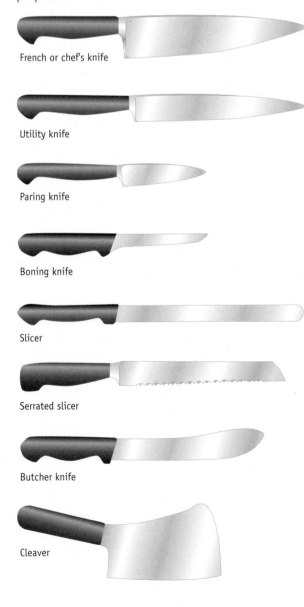

French or chef's knife

Utility knife

Paring knife

Boning knife

Slicer

Serrated slicer

Butcher knife

Cleaver

***Utility Knife***   The utility knife is geared toward lighter duties such as cutting tomatoes or carving meat.

***Paring Knife***   The smaller, shorter, 2- to 4-inch paring knife is used for more delicate jobs that require close control, such as the trimming of vegetables, fruits, and small pieces of meat like chicken breasts.

***Boning Knife***   The slightly curved boning knife is handy for separating meat from bone (e.g., deboning the breast of a chicken), disjointing poultry, cutting between the joints of larger pieces of meat, and dicing raw meats.

***Slicing Knife***   Slicing knives are long and flexible enough to portion off thin slices of meat or poultry. Serrated slicers are useful for cutting bread or angel food cake.

***Steak or Scimitar Knife***   These knives are used for cutting steaks from the appropriate parts of a carcass.

***Butcher Knife***   A variety of butcher knives are available for cutting raw meats.

***Heavy Cleavers***   Additional knives that are found in food service arenas include heavy cleavers for cutting through bone.

***Oyster and Clam Knives***   Oyster and clam knives are used for opening these shellfish.

## Purchasing Knives

Knives can range in price from a few to several hundred dollars. When selecting a knife, qualities to consider include size, weight, balance, the length of the tang, and the materials from which the blade and handle are made. Although the size selected will be determined by the use for which the knife is intended, the other factors depend on more qualitative assessments.

***Weight and Balance***   Sometimes the "balance" or the feel of the knife in the hand is a factor in selection. Some knives are blade-heavy, others handle-heavy, and some feel evenly divided between the two. A person should select the knife that feels "right" in his or her hand (2).

***Tang***   Another quality that varies among knives is the length of the tang, the part of the metal blade that extends into the handle. Better-quality knives have a tang that extends the full length of the handle.

***Blade***   Probably the most important factor in selecting a knife is the type of steel used for the blade: carbon, stainless steel, or high-carbon stainless steel. Carbon blades are almost obsolete because they are highly susceptible to rust and lose their edge quickly. Stainless steel, on the other hand, is rust-resistant but is difficult to sharpen and to keep sharp. This steel is actually a combination of metals, including chromium, which is added for its resistance to stains, corrosion, and heat. Despite its name, stainless steel is not stain-proof, but it does stain less than knives not made from stainless steel when it comes into contact with food and beverages, especially salad dressing, vinegar, salt, mustard, tea, and coffee.

High-carbon stainless-steel knives are usually preferred because, in addition to not staining, they keep a sharp edge and do not rust (2). High-carbon knives are further distinguished by whether they have a stamped or a forged blade. Stamped blades are cut from a sheet of steel, ground, and polished. Because they are mass produced, they are less expensive than forged knives. Forged blades are made from a single piece of steel that has been exposed to extremely high heat, submerged in a chemical bath, and set in a die before being hand-hammered into shape. The resultant blades are more costly, but they are also heavier, tougher, hold their edge longer, and require less pressure when cutting.

*Handle*　The knife's handle may be made of wood, carved bone, plastic, or metal. Wood is easier to hold, but water damage from frequent washing reduces the length of its life. Plastic or metal handles are more durable, but they are slippery. Bone-handled knives are both water- and wear-resistant and, when combined with a high-carbon stainless-steel knife, can last a lifetime or more.

## Care of Knives

Cutting knives should never go in a dishwasher. Strong detergents not only dull the blade but, when combined with hot water and air, can ruin wooden handles. Nicks can occur if the blade bumps against other metal utensils. Instead, knives should be washed immediately after use with soap or detergent, dried thoroughly, and then stored in such a way that their blades do not contact each other to prevent nicks. The blades may be kept separated by slipping them into a wooden knife block (blade turned upward) or a shallow knife block that fits inside a drawer, or by placing them along a magnetic strip.

Knives can be sharpened using one of the many electric and mechanical knife sharpeners on the market, but a better result can be achieved by hand. The two basic ways to hand-sharpen a knife are with a stone or a steel implement. The sharpening stone, also known as a whetstone, is used by rubbing a bit of moisture on the stone and sliding each side of the blade until the proper sharpness is acquired. A sharpening steel looks like a round sword and is held firmly in one hand while the knife, held in the other hand, is brandished against the steel (Figure A-17). Most chefs employ both implements,

**FIGURE A-17**　Sharpening knives using either a stone or steel.

| STONE | STEEL |
| --- | --- |
| Produces a sharp cutting edge* | Maintains a sharp cutting edge |

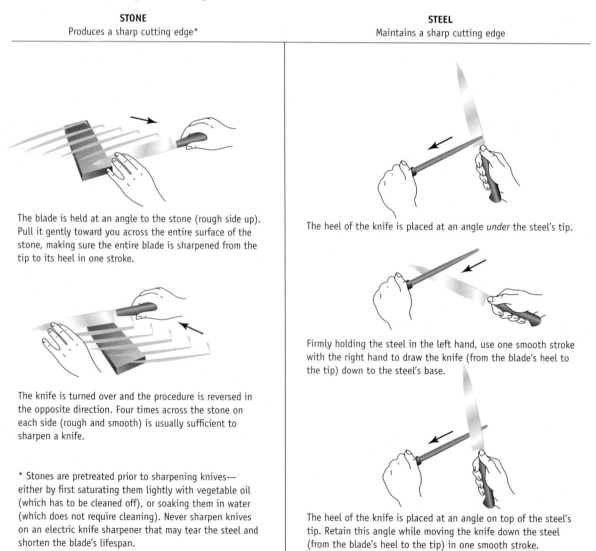

The blade is held at an angle to the stone (rough side up). Pull it gently toward you across the entire surface of the stone, making sure the entire blade is sharpened from the tip to its heel in one stroke.

The knife is turned over and the procedure is reversed in the opposite direction. Four times across the stone on each side (rough and smooth) is usually sufficient to sharpen a knife.

* Stones are pretreated prior to sharpening knives— either by first saturating them lightly with vegetable oil (which has to be cleaned off), or soaking them in water (which does not require cleaning). Never sharpen knives on an electric knife sharpener that may tear the steel and shorten the blade's lifespan.

The heel of the knife is placed at an angle *under* the steel's tip.

Firmly holding the steel in the left hand, use one smooth stroke with the right hand to draw the knife (from the blade's heel to the tip) down to the steel's base.

The heel of the knife is placed at an angle on top of the steel's tip. Retain this angle while moving the knife down the steel (from the blade's heel to the tip) in one smooth stroke.

for different purposes: one for getting an edge and the other for refining it.

## Cutting Boards

Cutting boards are used primarily for cutting meat, poultry, vegetables, and fruits, but they may also be used for kneading and rolling out dough. They may be made of wood, hard plastic, glass, or ceramic tiles; the latter three, however, are hard on knife blades (1). Cutting boards should be carefully scraped and thoroughly washed and dried every time they are used.

## Preparation Utensils

Figure A-18 shows some of the supporting utensils most commonly used in food preparation. Spoons, available in solid, slotted, or perforated versions, are used for mixing and serving. The holes in the slotted and perforated spoons allow liquids to drain. Wire whisks are used for mixing and are categorized by their shapes. Straight (French) whisks are ideal for general purposes or smooth sauces. The very thin wires of balloon whisks are designed for beating the maximum amount of air into thin liquids such as egg whites and cream. Flat whisks are fashioned for creating sauces and gravies when it is important to lift up materials from the corners of a pan (7).

Spatulas come in a variety of shapes for their many purposes. Rubber spatulas or scrapers are used to scrape bowls or to fold beaten egg whites or other ingredients into each other. The straight spatula or palette knife is used for measuring ingredients, and it is ideal for spreading icings onto cakes. The sandwich spreader, with a broader blade, is used, as the name implies, on sandwich fillings, butter, and jams. A pie server is an angled spatula used to lift pie, cake, or pizza wedges. Similar in design, but wider and with a larger bend, is the offset spatula, which is used to turn items such as hamburgers, eggs, and pancakes.

Other preparation utensils include the bench scraper, for scraping and for cutting dough; the pastry wheel, which is designed to cut pastry dough, but which can also be used to cut pizza; and the pastry brush, which is used to coat pastry with egg white or sugar glaze.

## Types of Measuring Utensils

About five different types of measuring utensils are frequently used in food preparation: liquid and dry measuring cups, measuring spoons, ladles, and scoops (Figure A-19).

**Liquid Measuring Cups**   Available in 1-cup, 2-cup (1 pint), and 4-cup (1 quart) capacities. Their volumes are divided into increments of ¼, 1/3, ½, 2/3, and ¾ cup. They are usually glass, have a pouring lip, and are all-purpose.

**Dry Measuring Cups**   Fractional, flat-topped (no pouring lip), single-volume cups (¼, 1/3, ½, and 1) are best because they can be leveled with a spatula for a more accurate result. Accuracy is also improved by using the 1-cup measure rather than four ¼ cups.

**Measuring Spoons**   Used to measure both liquid and dry ingredients requiring less than ¼ cup, they consist of 1 tablespoon, and 1,½, ¼, and occasionally 1/8 teaspoon. A tablespoon equals 3 teaspoons, and 2 tablespoons equal 1 fluid ounce.

**Ladles**   Liquids can be measured by ladles that are individually stamped with their capacity in ounces (Table A-1).

**Scoops or Dippers**   The various sizes are identified by a scoop number (Table A-2), which indicates the number of portions from a quart (e.g., a number 8 scoop yields eight ½-cup portions from 1 quart). The larger the scoop or dipper number, the smaller the serving. Measured scoops and dippers are used primarily by food service establishments for serving ice cream, mashed potatoes, and other soft foods.

**TABLE A-1**  Ladles—Approximate Measures and Their Uses

| Ladle Size | Measure | Use |
|---|---|---|
| 1 oz | 2 tbsp | Sauces, salad dressings, cream |
| 2 oz | ¼ C | Gravies, sauces |
| 3 oz | 1/3 C | Cereals, casseroles, meat sauces |
| 4 oz | ½ C | Puddings, creamed vegetables |
| 6 oz | ¾ C | Stews, creamed entrées, soup |
| 8 oz | 1 C | Soup |
| 12 oz | 1½ C | |
| 16 oz | 2 C (pt) | |
| 24 oz | 3 C | |
| 32 oz | 4 C (qt) | |

**TABLE A-2**  Scoops—Approximate Measures and Their Uses

| Scoop or Dipper Number* | Weight | Measure | Use |
|---|---|---|---|
| 6 | 6 oz | ¾ C | Soups |
| 8 | 4–5 oz | ½ C | Luncheon entrées, potatoes |
| 10 | 3–4 oz | 3/8 C | Desserts, meat patties, ice cream |
| 12 | 2½–3 oz | 1/3 C | Vegetables, desserts, puddings |
| 16 | 2–2¼ oz | ¼ C | Muffins, cottage cheese, croquettes, dessert |
| 20 | 1¾–2 oz | 3 tbs ¾ tsp | Muffins, cupcakes, meat salads |
| 24 | 1½–1¾ oz | 2 tbs 2 tsp | Cream puffs, ice cream |
| 30 | 1–1½ oz | 2 tbs ¾ tsp | Drop cookies |
| 40 | ¾ oz | 1 tbs 2¼ tsp | Whipped cream, toppings, gravy |
| 60 | ½ oz | 1 tbs | Salad dressings, toppings |
| 70 | 1/3 oz | 2¾ tsp | Cream cheese, salad dressing, jelly |
| 100 | ¼ oz | 2 tsp | Whipped butter |

*Dipper/Scoop = Servings/Quart

# FIGURE A-18 Food preparation utensils.

## Spatulas

Used to level off ingredients when measuring, remove food from flat pans, and spread frostings, butter, and other soft foods. Plastic or rubber spatulas are used for scraping bowls.

Larger offset spatulas or turners—Used to turn meat, pancakes, potatoes, and other foods while browning. The offset or bend keeps the hand away from the hot surface. Turner's blade—solid or perforated and used as a scraper to remove grease and other materials.

## Tongs

Used to turn meat while broiling, lift vegetables like corn-on-the-cob from a steamer, serve food, and serve ice cubes.

## Forks

Used to turn meat while cooking or to hold meat and other foods while being sliced.

## Whisks

**Straight whisk**—For general purposes and smooth sauces.

**Balloon whisk**—For incorporating air into egg whites and whipped cream.

**Flat whisk**—For sauces and gravies.

## Molds

**Molds** in a variety of shapes are used for gelatins and desserts.

## Brushes

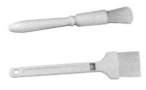

**Pastry brushes**—To spread melted butter or thin mixtures like icings or eggs and water.

**Grease brushes**—To remove grease from soups, stocks, and sauces.

**Vegetable brushes**—To clean vegetables and fruits.

## Spoons, skimmers, and strainers

**Wooden spoons**—For stirring, mixing, creaming, tossing, folding, and serving.

**Solid spoons**—To lift foods, including the liquid, out of the pot. They are also used to spoon liquids over foods.

**Slotted or perforated spoons**—To lift foods out of the liquid in which they were cooked.

**Skimmers**—Used to remove scum and grease from the top of stocks, gravies, and other liquids; also used to lift food out of hot liquid.

**Strainers** in a variety of sizes and shapes are used for separating solids and liquid. The mesh varies from fine to coarse. A sieve has similar uses, but is a stainless steel perforated cup with a handle.

**Colander**—Used to drain cooked foods like pasta and to rinse salad greens and berries.

**China cap**—Used to strain liquids from solids when making soups and gravies; also used to purée foods. A pointed wooden mallet is used to force food through the strainer.

## Flour and dough utensils

**Sifter**—Used to sift flour or powdered sugar, and to blend dry ingredients.

**Pastry blender**—Used to cut shortening into flour.

**Rolling pin**—Used to roll out pastry, rolls, and cookies.

**Dough scraper**—Used to scrape the dough from the board.

**Pastry bags**—Used to make shaped pastries and decorations. The bag is used with a variety of tips or tubes designed to create different shapes when a soft food like icing is squeezed through.

**FIGURE A-19**    Measuring utensils.

Liquid measuring cup

Dry measuring cups

Measuring spoons

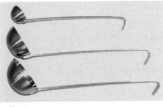
Ladles

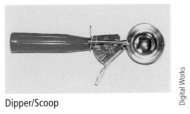
Dipper/Scoop

Digital Works

depending on the density, or weight per volume, of the object being measured. For example, half a cup of marshmallows weighs less than half a cup of vegetable oil.

Many countries measure ingredients by weight. In the United States, Americans tend to measure using volume measurements such as teaspoons, tablespoons, cups, pints, quarts, and gallons. Weight can be measured by a number of different types of scales: spring-type scales, used principally for weighing dry ingredients like grains, beans, dried pasta, vegetables, fruits, and cheese; portion scales and balance scales for weighing ingredients; and the baker's scale, used primarily for measuring dough ingredients (Figure A-20).

**FIGURE A-20**    Various scales used for weighing ingredients.

Spring-type scale

Balance scale

Portion scale

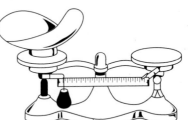
Baker's scale

Source: Texas Tech University

## Measuring Terms

***Mass vs. Volume***    Weight, commonly used to mean mass, is a much more accurate measurement than volume. As a result, many food service operations use weight rather than volume to measure recipe ingredients. Confusion between the two methods of measuring ingredients occurs because ounces can be measured either by volume, known as fluid ounces (fl), or by mass (weight), known as avoirdupois ounces (av). Water is the only substance whose fluid ounce is equal to its avoirdupois ounce. The mass of other substances will vary

***Metric vs. Nonmetric***   Metric measurements of volume are expressed in milliliters (mL). Metric cups come in sizes of 25, 50, 125, and 250 mL, and measuring spoons are divided into 1, 2, 5, 15, and 25 mL. A 250-mL metric cup is close to a nonmetric cup, which holds 236.59 mL. The 15 and 5 mL metric measures are almost equal to the nonmetric tablespoon and teaspoon, respectively. The inside back cover of this book lists the conversions between nonmetric and metric measuring units for volume and mass.

## Accuracy of Measuring Utensils

The American Association of Family and Consumer Sciences (AAFCS) has set certain tolerances for measuring the precise volume of household measuring utensils. One way to determine a cup's precise volume is to fill it with tap water and then pour it into a graduated cylinder. Both the utensil and the graduated cylinder should be on a level surface and the milliliters of water should be read at eye level at the bottom of the **meniscus** (Figure A-21). Any measurement evaluating accuracy should be done three times and then averaged to eliminate error. The resulting number should not deviate more than 5% from the standard set by the AAFCS. According to these standards, a metric cup of 250 mL can deviate 5 percent,

**FIGURE A-21**   Read the meniscus at eye level.

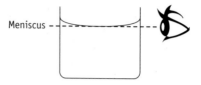

to 237.5 or 262.5 mL, and still be acceptable (3). Variations within the 5% range do not make any appreciable differences in ingredient proportions or in the quality of the final product.

## Serving Utensils

At last the meal is ready to be eaten, and serving utensils enter the picture. Basic tableware includes salad forks, dinner forks, regular knives, steak knives (optional), soup spoons, and teaspoons. A more extensive "wardrobe" of tableware might include butter knives, small two-tined forks known as seafood forks, dessert spoons, luncheon knives and forks (which are slightly smaller than standard knives and forks), iced-tea spoons, and grapefruit spoons. Eating utensils, for sanitary reasons, should always be touched by the handles.

# REFERENCES

1. Abrisham SH, et al. Bacterial adherence and viability on cutting board surfaces. *Journal of Food Safety* 14:153–172, 1994.

2. Albert A. Choosing great knives for confident, skillful cooking. *Fine Cooking* 24:50–53, 1998.

3. Birchfield JC. *Design and Layout of Foodservice Facilities.* Van Nostrand Reinhold, 1988.

4. Middleton S. Pros pick the best baking sheets. *Fine Cooking* 26:55–57, 1998.

5. Ovadia DZ, and C Walker. Impingement in food processing. *Food Technology* 52(4):46–50, 1998.

6. Pickett MS, MG Arnold, and LE Ketterer. *Household Equipment in Residential Design.* Waveland, 1986.

7. Stevens M. Choosing the best whisk. *Fine Cooking* 19:72, 1997.

**Meniscus** The imaginary line read at the bottom of the concave arc at the water's surface.

# Appendix B: Approximate Food Measurements

The quantity to purchase for an approximate yield.

| Food | Quantity to Purchase | Approximate Yield |
| --- | --- | --- |
| **DAIRY** | | |
| Cheese | | |
|     Cheddar | 1 lb | 2 C/4 C (grated) |
|     Cottage | 1 lb | 2 C |
|     Cream | 1 lb | 2 C |
| Cream | 1 C (½ pt) | 2 C |
| **EGGS** | | |
| Whole | 1 lb | 1¾ C |
| Whites (fresh) | 8–11 | 1 C |
| Yolks (fresh) | 12–14 | 1 C |
| **FATS AND OILS** | | |
| Butter/Margarine | 1 lb | 2 C |
| Vegetable Oil | 1 lb | 21/6 C |
| Vegetable Shortening | 1 lb | 2 1/3 C |
| **FLOUR** | | |
| All-Purpose | 1 lb | 4 C (sifted) |
| Cake | 1 lb | 4½ C (sifted) |
| Cornmeal | 1 lb | 3½ C (sifted) |
| Rye | 1 lb | 3½–5 C |
| Whole Wheat | 1 lb | 3 1/3 C (sifted) |
| **FRUIT** | | |
| Apples | 1 lb/3 med | 3 C (sliced) |
| Bananas | 1 lb/3 med | 2½ C (sliced) |
| Berries | 1 quart | 3½–4 C (sliced) |
| Coconut | 1 lb shredded | 5 C |
| Dates | 1 lb whole | 2¼ C or 2 C (pitted) |
| Lemon | 1 med | 1/3–½C juice |
| | | 1½–3 tsp (grated) |
| Orange | 1 med | 1/3–½ C juice |
| | | 1–2 tbs (grated) |
| Peaches | 1 lb | 4 C (sliced) |
| Prunes | 1 lb | 2 1/3 C |
| Raisins | 1 lb | 3 C |
| **NUTS** | | |
| Almonds | 1 lb shelled | 3 C |
| Pecans | 1 lb shelled | 4 C |
| Peanuts | 1 lb shelled | 3 C |
| Walnuts | 1 lb shelled | 4 C |
| **SUGAR/SALT** | | |
| Brown | 1 lb | 2¼–2½ C (firmly packed) |
| Confectioners | 1 lb | 4–4½ C (sifted) |
| Granulated | 1 lb | 2–2¼ C |
| Honey | 1 lb | 1–1¼ C |
| Salt | 1 lb | 1½ C |

*(Continued)*

# Appendix B (continued)

The quantity to purchase for an approximate yield.

| Food | Quantity to Purchase | Approximate Yield |
|------|----------------------|-------------------|
| VEGETABLES | | |
| Beets | 1 lb/4 med | 2 C |
| Cabbage | 1 lb | 4 C (shredded) |
| Carrots | 1 lb/4 med | 3 C (diced) |
| Celery | 1 lb/½ bunch | 4 C (diced) |
| Corn | 3 ears | 1 C (kernels) |
| Dried Beans | 1 lb/2 C | 5–6 C (cooked) |
| Green Beans | 1 lb | 3 C (chopped) |
| Lettuce | 1 lb/med | 6 C |
| Onion | 1 med | ½ C (diced) |
| Parsley | 1 med bunch | ½–1 C (finely chopped) |
| Potatoes | 1 lb/3 med | 2½ C (diced) |
| | | 3 C (peeled and sliced) |
| | | 2 C (mashed) |
| | | 2 C (French fries) |
| Tomatoes | 1 med | 1 C (chopped) |
| MISCELLANEOUS | | |
| Bread Crumbs (fresh) | 2 slices | 1 C |
| | 1 lb loaf | 10 C |
| Chocolate | | |
| Baking | 8-ounce pkg | 2 C (grated) |
| Cocoa | 1 lb | 4 C |
| Unsweetened | 8-ounce pkg | 8 1-ounce squares |
| Coffee | 1 lb ground | 5 C (about 2½ gallons) |
| | ½ C | 10 C |
| Crackers | | |
| Graham | 12 | 1 C (fine crumbs) |
| Saltines | 18 | 1 C (coarse crumbs) |
| | 24 | 1 C (fine crumbs) |
| Gelatin | 1 envelope | 1 T gelatin powder |
| Rice | 1 C uncooked | 3–4 C cooked |

# Appendix C: Substitution of Ingredients

| If Missing | Measurement | Substitute |
|---|---|---|
| **DAIRY** | | |
| Whole milk | 1 C | = ½ C evaporated milk + ½ C water |
| | | = 1/3 C nonfat dry milk + water to make one C + 2 T fat |
| | 1 quart | = 4 oz nonfat dry milk + water to make 1 qt + 1¼ oz fat |
| | | = ½ C heavy cream + ½ C cold water |
| Sweetened Condensed | 1 C | 1 C = ¾ C sugar + 1/3 evaporated milk + 2 T butter |
| Buttermilk/Sour Milk | 1 C | = 1 C fresh milk + 1 T fat vinegar or lemon juice (let stand for 5 minutes) |
| | 1 C | = 1 C unflavored plain yogurt |
| Cream | | |
|   Half & Half | 1 C | = ¾ cup milk + 2 T fat |
| | | = ½ C milk + ½ C light cream |
|   Heavy (Whipping) | 1 C | = 3/4 C milk + 1/3 C butter or margarine |
| Sour Cream | 1 C | = 1 C yogurt |
| **EGGS** | | |
| Whole | one | = 2 egg yolks + 1 T water |
| | | = 2 T dried whole eggs + 2½ T water |
| Whites, fresh | 1 white | = 2 T thawed frozen egg white or 2 tsp dry egg white + 2 T water |
| Yolks, fresh | 1 yolk | = 3½ T thawed frozen egg yolk or 2 T dry egg yolk + 2 tsp water |
| **FATS AND OILS** | | |
| Butter/margarine | 1 C | = 1 C margarine/butter |
| | | = 7/8 to 1 C hydrogenated fat + ½ tsp salt |
| | | = 7/8 C lard + ½ tsp salt |
| | | = 7/8 C vegetable oil |
| **FLOUR** | | |
| All-purpose | 1 C sifted | = 1 C unsifted all-purpose flour minus 2 T |
| | | = 1½ C bread flour |
| | | = 1 C rye |
| | | = 1 C + 2 T cake flour |
| | | = 1 C minus 2 T cornmeal |
| | | = 1 C graham flour |
| | | = 1 C minus 2 T rice flour |
| | | = 1 C rolled oats |
| | | = 1 C + 2 T coarsely ground whole wheat or graham flour or 13 T gluten flour |
| | | = 1¼ C rye flour |
| | | = ½ C barley flour |
| | 1 T (as thickener) | = ½ T cornstarch, potato starch, rice starch, arrowroot starch |
| | | = 1 T quick-cooking tapioca, waxy rice flour, waxy corn flour |
| Self-rising Flour | 1 lb | = 4 C of all-purpose flour + 2 T baking powder, 2 t salt |
| | | = 1 C = (1 C of all-purpose flour minus 2 t) + 1½ t baking powder + ½ t salt |
| Cake Flour | 1 C sifted | = 7/8 C sifted all-purpose flour or 1 C minus 2 tablespoons sifted all-purpose flour |

*(Continued)*

# Appendix C *(continued)*

| If Missing | Measurement | Substitute |
|---|---|---|
| **MISCELLANEOUS** | | |
| Allspice | 1 T | = ½ t cinnamon + ½ t ground cloves |
| Baking Powder | 1 t | = ¼ t baking soda + ½ t cream of tartar |
| | | = ¼ t baking soda + ½ C buttermilk or sour milk (replaces ½ C of liquid used in recipe) |
| Broth | 1 C | = 1 bouillon C (or 1 envelope powdered broth or 1 t powdered broth) + 1 C boiling water |
| Catsup | 1 C | = 1 C tomato sauce + ½ C sugar + 2 T vinegar |
| Chili Sauce | 1 C | = 1 C tomato sauce + ¼ C brown sugar + 2 T vinegar + ¼ t cinnamon + dash allspice/ ground cloves |
| Chives | | Scallion greens |
| Chocolate | | |
|   unsweetened | 1 ounce | = 3 T cocoa + 1 T fat |
|   baking | 1 square | = 3 T carob powder + 2 T water |
|   semisweet | 2 ounces | = 1 ounce unsweetened chocolate + 2 t sugar |
| Cocoa | 3 T | = 1 oz chocolate if recipe reduced by 1 T of fat |
| | | = 3 T carob powder |
| Cornstarch | 1 T | 2 T all-purpose flour |
| | 1 ounce | = 2 oz all-purpose flour |
| Garlic | 1 medium | = ½ t garlic salt |
| | clove | = 1/8 t garlic powder |
| Herbs | 1 T (fresh) | = ¼ t dried ground |
| | | = 1 t dried leaf |
| Lemon Juice | 1 T | = ½ T vinegar |
| Mayonnaise | 1 C | = ½ C yogurt + ½ C mayonnaise |
| | | = 1 C sour cream |
| | | = 1 C cottage cheese (pureed) |
| Pumpkin Pie Spice | 1 t | = ½ t cinnamon + ¼ t nutmeg + 1/8 t allspice + 1/8 t cardamom |
| Tomatoes (canned) | 1 C | = ½ C tomato puree + ½ C water |
| Tomato Juice | 1 C | = ½ C tomato puree or sauce + ½ up water |
| Tomato Purée | 2 C | = 1¼ C water + ¾ C tomato paste |
| Tomato Sauce | 2 C | = 1¼ C water + ¾ C tomato paste |
| **SUGAR/SWEETENERS** | | |
| Granulated | 1 C | = 1 1/3 C brown sugar |
| | | = 1½ C Confectioner's sugar |
| | | = 1 C honey minus ¼ to 1/3 liquid in recipe |
| | | = 1¼ to 1½ C corn syrup minus ¼ to ½ liquid in recipe |
| | | = 1 1/3 C molasses minus 1/3 C liquid in recipe |
| Brown | 1 C | = ½ C granulated sugar + ½ C liquid brown sugar |
| | | = 1 C granulated sugar + 2 T molasses or dark corn syrup |
| Confectioners | 1 C | = made by grinding 2 C granulated sugar in a processor |
| Honey | 1 C | = 1¼ C sugar + ¼ C liquid |
| Corn Syrup | 1 C | = 1 C sugar + ¼ C liquid |
| Molasses | 1 C | = ½ C honey 1¼ C melted brown sugar |

# Appendix D: Flavorings and Seasonings

| Name | Uses |
|------|------|
| Allspice | Allspice combines the flavors of cloves, cinnamon, and nutmeg. Whole allspice is used for pickling, gravies, broiled fish, and meats. Ground allspice is used for baked goods, fruit preserves, puddings, and relishes. |
| Almond | Almonds can be used in every dish from soup to dessert. Almond extract is used in cookies, confections, and Chinese cuisine. |
| Anise | Anise is a popular favorite for a few gourmet dishes such as Oysters Rockefeller. Also, it is used in bakery products, candies (especially licorice candy), certain kinds of cheese, pickles, and many liqueurs and cordials, including anisette and absinthe. |
| Anise-pepper | It is one of the ingredients in Chinese Five Spices and is commonly used for fish and strongly flavored foods. |
| Balm | Balm has a pleasant lemon scent and can be chopped and combined with other herbs for use in omelets and salads, and in the production of several liqueurs. Also, balm leaves are used to flavor soups and dressings. |
| Bay leaves | Bay leaves can be used either fresh or dried. They are one of the ingredients in bouquet garni, and are used in bouillon, marinades, olives, and pickles. They combine well with fish, potatoes, or tomatoes. |
| Bouquet garni | This is a French term meaning "bundle of sweet herbs." The bouquet garni is used in soups and stews, or any dish in which there is sufficient liquid to absorb the flavors. |
| Caper | Capers are much used in European cuisine. They are commonly used in making caper sauce, which is usually eaten with boiled lamb. They also go well with fish dishes and with casseroles of chicken and rabbit. |
| Caraway | The seeds (actually the dried whole fruits) are used in cakes, cheeses, confections, fresh cabbage, meat dishes, rye bread, salads, and sauerkraut. The chopped green leaves can be used in soups and salads. The roots can be cooked and eaten as a vegetable. |
| Cardamom seed | Freshly ground cardamom has many uses including: breads, cakes, cookies, cheese, curries, custard, liver sausage, meat dishes, pilaus, pork sausage, and punches. |
| Cassia | The stick cinnamon can be used in dishes to impart flavor, and then removed before serving; for example, some punches are flavored in this manner. Powdered cassia is used in combination with allspice, nutmeg, and cloves for spicing mincemeat, curries, pilaus, meat dishes, desserts and cakes. It is one of the ingredients of the famous Chinese Five Spices. |
| Cayenne pepper | A little goes a long way, but it is a spice that adds considerable interest to egg dishes, fish, and meat recipes. |
| Celery salt | This spice is slightly bitter, but it combines well with bouillon, eggs, fish, potato salad, and salad dressing. |
| Celery seed | Celery seeds have a slightly bitter taste, but they contribute a useful flavoring. They add special interest to many salads and salad dressings. |
| Chervil, garden | Chervil, which has a mild anise-caraway flavor, is one of the ingredients of Fines herbes, a mixture of chopped fresh herbs extensively used in French recipes. Chervil is used in omelets, soups, salads, sauces, and white wine vinegar. It should not be cooked, but must be added at the last minute; otherwise, it loses its flavor. |
| Chinese Five Spices | Chinese Five Spices (a blend of anise-pepper, star anise, cassia, cloves, and fennel seed) is an integral part of some of the recipes from the Far East. Also, it can be used to good advantage in flavoring pork dishes. |
| Chives | Chives are ideal as a garnish because of their delicate onion flavor and bright green color. Chives add interest and flavor to buttered beets, eggs, cottage or cream cheese, potato and other salads, sliced tomatoes, and soups. |
| Cinnamon | Cinnamon has a more delicate flavor than cassia and is more suitable for sweet dishes, cakes, and cookies. |
| Citron | Citron peel has a peculiar taste, quite different from other citrus. Is used in the U.S. as candied peel to be added to cakes, cookies, candies, and desserts. |
| Cloves | Whole cloves are used in many meat dishes, but a little goes a long way. Cloves are stuck into lemon slices for tea, into onions, and into hams for baking; they are also popular for apple cookery and pickle making. In the East, they go into many of the curry dishes. Whole cloves are also included in recipes for spiced wine and some liqueurs. Ground cloves are used in baked goods, borscht (beet soup), chocolate puddings, potato soup, and stews. |
| Cola | Cola is used in many soft drinks, and for coloring and flavoring some wines. |
| Coriander (Cilantro) | Coriander leaves are popular in Near, Middle, or Far East recipes, as well as Mexico and South America. The seeds are a principal ingredient of curry. Whole coriander seeds can be used in cakes, cookies, biscuits, gingerbread, green salads, pickles, and poultry stuffing. Ground seeds are added to many meat and sweet dishes. |

*(Continued)*

# Appendix D *(continued)*

| Name | Uses |
|---|---|
| Cress<br>Watercress<br>Garden Cress | The cresses are primarily used in salads and sandwiches, but they can be used to flavor soup, cooked greens, or sauces for fish dishes, and to garnish meals. |
| Cumin | Cumin's principal use is in curry powder. It is also used to flavor bread, stuffed eggs, meats, rice dishes, and soups. Commercially, it may be found in cheese, chutney, pickles, meats, and sausage. |
| Curry (powder) | Curry powder may be added to eggs, chicken, fish, meats, rice, soups, or a salad made of sweet potatoes and pineapple. |
| Dill | Dill loses its flavor when cooked, so it should be added at the last minute. Fresh dill leaves can be used for dishes containing chicken, mushrooms, or spinach. The seeds are used in dill pickles and dill vinegar, but they can be added to meat dishes, meat and fish sauces, sauerkraut, salads, and borscht (beet soup). |
| Fennel | Fennel has an anise-like flavor and is good with many foods: apple pie, candies, fish, liqueurs, pastries, pork, soups, and sweet pickles. |
| Fenugreek seed | Fenugreek seeds are usually used in Indian curries and chutneys. |
| Fines herbes | A combination of several herbs such as basil, chervil, chives, marjoram, oregano, parsley, rosemary, sage, tarragon, and thyme. Fines herbes can be used in many dishes such as fish sauces, meat stuffings, omelets, salads, salad dressing, and soups. |
| Garlic | Garlic blends with a wide range of dishes such as fish, game, meats, and vegetables. |
| Ginger | Ginger is used in numerous foods including beverages, biscuits, cakes, cookies, fish, gingerbread, ginger beer, ginger wine and cordials, puddings, sauces, and spice mixtures. It is used mostly in sweet preparation in European and North American cooking, but the Orient uses it extensively for chutney, fish, meat, and pickles. |
| Horseradish | Many cooks limit the use of horseradish to a sauce used on meats, but it can be added to chicken salads, egg dishes, and mayonnaise for use on fish dishes, or tomato combinations. |
| Leek | The leek is rather like a very mild onion. It is used mostly in soups and chowders. However, the leek may also be used as a bouquet for pork or lamb. |
| Lemon | Lemon juice can be used on salads instead of vinegar, and it is the predominant favorite for serving with most fishes. Grated lemon rind is added to cakes, cookies, desserts, and sauces, to give an added taste dimension. |
| Licorice | Licorice is used to flavor candy, chewing gum, and soft drinks. CAUTION: Licorice raises the blood pressure of some people dangerously high, due to retention of sodium. |
| Lime | Limes impart a unique taste to dishes, which cannot be replaced by lemons. Fish is often marinated in lime juice before cooking. |
| Mace | Mace can be added to apple dishes, beets, cakes, hot chocolate, coffee cakes, cookies, custards, eggnog, gingerbread, and muffins. |
| Marjoram | Is related to thyme; hence, they are often used together or to replace each other. It can be added to almost every dish to advantage. It should be added immediately before serving as the flavor is easily lost in cooking. Marjoram is used with egg dishes, lamb, poultry, sausage, soups, stews, and vegetables. |
| Mint,<br>Peppermint,<br>or Spearmint | Peppermint flavoring is used mostly for candies, cordials, desserts, icings, and liqueurs. Spearmint is the preferred mint for lamb as well as for iced tea and mint juleps. It can also be used in soups, stews, fish, and meat sauces. |
| Monosodium<br>Glutamate<br>(MSG) | MSG does not have any flavor of its own, but it intensifies and enhances the flavor in other foods, especially meat and fish. |
| Mustard<br>Black mustard<br>Brown<br>mustard<br>White or<br>Yellow<br>mustard | Whole mustard seeds add pungency to many foods, including pickles, meats, and salads. Powdered dry mustard is a common kitchen spice. Its sharp, hot flavor develops when the powder is moistened. It is used for roast beef, mustard pickles, sauces, and gravies. Prepared mustard is a mixture of powdered mustard with salt, spices, and lemon juice, with wine or vinegar to preserve the mustard's pungency. It may be used with ham, hamburgers, hot dogs, and sandwich spreads. |
| Nutmeg | Nutmeg is traditionally used in sweet foods such as cakes, custards, doughnuts, eggnog, pies, and puddings, but it goes very well with meat, sausage, spinach, sweet potatoes, and vegetables. |

*(Continued)*

# Appendix D (*continued*)

| Name | Uses |
| --- | --- |
| Onion | Onions are used either as a separate vegetable or as a flavoring for other foods. The leaves of the onion, along with the bulb, are used in salad. |
| Oregano | Oregano is used extensively in Italian cooking and can be added to cheese dishes, chili beans, fish, gravies, meats, sauces, sausage, salads, and soups. |
| Paprika | Paprika is used in many dishes both for its flavor and as a garnish. It can be added to chicken, sweet corn, fish, meats, sausages, tomato catsup, and tomato juice. |
| Parsley | Parsley can be added to fish and fish sauces, meats, sauces, soups, and vegetables. It is commonly used as a decoration for buffet dishes. |
| Pepper | Pepper loses much in aroma when ground or cooked, so freshly ground pepper should be used whenever possible. Whole peppercorn can be purchased as well as cracked, and coarsely or finely ground. Except for sweet dishes, pepper can be added to all other dishes. |
| Poppy seed | Poppy seeds have a pleasant nutlike flavor and aroma and are used primarily in baked goods, on the tops of rolls and bread, and in cakes and pastries. However, they are also used in confections, fruit salad dressings, and curries. |
| Rosemary | Rosemary is good with soups, on broiled steaks, or with other meat dishes, sauces, and vegetables. The taste is aromatic, pungent, and slightly bitter. |
| Saffron | Saffron is used as a flavoring and coloring (yellow) spice in biscuits, confections, boiled fish, fish soup, fancy rolls, and rice, and in some European dishes. |
| Sage | Sage is available whole, rubbed, or ground. It is used for baked fish, meats, and meat stuffings, sausages, cheeses, and sauces. |
| Savory | Savory is available whole or ground, and is often combined with other herbs to flavor meats. Also, it can be used in beans, scrambled eggs, peas, salads, sauces, and sausages. |
| Sesame seed | Sesame seeds develop a beautiful nutty taste when sprinkled on buns, rolls, or cakes, and then baked. They are also used in confections. |
| Shallot | Shallots can be used in the same way as the onion, although the flavor is much more subtle. Shallots should never be browned, as they turn bitter. |
| Soy sauce | Soy sauce can be used in a wide array of dishes, especially with beef, chicken, fish, soups, turkey, and vegetable dishes. |
| Star anise | Star anise has a strong flavor similar to anise, but slightly more bitter and pungent. In Chinese cooking, it is used for duck and pork recipes. |
| Sweet basil | Basil can be used for green beans, fish, soups, squashes, stews, tomatoes, and vinegar. |
| Sweet cicely | The plant smells and tastes somewhere between anise and licorice. The taproot can be boiled and used for salads, and the green fruit can be served with salad dressing. Europeans use the leaves in soups and salads. The plant is also used for flavoring desserts and liqueurs. |
| Tarragon | Tarragon is best known for flavoring vinegar, but it is also used for beef, chicken, eggs, fish, pickles, cookies, salads, and tartar sauce. It has a slightly anise flavor. |
| Thyme | Thyme is used with fish dishes, meats, poultry, sauces, tomato dishes, and vegetables. |
| Turmeric | Turmeric and mustard are inseparable partners (it is used to color mustard); and turmeric is superb for almost every meat and egg dish, for pickles, and for curries. It adds yellow color. |
| Vanilla | Vanilla is almost always used in sweet dishes such as bakery products and desserts. |
| Wintergreen | Wintergreen is used mainly for candies and lozenges. |

# Appendix E: Cheeses

| Name | Origin | Consistency | Flavor | Normal Ripening Period |
|------|--------|-------------|--------|------------------------|
| American pasteurized process | United States | Semisoft to soft; smooth, plastic body | Mild | Unripened after cheese(s) heated to blend |
| Asiago, fresh, medium, old | Italy | Semisoft (fresh), medium, or hard (old); tiny gas holes or eyes | Piquant, sharp in aged cheese | 60 days minimum for fresh (semisoft), 6 months minimum for medium, 12 months minimum for old (grating) |
| Bel paese | Italy | Soft; smooth, waxy body | Moderately robust | 6–8 weeks |
| Blue, Bleu | France | Semisoft; visible veins of mold on white cheese; pasty, sometimes crumbly | Piquant, tangy, spicy, peppery | 60 days minimum; 3–4 months usually; 9 months for more flavor |
| Breakfast, Frühstück | Germany | Soft; smooth, waxy body | Strong, aromatic | Little or none (either) |
| Brick | United States | Semisoft; smooth, open texture; numerous round and irregular-shaped eyes | Mild but pungent and sweet | 2–3 months |
| Brie | France | Soft, thin edible crust, creamy interior | Mild to pungent | 4–8 weeks |
| Caciocavallo | Italy | Hard, firm body; stringy texture | Sharp, similar to provolone | 3 months minimum for table use, 12 months or longer for grating |
| Camembert | France | Soft, almost fluid in consistency; thin edible crust, creamy interior | Mild to pungent | 4–5 weeks |
| Cheddar | England | Hard; smooth, firm body, can be crumbly | Mild to sharp | 60 days minimum; 3–6 months usually; 12 or longer for sharp flavor |
| Colby | United States | Hard but softer and more open in texture than Cheddar | Mild to mellow | 1–3 months |
| Cottage, Dutch, Farmers, Pot | Uncertain | Soft; moist, delicate, large or small curds | Mild, slightly acidic, flavoring may be added | Unripened |
| Cream | United States | Soft; smooth, buttery | Mild, slightly acid, flavoring may be added | Unripened |
| Edam | Holland | Semisoft to hard; firm, crumbly body; small eyes | Mild, sometimes salty | 2 months or longer |
| Feta | Greece | Soft, flaky; similar to very dry, high-acid cottage cheese | Salty | 4–5 days to 1 month |
| Gammelost | Norway | Semisoft | Sharp, aromatic | 4 weeks or longer |
| Gjetost | Norway | Hard; buttery | Sweet, caramel | Unripened |
| Gorgonzola | Italy | Semisoft; less moist than blue | Piquant, spicy, similar to blue | 3 months minimum, frequently 6 months to 1 year |
| Gouda | Holland | Hard, but softer than Cheddar; more open mealy body like Edam, small eyes | Mild, nutlike, similar to Edam | 2–6 months |
| Gruyère | Switzerland | Hard, tiny gas holes or eyes | Mild, sweet | 3 months minimum |
| Limburger | Belgium | Soft; smooth, waxy body | Strong, robust, highly aromatic | 1–2 months |

*(Continued)*

# Appendix E *(continued)*

| Name | Origin | Consistency | Flavor | Normal Ripening Period |
|------|--------|-------------|--------|------------------------|
| Monterey Jack | United States | Semisoft (whole milk), hard (low-fat or skim milk); smooth texture with small openings throughout | Mild to mellow | 3–6 weeks for table use, 6 months minimum for grating |
| Mozzarella | Italy | Semisoft; plastic | Mild, delicate | Unripened to 2 months |
| Muenster | Germany | Semisoft; smooth, waxy body, numerous small mechanical openings | Mild to mellow, between brick and Limburger | 2–8 weeks |
| Neufchatel | France | Soft; smooth, creamy | Mild | 3–4 weeks or unripened |
| Parmesan, Reggiano | Italy | Very hard (grating), granular, hard brittle rind | Sharp, piquant | 10 months minimum |
| Port du Salut, Oka | Trappist Monasteries | Semisoft; smooth, buttery | Mellow or mild to robust, similar to Gouda | 6–8 weeks |
| Primost | Norway | Semisoft | Mild, sweet, caramel | Unripened |
| Provolone | Italy | Hard, stringy texture; cuts without crumbling, plastic | Bland acid flavor to sharp and piquant, usually smoked | 6–14 months |
| Queso blanco, White cheese | Latin America | Soft, dry and granular if not pressed; hard open or crumbly if pressed | Salty, strong, may be smoked | Eaten within 2 days to 2 months or more; generally unripened if pressed |
| Ricotta | Italy | Soft, moist and grainy, or dry | Bland but semisweet | Unripened |
| Romano | Italy | Very hard, granular interior, hard brittle rind | Sharp, piquant if aged | 5 months minimum; usually 5–8 months for table cheese; 12 months minimum for grating cheese |
| Roquefort | France | Semisoft, pasty and sometimes crumbly | Sharp, spicy (pepper), piquant | 2 months minimum; usually 2–5 months or longer |
| Sap Sago | Switzerland | Very hard (grating), granular, frequently dried | Sharp, pungent, flavored with leaves; sweet | 5 months minimum |
| Schloss, Castle cheese | Germany, Northern Austria | Soft; small, ripened | Similar to, but milder than Limburger | Less than 1 month; less intensively than Limburger |
| Stirred curd, granular | United States | Semisoft to hard | Similar to mild Cheddar | 1–3 months |
| Stilton | England | Semisoft to hard; open flaky texture, more crumbly than blue | Piquant, spicy, but milder than Roquefort | 4–6 months or longer |
| Swiss, Emmentaler | Switzerland | Hard; smooth with large gas holes or eyes | Mild, sweet, nutty | 2 months minimum, 2–9 months usually |
| Washed curd | United States | Semisoft to hard | Similar to mild Cheddar | 1–3 months |

# Appendix F: Common Food Additives

| Name | Function | Food Use and Comments |
| --- | --- | --- |
| Acetic acid | pH control; preservative | Acid or vinegar is acetic acid; many food uses. |
| Adipic acid | pH control | Buffer and neutralizing agent; used in confectionery. |
| Ammonium alginate | Stabilizer and thickener; texturizer | Extracted from seaweed. Widespread food use. |
| Annatto | Color | Extracted from seeds of *Bixa orellana*. Butter, cheese, margarine, shortening, and sausage casings; coloring foods in general. |
| Arabinogalactan | Stabilizer and thickener; texturizer | Extracted from Western larch. Widespread food use; bodying agent in essential oils, nonnutritive sweeteners, flavor bases, nonstandardized dressings and pudding mixes. |
| Ascorbic acid (vitamin C) | Nutrient; antioxidant; preservative | Widespread use in foods to prevent rancidity, browning; used in meat curing; GRAS additive. |
| Aspartame | Sweetener, low calorie | Soft drinks, chewing gum, powdered beverages, whipping toppings, puddings, gelatin, tabletop sweetener. |
| Azodicarbonamide | Flour treating agent | Aging and bleaching ingredient in cereal flour. |
| Benzoic acid | Preservative | Widespread food use. |
| Benzoyl peroxide | Flour treating agent | Bleaching agent in flour; may be used in some cheeses. |
| Beta-apo-8 carotenol | Color | Natural food color. General use not to exceed 30 mg per lb or pt of food. |
| BHA (butylated hydroxyanisole) | Antioxidant; preservative | Fats, oils, dry yeast, beverages, breakfast cereals, dry mixes, shortening, potato flakes, chewing gum, sausage; often used in combination with BHT. |
| BHT (butylated hydroxytoluene) | Antioxidant; preservative | Rice, fats, oils, potato granules, breakfast cereals, potato flakes, shortening, chewing gum, sausage; often used in combination with BHA. |
| Biotin | Nutrient | Rich natural sources are liver, kidney, pancreas, yeast, milk; vitamin supplement. |
| Calcium alginate | Stabilizer and thickener; texturizer | Extracted from seaweeds. Widespread food use. |
| Calcium carbonate | Nutrient | Mineral supplement. |
| Calcium lactate | Preservative | General purpose and/or miscellaneous use. |
| Calcium phosphate | Leavening agent; sequestrant, nutrient | General purpose and/or miscellaneous use; mineral supplement. |
| Calcium propionate | Preservative | Bakery products, alone or with sodium propionate; inhibits mold and other microorganisms. |
| Calcium silicate | Anticaking agent | Used in baking powder, salt, and food; GRAS for use in baking powder and salt. |
| Canthaxanthin | Color | Widely distributed in nature. Color for foods; more red than carotene. |
| Caramel | Color | Miscellaneous and color for foods. |
| Carob bean gum | Stabilizer and thickener | Extracted from bean of carob tree (*Locust bena*). Numerous foods like confections, syrups, cheese spreads, frozen desserts, and salad dressings. |
| Carrageenan | Emulsifier; stabilizer and thickener | Extracted from seaweed. A variety of foods, primarily those with a water or milk base, especially ice cream. |
| Cellulose | Emulsifier; stabilizer and thickener | Component of all plants. Inert bulking agent in foods; may be used to reduce caloric content of food. |

*(Continued)*

# Appendix F (continued)

| Name | Function | Food Use and Comments |
|------|----------|----------------------|
| Citric acid | Preservative; antioxidant: pH control agent; sequestrant | Widely distributed in nature in both plants and animals. Miscellaneous use; used in lard, shortening, sausage, margarine, chili con carne, cured meats, and freeze-dried meats. |
| Citrus Red No. 2 | Color | Coloring skins of oranges. |
| Cochineal | Color | Derived from the dried female insect, *Coccus cacti*. Provides red color for such foods as meat products and beverages. |
| Corn endosperm oil | Color | Source of xanthophyll for yellow color. Used in chicken feed to color yolks of eggs and chicken skin. |
| Cornstarch | Anticaking agent; drying agent; formulation aid; processing aid; surface-finishing agent | Digestible polysaccharide used in many foods often in a modified form; example foods include baking powder, baby foods, soups, sauces, pie fillings, imitation jellies, custards, and candies. |
| Corn syrup | Flavoring agent; humectant; nutritive sweetener; preservative | Derived from hydrolysis of cornstarch. Employed in numerous foods, such as baby foods, bakery products, toppings, meat products, beverages, condiments, and confections. |
| Dextrose (glucose) | Flavoring agent; humectant; nutritive sweetener; synergist | Derived from cornstarch. Major users of dextrose are confection, wine, and canning industries; used to flavor meat products; used in production of caramel. |
| Diglycerides | Emulsifiers | Uses include frozen desserts, lard, shortening, and margarine. |
| Dioctyl sodium sulfosuccinate | Emulsifier; processing aid; surface active agent | Employed in gelatin dessert, dry beverages, fruit juice drinks, and noncarbonated beverages with cocoa fat; used in production of cane sugar and in canning. |
| Disodium guanylate | Flavor enhancer | Derived from dried fish or seaweed. Variety of uses. |
| Disodium inosinate | Flavor adjuvant | Derived from dried fish or seaweed; sodium guanylate a by-product. Variety of uses. |
| EDTA (ethylenedi- aminetetraacetic acid) | Antioxidant; sequestrant | Calcium disodium and disodium salt of EDTA employed in a variety of foods including soft drinks, alcoholic beverages, dressings, canned vegetables margarine, pickles, sandwich spreads, and sausage. |
| FD&C colors: Blue No. 1, Red No. 40, Yellow No. 5 | Color | Coloring foods in general. |
| Gelatin | Stabilizer and thickener; texturizer | Derived from collagen. Employed in many foods including confectionery, jellies, and ice cream. |
| Glycerine (glycerol) | Humectant | Miscellaneous and general purpose additive. |
| Grape skin extract | Color | Colorings for carbonated drinks, beverage bases, ades, and alcoholic beverages. |
| Guar gum | Stabilizer and thickener; texturizer | Extracted from seeds of the guar plant. Employed in such foods as cheese, salad dressings, ice cream, and soups. |
| Gum arabic | Stabilizer and thickener; texturizer | Gummy exudate of Acacia plants. Used in a variety of foods. |
| Gum ghatti | Stabilizer and thickener; texturizer | Gummy exudate of plant growing in India and Ceylon. A variety of food uses. |
| Hydrogen peroxide | Bleaching agent | Modification of starch and bleaching tripe; bleaching agent. |
| Hydrolyzed vegetable (plant) protein | Flavor enhancer | To flavor various meat products. |
| Invert sugar | Humectant; nutritive sweetener | Primarily used in confectionery and brewing industry. |

*(Continued)*

# Appendix F *(continued)*

| Name | Function | Food Use and Comments |
|------|----------|----------------------|
| Iron | Nutrient | Dietary supplements and foods. |
| Iron-ammonium citrate | Anticaking agent | Used in salt. |
| Karaya gum | Stabilizer and thickener | Derived from dried extract of *Sterculia urens*. Variety of food uses; a substitute for tragacanth gum. |
| Lactic acid | Preservative, pH control | Normal product of human metabolism. Numerous uses in foods and beverages; a miscellaneous general purpose additive. |
| Lecithin (phospha-tidylcholine) | Emulsifier; surface active agent | Normal tissue component of the body; naturally occurring in eggs; commercially derived from soybeans. Margarine, chocolate and wide variety of other uses. |
| Mannitol | Anticaking; nutritive sweetener; stabilizer and thickener; texturizer | Special dietary foods. A sugar alcohol. |
| Methylparaben | Preservative | Food and beverages. |
| Modified food starch | Drying agent; formulation aid; processing aid; surface finishing agent | Digestible polysaccharide used in many foods and stages of food processing; examples include baking powder, puddings, pie fillings, baby foods, soups, sauces, candies, etc. |
| Monoglycerides | Emulsifiers | Widely used in foods such as frozen desserts, lard, shortening and margarine. |
| MSG (monosodium glutamate) | Flavor enhancer | To enhance the flavor of a variety of foods including various meat products. |
| Papain | Texturizer | Used as a meat tenderizer. Achieves results through enzymatic action. |
| Paprika | Color; flavoring agent | To provide coloring and/or flavor to foods. |
| Pectin | Stabilizer and thickener; texturizer | Richest source of pectin is lemon and orange rind; present in cell walls of all plant tissues. Used to prepare jellies and similar foods. |
| Phosphoric acid | pH control | Used to increase effectiveness of antioxidants in lard and shortening. |
| Polyphosphates | Nutrient; flavor improver; sequestrant; pH control | Numerous food uses. |
| Polysorbates | Emulsifiers; surface active agent | Polysorbates designated by numbers such as 60, 65, and 80; variety of food uses including baking mixes, frozen custards, pickles, sherbets, ice creams, and shortening. |
| Potassium alginate | Stabilizer and thickener; texturizer | Extracted from seaweed. Wide usage. |
| Potassium bromate | Flour treating agent | Employed in flour, whole wheat flour, and fermented malt beverages, and to treat malt. |
| Potassium iodide | Nutrient | Added to table salt or used in mineral preparations as a source of dietary iodine. |
| Potassium nitrite | Curing and pickling agent | To fix color in cured products such as meats. |
| Potassium sorbate | Preservative | Inhibits mold and yeast growth in foods such as wines, sausage casings, and margarine. |
| Proplonic acid | Preservative | Mold inhibitor in breads and general fungicide; used in manufacture of fruit flavors. |
| Proply gallate | Antioxidant; preservative | Used in products containing oil or fat; employed in chewing gum; used to retard rancidity in frozen fresh pork sausage. |
| Propylene glycol | Emulsifier; humectant; stabilizer and thickener; texturizer | Miscellaneous and/or general purpose additive; uses include salad dressings, ice cream, ice milk, custards, and a variety of other foods. |

*(Continued)*

# Appendix F *(continued)*

| Name | Function | Food Use and Comments |
|------|----------|----------------------|
| Propylparaben | Preservative | Fungicide; controls mold in sausage casings; GRAS additive. |
| Saccharin | Nonnutritive sweetener | Special dietary foods and a variety of beverages; baked products; tabletop sweeteners. |
| Saffron | Color; flavoring agent | Derived from plant of western Asia and southern Europe. Used to color sausage casings, margarine, or product branding inks. |
| Silicon dioxide | Anticaking agent | Used in feed or feed components, beer production, production of special dietary foods; ink dilutent for marking fruits and vegetables. |
| Sodium acetate | pH control; preservative | Miscellaneous and/or general purpose use; meat preservation. |
| Sodium alginate | Stabilizer and thickener; texturizer | Extracted from seaweed; widespread food use. |
| Sodium aluminum sulfate | Leavening agent | Baking powders, confectionery; sugar refining. |
| Sodium benzoate | Preservative | To retard flavor reversion (i.e., margarine). |
| Sodium bicarbonate | Leavening agent; pH control | Separation of fatty acids and glyceroil on rendered fats; neutralize excess and clean vegetables in rendered fats, soups, and curing pickles. |
| Sodium chloride (salt) | Flavor enhancer; formulation acid; preservation | Widespread use of salt in many foods. |
| Sodium citrate | pH control; curing and pickling agent; sequestrant | Evaporated milk; miscellaneous and/or general purpose food use; accelerate color fixing in baking products. |
| Sodium diacetate | Preservative; sequestrant | An inhibitor of molds and rope-forming bacteria in baking products. |
| Sodium nitrate (Chile Saltpeter) | Curing and pickling agent; preservative | Used with or without sodium nitrite in smoked, cured fish; cured meat products. |
| Sodium nitrite | Curing and pickling agent; preservative | May be used with sodium nitrate in smoked, cured fish, cured meat products, and pet foods. |
| Sodium propionate | Preservative | A fungicide and mold preventative in bakery products, alone or with calcium propionate. |
| Sorbic acid | Preservative | Fungistatic agent for foods, especially cheeses; other uses include baked goods, beverages, dried fruits, fish, jams, jellies, meats, pickled products, and wines. |
| Sorbitan monostearate | Emulsifier; stabilizer and thickener | Widespread food usage such as whipped toppings, cakes, cake mixes, confectionery, icings, and shortenings; also many nonfood uses. |
| Sorbitol | Humectant; nutritive sweetener; stabilizer and thickener, sequestrant | Occurs naturally in berries, cherries, plums, pears, and apples; a sugar alcohol. Examples of use include chewing gum, meat products, icings, dairy products, beverages, and pet foods. |
| Sucrose (table sugar) | Nutritive sweetener; preservative | Sugar occurs naturally in some fruits and vegetables. The most widely used additive; used in beverages, baked goods, candies, jams and jellies—an endless list including meat products. |
| Tagetes (Aztec marigold) | Color | Source is flower petals of Aztec marigold. To enhance yellow color of chicken skin and eggs, incorporated in chicken feed. |
| Tartaric acid | pH control | Occurs free in many fruits, free or combined with calcium, magnesium, or potassium. In the soft drink industry, confectionery products, bakery products, and gelatin desserts. |
| Titanium dioxide | Color | For coloring foods generally, except standardized foods; used for coloring ingested and applied drugs. |

*(Continued)*

# Appendix F (*continued*)

| Name | Function | Food Use and Comments |
|------|----------|------------------------|
| Tocopherols (vitamin E) | Antioxidant; nutrient | To retard rancidity in foods containing fat; used as a supplement. |
| Tragacanth gum | Stabilizer and thickener; texturizer | Derived from the plant *Astragalus gummifier.* |
| Turmeric | Color | Derived from rhizome of *Curcuma longa.* Food use in general, except standardized foods; to color sausage casings, margarine or shortening; ink for branding or marking products. |
| Vanilla | Flavoring agent | Used in various bakery products, confectionery and beverages; natural flavoring extracted from cured, full-grown unripe fruit of *Vanilla panifolia.* |
| Vanillin | Flavoring agent and adjuvant | Widespread confectionery, beverage, and food use; synthetic form of vanilla. |
| Yellow prussiate of soda | Anticaking agent | Employed in salt. |

*A mole is a certain number (about $6 \times 10^{23}$) of molecules. The pH of a solution is defined as the negative logarithm of the hydrogen ion concentration of the solution. Thus, if the concentration is $10^{-2}$ (moles per liter), the pH is 2; if $10^{-8}$, the pH is 8; and so on.

# Glossary

**Acceptable Daily Intake (ADI)** The amount of food additive that can be safely ingested daily over a person's lifetime.

**Adenosine triphosphate (ATP)** Adenosine triphosphate is a universal energy compound in cells obtained from the metabolism of carbohydrate, fat, or protein. The energy of ATP, which is located in high-energy phosphate bonds, fuels chemical work at the cellular level.

**Agglomerate** A process in which small particles gather into a mass or ball. In the case of milk, the protein particles regroup into larger, more porous particles.

**Aging** Holding meat after slaughter to improve texture and tenderness. A ripening that occurs when carcasses are hung in refrigeration units for longer periods than that required for the reversal of rigor mortis.

**À la meunière** Fish seasoned, lightly floured, and sautéed in clarified butter or oil and served with a sauce made with butter and parsley.

**Albedo** The white, inner rind of citrus fruits, which is rich in pectin and aromatic oils.

**Al dente** Meaning "to the tooth" in Italian, it refers to pasta that is tender, yet firm enough to offer some resistance to the teeth.

**Amphoteric** Capable of acting chemically as either acid or base.

**Antibiotic** A substance used to prevent or treat infectious diseases by inhibiting or destroying the responsible microorganism.

**Antioxidant** A compound that inhibits oxidation, which can cause deterioration and rancidity.

**Aromatic compound** A compound that has a chemical configuration of a hexagon.

**Artesian water** Water that has surfaced on its own from an aquifer, rather than being pumped.

**Aspic** A clear gel prepared from stock or fruit or vegetable juices.

**As purchased (AP)** The total amount of food purchased prior to any preparation.

**Astringency** A sensory phenomenon characterized by a dry, puckery feeling in the mouth.

**Atoms** The basic building blocks of matter; individual elements found on the periodic table.

**Au gratin** Food prepared with a browned or crusted top. A common technique is to cover the food with a bread crumb/sauce mixture and pass it under a broiler.

**Au jus** Served with its own natural juices; a term usually used in reference to roasts.

**Bacteria** One-celled microorganisms abundant in the air, soil, water, and/or organic matter (i.e., the bodies of plants and animals).

**Baking powder** A chemical leavener consisting of a mixture of baking soda, acid(s), and an inert filler such as cornstarch.

**Baking soda** A white chemical leavening powder consisting of sodium bicarbonate.

**Barding** Tying thin sheets of fat or bacon over lean meat to keep the meat moist during roasting. The sheets of fat are often removed before serving.

**Baste** To add a liquid, such as drippings, melted fat, sauce, fruit juice, or water, to the surface of food (usually roasting meat) to help prevent drying.

**Batter** A flour mixture that contains more water than a dough does and whose consistency ranges from pourable to sticky.

**Beading** The formation of tiny syrup droplets on the surface of a baked meringue.

**Beurre manié** (pronounced *burr mahn-YAY*) A thickener that is a soft paste made from equal parts of soft butter and flour blended together.

**Bile** A digestive juice made by the liver from cholesterol and stored in the gall bladder.

**Biological value** The percentage of protein in food that can be utilized by an animal for growth and maintenance. High-quality, complete proteins are considered to have a high biological value.

**Biotechnology** Previously called genetic engineering, this term describes the alteration of a gene in a bacterium, plant, or animal for the purpose of changing one or more of its characteristics.

**Bisque** Traditionally, a cream soup made from shellfish. Marketers sometimes label creamed vegetable soups as bisques.

**Blanch** To dip a food briefly into boiling water.

**Blind bake** To bake an unfilled piecrust.

**Bloom** Cottony, fuzzy growth of molds.

**Body** The consistency of frozen desserts as measured by their firmness, richness, viscosity, and resistance to melting.

**Boiling point** The temperature at which a heated liquid begins to boil and changes to a gas.

**Bouillon** A broth made from meat and vegetables and then strained to remove any solid ingredients.

**Bouquet garni** A bundle of parsley, thyme, bay leaf, and whole black pepper rolled in a leek and tied together with twine.

**Bran** The hard outer covering just under the husk that protects the grain's soft endosperm.

**Broth** Stock made from meat or meat/bone combinations and some water with little or no flavoring. Broths are seldom reduced (simmered until much of the water evaporates) and therefore not as strong-flavored as stocks.

**Brown stock** The stock resulting from browning bones and/or meat prior to simmering them.

**Calorie (kcal)** The amount of energy required to raise 1 kilogram of water 1°C (measured between 14.5°C and 15.5°C at normal atmospheric pressure). (Small "c" calorie is defined by the amount of energy required to heat 1 "gram" of water.)

**Candling** A method of determining egg quality based on observing eggs held against a light.

**Caramelization** A process in which dry sugar, or sugar solution with most of its water evaporated, is heated until it melts into a clear, viscous liquid and, as heating continues, turns into a smooth, brown mixture.

**Carryover cooking**   The phenomenon in which food continues to cook after it has been removed from the heat source as the heat is distributed more evenly from the outer to the inner portion of the food.

**Casein**   The primary protein (80%) found in milk; it can be precipitated (solidified out of solution) with acid or certain enzymes.

**Catechins**   Flavonoid pigments that are a subgroup of the flavonol pigments.

**Chalaza**   (pl. *chalazae*) The ropy, twisted strands of albumen that anchor the yolk to the center of the thick egg white.

**Chemethesis**   The ability to feel a food's chemical properties (e.g., cool mints or hot chili peppers).

**Chiffon cake**   A cake made by combining the characteristics found in both shortened and unshortened cakes.

**Clarified butter**   Butter that will not burn because its milk solids and water have been removed.

**Clarify**   To make or become clear or pure.

**Coagulate**   To clot or become semisolid. In milk, denatured proteins often separate from the liquid by coagulation.

**Coagulation**   The clotting or precipitation of protein in a liquid into a semisolid compound.

**Codex Alimentarius Commission**   The international organization that develops international food standards, codes of practice, and other guidelines to protect consumers' health.

**Collagen**   A pearly white, tough, and fibrous protein that provides support to muscle and prevents it from overstretching. It is the primary protein in connective tissue.

**Colloidal dispersion**   A solvent containing particles that are too large to go into solution, but not large enough to precipitate out.

**Complete protein**   A protein, usually from animal sources, that contains all the essential amino acids in sufficient amounts for the body's maintenance and growth.

**Compound**   A substance whose molecules consist of unlike atoms.

**Conduction**   The direct transfer of heat from one substance to another that it is contacting.

**Congener**   Alcohol by-product such as methanol or wood alcohol.

**Connective tissue**   A protein structure that surrounds living cells, giving them structure and adhesiveness within themselves and to adjacent tissues.

**Consistency**   Describes a food's firmness or thickness.

**Consommé**   A richly flavored soup stock that has been clarified and made transparent by the use of egg whites.

**Convection**   The transfer of heat by moving air or liquid (water/fat) currents through and/or around food.

**Co-op**   Work-study program with a corporation that is often customized for the student.

**Country of Origin labeling**   The required identification of the country of origin on the label for fresh red meat (beef, pork, lamb, veal), marinated products (marinated meats), seafood, produce, and peanuts.

**Court bouillon**   Seasoned stock containing white wine and/or vinegar.

**Cover**   The table setting, including the place mat, flatware, dishes, and glasses.

**Creaming Method**   A procedure of cake batter mixing in which the shortening and sugar are first combined at slow or medium speed until the mixture becomes aerated, followed by the addition of eggs and, in alternate small portions, of milk and flour while mixing continues.

**Critical control point (CCP)**   A point in the HACCP process that must be controlled to ensure the safety of the food.

**Cross-contamination**   The transfer of bacteria or other microorganisms from one food to another.

**Cruciferous**   A group of indole-containing vegetables named for their cross-shaped blossoms; they are reported to have a protective effect against cancer in laboratory animals. Examples include broccoli, brussels sprouts, cabbage, cauliflower, kale, mustard greens, rutabaga, kohlrabi, and turnips.

**Crumb**   The cell structure appearing when a baked product is sliced. Evaluation is based on cell size (called *open* if medium to large, or *closed* if small), cell shape, and cell thickness (thin walls occur in fine crumb, whereas thick walls predominate in a coarse crumb).

**Crumbing**   A ceremonious procedure of Russian service in which a waiter, using a napkin or silver crumber, brushes crumbs off the tablecloth into a small container resembling a tiny dust pan.

**Crustacean**   An invertebrate animal with a segmented body covered by an exoskeleton consisting of a hard upper shell and a soft under shell.

**Crystalline candy**   Candies formed from sugar solutions yielding many fine, small crystals.

**Crystallization**   The precipitation of crystals from a solution into a solid, geometric network.

**Culture**   The ideas, customs, skills, and art of a group of people in a given period of civilization.

**Curd**   The coagulated or thickened part of milk.

**Cure**   To preserve food through the use of salt and drying. Sugar, spices, or nitrates may also be added.

**Curing**   To expose cheese to controlled temperature and humidity during aging.

**Cuticle (bloom)**   A waxy coating on an eggshell that seals the pores from bacterial contamination and moisture loss.

**Cycle menu**   A menu that consists of 2 or more weeks, usually 3 or 4, that cycles through a certain order of meals. Cycle menus offer a combination of variety and controlled costs.

**Deglaze**   To add liquid to pan drippings followed by simmering/stirring to dissolve and loosen cooked-on particles sticking to the bottom of the pan.

**Degorge**   To peel and slice vegetables, sprinkle them with salt, and allow them to stand at room temperature until droplets containing bitter substances form on the surface; the moisture is then removed.

**Dehydrate**   To remove at least 95% of the water from foods through exposure to high temperatures.

**Delaney Clause**   A clause added to the Food, Drug, and Cosmetic Act of 1938 stipulating that "no additive shall be deemed to be safe if it is found to induce cancer when ingested by man or animal."

**Denaturation**   The irreversible process in which the structure of a protein is disrupted, resulting in partial or complete loss of function.

**Density**   The concentration of matter measured by the amount of mass per unit volume. Objects with a higher density weigh more for their size.

**Dextrinization**   The breakdown of starch molecules to smaller, sweeter-tasting dextrin molecules in the presence of dry heat.

**Dextrose equivalent (DE)**   A measurement of dextrose concentration. A DE of 50 means the syrup contains 50% dextrose.

**Dietary fiber**   The undigested portion of carbohydrates remaining in a food sample after exposure to digestive enzymes.

**Dietitian (registered dietitian or RD)**   A health professional who counsels people about their medical nutrition therapy (diabetic, low-cholesterol, low-sodium, etc.). Registration requirements consist of completing an approved 4-year college degree, exam, internship, and ongoing continuing education.

**Distillation**   A procedure in which pure liquid is obtained from a solution by boiling, condensation, and collection of the condensed liquid in a separate container.

**Distilled water**   Water that has been purified through distillation to remove minerals, pathogens, and other substances.

**Diverticulosis**   An intestinal disorder characterized by pockets forming out from the digestive tract, especially the colon.

**Dough**   A flour mixture that is dry enough to be handled and kneaded.

**Drug**   A product able to treat, prevent, cure, mitigate, or diagnose a disease or disease symptom.

**Drupes**   Fruit with seeds encased in a pit. Examples are apricots, cherries, peaches, and plums.

**Dry-heat preparation**   A method of cooking in which heat is transferred by air, radiation, fat, or metal.

**Edible coating**   A thin layer of edible material, such as natural wax, oil, or petroleum-based wax, that serves as a barrier to gas and moisture.

**Edible portion (EP)**   Food in its raw state, minus that which is discarded—bones, fat, skins, and/or seeds.

**Electrolyte**   An electrically charged ion in a solution.

**Emulsifier**   A compound that possesses both water-loving (hydrophilic) and water-fearing (hydrophobic) properties so that it disperses in either water or oil.

**Emulsion**   A liquid dispersed in another liquid with which it is usually immiscible (incapable of being mixed).

**Endosperm**   The largest portion of the grain, containing all of the grain's starch.

**Enriched**   Foods that have had certain nutrients, which were lost through processing, added back to levels established by federal standards.

**Enrobe**   To coat food with melted chocolate that hardens to form a solid casing.

**Enzymatic browning**   A reaction in which an enzyme acts on a phenolic compound in the presence of oxygen to produce brown-colored products.

**Enzyme**   A protein that catalyzes (causes) a chemical reaction without itself being altered in the process.

**Essential nutrients**   Nutrients that the body cannot synthesize at all or in amounts necessary to meet the body's needs.

**Essential oil**   An oily substance that is volatile (easily vaporized), with 100 times the flavoring power of the material from which it originated.

**Eviscerate**   To remove the entrails from the body cavity.

**Extractives**   Flavor compounds consisting of nonprotein, nitrogen substances that are end products of protein metabolism.

**Fermentation**   The conversion of carbohydrates to carbon dioxide and alcohol by yeast or bacteria.

**Finfish**   Fish that have fins and internal skeletons.

**Fire point**   The temperature at which a heated substance (such as oil) bursts into flames and burns for at least 5 seconds.

**Flash point**   The temperature at which tiny wisps of fire streak to the surface of a heated substance (such as oil).

**Flatware**   Eating and serving utensils (e.g., knives, forks, and spoons).

**Flavor**   The combined sense of taste, odor, and mouthfeel.

**Flavoring**   Substance that adds a new flavor to food.

**Flavor reversion**   The breakdown (oxidation) of an essential fatty acid, linolenic acid, found in certain vegetable oils, leading to an undesirable flavor change prior to the start of actual rancidity.

**Flocculation**   A partial gel in which only some of the solid particles colloidally dispersed in a liquid have solidified.

**Foam**   A colloidal dispersion of a gas in a liquid.

**Food additive**   A substance added intentionally or unintentionally to food that becomes part of the food and affects its character.

**Foodborne illness**   An illness transmitted to humans by food.

**Food Code**   An FDA publication updated every two years that shows food service organizations how to prevent foodborne illness while preparing food.

**Food cost**   Often expressed as a percentage obtained by dividing the raw food cost by the menu price.

**Food group plan**   A diet-planning tool that "groups" foods together based on nutrient and calorie (kcal) content and then specifies the amount of servings a person should have based on their recommended calorie (kcal) intake.

**Food infection**   An illness resulting from ingestion of food containing large numbers of living bacteria or other microorganisms.

**Food intoxication**   An illness resulting from ingestion of food containing a toxin.

**Forecast**   A predicted amount of food that will be needed for a food service operation within a given time period.

**Fortified**   Foods that have had nutrients added that were not present in the original food.

**Free radical**   An unstable molecule that is extremely reactive and that can damage cells.

**Freeze-dry**   To remove water from food when it is in a frozen state, usually under a vacuum.

**Freezer burn**   White or grayish patches on frozen food caused by water evaporating into the package's air spaces.

**Freezing point**   The temperature at which a liquid changes to a solid.

**Fruit**   The edible part of a plant developed from a flower.

**Fumet**   A flavorful fish stock made with white wine.

**Functional food**   A food or beverage that imparts a physiological benefit that enhances overall health, helps prevent or treat a disease or condition, or improves physical/mental performance.

**Gaping**   The separation of fish flesh into flakes that occurs as the steak or fillet ages.

**Gelatinization**   The increase in volume, viscosity, and translucency of starch granules when they are heated in a liquid.

**Gene**   A unit of genetic information in the chromosome.

**Generally Recognized as Safe (GRAS) list**   A list of compounds exempt from the food additive definition because they are generally recognized as safe based on "a reasonable certainty of no harm from a product under the intended conditions of use."

**Genetically modified organisms (GMOs)**   Plants, animals, or microorganisms that have had their genes altered through genetic engineering using the application of recombinant deoxyribonucleic acid (rDNA) technology.

**Germ**   The smallest portion of the grain, and the embryo for a future plant.

**Glaze**   A sugar-coated icing poured over pies or pastries that hardens to provide flavor and structure. The word *glaze* is used both for pastries and soup stocks, but they have different meanings.

**Glaze**   A flavoring obtained from soup stock that has been concentrated by evaporation until it attains a syrupy consistency with a highly concentrated flavor.

**Gluten**   The protein portion of wheat flour with the elastic characteristics necessary for the structure of most baked products.

**Good Manufacturing Practices**   A set of regulations, codes, and guidelines for the manufacture of food products, drugs, medical devices, diagnostic products, and active pharmaceutical ingredients (APIs).

**Grading**   The voluntary process in which foods are evaluated for yield (a 1 to 5 grading for meats only) and quality (Prime, Choice, AA, A, Fancy, etc.).

**Gram**   A metric unit of weight. One gram (g) is equal to the weight of 1 cubic centimeter (cc) or milliliter (mL) of water (under a specific temperature and pressure).

**GRE**   The **G**raduate **R**ecord **E**xamination®, a general test that measures verbal reasoning, quantitative reasoning, critical thinking, and analytical writing skills.

**Gustatory**   Relating to the sense of taste.

**HACCP**   Hazard Analysis and Critical Control Point System, a systematized approach to preventing foodborne illness during the production and preparation of food.

**Halal**   An Arabic word meaning "permissible." Usually refers to permissible foods under Islamic law.

**Heat of solidification**   The temperature at which a substance converts from a liquid to a solid state.

**Heat of vaporization**   The amount of heat required to convert a liquid to a gas.

**Heat shock**   Repeated cycles of temperature fluctuations from cold to warm and back that cause larger ice crystal growth, reducing frozen dessert quality.

**Herb**   A plant leaf valued for its flavor or scent.

**Hermetically sealed**   Refers to foods that have been packaged airtight by a commercial sealing process.

**High-conversion corn syrups**   Corn syrups with a dextrose equivalent over 58.

**Homogenization**   A mechanical process that breaks up the fat globules in milk into much smaller globules that do not clump together and are permanently dispersed in a very fine emulsion.

**Hops**   The dried fruit of the *Humulus lupulus* plant, which grows in the Pacific Northwest of the United States.

**Hormone**   A substance (usually a peptide or steroid) produced by one tissue and sent through the bloodstream to another tissue site to act physiologically (growth or metabolism).

**Humectant**   A substance that attracts water to itself. If added to food, it increases the water-holding capacity of the food and helps to prevent it from drying out by lowering the water activity.

**Husk**   The rough outer covering protecting a grain.

**Hydrogenation**   A commercial process in which hydrogen atoms are added to the double bonds in monounsaturated or polyunsaturated fatty acids to make them more saturated.

**Hydrolysis**   A chemical reaction in which water (*hydro*) breaks (*lysis*) a chemical bond in another substance, splitting it into two or more new substances.

**Hydrophilic**   A term describing "water-loving" or water-soluble substances.

**Hydrophobic**   A term describing "water-fearing" or nonwater-soluble substances.

**Hygroscopic**   Having the ability to attract and retain moisture.

**Imitation milk**   A product defined by the FDA as having the appearance, taste, and function of its original counterpart but as being nutritionally inferior.

**Incomplete protein**   A protein, usually from plant sources, that does not provide all the essential amino acids.

**Induction**   The transfer of heat energy to a neighboring material without contact.

**Interesterification**   A commercial process that rearranges fatty acids on the glycerol molecule in order to produce fat with a smoother consistency.

**Interfering agent**   A substance added to the sugar syrup to prevent the formation of large crystals, resulting in a candy with a waxy, chewy texture.

**Internships**   Corporate internships are temporary job positions (usually for 3 months during the summer or for 6 months, which includes a summer and a semester) in industry, government, or academia. Dietetic internships are supervised practice experiences that average 12 months (6 to 24 months).

**Invert sugar**   An equal mixture of glucose and fructose, created by hydrolyzing sucrose.

**Ionize**   To separate a neutral molecule into electrically charged ions.

**Irradiation**   A food preservation process in which foods are treated with low doses of gamma rays, X-rays, or electrons.

**Job description**   An organized list of duties used for finding qualified applicants, training, performance appraisal, defining authority and responsibility, and determining salary.

**Julienne**   To cut food lengthwise into very thin, stick-like shapes.

**Kinetic energy**   Energy associated with motion.

**Knead**   To work the dough into an elastic mass by pushing, stretching, and folding it.

**Kosher**   From Hebrew, food that is "fit, right, proper" to be eaten according to Jewish dietary laws.

**Lamination**   The arrangement of alternating layers of fat and flour in rolled pastry dough. During baking, the fat melts and leaves empty spaces for steam to lift the layers of flour, resulting in a flaky pastry.

**Larding**   Inserting strips of bacon, salt pork, or other fat into slits in the meat with a large needle.

**Latent heat**   The amount of energy in calories (kcal) per gram absorbed or emitted as a substance undergoes a change in state (liquid/solid/gas).

**Legumes**   Members of the plant family *Leguminosae* that are characterized by growing in pods. Vegetable legumes include beans, peas, and lentils.

**Maillard reaction**   The reaction between a sugar (typically reducing sugars such as glucose/dextrose, fructose, lactose, or maltose) and a protein (specifically the nitrogen in an amino acid), resulting in the formation of brown complexes.

**Marbling**   Fat deposited in the muscle that can be seen as little white streaks or drops.

**Masa**   A word that means "dough" in Mexico. It is made by cooking corn in water, after which it is ground into a pourable slurry. Masa is often used to make tortillas, tamales, and many commercial corn-based snacks.

**Mealy**   A pastry with a grainy or less flaky texture, created by coating all of the flour with fat.

**Medical foods**   A food to be taken under the supervision of a physician and intended for the dietary management of a disease/condition for which distinctive nutritional requirements are established by scientific evaluation.

**Medical nutrition therapy**   Dietary therapy that applies the nutrition care process (NCP) in clinical settings to manage nutrition-related diseases such as heart disease, diabetes, cancer, renal disease, liver failure, and others. Registered dietitians, as part of the health care team, contribute to the patient's care by providing this dietary therapy often referred by a physician.

**Melting point**   The temperature at which a solid changes to a liquid state (liquid/solid/gas).

**Meniscus**   The imaginary line read at the bottom of the concave arc at the water's surface.

**Methylxanthine**   A compound that stimulates the central nervous system.

**Microorganism**   Plant or animal organism that can only be observed under the microscope—bacteria, mold, yeast, virus, or animal parasite.

**Milk solids-not-fat (MSNF)**   Federal standard identifying the total solids, primarily proteins and lactose, found in milk, minus the fat.

**Mineral water**   Water from natural springs having a strong taste or odor due to small amounts of salts of calcium, magnesium, and sodium (sodium bicarbonate, sodium carbonate, sodium chloride), and sometimes iron or hydrogen sulfide.

**Mirepoix**   A collection of lightly sautéed, chopped vegetables (a 2:1:1 ratio by weight of onions, celery, and carrots) flavored with spices and herbs (sage, thyme, marjoram, and chopped parsley are the most common).

**Modified starch**   A starch that has been chemically or physically modified to create unique functional characteristics.

**Moist-heat preparation**   A method of cooking in which heat is transferred by water, any water-based liquid, or steam.

**Mold**   A fungus (a plant that lacks chlorophyll) that produces a furry growth on organic matter.

**Molecule**   A unit composed of one or more types of atoms held together by chemical bonds.

**Mollusk**   An invertebrate animal with a soft unsegmented body usually enclosed in a shell.

**Monograph**   A summary sheet (fact sheet) describing a substance in terms of name (common and scientific), chemical constituents, functional uses (medical and common), dosage, side effects, drug interactions, and references.

**Mother sauce**   A sauce that serves as the springboard from which other sauces are prepared.

**Mycotoxin**   A toxin produced by a mold.

**Myocommata**   Large sheets of very thin connective tissue separating the myotomes.

**Myotomes**   Layers of short fibers in fish muscle.

**Noncrystalline (amorphous) candy**   Candies formed from sugar solutions that did not crystallize.

**Nonnutritive sweeteners**   Food additives requiring FDA approval that provide sweetness with no or insignificant amounts of energy (calories/kcal). Also known as alternative sweeteners, sugar substitutes, sugar replacers, and macronutrient substitutes.

**Nontempered coating**   A coating resembling chocolate that is not subject to bloom because it is made with fats other than cocoa butter.

**No-observed-effect level (NOEL)**   The no-observed-effect level is the level or dose at which an additive is fed to laboratory animals without any negative side effects.

**Nuclei**   Small aggregates of molecules serving as the starting point of crystal formation.

**Nutraceutical**   A bioactive compound (nutrients and nonnutrients) that has health benefits.

**Nutrition Care Process and Model**   A standardized model to guide registered dietitians and dietetic technicians, in providing high quality nutrition care.

**Nutrient content claims**   Food label descriptions communicating the amount of a nutrient or dietary substance contained in a food or beverage.

**Nutrients**   Food components that nourish the body to provide growth, maintenance, and repair.

**Nutrigenomics**   A field of study focusing on genetically determined, biochemical pathways linking specific dietary substances with health and disease.

**Objective tests**   Evaluations of food quality that rely on numbers generated by laboratory instruments, which are used to quantify the physical and chemical differences in foods.

**Ohmic heating**   A food preservation process in which an electrical current is passed through food, generating enough heat to destroy microorganisms.

**Olfactory**   Relating to the sense of smell.

**Omega-3 fatty acids**   A category of polyunsaturated fatty acids in which the first double bond is three carbons from the methyl ($CH_3$) end; examples are eicosapentaenoic acid (EPA) and docosahexaenoic acid (DHA).

**Organizational chart**   A descriptive diagram showing the administrative structure of an organization.

**Osmosis**   The movement of a solvent through a semipermeable membrane to the side with the higher solute concentration, equalizing solute concentration on both sides of the membrane.

**Osmotic pressure**   The pressure or pull that develops when two solutions of different solute concentration are on either side of a permeable membrane.

**Outbreak**   Defined by the CDC as the occurrence of two or more cases of a similar illness resulting from the ingestion of a common food.

**Oven spring**   The quick expansion of dough during the first 10 minutes of baking, caused by expanding gases.

**Overrun** The volume over and above the volume of the original frozen dessert mix, caused by the incorporation of air during freezing.

**Parasite** An organism that lives on or within another organism at the host's expense without any useful return.

**Parboil** To partially boil, but not fully cook, a food.

**Pascalization** A food preservation process utilizing ultrahigh pressures to inhibit the chemical processes of food deterioration.

**Pasteurization** A food preservation process that heats liquids to 161°F (72°C) for 15 seconds, or 143°F (62°C) for 30 minutes, in order to kill bacteria, yeasts, and molds.

**Patent flour** The finest streams of flour.

**Pathogenic** Causing or capable of causing disease.

**Peptide bond** The chemical bond between two amino acids.

**Percentage yield** The ratio of edible to inedible or wasted food.

**Phenolic** A chemical term to describe an aromatic (circular) ring attached to one or more hydroxyl (–OH) groups.

**pH scale** Measures the degree of acidity or alkalinity of a substance, with 1 the most acidic, 14 the most alkaline, and 7 neutral.

**Plain pastry** Pastry made for producing piecrusts, quiches, and main-dish pies.

**Plant stanol esters** Naturally occurring substances in plants that help block absorption of cholesterol from the digestive tract.

**Plasticity** The ability of a fat to be shaped or molded.

**Polymerization** A process in which free fatty acids link together, especially when overheated, resulting in a gummy, dark residue and an oil that is more viscous and prone to foaming.

**Polymorphism** The capability of solid fats to change into several crystalline forms, each with its own melting point, crystal structure, and solubility.

**Polyphenol** An organic compound with two or more phenols—carbon atoms structured into an aromatic ring with one or more hydroxyl (–OH) groups.

**Pomes** Fruit with seeds contained in a central core. Examples are apples and pears.

**Prawn** A large crustacean that resembles shrimp but is biologically different. Large shrimp are often called by this name.

**Prebiotics** Nondigestible food ingredients [generally fibers such as fructooligosaccharides (FOS) and inulin] that support the growth of probiotics.

**Precipitate** To separate or settle out of a solution.

**Prime (season)** To seal the pores of a pan's metal surface with a layer of heated-on oil.

**Prion** An infectious protein particle that does not contain DNA or RNA.

**Probiotics** Live microbial food ingredients (i.e., bacteria) that have a beneficial effect on human health.

**Process (processed) cheese** A cheese made from blending one or more varieties of cheese, with or without heat, and mixing the result with other ingredients.

**Product recall** Civil court action to seize or confiscate a product that is defective, unsafe, filthy, or produced under unsanitary conditions.

**Proof (alcohol)** Alcoholic strength indicated by a number that is twice the percent by volume of alcohol present.

**Proof (baking)** To increase the volume of shaped dough through continued fermentation.

**Proof box** A large, specially designed container that maintains optimal temperatures and humidity for the fermentation and rising of dough.

**Protein complementation** Two incomplete-protein foods, each of which supplies the amino acids missing in the other, combined to yield a complete protein profile.

**P/S ratio** The ratio of polyunsaturated fats to saturated fats. The higher the P/S ratio, the more polyunsaturated fats the food contains.

**Puff pastry** A delicate pastry that puffs up in size during baking because of many alternating layers of fat and flour.

**Purified water** Water that has undergone deionization, distillation, reverse osmosis, or any other method that removes minerals, chemicals, and flavor.

**Qualified health claims** An FDA term describing a relationship between a food, food component, or dietary supplement ingredient and reduced risk of a disease or health-related condition. Although the "weight of the evidence" qualifies them as a health claim, these claims are not held to the standard of significant scientific agreement.

**Quality grades** The USDA standards for beef, veal, lamb, and mutton.

**Quick bread** Bread leavened with air, steam, and/or carbon dioxide from baking soda or baking powder.

**Radiation** The transfer of heat energy in the form of waves of particles moving outward from their source.

**Rancid** The breakdown of the polyunsaturated fatty acids in fats that results in disagreeable odors and flavors.

**Reducing sugars** Sugars such as glucose, fructose, maltose, and others that have a reactive aldehyde or ketone group. Sucrose is not a reducing sugar.

**Reduction** The process in which a liquid is simmered or boiled until the volume is reduced through evaporation, leaving a thicker, more concentrated, flavorful mass; or the product of this process.

**Reference protein** A standard against which to measure the quality of other proteins. Registered dietitians, as part of the health care team, contribute to the patient's care by providing dietary therapy often pursuant to a physician's referral. The patient's nutrition status is assessed prior to recommending a dietary plan to treat medical conditions such as heart disease, diabetes, cancer, renal disease, liver failure, and others.

**Rennin** An enzyme obtained from the inner lining of a calf's stomach and sold commercially as rennet.

**Respiration rate** The rate of carbon dioxide produced from a given amount of produce over a certain unit of time.

**Retail cuts** Smaller cuts of meat obtained from wholesale cuts and sold to the consumer.

**Retrogradation** The seepage of water out of an aging gel because of the contraction of the gel (bonds tighten between the amylose molecules). Also known as *syneresis* or *weeping*.

**Rhizome** An underground (usually) stem that generates (1) shoots that rise up and/or horizontally to propagate new plants and (2) roots that grow down to the ground.

**Rigor mortis** Latin for "stiffness of death," the temporary stiff state following death as muscles contract.

**Ripening**   The chemical and physical changes that occur during the curing period.

**Rope**   The sticky, moist texture of breads resulting from contamination by *Bacillus mesentericus* bacteria.

**Roux**   A thickener made by cooking equal parts of flour and fat.

**Saturated solution**   A solution holding the maximum amount of dissolved solute at room temperature.

**Scalloped**   Baked with milk sauce and bread crumbs.

**Scampi**   A crustacean found in Italy and not generally available in North America. The term is often used incorrectly to describe a popular shrimp dish.

**Score**   The technique of taking a sharp knife or a special blade called a lame and creating ¼- to ½-inch-deep slashes on the risen dough's top surface just prior to baking.

**Sear**   To brown the surface of meat by brief exposure to high heat.

**Searing**   Cooking that exposes a cut of meat to very high initial temperatures; this is intended to seal the pores, increase flavor, and enhance color by browning.

**Seasoning**   Any compound that enhances the flavor already found naturally in a food.

**Seed**   To create nuclei or starting points from which additional crystals can form.

**Sensory or subjective tests**   Evaluations of food quality based on sensory characteristics and personal preferences as perceived by the five senses.

**Shortened cake**   A cake made with fat.

**Shortening**   A fat that tenderizes, or shortens, the texture of baked products by impeding gluten development, making them softer and easier to chew.

**Silence cloth**   A piece of fabric placed between the table and the tablecloth to protect the table, quiet the placement of dishes and utensils, and keep the tablecloth from slipping.

**Simple syrup**   A basic mixture of boiled sugar and water.

**Slurry**   A thickener made by combining starch and a cool liquid.

**Small sauce**   A secondary sauce created when a flavor is added to a mother sauce.

**Smoke point**   The temperature at which fat or oil begins to smoke.

**Sol**   A colloidal dispersion of a solid dispersed in a liquid.

**Solubility**   The ability of one substance to blend uniformly with another.

**Solute**   Solid, liquid, or gas compounds dissolved in another substance.

**Solution**   A completely homogeneous mixture of a solute (usually a solid) dissolved in a solvent (usually a liquid).

**Solvent**   A substance, usually a liquid, in which another substance is dissolved.

**Specifications**   Descriptive information used in food purchasing that defines the minimum and maximum levels of acceptable quality or quantity (i.e., U.S. grade, weight, size, fresh or frozen).

**Specific gravity**   The density of a substance compared to another substance (usually water).

**Specific heat**   The amount of heat required to raise the temperature of 1 gram of a substance 1°C.

**Spice**   A seasoning or flavoring added to food that is derived from the fruit, flowers, bark, seeds, or roots of a plant.

**Spore**   Encapsulated, dormant form assumed by some microorganisms that is resistant to environmental factors that would normally result in its death.

**Spring water**   Water that, according to the FDA requirements, flows from its source without being pumped and contains at least 250 parts per million of dissolved solids.

**Stabilizer**   A compound such as vegetable gum that attracts water and interferes with frozen ice crystal formation, resulting in a smoother consistency in frozen desserts.

**Standard Operating Procedures (SOPs)**   Established written procedures serving as compulsory instructions to be followed out exactly in carrying out a given operation. SOPs ensure quality control through always carrying out operations in the same correct manner.

**Standardized recipe**   A food service recipe that is a set of instructions describing how a particular dish is prepared by a specific establishment. It ensures consistent food quality and quantity, the latter of which provides portion/cost control.

**Standards of Fill**   The amount of raw product that must be put into a container before liquid (brine or syrup) is added.

**Standards of Identity**   Requirements for the type and amount of ingredients a food should contain in order to be labeled as that food.

**Standards of Minimum Quality**   Minimum quality requirements for tenderness, color, and freedom from defects in canned fruits and vegetables.

**Starter**   A culture of microorganisms, usually bacteria and/or yeasts, used in the production of certain foods such as sourdough bread, cheese, and alcoholic beverages.

**Sterilization**   The elimination of all microorganisms through extended boiling/heating to temperatures much higher than boiling or through the use of certain chemicals.

**Stock**   The foundational thin liquid of many soups, produced when meat, poultry, seafood, and/or their bones, or vegetables are reduced (simmered) and strained.

**Storage eggs**   Eggs that are treated with a light coat of oil or plastic and stored in high humidity at low refrigerator temperatures very close to the egg's freezing point (29°F–32°F/ −1.5°C–0°C).

**Straight flour**   Flour containing all the different types of streams produced during milling.

**Stream**   A division of milled flour based on particle size.

**Streusel topping**   A crunchy, flavorful topping that can be strewn over the top of pies; it is made by combining flour, butter or margarine, brown sugar, and possibly spices (cinnamon) and chopped nuts (pecans, walnuts, or almonds).

**Structure/function claims**   Statements identifying relationships between nutrients or dietary ingredients and body functions.

**Sublimation**   The process in which a solid changes directly to a vapor without passing through the liquid phase.

**Substrate**   A substance that is acted upon, such as by an enzyme.

**Superglycerinated**   Describes a shortening that has had mono- and diglycerides added for increased plasticity.

**Supersaturated solution**   An unstable solution created when more than the maximum solute is dissolved in solution.

**Surfactant**   Surface-active agent that reduces a liquid's surface tension to increase its wetting and blending ability.

**Surimi**   Japanese for "minced meat," a fabricated fish product usually made from Alaskan pollack, a deep-sea whitefish, which is skinned, deboned, minced, washed, strained, and shaped into pieces to resemble crab, shrimp, or scallops.

**Suspension**   A mixture in which particles too large to go into solution remain suspended in the solvent.

**Sweat**   The stage of cooking in which food, especially vegetables, becomes soft and translucent.

**Syneresis**   The oozing out of the liquid component of a gel.

**Tannins**   Polymers of various flavonoid compounds, of which some of the larger ones yield reddish and brown pigments.

**Temperature danger zone**   The temperature range that is ideal for bacterial growth; it is 40°F–140°F (4°C–60°C) for consumers and 41°F–135°F (5°C–57°C) for retailers.

**Tempering**   The process of heating and cooling chocolate to specific temperatures, making it more resistant to melting and resulting in a smooth, glossy, hard finish.

**Tenure track**   The process in which new faculty members are hired as assistant professors and are on probation for approximately 5 years. If their yearly contract is renewed for 5 years in a row based on satisfactory performance, they may apply for tenure (permanent hire) and promotion to associate professor. The next and final step is promotion to full professor.

**Three-compartment sink**   A sink divided into three sections, the first for soaking and washing, the second for rinsing, and the third for sanitizing.

**TOEFL**   The Test of English as a Foreign Language™ measures the ability of non-native speakers to understand English as it is spoken, written, and heard in colleges and universities.

**Truss**   To tie the legs and wings against the body of the bird to prevent it from overcooking before the breast is done.

**Turgor**   The rigid firmness of a plant cell resulting from being filled with water.

**Ultrahigh-temperature (UHT) milk**   Milk that has been pasteurized using very high temperatures, is aseptically sealed, and is capable of being stored unrefrigerated for up to 3 months.

**Ultrapasteurization**   A process in which a milk product is heated at or above 280°F (138°C) for at least 2 seconds.

**Unqualified health claim**   An FDA term describing a relationship between a food, food component, or dietary supplement ingredient and reducing risk of a disease or health-related condition. Significant scientific agreement supports these authorized claims.

**Unshortened cake**   A cake made without added fat.

**Variety meats**   The liver, sweetbreads (thymus), brain, kidneys, heart, tongue, tripe (stomach lining), and oxtail (tail of cattle).

**Verification form**   Documentation provided by the dietetic program director to an individual who has completed the undergraduate ADA course requirements approved by the ADA for that particular academic institution. This form is required for (1) acceptance into an ADA internship and (2) taking the ADA examination.

**Vinaigrette**   A salad dressing consisting only of oil, vinegar, and seasoning.

**Vintage**   The year in which a wine was bottled; especially, an exceptionally fine wine from a year with a good crop.

**Virus**   An infectious microorganism consisting of RNA or DNA that reproduces only in living cells.

**Viscosity**   The resistance of a fluid to flowing freely, caused by the friction of its molecules against a surface.

**Vitelline membrane**   The membrane surrounding the egg yolk and attached to the chalazae.

**Volatile molecules**   Molecules capable of evaporating like a gas into the air.

**Volume**   A measurement of three-dimensional space that is often used to measure liquids.

**Water activity (aw)**   Measures the amount of available (free) water in foods. Water activity ranges from 0 to the highest value of 1.00, which is pure water.

**Weeping (syneresis)**   The escape of liquid to the bottom of a meringue or the formation of pores filled with liquid.

**Well water**   Water pumped from an aquifer, an underground source of water.

**Whey**   The liquid portion of milk, consisting primarily of 93% water, lactose, and whey proteins (primarily lactalbumin and lactoglobulin). It is the watery component removed from the curd in cheese manufacture.

**White sauce**   A mixture of flour, milk, and usually fat.

**White stock**   The flavored liquid obtained by simmering the bones of beef, veal, chicken, or pork.

**Whole grains**   Defined by the American Association of Cereal Chemists (AACC) as "foods made from the entire grain seed, usually called the kernel, which consists of the bran, germ, and endosperm."

**Wholesale (primal) cuts**   The large cuts of an animal carcass, which are further divided into retail cuts.

**Winterizing**   A commercial process that removes from vegetable oils the fatty acids that have a tendency to crystallize and make the oils appear cloudy.

**Yeast**   A fungus (a plant that lacks chlorophyll) that is able to ferment sugars and that is used for producing food products such as bread and alcohol.

**Yeast bread**   Bread made with yeast, which produces carbon dioxide gas through the process of fermentation, causing the bread to rise.

**Yield grade**   The amount of lean meat on the carcass in proportion to fat, bone, and other inedible parts.

# Answers to Multiple Choice Questions

**Chapter 1**
1. b
2. b
3. d
4. b
5. d
6. d
7. d

**Chapter 2**
1. a
2. c
3. d
4. c
5. c
6. c
7. a

**Chapter 3**
1. a
2. b
3. d
4. b
5. d
6. a
7. d

**Chapter 4**
1. c
2. d
3. c
4. d
5. c
6. c
7. b

**Chapter 5**
1. a
2. d
3. a
4. b
5. c
6. a
7. d

**Chapter 6**
1. c
2. a
3. c
4. a
5. d
6. d
7. b

**Chapter 7**
1. c
2. d
3. a
4. a
5. d
6. c
7. b

**Chapter 8**
1. b
2. c
3. d
4. d
5. c
6. b
7. a

**Chapter 9**
1. d
2. c
3. b
4. a
5. d
6. b
7. c

**Chapter 10**
1. c
2. c
3. a
4. d
5. d
6. b
7. c

**Chapter 11**
1. d
2. b
3. c
4. d
5. c
6. b
7. d

**Chapter 12**
1. a
2. c
3. d
4. a
5. d
6. d
7. b

**Chapter 13**
1. c
2. d
3. b
4. d
5. b
6. d
7. a

**Chapter 14**
1. d
2. a
3. b
4. a
5. b
6. a
7. d

**Chapter 15**
1. d
2. b
3. d
4. d
5. a
6. a
7. c

**Chapter 16**
1. b
2. a
3. c
4. c
5. c
6. d
7. a

**Chapter 17**
1. b
2. c
3. a
4. b
5. d
6. a
7. c

**Chapter 18**
1. b
2. b
3. d
4. a
5. c
6. a
7. c

**Chapter 19**
1. c
2. b
3. d
4. d
5. b
6. b
7. a

**Chapter 20**
1. c
2. d
3. b
4. a
5. c
6. c
7. b

**Chapter 21**
1. b
2. a
3. a
4. d
5. b
6. c
7. b

**Chapter 22**
1. b
2. a
3. d
4. b
5. b
6. d
7. d

**Chapter 23**
1. b
2. d
3. b
4. b
5. a
6. c
7. b

**Chapter 24**
1. d
2. d
3. d
4. b
5. a
6. c
7. a

**Chapter 25**
1. b
2. a
3. b
4. c
5. d
6. c
7. c

**Chapter 26**
1. a
2. d
3. c
4. a
5. d
6. a
7. d

**Chapter 27**
1. b
2. c
3. b
4. a
5. d
6. a
7. c

**Chapter 28**
1. d
2. a
3. a
4. b
5. c
6. b
7. c

**Chapter 29**
1. b
2. a
3. b
4. c
5. d
6. b
7. d

**Chapter 30**
1. d
2. b
3. b
4. b
5. a
6. c
7. c

# Index

Note: Websites are listed at the end of each text chapter.

## A